Teacher, Student, and Parent One-Stop Internet Resources

Log on to gpescience.com

STUDY TOOLS

- Self-Check Quizzes
- Math Practice
- Interactive Tutor
- Vocabulary PuzzleMaker
- Chapter Review Tests
- Standardized Test Practice
- Key-Concept animations
- BrainPOP Movies
- Multilingual Glossary

EXTENSIONS

- WebQuest Projects
- Prescreened Web Links
- Unit Projects
- Internet Labs

INTERACTIVE STUDENT EDITION

- Complete Interactive Student Edition available at mhln.com

FOR TEACHERS

- Teacher Forum
- Teaching Today—Professional Development

SAFETY SYMBOLS

SAFETY SYMBOLS	HAZARD	EXAMPLES	PRECAUTION	REMEDY
DISPOSAL	Special disposal procedures need to be followed.	certain chemicals, living organisms	Do not dispose of these materials in the sink or trash can.	Dispose of wastes as directed by your teacher.
BIOLOGICAL	Organisms or other biological materials that might be harmful to humans	bacteria, fungi, blood, unpreserved tissues, plant materials	Avoid skin contact with these materials. Wear mask or gloves.	Notify your teacher if you suspect contact with material. Wash hands thoroughly.
EXTREME TEMPERATURE	Objects that can burn skin by being too cold or too hot	boiling liquids, hot plates, dry ice, liquid nitrogen	Use proper protection when handling.	Go to your teacher for first aid.
SHARP OBJECT	Use of tools or glassware that can easily puncture or slice skin	razor blades, pins, scalpels, pointed tools, dissecting probes, broken glass	Practice common-sense behavior and follow guidelines for use of the tool.	Go to your teacher for first aid.
FUME	Possible danger to respiratory tract from fumes	ammonia, acetone, nail polish remover, heated sulfur, moth balls	Make sure there is good ventilation. Never smell fumes directly. Wear a mask.	Leave foul area and notify your teacher immediately.
ELECTRICAL	Possible danger from electrical shock or burn	improper grounding, liquid spills, short circuits, exposed wires	Double-check setup with teacher. Check condition of wires and apparatus.	Do not attempt to fix electrical problems. Notify your teacher immediately.
IRRITANT	Substances that can irritate the skin or mucous membranes of the respiratory tract	pollen, moth balls, steel wool, fiberglass, potassium permanganate	Wear dust mask and gloves. Practice extra care when handling these materials.	Go to your teacher for first aid.
CHEMICAL	Chemicals can react with and destroy tissue and other materials	bleaches such as hydrogen peroxide; acids such as sulfuric acid, hydrochloric acid; bases such as ammonia, sodium hydroxide	Wear goggles, gloves, and an apron.	Immediately flush the affected area with water and notify your teacher.
TOXIC	Substance may be poisonous if touched, inhaled, or swallowed.	mercury, many metal compounds, iodine, poinsettia plant parts	Follow your teacher's instructions.	Always wash hands thoroughly after use. Go to your teacher for first aid.
FLAMMABLE	Flammable chemicals may be ignited by open flame, spark, or exposed heat.	alcohol, kerosene, potassium permanganate	Avoid open flames and heat when using flammable chemicals.	Notify your teacher immediately. Use fire safety equipment if applicable.
OPEN FLAME	Open flame in use, may cause fire.	hair, clothing, paper, synthetic materials	Tie back hair and loose clothing. Follow teacher's instruction on lighting and extinguishing flames.	Notify your teacher immediately. Use fire safety equipment if applicable.

Eye Safety
Proper eye protection should be worn at all times by anyone performing or observing science activities.

Clothing Protection
This symbol appears when substances could stain or burn clothing.

Animal Safety
This symbol appears when safety of animals and students must be ensured.

Handwashing
After the lab, wash hands with soap and water before removing goggles.

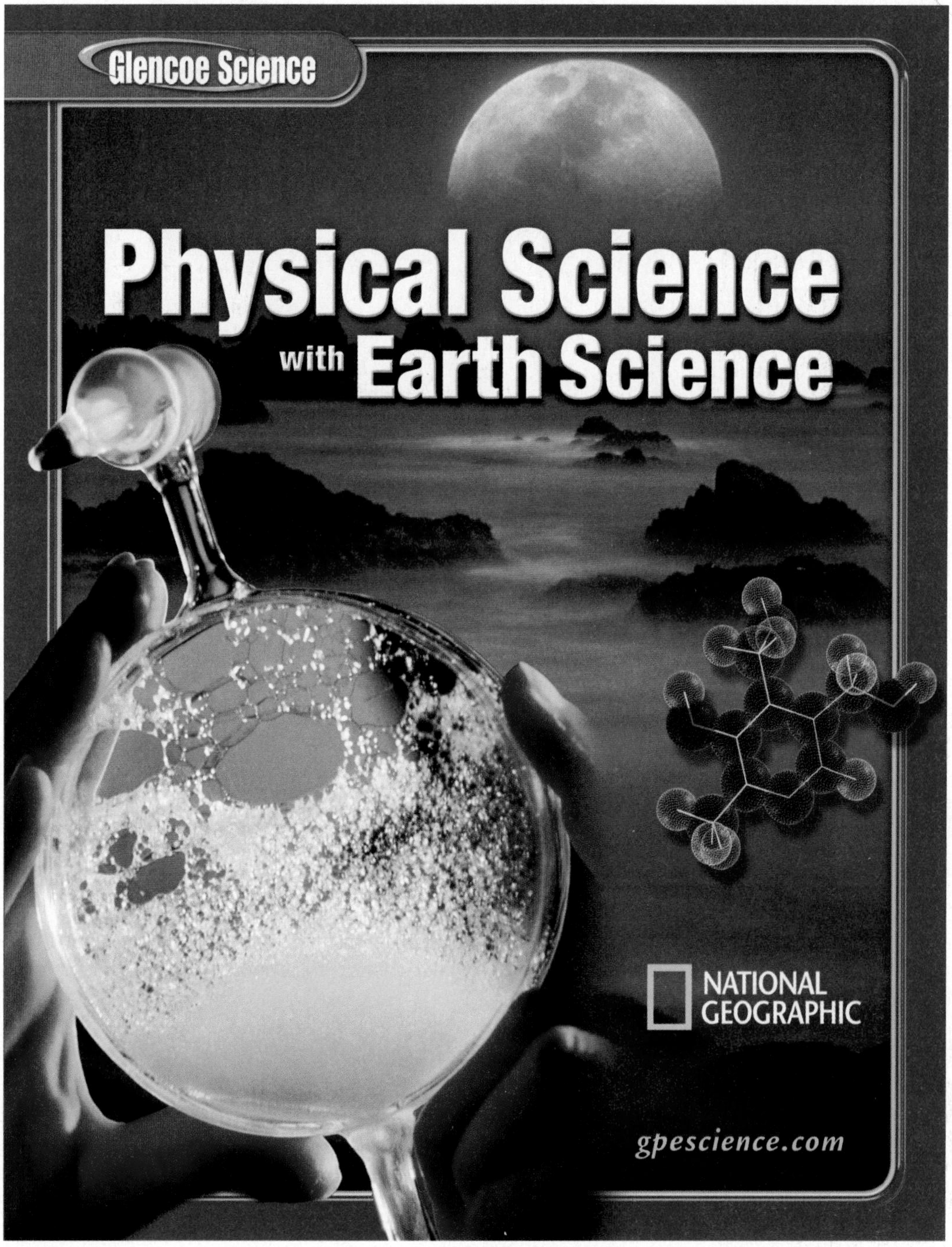

Glencoe Science

Physical Science with Earth Science

NATIONAL GEOGRAPHIC

gpescience.com

New York, New York Columbus, Ohio Chicago, Illinois Peoria, Illinois Woodland Hills, California

Glencoe Science

Physical Science with Earth Science

Moonrise over a rocky Pacific coast serves as the background for a color-enhanced image that shows the emulsifying property of a detergent and a computer-generated model of the chemical structure of vitamin B_6.

The McGraw·Hill Companies

Send all inquiries to:
Glencoe/McGraw-Hill
8787 Orion Place
Columbus, OH 43240-4027

ISBN: 0-07-868554-0

Printed in the United States of America.

4 5 6 7 8 9 10 071/043 09 08 07 06

Contents In Brief

Authors

Education Division
Washington, D.C.

Ralph M. Feather, Jr., PhD
Assistant Professor, Physics Department
Indiana University of Pennsylvania
Indiana, PA

Charles William McLaughlin, PhD
Senior Lecturer
University of Nebraska
Lincoln, NE

Marilyn Thompson, PhD
Assistant Professor, College of Education
Arizona State University
Tempe, AZ

Dinah Zike
Educational Consultant
Dinah-Might Activities, Inc.
San Antonio, TX

Teacher Advisory Board

Glencoe's National Teacher Advisory Board gave the authors, editorial staff, and the design team feedback on the content and design of the Student Edition. They were instrumental in providing valuable input toward the development of the 2006 edition of ***Glencoe Physical Science with Earth Science.*** We thank these teachers for their hard work and creative suggestions.

Melissa Shirley
Olentangy Liberty High School
Powell, OH

JoAnn Alverson
Pipestone Senior High
Pipestone, MN

Chuck Cambria
Springfield North High School
Springfield, OH

Susan Gleason
Middletown High School
Middletown, DE

Christy Saddic
Conestoga High School
Berwyn, PA

Laura Schmitmeyer
Memorial High School
St. Marys, OH

Jackie Barge
Walter Payton High School
Chicago, IL

Contributing Authors

Nancy Ross-Flanigan
Science Writer
Detroit, MI

Tina C. Hopper
Science and Health Writer
Rockwell, TX

Steve Hardesty
Adjunct Professor
Valencia Community College
Orlando, FL

Sharon Nicholson, PhD
Heinz and Katharina Lettau Professor of Climatology
Meteorology Department
Florida State University
Tallahassee, FL

Margaret K. Zorn
Science Writer
Yorktown, VA

Consultants

SCIENCE

Jack Cooper
Ennis High School
Ennis, TX

Elle Feth, MS
Columbus State Community College
Columbus, OH

David G. Haase, PhD
North Carolina State University
Raleigh, NC

William C. Keel, PhD
University of Alabama
Tuscaloosa, AL

Steve Letro
Meteorologist in Charge
WFO, National Weather Service
Jacksonville, FL

Madelaine Meek
Physics Consultant Editor
Lebanon, OH

Cheryl Wistrom, PhD
St. Joseph's College
Rensselaer, IN

Carl Zorn, PhD
Staff Scientist
Jefferson Laboratory
Newport News, VA

MATH

Michael Hopper, DEng.
Manager of Aircraft Certification
L-3 Communications
Greenville, TX

READING

Constance Cain, EdD
Literacy Coordinator
University of Central Florida
Tallahassee, FL

Maria Grant, EdD
Hoover High School
San Diego State University
San Diego, CA

ReLeah Lent, MS
Literacy Coordinator
University of Central Florida
Tallahassee, FL

SAFETY

Jack A. Gerlovich
Professor Science Education Safety
Drake University
Des Moines, IA

Denis McElroy
Assistant Professor Technology Education
Graceland University
Independence, MO

LAB TESTERS

Timothy Eimer
Phil-Mont Christina Academy
Erdenheim, PA

Nicholas Hainen
Worthington High School, Retired
Worthington, OH

James Marshall
Ohio State University
Columbus, OH

Jack Minot
Bexley High School
Bexley, OH

Reviewers

Sharla Adams
IPC Teacher
Allen High School
Allen, TX

Tom Bright
Concord High School
Charlotte, NC

Nora M. Prestinari Burchett
Saint Luke School
McLean, VA

Louise Chapman
Mainland High School
Daytona Beach, FL

Obioma Chukwu
J.H. Rose High School
Greenville, NC

Sandra Curtis
Atlantic High School
Port Orange, FL

Inga Dainton
Merrilville High School
Merrilville, IN

Robin Dillon
Hanover Central High School
Cedar Lake, IN

Dwight Dutton
East Chapel Hill High School
Chapel Hill, NC

Tracy Ebert
Robert Hungerford Preparatory High School
Eatonville, FL

Carolyn Elliott
South Iredell High School
Statesville, NC

Sueanne Esposito
Tipton High School
Tipton, IN

Kye Ewing
The Star Farm
Venus, FL

George Gabb
Great Bridge Middle School
Chesapeake Public Schools
Chesapeake, VA

Jan Gilliland
Alonso High School
Tampa, FL

Erick Hueck
Miami Senior High School
Miami, FL

Felecia Joiner
Judson High School
Converse, TX

Joe Kowalski
James "Nikki" Rowe High School
McAllen, TX

Bernie Leverett
Sublette High School
Sublette, KS

Jeff McKay
Spruce Creek High School
Port Orange, FL

Mike McKee
Cypress Creek High School
Orlando, FL

Armando S Miccoli, III
Adairsville High School
Adairsville, GA

Jane Nelson
University High School
Orlando, FL

Jim Nelson
University High School
Orlando, FL

Alicia K. Parker
New Smyrna Beach High School
New Smyrna Beach, FL

Annette Parrott
Lakeside High School
Atlanta, GA

Tom Schultz
Charlevoix High School
Charlevoix, MI

Ben Stofcheck
Citrus County Schools
Inverness, FL

Clabe Webb
Permian High School
Ector County ISD
Odessa, TX

Kim Wimpey
North Gwinnett High School
Suwanee, GA

Contents

Science and Technology—2

The Nature of Science—4

Science, Technology, and Society—36

Motion, Forces, and Energy—66

Motion, Acceleration, and Forces—68

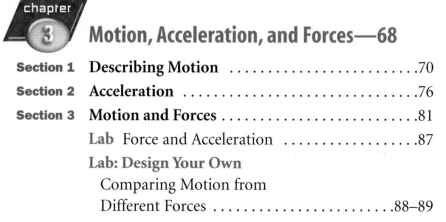

In each chapter, look for these opportunities for review and assessment:

- Reading Checks
- Caption Questions
- Section Review
- Chapter Study Guide
- Chapter Review
- Standardized Test Practice
- Online practice at **gpescience.com**

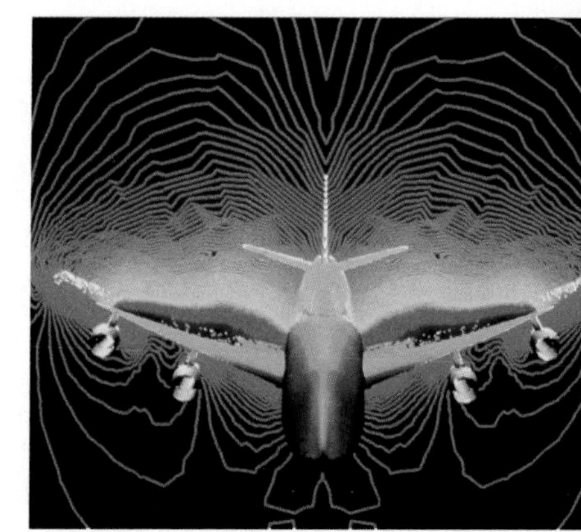

Contents

Contents

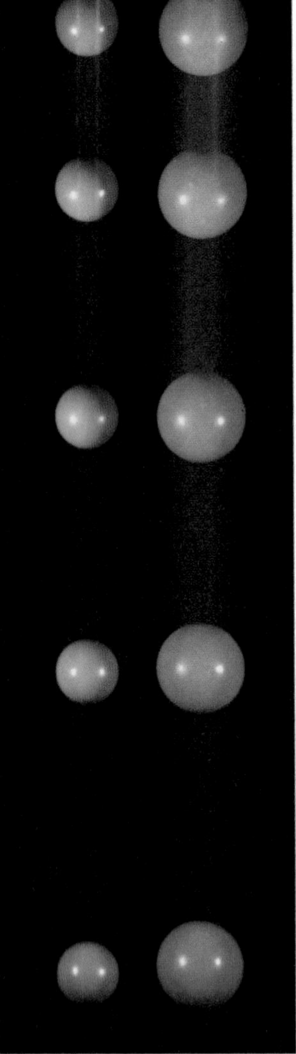

chapter 8 The Solar System—216

In each chapter, look for these opportunities for review and assessment:

- Reading Checks
- Caption Questions
- Section Review
- Chapter Study Guide
- Chapter Review
- Standardized Test Practice
- Online practice at **gpescience.com**

unit 3 Energy in Motion—250

chapter 9 Heat and States of Matter—252

chapter 10 Waves—286

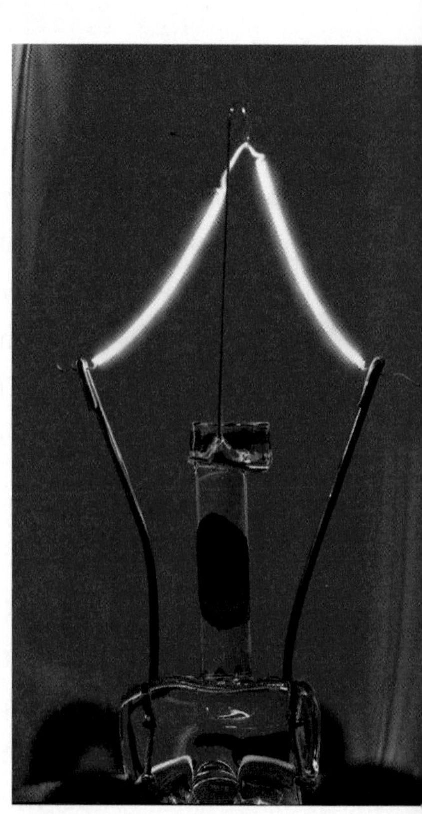

In each chapter, look for these opportunities for review and assessment:

- Reading Checks
- Caption Questions
- Section Review
- Chapter Study Guide
- Chapter Review
- Standardized Test Practice
- Online practice at gpescience.com

Matter and Earth Materials—548

Classification of Matter—550

Properties of Atoms and the Periodic Table—576

Earth Materials—606

In each chapter, look for these opportunities for review and assessment:

- Reading Checks
- Caption Questions
- Section Review
- Chapter Study Guide
- Chapter Review
- Standardized Test Practice
- Online practice at gpescience.com

Student Resources—848

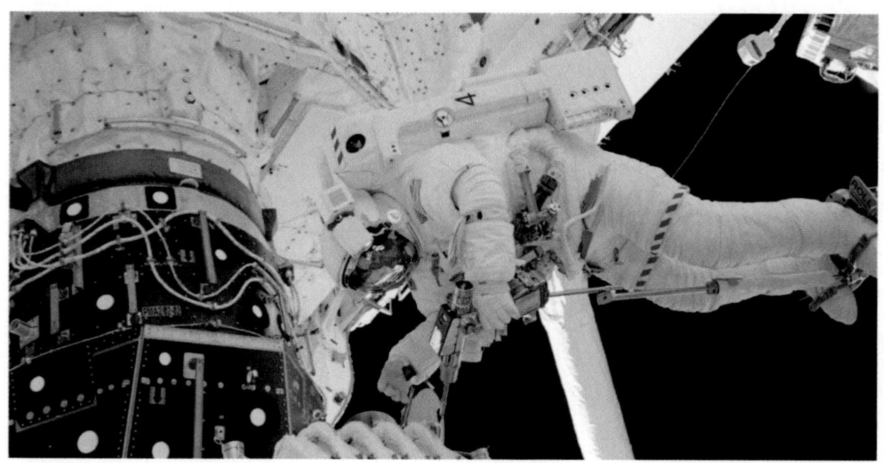

In each chapter, look for these opportunities for review and assessment:

- **Reading Checks**
- **Caption Questions**
- **Section Review**
- **Chapter Study Guide**
- **Chapter Review**
- **Standardized Test Practice**
- **Online practice at gpescience.com**

Cross-Curricular Readings

Content Details

NATIONAL GEOGRAPHIC Unit Openers

NATIONAL GEOGRAPHIC VISUALIZING

Cross-Curricular Readings

TIME Science and Society

TIME Science and History

Oops! Accidents in Science

Science and Language Arts

Science Stats

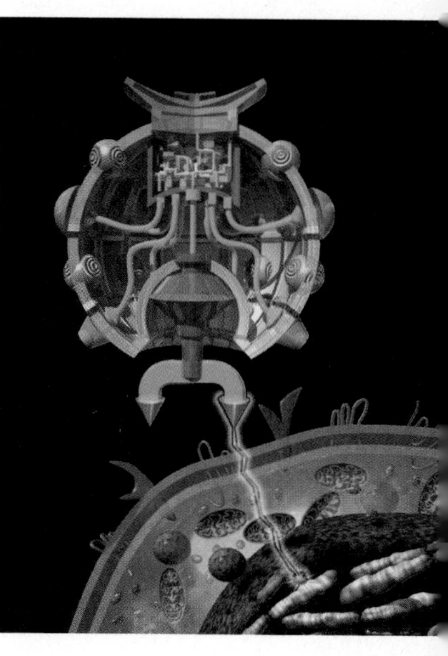

[video lab icon] available as a video lab

Mini LAB

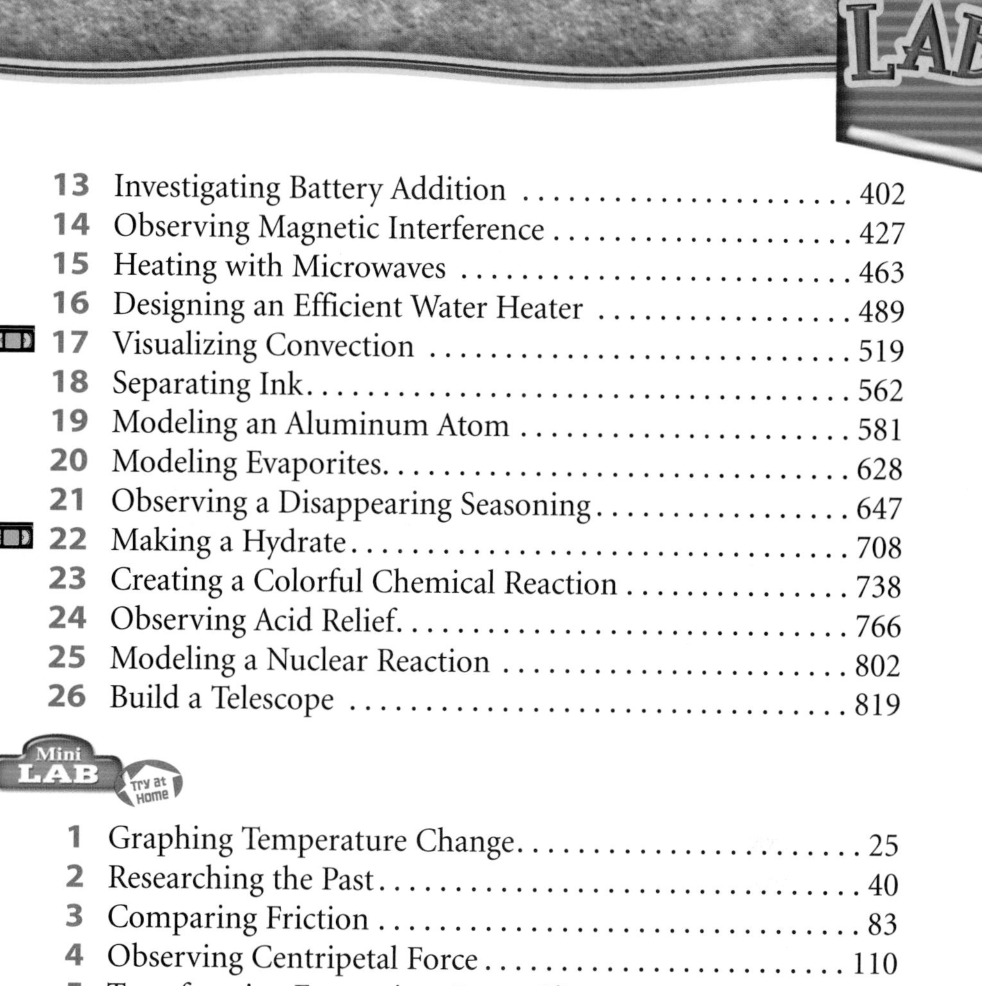

Content Details

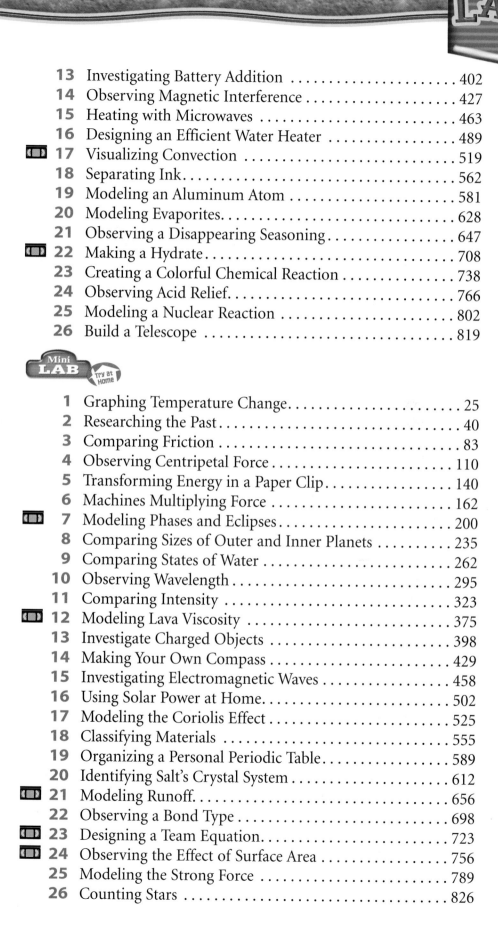

Mini LAB Try at Home

available as a video lab

Content Details

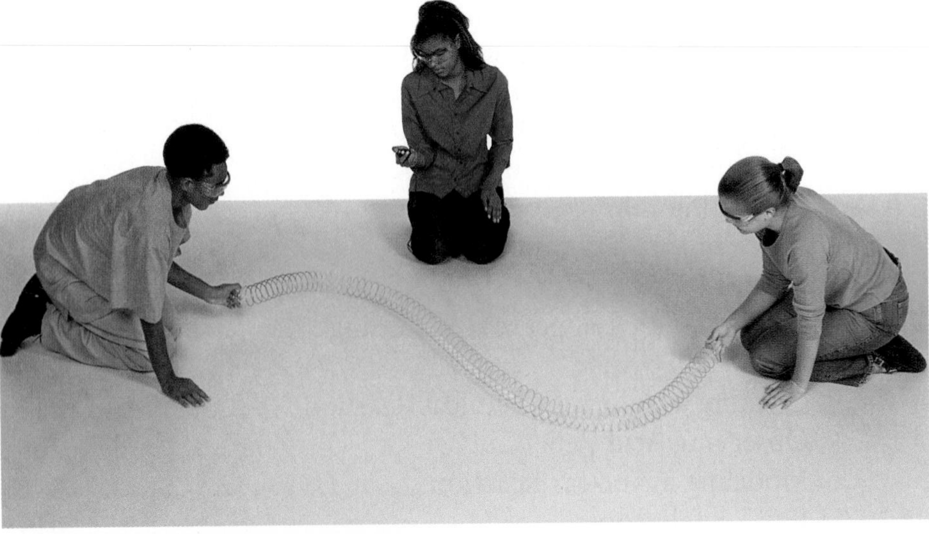

Traditional Labs

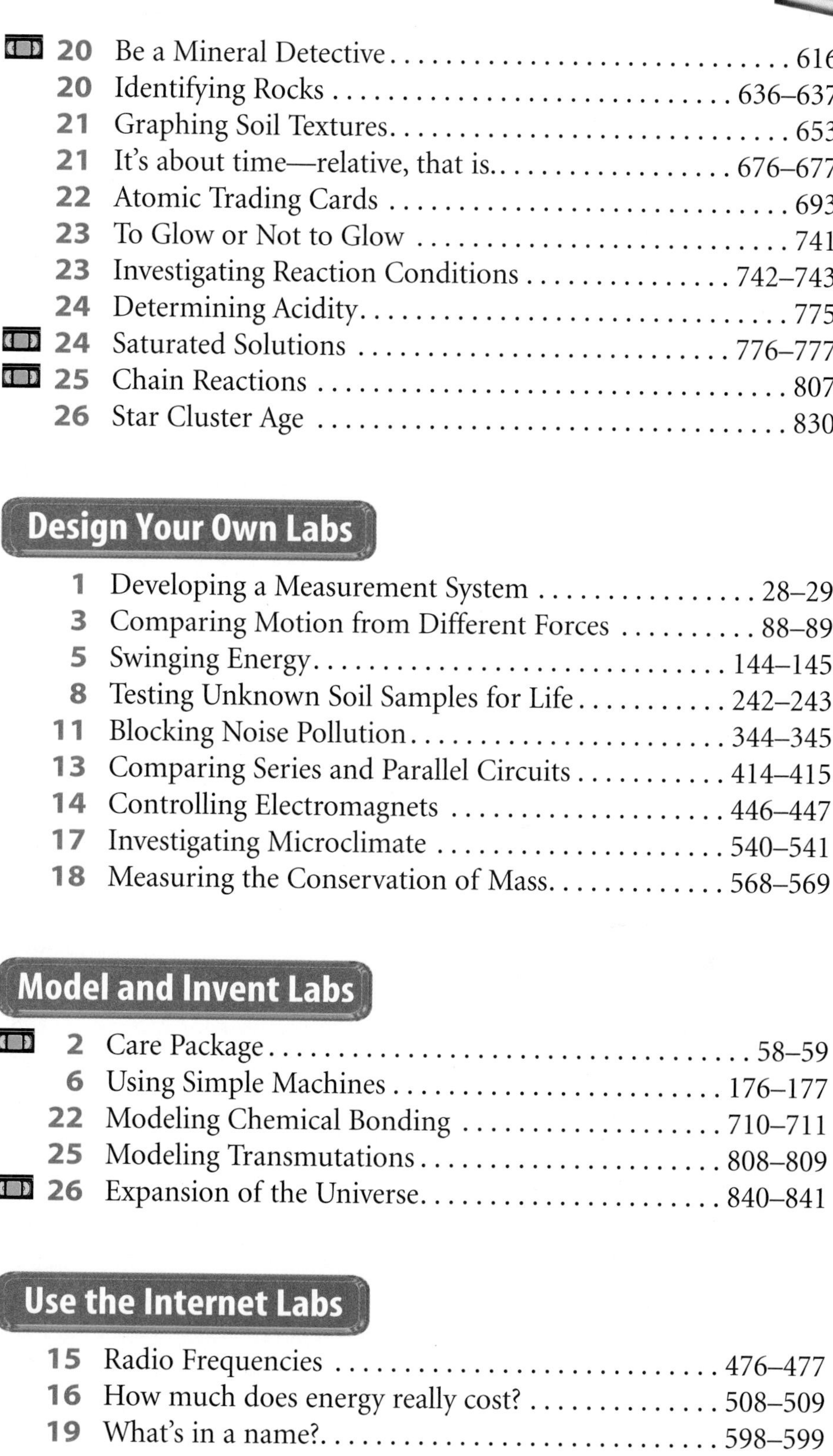

Design Your Own Labs

Model and Invent Labs

Use the Internet Labs

Activities

Content Details

Applying Math

Applying Science

INTEGRATE

Science Online

Standardized Test Practice

How Are Billiards & Bottles Connected?

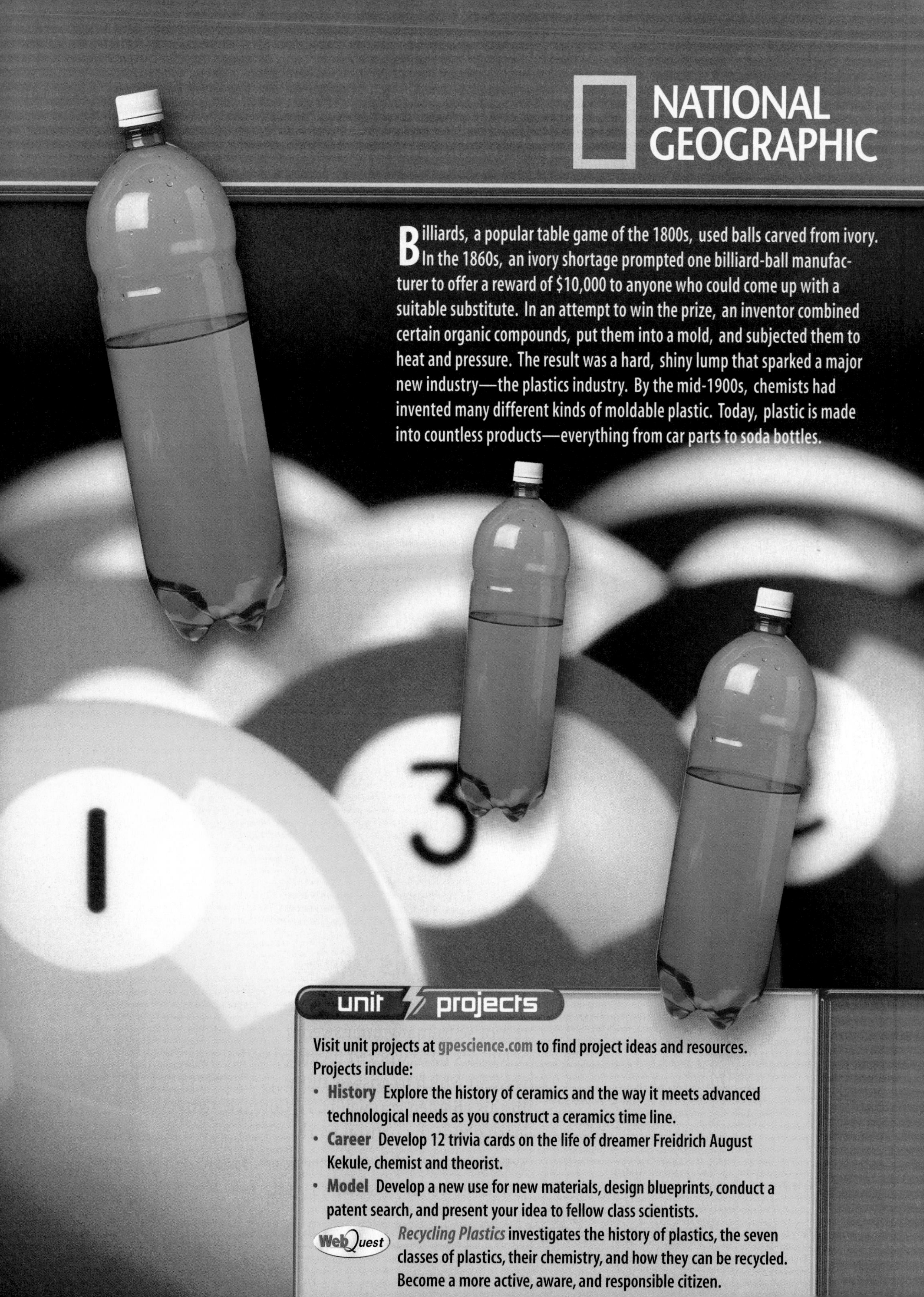

Billiards, a popular table game of the 1800s, used balls carved from ivory. In the 1860s, an ivory shortage prompted one billiard-ball manufacturer to offer a reward of $10,000 to anyone who could come up with a suitable substitute. In an attempt to win the prize, an inventor combined certain organic compounds, put them into a mold, and subjected them to heat and pressure. The result was a hard, shiny lump that sparked a major new industry—the plastics industry. By the mid-1900s, chemists had invented many different kinds of moldable plastic. Today, plastic is made into countless products—everything from car parts to soda bottles.

unit projects

Visit unit projects at gpescience.com to find project ideas and resources. Projects include:

- **History** Explore the history of ceramics and the way it meets advanced technological needs as you construct a ceramics time line.
- **Career** Develop 12 trivia cards on the life of dreamer Friedrich August Kekule, chemist and theorist.
- **Model** Develop a new use for new materials, design blueprints, conduct a patent search, and present your idea to fellow class scientists.

WebQuest *Recycling Plastics* investigates the history of plastics, the seven classes of plastics, their chemistry, and how they can be recycled. Become a more active, aware, and responsible citizen.

chapter 1

The Nature of Science

chapter preview

sections

1. The Methods of Science
2. Standards of Measurement
3. Communicating with Graphs

Lab *What's my graph?*

Lab *Developing a Measurement System*

Virtual Lab *How is a controlled experiment performed?*

Out of This World

The space program was developed in response to many unanswered questions. Scientists have worked together to develop ways in which to answer those questions. In this chapter, you will learn how scientists learn about the natural world.

Science Journal Look at the picture above. Write in your Science Journal why scientists study space.

Start-Up Activities

Understanding Measurements

Before there were measurement standards, people used parts of the body for measuring items. The length from the tip of the elbow to the end of the middle finger was a *cubit*. The *foot* was the length of a man's foot and was used to measure distance. How does the length of your classroom differ when measured using several students' feet?

1. Complete the safety form.
2. Estimate the distance in feet across the classroom. Record this number in your Science Journal.
3. Walk across your classroom by placing the heel of one foot against the toes of your other foot. Count and record the number of steps in your Science Journal.
4. Measure the lengths of your two feet and average them. Calculate the distance of the classroom by multiplying the average length of your feet by the number of steps.
5. Measure the distance across the classroom using a tape measure.
6. **Think Critically** Compare the distance across the classroom using your foot with the distance using other classmates' feet. How does your calculated measurement compare to your classmates' calculated measurements?

Scientific Processes Make the following Foldable to help identify what you already know, what you want to know, and what you learned about science.

STEP 1 **Fold** a sheet of paper vertically from side to side. Make the front edge about 1.25 cm shorter than the back edge.

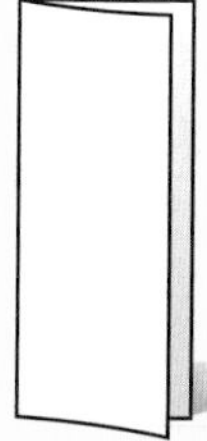

STEP 2 **Turn** lengthwise and **fold** into thirds.

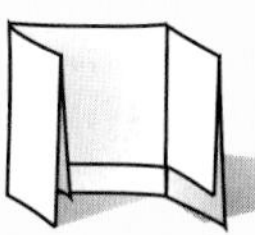

STEP 3 **Unfold and cut** only the top layer along both folds to make three tabs. **Label** each tab.

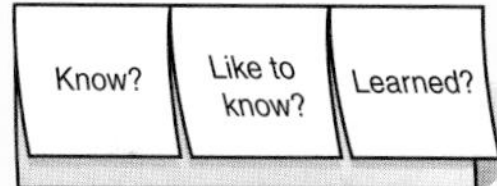

Identify Questions Before you read the chapter, write what you already know about science under the left tab of your Foldable, and write questions about what you'd like to know under the center tab. After you read the chapter, list what you learned under the right tab.

Preview this chapter's content and activities at gpescience.com

section 1

The Methods of Science

Reading Guide

What You'll Learn

- **Identify** the steps scientists often use to solve problems.
- **Describe** why scientists use variables.
- **Compare and contrast** science and technology.

Why It's Important

Using scientific methods will help you solve problems.

Review Vocabulary

investigation: to observe or study by close examination

New Vocabulary

- scientific method
- hypothesis
- experiment
- variable
- dependent variable
- independent variable
- constant
- control
- bias
- model
- theory
- scientific law
- technology

What is science?

Science is not just a subject in school. It is a method for studying the natural world. After all, science comes from the Latin word *scientia,* which means "knowledge." Science is a process that uses observation and investigation to gain knowledge about events in nature.

Nature follows a set of rules. Many rules, such as those concerning how the human body works, are complex. Other rules, such as the fact that Earth rotates about once every 24 h, are much simpler. Scientists ask questions to learn about the natural world.

Major Categories of Science Science covers many different topics that can be classified according to three main categories. (1) Life science deals with living things. (2) Earth science investigates Earth and space. (3) Physical science deals with matter and energy. In this textbook, you will study mainly physical science. Sometimes, though, a scientific study will overlap the categories. One scientist, for example, might study the motions of the human body to understand how to build better artificial limbs. Is this scientist studying energy and matter or how muscles operate? She is studying both life science and physical science. It is not always clear what kind of science you are using, as shown in **Figure 1.**

Figure 1 Astronaut Michael Lopez-Alegria uses a pistol-grip tool on the *International Space Station.*
Observe *What evidence do you see of the three main branches of science in the photograph?*

Science Explains Nature Scientific explanations help you understand the natural world. Sometimes these explanations must be modified. As more is learned about the natural world, some of the earlier explanations might be found to be incomplete or new technology might provide more accurate answers.

For example, look at **Figure 2.** In the late eighteenth century, most scientists thought that heat was an invisible fluid with no mass. Scientists observed that heat seems to flow like a fluid. It also moves away from a warm body in all directions, just as a fluid moves outward when you spill it on the floor.

However, the heat fluid idea did not explain everything. If heat was an actual fluid, an iron bar at a temperature of 1,000°C should have more mass than it does at 100°C because it would have more of the heat fluid in it. Eighteenth-century scientists thought they were not able to measure the small mass of the heat fluid on the balances they had. When additional investigations showed no difference in mass, scientists had to change the explanation.

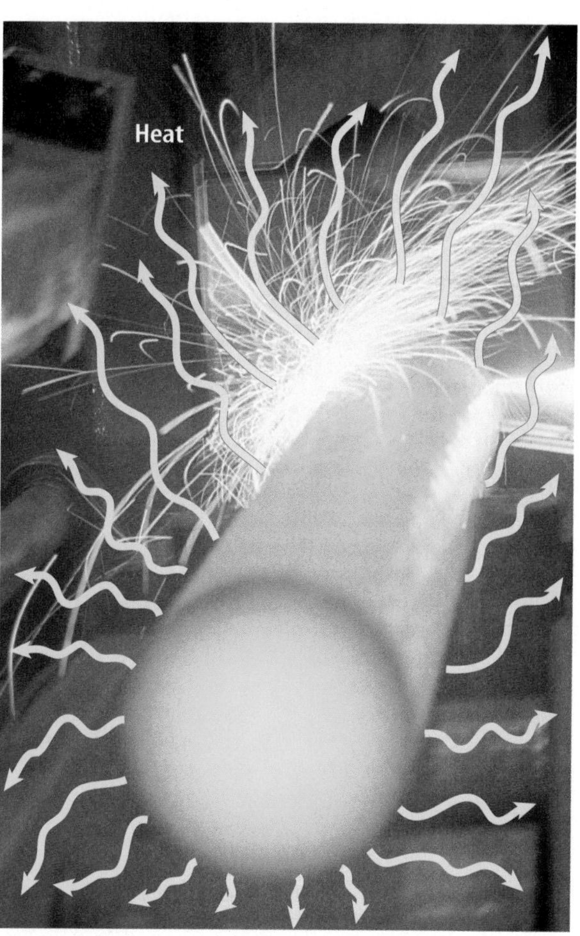

Figure 2 Many years ago, scientists thought that heat, such as in this metal rod, was a fluid.
Infer *how heat acts like a fluid.*

Investigations Scientists learn new information about the natural world by performing investigations, which can be done many different ways. Some investigations involve simply observing something that occurs and recording the observations, such as in a journal. Other investigations involve setting up experiments that test the effect of one thing on another. Some investigations involve building a model that resembles something in the natural world and then testing the model to see how it acts. Often, a scientist will use something from all three types of investigations when attempting to learn about the natural world.

Reading Check *Why do scientific explanations change?*

Scientific Methods

Although scientists do not always follow a rigid set of steps, investigations often follow a general pattern. An organized set of investigation procedures is called a **scientific method.** Six common steps found in scientific methods are shown in **Figure 3.** A scientist might add new steps, repeat some steps many times, or skip steps altogether when doing an investigation.

Topic: Prediction
Visit gpescience.com for Web links to information about why leaves change color in the autumn.

Activity Fill a glass with cold water and add a few drops of blue or red food coloring. Cut a piece of celery and place it in the glass. Over the next few days, observe what happens to the celery. Make a prediction about why this occurs. Support your answer with evidence.

Figure 3 The series of procedures shown here is one way to use scientific methods to solve a problem.
Explain *what should be done if your hypothesis is not supported.*

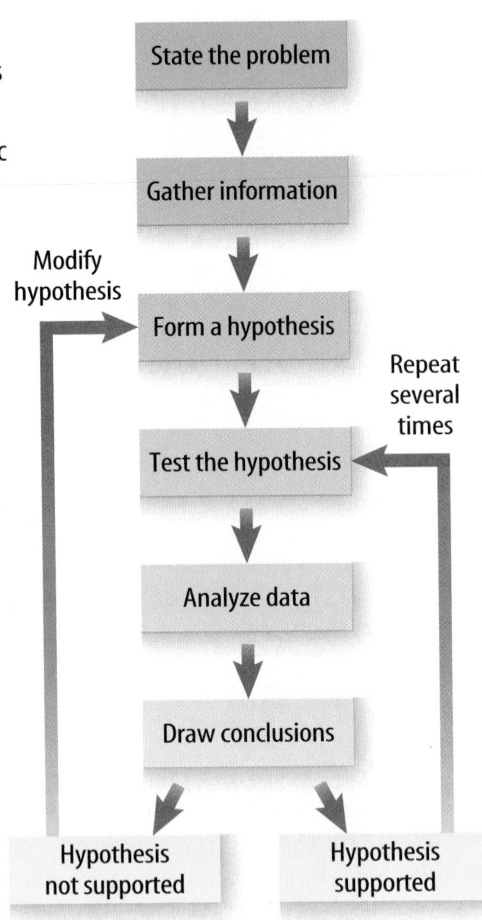

Stating a Problem Many scientific investigations begin when someone observes an event in nature and wonders why or how it occurs. Then the question of "why" or "how" is the problem. Sometimes a statement of a problem arises from an activity that is not working. Some early work on guided missiles showed that the instruments in the nose of the missiles did not always work. The problem statement involved finding a material to protect the instruments from the harsh conditions of flight.

Later, National Aeronautics and Space Administration (NASA) scientists made a similar problem statement. They wanted to build a new vehicle—the space shuttle—that could carry people to outer space and back again. Guided missiles did not have this capability. NASA needed to find a material for the outer surface of the space shuttle that could withstand the heat and forces of re-entry into Earth's atmosphere.

Researching and Gathering Information Before testing a hypothesis, it is useful to learn as much as possible about the background of the problem. Have others found information that will help determine what tests to do and what tests will not be helpful? The NASA scientists gathered information about melting points and other properties of the various materials that might be used. In many cases, tests had to be performed to learn the properties of new, recently created materials.

Forming a Hypothesis A **hypothesis** is a possible explanation for a problem using what you know and what you observe. NASA scientists knew that a ceramic coating had been found to solve the guided missile problem. They hypothesized that a ceramic material also might work on the space shuttle.

Testing a Hypothesis Some hypotheses can be tested by making observations. Others can be tested by building a model and relating it to real-life situations. One common way to test a hypothesis is to perform an experiment. An **experiment** tests the effect of one thing on another using controlled conditions.

Variables An experiment usually contains at least two variables. A **variable** is a factor that can cause a change in the results of an experiment. You might set up an experiment to determine the amount of fertilizer that will help plants grow the biggest. Before you begin your tests, you would need to think of all the factors that might cause the plants to grow bigger. Possible factors include plant type, amount of sunlight, amount of water, room temperature, type of soil, and type of fertilizer.

In this experiment, the amount of growth is the **dependent variable** because its value changes according to the changes in the other variables. The variable you change to see how it will affect the dependent variable is called the **independent variable.**

Constants and Controls To be sure you are testing to see how fertilizer affects growth, you must keep the other possible factors the same. A factor that does not change when other variables change is called a **constant.** You might set up one trial, using the same soil and type of plant. Each plant is given the same amount of sunlight and water and is kept at the same temperature. These are constants. Three of the plants receive a different amount of fertilizer, which is the independent variable.

The fourth plant is not fertilized. This plant is a control. A **control** is the standard by which the test results can be compared. Suppose after several days, the three fertilized plants grow between 2 and 3 cm. If the unfertilized plant grows 1.5 cm, you might infer that the growth of the fertilized plants was due to the fertilizers.

How might the NASA scientists set up an experiment to solve the problem of the damaged tiles shown in **Figure 4?** What are possible variables, constants, and controls?

Reading Check *Why is a control used in an experiment?*

Classification Systems Through observations of living organisms, Aristotle designed a classification system. Systems used today group organisms according to variables such as habits and physical and chemical features. Research to learn recent reclassifications of organisms. Share your findings with your class.

Figure 4 NASA has had an ongoing mission to improve the space shuttle. A technician is replacing tiles damaged upon re-entry into Earth's atmosphere.

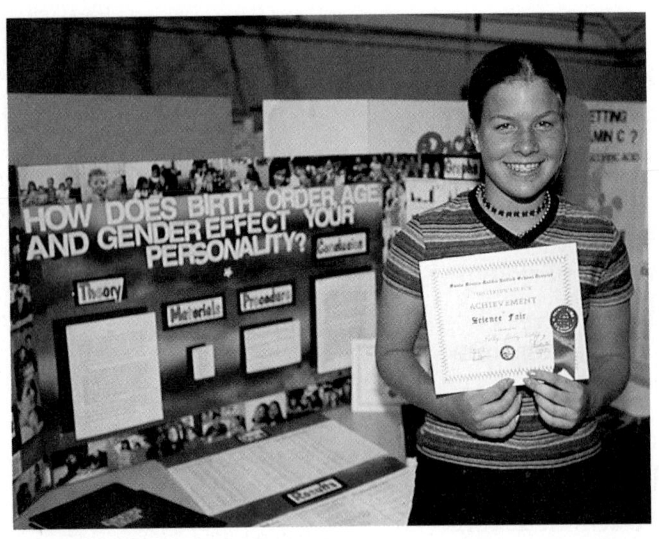

Figure 5 An exciting and important part of investigating something is sharing your ideas with others, as this student is doing at a science fair.
Identify *possible methods of how scientists can share their research data.*

Analyzing the Data An important part of every experiment includes recording observations and organizing the test data into easy-to-read tables and graphs. Later in this chapter you will study ways to display data. When you are making and recording observations, you should include all results, even unexpected ones. Many important discoveries have been made from unexpected occurrences.

Interpreting the data and analyzing the observations is an important step. If the data are not organized in a logical manner, wrong conclusions can be drawn. No matter how well a scientist communicates and shares that data, someone else might not agree with the data. Scientists share their data through reports and conferences. In **Figure 5,** a student is displaying her data.

Drawing Conclusions Based on the analysis of your data, you decide whether or not your hypothesis is supported. When lives are at stake, such as with the space shuttle, you must be very sure of your results. For the hypothesis to be considered valid and widely accepted, the experiment must result in the exact same data every time it is repeated. If your experiment does not support your hypothesis, you must reconsider the hypothesis. Perhaps it needs to be revised or your experiment needs to be conducted differently.

Being Objective Scientists also should be careful to reduce bias in their experiments. A **bias** occurs when what the scientist expects changes how the results are viewed. This expectation might cause a scientist to select a result from one trial over those from other trials. Bias also might be found if the advantages of a product being tested are used in a promotion and the drawbacks are not presented.

Scientists can lessen bias by running as many trials as possible and by keeping accurate notes of each observation made. Valid experiments also must have data that are measurable. For example, a scientist performing a global warming study must base his or her data on accurate measures of global temperature. This allows others to compare the results to data they obtain from a similar experiment. Most importantly, the experiment must be repeatable. Findings are supportable when other scientists perform the same experiment and get the same results.

Reading Check *What is bias in science?*

Visualizing with Models

Sometimes, scientists cannot see everything that they are testing. They might be observing something that is too large, too small, or takes too much time to see completely. In these cases, scientists use models. A **model** represents an idea, event, or object to help people better understand it.

Models in History Models have been used throughout history. One scientist, Lord Kelvin, who lived in England in the 1800s, was famous for making models. To model his idea of how light moves through space, he put balls into a bowl of jelly and encouraged people to move the balls around with their hands. Kelvin's work to explain the nature of temperature and heat still is used today.

High-Tech Models Scientific models don't always have to be something you can touch. Today, many scientists use computers to build models. NASA experiments involving space flight would not be practical without computers. The complex equations would take far too long to calculate by hand, and errors could be introduced much too easily.

Another type of model is a simulator, like the one shown in **Figure 6.** An airplane simulator enables pilots to practice problem solving with various situations and conditions they might encounter when in the air. This model will react the way a plane does when it flies. It gives pilots a safe way to test different reactions and to practice certain procedures before they fly a real plane.

Computer Models
Meteorology has changed greatly due to computer modeling. Using special computer programs, meteorologists now are able to more accurately predict disastrous weather. In your Science Journal, describe how computer models might help save lives.

Figure 6 Pilots and astronauts use flight simulators for training. **Explain** *how these models differ from actual airplanes and spacecraft.*

Topic: Archimedes' Principle

Visit gpescience.com for Web links to information about Archimedes' principle.

Activity Place a full soft-drink bottle, water bottle, or container of milk in a tub of water. What happens to the pop bottle or milk container? Would you classify Archimedes' principle as a scientific theory or scientific law?

Scientific Theories and Laws

A scientific **theory** is an explanation of things or events based on knowledge gained from many observations and investigations. It is not a guess. If scientists repeat an investigation and the results always support the hypothesis, the hypothesis can be called a theory. Just because a scientific theory has data supporting it does not mean it will never change. Recall that the theory about heat being a fluid was discarded after further experiments. As new information becomes available, theories can be modified. A theory accepted today might at some time in the future also be discarded.

A **scientific law** is a statement about what happens in nature and that seems to be true all the time. Laws tell you what will happen under certain conditions, but they don't explain why or how something happens. Gravity is an example of a scientific law. The law of gravity says that any one mass will attract another mass. To date, no experiments have been performed that disprove the law of gravity.

A theory can be used to explain a law. For example, many theories have been proposed to explain how the law of gravity works. Even so, there are few theories in science and even fewer laws.

What is the difference between a scientific theory and a scientific law?

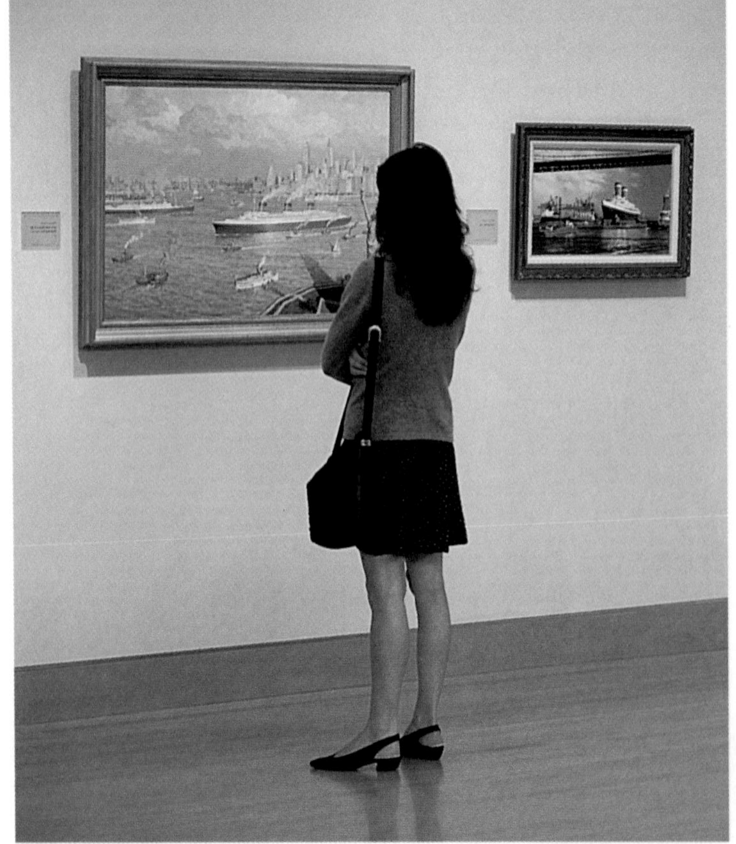

Figure 7 Science can't answer all questions.
Analyze *Can anyone prove that you like artwork? Explain.*

The Limitations of Science

Science can help you explain many things about the world, but science cannot explain or solve everything. Although it's the scientist's job to make hypotheses, the scientist also has to make sure his or her hypotheses can be tested and verified. How do you prove that people will like a play or a piece of music? You cannot and science cannot.

Most questions about emotions and values are not scientific questions. They cannot be tested. You might take a survey to get people's opinions about such questions, but that would not prove that the opinions are true for everyone. A survey might predict that you will like the art in **Figure 7,** but science cannot prove that you or others will.

Using Science—Technology

Many people use the terms *science* and *technology* interchangeably, but they are not the same. **Technology** is the application of science to help people. For example, when a chemist develops a new, lightweight material that can withstand great amounts of heat, science is used. When that material is used on the space shuttle, technology is applied. **Figure 8** shows other examples of technology.

Technology doesn't always follow science, however. Sometimes the process of discovery can be reversed. One important historic example of science following technology is the development of the steam engine. The inventors of the steam engine had little idea of how it worked. They just knew that steam from boiling water could move the engine. Because the steam engine became so important to industry, scientists began analyzing how it worked. Lord Kelvin, James Prescott Joule, and Sadi Carnot, who lived in the 1800s, learned so much from the steam engine that they developed revolutionary ideas about the nature of heat.

Science and technology do not always produce positive results. The benefits of some technological advances, such as nuclear technology and genetic engineering, are subjects of debate. Being more knowledgeable about science can help society address these issues as they arise.

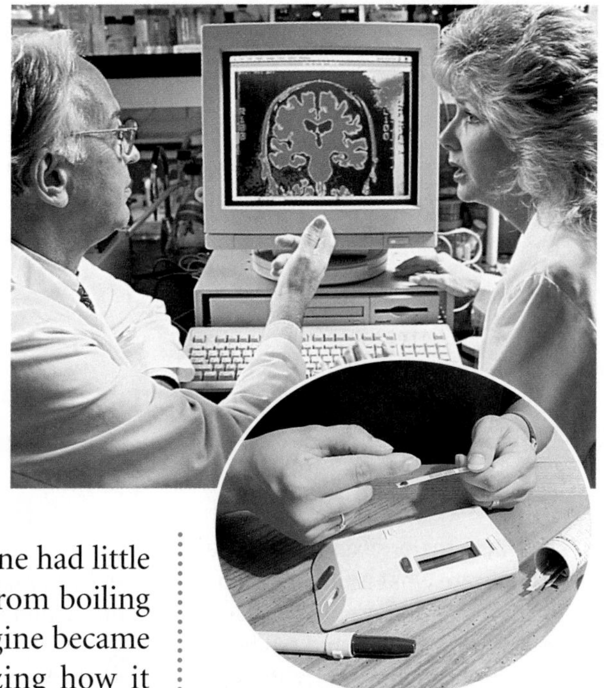

Figure 8 Technology is the application of science.
Identify *the type of science (life, Earth, or physical) that is applied in these examples of technology.*

section 1 review

Summary

What is science?

- Scientists ask questions and perform investigations to learn more about the natural world.

Scientific Methods

- Scientists perform the six-step scientific method to test their hypotheses.

Visualizing with Models

- Models help scientists visualize concepts.

Scientific Theories and Laws

- A theory is a possible explanation for observations, while a scientific law describes a pattern but does not explain why things happen.

Using Science—Technology

- Technology is the application of science in our everyday lives.

Self Check

1. **Define** the first step a scientist usually takes to solve a problem.
2. **Explain** why a control is needed in a valid experiment.
3. **Think Critically** What is the dependent variable in an experiment that shows how the volume of gas changes with changes in temperature?

Applying Math

4. **Find the Average** You perform an experiment to determine how many breaths a fish takes per minute. Your experiment yields the following data: minute 1: 65 breaths; minute 2: 73 breaths; minute 3: 67 breaths; minute 4: 71 breaths; minute 5: 62 breaths. Calculate the average number of breaths that the fish takes per minute.

More Section Review gpescience.com

section 2

Standards of Measurement

Reading Guide

What You'll Learn

- **Name** the prefixes used in SI and indicate what multiple of ten each one represents.
- **Identify** SI units and symbols for length, volume, mass, density, time, and temperature.
- **Convert** related SI units.

Why It's Important

By using uniform standards, nations can exchange goods and compare information easily.

Review Vocabulary

measurement: the dimensions, capacity, or amount of something

New Vocabulary

- precision
- accuracy
- volume
- mass
- density

Units and Standards

A standard is an exact quantity that people agree to use to compare measurements. Look at **Figure 9.** Suppose you and a friend want to make some measurements to find out whether a desk will fit through a doorway. You have no ruler, so you decide to use your hands as measuring tools. Using the width of his hands, your friend measures the doorway and says it is 8 hands wide. Using the width of your hands, you measure the desk and find it is $7\frac{3}{4}$ hands wide. Will the desk fit through the doorway? You can't be sure. What went wrong? Even though you both used hands to measure, you didn't check to see whether your hands were the same width as your friend's.

Figure 9 Hands are a convenient measuring tool, but using them can lead to misunderstanding.

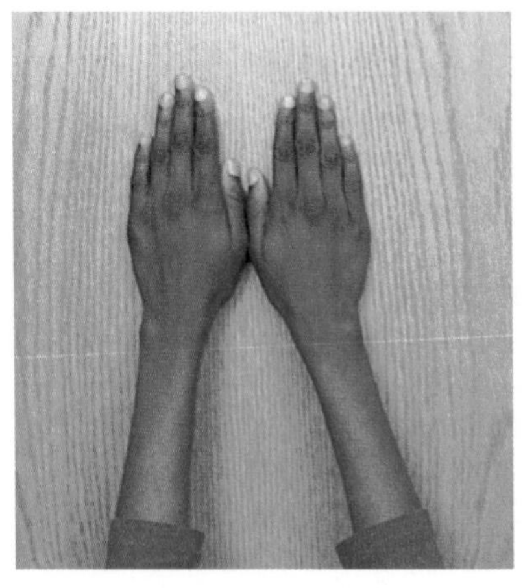

Precision and Accuracy

You are watching an archery event. The first person shoots five bull's-eyes in a row. The second person does not hit the bull's-eye at all, but the arrows all are in a similar location. What can be said about these two participants? The first person's aim was both precise and accurate. The second person's aim was only precise. **Precision** describes how closely measurements are to each other and how carefully measurements were made. **Accuracy** compares a measurement to the real or accepted value. When taking measurements, it is important to be precise and accurate.

International System of Units

In 1960, an improved version of the metric system was devised. Known as the International System of Units, this system is often abbreviated SI, from the French *Le Systeme Internationale d'Unites.* SI is an improved, accepted version of the metric system that is based on multiples of ten. It is understood by scientists throughout the world. The standard kilogram, which is kept in Sèvres, France, is shown in **Figure 10.** All kilograms used throughout the world must be exactly the same as the kilogram kept in France.

Each type of SI measurement has a base unit. The meter is the base unit of length. Every type of quantity measured in SI has a symbol for that unit. These names and symbols for the seven base units are shown in **Table 1.** All other SI units are obtained from these seven units.

Figure 10 The standard for mass, the kilogram, and other standards are kept at the International Bureau of Weights and Measures in Sèvres, France.
Explain *the purpose of a standard.*

SI Prefixes The SI system is easy to use because it is based on multiples of ten. Prefixes are used with the names of the units to indicate what multiple of ten should be used with the units. For example, the prefix *kilo-* means "1,000." This means that one kilometer equals 1,000 meters. Likewise, one kilogram equals 1,000 grams. Because *deci-* means "one-tenth," one decimeter equals one-tenth of a meter. A decigram equals one-tenth of a gram. The most frequently used prefixes are shown in **Table 2.**

Reading Check *How many meters are in 1 km? How many grams are in 1 dg?*

Table 1 SI Base Units

Quantity Measured	Unit	Symbol
Length	meter	m
Mass	kilogram	kg
Time	second	s
Electric current	ampere	A
Temperature	kelvin	K
Amount of substance	mole	mol
Intensity of light	candela	cd

Table 2 Common SI Prefixes

Prefix	Symbol	Multiplying Factor
Kilo-	k	1,000
Deci-	d	0.1
Centi-	c	0.01
Milli-	m	0.001
Micro-	μ	0.000 001
Nano-	n	0.000 000 001

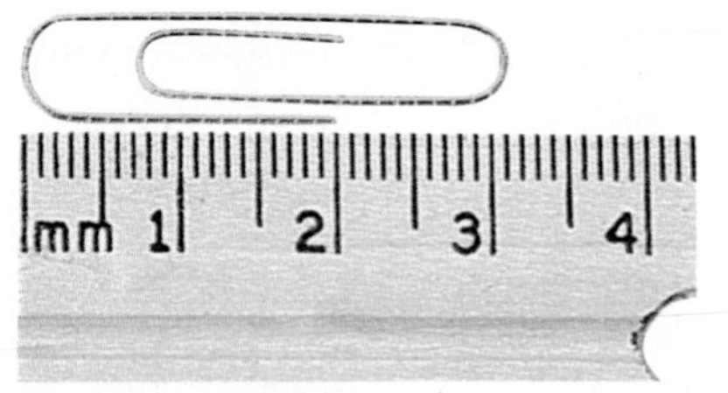

Figure 11 One centimeter contains 10 mm.
Determine *the length of the paper clip in centimeters and in millimeters.*

Converting Between SI Units Sometimes quantities are measured using different units, as shown in **Figure 11.** A conversion factor is a ratio that is equal to one and is used to change one unit to another. For example, there are 1,000 mL in 1 L, so 1,000 mL = 1 L. If both sides in this equation are divided by 1 L, the equation becomes:

$$\frac{1{,}000 \text{ mL}}{1 \text{ L}} = 1$$

To convert units, you multiply by the appropriate conversion factor. For example, to convert 1.255 L to mL, multiply 1.255 L by a conversion factor. Use the conversion factor with new units (mL) in the numerator and the old units (L) in the denominator.

$$1.255 \cancel{\text{L}} \times \frac{1{,}000 \text{ mL}}{1 \cancel{\text{L}}} = 1{,}255 \text{ mL}$$

Applying Math — Convert Units

CENTIMETERS How long in centimeters is a 3,075 mm rope?

IDENTIFY known values and the unknown value

Identify the known values:

The rope measures 3,075 mm; 1 m = 100 cm = 1,000 mm

Identify the unknown value:

How long is the rope in cm?

SOLVE the problem

This is the equation you need to use:

$$? \text{ cm} = 3{,}075 \text{ mm} \times \frac{100 \text{ cm}}{1{,}000 \text{ mm}}$$

Cancel units and multiply:

$$3{,}075 \cancel{\text{mm}} \times \frac{1\cancel{00} \text{ cm}}{1{,}0\cancel{00} \cancel{\text{mm}}} = 307.5 \text{ cm}$$

CHECK your answer

Does your answer seem reasonable? Check your answer by multiplying the answer by $\frac{1{,}000 \text{ mm}}{100 \text{ cm}}$. Did you calculate the original length in millimeters?

Practice Problems

1. Your pencil is 11 cm long. How long is it in millimeters?
2. The Bering Land Bridge National Preserve is a summer home to birds. Some birds migrate 20,000 miles. Assume 1 mile equals 1.6 kilometers. Calculate the distance birds fly in kilometers.

For more practice problems, go to page 879 and visit Math Practice at gpescience.com.

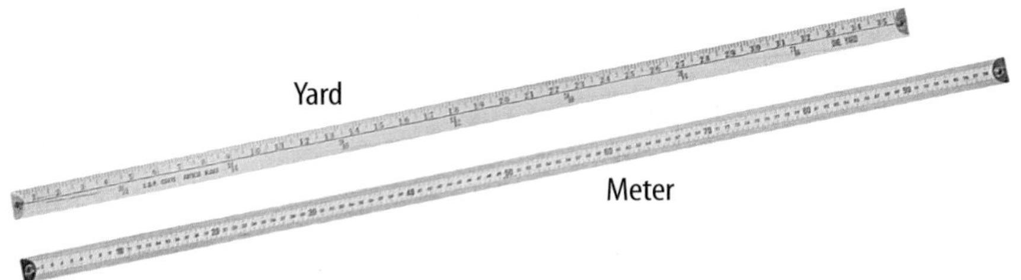

Figure 12 One meter is slightly longer than 1 yard, and 100 m is slightly longer than a football field. **Predict** *whether your time for a 100-m dash would be slightly more or less than your time for a 100-yard dash.*

Measuring Distance

The word *length* is used in many different ways. For example, the length of a novel is the number of pages or words it contains. In scientific measurement, length is the distance between two points. That distance might be the diameter of a hair or the distance from Earth to the Moon. The SI base unit of length is the meter, m. A baseball bat is about 1 m long. Metric rulers and metersticks are used to measure length. **Figure 12** compares a meter and a yard.

INTEGRATE Earth Science

Astronomical Units The standard measurement for the distance from Earth to the Sun is called the astronomical unit, AU. The distance is about 150 billion (1.5×10^{11}) m. In your Science Journal, calculate what 1 AU would equal in kilometers.

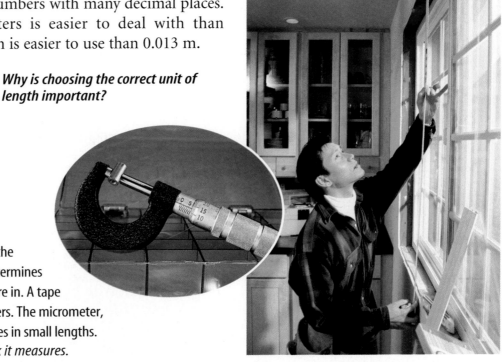

Choosing a Unit of Length As shown in **Figure 13,** the size of the unit you measure with will depend on the size of the object being measured. For example, the diameter of a shirt button is about 1 cm. You probably also would use the centimeter to measure the length of your pencil and the meter to measure the length of your classroom. What unit would you use to measure the distance from your home to school? You probably would want to use a unit larger than a meter. The kilometer, km, which is 1,000 m, is used to measure these kinds of distances.

By choosing an appropriate unit, you avoid large-digit numbers and numbers with many decimal places. Twenty-one kilometers is easier to deal with than 21,000 m, and 13 mm is easier to use than 0.013 m.

Reading Check ***Why is choosing the correct unit of length important?***

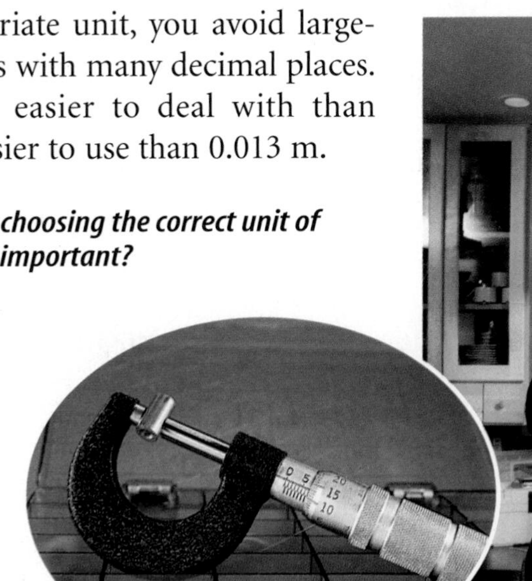

Figure 13 The size of the object being measured determines which unit you will measure in. A tape measure measures in meters. The micrometer, shown on the left, measures in small lengths. **State** *what unit you think it measures.*

Measuring Volume

The amount of space occupied by an object is called its **volume.** If you want to know the volume of a solid rectangle, such as a brick, you measure its length, width, and height and multiply the three numbers and their units together ($V = l \times w \times h$). For a brick, your measurements probably would be in centimeters. The volume would then be expressed in cubic centimeters, cm^3. To find out how much a moving van can carry, your measurements probably would be in meters, and the volume would be expressed in cubic meters, m^3, because when you multiply, you add exponents.

Topic: International System of Units

Visit gpescience.com for Web links to information about the International System of Units.

Activity Measure four different things in your classroom using a different unit—distance, volume, time, or temperature—for each. Use each unit only once. Write the measurements down and using the common prefixes in **Table 2,** convert them to different units.

Measuring Liquid Volume

How do you measure the volume of a liquid? A liquid has no sides to measure. In measuring a liquid's volume, you are indicating the capacity of the container that holds that amount of liquid. The most common units for expressing liquid volumes are liters and milliliters. These are measurements used in canned and bottled foods. A liter occupies the same volume as a cubic decimeter, dm^3. A cubic decimeter is a cube that is 1 dm, or 10 cm, on each side, as in **Figure 14.**

Look at **Figure 14.** One liter is equal to 1,000 mL. A cubic decimeter, dm^3, is equal to 1,000 cm^3. Because 1 L = 1 dm^3, it follows that

$$1 \text{ mL} = 1 \text{ cm}^3$$

Sometimes, liquid volumes such as doses of medicine are expressed in cubic centimeters.

Suppose you wanted to convert a measurement in liters to cubic centimeters. You would use conversion factors to convert L to mL and then mL to cm^3.

$$1.5 \cancel{\text{L}} \times \frac{1{,}000 \cancel{\text{mL}}}{1 \cancel{\text{L}}} \times \frac{1 \text{ cm}^3}{1 \cancel{\text{mL}}} = 1{,}500 \text{ cm}^3$$

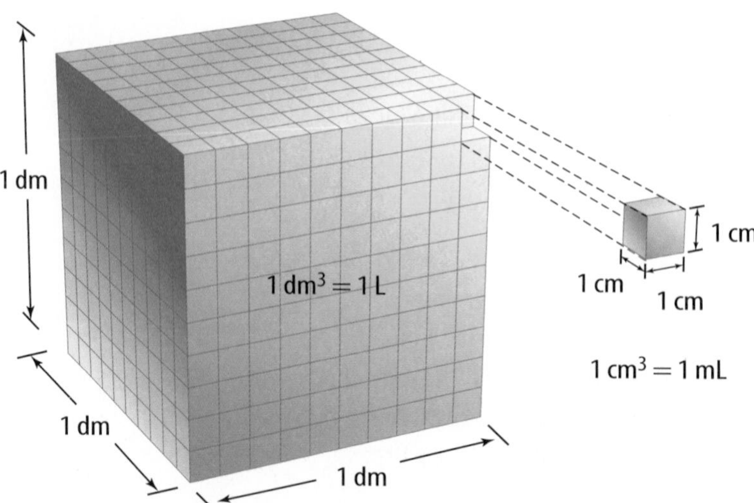

Figure 14 The large cube has a volume of 1 dm^3, which is equivalent to 1 L. **Calculate** *the cubic centimeters (cm^3) in the large cube.*

Table 3 Densities of Some Materials at 20°C

Material	Density (g/cm^3)	Material	Density (g/cm^3)
Hydrogen	0.000 09	Aluminum	2.7
Oxygen	0.001 4	Iron	7.9
Water	1.0	Gold	19.3

Measuring Matter

A table-tennis ball and a golf ball have about the same volume. But if you pick them up, you will notice a difference. The golf ball has more mass. **Mass** is a measurement of the quantity of matter in an object. The mass of the golf ball, which is about 45 g, is almost 18 times the mass of the table-tennis ball, which is about 2.5 g. To visualize SI units, see **Figure 15** on the following page.

Density A cube of polished aluminum and a cube of silver that are the same size not only look similar but also have the same volume. The mass and volume of an object can be used to find the density of the material the object is made of. **Density** is the mass per unit volume of a material. You find density by dividing an object's mass by the object's volume. For example, the density of an object having a mass of 10 g and a volume of 2 cm^3 is 5 g/cm^3. **Table 3** lists the densities of some familiar materials.

Reading Check *How is density determined?*

Derived Units The measurement unit for density, g/cm^3, is a combination of SI units. A unit obtained by combining different SI units is called a derived unit. An SI unit multiplied by itself also is a derived unit. Thus the liter, which is based on the cubic decimeter, is a derived unit. A meter cubed, expressed with an exponent—m^3—is a derived unit.

Measuring Time and Temperature

It is often necessary to keep track of how long it takes for something to happen, or whether something heats up or cools down. These measurements involve time and temperature.

Time is the interval between two events. The SI unit for time is the second. In the laboratory, you will use a stopwatch or a clock with a second hand to measure time.

Determining the Density of a Pencil

Procedure

1. Complete the safety form.
2. Find a **pencil** that will fit in a **100-mL graduated cylinder** below the 90-mL mark.
3. Measure the mass of the pencil in grams.
4. Put 90 mL of **water** (initial volume) into the 100-mL graduated cylinder. Lower the pencil, eraser first, into the cylinder. Push the pencil down until it is just submerged. Hold it there and record the final volume to the nearest tenth of a milliliter.

Analysis

1. Determine the water displaced by the pencil by subtracting the initial volume from the final volume.
2. Calculate the pencil's density by dividing its mass by the volume of water displaced.
3. Is the density of the pencil greater than or less than the density of water? How do you know?

NATIONAL GEOGRAPHIC

VISUALIZING SI DIMENSIONS

Figure 15

The characteristics of most of these everyday objects are measured using an international system known as SI dimensions. These dimensions measure length, volume, mass, density, and time. Celsius is not an SI unit but is widely used in scientific work.

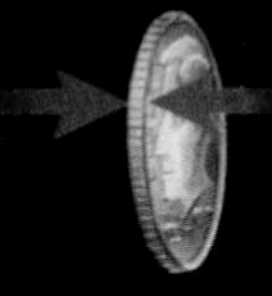

MILLIMETERS A dime is about 1 mm thick.

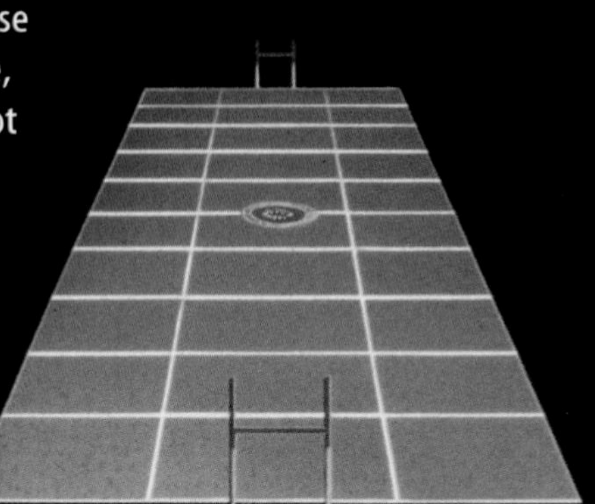

METERS A football field is about 91 m long.

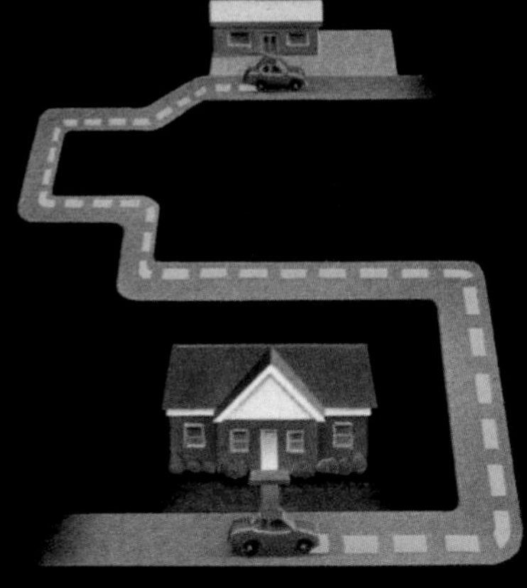

KILOMETERS The distance from your house to a store can be measured in kilometers.

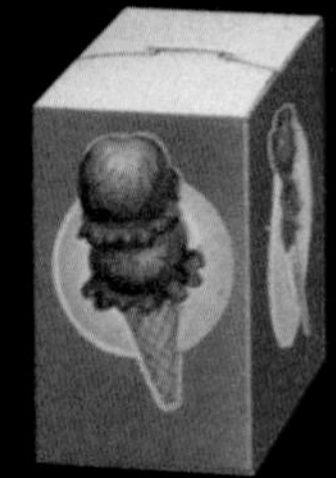

LITERS This carton holds 1.98 L of frozen yogurt.

MILLILITERS A teaspoonful of medicine is about 5 mL.

GRAMS/METER This stone sinks because it is denser—has more grams per cubic meter—than water.

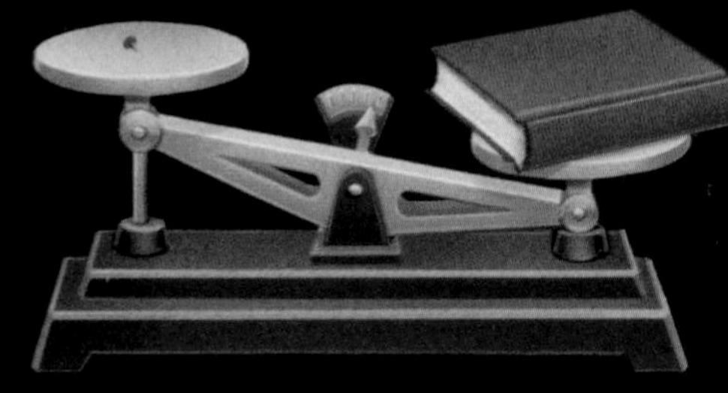

GRAMS The mass of a thumbtack and the mass of a textbook can be expressed in grams.

METERS/SECOND The speed of a roller-coaster car can be measured in meters per second.

CELSIUS Water boils at 100°C and freezes at 0°C.

What's Hot and What's Not You will learn the scientific meaning of the word *temperature* in a later chapter. For now, think of temperature as a measure of how hot or how cold something is.

Look at **Figure 16.** For most scientific work, temperature is measured on the Celsius (C) scale. On this scale, the freezing point of water is 0°C, and the boiling point of water is 100°C. Between these points, the scale is divided into 100 equal divisions. Each one represents 1°C. On the Celsius scale, the average human body temperature is 37°C, and a typical room temperature is between 20°C and 25°C.

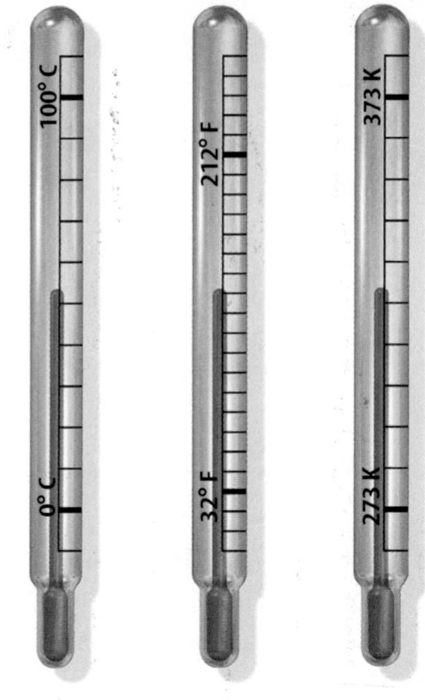

Figure 16 These three thermometers illustrate the scales of temperature between the freezing and boiling points of water. **Compare** *the boiling points of the three scales.*

Kelvin and Fahrenheit The SI unit of temperature is the kelvin (K). Zero on the Kelvin scale (0 K) is the coldest possible temperature, also known as absolute zero. Absolute zero is equal to −273°C, which is 273° below the freezing point of water.

Most laboratory thermometers are marked only with the Celsius scale. Because the divisions on the two scales are the same size, the Kelvin temperature can be found by adding 273 to the Celsius reading. So, on the Kelvin scale, water freezes at 273 K and boils at 373 K. Notice that degree symbols are not used with the Kelvin scale.

The temperature measurement you are probably most familiar with is the Fahrenheit scale, which was based roughly on the temperature of the human body, 98.6°.

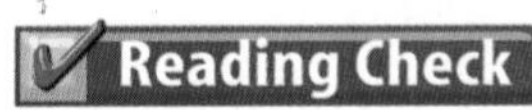

What is the relationship between the Celsius scale and the Kelvin scale?

section 2 review

Summary

Precision and Accuracy

- Precision is the description of how close measurements are to each other.

International System of Units

- The International System of Units, or SI, was established to provide a standard of measurement and reduce confusion.
- Conversion factors are used to change one unit to another and involve using a ratio equal to 1.

Measuring

- The size of an object being measured determines which unit you will measure in.

Self Check

1. **Explain** why it is important to have exact standards of measurement.
2. **Explain** why density is a derived unit.
3. **Think Critically** Using a metric ruler, measure a shoe box and a pad of paper. Find the volume of each in cubic centimeters. Then convert the units to mL.

Applying Math

4. **Convert Units** Make the following conversions: 27°C to Kelvin, 20 dg to milligrams, and 3 m to decimeters.
5. **Calculate Density** What is the density of an unknown metal that has a mass of 158 g and a volume of 20 mL? Use **Table 3** to identify this metal.

section 3

Communicating with Graphs

Reading Guide

What You'll Learn

- **Identify** three types of graphs and explain the ways they are used.
- **Distinguish** between dependent and independent variables.
- **Analyze** data using the various types of graphs.

Why It's Important

Graphs are a quick way to communicate a lot of information in a small amount of space.

Review Vocabulary

data: information gathered during an investigation or observation

New Vocabulary

- graph

A Visual Display

Scientists often graph the results of their experiments because they can detect patterns in the data easier in a graph than in a table. A **graph** is a visual display of information or data. **Figure 17** is a graph that shows a girl walking her dog. The horizontal axis, or the *x*-axis, measures time. Time is the independent variable because as it changes, it affects the measure of another variable. The distance from home that the girl and the dog walk is the other variable. It is the dependent variable and is measured on the vertical axis, or *y*-axis.

Figure 17 This graph tells the story of the motion that takes place when a girl takes her dog for an 8-min walk.

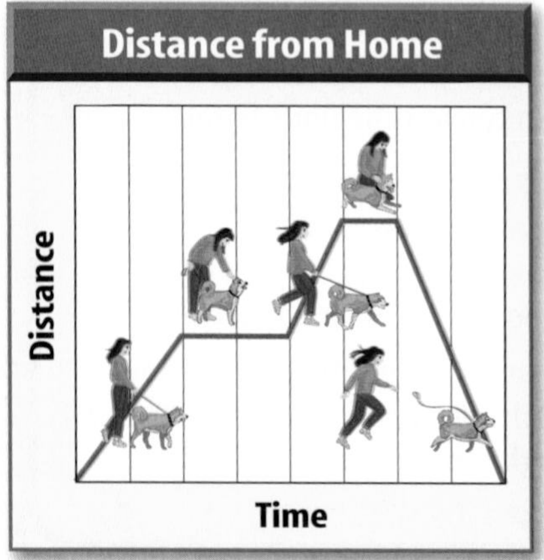

Graphs are useful for displaying numerical information in business, science, sports, advertising, and many everyday situations. Different kinds of graphs—line, bar, and circle—are appropriate for displaying different types of information.

Reading Check *What are three common types of graphs?*

Business people, as well as scientists, need an organized method to display data. Graphs make it easier to understand patterns by displaying data in a visual manner. Scientists often graph their data to detect patterns that would not have been evident in a table. Business people might graph sales dollars to determine trends. Different graphs display information by different methods. The conclusions drawn from graphs must be based on accurate information.

Line Graphs

A line graph can show any relationship where the dependent variable changes due to a change in the independent variable. Line graphs often show how a relationship between variables changes over time. You can use a line graph to track many things, such as how certain stocks perform or how the population changes over any period of time—a month, a week, or a year.

You can show more than one event on the same graph as long as the relationship between the variables is identical. Suppose a builder had three choices of thermostats for a new school. He wanted to test them to know which was the best brand to install throughout the building. He installed a different thermostat in classrooms A, B, and C. He set each thermostat at 20°C. He turned the furnace on and checked the temperatures in the three rooms every 5 min for 25 min. He recorded his data in **Table 4.**

Table 4 Room Temperature

Time*	Classroom Temperature (C°)		
	A	B	C
0	16	16	16
5	17	17	16.5
10	19	19	17
15	20	21	17.5
20	20	23	18
25	20	25	18.5

*minutes after turning on heat

The builder then plotted the data on a graph. He could see from the table that the data did not vary much for the three classrooms. So he chose small intervals for the *y*-axis and left part of the scale out (the part between 0° and 15°), as shown in **Figure 18.** This allowed him to spread out the area on the graph where the data points lie. You easily can see the contrast in the colors of the three lines and their relationship to the black horizontal line. The black line represents the thermostat setting and is the control. The control is what the resulting room temperature of the classrooms should be if the thermostats are working efficiently.

Figure 18 The room temperatures of classrooms A, B, and C are shown in contrast to the thermostat setting of 20°C.
Identify *the thermostat that achieved its temperature setting the quickest.*

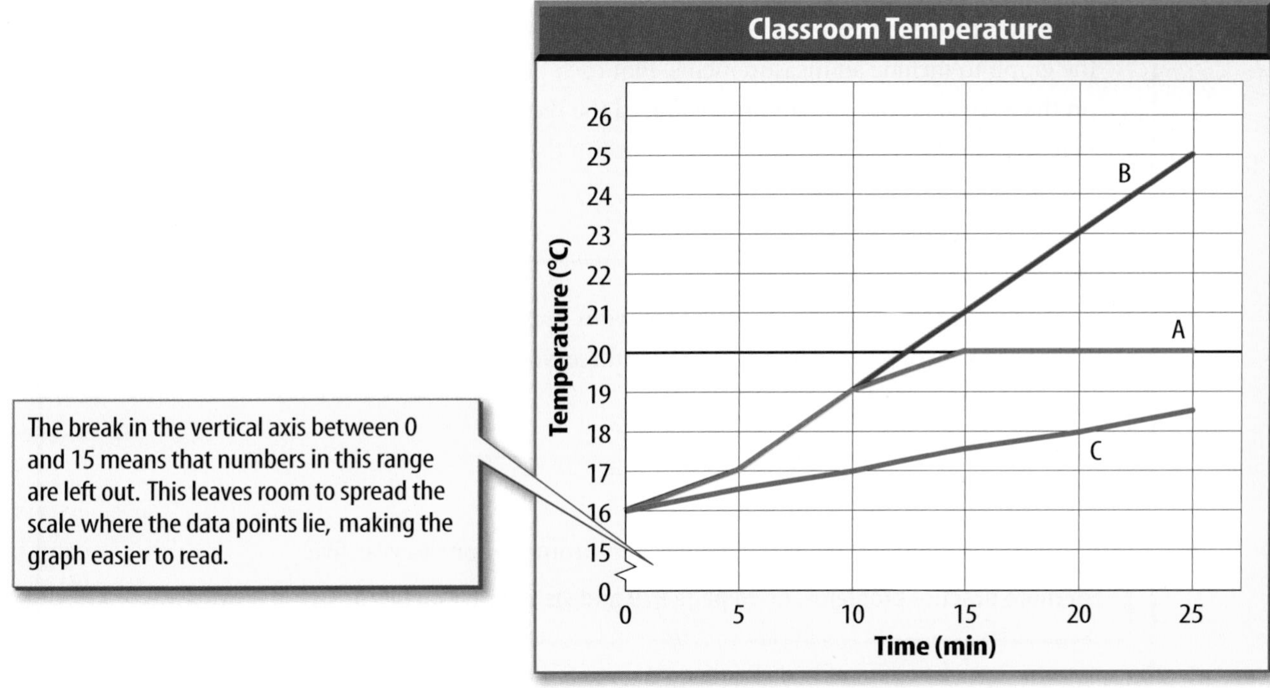

Figure 19 Graphing calculators are valuable tools for making graphs.

Constructing Line Graphs Besides choosing a scale that makes a graph readable, as illustrated in **Figure 18,** other factors are involved in constructing useful graphs. The most important factor in making a line graph is always using the *x*-axis for the independent variable. The *y*-axis always is used for the dependent variable. Because the points in a line graph are related, you connect the points.

Another factor in constructing a graph involves units of measurement. For example, you might use a Celsius thermometer for one part of your experiment and a Fahrenheit thermometer for another. But you must first convert your temperature readings to the same unit of measurement before you make your graph.

In the past, graphs had to be made by hand, with each point plotted individually. Today, scientists use a variety of tools, such as computers and graphing calculators like the one shown in **Figure 19,** to help them draw graphs.

Applying Math — Make and Use Graphs

TEMPERATURE In an experiment, you checked the air temperature at certain hours of the day. At 8 A.M., the temperature was 27°C; at noon, the temperature was 32°C; and at 4 P.M., the temperature was 30°C. Graph the results of your experiment.

IDENTIFY known values

time = independent variable, which is the *x*-axis

temperature = dependent variable, which is the *y*-axis

GRAPH the problem

Graph time on the *x*-axis and temperature on the *y*-axis. Mark the equal increments on the graph to include all measurements. Plot each point on the graph by finding the time on the *x*-axis and moving up until you find the recorded temperature on the *y*-axis. Place a point there. Continue placing points on the graph. Then connect the points from left to right.

Practice Problems

As you train for a marathon, you compare your previous times. In year one, you ran it in 5.2 h; in year two, you ran it in 5 h; in year three, you ran it in 4.8 h; in year four, you ran it in 4.3 h; and in year five, you ran it in 4 h.

1. Make a table of your data.
2. Graph the results of your marathon races.
3. Calculate your percentage of improvement from year one to year five.

For more practice problems, go to page 879 and visit Math Practice at gpescience.com.

Bar Graphs

A bar graph is useful for comparing information collected by counting. For example, suppose you counted the number of students in every classroom in your school on a particular day and organized your data as in **Table 5.** You could show these data in a bar graph like the one shown in **Figure 20.** Uses for bar graphs include comparisons of oil or crop productions, costs, or data in promotional materials. Each bar represents a quantity counted at a particular time, which should be stated on the graph. As on a line graph, the independent variable is plotted on the *x*-axis and the dependent variable is plotted on the *y*-axis.

Recall that you might need to place a break in the scale of the graph to better illustrate your results. For example, if your data were 1,002, 1,010, 1,030, and 1,040 and the intervals on the scale were every 100 units, you might not be able to see the difference from one bar to another. If you had a break in the scale and started your data range at 1,000 with intervals of ten units, you could make a more accurate comparison.

Reading Check ***Describe possible data where using a bar graph would be better than using a line graph.***

Graphing Temperature Change

Procedure

1. Complete the safety form.
2. Add 1 c of **cold water** to a **medium-sized plastic bowl.** Add 1/2 c of **ice** and 2 tbs of **table salt** to the water.
3. Fill a **clear-plastic cup** 2/3 full with **room temperature water.** Measure the temperature of the water using a **thermometer.**
4. Place the plastic cup with the thermometer into the iced water. Make sure the iced water surrounds the water in the cup but does not enter the cup.
5. Measure and record the temperature every 30 s for 5 min.

Analysis

1. Identify the dependent and independent variables.
2. Make a line graph of the data recorded in step 5.

Try at Home

Table 5 Classroom Size

Number of Students	Number of Classrooms
20	1
21	3
22	3
23	2
24	3
25	5
26	5
27	3

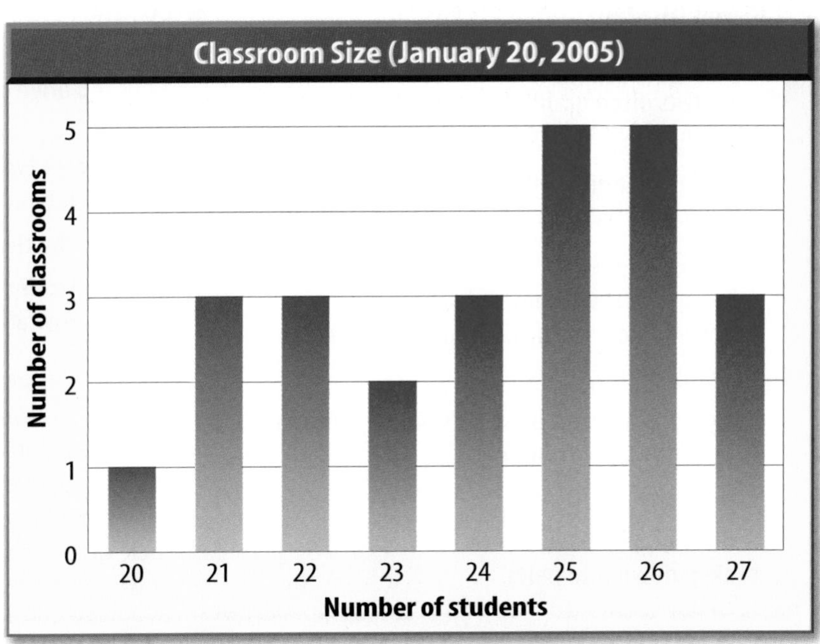

Figure 20 The height of each bar corresponds to the number of classrooms having a particular number of students.

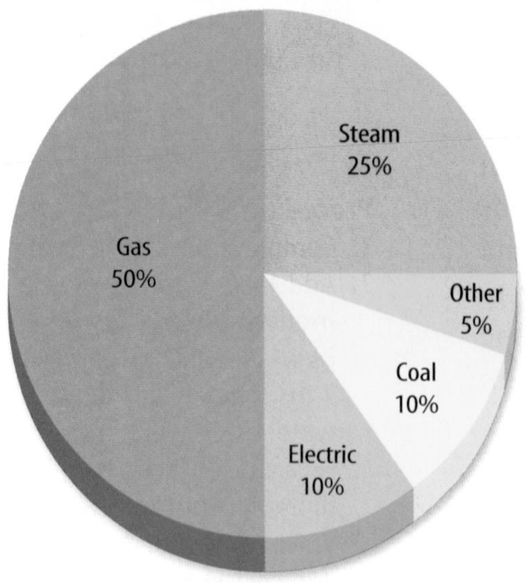

Figure 21 A circle graph shows the different parts of a whole quantity.
Calculate *the angle of gas usage.*

Circle Graphs

A circle graph, or pie graph, is used to show how some fixed quantity is broken down into parts. The circular pie represents the total. The slices represent the parts and usually are represented as percentages of the total.

Figure 21 illustrates how a circle graph could be used to show the percentage of buildings in a neighborhood using each of a variety of heating fuels. You easily can see that more buildings use gas heat than any other kind of heating fuel system. What else does the graph tell you?

To create a circle graph, you start with the total of what you are analyzing. There are 72 buildings in the neighborhood. For each type of heating fuel, you divide the number of buildings using each type of fuel by the total (72). You then multiply that decimal by 360° to determine the angle that the decimal makes in the circle. Eighteen buildings use steam. Therefore, $(18 \div 72) \times 360° = 90°$ on the circle graph. You then would measure 90° on the circle with your protractor to show 25 percent.

When you use graphs, think carefully about the conclusions you can draw from them. You want to make sure your conclusions are based on accurate information and that you use scales that help make your graph easy to read.

section 3 review

Summary

A Visual Display

- Graphs are a visual representation of data.
- Scientists often graph their data to detect patterns.
- The type of graph used is based on the conclusions you want to identify.

Line Graphs

- A line graph shows how a relationship between two variables changes over time.

Bar Graphs

- Bar graphs are best used to compare information collected by counting.

Circle Graphs

- A circle graph shows how a fixed quantity is broken down into parts.

Self Check

1. **Identify** the kind of graph that would best show the results of a survey of 144 people where 75 ride a bus, 45 drive cars, 15 carpool, and 9 walk to work.
2. **State** which type of variable is plotted on the *x*-axis and which type is plotted on the *y*-axis.
3. **Explain** why the points in a line graph are connected.
4. **Think Critically** How are line, bar, and circle graphs similar? How are they different?

Applying Math

5. **Percentage** In a survey, it was reported that 56 out of 245 people would rather drink orange juice in the morning than coffee. Calculate what percentage of a circle graph this data would occupy.

More Section Review gpescience.com

What's my graph?

You have heard that a picture is worth a thousand words. For scientists, it is also true that a graph is worth a thousand numbers. Graphs give us a visual display of data collected during experiments.

Real-World Problem

How are line, bar, and circle graphs used for analyzing different kinds of data?

Goals

- **Compare and contrast** the three different types of graphs and how they are used.

Materials

Science Journal
small ruler
protractor
pencil
compass
*circle template

Alternate material

Safety Precautions

Procedure

1. Complete the safety form.
2. Copy the data tables into your Science Journal. Examine the data listed in the tables.
3. **Discuss** with other students the type of graph to be used for each data table.
4. **Graph** the data for each table in your Science Journal.

Data Table 1 Home Energy Use

Type of Energy Use	Percentage
Heating and Cooling	44%
Water Heating	14%
Refrigerator	9%
Light Cooking and Other	33%

Data Table 2 Motion of an Object

Time (s)	Distance (m)
0	0
5	3
10	6
15	9

Data Table 3 Average Number of tornadoes by month

Month	Average # of Tornadoes
March	53
April	107
May	176
June	168
July	94

Conclude and Apply

1. **Explain** why you chose the type of graph you made for each data table.
2. **Discuss** the advantages of looking at a graph instead of just looking at numbers in a data table.

As a class, compare the type of graph made for each data table.

Design Your Own

Developing a Measurement System

Goals

- **Design** an experiment that involves devising and testing your own measurement system for length.
- **Measure** various objects with the string measurement system.

Possible Materials

string
scissors
marking pen
masking tape
miscellaneous objects for standards

Safety Precautions

Real-World Problem

To develop the International System of Units, people had to agree on set standards and basic definitions of scale. If you had to develop a new measurement system, people would have to agree with your new standards and definitions. In this lab, your team will use string to devise and test its own SI (String International) system for measuring length. What are the requirements for designing a new measurement system using string?

Form a Hypothesis

Based on your knowledge of measurement standards and systems, form a hypothesis that explains how exact units help keep measuring consistent.

Test Your Hypothesis

Make a Plan

1. Complete the safety form before you begin.
2. As a group, agree upon and write out the hypothesis statement.
3. As a group, list the steps that you need to take to test your hypothesis. Be specific, describing exactly what you will do at each step.
4. Make a list of the materials that you will need.

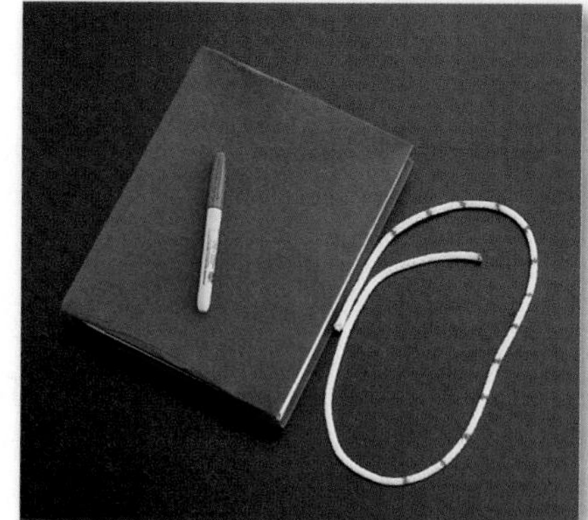

5. **Design** a data table in your Science Journal so it is ready to use as your group collects data.
6. As you read over your plan, be sure you have chosen an object in your classroom to serve as a standard. It should be in the same size range as what you will measure.
7. Consider how you will mark scale divisions on your string. Plan to use different pieces of string to try different-sized scale divisions.
8. What is your new unit of measurement called? Come up with an abbreviation for your unit. What will you name the smaller scale divisions?
9. What objects will you measure with your new unit? Be sure to include objects longer and shorter than your string. Will you measure each object more than once to test consistency? Will you measure the same object as another group and compare your findings?

Follow Your Plan

1. Make sure your teacher approves your plan before you start.
2. Carry out the experiment as it has been planned.
3. **Record** observations that you make and complete the data table in your Science Journal.

Analyze Your Data

1. **Explain** which of your string scale systems will provide the most accurate measurement of small objects.
2. **Describe** how you recorded measurements that were between two whole numbers of your units.

Conclude and Apply

1. **Explain** why, when sharing your results with other groups, it is important for them to know what you used as a standard.
2. **Infer** how it is possible for different numbers to represent the same length of an object.

Communicating Your Data

Compare your conclusions with other students' conclusions. Are there differences? Explain how these may have occurred.

Science and Language Arts

Thinking in Pictures: and other reports from my life with autism[1]

By Temple Grandin

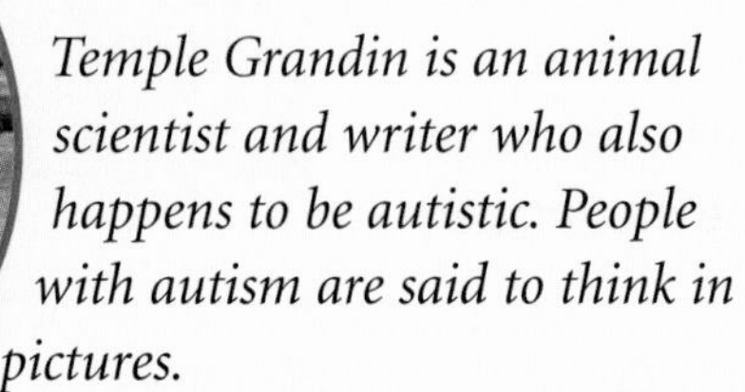

Temple Grandin is an animal scientist and writer who also happens to be autistic. People with autism are said to think in pictures.

I think in pictures. Words are like a second language to me. I translate both spoken and written words into full-color movies, complete with sound, which run like a VCR tape in my head. When somebody speaks to me, his words are instantly translated into pictures. Language-based thinkers often find this phenomenon difficult to understand, but in my job as equipment designer for the livestock industry, visual thinking is a tremendous advantage.

. . . I credit my visualization abilities with helping me understand the animals I work with. Early in my career I used a camera to help give me the animals' perspective as they walked through a chute for their veterinary treatment. I would kneel down and take pictures through the chute from the cow's eye level. Using the photos, I was able to figure out which things scared the cattle.

Every design problem I've ever solved started with my ability to visualize and see the world in pictures. I started designing things as a child, when I was always experimenting with new kinds of kites and model airplanes.

1 Autism is a complex developmental disability that usually appears during the first three years of life. Children and adults with autism typically have difficulties in communicating with others and relating to the outside world.

Understanding Literature

Identifying the Main Idea The most important idea expressed in a paragraph or essay is the main idea. The main idea in a reading might be clearly stated, but sometimes the reader has to summarize the contents of a reading in order to determine its main idea. What do you think is the main idea of the passage?

Respond to the Reading

1. How do people with autism think differently from other people?
2. What did the author use to see from a cow's point of view?
3. What did the author use for models to design things when she was a child?
4. **Linking Science and Writing** Research the use of a scientific model. Write a paragraph stating the main ideas and listing supporting details.

Models enable scientists to see things that are too big, too small, or too complex. Scientists might build models of DNA, airplanes, or other equipment. Temple Grandin's visual thinking and ability to make models enable her to predict how things will work when they are put together.

Reviewing Main Ideas

Section 1 The Methods of Science

1. Science is a way of learning about the natural world, such as the hurricane shown below, through investigation.

2. Scientific investigations can involve making observations, testing models, or conducting experiments.
3. Scientific experiments investigate the effect of one variable on another. All other variables are kept constant.
4. Scientific laws are repeated patterns in nature. Theories attempt to explain how and why these patterns develop.

Section 2 Standards of Measurement

1. A standard of measurement is an exact quantity that people agree to use as a basis of comparison. The International System of Units, or SI, was established to provide a standard and reduce confusion.
2. When a standard of measurement is established, all measurements are compared to the same exact quantity—the standard. Therefore, all measurements can be compared with one another.
3. The most commonly used SI units include length—meter, volume—liter, mass—kilogram, and time—second.
4. Any SI unit can be converted to any other related SI unit by multiplying by the appropriate conversion factor. These towers are 45,190 cm in height, which is equal to 451.9 m.

5. Precision is the description of how close measurements are to each other. Accuracy is comparing a measurement to the real or accepted value.

Section 3 Communicating with Graphs

1. Graphs are visual representations of data that make it easier for scientists to detect patterns.
2. Line graphs show continuous changes among related variables. Bar graphs are used to show data collected by counting. Circle graphs show how a fixed quantity can be broken into parts.
3. To create a circle graph, you have to determine the angles for your data.
4. In a line graph, the independent variable is always plotted on the horizontal x-axis. The dependent variable is always plotted on the vertical y-axis.

FOLDABLES Use the Foldable that you made at the beginning of this chapter to help you review scientific processes.

chapter 1 Review

Using Vocabulary

accuracy p. 14	mass p. 19
bias p. 10	model p. 11
constant p. 9	precision p. 14
control p. 9	scientific law p. 12
density p. 19	scientific method p. 7
dependent variable p. 9	technology p. 13
experiment p. 8	theory p. 12
graph p. 22	variable p. 9
hypothesis p. 8	volume p. 18
independent variable p. 9	

Match each phrase with the correct term from the list of vocabulary words.

1. comparing a measurement to the real or accepted value
2. the amount of space occupied by an object
3. application of science to help people
4. the amount of matter in an object
5. a variable that changes as another variable changes
6. a visual display of data
7. a test set up under controlled conditions
8. a variable that does **NOT** change as another variable changes
9. mass per unit volume
10. an educated guess using what you know and observe

Checking Concepts

Choose the word or phrase that best answers the question.

11. Which question **CANNOT** be answered by science?
    A) How do birds fly?
    B) How does a clock work?
    C) Is this a good song?
    D) What is an atom?

12. Which is an example of an SI unit?
    A) foot **C)** pound
    B) gallon **D)** second

13. One one-thousandth is expressed by which prefix?
    A) centi- **C)** milli-
    B) kilo- **D)** nano-

14. What is SI based on?
    A) English units **C)** powers of five
    B) inches **D)** powers of ten

15. What is the symbol for deciliter?
    A) dL **C)** dkL
    B) dcL **D)** Ld

16. Which is **NOT** a derived unit?
    A) dm^3 **C)** cm^3
    B) m **D)** g/mL

17. Which is **NOT** equal to 1,000 mL?
    A) 1 L **C)** 1 dm^3
    B) 100 cL **D)** 1 cm^3

Interpreting Graphics

Use the photo below to answer question 18.

18. **Define** The illustrations above show the items needed for an investigation. Which item is the independent variable? Which items are the constants? What might a dependent variable be?

Vocabulary PuzzleMaker gpescience.com

19. **Concept Map** Copy and complete this concept map on scientific methods.

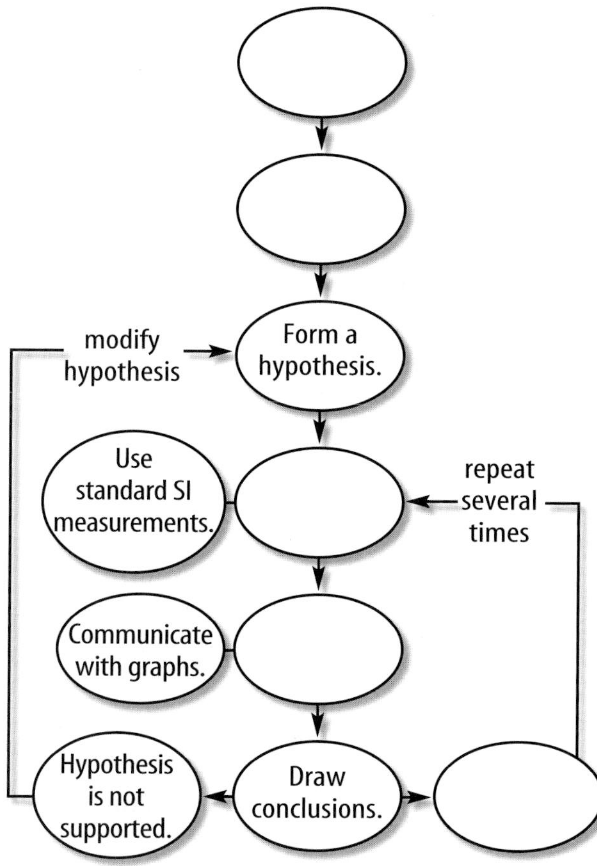

Thinking Critically

20. **Communicate** Standards of measurement used during the Middle Ages often were based on such things as the length of the king's arm. How would you go about convincing people to use a different system of standard units?

21. **Analyze** What are some advantages and disadvantages of adopting SI in the United States?

22. **Identify** when bias occurs in scientific experimentation. Describe steps scientists can take to reduce bias and validate experimental data.

23. **Demonstrate** Not all objects have a volume that is measured easily. If you were to determine the mass, volume, and density of your textbook, a container of milk, and an air-filled balloon, how would you do it?

24. **Apply** Suppose you set a glass of water in direct sunlight for 2 h and measure its temperature every 10 min. What type of graph would you use to display your data? What would the dependent variable be? What would the independent variable be?

25. **Form a Hypothesis** A metal sphere is found to have a density of 5.2 g/cm^3 at 25°C and a density of 5.1 g/cm^3 at 50°C. Form a hypothesis to explain this observation. How could you test your hypothesis?

26. **Compare and contrast** the ease with which conversions can be made among SI units versus conversions among units in the English system.

Applying Math

27. **Convert Units** Make the following conversions.

A) 1,500 mL to L	**C)** 5.8 dg to mg
B) 2 km to cm	**D)** 22°C to K

28. **Calculate the density** of an object having a mass of 17 g and a volume of 3 cm^3.

Use the illustration below to answer question 29.

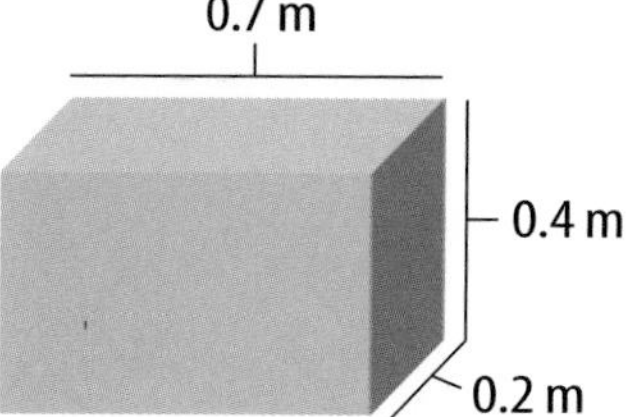

29. **Solve** Find the dimensions of the box in centimeters. Then find its volume in cubic centimeters.

chapter 1 Standardized Test Practice

Record your answers on the answer sheet provided by your teacher or on a sheet of paper.

Multiple Choice

Students drop objects from a height and measure the time it takes each to reach the ground.

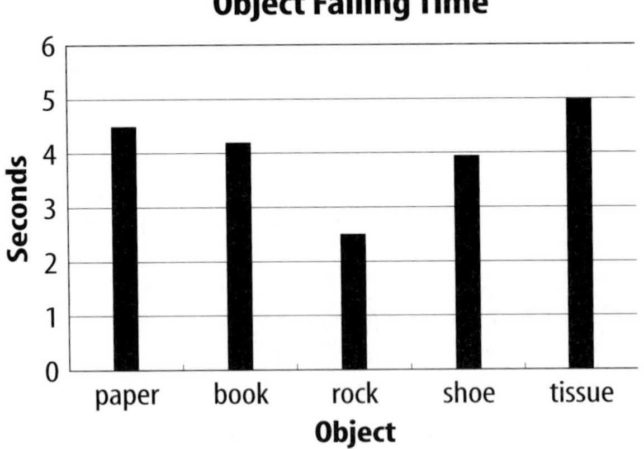

1. What is the dependent variable in this experiment?
   A. drop height
   B. falling time
   C. paper
   D. shoe

2. Which graph is most useful for showing how the relationship between independent and dependent variables changes over time?
   A. bar graph
   B. circle graph
   C. line graph
   D. pictograph

Test-Taking Tip

Recheck Your Answers Double check your answers before turning in the test.

3. Which is a statement about something that happens in nature which seems to be true all the time?
   A. theory
   B. scientific law
   C. hypothesis
   D. conclusion

4. Which best defines mass?
   A. the amount of space occupied by an object
   B. the distance between two points
   C. the quantity of matter in an object
   D. the interval between two events

Use the graph below to answer questions 5 and 6.

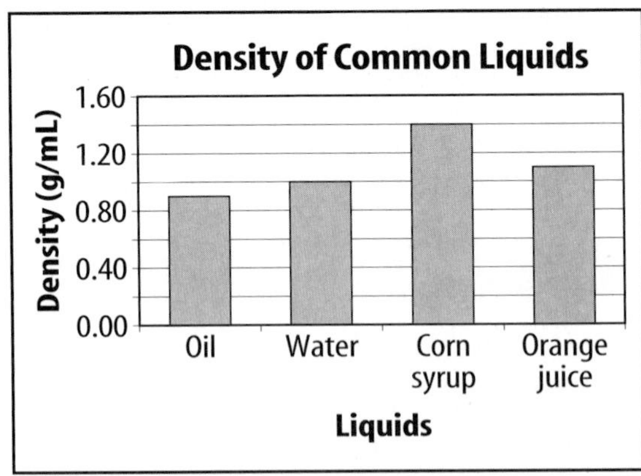

5. Which two liquids have the highest and the lowest densities?
   A. oil and water
   B. oil and corn syrup
   C. orange juice and water
   D. corn syrup and orange juice

6. Convert 615 mg to grams.
   - **A.** 0.00615 g
   - **B.** 0.615 g
   - **C.** 6.15 g
   - **D.** 61.5 g

7. What does the symbol *ns* represent?
   - **A.** microsecond
   - **B.** millisecond
   - **C.** nanosecond
   - **D.** picosecond

Gridded Response

8. Calculate the volume of the cube shown.

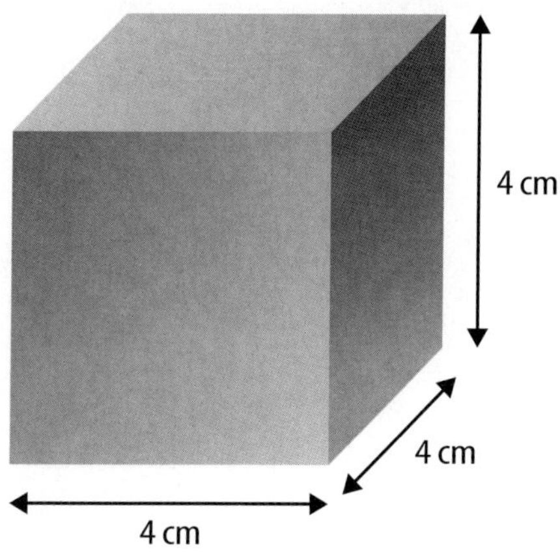

Short Response

9. Describe several ways scientists use investigations to learn about the natural world.

10. Why do scientists use graphs when analyzing data?

Extended Response

11. You are going on a hiking and camping trip. Space is limited, and you must carry your items during hikes. What measurements are important in your preparation?

Use the illustration below to answer question 12.

12. What is the standard unit shown in this illustration? Why is it kept under cover in a vacuum-sealed container?

13. Define the term *technology.* Identify three ways that technology makes your life easier, safer, or more enjoyable.

14. Describe the three major categories into which science is classified. Which branches of science would be most important to an environmental engineer? Why?

15. A friend frequently misses the morning school bus. Use the scientific method to address this problem.

chapter 2

Science, Technology, and Society

chapter preview

sections

Peas in a Pod

What looks like peas in a pod are actually carbon atoms inside a tube also made of carbon atoms. The width of the tube is less than one billionth the width of a human hair! Advancements in technology allow scientists to view and create these small-scale systems.

Science Journal List 10 types of technology you have used today.

Start-Up Activities

Technology in Your Life

You probably do not think about how technology affects your life. You use many types of technology every day without realizing it. How much technology do you use?

1. Complete the safety form.
2. Make a list of everything that you do from the time that you get home from school until you go to bed.
3. Circle the items on the list that involve some type of technology.
4. Select two of the items and identify the technologies involved in the item. For example, the technologies involved in a television may be the wiring or the wave that transports the signal.
5. **Thinking Critically** Write a list of activities that you could do after school if none of the technologies that you listed were available.

Preview this chapter's content and activities at gpescience.com

FOLDABLES™ Study Organizer

Science and Technology Make the following Foldable to help organize information about science and technology.

STEP 1 **Fold** a sheet of paper vertically in half from top to bottom.

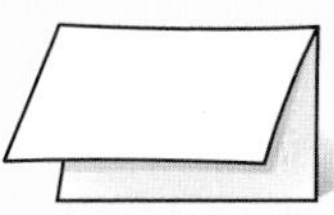

STEP 2 **Fold** in half from side to side with the fold at the top.

STEP 3 **Unfold** the paper once. **Cut** only the fold of the top flap to make two tabs.

STEP 4 **Turn** the paper vertically and label the flaps *Science* and *Technology*.

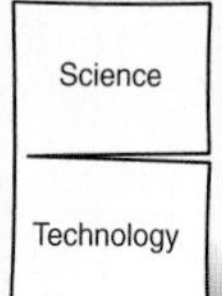

Read for Main Ideas As you read the chapter, list the characteristics of science and technology under the appropriate flap.

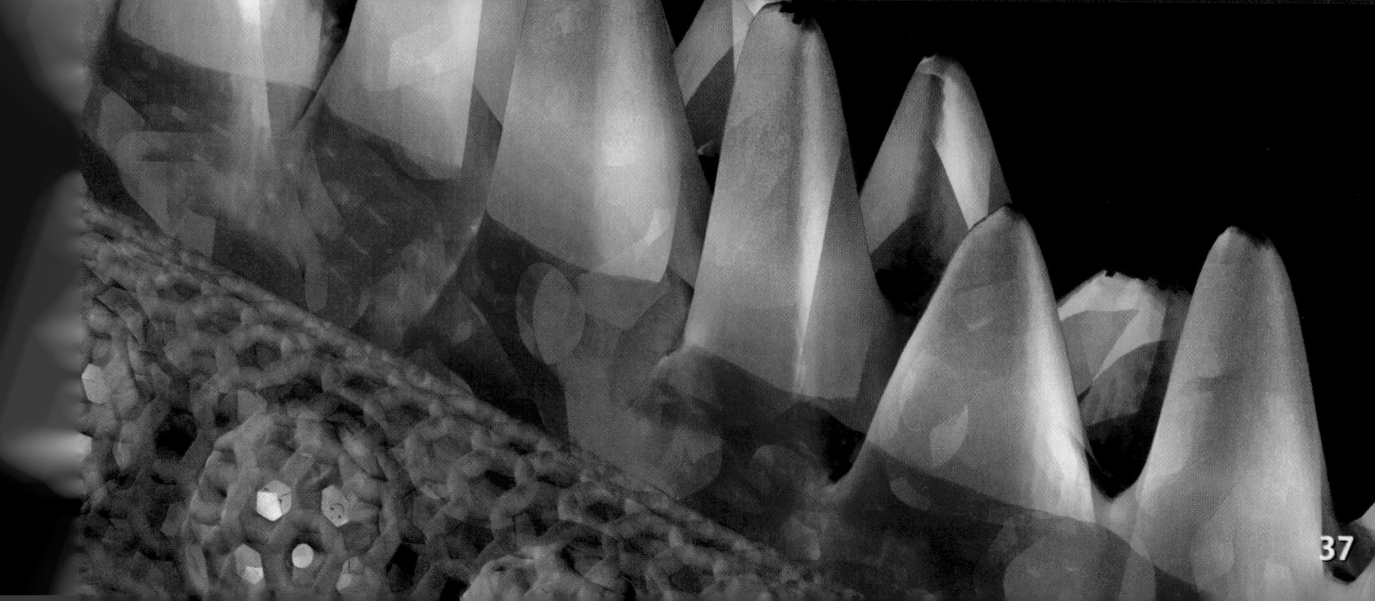

section 1

Science and Technology

Reading Guide

What You'll Learn

- **Discuss** how the study of science leads to a better understanding of natural events.
- **Describe** the different types of technology.
- **Explain** why the value of technology may differ for different people and at different times.

Why It's Important

Science provides information, insight, and analytical skills to help people understand the possible causes and effects of natural events.

Review Vocabulary

technology: the application of science to help people

New Vocabulary

- agricultural biotechnology

Figure 1 Physicians in the Middle Ages treated people who had the plague. The beak in the protective mask contains a mixture of materials to filter the "bad air."
Contrast *the methods used to help prevent the spread of diseases during the Middle Ages against methods used today.*

Scientific Discovery

The study of science usually leads to a better understanding of the world around you. For example, some scientists in the Middle Ages believed that the plague was spread by breathing "bad air". Doctors wore protective clothing to protect themselves, as shown in **Figure 1.** Various measures were taken by local doctors to rid homes of this "bad air" in an attempt to stop the plague. History is full of breakthroughs in science that have changed the course of human history. One such breakthrough is the discovery that microorganisms cause illnesses and disease. Before the mid-1800s, people had no idea that microorganisms caused many human illnesses. Many diseases were easily passed from person to person because people had no idea why the disease was spreading.

In the mid-1800s, Louis Pasteur, a French scientist, discovered the cause of most infectious diseases. He informed the public that microorganisms brought about infections and diseases. Knowing how infectious diseases spread, Alexandre Yersin, a bacteriologist, traveled to Hong Kong. A plague epidemic was underway. He was hoping to isolate the bacteria that caused this disease. In 1894, Yersin was able to isolate the microbe that caused the disease. Once scientists discovered what was causing the disease, they could focus on how the disease spread through the population. It took another four years before scientists discovered that fleas spread this microbe to humans.

Scientific Insight

Once scientists found the source of the plague, scientists brought the information to the public's attention. The information provided an explanation of why the plague occurred. Outbreaks of the plague took place throughout history. One outbreak in the 14th century killed approximately one-fourth of the population of Europe. When people clearly understood why the disease was spreading, actions could be taken to stop the disease.

Disease Prevention Today The fact that microbes spread illnesses is not news today. Young children are taught to wash their hands before eating. Restrooms contain water, soap, and towels to use to prevent the spread of microbes. Sterile bandages, cleansers, and medicines are sold in stores for use on injuries to prevent infection. Doctors and hospitals use sterile instruments for each patient to prevent the spread of disease. These measures are taken now because we know how contagious diseases are spread from person to person and what causes infections in wounds.

Weather Forecasting Scientific insight also is used in many other ways, including weather forecasting. If you had lived in the 1800s, you only could have guessed the type of weather that was approaching by looking at the sky. Today, meteorologists use a variety of specialized instruments to predict the weather. Their instruments can detect approaching conditions that are dangerous. With advance warning, people can seek shelter or leave areas that are threatened by violent weather. **Figure 2** demonstrates the importance of forecasting dangerous weather conditions.

Science has changed the way people respond to natural events, such as the spread of microbes and hazardous weather. The study of science can lead to an understanding of why a natural process occurs. Once there is a clear understanding, people learn how to respond to the event and try to control the outcome.

Reading Check ***What is another example of science being used to change people's behavior?***

Figure 2 In 1900, a hurricane struck Galveston, Texas, killing about 8,000 people. Today early-warning systems alert people of an approaching storm, which saves many lives.
Identify *a type of hazardous weather that occurs in your area that requires an early-warning system.*

Researching the Past

Procedure

1. Complete the safety form.
2. Prepare at least five questions to ask a senior citizen about technology when he or she was a teenager.
3. Perform the interview.
4. Write the responses to your questions in your **Science Journal**.

Analysis

Compare and contrast at least five differences in technology between then and now.

Try at Home

What is technology?

The terms *science* and *technology* often are used interchangeably. However, these terms have very different definitions. Science is an exploration process. Scientific processes are used to gain knowledge to explain and predict natural occurrences. Scientists often pursue scientific knowledge for the sake of learning new information. There may or may not be a plan to use the knowledge.

When scientific knowledge is used to solve a human need or problem, as shown in **Figure 3,** the result is referred to as technology. Technology is the application of scientific knowledge of materials and processes to benefit people. Given this definition, is an aspirin tablet technology? Is a car technology? What about the national highway system? Although these examples appear to be very different, they all represent examples of technology. Technology can be:

- any human-made object (such as a radio, computer, or pen),
- methods or techniques for making any object or tool (such as the process for making glass or ceramics),
- knowledge or skills needed to operate a human-made object (such as the skills needed to pilot an airplane), or
- a system of people and objects used to do a particular task (such as the Internet, which is a system to share information).

Technological Objects? The value of a technological object changes through time. What is considered technology today might be considered an antique tomorrow. For example, special feathers called quills were used long ago to write with ink. The quill pen was the height of technology for over one thousand years. Then, in the middle of the nineteenth century, metallic pens and writing points came into use. The modern ballpoint pen was not widely used until the 1940s.

What are five additional objects that are considered technology?

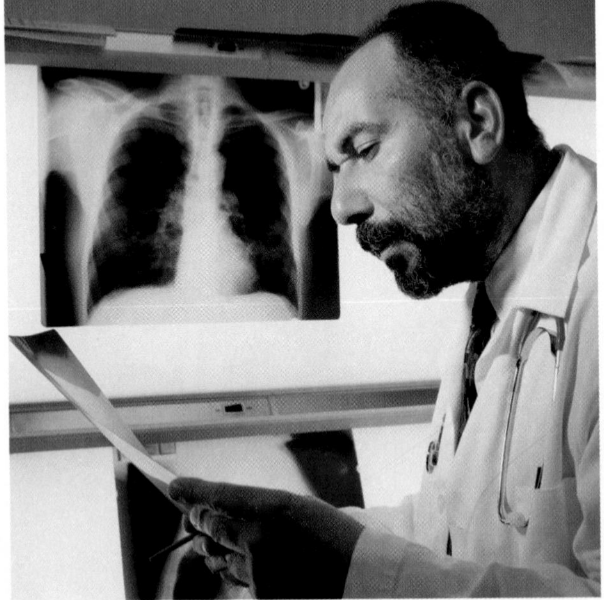

Figure 3 This figure shows technology. This image is a technological object. The knowledge needed to create and interpret this image is technology.
Identify *the technologies used to make a car.*

Figure 4 Technology changes through time. As new scientific knowledge and techniques are learned, the processes used to create objects change.
Identify *another process used to create objects that has changed through time.*

Technological Methods or Techniques Just as writing instruments have changed through time, so have techniques for performing various tasks. **Figure 4** shows the results of how the techniques for making books have changed. Long ago, people would sit for hours copying each page of a book by hand. They used whatever writing materials were available during that time. Books were expensive and bought only by the very rich. Today, books can be created in different ways. They can be made on a computer, printed with a computer printer, and bound with a simple machine. Modern printing presses are used to produce the majority of books used today, including your textbook.

The methods used to print books have changed through time. Each technique is technology. Other techniques that characterize technology are using a compact disc to store information, using a refrigerator to preserve food, and using email to correspond with friends.

Technological Knowledge or Skills Technology also can be the knowledge or skills needed to perform a task. For example, computer skills are needed to use the software that is used to make books and other documents. Printing press operators must use their skills to print books successfully. Anytime a complex machine is used to perform a task, technological skills must be used by the operator.

Technological Systems A network of people and objects that work together to perform a task also is technology. One example of this technology is the Internet. The Internet is a collection of computers and software that is used to exchange information. A technological system is a collection of the other types of technology that are combined to perform a specific function. The airline industry is an example of technological systems. This industry is a collection of objects, methods, systems, knowledge, and procedures. The airports, pilots, fuel, and ticketing process form a technology system that is used to move people and goods.

Global Technological Needs

The example of how books were made explained how technology has changed through the ages. The value of technology may differ for different people and at different times. The technology that is valued in the United States is not necessarily valued in other parts of the world. The needs for technology are different in developing and industrialized countries.

Developing Countries The people of some countries work hard for basic needs such as food, shelter, clothing, safe drinking water, and health care. For example, the Kenyan family shown in **Figure 5** lives without electricity and running water in their home. Health care is limited for many families in Africa. The lack of health care results in a life expectancy for rural Kenyans to be in the mid-forties. Tropical diseases, as well as infection by the human immunodeficiency virus (HIV) are a severe problem. In addition, droughts often cause shortages of food supplies in some parts of this country.

Technological solutions in developing countries focus on supplying basic needs for these families. Technology that would supply adequate and safe drinking water and food supplies would be valued. Increasing the accessibility of basic health care would improve the quality of life and increase the life expectancy in developing countries. The technology valued by rural people contrast with the technology valued by people in industrialized Kenyan cities.

Reading Check *What are two examples of technology that would be valued in developing countries?*

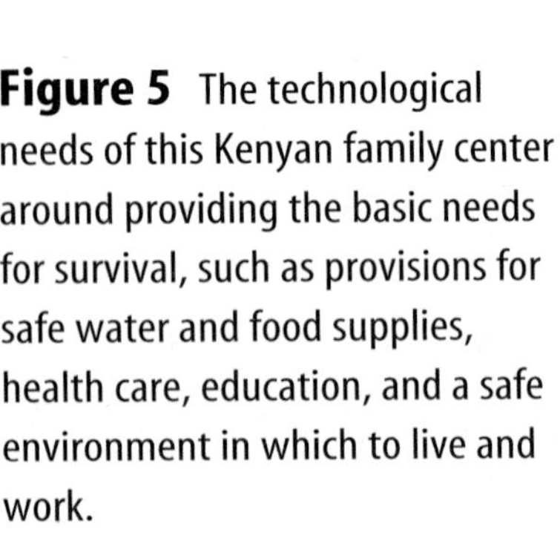

Figure 5 The technological needs of this Kenyan family center around providing the basic needs for survival, such as provisions for safe water and food supplies, health care, education, and a safe environment in which to live and work.

Industrialized Countries The United States is considered an industrialized country. Technology gives people the ability to clean polluted waters. Improving the quality of the food supply is also valued. But the level of urgency and focus appears to differ in developing countries. Most areas of the United States have adequate and safe water and food supplies. Most homes have electricity and running water. Quality health care is available to many people. The life expectancy of Americans is the late-seventies.

Look at the home in **Figure 6** and compare it to the home shown on the previous page. Does one of the homes have more technology? Look at the materials used to build each home. The materials used to build the home in **Figure 6** are technology, too.

Because the needs for survival are met in industrialized nations, money often is spent on technology. Technology is designed to improve the quality of life of individuals. Most homes in the United States contain many different types of technology, including computers, telephones, and televisions. People living in industrialized countries may value faster and faster computers. They may value small devices such as compact discs that store large amounts of information. Advances in health care can improve the quality of life due to the ability to cure diseases. Money is spent on such medical procedures as cosmetic surgery to remove wrinkles, and eye surgery so that a person no longer has to wear glasses.

Figure 6 The basic needs for survival are available in industrialized nations. In these countries, value is placed on technology that improves the quality of life such as devices that make tasks easier and devices that provide entertainment.
Describe *three technological objects that you value that would be of less value to the Kenyan family on the previous page. Explain your answers.*

Contrasting Needs As you can see, the human needs for developing and industrialized countries are very different. Both developing and industrialized countries value technologies that supply basic human needs as discussed in **Figure 7.** The actual technology required in each country may be very different.

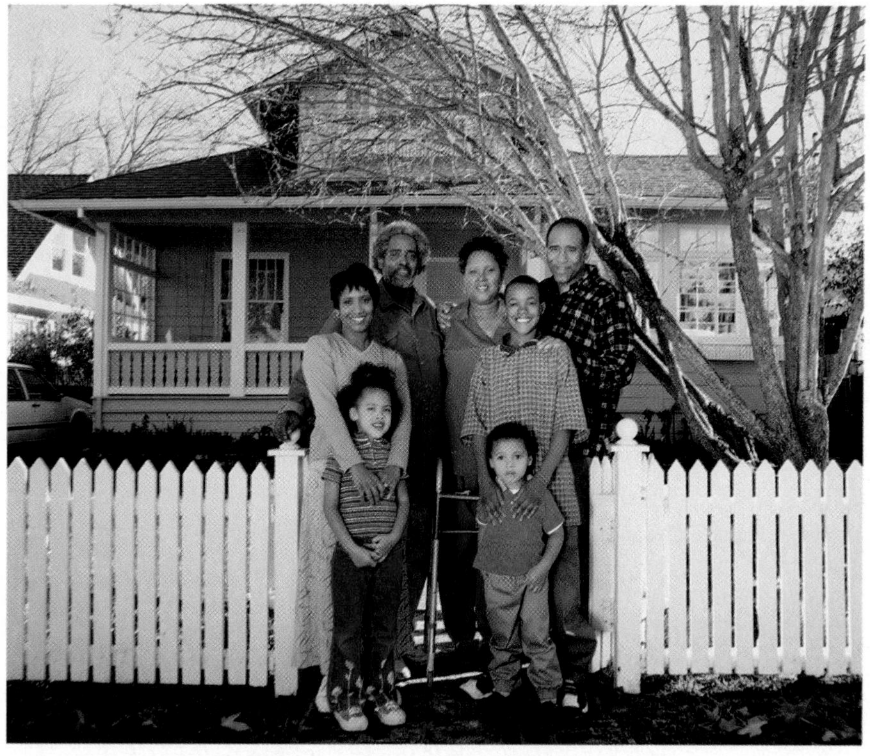

NATIONAL GEOGRAPHIC

VISUALIZING TECHNOLOGY SUPPLYING BASIC NEEDS

Figure 7

Technology is used to help supply the basic requirements of people across the globe. Many of life's necessities that we take for granted in the United States are lacking or inadequate in other countries.

▲ Modern Agriculture Farming with tractors on vast fields and using pesticides and specially engineered seed can increase the amount of food a country produces.

▲ Water Treatment Polluted water can cause diseases. Treating the water with chlorine kills germs.

◀ Vaccines Diseases like polio, measles, and tetanus kill hundreds of thousands of people a year. Vaccines cause the human body to make antibodies for a specific bacteria or virus. If this bacteria or virus enters the body, the antibodies destroy it.

▲ Power Plants Electricity is necessary for many technologies. Electricity can be produced in many ways including burning coal, using moving water, absorbing the heat of solar rays, and using heat deep in the earth.

▲ Transportation Systems Technology allows countries to build roads, airports, and canals that let food, medicine, and information travel quickly to where it is needed.

Bioengineering and Food Can technology stop world hunger and starvation? Agricultural biotechnology can increase the yields of farm crops. It also can increase the nutritional value of the food produced. **Agricultural biotechnology** is a collection of scientific techniques, including genetic engineering, that are used to create, improve, or modify plants, animals, and microorganisms. For example, changing the DNA of a plant such as corn can improve its tolerance to drought. Other benefits could make it disease and insect resistant, or improve its nutritional value. The end result may be a higher crop yield and a product that provides more nutritional value per serving.

Genetically altering plant species is a highly controversial practice. Some people argue that the practice is too dangerous. They argue that humans should not alter species that occur naturally on Earth. Others argue that new species could harm natural ecosystems. New crops could pose human health risks such as food allergies or toxins.

A discussion also exists over whether more food really is needed to feed the world population. One argument is that enough food is produced, but it is not equally distributed throughout the world. As you can see, using technology to solve human problems is not always an easy task. Often there are many obstacles and issues involved.

Topic: Biotech Foods
Visit gpescience.com for Web links to information about agricultural biotechnology and biotech foods.

Activity Create a pamphlet explaining the risks and benefits of biotech foods.

section 1 review

Summary

Scientific Discovery

- The study of science leads to a better understanding of natural events that occur.

Scientific Insight

- Once there is an understanding of why a natural process occurs, people learn how to respond to the event and try to control the outcome.
- Scientific insight changes human behavior.

What is technology?

- Technology can be an object, a technique, a skill, or a system.
- The value of technological objects varies over time.

Global Technological Needs

- Technological needs vary around the world.

Self Check

1. **Describe** a situation in which scientific insight changed people's understanding, resulting in behavior change in response to a natural event or process.
2. **Classify** List the types of technology and give at least two examples of each type.
3. **Explain** why the types of technology valued varies from location to location.
4. **Think Critically** Cell phones are a technology that is popular throughout the world. Would this technology be of use in rural areas of the world? Explain your answer.

Applying Math

5. **Calculate** Approximately 38 million people are infected with HIV worldwide. Two-thirds of this group lives in sub-Saharan Africa. Calculate how many sub-Saharans are affected.

section 2

Forces that Shape Technology

Reading Guide

What You'll Learn

- **Discuss** how funds for science research come from the federal government, private industries, and private foundations.
- **Explain** how consumers affect technological development.
- **Describe** the importance of responsible technological development.

Why It's Important

Society and economic forces influence which technologies will be developed and used.

Review Vocabulary

ecosystem: a complex community of organisms that function together in a particular environment

New Vocabulary

- society

Social Forces that Shape Technology

Science and society are closely connected. **Society** is a group of people that share similar values and beliefs. Discoveries in science and technology bring about changes in society. In turn, society affects how new technologies develop. The development of technology is affected by society and its changing values, politics, and economics.

In the past 100 years, attitudes in the United States have changed toward automobiles. Many people were able to own cars due to the changes in technology and manufacturing. As car ownership increased, so did fossil-fuel consumption. With rising gasoline prices, some consumers began buying more fuel-efficient cars. The automotive industry has researched and developed technologies that make cars more fuel-efficient. Today we have hybrid cars that use both gasoline and electricity.

Figure 8 Consumers often decide which technologies will be developed. If consumers do not purchase a product, additional money usually will not be spent on the production or improvement of the product.

Consumer Acceptance Purchasing technology is a direct way in which people support the development of technology, as shown in **Figure 8.** For example, if consumers continue to purchase fuel-efficient cars, additional money will be spent on improving the technology. If consumers fail to buy a product, companies usually will not spend additional money on that type of technology.

Personal Values People will support the development of technologies that agree with their personal values, directly and indirectly. For example, people vote for a congressional candidate based upon the candidate's views on various issues. This is an indirect way in which people's personal values influence how technology projects receive funding. People support the development of technology directly when they give their money to organizations committed to a specific project, such as cancer research.

Reading Check *How do the personal values of consumers influence which technologies will be developed?*

Economic Forces that Shape Technology

Many factors influence how much money is spent on technology. Before funding is given for a project, several questions should be answered. What is the benefit for this product? What is the cost? Who will buy this product? All these questions should be answered before money is given for a project. Various methods exist to fund new and existing technology.

Federal Government One way in which funds are allocated for research and development of technology is through the federal government. Every year, Congress and the president place large amounts of money in the federal budget for scientific research and development. These funds are reserved for specific types of research, such as agriculture, defense, energy, and transportation. This money is given to companies and institutions in the form of contracts and grants to do specific types of research. Citizens can influence elected officials by telling them how they want the official to vote, as shown in **Figure 9.**

Private Foundations Some scientific research is funded using money from private foundations. A private foundation, which is an organization not associated with the government, is a group of people who work together for a common goal. Funds are raised for various types of disease research, such as breast cancer and muscular dystrophy. Money is raised by events such as races and telethons. Many private foundations focus on research for a specific cause.

Evaluating the Benefits and Consequences of Technology

Procedure

1. Complete the safety form.
2. Choose a technology that you value and do not want to live without.
3. Make a two-column table. Label one column *Benefits* and the other column *Consequences.*
4. Provide at least three entries for each column.

Analysis

1. Do the benefits of the technology outweigh the consequences? Explain.
2. Infer who funded the development of your technology. Explain.
3. Infer how you may influence future funding on this type of technology.

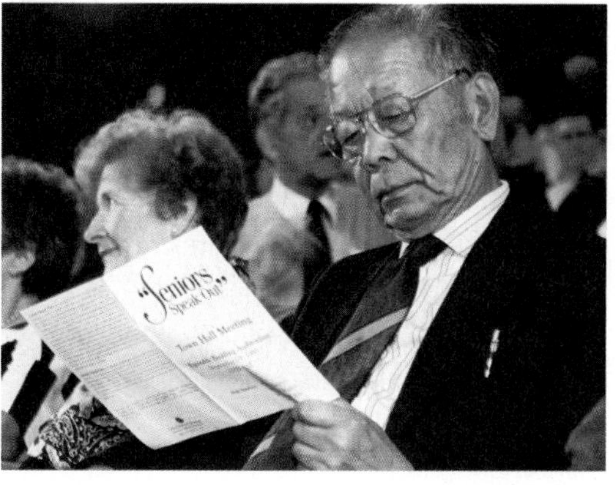

Figure 9 Participating in the political process is one way in which people can influence which technologies are developed and which are not.
Explain *how voting for someone in the Congress influences which technologies will be developed.*

Environmental Protection Agency The purpose of the EPA is to protect human health and the environment. The EPA was established in 1970 by the White House and Congress. Cleaner air and water, along with cleaner land, were things that U.S. citizens have demanded. Prior to the establishment of the EPA, the government had no organized method to reduce pollutants. Research the water quality for your city or state.

Private Industries Research and development also is funded by private industries. Industries budget a portion of their profits for research and development. Investing in research and development often makes money for the company. Bringing new products to the marketplace is one way companies make profits.

Responsible Technology

Humans have the ability to invent tools and processes that can have an impact on other living things. Humans are part of many ecosystems on Earth. It is important that any technology that is used does not destroy the environment, as shown in **Figure 10.** It is important that both positive and negative consequences of a particular technology be considered.

Environmental Issues Sometimes the consequences of technology are known, but the benefits are perceived to outweigh the risks. One example of technology that has both positive and negative consequences is gasoline-powered cars. Gasoline-powered cars give the benefit of reliable transportation. The negative consequences are the environmental problems created by drilling for oil, air pollution, and disposal of unwanted cars. The disposal of used petroleum products, such as motor oil, also is a problem.

Sometimes the benefits of technology are known immediately, but the consequences are not known for a period of time. For example, at one time, the pesticide DDT was eagerly accepted by people. It killed insects that were damaging agricultural crops. The fact that DDT was damaging the environment was not known immediately. When the damage was discovered, DDT was banned and other pesticides were developed. Consumers and voters have a responsibility to weigh the benefits and consequences of technology. They play a major role in determining which technologies are used and which are rejected.

Figure 10 In March 1989, approximately 42,000 kiloliters of oil were spilled in Prince William Sound, Alaska. Animals, such as ducks and birds, are affected by human-made environmental disasters.

The Chernobyl Accident Possibly the worst technological accident in modern times occurred on April 26, 1986, at the Chernobyl nuclear power plant in the Ukraine. Large areas of land were contaminated and more than 400,000 people had to be relocated. The city of Pripyat had to be abandoned, as shown in **Figure 11.** An accumulation of radioactive fallout in the upper layers of soil has destroyed important farmland. Groundwater and surface waters were contaminated. Some water wells had to be sealed. Structures had to be built to prevent rivers and streams in the area from contaminating additional rivers in the watershed, which led to the Black Sea.

The radiation also affected many animals and plants around Chernobyl. As the radiation diminished, plants and animals started to repopulate the area.

Why did the residents near the accident have to move?

Figure 11 This town once was home to about 45,000 residents. It is now abandoned because radioactive contamination has made it unsafe to live in.
Identify *possible health effects that Ukrainians faced.*

Applying Science

What impact does the accident at Chernobyl have on the future of nuclear power?

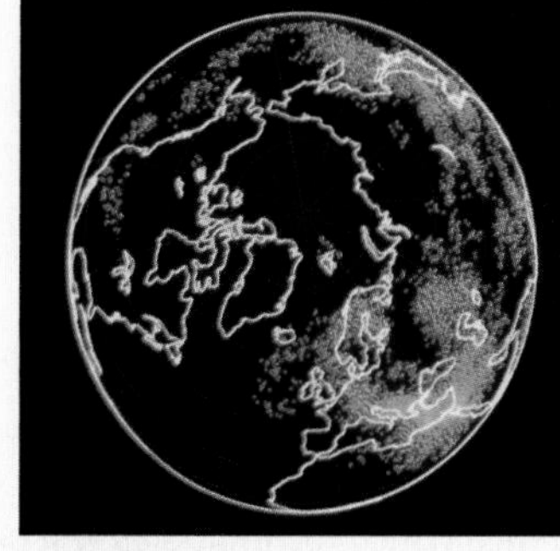

The nuclear accident at Chernobyl in 1986 demonstrates the far-reaching impact of a nuclear disaster. Not only were the residents and environment surrounding the facility influenced, but nuclear fallout was spread across most of the northern hemisphere. Parts of the USSR and Europe were seriously affected.

Identifying the Problem

Many countries, including the United States, use nuclear power as a source of electric power. Is it important for industrialized countries, such as the United States, that have vast resources and knowledge about nuclear power generation, to assist developing countries in their nuclear power generation programs?

Solving the Problem

1. What are the benefits and consequences of allowing developing countries that do not have the experience that developed countries have to develop nuclear power on their own?
2. What are the benefits and consequences of assisting developing countries?

Topic: Morals and Ethics in Science

Visit gpescience.com for Web links to information about morals and ethics in science.

Activity Choose a topic and write an essay describing the moral and ethical arguments surrounding the topic.

Moral and Ethical Issues When people need to distinguish between right and wrong, what is fair, and what is in the best interest of all people, moral and ethical issues are raised. Ethical issues in science pose questions and establish rules about how scientific hypotheses should be tested and how society should use scientific knowledge.

Ethics help scientists establish standards that they agree to follow when they collect, analyze, and report data. Scientists are expected to conduct investigations honestly and openly. Honesty is important because scientists share the results of their investigations with each other. As a result, one scientist's dishonesty can harm the investigations of many scientists.

Other ethical questions in science concern the use of animals and humans in scientific investigations. In the past, the inhumane treatment of humans and animals in experiments has led to public outcry. For example, human test subjects have been subjected to experiments against their will, or without informing them of risks associated with the research or the true nature of the experiments. Ethical questions about these practices helped to create laws and guidelines to prevent unethical treatment of both humans and animals in scientific research.

section 2 review

Summary

Social and Economic Forces that Shape Technology

- Societal and economic forces influence which technologies will be developed and used.
- The federal government, private foundations, and private industries fund research and development of technology.
- Consumers influence which technologies will be developed by their voting and buying habits.

Responsible Technology

- Humans must be responsible in their technological choices because the technology may have an enormous impact on other living things and the environment.
- The risks and benefits of technology must be considered before developing and using it.
- Consumers and scientists must consider moral and ethical issues when choosing which technologies should be developed.

Self Check

1. **Describe** how private citizens have a voice in which projects the federal government will fund.
2. **Explain** how you can influence whether or not funding will be spent on developing certain technology.
3. **Describe** a situation in which technology had unexpected consequences.
4. **Think Critically** Many products carry a message on their label that no animals or humans were used in the testing of the product. Is this good or bad for the consumer? Explain your answer.

Applying Math

5. **Use Percentages** In 2001, approximately \$41 million was spent by the Department of Defense for Research, Development, Test, and Evaluation. If the total amount of money spent by the department was over \$305 million, what percent of the defense budget does this represent?

More Section Review gpescience.com

Who contributes CO_2?

Many scientists believe that Earth's surface temperature is rising because of increasing levels of heat-trapping carbon dioxide, CO_2, in Earth's atmosphere. The combustion of fossil-fuels contributes CO_2 to Earth's atmosphere.

Real-World Problem

Which countries produce the most carbon dioxide from the combustion of fossil fuels?

Goals

- **Compare and contrast** the total carbon dioxide emissions from selected countries.
- **Calculate** the average annual per capita amount of carbon dioxide for each country.
- **Form a hypothesis** that explains the amount of carbon dioxide produced by each country.

Materials

Science Journal
pen or pencil

Procedure

1. Complete the safety form before you begin.
2. Copy and complete the data table in your Science Journal by calculating the amount of carbon dioxide produced per person for each country.
3. **Formulate** a hypothesis that explains the amount of carbon dioxide produced by the inhabitants of the countries listed in the table. For each country, consider the following factors: size; population density; transportation requirements; heating and cooling requirements; development and industry; means of generating electricity.

Countries CO_2 Emissions

Country	CO_2 Emissions (millions of tons of carbon)	Total Population (millions of) persons)	CO_2 Produced (tons/person)
United States	1,446.8	263.8	
China	918.0	1,210.0	
Russian Federation	431.1	149.9	
Japan	318.7	125.5	
India	272.2	936.5	
Canada	111.7	28.4	
France	98.8	58.1	

Conclude and Apply

1. **Graph** the data of carbon dioxide produced per person using a bar graph.
2. **Think Critically** Why might more developed, industrial countries have more carbon dioxide produced per person?
3. **Evaluate** Did the results obtained from your calculations support your hypothesis?

Communicating Your Data

Compare your bar graphs with those of other students. Discuss any differences, particularly if some students used computer graphing programs to prepare their graphs.

section 3

Developing Technology

Reading Guide

What You'll Learn

- **Discuss** how the roles of a scientist and an engineer differ.
- **Explain** how technological problems often create a demand for new scientific knowledge.
- **Describe** the general process used to find technical solutions to problems or human needs.

Why It's Important

Understanding how technology is developed can help you make better decisions as a consumer.

Review Vocabulary

system: a group of devices that work together to perform a specific function

New Vocabulary

- engineer
- constraints
- computer simulation
- prototype
- pilot plant
- control system

Scientists and Engineers

How do a scientist and an engineer differ? This may seem to be a very odd question, but it is important to learn the roles of these two types of professionals. Recall that a scientist is someone who has knowledge about science. He or she works to learn more information about science. Scientists may not know how or if the results of their work will be used, but they pursue knowledge for the sake of learning new things.

Scientists have knowledge of scientific principles. Most scientists have some knowledge of basic chemistry, physics, and biology. Many scientists specialize in one area of science. For example, a scientist may have a chemistry background but specialize in Earth science. This scientist will study the chemical and physical changes that occur on Earth. An Earth scientist may specialize even further and study topics such as glaciers, earthquakes, or volcanoes.

Scientists often work in research laboratories doing research. In addition, scientists also work in the field, as shown in **Figure 12.** An Earth scientist is an example of a scientist that would conduct investigations and research in the field, as well as in the laboratory.

Figure 12 This meteorologist is collecting data on an approaching tornado. This information can be used in the study of tornadoes and how to protect people and property from damage.

Reading Check *What is a scientist?*

Engineers A researcher who is responsible for bringing technology to the consumer is called an **engineer.** Using scientific information or an idea, an engineer creates a way to solve a problem or to produce a product. For example, decreasing petroleum deposits and high prices for gasoline have created a demand for cars that get better gas mileage. Engineers are working to improve an engine called a hybrid. The hybrid engine operates on electricity and gasoline or diesel to power the car. Fuel efficiency increases by combining the two power sources. Cars currently are available using this technology. As this example demonstrates, consumers want cars with better gas mileage. Engineers worked to create technology that would satisfy this consumer demand.

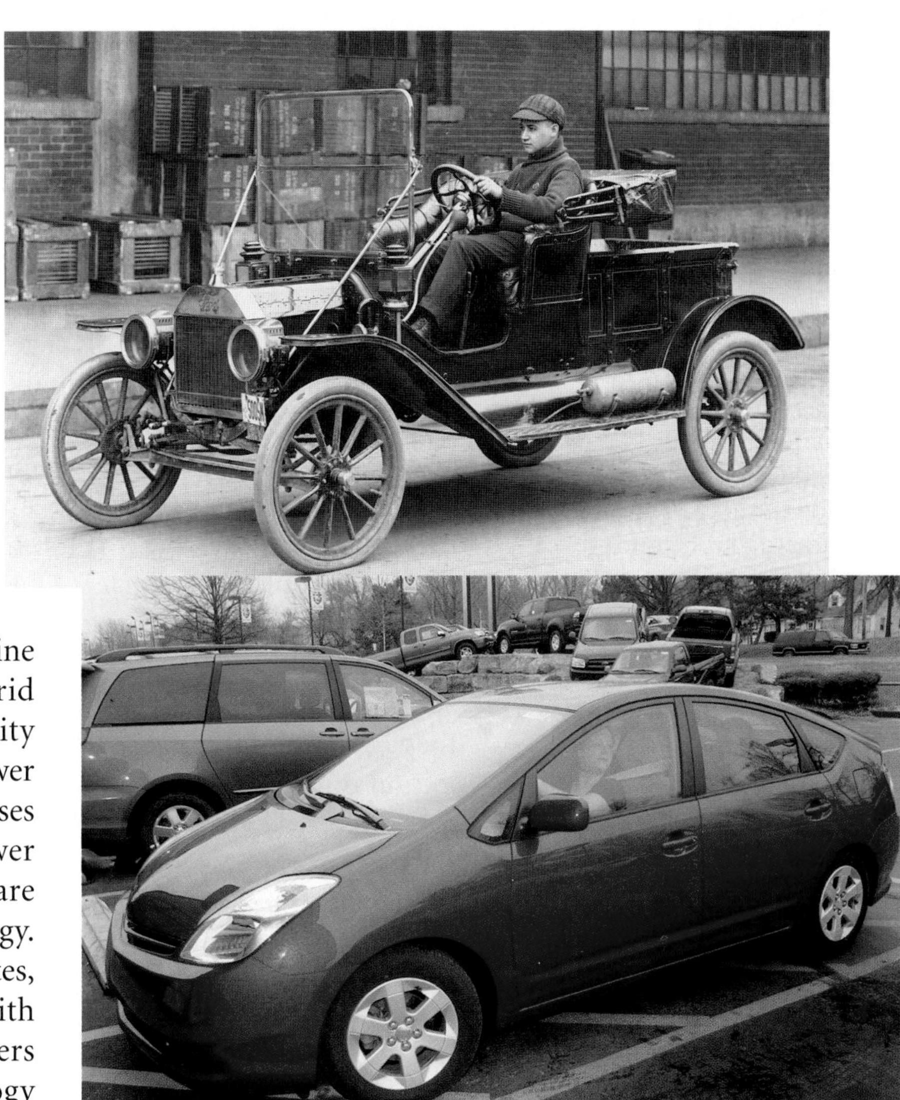

Figure 13 New technology often creates a demand for even more technology. Many technological advances have been made in the automobile since 1950.
Identify *at least three different technological advances that have been made in the automobile since 1950.*

Technological problems often create a demand for new scientific knowledge. Time and money are devoted to finding new scientific information, which makes it possible for scientists to extend their research in a way that advances science. Many unexpected discoveries happened when scientists were looking for a technical solution. In addition, new technology often creates a need for even more technology, as shown in **Figure 13.**

Engineers not only design automobiles, they work in many areas of science. There are aeronautical, aerospace, biomedical, chemical, computer, electrical, mechanical, and many other types of engineers. At some point of the manufacturing process, engineers are involved in the development of the products you use every day.

Reading Check *How do scientists and engineers differ?*

Young Owl Gets Help Technology can be used in unusual ways. A young, great-horned owl was observed sitting on a fence for several days without moving. It seems the owl could not see and could not search for food. Veterinarians removed the owl's cataracts and fitted the owl with new artificial lenses. The owl recovered and was returned to the wild. Research other medical procedures that can be used to help animals.

Finding Solutions

Scientists and engineers work together to find technological solutions, as shown in **Figure 14.** They often use a system much like the scientific method. The general process is outlined in this section. However, there are many variations to this procedure, just as there are many kinds of technical problems.

Identifying the Problem The first step in finding a technical solution is to define clearly the problem that you are trying to solve. The problem must have a narrow enough focus so that a solution is possible. For example, it would not be helpful to define the problem as, "Create and build a new vehicle." This definition of the problem is too broad. What type of vehicle is needed? Should the vehicle be designed to carry passengers or heavy loads or both? What type of fuel should the new vehicle use? What types of materials should be used to construct the vehicle? Is there a market for the vehicle once it is built? All of these questions and many more should be answered in the definition of the problem. A better statement of the problem may be, "Build a new passenger vehicle from readily available materials that has an improved gas mileage of at least 20 percent over current, similar-size models and is competitively priced."

Proposing Solutions Once the problem is clearly defined, the search for the solution can begin. This is not always an easy process. There are many factors to consider when searching for the best solution to a problem. The best solution must be found while working with many limitations.

Reading Check *Why is it important to have a clearly defined problem?*

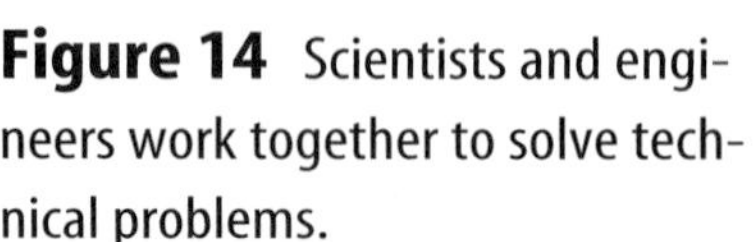

Figure 14 Scientists and engineers work together to solve technical problems.
Identify *the type of system scientists and engineers use to find technological solutions.*

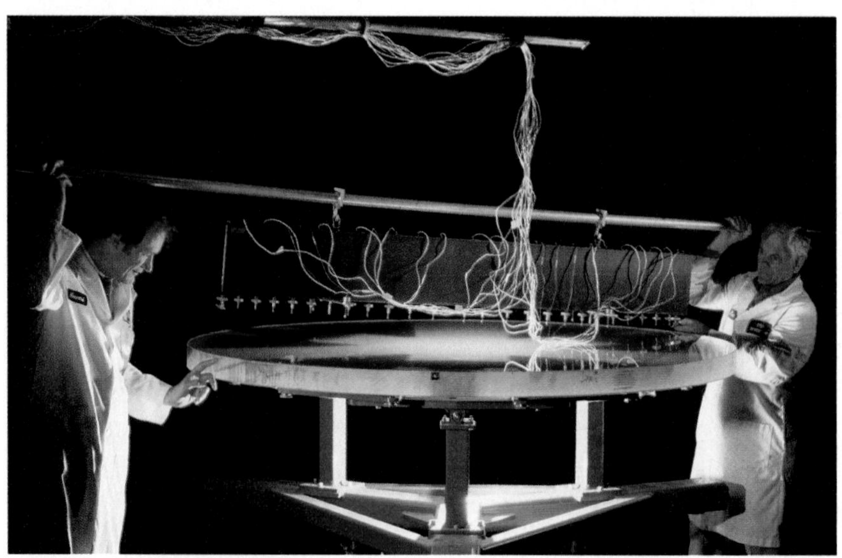

Constraints Design restrictions for products from outside factors are known as **constraints.** Examples of constraints are cost, environmental impact, and available materials. Constraints also include limited time to complete the task, limited funding, and design changes that must be made to satisfy laws and regulations. In addition, working with too little information or technical data can be constraints. Some designs may be perfect solutions but they cannot be built. Some materials may not be available or do not exist. Money may not be available. The solution to the problem may be a compromise because of many design constraints.

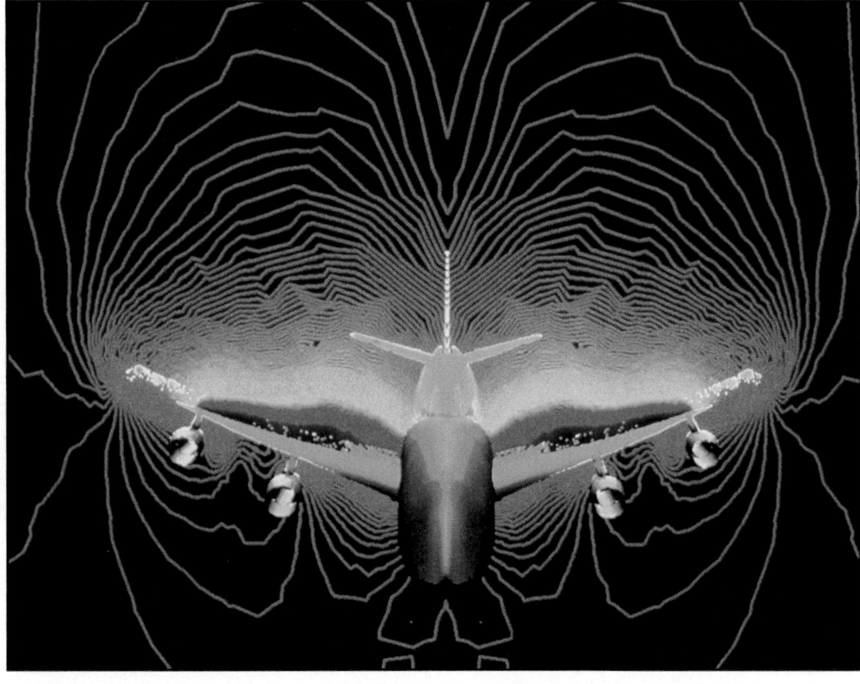

Figure 15 This is a computer simulation of an aircraft in a wind tunnel. The lines and colors surrounding the aircraft show airflow and turbulence. Scientists who use computer simulation can work to improve the aircraft design.
List *other ways that a computer simulation can be used.*

Performance Testing Once design ideas are accepted, drawings or models usually are constructed. These are carefully evaluated to identify possible design flaws. It is important to find design flaws as early in the process as possible. The later a design flaw is found in the process, the more expensive it is to correct.

After the drawing or model has been thoroughly evaluated, the design must be tested to make sure everything operates as planned. Many ways exist to performance test a design. The type of performance test used depends upon the design. For example, a chemical process may be tested initially by using a computer simulation. A **computer simulation** uses a computer to imitate the process to collect data or to test a process or procedure. Computer software is designed to mimic the processes in the design, as shown in **Figure 15.**

Complying with Laws and Regulations Local communities, states, and the federal government have laws and regulations regarding manufacturing processes, products, or buildings. These regulations often cover topics such as worker safety, environmental protection, product transportation, and how pleasing the manufacturing facility looks in the community. Products also must meet ethical or moral standards in their production, testing, and use. All of these factors must be considered when designing technology.

Figure 16 This robot, which is 39 cm tall, can be controlled by a cellular phone or by voice commands. The robot can be used in schools to teach about engineering and computer programming. **Explain** *why it is important to make a prototype of a new product.*

Prototypes and Pilot Plants Constructing a prototype or manufacturing in a pilot plant are methods of performance testing. A **prototype** is a full-scale model that is used to test a new product such as a new car design or a new airplane product. A prototype, as shown in **Figure 16,** is the first full-scale product that is built. A **pilot plant** is a smaller version of the real production equipment that closely models actual manufacturing conditions. A pilot plant would be used to test a new manufacturing process, such as a production line to produce a new medicine.

Flight-Test Engineer
One of the crew members responsible for flight-testing an airplane or helicopter prototype is a flight-test engineer. The flight-test engineer is responsible for planning the test, acquiring the test equipment, writing the test plan, conducting the test, and then writing the test report. Research the educational requirements of a flight-test engineer.

Limiting System Failure Performance testing is one way to limit system failures. A system is a group of devices that work together to perform a specific function. Other ways to limit system failures are to put redundant systems within a design. A redundant system performs the same function within a design. These systems are used for important processes that would force the entire system to shut down if there is a failure. When redundancy is built into the design, if one part fails the other system will take over, keeping the entire system from failing.

Control systems, such as a fail safe system, also are incorporated into designs to keep the system from failing. A **control system** is a device or collection of devices that monitors a system. This device makes corrections to keep the system operating at preset conditions. A simple control system is the thermostat on an air conditioner or heater. An air conditioner thermostat turns the unit on when the air temperature rises to a preset high temperature. When the air temperature drops to the preset low temperature, the unit is turned off. Control systems can be simple, like a thermostat, or very complicated, like an elaborate computer system that controls heating and cooling. The purpose of control systems is to keep the system operating at desired conditions.

Intellectual Property Once a new product is introduced, it is important that a company protects its investment. Research and development of a new product is costly. Often, it takes years to bring a new product to the consumer. Companies protect their rights to sell their new product or process by applying for a patent. A patent is a legal document granted by the government giving an inventor the exclusive right to make, use, and sell an invention for a specific number of years. In the United States, a new patent lasts for 20 years. A patent allows a company to get the money back that they have spent on creating the new product or process.

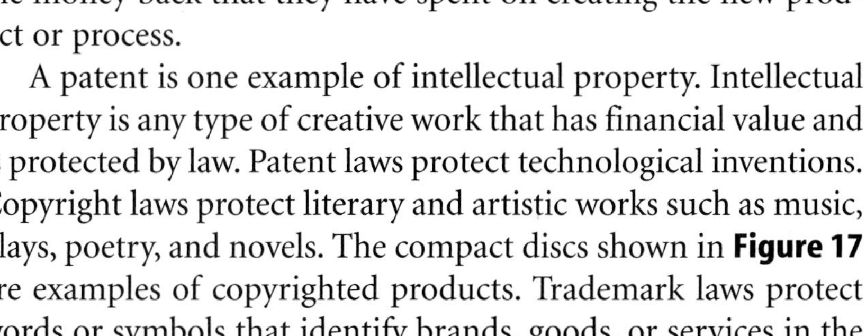

A patent is one example of intellectual property. Intellectual property is any type of creative work that has financial value and is protected by law. Patent laws protect technological inventions. Copyright laws protect literary and artistic works such as music, plays, poetry, and novels. The compact discs shown in **Figure 17** are examples of copyrighted products. Trademark laws protect words or symbols that identify brands, goods, or services in the marketplace.

Reading Check *What are three types of intellectual property, and can you give an example of each type?*

Figure 17 Music compact discs are copyrighted. Copyright protection lasts for the author's life plus an additional seventy years after the author's death.

section 3 review

Summary

Scientists and Engineers

- Scientists seek new scientific knowledge, but they may not have a plan to use that knowledge.
- Engineers use science to find solutions to problems or human needs.
- Technical problems often create a demand for new scientific knowledge.

Finding Solutions

- Scientists and engineers often work together to solve technical problems.
- Performance testing often is conducted using models and computer simulations to reduce the chance of system failure.
- Intellectual property such as technology, music, and symbols that identify products is protected by law.
- Companies protect their investment in technology with a patent.

Self Check

1. **Compare and contrast** the roles of scientists and engineers.
2. **Describe** a specific situation in which new technology created a demand for even more technology.
3. **Describe** Briefly describe the general process that is used to find technological solutions.
4. **Think Critically** Some foreign countries do not have intellectual property laws. It is common for people in these countries to illegally manufacture goods that are protected in the United States and sell them to consumers. Explain how this financially harms companies in the United States as well as the consumer.

Applying Math

5. **Use Percentages** The U.S. Customs and Border Protection made 6,500 seizures of merchandise in violation of intellectual property rights in the year 2003. This was an increase of about 700 seizures over the previous year. What was the percentage increase in the number of seizures from 2002 to 2003?

More Section Review gpescience.com

Model and Invent

Care Package

Goal

- **Model** packaging and shipping a product for consumer interest.
- **Calculate** the cost to produce a consumer product.
- **Test** your packaging design to see if it keeps the snacks from being damaged.

Possible Materials

shoe box
assorted snack foods
packing peanuts
shredded newspaper
packing tape
wrapping paper
markers

Safety Precautions

Real-World Problem

Suppose you are a young businessperson with a new idea for a product. You are going to create a care package for students to use during exams week. You must select the items for your care package, pack the items for display and shipping, and calculate the cost of the care package. How can you package foods to survive the hazards of shipping and keep your costs to a minimum?

Make a Model

1. Complete the safety form before you begin.
2. **Choose** the snack items that you would like to include in your care package.
3. **Plan** how you are going to pack your items so that they are nicely displayed and safe for shipping.
4. **Plan** the design of your decorative wrapper and labels. Remember to make the wrapper attractive so that customers will want to purchase your product.
5. **Design** a data table in your Science Journal to keep track of all your costs to make the care package.
6. **Evaluate** your design plan. Determine if there are ways to cut costs to increase your profit.
7. Make sure your teacher approves your plan before you start.

Test Your Model

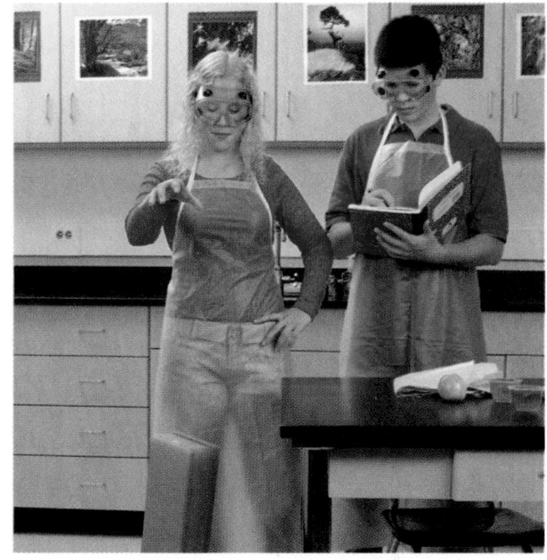

1. Assemble the items that you need to carry out your plan.
2. Pack the materials according to your plan. Do not put on the decorative wrapper until after you have tested your package.
3. After you have carefully packed your box, test to make sure that everything is secure. Drop the box from the height of your lab table five times.
4. Unwrap your box to see if anything is damaged. If the contents are broken, determine changes that can be made to improve your design.
5. If needed, retest your new design. After the drop tests are successful, apply your wrapper and labels.

Analyze Your Data

1. **Calculate** the cost of your package.
2. **Determine** the constraints of your design.

Conclude and Apply

1. **Explain** how you can improve your design. Consider cost, contents of package, and packaging.
2. **Determine** under what conditions the contents of your package may be damaged. For example, if your package contains chocolate, the package cannot be shipped during hot summer months. What changes can you make to your design to remove the shipping constraints?
3. **Infer** what would happen to the contents of your package if the package were dropped from heights taller than your lab table.

Communicating Your Data

Present your design to your classmates. Compare your design to those of your classmates. Look for ways that you could improve your design.

A Slippery Situation

Oddly enough, one of the slipperiest substances, Teflon®, was discovered by accident. Teflon was discovered in 1938 by Roy Plunkett, a research scientist. He was trying to improve the chemicals that refrigerators use to stay cool. He experimented with dozens of chemicals. One day, he noticed a gas cylinder that wouldn't operate. When he took the cylinder apart, he found that a white, slippery powder had clogged it.

Instead of being frustrated by a glitch in his experiment, Plunkett took the time to test this strange substance. What he found was amazing. This substance turned out to be extremely resistant to heat and highly inert. A substance that is inert will not react with other substances.

Why is Teflon resistant to heat and corrosion? Because it is a polymer.

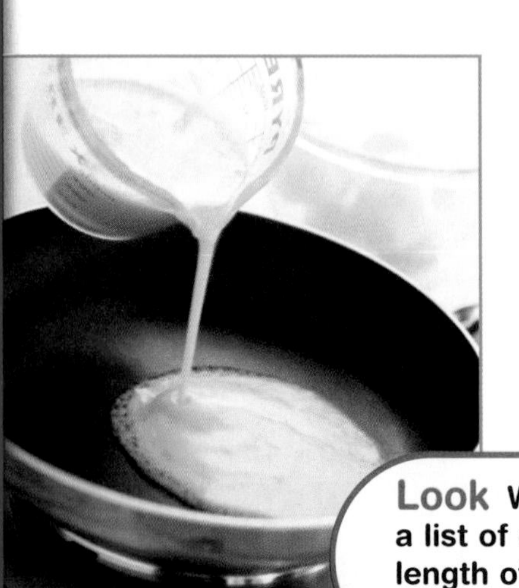

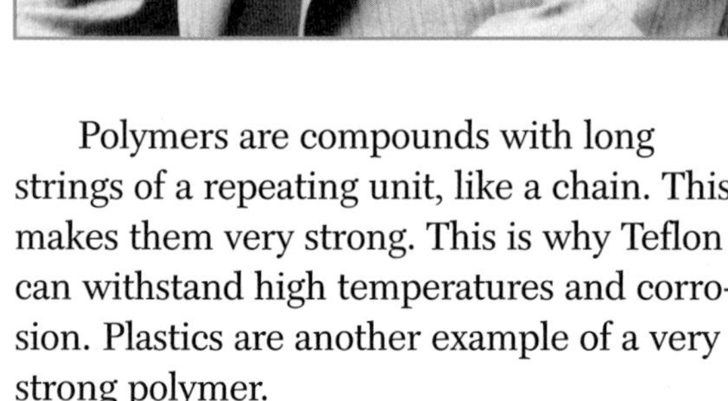

Polymers are compounds with long strings of a repeating unit, like a chain. This makes them very strong. This is why Teflon can withstand high temperatures and corrosion. Plastics are another example of a very strong polymer.

At first, scientists were skeptical about any practical uses for Teflon because it was so difficult to manufacture and therefore, very expensive. However, that changed when the Manhattan Project (the group of scientists who developed the atomic bomb) needed a substance that was heat resistant and chemically inert. This was the perfect job for Teflon. Since then, Teflon has found its way into many items, including the space shuttle, surgical implants, stain guards, and, of course, easy-to-clean pots and pans.

Look **With a partner, walk around the room and make a list of objects that are made from polymers. The length of the list will surprise you. Your teacher can help you find many polymers.**

For more information, visit gpescience.com

Reviewing Main Ideas

Section 1 Science and Technology

1. The study of science leads to a better understanding of the world around you.
2. Scientists bring information and insight to the public's attention and provide an understanding of why an event occurs. The knowledge that microbes cause illness and diseases changed people's behavior. Today we take precautions to prevent the spread of microbes by washing our hands often and covering our mouth when we cough.

3. Once people understand why an event or natural process occurs, they can respond and try to control the outcome.
4. Technology can be an object, a method or technique for making an object or tool, the knowledge or skill needed to operate a human-made object, or a system of people or objects that perform a task.
5. The value of technology varies over time and from location to location. This electric guitar may be valued by some in industrialized nations, but people in developing countries do not have electricity in their homes. This technology would be useless to them.

Section 2 Forces that Shape Technology

1. Funding for scientific research and technology comes from the federal government, private industries, and private foundations.
2. Consumers affect which technologies will be developed by their voting and buying habits.
3. Humans can have an enormous impact on other living things and must practice responsible technology.
4. The consequences of technology may be known immediately, but sometimes they are not known for a period of time.
5. Moral and ethical issues should be considered when doing scientific research or developing technology.

Section 3 Developing Technology

1. Scientists and engineers have different roles in developing technology.
2. Technological problems often create a demand for new scientific knowledge.

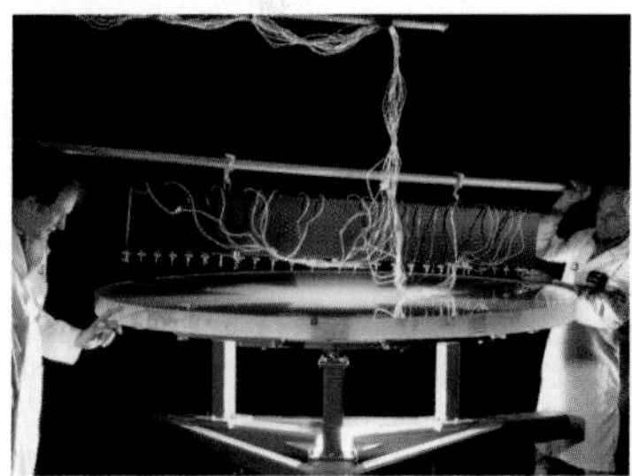

3. Scientists and engineers use a methodical process to develop technology. Performance testing is an important part of this process that reduces or prevents product failure.

FOLDABLES™ Use the Foldable that you made at the beginning of this chapter to help you review science, technology, and society.

Using Vocabulary

agricultural biotechnology p. 45
computer simulation p. 55
constraints p. 55
control system p. 56
engineer p. 53
pilot plant p. 56
prototype p. 56
society p. 46

Answer each question using the correct vocabulary word(s).

1. Which professional in science devises a way to use scientific knowledge to solve a problem or mass-produce a product?
2. What is performance-testing using a computer to imitate the process called?
3. What is the name of a group of people that share similar values and beliefs?
4. What is the term for design limitations put on the design by outside factors?
5. What is the full-scale model that is used to performance-test a new car design called?
6. What is the name of a scaled-down version of real production equipment that closely models actual manufacturing conditions?
7. What is a device that monitors a system to keep it operating at preset conditions called?

Checking Concepts

Choose the word or phrase that best answers the question.

8. Which represents technology?
   A) canary
   B) digital camera
   C) eating an apple
   D) native grass
9. Which technological object would be most valued in a developing country?
   A) compact discs
   B) electric can opener
   C) portable water purifier
   D) video recorder
10. What is a collection of scientific techniques that are used to create, improve, or modify plants, animals, and microorganisms?
    A) agricultural biotechnology
    B) agricultural cloning
    C) agricultural prototype
    D) agricultural simulation
11. Which is **NOT** normally a source of funding for technology?
    A) federal government
    B) local government
    C) private foundation
    D) private industry
12. Which is **NOT** considered part of moral and ethical considerations?
    A) doing what is cost effective
    B) doing what is fair
    C) doing what is honest
    D) doing what is right
13. What is the term for limitations put on the design of a product from outside factors?
    A) conflicts
    B) constraints
    C) parameters
    D) specifications
14. A thermostat on an air conditioner is an example of which of the following?
    A) control system
    B) environmental system
    C) monitoring system
    D) on/off system
15. What are patents legal protection for?
    A) new medicine
    B) new novels
    C) new product symbol
    D) new songs
16. Which does NOT represent technology?
    A) calculator
    B) computer program
    C) granite rock
    D) compact discs

Interpreting Graphics

17. Copy and complete the following concept map on technology.

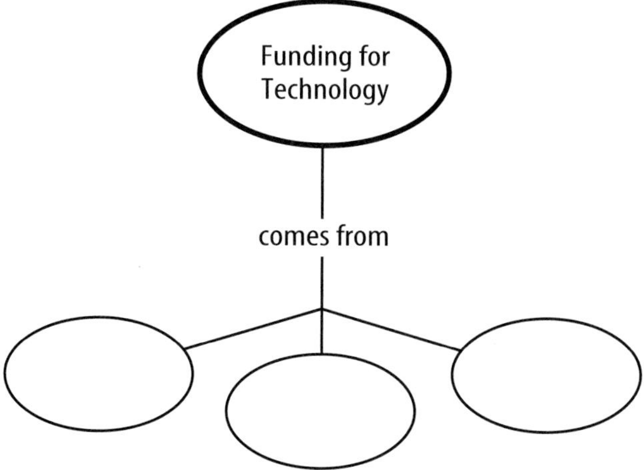

18. **Make a table** classifying each of the following actions as science or technology: finding the composition of Saturn's rings, developing the vehicle that can travel to Saturn, finding the human genes that cause a birth defect, and producing a medicine that can cure a disease.

Thinking Critically

19. **Draw Conclusions** Explain why technology of the past, such as the protective clothing shown in **Figure 1,** seems ridiculous to us today even though the people of the time accepted the technology as reasonable.

20. **Describe** one way in which scientific knowledge has changed people's understanding of a natural process or event.

21. **Describe** each type of technology and give a new example of each type.

22. **Explain** why the aircraft radar used by air-traffic controllers represents all types of technology.

23. **Explain** why the value of technology may differ for people living in different parts of the world.

24. **List** three technological items that you value that would not be important in a developing country such as Kenya.

25. **Describe** how technology can be used to supply basic human needs for survival to those who are in need.

26. **Explain** why ethics are important to scientists.

27. **Explain** how citizens of a local community can influence the development of new technology.

28. **Explain** During the mid 1970s, gasoline was in short supply. The price for gasoline increased. Automobile manufacturers produced smaller, fuel-efficient cars. The crisis soon was over and the smaller cars did not sell well. Explain how consumers' actions determined where the manufacturers spent their research funds.

29. **Explain** What type of intellectual property is protected by copyright? How does a copyright differ from a patent?

Applying Math

30. **Calculate** A hybrid car gets 20 km/L in city driving. The same model that is gasoline-powered gets 14 km/L in city driving. If gasoline costs $0.53 per liter, how much money would be saved per year if you drove 19,000 km of city driving?

31. **Calculate** How many liters of gasoline would be saved using the hybrid car instead of the gasoline-powered car?

chapter 2 Standardized Test Practice

Record your answers on the answer sheet provided by your teacher or on a sheet of paper.

Multiple Choice

1. What did Louis Pasteur discover as the cause of infections and diseases?

A. bad air

B. fleas

C. microbes

D. rats

Use the photo below to answer question 2.

2. Which technological object would be more valuable to the family shown?

A. airline industry

B. automobile industry

C. the Internet

D. medical center

3. Which is a technological system for sharing information?

A. airline industry

B. automobile industry

C. the Internet

D. medical center

4. What does a society share?

A. biotechnology

B. consumed goods

C. new technologies

D. similar values and beliefs

Use the photo below to answer question 5.

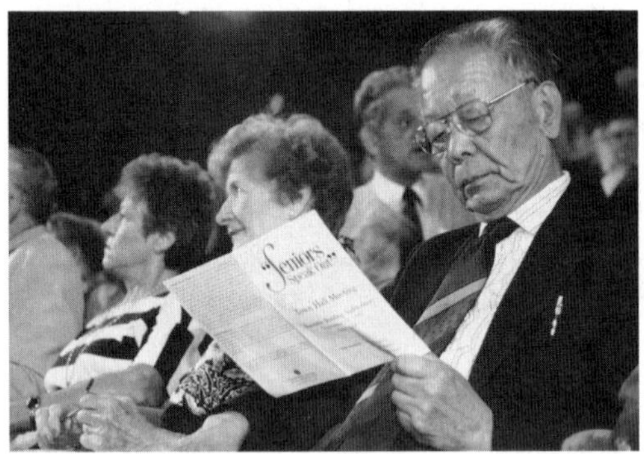

5. Which process used by people to influence technological development is shown?

A. consumer purchase

B. federal spending

C. political discussion

D. private funding

Gridded Response

6. If a liter of gasoline costs $0.58 and a car gets 11 km/L, what would it cost to travel 100 km?

7. Assuming that gasoline costs the same as in the previous question, how much would it cost to travel 100 km in a car that gets 6 km/L?

Short Response

Use photo below to answer question 8.

8. Identify one positive and one negative consequence from the use of gasoline-powered cars.

9. How would technological advances in health care benefit people in developing countries?

10. Why were people less concerned about fuel-efficient cars in the past than they are today?

11. What led to laws being created to prevent unethical treatment of animals and humans?

12. How does consumer acceptance affect the development of technology?

Extended Response

Use photo below to answer question 13.

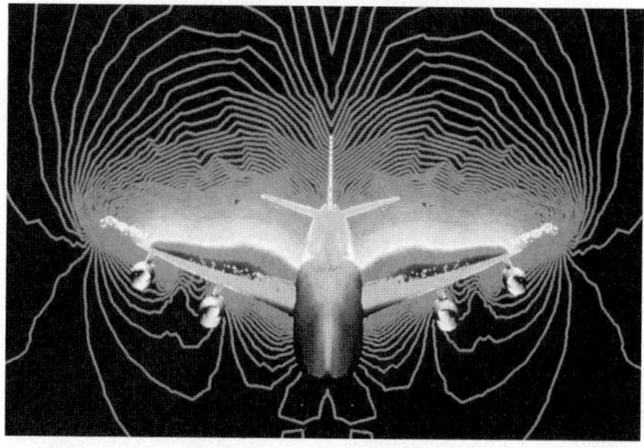

13. **Part A** In general, how do scientists and engineers work together to find a technological solution?
    Part B How would scientists and engineers determine if a new airplane design were more aerodynamic?

Test-Taking Tip

Get plenty of sleep—at least 8 hours every night—during test week and the week before the test.

unit 2 Motion, Forces, and Energy

How Are Waffles & Running Shoes Connected?

For centuries, shoes were made mostly of leather, cloth, or wood. These shoes helped protect feet, but they didn't provide much traction on slippery surfaces. In the early twentieth century, manufacturers began putting rubber on the bottom of canvas shoes, creating the first "sneakers." Sneakers provided good traction, but the rubber soles could be heavy—especially for athletes. One morning in the 1970s, an athletic coach stared at the waffles on the breakfast table and had an idea for a rubber sole that would be lighter in weight but would still provide traction. That's how the first waffle soles were born. Waffle soles soon became a world standard for running shoes.

unit projects

Visit unit projects at gpescience.com to find project ideas and resources.
Projects include:

- **Career** Explore the field of mechanical engineering through sports. Create an advertisement for your aerodynamic sports equipment.
- **Technology** Create a brochure showing five different shoe treads with reference to particular sports and environmental conditions.
- **Model** Design and build a model of a space station on the Moon or on a planet in our solar system.

Using virtual programming, *Roller Coaster Physics* provides an opportunity to engineer, test, and evaluate roller coaster design, and then build your own three-dimensional coaster.

chapter 3

Motion, Acceleration, and Forces

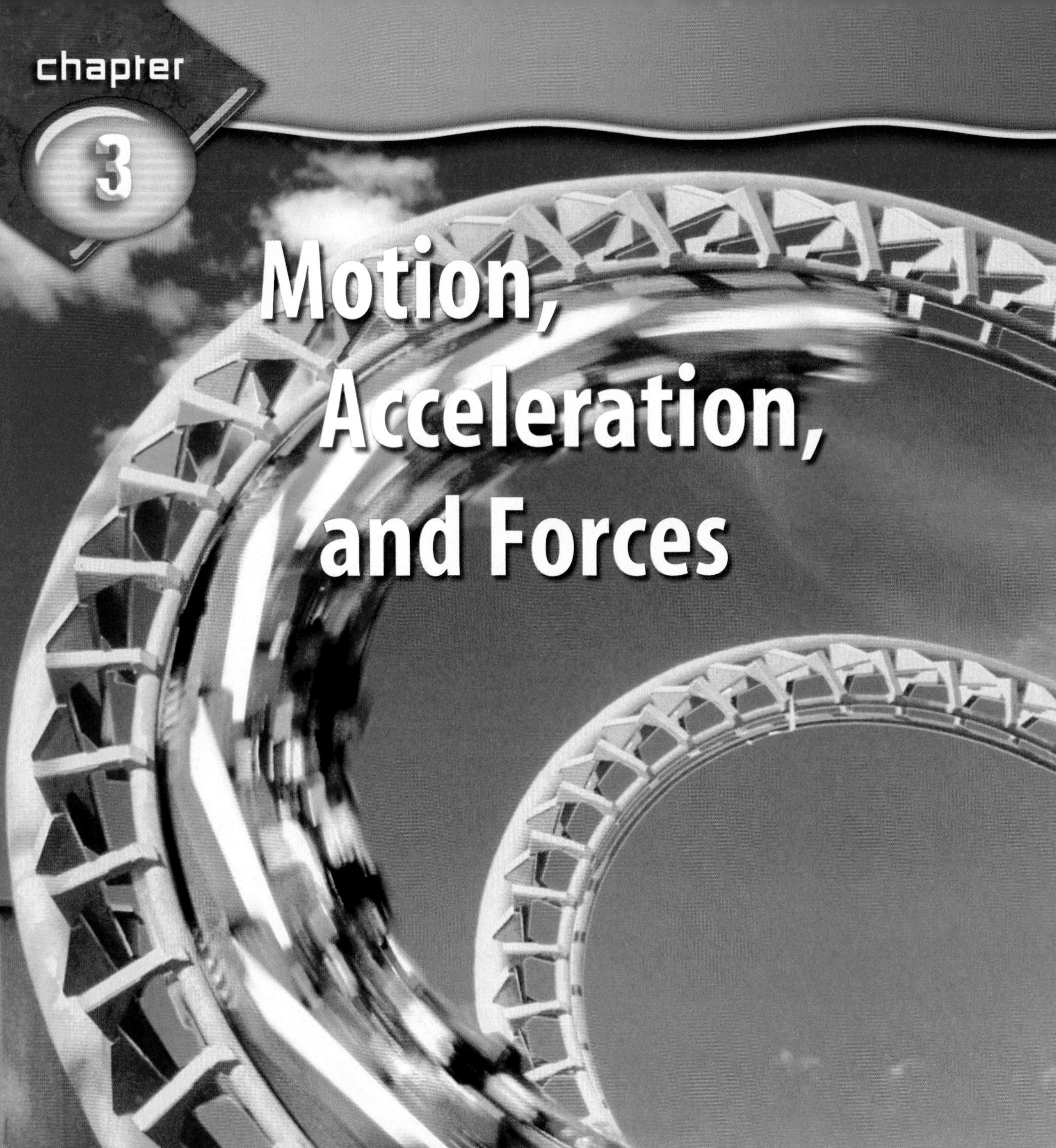

chapter preview

sections

Taking the Plunge

You probably don't think of an amusement park as a physics laboratory. The fast speeds, quick turns, and plunging falls are a great place to study the laws of gravity and motion. In fact, engineers use these laws when they are designing and building amusement park rides.

Science Journal Write a paragraph describing three rides in an amusement park and how the rides cause you to move.

Start-Up Activities

Compare Speeds

A cheetah can run at a speed of almost 120 km/h and is the fastest runner in the world. A horse can reach a speed of 64 km/h; and the fastest snake slithers at a speed of about 3 km/h. The speed of an object is calculated by dividing the distance the object travels by the time it takes it to move that distance. How does your speed compare to the speeds of these animals?

1. Complete the safety form.
2. Use a meterstick to mark off 10 m.
3. Have your partner use a stopwatch to determine how fast you run 10 m.
4. Divide 10 m by your time in seconds to calculate your speed in m/s.
5. Multiply your answer by 3.6 to determine your speed in km/h.
6. **Think Critically** Write a paragraph in your Science Journal comparing your speed with the maximum speed of a cheetah, horse, and snake. Could you win a race with any of them?

Preview this chapter's content and activities at gpescience.com

Motion Many things are in motion in your everyday life. Make the following Foldable to help you better understand motion as you read the chapter.

STEP 1 **Fold** a sheet of paper in half lengthwise. Make the back edge about 1.25 cm longer than the front edge.

STEP 2 **Fold** in half, then fold in half again to make three folds.

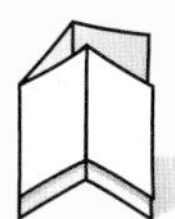

STEP 3 **Unfold and cut** only the top layer along the three folds to make four tabs.

STEP 4 **Label** the tabs.

Identify Questions Before you read the chapter, select a motion you can observe and write it under the left tab. As you read the chapter, write answers to the other questions under the appropriate tabs.

section 1

Describing Motion

Reading Guide

What You'll Learn

- **Distinguish** between distance and displacement.
- **Calculate** average speed.
- **Explain** the difference between speed and velocity.
- **Interpret** motion graphs.

Why It's Important

Understanding the nature of motion and how to describe it helps you understand why motion occurs.

Review Vocabulary

instantaneous: occurring at a particular instant of time

New Vocabulary

- displacement
- vector
- speed
- instantaneous speed
- average speed
- velocity

Figure 1 This mail truck is in motion.
Infer *How do you know the mail truck has moved?*

Motion

Every day you see objects and people in motion. Cars, buses, and trucks move along streets and highways. People around you walk and run and move in different directions. How would you describe the motion of the people and objects around you? One way to describe motion is to say that it is fast or slow. However, sometimes you might need to know the speed of an object, as well as the direction in which it is moving. It also might be important to know how the motion of something is changing.

Motion is a Change in Position You don't always need to see something move to know that motion has taken place. For example, suppose you see a mail truck stopped next to a mailbox, as in **Figure 1.** A little while later, you see the same truck stopped farther along the street. Although you didn't see the truck move, you know it moved because its position relative to the mailbox changed. Motion occurs when an object changes position.

Position Depends on a Reference Frame To measure the position of an object, a reference frame must be chosen. A reference frame is a group of objects that are not moving relative to each other. One point in the reference frame is chosen as the reference point. Then the position of an object is the distance and the direction of the object from the reference point. For example, in **Figure 1** the mailbox, the tree, and the houses can form a reference frame because they aren't moving relative to each other.

Motion is Relative The motion of an object depends on the reference frame that is chosen. For example, suppose you are sitting in a moving car. You are not moving if the car is the reference frame, but you are moving relative to the ground. Using the Sun as a reference frame, Earth is moving at a speed of 30 km/s. However, the Sun also is moving relative to a reference frame centered on the Milky Way Galaxy. Your motion would be different in each of these reference frames. However, there is no special reference frame that really is at rest compared to all other reference frames. As a result, an object does not have absolute motion, but only motion relative to a chosen reference frame.

Distance and Displacement Suppose a runner jogs to the 50-m mark and then turns around and runs back to the 20-m mark, as shown in **Figure 2.** How far did the runner travel? For a moving object, distance is the length of the path the object travels. The runner traveled a distance of 50 m plus 30 m, or 80 m. However, even though she ran 80 m, she is only 20 m from the starting line. Her final position is 20 m north of her initial position. **Displacement** is the distance and direction of an object's final position from its initial position. The runner's displacement is 20 m north.

Displacement includes both a size and a direction. The size of the displacement is the distance between the initial and final position. A quantity that is specified by both a size and a direction is a **vector.** Displacement includes both a size and a direction and is an example of a vector. However, distance is a physical quantity that does not include a direction and isn't a vector.

Measuring Average Speed

Procedure

1. Complete the safety form.
2. Mark a point on the floor with **tape.** Place a **toy car** at this point.
3. Give the toy car a gentle push forward. At the same time, start a **stopwatch.**
4. Record the time it takes for the car to come to a stop.
5. Use a **meterstick** to measure the distance the car traveled.

Analysis

Calculate the average speed of the car in m/s. How does the average speed depend on the direction of motion?

How do distance and displacement differ?

Speed

To describe how fast something is moving, you need to know how far it travels in a given amount of time. **Speed** is the distance an object travels per unit of time. In SI units, the unit of speed is meters per second (m/s). Usually the speed of an object changes as it moves from one place to another.

One way to describe the speed of an object is to measure its speed at a single instant of time. The **instantaneous speed** of an object is the speed at a single instant of time. A car's speedometer measures the instantaneous speed of the car.

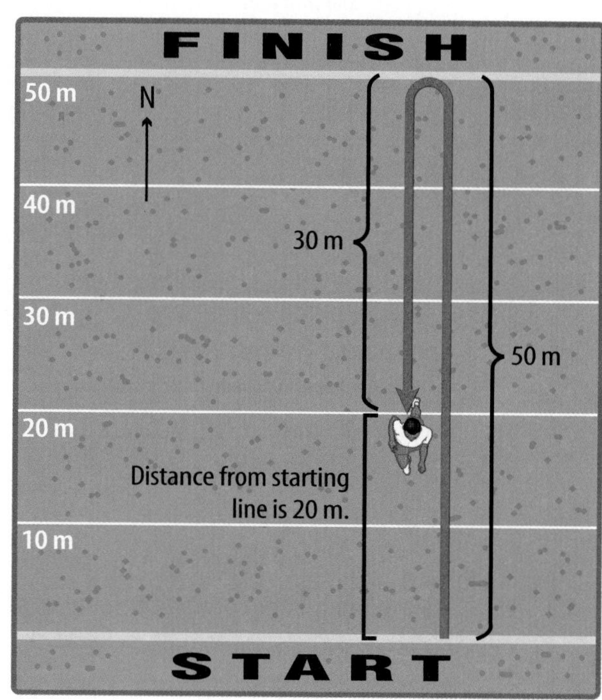

Figure 2 The runner's displacement is 20 m north of the starting line. However, the total distance traveled is 80 m.

Topic: Running Speeds
Visit gpescience.com for Web links to information about the speeds of various animals.

Activity Make a table comparing the top speed of various animals.

Average Speed Another way to describe the motion of an object is to determine the object's average speed. The average speed describes how quickly an object moved over the entire distance it traveled. The **average speed** of any object is the total distance traveled divided by the total travel time:

Average Speed Equation

$$\text{average speed (in meters/second)} = \frac{\text{total distance (in meters)}}{\text{total time (in seconds)}}$$

$$\bar{v} = \frac{d}{t}$$

The units of average speed are always a distance unit divided by a time unit. Sometimes it is more convenient to express average speed in units other than m/s, such as kilometers per hour (km/h).

Applying Math — Solve a One-Step Equation

AVERAGE SPEED What is the average speed of a car that travels a distance of 750 m in 25 s?

IDENTIFY known values and the unknown value

Identify the known values:

travels a distance of 750 m → means → $d = 750$ m

in 25 s → means → $t = 25$ s

Identify the unknown value:

What is the average speed? → means → $\bar{v} = ?$ m/s

SOLVE the problem

Substitute the given values of distance and time into the average speed equation:

$$\bar{v} = \frac{d}{t} = \frac{750\text{ m}}{25\text{ s}} = 30\text{ m/s}$$

CHECK your answer:

Does your answer seem reasonable? Check your answer by multiplying the time by the average speed. The result should be the distance given in the problem.

Practice Problems

1. An elevator travels a distance of 220 m from the first floor to the 60th floor in 27.5 s. What is the elevator's average speed?
2. A motorcyclist travels with an average speed of 20 km/h. If the cyclist is going to a friend's house 5 km away, how long does it take the cyclist to make the trip?

For more practice problems, go to page 879 and visit Math Practice at gpescience.com.

Velocity

You turn on the TV and see a news story about a hurricane, such as the one in **Figure 3,** that is approaching land with a speed of 10 km/h. The eye of the hurricane is located 450 km east of your location. Should you take shelter?

If you know only the hurricane's speed and location, you don't have enough information to answer the question. To decide whether the hurricane will reach your location, you need to know the direction in which the storm is moving. In other words, you need to know the velocity of the hurricane. The **velocity** of an object is the speed of the object and its direction of motion. Just like displacement, velocity is a vector that has a size and a direction. The size of an object's velocity is the object's speed.

Figure 3 The speed of a hurricane is not enough information to plot the hurricane's path. The direction of its motion also must be known.

Objects have different velocities if they are moving at different speeds, or in different directions. The people on the up escalator and the down escalator in **Figure 4** are moving with the same speed, but in opposite directions. As a result, the people on the up escalator and the down escalator have a different velocity.

The velocity of an object can change even if the speed of the object remains constant. For example, look at **Figure 4.** The race car has a constant speed as it goes around an oval track. Even though the speed remains constant, the velocity changes because the direction of the car's motion is changing.

Reading Check *How are velocity and speed different?*

Figure 4 The velocity of an object depends on both its speed and its direction of motion.
Describe *how the car's velocity changes as the car travels from the starting line to the top of the figure.*

The people on these two escalators have the same speed. However, their velocities are different because they are traveling in opposite directions.

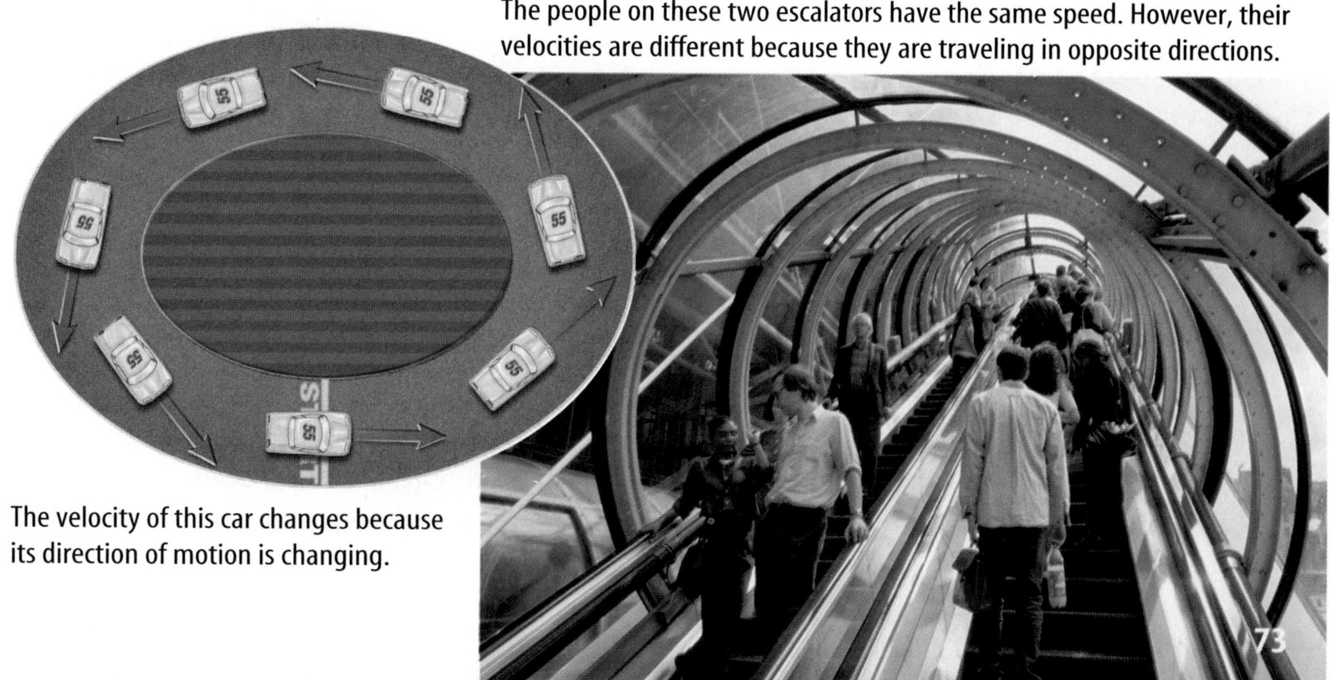

The velocity of this car changes because its direction of motion is changing.

Figure 5 The slope of a line on a distance-time graph gives the speed of an object in motion. **Identify** *the part of the graph that shows one of the swimmers resting for 10 min.*

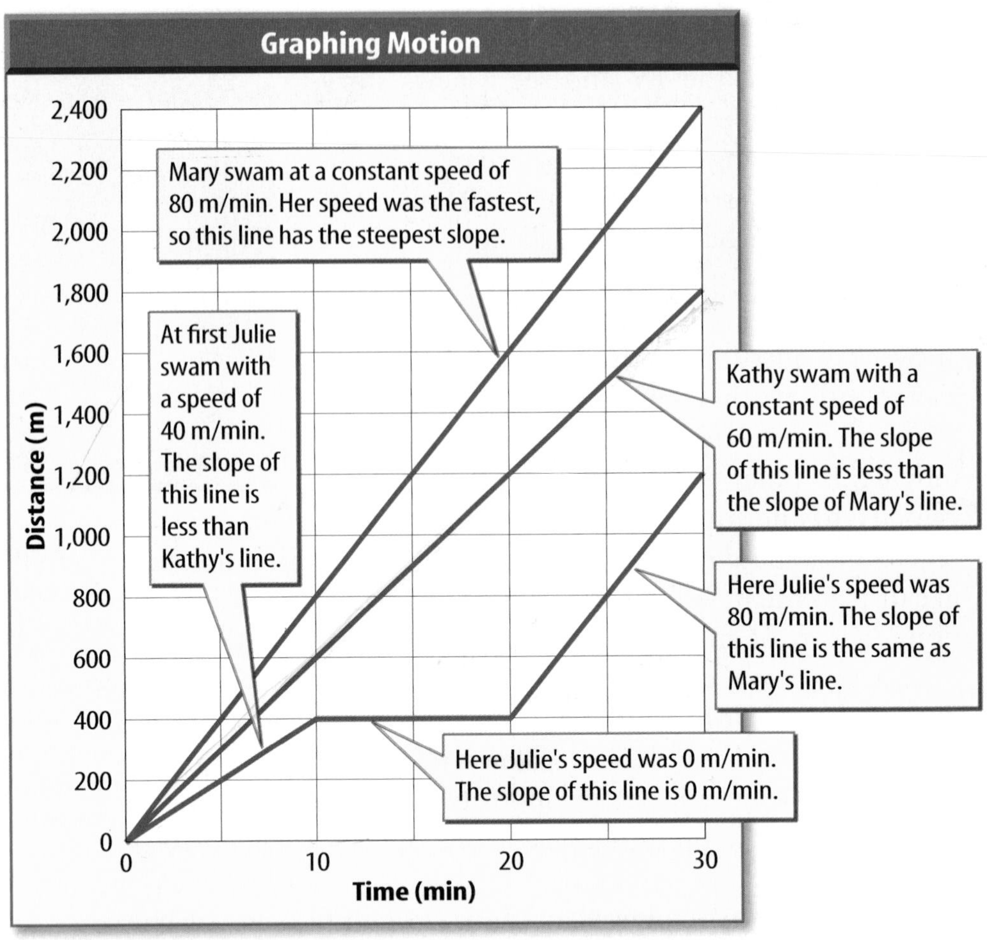

Graphing Motion
Mary swam at a constant speed of 80 m/min. Her speed was the fastest, so this line has the steepest slope.
At first Julie swam with a speed of 40 m/min. The slope of this line is less than Kathy's line.
Kathy swam with a constant speed of 60 m/min. The slope of this line is less than the slope of Mary's line.
Here Julie's speed was 80 m/min. The slope of this line is the same as Mary's line.
Here Julie's speed was 0 m/min. The slope of this line is 0 m/min.
Distance (m)
Time (min)
0
200
400
600
800
1,000
1,200
1,400
1,600
1,800
2,000
2,200
2,400
10
20
30

Graphing Motion

The motion of an object over a period of time can be shown on a distance-time graph. Time is plotted along the horizontal axis of the graph and the distance traveled is plotted along the vertical axis of the graph. If the object moves with constant speed, the increase in distance over equal time intervals is the same. As a result, the line representing the object's motion is a straight line.

For example, the graph shown in **Figure 5** represents the motion of three swimmers during a 30-min workout. The straight red line represents the motion of Mary, who swam with a constant speed of 80 m/min over the 30-min workout. The straight blue line represents the motion of Kathy, who swam with a constant speed of 60 m/min during the workout.

The graph shows that the line representing the motion of the faster swimmer is steeper. The steepness of a line on a graph is the slope of the line. The slope of a line on a distance-time graph equals the speed. A horizontal line on a distance-time graph has zero slope, and represents an object at rest. Because Mary has a larger speed than Kathy, the line representing her motion has a larger slope.

INTEGRATE
Astronomy

Orbital Speeds Size, mass, composition, and orbital radius are some of the ways astronomers classify planets. Astronomers also measure the speed of a planet relative to the Sun. Make a table of the average orbital speeds and orbital radii of the nine planets in the solar system. How are distance from the Sun and orbital speed related?

Changing Speed The green line represents the motion of Julie, who did not swim at a constant speed. She covered 400 m at a constant speed during the first 10 min, rested for the next 10 min, and then covered 800 m during the final 10 min. During the first 10 min, her speed was less than Mary's or Kathy's, so her line has a smaller slope. During the middle period her speed is zero, so her line over this interval is horizontal and has zero slope. During the last time interval she swam as fast as Mary, so that part of her line has the same slope.

Plotting a Distance-Time Graph On a distance-time graph, the distance is plotted on the vertical axis and the time on the horizontal axis. Each axis must have a scale that covers the range of numbers to be plotted. In **Figure 5** the distance scale must range from 0 to 2,400 m and the time scale must range from 0 to 30 min. Then, each axis can be divided into equal time intervals to represent the data. Once the scales for each axis are in place, the data points can be plotted. After plotting the data points, draw a line connecting the points.

Topic: Olympic Swimming Speeds
Visit gpescience.com for Web links to information about the speeds of Olympic swimmers over the past 60 years.

Activity Make a speed-year graph showing the swimming speeds over time. Are there any trends in the speed data?

section 1 review

Summary

Position and Motion

- The position of an object is determined relative to a reference point.
- Motion occurs when an object changes its position relative to a reference point.
- Distance is the length of the path an object has traveled. Displacement is the distance and direction of a change in position.

Average Speed and Velocity

- Average speed is the total distance traveled divided by the total time:

$$\bar{v} = \frac{d}{t}$$

- The velocity of an object includes the object's speed and its direction of motion relative to a reference point.

Graphing Motion

- On a distance-time graph, time is the horizontal axis and distance is the vertical axis.
- The slope of a line plotted on a distance-time graph is the speed.

Self Check

1. **Infer** whether the size of an object's displacement could be greater than the distance the object travels.
2. **Describe** the motion represented by a horizontal line on a distance-time graph.
3. **Explain** whether, during a trip, a car's instantaneous speed can ever be greater than its average speed.
4. **Describe** the difference between average speed and constant speed.
5. **Think Critically** You are walking toward the back of a bus that is moving forward with a constant velocity. Describe your motion relative to the bus and relative to a point on the ground.

Applying Math

6. **Calculate Average Speed** Michiko walked a distance of 1.60 km in 30 min. Find her average speed in m/s.
7. **Calculate Distance** A car travels at an average speed of 30.0 m/s for 0.8 h. Find the total distance traveled in km.

section 2

Acceleration

Reading Guide

What You'll Learn

- **Identify** how acceleration, time, and velocity are related.
- **Describe** how to calculate the average acceleration of an object.
- **Explain** how positive and negative acceleration affect motion.

Why It's Important

Acceleration occurs all around you as objects speed up, slow down, or change direction.

Review Vocabulary

velocity: a vector with a size that is an object's speed and a direction that is the direction of an object's motion

New Vocabulary

- acceleration

Acceleration, Speed, and Velocity

A car at a stoplight starts moving when the light finally turns green. When the car started moving, the car's velocity increased and the car was accelerating. **Acceleration** is the change in velocity divided by the time for the change to occur.

Remember that velocity is a vector that includes both the speed and direction of motion. So a change in velocity can either be a change in speed or a change in direction of motion. Acceleration occurs when the speed of an object changes or its direction of motion changes.

Reading Check *When does acceleration occur?*

Just as velocity has a size and direction, acceleration also is a vector that has a size and direction. **Figure 6** shows how acceleration can cause an object to speed up or slow down.

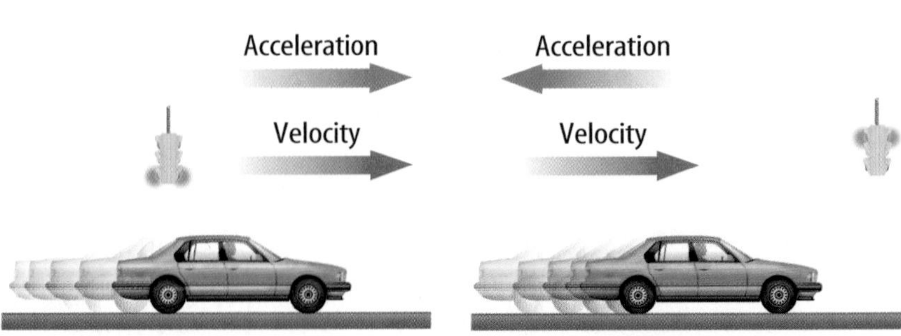

The car speeds up when the acceleration and velocity are in the same direction.

The car slows down when the acceleration and velocity are in opposite directions.

Figure 6 These cars are both accelerating because their speed is changing.
Infer *how the car would move if its acceleration were zero.*

Changing Direction A change in velocity can be either a change in how fast something is moving or a change in the direction of movement. Any time a moving object changes direction, its velocity changes and it is accelerating. Think about a horse on a carousel. Although the horse's speed remains constant, the horse is accelerating because it is changing direction constantly as it travels in a circular path, as shown in **Figure 7.** Another example is the motion of Earth around the Sun in a nearly circular path. As a result, the direction of Earth's velocity is continually changing. This means that Earth is accelerating as it orbits the Sun.

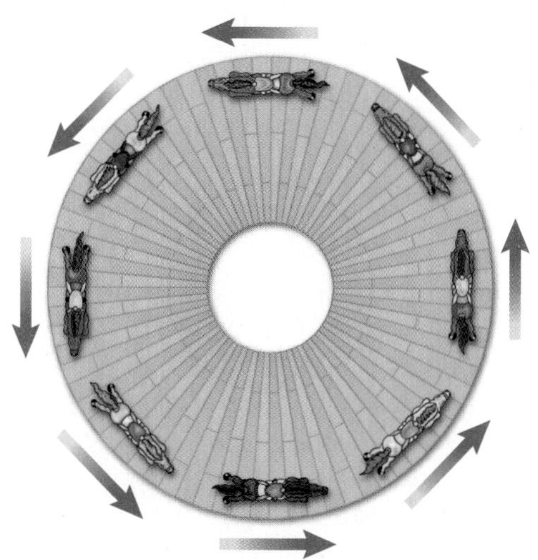

Figure 7 The speed of each horse on this carousel is constant but each is accelerating because the direction of its velocity is changing.

Calculating Acceleration

When an object moves from one place to another, it might speed up, slow down, and change direction many times. Each change in velocity causes the acceleration of the object to change. When acceleration is changing, the size of the average acceleration over some period of time can be calculated. If the direction of motion of an object doesn't change, the size of its average acceleration can be calculated using the following equation:

Acceleration Equation

average acceleration (in m/s^2) =

$$\frac{(\text{final velocity (in m/s)} - \text{initial velocity (in m/s)})}{(\text{final time (s)} - \text{initial time (s)})}$$

$$\bar{a} = \frac{(v_f - v_i)}{(t_f - t_i)}$$

In this equation, v_i is the size of the velocity at the start of the time period and v_f is the size of the velocity at the end of the time period. For motion in a single direction, v_i is the same as the initial speed, and v_f is the same as the final speed. In SI units, acceleration has units of m/s^2—meters per second squared. An acceleration of 1 m/s^2 means that the velocity of the object increases by 1 m/s each second. If the object was initially at rest, its velocity would be 1 m/s after 1 s, 2 m/s after 2 s, and so on.

Speed-Time Graphs For an object moving in one direction, the acceleration can be found from a speed-time graph. For this type of graph, the vertical axis is the object's speed and the horizontal axis is the time. Then the slope of the plotted line is the size of the object's acceleration. **Figure 8** shows some examples of speed-time graphs.

Aircraft Carriers In 1911, American pilot Eugene Ely successfully landed on a specially equipped deck on the battleship *Pennsylvania.* Today, huge aircraft carriers provide floating landing strips for military airplanes. Carriers must be equipped to provide large accelerations, both positive and negative, to the jets launched from their decks. Research the accelerations provided by modern carriers as they launch and land aircraft.

NATIONAL GEOGRAPHIC

VISUALIZING ACCELERATION

Figure 8

Acceleration can be positive, negative, or zero depending on whether an object is speeding up, slowing down, or moving at a constant speed. If the speed of an object is plotted on a graph, with time along the horizontal axis, the slope of the line is related to the acceleration.

Ⓐ The car in the photograph on the right is maintaining a constant speed of about 90 km/h. Because the speed is constant, the car's acceleration is zero. A graph of the car's speed with time is a horizontal line.

Ⓑ The green graph shows how the speed of a bouncing ball changes with time as it falls from the top of a bounce. The ball speeds up as gravity pulls the ball downward, so the acceleration is positive. For positive acceleration, the plotted line slopes upward to the right.

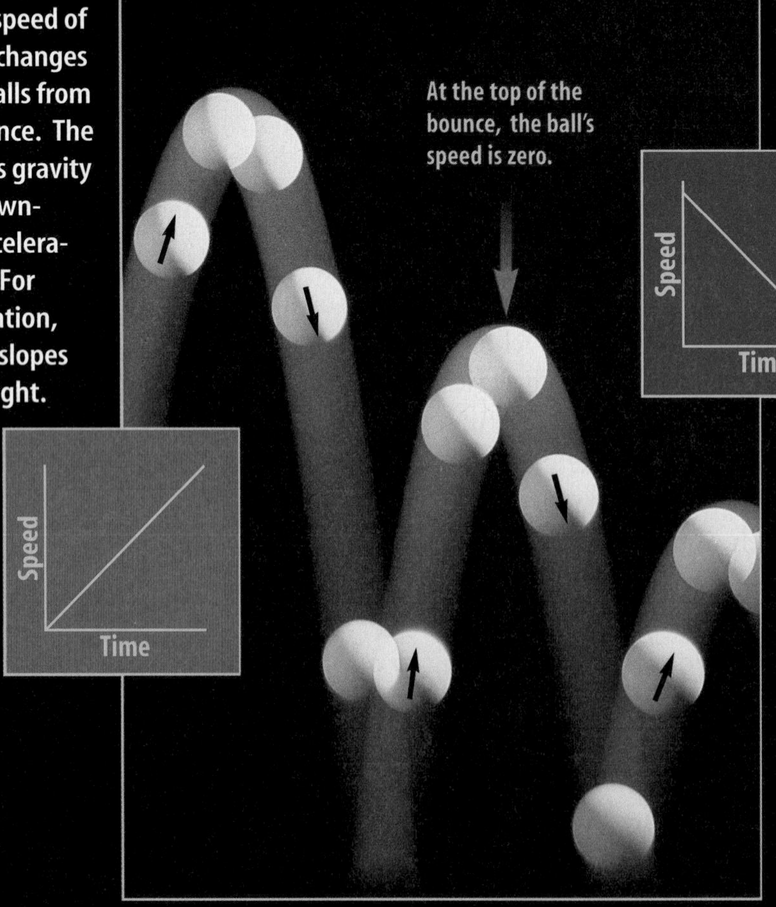

Ⓒ The blue graph shows the change with time in the speed of a ball after it hits the ground and bounces upward. The climbing ball slows as gravity pulls it downward, so the acceleration is negative. For negative acceleration, the plotted line slopes downward to the right.

Calculating Positive Acceleration How is the acceleration for an object that is speeding up different from that of an object that is slowing down? Suppose the jet airliner in **Figure 9** starts at rest and moves down the runway in a single direction. After accelerating for 20 s, it reaches a speed of 80 m/s. Because it started from rest, its initial speed was zero. The jetliner's average acceleration can be calculated as follows:

$$\bar{a} = \frac{(v_f - v_i)}{(t_f - t_i)} = \frac{(80 \text{ m/s} - 0 \text{ m/s})}{20 \text{ s}} = 4 \text{ m/s}^2$$

The airliner is speeding up, so the final speed is greater than the initial speed and the acceleration is a positive number.

Calculating Negative Acceleration Now imagine that the skateboarder in **Figure 9** is moving in a straight line at a constant speed of 3 m/s and comes to a stop in 2 s. The final speed is zero and the initial speed was 3 m/s. The skateboarder's acceleration is calculated as follows:

$$\bar{a} = \frac{(v_f - v_i)}{(t_f - t_i)} = \frac{(0 \text{ m/s} - 3 \text{ m/s})}{2 \text{ s}} = -1.5 \text{ m/s}^2$$

The skateboarder is slowing down, so the final speed is less than the initial speed and the acceleration is a negative number. For an object moving in one direction, the acceleration will be a positive number if an object is speeding up and a negative number if the object is slowing down.

Figure 9 A speed-time graph tells you if acceleration is a positive or negative number.
Infer *what a steeper slope indicates on a speed-time graph.*

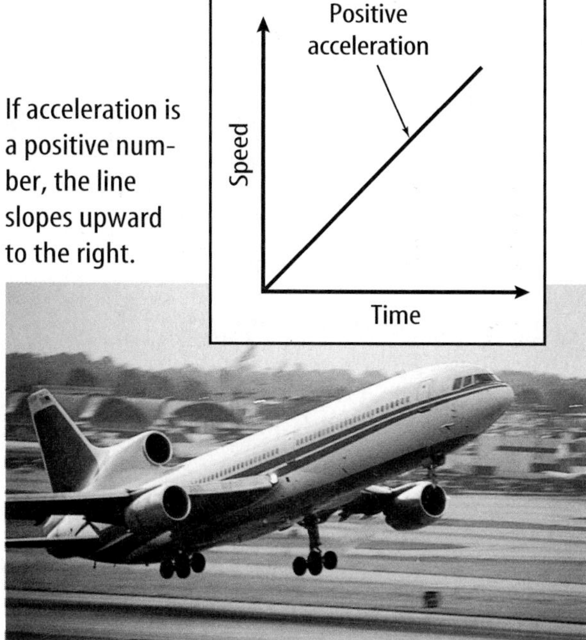
If acceleration is a positive number, the line slopes upward to the right.
Positive acceleration
Speed
Time

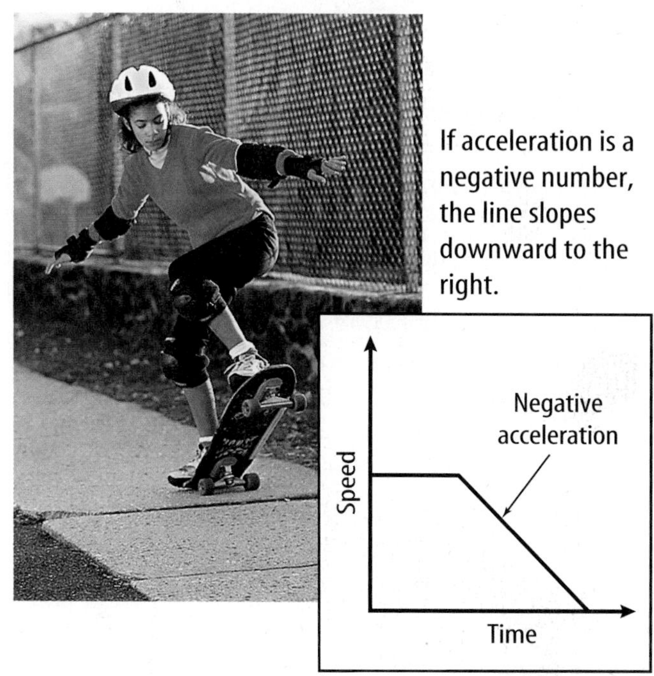
If acceleration is a negative number, the line slopes downward to the right.
Negative acceleration
Speed
Time

Figure 10 Cars on this roller coaster can reach a speed of about 150 km/h in 4 s.

Amusement Park Acceleration

Riding roller coasters in amusement parks can give you the feeling of danger, but these rides are designed to be safe. Engineers use the laws of physics to design amusement park rides that are thrilling, but harmless. Roller coasters are constructed of steel or wood. Because wood is not as rigid as steel, wooden roller coasters do not have hills that are as high and steep as some steel roller coasters have. As a result, the highest speeds and accelerations usually are produced on steel roller coasters.

Steel roller coasters can offer multiple steep drops and inversion loops, which give the rider large accelerations. As the rider moves down a steep hill or an inversion loop, he or she will accelerate toward the ground due to gravity. When riders go around a sharp turn, they also are accelerated. This acceleration makes them feel as if a force is pushing them toward the side of the car. **Figure 10** shows the fastest roller coaster in the United States.

Reading Check *What happens when riders on a roller coaster go around a sharp turn?*

section 2 review

Summary

Acceleration, Speed, and Velocity

- Acceleration is the change in velocity divided by the time needed for the change to occur.
- A change in velocity occurs when an object changes its speed or direction of motion.
- The speed of an object increases if the acceleration is in the same direction as the velocity.
- The speed of an object decreases if the acceleration and the velocity of the object are in opposite directions.

Calculating Acceleration

- The average acceleration of an object can be calculated using this equation:

$$\bar{a} = \frac{(v_f - v_i)}{(t_f - t_i)}$$

- When an object moving in a straight line speeds up, its acceleration is positive
- When an object moving in a straight line slows down, its acceleration is negative.

Self Check

1. **Determine** whether you are accelerating as Earth rotates once every 24 h.
2. **Determine** the change in velocity of a car that starts at rest and has a final velocity of 20 m/s north.
3. **Explain** why streets and highways have speed limits rather than velocity limits.
4. **Describe** the motion of an object that has an acceleration of 0 m/s^2.
5. **Think Critically** Suppose a car is accelerating so that its speed is increasing. Describe the plotted line of the motion of the car on a distance-time graph.

Applying Math

6. **Calculate Acceleration** A ball is dropped from a cliff and reaches a speed of 29.4 m/s after 3.0 s. What is the ball's average acceleration?
7. **Calculate Speed** A sprinter runs with an average acceleration of 4.5 m/s^2. What is the sprinter's speed 2 s after leaving the starting blocks?

More Section Review gpescience.com

section 3 Motion and Forces

Reading Guide

What You'll Learn

- **Explain** how forces and motion are related.
- **Compare and contrast** static friction and sliding friction.
- **Describe** the effects of air resistance on falling objects.

Why It's Important

All changes in motion are caused by the forces that act on objects.

Review Vocabulary

vector: a quantity that includes both a size and a direction

New Vocabulary

- force
- net force
- balanced forces
- unbalanced forces
- friction
- static friction
- sliding friction
- air resistance

What is force?

When you shoot a basketball or kick a soccer ball, you are exerting a force on an object. In fact, every push or pull you exert results in a force being applied to some object. A **force** is a push or pull that one object exerts on another. Just like velocity and acceleration, force also is a vector that has a size and a direction. The size of a force often is called the strength of the force. The direction of a force is the direction in which the push or pull is applied. For example, when you lift your backpack, you apply an upward force. In SI units, force is measured in newtons (N). You have to apply a force of about 3 N to lift a full can of soft drink.

Changing Motion What happens to the motion of an object when you exert a force on it? A force can cause the motion of an object to change, as in **Figure 11.** The racket strikes the ball with a force that causes the ball to stop and then move in the opposite direction. If you have played billiards, you know that you can cause a ball at rest to roll into a pocket by striking it with another ball. The force applied by the moving ball causes the ball at rest to move in the direction of the force. In these cases, the velocities of the tennis ball and the billiard ball were changed by a force.

Figure 11 This ball is hit with a force. The racket strikes the ball with a force in the opposite direction of its motion. As a result, the ball changes the direction it is moving.

Figure 12 Forces can be balanced and unbalanced.

→ + ← = →

Net Force = →

B These students are pushing on the box with unequal forces in opposite directions. The box will be moved in the direction of the larger force.

→ + ← = 0

Net Force = 0

A These students are pushing on the box with an equal force but in opposite directions. Because the forces are balanced, the box does not move.

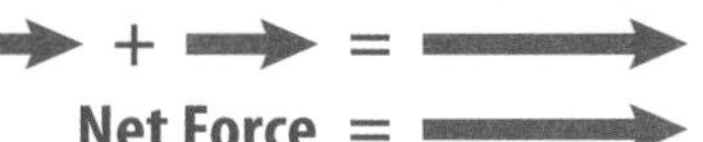

Net Force = →

C These students are pushing on the box in the same direction. The combined forces will cause the box to move.

Balanced Forces Force does not always change velocity. In **Figure 12A,** two students are pushing on opposite sides of a box. Both students are pushing with an equal force but in opposite directions. When two or more forces act on an object at the same time, the forces combine to form the **net force.** The net force on the box in **Figure 12A** is zero because the two forces cancel each other. Forces on an object that are equal in size and opposite in direction are called **balanced forces.**

Unbalanced Forces Another example of how forces combine is shown in **Figure 12B.** When two students are pushing with unequal forces in opposite directions, a net force occurs in the direction of the larger force. When forces combine to produce a net force that is not zero, the forces acting on the object are **unbalanced forces.** The net force that causes the box to accelerate will be the difference between the two forces because they are in opposite directions.

In **Figure 12C,** the students are pushing on the box in the same direction. These forces are combined, or added together, because they are exerted on the box in the same direction. The net force that acts on this box is found by adding the two forces together.

Reading Check *Give another example of an unbalanced force.*

Topic: Forces and Fault Lines

Visit gpescience.com for Web links to information about the unbalanced forces that occur along Earth's fault lines.

Activity Use inexpensive materials such as bars of soap to model the forces and movements along the fault lines. Share your models and demonstrations with your class.

Unbalanced Forces Change Velocity When the forces acting on an object are balanced, the velocity of an object doesn't change. If you and a friend push on a door from opposite sides with the same size force, the door doesn't move. The net force is zero and the forces are balanced. But if you push harder, the door moves in the direction of your push. The velocity the door, or any object, changes only when the forces on it are unbalanced.

Friction

Suppose you give a skateboard a push with your hand. The skateboard speeds up as you push it and then keeps moving after it leaves your hand. What happens to the skateboard's speed if it is moving on a flat, level surface? You know the answer. The skateboard slows down and finally stops.

After it left your hand, the skateboard's velocity changed because the forces acting on it were unbalanced. The unbalanced force that slowed the skateboard was friction. **Friction** is the force that opposes the sliding motion of two surfaces that are in contact.

What causes friction? The size of the frictional force exerted by one surface on another depends on the materials the surfaces are made from and the roughness of the surfaces. All surfaces have bumps and dips, including highly-polished metal surfaces that seem very smooth. When two surfaces are in contact, sticking occurs where the bumps and dips touch each other. This causes microwelds to form between the two surface. These microwelds tend to make the surfaces stick together and cause friction to occur.

The frictional force between two surfaces increases when the force pushing the surfaces together increases, as shown in **Figure 13.** When the surfaces are pushed together with more force, more of the bumps and dips come into contact. This increases the strength of the microwelds.

Mini
LAB

Comparing Friction

Procedure

1. Complete the safety form.
2. Place an **ice cube,** a **rock,** an **eraser,** a **wood block,** and a square of **aluminum foil** at one end of a **metal or plastic tray.**
3. Slowly lift the end of the tray with the items.
4. Have a partner use a **metric ruler** to measure the height of the raised end of the tray at which each object begins to slide. Record your measurements.

Analysis

1. List the height at which each object began to slide.
2. How did the height at which objects began to slide depend on the roughness of the objects?
3. How did the forces due to static and sliding friction compare for each object?

Try at Home

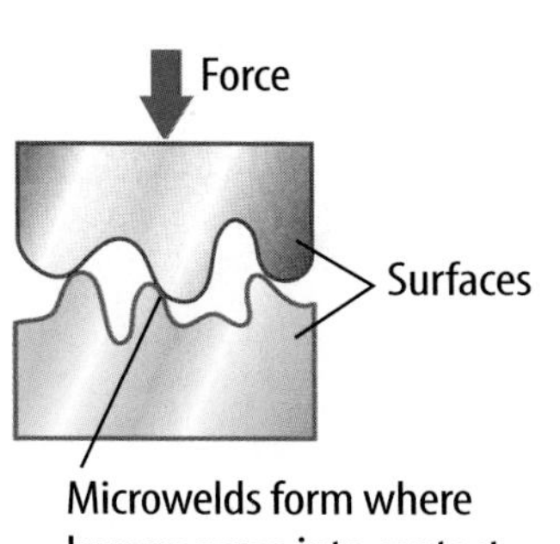
Force
Surfaces
Microwelds form where bumps come into contact.

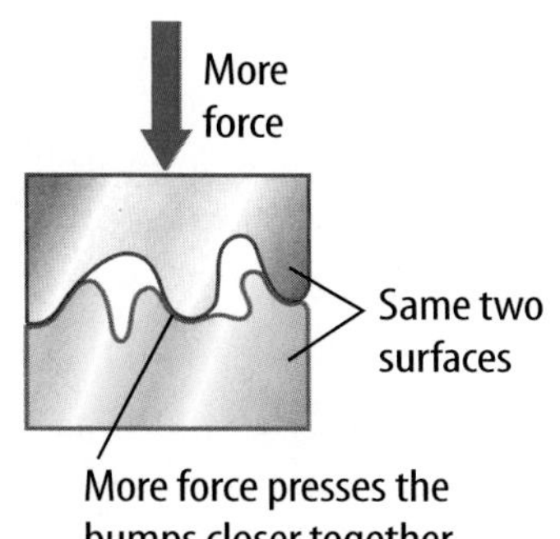
More force
Same two surfaces
More force presses the bumps closer together.

Figure 13 Friction is due to the microwelds that form between two surfaces in contact. These microwelds become stronger when the force pushing the surfaces together increases.
Explain *how the area of contact between the surfaces changes when they are pushed together with a larger force.*

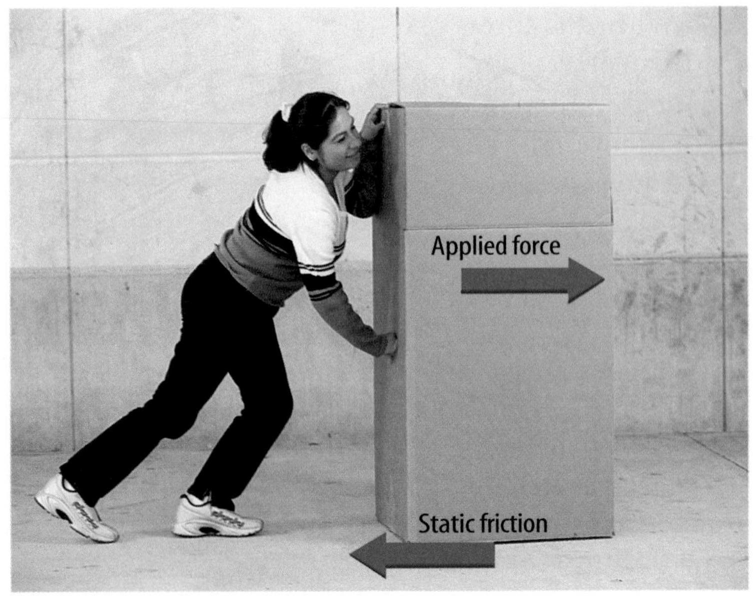

Figure 14 The static friction force balances the applied force so the box doesn't move.
Infer *the net force on the box.*

Static Friction Suppose you push on a heavy box, like the one in **Figure 14,** and it doesn't budge. The velocity of the box didn't change, so another force is acting on the box that balances your push. This force is a frictional force called static friction that is due to the microwelds that have formed between the box and the floor. **Static friction** is the frictional force that prevents two surfaces in contact from sliding past each other. In this case, the box didn't move because your push is not large enough to break the microwelds between the two surfaces.

Sliding Friction You and a friend push together on the box and the box slides along the floor, as shown in **Figure 15.** As the box slides on the floor, another frictional force—sliding friction—opposes the motion of the box. **Sliding friction** is the force that acts in the opposite direction to the motion of a surface sliding on another surface. If you stop pushing, sliding friction causes the box to slow down and stop. The force due to sliding friction between two surfaces is smaller than the force due to static friction.

Reading Check *Sliding friction always is in what direction?*

Rolling Friction When an object rolls over a surface, a frictional force due to rolling friction slows the object down. Rolling friction usually is much less than sliding friction. This is why it is easier to move a heavy object if it is on wheels.

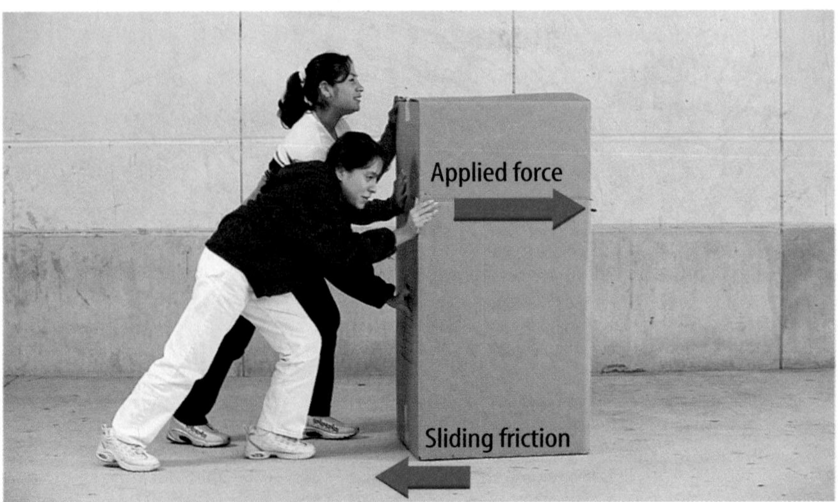

Figure 15 Sliding friction acts on the sliding box in the direction opposite to the motion of the box.

Air Resistance

When an object falls toward Earth, it is pulled downward by the force due to gravity. However, a type of frictional force called **air resistance** opposes the motion of objects that move through the air. Air resistance causes objects to fall with different accelerations and different speeds. If there were no air resistance, then all objects, like the apple and feather shown in **Figure 16,** would fall with the same acceleration.

Air resistance acts in the direction opposite to the velocity of an object moving in air. If an object is falling downward, air resistance acts upward on the object. The size of the air resistance force depends on the size and shape of an object. Imagine dropping two identical plastic bags. One is crumpled into a ball and the other is spread out. When the bags are dropped, the crumpled bag falls faster than the spread-out bag. The downward force of gravity on both bags is the same, but the upward force of air resistance on the crumpled bag is less. As a result, the net downward force on the crumpled bag is greater, as shown in **Figure 17.**

The amount of air resistance on an object depends on the speed, size, and shape of the object. Air resistance, not the object's mass, is why feathers, leaves, and sheets of paper fall more slowly than pennies, acorns, and apples.

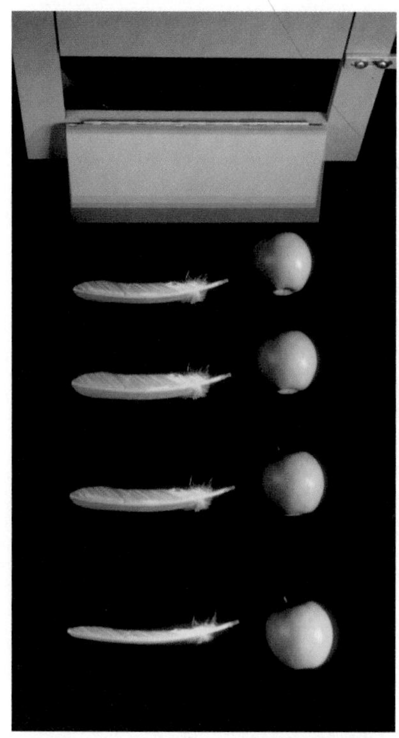

Figure 16 This photograph shows an apple and a feather falling in a vacuum. The photograph was taken with a strobe light that flashes on and off at a steady rate. Because there is no air resistance in a vacuum, the feather and the apple fall with the same acceleration.

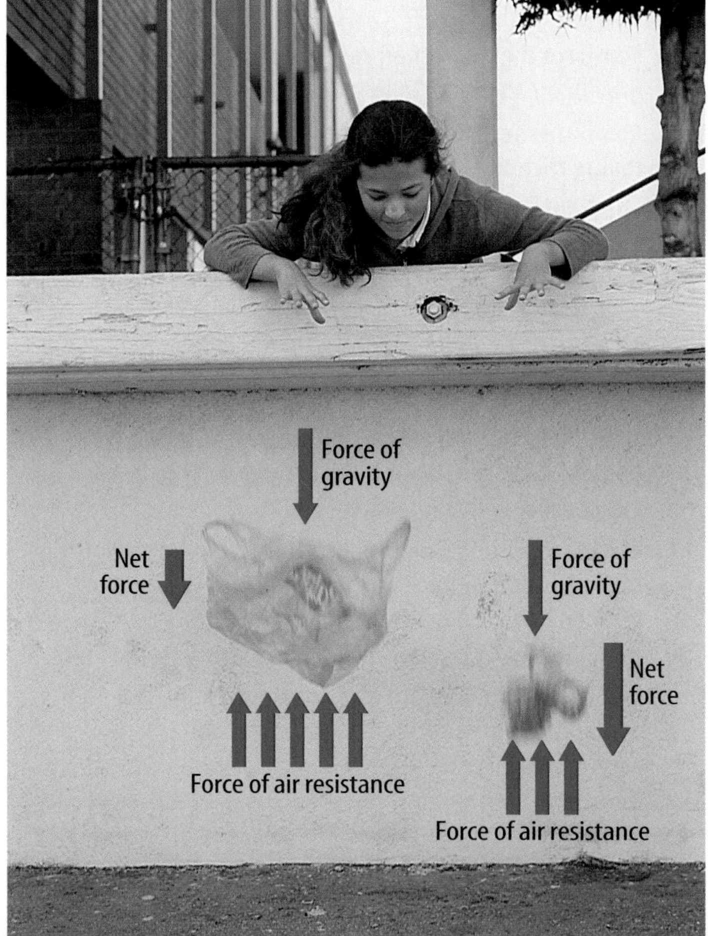

Figure 17 Because the bag on the left has a greater surface area, the bag on the left has more air resistance acting on it as it falls.

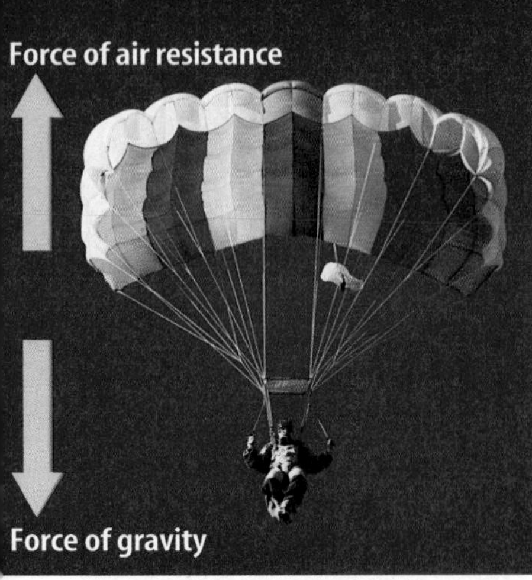

Figure 18 When the sky diver reaches terminal velocity, the force due to air resistance balances the force due to gravity.

Terminal Velocity As an object falls, the downward force of gravity causes the object to accelerate. For example, after falling 2,000 m, without the effects of air resistance a skydiver's speed would be almost 200 m/s, or more than 700 km/h.

However, as the speed of a falling object increases, the upward force of air resistance also increases. This causes the net force on a sky diver to decrease as the sky diver falls. Finally, the upward air resistance force becomes large enough to balance the downward force of gravity. The net force on the object is zero, and the velocity of the object doesn't change. The object falls with a constant velocity called the terminal velocity. The terminal velocity is the highest velocity the falling object will reach.

The terminal velocity depends on the size, shape, and mass of the falling object. The air resistance force on an open parachute, like the one in **Figure 18,** is much larger than the air resistance on the a sky diver with a closed parachute. With the parachute open, the terminal velocity of the sky diver becomes small enough that the sky diver can land safely.

section 3 review

Summary

Balanced and Unbalanced Forces

- A force is a push or a pull that one object exerts on another object.
- The net force on an object is the combination of all the forces acting on the object.
- When the net force on an object is zero, the forces acting on the object are balanced.
- When the net force on an object is not zero, the forces acting on the object are unbalanced.
- Unbalanced forces cause the velocity of an object to change.

Friction

- Friction is the force that opposes motion between two surfaces that are touching each other.
- Friction depends on the types of surfaces and the force pressing the surfaces together.
- Static friction prevents surfaces that are in contact from sliding past each other.

Air Resistance

- Air resistance is a force that acts on objects that move through the air.

Self Check

1. **Explain** why the frictional force between two surfaces increases when the force pushing the surfaces together increases.
2. **Compare** the force of air resistance and the force of gravity on an object falling at its terminal velocity.
3. **Compare** the size of forces due to static, sliding, and rolling friction between two surfaces.
4. **Think Critically** What is the net force on a car stopped at a stop sign? What is the net force on a car moving in a straight line with a constant speed?

Applying Math

5. **Calculate Net Force** Two students push on a box in the same direction, and another pushes on the box in the opposite direction. What is the net force on the box if each pushes with a force of 50 N?
6. **Calculate Sliding Friction** You push a box with a force of 80 N. If the net force on the box is 50 N, what is the force on the box due to sliding friction?
7. **Calculate Acceleration** The downward force of gravity and the upward force of air resistance on a falling ball are both 5 N. What is the ball's acceleration?

More Section Review gpescience.com

Force and Acceleration

If you stand at a stoplight, you will see cars stopping for red lights and then taking off when the light turns green. What makes the cars slow down? What makes them speed up? How do balanced and unbalanced forces affect the acceleration of objects?

Real-World Problem

How does an unbalanced force affect the motion of an object?

Goals

- **Observe** how changing the net force on an object affects its acceleration.
- **Interpret** data collected for several trials.

Materials

tape	this science book
paper clip	triple-beam balance
10-N spring scale	**electronic balance*
large book	**Alternate materials*

Safety Precautions

Proper eye protection should be worn at all times while performing this lab.

Procedure

1. Complete the safety form.
2. Tie the string around the book and attach the paper clip to the string.
3. Prepare a data table with the following headings: *Force, Mass.*
4. If available, use a large balance to find the mass of the two books.
5. Place the book on the floor or on the surface of a long table. Use the paper clip to hook the spring scale to the book.

6. Pull the book across the floor or table at a slow but constant velocity. While pulling, read the force measured by the spring scale and record it in your data table.
7. Repeat step 5 two more times, once accelerating slowly and once accelerating quickly. Be careful not to pull too hard. Your spring scale will read only up to 10 N.
8. Place a second book on top of the first book and repeat steps 3 through 6.

Conclude and Apply

1. **Organize** the pulling forces from greatest to least for each set of trials. What is the relationship between the net force and the acceleration of the book?
2. **Explain** how adding the second book changed your results.

Communicating Your Data

Compare your conclusions with those of other students in your class.

Design Your Own

Comparing Motion from Different Forces

Goals

- **Identify** several forces that you can use to propel a small toy car across the floor.
- **Demonstrate** the motion of the toy car using each of the forces.
- **Graph** the position versus time for each force.
- **Compare** the motion of the toy car resulting from each force.

Possible Materials

small toy car
ramps or boards of different lengths
springs or rubber bands
string
stopwatch
meterstick or tape measure
graph paper

Safety Precautions

Real-World Problem

Think about a small ball. How many ways could you exert a force on the ball to make it move? You could throw it, kick it, roll it down a ramp, blow it with a large fan, etc. Do you think the distance and speed of the ball's motion will be the same for all of these forces? Do you think the acceleration of the ball would be the same for all of these types of forces?

Form a Hypothesis

Based on your reading and observations, state a hypothesis about how a force can be applied that will cause a toy car to go fastest.

Test Your Hypothesis

Make a Plan

1. Complete the safety form.
2. As a group, agree upon the hypothesis and decide how you will test it. Identify which results will confirm your hypothesis.
3. **List** the steps you will need to test your hypothesis. Be sure to include a control run. Be specific. Describe exactly what you will do in each step. List your materials.
4. **Prepare** a data table in your Science Journal to record your observations.

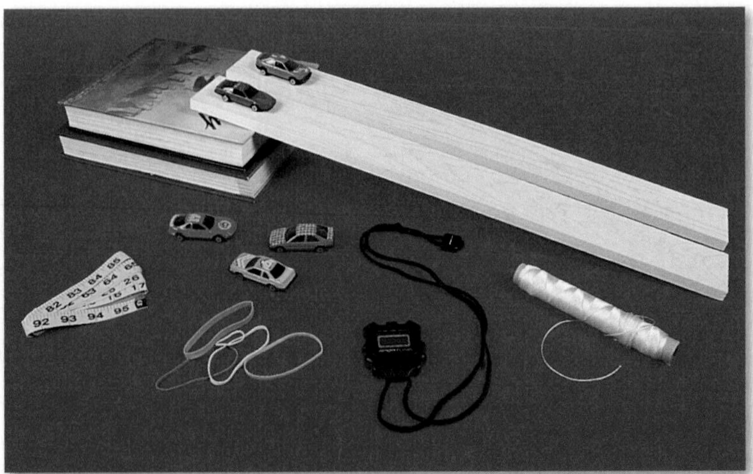

5. **Read** the entire experiment to make sure all steps are in logical order and will lead to a useful conclusion.
6. **Identify** all constants, variables, and controls of the experiment. Keep in mind that you will need to have measurements at multiple points. These points are needed to graph your results. You should make sure to have several data points taken after you stop applying the force and before the car starts to slow down. It might be useful to have several students taking measurements, making each responsible for one or two points.

Follow Your Plan

1. Make sure your teacher approves your plan before you start.
2. Carry out the experiment as planned.
3. While doing the experiment, record your observations and complete the data tables in your Science Journal.

Analyze Your Data

1. **Graph** the position of the car versus time for each of the forces you applied. How can you use the graphs to compare the speeds of the toy car?
2. **Calculate** the speed of the toy car over the same time interval for each of the forces that you applied. How do the speeds compare?

Conclude and Apply

1. **Evaluate** Did the speed of the toy car vary depending upon the force applied to it?
2. **Determine** For any particular force, did the speed of the toy car change over time? If so, how did the speed change? Describe how you can use your graphs to answer these questions.
3. **Draw Conclusions** Did your results support your hypothesis? Why or why not?

Communicating Your Data

Compare your data with those of other students. **Discuss** how the forces you applied might be different from those others applied and how that affected your results.

Science and Language Arts

"A Brave and Startling Truth"

by Maya Angelou

We, this people, on a small and lonely planet
Traveling through casual space
Past aloof stars, across the way of indifferent suns
To a destination where all signs tell us
It is possible and imperative that we learn
A brave and startling truth ...

When we come to it
Then we will confess that not the Pyramids
With their stones set in mysterious perfection ...
Not the Grand Canyon
Kindled into delicious color
By Western sunsets
These are not the only wonders of the world ...

When we come to it
We, this people, on this minuscule and kithless[1]
globe ...
We this people on this mote[2] of matter

When we come to it
We, this people, on this wayward[3], floating body
Created on this earth, of this earth
Have the power to fashion for this earth
A climate where every man and every woman
Can live freely without sanctimonious piety[4]
Without crippling fear

When we come to it
We must confess that we
are the possible
We are the miraculous, the
true wonder of the world
That is when, and only
when
We come to it.

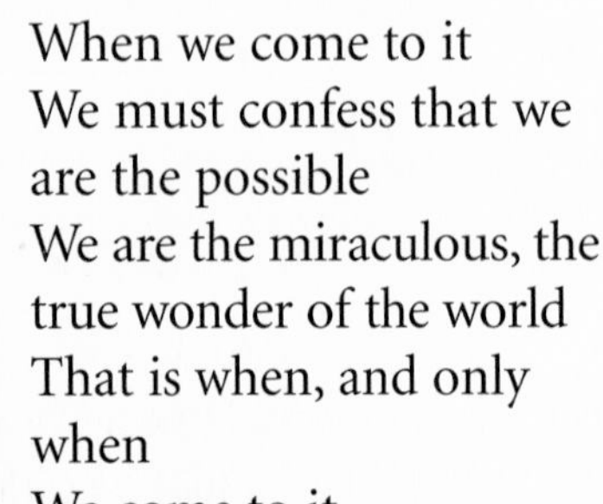

Understanding Literature

Descriptive Writing The poet names some special places on Earth. These places, although marvelous, fall short of being really wonderful. How does Angelou contrast Earth's position within the universe to emphasize the importance of people?

Respond to the Reading

1. What adjectives does the poet use to describe Earth?
2. What does the poet believe are the true wonders of the world?
3. **Linking Science and Writing** Write a six-line poem that describes Earth's movement from the point of view of the Moon.

Sometimes a person doesn't need to see movement to know that something has moved. Even though we don't necessarily see Earth's movement, we know Earth moves relative to a reference point such as the Sun. If the Sun is the reference point, Earth moves because the Sun appears to change its position in the sky. The poem describes Earth's movement from a reference point outside of Earth, somewhere in space.

1 to be without friends or neighbors
2 small particle
3 wanting one's own way in spite of the advice or wishes of another
4 a self-important show of being religious

Reviewing Main Ideas

Section 1 Describing Motion

1. Motion is a change of position of a body. Distance is the measure of how far an object moved. Displacement is the distance and direction of an object's change in position from the starting point.
2. A reference point must be specified in order to determine an object's position.
3. The average speed of an object can be calculated from this equation:

$$\bar{v} = \frac{d}{t}$$

4. Velocity is a vector that includes the speed and direction of a moving object.
5. The slope of a line on a distance-time graph is equal to the speed.

Section 2 Acceleration

1. Acceleration occurs when an object changes speed or changes direction.
2. An object speeds up if its acceleration is in the direction of its motion.

3. An object slows down if its acceleration is opposite to the direction of its motion.
4. Acceleration is the rate of change of velocity, and is calculated from this equation:

$$\bar{a} = \frac{v_f - v_i}{(t_f - t_i)}$$

Section 3 Motion and Forces

1. A force is a push or a pull.
2. The net force acting on an object is the combination of all the forces acting on the object.
3. The forces on an object are balanced if the net force is zero.
4. Sliding friction is a force opposing the sliding motion of two surfaces in contact. Static friction prevents surfaces in contact from sliding past each other.
5. Air resistance opposes the motion of objects that move in the air.

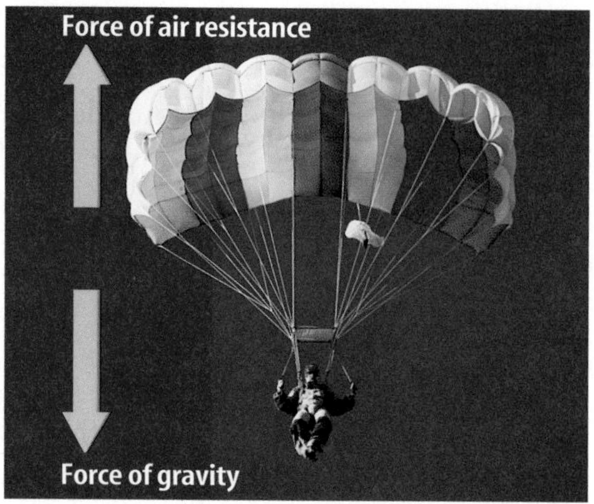

FOLDABLES™ Use the Foldable that you made at the beginning of this chapter to help you review motion.

Using Vocabulary

acceleration p. 76
air resistance p. 85
average speed p. 72
balanced forces p. 82
displacement p. 71
force p. 81
friction p. 83
instantaneous speed p. 71
net force p. 82
sliding friction p. 84
speed p. 71
static friction p. 84
unbalanced forces p. 82
vector p. 71
velocity p. 72

Compare and contrast the following pairs of vocabulary words.

1. speed—velocity
2. distance—displacement
3. average speed—instantaneous speed
4. balanced force—net force
5. acceleration—velocity
6. velocity—instantaneous speed
7. static friction—sliding friction
8. friction—air resistance

Checking Concepts

Choose the word or phrase that best answers the question.

9. Which of the following do you calculate when you divide the total distance traveled by the total travel time?
   A) average speed
   B) constant speed
   C) variable speed
   D) instantaneous speed

10. Which term below best describes the forces on an object with a net force of zero?
    A) inertia
    B) balanced forces
    C) acceleration
    D) unbalanced forces

11. Which of the following is a proper unit of acceleration?
    A) s/km^2 **C)** m/s^2
    B) km/h **D)** cm/s

12. Which of the following is not used in calculating acceleration?
    A) initial velocity **C)** time interval
    B) average speed **D)** final velocity

13. In which of the following conditions does the car NOT accelerate?
    A) A car moves at 80 km/h on a flat, straight highway.
    B) The car slows from 80 km/h to 35 km/h.
    C) The car turns a corner.
    D) The car speeds up from 35 km/h to 80 km/h.

14. What is the slope of a line on a distance-time graph?
    A) displacement **C)** speed
    B) force **D)** acceleration

15. How can speed be defined?
    A) acceleration/time
    B) change in velocity/time
    C) distance/time
    D) displacement/time

Interpreting Graphics

Use the table below to answer question 16.

Distance-Time for Runners				
Time (s)	1	2	3	4
Sally's distance (m)	2	4	6	8
Alonzo's distance (m)	1	2	2	4

16. **Graph** the motion of both runners on a distance-time graph. What is the average speed of each runner? Which runner stops briefly? Over what time interval do they both have the same speed?

17. Copy and complete this concept map on motion.

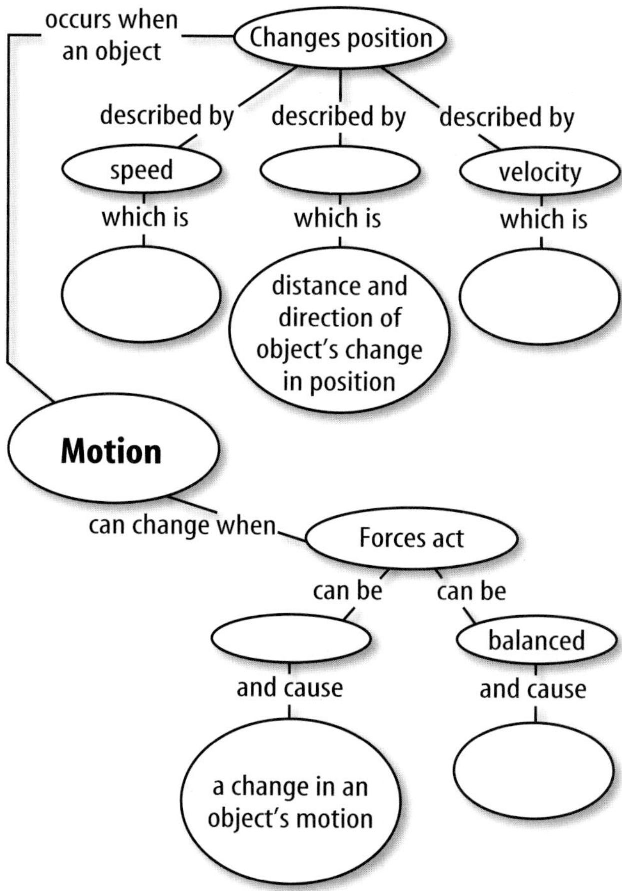

Thinking Critically

18. **Evaluate** Which of the following represents the greatest speed: 20 m/s, 200 cm/s, or 0.2 km/s?

19. **Describe** a line on a distance-time graph representing an object that is not moving.

20. **Infer** You push a book on a table so that its velocity is constant. If the force you apply is 15 N, what is the force due to sliding friction?

21. **Determine** If you walked 20 m, took a book from a library table, turned around and walked back to your seat, what are the distance traveled and displacement?

22. **Explain** When you are describing the rate that a race car goes around a track, should you use the term *speed* or *velocity* to describe the motion?

Applying Math

23. **Calculate Speed** A cyclist must travel 800 km. How many days will the trip take if the cyclist travels 8 h/day at an average speed of 16 km/h?

24. **Calculate Acceleration** A satellite's speed is 10,000 m/s. After 1 min, it is 5,000 m/s. What is the satellite's acceleration?

25. **Calculate Displacement** A cyclist leaves home and rides due east for a distance of 45 km. She returns home on the same bike path. If the entire trip takes 4 h, what is her average speed? What is her displacement?

26. **Calculate Velocity** The return trip of the cyclist in question 13 took 30 min longer than her trip east, although her total time was still 4 h. What was her velocity in each direction?

Use the graph below to answer question 27.

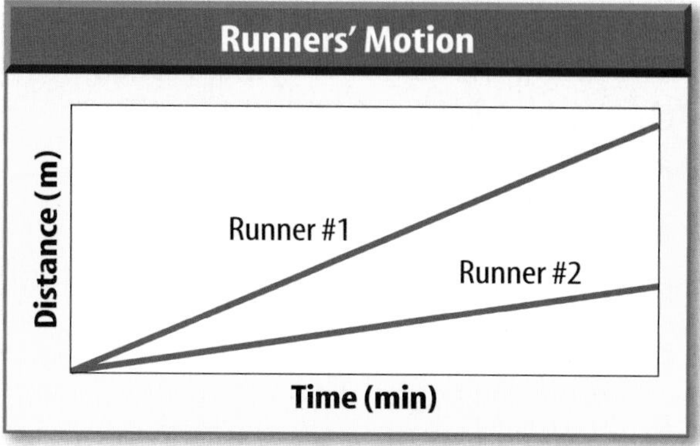

27. **Interpret a Graph** Use the graph to determine which runner had the greatest speed.

Standardized Test Practice

Record your answers on the answer sheet provided by your teacher or on a sheet of paper.

Multiple Choice

1. Sound travels at a speed of 330 m/s. How long does it take for the sound of thunder to travel 1,485 m?

   A. 45 s

   B. 4.5 s

   C. 4,900 s

   D. 0.22 s

Use the graph below to answer questions 2–4.

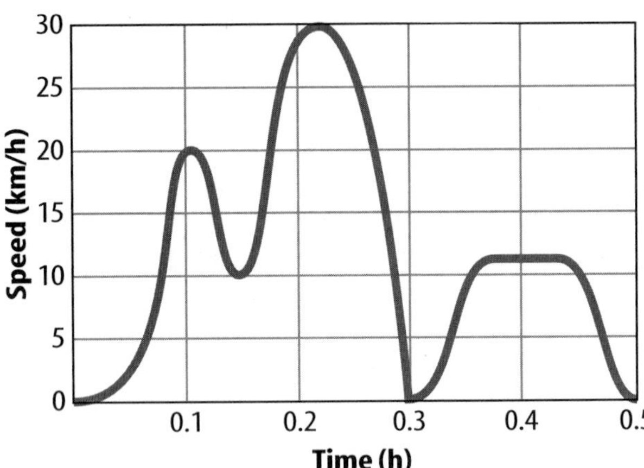

2. The graph shows how a cyclist's speed changed over 0.5 h. What is the cyclist's average speed if the trip was 6 km?

   A. 3 km/h

   B. 10 km/h

   C. 12 km/h

   D. 20 km/h

Test-Taking Tip

Read Carefully Read each question carefully for full understanding.

3. Once the trip was started, how many times did the cyclist stop?

   A. 0

   B. 4

   C. 2

   D. 5

4. What was the fastest speed the cyclist traveled?

   A. 20 km/h

   B. 30 km/h

   C. 12 km/h

   D. 10 km/h

Use the table below to answer questions 5 and 6.

Runner	Distance covered (km)	Time (min)
Daisy	12.5	42
Jane	7.8	38
Bill	10.5	32
Joe	8.9	30

5. What is Daisy's average speed?

   A. 0.30 km/min

   B. 530 km/min

   C. 3.0 km/min

   D. 3.4 km/min

6. Which runner has the fastest average speed?

   A. Daisy

   B. Jane

   C. Bill

   D. Joe

7. A skier is going down a hill at a speed of 9 m/s. The hill gets steeper and her speed increases to 18 m/s in 3 s. What is her acceleration?

   A. 9 m/s^2

   B. 3 m/s^2

   C. 27 m/s^2

   D. 6 m/s^2

8. Which of the following best describes an object with constant velocity?

   A. It is changing direction.

   B. Its acceleration is increasing.

   C. Its acceleration is zero.

   D. Its acceleration is negative.

Gridded Response

9. If a car is traveling at a speed of 40 km/h and then comes to a stop in 5 s, what is its acceleration in m/s^2?

Short Response

Use the table below to answer question10.

Car	Mass (kg)	Stopping Distance(m)
A	1,000	80
B	1,250	100
C	1,500	120
D	2,000	160

10. What is the relationship between a car's mass and its stopping distance?

11. If the speedometer on a car indicates a constant speed, can you be sure the car is not accelerating? Explain.

Extended Response

Use the graph below to answer question 12.

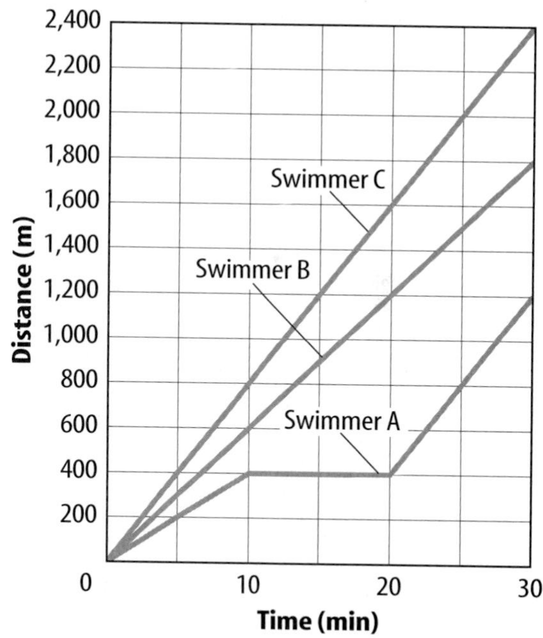

12. The graph shows the motion of three swimmers during a 30-min workout. Describe the motion of the three swimmers.

13. Where would you place the location of a reference point in order to describe the motion of a space probe traveling from Earth to Jupiter? Explain your choice.

chapter 4

The Laws of Motion

chapter preview

sections

Who's a dummy?

Crunch! This test dummy would have some explaining to do if this were a traffic accident. But in a test crash, the dummy plays an important role. The forces acting on it during a crash are measured and analyzed in order to learn how to make cars safer.

Science Journal Explain which would be a safer car—a car with a front that crumples in a crash, or one with a front that doesn't crumple.

Start-Up Activities

The Force of Gravity

The force of Earth's gravity pulls all objects downward. However, objects such as rocks seem to fall faster than feathers or leaves. Do objects with more mass fall faster?

1. Complete the safety form.
2. Measure the mass of a softball, a tennis ball, and a flat sheet of paper. Copy the data table below and record the masses.
3. Drop the softball from a height of 2.5 m and use a stopwatch to measure the time it takes for the softball to hit the floor. Record the time in your data table.
4. Repeat step 2 using the tennis ball and the flat sheet of paper. Record the times in your data table.
5. Crumple the flat sheet of paper into a ball, and measure and record the time for it to fall 2.5 m.
6. **Think Critically** Write a paragraph comparing the times it took each item to fall 2.5 m. From your data, infer if the speed of a falling object depends on the object's mass.

Falling Object Data

Object	Mass	Time
Softball		
Tennis ball		
Flat paper		
Crumpled paper		

The Laws of Motion Make the following Foldable to help you better understand Newton's laws of motion as you read the chapter.

STEP 1 **Fold** the top of a vertical piece of paper down and the bottom up to divide the paper into thirds.

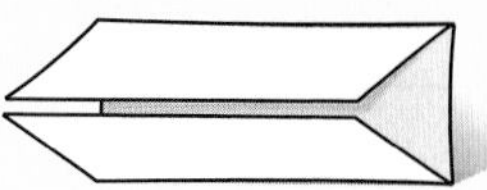

STEP 2 **Unfold and label** the rows *First Law, Second Law, Third Law.*

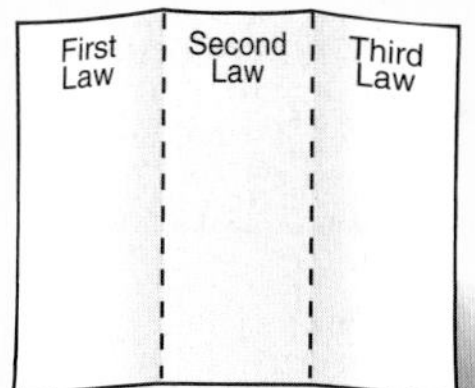

Find Main Ideas As you read the chapter, list the main ideas in the sections on the laws of motion under the appropriate column.

Preview this chapter's content and activities at gpescience.com

section 1

The First Two Laws of Motion

Reading Guide

What You'll Learn

- **Define** Newton's first law of motion.
- **Explain** how inertia and mass are related.
- **Define** Newton's second law of motion.
- **Apply** Newton's second law of motion.

Why It's Important

Newton's first two laws of motion explain how forces cause the motion of objects to change.

Review Vocabulary

net force: the combination of all forces acting on an object

New Vocabulary

- first law of motion
- inertia
- second law of motion

Newton's Laws of Motion

When you lift a backpack, you exert a force on an object and cause it to move. The backpack initially was at rest. The force you exerted caused its velocity to change. But if you push downward on a table, the table doesn't move. The force you applied did not cause the velocity of the table to change. How are forces and motion related?

The British scientist Isaac Newton, who lived from 1642 to 1727, published a set of three rules in 1687 that explained how forces and motion are related. These three rules are called Newton's laws of motion. They apply to the motion of all objects you encounter every day, such as those in **Figure 1,** as well as to the motion of planets, stars, and galaxies.

Figure 1 Newton's laws of motion apply to all objects, including the volleyball and the volleyball players.

The First Law of Motion

Newton's first law of motion describes how an object moves when the net force acting on it is zero. According to Newton's **first law of motion,** if the net force acting on an object is zero, the object remains at rest, or if the object is moving, it continues moving in a straight line with constant speed. The first law of motion means that if the forces acting on an object are balanced, then the velocity of the object doesn't change.

Only Unbalanced Forces Change Velocity The first law of motion connects forces with changes in motion. For example, suppose a skateboard is at rest. The skateboard doesn't move until you give it a push. Then the velocity of the skateboard increases while you are pushing it. The forces on the skateboard are unbalanced while you are pushing on it. As a result, the velocity of the skateboard changes as its speed increases.

However, after the skateboard leaves your hand, it slows down and stops. It might seem that the skateboard stops moving because there are no forces acting on it. However, there is an unbalanced frictional force that acts on the skateboard as it rolls. This is the force that causes the skateboard to stop. The unbalanced frictional force causes the velocity of the skateboard to change as its speed decreases. The velocity of an object changes only when there are unbalanced forces acting on it.

Inertia and Mass

All objects have a property called inertia. **Inertia** is the tendency of an object to resist a change in its motion. The dirt bike in **Figure 2** has inertia that tends to keep it moving with a constant velocity when its rider tries to steer around a curve. The inertia of an object depends on the object's mass. The greater the mass of an object, the greater its inertia.

For example, a bowling ball has more mass than a volleyball. Which would be harder to stop—a volleyball or a bowling ball rolling at the same speed? You would have to exert a greater force on the bowling ball to make it stop. The bowling ball has more inertia than the volleyball because it has more mass.

Reading Check ***How does the inertia of an object depend on the object's mass?***

Mini LAB

Observing Inertia

Procedure

1. Complete the safety form.
2. Create an inclined plane using a **board** and **textbooks.**
3. Place a small object in a **cart** and let the cart roll down the plane. Record the results.
4. Secure the object in the cart with rubber bands. Let the cart roll down the plane and record your results.

Analysis

1. Identify the forces acting on the object in the cart in both runs.
2. Explain how safety belts reduce the risk of injury in a car crash.

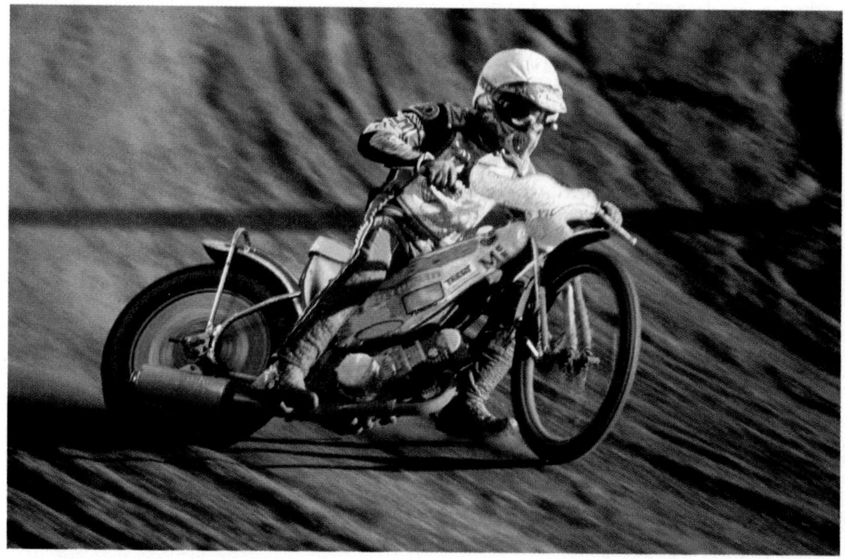

Figure 2 Inertia causes the bike to move in a straight line despite the efforts of the rider to steer the bike around the curve.

Figure 3 The inertia of the boxes causes them to keep moving after the cart has been brought to a sudden stop.
Describe *the velocity of the boxes just after the cart stops.*

Inertia and the First Law of Motion According to the first law of motion, the velocity of an object doesn't change unless the forces acting on the object are unbalanced. In other words, if you observe a change in an object's velocity, you know an unbalanced force acted on it. This is similar to the definition of inertia—an object resists a change in its motion. As a result, the first law of motion sometimes is called the law of inertia.

Inertia and the first law of motion explain why the boxes in **Figure 3** slide off the cart when the cart comes to a sudden stop. The student applies an unbalanced force to the cart that makes it stop. However, the inertia of the boxes causes them to keep moving even after the cart stops.

What happens in a crash?

The first law of motion can explain what happens in a car crash. When a car traveling 50 km/h collides head-on with something solid, the car crumples, slows down, and stops within about 0.1 s. Any passenger not wearing a safety belt continues to move forward at the same speed the car was traveling. Within about 0.02 s after the car stops, unbelted passengers slam into the dashboard, windshield, or steering wheel, as shown in **Figure 4.** Passengers in the back seat might slam into the backs of the front seats. Because of their inertia, they are traveling at the car's original speed of 50 km/h—about the same speed they would reach falling from a three-story building.

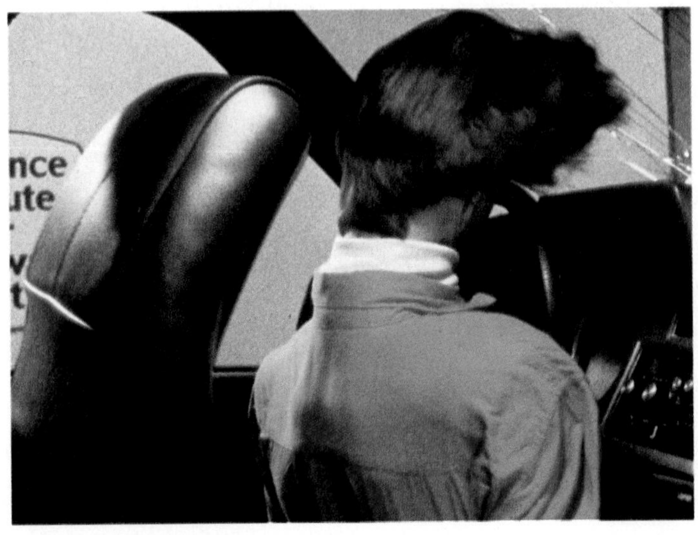

Figure 4 This crash test dummy is not restrained in this low-speed crash. Inertia causes the dummy to slam into the steering wheel.

Safety Belts In **Figure 5** the crash test dummies are restrained with safety belts. As a result, they slow down at the same rate that the car slows down. However, the force exerted on a person that slows from 50 km/h to 0 km/h in 0.1 s is about 14 times as large as the force of gravity on the person. To further reduce the force acting on a person, the seat belt loosens slightly, increasing the time it takes to slow the person. Increasing the time needed to stop reduces the force exerted on the person. The safety belt also prevents the person from being thrown out of the car. Car-safety experts say that about half the people who die in car crashes would survive if they wore safety belts.

Figure 5 These crash test dummies were restrained with safety belts in this crash. This prevented the dummies from hitting the dashboard or windshield.

Reading Check *What are two advantages of wearing a safety belt in a crash?*

The Second Law of Motion

According to Newton's first law of motion, unbalanced forces cause the velocity of an object to change. Newton's second law of motion describes how the net force on an object, its mass, and its acceleration are related.

Force and Acceleration What's different about throwing a ball horizontally as hard as you can and tossing it gently? When you throw hard, the net force on the ball is greater than when you toss it. Also, the ball has a greater velocity when it leaves your hand. As a result, the hard-thrown ball has a greater change in velocity, and this change can occur over a shorter period of time. Recall that acceleration is the change in velocity divided by the time needed for the change to occur. So, a hard-thrown ball has a greater acceleration than a gently-thrown ball.

Mass and Acceleration If you throw a baseball and a softball with the same amount of force, why does the baseball move faster? A softball has more mass than a baseball. Even though the net force exerted on both balls is the same, the softball has less velocity when it leaves your hand than the baseball does. If it takes the same amount of time to throw both balls, the acceleration of the softball is less than that of the baseball. So, the acceleration of an object depends on its mass, as well as the net force exerted on it. Net force, mass, and acceleration are related.

Topic: Motion in Sports
Visit gpescience.com for Web links to information about methods used to analyze the motions of athletes.

Activity Choose a sport and write a report on how analyzing the motions involved in the sport can improve performance and reduce injuries.

The Second Law of Motion

Newton's **second law of motion** states that the acceleration of an object is in the same direction as the net force on the object, and that the acceleration can be calculated from the following equation:

Newton's Second Law of Motion

$$\text{acceleration (in meters/second}^2) = \frac{\text{net force (in newtons)}}{\text{mass (in kilograms)}}$$

$$a = \frac{F_{net}}{m}$$

Applying Math — Solve a Simple Equation

THE ACCELERATION OF A SLED You push a friend on a sled. Your friend and the sled together have a mass of 70 kg. If the net force on the sled is 35 N, what is the sled's acceleration?

IDENTIFY known values and the unknown value

Identify the known values:

The net force on the sled is 35 N means $F_{net} = 35$ N

Your friend and the sled together have a mass of 70 kg means $m = 70$ kg

Identify the unknown value:

What is the sled's acceleration? means $a = ?$ m/s^2

SOLVE the problem

Substitute the known values $F_{net} = 35$ N and $m = 70$ kg into the equation for Newton's second law of motion:

$$a = \frac{F_{net}}{m} = \frac{35\text{ N}}{70\text{ kg}} = 0.5\,\frac{\text{N}}{\text{kg}} = 0.5\,\frac{\cancel{\text{kg}}\,\text{m}}{\text{s}^2} \times \frac{1}{\cancel{\text{kg}}} = 0.5\text{ m/s}^2$$

CHECK your answer

Does your answer seem reasonable? Check your answer by multiplying the acceleration you calculated by the mass given in the problem. The result should be the net force given in the problem.

Practice Problems

1. If the mass of a helicopter is 4,500 kg, and the net force on it is 18,000 N, what is the helicopter's acceleration?
2. What is the net force on a dragster with a mass of 900 kg if its acceleration is 32.0 m/s^2?
3. A car is being pulled by a tow truck. What is the car's mass if the net force on the car is 3,000 N and it has an acceleration of 2.0 m/s^2?

For more practice problems go to page 879, and visit Math Practice at gpescience.com.

Calculating Net Force with the Second Law Newton's second law also can be used to calculate the net force if mass and acceleration are known. To do this, the equation for Newton's second law must be solved for the net force, *F*. To solve for the net force, multiply both sides of the above equation by the mass:

$$\cancel{m} \times \frac{F_{net}}{\cancel{m}} = ma$$

The mass, *m*, on the left side cancels, giving the equation:

$$F_{net} = ma$$

For example, when the tennis player in **Figure 6** hits a ball, the ball might be in contact with the racket for only a few thousandths of a second. Because the ball's velocity changes over such a short period of time, the ball's acceleration could be as high as 5,000 m/s^2. The ball's mass is 0.06 kg, so the net force exerted on the ball would be:

$$F_{net} = ma = (0.06 \text{ kg})(5{,}000 \text{ m/s}^2) = 300 \text{ kg m/s}^2 = 300 \text{ N}$$

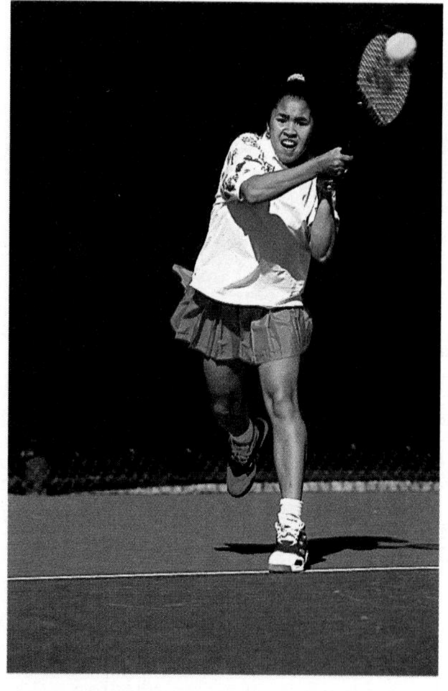

Figure 6 The tennis racket exerts a force on the ball that causes it to accelerate.

section 1 review

Summary

Inertia and the First Law of Motion

- According to Newton's first law of motion, the velocity of an object does not change unless an unbalanced net force acts on the object.
- The inertia of an object is the tendency of the object to resist a change in motion.
- The larger the mass of an object, the greater its inertia.
- In a car crash, inertia causes an unrestrained passenger to continue moving at the speed of the car before the crash.

The Second Law of Motion

- The acceleration of an object depends on the mass of the object and the net force exerted on the object.
- According to Newton's second law of motion, the acceleration of an object is in the direction of the net force on the object, and can be calculated from this equation:

$$a = \frac{F_{net}}{m}$$

Self Check

1. **Infer** whether the inertia of an object changes as the object's velocity changes.
2. **Determine** whether or not there must be an unbalanced force acting on any object that is moving. Explain.
3. **Determine** whether or not there can be forces acting on an object that is at rest. Explain.
4. **Infer** the net force on a refrigerator if you push on the refrigerator and it doesn't move.
5. **Think Critically** Describe three situations in which a force changes the velocity of an object. Let at least one of the situations involve the force due to gravity.

Applying Math

6. **Calculate Mass** You push yourself on a skateboard with a force of 30 N and accelerate at 0.5 m/s^2. Find the mass of you and the skateboard together.
7. **Calculate Net Force** You push a book that has a mass of 2.0 kg so that it has an acceleration of 1.0 m/s^2. What is the net force on the book?

More Section Review gpescience.com

section 2

Gravity

Reading Guide

What You'll Learn

- **Describe** the gravitational force.
- **Distinguish** between mass and weight.
- **Explain** why objects that are thrown will follow a curved path.
- **Compare** circular motion with motion in a straight line.

Why It's Important

There is a gravitational force between you and every other object in the universe.

Review Vocabulary

acceleration: the rate of change of velocity which occurs when an object changes speed or direction

New Vocabulary

- gravity
- weight
- centripetal acceleration
- centripetal force

Figure 7 The gravitational force between two objects depends on their masses and the distance between them.

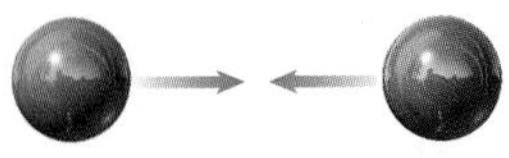

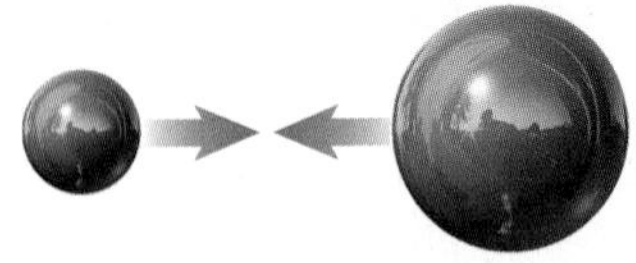

Increasing the mass of either object increases the gravitational force between them.

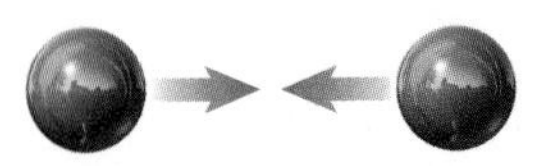

If the objects are closer together, the gravitational force between them increases.

What is gravity?

There's a lot about you that's attractive. At this moment, you are exerting an attractive force on everything around you—your desk, your classmates, even the planet Jupiter millions of kilometers away. It's the attractive force of gravity.

Anything that has mass is attracted by the force of gravity. **Gravity** is an attractive force between any two objects that depends on the masses of the objects and the distance between them. This force increases as the mass of either object increases, or as the objects move closer, as shown in **Figure 7.**

You can't feel any gravitational attraction between you and this book because the force is weak. Only Earth is both close enough and has a large enough mass that you can feel its gravitational attraction. While the Sun has much more mass than Earth, the Sun is too far away to exert a noticeable gravitational attraction on you. And while this book is close, it doesn't have enough mass to exert an attraction you can feel.

Gravity—A Basic Force Gravity is one of the four basic forces. The other basic forces are the electromagnetic force, the strong nuclear force, and the weak nuclear force. The two nuclear forces only act on particles in the nuclei of atoms. Electricity and magnetism are caused by the electromagnetic force. Chemical interactions between atoms and molecules also are due to the electromagnetic force.

The Law of Universal Gravitation

For thousands of years, people have observed the night sky and collected data on the motions of the stars and planets. Isaac Newton used some of these data on the motions of planets to formulate the law of universal gravitation, which he published in 1687. This law can be written as the following equation.

The Law of Universal Gravitation

$$\text{gravitational force} = (\text{constant}) \times \frac{(\text{mass 1}) \times (\text{mass 2})}{(\text{distance})^2}$$

$$F = G\frac{m_1 m_2}{d^2}$$

In this equation G is a constant called the universal gravitational constant, and d is the distance between the two masses, m_1 and m_2. The law of universal gravitation enables the force of gravity to be calculated between any two objects if their masses and the distance between them are known.

The Range of Gravity According to the law of universal gravitation, the gravitational force between two masses decreases rapidly as the distance between the masses increases. For example, if the distance between two objects increases from 1 m to 2 m, the gravitational force between them becomes one fourth as large. If the distance increases from 1 m to 10 m, the gravitational force between the objects is one hundredth as large. However, no matter how far apart two objects are, the gravitational force between them never completely disappears. As a result, gravity is called a long-range force.

Reading Check *Why is gravity called a long-range force?*

Finding Other Planets Earth's motion around the Sun is affected by the gravitational pulls of the other planets in the solar system. In the same way, the motion of every planet in the solar system is affected by the gravitational pulls of all the other planets.

In the 1840s the most distant planet known was Uranus. The motion of Uranus calculated from the law of universal gravitation disagreed slightly with its observed motion. Some astronomers suggested that there must be an undiscovered planet affecting the motion of Uranus. Using the law of universal gravitation and Newton's laws of motion, two astronomers independently calculated the orbit of this planet. As a result of these calculations, the planet Neptune, shown in **Figure 8,** was found in 1846.

Topic: Gravity on Other Planets
Visit gpescience.com for Web links to information about the gravitational acceleration near the surface of different planets in the solar system.

Activity Make a graph with the gravitational acceleration on the *y*-axis, and the planet's mass on the *x*-axis. Infer from your graph how the gravitational acceleration depends on a planet's mass.

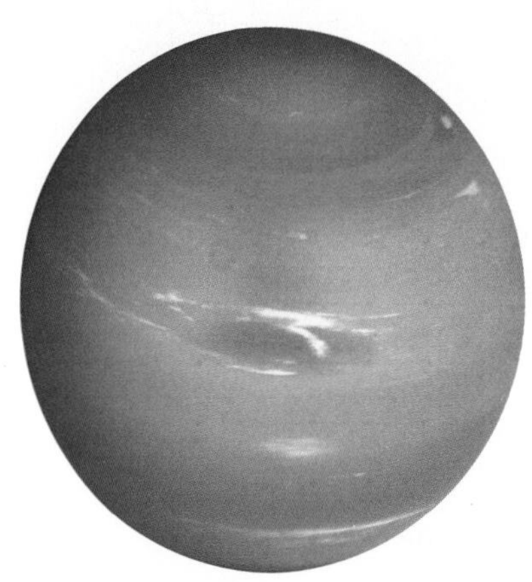

Figure 8 The location of the planet Neptune in the night sky was correctly predicted using Newton's laws of motion and the law of universal gravitation.

Earth's Gravitational Acceleration

If you dropped a bowling ball and a marble at the same time, which would hit the ground first? Suppose the effects of air resistance are small enough to be ignored. When all forces except gravity acting on an a falling object can be ignored, the object is said to be in free fall. Then all objects near Earth's surface would fall with the same acceleration, just like the two balls in **Figure 9.**

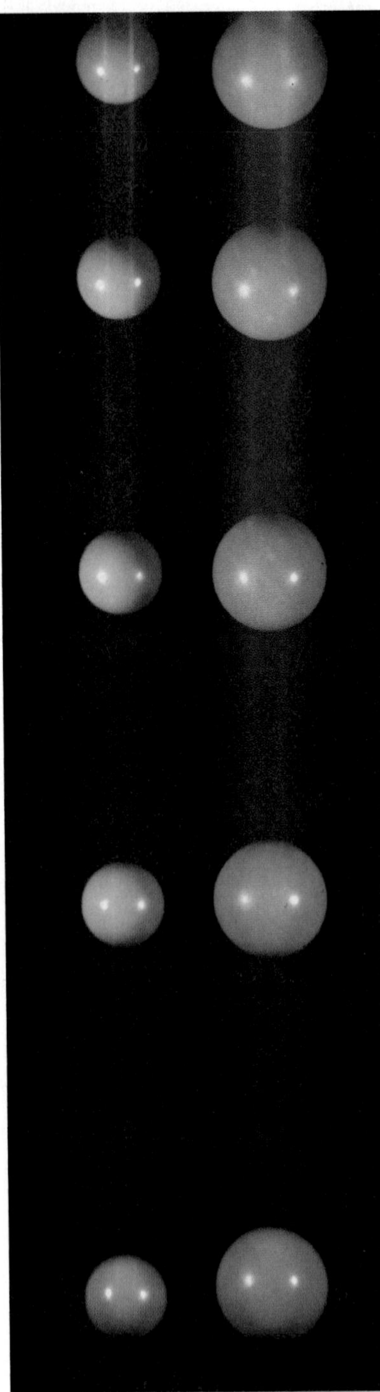

Figure 9 Time-lapse photography shows that two balls of different masses fall at the same rate.

Close to Earth's surface, the acceleration of an object in free fall is 9.8 m/s^2. This acceleration is the gravitational acceleration and is given the symbol g. By the second law of motion, the gravitational force exerted by Earth on a falling object is the object's mass times the gravitational acceleration. The gravitational force can be calculated from this equation:

Gravitational Force Equation

gravitational force (N) = mass (kg) × gravitational acceleration (m/s^2)

$$F = mg$$

For example, the gravitational force on a skydiver with a mass of 60 kg would be

$$F = mg = (60 \text{ kg})\ (9.8 \text{ m/s}^2) = 588 \text{ N}$$

Weight Even if you are not falling, the force due to Earth's gravity still is pulling you downward. If you are standing on a floor, the net force on you is zero. The force due to Earth's gravity pulls you downward, but the floor exerts an upward force on you that balances gravity's downward pull.

Whether you are standing, jumping, or falling, Earth exerts a gravitational force on you. The gravitational force exerted on an object is called the object's **weight.** The weight of an object on Earth is equal to the gravitational force exerted by Earth on the object. As a result, near Earth's surface an object's weight can be calculated from this equation:

Weight Equation

weight (N) = mass (kg) × gravitational acceleration (m/s^2)

$$W = mg$$

On Earth, where g equals 9.8 m/s^2, a cassette tape weighs about 0.5 N, a backpack full of books could weigh 100 N, and a jumbo jet weighs about 3.4 million N.

Weight and Mass Weight and mass are not the same. Weight is a force and mass is a measure of the amount of matter an object contains. However, according to the weight equation on the previous page, weight and mass are related. Weight increases as mass increases.

The weight of an object usually is the gravitational force between the object and Earth. But the weight of an object can change, depending on the gravitational force on the object. For example, the acceleration due to gravity on the Moon is 1.6 m/s^2, about one sixth as large as Earth's gravitational acceleration. As a result, a person, like the astronaut in **Figure 10,** would weigh only about one sixth as much on the Moon as on Earth. **Table 1** shows how various weights on Earth would be different on the Moon and some of the planets.

Reading Check *How are weight and mass related?*

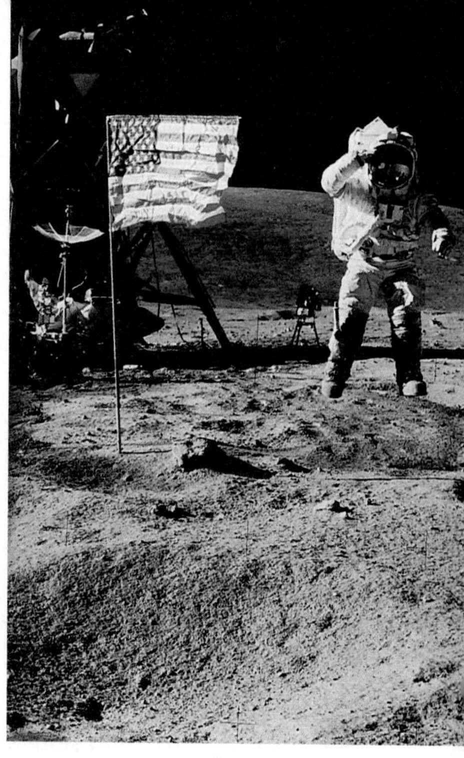

Figure 10 On the Moon, the gravitational force on the astronaut is less than it is on Earth. As a result, the astronaut can take longer steps and jump higher than on Earth.
Describe *how the astronaut's mass and weight on the Moon compare to his mass and weight on Earth.*

Weightlessness and Free Fall

You've probably seen pictures of astronauts and equipment floating inside the space shuttle. Any item that is not fastened down seems to float throughout the cabin. They are said to be experiencing the sensation of weightlessness.

However, for a typical mission, the shuttle orbits Earth at an altitude of about 400 km. According the law of universal gravitation, at 400-km altitude the force of Earth's gravity is about 90 percent as strong as it is at Earth's surface. So an astronaut with a mass of 80 kg still would weigh about 700 N in orbit, compared with a weight of about 780 N at Earth's surface.

Table 1 Weight Comparison Table

Weight on Earth (N)	Weight on Other Bodies in the Solar System (N)				
	Moon	Venus	Mars	Jupiter	Saturn
75	12	68	28	190	87
100	17	90	38	254	116
150	25	135	57	381	174
500	84	450	190	1,270	580
700	119	630	266	1,778	812
2,000	333	1,800	760	5,080	2,320

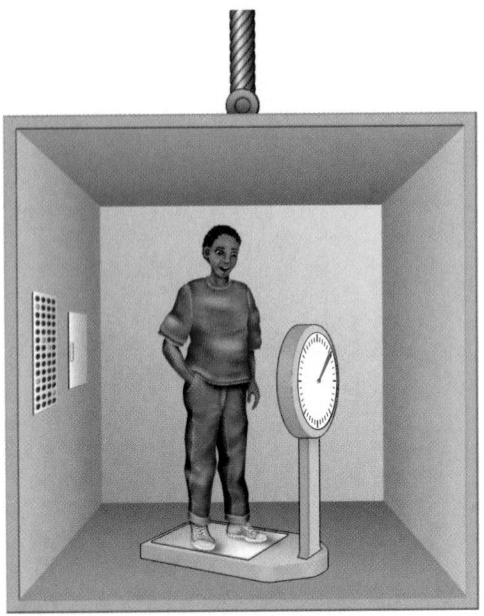

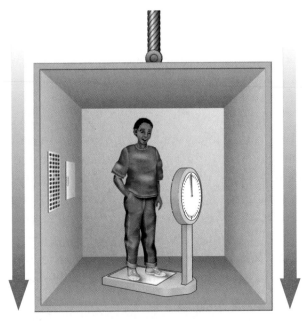

A When the elevator is stationary, the scale shows the boy's weight.

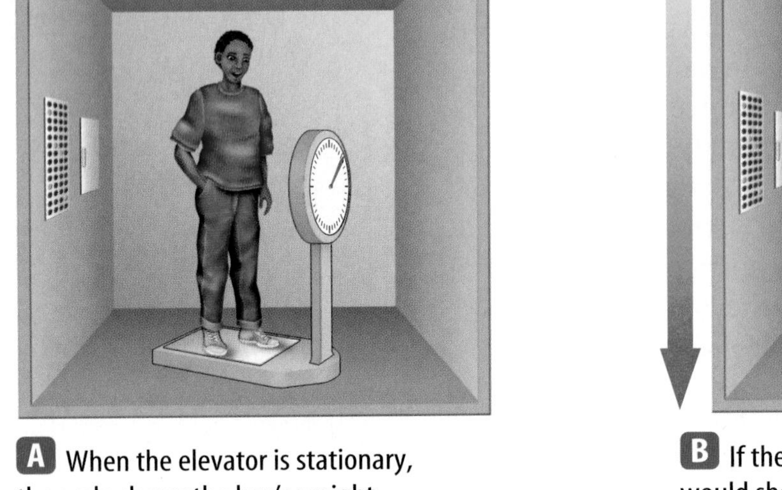

B If the elevator were in free fall, the scale would show a zero weight.

Figure 11 The boy and the scale no longer exert forces on each other when they both are in free fall. As a result, the boy seems to be weightless.

Floating in Space Why do objects seem weightless in orbit? Think about measuring your weight. When you stand on a scale, as in **Figure 11A,** the net force on you is zero. The scale exerts an upward force on you equal to your weight. Now suppose you stand on a scale in an elevator that is in free fall, as shown in **Figure 11B.** Then the only force acting on you is the gravitational force. You and the scale no longer exert forces on each other. The scale would show zero weight, even though the gravitational force on you hasn't changed.

Reading Check *What is the only force on an object in free fall?*

A space shuttle in orbit is in free fall, but it is falling around Earth, rather than straight downward. Everything in the orbiting space shuttle is falling around Earth at the same rate, in the same way you and the scale were falling in the elevator. Objects in the shuttle seem to be floating because they are all falling with the same acceleration.

INTEGRATE Earth Science

Gravity and Earth's Atmosphere Apart from simply keeping your feet on the ground, gravity is important for life on Earth for other reasons, too. Because Earth has a sufficient gravitational pull, it can hold around it the oxygen/nitrogen atmosphere necessary for sustaining life. Research other ways in which gravity has played a role in the formation of Earth.

Projectile Motion

If you've tossed a ball to someone, you've probably noticed that thrown objects don't always travel in straight lines. They curve downward. Anything that's thrown or shot through the air is called a projectile. Earth's gravity causes projectiles to follow a curved path.

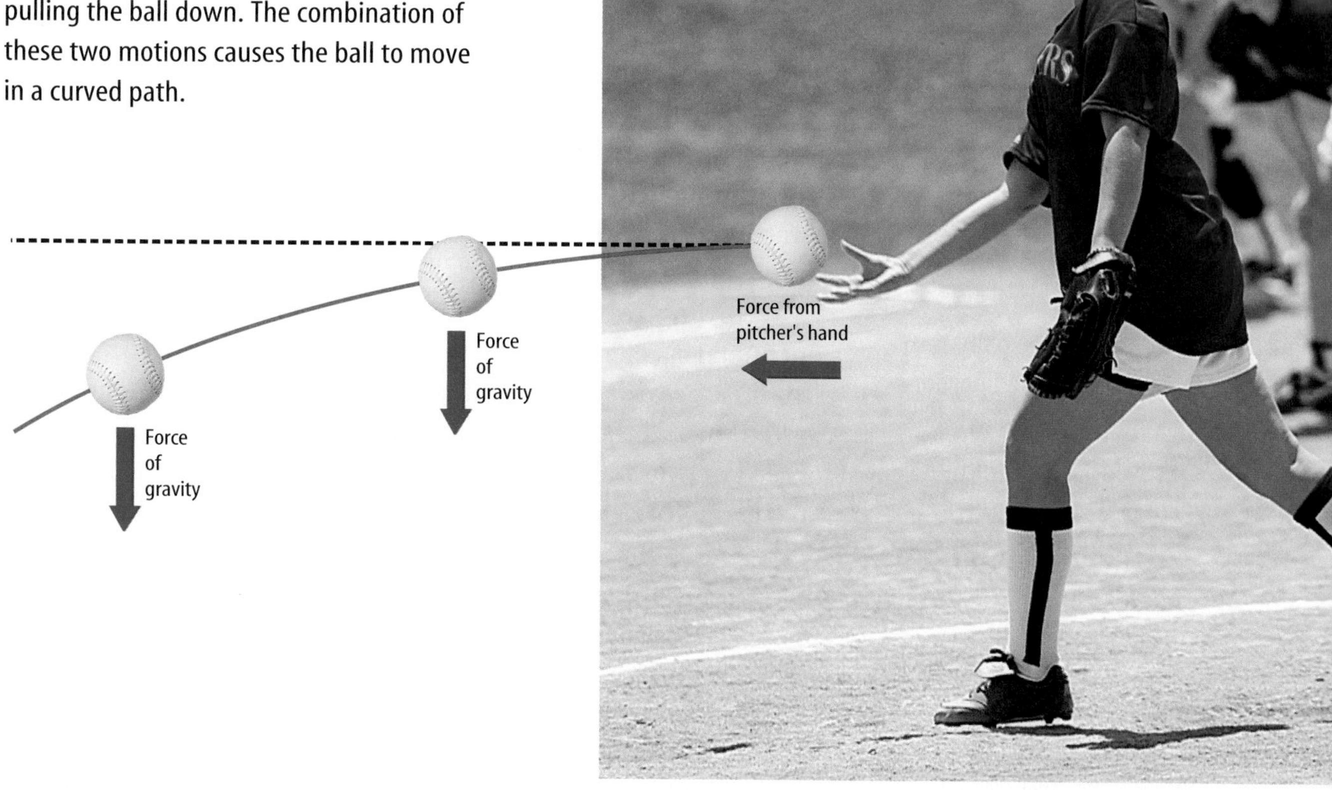

Figure 12 The pitcher gives the ball a horizontal motion. Gravity, however, is pulling the ball down. The combination of these two motions causes the ball to move in a curved path.

Horizontal and Vertical Motions When you throw a ball, like the pitcher in **Figure 12,** the force exerted by your hand pushes the ball forward. This force gives the ball horizontal motion. After you let go of the ball, no force accelerates it forward, so its horizontal velocity is constant, if you ignore air resistance.

However, when you let go of the ball, gravity can accelerate it downward, giving it vertical motion. Now the ball has constant horizontal velocity but increasing vertical velocity. Gravity exerts an unbalanced force on the ball, changing the direction of its path from only forward to forward and downward. The result of these two motions is that the ball appears to travel in a curve, even though its horizontal and vertical motions are completely independent of each other.

Horizontal and Vertical Distance If you were to throw a ball as hard as you could from shoulder height in a perfectly horizontal direction, would it take longer to reach the ground than if you dropped a ball from the same height? Surprisingly, it won't. A thrown ball and one dropped will hit the ground at the same time. Both balls in **Figure 13** travel the same vertical distance in the same amount of time. However, the ball thrown horizontally travels a greater horizontal distance than the ball that is dropped.

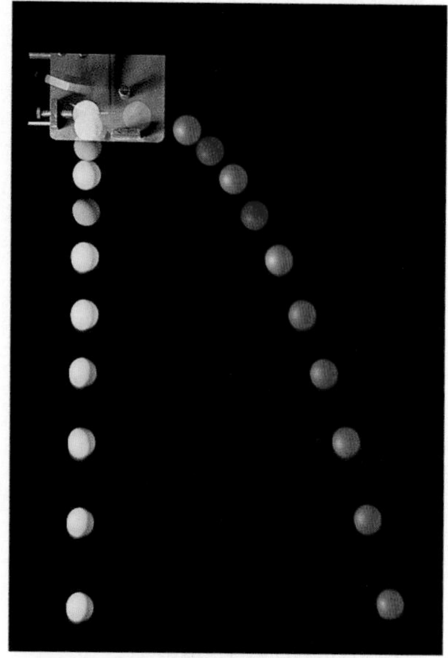

Figure 13 Multiflash photography shows that each ball has the same acceleration downward, whether it's thrown or dropped.

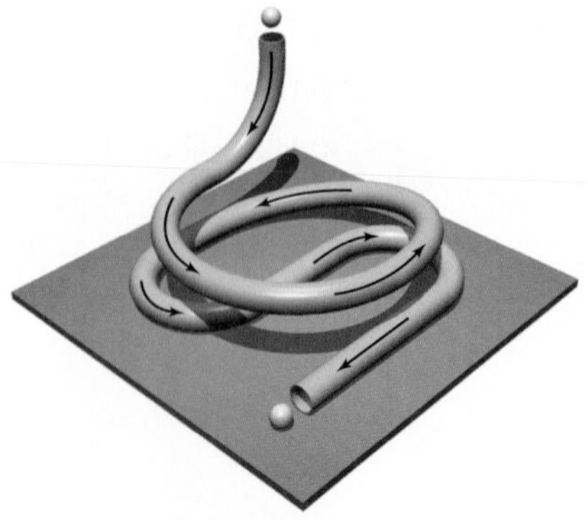

Figure 14 When the ball moves through the curved portions of the tube, it is accelerating because its velocity is changing. **Identify** *the forces acting on the ball as it falls through the tube.*

Centripetal Force

Look at the path the ball follows as it travels through the curved tube in **Figure 14.** The ball may accelerate in the straight sections of the pipe maze if it speeds up or slows down. However, when the ball enters a curve, even if its speed does not change, it is accelerating because its direction is changing. When the ball goes around a curve, the change in the direction of the velocity is toward the center of the curve. Acceleration toward the center of a curved or circular path is called **centripetal acceleration.**

According to the second law of motion, when the ball has centripetal acceleration, the direction of the net force on the ball also must be toward the center of the curved path. The net force exerted toward the center of a curved path is called a **centripetal force.** For the ball moving through the tube, the centripetal force is the force exerted by the walls of the tube on the ball.

Centripetal Force and Traction When a car rounds a curve on a highway, a centripetal force must be acting on the car to keep it moving in a curved path. This centripetal force is the frictional force, or the traction, between the tires and the road surface. If the road is slippery and the frictional force is small, the centripetal force might not be large enough to keep the car moving around the curve. Then the car will slide in a straight line. Anything that moves in a circle, such as the people on the amusement park ride in **Figure 15,** is doing so because a centripetal force is accelerating it toward the center.

Observing Centripetal Force

Procedure 

1. Complete the safety form.
2. Thread a **string** about 1 m long through the holes of a **plastic, slotted ball.**
3. Swing the ball in a vertical circle.
4. Swing the ball at different speeds and observe the motion of the ball and the tension in the string.

Analysis

1. What force does the string exert on the ball when the ball is at the top, sides, and bottom of the swing?
2. How does the tension in the string depend on the speed of the ball?

Try at Home

Figure 15 Centripetal force keeps these riders moving in a circle.

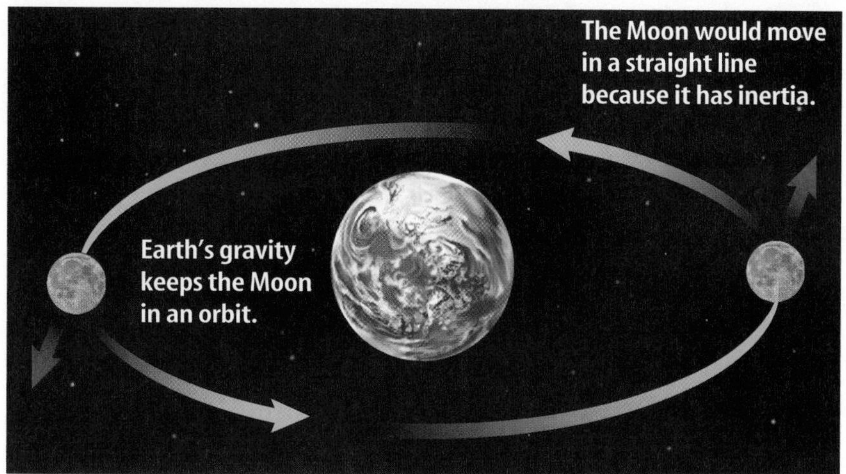
The Moon would move in a straight line because it has inertia.
Earth's gravity keeps the Moon in an orbit.

Figure 16 The Moon would move in a straight line except that Earth's gravity keeps pulling it toward Earth. This gives the Moon a nearly circular orbit.

Gravity Can Be a Centripetal Force Imagine whirling an object tied to a string above your head. The string exerts a centripetal force on the object that keeps it moving in a circular path. In the same way, Earth's gravity exerts a centripetal force on the Moon that keeps it moving in a nearly circular orbit, as shown in **Figure 16.**

section 2 review

Summary

Gravity

- According to the law of universal gravitation, the gravitational force between two objects depends on the masses of the objects and the distance between them.
- The acceleration due to gravity near Earth's surface has the value 9.8 m/s^2.
- Near Earth's surface, the gravitational force on an object with mass, m, is given by:

$$F = mg$$

Weight

- The weight of an object is related to its mass according to the equation:

$$W = mg$$

- An object in orbit seems to be weightless because it is falling around Earth.

Projectile Motion and Centripetal Force

- Projectiles follow a curved path because their horizontal motion is constant, but gravity causes the vertical motion to be changing.
- The net force on an object moving in a circular path is called the centripetal force.

Self Check

1. **Describe** how the gravitational force between two objects depends on the mass of the objects and the distance between them.
2. **Distinguish** between the mass of an object and the object's weight.
3. **Explain** what causes the path of a projectile to be curved.
4. **Describe** the force that causes the planets to stay in orbit around the Sun.
5. **Think Critically** Suppose Earth's mass increased, but its diameter didn't change. How would the acceleration due to gravity near Earth's surface change?

Applying Math

6. **Calculate Weight** On Earth, what is the weight of a large-screen TV that has a mass of 75 kg?
7. **Calculate Gravity on Mars** Find the acceleration due to gravity on Mars if a person with a mass of 60.0 kg weighs 222 N on Mars.
8. **Calculate Force** Find the force exerted by a rope on a 10-kg mass that is hanging from the rope.

GRAVITY AND AIR RESISTANCE

If you dropped a bowling ball and a feather from the same height on the Moon, they would both hit the surface at the same time. All objects dropped on Earth are attracted to Earth with the same acceleration. But on Earth, a bowling ball and a feather will not hit the ground at the same time. Air resistance slows the feather down.

Real-World Problem

How does air resistance affect the acceleration of falling objects?

Goals

- **Measure** the effect of air resistance on sheets of paper with different shapes.
- **Design** and create a shape from a piece of paper that maximizes air resistance.

Materials

paper (4 sheets of equal size)
scissors
meterstick
stopwatch
masking tape

Safety Precautions

Procedure

1. Complete the safety form.
2. Copy the data table above in your Science Journal, or create it on a computer.
3. Measure a height of 2.5m on the wall and mark the height with a piece of masking tape.
4. Have one group member drop the flat sheet of paper from the 2.5-m mark. Use the stopwatch to time how long it takes for the paper to reach the ground. Record your time in your data table.

Effects of Air Resistance

Paper Type	Time
Flat paper	Do not write in this book.
Loosely crumpled paper	
Tightly crumpled paper	
Your paper design	

5. Crumple a sheet of paper into a loose ball and repeat step 4.
6. Crumple a sheet of paper into a tight ball and repeat step 4.
7. Use scissors to shape a piece of paper so that it will fall slowly. You may cut, tear, or fold your paper into any design you choose.

Conclude and Apply

1. **Compare** the falling times of the different sheets of paper.
2. **Explain** why the different-shaped papers fell at different speeds.
3. **Explain** how your design caused the force of air resistance on the paper to be greater than the air resistance on the other paper shapes.

Communicating Your Data

Compare your paper design with the designs created by your classmates. As a class, compile a list of characteristics that increase air resistance.

The Third Law of Motion

Reading Guide

What You'll Learn

- **State** Newton's third law of motion.
- **Identify** action and reaction forces.
- **Calculate** momentum.
- **Recognize** when momentum is conserved.

Why It's Important

The third law of motion explains how you affect Earth when you walk, and how Earth affects you.

Review Vocabulary

velocity: describes the speed and direction of a moving object

New Vocabulary

- third law of motion
- momentum
- law of conservation of momentum

Newton's Third Law

Push against a wall and what happens? If the wall is sturdy enough, usually nothing happens. If you pushed against a wall while wearing roller skates, you would go rolling backward. Your action on the wall produced a reaction—movement backward. This is a demonstration of Newton's third law of motion.

The **third law of motion** describes action-reaction pairs this way: When one object exerts a force on a second object, the second one exerts a force on the first that is equal in strength and opposite in direction. Another way to say this is "to every action force there is an equal and opposite reaction force."

Action and Reaction When a force is applied in nature, a reaction force occurs at the same time. When you jump on a trampoline, for example, you exert a downward force on the trampoline. Simultaneously, the trampoline exerts an equal force upward, sending you high into the air.

Action and reaction forces are acting on the two skaters in **Figure 17.** The male skater is pulling upward on the female skater, while the female skater is pulling downward on the male skater. The two forces are equal, but in opposite directions.

Figure 17 According to Newton's third law of motion, the two skaters exert forces on each other. The two forces are equal, but in opposite directions.

Space Medicine Astronauts who stay in outer space for extended periods of time may develop health problems. Their muscles, for example, may begin to weaken because they don't have to exert as much force to get the same reaction as they do on Earth. A branch of medicine called space medicine deals with the possible health problems that astronauts may experience. Research some other health risks that are involved in going into outer space.

Action and Reaction Forces Don't Cancel If action and reaction forces are equal, you might wonder how some things ever happen. For example, how does a swimmer move through the water in a pool if each time she pushes on the water, the water pushes back on her? According to the third law of motion, action and reaction forces act on different objects. Thus, even though the forces are equal, they are not balanced because they act on different objects. In the case of the swimmer, as she "acts" on the water, the "reaction" of the water pushes her forward. Thus, a net force, or unbalanced force, acts on her so a change in her motion occurs. As the swimmer moves forward in the water, how does she make the water move?

Reading Check *Why don't action and reaction forces cancel?*

Rocket Propulsion Suppose you are standing on skates holding a softball. You exert a force on the softball when you throw the softball. According to Newton's third law, the softball exerts a force on you. This force pushes you in the direction opposite the softball's motion. Rockets use the same principle to move even in the vacuum of outer space. In the rocket engine, burning fuel produces hot gases. The rocket engine exerts a force on these gases and causes them to escape out the back of the rocket. By Newton's third law, the gases exert a force on the rocket and push it forward. The car in **Figure 18** uses a rocket engine to propel it forward. **Figure 19** shows how rockets were used to travel to the Moon.

Figure 18 Newton's third law enables the rocket engine to push the car forward. The rocket engine pushes the hot gases produced by burning fuel backward out of the engine's nozzle. By the third law, the gases exert a force of equal size on the engine in the forward direction.
Infer *the direction of the car's acceleration when the rocket engine is fired.*

NATIONAL GEOGRAPHIC

VISUALIZING ROCKET MOTION

Figure 19

On the afternoon of July 16, 1969, *Apollo 11* lifted off from Cape Kennedy, Florida, bound for the Moon. Eight days later, the spacecraft returned to Earth, splashing down safely in the Pacific Ocean. The motion of the spacecraft to the Moon and back is governed by Newton's laws of motion.

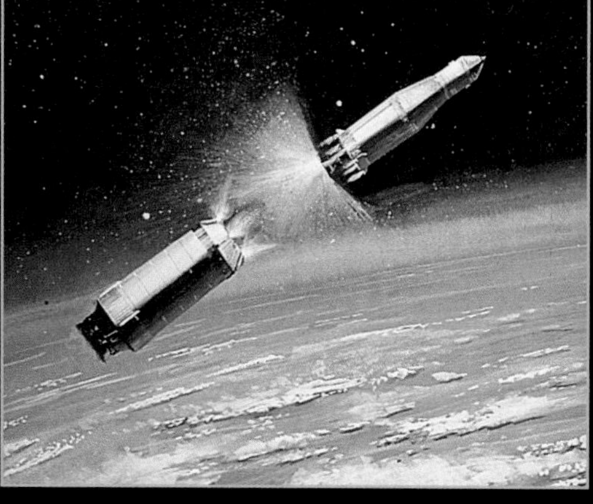

▲ As *Apollo* rises, it burns fuel and ejects its rocket booster engines. This decreases its mass, and helps *Apollo* move faster. This is Newton's second law in action: As mass decreases, acceleration can increase.

◀ *Apollo 11* roars toward the Moon. At launch, a rocket's engines must produce enough force and acceleration to overcome the pull of Earth's gravity. A rocket's liftoff is an illustration of Newton's third law: For every action there is an equal and opposite reaction.

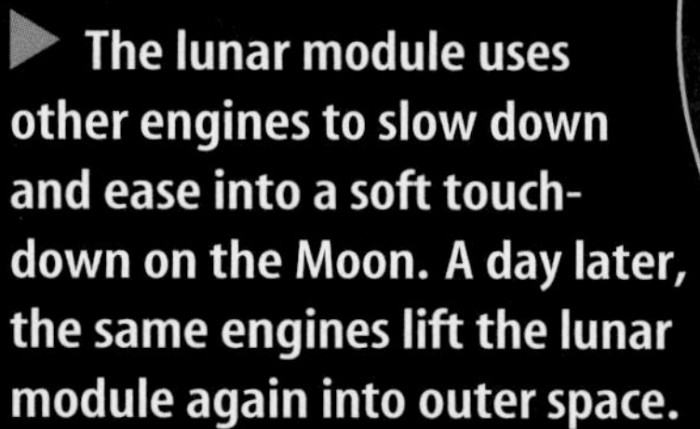

▶ The lunar module uses other engines to slow down and ease into a soft touchdown on the Moon. A day later, the same engines lift the lunar module again into outer space.

▲ The lunar module returns to the *Apollo* spacecraft, which fires its engines to start it moving toward Earth. As the spacecraft gets closer to Earth, the gravitational force exerted by Earth becomes larger, continually increasing the acceleration of the spacecraft.

Momentum

A moving object has a property called momentum that is related to how much force is needed to change its motion. The **momentum** of an object is the product of its mass and velocity. Momentum is given the symbol p and can be calculated with the following equation:

Momentum Equation

momentum (kg m/s) = **mass** (kg) × **velocity** (m/s)

$$p = mv$$

The unit for momentum is kg·m/s. Notice that momentum has a direction because velocity has a direction.

Applying Math — Solve a Simple Equation

THE MOMENTUM OF A SPRINTER At the end of a race, a sprinter with a mass of 80 kg has a speed of 10 m/s. What is the sprinter's momentum?

IDENTIFY known values and the unknown value

Identify the known values:

a sprinter with a mass of 80 kg → means → $m = 80$ kg

has a speed of 10 m/s → means → $v = 10$ m/s

Identify the unknown value:

What is the sprinter's momentum? → means → $p = ?$ kg•m/s

SOLVE the problem

Substitute the known values $m = 80$ kg and $v = 10$ m/s into the momentum equation:

$$p = mv = (80 \text{ kg})(10 \text{ m/s}) = 800 \text{ kg•m/s}$$

CHECK your answer

Does your answer seem reasonable? Check your answer by dividing the momentum you calculated by the mass given in the problem. The result should be the speed given in the problem.

Practice Problems

1. What is the momentum of a car with a mass of 1,300 kg traveling at a speed of 28 m/s?
2. A baseball thrown by a pitcher has a momentum of 6.0 kg•m/s. If the baseball's mass is 0.15 kg, what is the baseball's speed?
3. What is the mass of a person walking at a speed of 0.8 m/s if their momentum is 52.0 kg•m/s?

For more practice problems go to page 879, and visit Math Practice at gpescience.com.

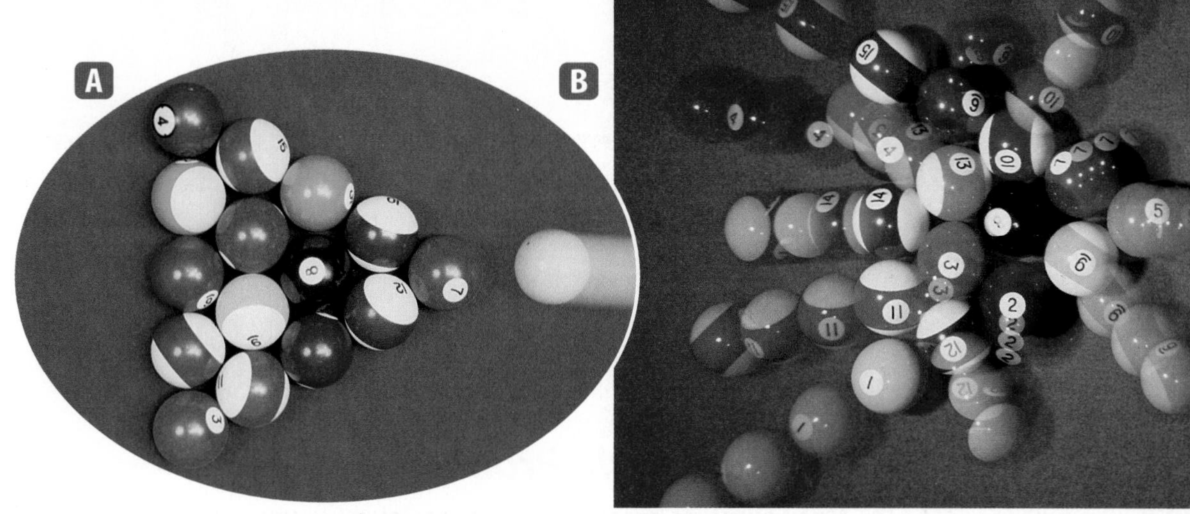

Law of Conservation of Momentum The momentum of an object doesn't change unless its mass, velocity, or both change. Momentum, however, can be transferred from one object to another. Consider the game of pool shown in **Figure 20.**

When the cue ball hits the group of balls that are motionless, the cue ball slows down and the rest of the balls begin to move. The momentum the group of balls gained is equal to the momentum that the cue ball lost. The total momentum of all the balls just before and after the collision would be the same. If no other forces act on the balls, their total momentum is conserved—it isn't lost or created. This is the **law of conservation of momentum**—if a group of objects exerts forces only on each other, their total momentum doesn't change.

Figure 20 Momentum is transferred in collisions. **A** Before the collision, only the cue ball has momentum. **B** When the cue ball strikes the other balls, it transfers some of its momentum to them.

section 3 review

Summary

Newton's Third Law

- According to Newton's third law of motion, for every action force, there is an equal and opposite reaction force.
- Action and reaction forces act on different objects.

Momentum

- The momentum of an object is the product of its mass and velocity:

$$p = mv$$

The Law of Conservation of Momentum

- According to the law of conservation of momentum, if objects exert forces only on each other, their total momentum is conserved.
- In a collision, momentum is transferred from one object to another.

Self Check

1. **Determine** You push against a wall with a force of 50 N. If the wall doesn't move, what is the net force on you?
2. **Explain** how a rocket can move through outer space where there is no matter for it to push on.
3. **Compare** the momentum of a 6,300-kg elephant walking 0.11 m/s and a 50-kg dolphin swimming 10.4 m/s.
4. **Describe** what happens to the momentum of two billiard balls that collide.
5. **Think Critically** A ballet director assigns slow, graceful steps to larger dancers, and quick movements to smaller dancers. Why is this plan successful?

Applying Math

6. **Calculate Momentum** What is the momentum of a 100-kg football player running at a speed of 4 m/s?

The Momentum of Colliding Objects

Goals

- **Observe** and calculate the momentum of different balls.
- **Compare** the results of collisions involving different amounts of momentum.

Materials

meterstick
softball
racquetball
tennis ball
baseball
stopwatch
masking tape
balance

Safety Precautions

Real-World Problem

Many scientists hypothesize that dinosaurs became extinct 65 million years ago when an asteroid collided with Earth. The asteroid's diameter was probably no more than 10 km. Earth's diameter is more than 12,700 km. How could an object that size change Earth's climate enough to cause the extinction of animals that had dominated life on Earth for 140 million years? The asteroid could have caused such damage because it may have been traveling at a velocity of 50 m/s, and had a huge amount of momentum. The combination of an object's velocity and mass will determine how much force it can exert. How do the mass and velocity of a moving object affect its momentum?

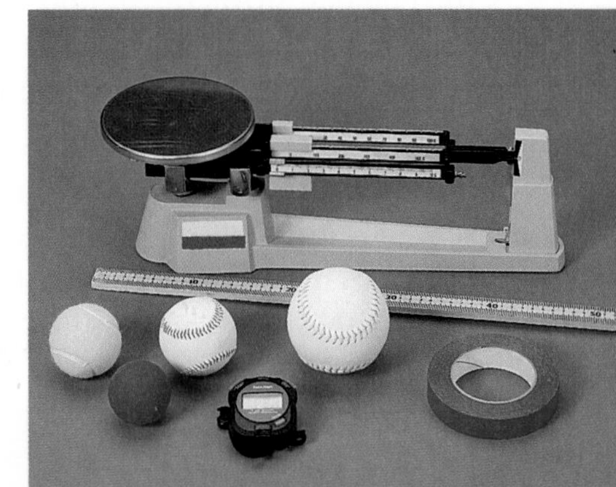

Momentum of Colliding Balls

Action	Time	Velocity	Mass	Momentum	Distance Softball Moved
Racquetball rolled slowly					
Racquetball rolled quickly					
Tennis ball rolled slowly		Do not write in this book.			
Tennis ball rolled quickly					
Baseball rolled slowly					
Baseball rolled quickly					

Procedure

1. Complete the safety form.
2. Copy the data table on the previous page in your Science Journal.
3. Use the balance to measure the mass of the racquetball, tennis ball, and baseball. Record these masses in your data table.
4. Measure a 2-m distance on the floor and mark it with two pieces of masking tape.
5. Place the softball on one piece of tape.
6. Use a stopwatch to time how long it takes the racquetball to slowly roll from the other piece of tape and hit the softball. Record this time in your data table.
7. Measure and record the distance the racquetball moved the softball.
8. Repeat steps 6–7, rolling the racquetball quickly.
9. Repeat steps 6–7, rolling the tennis ball slowly, then quickly.
10. Repeat steps 6–7, rolling the baseball slowly, then quickly.

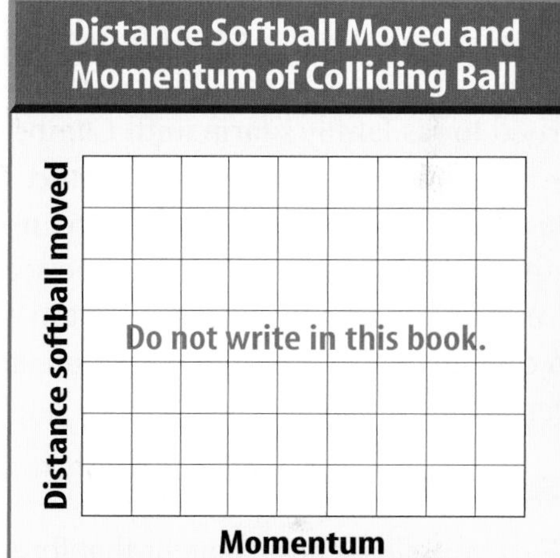

Analyze Your Data

1. **Calculate** the momentum for each type of ball and action using the formula $p = mv$. Record your calculations in the data table.
2. **Graph** the relationship between the momentum of each ball and the distance the softball was moved using a graph like the one shown.

Conclude and Apply

1. **Infer** from your graph how the distance the softball moves after each collision depends on the momentum of the ball that hits it.
2. **Explain** how the motion of the balls after they collide obeys Newton's laws of motion.

Communicating Your Data

Compare your graph with the graphs made by other students in your class. **Discuss** why the graphs might look different.

Newton and the Plague

In Einstein's theory of general relativity, gravity is due to a distortion in space-time.

In 1665, the bubonic plague swept through England and other parts of Europe. Isaac Newton, then a 23-year-old university student, returned to his family's farm until Cambridge university reopened. To occupy his time, Newton made a list of 22 questions. During the next 18 months, Newton buried himself in the search for answers. And in that brief time, Newton developed calculus, the three laws of motion, and the universal law of gravitation!

The Laws of Motion

Earlier philosophers thought that force was necessary to keep an object moving. By analyzing the data collected by Galileo and others, Newton realized that forces did not cause motion. Instead, forces cause a change in motion. Newton came to understand that force and acceleration were related and that objects exert equal and opposite forces on each other. Newton's three laws of motion were able to explain how all things moved, from an apple falling from a tree to the motions of the moon and the planets, in terms of force, mass, and acceleration.

Isaac Newton was a university student when he developed the laws of motion.

What is gravity?

Using the calculus and data on the motion of the planets, Newton deduced the law of universal gravitation. This law enabled the force of gravity between any two objects to be calculated, if their masses and the distance between them were known.

Newton was able to show mathematically that the law of universal gravitation predicted that the planets' orbits should be ellipses, just as Johannes Kepler had discovered two generations earlier. The application of Newton's laws of motion and the law of universal gravitation also were able to explain phenomena such as tides, the motion of the moon and the planets, and the bulge at the Earth's equator.

A Different View of Gravity

In 1916, Albert Einstein proposed a different model for gravity called the general theory of relativity. In Einstein's model, objects create distortions in space-time, like a bowling ball dropped on a sheet. What we see as the force of gravity is the motion of an object on distorted space-time. Today, Einstein's theory is used to help explain the nature of the big bang and the structure of the universe.

Investigate Research how both Newton's law of gravitation and Einstein's general theory of relativity have been used to develop the current model of the universe.

TIME

For more information, visit gpescience.com

Reviewing Main Ideas

Section 1 The First Two Laws of Motion

1. According to the first law of motion, the velocity of an object doesn't change unless an unbalanced net force acts on the object.
2. Inertia is the tendency of an object to resist a change in its motion.
3. The second law of motion states that a net force causes an object to accelerate in the direction of the net force and that the acceleration is given by

$$a = \frac{F_{net}}{m}$$

Section 2 Gravity

1. Gravity is an attractive force between any two objects with mass. The gravitational force depends on the masses of the objects and the distance between them.
2. The gravitational acceleration, g, near Earth's surface equals 9.8 m/s^2. The force of gravity on an object with mass, m, is:

$$F = mg$$

3. The weight of an object near Earth's surface is:

$$W = mg$$

4. Projectiles travel in a curved path because of their horizontal motion and vertical acceleration due to gravity.
5. The centripetal force is the net force on an object in circular motion and is directed toward the center of the circular path.

Section 3 The Third Law of Motion

1. Newton's third law of motion states that for every action there is an equal and opposite reaction.
2. The momentum of an object can be calculated by the equation $p = mv$.

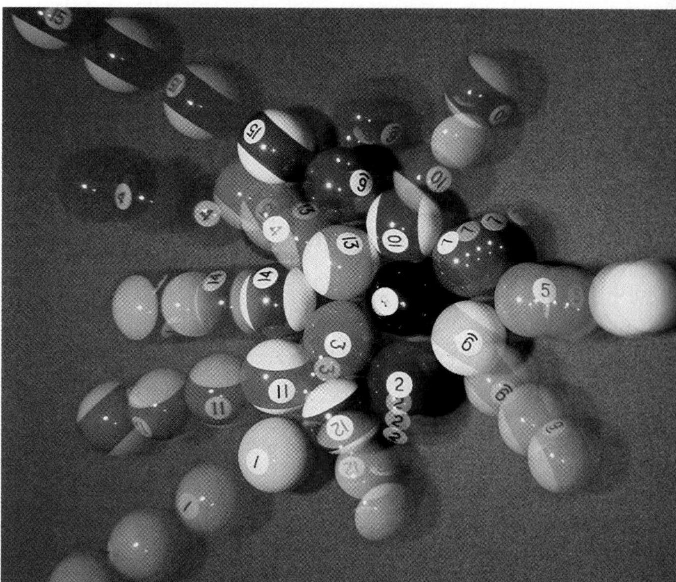

3. When two objects collide, momentum can be conserved. Some of the momentum from one object is transferred to the other.

FOLDABLES Use the Foldable that you made at the beginning of this chapter to help you review the laws of motion.

Using Vocabulary

centripetal acceleration p. 110
centripetal force p. 110
first law of motion p. 98
gravity p. 104
inertia p. 99
law of conservation of momentum p. 117
momentum p. 116
second law of motion p. 102
third law of motion p. 113
weight p. 106

Complete each statement using a word(s) from the vocabulary list above.

1. The way in which objects exert forces on each other is described by ________.
2. The ________ of an object depends on its mass and velocity.
3. The ________ of an object is different on other planets in the solar system.
4. When an object moves in a circular path, the net force is called a(n) ________.
5. The attractive force between two objects that depends on their masses and the distance between them is ________.
6. ________ relates the net force exerted on an object to its mass and acceleration.

Checking Concepts

Choose the best answer for each question.

7. What is the gravitational force exerted on an object called?
   A) centripetal force **C)** momentum
   B) friction **D)** weight
8. Which of the following best describes why projectiles move in a curved path?
   A) They have horizontal velocity and vertical acceleration.
   B) They have momentum.
   C) They have mass.
   D) They have weight.
9. Which of the following explains why astronauts seem weightless in orbit?
   A) Earth's gravity is much less in orbit.
   B) The space shuttle is in free fall.
   C) The gravity of Earth and the Sun cancel.
   D) The centripetal force on the shuttle balances Earth's gravity.
10. Which of the following exerts the strongest gravitational force on you?
   A) the Moon **C)** the Sun
   B) Earth **D)** this book

Use the graph below to answer question 11.

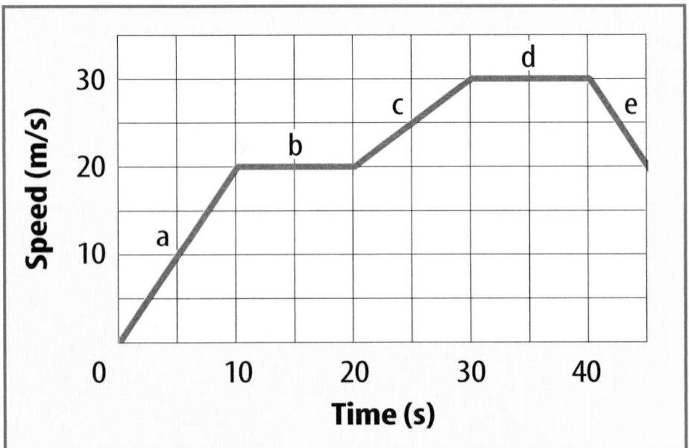

11. The graph shows the speed of a car moving in a straight line. Over which segments are the forces on the car balanced?
   A) A and C **C)** C and E
   B) B and D **D)** D only
12. Which of the following is true about an object in free fall?
   A) Its acceleration depends on its mass.
   B) It has no inertia.
   C) It pulls on Earth, and Earth pulls on it.
   D) Its momentum is constant.
13. The acceleration of an object is in the same direction as which of the following?
   A) net force **C)** static friction
   B) air resistance **D)** gravity
14. Which of the following is NOT a force?
   A) weight **C)** momentum
   B) friction **D)** air resistance

Interpreting Graphics

15. Copy and complete the following concept map on forces.

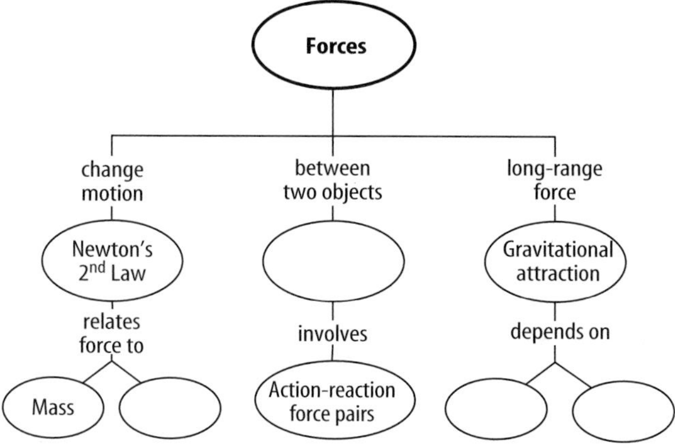

Use the table below to answer questions 16–18.

Time of Fall for Dropped Objects		
Object	Mass (g)	Time of Fall (s)
A	5.0	2.0
B	5.0	1.0
C	30.0	0.5
D	35.0	1.5

16. If the objects in the data table above all fell the same distance, which object fell with the greatest average speed?

Thinking Critically

17. **Describe** the forces acting on a softball after it leaves the pitcher's hand. Ignore the effects of air resistance.

18. **Determine** the direction of the net force on a book sliding on a table if the book is slowing down.

19. **Explain** whether there can be any forces acting on a car moving in a straight line with constant speed.

20. **Explain** You pull a door open. If the force the door exerts on you is equal to the force you exert on the door, why don't you move?

21. **Predict** Suppose you are standing on a bathroom scale next to a sink. How does the reading on the scale change if you push down on the sink?

22. **Describe** the action and reaction force pairs involved when an object falls toward Earth. Ignore the effects of air resistance.

Applying Math

23. **Calculate Mass** Find your mass if a scale on Earth reads 650 N when you stand on it.

24. **Calculate Acceleration of Gravity** You weigh yourself at the top of a high mountain and the scale reads 720 N. If your mass is 75 kg, what is the acceleration of gravity at your location?

Use the figure below to answer question 25.

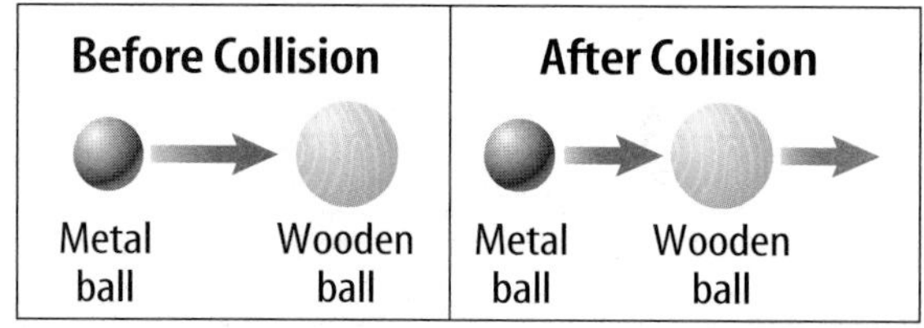

25. **Calculate Speed** The 2-kg metal ball moving at a speed of 3 m/s strikes a 1-kg wooden ball that is at rest. After the collision, the speed of the metal ball is 1 m/s. Assuming momentum is conserved, what is the speed of the wooden ball?

26. **Calculate Mass** Find the mass of a car that has a speed of 30 m/s and a momentum of 45,000 kg,m/s.

27. **Calculate Sliding Friction** A box being pushed with a force of 85 N slides along the floor with a constant speed. What is the force of sliding friction on the box?

chapter 4 Standardized Test Practice

Record your answers on the answer sheet provided by your teacher or on a sheet of paper.

Multiple Choice

1. The net force on an object moving with constant speed in circular motion is in which direction?
   A. downward
   B. opposite to the object's motion
   C. toward the center of the circle
   D. in the direction of the object's velocity

Use the table below to answer questions 2–4.

Object in Free Fall with Air Resistance	
Time (s)	Speed (m/s)
0	0
1	9.1
2	15.1
3	18.1
4	19.3
5	19.9

2. Over which of the following time intervals is the acceleration of the object the greatest?
   A. 0 s to 1 s
   B. 1 s to 2 s
   C. 4 s to 5 s
   D. The acceleration is constant.

Test-Taking Tip

Keep Track of Time If you are taking a timed test, keep track of time. If you find you are spending too much time on a multiple-choice question, mark your best guess and move on.

3. Over which of the following time intervals is the net force on the object the smallest?
   A. 0 s to 1 s
   B. 1 s to 2 s
   C. 4 s to 5 s
   D. The force is constant.

4. According to the trend in these data, which of the following values is most likely the speed of the object after falling for 6 s?
   A. 26.7 m/s
   B. 15.1 m/s
   C. 20.1 m/s
   D. 0 m/s

5. Which of the following would cause the gravitational force between object A and object B to increase?
   A. Decrease the distance between them.
   B. Increase the distance between them.
   C. Decrease the mass of object A.
   D. Decrease the mass of both objects.

Use the graphs below to answer question 6.

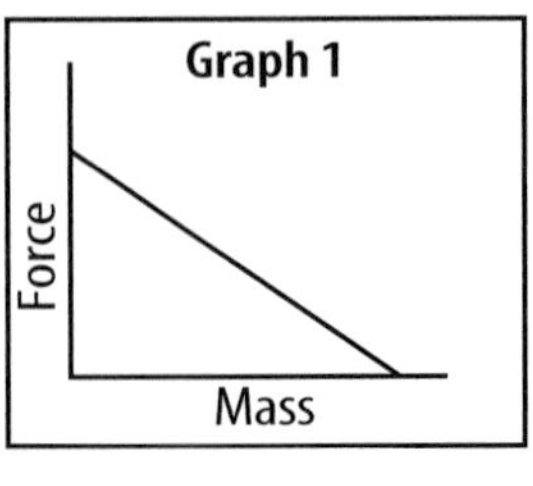

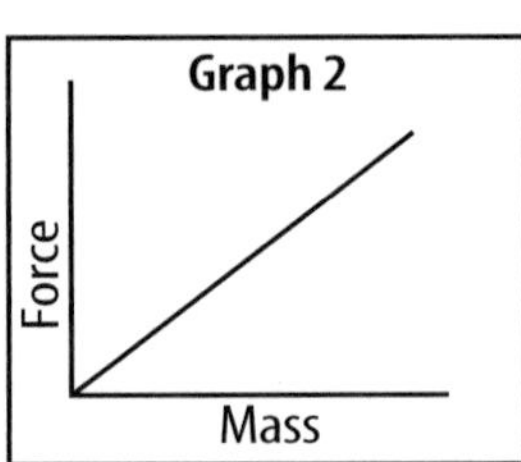

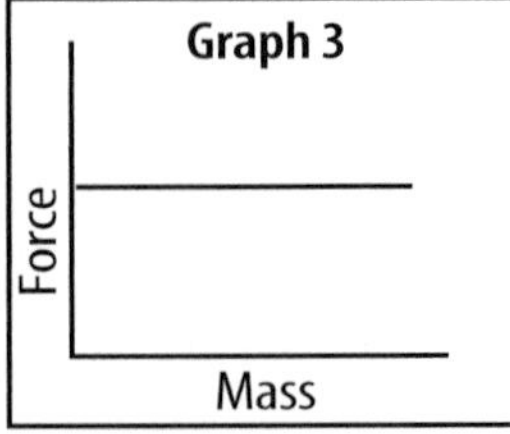

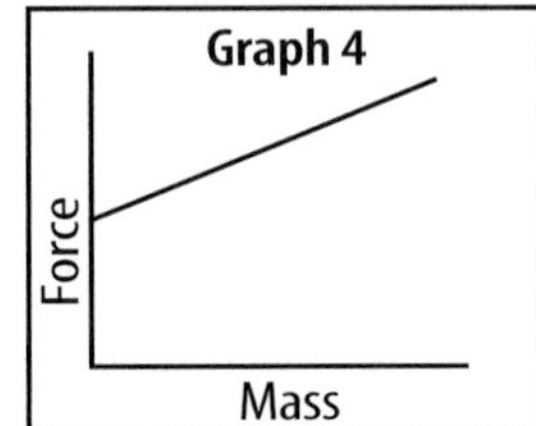

6. Which of the graphs on the previous page shows how the force on an object changes if the mass increases and the acceleration stays constant?

   A. Graph 1

   B. Graph 2

   C. Graph 3

   D. Graph 4

Gridded Response

7. You are pushing a 30-kg wooden crate across the floor. The force of sliding friction on the crate is 90 N. How much force, in newtons, must you exert on the crate to keep it moving with a constant velocity?

8. A skydiver with a mass of 60 kg jumps from an airplane. Five seconds after jumping, the force of air resistance on the skydiver is 300 N. What is the skydiver's acceleration in m/s^2 five seconds after jumping?

Short Response

Use the figure below to answer question 9.

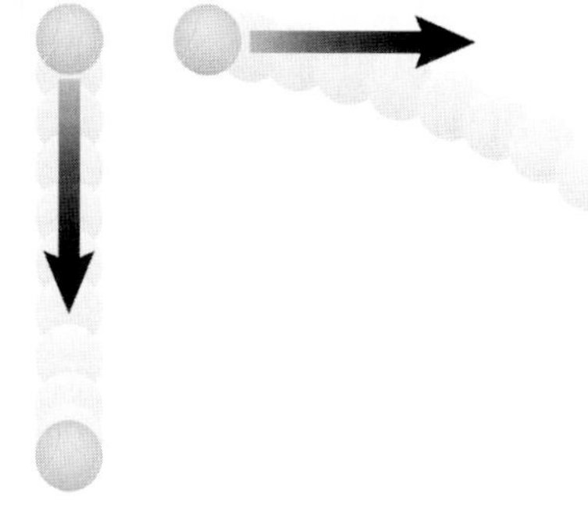

9. Two balls are at the same height. One ball is dropped and the other initially moves horizontally, as shown in the figure. After 1 s, which ball has fallen the greatest vertical distance?

10. How does the acceleration of gravity 5,000 km above Earth's surface compare with the acceleration of gravity at Earth's surface?

Extended Response

Use the graph below to answer question 11.

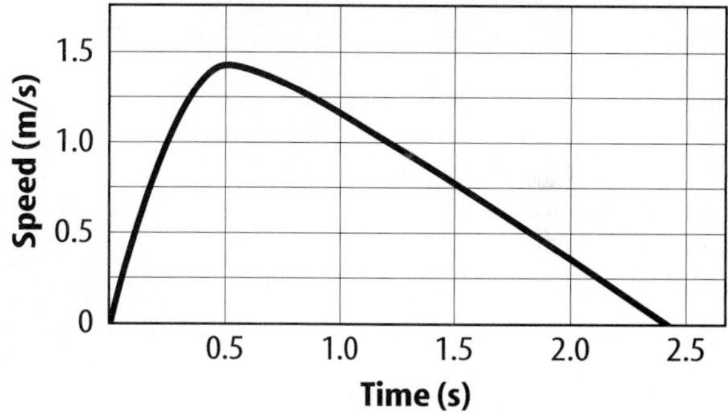

11. The graph above shows the how the speed of a book changes as it slides across a table. Over what time interval is the net force on the book in the opposite direction of the book's motion?

12. When the space shuttle is launched from Earth, the rocket engines burn fuel and apply a constant force on the shuttle until they burn out. Explain why the shuttles acceleration increases while the rocket engines are burning fuel.

chapter 5

Energy

chapter preview

sections

A Big Lift

How does this pole vaulter go from standing at the end of a runway to climbing through the air? The answer is energy. During her vault, energy originally stored in her muscles is converted into other forms of energy that enable her to soar.

Science Journal Which takes more energy: walking up stairs, or taking an escalator? Explain your reasoning.

Start-Up Activities

Energy Conversions

One of the most useful inventions of the nineteenth century was the electric lightbulb. Being able to light up the dark has enabled people to work and play longer. A lightbulb converts electrical energy to heat energy and light, another form of energy. The following lab shows how electrical energy is converted into other forms of energy.

WARNING: *Steel wool can become hot—connect to battery only for a brief time.*

1. Complete the safety form.
2. Obtain two D-cell batteries, tape, metal tongs, two non-coated paper clips, and some steel wool. Separate the steel wool into thin strands and straighten the paper clips.
3. Tape the batteries together and then tape one end of each paper clip to the battery terminals.
4. While holding the steel wool with the tongs, briefly complete the circuit by placing the steel wool in contact with both the paper clip ends.
5. **Think Critically** In your Science Journal, describe what happened to the steel wool. What changes did you observe?

Preview this chapter's content and activities at gpescience.com

FOLDABLES™ Study Organizer

Energy Make the following Foldable to help you identify what you already know, what you want to know, and what you learned about energy.

STEP 1 **Fold** a sheet of paper vertically from side to side. Make the front edge about 1 cm shorter than the back edge.

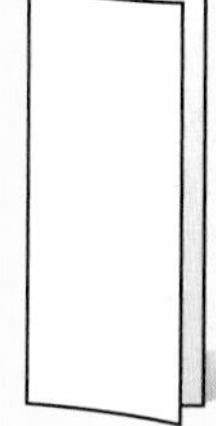

STEP 2 **Turn** lengthwise and **fold** into thirds.

STEP 3 **Unfold and cut** only the top layer along both folds to make three tabs.

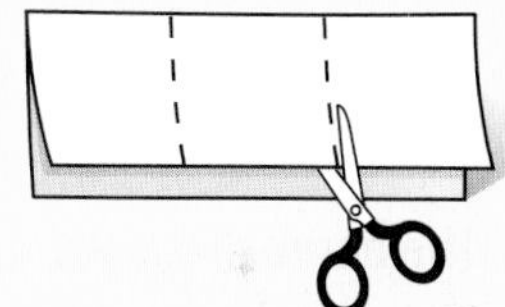

STEP 4 **Label** each tab as shown.

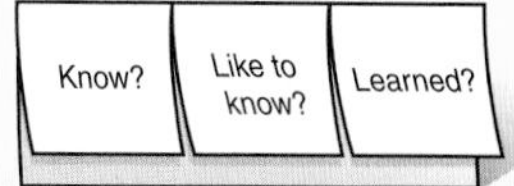

Question Before you read the chapter, write what you already know about energy under the left tab of your Foldable, and write questions about what you'd like to know under the center tab. After you read the chapter, list what you learned under the right tab.

section 1

The Nature of Energy

Reading Guide

What You'll Learn

- **Distinguish** between kinetic and potential energy.
- **Calculate** kinetic energy.
- **Describe** different forms of potential energy.
- **Calculate** gravitational potential energy.

Why It's Important

All of the changes that occur around you every day involve the conversion of energy from one form to another.

Review Vocabulary

gravity: the attractive force between any two objects that have mass

New Vocabulary

- energy
- kinetic energy
- joule
- potential energy
- elastic potential energy
- chemical potential energy
- gravitational potential energy

What is energy?

Wherever you are sitting as you read this, changes are taking place—lightbulbs are heating the air around them, the wind might be rustling leaves, or sunlight might be glaring off a nearby window. Even you are changing as you breathe, blink, or shift position in your seat.

Every change that occurs—large or small—involves energy. Imagine a baseball flying through the air. It hits a window, causing the glass to break as shown in **Figure 1.** The window changed from a solid sheet of glass to a number of broken pieces. The moving baseball caused this change—a moving baseball has energy. Even when you comb your hair or walk from one class to another, energy is involved.

Figure 1 The baseball caused changes to occur when it hit the window.
Describe *the changes that are occurring.*

Energy and Work In science, **energy** is defined as the ability to do work. Work is done when a force causes something to move. The ball in **Figure 1** does work on the window when it exerts a force on the glass and causes it to break. Energy also can be defined as the ability to cause change. When work is done, something moves and a change occurs. When work is done and a change occurs, energy moves from place to place or changes from one form to another.

Figure 2 Energy can be stored and it can move from place to place.
Infer *which materials are storing chemical energy.*

Different Forms of Energy

Turn on an electric light, and a dark room becomes bright. Turn on your CD player, and sound comes through your headphones. In both situations, energy moves from one place to another. These changes are different from each other, and differ from the baseball shattering the window in **Figure 1.** This is because energy has several different forms—electrical, chemical, radiant, and thermal.

Figure 2 shows some examples of everyday situations which involve forms of energy. Is the chemical energy stored in food the same as the energy that comes from the Sun or the energy stored in gasoline? Radiant energy from the Sun travels a vast distance through space to Earth, warming the planet and providing energy that enables green plants to grow. When you make toast in the morning, you are using electrical energy. In short, energy plays a role in every activity that you do.

Reading Check *What are some different forms of energy?*

An Energy Analogy Money can be used in an analogy to help you understand energy. If you have $100, you could store it in a variety of forms—cash in your wallet, a bank account, travelers' checks, or gold or silver coins. You could transfer that money to different forms. You could deposit your cash into a bank account or trade the cash for gold. Regardless of its form, money is money. The same is true for energy. Energy from the Sun that warms you and energy from the food that you eat are only different forms of the same thing.

Topic: Glacier Flow

Visit gpescience.com for Web links to information about the speeds at which glaciers flow.

Activity For the five fastest glaciers, calculate the kinetic energy a 1-kg block of ice in each glacier would have.

Kinetic Energy

When you think of energy, you might think of action—or objects in motion—like the baseball that shatters a window. An object in motion does have energy. **Kinetic energy** is the energy a moving object has because of its motion. The kinetic energy of a moving object depends on the object's mass and its speed.

Kinetic Energy Equation

$$\text{kinetic energy (in joules)} = \frac{1}{2}\,\text{mass (in kg)} \times [\text{speed (in m/s)}]^2$$

$$KE = \frac{1}{2}mv^2$$

The SI unit of energy is the **joule,** abbreviated J. If you dropped a softball from a height of about 0.5 m, it would have a kinetic energy of about one joule before it hit the floor.

Applying Math — Solve a Simple Equation

KINETIC ENERGY A jogger whose mass is 60 kg is moving at a speed of 3 m/s. What is the jogger's kinetic energy?

IDENTIFY known values and the unknown value

Identify the known values:

a jogger whose mass is 60 kg → means → $m = 60$ kg

is moving at a speed of 3 m/s → means → $v = 3$ m/s

Identify the unknown value:

What is the jogger's kinetic energy? → means → $KE = ?$ J

SOLVE the problem

Substitute the known values $m = 60$ kg and $v = 3$ m/s into the kinetic energy equation:

$$KE = \frac{1}{2}mv^2 = \frac{1}{2}(60\text{ kg})(3\text{ m/s})^2 = \frac{1}{2}(60)(9)\text{ kg m}^2/\text{s}^2 = 270\text{ J}$$

CHECK your answer

Does your answer seem reasonable? Check your answer by dividing the kinetic energy you calculate by the square of the given velocity, and then multiplying by 2. The result should be the mass given in the problem.

Practice Problems

1. What is the kinetic energy of a baseball moving at a speed of 40 m/s if the baseball has a mass of 0.15 kg?
2. A car moving at a speed of 20 m/s has a kinteic energy of 300,000 J. What is the car's mass?
3. A sprinter has a mass of 80 kg and a kinetic energy of 4,000 J. What is the sprinter's speed?

For more practice problems go to page 879, and visit Math Practice at gpescience.com.

Figure 3 As natural gas burns, it combines with oxygen to form carbon dioxide and water. In this chemical reaction, chemical potential energy is released.

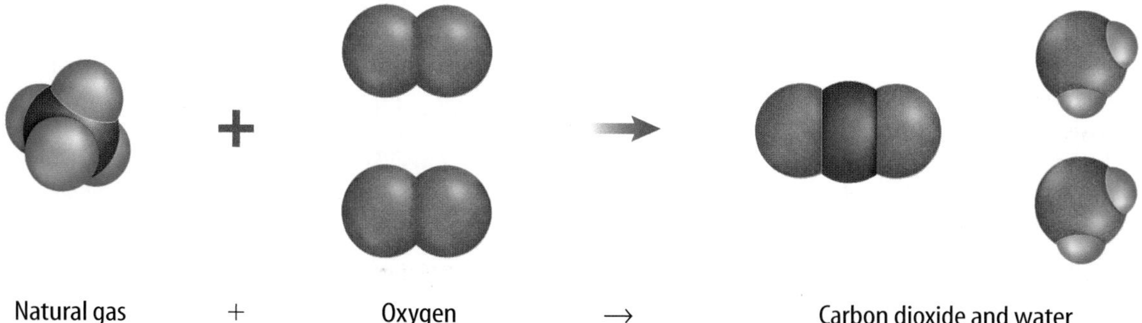

Potential Energy

Energy doesn't have to involve motion. Even motionless objects can have energy. This energy is stored in the object. Therefore, the object has potential to cause change. A hanging apple in a tree has stored energy. When the apple falls to the ground, a change occurs. Because the apple has the ability to cause change, it has energy. The hanging apple has energy because of its position above Earth's surface. Stored energy due to position is called **potential energy.** If the apple stays in the tree, it will keep the stored energy due to its height above the ground. If it falls, that stored energy of position is converted to energy of motion.

Elastic Potential Energy Energy can be stored in other ways, too. If you stretch a rubber band and let it go, it sails across the room. As it flies through the air, it has kinetic energy due to its motion. Where did this kinetic energy come from? Just as the apple hanging in the tree had potential energy, the stretched rubber band had energy stored as elastic potential energy. **Elastic potential energy** is energy stored by something that can stretch or compress, such as a rubber band or spring.

Chemical Potential Energy The cereal you eat for breakfast and the sandwich you eat at lunch also contain stored energy. Gasoline stores energy in the same way as food stores energy—in the chemical bonds between atoms. Energy stored in chemical bonds is **chemical potential energy. Figure 3** shows a molecule of natural gas. Energy is stored in the bonds that hold the carbon and hydrogen atoms together and is released when the gas is burned.

How is elastic potential energy different from chemical potential energy?

Interpreting Data from a Slingshot

Procedure

1. Complete the safety form.
2. Using two fingers, carefully stretch a **rubber band** on a table until it has no slack.
3. Place a **nickel** on the table, slightly touching the midpoint of the rubber band.
4. Push the nickel back 0.5 cm into the rubber band and release. Measure the distance the nickel travels.
5. Repeat step 4, each time pushing the nickel back an additional 0.5 cm.

Analysis

1. How did the takeoff speed of the nickel seem to change relative to the distance that you stretched the rubber band?
2. What does this imply about the kinetic energy of the nickel?

The Myth of Sisyphus In Greek mythology, a king named Sisyphus angered the gods by attempting to delay death. As punishment, he was doomed for eternity to endlessly roll a huge stone up a hill, only to have it roll back to the bottom again. Explain what caused the potential energy of the stone to change as it moved up and down the hill.

Gravitational Potential Energy Anything that can fall has stored energy called gravitational potential energy. **Gravitational potential energy** (GPE) is energy stored by objects due to their position above Earth's surface. The GPE of an object depends on the object's mass and height above the ground. Gravitational potential energy can be calculated from the following equation.

Gravitational Potential Energy Equation

gravitational potential energy (J) =
mass (kg) × acceleration due to gravity (m/s²) × **height** (m)

$$GPE = mgh$$

On Earth, the acceleration due to gravity is 9.8 m/s² and has the symbol *g*. Like all forms of energy, gravitational potential energy is measured in joules.

Applying Math — Solve a Simple Equation

GRAVITATIONAL POTENTIAL ENERGY What is the gravitational potential energy of a ceiling fan that has a mass of 7 kg and is 4 m above the ground?

IDENTIFY known values and the unknown value

Identify the known values:

has a mass of 7 kg → means → $m = 7$ kg

is 4 m above the ground → means → $h = 4$ m

Identify the unknown value:

what is the gravitational potential energy → means → $GPE = ?$ J

SOLVE the problem

Substitute the known values $m = 7$ kg, $h = 4$ m, and $g = 9.8$ m/s² into the gravitational potential energy equation:

$$GPE = mgh = (7 \text{ kg})(9.8 \text{ m/s}^2)(4 \text{ m}) = (274) \text{ kg m}^2/\text{s}^2 = 274 \text{ J}$$

CHECK your answer

Does your answer seem reasonable? Check your answer by dividing the gravitational potential energy you calculate by the given mass, and then divide by 9.8 m/s². The result should be the height given in the problem.

Practice Problems

1. Find the height of a baseball with a mass of 0.15 kg that has a GPE of 73.5 J.
2. Find the GPE of a coffee mug with a mass of 0.3 kg on a 1-m high counter top.
3. What is the mass of a hiker 200 m above the ground if her GPE is 117,600 J?

For more practice problems go to page 879, and visit Math Practice at gpescience.com.

Changing GPE Look at the objects in the bookcase in **Figure 4.** Which of these objects has the most gravitational potential energy? According to the equation for gravitational potential energy, the GPE of an object can be increased by increasing its height above the ground. If two objects are at the same height, then the object with the larger mass has more gravitational potential energy.

In **Figure 4,** suppose the green vase on the lower shelf and the blue vase on the upper shelf have the same mass. Then the blue vase on the upper shelf has more gravitational potential energy because it is higher above the ground.

Imagine what would happen if the two vases were to fall. As they fall and begin moving, they have kinetic energy as well as gravitational potential energy. As the vases get closer to the ground, their gravitational potential energy decreases. At the same time, they are moving faster, so their kinetic energy increases. The vase that was higher above the floor has fallen a greater distance. As a result, the vase that initially had more gravitational potential energy will be moving faster and have more kinetic energy when it hits the floor.

Figure 4 An object's gravitational potential energy increases as its height increases.

section 1 review

Summary

Energy

- Energy is the ability to cause change.
- Forms of energy include electrical, chemical, thermal, and radiant energy.

Kinetic Energy

- Kinetic energy is the energy a moving object has because of its motion.
- The kinetic energy of a moving object can be calculated from this equation:

$$KE = \frac{1}{2}mv^2$$

Potential Energy

- Potential energy is stored energy due to the position of an object.
- Different forms of potential energy include elastic potential energy, chemical potential energy, and gravitational potential energy.
- Gravitational potential energy can be calculated from this equation:

$$GPE = mgh$$

Self Check

1. **Explain** whether an object can have kinetic energy and potential energy at the same time.
2. **Describe** three situations in which the gravitational potential energy of an object changes.
3. **Explain** how the kinetic energy of a truck could be increased without increasing the truck's speed.
4. **Think Critically** The different molecules that make up the air in a room have, on average, the same kinetic energy. How does the speed of the different air molecules depend on their masses?

Applying Math

5. **Calculate Kinetic Energy** Find the kinetic energy of a ball with a mass of 0.06 kg moving at 50 m/s.
6. **Use Ratios** A boulder on top of a cliff has potential energy of 8,800 J and twice the mass of a boulder next to it. What is the GPE of the smaller boulder?
7. **Calculate GPE** An 80-kg diver jumps from a 10-m high platform. What is the gravitational potential energy of the diver halfway down?

Bouncing Balls

What happens when you drop a ball onto a hard, flat surface? It starts with potential energy. It bounces up and down until it finally comes to a rest. Where did the energy go?

Real-World Problem

Why do bouncing balls stop bouncing?

Goals

- **Identify** the forms of energy observed in a bouncing ball.
- **Infer** why the ball stops bouncing.

Materials

tennis ball	masking tape
rubber ball	cardboard box
balance	**shoe box*
meterstick	**Alternate materials*

Safety Precautions

Procedure

1. Complete the safety form.
2. **Measure** the mass of the two balls.
3. Have a partner drop one ball from 1 m. Measure how high the ball bounced. Repeat this two more times so you can calculate an average bounce height. Record your values on the data table.
4. Repeat step 3 for the other ball.
5. **Predict** whether the balls would bounce higher or lower if they were dropped onto the cardboard box. Design an experiment to measure how high the balls would bounce off the surface of a cardboard box.

Bounce Height

Type of Ball	Surface	Trial	Height (cm)
Tennis	Floor	1	
Tennis	Floor	2	
Tennis	Floor	3	
Rubber	Floor	1	
Rubber	Floor	2	
Rubber	Floor	3	
Tennis	Box	1	

Conclude and Apply

1. **Calculate** the gravitational potential energy of each ball before dropping it.
2. **Calculate** the average bounce height for the three trials under each condition. Describe your observations.
3. **Compare** the bounce heights of the balls dropped on a cardboard box with the bounce heights of the ballls dropped on the floor. *Hint: Did you observe any movement of the box when the balls bounced?*
4. **Explain** why the balls bounced to different heights, using the concept of elastic potential energy.

Communicating Your Data

Meet with three other lab teams and compare average bounce heights for the tennis ball on the floor. Discuss why your results might differ. **For more help, refer to the Science Skill Handbook.**

section 2 Conservation of Energy

Reading Guide

What You'll Learn

- **Describe** how energy can be transformed from one form to another.
- **Explain** how the mechanical energy of a system is the sum of the kinetic and potential energy.
- **Discuss** the law of conservation of energy.

Why It's Important

All the energy transformations that occur inside you and around you obey the law of conservation of energy.

Review Vocabulary

friction: a force that opposes the sliding motion of two surfaces that are touching each other

New Vocabulary

- mechanical energy
- law of conservation of energy

Changing Forms of Energy

Unless you were talking about potential energy, you probably wouldn't think of the book on top of a bookshelf as having much to do with energy—until it fell. You'd be more likely to think of energy as race cars roar past, or as your body uses energy from food to help it move, or as the Sun warms your skin on a summer day. These situations involve energy changing from one form to another form.

Transforming Electrical Energy You use many devices every day that convert one form of energy to other forms. For example, you might be reading this page in a room lit by lightbulbs. The lightbulbs transform electrical energy into light so you can see. The warmth you feel around the bulb is evidence that some of that electrical energy is transformed into thermal energy, as illustrated in **Figure 5.** What other devices have you used today that make use of electrical energy? You might have been awakened by an alarm clock, styled your hair, made toast, listened to music, or played a video game. What form or forms of energy is electrical energy converted to in these examples?

Figure 5 A lightbulb is a device that transforms electrical energy into light energy and thermal energy.
Identify *other devices that convert electrical energy to thermal energy.*

Spark plug fires

When a spark plug fires in an engine, chemical potential energy is converted into thermal energy.

Gases expand

As the hot gases expand, thermal energy is converted into kinetic energy.

Figure 6 In the engine of a car, several energy conversions occur.

Transforming Chemical Energy Fuel stores energy in the form of chemical potential energy. For example, the car or bus that might have brought you to school this morning probably uses gasoline as a fuel. The engine transforms the chemical potential energy stored in gasoline molecules into the kinetic energy of a moving car or bus. Several energy conversions occur in this process, as shown in **Figure 6.** An electric spark ignites a small amount of fuel. Igniting the fuel changes chemical energy to thermal energy. The thermal energy causes gases to expand and move parts of the engine, producing kinetic energy.

Some energy transformations are less obvious because they do not result in visible motion, sound, heat, or light. Every green plant you see converts light energy from the Sun into energy stored in chemical bonds in the plant. If you eat an ear of corn, the chemical potential energy in the corn is transformed into other forms of energy by your body.

Conversions Between Kinetic and Potential Energy

You have experienced many situations that involve conversions between potential and kinetic energy. Systems such as bicycles, roller coasters, and swings can be described in terms of potential and kinetic energy. Even launching a rubber band or using a bow and arrow involves converting potential energy into kinetic energy. To understand the energy conversions that occur, it is helpful to identify the mechanical energy of a system. **Mechanical energy** is the total amount of potential and kinetic energy in a system and can be expressed by this equation.

$$\text{mechanical energy} = \text{potential energy} + \text{kinetic energy}$$

In other words, mechanical energy is energy due to the position and the motion of an object or the objects in a system. What happens to the mechanical energy of an object as potential and kinetic energy are converted into each other?

Falling Objects Standing under an apple tree can be hazardous. An apple on a tree, like the one in **Figure 7,** has gravitational potential energy due to Earth pulling down on it. The apple does not have kinetic energy while it hangs from the tree. However, the instant the apple comes loose from the tree, it accelerates due to gravity. As it falls, it loses height so its gravitational potential energy decreases. This potential energy is transformed into kinetic energy as the speed of the apple increases.

Look back at the equation for mechanical energy. If the potential energy is being converted into kinetic energy, then the mechanical energy of the apple doesn't change as it falls. The potential energy that the apple loses is gained back as kinetic energy. The form of energy changes, but the total amount of energy remains the same.

What happens to the mechanical energy of the apple as it falls from the tree?

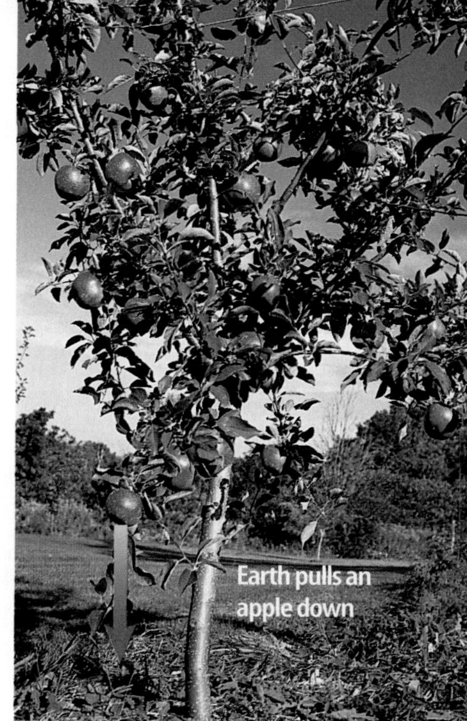

Figure 7 Objects that can fall have gravitational potential energy.
Apply *What objects around you have gravitational potential energy?*

Energy Transformations in Projectile Motion Energy transformations also occur during projectile motion when an object moves in a curved path. Look at **Figure 8.** When the ball leaves the bat, it has mostly kinetic energy. As the ball rises, its speed decreases, so its kinetic energy must decrease, too. However, the ball's gravitational potential energy increases as it goes higher. At its highest point, the baseball has the maximum amount of gravitational potential energy. The only kinetic energy it has at this point is due to its forward motion. Then, as the baseball falls, gravitational potential energy decreases while kinetic energy increases as the ball moves faster. However, the mechanical energy of the ball remains constant as it rises and falls.

Figure 8 Kinetic energy and gravitational potential energy are converted into each other as the ball rises and falls.

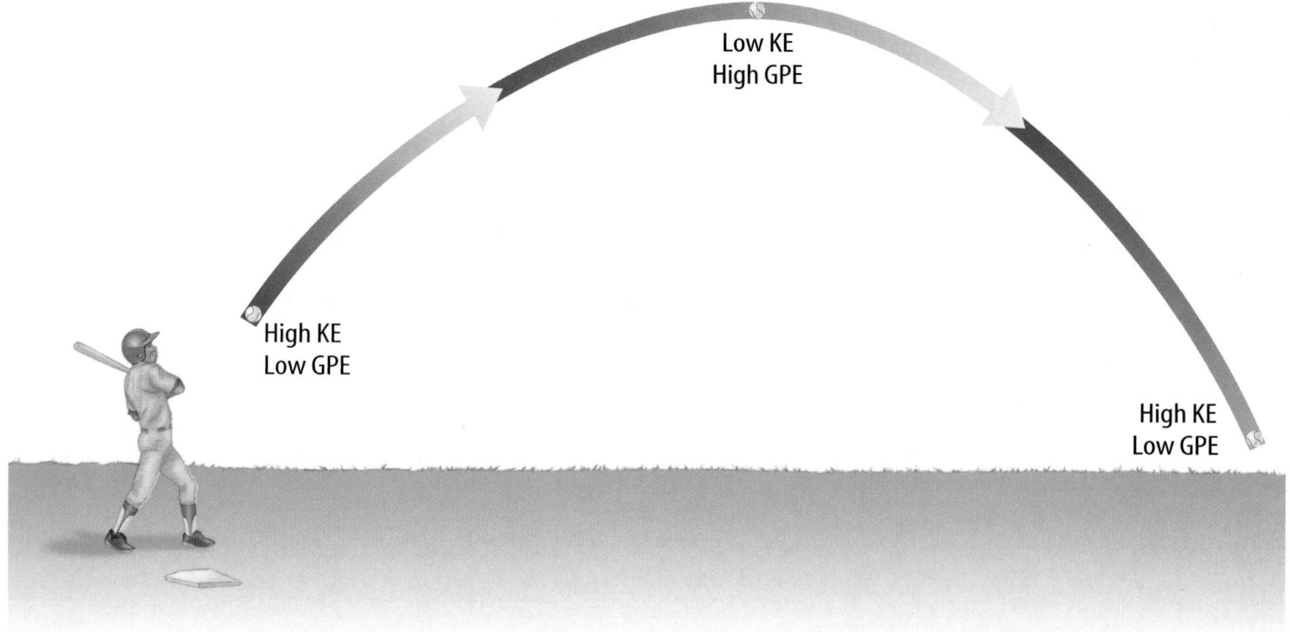

NATIONAL GEOGRAPHIC

VISUALIZING ENERGY TRANSFORMATIONS

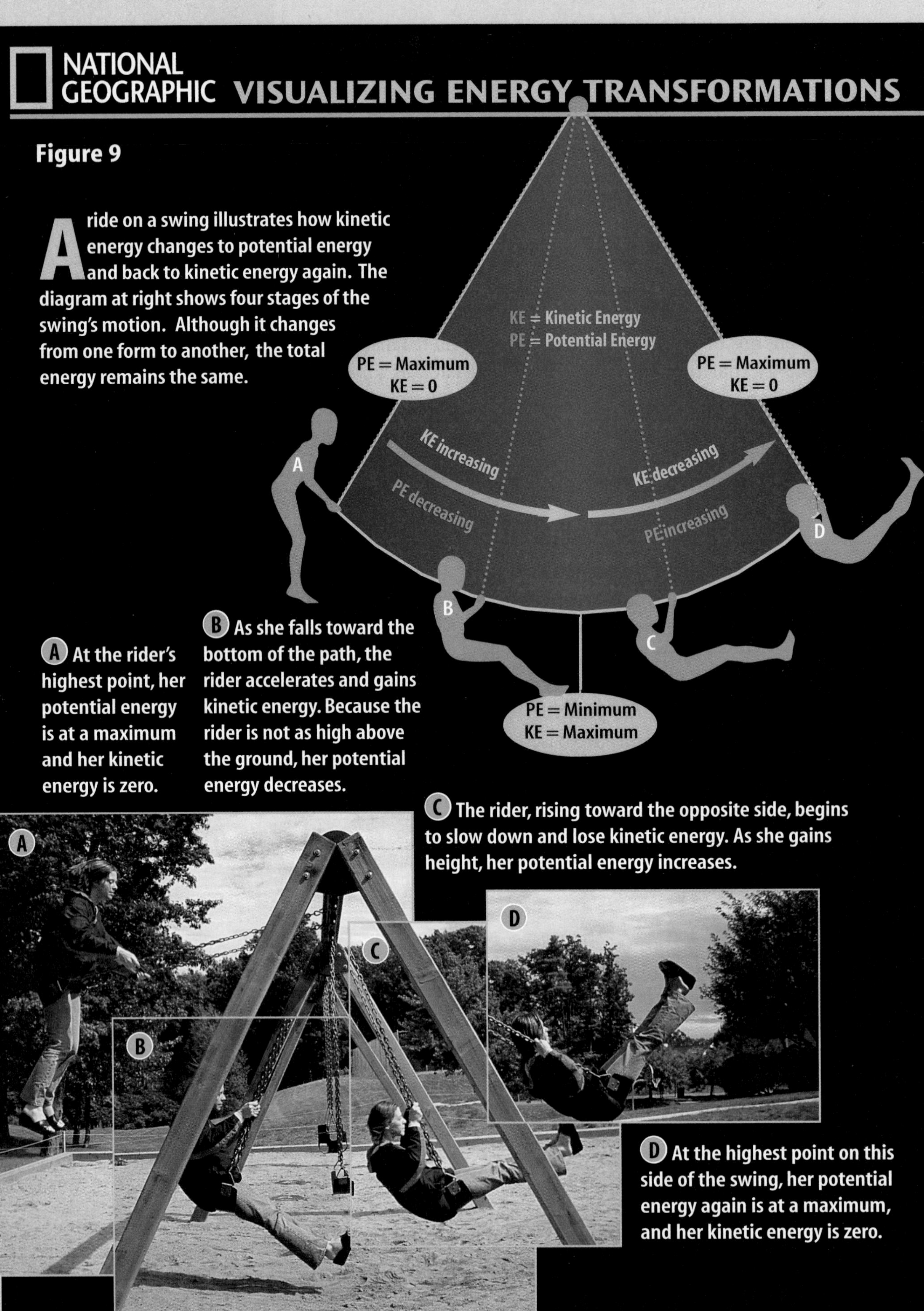

Figure 9

A ride on a swing illustrates how kinetic energy changes to potential energy and back to kinetic energy again. The diagram at right shows four stages of the swing's motion. Although it changes from one form to another, the total energy remains the same.

A At the rider's highest point, her potential energy is at a maximum and her kinetic energy is zero.

B As she falls toward the bottom of the path, the rider accelerates and gains kinetic energy. Because the rider is not as high above the ground, her potential energy decreases.

C The rider, rising toward the opposite side, begins to slow down and lose kinetic energy. As she gains height, her potential energy increases.

D At the highest point on this side of the swing, her potential energy again is at a maximum, and her kinetic energy is zero.

Energy Transformations in a Swing When you ride on a swing, like the one shown in **Figure 9,** part of the fun is the feeling of almost falling as you drop from the highest point to the lowest point of the swing's path. Think about energy conservation to analyze such a ride.

The ride starts with a push that gets you moving, giving you kinetic energy. As the swing rises, you lose speed but gain height. In energy terms, kinetic energy changes to gravitational potential energy. At the top of your path, potential energy is at its greatest. Then, as the swing accelerates downward, potential energy changes to kinetic energy. At the bottom of each swing, the kinetic energy is at its greatest and the potential energy is at its minimum. As you swing back and forth, energy continually converts from kinetic to potential and back to kinetic. What happens to your mechanical energy as you swing?

Energy and the Food Chain One way energy enters ecosystems is when green plants transform radiant energy from the Sun into chemical potential energy in the form of food. Energy moves through the food chain as animals that eat plants are eaten by other animals. Some energy leaves the food chain, such as when living organisms release thermal energy to the environment. Diagram a simple biological food chain showing energy conservation.

The Law of Conservation of Energy

When a ball is thrown into the air or a swing moves back and forth, kinetic and potential energy are constantly changing as the object speeds up and slows down. However, mechanical energy stays constant. Kinetic and potential energy simply change forms and no energy is destroyed.

This is always true. Energy can change from one form to another, but the total amount of energy never changes. Even when energy changes form from electrical to thermal and other energy forms as in the hair dryer shown in **Figure 10,** energy is never destroyed. Another way to say this is that energy is conserved. This principle is recognized as a law of nature. The **law of conservation of energy** states that energy cannot be created or destroyed. On a large scale, this law means that the total amount of energy in the universe does not change.

Reading Check *What law states that the total amount of energy never changes?*

Conserving Resources You might have heard about energy conservation or been asked to conserve energy. These ideas are related to reducing the demand for electricity and gasoline, which lowers the consumption of energy resources such as coal and fuel oil. The law of conservation of energy, on the other hand, is a universal principle that describes what happens to energy as it is transferred from one object to another or as it is transformed.

Figure 10 The law of conservation of energy requires that the total amount of energy going into a hair dryer must equal the total amount of energy coming out of the hair dryer.

Figure 11 In a swing, mechanical energy is transformed into thermal energy because of friction and air resistance.
Infer *how the kinetic and potential energy of the swing change with time.*

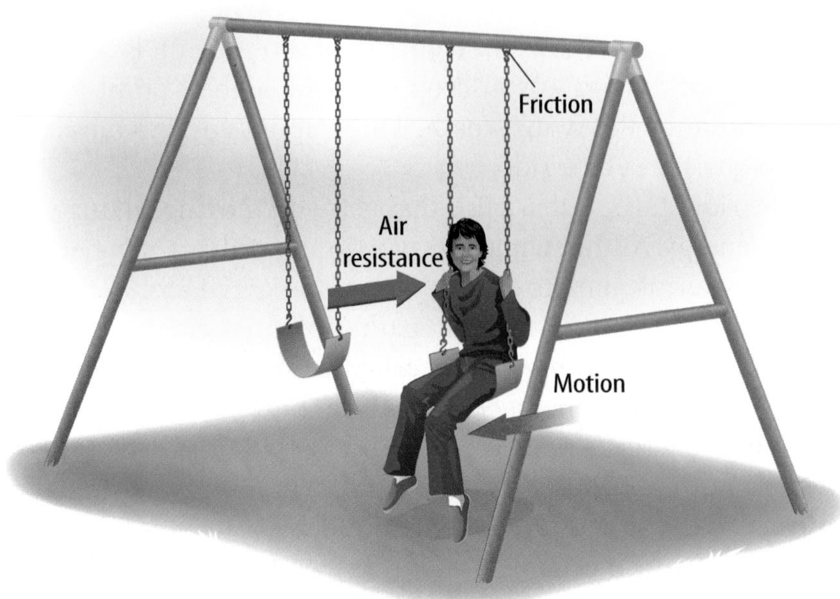

Transforming Energy in a Paper Clip

Procedure

1. Complete the safety form.
2. Straighten a **paper clip.** While holding the ends, touch the paper clip to the skin just below your lower lip. Note whether the paper clip feels warm, cool, or about room temperature.
3. Quickly bend the paper clip back and forth five times. Touch it below your lower lip again. Note whether the paper clip feels warmer or cooler than before.

Analysis

1. What happened to the temperature of the paper clip? Why?
2. Describe the energy conversions that take place as you bend the paper clip.

Try at Home

Is energy always conserved? You might be able to think of situations where it seems as though energy is not conserved. For example, while coasting along a flat road on a bicycle, you know that you will eventually stop if you don't pedal. If energy is conserved, why wouldn't your kinetic energy stay constant so that you would coast forever? In many situations, it might seem that energy is destroyed or created. Sometimes it is hard to see the law of conservation of energy at work.

The Effect of Friction You know from experience that if you don't continue to pump a swing or be pushed by somebody else, your arcs will become lower and you eventually will stop swinging. In other words, the mechanical (kinetic and potential) energy of the swing seems to decrease, as if the energy were being destroyed. Is this a violation of the law of conservation of energy?

It can't be—it's the law! If the energy of the swing decreases, then the energy of some other object must increase by an equal amount to keep the total amount of energy the same. What could this other object be that experiences an energy increase? To answer this, you need to think about friction. With every movement, the swing's ropes or chains rub on their hooks and air pushes on the rider, as illustrated in **Figure 11.** Friction and air resistance cause some of the mechanical energy of the swing to change to thermal energy. With every pass of the swing, the temperature of the hooks and the air increases a little, so the mechanical energy of the swing is not destroyed. Rather, it is transformed into thermal energy. The total amount of energy always stays the same.

Converting Mass into Energy You might have wondered how the Sun unleashes enough energy to light and warm Earth from so far away. A special kind of energy conversion—nuclear fusion—takes place in the Sun and other stars. During this process a small amount of mass is transformed into a tremendous amount of energy. An example of a nuclear fusion reaction is shown in **Figure 12.** In the reaction shown here, the nuclei of the hydrogen isotopes deuterium and tritium undergo fusion.

Nuclear Fission Another process involving the nuclei of atoms, called nuclear fission, converts a small amount of mass into enormous quantities of energy. In this process, nuclei do not fuse—they are broken apart, as shown in **Figure 12.** In either process, fusion or fission, mass is converted to energy. In processes involving nuclear fission and fusion, the total amount of energy is still conserved if the energy content of the masses involved are included. Then the total energy before the reaction is equal to the total energy after the reaction, as required by the law of conservation of energy. The process of nuclear fission is used by nuclear power plants to generate electrical energy.

Topic: Nuclear Fusion
Visit gpescience.com for Web links to information about using nuclear fusion as a source of energy in electric power plants.

Activity Make a table listing the advantages and disadvantages of using nuclear fusion as an energy source.

Figure 12 Mass is converted to energy in the processes of fusion and fission.

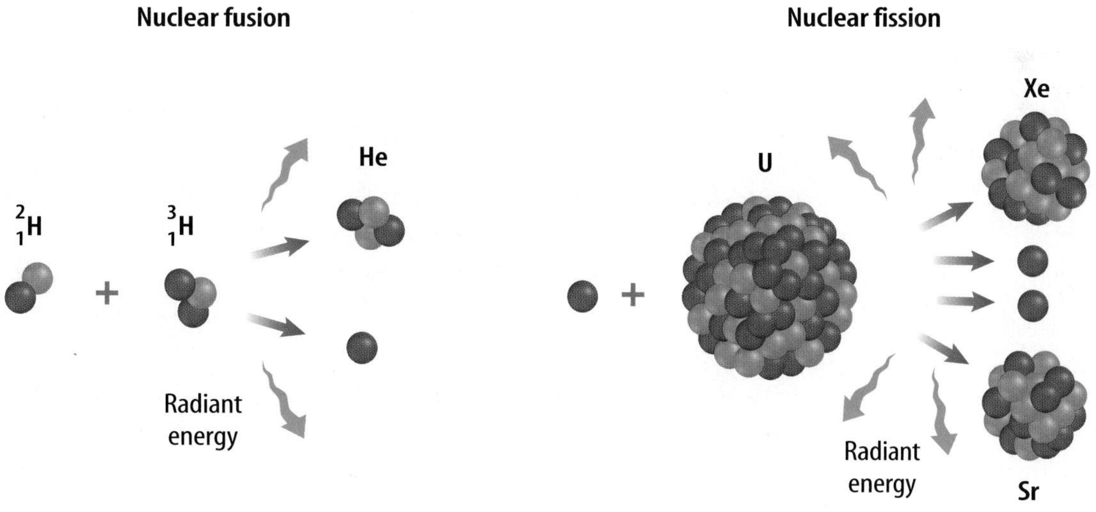

Nuclear fusion

$^{2}_{1}H$ + $^{3}_{1}H$ → He, Radiant energy

Mass $^{2}_{1}$**H** + Mass $^{3}_{1}$**H** > Mass **He** + Mass **neutron**

In this fusion reaction, the combined mass of the two hydrogen nuclei is greater than the mass of the helium nucleus, He, and the neutron.

Nuclear fission

U, Xe, Sr, Radiant energy

Mass **U** + Mass **neutron** > Mass **Xe** + Mass **Sr** + Mass **neutrons**

In nuclear fission, the mass of the large nucleus on the left is greater than the combined mass of the other two nuclei and the neutrons.

The Human Body—Balancing the Energy Equation

What forms of energy discussed in this chapter can you find in the human body? With your right hand, reach up and feel your left shoulder. With that simple action, stored potential energy within your body was converted to the kinetic energy of your moving arm. Did your shoulder feel warm to your hand? Some of the chemical potential energy stored in your body is used to maintain a nearly constant internal temperature. A portion of this energy also is converted to the excess heat that your body gives off to its surroundings. Even the people shown standing in **Figure 13** require energy conversions to stand still.

Energy Conversions in Your Body The complex chemical and physical processes going on in your body also obey the law of conservation of energy. Your body stores energy in the form of fat and other chemical compounds. This chemical potential energy is used to fuel the processes that keep you alive, such as making your heart beat and digesting the food you eat. Your body also converts this energy to heat that is transferred to your surroundings, and you use this energy to make your body move. **Table 1** shows the amount of energy used in doing various activities. To maintain a healthy weight, you must have a proper balance between energy contained in the food you eat and the energy your body uses.

Figure 13 The runners convert the energy stored in their bodies more rapidly than the spectators do.
Calculate *Use* ***Table 1*** *to calculate how long a person would need to stand to burn as much energy as a runner burns in 1 h.*

Food Energy Your body has been busy breaking down your breakfast into molecules that can be used as fuel. The chemical potential energy in these molecules supplies the cells in your body with the energy they need to function. Your body also can use the chemical potential energy stored in fat for its energy needs. The food Calorie (C) is a unit used by nutritionists to measure how much energy you get from various foods—1 C is equivalent to about 4,184 J. Every gram of fat a person consumes can supply 9 C of energy. Carbohydrates and proteins each supply about 4 C of energy per gram.

Look at the labels on food packages. They provide information about the Calories contained in a serving, as well as the amounts of protein, fat, and carbohydrates.

Table 1 Calories Used in One Hour

Type of Activity	Body Frames		
	Small	Medium	Large
Sleeping	48	56	64
Sitting	72	84	96
Eating	84	98	112
Standing	96	112	123
Walking	180	210	240
Playing tennis	380	420	460
Bicycling (fast)	500	600	700
Running	700	850	1,000

section 2 review

Summary

Energy Transformations

- Energy can be transformed from one form to another.
- Devices such as lightbulbs, hair dryers, and automobile engines convert one form of energy into other forms.
- The mechanical energy of a system is the sum of the kinetic and potential energy in the system:

  mechanical energy = *KE* + *PE*
- In falling, projectile motion, and swings, kinetic and potential energy are transformed into each other and the mechanical energy doesn't change.

The Law of Conservation of Energy

- According to the law of conservation of energy, energy cannot be created or destroyed.
- Friction converts mechanical energy into thermal energy.
- Fission and fusion are nuclear reactions that convert a small amount of mass in a nucleus into an enormous amount of energy.

Self Check

1. **Explain** how friction affects the mechanical energy of a system.
2. **Describe** the energy transformations that occur as you coast down a long hill on a bicycle and apply the brakes, causing the brake pads and bicycle rims to feel warm.
3. **Explain** how energy is conserved when nuclear fission or fusion occurs.
4. **Think Critically** A roller coaster is at the top of a hill and rolls to the top of a lower hill. If mechanical energy is conserved, on the top of which hill is the kinetic energy of the roller coaster larger?

Applying Math

5. **Calculate Kinetic Energy** The potential energy of a swing is 200 J at its highest point and 50 J at its lowest point. If mechanical energy is conserved, what is the kinetic energy of the swing at its lowest point?
6. **Calculate Thermal Energy** The mechanical energy of a bicycle at the top of a hill is 6,000 J. The bicycle stops at the bottom of the hill by applying the brakes. If the potential energy of the bicycle is 2,000 J at the bottom of the hill, how much thermal energy was produced?

Design Your Own

Swinging Energy

Real-World Problem

Imagine yourself swinging on a swing. What would happen if a friend grabbed the swing's chains as you passed the lowest point? Would you come to a complete stop or continue rising to your previous maximum height? How does the motion and maximum height reached by a swing change if the swing is interrupted?

Form a Hypothesis

Examine the diagram on this page. How is it similar to the situation in the introductory paragraph? An object that is suspended so that it can swing back and forth is called a pendulum. Hypothesize what will happen to the pendulum's motion and final height if its swing is interrupted.

Goals

- **Construct** a pendulum to compare the exchange of potential and kinetic energy when a swing is interrupted.
- **Measure** the starting and ending heights of the pendulum.

Possible Materials

ring stand
test-tube clamp
support-rod clamp, right angle
30-cm support rod
2-hole, medium rubber stopper
string (1 m)
metersticks (2)
graph paper

Safety Precautions

WARNING: *Be sure the base is heavy enough or well anchored so that the apparatus will not tip over.*

Test Your Hypothesis

Make a Plan

1. Complete the safety form.
2. As a group, write your hypothesis and list the steps that you will take to test it. Be specific. Also list the materials you will need.
3. **Design** a data table and place it in your Science Journal.
4. Set up an apparatus similar to the one shown in the diagram.
5. **Devise** a way to measure the starting and ending heights of the stopper. Record your starting and ending heights in a data table. This will be your control.
6. Be sure you test your swing, starting it above and below the height of the cross arm. How many times should you repeat each starting point?

Follow Your Plan

1. Make sure your teacher approves your plan before you start.
2. Carry out the approved experiment as planned.
3. While the experiment is going on, write any observations that you make and complete the data table in your Science Journal.

Analyze Your Data

1. When the stopper is released from the same height as the cross arm, is the ending height of the stopper exactly the same as its starting height? Use your data to support your answer.
2. **Analyze** the energy transfers. At what point along a single swing does the stopper have the greatest kinetic energy? The greatest potential energy?

Conclude and Apply

1. **Explain** Do the results support your hypothesis?
2. **Compare** the starting heights to the ending heights of the stopper. Is there a pattern? Can you account for the observed behavior?
3. **Discuss** Do your results support the law of conservation of energy? Why or why not?
4. **Infer** What happens if the mass of the stopper is increased? Test it.

Communicating Your Data

Compare your conclusions with those of the other lab teams in your class. **For more help, refer to the** Science Skill Handbook.

TIME **SCIENCE** AND HISTORY

SCIENCE CAN CHANGE THE COURSE OF HISTORY!

The Impossible Dream

A machine that keeps on going? It has been tried for hundreds of years.

Many people have tried throughout history—and failed—to build perpetual-motion machines. In theory, a perpetual-motion machine would run forever and do work without a continual source of energy. You can think of it as a car that you could fill up once with gas, and the car would run forever. Sound impossible? It is!

Science Puts Its Foot Down

For hundreds of years, people have tried to create perpetual-motion machines. But these machines won't work because they violate two of nature's laws. The first law is the law of conservation of energy, which states that energy cannot be created or destroyed. It can change form—say, from mechanical energy to electrical energy—but you always end up with the same amount of energy that you started with.

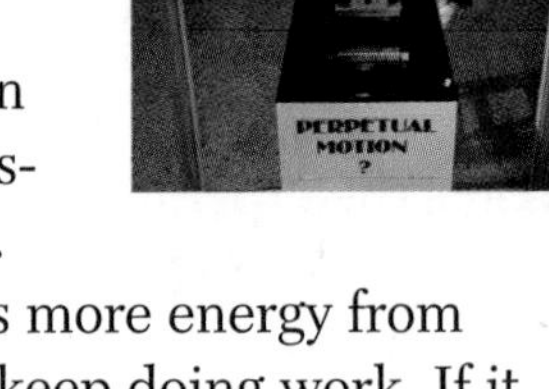

How does that apply to perpetual-motion machines? When a machine does work on an object, the machine transfers energy to the object. Unless that machine gets more energy from somewhere else, it can't keep doing work. If it did, it would be creating energy.

The second law states that heat by itself always flows from a warm object to a cold object. Heat will only flow from a cold object to a warm object if work is done. In the process, some heat always escapes.

To make up for these energy losses, energy constantly needs to be transferred to the machine. Otherwise, it stops. No perpetual motion. No free electricity. No devices that generate more energy than they use. No engine motors that run forever without refueling. Some laws just can't be broken.

Visitors look at the Keely Motor, the most famous perpetual-motion machine fraud of the late 1800s.

Analyze Using your school or public-library resources, locate a picture or diagram of a perpetual-motion machine. Figure out why it won't run forever. Explain to the class what the problem is.

For more information, visit gpescience.com

Reviewing Main Ideas

Section 1 The Nature of Energy

1. Energy is the ability to cause change.
2. Energy can have different forms, including kinetic, potential, and thermal energy.

3. Moving objects have kinetic energy that depends on the object's mass and velocity, and can be calculated from this equation:

$$KE = \frac{1}{2}mv^2$$

4. Potential energy is stored energy. An object can have gravitational potential energy that depends on its mass and its height, and is given by this equation:

$$GPE = mgh$$

Section 2 Conservation of Energy

1. Energy can change from one form to another. Devices you use every day transform one form of energy into other forms that are more useful.
2. Falling, swinging, and projectile motion all involve transformations between kinetic energy and gravitational potential energy.
3. The total amount of kinetic energy and gravitational potential energy in a system is the mechanical energy of the system:

$$\textit{mechanical energy} = KE + GPE$$

4. The law of conservation of energy states that energy never can be created or destroyed. The total amount of energy in the universe is constant.

5. Friction converts mechanical energy into thermal energy, causing the mechanical energy of a system to decrease.
6. Mass is converted into energy in nuclear fission and fusion reactions. Fusion and fission occur in the nuclei of certain atoms, and release tremendous amounts of energy.

FOLDABLES Use the Foldable you made at the beginning of this chapter to review what you learned about energy.

chapter 5 Review

Using Vocabulary

chemical potential energy p. 131
elastic potential energy p. 131
energy p. 128
gravitational potential energy p. 132
joule p. 130
kinetic energy p. 130
law of conservation of energy p. 139
mechanical energy p. 136
potential energy p. 131

Complete each statement using a word(s) from the vocabulary list above.

1. If friction can be ignored, the ________ of a system doesn't change.
2. The energy stored in a compressed spring is ________.
3. The ________ is the SI unit for energy.
4. When a book is moved from a higher shelf to a lower shelf, its ________ changes.
5. The muscles of a runner transform chemical potential energy into ________.
6. According to the ________, the amount of energy in the universe doesn't change.

Checking Concepts

Choose the word or phrase that best answers the question.

7. What occurs when energy is transferred from one object to another?
   A) an explosion
   B) a chemical reaction
   C) nuclear fusion
   D) a change
8. For which of the following is kinetic energy converted into potential energy?
   A) a boulder rolls down a hill
   B) a ball is thrown upward
   C) a swing comes to a stop
   D) a bowling ball rolls horizontally
9. The gravitational potential energy of an object changes when which of the following changes?
   A) the object's speed
   B) the object's mass
   C) the object's temperature
   D) the object's length
10. Friction causes mechanical energy to be transformed into which of these forms?
    A) thermal **C)** kinetic
    B) nuclear **D)** potential
11. The kinetic energy of an object changes when which of the following changes?
    A) the object's chemical potential energy
    B) the object's volume
    C) the object's direction of motion
    D) the object's speed
12. When an energy transformation occurs, which of the following is true?
    A) Mechanical energy doesn't change.
    B) Mechanical energy is lost.
    C) The total energy doesn't change.
    D) Mass is converted into energy.

Interpreting Graphics

13. Copy and complete the following concept map on energy.

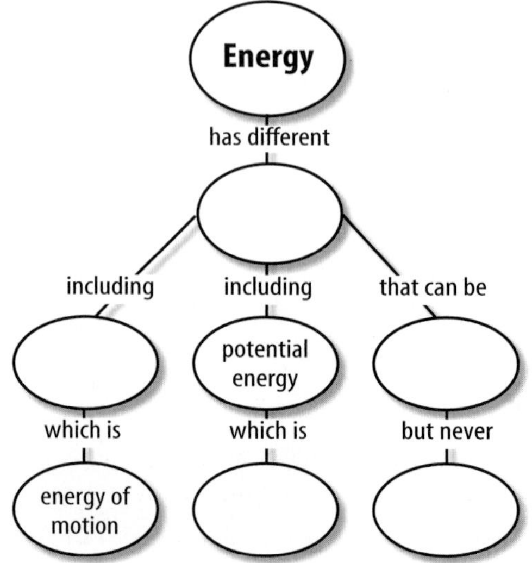

Use the table below to answer question 14.

Toy Cars Rolling Down Ramps			
Ramp Height (m)	Speed at Bottom (m/s)	GPE (J)	KE (J)
0.50	3.13	Do not write in this book.	
0.75	3.83		
1.00	4.43		

14. **Make and Use Tables** Three toy cars, each with a mass of 0.05 kg, roll down ramps with different heights. The height of each ramp and the speed of each car at the bottom of each ramp is given in the table. Copy and complete the table by calculating the GPE of each car at the top of the ramp and the KE for each car at the bottom of the ramp to two decimal places. How do the values of GPE and KE you calculate compare?

Use the graph below to answer questions 15–17.

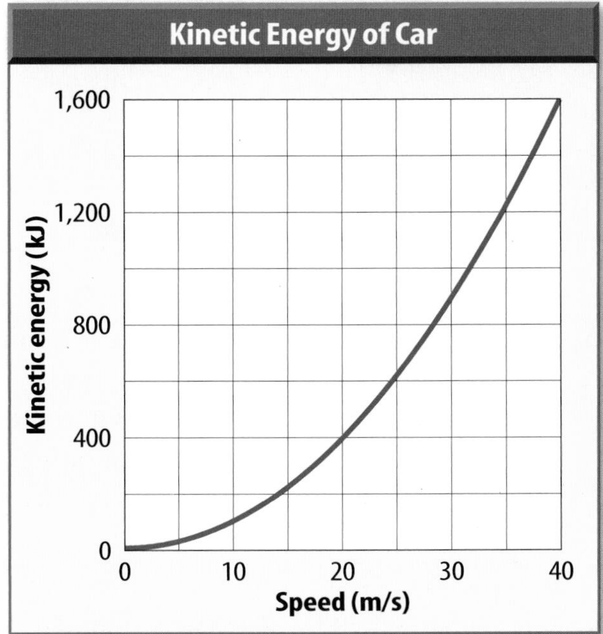

15. When the car's speed doubles from 20 m/s to 40 m/s, by how many times does the car's kinetic energy increase?

16. Using the graph, estimate the car's kinetic energy at a speed of 50 m/s.

17. If the car's kinetic energy at a speed of 20 m/s is 400 kJ, what is the car's kinetic energy at a speed of 10 m/s?

Thinking Critically

18. **Describe** the energy changes that occur in a swing. Explain how energy is conserved as the swing slows down and stops.

19. **Explain** why the law of conservation of energy must also include changes in mass.

20. **Infer** why the tires of a car get hot when the car is driven.

21. **Diagram** On a cold day you rub your hands together to make them warm. Diagram the energy transformations that occur, starting with the chemical potential energy stored in your muscles.

Applying Math

22. **Calculate Kinetic Energy** What is the kinetic energy of a 0.06-kg tennis ball traveling at a speed of 150 m/s?

23. **Calculate Potential Energy** A boulder with a mass of 2,500 kg rests on a ledge 200 m above the ground. What is the boulder's potential energy?

24. **Calculate Mechanical Energy** What is the mechanical energy of a 500-kg roller-coaster car moving with a speed of 3 m/s at the top of hill that is 30 m high?

25. **Calculate Speed** A boulder with a mass of 2,500 kg on a ledge 200 m above the ground falls. If the boulder's mechanical energy is conserved, what is the speed of the boulder just before it hits the ground?

chapter 5 Standardized Test Practice

Record your answers on the answer sheet provided by your teacher or on a sheet of paper.

Multiple Choice

1. What is the potential energy of a 5.0-kg object located 2.0 m above the ground?

A. 2.5 J

B. 10 J

C. 98 J

D. 196 J

Use the figure below to answer questions 2–4.

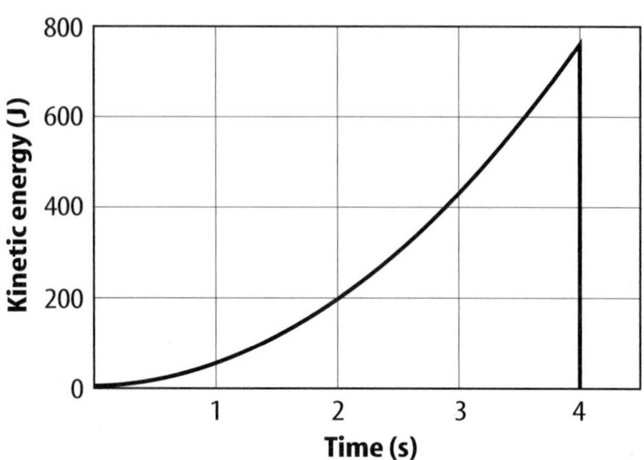

2. According to the graph, which of the following is the best estimate for the kinetic energy of the rock after it has fallen for 1 s?

A. 100 J

B. 50 J

C. 200 J

D. 0 J

3. According to the graph, which of the following is the best estimate for the potential energy of the rock before it fell?

A. 400 J

B. 750 J

C. 200 J

D. 0 J

4. If the rock has a mass of 1 kg, which of the following is the speed of the rock after it has fallen for 2 s?

A. 10 m/s

B. 100 m/s

C. 20 m/s

D. 200 m/s

Use the figure below to answer questions 5 and 6.

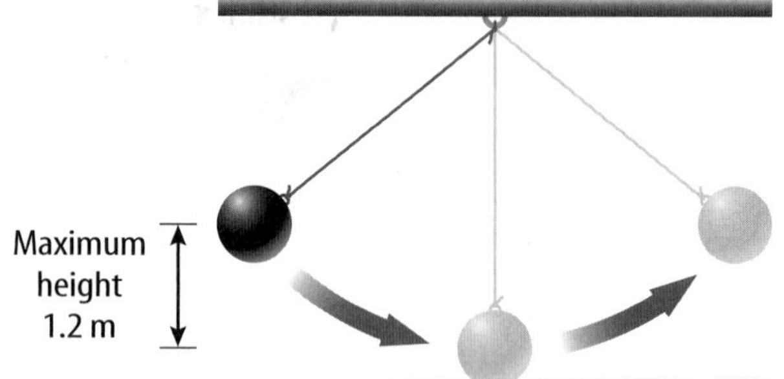

5. At its highest point, the pendulum is 1.2 m above the ground and has a gravitational potential energy of 62 J. If the gravitational potential energy is 10 J at its lowest point, what is the pendulum's kinetic energy at this point?

A. 0 J

B. 31 J

C. 62 J

D. 52 J

6. What is the mass of the pendulum bob?
   A. 2.7 kg
   B. 5.3 kg
   C. 6.3 kg
   D. 52 kg

Gridded Response

7. What is the difference, in J, in the gravitational potential energy of a 7.75-kg book that is 1.50 m above the ground and a 9.53-kg book that is 1.75 m above the ground?

Short Response

Use the figure below to answer question 8.

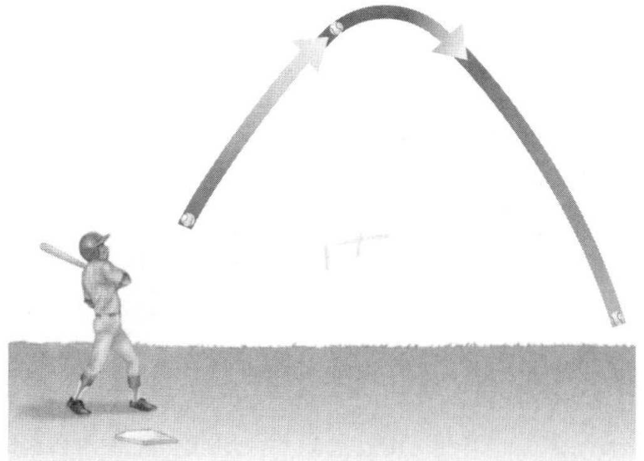

8. How does the mechanical energy of the ball change from the moment just after the batter hits it to the moment just before it touches the ground?

9. Explain why the law of conservation of energy also includes mass when applied to nuclear reactions.

Extended Response

Use the figure below to answer question 10.

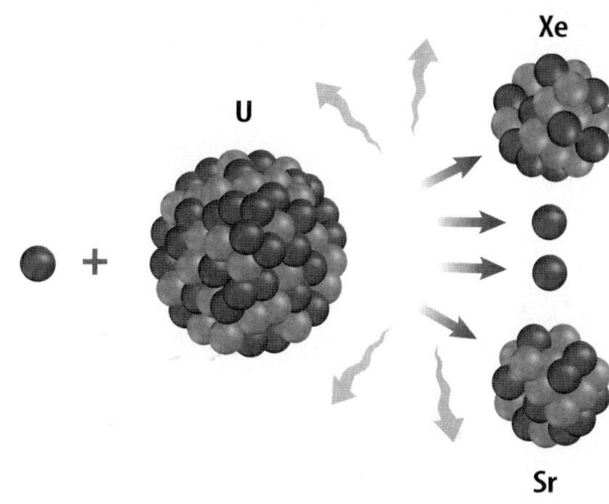

10. The figure shows a nuclear reaction.
    Part A Describe the process and explain how it obeys the law of conservation of energy.
    Part B Describe how the total mass of the particles before the reaction occurs compares to the total mass of the particles produced by the reaction.

11. Describe a process in which energy travels through the environment and changes from one form to another.

Test-Taking Tip

Show All Your Work For extended response questions, show all your work and any calculations on your answer sheet.

Question 10 On your answer sheet, list the energy changes that occur during each step.

chapter 6

Work and Machines

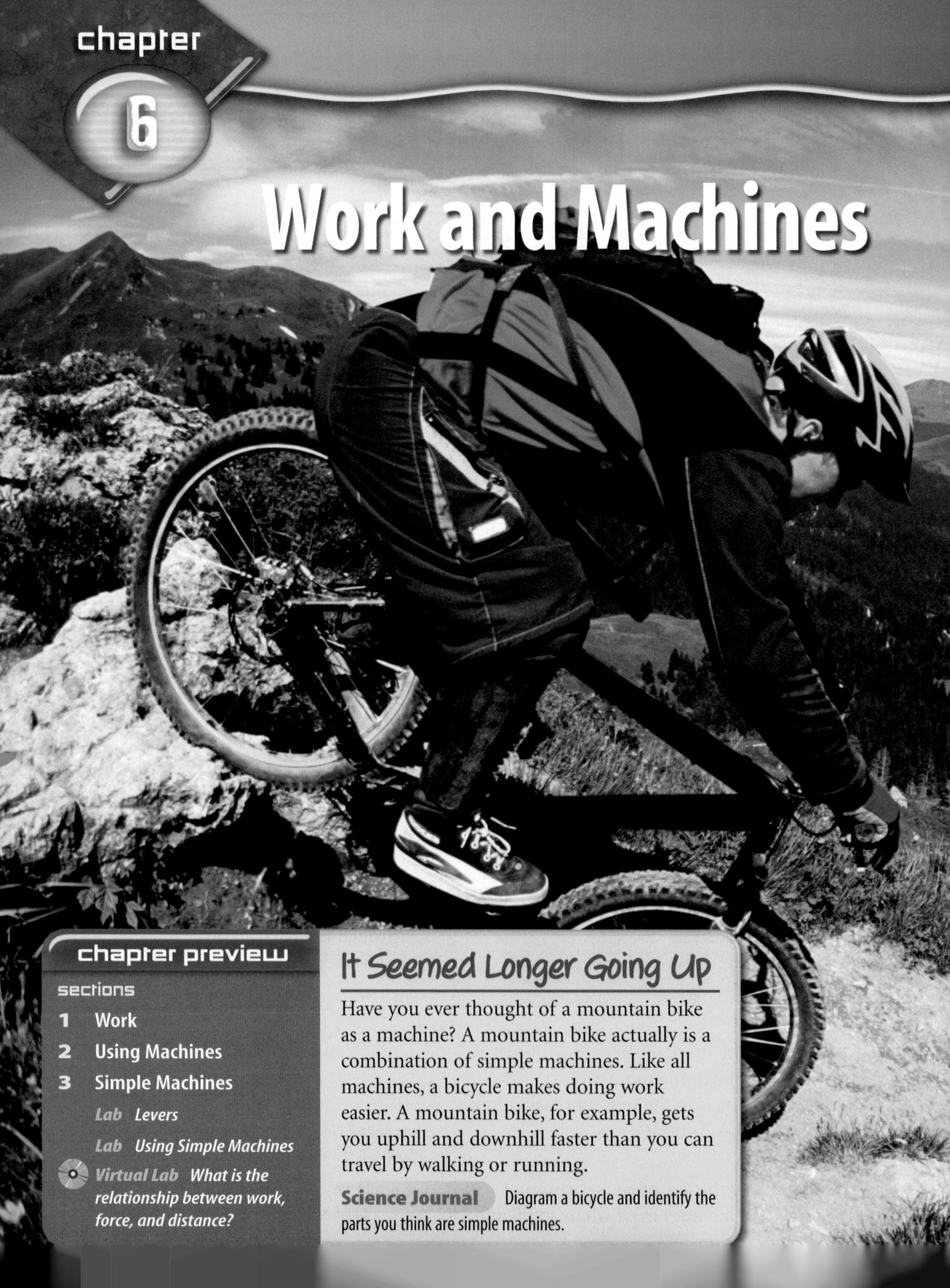

chapter preview

sections

1 Work
2 Using Machines
3 Simple Machines

Lab *Levers*

Lab *Using Simple Machines*

Virtual Lab *What is the relationship between work, force, and distance?*

It Seemed Longer Going Up

Have you ever thought of a mountain bike as a machine? A mountain bike actually is a combination of simple machines. Like all machines, a bicycle makes doing work easier. A mountain bike, for example, gets you uphill and downhill faster than you can travel by walking or running.

Science Journal Diagram a bicycle and identify the parts you think are simple machines.

Start-Up Activities

Doing Work with a Simple Machine

Did you know you can lift several times your own weight with the help of a pulley? Before the hydraulic lift was invented, a car mechanic used pulleys to raise a car off the ground. In this lab you'll see how a pulley can increase a force.

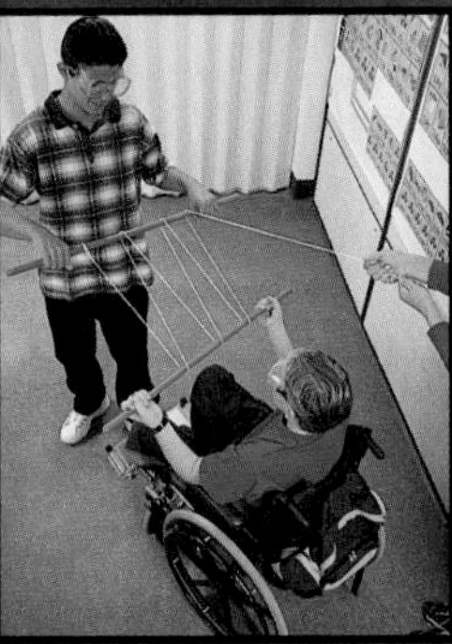

1. Complete the safety form.
2. Tie a rope several meters in length to the center of a broom handle. Have one student hold both ends of the handle.
3. Have another student hold the ends of a second broom handle and face the first student as shown in the photo.
4. Have a third student loop the free end of the rope around the second handle, making six or seven loops.
5. The third student should stand to the side of one of the handles and pull on the free end of the rope. The two students holding the broom handles should prevent the handles from coming together.
6. **Think Critically** Describe how the applied force was changed. What would happen if the number of rope loops were increased?

Work and Machines Make the following Foldable to help you understand how machines make doing work easier.

STEP 1 **Fold** a sheet of paper vertically in half from top to bottom.

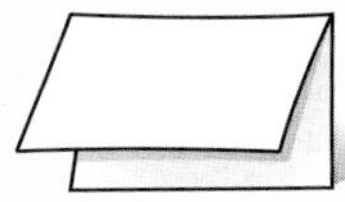

STEP 2 **Fold** in half from side to side with the fold at the top.

STEP 3 **Unfold** the paper once. **Cut** only the fold of the top flap to make two tabs.

STEP 4 **Turn** the paper vertically and **label** the front tabs as shown.

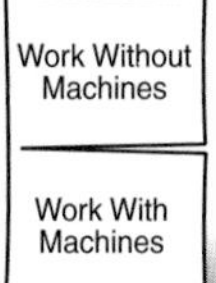

List Before you read the chapter, list five examples of work you do without machines, and five examples of work you do with machines. Next to each example, rate the effort needed to do the work on a scale of 1 (little effort) to 3 (much effort).

Preview this chapter's content and activities at gpescience.com

section 1

Work

Reading Guide

What You'll Learn

- **Explain** the meaning of work.
- **Describe** how work and energy are related.
- **Calculate** work.
- **Calculate** power.

Why It's Important

Doing work is another way of transferring energy from one place to another.

Review Vocabulary

energy: the ability to cause change

New Vocabulary

- work
- power

What is work?

Have you done any work today? To many people, the word *work* means something they do to earn money. In that sense, work can be anything from filling fast-food orders or loading trucks to teaching or doing word processing on a computer. The word *work* also means exerting a force with your muscles. A person might say he or she does work while pushing as hard as they can against a wall that doesn't move. However, in science the word *work* is used in a different way.

Work Makes Something Move Press your hand against the surface of your desk as hard as you can. Have you done any work? The answer is no, no matter how tired your effort makes you feel. Remember that a force is a push or a pull. For work to be done, a force must make something move. **Work** is the energy transferred when a force makes an object move. If you push against the desk and it doesn't move, then you haven't done any work on the desk.

Doing Work There are two conditions that have to be satisfied for a force to do work on an object. One is that the applied force must make the object move, and the other is that the movement must be in the same direction as the applied force.

For example, if you pick up a pile of books from the floor, as in **Figure 1,** you do work on the books. The books move upward, in the direction of the force you are applying. If you hold the books in your arms without moving the books, you are not doing work on the books. You're still applying an upward force to keep the books from falling, but no movement is taking place.

Figure 1 When you lift a stack of books, your arms apply a force upward and the books move upward. Because the force and distance are in the same direction, your arms have done work on the books.

Force and Direction of Motion When you carry books while walking, like the student in **Figure 2,** you might think that your arms are doing work. After all, you are exerting a force on the books with your arms, and the books are moving. Your arms might even feel tired. However, in this case the force exerted by your arms does no work on the books. The force exerted by your arms on the books is upward, but the books are moving horizontally. The force you exert upward is at right angles to the direction the books are moving. As a result, your arms exert no force in the direction the books are moving.

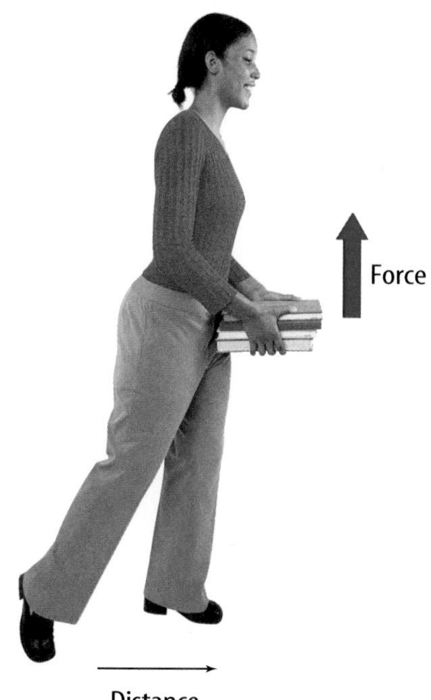

Figure 2 If you hold a stack of books and walk forward, your arms are exerting a force upward. However, the distance the books move is horizontal. Therefore, your arms are not doing work on the books.

Reading Check *How are an applied force and an object's motion related when work is done?*

Work and Energy

How are work and energy related? When work is done, a transfer of energy always occurs. This is easier to understand when you think about carrying a heavy box up a flight of stairs. Remember that when the height of an object above Earth's surface increases, the gravitational potential energy of the object increases. As you move up the stairs, you increase the height of the box above Earth's surface. This causes the gravitational potential energy of the box to increase.

You may recall that energy is the ability to cause change. If something has energy, it can transfer energy to another object by doing work on that object. When you do work on an object, you increase its energy. The student carrying the box in **Figure 3** transfers chemical energy in his muscles to the box. Energy is always transferred from the object that is doing the work to the object on which the work is done.

Figure 3 By carrying a box up the stairs, you are doing work. You transfer energy to the box. **Explain** *how the energy of the box changes as the student climbs the stairs.*

Calculating Work The amount of work done depends on the amount of force exerted and the distance over which the force is applied. When a force is exerted and an object moves in the direction of the force, the amount of work done can be calculated as follows.

Work Equation

work (in joules) = **applied force** (in newtons) × **distance** (in meters)

$$W = Fd$$

In this equation, force is measured in newtons and distance is measured in meters. Work, like energy, is measured in joules. One joule is about the amount of work required to lift a baseball a vertical distance of 0.7 m.

Applying Math Solve a Simple Equation

WORK You push a refrigerator with a force of 100 N. If you move the refrigerator a distance of 5 m, how much work do you do?

IDENTIFY known values and the unknown value

Identify the known values:

you push a refrigerator with a force of 100 N means $F = 100$ N

you move the refrigerator a distance of 5 m means $d = 5$ m

Identify the unknown value:

how much work you do means $W = ?$ J

SOLVE the problem

Substitute the known values $F = 100$ N and $d = 5$ m into the work equation.

$$W = Fd = (100 \text{ N})\ (5 \text{ m}) = 500 \text{ Nm} = 500 \text{ J}$$

CHECK your answer

Does your answer seem reasonable? Check your answer by dividing the work you calculated by the force given in the problem. The result should be the distance given in the problem.

Practice Problems

1. A force of 75 N is exerted on a 45-kg couch, and the couch is moved 5 m. How much work is done in moving the couch?
2. A lawn mower is pushed with a force of 80 N. If 12,000 J of work is done in mowing a lawn, what is the total distance the lawn mower was pushed?
3. The brakes on a car do 240,000 J of work in stopping the car. If the car travels a distance of 50 m while the brakes are being applied, what is the force the brakes exert on the car?

For more practice problems go to page 879, and visit Math Practice at gpescience.com.

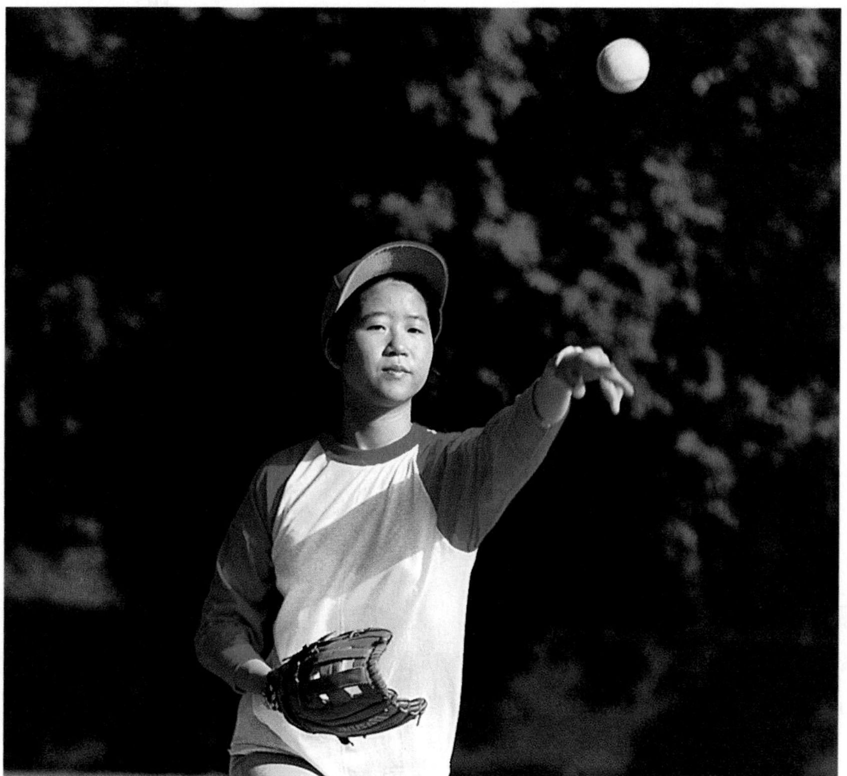

Figure 4 A pitcher exerts a force on the ball to throw it to the catcher. After the ball leaves her hand, the pitcher no longer is exerting any force on the ball. She does work on the ball only while it is in her hand.

When is work done? Suppose you give a book a push and it slides along a table for a distance of 1 m before it comes to a stop. The distance you use to calculate the work you did is how far the object moved while the force was being applied. Even though the book moved 1 m, you did work on the book only while your hand was in contact with it. The distance in the formula for work is the distance the book moved while your hand was pushing on the book. As **Figure 4** shows, work is done on an object only when a force is being applied to the object.

Power

Suppose you and another student are pushing boxes of books up a ramp to load them into a truck. To make the job fun, you make a game of it, racing to see who can push a box up the ramp faster. The boxes weigh the same, but your friend is able to push a box a little faster than you can. She moves a box up the ramp in 30 s. It takes you 45 s. You both do the same amount of work on the books because the boxes weigh the same and are moved the same distance. The only difference is the time it takes to do the work.

In this game, your friend has more power than you do. **Power** is the amount of work done in one second. It is a rate, the rate at which work is done.

Reading Check *How is power related to work?*

Calculating Your Work and Power

Procedure

1. Complete the safety form.
2. Find a set of **stairs** that you can safely walk and run up. Measure the vertical height of the stairs in meters.
3. Record how many seconds it takes you to walk and then run up the stairs.
4. Calculate the work you did in both walking and running up the stairs. The force is your weight in newtons (your weight in pounds times 4.5). The distance is the vertical height of the stairs.
5. Use the formula $P = W/t$ to calculate the power you needed to both walk and to run up the stairs.

Analysis

1. Is the work you did walking and running up the steps the same?
2. Which required more power, walking or running up the steps? Why?

Calculating Power Power is the rate at which work is done. To calculate power, divide the work done by the time that is required to do the work.

Power Equation

$$\text{power (in watts)} = \frac{\text{work (in joules)}}{\text{time (in seconds)}}$$

$$P = \frac{W}{t}$$

The SI unit for power is the watt (W). One watt equals one joule of work done in one second. Because the watt is a small unit, power often is expressed in kilowatts. One kilowatt (kW) equals 1,000 W.

Applying Math — Solve a Simple Equation

POWER You do 900 J of work in pushing a sofa. If it took 5 s to move the sofa, what was your power?

IDENTIFY known values and the unknown value

Identify the known values:

You do 900 J of work — means → $W = 900$ J

it took 5 s to move the sofa — means → $t = 5$ s

Identify the unknown value:

what was your power? — means → $P = ?$ W

SOLVE the problem

Substitute the known values $W = 900$ J and $t = 5$ s into the power equation.

$$P = \frac{W}{t} = \frac{900\ \text{J}}{5\ s} = 180\ \text{J/s} = 180\ \text{W}$$

The symbol for work, *W*, is usually italicized. However, the abbreviation for watt, W, is not italicized.

CHECK your answer

Does your answer seem reasonable? Check your answer by multiplying the power you calculated by the time given in the problem. The result should be the work given in the problem.

Practice Problems

1. To lift a baby from a crib, 50 J of work is done. How much power is needed if the baby is lifted in 0.5 s?
2. If a runner's power is 130 W, how much work is done by the runner in 10 minutes?
3. The power produced by an electric motor is 500 W. How long will it take the motor to do 10,000 J of work?

For more practice problems go to page 879, and visit Math Practice at gpescience.com.

Power and Energy Doing work is a way of transferring energy from one object to another. Just as power is the rate at which work is done, power is also the rate at which energy is transferred. When energy is transferred, the power involved can be calculated by dividing the energy transferred by the time needed for the transfer to occur.

Power Equation for Energy Transfer

$$\text{power (in watts)} = \frac{\text{energy transferred (in joules)}}{\text{time (in seconds)}}$$

$$P = \frac{E}{t}$$

For example, when the lightbulb in **Figure 5** is connected to an electric circuit, energy is transferred from the circuit to the lightbulb filament. The filament converts the electrical energy supplied to the lightbulb into heat and light. The power used by the lightbulb is the amount of electrical energy transferred to the lightbulb each second.

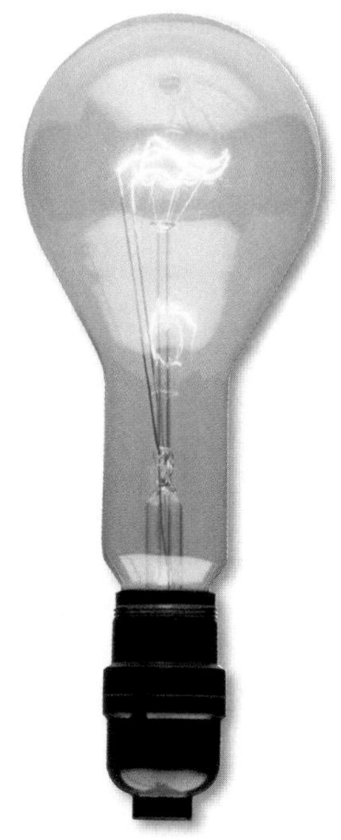

Figure 5 This 100-W lightbulb converts electrical energy into light and heat at a rate of 100 J/s.

section 1 review

Summary

Work and Energy

- Work is done on an object when a force is exerted on the object and it moves in the direction of the force.
- If a force, F, is exerted on an object while the object moves a distance, d, in the direction of the force, the work done is

$$W = Fd$$

- When work is done on an object, energy is transferred to the object.

Power

- Power is the rate at which work is done or energy is transferred.
- When work is done, power can be calculated from the equation

$$P = \frac{W}{t}$$

- When energy is transferred, power can be calculated from the equation

$$P = \frac{E}{t}$$

Self Check

1. **Explain** how the scientific definition of *work* is different from the everyday meaning.
2. **Describe** a situation in which a force is applied but no work is done.
3. **Explain** how work and energy are related.
4. **Think Critically** In which of the following situations is work being done?
   a. A person shovels snow off a sidewalk.
   b. A worker lifts bricks, one at a time, from the ground to the back of a truck.
   c. A roofer's assistant carries a bundle of shingles across a construction site.

Applying Math

5. **Calculate Force** Find the force a person exerts in pulling a wagon 20 m if 1,500 J of work is done.
6. **Calculate Work** A car's engine produces 100 kW of power. How much work does the engine do in 5 s?
7. **Calculate Energy** A color TV uses 120 W of power. How much energy does the TV use in 1 hour?

section 2

Using Machines

Reading Guide

What You'll Learn

- **Explain** how machines make doing work easier.
- **Calculate** the mechanical advantage of a machine.
- **Calculate** the efficiency of a machine.

Why It's Important

Cars, stairs, and teeth are all examples of machines that make your life easier.

Review Vocabulary

force: a push or a pull

New Vocabulary

- machine
- input force
- output force
- mechanical advantage
- efficiency

What is a machine?

A **machine** is a device that makes doing work easier. When you think of a machine, you may picture a device with an engine and many moving parts. However, machines can be simple. Some machines, such as knives, scissors, and doorknobs, are used every day to make doing work easier.

Making Work Easier

Machines can make work easier by increasing the force that can be applied to an object. A screwdriver increases the force you apply to a screw. A second way that machines can make work easier is by increasing the distance over which a force can be applied. A leaf rake is an example of this type of machine. Machines also can make work easier by changing the direction of an applied force. A simple pulley changes a downward force to an upward force.

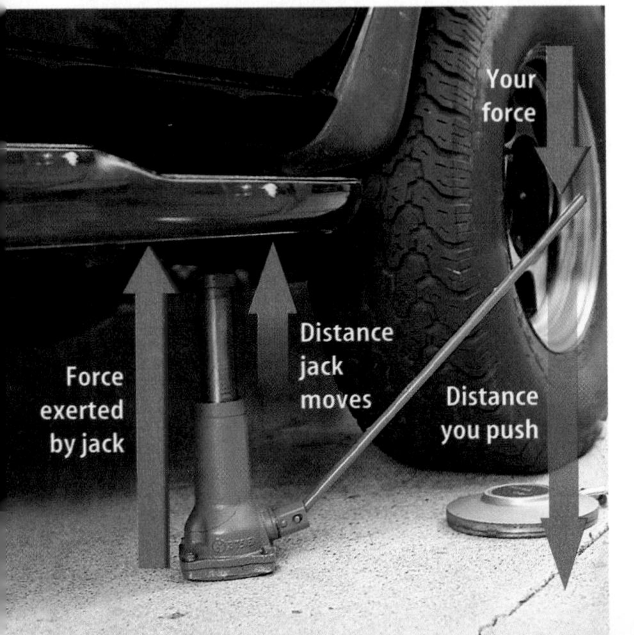

Figure 6 A car jack is an example of a machine that increases an applied force.

Increasing Force A car jack, such as the one in **Figure 6,** is an example of a machine that increases an applied force. The upward force exerted by the jack is greater than the downward force you exert on the handle. However, the distance you push the handle downward is greater than the distance the car is pushed upward. Because work is the product of force and distance, the work done by the jack is not greater than the work you do on the jack. The jack increases the applied force, but it doesn't increase the work done.

Figure 7 Whether the mover slides the chair up the ramp or lifts it directly into the truck, she will do the same amount of work. Doing the work over a longer distance allows her to use less force.

Force and Distance Why does the mover in **Figure 7** push the heavy furniture up the ramp instead of lifting it into the truck? It is easier because less force is needed to push the furniture up the ramp than is needed to lift it.

The work done in lifting an object depends on the change in height of the object. The same amount of work is done whether the mover pushes the furniture up the long ramp or lifts it straight up. If she uses a ramp to lift the furniture, she moves the furniture a longer distance than if she just raised it straight up. If the work stays the same and the distance is increased, then less force will be needed to do the work.

Reading Check *How does a ramp make lifting an object easier?*

Changing Direction Some machines change the direction of the force you apply. When you use the car jack, you exert a force downward on the jack handle. The force exerted by the jack on the car is upward. The direction of the force you apply is changed from downward to upward. Some machines change the direction of the force that is applied to them in another way. The wedge-shaped blade of an ax is one example. When you use an ax to split wood, you exert a downward force as you swing the ax toward the wood. As **Figure 8** shows, the blade changes the downward force into a horizontal force that splits the wood apart.

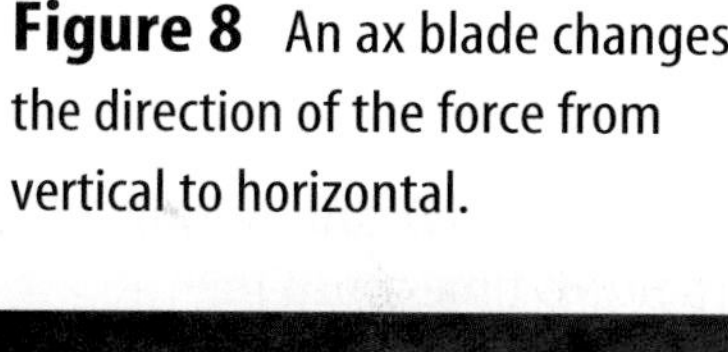

Figure 8 An ax blade changes the direction of the force from vertical to horizontal.

Figure 9 A crowbar increases the force you apply and changes its direction.

Machines Multiplying Force

Procedure

1. Complete the safety form.
2. Open a **can of food** using a **manual can opener** that has circular blades. ***WARNING:*** *Do not touch can opener's cutting blades or the cut edges of the cans' lids.*
3. Open another **can of food,** but this time grasp the handle as close as you can to the handle's pivot point.

Analysis

1. Compare how difficult it was to open the first can with how difficult it was to open the second can.
2. Infer why a can opener makes it easier to open a metal can. How would you design a can opener to make opening cans even easier?

Try at Home

The Work Done by Machines

To pry the lid off a wooden crate with a crowbar, you'd slip the end of the crowbar under the edge of the crate lid and push down on the handle. By moving the handle downward, you do work on the crowbar. As the crowbar moves, it does work on the lid, lifting it up. **Figure 9** shows how the crowbar increases the amount of force being applied and changes the direction of the force.

When you use a machine such as a crowbar, you are moving something that resists being moved. For example, if you use a crowbar to pry the lid off a crate, you are working against the friction between the nails in the lid and the crate. You also could use a crowbar to move a large rock. In this case, you would be working against gravity—the weight of the rock.

Input and Output Forces Two forces are involved when a machine is used to do work. You exert a force on the machine, such as a bottle opener, and the machine then exerts a force on the object you are trying to move, such as a bottle cap. The force that is applied to the machine is called the **input force.** F_{in} stands for the input force. The force applied by the machine is called the **output force,** symbolized by F_{out}. When you try to pull a nail out of wood with a hammer, as in **Figure 10,** you apply the input force on the handle. The output force is the force the claw applies to the nail.

Two kinds of work need to be considered when you use a machine: the work done by you on the machine and the work done by the machine. When you use a crowbar, you do work when you apply force to the crowbar handle and make it move. The work done by you on a machine is called the input work and is symbolized by W_{in}. The work done by the machine is called the output work and is abbreviated W_{out}.

Conserving Energy Remember that energy is always conserved. When you do work on a machine, you transfer energy to the machine. When the machine does work on an object, energy is transferred from the machine to the object. Because energy cannot be created or destroyed, the amount of energy the machine transfers to the object cannot be greater than the amount of energy you transfer to the machine. A machine cannot create energy, so W_{out} is never greater than W_{in}.

However, the machine does not transfer all of the energy it receives to the object. In fact, when a machine is used, some of the input energy changes to thermal energy due to friction. The energy that changes to thermal energy cannot be used to do work, so W_{out} is always smaller than W_{in}.

Figure 10 When prying a nail out of a piece of wood with a claw hammer, you exert the input force on the handle of the hammer, and the claw exerts the output force. **Describe** *how the hammer changes the input force.*

Ideal Machines Remember that work is calculated by multiplying force by distance. The input work is the product of the input force and the distance over which the input force is exerted. The output work is the product of the output force and the distance over which that force is exerted.

Suppose a perfect machine could be built in which there were no friction. None of the input work or output work would be converted to thermal energy. For such an ideal machine, the input work would equal the output work. For an ideal machine,

$$W_{in} = W_{out}$$

Suppose the ideal machine increases the force applied to it. This means that the output force, F_{out}, is greater than the input force, F_{in}. Recall that work is equal to force times distance. If F_{out} is greater than F_{in}, then W_{in} and W_{out} can be equal only if the input force is applied over a greater distance than the output force.

For example, suppose the hammer claw in **Figure 10** moves a distance of 1 cm to remove a nail. If an output force of 1,500 N is exerted by the claw of the hammer, and you move the handle of the hammer 5 cm, you can find the input force as follows.

$$\begin{aligned} W_{in} &= W_{out} \\ F_{in}\, d_{in} &= F_{out}\, d_{out} \\ F_{in}\ (0.05\ \text{m}) &= (1{,}500\ \text{N})\ (0.01\ \text{m}) \\ F_{in}\ (0.05\ \text{m}) &= 15\ \text{N}\cdot\text{m} \\ F_{in} &= 300\ \text{N} \end{aligned}$$

Because the distance you move the hammer is longer than the distance the hammer moves the nail, the input force is less than the output force.

Topic: Rube Goldberg
Visit gpescience.com for Web links to information about Rube Goldberg devices.

Activity Use the information that you find to sketch a Rube-Goldberg-like device designed to burst a balloon.

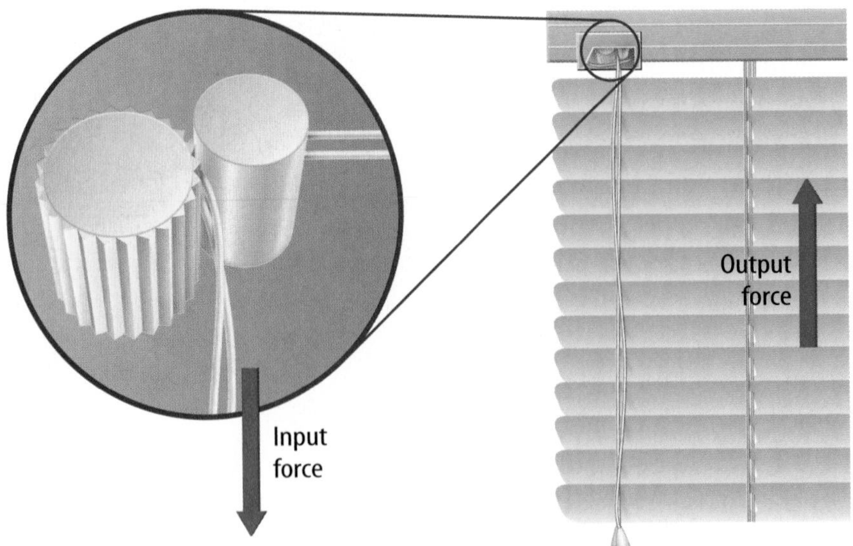

Figure 11 Window blinds use a machine that changes the direction of an input force. A downward pull on the cord is changed to an upward force on the blinds. The input and output forces are equal, so the *MA* is 1.

Mechanical Advantage

Machines such as the car jack, the ramp, the crow bar, and the claw hammer make work easier by making the output force greater than the input force. The ratio of the output force to the input force is the **mechanical advantage** of a machine. The mechanical advantage (*MA*) of a machine can be calculated with the following equation.

Mechanical Advantage Equation

$$\text{mechanical advantage} = \frac{\text{output force (in newtons)}}{\text{input force (in newtons)}}$$

$$MA = \frac{F_{out}}{F_{in}}$$

Figure 11 shows that the mechanical advantage equals one when only the direction of the input force changes.

Ideal Mechanical Advantage The mechanical advantage of a machine without friction is called the ideal mechanical advantage, or *IMA*. The *IMA* can be calculated by dividing the input distance by the output distance. For a real machine, the *IMA* would be the mechanical advantage of the machine if there were no friction.

INTEGRATE Chemistry

Graphite Lubricant Graphite is a solid that sometimes is used as a lubricant to increase the efficiency of machines. Find out what element graphite is made of and why graphite is a good lubricant.

Efficiency

For real machines, some of the energy put into a machine is always converted to thermal energy by frictional forces. For this reason, the output work of a machine is always less than the work put into the machine.

Efficiency is a measure of how much of the work put into a machine is changed into useful output work by the machine. A machine with high efficiency produces less thermal energy from friction, so more of the input work is changed to useful output work.

Why is the output work always less than the input work for a real machine?

Calculating Efficiency To calculate the efficiency of a machine, the output work is divided by the input work. Efficiency is usually expressed as a percentage by this equation:

Efficiency Equation

$$\text{efficiency (\%)} = \frac{\text{output work (in joules)}}{\text{input work (in joules)}} \times 100\%$$

$$efficiency = \frac{W_{out}}{W_{in}} \times 100\%$$

In an ideal machine, there is no friction and the output work equals the input work. So the efficiency of an ideal machine is 100 percent. In a real machine, friction causes the output work to always be less than the input work. So the efficiency of a real machine is always less than 100 percent.

Increasing Efficiency Machines can be made more efficient by reducing friction. This usually is done by adding a lubricant, such as oil or grease, to surfaces that rub together, as shown in **Figure 12.** A lubricant fills in the gaps between the surfaces, enabling the surfaces to slide past each other more easily.

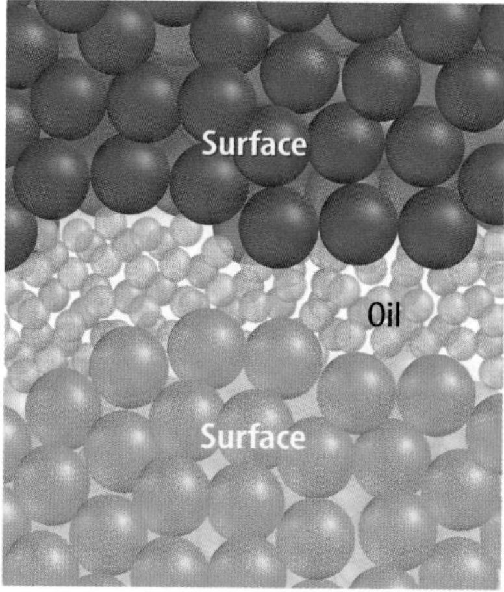

Figure 12 Oil reduces the friction between two surfaces. Oil fills the space between the surfaces so high spots don't rub against each other.

section 2 review

Summary

Work and Machines

- Machines make doing work easier by changing the applied force, changing the distance over which the force is applied, or changing the direction of the applied force.
- Because energy cannot be created or destroyed, the output work cannot be greater than the input work.
- In a real machine, some of the input work is converted into heat by friction.

Mechanical Advantage and Efficiency

- The mechanical advantage of a machine is the output force divided by the input force:

$$MA = \frac{F_{out}}{F_{in}}$$

- The efficiency of a machine is the output work divided by the input work times 100%:

$$efficiency = \frac{W_{out}}{W_{in}} \times 100\%$$

Self Check

1. **Describe** the circumstances for which the output work would equal the input work in a machine.
2. **Infer** how lubricating a machine affects the output force exerted by the machine.
3. **Explain** why, in a real machine, the output work is always less than the input work.
4. **Think Critically** The mechanical advantage of a machine is less than one. Compare the distances over which the input and output forces are applied.

Applying Math

5. **Calculate** the mechanical advantage of a hammer if the input force is 125 N and the output force is 2,000 N.
6. **Calculate Efficiency** Find the efficiency of a machine that does 800 J of work if the input work is 2,400 J.
7. **Calculate Force** Find the force needed to lift a 2,000-N weight using a machine with a mechanical advantage of 15.

section 3

Simple Machines

Reading Guide

What You'll Learn

- **Describe** the six types of simple machines.
- **Explain** how the different types of simple machines make doing work easier.
- **Calculate** the ideal mechanical advantage of the different types of simple machines.

Why It's Important

All complex machines, such as cars, are made of simple machines. Even your body contains simple machines.

Review Vocabulary

compound: composed of separate elements or parts

New Vocabulary

- simple machine
- lever
- pulley
- wheel and axle
- inclined plane
- screw
- wedge
- compound machine

Figure 13 There are three classes of levers.

First-Class Lever
The fulcrum is between the input force and the output force.

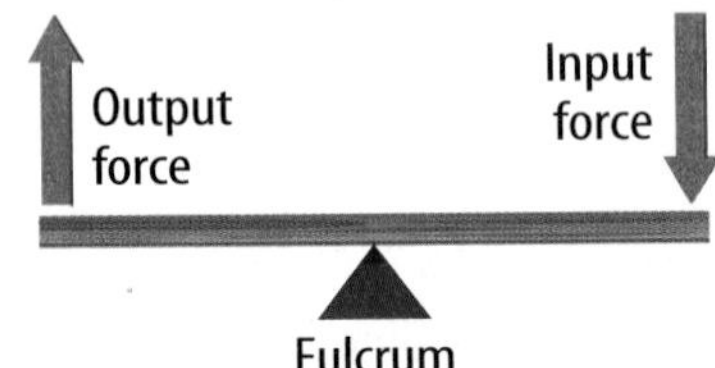

Second-Class Lever
The output force is between the fulcrum and the input force.

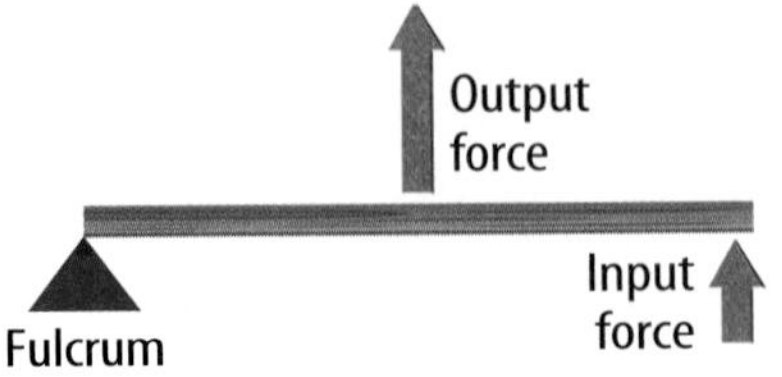

Third-Class Lever
The input force is between the fulcrum and the output force.

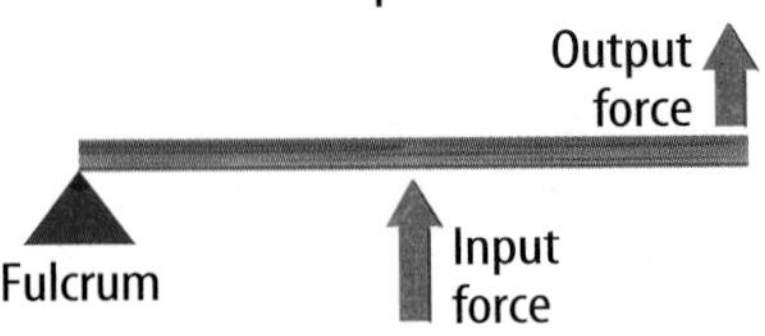

Types of Simple Machines

If you cut your food with a knife, use a screwdriver, or even chew your food, you are using a simple machine. A **simple machine** is a machine that does work with only one movement of the machine. There are six types of simple machines: lever, pulley, wheel and axle, inclined plane, screw, and wedge. The pulley and the wheel and axle are modified levers, and the screw and the wedge are modified inclined planes.

Levers

You've used a lever if you've used a wheelbarrow, a lawn rake, or swung a baseball bat. A **lever** is a bar that is free to pivot or turn around a fixed point. The fixed point the lever pivots on is called the fulcrum. The input arm of the lever is the distance from the fulcrum to the point where the input force is applied. The output arm is the distance from the fulcrum to the point where the output force is exerted by the lever.

The output force produced by a lever depends on the lengths of the input arm and the output arm. If the output arm is longer than the input arm, the law of conservation of energy requires that the output force be less than the input force. If the output arm is shorter than the input arm, then the output force is greater than the input force.

There are three classes of levers, as shown in **Figure 13.** The differences among the three classes of levers depend on the locations of the fulcrum, the input force, and the output force.

Figure 14 Levers are classified by the locations of the input force, the output force, and the fulcrum.

A The screwdriver is being used as a first-class lever. The fulcrum is the paint can rim.

B A wheelbarrow is a second-class lever. The fulcrum is the wheel.

C A baseball bat is a third-class lever. The fulcrum here is the batter's left hand.

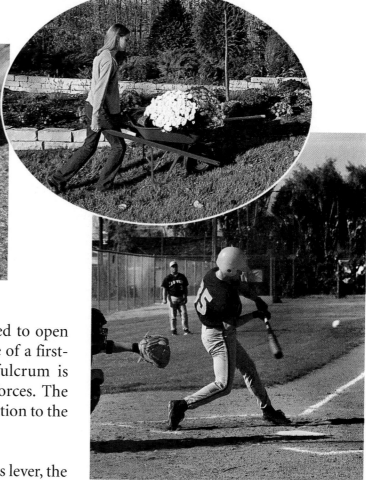

First-Class Lever The screwdriver used to open the paint can in **Figure 14A** is an example of a first-class lever. For a first-class lever, the fulcrum is located between the input and output forces. The output force is always in the opposite direction to the input force in a first-class lever.

Second-Class Lever For a second-class lever, the output force is located between the input force and the fulcrum. Look at the wheelbarrow in **Figure 14B.** The girl applies an upward input force on the handles, and the wheel is the fulcrum. The output force is exerted between the input force and the fulcrum. For a second-class lever, the output force is always greater than the input force.

Third-Class Lever Many pieces of sports equipment, such as a baseball bat, are third-class levers. For a third-class lever, the input force is applied between the output force and the fulcrum. The right-handed batter in **Figure 14C** applies the input force with the right hand, and the left hand is the fulcrum. The output force is exerted by the bat above the right hand. The output force is always less than the input force in a third-class lever. Instead, the distance over which the output force is applied is increased.

Every lever can be placed in one of these classes. Each class can be found in your body, as shown in **Figure 15** on the next page.

Archimedes and the Lever
The Greek mathematician Archimedes (287–212 B.C.) described how a lever could be used to increase force. He also supposedly said "Give me a place to stand and I will move the Earth." Research one of Archimedes' inventions and write a paragraph describing it.

NATIONAL GEOGRAPHIC

VISUALIZING LEVERS IN THE HUMAN BODY

Figure 15

▲ Fulcrum
⬇ Input force
⬆ Output force

All three types of levers—first-class, second-class, and third-class—are found in the human body. The forces exerted by muscles in your body can be increased by first-class and second-class levers, while third-class levers increase the range of movement of a body part. Examples of how the body uses levers to help it move are shown here.

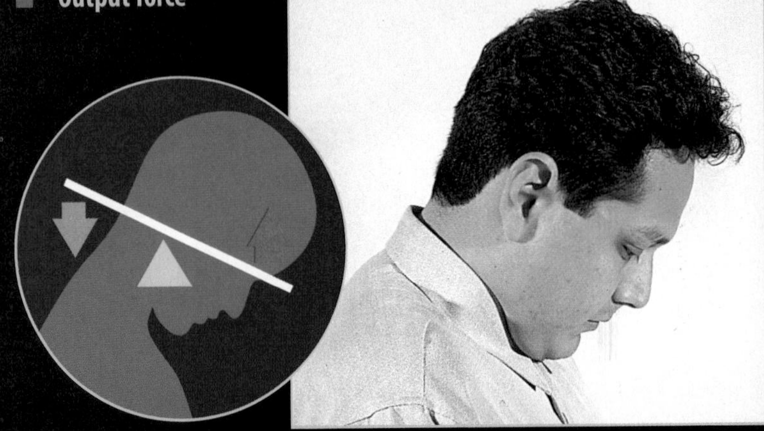

▲ FIRST-CLASS LEVER The fulcrum lies between the input force and the output force. Your head acts like a first-class lever. Your neck muscles provide the input force to support the weight of your head.

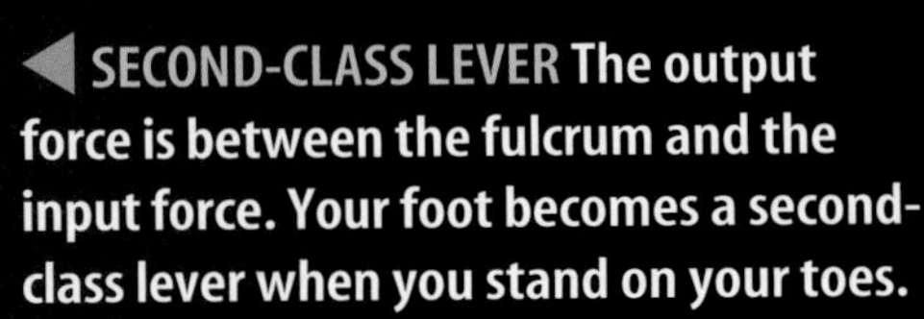

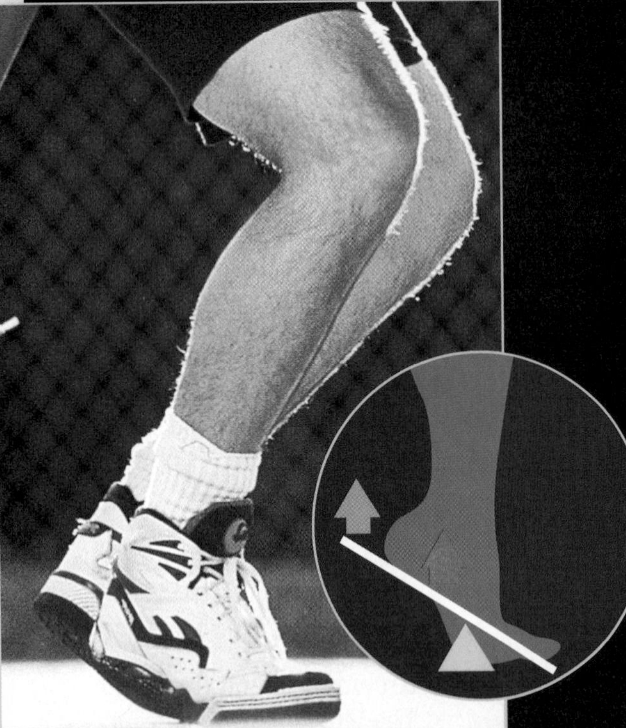

◀ SECOND-CLASS LEVER The output force is between the fulcrum and the input force. Your foot becomes a second-class lever when you stand on your toes.

▶ THIRD-CLASS LEVER The input force is between the fulcrum and the output force. A third-class lever increases the range of motion of the output force. When you do a curl with a dumbbell, your forearm is a third-class lever.

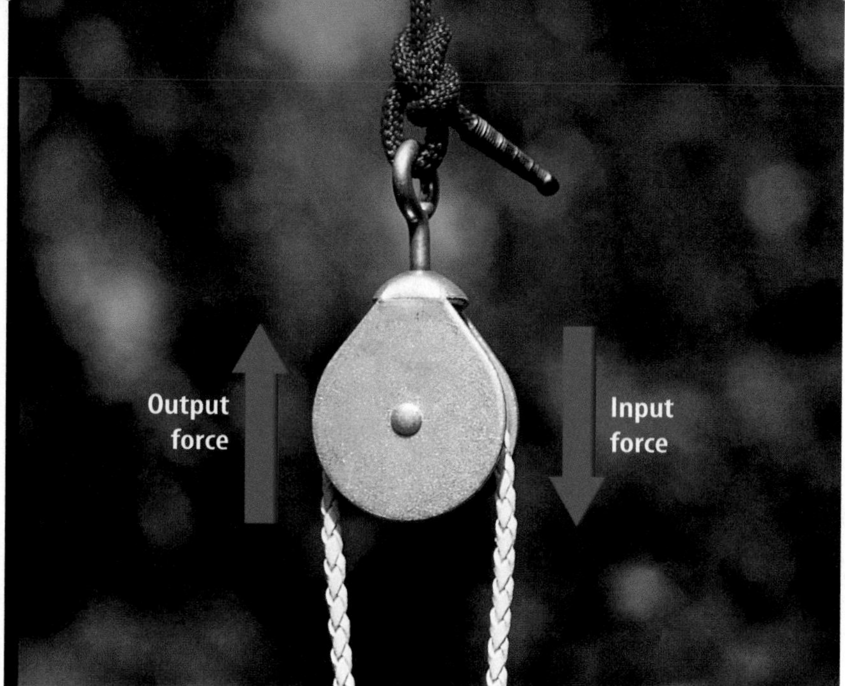

Figure 16 A fixed pulley is another form of the lever. **Infer** *the lengths of the input arm and output arm in a pulley.*

Ideal Mechanical Advantage of a Lever The ideal mechanical advantage, or *IMA*, for a lever can be calculated by dividing the length of the input arm by the length of the output arm:

IMA of a Lever

$$\text{ideal mechanical advantage} = \frac{\text{length of input arm (m)}}{\text{length of output arm (m)}}$$

$$IMA = \frac{L_{in}}{L_{out}}$$

Pulleys

A **pulley** is a grooved wheel with a rope, chain, or cable running along the groove. A fixed pulley, as shown in **Figure 16,** is a modified first-class lever. The axle of the pulley acts as the fulcrum. The two sides of the pulley are the input arm and output arm. A pulley can change the direction of the input force or increase the output force, depending on whether the pulley is fixed or movable. A system of pulleys can change the direction of the input force and make the output force larger.

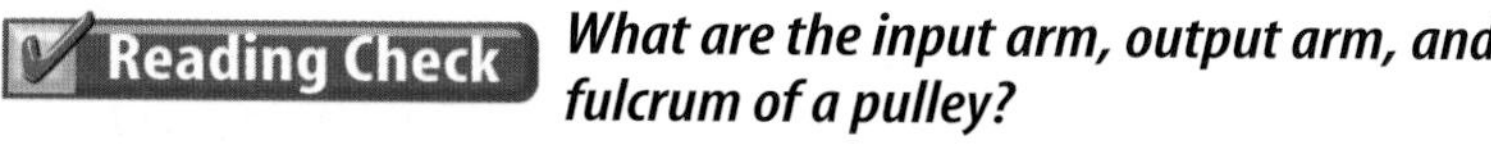

Reading Check *What are the input arm, output arm, and fulcrum of a pulley?*

Fixed Pulleys The cable attached to an elevator passes over a fixed pulley at the top of the elevator shaft. A fixed pulley, such as the one in **Figure 17,** is attached to something that doesn't move, such as a ceiling or wall. Because a fixed pulley changes only the direction of force, the IMA is 1.

Figure 17 A fixed pulley changes only the direction of a force. You need to apply an input force of 4 N to lift the 4-N weight.

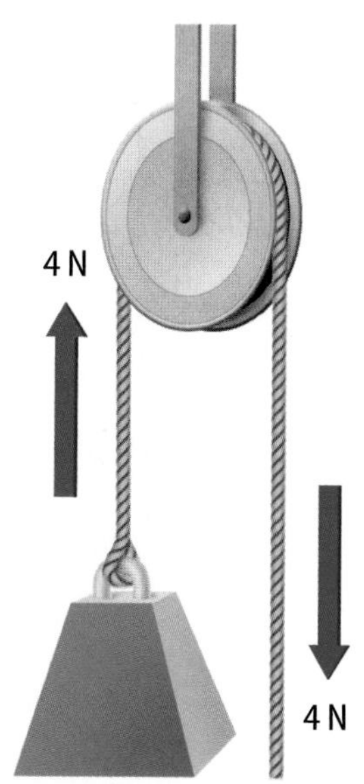

Figure 18 A movable pulley and a pulley system called a block and tackle reduce the force needed to lift a weight.

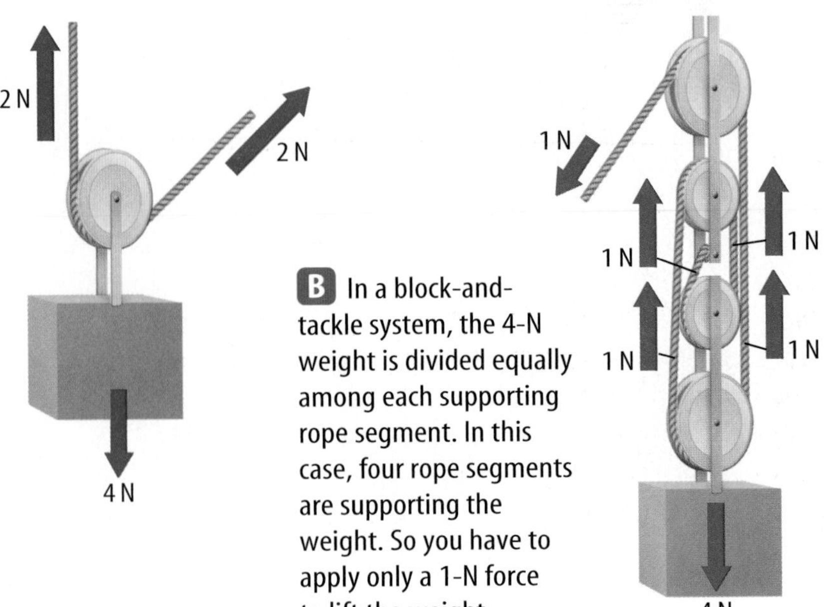

A With a movable pulley, the attached side of the rope supports half of the 4-N weight. You have to apply a 2-N force to lift the weight.

B In a block-and-tackle system, the 4-N weight is divided equally among each supporting rope segment. In this case, four rope segments are supporting the weight. So you have to apply only a 1-N force to lift the weight.

Movable Pulleys A pulley in which one end of the rope is fixed and the wheel is free to move is called a movable pulley. Unlike a fixed pulley, a movable pulley does multiply force. Suppose a 4-N weight is hung from the movable pulley in **Figure 18A.** The ceiling acts like someone helping you to lift the weight. The rope attached to the ceiling will support half of the weight—2 N. You need to exert only the other half of the weight—2 N—to support and lift the weight. The output force exerted on the weight is 4 N, and the applied input force is 2 N. Therefore the *IMA* of the movable pulley is 2.

For a fixed pulley, the distance you pull the rope downward equals the distance the weight moves upward. For a movable pulley, the distance you pull the rope upward is twice the distance the weight moves upward.

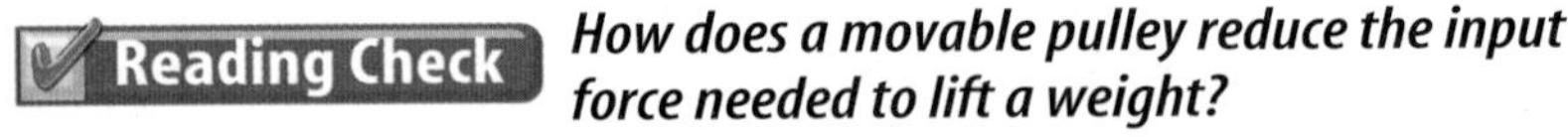

How does a movable pulley reduce the input force needed to lift a weight?

The Block and Tackle A system of pulleys consisting of fixed and movable pulleys is called a block and tackle. **Figure 18B** shows a block and tackle made up of two fixed pulleys and two movable pulleys. If a 4-N weight is suspended from the movable pulley, each rope segment supports one-fourth of the weight, reducing the input force to 1 N. The *IMA* of a pulley system is equal to the number of rope segments that support the weight. The block and tackle shown in **Figure 18B** has an *IMA* of 4. The *IMA* of a block and tackle can be increased by increasing the number of pulleys in the pulley system.

Figure 19 The handle on a pencil sharpener is part of a wheel and axle. You apply a force to the handle. This force is made larger by the wheel and axle, making it easy to turn the sharpening mechanism.

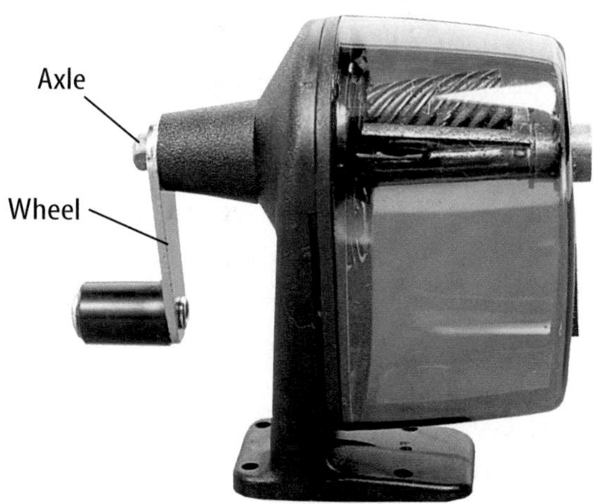

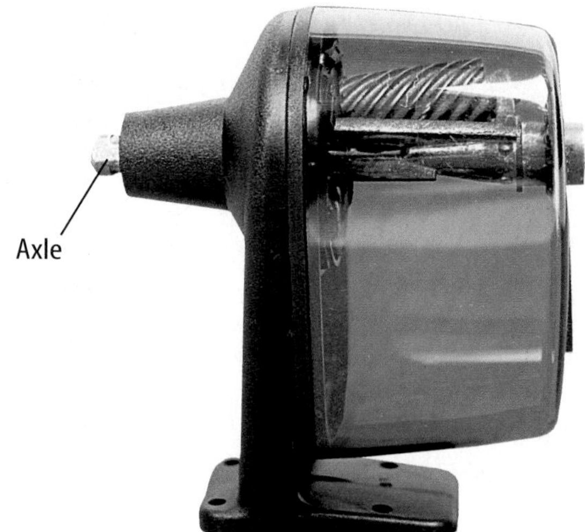

Wheel and Axle

Could you use the pencil sharpener in **Figure 19** if the handle weren't attached? The handle on the pencil sharpener is part of a wheel and axle. A **wheel and axle** is a simple machine consisting of a shaft or axle attached to the center of a larger wheel, so that the wheel and axle rotate together. Doorknobs, screwdrivers, and faucet handles are examples of wheel and axles. Usually, the input force is applied to the wheel, and the output force is exerted by the axle.

Mechanical Advantage of a Wheel and Axle

Recall that the mechanical advantage of a lever is the length of the input arm divided by the length of the output arm. A wheel and axle is another modified lever. The center of the axle is the fulcrum. The input force is applied at the rim of the wheel. So the length of the input arm is the radius of the wheel. The output force is exerted at the rim of the axle. So the length of the output arm is the radius of the axle. The *IMA* of a wheel and axle is:

IMA of a Wheel and Axle

$$\text{ideal mechanical advantage} = \frac{\text{radius of wheel (m)}}{\text{radius of axle (m)}}$$

$$IMA = \frac{r_w}{r_a}$$

According to this equation, the *IMA* of a wheel and axle can be increased by increasing the radius of the wheel.

Figure 20 If an input force is applied to the larger gear and it rotates clockwise, the smaller gear rotates counterclockwise. The output force exerted by the smaller gear is less than the input force applied to the larger gear.

Gears A gear is a wheel and axle with the wheel having teeth around its rim. When the teeth of two gears interlock, the turning of one gear causes the other gear to turn.

When two gears of different sizes are interlocked, they rotate at different rates. Each rotation of the larger gear causes the smaller gear to make more than one rotation. If the input force is applied to the larger gear, the output force exerted by the smaller gear is less than the input force.

Gears also may change the direction of the force, as shown in **Figure 20.** When the larger gear in is rotated clockwise, the smaller gear rotates counterclockwise.

Inclined Planes

Why do the roads and paths on mountains zigzag? Would it be easier to climb directly up a steep incline or walk a longer path gently sloped around the mountain? An **inclined plane** is a sloping surface, such as a ramp, that reduces the amount of force required to do work.

Mechanical Advantage of an Inclined Plane You do the same work by lifting a box straight up or pushing it up an inclined plane. By pushing the box up an inclined plane, however, the input force is exerted over a longer distance compared to lifting the box straight up. As a result, the input force is less than the force needed to lift the box straight upward. The *IMA* of an inclined plane can be calculated from the following equation:

IMA of an Inclined Plane

$$\text{ideal mechanical advantage} = \frac{\text{length of slope (m)}}{\text{height of slope (m)}}$$

$$IMA = \frac{l}{h}$$

The *IMA* of an inclined plane for a given height is increased by making the plane longer.

When you think of an inclined plane, you might think of moving an object up a ramp—you move and the inclined plane remains stationary. The screw and the wedge, however, are variations of the inclined plane in which the inclined plane moves and the object remains stationary.

Topic: Nanobots

Visit gpescience.com for Web links to information about how tiny robots called nanobots might be used as microsurgical instruments.

Activity Use the information you find to make a diagram of a nanobot that might be used to perform surgery.

Figure 21 A screw has an inclined plane that wraps around the post of the screw.

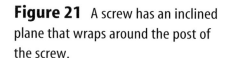

The thread gets thinner farther from the post. This helps the screw force its way into materials.

Many lids, such as those on peanut butter jars, also contain threads.

The Screw

A **screw** is an inclined plane wrapped in a spiral around a cylindrical post. If you look closely at the screw in **Figure 21,** you'll see that the threads form a tiny ramp that runs upward from its tip. You apply the input force by turning the screw. The output force is exerted along the threads of the screw. The *IMA* of a screw is related to the spacing of the threads. The *IMA* is larger when the threads are closer together. However, when the *IMA* is larger, more turns of the screw are needed to drive it into a material.

How do you remove the lid from a jar of peanut butter, such as the one in **Figure 21?** If you look closely, you will see threads similar to the ones on the screw in **Figure 21.** Where else can you find examples of a screw?

The Wedge

Like the screw, the wedge is also a simple machine in which the inclined plane moves through an object or material. A **wedge** is an inclined plane with one or two sloping sides. It changes the direction of the input force.

Look closely at the knife in **Figure 22.** One edge is sharp, and it slopes outward at both sides, forming an inclined plane. As it moves through the apple, the downward input force is changed to a horizontal force, forcing the apple apart.

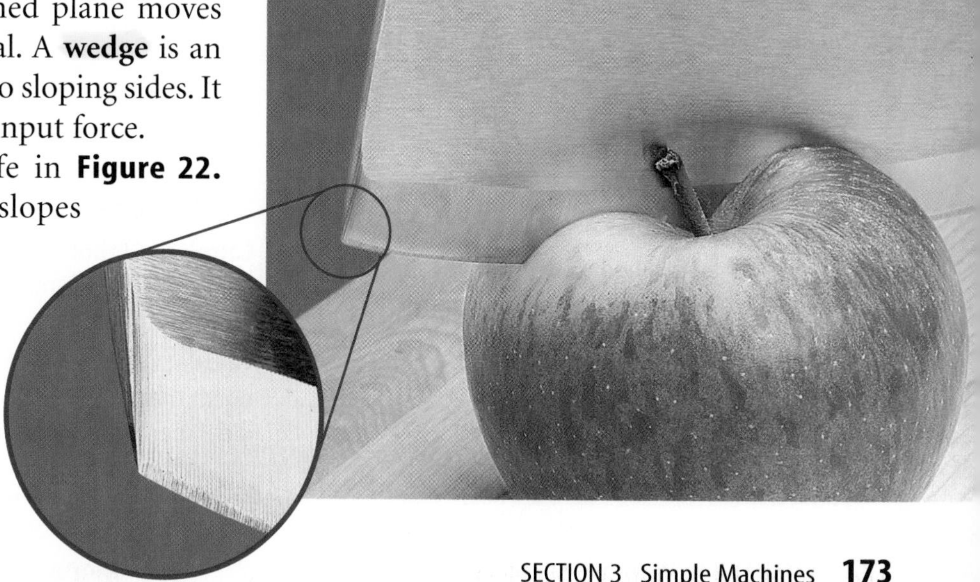

Figure 22 A knife blade is a wedge. As you cut through the apple, the blade pushes the halves of the apple apart.

Figure 23 A compound machine, such as a can opener, is made up of simple machines.

Compound Machines

Some of the machines you use every day are made up of several simple machines. Two or more simple machines that operate together form a **compound machine.**

Look at the can opener in **Figure 23.** To open the can, you first squeeze the handles together. The handles act as levers and increase the force applied on a wedge, which then pierces the can. You then turn the handle, a wheel and axle, to open the can.

A car is also a compound machine. Burning fuel in the cylinders of the engine causes the pistons to move up and down. This up-and-down motion makes the crankshaft rotate. The force exerted by the rotating crankshaft is transmitted to the wheels through other parts of the car, such as the transmission and the differential. Both of these parts contain gears which can change the rate at which the wheels rotate, the force exerted by the wheels, and even reverse the direction of rotation.

section 3 review

Summary

The Lever Family

- A lever is a bar that is free to pivot about a fixed point called the fulcrum.
- There are three classes of levers based on the relative locations of the input force, the output force, and the fulcrum.
- A pulley is a grooved wheel with a rope, chain, or cable placed in the groove and is a modified form of a lever.
- The *IMA* of a lever is the length of the input arm divided by the length of the output arm.
- A wheel and axle consists of a shaft or axle attached to the center of a larger wheel.

The Inclined Plane Family

- An inclined plane is a ramp or sloping surface that reduces the force needed to do work.
- The *IMA* of an inclined plane is the length of the plane divided by the height of the plane.
- A screw consists of an inclined plane wrapped around a shaft.
- A wedge is an inclined plane that moves and can have one or two sloping surfaces.

Self Check

1. **Classify** a screwdriver being used to turn a screw as one of the six types of simple machines. Explain how the *IMA* of a screwdriver could be increased.
2. **Determine** for which class of lever the output force is always greater than the input force. For which class is the output force always less than the input force?
3. **Make a diagram** of a bicycle and label the parts of a bicycle that are simple machines.
4. **Think Critically** Use the law of conservation of energy to explain why, in a second-class lever, the distance over which the input force is applied is always greater than the distance over which the output force is applied.

Applying Math

5. **Calculate *IMA*** What is the *IMA* of a car's steering wheel if the wheel has a diameter of 40 cm and the shaft it's attached to has a diameter of 4 cm?
6. **Calculate Output Arm Length** A lever has an *IMA* of 4. If the length of the input arm is 1.0 m, what is the length of the output arm?
7. **Calculate *IMA*** A 6.0-m ramp runs from a sidewalk to a porch that is 2.0 m above the sidewalk. What is the ideal mechanical advantage of this ramp?

More Section Review gpescience.com

Levers

Have you ever tried to balance a friend on a seesaw? If your friend was lighter, you had to move toward the fulcrum. In this lab, you will use the same method to measure the mass of a coin.

Real-World Problem

How can a lever be used to measure mass?

Goals

- **Calculate** the ideal mechanical advantage of a lever.
- **Determine** the mass of a coin.

Materials

stiff cardboard, 3 cm by 30 cm
coins (one quarter, one dime, one nickel)
balance
metric ruler

Safety Precautions

Procedure

1. Complete the safety form.
2. **Measure** the mass of each coin.
3. Mark a line 2 cm from one end of the cardboard strip. Label this line *Output.*
4. Slide the other end of the cardboard strip over the edge of a table until the strip begins to tip. Mark a line across the strip at the table edge and label this line *Input.*
5. **Measure** the mass of the strip to the nearest 0.1 g. Write this mass on the input line.
6. Center a dime on the output line. Slide the cardboard strip until it begins to tip. Mark the balance line. Label it *Fulcrum 1.*

7. **Measure** the lengths of the output and input arms to the nearest 0.1 cm.
8. Calculate the *IMA* of the lever. Multiply the *IMA* by the mass of the lever to find the approximate mass of the coin.
9. Repeat steps 6 through 8 with the nickel and the quarter. Mark the fulcrum line *Fulcrum 2* for the nickel and *Fulcrum 3* for the quarter.

Conclude and Apply

1. **Explain** why there might be a difference between the mass of each coin measured by the balance and the mass measured using the lever.
2. **Explain** what provides the input and output forces for the lever.
3. **Explain** why the *IMA* of the lever changes as the mass of the coin changes.

Compare your results with those of other students in your class. **For more help, refer to the Science Skill Handbook.**

Model and Invent

USING SIMPLE MACHINES

Goals

- **Model** lifting devices based on a block and tackle and on an inclined plane.
- **Calculate** the output work that will be accomplished.
- **Measure** the force needed by each machine to lift a weight.
- **Calculate** the input work and efficiency for each model machine.
- **Select** the best machine for your job based on the force required.

Possible Materials

spring scale, 0–10 N range
9.8-N weight (1-kg mass)
double pulley
single pulley
string for the pulleys
stand or support for the pulleys
wooden board, 40 cm long
support for board, 10 cm high

Safety Precautions

Real-World Problem

Suppose that you are the contractor on a one-story building with a large air conditioner. How can you get the air conditioner to the roof? How can you minimize the force needed to lift an object? What machines could you use? Consider a fixed pulley, a block and tackle system of pulleys, and an inclined plane. How can you find the efficiency of machines?

Make a Model

1. Complete the safety form.
2. Work in teams of at least two. **Collect** all the needed equipment.
3. Sketch a model for each lifting machine. Include a control in which the weight is lifted while being suspended directly from the spring scale. **Model** the inclined plane with a board 40-cm long and raised 10 cm at one end.
4. Make a data table like the one below.
5. Is the pulley support high enough that the block and tackle can lift a weight 10 cm?
6. Obtain your teacher's approval of your sketches and data table before proceeding.

Problem Data

	Control	Inclined Plane	Block and Tackle
Ideal Mechanical Advantage, *IMA*	1		
Input force, F_{in}, N		Do not write in this book.	
Input distance, d_{in}, m	0.10		
Output force, F_{out}, N	9.8	9.8	9.8
Output distance, d_{out}, m	0.10	0.10	0.10
$Work_{in} = F_{in}\ d_{in}$, Joules			
$Work_{out} = F_{out}\ d_{out}$, Joules	0.98	0.98	0.98
$Efficiency = (Work_{out}/Work_{in}) \times 100\%$			

Test Your Model

1. Tie the weight to the spring scale and measure the force required to lift it. Record the input force in your data table under *Control,* along with the 10-cm input distance.
2. Assemble the inclined plane so that the weight can be pulled up the ramp at a constant rate. The 40-cm board should be supported so that one end is 10 cm higher than the other end.
3. Tie the string to the spring scale and measure the force required to move the weight up the ramp at a constant speed. Record this input force under *Inclined Plane* in your data table. Record 40 cm as the input distance for the inclined plane.
4. Assemble the block and tackle using one fixed double pulley and one movable single pulley.
5. Tie the weight to the single pulley and tie the spring scale to the string at the top of the upper double pulley.
6. **Measure** the force required to lift the weight with the block and tackle. Record this input force.
7. **Measure** the length of string that must be pulled to raise the weight 10 cm. Record this input distance.

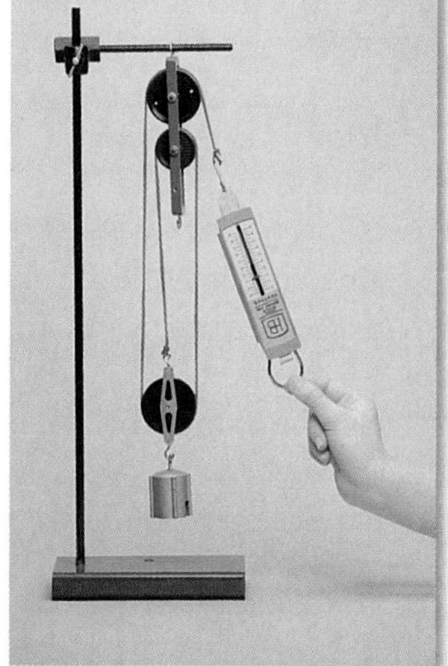

Analyze Your Data

1. **Calculate** the output work for all three methods of lifting the 9.8-N weight a distance of 10 cm.
2. **Calculate** the input work and the efficiency for the control, the inclined plane, and the block and tackle.
3. **Compare** the efficiencies of each of the three methods of lifting.

Conclude and Apply

1. **Explain** how you might improve the efficiency of the machine in each case.
2. **Infer** what types of situations would require the use of a ramp over a pulley to help lift something.
3. **Infer** which machines would be most likely to be affected by friction.

Communicating Your Data

Make a poster showing how the best machine would be used to lift the air conditioner to the roof of the building.

The Science of very, very small

Imagine an army of tiny robots, each no bigger than a bacterium swimming through your bloodstream.

Welcome to the world of nanotechnology, the science of creating molecular-sized machines. These machines are called nanobots.

This is the smallest guitar in the world. It is about as big as a human white blood cell. Each of its six silicon strings is 100 atoms wide. You can see the guitar only with an electron microscope.

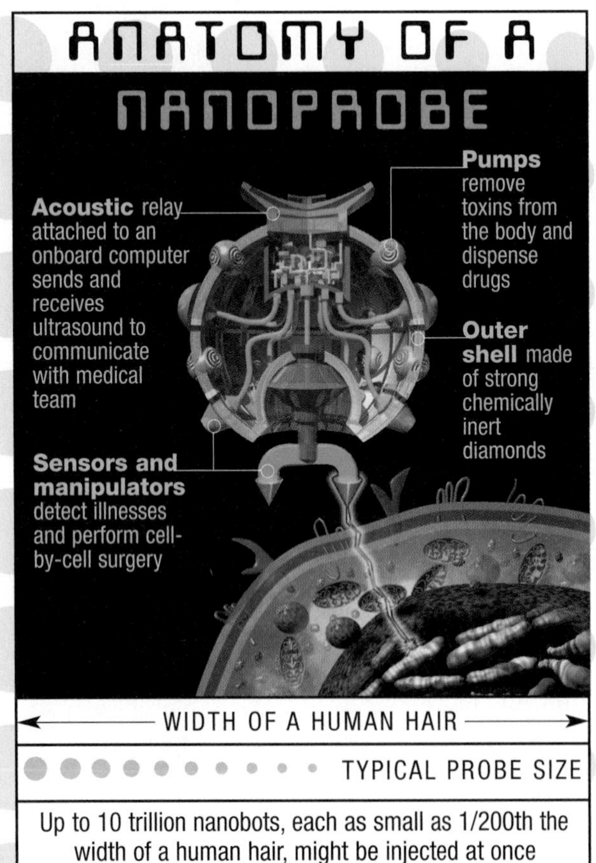
ANATOMY OF A NANOPROBE
Pumps remove toxins from the body and dispense drugs
Acoustic relay attached to an onboard computer sends and receives ultrasound to communicate with medical team
Outer shell made of strong chemically inert diamonds
Sensors and manipulators detect illnesses and perform cell-by-cell surgery
WIDTH OF A HUMAN HAIR
TYPICAL PROBE SIZE
Up to 10 trillion nanobots, each as small as 1/200th the width of a human hair, might be injected at once

The smallest of these machines are only billionths of a meter in size. They are so tiny that they can do work on the molecular scale.

Small, Smaller, Smallest

Nanotechnologists are predicting that within a few decades they will be creating nanobots that can do just about anything, as long as it's small. Already, nanotechnologists have built gears 10,000 times thinner than a human hair. They've also built tiny molecular "motors" only 50 atoms long. At Cornell University, nanotechnologists created the world's smallest guitar. It is appoximately the size of a white blood cell and it even has six strings.

In the future, nanobots might transmit your internal vital signs to a nanocomputer implanted under your skin. There the data could be analyzed for signs of disease. Other nanomachines then could be sent to scrub your arteries clean of dangerous blockages, or mop up cancer cells, or even vaporize blood clots with tiny lasers. These are just some of the possibilities in the imaginations of those studying the new science of nanotechnology.

Design **Think up a very small simple or complex machine that could go inside the body and do something. What would the machine do? Where would it go? Share your diagram or design with your classmates.**

TIME
For more information, visit gpescience.com

Reviewing Main Ideas

Section 1 Work

1. Work is the energy transferred when a force makes an object move.
2. Work is done only when a force produces motion in the direction of the force.
3. Power is the amount of work, or the amount of energy transferred, in a certain amount of time.

Section 2 Using Machines

1. A machine makes work easier by changing the size of the force applied, by increasing the distance an object is moved, or by changing the direction of the applied force.
2. The number of times a machine multiplies the force applied to it is the mechanical advantage of the machine. The actual mechanical advantage is always less than the ideal mechanical advantage.
3. The efficiency of a machine equals the output work divided by the input work.
4. Friction always causes the output work to be less than the input work, so no real machine can be 100 percent efficient.

Section 3 Simple Machines

1. A simple machine is a machine that can do work with a single movement.
2. A simple machine can increase an applied force, change its direction, or both.
3. A lever is a bar that is free to pivot about a fixed point called a fulcrum. A pulley is a grooved wheel with a rope running along the groove. A wheel and axle consists of two different-sized wheels that rotate together. An inclined plane is a sloping surface used to raise objects. The screw and the wedge are special types of inclined planes.
4. A combination of two or more simple machines is called a compound machine.

FOLDABLES™ Use the Foldable that you made at the beginning of this chapter to help you review how machines make doing work easier.

chapter 6 Review

Using Vocabulary

compound machine p. 174	output force p. 162
efficiency p. 164	power p. 157
inclined plane p. 172	pulley p. 169
input force p. 162	screw p. 173
lever p. 166	simple machine p. 166
machine p. 160	wedge p. 173
mechanical advantage p. 164	wheel and axle p. 171
	work p. 154

Complete each statement using a word(s) from the vocabulary list above.

1. A combination of two or more simple machines is a(n) ________.
2. A wedge is another form of a(n) ________.
3. The ratio of the output force to the input force is the ________ of a machine.
4. A(n) ________ is a grooved wheel with a rope, chain, or cable in the groove.
5. The force exerted by a machine is the ________.
6. Energy is transferred when ________ is done.
7. ________ is the rate at which work is done or energy is transferred.

Checking Concepts

Choose the word or phrase that best answers the question.

8. Which of these is not done by a machine?
   A) multiply force
   B) multiply energy
   C) change direction of a force
   D) work
9. Which of the following increases as the efficiency of a machine increases?
   A) work input
   B) work output
   C) friction
   D) *IMA*
10. In an ideal machine, which of the following is true?
    A) Work input is equal to work output.
    B) Work input is greater than work output.
    C) Work input is less than work output.
    D) The *IMA* is always equal to one.

Use the picture below to answer question 11.

11. What simple machines make up a pair of scissors?
    A) lever and pulley
    B) pulley and wheel and axle
    C) wedge and pulley
    D) wedge and lever
12. What term indicates the number of times a machine multiplies the input force?
    A) efficiency
    B) power
    C) mechanical advantage
    D) resistance
13. How could you increase the *IMA* of an inclined plane?
    A) increase its length
    B) increase its height
    C) decrease its length
    D) make its surface smoother
14. What is the *IMA* of a screwdriver with a shaft radius of 3 mm and a handle radius of 10 mm?
    A) 0.3
    B) 3.3
    C) 30
    D) 13
15. What is the *IMA* of an inclined plane that is 2.1 m long and 0.7 m high?
    A) 0.3
    B) 2.8
    C) 1.5
    D) 3.0

Vocabulary PuzzleMaker gpescience.com

Interpreting Graphics

16. Copy and complete the concept map of simple machines using the following terms: *compound machines, mechanical advantage, output force, work.*

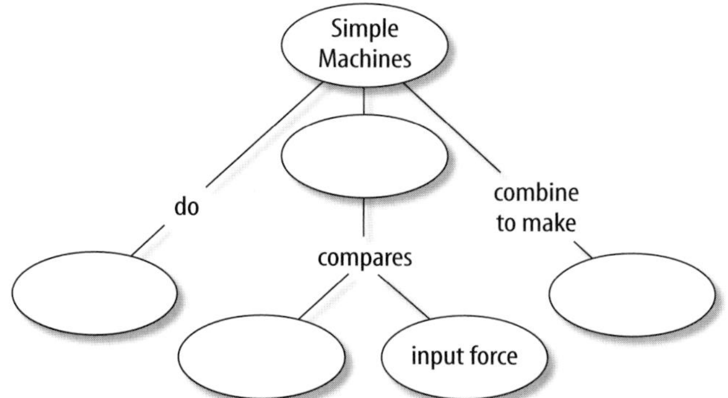

Use the table below to answer questions 17 and 18.

Lever Input and Output Arms

Lever	Input Arm (cm)	Output Arm (cm)
A	25	75
B	53	42
C	36	36
D	32	99
E	10	30

17. **Determine** which of the levers listed in the table above has the largest *IMA*.

18. **Calculate** An input force of 50 N is applied to lever B. If the lever is 100 percent efficient, what is the output force?

Thinking Critically

19. **Describe** how the input force, output force, and fulcrum should be arranged so that a child can lift an adult using his own body weight on a seesaw.

20. **Determine** what arrangement of movable and fixed pulleys would give a mechanical advantage of 3.

21. **Explain** which would give the best mechanical advantage for driving a screw down into a board: a screwdriver with a long, thin handle, or a screwdriver with a short, thick handle.

Applying Math

22. **Calculate Work** Find the work needed to lift a 20.0-N book 2.0 m.

23. **Calculate Axle Radius** A doorknob has an *IMA* equal to 8.5. If the diameter of the doorknob is 8.0 cm, what is the radius of the shaft the doorknob is connected to?

24. **Calculate Input Work** A machine has an efficiency of 61 percent. Find the input work if the output work is 140 J.

25. **Calculate Efficiency** Using a ramp 6 m long, workers apply an input force of 1,250 N to move a 2,000-N crate onto a platform 2 m high. What is the efficiency of the ramp?

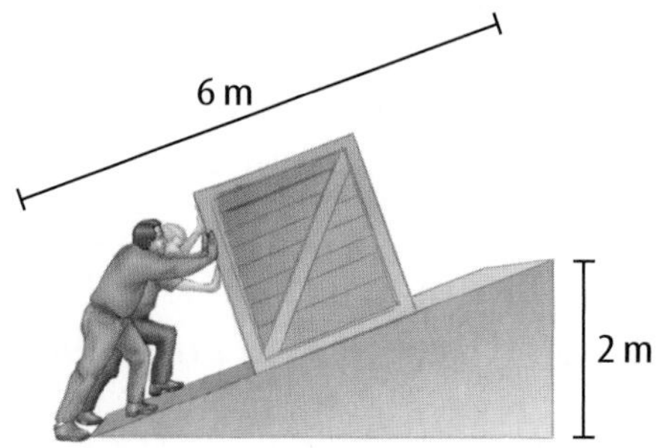

26. **Calculate Power** A person weighing 500 N climbs 3 m. How much power is needed to make the climb in 5 s?

chapter 6 Standardized Test Practice

Record your answers on the answer sheet provided by your teacher or on a sheet of paper.

Multiple Choice

Use the figure below to answer questions 1 and 2.

1. The figure above shows a doorknob with a radius of 4.8 cm and a mechanical advantage of 4.0. What is the radius of the inner rod that connects the knob to the door?

   A. 0.6 cm

   B. 1.2 cm

   C. 1.8 cm

   D. 2.4 cm

2. What would happen to the mechanical advantage if the radius of the doorknob were doubled?

   A. It would be multiplied by 4.

   B. It would be multiplied by 2.

   C. It would be divided by 4.

   D. It would be divided by 2.

3. A ramp is 2.8 m long and 1.2 m high. How much power is needed to push a box up the ramp in 4.6 s with a force of 96 N?

   A. 21 W

   B. 25 W

   C. 58 W

   D. 270 W

4. How much work is done in lifting a 9.10-kg box onto a shelf 1.80 m high?

   A. 5.06 J

   B. 16.4 J

   C. 49.5 J

   D. 161 J

5. An input force of 80 N is used to lift an object weighing 240 N with a system of pulleys. How far down must the rope around the pulleys be pulled to lift the object a distance of 1.4 m?

   A. 0.47 m

   B. 1.4 m

   C. 2.8 m

   D. 4.2 m

Use the figure below to answer questions 6 and 7.

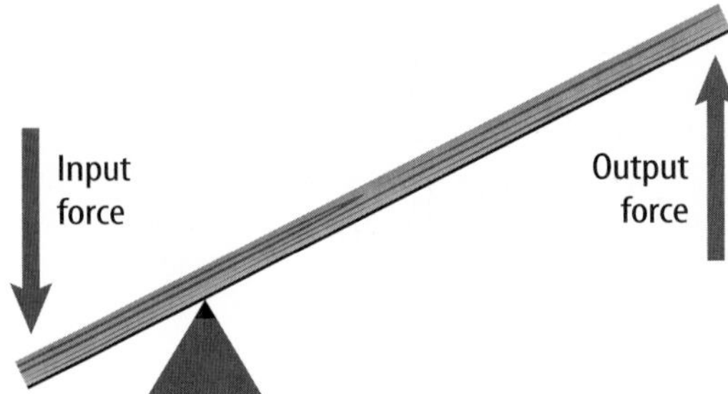

6. If a lever's input arm is 8 cm and its output arm is 24 cm, what is the ideal mechanical advantage of the lever?

   A. 4

   B. 3

   C. 0.33

   D. 0.25

Gridded Response

7. How much more work is done to push a box 2.5 m with a force of 30 N than to push a box 2.0 m with a force of 26 N?

8. As you throw a ball, you exert a force of 4.2 N on the ball. You exert this force on the ball while the ball moves a distance of 0.45 m. The ball leaves your hand and travels a horizontal distance of 8.5 m to your friend. How much work have you done on the ball?

Short Response

Use the illustration below to answer questions 9 and 10.

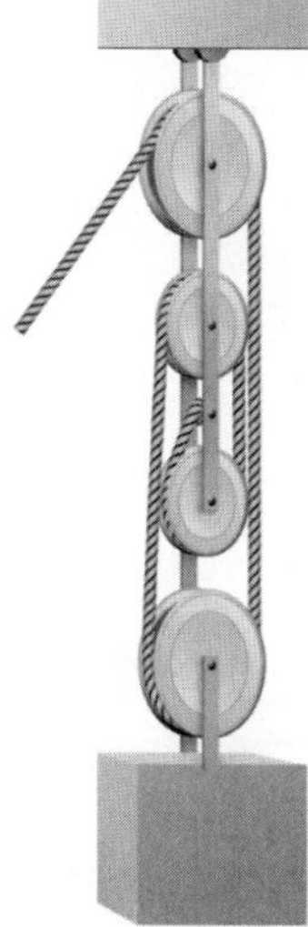

9. What is the ideal mechanical advantage of the pulley system shown in the figure above?

10. If the block supported by the pulley system has a weight of 20 N, what is the input force on the rope?

Extended Response

11. Use the law of conservation of energy to explain why it is impossible for the output work of a machine to be greater than the machine's input work.

Use the illustration below to answer question 12.

12. The boy in the figure below is carrying a box to the top of the stairs.

PART A Describe how the work that the boy does on the box is related to the energy transfer that occurs. How does the energy of the box change form as the boy carries the box up the stairs?

PART B Explain how the work, power, and energy would change if the boy walked faster. How would the work, power, and energy change if the steps were the same height but steeper?

Test-Taking Tip

Don't Panic Stay calm during the test. If you feel yourself getting nervous, close your eyes and take five slow, deep breaths.

chapter 7

The Earth-Moon-Sun System

chapter preview

sections

1 Earth in Space
2 Time and Seasons
Lab Comparing the Angle of Sunlight to Intensity
3 Earth's Moon
Lab Identifying the Moon's Surface Features and Apollo Landing Sites

Are there tides in a desert?

Did you know the Moon exerts gravitational force on Earth? On Earth, the Moon's gravitational attraction is evidenced as tides in oceans and seas. Even in the desert, the Moon's gravity influences the flexing of tectonic plates.

Science Journal Research to discover what landforms or events are affected by the Moon's gravitational force on Earth.

Start-Up Activities

Relative Sizes of Earth, the Moon, and the Sun

Can you picture the relative sizes of Earth, the Moon, and the Sun? Earth is about four times larger than the Moon in diameter, but the Sun is much larger than either. The Sun's diameter is about 100 times that of Earth and about 400 times that of the Moon. In this Lab, you'll investigate the relative sizes of all three objects.

1. Get permission to draw some circles on a sidewalk or paved area with chalk. You could also use a stick to draw circles on a dirt playing field.
2. Select a scale that will enable you to draw circles that will represent each object. *Hint: Using 1 cm for the Moon's diameter is a good start.*
3. Use a meterstick to draw a circle with a 1-cm diameter for the Moon.
4. Now draw two more circles to represent Earth and the Sun.
5. **Think Critically** In your Science Journal, explain how the Moon and the Sun can appear to be about the same size in the sky. Think about how things look smaller the farther they are from you.

Earth, Sun, and Moon Make the following Foldable to help organize what you learn about the Earth-Moon-Sun system.

STEP 1 **Fold** a sheet of paper vertically from side to side. Make the front edge about $\frac{1}{2}$ inch longer than the back edge.

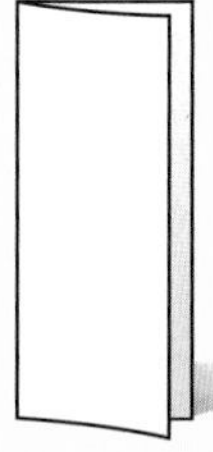

STEP 2 **Turn** lengthwise and fold into thirds.

STEP 3 **Unfold** and cut only the top layer along both folds to make three tabs.

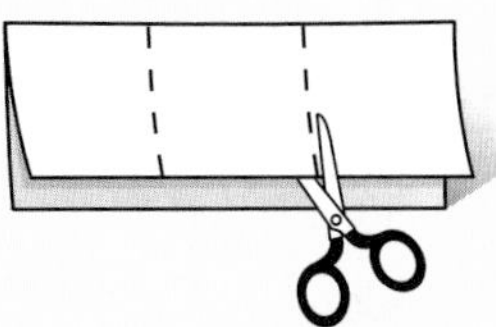

STEP 4 **Label** each tab.

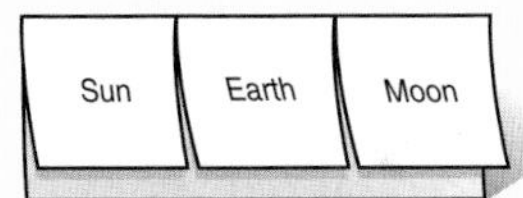

Questions As you read the chapter, write what you learn about each body under the correct tab of your Foldable. After you read the chapter, note the many ways that the three affect each other.

Preview this chapter's content and activities at gpescience.com

section 1

Earth in Space

Reading Guide

What You'll Learn

- **Compare and contrast** Earth's physical characteristics with those of other planets.
- **Explain** Earth's magnetic field.
- **Describe** Earth's movement in space and how eclipses occur.

Why It's Important

Gravity from the Earth-Moon-Sun system directly affects what it's like to live here on Earth.

Review Vocabulary

orbit: curved path of one object, such as the Moon, around another object, such as Earth

New Vocabulary

- sphere
- ellipse
- gravity

Figure 1 Objects fall toward Earth's center.
Infer *How would apples fall from trees if Earth were shaped like a cube?*

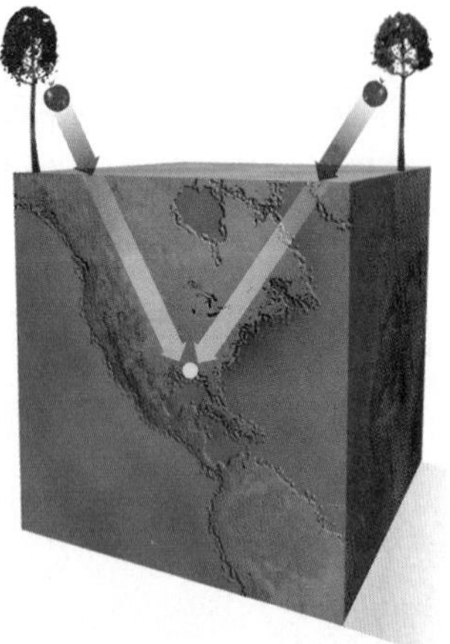

Earth's Size and Shape

Like most people, you are aware that Earth is round like a ball. But can you prove that this is true? If you jump up, you know that you'll come back down, but why is this so? What is the force that brings you down? You may have used a compass to tell directions, but do you know how a compass works? You will learn the answers to these questions and also about many physical characteristics of Earth in this section.

Ancient Measurements Earth's shape is similar to a sphere. A **sphere** is a round, three-dimensional object, the surface of which is the same distance from the center in all directions. Even ancient astronomers knew that Earth is spherical in shape. We have pictures of Earth from space that show us that it is spherical, but how could astronomers from long ago have learned this? They used evidence from observations.

Aristotle was one of these early astronomers. He made three different observations that indicated that Earth's shape is spherical. First, as shown in **Figure 1,** no matter where you are on Earth, objects fall straight down to the surface, as if they are falling toward the center of a sphere. Second, Earth's shadow on the Moon during a lunar eclipse is always curved. If Earth weren't spherical, this might not always be the case. For example, a flat disk casts a straight-edged shadow sometimes. Finally, people in different parts of the world see different stars above their horizons. More specifically, the pole star Polaris is lower in the sky at some locations on Earth than at others.

Everyday Evidence of Earth's Shape What have you seen, other than pictures from space, that indicates Earth's shape? Think about walking toward someone over a hill. First, you see the top of the person's head, and then you can see more and more of that person. Similarly, if you sail toward a lighthouse, you first see the top of the lighthouse and then see more and more of it as you move over Earth's curved surface.

You can see other evidence, too. Just like ancient astronomers, you can see for yourself that objects always fall straight down. Today, however, we know more about gravity. **Gravity** is the attractive force between two objects that depends on the masses of the objects and the distance between them. Astronomers think Earth formed by the accumulation of infalling objects toward a central mass. Energy released in the impacts kept the growing Earth molten. Gravity caused it to form into the most stable shape, a sphere. In this shape, the pull of gravity toward the center of the planet is the same in all directions. If a planet is massive enough, the pull of gravity could be so strong that even tall mountains would collapse under their own weight. **Table 1** lists some of Earth's other properties.

Reading Check *How does the pull of gravity indicate that Earth's shape is spherical?*

Topic: Comparing Earth to Other Planets

Visit gpescience.com for Web links to information about how Earth is similar to and different from other planets in the solar system.

Activity Make a table that lists the similarities and differences of Mercury, Venus, Earth, and Mars.

Table 1 Earth's Physical Properties

Diameter (pole to pole)	12,714 km
Diameter (through equator)	12,756 km
Circumference (poles)	40,008 km
Circumference (equator)	40,075 km
Mass	5.98×10^{24} kg
Average density	5.52 g/cm^3
Average distance to the Sun	149,600,000 km
Average distance to the Moon	384,400 km
Period of rotation	23 h, 56 min
Period of revolution	365 days, 6 h, 9 min

Van Allen Belts The magnetosphere lies above the outer layers of Earth's atmosphere. Within this magnetosphere are belts of charged particles known as the Van Allen belts. They contain thin plasma composed of protons (inner belt) and electrons (outer belt) that are trapped by Earth's magnetic field. Research how the magnetosphere protects Earth from the solar wind and why the Van Allen belts are hazardous to astronauts and satellites. Report your findings to the class.

Earth's Magnetic Field

Earth has a magnetic field that protects us from harmful radiation from the Sun. Scientists hypothesize that Earth's rotation and movement of matter in the core set up a strong magnetic field in and around Earth. This field resembles that surrounding a bar magnet, shown in **Figure 2.** Earth's magnetic field is concentrated at two ends of an imaginary magnetic axis running from Earth's north magnetic pole to its south magnetic pole. This axis is tilted about 11.5° from Earth's geographic axis of rotation.

Wandering Poles The locations of Earth's magnetic poles change slowly over time. Large-scale movements, called polar wandering, are thought to be caused by movements in Earth's crust and upper mantle. The magnetic north pole is carefully remapped periodically to pinpoint its location.

The Aurora An area within Earth's magnetic field, called the magnetosphere, deflects harmful radiation coming from the Sun, a stream of particles called solar wind. Some of these ejected particles from the Sun produce other charged particles in Earth's outer atmosphere. These charged particles spiral along Earth's magnetic field lines toward Earth's magnetic poles. There they collide with atoms in the atmosphere. These collisions cause the atoms to emit light. This light is called the aurora borealis (northern lights) in the northern hemisphere and the aurora australis (southern lights) in the southern hemisphere.

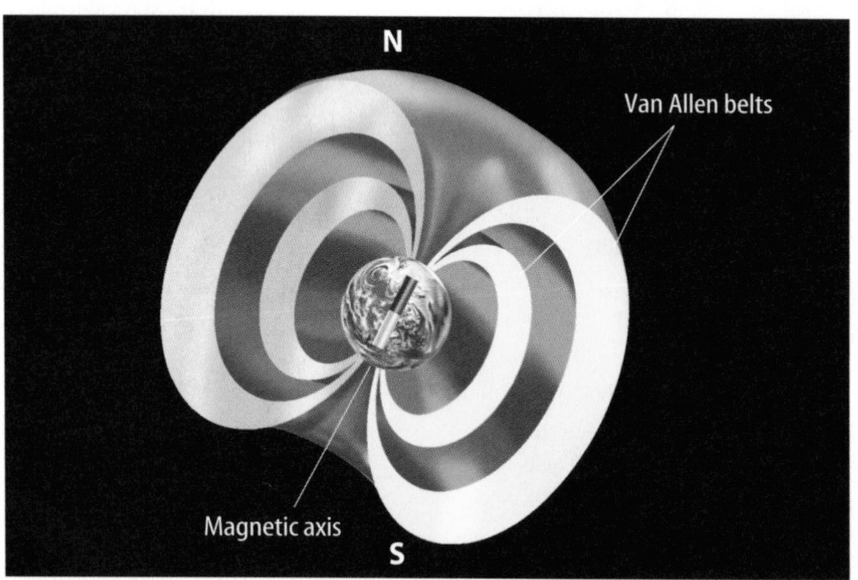

Figure 2 Like a common bar magnet, Earth also has north and south magnetic poles. The inner and outer gold shells respectively represent the positive and negative Van Allen Belts.
Explain *why the Van Allen Belts are shaped as they are.*

Earth Orbits the Sun

Earth orbits the Sun at an average distance of 149,600,000 km. Its orbit, like those of all the planets, moons, asteroids, and many comets, is shaped like an ellipse. An **ellipse** is an elongated, closed curve with two foci. The Sun is not located at the center of the ellipse, but at one of its two foci. This means the distance from Earth to the Sun varies during the year. Earth is closest to the Sun—about 147 million km away—around January 3 and is farthest from the Sun—about 152 million km away—around July 4 of each year.

Earth as a Planet Earth is a planet, just as Venus, Mars, and Jupiter are planets. However, Earth is the only planet whose characteristics make it possible for life as we know it to survive.

Earth resembles Venus more than any other planet. Earth and Venus are nearly the same size, and both have atmospheres that contain carbon dioxide, although in greatly different amounts. Earth's oceans, shown in **Figure 3,** absorbed much of the carbon dioxide in Earth's early atmosphere. Also, Venus's atmosphere is much denser than Earth's, with pressures as high as those encountered by submarines in Earth's oceans at depth of 900 m.

Another difference is the surface temperatures. On Earth, you can walk outside and feel how cold or warm it is. However, temperature on Venus is over 450°C. This high temperature is caused by the large amount of carbon dioxide in Venus's atmosphere, which traps heat energy and prevents it from escaping. Think about what life on Earth might be like with more carbon dioxide in the atmosphere.

Figure 3 By absorbing carbon dioxide, oceans protect Earth from experiencing a greenhouse effect like that on Venus.

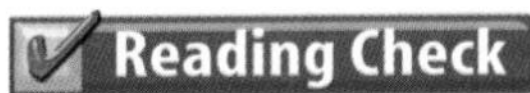

What feature of Earth's surface has led to such a difference between Earth and Venus?

Although Mars is almost half the size of Earth, its surface gravitational pull is less than two-fifths that of Earth's. Yet conditions there are more like those on Earth than any other planet in the solar system. Mars may even have frozen water near its surface. Mercury is very different from Earth; it has no atmosphere, and is cratered like Earth's moon.

section 1 review

Summary

Earth's Size and Shape

- Earth has a spherical shape.
- Gravity causes very large objects in space to form spheres.

Earth's Magnetic Field

- Earth's magnetic field protects life on Earth's surface from harmful radiation.
- Interaction of the solar wind with Earth's magnetic field produces the aurora.

Earth Orbits the Sun

- Earth orbits the Sun in an elliptical orbit.
- Earth and Venus are similar in size, and both have atmospheres that contain carbon dioxide.

Self Check

1. **Identify** two pieces of evidence that prove Earth's spherical shape.
2. **Define** the term *gravity.*
3. **Explain** what produces Earth's magnetic field.
4. **Describe** how Venus and Earth are similar.
5. **Think Critically** Evidence indicates that Mars once had liquid water on its surface. Discuss some possible reasons why it has none today.

Applying Math

6. **Calculate Speed** Earth's circumference at the equator is 40,075 km. If it spins once each day, what is the speed of spinning in km/h?

More Section Review gpescience.com

section 2

Time and Seasons

Reading Guide

What You'll Learn

- **Calculate** the time and date in different time zones.
- **Differentiate** between revolution and rotation.
- **Discuss** what causes seasons to change.

Why It's Important

Days and years are measurements of Earth's movements.

Review Vocabulary

latitude: angular distance north or south of Earth's equator

New Vocabulary

- time zone
- rotation
- revolution
- ecliptic
- equinox
- solstice

Measuring Time on Earth

People can determine the approximate time of day by determining where the Sun is in the sky. If the Sun is near an imaginary line drawn from due north to due south, it is about 12:00 noon. Humans have used movements of Earth, the Moon, and the Sun to measure time for thousands of years.

Around 3000 B.C., the Babylonians devised a method of timekeeping using their counting methods, which were based on 60. They noticed that the Sun appeared to take a circular path through the sky. Because their counting methods were based on 60, they divided a circle into 360 parts called degrees. The symbol for degree (°) was taken from their symbol for the Sun.

Earth Movements Measure Time Earth spins and makes one complete turn in about 24 hours, as shown in **Figure 4.** This spinning causes the Sun to appear to move across the sky from east to west. It takes 24 hours from when the Sun is highest in the sky (noon) until it is highest in the sky again (noon the next day).

If Earth spins approximately 360° in 24 hours, then it spins through 15° in one hour. This led to the setting up of time zones on Earth that have the same time in minutes but vary in hours. A **time zone** is an area 15° wide in which the time is the same. **Figure 5** shows the time zones in the U.S. Ideally, time zones should be equal in size and follow lines drawn from the north pole to the south pole. However, for convenience, time zones are modified to fit around city, state, and country borders, and other key sites.

Figure 4 The sunlit side of Earth is day; the shadow side is night.

Reading Check *How many degrees does Earth spin in one hour?*

The Date Line You can see that a problem would quickly arise if you just kept dropping back an hour earlier for each 15°. Eventually, you would come around Earth and it would be 24 hours earlier. It cannot be two different days at the same spot, so a day is added to the time at the International Date Line. If it is Monday to the east of the date line, then it is the same hour on Tuesday to the west of the date line. This line is drawn down through the Pacific Ocean (around islands, such as New Zealand) directly opposite the Prime Meridian, the starting point for this worldwide system of measuring time. The Prime Meridian is an imaginary line drawn on Earth that passes through Greenwich, England. Time based on this method is called Coordinated Universal Time (UTC). In some areas, this time is modified in summer so that there are more hours of daylight in the evening. This is referred to as Daylight Saving Time (DST). Some areas apply local modifications to this system as well.

Figure 5 The globe is divided into 24 time zones. Lines of longitude roughly determine the locations of time zone boundaries. Notice that each successive time zone to the west is one hour earlier.
Think Critically *If you leave Russia at 12:30 A.M. on Tuesday and fly east for one hour across the Bering Strait, what time and day will you arrive in Alaska?*

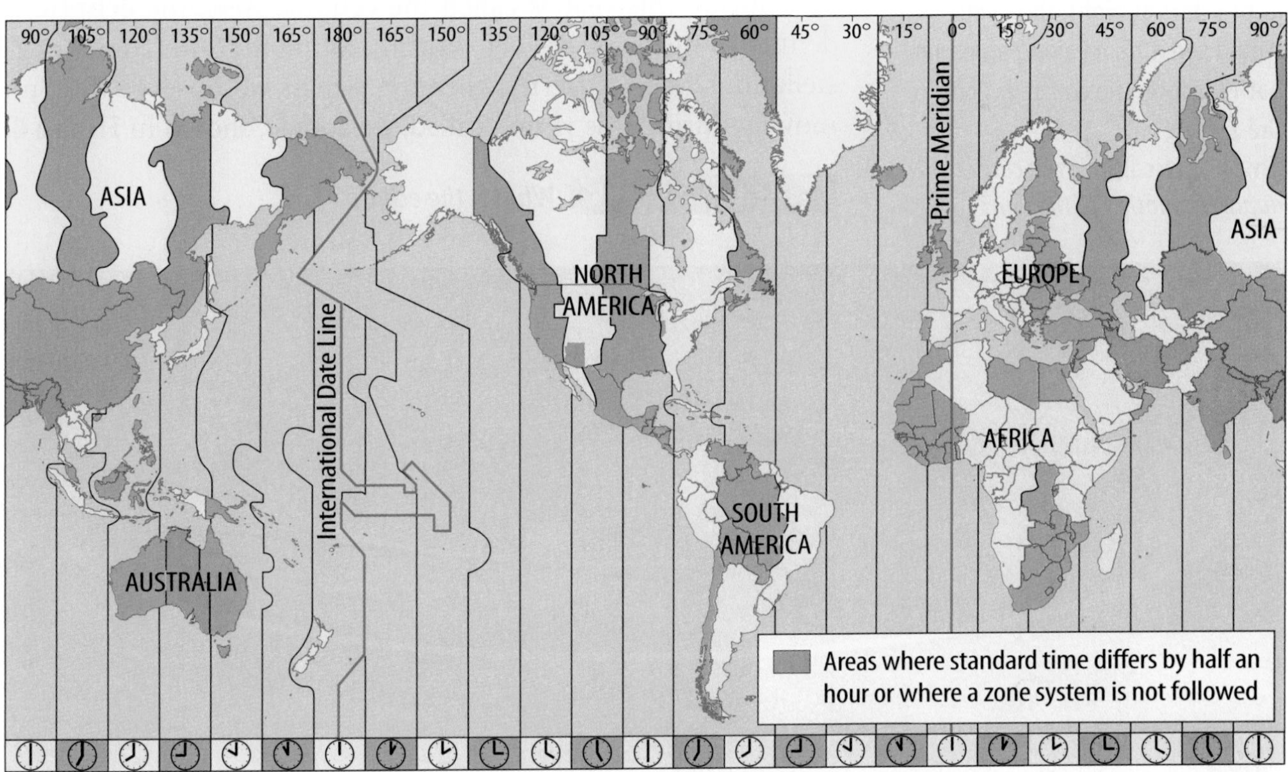

Rotation Measures Days The spinning motion of Earth enables you to measure the passing hours of the day. **Rotation** is the spinning of Earth on its axis, an imaginary line drawn through Earth from its rotational north pole to its rotational south pole.

The apparent movement of the Sun from noon one day until noon the next day is called a solar day. This period is a bit longer than the time it takes Earth to rotate on its axis, however. This is because while Earth rotates, it also moves in orbit around the Sun and must rotate a bit more each day to make the Sun reach noon. However, if you measure time based on when a certain star rises above the horizon until it rises again, you will see a slightly shorter time period (23 h 56 m 4 s). This is called a sidereal day and is the true measure of the time it takes for Earth to rotate once on its axis.

Revolution Measures Years The motion of Earth around the Sun enables you to measure the passing of years. **Revolution** is the motion of Earth in its orbit around the Sun. **Figure 6** shows Earth's orbit around the Sun. As Earth revolves in its orbit, the Sun appears to move through the skies compared to the seemingly fixed positions of the stars.

The time it takes for the Sun to make one complete trip through the sky in reference to the background of stars is the same amount of time it takes for Earth to complete one trip around the Sun, or one sidereal year. The apparent path of the Sun during this year is called the ecliptic. Also, the **ecliptic** is defined as the plane of Earth's orbit around the Sun. The 12 constellations (star patterns) through which we observe the Sun moving during this year is called the zodiac, shown in **Figure 6.**

Reading Check *What is the ecliptic?*

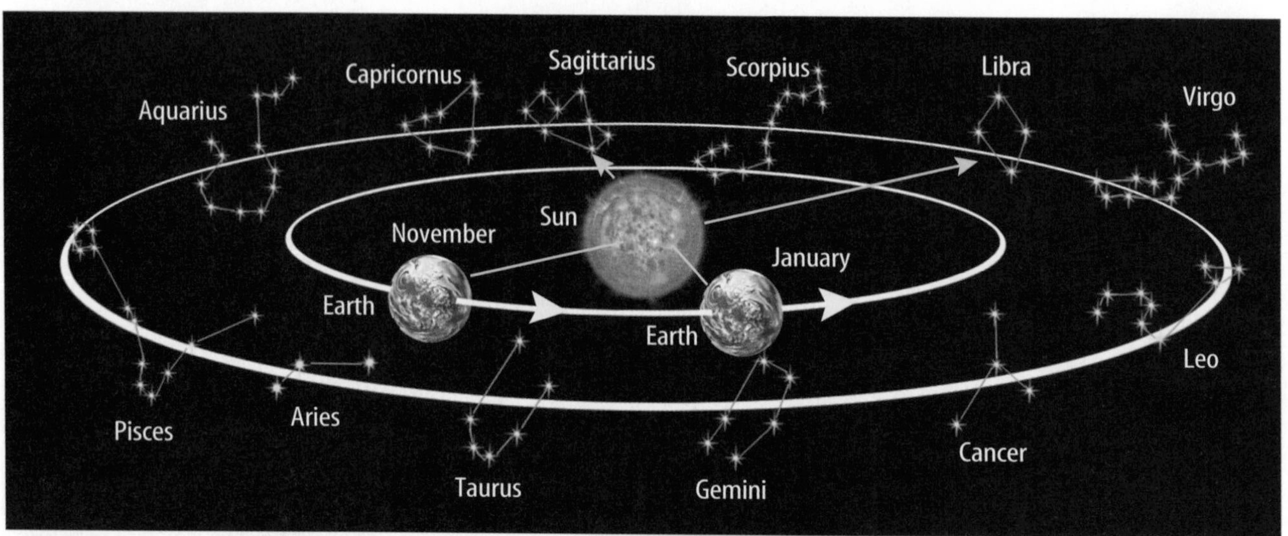

Figure 6 If the Sun were as faint as the stars at night, then you would see it travel along an annual path through the constellations of the Zodiac.
Think Critically *What causes this apparent motion of the Sun?*

Why do seasons change?

Recall that Earth's orbit around the Sun is an ellipse. This means that Earth is closer to the Sun at one time than it is at other times. Is this the cause of seasonal changes on Earth? Because Earth is closest to the Sun in January, you would expect this to be the warmest month. However, you know this isn't true in the northern hemisphere; something else must be causing the change. These seasonal changes are caused by Earth's rotation, its revolution, and the tilt of its axis.

Seasons change on Earth because the number of hours of daylight each day varies and also because the angle at which sunlight strikes Earth's surface varies at different times of the year. Earth's axis is tilted 23.5° from a line drawn perpendicular to the plane of its orbit, or ecliptic. Because of this tilt, Earth's north geographic pole points toward Polaris throughout the year. Later, you will learn how this tilt, along with Earth's revolution, causes the seasons.

Changing Angle of Sunlight During the summer, the Sun is higher in the sky, and sunlight hits Earth's surface at a higher angle. As the year progresses, the Sun is lower and lower in the sky, and sunlight strikes Earth's surface at lower angles. When striking Earth's surface at higher angles, approaching 90°, sunlight is more intense and warms Earth's surface more than when it strikes the surface at lower angles. Because Earth remains tilted in the same direction as it revolves, different hemispheres are tilted toward the Sun at different times of the year. As shown in **Figure 7,** the hemisphere tilted toward the Sun receives sunlight at higher angles than the hemisphere tilted away from the Sun. The greater intensity of sunlight is one reason why summer is warmer than winter, but it is not the only reason. Another factor is involved.

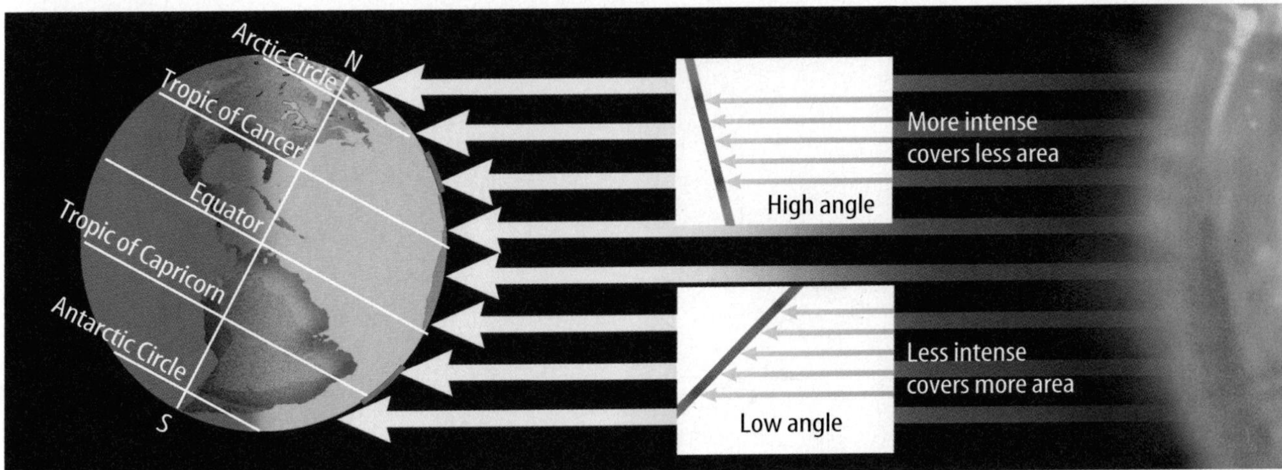

Figure 7 The Sun's rays strike Earth's surface at higher angles in the northern hemisphere when the north pole is tilted toward the Sun.

More Hours of Daylight in Summer During the summer, the Sun is above the horizon for more hours than it is when school begins in the fall. As the year progresses, the number of hours of daylight each day becomes fewer and fewer until it reaches a minimum around December 21 for the northern hemisphere. When do you think the number of hours of daylight would be at a maximum in the northern hemisphere? This happens six months later, around June 21. As shown in **Figure 8,** the hemisphere of Earth that is tilted toward the Sun receives more hours of daylight each day than the hemisphere tilted away from the Sun. This longer period of daylight is the second reason why summer is warmer than winter.

Reading Check ***During which month does Earth's northern hemisphere experience more hours of daylight?***

Figure 8 During the winter solstice for the northern hemisphere, the Sun's rays strike Earth perpendicular at the Tropic of Capricorn, while the area within the arctic circle remains in darkness. During the summer solstice, the Sun's rays are perpendicular to Earth at the Tropic of Cancer, while the area within the arctic circle remains in sunlight.

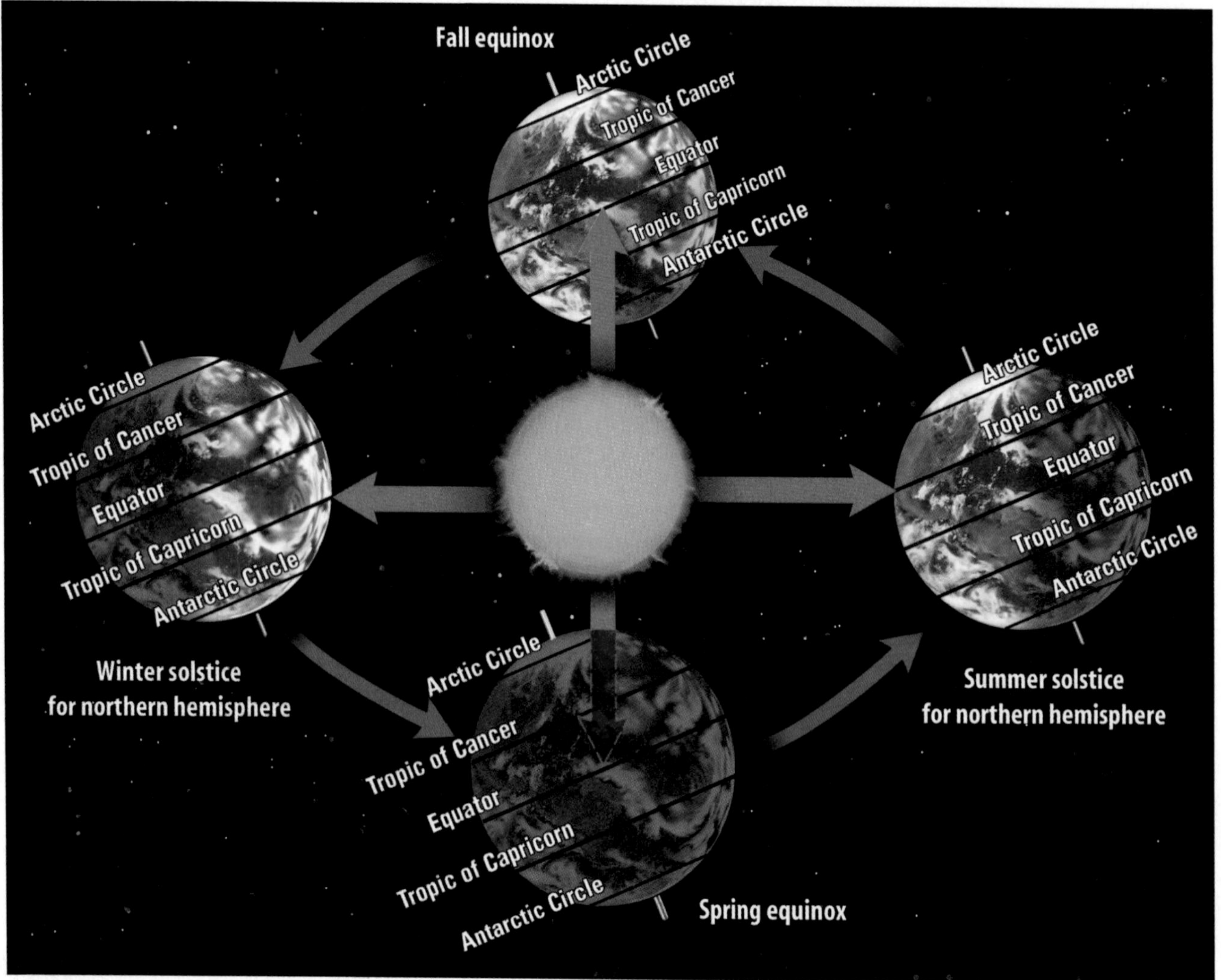

Equinoxes and Solstices Because of the tilt of Earth's axis, the Sun's position relative to Earth's equator constantly changes. Most of the time, the Sun is north or south of the equator, but two times during the year, the Sun is directly over the equator. **Figure 8** shows that the Sun reaches an **equinox** when it is directly above Earth's equator, and the number of daylight hours equals the number of nighttime hours all over the world. The term *equinox* is derived from two words meaning "equal" and "night." At that time, neither the northern nor the southern hemisphere is tilted toward the Sun. In the northern hemisphere, the Sun reaches the spring equinox on March 20 or 21, and the fall equinox on September 22 or 23. In the southern hemisphere, the equinoxes are reversed.

The **solstice** is the point at which the Sun reaches its greatest distance north or south of the equator the Tropic of Cancer and Tropic of Capricorn, respectively. The term *solstice* is derived from the Sun's name, *Sol*, and a Latin word meaning "standing"; that is, it appears to stand, or stop moving, north or south in the sky. In the northern hemisphere, the Sun reaches the summer solstice on June 21 or 22, and it reaches the winter solstice on December 21 or 22. Just the opposite is true for the southern hemisphere. When the Sun is at the summer solstice, there are more hours of daylight than during any other day of the year, and the Sun's rays strike at a higher angle. When it's at the winter solstice, on the shortest day of the year, the most nighttime hours occur and the Sun's rays strike at the lowest angle.

Modeling the Sun's Rays at Solstice

Procedure

1. Use a **globe** with the equator, the Tropic of Cancer, and the Tropic of Capricorn indicated.
2. Set up a light source, such as a **flashlight or gooseneck lamp,** so light shines vertically at Earth's equator.
3. Tilt the globe 23.5° from vertical so that first the northern and then the southern hemisphere is tilted toward the light.

Analysis

1. When the globe was tilted 23.5° toward and away from the light, what latitudes received vertical rays?
2. What areas of Earth never received vertical rays?

section 2 review

Summary

Measuring Time on Earth

- Humans use movements of Earth, the Moon, and the Sun to measure time.
- Earth's rotation is used to measure days.
- Earth's revolution is used to measure years.
- The ecliptic is the Sun's apparent yearly path through the zodiac.

Why do seasons change?

- Earth's seasonal changes are caused by its tilt, rotation, and revolution.
- Equinoxes occur when the Sun is directly over Earth's equator. Solstices occur when the Sun reaches it greatest distance north or south of the equator.

Self-Check

1. **Determine** how long it takes Earth to make one complete turn on its axis.
2. **Explain** why each time zone contains 15° of longitude. If it is 4:15 P.M. in one time zone, what is the time two time zones to the west?
3. **Explain** how a solar day differs from a sidereal day.
4. **Compare and contrast** rotation and revolution.
5. **Think Critically** Why does Earth's surface become warmer in summer than it does in winter?

Applying Math

6. **Use Decimals** It takes Earth about 365.25 days to make one trip around the Sun. As it does this, Earth travels 360° in its orbit. On average, how many degrees does Earth travel each day?

More Section Review gpescience.com

Comparing the Angle of Sunlight to Intensity

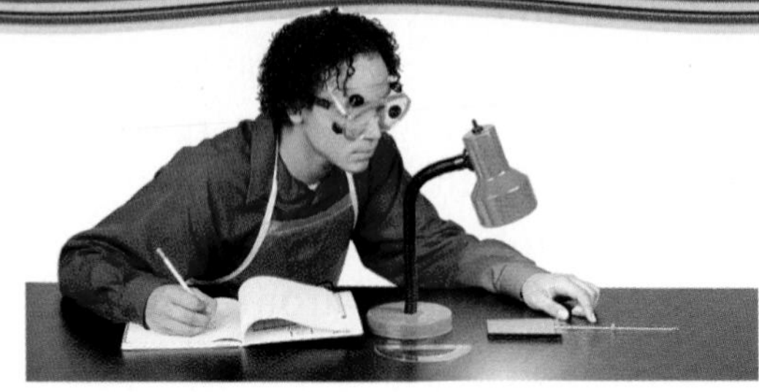

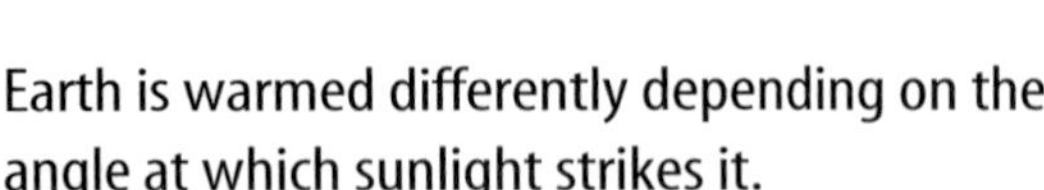

Earth is warmed differently depending on the angle at which sunlight strikes it.

Real-World Problem

How can you a model the angle at which sunlight strikes Earth's surface?

Goals

- **Model** different angles at which sunlight strikes Earth's surface.
- **Compare and contrast** the amount of heat generated by light striking at different angles.

Materials

75-W bulb in a gooseneck lamp
alcohol thermometers (2)
sheets of construction paper, one color (2)
protractor
** unshaded 75-W lamp*
** books to change the angle of the thermometers*
**Alternate materials*

Safety Precautions

WARNING: *Do not touch lamp or lightbulb without safety gloves. They stay hot after being turned off. Handle thermometers carefully.*

Procedure

1. Copy the data table shown on this page.
2. Label the thermometers *T-1* and *T-2* and record their temperatures in the data table.
3. Fold the construction paper to form a pocket that will conceal the thermometer's bulb.
4. Place *T-1* in the pocket and lay them on a desktop. Turn on the lamp. Position the lamp so that light strikes the pocket at an angle of 75°.
5. Record the temperature of *T-1* at 10 min and at 20 min.
6. Repeat steps 3, 4, and 5 using *T-2*, but aim the lamp at an angle of 20°.

Conclude and Apply

1. **Compare and contrast** the temperature readings of each thermometer.
2. **Infer** which angle models the Sun's position during the summer and during the winter.
3. **Explain** how changes in the angle at which sunlight strikes Earth's surface are one cause of Earth's changing seasons.

Data Table

Thermometer	Original Temperature	Temperature at 10 min	Temperature at 20 min
T-1			
T-2			

Communicating Your Data

Compile your classmates' data. Find the average temperatures—original, at 10 min, and at 20 min—for *T-1* and *T-2*. Compare and contrast your results with the class averages.

section 3

Earth's Moon

Reading Guide

What You'll Learn

- **Describe** how tides on Earth are caused by the Moon.
- **Explain** how the Moon's phases depend on the relative positions of the Sun, the Moon, and Earth.
- **Compare and contrast** solar and lunar eclipses.
- **Analyze** what surface features of the Moon reveal about its history.

Why It's Important

The Moon is our nearest neighbor in space and affects Earth in many ways.

Review Vocabulary

lava: molten rock

New Vocabulary

- tide
- moon phase
- solar eclipse
- lunar eclipse
- maria
- regolith

Movement of the Moon

You have seen the Moon move across the sky from east to west, just like the Sun. This is an apparent movement like the Sun's, caused by Earth's rotation. But, the Moon actually does move in another way. If you look at the Moon each day at the same time over a period of a few days, you will see that it moves toward the east.

Rotation and Revolution This eastward movement of the Moon is an actual movement that is caused by the Moon's revolution in its orbit. It takes 27.3 days (a sidereal month) for the Moon to revolve once around Earth and line up with the same star again. Because Earth also revolves around the Sun, it takes more than two more days for the Moon to line up with Earth and the Sun again. This means that a complete lunar phase cycle takes 29.5 days, known as a synodic month.

Many people think the Moon does not rotate because it always keeps the same side facing Earth. This is not true. As shown in **Figure 9,** the Moon keeps the same side facing Earth because it takes 27.3 days to rotate once on its axis—the same amount of time that it takes to revolve once around Earth. You can observe this by having a friend move the ball around you while keeping the same side of it facing you. You will see only one side.

Figure 9 The face of the "man in the moon" is always facing Earth. **Explain** *why the same side of the Moon always faces Earth.*

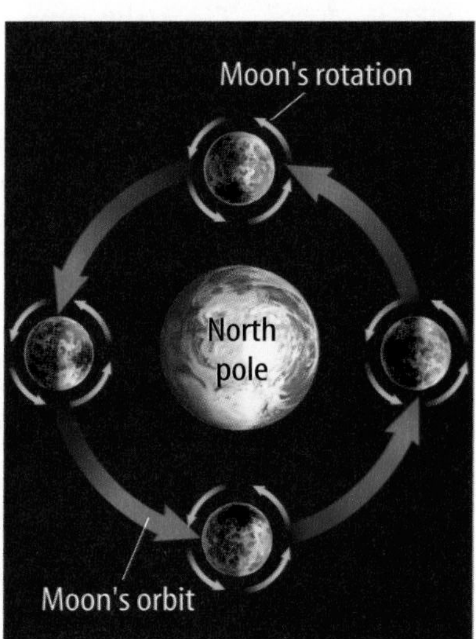

How does the Moon affect Earth?

The Moon affects Earth in many ways, some obvious and others less so. If you have ever been to a beach for vacation, you realized that the water is not always at the same location on the beach. Sometimes the water comes farther up the beach than at other times. Have you ever placed your towel on the beach and come back later to find it wet because the tide came in? You also may have noticed that the Moon doesn't look the same each evening. Sometimes you can see all of the side facing Earth, while at other times, you can barely see any of it. Let's take a look at these and some other effects of the Moon on Earth.

Applying Math — Solve a Simple Equation

TIDES The Moon rises an average of 52.7 min later each day. If the time of high tide is known for one day, this formula can be used to determine when high tide will occur on the next day or any successive day.

$$T_N = T_0 + N \times 52.7 \text{ min}$$

In this formula, T_0 is the original time of high tide on a given day and T_N is the time of high tide on any successive day. N is the number of days later for which you wish to determine the time of high tide. If the tide is high at 1:00 P.M., find out what time it will be high in 7 days.

IDENTIFY known values and unknown values

Identify the known values:

T_0 = original time of high tide = 1:00 P.M.

N = the number of days later for which you wish to determine high tide = 7

52.7 min = how much later the Moon rises each day

Identify the unknown value:

T_N = the time of high tide on N number of days

SOLVE the problem

Substitute the known values into the equation for time.

T_N = 1:00 P.M. + 7 × 52.7 min = 1:00 P.M. + 369 min

T_N = 1:00 P.M. + 6 h 09 min = 7:09 P.M.

Practice Problem

Low tide is the best time to hunt for seashells. If you see that the tide is low at noon on Thursday, when during the day will it be low on the following Sunday?

For more practice problems, go to page 879 and visit Math Practice at gpescience.com.

Tides Think again of the beach and how the level of the sea rose and fell during the day. This rise and fall in sea level is called a **tide**. A tide on Earth is caused by a giant wave produced by the gravitational pulls of the Sun and the Moon. This wave has a wave height of only 1 or 2 m, but it has a wavelength of thousands of kilometers. As the crest of this large wave approaches the shore, the level of the water in the ocean rises. This rise of sea level is called high tide. About six hours later, as the trough of the wave approaches, sea level drops, causing a low tide.

Earth and the Moon both revolve around their common center of mass located about 1,700 km below Earth's surface. Because Earth is much more massive, the Moon does most of the moving, and it seems to us as though the Moon were revolving around Earth. This center of mass, in turn, revolves around the Sun. This is why Earth, Moon, Sun are considered as a three-body system.

As Earth rotates and the Moon revolves, different locations on Earth's surface pass through the high and low tides. Although the Sun is much more massive than the Moon, it also is much farther away. Because of this, the Moon has a greater effect on Earth's tides than does the Sun. However, the Sun does affect Earth's tides: it can strengthen or weaken the tidal effect. When the Moon and the Sun pull together, when they are lined up, high tides are much higher and low tides are much lower. This is called a spring tide, as shown in **Figure 10.** However, when the two are at right angles to each other, the high tide is not as high and the low tide not as low, producing a neap tide.

Reading Check *What happens to the sea level at spring tide?*

Moonlight

The Moon shines because it reflects sunlight from its surface. Just as half of Earth experiences day as the other half experiences night, half of the Moon is lighted while the other half is dark. As the Moon revolves around Earth, different portions of the side facing Earth are lighted, causing the Moon's appearance to change. **Moon phases** are the changing appearances of the Moon as seen from Earth. The phase you see depends on the relative positions of the Moon, Earth, and the Sun.

Phases of the Moon A new moon occurs when the Moon is between Earth and the Sun. During a new moon, the side of the Moon facing away from Earth is lighted and the side of the Moon facing Earth receives no light from the Sun. The Moon is in the sky, but it cannot be seen, except for a special alignment you will learn about later.

Figure 10 Earth's tides are an example of how the Sun, Moon, and Earth pull on each other and operate as a three-body system.
A The Sun, Earth, and Moon are in alignment during spring tide.
B The Sun, Earth, and Moon form a right angle during neap tide.
Identify *whether high tide is higher during spring tide or neap tide.*

A

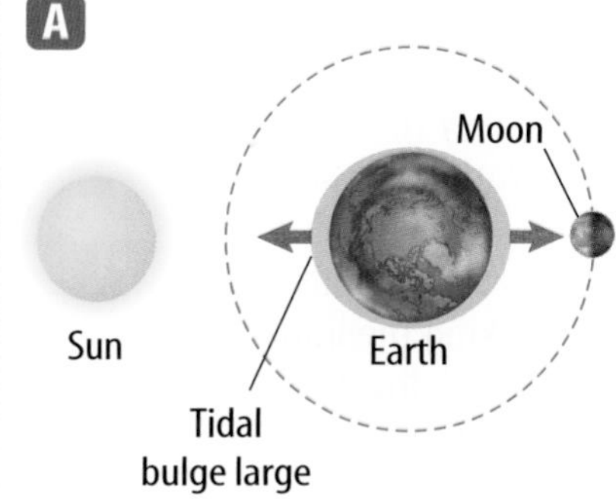

B

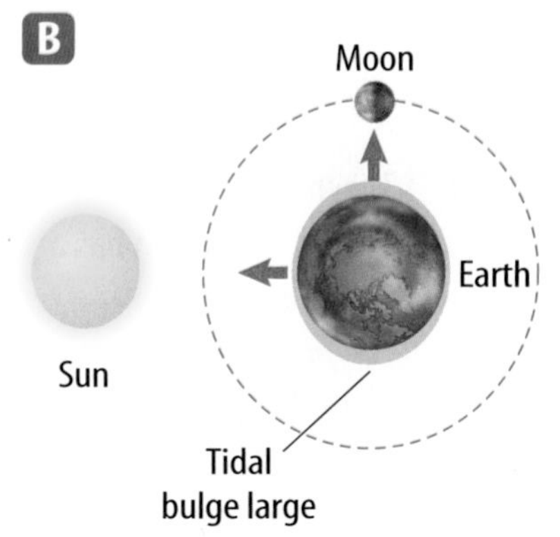

Modeling Phases and Eclipses

Procedure

1. Turn on a **lamp with no shade** and place a **small, white, plastic-foam ball** on the end of a **pencil.**
2. Stand facing the lamp and hold the white plastic-foam ball between your head and the lamp.
3. Slowly move the ball counterclockwise around your head and observe how much of the side of the ball facing you is lighted at different positions.
4. Note positions where the ball blocks light from the lamp—or moves into the shadow cast by your head.

Analysis

1. Describe what happened to the ball as it was moved around your head, and identify the moon phases at various positions.
2. At which positions (phases) was the ball blocking the light or falling into shadow? At which phase(s) can eclipses occur?

Try at Home

Waxing Phases After a new moon, the moon's phases are said to be waxing—the lighted portion that we see appears larger each night. The first phase we see after a new moon is called the waxing crescent. About a week after a new moon, we see one-half of the Moon's lighted side, or one-quarter of the Moon's surface. This phase is the first-quarter.

The moon is in the waxing gibbous phase from the first quarter up until full moon. A full moon occurs when we see all of the Moon's lighted side. At this time, the Moon is on the side of Earth opposite from the Sun.

Waning Phases After a full moon, the lighted portion that we see begins to appear smaller. The phases are said to be waning. When only half of the side of the Moon facing Earth is lighted, the third-quarter phase occurs. The waning crescent occurs before another new moon. Only a small slice of the side of the Moon facing Earth is lighted.

The word *month* is derived from the same root word as *Moon.* The complete cycle of the Moon's phases, shown in **Figure 11,** takes about 29.5 days, or one synodic month. Recall that it takes about 27.3 days for the Moon to revolve around Earth. The discrepancy between these two numbers is due to Earth's revolution around the Sun. It takes the Moon a little over two days to "catch up" with Earth's advancement around the Sun.

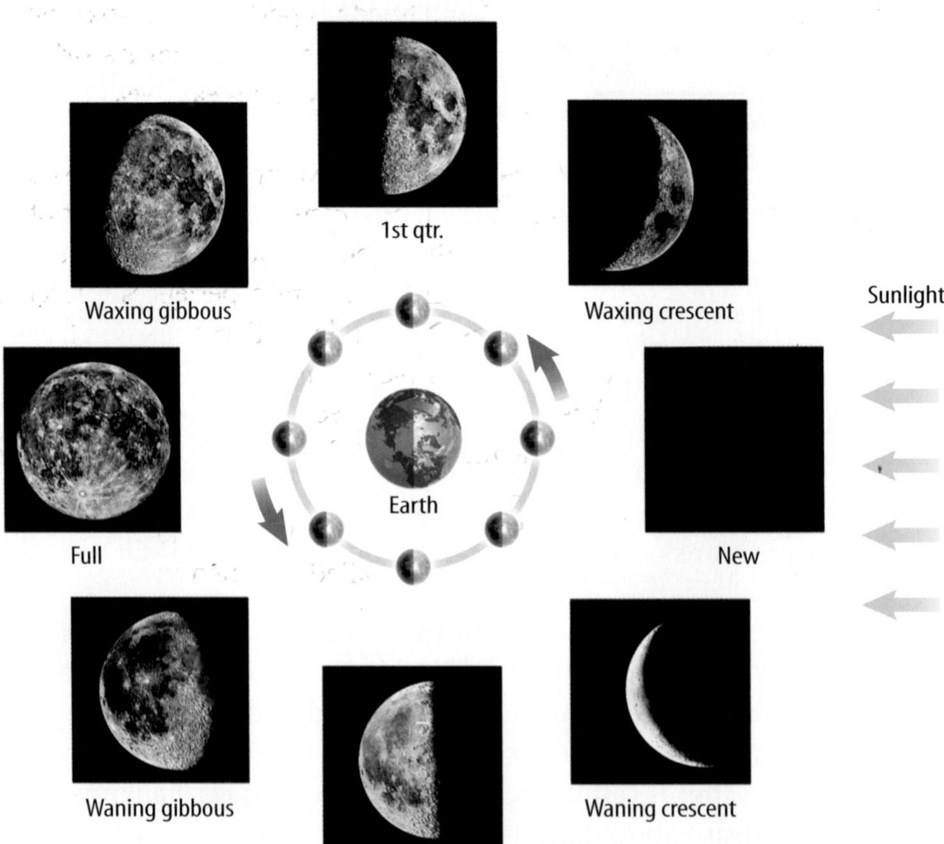

Figure 11 When viewed from Earth's north pole, the Moon has a counterclockwise orbit.
Infer *In this figure, the Sun's rays are coming from the right. Why are the Moon's waning phases showing sunlight on the left?*

Eclipses

If you knew nothing about what the Sun is or why it produces so much light and heat, wouldn't you be concerned if suddenly it darkened? Think of how humans from long ago must have reacted when the Moon passed in front of the Sun and the source of light and heat was blocked, as it is during an eclipse.

In the year 585 B.C., a battle was raging between the armies of the Lydians and the Persians when suddenly, the Sun was eclipsed by the Moon. The two armies were so stunned by the event that they put down their weapons and stopped fighting.

Now we understand what causes eclipses of both the Sun and the Moon. We know that for the Moon to block out the Sun, it must appear to be the same size. In fact, both the Sun and the Moon have apparent diameters that are almost the same, about 0.5°. If it weren't for this, a total eclipse of the Sun might never happen. Because the Sun is about 400 times larger than the Moon, it also must be about 400 times farther from Earth for a total solar eclipse to occur.

Eclipses occur when Earth or the Moon temporarily blocks sunlight from reaching the other object. Sometimes, during a new moon, a shadow cast by the Moon falls on Earth, causing a solar eclipse. During a full moon, a shadow of Earth can be cast on the Moon, resulting in a lunar eclipse. Eclipses can occur only when the Sun, the Moon, and Earth are lined up perfectly. Because the Moon's orbit is tilted about 5° from the plane of Earth's orbit around the Sun, eclipses happen only a few times each year.

Solar Eclipses A **solar eclipse** occurs when the Moon moves directly between the Sun and Earth and casts a shadow on part of Earth. The darkest portion of the Moon's shadow is called the umbra. A person standing within the umbra experiences a total solar eclipse. As shown in **Figure 12,** the only portion of the Sun that is visible during a total eclipse is part of its atmosphere, which appears as a pearly white glow around the edge of the eclipsing Moon. This is the only time the entire disk of the new moon phase can be photographed—it appears black against the Sun.

As shown in **Figure 13,** surrounding the umbra is a lighter shadow on Earth's surface called the penumbra. Persons standing in the penumbra experience a partial solar eclipse. **WARNING:** *Regardless of where you are standing, never look directly at a solar eclipse. The light can permanently damage your eyes.*

How are Earth, the Moon, and the Sun aligned during a solar eclipse?

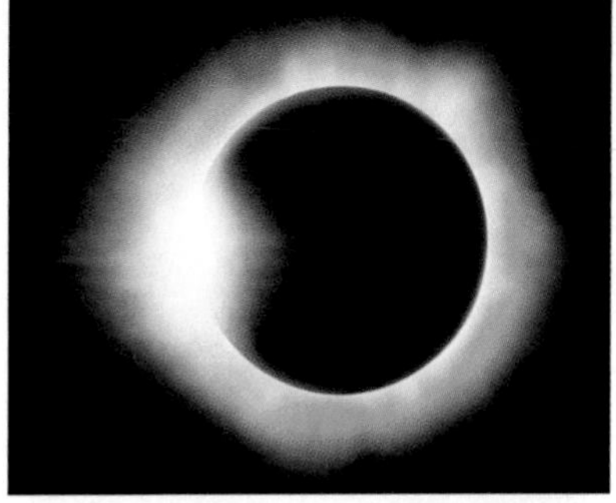

Figure 12 The solar corona can be seen as a pale glow around the lunar disk. Sunlight shining through lunar valleys produces a diamond-ring effect.

Early Civilizations
Celestial objects were studied by early civilizations. Some hypotheses proposed by the Babylonians and the early Greeks were close to reality, but other ideas were wrong. Research how some early civilizations explained eclipses and other astronomical observations and report your findings to your class using drawings and diagrams.

Figure 13 A total solar eclipse appears only within the Moon's umbra.

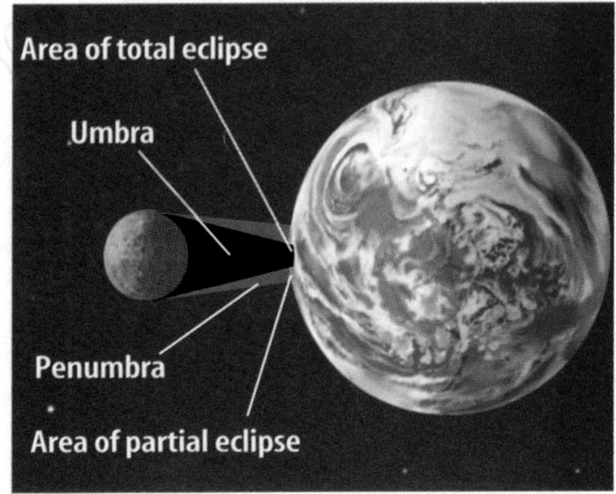

Figure 14 Sometimes during a partial lunar eclipse, Earth's curvature can be seen silhouetted on the Moon. The red coloration is caused by the refraction of sunlight passing through Earth's atmosphere before reaching the Moon's surface.

Lunar Eclipses When Earth's shadow falls on the Moon, a **lunar eclipse** occurs. A lunar eclipse begins when the Moon moves into Earth's penumbra. As the Moon continues to move, it enters Earth's umbra, and you see a curved shadow on the Moon's surface, as shown in **Figure 14.** It was this shadow that led Aristotle to conclude that Earth is spherical. When the Moon moves completely into Earth's umbra, a total lunar eclipse occurs, as shown in **Figure 15.** The Moon sometimes becomes red during an eclipse because light from the Sun is scattered and refracted by Earth's atmosphere. Longer wavelength red light is affected less than shorter wavelengths, so more red light falls on the Moon.

A partial lunar eclipse occurs when only a portion of the Moon moves into Earth's umbra. The remainder of the Moon is in Earth's penumbra and, therefore, receives some direct sunlight. A partial lunar eclipse also occurs when the Moon is partially or totally within Earth's penumbra. However, this can be difficult to see because some direct sunlight falls on the Moon, making it appear only slightly dimmer than usual.

A total solar eclipse can occur as often as twice a year, yet most people live their entire lives without witnessing one. You may never see a total solar eclipse, but it is almost certain you will have a chance to see a total lunar eclipse. The reason why it is so rare to view a total solar eclipse is that only those people in the small region where the Moon's umbra strikes Earth can see one and, even then, there must be clear skies. In contrast, the opportunities to witness lunar eclipses are much more frequent, and anyone on the night side of Earth can see them.

Reading Check *How are Earth, the Moon, and the Sun aligned during a lunar eclipse?*

Figure 15 A total lunar eclipse occurs when the Moon is entirely within Earth's umbra. *Umbra* is the Latin word for "shadow." Just as a peninsula is almost an island, a penumbra is almost a shadow.
Research *What does* umbrella *mean?*

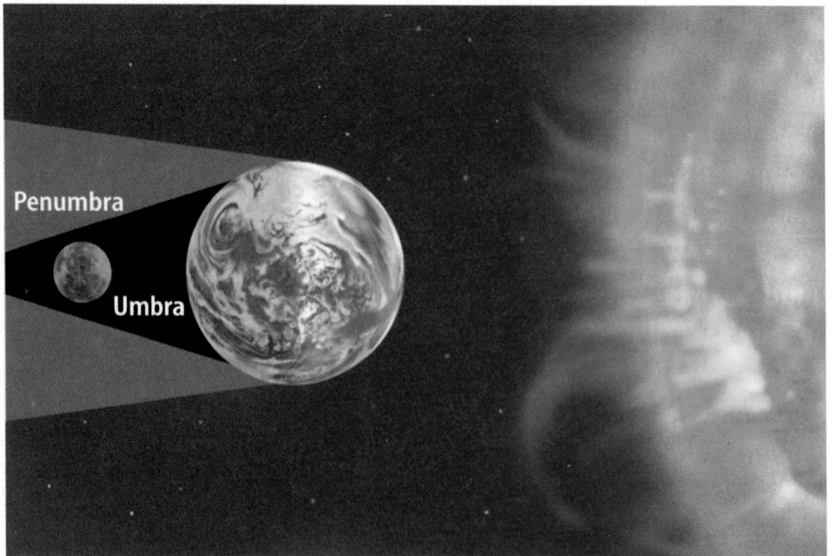

The Moon's Surface

When you look at the Moon, as shown in **Figure 16,** you can see many of its larger surface features. Craters, rays, mountains, and maria can easily be seen through a small telescope or a pair of binoculars. What are these different features, how did they form, and what do they tell us about the Moon's history and interior?

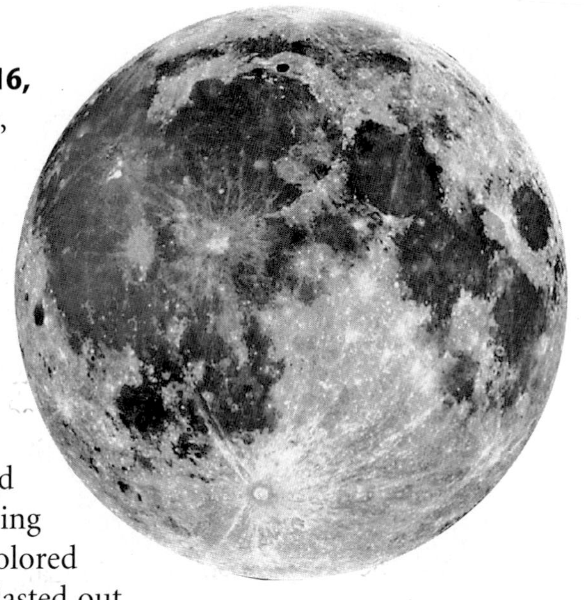

Figure 16 Notice the light-colored material radiating outward from the large crater near the base of the Moon.
Identify *What are the flat, dark-colored regions called? What are the light-colored areas radiating from the craters called?*

Craters, Maria, and Mountains Many depressions on the Moon were formed by meteorites, asteroids, and comets, which strike the surfaces of planets and their satellites. These depressions, which are called craters, formed early in the Moon's history. Surrounding many craters are ray patterns produced by lighter-colored material from just below the lunar surface that was blasted out on impact and settled on top of the darker surface material around the craters. During the impact, when these large basins formed, cracks may have formed in the Moon's crust, allowing lava from the still-molten interior to reach the surface and fill in the basins, forming maria.

Maria are the dark-colored, relatively flat regions on the Moon's surface, shown in **Figure 16.** The igneous rocks of the maria are 3 to 4 billion years old. They are the youngest rocks found on the Moon so far. This indicates that the craters formed after the Moon's surface originally cooled. However, the maria formed early enough in the Moon's history that molten rock material still remained in the Moon's interior.

Surrounding the large depressions that later filled with lava are areas that were thrown upward in the original collision and formed mountains. The largest mountain ranges on the Moon surround the large, flat, dark-colored maria.

Regolith When NASA scientists started to plan for crewed spacecraft to land on the Moon, they were concerned about whether the lunar surface would be able to support the craft? To find out, unmanned *Surveyor* spacecraft were landed on its surface. One *Surveyor* craft actually bounced a few times as it landed on the side of a crater. What was this material that the spacecraft had landed on?

Impacts on the Moon thoughout its history led to the accumulation of debris known as **regolith.** On some areas of the Moon, this regolith is almost 40 m thick, while in other locations, it is only a few centimeters thick. Some regolith is coarse, but some is a fine dust. If you watch astronauts walking on the Moon, you will notice that they often kick up a lot of dust.

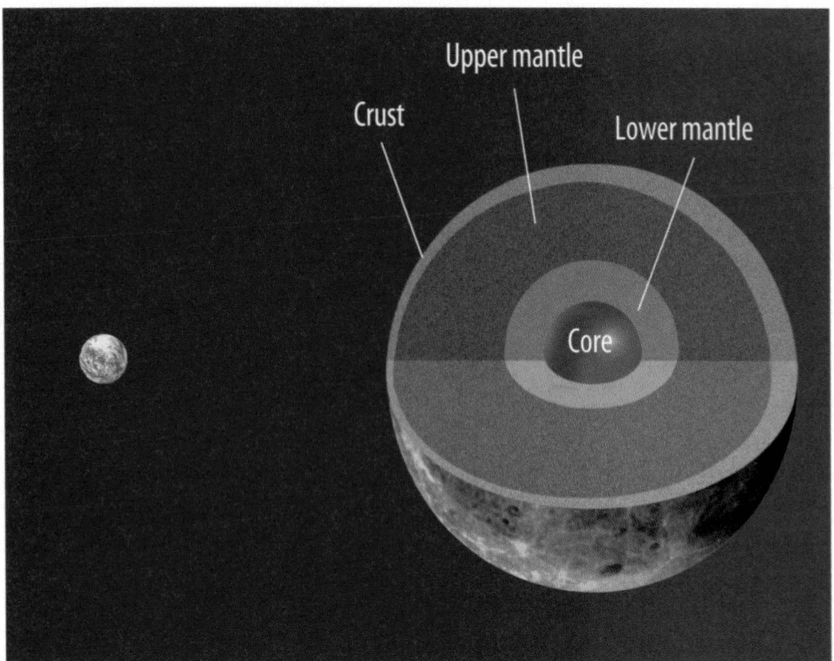

Figure 17 The Moon's crust is thinnest on the side nearest to Earth.
Infer *Why is the Moon's crust thinnest on the side facing Earth?*

The Moon's Interior

The presence of maria on the Moon's surface tells us something about its interior. If cracks did form when the large depressions were produced by impacts, and lava did flow onto the lunar surface, then the interior of the Moon just below its surface must have been molten at that time. It is believed that this was the case and that before the Moon cooled to what it is like today, its interior separated into layers.

Other information about the Moon's interior comes from seismographs left on the Moon by *Apollo* astronauts. Just as the study of earthquakes allows scientists to map Earth's interior, the study of moonquakes helps them study the Moon's interior and has led to the model shown in **Figure 17.** This model shows that the Moon's crust is about 60 km thick on the side facing Earth and about 150 km thick on the side facing away. Below the crust, a solid mantle may extend to a depth of 1,000 km. A partly molten zone of the mantle extends farther down. Below this is an iron-rich, solid core.

Reading Check *Where is the Moon's crust thickest?*

Exploring the Moon

More than 20 years after the *Apollo* program ended, the *Clementine* spacecraft was placed in lunar orbit. *Clementine* compiled a detailed map of the Moon's surface, including the South Pole-Aitken Basin. This is the oldest identifiable impact feature on the Moon's surface. It is also the largest and deepest impact basin or depression found so far anywhere in the solar system, measuring 12 km in depth and 2,500 km in diameter. Because the angle of sunlight is always low near the poles, much of this depression remains in shadow throughout the Moon's rotation. This location provides a cold area where ice deposits from impacting comets may have collected. The *Clementine* spacecraft, and later the *Lunar Prospector*, both collected evidence that supports the hypothesis that water-ice has accumulated in South Pole-Aitken Basin. See **Figure 18** to learn more about this.

Topic: *Clementine* and *Lunar Prospector* Missions
Visit gpescience.com for Web links to information about the missions of the *Clementine* and *Lunar Prospector* spacecraft.

Activity Make a table that lists the major goals of each mission and whether or not it succeeded.

VISUALIZING WATER ON THE MOON

Figure 18

Astronomers long believed that the Moon was a cold, dry place without atmosphere. But a few scientists hypothesized that water could exist on the Moon under certain conditions. This hypothesis was proven correct in the late 1990s by data from the spacecrafts *Clementine* and *Lunar Prospector*.

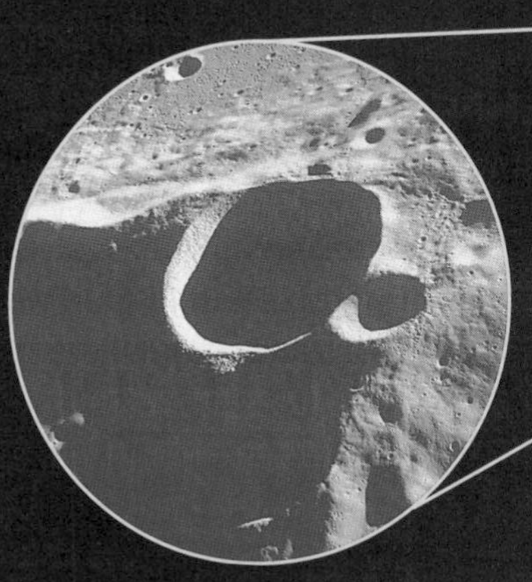

HOW DID IT GET THERE? Throughout its history, the Moon has been bombarded by comets and meteorites, most of which contain water-ice. Upon impact, some of the water would have quickly vaporized and be lost to space. However, some was deposited in the bottom of deep polar craters. Temperatures in these craters never exceed about −173°C. At these temperatures, ice could persist for billions of years.

HOW MUCH IS THERE? Estimates of how much water-ice exists vary, but some estimates are as high as 6 trillion kg. Ice might be buried several meters below the surface, either in solid blocks or as ice crystals mixed with lunar regolith. Water is important for many reasons. It would be needed for the survival of humans at any future lunar bases. Also, it could be split using solar power into hydrogen and oxygen to make rocket fuel.

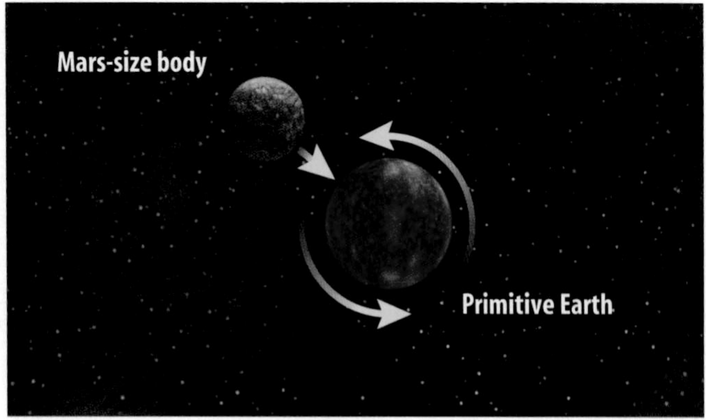

A large, Mars-sized body collided with primitive Earth approximately 4.6 billion years ago.

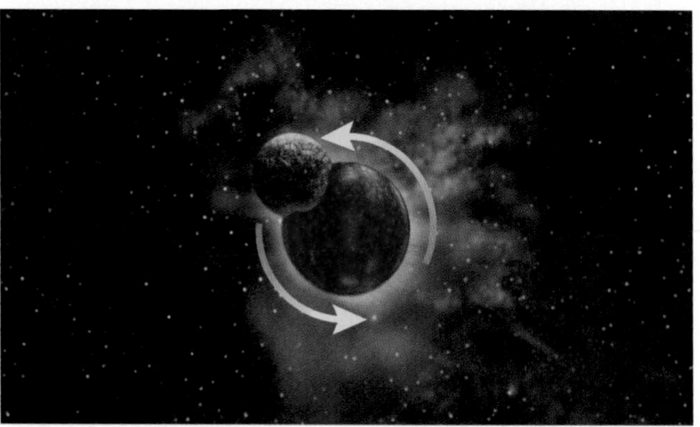

The violent collision melted and vaporized some of Earth's crust and mantle and hurled it into space.

Figure 19 According to the giant impact theory, the Moon formed after Earth was struck a glancing blow by a body that was more massive than Mars. For millions of years after the Moon's birth, stray rock fragments ejected by the original impact continued to pelt its surface, creating the craters that now blanket the Moon.

Clementine and Lunar Prospector Data from *Clementine* confirmed that the crust on the side facing Earth is much thinner than on the far side. Data also found that the crust thins under impact basins and showed the location of mascons, concentrations of mass that are located under impact basins. What do you think they might be? *Clementine* also provided information on the mineral content of Moon rocks. In fact, this part of its mission explains the spacecraft's name. Clementine was the daughter of a miner in the ballad "My Darlin' Clementine."

Reading Check *What are mascons?*

In 1998, the *Lunar Prospector* spacecraft orbited the Moon, taking photographs of the lunar surface. Maps made using these photographs confirmed the *Clementine* data. Also, data from *Lunar Prospector* confirmed that the Moon has a small, iron-rich core about 600 km in diameter. *Lunar Prospector* also conducted a detailed study of the Moon's surface searching for clues as to its origin and structure.

Origin of the Moon

Prior to the data obtained from the Apollo space missions, there were three theories about the Moon's origin. The first was that the Moon was captured by Earth's gravity (the capture theory). It had formed elsewhere and wandered near Earth. The second theory was that the Moon condensed from the same loose material that Earth formed from during the early formation of the solar system (the binary accretion theory). The third theory was that a glob of molten material was ejected from Earth while Earth was still molten (the fission theory). Ironically, the goal of one Apollo mission was to help determine which of these theories was correct. Instead, the mission showed that none of the three theories can explain the Moon's composition.

Some material fell back to Earth, some escaped into interplanetary space, and some orbited Earth as a ring of hot gas and debris.

Solid particles eventually condensed from the cooling gas and the Moon began to accumulate.

Giant Impact Theory Data gathered by the *Apollo* missions led many scientists to form a new giant impact theory, which has gained wide acceptance among astronomers. According to this theory, the Moon formed about 4.6 billion years ago when a Mars-sized object collided with Earth. After colliding, the cores of the two bodies combined and settled toward the center of the larger object. Gas and other debris were thrown into orbit. Some fell back to Earth, but the remainder condensed into a large mass, forming the Moon. This sequence is shown in **Figure 19.** This theory helps to explain how the Moon and Earth are similar, yet not similar enough to have formed from the same condensing mass. If the core of the Mars-sized body was added to the core of Earth, this explains why the Moon's composition is like Earth's mantle and why the Moon has a much smaller central core than expected.

section 3 review

Summary

How does the Moon affect Earth?

- The Moon and the Sun affect Earth's tides.
- As the Moon revolves, different amounts of the side facing Earth are lighted by the Sun, causing the changing phases.
- The alignment of Earth, the Moon, and the Sun produces eclipses.

The Moon's Surface and Interior

- The Moon's surface has craters, mountains, and maria.
- Maria are the dark-colored, relatively flat regions on the Moon.
- The *Clementine* and *Lunar Prospector* found evidence of water-ice on the Moon.

Self Check

1. **Compare and contrast** solar and lunar eclipses.
2. **Explain** tides in Earth's oceans.
3. **Diagram** the positions of Earth, the Moon, and the Sun during a full moon.
4. **Describe** how lunar maria might have formed.
5. **Think Critically** Why is it so important to future space exploration if the Moon has water-ice near its surface?

Applying Math

6. **Calculate** An estimate of the amount of water frozen at the Moon's south pole is 100,000 m^3. If this deposit was spread over an area measuring 160 m by 125 m, how many meters deep would the deposit be?

Identifying the Moon's Surface Features and *APOLLO* Landing Sites

Goals

- **Identify** prominent surface features on the Moon.
- **Determine** the relative ages of features on the Moon's surface.
- **Locate** the *Apollo* landing sites on the Moon.

Materials

large-scale maps or globes of the Moon
individual, smaller maps of the Moon

Real-World Problem

When you look at a full moon in the night sky, you can see light and dark areas. When you look through binoculars or a small telescope, you can see many craters and the large, dark-colored maria. Many craters are named after great philosophers and scientists. The maria are named for what early scientists thought they saw there; for example, *Oceanus Procellarum* means "ocean of storms." Can you tell the difference between an old crater and one that has formed more recently? If craters are seen on a maria, which of the features is older? If one crater partially covers another, which one formed first?

Procedure

1. Obtain a large-scale map or globe of the Moon.
2. Familiarize yourself with some of the more prominent surface features of the Moon.

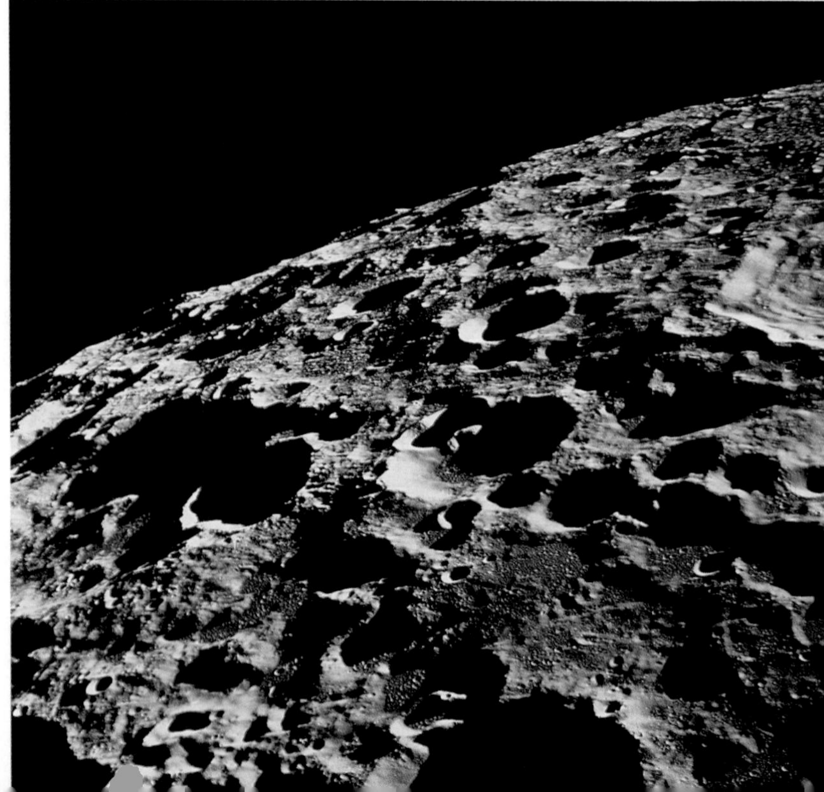

3. Look for examples of younger craters (those with sharp sides and peaks in the center) and examples of older craters (those whose sides are worn down or missing).
4. Using a large-scale, labeled map or globe of the Moon, locate, identify, and label the following prominent surface features of the Moon and *Apollo* landing sites on a copy of the Moon's surface. (If an unlabeled Moon map is not available, draw one and illustrate and label the features listed below.)

Features on the Moon

Maria	Craters	Mountain Ranges	Apollo Landing Sites
Mare Crisium	Alphonsus	Alps	11-Mare Tranquillitatus
Mare Frigoris	Aristarchus	Apennine	12-Oceanus Procellarum
Mare Imbrium	Clavius	Caucasus	14-Fra Mauro
Mare Serenitatis	Copernicus	Jura	15-Mt. Hadley
Mare Tranquillitatus	Fra Mauro	Mt. Hadley	16-Descartes
Oceanus Procellarum	Grimaldi		17-Taurus Littrow
	Kepler		
	Plato		
	Tycho		

Analyze Your Data

1. **Describe** the specific lunar features you studied and what you learned from them.
2. **Identify** one specific observation that helped you decide which of two features formed first.

Conclude and Apply

1. **Infer** from your study of the large-scale map or globe whether Copernicus Crater or Grimaldi Crater is older. Do the same for Fra Mauro Crater and Tycho Crater. **Explain** your answers.
2. **Research** the *Apollo* missions and **explain** why there is no landing site for *Apollo 13*.

Compare your map of the Moon with the maps labeled by other students in your class. **Discuss** why individual maps may be labeled differently or why map illustrations might look different.

Oops! Accidents in SCIENCE

SOMETIMES GREAT DISCOVERIES HAPPEN BY ACCIDENT!

Even Great Scientists Make Mistakes

"If I have been able to see further, it was only because I stood on the shoulders of giants."
—Sir Isaac Newton

Today, scientists know that sometimes light behaves like a wave and at other times it behaves like a particle. However, early scientists believed it had to be one way or the other. Sir Isaac Newton, 1642–1727, believed in the particle nature of light. Based on his observations and a few erroneous assumptions, he eventually invented what is now called the Newtonian reflecting telescope.

Where did Newton go wrong?

One erroneous assumption involved the behavior of light as it passes through matter such as glass, a property known as chromatic aberration. White light is composed of many different wavelengths that can disperse when refracted. This also happens with telescopes because lenses act like a combination of many prisms and disperse white light into the colors of the rainbow. We now know that the larger the lens, the less of a problem this dispersal of light causes. Newton thought the opposite was true. He tried to eliminate chromatic aberration by changing or adding lenses and using prisms, but was unable to correct the problem.

Another mistake made by Newton was that he assumed light particles begin to refract before they contact matter. Because of these assumptions, he concluded that there was no way to fix the problem. So, he gave up on refracting telescopes and used a curved mirror to focus light. This worked, because light does not disperse when it is reflected. His incorrect assumptions led to the invention of the most-used instrument in astronomy, the reflecting telescope.

Sometimes the greatest discoveries in science are based on incorrect assumptions. But, that's okay—it is called science.

Newtonian Telescope

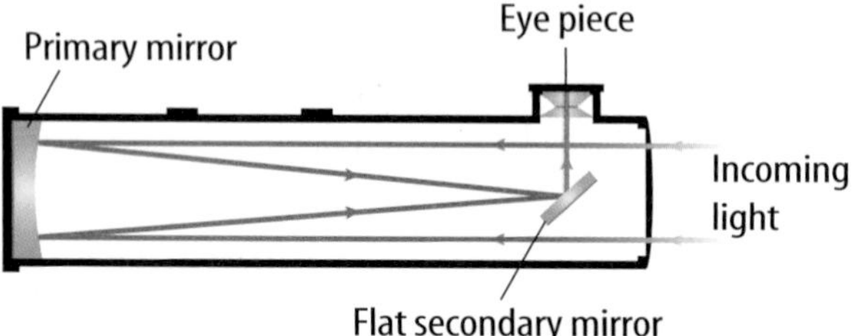

Try it yourself Experiment with shining light through a prism and reflecting it from a mirror. Try using several different prisms with different angles.

Reviewing Main Ideas

Section 1 Earth in Space

1. The fact that Earth always casts a curved shadow, as shown here, is evidence of its spherical shape.

2. When Earth was molten, the force of gravity pulling equally in all directions caused Earth to form into a spherical shape.
3. Earth's rotation and movement of matter in its core cause Earth's magnetic field.
4. Earth is different than other planets. These differences enable life to flourish on Earth.

Section 2 Time and Seasons

1. Earth's rotation and revolution are used to measure time in days and years.
2. During the year, the Sun appears to move on the ecliptic through a background of constellations called the zodiac.
3. Earth experiences changing seasons because locations on Earth receive sunlight for varying amounts of time each day and at varying angles throughout the year, as shown here.

4. The Sun reaches its greatest distance north of the equator on the summer solstice for the northern hemisphere and on the winter solstice for the southern hemisphere.

Section 3 Earth's Moon

1. Tides in Earth's oceans are affected by the gravity of the Moon and the Sun. The Moon's gravity has a greater effect because the Moon is much closer to Earth.
2. Changing positions of the Moon in relation to the Sun cause it to go through a lunar phase cycle every 29.5 days.
3. Solar eclipses occur during a new moon, and lunar eclipses occur during a full moon.
4. Craters, like this one, are depressions on the Moon's surface. Some large depressions may have filled with lava, forming maria.

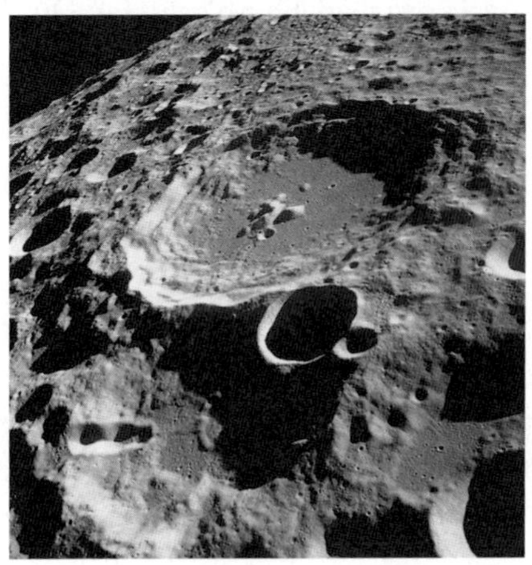

FOLDABLES™ Use the foldable that you made at the beginning of this chapter to review the Earth-Moon-Sun system.

chapter 7 Review

Using Vocabulary

ecliptic p. 192	revolution p. 192
ellipse p. 188	rotation p. 192
equinox p. 195	solar eclipse p. 201
gravity p. 187	solstice p. 195
lunar eclipse p. 202	sphere p. 186
maria p. 203	tide p. 199
moon phase p. 199	time zone p. 190
regolith p. 203	

Match the correct vocabulary word or phrase with each definition given below.

1. dark-colored, relatively flat areas on the Moon
2. Earth spinning on its axis
3. a large wave in Earth's oceans caused by the gravity of the Moon and the Sun
4. a round, three-dimensional object, the surface of which is the same distance from the center in all directions
5. Earth moving in orbit around the Sun
6. eclipse that occurs during a new moon
7. 15°-wide area on Earth's surface in which the time is the same
8. occurs when the Sun is directly above Earth's equator
9. attractive force between two objects
10. yearly path of Earth around the Sun

Checking Concepts

Choose the word or phrase that best answers each question.

11. How long is a month of lunar phases?
    A) 14 days
    B) 27.3 days
    C) 29.5 days
    D) 365 days

12. During winter solstice in the northern hemisphere, the Sun is directly over which part of Earth?
    A) equator
    B) pole
    C) Tropic of Cancer
    D) Tropic of Capricorn

13. Which movement causes lunar phases?
    A) Earth's revolution
    B) Earth's rotation
    C) the Moon's revolution
    D) the Moon's rotation

14. Which eclipse do you experience if you are standing in the Moon's umbra?
    A) partial lunar **C)** total lunar
    B) partial solar **D)** total solar

15. Which phase occurs when the Moon is on the opposite side of Earth from the Sun?

A) 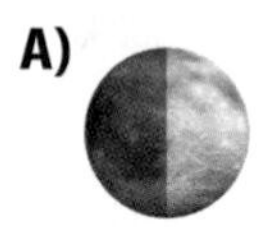 **B)** 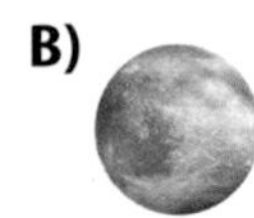 **C)** 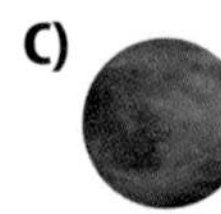 **D)** 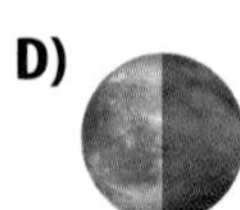

16. Which material may have been found on the Moon by the *Clementine* spacecraft?
    A) atmosphere
    B) dark-colored rocks
    C) light-colored rocks
    D) water-ice

17. On average, how many degrees of longitude are contained in one time zone?
    A) 0.5° **C)** 23.5°
    B) 15° **D)** 30°

18. Which occurs a few days after a full moon?
    A) waning crescent
    B) waxing crescent
    C) waning gibbous
    D) waxing gibbous

19. Which season begins around December 21 in Australia?

A) spring **C)** fall
B) summer **D)** winter

20. Which forms when small meteorites crash into the Moon?

A) craters **C)** mountains
B) maria **D)** cracks

Interpreting Graphics

21. Make an illustration showing how magnetic force lines surrounding Earth are similar to those surrounding a bar magnet.

Use the data in the table below to answer question 22.

Earth's Physical Properties	
Diameter (pole to pole)	12,714 km
Diameter (through equator)	12,756 km
Circumference (poles)	40,008 km
Circumference (equator)	40,075 km
Average distance to the Sun	149,600,000 km
Average distance to the Moon	384,400 km

22. What is the difference between the diameter of Earth through the poles compared to the equator? How many times farther from Earth is the Sun, compared to the Moon?

Thinking Critically

23. Infer why more craters are present on the Moon's surface than on Earth's. *Hint: Consider gravity and the presence of an atmosphere.*

24. Infer how seasons would be affected if Earth had no tilt instead of the 23.5° tilt that it has.

25. Explain why a lunar base would best be built on a plateau that is always in sunlight.

26. Form a hypothesis about why during crescent phases we can often see a dim image of the rest of that side of the Moon. *Hint: Recall the arrangement of Earth, the Moon, and the Sun during crescent phases.*

27. Form a hypothesis about how the thickness of the Moon's crust might play a part in the fact that the side of the Moon facing Earth has more maria than the side facing away.

Applying Math

Use the illustration below to answer questions 28–29.

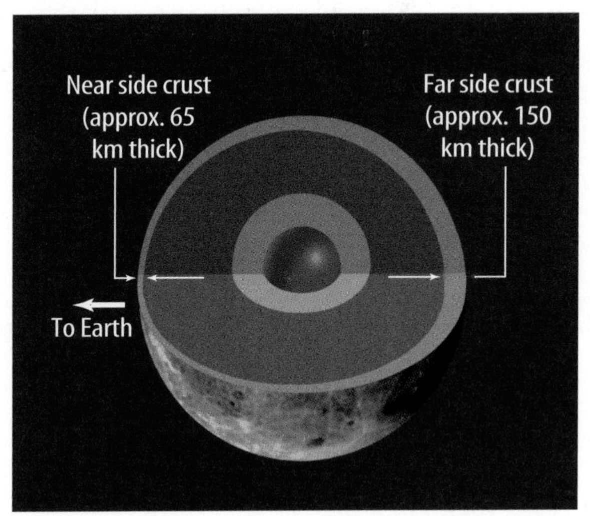

28. Model to Scale If you are making a scale model of the Moon (diameter approx. 3,500 km), what scale should you use to obtain a model that is about 35 cm in diameter?

29. Calculate Using your scale, what would the thicknesses be, in centimeters, for the near-side crust and the far-side crust?

chapter 7 Standardized Test Practice

Record your answers on the answer sheet provided by your teacher or on a sheet of paper.

Multiple Choice

Use the illustration below to answer question 1.

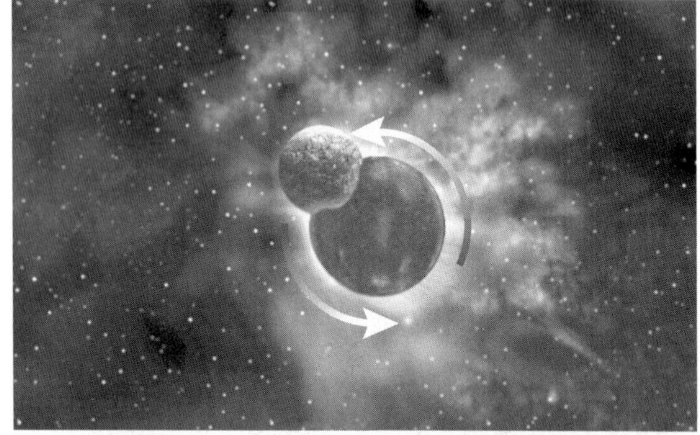

1. Which theory explaining the Moon's origin is widely accepted by astronomers?
   A. binary accretion theory
   B. capture theory
   C. fission theory
   D. giant impact theory

2. How far is Earth's magnetic axis tilted from its geographic axis?
   A. 5°
   B. 11.5°
   C. 15°
   D. 23.5°

3. On which number did the Babylonians base their counting methods?
   A. 10
   B. 60
   C. 100
   D. 360

4. Which is a way that Venus and Earth are similar?
   A. atmospheric density
   B. liquid water oceans
   C. rocky nature
   D. surface temperature

Use the illustration below to answer question 5.

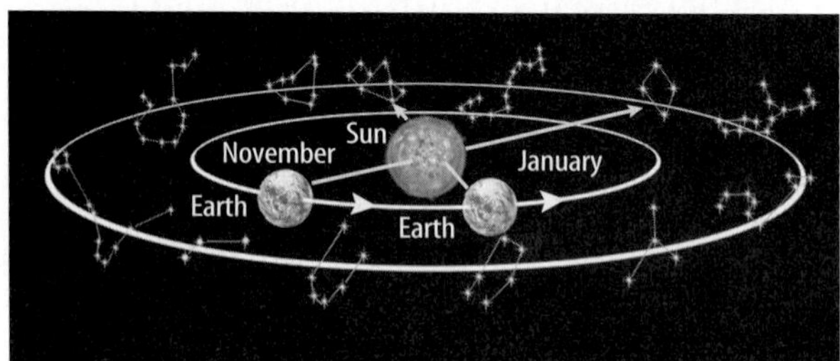

5. Which is a group of constellations through which the Sun appears to move?
   A. ecliptic
   B. equinox
   C. solstice
   D. zodiac

6. Which contains bands of charged particles known as the Van Allen belts?
   A. exosphere
   B. ionosphere
   C. magnetosphere
   D. stratosphere

Test-Taking Tip

Caution Read each question carefully for full understanding.

7. During which month of the year is Earth farthest from the Sun?

   A. January

   B. April

   C. July

   D. September

Gridded Response

8. If a synodic month is 29.5 days long and a sidereal month is 27.3 days long, how much longer is a synodic month?

Use the illustration below to answer question 9.

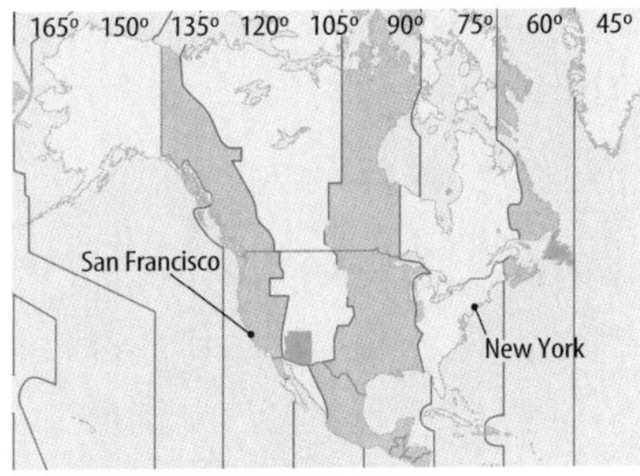

9. If it is 9:00 A.M. in New York city, what time is it in San Francisco?

10. If the collision of two planetary-sized objects (Earth and another object) formed the Moon, why is the Moon's iron core so small compared to Earth's?

Short Response

11. What may have caused the great difference in the percentage of CO_2 found in Earth's atmosphere compared to those of Venus and Mars?

12. What causes the auroras?

Extended Response

Use the illustration below to answer question 13.

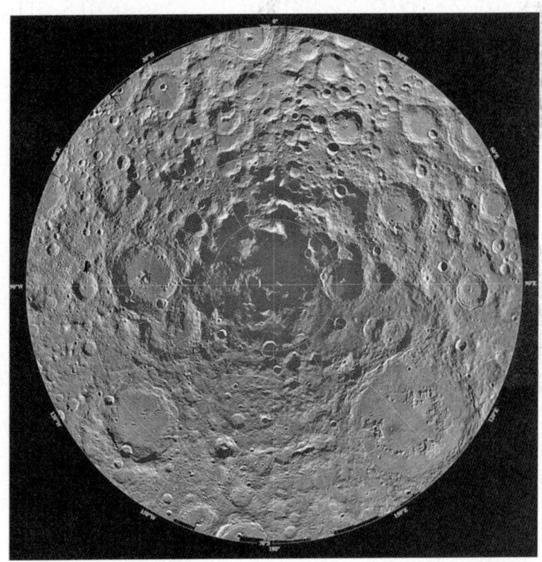

13. **PART A** The *Lunar Prospector* discovered possible concentrations of water-ice in the area of the South Pole-Aitken Basin on the Moon. How does location and geographic terrain affect the possibility of finding water there?

    PART B How might this discovery affect the future of spaceflight?

chapter 8

The Solar System

chapter preview

sections

Home Sweet Home

Nestled within one of the great spiral arms of the Milky Way lies our solar system. As large as the solar system is, it is still only a tiny fraction of our galaxy!

Science Journal In your Science Journal, write a hypothesis about whether life might exist beyond Earth, or even beyond the solar system, and how you would test for it.

Start-Up Activities

The Size of the Solar System

Even if you consider only the planets, the solar system is enormous compared to things you see in your daily life. You may be better able to understand the size of the solar system by using a scale model. Distances between the orbits of the nine planets usually are measured in astronomical units (AUs). One AU equals 150 million km, the average distance between Earth and the Sun. Using a scale of 1 AU = 10 cm and the data from the table below, draw a model of the solar system.

Planet	Distance to Sun (AU)
Mercury	0.39
Venus	0.72
Earth	1.00
Mars	1.52
Jupiter	5.20
Saturn	9.54
Uranus	19.19
Neptune	30.07
Pluto	39.48

1. Obtain a roll of adding-machine tape at least 4 m long. Place a symbol for the Sun on one end of the tape.
2. Using the scale above, measure and place the name of each planet at its proper distance from the Sun.
3. **Think Critically** In your Science Journal, explain how a scale model helped you to visualize the size of the solar system.

The Solar System Make the following Foldable to help you organize information about the planets.

STEP 1 **Collect** five sheets of paper and layer them about $\frac{1}{2}$ inch apart vertically. Keep the edge level.

STEP 2 **Fold** up the bottom edges of the paper to form ten equal tabs.

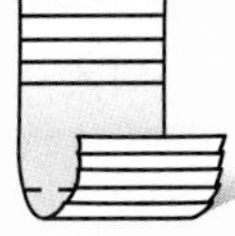

STEP 3 **Fold** the papers and crease well to hold the tabs in place. Staple along the fold. **Label** each tab with the names of the nine planets and Sedna.

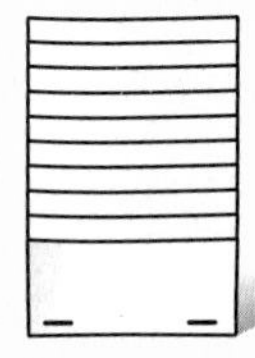

Organize As you read the chapter, write information about each planet on its sheet.

Preview this chapter's content and activities at gpescience.com

section 1

Planet Motion

Reading Guide

What You'll Learn

- **Compare and contrast** geocentric and heliocentric models of the solar system.
- **Describe** each planet's position in orbit around the Sun.
- **Classify** planets by location and characteristics.

Why It's Important

Understanding the solar system helps you understand Earth's position in space.

Review Vocabulary

ellipse: an oblong, closed curve drawn around two foci

New Vocabulary

- geocentric model
- heliocentric model
- astronomical unit
- extrasolar planet

Models of the Solar System

Think how difficult it would be to construct a model of the solar system if you didn't know that Earth is a rotating planet. All you can see is the nightly movement of planets across the sky and they all appear to move in a path around Earth. You can understand why many early scientists thought that Earth was the center around which everything they saw in the sky revolved.

Figure 1 Ptolemy presented his geocentric model of the solar system in 140 A.D.

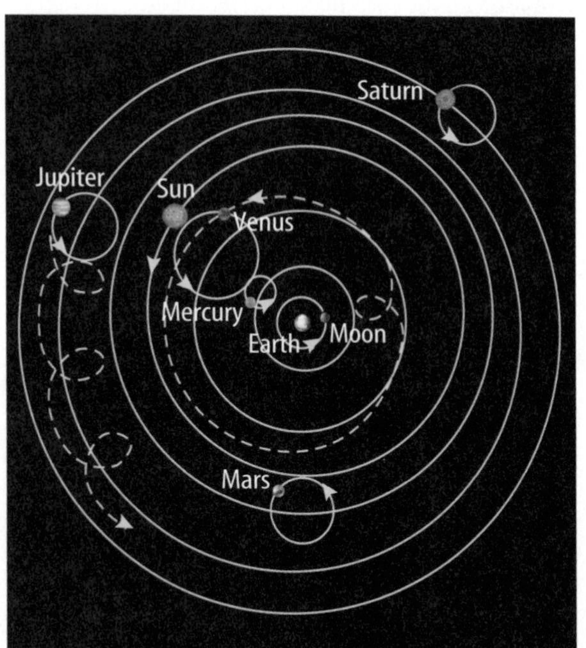

Geocentric Model In the **geocentric model** of the solar system, Earth is considered the center and everything else revolves around it. You have seen why ancient scientists thought this. Many early Greek scientists explained this by saying that the planets, the Sun, and the Moon were embedded in separate crystal spheres that rotated around Earth. They thought that the stars were imbedded in another sphere that also rotated around Earth. The sphere bearing the stars moved in a regular, predictable way, but those bearing the planets seemed often to move erratically against this background. For this reason, they were called *planasthai*, a Greek word that means "to wander." Our word *planet* comes from this. These ideas led to the model of the solar system devised by Ptolemy, shown in **Figure 1.**

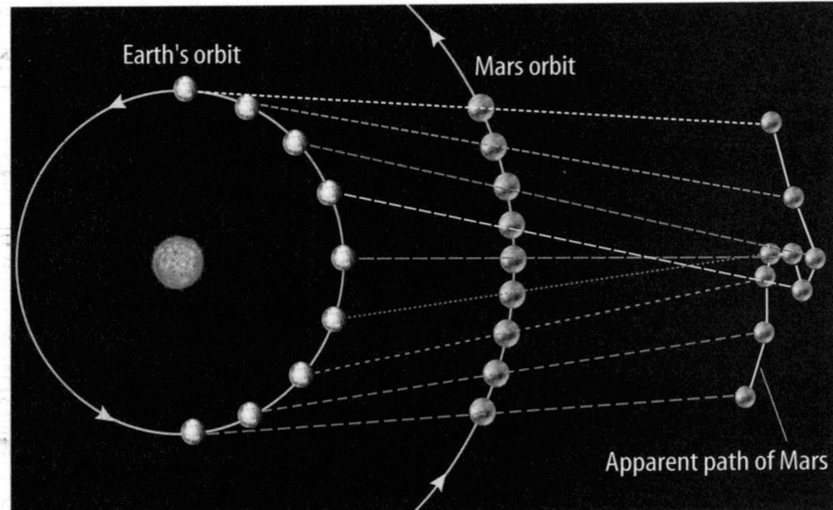

Figure 2 As Earth catches up and passes Mars, Mars appears to move backward, or retrograde.

Geocentric Modifications Although the Ptolemaic model of the solar system was accepted and used for centuries, there were many problems with it. One problem was the fact that planets periodically appear to move in a retrograde, or backward, direction when viewed against a background of stars, as shown in **Figure 2.** Modifications helped, but eventually the model grew very complex.

Heliocentric Model In spite of its complexity, the Earth-centered model held until 1543 when Polish astronomer Nicholas Copernicus published a different view. Using an idea proposed by an early Greek scholar, Copernicus stated that the Moon revolves around Earth, which is a planet, and that Earth and the other planets revolve around the Sun. He also stated that the apparent motion of the planets, the stars, and the Sun is due to Earth's rotation. This is the **heliocentric model,** or Sun-centered model of the solar system, shown in **Figure 3.**

Galileo Using his telescope, Italian astronomer Galileo Galilei found evidence that supported the ideas of Copernicus. He observed that Venus went through phases like the Moon's. These phases could be explained only if Venus orbited the Sun and passed between Earth and the Sun on each orbit. He also saw moons in orbit around Jupiter. These observations convinced him that everything didn't move around Earth, and he concluded that Earth and Venus revolve around the Sun and that the Sun is the center of the solar system.

Plotting Retrograde Motion

Procedure

1. Using **graph paper,** draw and label the following: *x*-axis, Right ascension, from 12 hours to ten hours; *y*-axis, Declination, from 0° to 20°.
2. Plot and label these positions of Jupiter:
   Oct. 1, 2003 (10h 38m, 10°);
   Jan. 1, 2004 (11h 21m, 6°);
   April 1, 2004 (10h 51m, 9°);
   May 1, 2004 (10h 44m, 9°);
   Aug. 1, 2004 (11h 20m, 6°).
3. Connect the plotted positions in chronological order with a solid line.

Analysis

1. What do you notice about Jupiter's direction of motion?
2. Using a heliocentric model of the solar system, explain why Jupiter appears to move as it does.

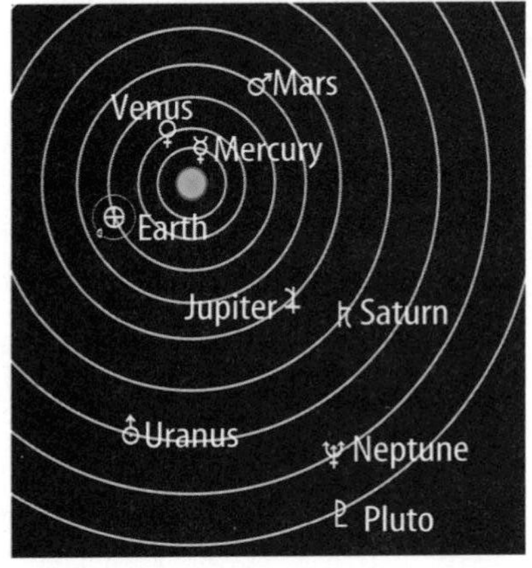

Figure 3 The heliocentric model of the solar system states that Earth and the other planets revolve around the Sun.

Planet Motion Johannes Kepler devised a system of equations that determine a planet's motion based on its position in the solar system. Use this equation to calculate the orbital period (in years) for Jupiter, which is 5.2 AU from the Sun.

$$(\text{planet's distance in AUs})^3 = (\text{planet's period in years})^2$$

Check your result against the known value.

Understanding the Solar System

Copernicus and Galileo led the way, but many scientists contributed to our understanding of the solar system. Foremost among them is German mathematician Johannes Kepler. In the early 1600s, Kepler discovered that the planets travel around the Sun in ellipses, not circles, as Copernicus had believed. Kepler also learned that the planets travel at different speeds in their orbits. The closer a planet is to the Sun, the faster it travels. This means that the outer planets take more time to orbit than the inner planets. Mercury takes just 88 days to complete one orbit, or period, but Pluto takes over 248 years to complete an orbital period.

To measure the large distances within the solar system, astronomers use a different unit. This is the **astronomical unit** (AU), which equals the average distance from Earth to the Sun, about 150 million km.

Applying Math — Use Geometry

HELIOCENTRIC LONGITUDE (H.L.) This is the position of a planet on any day of any year. It is plotted by measuring the angle between a line of the spring equinox in the northern hemisphere and the position of the planet on its orbit. This measurement is done counterclockwise because that is how planets appear to move when viewed from above the north pole of Earth. Plot the heliocentric longitudes given below.

IDENTIFY known values and the unknown value

Identify known values:

H.L. on March 4, 2004 for Mercury, Venus, Earth, Mars, and Jupiter are 344°, 103°, 164°, 84°, and 164° respectively.

Identify unknown values:

the position of each planet marked below

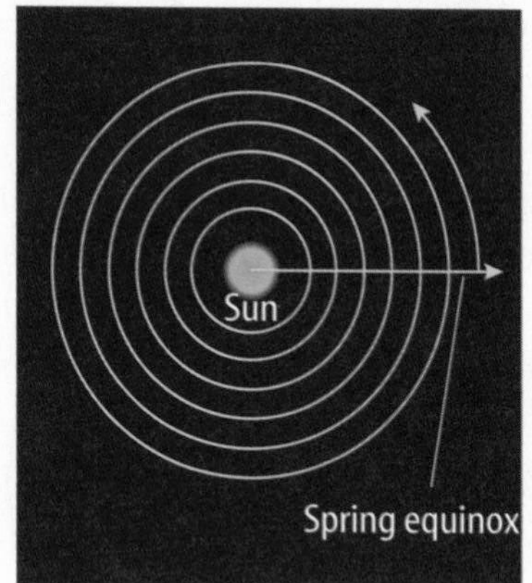

SOLVE the problem

Using this diagram as a model, make a larger drawing on a piece of paper. Use a protractor to measure the H.L. angles and mark the position of each planet on its orbit in the diagram and label it with its name or planet symbol: Mercury ☿, Venus ♀, Earth ⊕, Mars ♂, and Jupiter ♃

CHECK the answers

Switch papers with another student and use a protractor to check each other's plots.

Practice Problems

For more practice problems, go to p. 879 and visit Math Practice at gpescience.com.

Classifying Planets The nine planets can be classified in several ways. One system uses size and other characteristics. Those similar to Earth are called terrestrial planets, and the giant planets are jovian planets. Pluto is sometimes called terrestrial, but does not perfectly fit either category.

Two other systems classify planets by location. The system used most often classifies planets whose orbits are between the Sun and the asteroid belt as inner planets and those beyond the asteroid belt as outer planets. The other system classifies planets whose orbits are between Earth's orbit and the Sun as inferior planets, and those whose orbits are beyond Earth's orbit as superior planets.

What separates the inner planets from the outer planets?

Origin of the Solar System Evidence suggests that the Sun probably formed as part of a multiple star system or star cluster. Scientists hypothesize that a huge cloud of gas, ice, and dust began forming into the present solar system about 4.6 billion years ago. This cloud, called a nebula, might have been as large as a million billion kilometers across. When something in nearby space produced an immense shock wave, the nebula began to condense and fragment because of gravitational instability.

Eventually fragmentation stopped, but contraction continued. The cloud fragment from which our Sun formed was about the size of the entire solar system and was slowly rotating in space. As it continued to contract, the matter in the cloud fragment condensed into less space. As its density became greater, gravity pulled more gas, ice, and dust toward the center. This caused the cloud fragment to rotate faster, which in turn caused it to flatten into a disk with a dense center, as shown in **Figure 4.**

Temperature inside the cloud fragment increased. When it reached about ten million degrees Celsius, hydrogen fused into helium, and the Sun was born. Viewed from above the north pole of the Sun, the nebular disk rotated in a counterclockwise direction. Leftover mass in the outer portions of the disk condensed and formed planets and other objects that continue to revolve around the Sun in a counterclockwise direction.

Figure 4 Our solar system formed from one fragment of a giant nebula in a sequence of contractions and condensations.

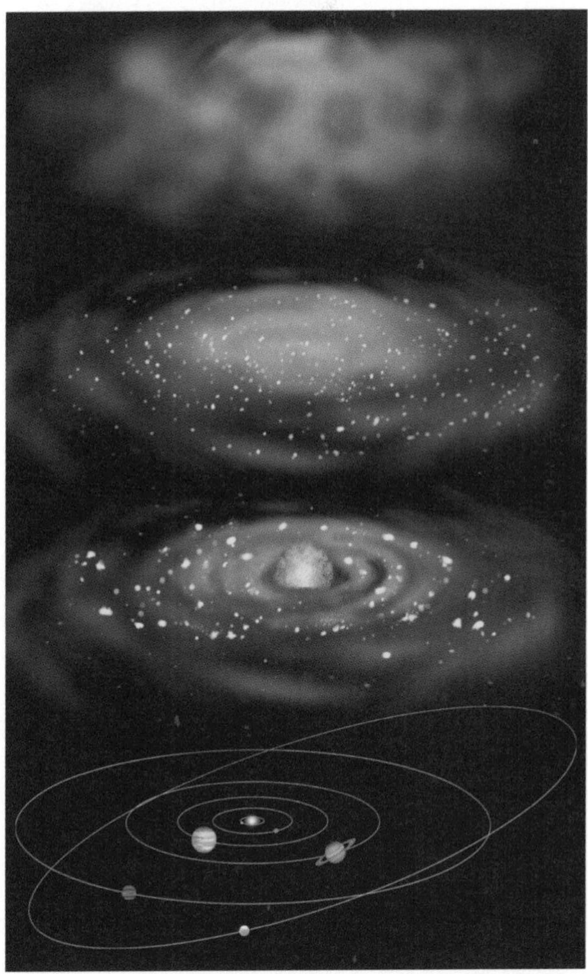

Figure 5 The planetary system around Upsilon Andromedae is shown below. This star is barely visible to the unaided eye.

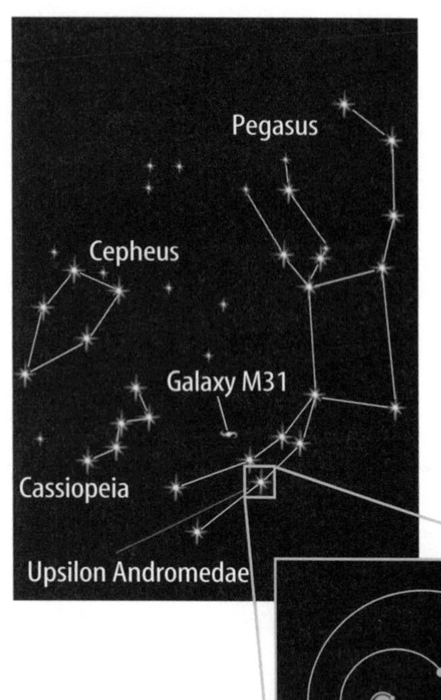

Other Solar Systems

Until recently, our solar system was the only one known. Now we know that many other stars also have planets around them. **Figure 5** shows the orbits of three planets around the star Upsilon Andromedae, a yellow star about 44 light-years away. These **extrasolar planets**—planets in orbit around other stars—are helping astronomers learn how planetary systems form.

Astronomers have devised new techniques and instruments to find planets around other stars. So far, they have found over 100 stars that have planetary systems, and many more probably exist. Some of these planetary systems have been found around stars like the Sun, but so far only giant planets similar to Jupiter have been detected. If these large planets exist, scientists say the chances are good that planets of lower mass exist also.

New instruments developed by NASA will look for Earthlike planets around other stars. It is difficult because the low mass of planets like Earth make them harder to find than giant planets. Even so, some of the planets found so far are comparatively low in mass and a few of these have a mass very similar to Neptune's. Scientists hypothesize this indicates that many more planets of even lower mass will be found.

What is an extrasolar planet?

section 1 review

Summary

Models of the Solar System

- Geocentric models of the solar system place Earth in the center.
- Heliocentric models of the solar system place the Sun in the center.

Understanding the Solar System

- Planet orbits are ellipses.
- Planets are classified by their positions in the solar system and by their characteristics.
- Planets move around the Sun in a counter-clockwise direction.

Other Solar Systems

- Extrasolar planetary systems have been discovered.

Self Check

1. **Compare and contrast** the geocentric and heliocentric models of the solar system.
2. **Explain** how heliocentric longitude is used to show the position of planets in orbit around the Sun.
3. **Classify** planets two ways using their positions in the solar system.
4. **Define** terrestrial planets.
5. **Think Critically** Would a year on Venus be longer or shorter than an Earth year? Explain.

Applying Math

6. **Using Equations** Use the equation (planet's period in years)^2 = (planet's distance in AUs)^3 to find the orbital period of a planet that is 19.2 AU from the Sun.

More Section Review gpescience.com

section 2

The Inner Planets

Reading Guide

What You'll Learn

- **Compare and contrast** other inner planets with Earth.
- **Describe** important characteristics for each inner planet.
- **Evaluate** the success of various missions to Mars.

Why It's Important

Answering questions about life on other planets helps you better appreciate the unique qualities of Earth.

Review Vocabulary

robot lander: part of a spacecraft that lands on another planet or other object

New Vocabulary

- Mercury
- Venus
- Earth
- Mars

Planets Near the Sun

Recall that not all the gas, ice, and dust in the nebula fragment that formed the Sun was drawn into its core. Remaining gas, ice, and dust particles collided and stuck together, forming solid objects that in turn attracted more particles because of the stronger pull of gravity. Close to the Sun, high temperatures vaporized lighter elements so they could not condense. This explains why light elements are scarcer in planets closer to the Sun than in those that formed farther away. Also, any lighter elements surviving were probably driven from the surface by the solar wind. As a result, small, rocky planets with iron cores formed in the inner solar system.

Mercury

The second-smallest and closest planet to the Sun is **Mercury.** *Mariner 10* was the first American space probe sent to Mercury. During 1974 and 1975, it flew by the planet and sent pictures back to Earth, but only 45 percent of Mercury's surface was photographed. What we can see of its surface looks very much like our Moon's. Mercury is covered by craters, some of which have double rings, as shown in **Figure 6.** Scientists do not yet understand why this is true.

The magnetic field around Mercury, discovered by *Mariner 10,* indicates that Mercury has a much larger iron core than would be expected and is missing some lighter materials you would expect to find in its mantle. One theory is that two similarly sized bodies collided. Their cores merged to form one large iron core, and some of the lighter material vaporized and was lost to space.

Figure 6 The outer ring of this double-ringed crater on Mercury is 170 km across.

Figure 7 Thick clouds obscure the surface of Venus. This view of the surface was assembled from radar data of *Magellan* and *Pioneer Venus Orbiter* and then colorized.

Mercury's Surface Mercury's relatively large core and thin outer layers resulted in some extreme differences in its surface terrain. Mercury's large, solid core shrank much more rapidly than its thin outer layers. As the outer layers adjusted, they wrinkled, forming dramatic cliffs as high as 3 km. This is similar to what happens when an apple dries and shrivels up.

Is there an atmosphere around Mercury? Mercury is small, has a comparatively low gravitational pull, and its surface experiences extremes in temperature. These conditions indicate that the planet has no atmosphere. *Mariner 10* and Earth-based observations did find traces of what was thought to be an atmosphere, but turned out to be gases that remain for a while and then leak into space. They are thought to be hydrogen and helium deposited by the solar wind and sodium and potassium vapors that diffuse upward through the crust. Mercury has no true atmosphere. This, along with its closeness to the Sun, causes surface temperatures that vary from 427°C to –170°C.

Venus

Venus, shown in **Figure 7,** is the second planet from the Sun. Venus and Earth are similar in some respects; their sizes and masses are almost the same. However, one major difference is that the entire surface of Venus is blanketed by a dense atmosphere. The atmosphere of Venus, which has 92 times the surface pressure of Earth's at sea level, is mostly carbon dioxide. The clouds in the atmosphere also contain droplets of sulfuric acid, which gives them a slightly yellow color.

Clouds on Venus are so dense that only about two percent of sunlight reaches the planet's surface. However, the solar energy that reaches warms Venus's surface. Then, heat radiated from Venus's surface is absorbed by the carbon dioxide gas, causing what is called a greenhouse effect. Due to this intense greenhouse effect, temperatures on the surface of Venus are between 450°C and 475°C.

Why is Venus's surface temperature so high?

Figure 8 Radar images of volcanoes with visible lava flows on Venus were captured by the *Magellan* space probe.

In the 1970s, the former Soviet Union sent the *Venera* lander probes to Venus. One photographed and mapped the surface before landing, but after about an hour, was destroyed by the intense heat. In 1995, the U. S. *Magellan* probe sent back radar images that revealed features of Venus, as shown in **Figure 8.**

Earth

The third planet from the Sun is **Earth,** shown in **Figure 9.** Unlike other planets, surface temperatures on Earth allow water to exist as a solid, a liquid, and a gas. Earth's atmosphere causes most meteors to burn up before they reach the surface. Ozone in Earth's atmosphere also protects life from the Sun's intense radiation. Life exists all over the planet. Life has been found on Earth in extreme conditions of temperature and pressure in which scientists would never have guessed it could survive. These finds are encouraging for the possibility of life existing elsewhere in the solar system. You will learn more about this in Section 4.

Figure 9 Seen from space, Earth's white clouds and blue oceans testify to the abundant water that makes life possible.

Mars

The fourth planet from the Sun, **Mars,** is called the red planet because iron oxide in some of the weathered rocks on its surface gives it a reddish color, as shown in **Figure 10.** Other Martian features visible from Earth are the polar ice caps and changes in the coloring of Mars's surface.

Mars is tilted 25° on its axis, which is very close to Earth's tilt. Because of this, Mars goes through seasons as it orbits the Sun, just like Earth does. The polar ice caps, which are made of frozen carbon dioxide and frozen water, get larger during the Martian winter and shrink during the summer. The color of Mars's surface changes according to its seasons.

In fact, both the changing sizes of the ice caps and changes in Mars's surface coloration were observed from Earth long before we ever sent a probe to the planet. Some people saw these changes and thought that life forms such as lichens might exist on Mars and grew more during the summer, thus making the area look darker. We know now that what actually causes this seasonal change in coloration of the Martian surface is wind, not the cyclic nature of plant life. As the seasons change during the Martian year, winds blow the dust around on the planet's surface. When the wind blows dust off one area, it may appear darker.

Figure 10 The dark region below Mars's polar cap in this *Hubble* photograph is composed of pulverized volcanic rock. A swirling storm of water-ice is visible to the left of the polar cap.

Figure 11 These spherical concretions found on Mars provide evidence of Mar's wet past.

Mars's Atmosphere The Martian atmosphere is much thinner than Earth's and is composed mostly of carbon dioxide with some nitrogen and argon. The thin atmosphere does not filter out harmful rays from the Sun as Earth's atmosphere does. Surface temperatures range from 37°C to −123°C. This temperature difference sets up strong winds on the planet, which can cause global dust storms.

Martian Moons Mars has two small, heavily cratered moons called Phobos and Deimos. Phobos is slowly spiraling inward toward Mars, and is expected to hit the Martian surface in about 50 million years. Deimos, which is smaller, orbits about 23,500 km above the surface.

Was Mars once wet? Currently, Mars appears to be a very dry world, but has it always been dry? Does it have any water now? The *Spirit* and *Opportunity* probes sent to Mars seem to indicate that liquid water was once on the planet and may still be there. **Figure 11,** a photo from *Opportunity,* shows deposits on Mars that have been labeled "blueberries" because of their round shape. These are concretions, most likely composed of hematite. Similar concretions form on Earth in the presence of water. **Figure 12** shows photographs taken in 2000 by the *Mars Global Surveyor* spacecraft. You can see the layering and gullies that look as if they were made by running water. Wind might have formed the layering of sediment and gullies, but more and more evidence indicates that water was once present on Mars. Other formations at the edges of craters suggest that frozen water in Mars's surface may flow outward from time to time, forming the channels. NASA has sent many spacecraft to Mars to find out if Mars is or was once wet.

Figure 12 These photographs taken by *Mars Global Surveyor* show gullies and channels inside the Newton Basin.

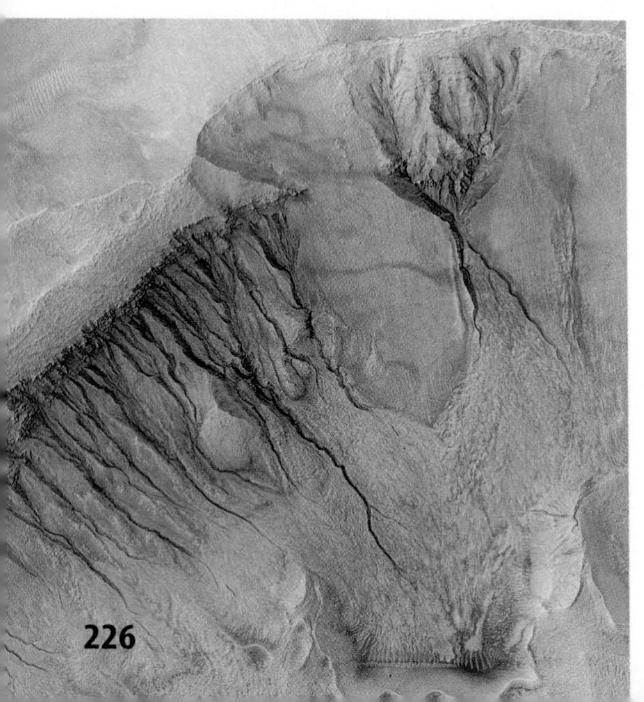

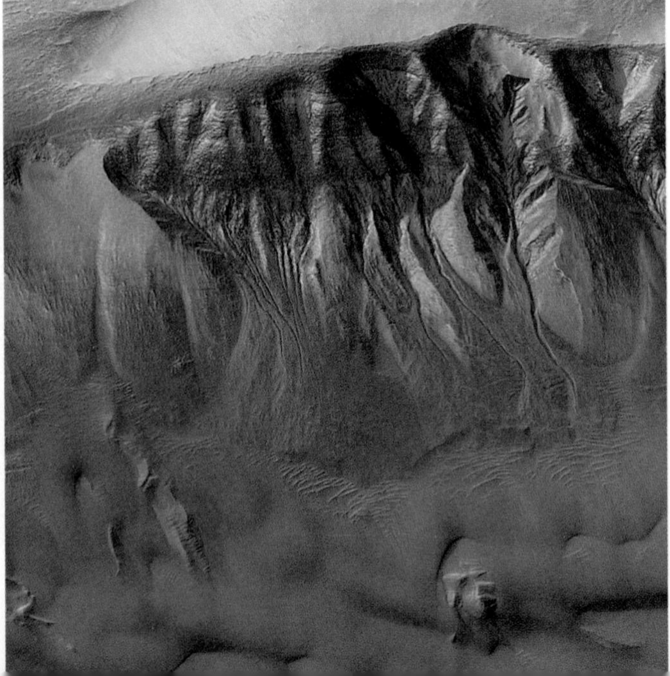

NASA on Mars

Mars has been a major focal point of NASA planetary exploration for many years. Although not all missions to Mars have been successful, several missions have returned a great amount of data about our reddish-colored neighbor. One of the earliest missions to Mars was the *Mariner 9* space probe that orbited Mars in 1971–1972. It revealed long channels on the planet that may have been carved by flowing water. *Valles Marineris* (the valley of the Mariners), shown in **Figure 13,** is a large canyon that was discovered by this early mission. It is so large that, if it were on Earth, this canyon would reach from San Francisco to New York City. It probably formed when Mars was still geologically active. *Mariner 9* also found large, extinct volcanoes. One of these, Olympus Mons, is the largest volcano found in the solar system.

Topic: NASA Missions to Mars

Visit gpescience.com for Web links to information from recent studies of missions to Mars.

Activity List new information you learned about Mars and write about the significance of each discovery in your Science Journal.

Reading Check *What is* Valles Marineris?

Until 2003, most of the information about Mars came from the *Mariner 9, Viking probes, Mars Global Surveyor, Mars Pathfinder,* and *Odyssey.* The most recent data come from the *Mars Exploration Rover Mission* and its two landers, *Spirit* and *Opportunity.*

The Viking Probes

In 1976, the *Viking 1* and *Viking 2* probes landed on Mars at the locations shown in **Figure 13.** Each spacecraft consisted of an orbiter and a lander. The *Viking 1* and *Viking 2* orbiters photographed the entire surface of Mars from orbit, while their landers conducted meteorological, chemical, and biological experiments on the planet's surface. Although initial results of the biological tests provided data that seemed to indicate the presence of life, scientists soon realized that the same data also could indicate chemical reactions. The biological experiments found no conclusive evidence of life in the soil.

The *Viking* landers also sent back pictures of Mars's reddish-colored, barren, rocky, and windswept surface.

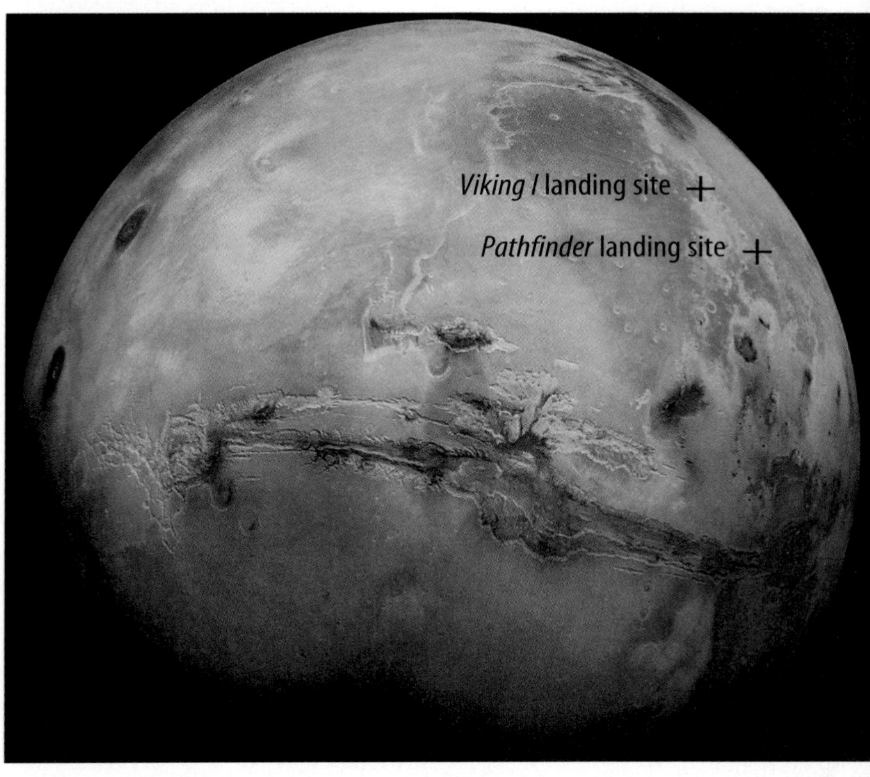

Figure 13 Mars's *Valles Marineris* is six to seven times deeper than Earth's Grand Canyon.

Martian Minerals In addition to "blueberries," *Opportunity* discovered evidence that water once existed on Mars. It found deposits of salts and high concentrations of sulfur. It also found holes in Martian deposits left behind when a mineral is eroded. Research evidence that indicates water was in *Meridiani Planum.* Write about it in your Science Journal.

Global Surveyor, Pathfinder, and _Odyssey_ Cameras on board *Global Surveyor* showed that the walls of *Valles Marineris,* shown in **Figure 13,** have distinct layers similar to those of the Grand Canyon. They also showed that a vast flat region, resembling a dried-up seabed or mudflat, covers a large area of Mars's northern hemisphere. Many believe the water is now frozen into Mars's crust at the poles or has soaked into the ground. In fact, a more recent probe, *Mars Odyssey,* provided evidence for water as frost beneath a thin layer of soil in the far northern and southern parts of Mars.

The *Mars Pathfinder* and its rover, *Sojourner,* gathered data that indicated that iron in Mar's crust may have been leached out by groundwater.

Mars Exploration Rover Mission Recall the round concretions, called "blueberries," that indicate water may once have existed on the surface of Mars. These were discovered by the *Opportunity* rover in *Meridiani Planum.* The rover confirmed that the concretions were hematite deposits. The fact that the hematite was deposited on top of other features suggests that it was deposited in a standing body of water. Other deposits of hematite have been found by the *Spirit* rover in *Gusev Crater.* The *Mars Exploration Rover* mission that placed these two landers on Mars provided a great deal of evidence that water existed on Mars in the past. Photographs of the Martian surface taken by the two landers *Spirit* and *Opportunity* are shown in **Figure 14.**

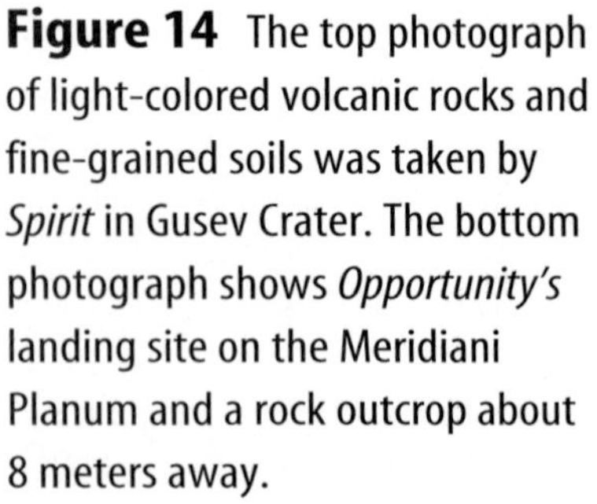

Figure 14 The top photograph of light-colored volcanic rocks and fine-grained soils was taken by *Spirit* in Gusev Crater. The bottom photograph shows *Opportunity's* landing site on the Meridiani Planum and a rock outcrop about 8 meters away.

Martian Meteorites Space probes sent to Mars are not the only way we have learned about our planetary neighbor. Meteorites have been found on Earth that are believed to have been blasted into space from Mars. An interesting discovery was reported in 1996 that caused a lot of debate over the possibility of life on Mars. A meteorite from Mars (ALH84001) that had been found in the Antarctic in 1984 was studied by a group of scientists. They found long, egg-shaped microscopic structures that they thought were fossilized evidence of ancient life on Mars. The meteorite, shown in **Figure 15,** is thought to have been ejected by an impact on Mars about 16 million years ago.

Figure 15 This meteorite fragment found in Antarctica is believed to have been dislodged from Mars by a huge impact about 16 million years ago.

How was a meteorite that originated on Mars found on Earth?

A debate began in which many scientists suggested other possible reasons for the egg-shaped structures. They pointed out that processes not related to life might have formed these structures. Although no firm conclusions can be drawn from this evidence, research will continue until data are found to determine whether life exists, or may once have existed, on Mars.

section 2 review

Summary

Mercury

- Mercury has an extreme range of temperatures, many craters, and long, steep cliffs.

Venus

- The sizes and masses of Venus and Earth are similar.
- The carbon dioxide in Venus's atmosphere has caused an extreme greenhouse effect, creating high surface temperatures.

Earth

- Surface temperatures on Earth allow water to exist as a solid, a liquid, and a gas.
- Earth is the only planet definitely known to support life.

Mars

- Mars has ice caps and surface coloring that experience seasonal changes.
- There is much evidence of water on Mars in both past and present.

Self Check

1. **Compare and contrast** Mercury, Venus, Earth, and Mars.
2. **Describe** the atmosphere of Venus.
3. **Explain** why scientists believe Mars was once a wetter planet.
4. **Explain** why the *Mars Exploration Rover* mission is considered a success.
5. **Think Critically** Explain why Mars is a more likely planet for humans to visit than Venus.

Applying Math

6. **Using Percentages** If the diameter of Mars is 53.3 percent of Earth's diameter (12,756 km), find its diameter in kilometers.
7. **Converting Units** If Earth takes 365.26 days to orbit the Sun and Mars takes 686.98 days, how many Earth years does Mars take to orbit once around the Sun?

More Section Review gpescience.com

Interpreting Features of Mars's Surface

Much of what scientists have learned about Mars involves interpreting data and photographs obtained from spacecraft that explored Mars. Comparing features found on the surface of Mars to similar features on Earth enables scientists to hypothesize about what may have caused these features. How can you interpret Martian surface features in the classroom?

Real-World Problem

How can you interpret the surface features shown on a photograph of the surface of Mars?

Goals

- **Identify** surface features shown on a photograph of Mars.
- **Interpret** what processes were involved in the formation of these surface features.

Materials

photograph of the Martian surface from NASA
clear-plastic sheet
scissors
erasable marker
transparent removable tape

Alternate Materials

additional laminated photographs of the Martian surface

Safety Precautions 

Procedure

1. Cut the clear-plastic sheet to fit, and tape it over the photograph.
2. Use the erasable marker to mark, trace, or circle the features you can recognize, such as craters, lava flows, flow structures (channels), terraces, and sediment layering.
3. Indicate the direction that a flow of lava or water may have traveled.

Conclude and Apply

1. **Make a table** to show the data you will obtain in this Lab.
2. **List** in your table the features you were able to recognize in the photograph.
3. **Identify** and **list** in your table a process that might have formed each feature.

NASA Photograph MOC2-102

Surface Feature	Formation Process
Do not write in this book.	

4. **Recognize cause and effect** for the flow structures you identified. Record possible flow material and the direction for flow.
5. **Infer** from which direction sunlight was shining on the Martian surface when this photograph was taken.

Communicating Your Data

Explain to friends, classmates, or your family that this is what NASA scientists do with data obtained from spacecraft.

section 3

The Outer Planets

Reading Guide

What You'll Learn

- **Compare and contrast** the outer planets.
- **Describe** important characteristics for each outer planet.
- **Evaluate** the success of NASA missions *Galileo* and *Cassini*.

Why It's Important

Studying the outer planets helps scientists to understand the formation of planets orbiting other stars.

Review Vocabulary

space probe: an instrument that is sent to space to gather information

New Vocabulary

- Jupiter
- Saturn
- Uranus
- Neptune
- Pluto
- comet
- asteroid
- meteoroid
- Sedna

Why are the outer planets so different?

Recall how the small, rocky inner planets formed. High temperatures near the Sun prevented the lighter elements from condensing readily, so these elements are scarcer in the planets closer to the Sun. Things were different in the outer portion of the cloud fragment that formed our solar system. There, lighter elements condensed, collided, and stuck together. The growing masses collected still more particles because of the stronger pull of gravity. This produced the giant planets found in the outer part of our solar system. Only Pluto does not fit this description.

Jupiter

The largest and fifth planet from the Sun is **Jupiter.** It is composed mostly of hydrogen and helium, with some ammonia, methane, and water vapor. Scientists theorize that the atmosphere of hydrogen and helium gradually changes to a planetwide ocean of liquid metallic hydrogen toward the middle of the planet. Below this liquid layer may be a solid rocky core much larger than Earth. However, the extreme pressure (about 50 million Earth atmospheres) and temperature (40,000°C), make the core different from any rock on Earth.

You've probably seen pictures of Jupiter's colorful clouds. Its atmosphere has bands of white, red, tan, and brown clouds, as shown in **Figure 16.** Continuous storms of swirling, high-pressure gas have been observed on Jupiter. The Great Red Spot is the most spectacular of these storms. Lightning also has been observed within Jupiter's clouds.

Figure 16 The Great Red Spot on Jupiter in the lower right is a storm that has raged for more than 300 years.

Figure 17 Jupiter's gravity holds more than sixty moons in orbit. Two of them are shown above.

Space Probes to Jupiter

In 1979, *Voyager 1* and *Voyager 2* flew past Jupiter, and the *Galileo* space probe reached Jupiter in 1995. The major discoveries of these probes include information about the composition and motion of Jupiter's atmosphere, characteristics of some of its moons (including volcanoes on one of them), and the discovery of new moons. The *Voyager* probes also discovered that Jupiter has faint dust rings around it.

Jupiter's Moons Jupiter has more than 60 moons. You can see some of them orbiting the planet in **Figure 17.** Many of these are small, rocky bodies that may be captured asteroids. However, four are large enough to be considered small planets. Galileo was the first to discover these moons when he looked through his telescope and he drew a picture of what he saw. These Galilean moons of Jupiter are Io, Europa, Ganymede, and Callisto, and are described in more detail in **Figure 18.** One of these, Ganymede, is larger than the planet Mercury and is the largest moon in the solar system.

Io is under a constant tug-of-war between the gravities of Jupiter and Europa. This heats up the interior of Io and causes it to be the most volcanically active body in the solar system. The volcanoes on Io were first seen in photographs from the *Voyager* probes.

The most recent interpretation of the data sent back by the *Galileo* space probe about the other three Galilean moons is that they all most likely have an ocean of water underneath an ice-rock crust. Some scientists speculate that Europa's ocean is very large and could possibly harbor some kind of life. Using gravity and magnetic field measurements, NASA scientists generated a model that shows the existence of water beneath Europa, Ganymede, and Callisto. When *Galileo's* mission ended, NASA chose to send the probe into Jupiter's atmosphere, where it was crushed and destroyed. They didn't want to risk having it crash on Europa and contaminate any life that might exist there.

Reading Check *Why do scientists think Europa is a good candidate for supporting life?*

Naming Moons You may know that the names of all the planets except Earth come from names of Greek and Roman gods. But did you realize that many of the names chosen for moons come from either mythology or literature? For example, Charon was the name of the boatman who ferried the dead to the underworld kingdom of Pluto. Research the moons of Mars and find out why their names are appropriate.

NATIONAL GEOGRAPHIC

VISUALIZING WHAT GALILEO SAW

Figure 18

Galileo Galilei, 1564–1642 In 1610, Italian scientist Galileo Galilei did something revolutionary for his time. Galileo used a telescope to look at the planet Jupiter and saw four objects revolving around it. This and other observations convinced him that our solar system is heliocentric.

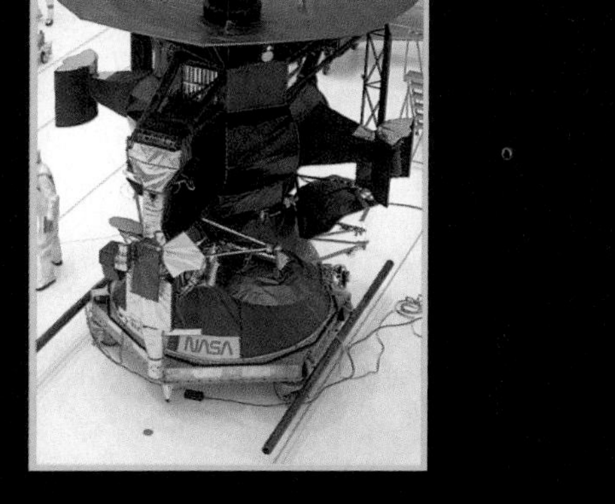

Like its namesake, the *Galileo* spacecraft changed our understanding of the solar system. *Galileo* launched a probe into Jupiter's atmosphere, which gathered data about cloud layers, chemical composition, and lightning.

Ganymede, the largest moon in the solar system, is larger than the planet Mercury. It has grooves covering its surface that indicate a violent past. These grooves most likely resurfaced over time through volcanic activity.

Callisto is marked by one giant crater called Valhalla, which was formed by an impact over four billion years ago. Callisto has a thin atmosphere of carbon dioxide and most likely has an ocean of salty water surrounding its interior of ice and rock.

Io is caught between Jupiter's tremendous gravitational force and the gravity of neighbor Europa. It is the most volcanically active object in the solar system.

Europa is mostly rock with a thick, cracked but smooth crust of ice.

Figure 19 *Cassini-Huygens* provided detailed views of Saturn and its rings.

Saturn

NASA began a new study of Saturn when the *Cassini-Huygens* spacecraft reached Saturn in July 2004. The spacecraft's arrival, insertion into orbit, and initial operation went flawlessly. Many data about Saturn, its rings, and its moons were collected during its approach. **Saturn,** shown in **Figure 19,** is the sixth planet from the Sun and is known as the ringed planet. It is not the only ringed planet, but it has the largest and most complex ring system. Based on *Cassini-Huygens* data, Saturn could have at least 34 moons. Although Saturn is the second-largest planet in the solar system, it has the lowest density. This is because Saturn's gravitational pull is weaker than Jupiter's—Saturn's cloud layers are not squeezed as close together as cloud layers in Jupiter's atmosphere are.

Like Jupiter, Saturn is a large planet with a thick outer atmosphere composed mostly of hydrogen and helium with some ammonia, methane, and water vapor. As you go deeper into Saturn's atmosphere, the gases gradually change to liquid hydrogen and helium. Below its atmosphere and liquid ocean, Saturn may have a small rocky core, but no rocky surface like Earth's. When compared to Earth, the pressures and temperatures at the core must be tremendous.

Topic: Discoveries of *Cassini-Huygens*
Visit gpescience.com for Web links to information about discoveries made by the *Cassini* and *Huygens* probes at Saturn.

Activity Write a paragraph describing new discoveries made by *Cassini* space probe about Saturn.

Saturn's Rings Saturn's rings are composed of countless ice and rock particles ranging in size from a speck of dust to tens of meters across. The rings appear as several broad bands, each of which is composed of thousands of thin ringlets.

As *Cassini-Huygens* entered its orbit around Saturn, it took close-up pictures of these rings that showed a density wave, just as had been predicted by physicists, as shown in **Figure 20.** Also shown in this figure, are areas where the ring material bends up and down. Other *Cassini-Huygens* photographs showed that the gravity of Saturn's moons affects its rings and the presence of a new ring associated with the moon, Atlas.

Figure 20 *Cassini* produced this image by filtering starlight through Saturn's rings. The bands of color represent changes in ring density as rings are affected by the gravity of two of Saturn's moons.

***Cassini-Huygens* at Saturn** *Cassini-Huygens* is actually two spacecraft in one. *Cassini* was programmed to orbit Saturn for over four years and to release a probe named *Huygens* when it passed Saturn's moon Titan. *Huygens* was designed to descend through the thick atmosphere of Titan and land on its surface. You will learn more about Titan later.

Uranus

After touring Saturn, *Voyager 2* flew by Uranus in 1986. **Uranus,** shown in **Figure 21,** is the seventh planet from the Sun and wasn't discovered until 1781. It is a large planet with 27 moons.

Uranus's axis of rotation is tilted on its side, so that it is nearly perpendicular to the plane of its orbit. This may have happened as the result of a collision with another object. *Voyager* also revealed numerous thin rings, and detected that the planet's magnetic field is tilted 55 degrees from its rotational poles.

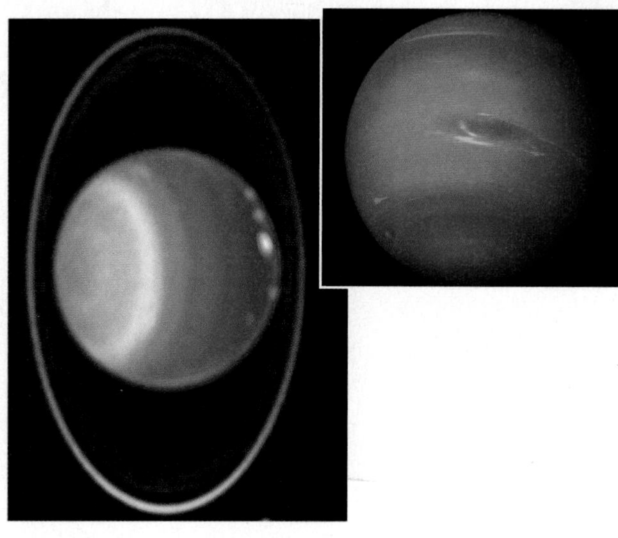

Figure 21 Uranus, left, and Neptune, right, are about the same size. Uranus has faint cloud bands and a ring system that are visible only in false color.

The atmosphere of Uranus contains hydrogen, helium, and about two percent methane. The methane gives the planet its blue-green color because methane absorbs red and yellow light, and the clouds reflect green and blue light. No cloud bands and few storm systems are seen on Uranus because of its cold upper atmosphere. This lack of bands and storms may be connected to the fact that Uranus has less internal heat than Neptune. Evidence suggests that under its atmosphere, Uranus has a mantle of liquid water, methane, and ammonia surrounding a rocky core.

Reading Check *What causes Uranus to have a blue-green color?*

Neptune

From Uranus, *Voyager 2* traveled on to Neptune. **Neptune,** another large planet similar in size to Uranus, is the eighth planet from the Sun. Neptune, also shown in **Figure 21,** was discovered in 1846 when studies of Uranus's orbit indicated that the gravity of another unseen planet must be pulling on it. Under its atmosphere, Neptune is thought to have liquid water, methane, and ammonia covering a rocky core. *Voyager* also detected that Neptune has rings that are thin in some places and thick in others.

Neptune's atmosphere is similar to that of Uranus, but has a little more methane—about three percent—causing it to look bluer. Neptune's atmosphere has dark spots similar to the Great Red Spot on Jupiter. With its dynamic atmosphere, Neptune's internal heat must be greater than Uranus's.

Neptune has at least 13 moons. Triton, the largest, has a diameter of 2,700 km and a thin atmosphere composed mostly of nitrogen. Some methane is also present; *Voyager* detected methane geysers erupting on Triton.

Comparing Sizes of Outer and Inner Planets

Procedure

1. Research the planets to find their diameters in kilometers.
2. Select a scale to draw the planets. The scale must be such that the small planets can be seen and the larger planets can fit your paper.
3. Using your scale, draw a circle with the diameter of the smallest planet.
4. Using the same scale, draw circles representing each of the other planets.

Analysis

1. How do the model sizes of the inner planets compare to those of the outer planets?
2. Which planet is smallest? Largest?

Try at Home

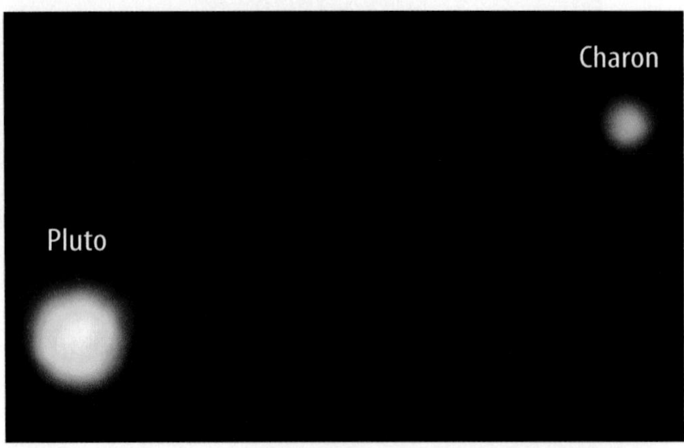

Figure 22 Pluto's moon, Charon, is about half the size of Pluto.

Pluto

The smallest planet in our solar system and the one we know the least about is Pluto. Because **Pluto** is farther from the Sun than Neptune during most of its orbit, it is considered the ninth planet from the Sun. Although Pluto comes closer to the Sun than Neptune for part of its orbit, the two planets will never collide because Pluto's orbit is inclined farther from the ecliptic plane. Because of this angle, the two orbits cannot intersect.

Pluto is unlike the other outer planets. It has a thin atmosphere and is the only outer planet with a solid, ice-rock surface. Pluto's only moon, Charon, is large compared to Pluto and orbits close to Pluto, as shown in **Figure 22.**

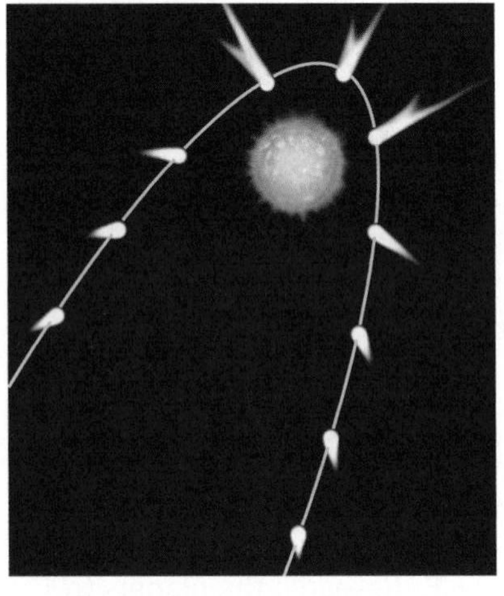

Figure 23 Comets often have two tails—one made of dust and one made of ions.

Comets and Other Objects

A **comet** is composed of dust and rock particles mixed with frozen water, methane, and ammonia. As a comet approaches the Sun, it begins to vaporize because of heat from the Sun. The released dust and gases form a bright cloud called a coma around the nucleus, the solid part of the comet. The solar wind pushes on the vaporized coma, forming a tail that always points away from the Sun, as shown in **Figure 23.**

Most comets come from two places—a vast disk of icy comets called the Kuiper Belt near Neptune's orbit and the Oort Cloud, which completely surrounds the solar system. Gravity from the Sun or from passing stars may periodically disturb these areas, sending comets on long journeys toward the Sun. Once in orbit around the Sun, comets reappear at predictable times. One of the most famous of these is Halley's comet, which returns every 76 years.

Figure 24 Asteroid 433 Eros has an irregular shape and is 23 km long. Spacecraft *NEAR-Shoemaker* landed on its surface in 2001.

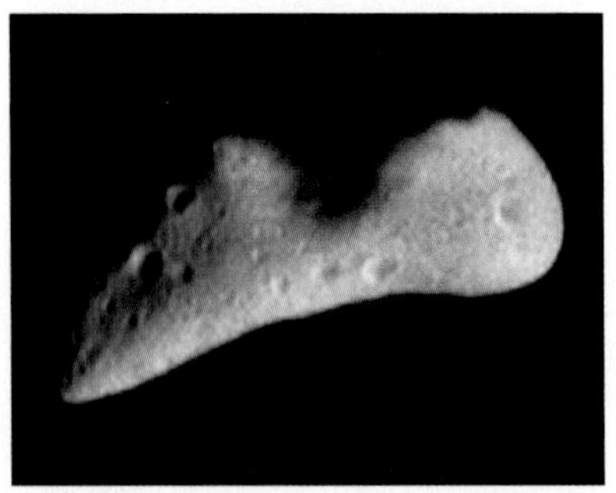

Asteroids Rocky objects formed from material similar to that of the planets are called **asteroids.** One of these is shown in **Figure 24.** Most asteroids are found in a belt between the orbits of Mars and Jupiter. Asteroids range in size from tiny particles to objects 940 km in diameter.

Meteoroids Other rocky objects orbiting within the solar system are **meteoroids.** They may have formed when asteroids collided or when comets passed close to the Sun, leaving a trail of debris. When Earth passes through these trails, meteoroids may enter the atmosphere. Most burn up completely and we see them as meteors or "shooting stars." Others do not burn completely and strike Earth. These are called meteorites.

Sedna Another object in the solar system that has puzzled astronomers is Sedna, shown in **Figure 25. Sedna** has been labeled a distant planetoid and with a diameter of 1,200 to 1,700 km, it is smaller than Pluto, but larger than comets in the Kuiper Belt. Also, it has a very elliptical orbit. Sedna comes as close as 76 AU from the Sun, but travels to a distance of 950 AU—far beyond the Kuiper Belt, but much closer than the Oort Cloud.

Some suggest that the Sedna's orbit was severely affected by a passing star less than 100 million years after the Sun formed. This supports the idea that our Sun formed as part of a star cluster. A close encounter with another star so soon after the Sun's formation and so nearby (800 AU) could be explained if the Sun formed within a cluster.

Figure 25 Sedna is larger than an asteroid or comet, but smaller than a planet.

Another puzzle is Sedna's apparent rate of rotation of 40 days. This is best explained if the object had another object in orbit around it. Even if Sedna has no companion, there may be other similar objects, some even larger than Pluto, out there. At present, objects this far away and having orbits this elliptical can be detected only when they are closest to the Sun.

section 3 review

Summary

Jupiter

- Jupiter is the largest planet in the solar system.
- The four largest moons of Jupiter are Io, Europa, Ganymede, and Callisto.

Saturn

- Saturn has the most complex system of rings.

Uranus, Neptune, and Pluto

- Uranus's axis of rotation is tilted on its side.
- Neptune's atmosphere has storms similar to those on Jupiter.
- Pluto has a solid, icy-rock surface.

Comets, Asteroids, and Meteoroids

- The asteroid belt lies between the orbits of Mars and Jupiter.
- Meteoroids form from pieces of comets and asteroids.

Self Check

1. **Compare and contrast** Jupiter, Saturn, Uranus, Neptune, and Pluto.
2. **Explain** what Saturn's rings are composed of.
3. **List** three things scientists learned about Jupiter and its moons from the *Galileo* space probe.
4. **Think Critically** Explain what would happen to the density of Saturn if its atmospheric layers squeezed inward toward the center of the planet.

Applying Math

5. **Use Numbers** If the diameters of Earth and Jupiter are 12,756 km and 142,984 km, respectively, how many Earths could fit across Jupiter's equator?
6. **Use Decimals** On average, if Earth is 150 million km, or 1 AU, from the Sun and Venus is 0.71 AU from the Sun, what is the distance to Venus in kilometers?

section 4

Life in the Solar System

Reading Guide

What You'll Learn

- **Evaluate** other planets in the solar system for the possibility of containing life.
- **Examine** life forms that exist in exotic locations on Earth.
- **Discuss** locations in which life might exist on other planets.

Why It's Important

The lack of evidence showing that life exists elsewhere helps you realize how important it is to take care of Earth.

Review Vocabulary

fossil: remains, imprints, or traces of past life

New Vocabulary

- extraterrestrial life

Life As We Know It

We know life exists in our solar system. It's all around us; in the air, on rocks, and in the sea, as shown in **Figure 26.** It thrives even in harsh environments on Earth, but does it exist anywhere else in the solar system and, if so, where? What does life need to survive? Life on Earth is carbon-based and requires water to survive. Does water exist elsewhere in sufficient quantities to make life possible?

Think about what you consider to be life. You might think of yourself, your friends, your pet, and maybe birds flying by the window. There is so much diversity to life on Earth. There is life we easily can see, such as grass, trees, and animals, but there is also a multitude of life that we cannot see. For example, microorganisms are so small you need a microscope to see them. When we define life, we might forget about microorganisms but they are very important to us.

Figure 26 Tube worms and mussels at this hydrocarbon seep on the floor of the Gulf of Mexico get their energy from bacteria that consume methane. Predators, such as crabs feed on the community.

If you were asked to check a sample of soil to see if there was or ever had been life in it, what would you do? You might look for moving things, but motion doesn't always mean something is alive. You might look for residues or traces of life that once lived in the soil, such as fossils. You know that you release gases as you breathe. You might check to see if gases are being emitted from the soil. Think about what type of tests you could perform on Martian soil to check for life, and remember that these tests would have to be done by a robot lander.

Exotic Life on Earth Before we rule out the possibility of life on other worlds, we should take a closer look at life on Earth. Some of the exotic life-forms that exist on Earth are strange, indeed.

In the 1970s, a research submarine, the *Alvin,* found some interesting life-forms while exploring hot, volcanic vents on the ocean floor at depths where light could not penetrate. A variety of life was living where scientists didn't think it could. Life-forms found included crabs, clams, and tubeworms. They were living and thriving in a dark environment under tremendous pressures. *Alvin* also discovered colonies of bacteria living off the extremely hot material spewing from volcanic vents. These organisms use thermal and chemical energy to survive and are the basis of food webs there.

A surprising amount of life has been found elsewhere on Earth in extreme conditions—scalding heat, freezing cold, salt, lye, and darkness. For example, a moss shown in **Figure 27** was dormant for 40,000 years in permafrost in northeastern Siberia. Three forms of bacteria thrive in the mud of Mono Lake, California, where the environment is dark, lacks oxygen, and contains a large amount of salt and lye. What does this tell us about the ability of life to exist in the harsh environments of other planets?

Figure 27 The moss shown to the left is growing again after being dormant for 40,000 years. The photograph above shows a researcher collecting mud samples from Mono Lake, California. One bacteria in this mud obtains energy from chemicals rather than sunlight.

Can life exist on other worlds?

Think about what you have learned about Earth compared to other planets. None of the planets you have studied has characteristics so close to Earth's that you could definitely say life as we know it could exist there. But can you say for certain that life cannot exist on other planets in the solar system?

The high temperatures found on Mercury and Venus and the lack of a surface on Jupiter and the other gaseous giants make finding life as we know it unlikely. But such facts do not rule out the possibility of some form of **extraterrestrial life,** that is, life on other worlds. Let's take a look at some places where scientists are searching for extraterrestrial life.

Reading Check ***What is extraterrestrial life?***

Figure 28 This crater on Mars is called a splosh crater. It is hypothesized that a splosh crater forms from the impact of an object on a water-soaked surface.

Mars During most of Earth's history, one-celled organisms were the major form of life on Earth. Could such life-forms be present on Mars today or could they have existed on Mars in the past when the planet was wetter? Evidence from space probes has shown that Mars probably had large amounts of water on its surface, as shown in **Figure 28.** If life existed and left evidence, future astronauts will find it. Such a discovery would be almost as exciting as finding living life-forms.

Actually, searches for life on Mars already have been done. *Viking* landers on Mars performed some tests designed to detect life. The first tests showed positive results, but additional study showed that these results could be explained just as well by chemical reactions that did not require life. Ultimately, the *Viking* experiments found no evidence of life. However, the tests used only would detect life similar to that which exists on Earth. What if life on Mars is totally different from life on Earth?

Europa Data from the *Galileo* probe of Jupiter and its moons indicate that three Galilean moons have ice-covered oceans. The oceans under the ice of Ganymede and Callisto might be partially frozen. However, data related to surface cracking, as shown in **Figure 29,** tidal bulges, and movement of the surface over some lubricating substance below, support the idea of a liquid ocean about 25 km deep in Europa. Europa's ocean could hold more than twice the amount of water that Earth's oceans hold.

Some scientists speculate there is a chance that Europa's ocean could harbor life. One reason is that it might be deep and long-lasting, possibly even liquid water that is warmed by Jupiter's gravitational pull on Europa.

On Earth, we find life everywhere there is water. If life can exist in extremely hot liquid flowing from volcanic vents on Earth's ocean floor, could it not exist in a similar environment if it exists on Europa? In fact, ocean currents have been detected on Europa. Today, some scientists consider Europa to be the most promising location to look for extraterrestrial life.

Figure 29 Surface cracking on Europa may indicate that a liquid ocean lies beneath.

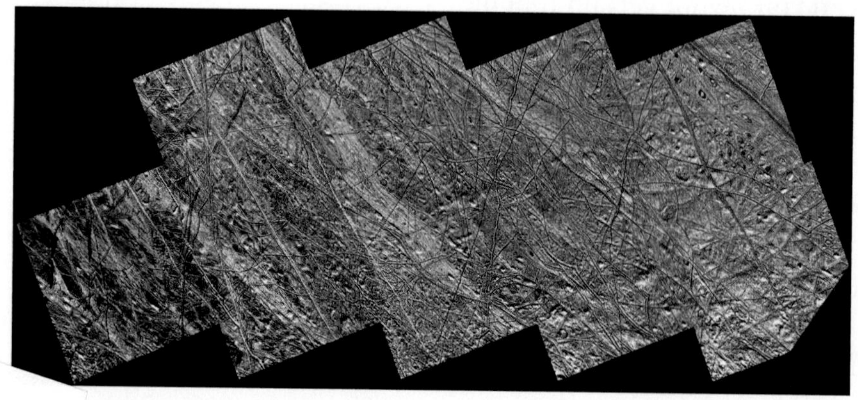

Titan Saturn's moon, Titan, shown in **Figure 30,** is larger than Mercury and has an atmosphere composed mostly of nitrogen, with some argon and methane. Certain features on Titan's surface indicate this moon has a long-term, interesting geologic history. Some darker areas contain relatively pure water-ice, while other, brighter areas are composed of hydrocarbons, organic compounds containing hydrogen and carbon. Recall that life on Earth is carbon-based. A bright cloud of the hydrocarbon methane is visible near the south pole. Large particles in this area indicate that a dynamically active atmosphere exists there.

The presence of hydrocarbons on the surface of Titan interests exobiologists, scientists who search for evidence of life on other worlds. Also, the apparent absence of large impact craters indicates that Titan has experienced internal geologic activity. This activity could provide the energy needed for organic molecules to develop into the building blocks of life.

A more detailed study of Titan is programmed for the *Huygens* probe. Data obtained from the *Huygens* probe might help solve the mystery of whether life exists or may have existed on Titan.

Figure 30 This image of Titan taken by *Cassini* shows two thin layers of hazy atmosphere that are thought to contain nitrogen and methane.

What fact about Titan particularly interests exobiologists?

section 4 review

Summary

Life As We Know It

- Life as we know it is carbon-based and needs liquid water to survive.

Exotic Life On Earth

- Life exists in extreme environments on Earth.
- Life can remain dormant for long periods of time and still survive.

Mars

- *Viking's* tests for life on Mars were inconclusive.

Europa

- There is strong evidence for a warm, liquid ocean on Europa.
- Europa is considered the most promising location to look for extraterrestrial life.

Titan

- Organic compounds exist on Titan.

Self Check

1. **Define** extraterrestrial life.
2. **Describe** how finding life in extreme environments on Earth supports the idea that life may exist elsewhere.
3. **Explain** how hot fluids from volcanic vents on Earth's ocean floor support energy for life.
4. **Discuss** other locations in the solar system where life may once have existed.
5. **Think Critically** Explain why scientists consider Europa to be a good place to search for extraterrestrial life.

Applying Math

6. **Use Numbers** Atmospheric pressure at the surface of Venus is 92 times Earth's at sea level. If atmospheric pressure on Earth at sea level is 101.3 kPa, find the atmospheric pressure on the surface of Venus in kPa.
7. **Use Percentages** The average composition of Mars's atmosphere is 95% CO_2, 3% N, and 2% Ar. Approximately what fraction of Mars's atmosphere is nitrogen?

Design Your Own

Testing Unknown Soil Samples for Life

Goals

- **Identify** several things that you might see if life existed in an unknown sand or soil sample.
- **Demonstrate** processes you could use to search for life in the unknown sample.
- **Illustrate** evidence of life that you observe in each of your samples.
- **Compare and contrast** observations that may or may not indicate that life exists in your samples.

Possible Materials

clear plastic cups (3)
sand or soil
warm tap water
magnifying glass
thermometer

Safety Precautions

Real-World Problem

What is life? You know that life needs water, air, and food to survive. The type of food depends on the life form. You eat certain foods to provide you with the energy to live. Other forms of life need different foods for energy. You know that many plants make their own food using photosynthesis, and you have learned about life-forms on Earth that obtain their energy from chemical reactions. How might you find out if life is present in an unknown sample? Perhaps you might look for evidence that an organism was taking in one gas and expelling others. Perhaps you might look for energy being produced. How could you determine if an unknown sample of sand or soil contained life?

Form a Hypothesis

Form a hypothesis about what evidence might indicate that a sand or soil sample contained life. Have your teacher approve your hypothesis before you begin your experiment.

Test Your Hypothesis

Make a Plan

1. As a group, agree upon the hypothesis and decide how to test it. Identify which results will confirm your hypothesis.
2. **List** the steps you will need to test your hypothesis. Be sure to include a control run. Describe exactly what you will do in each step.

3. **List** the materials you will need to perform the experiment. Determine who will prepare your unknown samples from your materials.
4. **Prepare** an illustration in your Science Journal where you will record your observations.
5. **Read** the entire experiment to make sure all steps will be performed in a logical order.
6. **Identify** all constants, variables, and controls of the experiment. Keep in mind that the person who prepares your unknowns must follow your instructions carefully and not divulge which sample(s) contain life.

Follow Your Plan

1. Make sure your teacher approves your plan and oversees the preparation of your unknown samples before you start.
2. Carry out your experiment as planned.
3. While doing the experiment, record your observations in your illustration in your Science Journal.

Analyze Your Data

1. **Illustrate** your visual observations of unknown sand or soil samples.
2. Label your drawings with other observations that you make, such as production of heat or change in odor.

Conclude and Apply

1. **Examine** Do the illustrations show the observations you expected to make? Describe any unexpected observations noted.
2. **Determine** Do any of your observations indicate that life exists in any of your unknown samples? Describe any inconclusive observations.
3. **Draw Conclusions** Did your results support your hypothesis? Explain why or why not.

Communicating Your Data

Compare your data with those of other students. Discuss how your experiments may have differed, and how this could have changed the observations.

SCIENCE Stats

Night Sky Wonders

Did You Know...

. . . Five of the nine planets can be seen from Earth without the aid of a telescope or binoculars. Mars, Venus, Mercury, Jupiter, and Saturn can be seen at various times throughout the year in the night or early morning sky. Mercury is visible for only a few days, but the other planets are often clearly visible, such as Mars and Venus shown here.

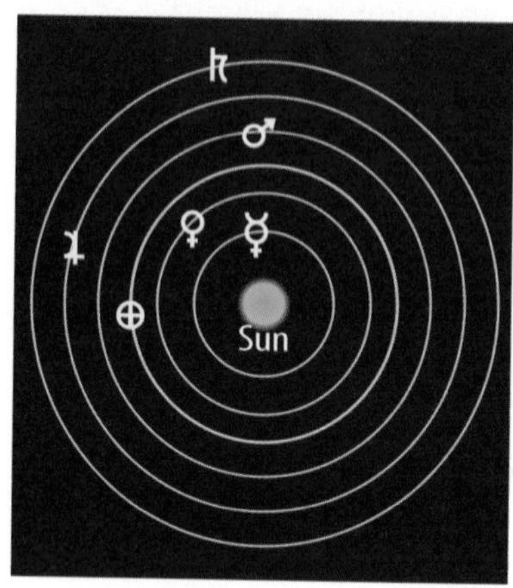

. . . All five planets were visible as a group in the evening sky in the spring of 2004. All five planets were briefly positioned on the same side of the Sun at the same time. This stunning sight will not be seen in the evening skies again until 2036. Because they orbit the Sun at different rates, it is rare for all five visible planets to be as close together in the sky as they were in 2004.

Applying Math

1. How old will you be when the next planetary alignment like the one that was visible in 2004 can be seen from Earth again?
2. The reflected light we see when observing planets travels through space at 300,000 km/s. If Jupiter is 588.5 million km from Earth at its closest approach, what is the minimum amount of time required for light to reach Earth from Jupiter?
3. Using the heliocentric longitude diagram you made in the Applying Math feature, plot the positions of the planets using the following longitudes: Mercury 96°, Venus 135°, Earth 184°, Mars 94°, Jupiter 165°, and Saturn 103°.

Reviewing Main Ideas

Section 1 Planet Motion

1. In the geocentric model of the solar system, Earth is in the center. In the heliocentric model, the Sun is in the center.
2. The position of a planet in orbit around the Sun can be shown by heliocentric longitude.
3. Inner planets have orbits closer to the Sun than the asteroid belt; outer planets have orbits beyond the asteroid belt.
4. A large nebula, like the one shown below, fragmented into smaller ones, one of which later condensed and flattened to form the solar system.

Section 2 The Inner Planets

1. Mercury has a larger iron core than expected for a planet its size.
2. The carbon dioxide in Venus's dense atmosphere has caused a strong greenhouse effect that produces very high surface temperatures.
3. Mars landers like the one shown in the next column have provided evidence that Mars, once had more water on its surface than it does today.

Section 3 The Outer Planets

1. Jupiter is the largest planet in the solar system and has the most moons. Four of these moons were first seen by Galileo.
2. Saturn has the most developed ring system of all outer planets.
3. The green-blue color of Uranus and the blue color of Neptune are caused by methane in their atmospheres.
4. Pluto is the smallest planet, and unlike other outer planets, it has a thin atmosphere and an icy rock surface.

Section 4 Life in the Solar System

1. Life exists in extreme environments on Earth and may be able to survive in extreme environments on other planets.
2. Europa is considered the most promising location in the solar system to look for extraterrestrial life.

FOLDABLES™ Use the Foldable that you made at the beginning of this chapter to help you review details about each of the planets.

chapter 8 Review

Using Vocabulary

asteroid p. 236	Mars p. 225
astronomical unit p. 220	Mercury p. 223
comet p. 236	meteoroid p. 236
Earth p. 225	Neptune p. 235
extrasolar planet p. 222	Pluto p. 236
extraterrestrial life p. 239	Saturn p. 234
geocentric model p. 218	Sedna p. 237
heliocentric model p. 219	Uranus p. 235
Jupiter p. 231	Venus p. 224

Use what you know about the vocabulary terms to explain the differences in the following sets of words. Then explain how they are related.

1. Uranus—Neptune
2. extrasolar planet—extraterrestrial life
3. Pluto—Sedna
4. Venus—Earth
5. Mercury—Jupiter
6. asteroid—Sedna
7. geocentric model—heliocentric model
8. astronomical unit—Earth
9. Jupiter—Saturn
10. Earth—Mars

Checking Concepts

Choose the word or phrase that best answers the question.

11. Who proposed the heliocentric model of the solar system?
    A) Copernicus **C)** Kepler
    B) Galileo **D)** Ptolemy

12. Which planet has the most complex ring system?
    A) Jupiter **C)** Saturn
    B) Neptune **D)** Uranus

13. Which planet is the red planet?
    A) Jupiter **C)** Neptune
    B) Mars **D)** Venus

14. Which chemical or gas produces a green-blue to blue color in Uranus and Neptune?
    A) carbon dioxide **C)** oxygen
    B) methane **D)** sulfur dioxide

15. Which planet has the highest surface temperature?
    A) Earth **C)** Mercury
    B) Mars **D)** Venus

16. Which fact is true of superior planets?
    A) Earth-sized
    B) Jupiter-sized
    C) lie beyond Earth's orbit
    D) lie beyond the asteroid belt

17. Which Galilean satellite has active volcanoes?
    A) Callisto **C)** Ganymede
    B) Europa **D)** Io

Use the illustration below to answer question 18.

18. Which planet is shown?
    A) Jupiter
    B) Mars
    C) Saturn
    D) Venus

19. Which part of a comet is on the opposite side of a comet from the Sun?
    A) coma **C)** nucleus
    B) head **D)** tail

20. What is a reason Europa might be a good place to look for extraterrestrial life?
    A) liquid methane
    B) presence of organic molecules
    C) rust-colored rocks
    D) water ocean and tidal energy

Interpreting Graphics

21. **Make a table** summarizing the gases contained in the atmospheres of the nine planets.
22. **Make a model** of how you think the geocentric model of the solar system was organized.

Use the data in the table below to answer questions 23–25.

Planet Orbits

Planet	Distance to the Sun (AU)
Mercury	0.39
Venus	0.72
Earth	1.00
Mars	1.52
Jupiter	5.20
Saturn	9.54
Uranus	19.19
Neptune	30.07
Pluto	39.48

23. Which planet is a little over nine times farther from the Sun than Earth?
24. How far is the orbit of Mars from that of Jupiter in AU's?
25. How much closer to Earth is Venus than is Mars?

Thinking Critically

26. **Explain** why Venus has a higher surface temperature than Mercury even though Mercury is closer to the Sun.
27. **Form a hypothesis** about where the oxygen that combined with the iron in rocks on Mars came from.
28. **Classify** a newly discovered planet, by using three classification systems, with the following characteristics: 10,000 km in diameter, rocky surface, an orbit beyond Pluto's.
29. **Explain** why Mars, Europa, and Titan are of interest to scientists searching for extraterrestrial life.
30. **Explain** several reasons why Earth is the only planet in the solar system that appears capable of supporting life as we know it.

Applying Math

31. **Use Angles** How far around in its orbit would a planet be if its heliocentric longitude was 90 degrees?
32. **Calculate** Jupiter's distance from the Sun is 5.2 AUs. Use the equation (planet's distancein AUs)^3 = (planet's period in years)^2 to calculate Jupiter's period of revolution.

Use the diagram below to answer No. 33.

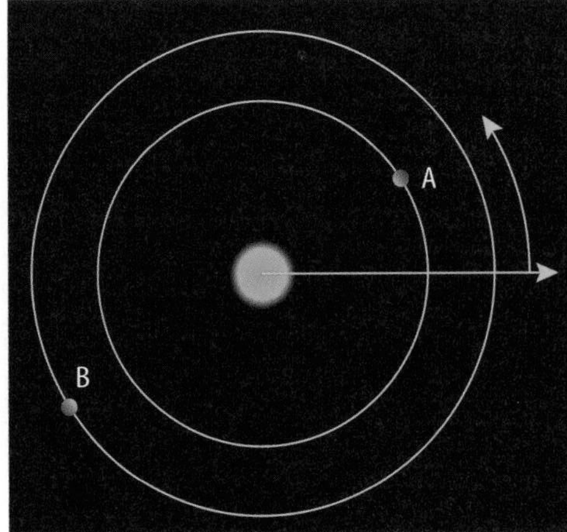

33. **Use Angles** If the Heliocentric Longitude (H. L.) of planet A is 33 degrees, what is the approximate H.L. of planet B?
34. **Using Equations** Use the equation $V = \frac{4}{3}(\pi r^3)$ to determine the volume of Venus, which has a radius of 6052 km.

chapter 8 Standardized Test Practice

Record your answers on the answer sheet provided by your teacher or on a sheet of paper.

Multiple Choice

Use the illustration below to answer question 1.

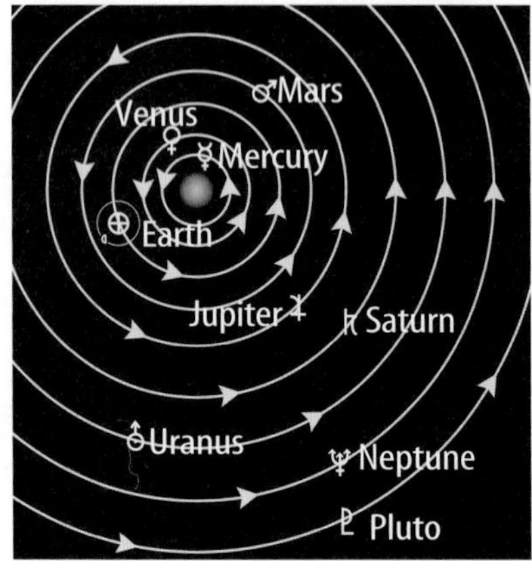

1. As seen from above their north poles, in which direction do planets move around the Sun?

A. clockwise

B. counterclockwise

C. geocentric

D. heliocentric

2. Which model of the solar system places the Sun in the center?

A. geocentric

B. heliocentric

C. inferior

D. superior

Test-Taking Tip

Practice Remember that test-taking skills can improve with practice. If possible, take at least one practice test and familiarize yourself with the test format and instructions.

3. What did Galileo observe about Venus?

A. It goes through phases.

B. It has two moons.

C. It is covered by clouds.

D. Its surface temperature is over 450°C.

4. Who determined that planets orbit the Sun in elliptical orbits?

A. Brahe

B. Copernicus

C. Kepler

D. Ptolemy

Use the photo below to answer question 5.

5. Which moon of Jupiter, pictured here, is most likely to have liquid water under its ice crust?

A. Callisto

B. Europa

C. Ganymede

D. Io

6. Which is composed of dust and rock particles mixed in with frozen water, methane, and ammonia?

   A. asteroid

   B. comet

   C. meteoroid

   D. meteorite

Gridded Response

7. Saturn travels 360° around the Sun for 29.5 years to complete one orbit. How many years will it take Saturn to complete 195° of its orbit?

8. If the orbital radius of Jupiter is 5.2 AU, what fraction of that distance is Earth's orbital radius?

Short Response

Use the illustration below to answer question 9.

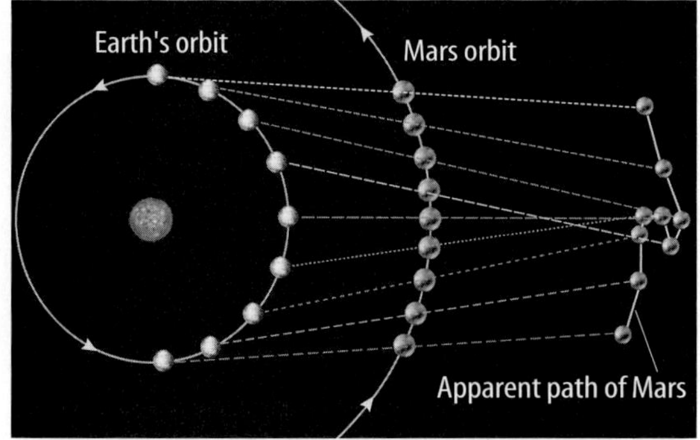

9. How does retrograde motion of planets, such as Mars shown in the illustration, provide support for the heliocentric model of the solar system?

10. What causes Mars to have a reddish-orange color?

11. Why was the *Galileo* spacecraft given this name?

12. What is generally true of all jovian planets?

13. Compare and contrast Venus to Earth. Why does it appear unlikely that Venus is capable of supporting life as we know it?

14. Compare and contrast Jupiter to Earth. Why does it appear unlikely that Jupiter is capable of supporting life as we know it?

15. Why would the discovery of oceans existing under the surface ice of Europa be so interesting to astronomers?

Extended Response

Use the photo below to answer question 16.

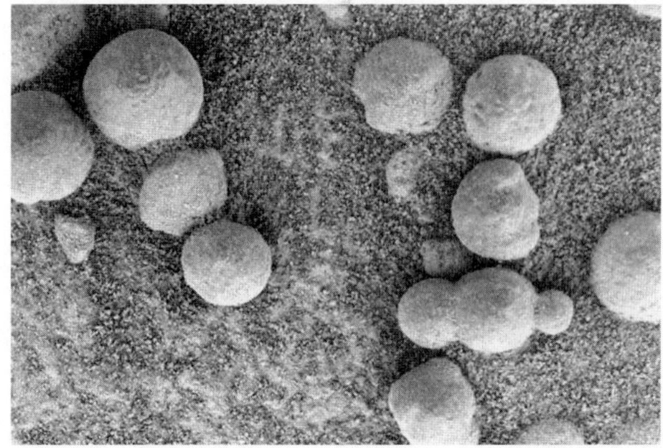

16. **PART A** What, shown in the photograph, did the NASA spacecraft, *Opportunity*, discover on Mars?

    PART B Why is this discovery so important?

How Are Glassblowing & X Rays Connected?

Glassblowing (far left) is an art in which air is blown through a tube to shape melted glass. In the mid-1800s, a glassblower created a glass tube, sealed metal electrodes into the ends, and removed most of the air from inside. When electricity was passed through the tube, it glowed. The glow aroused the curiosity of scientists, who began experimenting with similar tubes. In order to observe the glow more closely, one physicist surrounded a tube with black cardboard and darkened the laboratory. When the electric current was turned on, the tube glowed—but so did an object across the room! Apparently the tube was emitting some kind of radiation that could pass through cardboard. The mysterious radiation became known as X rays. Scientists eventually learned that X rays are a form of electromagnetic radiation, similar to visible light but with shorter wavelengths and higher energy. Because X rays pass through many substances, they have become important in medicine and science, making it possible to "see" structures inside the bodies of people—and also fish.

unit projects

Visit unit projects at gpescience.com to find project ideas and resources. Projects include:

- **History** Adopt a landform to research and record its geological and environmental changes throughout history.
- **Technology** Research medical procedures that use waves, sound, light, or heat to form diagnostic images.
- **Model** Design and construct your own creative musical instrument. Participate in a class orchestra of homemade instruments.

WebQuest *Created Gemstones: The Real Thing* offers an opportunity to evaluate the process and value of manufactured gems.

chapter 9

Heat and States of Matter

chapter preview

sections

Hot Stuff

This hot, glowing liquid will become solid steel when it cools. The difference between liquid and solid steel is energy—thermal energy. Increasing the thermal energy of solid steel can cause it to melt and change into a fiery liquid.

Science Journal Describe things you do to make yourself feel warmer or cooler.

Start-Up Activities

Expansion of a Gas

Why does mercury in a thermometer rise? Why do sidewalks, streets, and bridges have cracks? Many substances expand when heated and contract when cooled, as you will see during this lab.

1. Blow up a balloon until it is half filled. Use a tape measure to measure the circumference of the balloon.
2. Pour water into a large beaker until it is half full. Place the beaker on a hot plate and wait for the water to boil.
3. Set the balloon on the mouth of the beaker and observe for 5 min. Be careful not to allow the balloon to touch the hot plate. Again measure the circumference of the balloon.
4. **Think Critically** Write a paragraph in your Science Journal describing the changing size of the balloon's circumference. Infer why the balloon's circumference changed.

Thermal Energy and States of Matter Make the following Foldable to help you understand thermal energy and states of matter.

STEP 1 **Fold** a vertical sheet of paper in half from top to bottom.

STEP 2 **Fold** in half from side to side with the fold at the top.

STEP 3 **Unfold** the paper once. **Cut** only the fold of the top flap to make two tabs.

STEP 4 **Turn** the paper vertically and **label** the front tabs as shown.

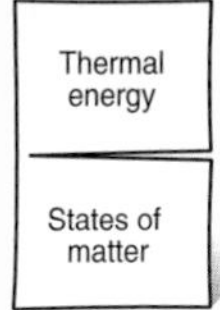

Find Main Ideas As you read the chapter, write the main ideas you find about thermal energy and states of matter under the appropriate tab.

Preview this chapter's content and activities at gpescience.com

section 1

Temperature and Thermal Energy

Reading Guide

What You'll Learn

- **Define** temperature.
- **Explain** how thermal energy and temperature are related.
- **Calculate** the change in thermal energy of an object due to a temperature change.

Why It's Important

Cars, buses, trucks, and airplanes could not operate without thermal energy.

Review Vocabulary

kinetic energy: energy an object has due to its motion

New Vocabulary

- kinetic theory
- Kelvin
- temperature
- thermal energy
- heat
- specific heat

Kinetic Theory of Matter

All around you objects are warming and cooling. What is temperature and what causes the temperature of an object to change? The temperature of an object is related to the motion of the particles in that object. The motion of the particles in matter is described by the **kinetic theory** of matter.

According to the kinetic theory, matter is composed of particles that are atoms, molecules, or ions that always are in random motion. This means that the particles in matter are moving at different speeds in all directions. Because these particles are in motion, they have kinetic energy. These particles also can collide and kinetic energy can be transferred from one particle to another particle.

Reading Check *How do the particles in matter move?*

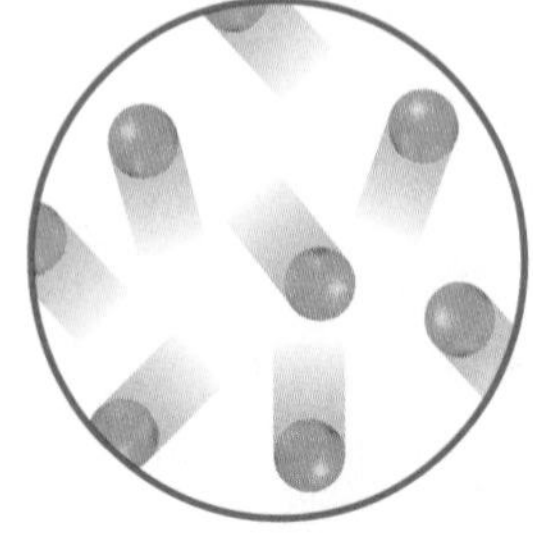

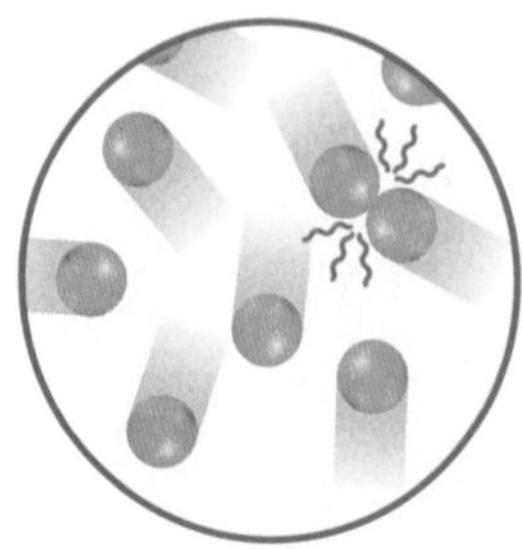

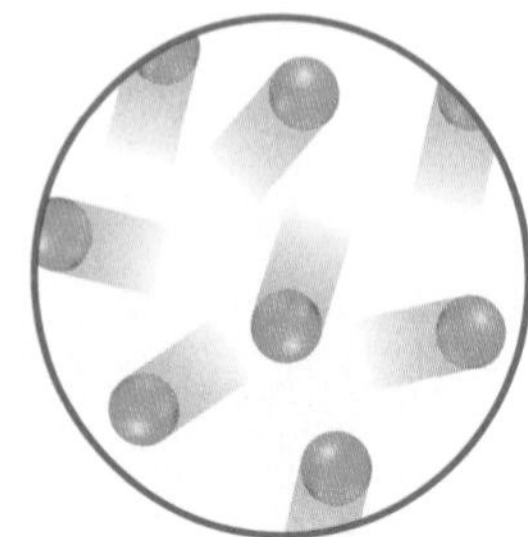

Figure 1 When particles collide, kinetic energy can be transferred from one particle to another and the direction of particle motion changes.

Temperature

The particles in matter are moving at various speeds and have a range of kinetic energies. The **temperature** of a substance is a measure of the average kinetic energy of its particles. As the average speed of the particles in random motion increases, the temperature of the substance increases. For example, the particles in hot tea move faster on average than those in iced tea.

Temperature Scales The SI unit for temperature is the **Kelvin** (K). More commonly used temperature scales are the Celsius scale and the Fahrenheit scale. Subtracting 273 from a temperature in Kelvin converts the temperature to degrees Celsius. Temperatures measured on the Celsius and Fahrenheit scales can be converted into one another using these equations:

Celsius-Fahrenheit Conversion Equations

To convert temperature in °F to °C: $°C = \frac{5}{9}(°F - 32)$

To convert temperature in °C to °F: $°F = \frac{9}{5}°C + 32$

Applying Math — Solve a Simple Equation

CONVERTING TO CELSIUS What is a temperature of 86°F on the Celsius scale?

IDENTIFY known values and the unknown value

Identify the known values:

the temperature is 86°F → means → $°F = 86$

Identify the unknown values:

What is this temperature on the Celsius scale? → means → $°C = ?$

SOLVE the problem

Substitute the temperature 86°F into the equation that converts °F to °C:

$$°C = \frac{5}{9}(°F - 32) = \frac{5}{9}(54) = 30°C$$

CHECK your answer

Does your answer seem reasonable? Multiply your answer by $\frac{9}{5}$ and add 32. The result should be the given Fahrenheit temperature.

Practice Problems

1. What is –40°F in degrees Celsius?
2. What is a temperature of 22°C in degrees Fahrenheit?

For more practice problems, go to page 879 and visit Math Practice at gpescience.com.

Topic: Weather Satellites
Visit gpescience.com for Web links to information about how weather satellites use thermal energy to acquire weather data.

Activity Draw a diagram showing how a satellite uses thermal energy to produce images.

Thermal Energy

According to the kinetic theory, atoms and molecules that make up matter are in constant motion. However, not only do these particles have kinetic energy due to their motion, they also have potential energy. This is because atoms and molecules exert attractive electric forces on each other. Recall that Earth exerts an attractive gravitational force on a ball. When the ball is above the ground, the ball and Earth are separated, and the ball has potential energy. In the same way, particles that exert attractive forces on each other have potential energy that increases as they get farther apart. The sum of the kinetic and potential energies of all the particles in an object is the **thermal energy** of the object. **Figure 2** shows how thermal energy depends on the kinetic and potential energies of the particles in a material.

What type of energy results from the attractive forces between particles?

Thermal Energy of a Moving Object You've learned that the faster an object moves, the greater its kinetic energy. A ball rolling across the floor has greater kinetic energy than if it were sitting still. Does this mean that the rolling ball also has greater thermal energy? At first, the answer might seem to be yes, but in fact, the thermal energy is the same for both of the balls, if they are at the same temperature. Thermal energy is the sum of the kinetic energy and potential energy of the particles that make up the ball, not of the ball itself. The increase in the ball's speed doesn't affect the random motion of its particles, so it doesn't affect its thermal energy.

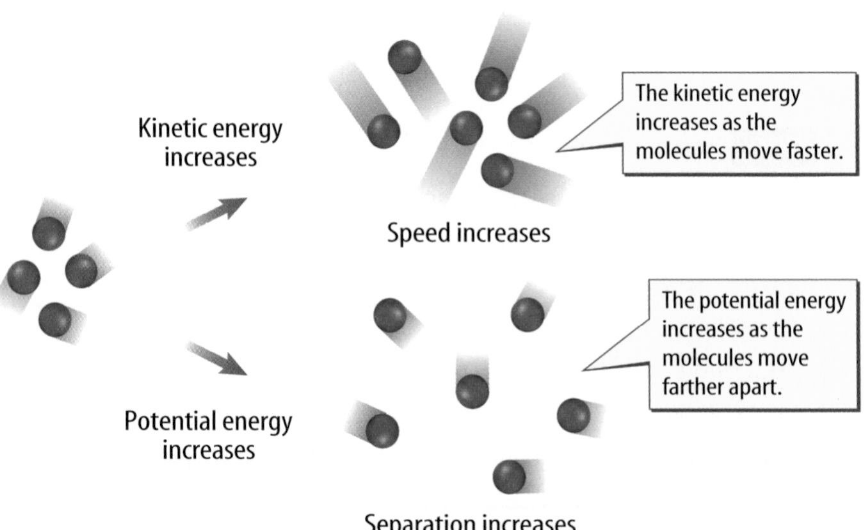
Kinetic energy increases
The kinetic energy increases as the molecules move faster.
Speed increases
Potential energy increases
The potential energy increases as the molecules move farther apart.
Separation increases

Figure 2 The thermal energy of a substance is the sum of the kinetic and potential energies of its molecules.
Infer *why increasing the temperature of an object increases its thermal energy.*

Heat

Can you tell if someone has been sitting in your chair? Perhaps you've noticed that your chair felt warm, and maybe you concluded that someone was sitting in it recently. The chair felt warmer because thermal energy from the person's body flowed to the chair, increasing the chair's temperature.

Heat is thermal energy that flows from something at a higher temperature to something at a lower temperature. Heat is a form of energy, so it is measured in joules. Thermal energy always flows from a warmer to a cooler material. How did the ice cream in **Figure 3** become cold? Thermal energy flowed from the warmer liquid ingredients to the cooler ice-and-salt mixture. The liquid ingredients lost enough thermal energy to become cold enough to freeze. Meanwhile, the ice-and-salt solution absorbed thermal energy, causing some of the ice to melt.

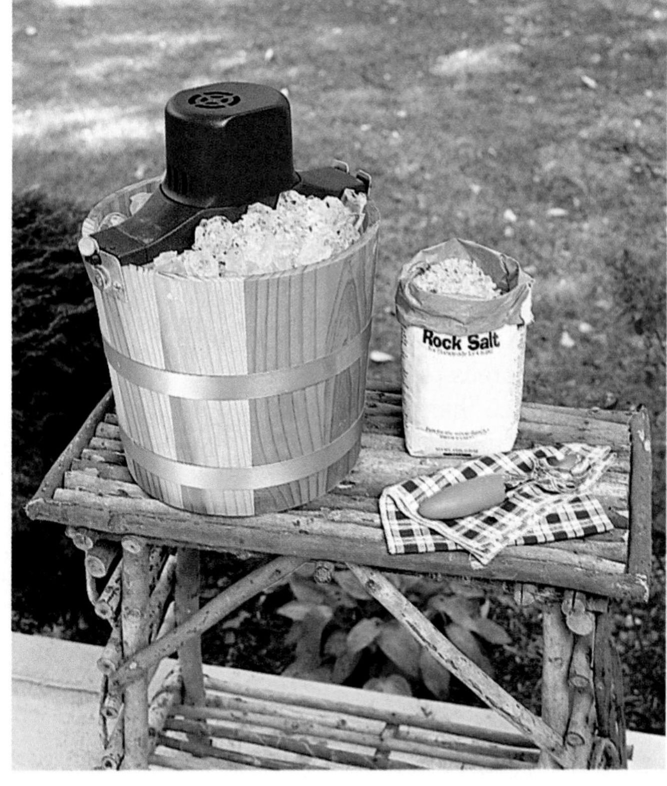

Figure 3 Thermal energy flows from the warmer ingredients inside the container to the ice-and-salt mixture around it.

Reading Check *How are heat and thermal energy related?*

Specific Heat

If you are at the beach in the summer, you might notice that the ocean seems cooler than the air or sand. Even though energy from the Sun goes to the air, sand, and water at the same rate, the water's temperature changes less than that of the air or sand.

As a substance is heated, its temperature change depends both on the amount of thermal energy added and the nature of the substance. For an equal mass of water and sand, about six times as much thermal energy is needed to raise the temperature of the water than the sand by 1°C. It's not surprising, then, that the sand feels much warmer than the ocean. The amount of thermal energy needed to raise the temperature of 1 kg of some material by 1°C is called the **specific heat** of the material. Specific heat is measured in joules per kilogram per degree Celsius, which is written as J/(kg °C). **Table 1** shows the specific heats of some familiar materials.

Table 1 Specific Heat of Some Common Materials

Substance	Specific Heat [J/(kg°C)]
Water	4,184
Ice	2,110
Asphalt	920
Glass	800
Iron	450

Reading Check *What is the specific heat of a material?*

Coastal Climates The high specific heat of water causes large bodies of water to heat up and cool down more slowly than land masses. As a result, the temperature changes in coastal areas tend to be less extreme than they are farther inland.

Changes in Thermal Energy The thermal energy of an object changes when its temperature changes. If Q is the change in thermal energy and C is the specific heat, the change in thermal energy can be calculated from the following equation:

Thermal Energy Equation

change in thermal energy (J) =
mass (kg) × change in temperature (°C) × **specific heat** $\left(\frac{\text{J}}{\text{kg°C}}\right)$

$$Q = m(T_f - T_i)C$$

Applying Math

Solve a Simple Equation

CHANGE IN THERMAL ENERGY Find the change in thermal energy of a 20-kg wooden chair that warms from 15°C to 25°C if the specific heat of wood is 1,700 J/(kg °C).

IDENTIFY known values and the unknown value

Identify the known values:

A wooden chair with a mass of 20 kg → means → $m = 20$ kg

warms from 15°C to 25°C → means → $T_i = 15$°C and $T_f = 25$°C

the specific heat of wood is 1,700 J/(kg °C) → means → $C = 1{,}700$ J/(kg °C)

Identify the unknown value:

What is the change in thermal energy → means → $Q = ?$ J

SOLVE the problem

Substitute the known values into the thermal energy equation:

$$Q = m(T_f - T_i)C = (20\text{ kg})(25°\text{C} - 15°\text{C})\left(1{,}700\ \frac{\text{J}}{\text{kg °C}}\right)$$

$$= (20\text{ kg})(10°\text{C})\left(1{,}700\ \frac{\text{J}}{\text{kg °C}}\right) = 340{,}000\text{ kg °C}\ \frac{\text{J}}{\text{kg °C}} = 340{,}000\text{ J}$$

CHECK your answer

Does your answer seem reasonable? Check your answer by dividing the change in thermal energy you calculated by the mass and the specific heat given in the problem. The result should be the difference in temperature given in the problem.

Practice Problems

The air in a living room has a mass of 72 kg and a specific heat of 1,010 J/(kg °C). What is the change in thermal energy of the air when it warms from 20°C to 25°C?

For more practice problems, go to page 879 and visit Math Practice at gpescience.com.

Measuring Specific Heat

The specific heat of a material can be measured using a device called a calorimeter, shown in **Figure 4.** The specific heat of a material can be determined if the mass of the material, its change in temperature, and the amount of thermal energy absorbed or released are known. In a calorimeter, a heated sample transfers thermal energy to a known mass of water. The energy absorbed by the water can be calculated by measuring the water's temperature change. The thermal energy released by the sample is equal to the thermal energy absorbed by the water.

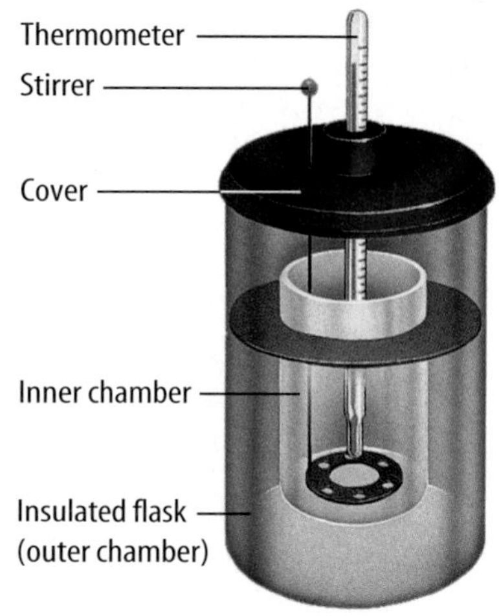

Figure 4 A calorimeter can be used to measure the specific heat of materials. The sample is placed in the inner chamber.

Using a Calorimeter To measure the specific heat of a material, the mass of a sample of the material is measured, as is the initial temperature of the water in the calorimeter. The material then is heated, its temperature is measured, and the sample is placed in the water in the inner chamber of the calorimeter. The sample cools as thermal energy is transferred to the water, and the temperature of the water increases. The transfer continues until the sample and the water are at the same temperature. Then the initial and final temperatures of the water are known, and the thermal energy gained by the water can be calculated.

section 1 review

Summary

Temperature

- The kinetic theory is a description of how the particles in a material are moving.
- The temperature of an object is a measure of the average kinetic energy of the particles that make up the object.

Thermal Energy and Heat

- Thermal energy is the sum of the kinetic and potential energies of the particles in an object.
- Heat is thermal energy that is transferred from an object at a higher temperature to an object at a lower temperature.

Specific Heat

- The specific heat of a material is the amount of thermal energy needed to raise the temperature of 1 kg of the material by 1°C.
- The change in thermal energy of an object can be calculated from this equation:

$$Q = m(T_f - T_i)C$$

Self Check

1. **Explain** how energy moves when you touch a block of ice with your hand.
2. **Describe** how the thermal energy of an object changes when the object's temperature changes.
3. **Infer** When thermal energy flows between two objects, does the temperature increase of one object always equal the temperature decrease of the other? Explain.
4. **Describe** how the particles in matter move, using the assumptions of the kinetic theory.
5. **Think Critically** Explain whether or not the following statement is true: For any two objects, the one with the higher temperature always has more thermal energy.

Applying Math

6. **Calculate** the change in thermal energy of the water in a pond with a mass of 1,000 kg and a specific heat of 4,184 J/(kg °C) if the water cools by 1°C.
7. **Calculate** the specific heat of a metal if 0.5 kg absorbs 9,000 J of thermal energy as it warms by 10°C.

section 2

States of Matter

Reading Guide

What You'll Learn

- **Describe** the differences in the four states of matter.
- **Explain** how the motion of particles in a material changes at the material's melting point and boiling point.
- **Explain** why thermal expansion occurs.

Why It's Important

You can use energy that is lost or gained when a substance changes from one state to another.

Review Vocabulary

force: a push or a pull that one object exerts on another

New Vocabulary

- plasma
- heat of fusion
- heat of vaporization

Four States of Matter

Every day you encounter different states of matter. Your lunch probably includes a solid and a liquid, such as pizza and a drink. The air you breathe is a gas. The Sun, lightning, and fluorescent lights contain a fourth state of matter called plasma. The differences between solids, liquids and gases are due to differences in the attraction between particles.

Solid State The spacing between particles is different for solids, liquids, and gases, as shown in **Figure 5.** The particles of a solid are packed closely together and are constantly vibrating in place. Because the attractions between particles are strong, solids have a fixed volume and shape.

Figure 5 In solids, strong attractive forces hold particles in place. Particles in liquids and gases do not have fixed arrangements. Particles are farther apart in gases than in solids and liquids.

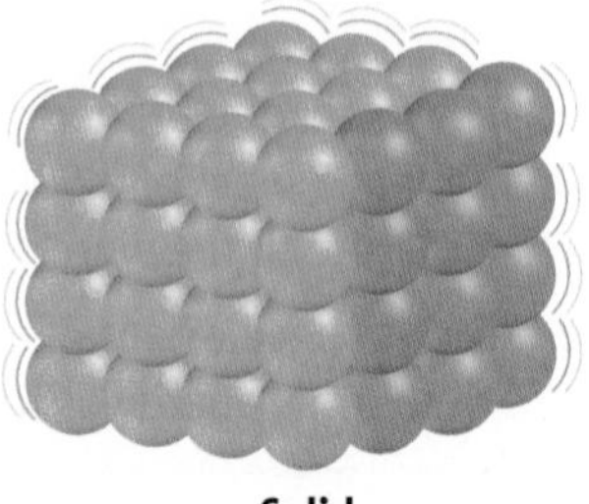

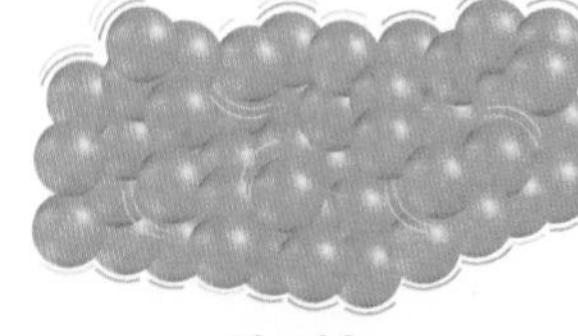

Liquid State In a liquid the attractive forces between particles are weaker than in a solid. The particles can slide past each other, allowing liquids to flow and take the shape of their container. However, the attractive forces are strong enough to cause particles to cling together. As a result, liquids have a definite volume, but not a definite shape.

Why do liquids flow?

Gas State In a gas, the particles are much farther apart than in a liquid or a solid. Because the particles are so far apart, the attractive forces between them are weak. In a gas the forces between particles are so weak that the particles no longer cling together. As a result, gases do not have a definite shape or volume.

Because the particles in a gas are so far apart, a gas contains mostly empty space. This means that the density of gases is much less than the density of solids and liquids. Also, the particles in a gas will spread throughout a given volume until they are distributed evenly. This process is called diffusion. Diffusion also occurs in solids and liquids, but occurs much more rapidly in gases.

INTEGRATE Astronomy

Plasma State The most common state of matter in the universe is the plasma state. **Plasma** is matter consisting of positively and negatively charged particles and does not have a definite shape or volume. It results from collisions between particles moving at such high speeds that electrons are knocked from the atoms, leaving only charged particles. All of the observed stars, including the Sun shown in **Figure 6,** contain plasma. Lightning bolts, neon and fluorescent tubes, and auroras also contain plasmas.

Changing States

If you hold an ice cube in your hand, you can see it change from a solid to a liquid. A liquid changes to a gas when a puddle of water disappears after a rain on a hot day. Adding thermal energy caused the solid ice to change to a liquid and the liquid water to become a gas. Changes in the thermal energy of a material can cause it to change from one state to another.

Topic: States of Matter
Visit gpescience.com for Web links to information about states of matter.

Activity Create a slide presentation about the states of matter using computer presentation software.

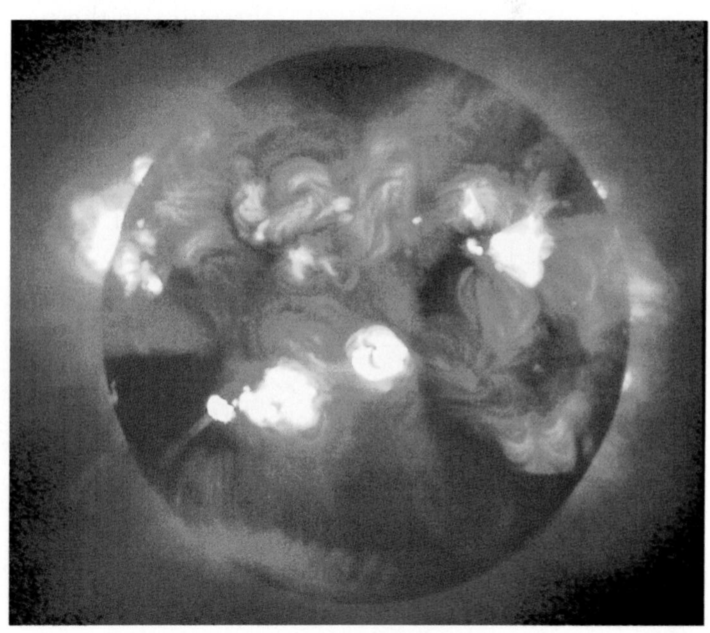

Figure 6 The Sun and other stars contain matter that is in the plasma state. Plasma usually exists where the temperature is extremely high.
Describe *the plasma state.*

Comparing States of Water

Procedure

1. Fill a **plastic bowl or ice tray** with **water**. Mark the level of the top surface of the water with a **marker** or piece of **masking tape**.
2. Place the container in a freezer overnight.
3. Remove the container from the freezer and observe the level of the ice that has formed.

Analysis

1. How does the level of ice compare with the level of water that you placed in the container?
2. Explain why the level of ice is different from the level of the water.

Try at Home

Melting What causes an ice cube to melt into liquid water? As the temperature of the ice increases, its particles move faster, and the attractive forces between them aren't strong enough to keep them in place. The particles can slip out of their ordered arrangement, and the ice melts. The temperature at which a solid begins to melt is its melting point. Energy is required for the particles to slip out of the ordered arrangement. The amount of energy required to change 1 kg of a substance from a solid to a liquid at its melting point is known as the **heat of fusion.**

Freezing The heat of fusion is also the energy released when a liquid freezes, or changes to a solid. If you lower a liquid's temperature, you decrease its particles' average kinetic energy, and they move slower. The attractive forces then are strong enough that the particles form an ordered arrangement.

Vaporization As the temperature of a liquid increases, the particles in the liquid move faster. As the particles move faster and become farther apart, the forces between them can become so weak that the particles no longer cling together. Then vaporization occurs as liquid changes into a gas. Vaporization can occur within the liquid or at the surface of the liquid. Vaporization that occurs at the surface of a liquid is called evaporation. Particles that enter the gas state have more energy than the particles left in the liquid. Evaporation causes the temperature of the liquid to decrease.

Boiling A second way that a liquid can vaporize is by boiling, shown in **Figure 7.** Unlike evaporation, boiling occurs throughout a liquid. It occurs only at a certain temperature depending on the pressure on the surface of the liquid. The boiling point of a liquid is the temperature at which the pressure of the vapor in the liquid is equal to the external pressure acting on its surface. This external pressure is a force pushing down upon a liquid, keeping particles from escaping. The **heat of vaporization** is the amount of energy required for 1 kg of the liquid at its boiling point to become a gas.

Figure 7 Boiling occurs throughout a liquid when the pressure of the pockets of gas forming in the liquid equals the pressure of the vapor at the surface of the liquid.
Explain *the difference between boiling and evaporation.*

Condensation The heat of vaporization is also the amount of energy released during condensation, when a gas changes into a liquid. You've probably seen condensation form on the sides of a glass of ice water on a hot day. The cold glass reduced the temperature of nearby water vapor. The average kinetic energy of the water particles decreased, and the water vapor changed to a liquid. As shown in **Figure 8,** dew also is the result of condensation.

What change of state occurs during condensation?

Figure 8 The formation of dew is an example of condensation. Dew forms when the air around matter, such as spider webs, cools enough so that water vapor in the air changes to a liquid.
Identify *the change in state that occurs when dew evaporates.*

Heating Curve of a Substance A graph representing water being heated from –20°C to 120°C is shown in **Figure 9.** This graph is called a heating curve because it shows the temperature change of water as thermal energy is added. There are two areas on the graph where the temperature does not change. One is at 0°C, where ice is melting. All of the energy absorbed by the ice at this temperature increases the potential energy of the water molecules. This causes water molecules to move farther apart and weakens the forces between molecules. However, the average kinetic energy of the water molecules doesn't change. As a result, the temperature remains constant during melting.

After the ice melts completely, adding additional energy causes the water molecules to move faster so that the temperature of the water increases. At 100°C, the water changes from a liquid to a gas. As the water absorbs energy, only the potential energy of the water molecules increases and the temperature of the water doesn't change. After the liquid water has changed completely into a gas, the temperature of the gas increases as energy is added.

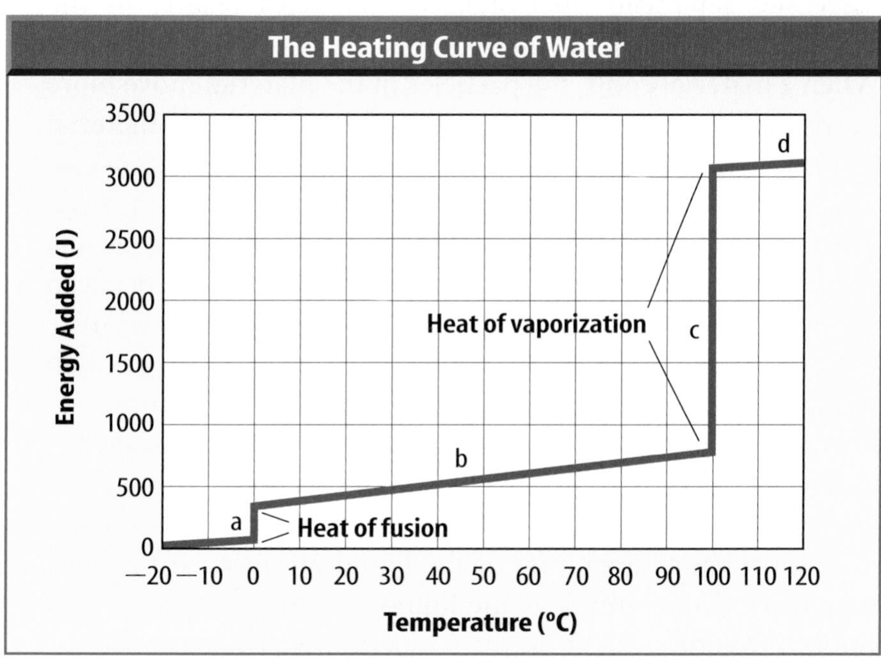

Figure 9 This graph shows the heating curve of one gram of water from −20°C to 120°C. At **a** and **c,** the energy absorbed causes water molecules to move faster and so the temperature increases. At **b** and **d,** the water changes from one state to another. The added energy increases the potential energy of the water molecules, but not the kinetic energy, so the temperature doesn't change.

Figure 10 This expansion joint enables the road surface on this bridge to expand without causing the road surface to crack or buckle.

Thermal Expansion

When you have gone over a highway bridge in a car, you might have noticed the gaps between sections of the bridge, such as the gap shown in **Figure 10.** This gap is called an expansion joint. The materials that make up the bridge and the road surface expand when they get warmer. Expansion joints provide space for these materials to expand on hot days without causing the materials to become warped or cracked.

The Thermal Expansion of Matter The kinetic theory explains why an object expands when its temperature increases. According to the kinetic theory, the particles in an object are in constant motion. The speed of these particles increases as the temperature of the object increases. As the particles move faster, the attractive forces between them become weaker and the distance between particles increases. The increased separation between the particles results in the expansion of the object and the size of the object increases. When a material cools, the particles in the material move more slowly and become closer together. As a result, the material contracts and its dimensions decrease.

Figure 11 The liquid in the thermometer tube expands as its temperature increases. Because the tube expands much less than the liquid, the height of the liquid column increases as the temperature increases.

Thermal Expansion of Liquids The forces between the particles in liquids are usually weaker than the forces between the particles in a solid. As a result, the same temperature increase usually causes liquids to expand much more than solids. For example, the liquid in the thermometer shown in **Figure 11** expands as its temperature increases. However, the dimensions of the glass tube containing the liquid hardly change at all. This causes the height of the liquid column in the tube to increase as the temperature of the liquid increases. If the glass tube expanded as much as the liquid, the height of the liquid column wouldn't change as the temperature changed.

Thermal Expansion of Gases In a gas, the forces between particles are much weaker than they are in liquids. As a result, gases expand even more than liquids for the same increase in temperature. When thermal expansion occurs, the volume of the material increases. Recall that the density of an object is the mass of the object divided by its volume. Because the material's volume has increased, but its mass hasn't changed, the material becomes less dense as it expands.

Reading Check *How does the density of a material change when the material expands?*

The decrease in density caused by thermal expansion enables the hot-air balloon in **Figure 12** to rise. The air inside the balloon is heated by a burner. The hot air in the balloon expands and becomes less dense than the cooler air outside the balloon. As a result, the hot-air balloon is pushed upward by the denser, cooler air around it.

Figure 12 The air inside the hot-air balloon expands as it is heated. Its density decreases and the balloon rises.

section 2 review

Summary

Four States of Matter

- In solids, particles are held in place by strong attractive forces. Solids have a definite volume and shape.
- In liquids, the attractive forces are weaker and particles can move past each other. Liquids have a definite volume, but no definite shape.
- In gases, the attractive forces between particles are weakest. Gases have neither a definite volume nor shape.
- A plasma is made of electrically charged particles.

Changing States

- The heat of fusion is the amount of energy needed to change 1 kg of a material at its melting point from a solid to a liquid.
- The heat of vaporization is the amount of energy needed to change 1 kg of material at its boiling point from a liquid to a gas.

Self Check

1. **Compare and contrast** solids, liquids, gases, and plasma.
2. **Explain** whether the temperature of a substance must increase when thermal energy is added to the substance.
3. **Explain** how the forces between water molecules in an ice cube change when the ice cube melts.
4. **Think Critically** The atmospheric pressure on a mountain peak is lower than the atmospheric pressure at sea level. Compare the boiling point of water at sea level to its boiling point on the mountain peak.

Applying Math

5. **Calculate** the amount of thermal energy required to change a 0.45-kg block of ice at 0°C to water at 0°C if the heat of fusion of water is 334,000 J/kg.
6. **Calculate** The heat of vaporization for water is 2.26×10^6 J/kg. What mass of liquid water at 100°C can be changed to water vapor at 100°C by adding 1.0×10^6 J of thermal energy to the liquid water?

section 3

Transferring Thermal Energy

Reading Guide

What You'll Learn

- **Compare and contrast** the transfer of thermal energy by conduction, convection, and radiation.
- **Compare and contrast** thermal conductors and insulators.
- **Explain** how thermal insulators are used to control the transfer of thermal energy.

Why It's Important

You must be able to control the flow of thermal energy to keep from being too hot or too cold.

Review Vocabulary

density: the mass per unit volume of a substance

New Vocabulary

- conduction
- convection
- radiation
- thermal insulator

Ways to Transfer Thermal Energy

When water is heated on a stove or when a hot slice of pizza becomes cool enough to eat, thermal energy is transferred from one place to another. Thermal energy can be transferred from one place to another by conduction, convection, or radiation.

Conduction

According to the kinetic theory, particles in matter are always in random motion. Thermal energy can be transferred when particles collide. This transfer of thermal energy between colliding particles is **conduction. Figure 13** shows a metal pot being heated over a fire. The metal atoms nearest the fire absorb the most thermal energy, causing their kinetic energy to increase. Kinetic energy is transferred when these faster-moving atoms collide with slower-moving particles. As these collisions continue, thermal energy is transferred throughout the pot. When thermal energy is transferred by conduction, it is transferred by the collisions between particles, not by the movement of matter.

Figure 13 Conduction occurs within the metal of this pot as faster-moving particles transfer thermal energy when they collide with slower-moving particles.

Thermal Conductors Thermal energy can be transferred by conduction in all materials. However, the rate at which thermal energy is transferred depends on the material. A material in which thermal energy is transferred easily is called a thermal conductor. The conduction of thermal energy in solids and liquids is faster than in gases. Particles in gases are farther apart than in solids or liquids and collide less frequently. Therefore, gases are poorer thermal conductors than solids or liquids. Solids usually are better thermal conductors than liquids or gases. The best thermal conductors are metals.

Convection

Unlike solids, liquids and gases can flow and are classified as fluids. In fluids, thermal energy can be transferred by convection. **Convection** is the transfer of thermal energy in a fluid by the movement of fluid from place to place. When convection occurs, particles with more energy transfer energy to other particles in the fluid as they move from place to place.

As the particles move faster, they tend to be farther apart. As a result, a fluid expands as its temperature increases. Recall that density is the mass of a material divided by its volume. When a fluid expands, its volume increases, but its mass doesn't change. The density of the fluid therefore decreases. The same is true for parts of a fluid that have been heated. The density of the warmer fluid is less than that of the surrounding cooler fluid.

Convection Currents The difference in densities between warmer and cooler fluids can cause convection currents to occur. **Figure 14** shows the formation of convection currents in a beaker of water that is being heated from the bottom. Warm water at the bottom of the beaker is less dense than the cooler water above it. Because the warmer water is less dense, it is forced upward by the sinking cooler water. As it rises, the warm water transfers thermal energy to the cooler water around it. When the warm water cools, it becomes denser than the surrounding warmer water and sinks to the bottom of the beaker. The rising and sinking water forms a convection current. In this way, convection currents transfer thermal energy by the movement of water from place to place.

Comparing Thermal Conductors

Procedure

1. Obtain a **plastic spoon,** a **metal spoon,** and a **wooden spoon** with similar lengths.
2. Stick a small **plastic bead** to the handle of each spoon with a dab of **butter or wax.** Each bead should be the same distance from the tip of the spoon.
3. Stand the spoons in a **beaker,** with the beads hanging over the edge of the beaker.
4. Carefully pour about 5 cm of **boiling water** in the beaker holding the spoons.

Analysis

1. Describe how heat was transferred from the water to the beads.
2. Rank the spoons in their ability to conduct heat.

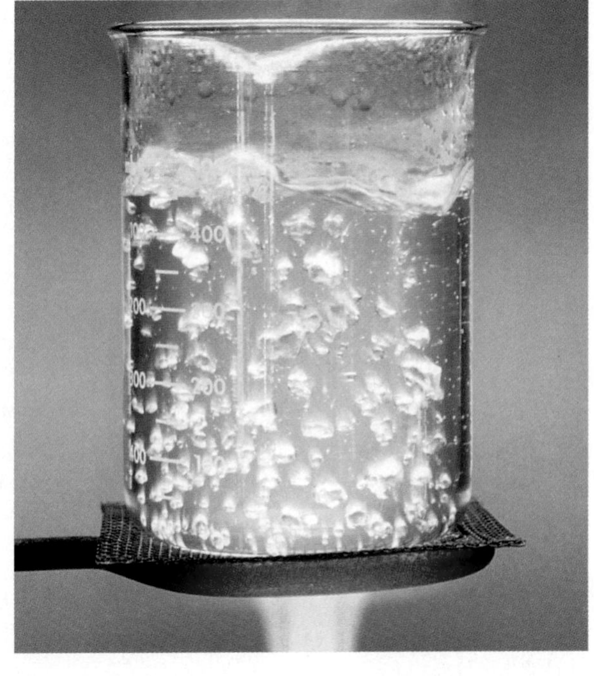

Figure 14 Convection currents result when heated fluid rises and cooler fluid sinks.

NATIONAL GEOGRAPHIC

VISUALIZING CONVECTION CURRENTS

Figure 15

When the Sun beats down on the equator, warm, moist air begins to rise. As it rises, the air cools and loses its moisture as rain that sustains rain forests near the equator. Convection currents carry the now dry air farther north and south. Some of this dry air descends at the tropics, where it creates a zone of deserts.

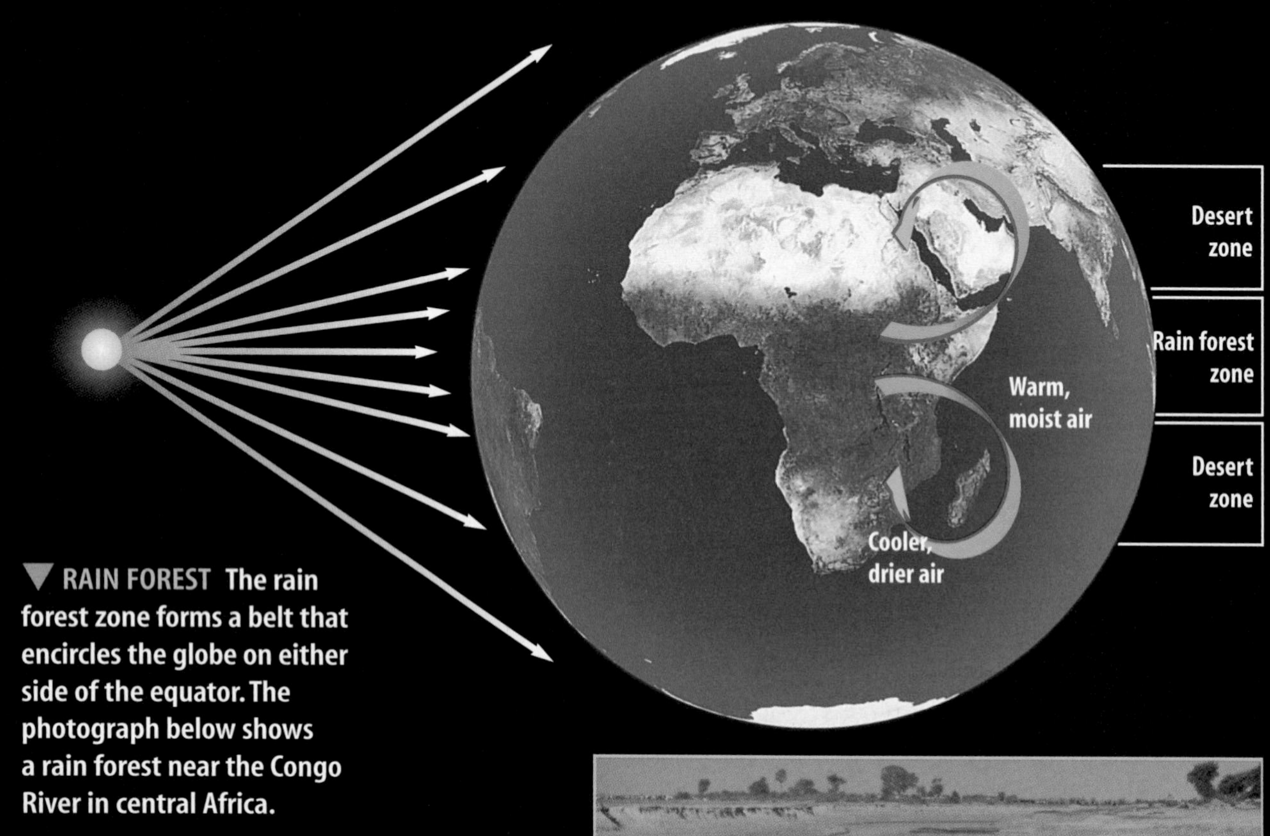

▼ RAIN FOREST The rain forest zone forms a belt that encircles the globe on either side of the equator. The photograph below shows a rain forest near the Congo River in central Africa.

▲ DESERT Like many of the great desert regions of the world, the Sahara, in northern Africa, is largely a result of atmospheric convection currents. Here, a group of nomads gather near a dried-up river in Mali.

Deserts and Rain Forests Earth's atmosphere is made of various gases and is a fluid. The atmosphere is warmer at the equator than it is at the north and south poles. Also, the atmosphere is warmer at Earth's surface than at higher altitudes. These temperature differences create convection currents that transfer thermal energy to cooler regions. **Figure 15** shows how these currents create rain forests and deserts in different regions.

Radiation

Earth is warmed by energy from the Sun, but how does that energy travel through space? Almost no matter exists in the space between Earth and the Sun, so thermal energy cannot be transferred by conduction or convection. Instead, the Sun's energy reaches Earth by radiation.

Radiation is the transfer of energy by electromagnetic waves. These waves travel through space even when no matter is present. Energy transferred by radiation often is called radiant energy. When you stand near a fire, much of the warmth you feel is caused by radiation transferring thermal energy to your skin. Radiation can pass through solids, liquids, and gases, but the transfer of energy by radiation is most important in gases.

Radiant Energy and Matter When radiation strikes a material, some energy is absorbed, some is reflected, and some may be transmitted through the material. The amount of energy that does each depends on the material. Light-colored materials reflect more radiant energy, while dark-colored materials are more absorbent. When a material absorbs radiant energy, its thermal energy increases. **Figure 16** shows what happens to radiant energy from the Sun as it reaches Earth.

For example, when a car sits out in the Sun, some of the Sun's radiation passes through the car windows. Materials inside the car absorb some of this radiation and become hot.

The Flow of Thermal Energy

You do a number of things every day to control the flow of thermal energy. When it's cold outside you wear a coat. You use an oven mitt when handling a hot dish. In each case, you use an insulating material to help reduce the flow of thermal energy. Your coat keeps you warm by reducing the flow of thermal energy from your body to the surrounding air. The oven mitt keeps your hand from being burned by reducing the flow of thermal energy from the hot dish to your hand.

Figure 16 Not all of the Sun's radiant energy reaches Earth's surface. Some of it is reflected or absorbed by the atmosphere. Even some of the radiation that reaches Earth's surface is reflected.

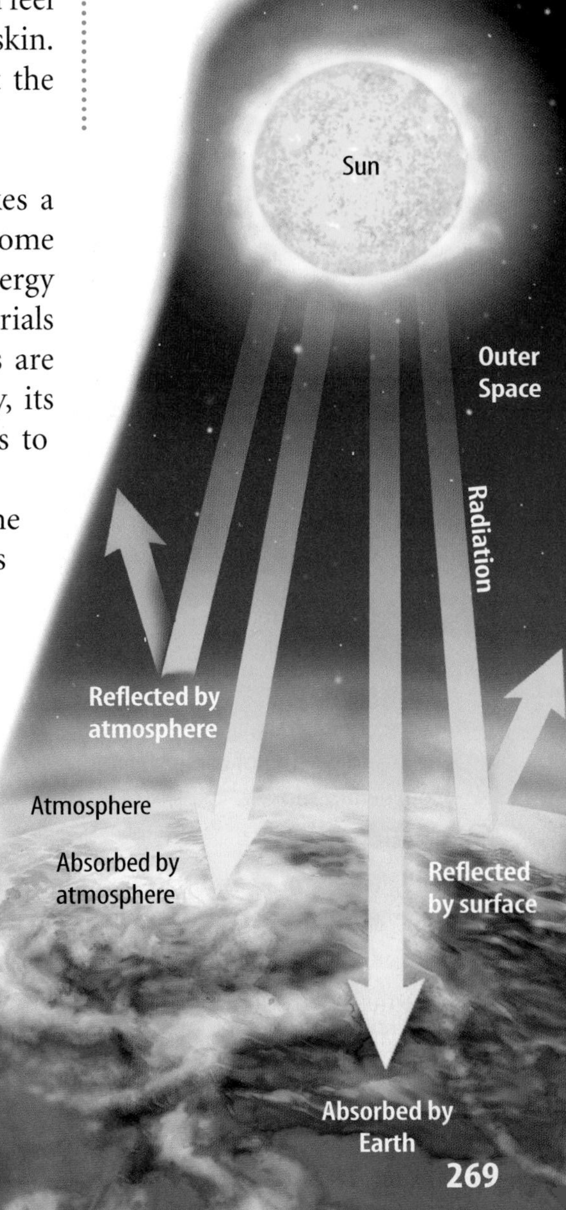

Thermal Insulators

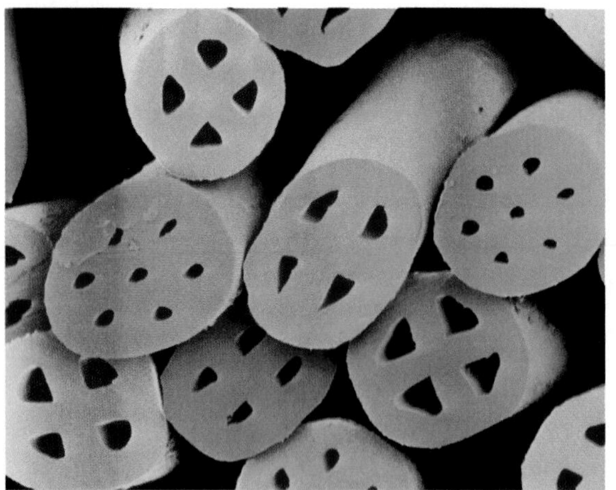

Figure 17 The fibers of this insulating material contain air spaces that are good thermal insulators. They reduce the rate at which thermal energy flows in the material.

A jacket and an oven mitt are examples of thermal insulators. A material in which thermal energy moves slowly is a **thermal insulator.** Examples of materials that are thermal insulators are wood, some plastics, fiberglass, and air. Materials that are good conductors of thermal energy, such as metals, are poor thermal insulators.

Gases, such as air, are usually much better thermal insulators than solids or liquids. Some types of insulating material contain many pockets of trapped air. These air pockets are poor conductors of thermal energy and also keep convection currents from forming. Winter jackets often are made of insulating materials consisting of fibers, like the ones shown in **Figure 17,** that trap air and hold it next to you. This trapped air slows the flow of your body heat to the colder air outside the jacket. Gradually, the air trapped by the material is warmed by your body heat. Underneath the jacket you are wrapped in a blanket of warm air.

Reading Check *Why is a material that traps air a good insulator?*

section 3 review

Summary

Conduction

- Conduction is the transfer of thermal energy by collisions between particles.
- Conduction occurs in solids, liquids, and gases. Metals are the best conductors of heat.

Convection

- Convection is the transfer of thermal energy by the movement of warmer and cooler material.
- Convection occurs in fluids. Rising of warmer fluid and sinking of cooler fluid forms a convection current.

Radiation

- Radiation is the transfer of energy by electromagnetic waves.

Thermal Insulators

- Thermal energy moves slowly in thermal insulators. Thermal insulators are used to control the flow of thermal energy.

Self Check

1. **Explain** why convection can occur in fluids but not in solids.
2. **Explain** why the air temperature near the ceiling of a room tends to be warmer than near the floor.
3. **Predict** whether plastic foam, which contains pockets of air, would be a good conductor or a good insulator.
4. **Describe** how a convection current occurs.
5. **Think Critically** Several days after a snowfall, the roofs of some homes on a street have almost no snow on them, while the roofs of other houses are still snow covered. Describe what would cause this difference.

Applying Math

6. **Calculate Solar Radiation** Averaged over a year in the central United States, radiation from the Sun transfers about 200 W to each square meter of Earth's surface. If a house is 10 m long by 10 m wide, how much solar energy falls on the house each second?

More Section Review gpescience.com

Convection in Gases and Liquids

A hawk gliding through the sky will rarely flap its wings. Hawks and some other birds conserve energy by gliding on columns of warm air rising up from the ground. These convection currents form when gases or liquids are heated unevenly and the warmer, less dense fluid is forced upward.

Real-World Problem

How can convection currents be modeled and observed?

Goals

- **Model** the formation of convection currents in water.
- **Observe** convection currents formed in water.
- **Observe** convection currents formed in air.

Materials

burner or hot plate
water
candle
500-mL beaker
black pepper

Safety Precautions 

WARNING: *Use care when working with hot materials. Remember that hot and cold glass appear the same.*

Procedure

1. Pour 450 mL of water into the beaker.
2. Use a balance to measure 1 g of black pepper.
3. Sprinkle the pepper into the beaker of water and let it settle to the bottom of the beaker.
4. Heat the bottom of the beaker using the burner or the hotplate.
5. **Observe** how the particles of pepper move as the water is heated and make a drawing showing their motion in your Science Journal.
6. Turn off the hot plate or burner. Light the candle and let it burn for a few minutes.
7. Blow out the candle and observe the motion of the smoke.
8. Make a drawing of the movement of the smoke in your Science Journal.

Conclude and Apply

1. **Describe** how the particles of pepper moved as the water became hotter.
2. **Explain** how the motion of the pepper particles is related to the motion of the water.
3. **Explain** how a convection current formed in the beaker.
4. **Explain** why the motion of the pepper changed when the heat was turned off.
5. **Predict** how the pepper would move if the water were heated from the top.
6. **Describe** how the smoke particles moved when the candle was blown out.
7. **Explain** why the smoke moved as it did.

Communicating Your Data

Compare your conclusions with other students in your class.

section 4

Using Thermal Energy

Reading Guide

What You'll Learn

- **Describe** the first and second laws of thermodynamics.
- **Explain** how an internal combustion engine converts thermal energy to mechanical energy.
- **Describe** how the entropy of the universe changes when any event occurs.

Why It's Important

Imagine your life without heating systems, cooling systems, or cars.

Review Vocabulary

work: the product of the force exerted on an object and the distance the object moves in the direction of the force

New Vocabulary

- first law of thermodynamics
- second law of thermodynamics
- entropy

How is thermal energy used?

If you've ever lit a campfire to keep warm or turned on a stove to boil water, you've used thermal energy. Heating a house and using a gasoline engine to mow a lawn are other ways that thermal energy is used. Sometimes thermal energy is transferred to an object to heat it, as when water is boiled or a room is heated. Sometimes thermal energy is converted into work.

Heating Systems

Almost everywhere in the United States a heating system is needed to make indoor temperatures comfortable on cold days. Heating systems add thermal energy to the interiors of rooms. The best heating system for a building depends on the local climate and how the building is constructed.

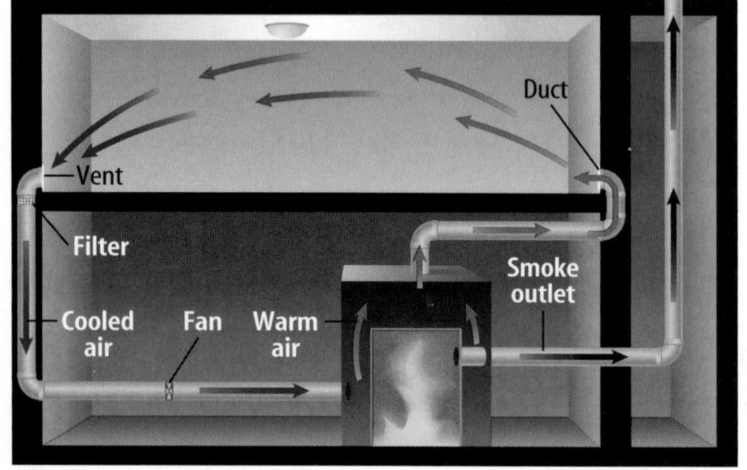

Figure 18 In a forced-air system, heated air is blown through ducts that lead to rooms in the building.

Forced-Air Systems The most common type of heating system in use today is the forced-air system, as shown in **Figure 18.** In this system, fuel is burned in a furnace and heats a volume of air. A fan blows the heated air through a series of large pipes called ducts. The ducts lead to openings called vents in each room. Cool air is returned through additional vents to the furnace, where it is reheated.

Radiator Systems Before forced-air systems were widely used, many homes and buildings were heated by radiators. A radiator is a closed metal container that contains hot water or steam. The thermal energy contained in the hot water or steam is transferred to the air surrounding the radiator by conduction. This warm air then moves through the room by convection.

In radiator heating systems, fuel burned in a central furnace heats a tank of water. A system of pipes carries the hot water to radiators in the rooms of the building. After the water cools, it flows through the pipes back to the water tank, and is reheated.

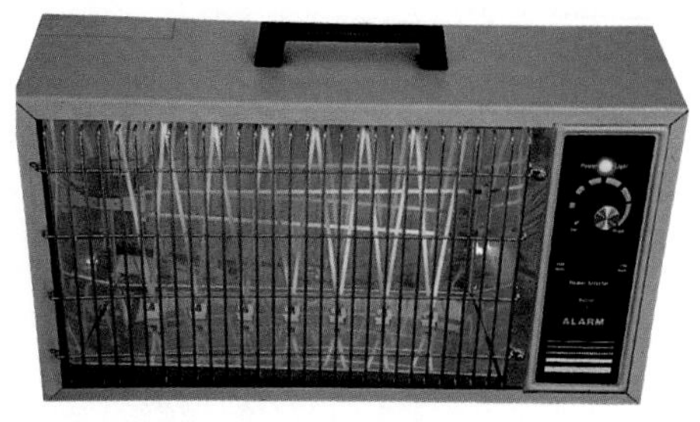

Figure 19 Electric heaters convert electrical energy to thermal energy. Thermal energy is transferred to the room mainly by radiation and convection.

Electric Heating Systems Unlike forced-air and radiator systems, electric heaters, like the one in **Figure 19,** convert electrical energy to thermal energy. In an electric heating system, electrically heated coils placed in floors and in walls heat the surrounding air by conduction and convection. Electric heating systems are not as widely used as forced-air systems. However, in warmer climates the walls and floors of some buildings may not be thick enough to contain pipes and ducts. Then an electric heating system might be the only practical way to provide heat.

Thermodynamics

There is another way to increase the thermal energy of an object besides heating it. Have you ever rubbed your hands together to warm them on a cold day? Your hands get warmer and their thermal energy and temperature increase. You did work on your hands by rubbing them together. The work you did caused the thermal energy of your hands to increase. Thermal energy, heat, and work are related. The study of the relationship among thermal energy, heat, and work is thermodynamics.

Heating and Work Increase Thermal Energy You can warm your hands by placing them near a fire. Thermal energy is added to your hands by radiation. You can make your hands even warmer by rubbing them together and holding them near a fire. Both the work you do and energy from the fire increase the thermal energy of your hands.

In the example above, your hands can be considered as a system. A system can be a group of objects, such as a galaxy or a car's engine, or something as simple as a ball. In fact, a system is anything you can draw a boundary around. The energy transferred to a system is the amount of energy flowing into the system across the boundary. Heating a system transfers energy to the system. The work done on a system is the work done by something outside the system's boundary.

Thermodynamics and the Steam Engine By the end of the eighteenth century, the steam engine was being used in manufacturing and transportation, leading to a wide-spread change in society called the industrial revolution. A young French engineer named Sadi Carnot (1796–1832) studied how the efficiency of steam engines could be improved. His work provided the basis for the development of thermodynamics by others in the nineteenth century. Research the life of Sadi Carnot and write a paragraph describing the results of his studies.

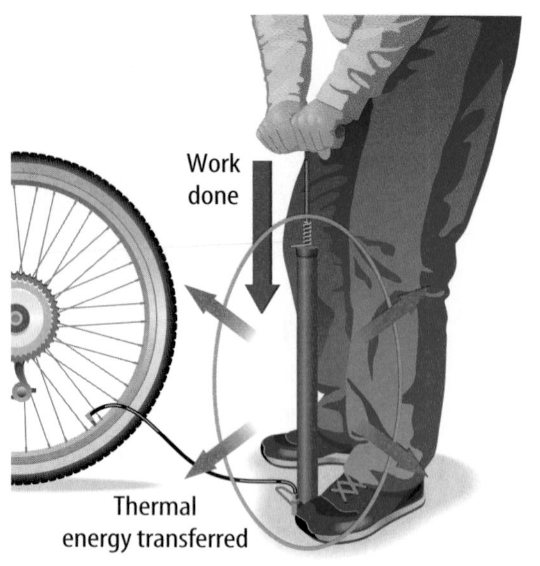

Figure 20 A bicycle air pump can be a system. Work is done on the system by pushing down on the handle. This causes the pump to become warm and thermal energy is transferred from the system to the air around it.
Infer *if the air pump is an open or a closed system.*

The First Law of Thermodynamics

According to the **first law of thermodynamics,** the increase in thermal energy of a system equals the work done on the system plus the thermal energy transferred to the system. Doing work on a system is a way of adding energy to a system, as shown in **Figure 20.** As a result, the temperature of a system can be increased by heating the system, doing work on the system, or both. The first law of thermodynamics is another way of stating the law of conservation of energy. The increase in energy of a system equals the energy added to the system.

Closed and Open Systems

A system is an open system if thermal energy flows across the boundary or if work is done across the boundary. The energy of an open system can change. If no thermal energy flows across the boundary and no outside work is done, the system is a closed system. According to the first law of thermodynamics, the thermal energy of a closed system doesn't change. Energy might change form within the system, but the total energy of the system doesn't change.

The Second Law of Thermodynamics

When thermal energy flows from a warm object to a cool object, the first law of thermodynamics is satisfied. The increase in thermal energy of the cool object equals the decrease in thermal energy of the warm object. Energy is neither created or destroyed.

Reading Check *How does the flow of thermal energy from a warm object to a cool object satisfy the first law of thermodynamics?*

Can thermal energy flow spontaneously from a cool object to a warm object? This never happens, but it doesn't violate the first law of thermodynamics. The first law requires only that energy be conserved. The first law would be satisfied if the decrease in thermal energy of the cool object equaled the increase in thermal energy of the warm object.

However, the flow of thermal energy spontaneously from a cool object to a warm object never happens because it violates another law—the second law of thermodynamics. One way to state the **second law of thermodynamics** is that it is impossible for thermal energy to flow from a cool object to a warmer object unless work is done. For example, if you hold an ice cube in your hand, no work is done. As a result, thermal energy flows only from your warmer hand to the colder ice.

Science Online
Topic: Solar Heating
Visit gpescience.com for Web links to information about heating systems that use solar energy to heat buildings.

Activity Draw a diagram showing how an active solar heating system is used to heat a home.

Converting Thermal Energy to Work

If you give a book sitting on a table a push, the book slides for a while and then stops. Friction between the book and the table converts all of the work you did on the book into thermal energy. Is it possible to convert thermal energy completely into work? Even though this process wouldn't violate the first law of thermodynamics, it violates the second law of thermodynamics. No device or process can convert thermal energy completely into work.

Heat Engines A device that converts thermal energy into work is called a heat engine. A car's engine is a heat engine. It converts some of the thermal energy produced by burning fuel into work that causes the car to move. However, only about 25 percent of the thermal energy in the burning fuel is converted into work. The rest of the thermal energy is transferred to the air surrounding the engine. When thermal energy is converted into work, some thermal energy always is transferred to the surroundings.

Internal Combustion Engines The heat engine in a car is an internal combustion engine, in which fuel is burned inside the engine in chambers or cylinders. Each cylinder contains a piston that moves up and down. Each up or down movement of the piston is called a stroke. Automobile and diesel engines make four different strokes, as shown in **Figure 21.**

Figure 21 The up-and-down movement of a piston in an automobile engine consists of four separate strokes. The four strokes form a cycle that is repeated many times each second by each piston.

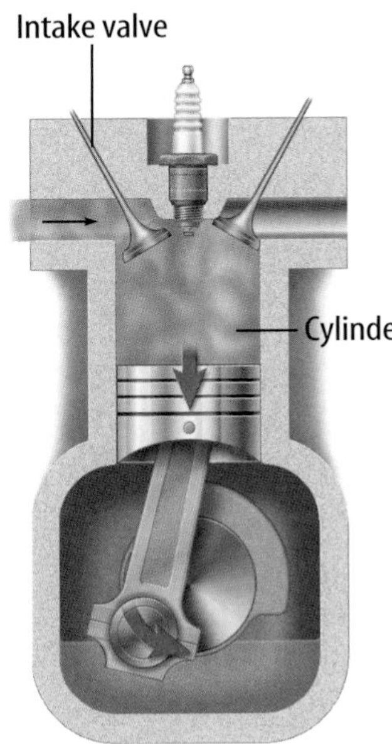

Intake stroke
The intake valve opens as the piston moves downward, drawing gasoline and air into the cylinder.

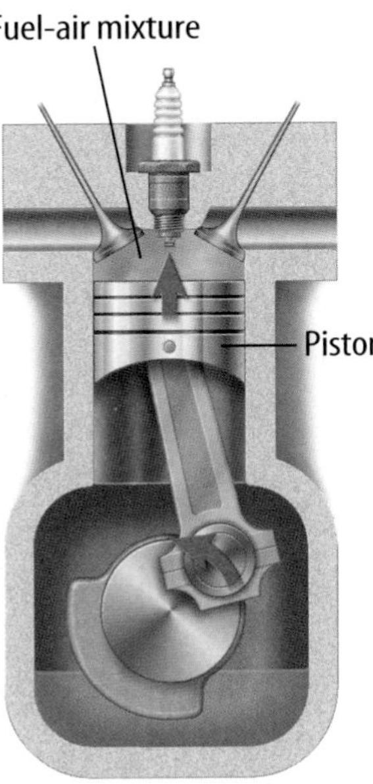

Compression stroke
The intake valve closes as the piston moves upward, compressing the fuel-air mixture.

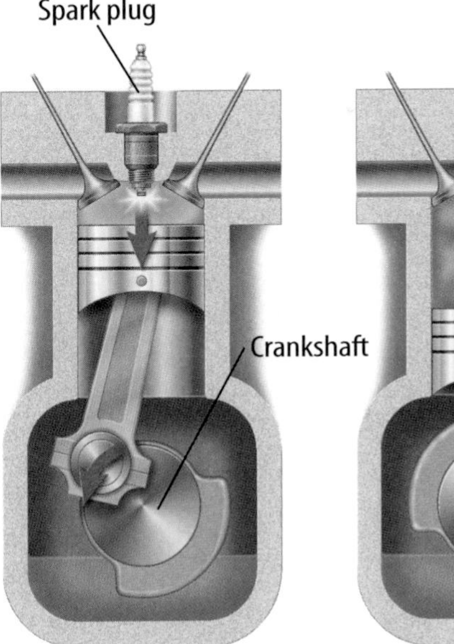

Power stroke
A spark plug ignites the fuel-air mixture. The hot gases expand, pushing the piston down.

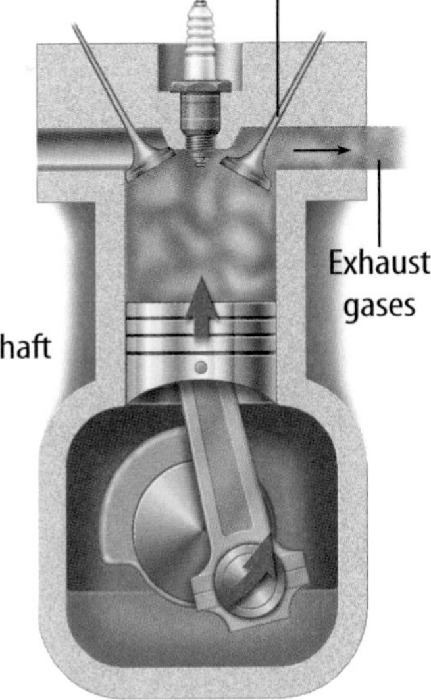

Exhaust stroke
As the piston moves up, the exhaust valve opens, and the hot gases are pushed out of the cylinder.

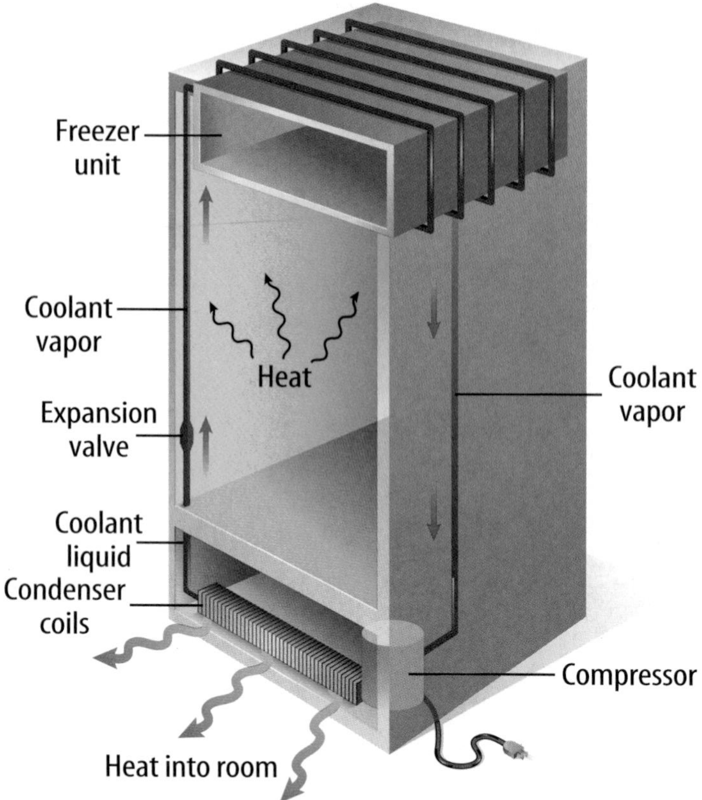

Figure 22 A refrigerator does work on the coolant in order to transfer thermal energy from inside the refrigerator to the warmer air outside. Work is done when the compressor compresses the coolant, causing its temperature to increase.

Moving Thermal Energy

How do refrigerators stay cold? The second law of thermodynamics prevents thermal energy from spontaneously flowing from inside the refrigerator to the warmer room. However, the second law of thermodynamics allows thermal energy to be moved from a cold area to a warmer area if work is done in the process. A refrigerator does work as it moves thermal energy from inside the refrigerator to the warmer room. The energy to do the work comes from the electrical energy the refrigerator obtains from an electric outlet.

Refrigerators A refrigerator contains a coolant that is pumped through pipes in the refrigerator. The coolant is a special substance that evaporates at a low temperature. **Figure 22** shows how a refrigerator operates. Liquid coolant is pumped through an expansion valve and changes into a gas. When the coolant changes to a gas, it cools. The cold gas is pumped through pipes inside the refrigerator. The coolant absorbs thermal energy and the inside of the refrigerator cools.

The gas then is pumped to a compressor that does work by compressing the gas. This makes the coolant gas warmer than the room. The warm gas then flows through condenser coils. There thermal energy flows from the gas to the air in the room. Some of this thermal energy is the thermal energy the coolant gas absorbed from inside the refrigerator. As the gas gives off thermal energy, it cools and changes into a liquid. Then liquid coolant is changed back into a gas, and the cycle is repeated.

Entropy

According to the laws of thermodynamics, work can be converted completely into thermal energy, but thermal energy can't be converted completely into work. What prevents thermal energy from being converted completely into work? The answer has to do with entropy. **Entropy** is a measure of how spread out, or dispersed, energy is. Entropy increases when energy becomes more spread out and less concentrated. Converting thermal energy completely into work causes energy to become less spread out and more concentrated. As a result, entropy decreases.

Entropy Always Increases All events that occur, such as the production of work in an automobile engine, must obey the entropy principle. According to the entropy principle, all events that occur cause the entropy of the universe to increase. This means that all events cause energy to spread out and become less concentrated. **Figure 23** shows how the potential energy stored in a ball becomes less concentrated when the ball is dropped and hits the ground.

How does an automobile engine cause entropy to increase? The energy source in an automobile engine is the chemical potential energy concentrated in the gasoline in the fuel tank. When gasoline is burned in the cylinders, 75 percent of the thermal energy produced is not converted into work, but spreads out into the air surrounding the car. Because energy becomes more spread out, entropy increases. However, the energy that becomes spread out is no longer useable. Any event that occurs causes the amount of useable energy to decrease.

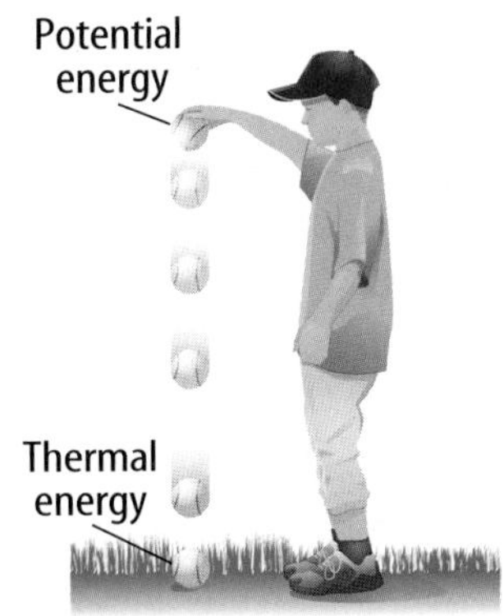

Figure 23 When the ball hits the ground, the potential energy it initially had is converted into thermal energy. This thermal energy spreads out and becomes less concentrated. This causes entropy to increase.

section 4 review

Summary

Heating Systems

- A forced-air heating system uses a fan to force air heated by a furnace through a system of ducts.
- Heating systems transfer thermal energy to rooms by conduction and convection.

Thermodynamics

- The first law of thermodynamics states that the increase in thermal energy of a system equals the work done on the system plus the thermal energy added to the system.
- One way to state the second law of thermodynamics is that thermal energy will not flow from a hot to a cold object unless work is done.

Converting Thermal Energy to Work

- The second law of thermodynamics states that thermal energy cannot be converted completely into work.
- Entropy is a measure of how spread out or dispersed energy is. Entropy increases when thermal energy is produced in an energy transformation.
- According to the entropy principle, all events that occur cause the entropy of the universe to increase. As a result, the amount of usable energy decreases.

Self Check

1. **Explain** how the thermal energy of a closed system changes with time.
2. **Compare and contrast** a forced-air heating system with a radiator system.
3. **Explain** whether or not a heat engine could be made 100 percent efficient by eliminating friction.
4. **Explain** Use the entropy principle to explain why thermal energy can never be converted completely into work.
5. **Think Critically** Suppose you vigorously shake a bottle of fruit juice. Predict how the temperature of the juice will change. Explain your reasoning.

Applying Math

6. **Calculate Change in Thermal Energy** You push down on the handle of a bicycle pump with a force of 20 N. The handle moves 0.3 m, and the pump does not absorb or release any thermal energy. What is the change in thermal energy of the bicycle pump as a result of the work you did on it?
7. **Calculate Work** The thermal energy released when a gallon of gasoline is burned in a car's engine is 140 million J. If the engine is 25 percent efficient, how much work does it do when one gallon of gasoline is burned?

Conduction in Gases

Goals

- **Measure** temperature changes in air near a heat source.
- **Observe** conduction of heat in air.

Materials

thermometers (3)
foam cups (2)
400-mL beakers (2)
burner or hot plate
paring knife
thermal mitts (2)

Safety Precautions

WARNING: *Use care when handling hot water. Pour hot water using both hands.*

Real-World Problem

Does smog occur where you live? If so, you may have experienced a temperature inversion. Usually the Sun warms the ground and the air above it. When the air near the ground is warmer than the air above, convection occurs. This convection also carries smoke and other gases emitted by cars, chimneys, and smokestacks upward into the atmosphere. If the air near the ground is colder than the air above, convection does not occur. Then smoke and other pollutants can be trapped near the ground, sometimes forming smog. How do the insulating properties of air cause a temperature inversion to occur?

Procedure

1. Using the paring knife, carefully cut the bottom from one foam cup.
2. Use a pencil or pen to poke holes about 2 cm from the top and bottom of each foam cup, as shown in the photo.
3. Turn both cups upside down, and poke the ends of the thermometers through the upper holes and lower holes, so both thermometers are supported horizontally. The bulb end of each thermometer should extend into the middle of the bottomless cup.

4. Heat about 350 mL of water to about 80°C in one of the beakers.
5. Place an empty 400-mL beaker on top of the bottomless cup. Record the temperatures of the two thermometers in a data table like the one shown here.
6. Add about 100 mL of hot water to the empty beaker. After one minute, record the temperatures of the thermometers in your data table.
7. Continue to record the temperatures every minute for 10 min. Add hot water as needed to keep the temperature of the water at about 80°C.

Air Temperatures in Foam Cup

Time (min)	Upper Thermometer (°C)	Lower Thermometer (°C)
0		
1	Do not write in this book.	
2		
3		
4		
5		

Analyze Your Data

1. **Graph** the temperatures measured by the upper and lower thermometers on the same graph. Make the vertical axis the temperature and the horizontal axis the time.
2. **Calculate** the total temperature change for each thermometer by subtracting the initial temperature from the final temperature.
3. **Calculate** the average rate of temperature change for each thermometer by dividing the total temperature change by 10 min.

Conclude and Apply

1. **Explain** whether convection can occur in the foam cup if it's being heated from the top.
2. **Describe** how heat was transferred through the air in the foam cup.
3. **Explain** why the average rates of temperature change were different for the two thermometers.

Compare your results with other students in your class. **Identify** the factors that caused the average rate of temperature change to be different for different groups.

SCIENCE Stats

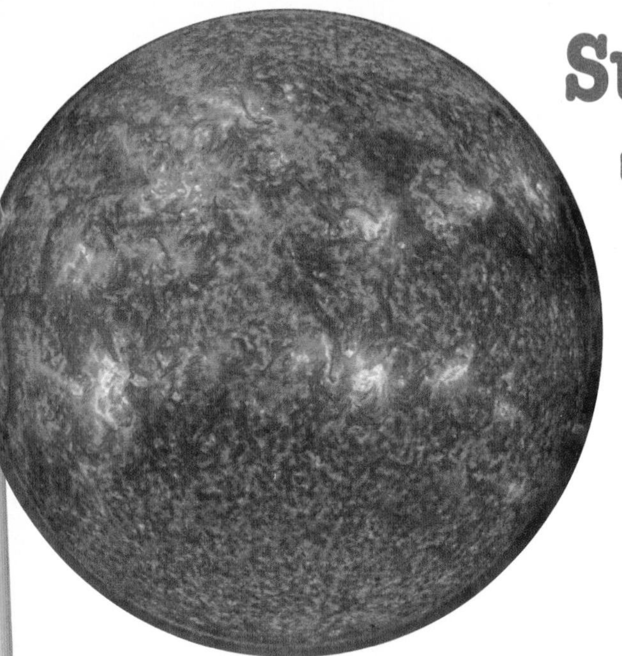

Surprising Thermal Energy

Did you know...

. . . The average amount of solar energy that reaches the United States each year is about 600 times greater than the nation's annual energy demands.

. . . When a space shuttle reenters Earth's atmosphere at more than 28,000 km/h, its outer surface is heated by friction to nearly 1,650°C. This temperature is high enough to melt steel.

. . . A lightning bolt heats the air in its path to temperatures of about 25,000°C. That's about 4 times hotter than the average temperature on the surface of the Sun.

Applying Math

1. The highest recorded temperature on Earth is 58°C and the lowest is −89°C. What is the range between the highest and lowest recorded temperatures?
2. What is the average temperature of the surface of the Sun? Draw a bar graph comparing the temperature of a lightning bolt to the temperature of the surface of the Sun.
3. The Sun is almost 150 million km from Earth. How long does it take solar energy to reach Earth if the energy travels at 300,000 km/s?

Reviewing Main Ideas

Section 1 Temperature and Thermal Energy

1. According to the kinetic theory, all matter is made of constantly moving particles that collide without losing energy.
2. The temperature of a material is a measure of the average kinetic energy of the particles in the material.
3. The thermal energy of an object is the total kinetic and potential energies of the particles in the object.
4. Heat is thermal energy that flows from a higher to a lower temperature.
5. The specific heat is the amount of energy needed to raise the temperature of 1 kg of a substance by 1°C.

Section 2 States of Matter

1. The states of matter are solid, liquid, gas, and plasma.
2. Changes of state can be interpreted in terms of the kinetic theory of matter.
3. Most matter expands when heated and contracts when cooled. An expansion joint allows concrete to expand and contract without damage.

Section 3 Transferring Thermal Energy

1. Conduction occurs when thermal energy is transferred by collisions between particles. Matter is not transferred.
2. Convection occurs in a fluid as warmer and cooler fluid move from place to place.
3. Radiation is the transfer of energy by electromagnetic waves. Radiation can transfer energy through empty space.
4. Thermal energy flows more quickly in materials that are conductors than in insulators.
5. Some insulating materials contain pockets of air that reduce the flow of thermal energy.

Section 4 Using Thermal Energy

1. According to the first law of thermodynamics, the increase in the thermal energy of a system equals the work done and the amount of thermal energy added to the system.

2. The second law of thermodynamics states that thermal energy cannot flow from a colder to a hotter temperature unless work is done.
3. Any event that occurs decreases the amount of usable energy and increases the entropy of the universe.

FOLDABLES™ Use the Foldable that you made at the beginning of this chapter to help you review thermal energy.

chapter 9 Review

Using Vocabulary

conduction p. 266	kinetic theory p. 254
convection p. 267	plasma p. 261
entropy p. 276	radiation p. 269
first law of thermodynamics p. 274	second law of thermodynamics p. 274
heat p. 257	specific heat p. 257
heat of fusion p. 262	temperature p. 255
heat of vaporization p. 262	thermal energy p. 256
Kelvin p. 255	thermal insulator p. 270

Complete each statement using a word or phrase from the vocabulary list above.

1. Thermal energy that is transferred from warmer to cooler materials is ________.
2. ________ is a measure of the average kinetic energy of the particles in a material.
3. The energy required to raise the temperature of 1 kg of a material 1°C is a material's ________.
4. In ________, thermal energy is transferred by the collisions between particles, not by the transfer of matter.
5. ________ is the theory used to explain the behavior of particles in matter.
6. The transfer of energy by electromagnetic waves is ________.

Checking Concepts

Choose the word or phrase that best answers the question.

7. Which is NOT a method of heat transfer?
   A) conduction **C)** radiation
   B) convection **D)** specific heat
8. Which state of matter is made of electrically charged particles?
   A) gas **C)** plasma
   B) liquid **D)** solid
9. During which phase of a four-stroke engine are waste gases removed?
   A) combustion stroke
   B) exhaust stroke
   C) intake stroke
   D) power stroke

Use the table below to answer question 10.

Latent Heats

Substance	Heat of Fusion (J/g)	Heat of Vaporization (J/g)
Ethanol	109	838
Water	334	2,258
Aluminum	397	10,900
Propane	61	387

10. Which material would require the least amount of energy to change 1 kg from a solid to a liquid?
    A) aluminum **C)** propane
    B) ethanol **D)** water
11. Which of the following is a measure of the average kinetic energy of the particles in an object?
    A) potential energy
    B) specific heat
    C) temperature
    D) thermal energy
12. Which of these is NOT used to calculate change in thermal energy?
    A) mass
    B) specific heat
    C) temperature change
    D) volume
13. Which of the following processes does NOT require the presence of particles of matter?
    A) combustion
    B) conduction
    C) convection
    D) radiation

Vocabulary PuzzleMaker gpescience.com

14. Which of the following is the name for thermal energy that is transferred only from a higher temperature to a lower temperature?
   A) heat **C)** potential energy
   B) kinetic energy **D)** solar energy

15. According to the entropy principle, any event that occurs in the universe must cause which of the following to happen?
   A) energy to become more concentrated
   B) energy to become less concentrated
   C) total amount of energy to increase
   D) total amount of energy to decrease

Interpreting Graphics

16. Copy and complete this concept map.

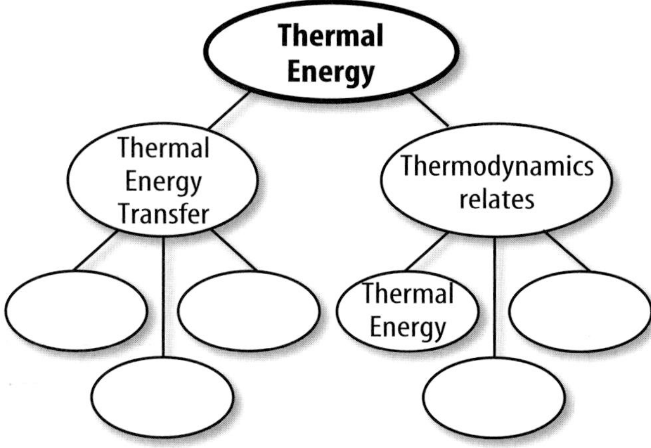

Thinking Critically

17. **Explain** On a hot day a friend suggests that you can make your kitchen cooler by leaving the refrigerator door open. Explain whether leaving the refrigerator door open would cause the air temperature in the kitchen to decrease.

18. **Explain** Which has the greater amount of thermal energy, one liter of water at 50°C or two liters of water at 50°C?

19. **Explain** whether or not the following statement is true: If the thermal energy of an object increases, the temperature of the object must also increase.

20. **Predict** Suppose a beaker of water is heated from the top. Explain which is more likely to occur in the water—thermal energy transfer by conduction or convection.

21. **Classify** Order the events that occur in the removal of thermal energy from an object by a refrigerator. Draw the complete cycle, from the placing of a warm object in the refrigerator to the changes in the coolant.

Applying Math

Use the table below to answer questions 22 to 24.

Specific Heat of Materials

Material	Specific Heat (J/kg°C)
Water	4,184
Copper	385
Silver	235
Graphite	710
Iron	450

22. **Calculate Thermal Energy** How much thermal energy is needed to raise the temperature of 4.0 kg of water from 25°C to 75°C?

23. **Calculate Temperature Change** How does the temperature of 33.0 g of graphite change when it absorbs 350 J of thermal energy?

24. **Calculate Mass** A certain mass of water gains 1,673 J of thermal energy when its temperature increases by 2.0°C. What is the mass of the water?

Standardized Test Practice

Record your answers on the answer sheet provided by your teacher or on a sheet of paper.

Multiple Choice

Use the graph below to answer questions 1 and 2.

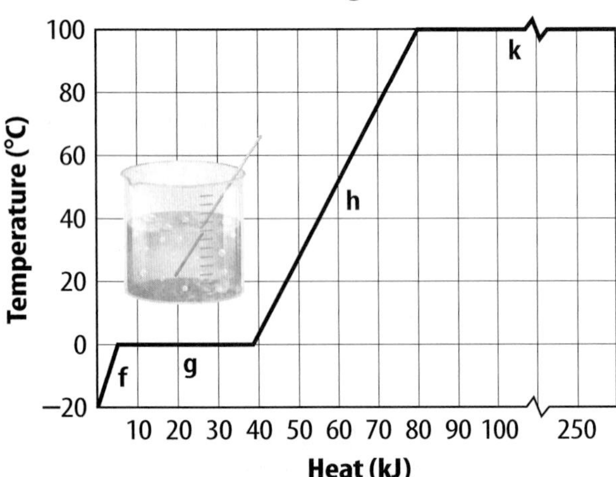

1. At which points on the graph is the average kinetic energy of the water molecules increasing?
   - **A.** F and G
   - **B.** G and K
   - **C.** F and H
   - **D.** H and K

2. At which points on the graph is the potential energy of the water molecules increasing, but the average kinetic energy is not?
   - **A.** F and G
   - **B.** G and K
   - **C.** F and H
   - **D.** H and K

Test-Taking Tip

Read Carefully Read all choices before answering the question.

3. Which is the term for the amount of energy required for 1 kg of a liquid at its boiling point to become a gas?
   - **A.** heat of vaporization
   - **B.** diffusion
   - **C.** heat of fusion
   - **D.** thermal energy

Use the table below to answer questions 4 and 5.

Material	Specific Heat [J/(kg °C)]
Copper	385
Gold	449
Lead	129
Tin	228
Zinc	388

4. According to the table above, a 2-kg block of which of the following materials would require 898 J of heat to increase its temperature by 1°C?
   - **A.** gold
   - **B.** lead
   - **C.** tin
   - **D.** zinc

5. Which of the following materials would require the most heat to raise a 5-kg sample of the material from 10°C to 50°C?
   - **A.** gold
   - **B.** lead
   - **C.** tin
   - **D.** zinc

Use the figure below to answer question 6.

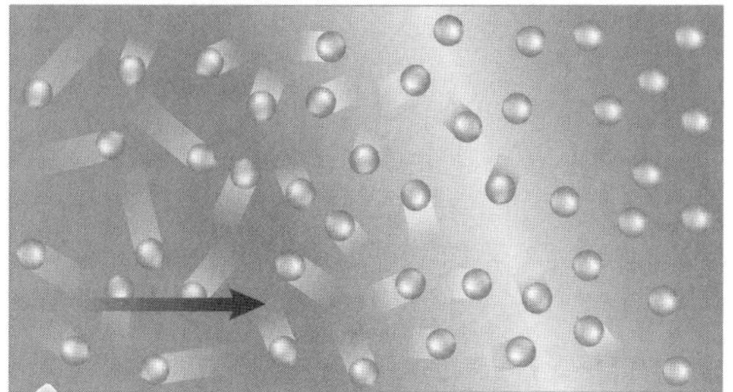

6. The illustration above shows the particles in a material being heated at one end. Thermal energy is transferred from the warmer part of the material to the cooler part. What heat transfer process is occurring in the material?

   A. conduction

   B. convection

   C. radiation

   D. entropy

Gridded Response

7. The temperature of a 24.5-g block of aluminum decreases from 30.0°C to 21.5°C. If aluminum has a specific heat of 897 J/(kg °C), what is the change in thermal energy, measured in joules, of the block of aluminum?

8. The difference between the boiling point and the freezing point of potassium is 695.72 K. What is this difference in degrees on the Celsius temperature scale?

Short Response

9. Explain how the temperature and thermal energy of an object are related.

10. What property of water makes it useful as a coolant?

Extended Response

Use the figure below to answer question 11.

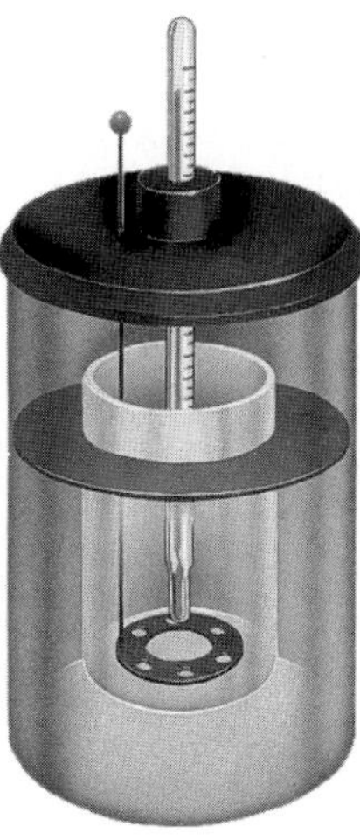

11. The figure shows a calorimeter that is used to measure the specific heat of materials.

    PART A Explain why it is necessary to measure the temperature change of the water when using a calorimeter to measure the specific heat of a material.

    PART B Explain why the calorimeter must be insulated from its surroundings.

12. Explain how changes in a fluid's density enable convection currents to occur.

chapter 10

Waves

chapter preview

sections

Waves, anyone?

This surfer in Hawaii is surrounded by an ocean wave that forms a huge wall of water. You are surrounded by waves, too. Everything you see or hear is brought to you by waves. Easy to see—like this ocean wave—or invisible, all waves transfer energy.

Science Journal Write down three things you already know about waves and one thing you would like to learn about waves.

Start-Up Activities

How do waves transfer energy?

Light enters your eyes and sound strikes your ears, enabling you to sense the world around you. Light and sound are waves that transfer energy from one place to another. Do waves transfer anything else along with their energy? Does a wave transfer matter, too? In this activity, you'll observe one way that waves can transfer energy.

1. Complete the safety form.
2. Place your textbook flat on your desk. Line up four marbles on the groove at the edge of the textbook so that the marbles are touching each other.
3. Hold the first three marbles in place using three fingers of one hand.
4. Tap the first marble with a pen or pencil.
5. Observe the behavior of the fourth marble.
6. **Think Critically** Write a paragraph explaining how the fourth marble reacted to the pen tap. Diagram how energy was transferred through the marbles.

Preview this chapter's content and activities at gpescience.com

Types of Waves Make the following Foldable to compare and contrast two types of waves.

STEP 1 **Fold** one sheet of paper lengthwise.

STEP 2 **Fold** into thirds.

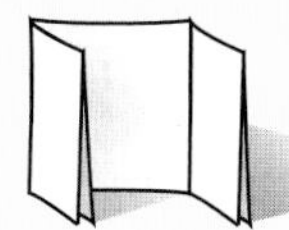

STEP 3 **Unfold and draw** overlapping ovals. Cut the top sheet along the folds.

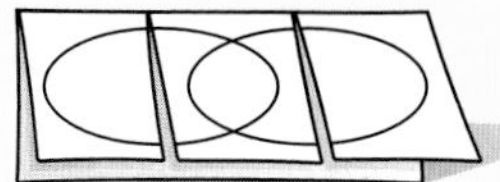

STEP 4 **Label** the ovals as shown.

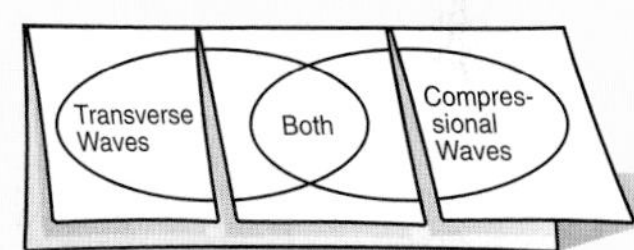

Construct a Venn Diagram As you read this chapter, list properties and characteristics unique to transverse waves under the left tab, those unique to compressional waves under the right tab, and those common to both under the middle tab.

section 1

The Nature of Waves

Reading Guide

What You'll Learn

- **Recognize** that waves transfer energy but not matter.
- **Define** mechanical waves.
- **Compare and contrast** transverse waves and compressional waves.

Why It's Important

You can see and hear the world around you because of the energy transferred by waves.

Review Vocabulary

energy: the ability to do work

New Vocabulary

- wave
- medium
- transverse wave
- compressional wave

What's in a wave?

A surfer bobs in the ocean waiting for the perfect wave, microwaves warm up your leftover pizza, and sound waves from your CD player bring music to your ears. Do these and other types of waves have anything in common with one another?

A **wave** is a repeating disturbance or movement that transfers energy through matter or space. For example, ocean waves disturb the water and transfer energy through it. During earthquakes, energy is transferred in powerful waves that travel through Earth. Light is a type of wave that can travel through empty space to transfer energy from one place to another, such as from the Sun to Earth.

Figure 1 Falling pebbles transfer their kinetic energy to the particles of water in a pond, forming waves.

Waves and Energy

Kerplop! A pebble falls into a pool of water and ripples form. As **Figure 1** shows, the pebble causes a disturbance that moves outward in the form of a wave. Because it is moving, the falling pebble has energy. As it splashes into the pool, the pebble transfers some of its energy to nearby water molecules, causing them to move. These molecules then pass the energy along to neighboring water molecules, which, in turn, transfer it to their neighbors. The energy moves farther and farther from the source of the disturbance. What you see is energy traveling in the form of a wave on the surface of the water.

Waves and Matter Imagine that you're in a boat on a lake. Approaching waves bump against your boat, but they don't carry it along with them as they pass. The boat does move up and down and maybe even a short distance back and forth because the waves transfer some of their energy to it. However, after the waves have moved on, the boat is still in nearly the same place. The waves don't even carry the water along with them. Only the energy transferred by the waves moves forward. All waves have this property—they transfer energy without transporting matter from place to place.

Reading Check *What do waves transfer?*

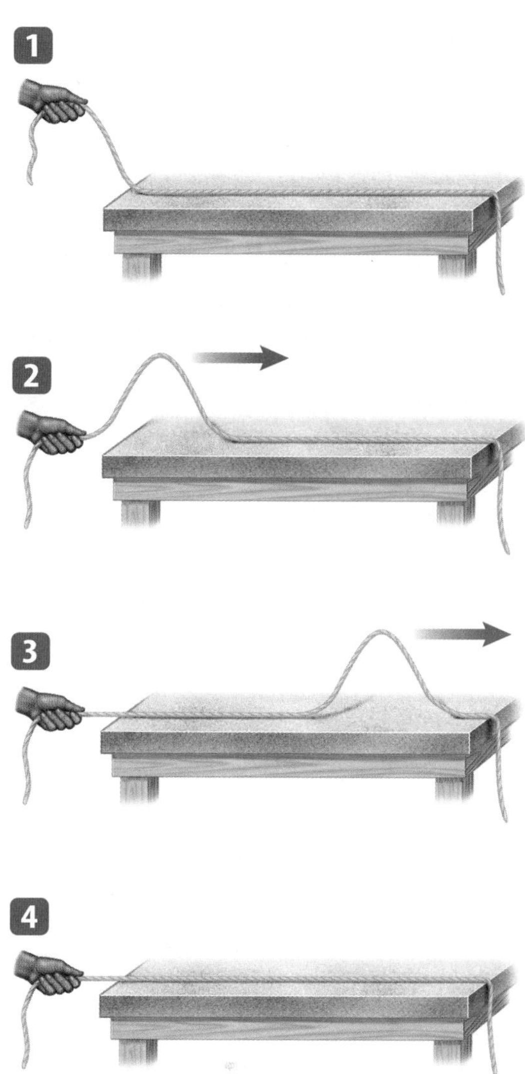

Figure 2 A wave will exist only as long as it has energy to transfer. **Explain** *what happened to the energy that was transferred by the wave in this rope.*

Making Waves A wave will travel only as long as it has energy to transfer. For example, when you drop a pebble into a puddle, the ripples soon die out and the surface of the water becomes still again.

Suppose you are holding a rope at one end, and you give it a shake. You would create a pulse that would travel along the rope to the other end, and then the rope would be still again, as **Figure 2** shows. Now suppose you shake your end of the rope up and down for a while. You would make a wave that would travel along the rope. When you stop shaking your hand up and down, the rope will be still again. It is the up-and-down motion of your hand that creates the wave.

Anything that moves up and down or back and forth in a rhythmic way is vibrating. The vibrating movement of your hand at the end of the rope created the wave. In fact, all waves are produced by something that vibrates.

Mechanical Waves

Sound waves travel through the air to reach your ears. Ocean waves move through water to reach the shore. In both cases, the matter the waves travel through is called a **medium.** A medium can be a solid, a liquid, a gas, or a combination of these. For sound waves, the medium is air, and for ocean waves, the medium is water. Not all waves need a medium. Some waves, such as light and radio waves, can travel through space. Waves that can travel only through matter are called mechanical waves. The two types of mechanical waves are transverse waves and compressional waves.

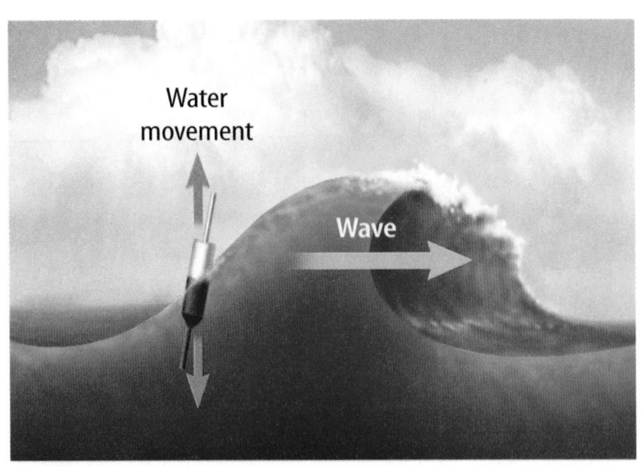

Figure 3 A water wave travels horizontally as the water moves vertically up and down.

Transverse Waves In a **transverse wave,** matter in the medium moves back and forth at right angles to the direction that the wave travels. For example, **Figure 3** shows how a wave in the ocean moves horizontally, but the water that the wave passes through moves up and down. When you shake one end of a rope while your friend holds the other end, you are making transverse waves. The wave and its energy travel from you to your friend as the rope moves up and down.

Compressional Waves In a **compressional wave,** matter in the medium moves back and forth along the same direction that the wave travels. You can model compressional waves with a coiled-spring toy, as shown in **Figure 4.** Squeeze several coils together at one end of the spring. Then let go of the coils, still holding onto coils at both ends of the spring. A wave will travel along the spring. As the wave moves, it looks as if the whole spring is moving toward one end. Suppose you watched the coil with yarn tied to it, as in **Figure 4.** You would see that the yarn moves back and forth as the wave passes, and then it stops moving after the wave has passed. The wave transfers energy, but not matter, forward along the spring. Compressional waves also are called longitudinal waves.

Sound Waves Sound waves are compressional waves. When a noise is made, such as when a locker door slams shut and vibrates, nearby air molecules are pushed together by the vibrations. The air molecules are squeezed together like the coils in a coiled-spring toy are when you make a compressional wave with it. The compressions travel through the air to make a wave.

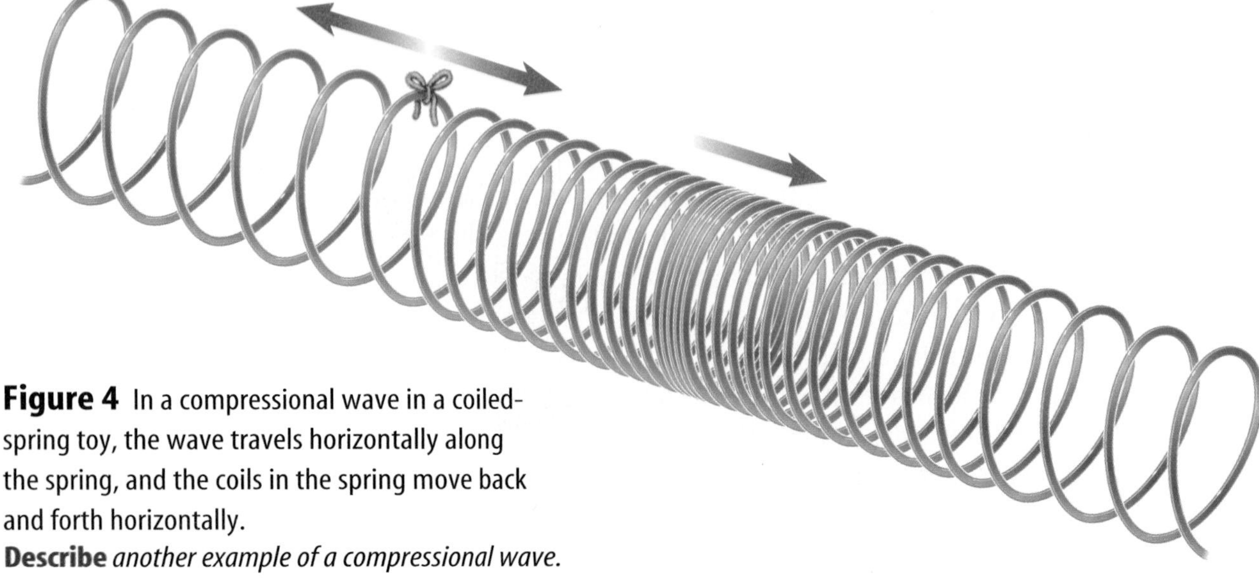

Figure 4 In a compressional wave in a coiled-spring toy, the wave travels horizontally along the spring, and the coils in the spring move back and forth horizontally.
Describe *another example of a compressional wave.*

Sound in Other Materials Sound waves also can travel through other mediums, such as water and wood. Particles in these mediums also are pushed together and move apart as the sound waves travel through them. When a sound wave reaches your ear, it causes your eardrum to vibrate. Your inner ear then sends signals to your brain, and your brain interprets the signals as sound.

Reading Check *How do sound waves travel in solids?*

Water Waves Water waves are not purely transverse waves. The water moves up and down as the waves go by. However, the water also moves a short distance back and forth along the direction the wave is moving. This movement happens because the low part of the wave can be formed only by pushing water forward or backward toward the high part of the wave, as in **Figure 5A.** Then, as the wave passes, the water that was pushed aside moves back to its initial position, as in **Figure 5B.** In fact, if you looked closely, you would see that the combination of this up-and-down and back-and-forth motion causes water to move in circles. Anything floating on the surface of the water absorbs some of the waves' energy and bobs in a circular motion.

Ocean waves are formed most often by wind blowing across the ocean surface. As the wind blows faster and slower, the changing wind speed is like a vibration. The size of the waves that are formed depends on the wind speed, the distance over which the wind blows, and how long the wind blows. **Figure 6** on the next page shows how ocean waves are formed.

Figure 5 A water wave causes water to move back and forth, as well as up and down. Water is pushed back and forth to form the crests and troughs.

A The low point of a water wave is formed when water is pushed aside and up to the high point of the wave.

B The water that is pushed aside returns to its initial position.

NATIONAL GEOGRAPHIC VISUALIZING FORMATION OF OCEAN WAVES

Figure 6

When wind blows across an ocean, friction between the moving air and the water causes the water to move. As a result, energy is transferred from the wind to the surface of the water. The waves that are produced depend on the length of time and the distance over which the wind blows, as well as the wind speed.

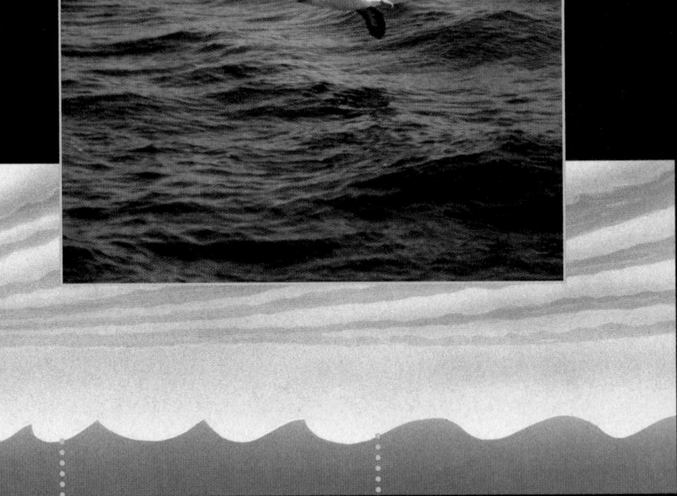

▲ Wind causes ripples to form on the surface of the water. As ripples form, they provide an even larger surface area for the wind to strike, and the ripples increase in size.

▲ Waves that are higher and have longer wavelengths grow faster as the wind continues to blow, but the steepest waves break up, forming whitecaps. The surface is said to be choppy.

▲ The shortest-wavelength waves break up, while the longest-wavelength waves continue to grow. When these waves have reached their maximum height, they form fully developed seas.

▲ After the wind dies down, the waves lose energy and become lower and smoother. These smooth, long-wavelength ocean waves are called swells.

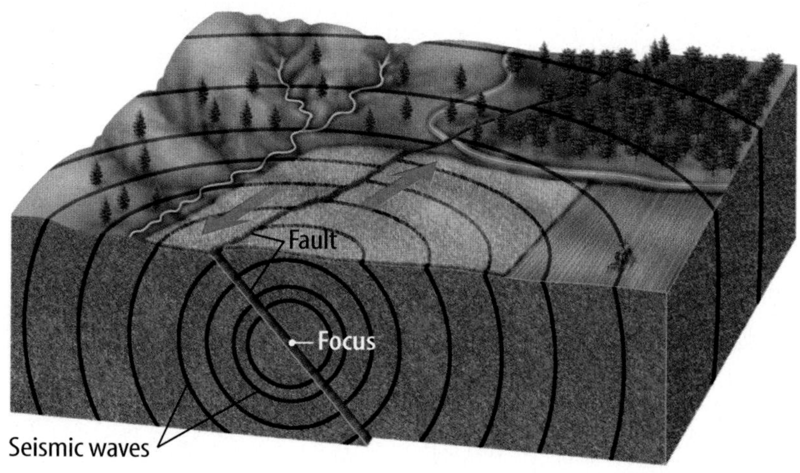

Figure 7 When Earth's crust shifts or breaks, the energy that is released is transmitted outward, causing an earthquake.
Explain *why earthquakes are mechanical waves.*

INTEGRATE Earth Science

Seismic Waves A guitar string makes a sound when it breaks. The string vibrates for a short time after it breaks and produces sound waves. In a similar way, forces in Earth's crust can cause regions of the crust to shift, bend, or even break. The breaking crust vibrates, creating seismic waves that transfer energy outward, as shown in **Figure 7.** Seismic waves are a combination of compressional and transverse waves. They can travel through Earth and along Earth's surface. When objects on Earth's surface absorb some of the energy transferred by seismic waves, they move and shake. The more the crust moves during an earthquake, the more energy is released.

Topic: Seismic Waves
Visit gpescience.com for Web links to information about seismic waves.

Activity Write a summary of how seismic waves are used to map Earth's interior.

section 1 review

Summary

Waves and Energy

- A wave is a repeating disturbance or movement that transfers energy through matter or space.
- Waves transfer energy without transporting matter.
- Waves are produced by something that is vibrating.

Mechanical Waves

- Mechanical waves must travel in matter.
- Mechanical waves can be transverse waves or compressional waves.
- In a transverse wave, matter in the medium moves at right angles to the wave motion.
- In a compressional wave, matter in the medium moves back and forth along the direction of the wave motion.

Self Check

1. **Compare and contrast** a transverse wave and a compressional wave. Give an example of each type.
2. **Describe** the motion of a buoy when a water wave passes. Does it move the buoy forward?
3. **Explain** how you could model a compressional wave using a coiled-spring toy.
4. **List** the characteristics of a mechanical wave.
5. **Think Critically** Why do boats need anchors if ocean waves do not carry matter forward?

Applying Math

6. **Calculate Time** The average speed of sound in water is 1,500 m/s. How long would it take a sound wave to travel 9,000 m?

Wave Properties

Reading Guide

What You'll Learn

- **Define** wavelength, frequency, period, and amplitude.
- **Describe** the relationship between frequency and wavelength.
- **Explain** how a wave's energy and amplitude are related.
- **Calculate** wave speed.

Why It's Important

Waves with different properties can be used in different ways.

Review Vocabulary

vibration: a back and forth movement

New Vocabulary

- crest
- trough
- rarefaction
- wavelength
- frequency
- period
- amplitude

The Parts of a Wave

What makes sound waves, water waves, and seismic waves different from each other? Waves can differ in how much energy they transfer and in how fast they travel. Waves also have other characteristics that make them different from each other.

Suppose you shake the end of a rope and make a transverse wave. The transverse wave in **Figure 8** has alternating high points, called **crests,** and low points, called **troughs.**

On the other hand, a compressional wave has no crests and troughs. When a compressional wave passes through a medium, it creates regions where the medium becomes crowded together and more dense, as in **Figure 8.** These regions are compressions. When you make compressional waves in a coiled spring, the compressions are regions where the coils are close together. **Figure 8** also shows that the coils in the regions next to a compression are spread apart, or less dense. These less-dense regions of a compressional wave are called **rarefactions.**

Figure 8 Transverse and compressional waves have different features that travel through a medium and form the wave.

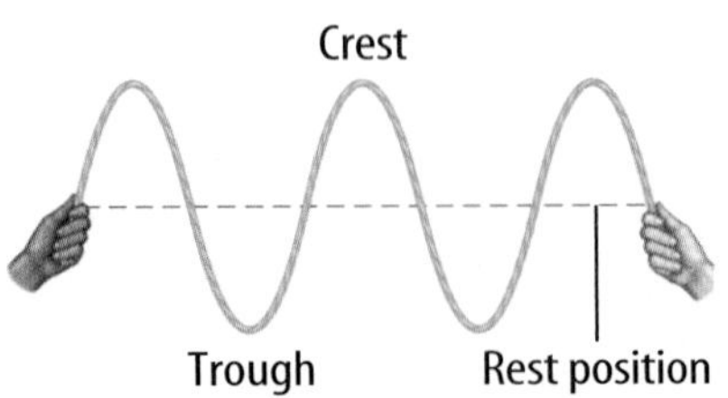

A transverse wave is made of crests and troughs that travel through the medium.

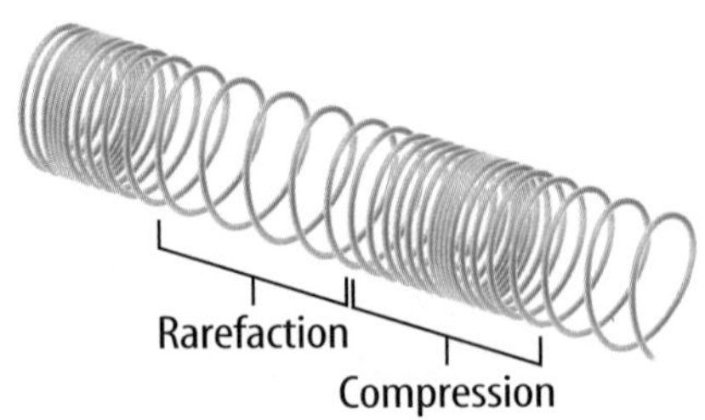

A compressional wave is made of compressions and rarefactions that travel through the medium.

Figure 9 One wavelength starts at any point on a wave and ends at the nearest point just like it.

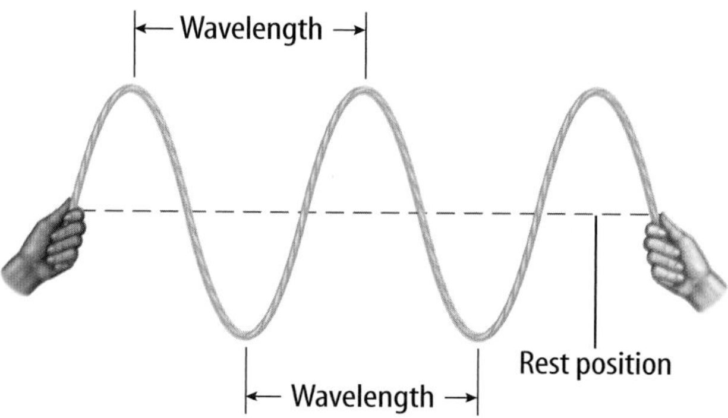

For transverse waves, a wavelength can be measured from crest to crest or trough to trough.

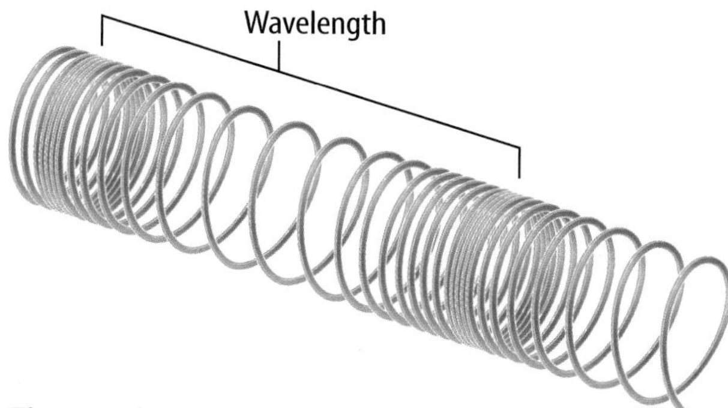

The wavelength of a compressional wave can be measured from compression to compression or from rarefaction to rarefaction.

Wavelength

Waves also have a property called wavelength. A **wavelength** is the distance between one point on a wave and the nearest point just like it. **Figure 9** shows that for transverse waves, the wavelength is the distance from crest to crest or trough to trough.

A wavelength in a compressional wave is the distance between two neighboring compressions or two neighboring rarefactions, as shown in **Figure 9.** You can measure from the start of one compression to the start of the next compression or from the start of one rarefaction to the start of the next rarefaction. The wavelengths of sound waves that you can hear range from a few centimeters for the highest-pitched sounds to about 15 m for the deepest sounds.

Reading Check *How is wavelength measured in transverse and compressional waves?*

Frequency and Period

When you tune your radio to a station, you are choosing radio waves of a certain frequency. The **frequency** of a wave is the number of wavelengths that pass a fixed point each second. You can find the frequency of a transverse wave by counting the number of crests or troughs that pass by a point each second. The frequency of a compressional wave is the number of compressions or rarefactions that pass a point every second. Frequency is expressed in hertz (Hz). A frequency of 1 Hz means that one wavelength passes by in 1 s. In SI units, 1 Hz is the same as 1/s. The **period** of a wave is the amount of time it takes one wavelength to pass a point. As the frequency of a wave increases, the period decreases. Periods are measured in units of seconds.

Observing Wavelength

Procedure

1. Complete the safety form.
2. Fill a **pie plate or other wide pan** with **water** about 2 cm deep.
3. Lightly tap your finger once per second on the surface of the water and observe the spacing of the water waves.
4. Increase the rate of your tapping, and observe the spacing of the water waves.

Analysis

1. How is the spacing of the water waves related to their wavelength?
2. How does the spacing of the water waves change when the rate of tapping increases?

Try at Home

Wavelength Is Related to Frequency If you make transverse waves with a rope, you increase the frequency by moving the rope up and down faster. Moving the rope faster also makes the wavelength shorter. This relationship is always true: as frequency increases, wavelength decreases. **Figure 10** compares the wavelengths and frequencies of two different waves.

The frequency of a wave is always equal to the rate of vibration of the source that creates it. If you move the rope up, down, and back up in 1 s, the frequency of the wave you generate is 1 Hz. If you move the rope up, down, and back up five times in 1 s, the resulting wave has a frequency of 5 Hz.

How are the wavelength and frequency of a wave related?

Wave Speed

You're at a large stadium watching a baseball game, but you're high up in the bleachers, far from the action. The batter swings and you see the ball rise. An instant later, you hear the crack of the bat hitting the ball. You see the impact before you hear it because light waves travel much faster than sound waves. Therefore, the light waves reflected from the flying ball reach your eyes before the sound waves created by the crack of the bat reach your ears.

The speed of a wave depends on the medium it is traveling through. Sound waves usually travel faster in liquids and solids than they do in gases. However, light waves travel more slowly in liquids and solids than they do in gases or in empty space. Also, sound waves usually travel faster in a material when the temperature of the material is increased. For example, sound waves travel faster in air at 20°C than in air at 0°C.

Figure 10 The wavelength of a wave decreases as the frequency increases.

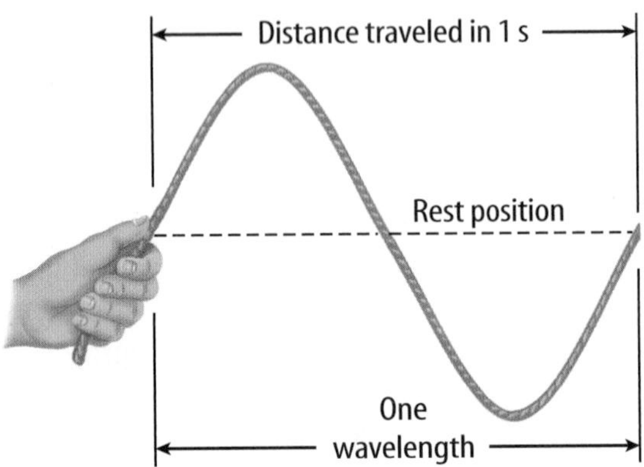

The rope is moved down, up, and down again one time in 1 s. One wavelength is created on the rope.

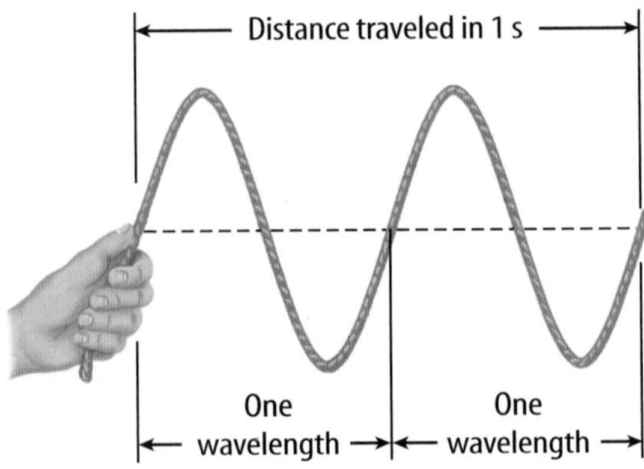

The rope is shaken down, up, and down again twice in 1 s. Two wavelengths are created on the rope.

Calculating Wave Speed You can calculate the speed of a wave, represented by v, by multiplying its frequency times its wavelength. Wavelength is represented by the Greek letter *lambda* (λ), and frequency is represented by f.

Wave Speed Equation

speed (in m/s) = frequency (in Hz) × wavelength (in m)

$$v = f\lambda$$

Why does multiplying the frequency unit Hz by the distance unit m give the unit for speed, m/s? Recall that the SI unit Hz is the same as 1/s. Multiplying m × Hz equals m × 1/s, which equals m/s.

Deadly Ocean Waves
Tsunamis can cause serious damage when they hit land. These waves can measure up to 30 m tall and can travel faster than 700 km/h. Research which areas of the world are most vulnerable to tsunamis. Describe the effects of a tsunami that has occurred in these areas.

Applying Math — Solve a Simple Equation

THE SPEED OF SOUND What is the speed of a sound wave that has a wavelength of 2.00 m and a frequency of 170.5 Hz?

IDENTIFY known values and the unknown value

Identify the known values:

wavelength of 2.0 m → means → $\lambda = 2.00$ m

frequency of 170.5 Hz → means → $f = 170.5$ Hz

Identify the unknown value:

the speed of a sound wave → means → $v = ?$ m/s

SOLVE the problem

Substitute the known values $\lambda = 2.00$ m and $f = 170.5$ Hz into the wave speed equation:

$$v = f\lambda = (170.5 \text{ Hz})(2.00 \text{ m}) = 341 \text{ m/s}$$

CHECK the answer

Does your answer seem reasonable? Check your answer by dividing the wave speed you calculated by the wavelength given in the problem. The result should be the frequency given in the problem.

Practice Problems

1. A wave traveling in water has a frequency of 500.0 Hz and a wavelength of 3.0 m. What is the speed of the wave?
2. The lowest-pitched sounds humans can hear have a frequency of 20.0 Hz. What is the wavelength of these sound waves if their wave speed is 340.0 m/s?
3. The highest-pitched sound humans can hear have a wavelength of 0.017 m in air. What is the frequency of these sound waves if their wave speed is 340.0 m/s?

For more practice problems go to page 879, and visit Math Practice at gpescience.com.

Amplitude and Energy

Why do some earthquakes cause terrible damage, while others are hardly felt? This is because the amount of energy a wave transfers can vary. **Amplitude** is related to the energy transferred by a wave. The greater the wave's amplitude, the more energy the wave transfers. Amplitude is measured differently for compressional and transverse waves.

Amplitude of Compressional Waves The amplitude of a compressional wave is related to how tightly the medium is pushed together at the compressions. The denser the medium is at the compressions, the larger its amplitude is and the more energy the wave transfers. For example, it takes more energy to push the coils in a coiled-spring toy tightly together than to barely move them. The closer the coils are in a compression, the farther apart they are in a rarefaction. So, the less dense the medium is at the rarefactions, the more energy the wave transfers. **Figure 11** shows compressional waves with different amplitudes.

Figure 11 The amplitude of a compressional wave depends on the density of the medium in the compressions and rarefactions.

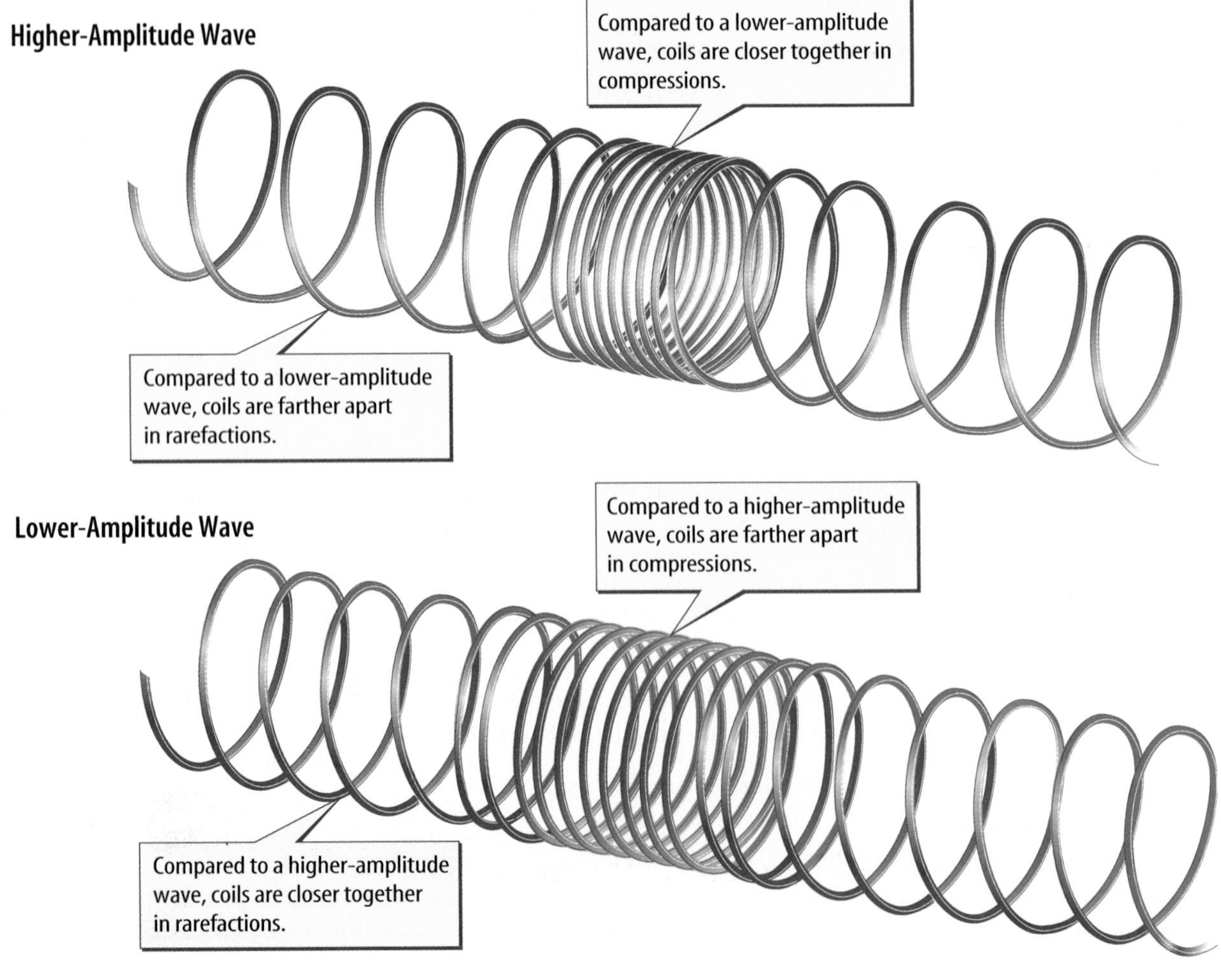
Higher-Amplitude Wave
Compared to a lower-amplitude wave, coils are closer together in compressions.
Compared to a lower-amplitude wave, coils are farther apart in rarefactions.
Lower-Amplitude Wave
Compared to a higher-amplitude wave, coils are farther apart in compressions.
Compared to a higher-amplitude wave, coils are closer together in rarefactions.

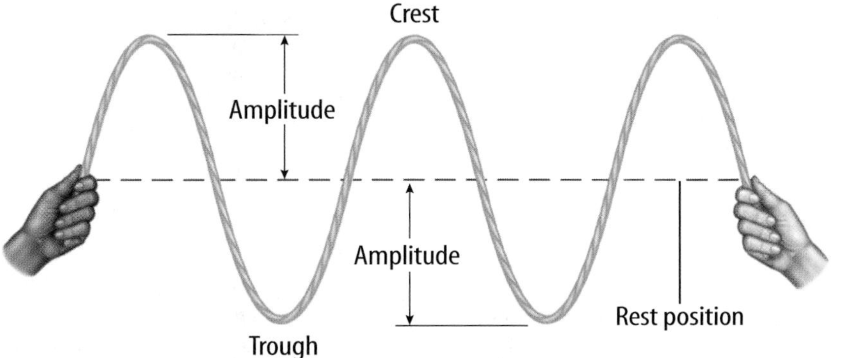

Figure 12 The amplitude of a transverse wave is the distance between a crest or a trough and the position of the medium at rest. **Describe** *how you could create waves with different amplitudes in a piece of rope.*

Amplitude of Transverse Waves If you've ever been knocked over by an ocean wave, you know that the higher the wave, the more energy it transfers. Remember that the amplitude of a wave increases as the energy transferred by the wave increases. So a tall ocean wave has a greater amplitude than a short ocean wave does. The amplitude of any transverse wave is the distance from the crest or trough of the wave to the rest position of the medium, as shown in **Figure 12.**

section 2 review

Summary

The Parts of a Wave

- Transverse waves have repeating high points called crests and low points called troughs.
- Compressional waves have repeating high-density regions called compressions, and low-density regions called rarefactions.

Wavelength, Frequency, and Period

- Wavelength is the distance between a point on a wave and the nearest point just like it.
- A wave's frequency is the number of wavelengths passing a fixed point each second.
- A wave's period is the amount of time it takes one wavelength to pass a fixed point.

Wave Speed and Amplitude

- The speed of a wave depends on the material it is traveling in, and on the temperature.
- The speed of a wave is the product of its frequency and its wavelength:

$$v = f\lambda$$

- As the amplitude of a wave increases, the energy transferred by the wave increases.

Self Check

1. **Describe** the difference between a compressional wave with a large amplitude and one with a small amplitude.
2. **Describe** how the wavelength of a wave changes when the wave slows down but its frequency doesn't change.
3. **Explain** how the frequency of a wave changes when the period of the wave increases.
4. **Form a hypothesis** to explain why a sound wave travels faster in a solid than in a gas.
5. **Think Critically** You make a transverse wave by shaking the end of a long rope up and down. Explain how you would shake the end of the rope to make the wavelength shorter. How would you shake the end of the rope to increase the energy transferred by the wave?

Applying Math

6. **Calculate** the frequency of a water wave that has a wavelength of 0.5 m and a speed of 4.0 m/s.
7. **Calculate Wavelength** An FM radio station broadcasts radio waves with a frequency of 100,000,000 Hz. What is the wavelength of these radio waves if they travel at a speed of 300,000 km/s?

Wave Speed and Tension

Before playing her violin, a violinist must adjust the tension, or the amount of force pulling on each string, to tune the violin.

Real-World Problem

How does the tension in a material affect the waves traveling in the material?

Goal

- **Determine** the relationship between tension and wave speed.

Materials

coiled-spring toy
meterstick
stopwatch

Safety Precautions

Procedure

1. Complete the safety form.
2. Attach one end of the spring to a chair leg so that the spring rests on a smooth floor.
3. Stretch the spring along the floor to a length of 1 m.
4. Make a compressional wave by squeezing together several coils and then releasing them.
5. Have your partner time how long the wave takes to travel two or three lengths of the spring. Record the time in your data table. Record the distance the wave traveled in your data table.
6. Repeat steps 4 and 5 two more times for waves 2 and 3.
7. Stretch the spring to a length of 1.5 m.
8. Repeat steps 4 and 5 for waves 4, 5, and 6.

Data Table

	Distance (m)	Wave Time (s)	Wave Speed (m/s)
Wave 1			
Wave 2			
Wave 3	Do not write in this book.		
Wave 4			
Wave 5			
Wave 6			

Analyze Your Data

1. **Calculate** the wave speed for each wave using the formula:

   speed = distance/time

2. Calculate the average speed of the waves on the spring with a length of 1.0 m by adding the measured wave speeds and dividing by 3.
3. Calculate the average speed of the waves on the spring with a length of 1.5 m.

Conclude and Apply

1. How did the tension in the spring change as the length of the spring increased?
2. How did the wave speed depend on the tension? How could you make the waves travel even faster? Test your prediction.

Communicating Your Data

Compare your results with those of other students in your class. Discuss why results might be different.

section 3

The Behavior of Waves

Reading Guide

What You'll Learn

- **State** the law of reflection.
- **Explain** why waves change direction when they travel from one material to another.
- **Compare and contrast** refraction and diffraction.
- **Describe** how waves interfere with each other.

Why It's Important

You can hear an echo, see shadows, and check your reflection in a mirror because of how waves behave.

Review Vocabulary

perpendicular: a line that forms a 90-degree angle with another line

New Vocabulary

- refraction
- diffraction
- interference
- standing wave
- resonance

Reflection

If you are one of the last people to leave your school building at the end of the day, you'll probably find the hallways quiet and empty. When you close your locker door, the sound echoes down the empty hall. Your footsteps also make a hollow sound. Thinking you're alone, you may be startled by your own reflection in a classroom window. The echoes and your image looking back at you from the window are caused by wave reflection.

Reflection occurs when a wave strikes an object and bounces off it. All types of waves—including sound, water, and light waves—can be reflected. How does the reflection of light allow the boy in **Figure 13** to see himself in the mirror? It happens in two steps. First, light strikes his face and bounces off. Then, the light reflected off his face strikes the mirror and is reflected into his eyes.

Figure 13 The light that strikes the boy's face is reflected into the mirror. The light then reflects off the mirror into his eyes.
List *examples of waves that can be reflected.*

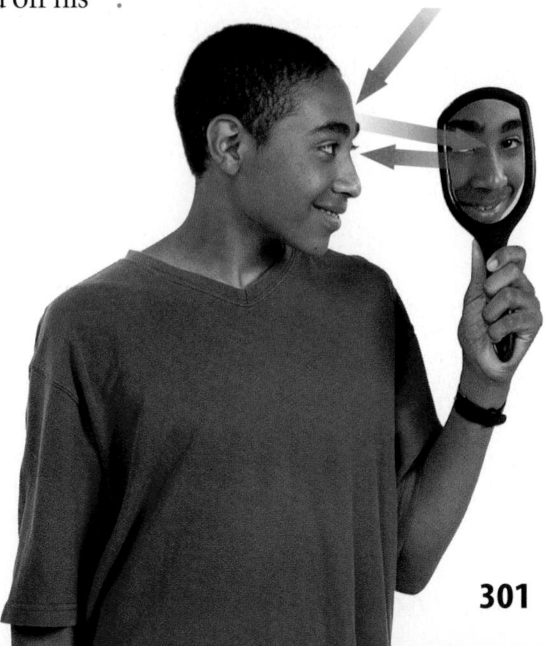

Echoes A similar thing happens to sound waves when your footsteps echo. Sound waves form when your foot hits the floor and the waves travel through the air to both your ears and other objects. Sometimes, when the sound waves hit another object, they reflect off it and come back to you. Your ears hear the sound again, a few seconds after you first heard your footstep.

Bats and dolphins use echoes to learn about their surroundings. A dolphin makes a clicking sound and listens to the echoes. These echoes enable the dolphin to locate nearby objects.

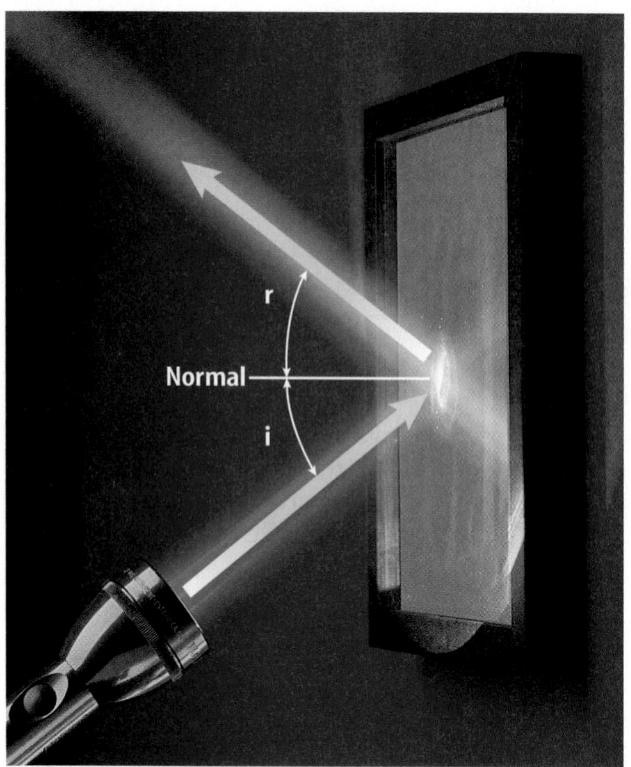

Figure 14 A flashlight beam is made of light waves. When any wave is reflected, the angle of incidence, *i*, equals the angle of reflection, *r*.

The Law of Reflection Look at the two light beams in **Figure 14.** The beam striking the mirror is called the incident beam. The beam that bounces off the mirror is called the reflected beam. The line drawn perpendicular to the surface of the mirror is called the normal. The angle formed by the incident beam and the normal is the angle of incidence, labeled *i*. The angle formed by the reflected beam and the normal is the angle of reflection, labeled *r*. According to the law of reflection, the angle of incidence is equal to the angle of reflection. All reflected waves obey this law. Objects that bounce from a surface sometimes behave like waves that are reflected from a surface. For example, suppose you throw a bounce pass while playing basketball. The angle between the ball's direction and the normal to the floor is the same before and after it bounces.

Refraction

Do you notice anything unusual in **Figure 15?** The pencil looks as if it is broken into two pieces. But if you pulled the pencil out of the water, you would see that it is unbroken. This illusion is caused by refraction. How does it work?

Remember that a wave's speed depends on the medium it is moving through. When a wave passes from one medium to another—such as when a light wave passes from air to water—it changes speed. If the wave is traveling at an angle when it passes from one medium to another, it changes direction, or bends, as it changes speed. **Refraction** is the bending of a wave caused by a change in its speed as it moves from one medium to another. The greater the change in speed is, the more the wave bends.

Reading Check *When does refraction occur?*

Figure 16A on the next page shows what happens when a wave passes into a material in which it slows down. The wave is refracted (bent) toward the normal. **Figure 16B** shows what happens when a wave passes into a medium in which it speeds up. Then the wave is refracted away from the normal.

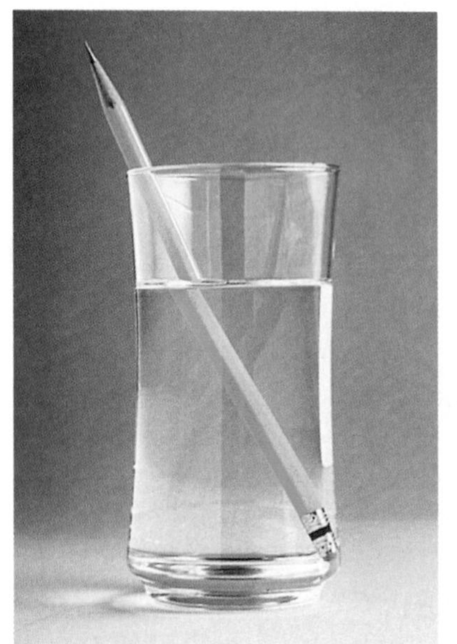

Figure 15 The pencil looks like it is broken at the surface of the water because of refraction.

Figure 16 Light waves travel more slowly in water than in air. This causes light waves to change direction when they move from water to air or air to water. **Predict** *how the beam would bend if the speed were the same in both air and water.*

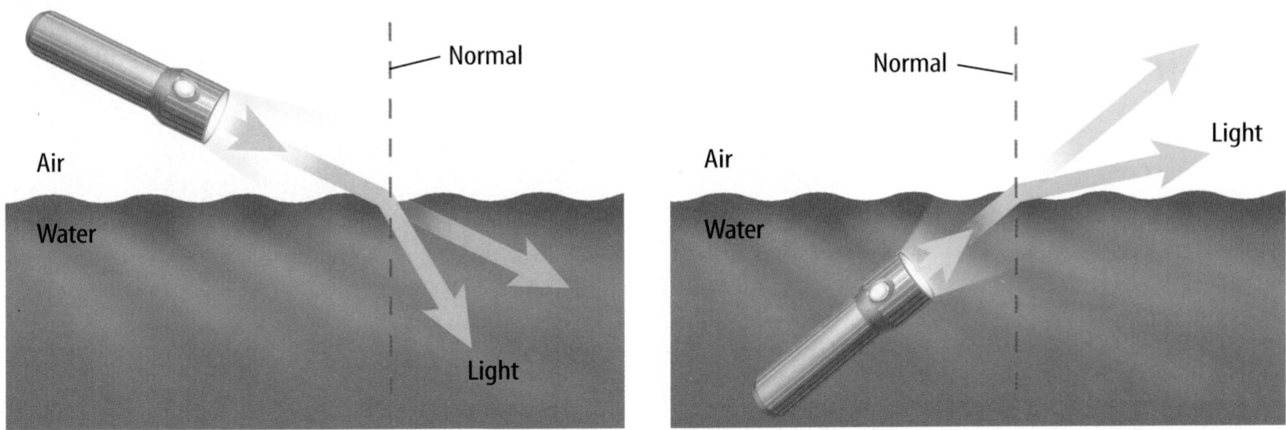

A When light waves travel from air to water, they slow down and bend toward the normal.

B When light waves travel from water to air, they speed up and bend away from the normal.

Refraction of Light in Water You may have noticed that objects that are under water seem closer to the surface than they really are. **Figure 17** shows how refraction causes this illusion. In the figure, the light waves reflected from the boy's foot are refracted away from the normal and enter your eyes. However, your brain assumes that all light waves have traveled in a straight line. The light waves that enter your eyes seem to have come from a foot that was higher in the water. This is also why the pencil in **Figure 15** seems broken. The light waves coming from the part of the pencil that is under water are refracted, but your brain interprets them as if they have traveled in a straight line. However, the light waves coming from the part of the pencil above the water are not refracted. So, the part of the pencil that is under water looks as if it has shifted.

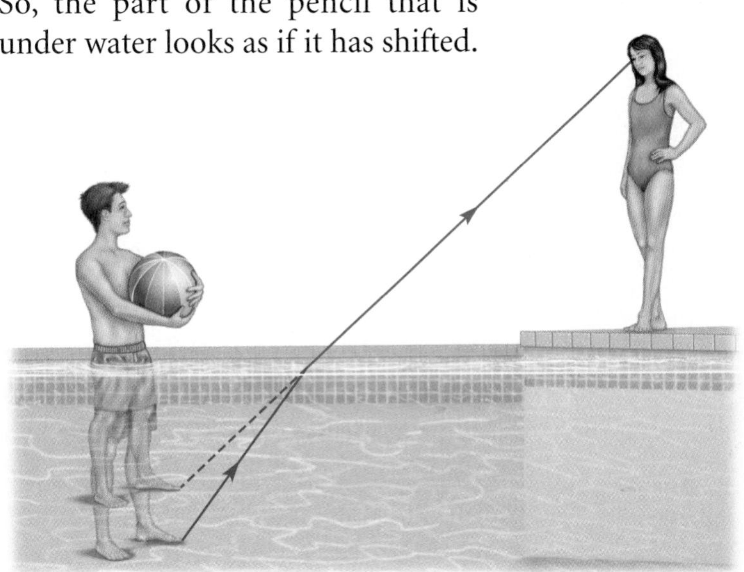

Figure 17 Light waves from the boy's foot bend away from the normal as they pass from water to air. This makes his foot look closer to the surface than it really is. **Infer** *whether the boy's knee would seem closer to the surface than it is.*

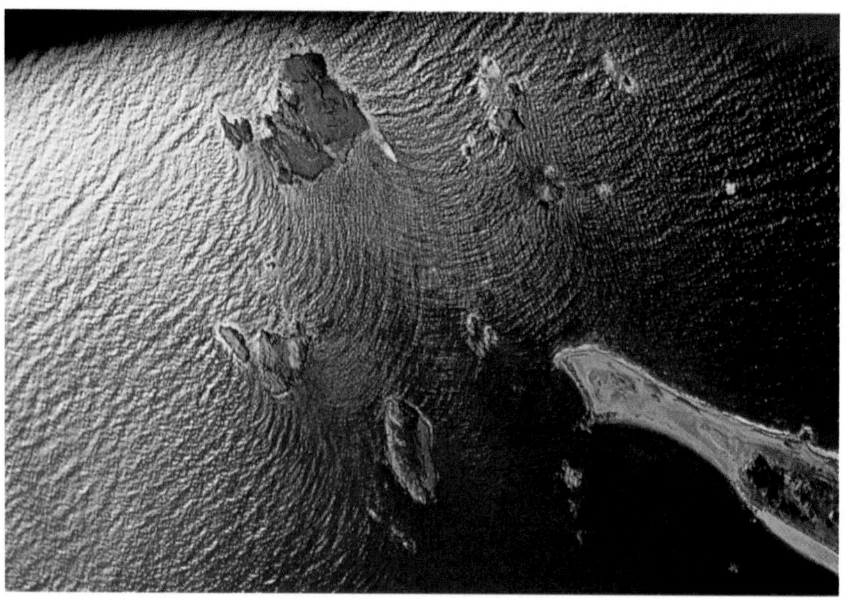

Figure 18 Diffraction causes ocean waves to change direction as they pass a group of islands.

Diffraction

When waves strike an object, several things can happen. Some of the wave's energy can be absorbed by the object. The waves can also be reflected. If the object is transparent, light waves can be refracted as they pass through it. Sometimes, the waves may be both reflected and refracted. If you look into a glass window, sometimes you can see your reflection in the window, as well as objects behind it.

Waves also can behave another way when they strike an object. The waves can bend around the object. **Figure 18** shows how ocean waves change direction and bend after they strike an island. **Diffraction** occurs when an object causes a wave to change direction and bend around it. Diffraction and refraction both cause waves to bend. The difference is that refraction occurs when waves pass through an object, while diffraction occurs when waves pass around an object.

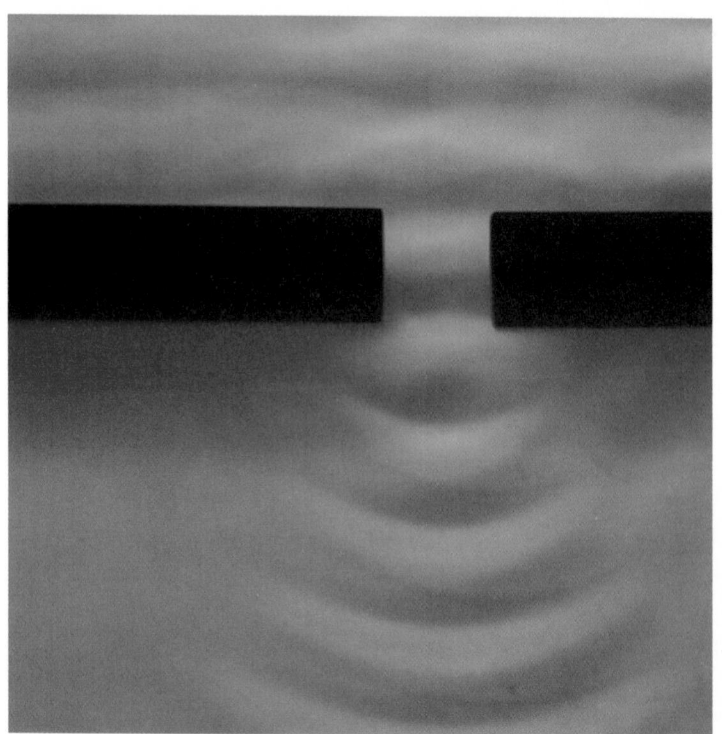

Figure 19 When water waves pass through a small opening in a barrier, they diffract and spread out after they pass through the hole.

Reading Check *How do diffraction and refraction differ?*

Waves also can be diffracted when they pass through a narrow opening, as shown in **Figure 19.** After they pass through the opening, the waves spread out. In this case, the waves are bending around the corners of the opening.

Diffraction and Wavelength How much does a wave bend when it strikes an object or an opening? The amount of diffraction that occurs depends on how big the obstacle or opening is compared to the wavelength, as shown in **Figure 20.** When an obstacle is smaller than the wavelength, the waves bend around it. But if the obstacle is larger than the wavelength, the waves do not diffract as much. In fact, if the obstacle is much larger than the wavelength, almost no diffraction occurs. The obstacle casts a shadow because almost no waves bend around it.

Hearing Around Corners For example, suppose that you're walking down a hallway and hear sounds coming from the lunchroom before you reach the open lunchroom door. However, you can't see into the room until you reach the doorway. Why can you hear the sound waves but not see the light waves while you're still in the hallway? The wavelengths of sound waves are similar in size to a door opening. Sound waves diffract around the door and spread out down the hallway. Light waves have a much shorter wavelength. They are hardly diffracted at all by the door. This is why you can't see into the room until you get close to the door.

Diffraction of Radio Waves Diffraction also affects your radio's reception. AM radio waves have longer wavelengths than FM radio waves do. Because of their longer wavelengths, AM radio waves diffract around obstacles such as buildings and mountains. The FM waves with their short wavelengths do not diffract as much. As a result, AM radio reception is often better than FM reception around tall buildings and natural barriers such as hills.

Topic: Diffraction
Visit gpescience.com for Web links to information about diffraction.

Activity Research the wave lengths of several types of waves. For each wave type, give an example of an object that could cause diffraction to occur.

Figure 20 The diffraction of waves around an obstacle depends on the wavelength and the size of the obstacle.

More diffraction occurs when the wavelength is the same size as the obstacle.

Less diffraction occurs when the wavelength is smaller than the obstacle.

Interference

Suppose two waves are traveling toward each other on a long rope, as in **Figure 21A.** What will happen when the two waves meet? If you did this experiment, you would find that the two waves pass right through each other, and each one continues to travel in its original direction, as shown in **Figure 21B** and **Figure 21C.** If you look closely at the waves when they meet each other in **Figure 21B,** you will see a wave that looks different from either of the two original waves. When the two waves arrive at the same place at the same time, they combine to form a new wave. When two or more waves overlap and combine to form a new wave, the process is called **interference.** This new wave exists only while the two original waves continue to overlap. The two ways that the waves can combine are called constructive interference and destructive interference.

Figure 21 Interference occurs while two waves are overlapping. Then the waves combine to form a new wave. Two waves traveling on a rope can interfere with each other.

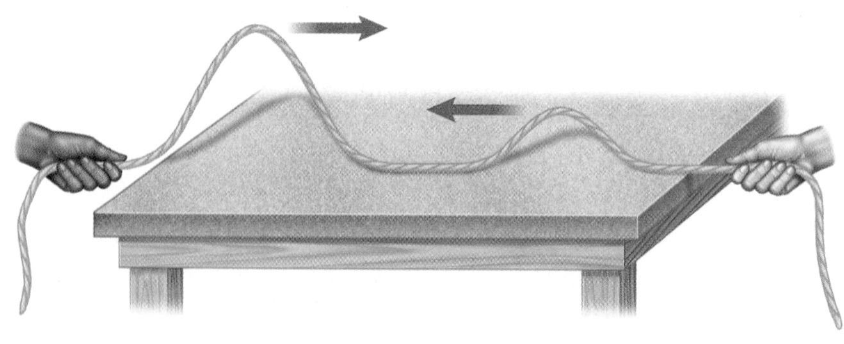

A Two waves move toward each other on a rope.

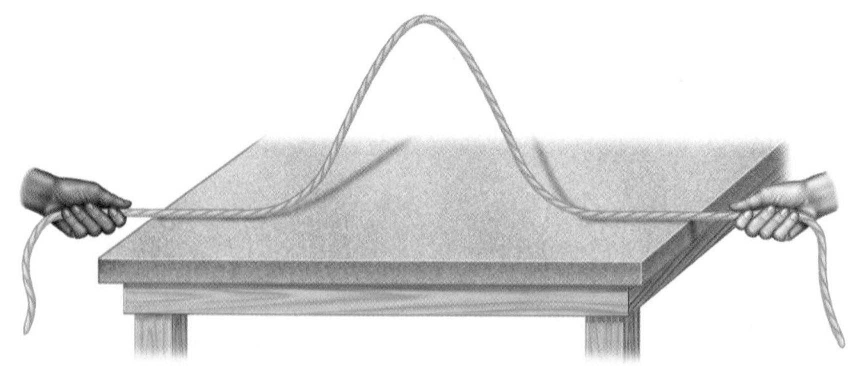

B As the waves overlap, they interfere to form a new wave.
Identify *What is the amplitude of the new wave?*

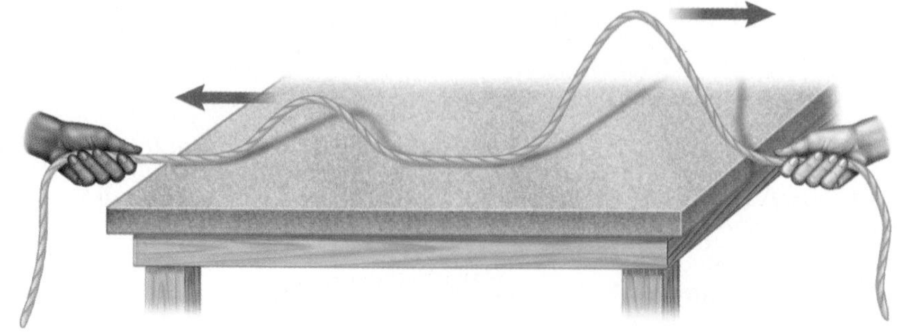

C While the two waves overlap, they continue to move right through each other. Afterward, they continue moving unchanged, as if they had never met.

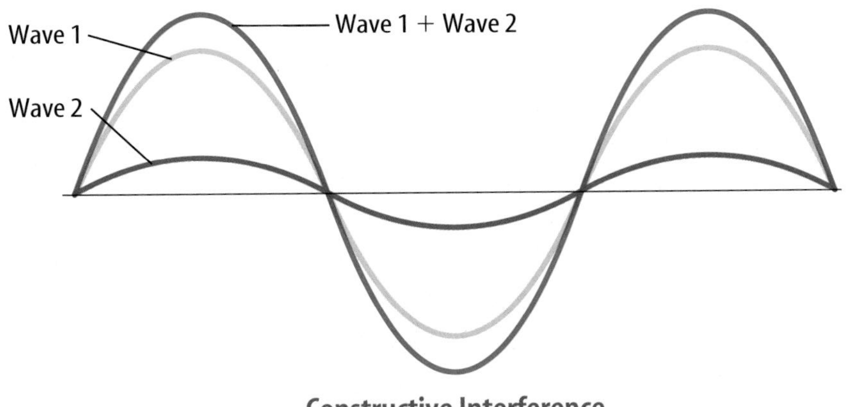

A If Wave 1 and Wave 2 overlap, they constructively interfere and form the green wave. Wave 1 and Wave 2 are in phase.

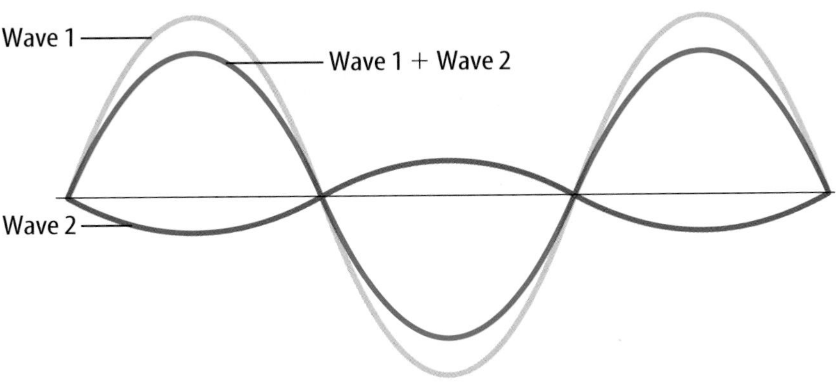

B If Wave 1 and Wave 2 overlap, they destructively interfere and form the green wave. Wave 1 and Wave 2 are out of phase.

Figure 22 When waves interfere with each other, constructive and destructive interference can occur. **Infer** *how the energy transferred by each wave changes when interference occurs.*

Constructive Interference In constructive interference, shown in **Figure 22A,** the waves add together. This happens when the crests of two or more transverse waves arrive at the same place at the same time and overlap. The amplitude of the new wave that forms is equal to the sum of the amplitudes of the original waves. Constructive interference also occurs when the compressions of different compressional waves overlap. If the waves are sound waves, for example, constructive interference produces a louder sound. Waves undergoing constructive interference are said to be in phase.

Destructive Interference In destructive interference, the waves subtract from each other as they overlap. This happens when the crests of one transverse wave meet the troughs of another transverse wave, as shown in **Figure 22B.** The amplitude of the new wave is the difference between the amplitudes of the waves that overlapped. With compressional waves, destructive interference occurs when the compression of one wave overlaps with the rarefaction of another wave. The compressions and rarefactions combine and form a wave with reduced amplitude. When destructive interference happens with sound waves, it causes a decrease in loudness. Waves undergoing destructive interference are said to be out of phase.

Noise Damage People who are exposed to constant loud noises, such as those made by airplane engines, can suffer hearing damage. Special ear protectors have been developed that use destructive interference to cancel damaging noise. With a classmate, list all the jobs you can think of that require ear protectors.

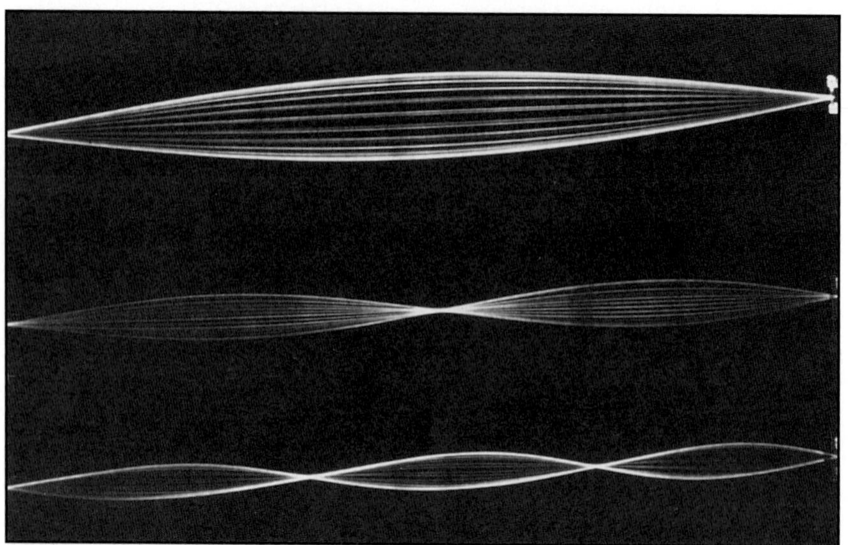

Figure 23 Standing waves form a wave pattern that seems to stay in the same place.
Explain *how nodes form in a standing wave.*

Standing Waves

When you make transverse waves with a rope, you might shake one end while your friend holds the other end still. What would happen if you both shook the rope continuously to create identical waves moving toward each other? As the two waves travel in opposite directions down the rope, they continually pass through each other. Interference takes place as the waves from each end overlap along the rope. At any point where a crest meets a crest, a new wave with a larger amplitude forms. However, at points where crests meet troughs, the waves cancel each other and no motion occurs.

The interference of the two identical waves makes the rope vibrate in a special way, as shown in **Figure 23.** The waves create a pattern of crests and troughs that do not seem to be moving. Because the wave pattern stays in one place, it is called a standing wave. A **standing wave** is a special type of wave pattern that forms when waves equal in wavelength and amplitude, but traveling in opposite directions, continuously interfere with each other. The places where the two waves always cancel are called nodes. The nodes always stay in the same place on the rope. Meanwhile, the wave pattern vibrates between the nodes.

Standing Waves in Music When the string of a violin is played with a bow, it vibrates and creates standing waves. The standing waves in the string help produce a rich, musical tone. Other instruments also rely on standing waves to produce music. Some instruments, such as flutes, create standing waves in a column of air. In other instruments, such as drums, a tightly stretched piece of material vibrates in a special way to create standing waves. As the material in a drum vibrates, nodes are created on the surface of the drum.

Resonance

You might have noticed that bells of different sizes and shapes create different notes. When you strike a bell, the bell vibrates at certain frequencies called the natural frequencies. All objects have their own natural frequencies of vibration, which depend on the object's size, shape, and the material it is made from.

There is another way to make something vibrate at its natural frequencies. Suppose you have a tuning fork that has a single natural frequency of 440 Hz. Imagine that a sound wave of the same frequency strikes the tuning fork. Because the sound wave has the same frequency as the natural frequency of the tuning fork, the tuning fork will vibrate. The process by which an object is made to vibrate by absorbing energy at its natural frequencies is called **resonance.**

Sometimes, resonance can cause an object to absorb a large amount of energy. Recall that the amplitude of a wave increases as the energy it transfers increases. In the same way, an object vibrates more strongly as it continues to absorb energy at its natural frequencies. If enough energy is absorbed, the object can vibrate so strongly that it breaks apart.

Experimenting with Resonance

Procedure

1. Complete the safety form.
2. Strike a **tuning fork** with a **mallet.**
3. Hold the vibrating tuning fork 1 cm from a **second tuning fork** that has the same frequency.
4. Strike the tuning fork again. Hold it 1 cm from a **third tuning fork** that has a different frequency.

Analysis

How did the vibrating tuning fork affect the other tuning forks? Explain.

section 3 review

Summary

Reflection and Refraction

- When reflection of a wave occurs, the angle of incidence equals the angle of reflection.
- Refraction occurs when a wave changes direction as it moves from one medium to another.

Diffraction

- Diffraction occurs when a wave changes direction by bending around an obstacle.
- The effects of diffraction are greatest when the wavelength is nearly the obstacle's size.

Interference and Resonance

- Interference occurs when two or more waves overlap and form a new wave.
- Interference between two waves with the same wavelength and amplitude, but moving in opposite directions, produces a standing wave.
- Resonance occurs when an object is made to vibrate by absorbing energy from vibrations at its natural frequencies.

Self Check

1. **Compare** the loudness of sound waves that are in phase when they interfere with the loudness of sound waves that are out of phase when they interfere.
2. **Describe** how the reflection of light waves enables you to see your image in a mirror.
3. **Describe** the energy transformations that occur when one tuning fork makes another tuning fork resonate.
4. **Think Critically** Suppose the speed of light were greater in water than in air. Draw a diagram to show whether an object under water would seem deeper or closer to the surface than it really is.

Applying Math

5. **Use Percentages** You aim a flashlight at a window. The radiant energy in the reflected beam is two-fifths of the energy in the incident beam. What percentage of the incident beam energy passed through the window?
6. **Calculate Angle of Incidence** The angle between a flashlight beam that strikes a mirror and the reflected beam is 80°. What is the angle of incidence?

Measuring Wave Properties

Goals

- **Measure** the speed of a transverse wave.
- **Create** waves with different amplitudes.
- **Measure** the wavelength of a transverse wave.

Materials

long spring, rope, or hose
meterstick
stopwatch

Safety Precautions

Real-World Problem

Some waves travel through space; others pass through a medium such as air, water, or earth. Each wave has a wavelength, speed, frequency, and amplitude. How can the speed of a wave be measured? How can the wavelength be determined from the frequency?

Procedure

1. Complete the safety form before you begin.
2. With a partner, stretch your spring across an open floor and measure the length of the spring. Record this measurement in the data table. Make sure the spring is stretched to the same length for each step.
3. Have your partner hold one end of the spring. Create a single wave pulse by shaking the other end of the spring back and forth.
4. Have a third person use a stopwatch to measure the time needed for the pulse to travel the length of the spring. Record this measurement in the *Wave Time* column of your data table.
5. Repeat steps 3 and 4 two more times.
6. **Calculate** the speed of waves 1, 2, and 3 in your data table by using the formula speed = distance/time. Average the speeds of waves 1, 2, and 3 to find the speed of waves on your spring.
7. **Create** a wave with several wavelengths. You make one wavelength when your hand moves left, right, and left again. Count the number of wavelengths that you generate in 10 s. Record this measurement for wave 4 in the *Wavelength Count* column in your data table.
8. **Repeat** step 7 two more times. Each time, create a wave with a different wavelength by shaking the spring faster or slower.

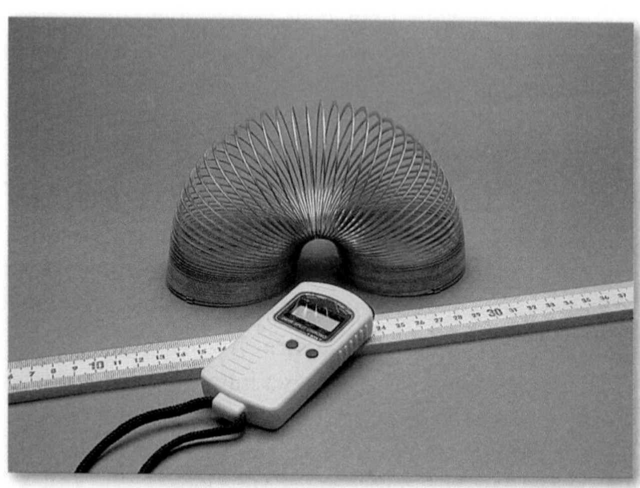

Analyze Your Data

1. **Calculate** the frequency of waves 4, 5, and 6 by dividing the number of wavelengths by 10 s.
2. Calculate the wavelength of waves 4, 5, and 6 using the formula wavelength = wave speed/frequency.

   Use the average speed calculated in step 5 for the wave speed.

Wave Property Measurement

	Spring Length	Wave Time	Wave Speed
Wave 1			
Wave 2			
Wave 3	Do not write in this book.		
	Wavelength Count	**Frequency**	**Wavelength**
Wave 4			
Wave 5			
Wave 6			

Conclude and Apply

1. Was the wave speed different for the three different pulses you created? Why or why not?
2. Why would you average the speeds of the three different pulses to calculate the speed of waves on your spring?
3. How did the wavelength of the waves you created depend on the frequency of the waves?

Ask your teacher to set up a contest between the groups in your class. Have each group compete to determine who can create waves with the longest wavelength, the highest frequency, and the largest wave speed. Record the measurements of each group's efforts on the board.

TIME **SCIENCE** AND **HISTORY**

SCIENCE CAN CHANGE THE COURSE OF HISTORY!

MAKING WAVES

Sonar Helps Create Deep-Sea Pictures and Save Lives

This machine houses side-scan sonar. It was used to help locate the wreck of the *Titanic*.

What is sonar?

Sonar is a device that uses sound waves to locate and measure the distance to underwater objects. Its name is a shortened version of SOund NAvigation and Ranging.

How does sonar work?

Sonar sends out a ping sound that reflects back when it hits an underwater object. Because sound travels through water at a known speed (about 1,500 m/s), scientists measure how long the sound takes to return, then they calculate the distance.

Why was it invented?

Sonar was developed by scientists in the early twentieth century as a way to detect icebergs and prevent boating disasters. Its technical advancement was hurried by the Allies' need to detect German submarines in World War I. By 1918, the United States and Great Britain had developed an active sonar system placed in submarines sent to attack other subs.

By World War II, sonar allowed ships to defend themselves effectively from enemy subs. The strategy was to use sonar to find subs and then fire depth charges at them from a safe distance. After the war, sonar-absorbing hulls and quiet engines and machinery ensured that subs could partly shield themselves from sonar.

Sonar is now used to help find schools of fish. Oceanographers also use it to map ocean and lake floors. Sonar has been vital, too, in the discovery of sunken airplanes and ships.

In 1985, a French and American team used a new type of sonar device called side-scan sonar to locate the *Titanic*, the passenger liner that sank in 1912. This kind of sonar projects a tight beam of sound to create detailed images of the sea bed. Members of the expedition towed this sonar device about 170 m above the seabed across a section of the Atlantic Ocean where the *Titanic* went down. Weeks later, video cameras finally spotted the wreck.

The *Titanic* was found thanks to sonar.

Report Research how sonar was used by navies in World War I and World War II. Did sonar affect each war's outcome? How did it save lives? What uses can you think of for sonar if it could be used in everyday life?

TIME

For more information, visit gpescience.com

Reviewing Main Ideas

Section 1 The Nature of Waves

1. Waves are rhythmic disturbances that transfer energy through matter or space.
2. Waves transfer only energy, not matter.

3. Mechanical waves need matter to travel through. Mechanical waves can be compressional or transverse.
4. When a transverse wave travels in a medium, matter in the medium moves at right angles to the direction the wave travels.
5. When a compressional wave travels in a medium, matter moves back and forth along the same direction that the wave travels.

Section 2 Wave Properties

1. The movement of high points in a medium, called crests, and low points, called troughs, forms a transverse wave.
2. The movement of more-dense regions, called compressions, and less-dense regions, called rarefactions, forms a compressional wave.
3. Transverse and compressional waves can be described by their wavelengths, frequencies, periods, and amplitudes. As frequency increases, wavelength always decreases.
4. The greater a wave's amplitude is, the more energy it transfers.

5. A wave's velocity can be calculated by multiplying its frequency times its wavelength.

Section 3 The Behavior of Waves

1. For all waves, the angle of incidence equals the angle of reflection.
2. A wave is bent, or refracted, when it changes speed as it enters a new medium.

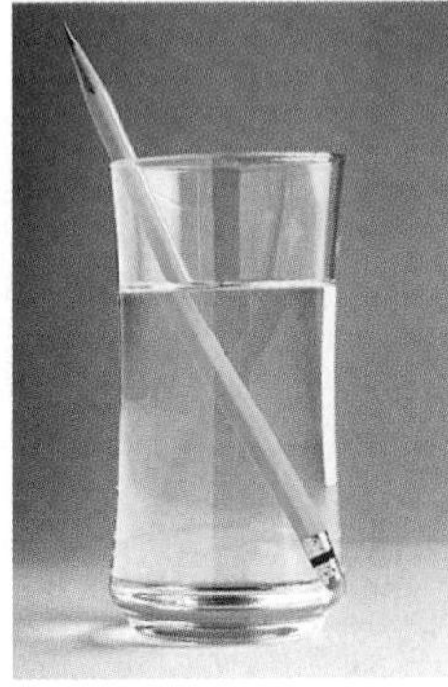

3. When two or more waves overlap, they combine to form a new wave. This process is called interference.

FOLDABLES™ Use the Foldable that you made at the beginning of this chapter to help you review waves.

chapter 10 Review

Using Vocabulary

amplitude p. 288	rarefaction p. 294
compressional wave p.290	refraction p. 302
crest p. 294	resonance p. 309
diffraction p. 304	standing wave p. 308
frequency p. 295	transverse wave p. 290
interference p. 306	trough p. 294
medium p. 289	wave p. 298
period p. 295	wavelength p. 295

Answer the following questions using complete sentences.

1. Compare and contrast reflection and refraction.
2. Which type of wave has points called nodes that do not move?
3. Which part of a compressional wave has the lowest density?
4. Find two words in the vocabulary list that describe the bending of a wave.
5. What occurs when waves overlap?
6. What is the relationship among amplitude, crest, and trough?
7. What does frequency measure?
8. What does a mechanical wave always travel through?

Checking Concepts

Choose the word or phrase that best answers each question.

9. Which of the following do waves transfer?
   A) matter **C)** matter and energy
   B) energy **D)** the medium
10. What is the formula for calculating wave speed?
    A) $v = \lambda f$ **C)** $v = \lambda / f$
    B) $v = f - \lambda$ **D)** $v = \lambda + f$
11. When a compressional wave travels through a medium, which way does matter in the medium move?
    A) backward
    B) forward
    C) perpendicular to the rest position
    D) along the same direction the wave travels
12. What is the highest point of a transverse wave called?
    A) crest **C)** wavelength
    B) compression **D)** trough
13. If the frequency of the waves produced by a vibrating object increases, how does the wavelength of the waves produced change?
    A) It stays the same. **C)** It vibrates.
    B) It decreases. **D)** It increases.
14. If the amplitude of a wave changes, which of the following changes?
    A) wave energy **C)** wave speed
    B) frequency **D)** refraction
15. Which term describes the bending of a wave around an object?
    A) resonance **C)** diffraction
    B) interference **D)** reflection
16. What is equal to the angle of reflection?
    A) refraction angle **C)** bouncing angle
    B) normal angle **D)** angle of incidence

Use the table below to answer question 17.

Speed of Sound in Air

Temperature (°C)	Sound Speed (m/s)
0	331.4
10	337.4
20	343.4

17. Based on the data in the table above, which of the following would be the speed of sound in air at 30°C?
    A) 340.4 m/s **C)** 353.4 m/s
    B) 346.4 m/s **D)** 349.4 m/s

Vocabulary PuzzleMaker gpescience.com

Interpreting Graphics

18. Copy and complete the following concept map.

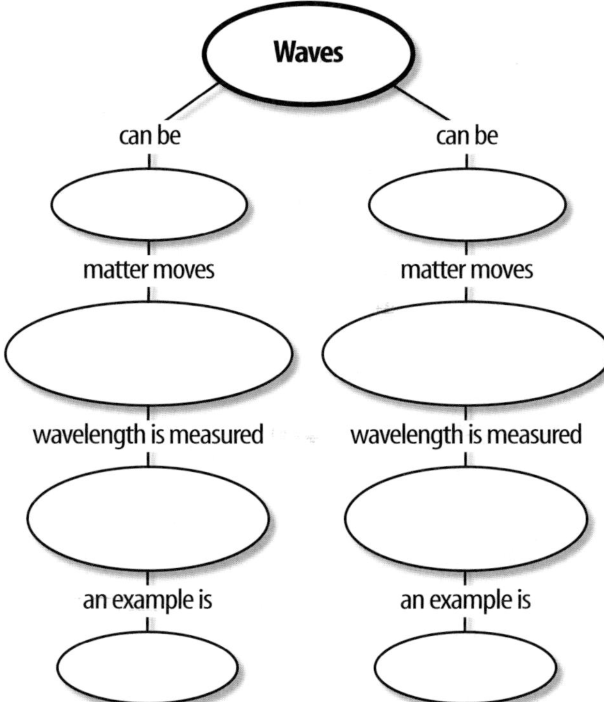

Thinking Critically

19. Explain An earthquake on the ocean floor produces a tsunami that hits a remote island. Is the water that hits the island the same water that was above the earthquake on the ocean floor?

20. Compare Suppose waves with different amplitudes are produced by a vibrating object. How do the frequencies of the waves with different amplitudes compare?

21. Explain Use the law of reflection to explain why you see only a portion of the area behind you when you look in a mirror.

22. Explain why you can hear a fire engine coming around a street corner before you can see it.

23. Describe the objects or materials that vibrated to produce three of the sounds you've heard today.

24. Form a Hypothesis In 1981, people dancing on the balconies of a Kansas City, Missouri, hotel caused the balconies to collapse. Use what have you learned about wave behavior to form a hypothesis that explains why this happened.

25. Make and Use Tables Find information in newspaper articles or magazines describing five recent earthquakes. Construct a table that shows for each earthquake the date, location, magnitude, and whether the damage caused by the earthquake was light, moderate, or heavy.

26. Concept Map Design a concept map that shows the characteristics of transverse waves. Include the terms *crest, trough, medium, wavelength, frequency, period,* and *amplitude.*

Applying Math

27. Calculate Wavelength Calculate the wavelength of a wave traveling on a spring if the wave moves at 0.2 m/s and has a period of 0.5 s.

28. Calculate Wavespeed The microwaves produced inside a microwave oven have a wavelength of 12.0 cm and a frequency of 2,500,000,000 Hz. At what speed do the microwaves travel in units of m/s?

29. Calculate Frequency Water waves on a lake travel toward a dock with a speed of 2.0 m/s and a wavelength of 0.5 m. How many wave crests strike the dock each second?

Standardized Test Practice

Record your answers on the answer sheet provided by your teacher or on a sheet of paper.

Multiple Choice

1. When a transverse wave travels through a medium, which way does matter in the medium move?

A. backward

B. in all directions

C. at right angles to the direction the wave travels

D. in the same direction the wave travels

Use the illustration below to answer questions 2 and 3.

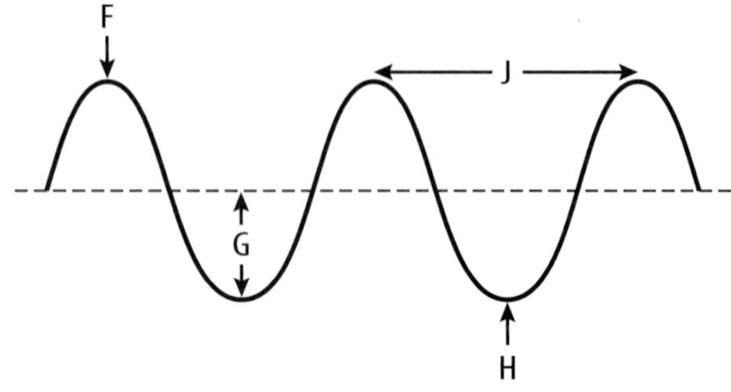

2. What wave property is shown at G?

A. amplitude

B. wavelength

C. crest

D. trough

3. What property of the wave is shown at H?

A. amplitude

B. wavelength

C. crest

D. trough

Use the illustration below to answer questions 4 and 5.

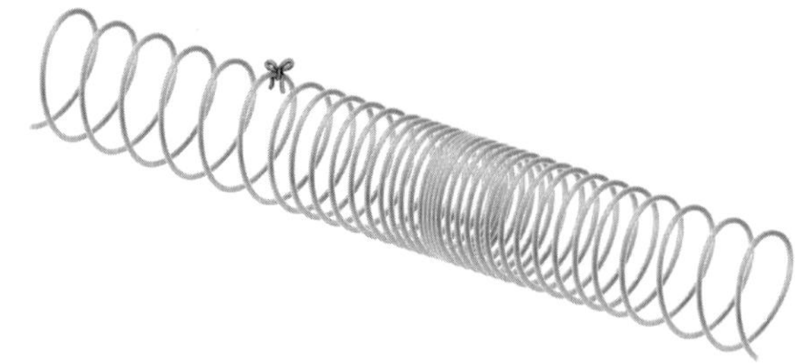

4. What kind of wave is shown?

A. mechanical

B. compressional

C. transverse

D. both A and B

5. What happens to the yarn tied to the coil?

A. It moves back and forth as the wave passes.

B. It moves up and down as the wave passes.

C. It does not move as the wave passes.

D. It moves to the next coil as the wave passes.

6. What kind of waves are the seismic waves produced during an earthquake?

A. transverse

B. compressional

C. combination of transverse and compressional

D. electromagnetic

7. What is the bending of a wave as it enters a new material called?

   A. refraction

   B. diffraction

   C. reflection

   D. interference

Gridded Response

8. A tuning fork vibrates at a frequency of 256 Hz. The wavelength of the sound produced by the tuning fork is 1.32 m. What is the speed of the wave in m/s?

9. A wave has a speed of 345 m/s and its frequency is 2050 Hz. What is its wavelength in m?

Short Response

Use the illustration below to answer questions 10 and 11.

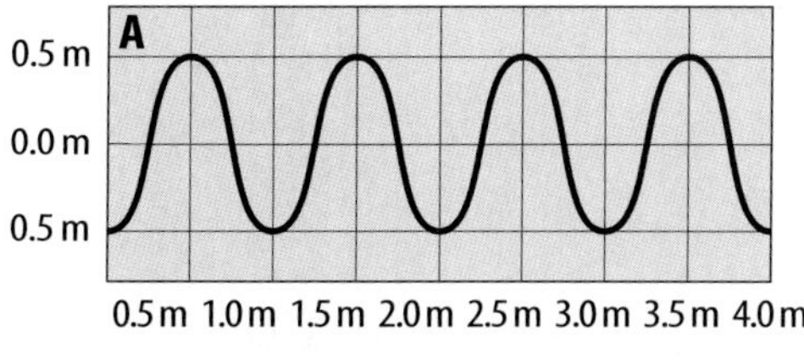

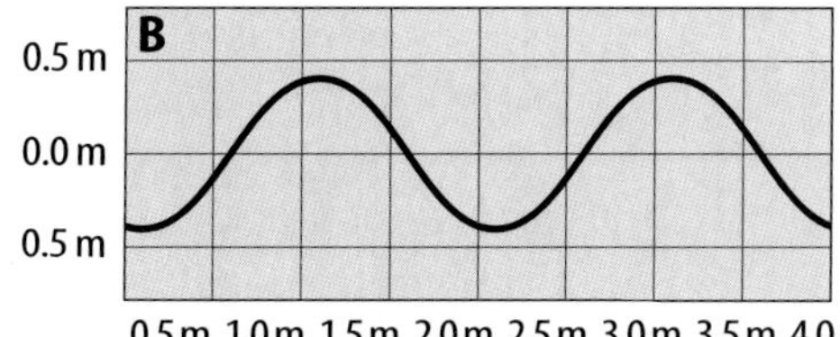

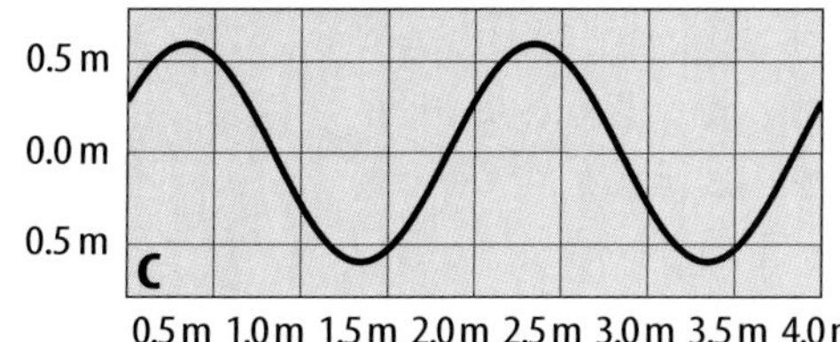

10. Determine the amplitudes and the wavelengths of each of the three waves.

11. If the length of the x-axis on each diagram represents 2 s of time, what is the frequency of each wave?

Extended Response

12. In a science fiction movie, a huge explosion occurs on the surface of a planet. People in a spaceship heading toward the planet see and hear the explosion. Is this realistic? Explain.

Use the illustrations below to answer question 13.

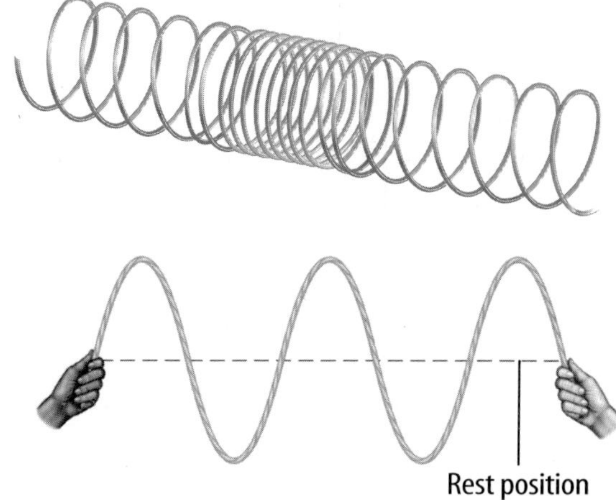

13. The illustrations show two mechanical waves.

    Part A Describe how the amplitude of each of the two waves shown is defined.

    Part B Describe how you would change both drawings to show waves that transfer more energy.

Test-Taking Tip

Answer Every Question Never leave any open-ended answer blank. Answer each question as best you can. You can receive partial credit for partially correct answers.

Question 12 Before you answer the question, list what you know about light waves and sound waves.

chapter 11

Sound and Light

chapter preview

sections

Cracking the Sound Barrier

Have you ever heard a sonic boom? A plane moving faster than the speed of sound makes a sonic boom. The sound waves from the plane add together to form a cone-shaped shock wave. Behind the cone are low-pressure regions that can cause water vapor to condense, forming the cloud you see here.

Science Journal Write three things that you would like to learn about sound.

Start-Up Activities

What sound does a ruler make?

Musical instruments come in many shapes and sizes. Some have strings, some have hollow tubes, and others have keys or pedals. They are played with various techniques. These differences give each instrument a unique sound. What would an instrument made from a ruler sound like?

1. Complete the safety form.
2. Firmly hold one end of a thin ruler lying flat on a desk, allowing the free end to extend beyond the edge of the desk.
3. Gently pull up on and release the end of the ruler. What do you see and hear?
4. Vary the length of the overhanging portion and repeat the experiment several times.
5. **Think Critically** In your Science Journal, write instructions for playing a song with the ruler. Explain how the length of the overhanging part of the ruler affects the sound.

Preview this chapter's content and activities at gpescience.com

Sound and Light Make the following Foldable to help you identify ways in which sound and light are different and alike.

STEP 1 **Fold** one sheet of paper lengthwise.

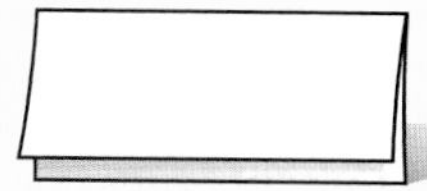

STEP 2 **Fold** into thirds.

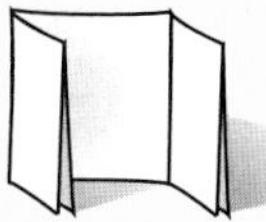

STEP 3 **Unfold and draw** overlapping ovals. **Cut** the top sheet along the folds.

STEP 4 **Label** the ovals as shown.

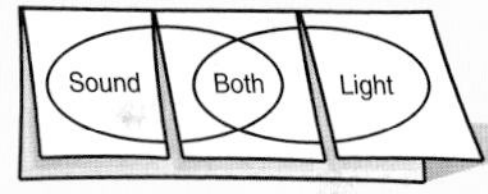

Constructing a Venn Diagram As you read the chapter, list the characteristics unique to sound under the left tab, the characteristics unique to light under the right tab, and those common to both under the middle tab.

section 1

Sound

Reading Guide

What You'll Learn
- **Explain** how sound travels through different mediums.
- **Explain** different properties of sound.
- **Discuss** the Doppler effect.
- **Identify** ways sound can be used.

Why It's Important
Sounds are one way that you learn about the world around you.

Review Vocabulary
vibration: rhythmic back-and-forth motion

New Vocabulary
- intensity
- loudness
- decibel
- pitch
- Doppler effect

Sound Waves

An amusement park can be a noisy place. With all the racket of carousel music and booming loudspeakers, it can be hard to hear what your friends say. The many sounds you hear every day are different, but they do have something in common—each sound is produced by an object that vibrates.

Making Sound Waves When an object like the tuning fork in **Figure 1** vibrates, it creates sound waves. Sound waves are compressional waves. Recall that a compressional wave is made of compressions and rarefactions. When an end of the tuning fork moves outward, it forms a compression on that side by pushing the molecules in air together. The compression moves away from the tuning fork as these molecules collide with other molecules in air. When the end of the tuning fork moves back, a rarefaction is formed where the molecules are farther apart. As the tuning fork vibrates, it produces a series of compressions and rarefactions that travels outward. This series of compressions and rarefactions is the sound wave that you hear.

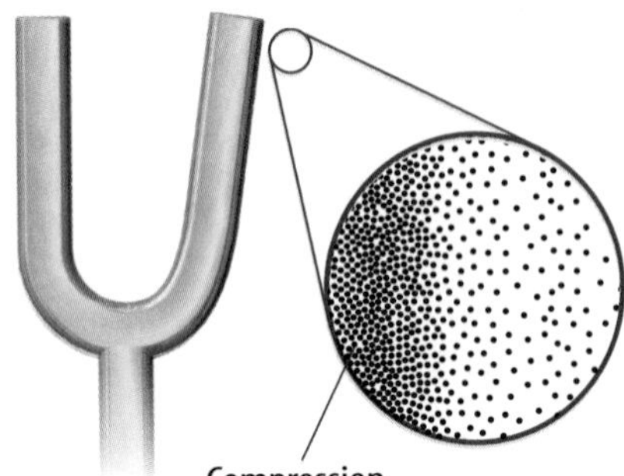

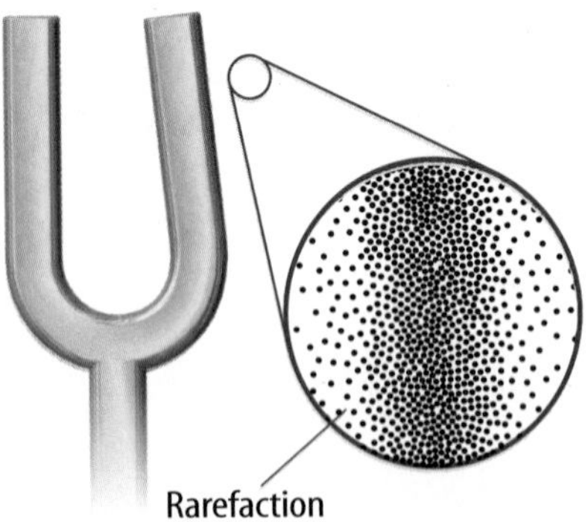

Figure 1 A vibrating tuning fork produces compressions and rarefactions that travel outward from the tuning fork. These compressions and rarefactions form a sound wave.

The Speed of Sound

A sound wave moves in air as collisions between the molecules in air transfer energy from place to place. Sound waves also can move in other materials. The material in which a sound wave moves is called a medium. Just as in air, sound waves travel in solids, liquids, and other gases as a vibrating object transfers energy to the particles in the material. However, sound waves cannot travel in empty space where there are no particles.

The Speed of Sound in Different Materials The speed of a sound wave in a medium depends on the type of substance and whether it is a solid, liquid, or gas. For example, **Table 1** shows the speed of sound in different mediums at room temperature. In general, sound travels slowest in gases and fastest in solids.

Temperature and the Speed of Sound The speed of sound waves also depends on the temperature of the medium. As the temperature of a substance increases, its atoms and molecules move faster. At a higher temperature, atoms and molecules collide with each other more frequently. As a result, sound waves move faster. For example, in air at a temperature of 0°C, sound travels at a speed of 331 m/s, but in air at a temperature of 20°C, sound travels at 343 m/s.

Table 1 Speed of Sound in Different Mediums

Medium	Speed of Sound (m/s)
Air (20°C)	343
Cork	500
Water	1,498
Brick	3,650
Aluminum	5,000

Amplitude and Energy of Sound Waves

Think about the differences among sounds you hear. Some are loud; others are soft. The flute plays high notes, whereas the tuba is much lower. What properties of sound waves cause these differences in the sounds you hear?

Recall that the amount of energy a wave carries corresponds to its amplitude. For a compressional wave, amplitude is related to the density of the particles in the compressions and rarefactions. A vibrating object makes a wave by transferring energy to the medium. More energy is transferred to the medium when the particles of the medium are forced closer together in the compressions and spread farther apart in the rarefactions. Look at **Figure 2.** A sound wave has a higher amplitude and carries more energy when particles in the medium are closer together in the compressions and more spread out in the rarefactions.

Figure 2 The amplitude of a sound wave depends on the density of the particles in the compressions and rarefactions.

In a high-amplitude sound wave, particles are tightly packed in the compressions and far apart in the rarefactions.

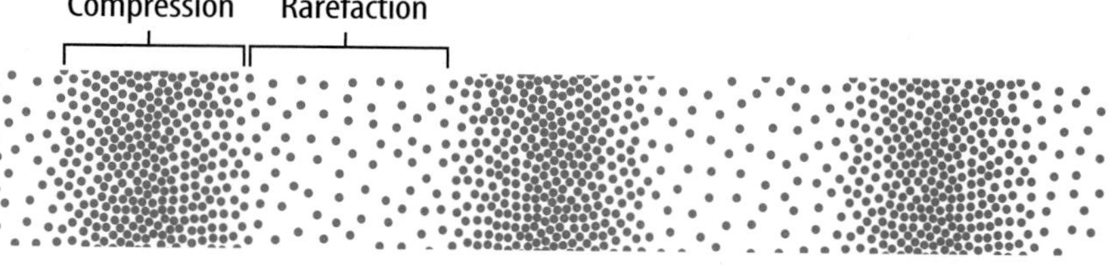

High-amplitude sound wave

In a low-amplitude sound wave, particles are less tightly packed in the compressions and not as far apart in the rarefactions.

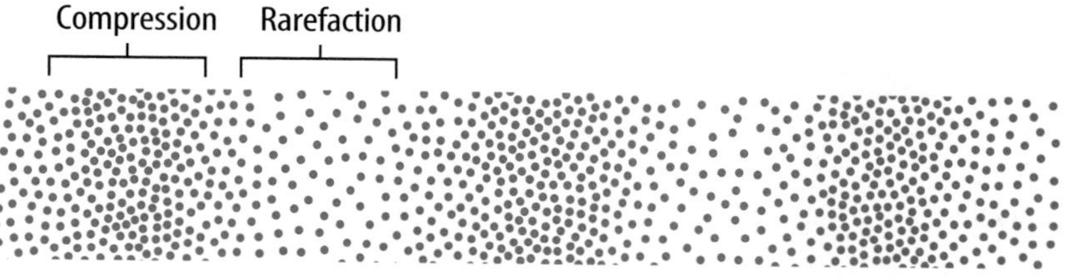

Low-amplitude sound wave

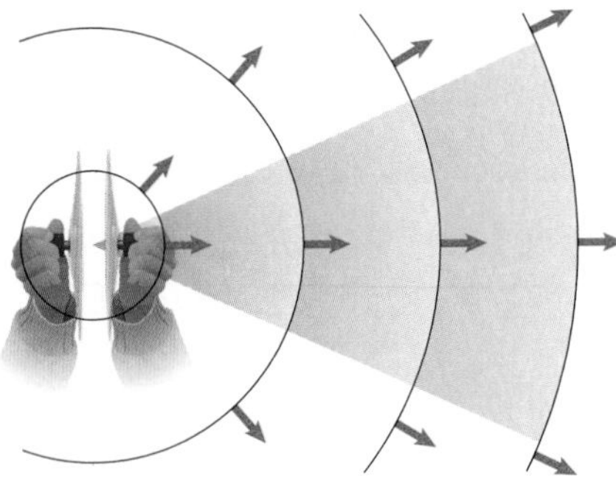

Figure 3 The intensity of a sound wave decreases as the wave spreads out from the source of the sound. The energy of the wave is spread over a larger area as the wave spreads out.

Intensity and Loudness

Imagine you are listening to music from your CD player. Sound waves produced by the CD player travel through the air and transfer energy from the CD player to your ears. The energy that reaches your ears depends not only on the amplitude of the sound waves, but also on how close you are to the CD player. **Figure 3** shows that as you get farther from a source of sound, the energy of the sound waves is spread over a greater area. The amount of energy transferred by a sound wave through a certain area each second is the **intensity** of the sound wave. As sound waves travel away from the source of the sound, the intensity of the waves decreases as the waves spread out. This means that as you get farther from the source, less energy reaches your ears each second.

Loudness When you hear different sounds, you do not need special equipment to know which sounds have greater intensity. Your ears and brain can tell the difference. **Loudness** is the human perception of sound intensity. As the intensity of a sound wave increases, the loudness of the sound also increases.

Reading Check *How are intensity and loudness related?*

A Scale for Sound Intensity The intensity of sound can be described using a measurement scale. Each unit on the scale for sound intensity is called a **decibel,** abbreviated dB. On this scale, the faintest sound that most people can hear is 0 dB. Sounds with intensity levels above 120 dB may cause pain and permanent hearing loss. During some rock concerts, sounds reach this damaging intensity level. Wearing ear protection, such as earplugs, around loud sounds can help protect against hearing loss. **Figure 4** shows some sounds and their intensity levels in decibels.

Figure 4 The decibel scale measures the intensity of sound. **Identify** *where a normal speaking voice would fall on the scale.*

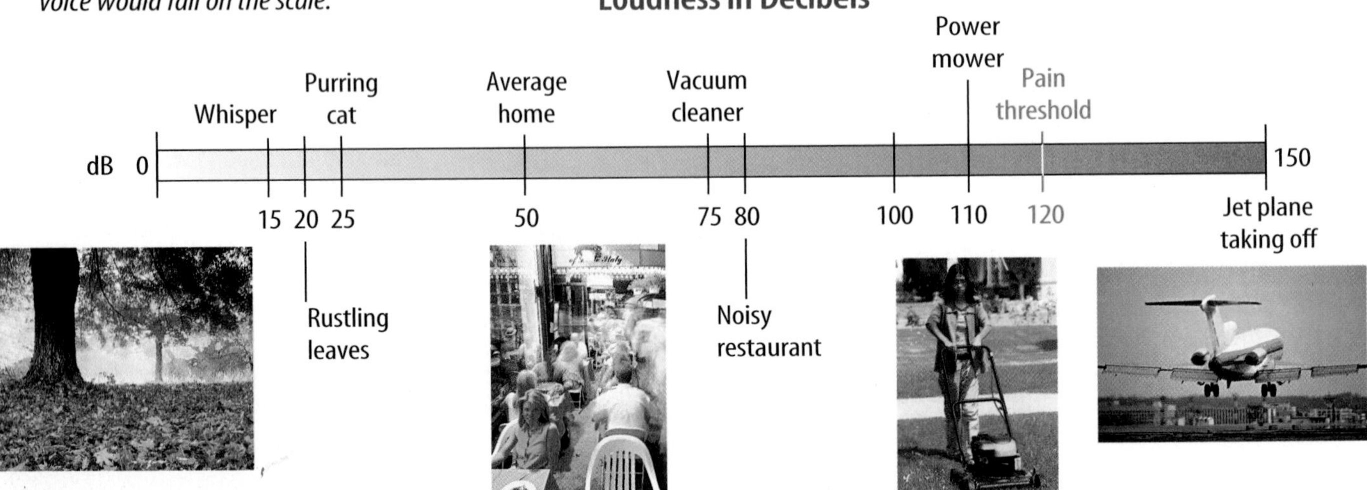

Pitch and Frequency

You might be familiar with the music scale do, re, mi, fa, sol, la, ti, do shown in **Figure 5.** If you were to sing this scale, you would hear the pitch of your voice get higher. **Pitch** is the human perception of the frequency of sound waves. Pitch gets higher as the frequency of the sound waves increases.

C	D	E	F	G	A	B	C
do	re	mi	fa	sol	la	ti	do
262 Hz	294 Hz	330 Hz	349 Hz	392 Hz	440 Hz	494 Hz	523 Hz

Figure 5 Every note has a different frequency, which gives it a distinct pitch.
Describe *how pitch changes when frequency increases.*

Frequency Frequency is a measure of how many wavelengths pass a particular point each second. For a compressional wave, such as sound, the frequency is the number of compressions or the number of rarefactions that pass by each second. Frequency is measured in hertz (Hz). A frequency of 1 Hz means that one complete wavelength passes by in 1 s.

A healthy human ear can hear sound waves with frequencies from about 20 Hz to 20,000 Hz. The human ear is most sensitive to sounds in the range of 440 Hz to about 7,000 Hz. In this range, most people can hear much fainter sounds than they can at higher or lower frequencies.

Ultrasonic and Infrasonic Waves Most people can't hear sound frequencies above 20,000 Hz, which are called ultrasonic waves. Dogs can hear sounds with frequencies up to about 35,000 Hz, and bats can detect frequencies higher than 100,000 Hz. Even though humans can't hear ultrasonic waves, they have a number of uses. For example, ultrasonic waves directed into the human body are used in medical diagnosis and treatment. Ultrasonic waves are reflected by objects underwater, and can be used to determine the size, shape, and depth of underwater objects.

Infrasonic, or subsonic, waves have frequencies below 20 Hz—too low for most people to hear. These waves are produced by sources that vibrate slowly, such as wind, heavy machinery, and earthquakes. Although you can't hear infrasonic waves, you might feel them as a rumble inside your body.

The Doppler Effect

Imagine that you are standing at the side of a racetrack with race cars zooming past. As they move toward you, the pitches of their engines become higher. As they move away, the pitches become lower. The change in pitch or frequency due to the relative motion of a wave source is called the **Doppler effect.**

Comparing Intensity

Procedure

1. Complete the safety form.
2. Tie the middle of a 50-cm length of **string** to a metal object such as a **spoon.**
3. Wrap each end of the string around a finger on each hand. Place one of these fingers in each ear.
4. Swing the object so it bumps against a **table** or a **chair.** Note the sound.
5. Take your fingers out of your ears and repeat step 4.

Analysis

1. Compare the sounds you heard in both trials.
2. Compare the intensity of the sound waves that reached your ears in both trials.

Try at Home

Figure 6 The Doppler effect occurs when the source of a sound wave is moving relative to a listener.

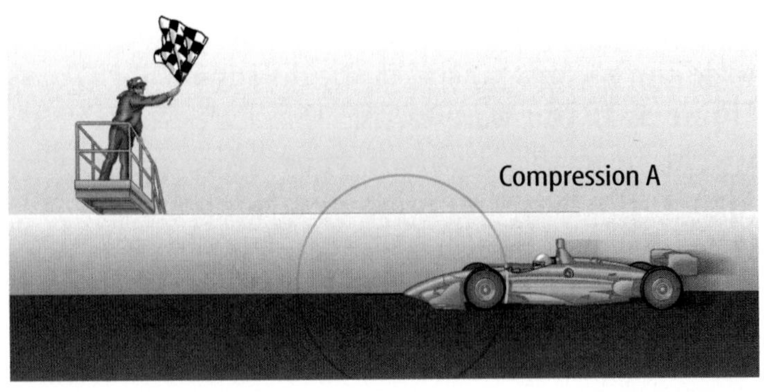

A The race car creates compression A, which spreads through the air in all directions from the point where it was created.

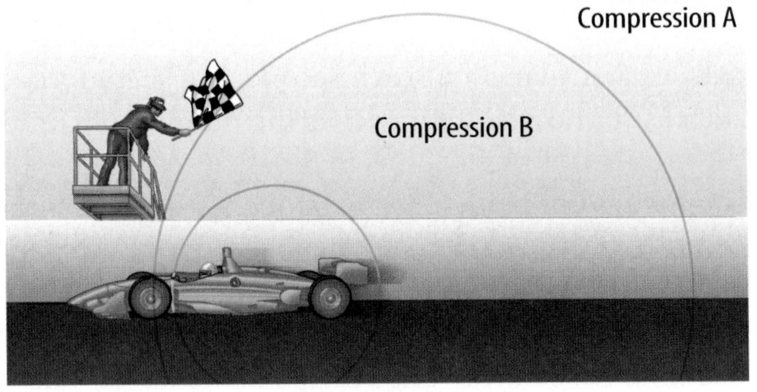

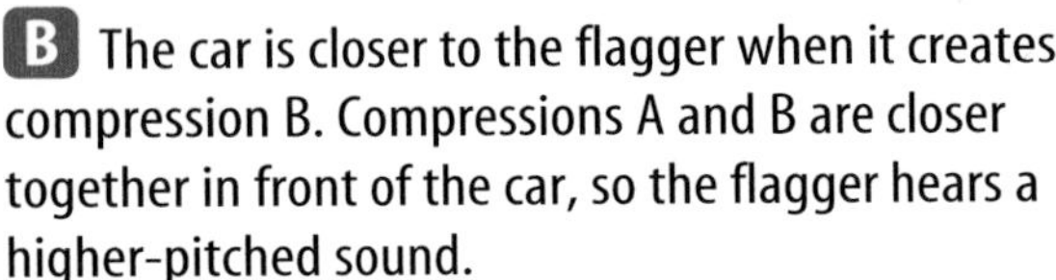

B The car is closer to the flagger when it creates compression B. Compressions A and B are closer together in front of the car, so the flagger hears a higher-pitched sound.

A Moving Source of Sound As a race car moves, it produces sound waves in the form of compressions and rarefactions. In **Figure 6A,** the race car creates a compression, labeled A. Compression A moves through the air toward the flagger. By the time compression B leaves the race car in **Figure 6B,** the car has moved forward. Because the car has moved since the time it created compression A, compressions A and B are closer together in front of the car than they would be if the car had stayed still. Because the compressions are closer together, more compressions pass by the flagger each second than if the car had been at rest. As a result, the flagger hears a higher pitch. **Figure 6B** also shows that the compressions behind the moving car are farther apart, resulting in a lower frequency. The flagger hears a lower pitch after the car passes.

Red Shift The Doppler effect can also be observed in light waves emanating from moving sources—although the sources must be moving at tremendous speeds. Astronomers have learned that the universe is expanding by observing the Doppler effect in light waves. Research the phenomenon known as red shift and explain in your Science Journal how it relates to the Doppler effect.

A Moving Listener You also can hear the Doppler effect when you are moving past a sound source that is standing still. Suppose you were riding in a school bus and passed a building with a ringing bell. The pitch would become higher as you approached the building and lower as you rode away from it. The Doppler effect happens any time the source of a sound is changing position relative to the listener. It occurs no matter whether it is the sound source or the listener that is moving. The faster the change in position, the greater the change in frequency and pitch.

How does pitch change if you are moving away from a sound source?

Using Sound

When sound waves strike an object, they can be absorbed by the object, transmitted through the object, or reflected from the object. By detecting the sound waves reflected from an object, the size, shape, and location of an object can be determined.

Echolocation and Sonar Some species of bats, as well as dolphins, whales, and other animals, use sound waves to detect their prey. Echolocation is the process of locating objects by emitting sounds and detecting the sound waves that reflect back.

Sonar is a system that uses the reflection of underwater sound waves to detect objects. First, a sound pulse is emitted underwater. The sound waves travel in the water and are reflected when they strike an object, such as a fish or a ship. An underwater microphone detects the reflected waves. Because the speed of sound in water is known, the distance to the object can be calculated by measuring how much time passes between when the sound pulse is emitted and when the reflected signal is received.

Topic: Sonar

Visit gpescience.com for Web links to information about ways sonar is used.

Activity Research to find out the different types of sonar. Make a poster that describes active and passive sonar.

Applying Math — Solve a Simple Equation

USING SONAR A sonar pulse returns in 3.00 s from a sunken ship that is directly below. Find the depth of the ship if the speed of the pulse is 1,500 m/s. *Hint:* The sonar pulse travels a distance equal to twice the depth of the ship, so use the equation $d = st/2$ to find the depth.

IDENTIFY known values and the unknown value

Identify the known values:

speed of sound in water → means → $s = 1{,}500$ m/s

time for round trip → means → $t = 3.00$ s

Identify the unknown value:

depth of the ship → means → $d = ?$ m

SOLVE the problem

Substitute the known values $s = 1{,}500$ m/s and $t = 3.00$ s into the equation for depth:

$$d = \frac{st}{2} = \frac{(1{,}500 \text{ m/s})(3.00 \text{ s})}{2} = \frac{1}{2}(4{,}500 \text{ m}) = 2{,}250 \text{ m}$$

CHECK your answer

Check your answer by dividing the distance by the time and multiplying by 2. The result should be the speed of sound given in the problem.

Practice Problems

Find the speed of a sonar pulse that returns in 2.0 s from a ship at a depth of 1,500 m.

For more practice problems go to page 879, and visit Math Practice at gpescience.com.

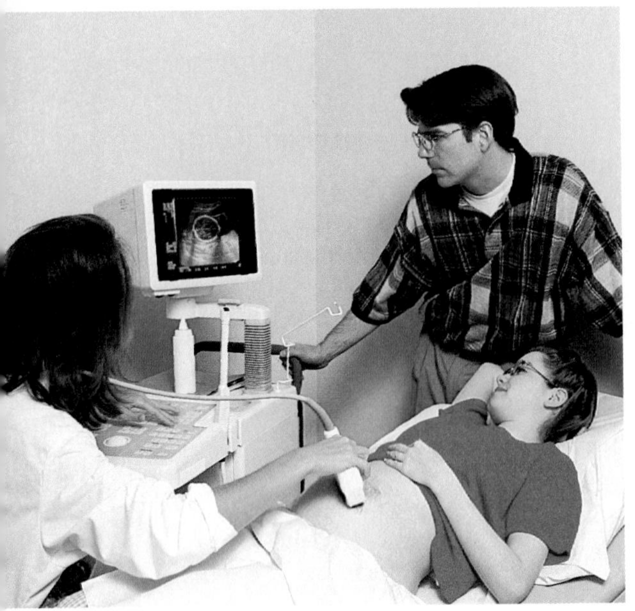

Figure 7 Ultrasonic waves are directed into a pregnant woman's uterus to form images of her fetus.

Ultrasound in Medicine High-frequency sound waves can also be used to remove dirt buildup on jewelry and glassware. One of the most important uses of ultrasonic waves, though, is in medicine. Using special instruments, medical professionals can send ultrasonic waves into a specific part of a patient's body. Reflected ultrasonic waves are used to examine different body parts and to detect and monitor certain types of heart disease and cancer. Ultrasound imaging also is used to monitor the development of a fetus, as shown in **Figure 7.** However, ultrasound does not produce good images of the bones and lungs, because hard tissues and air absorb the ultrasonic waves instead of reflecting them.

High-frequency sound waves can be used to treat certain medical problems. For example, sometimes small, hard deposits of calcium compounds or other minerals form in the kidneys, making kidney stones. In the past, physicians had to perform surgery to remove kidney stones. Now ultrasonic treatments are commonly used to break them up instead. Bursts of ultrasound create vibrations that cause the stones to break into small pieces that can easily pass out of the body with the urine. A similar treatment is available for gallstones.

section 1 review

Summary

Sound Waves

- Sound waves are compressional waves that are produced by vibrating objects.
- Sound usually travels fastest in solids and slowest in gases.
- Sound travels faster as the temperature of the medium increases.

Properties of Sound

- The amplitude of sound waves increases as the density of particles increases in the compressions and decreases in rarefactions
- Loudness is the human perception of the intensity of sound waves. Pitch is the human perception of the frequency of sound waves.
- The Doppler effect is the change in the frequency of a sound wave when the source of the sound is moving relative to a listener.

Self Check

1. **Explain** how sound travels from your vocal cords to your friend's ears when you talk.
2. **Explain** how the loudness of a sound wave depends on the intensity of the sound wave.
3. **Draw and label** a diagram that explains the Doppler effect.
4. **Describe** at least three uses of ultrasonic technology in medicine.
5. **Think Critically** How could sonar technology be used to locate deposits of oil and minerals?

Applying Math

6. **Calculate Time** How long would it take a sound wave from a car alarm to travel 1.0 km in air if the air temperature were 0°C?
7. **Calculate Distance** If sound travels in the ocean with a speed of 1,500 m/s, how far will a sonar pulse travel in 45 s?

More Section Review gpescience.com

section 2

Reflection and Refraction of Light

Reading Guide

What You'll Learn

- **Describe** how light waves interact with matter.
- **Explain** the difference between regular and diffuse reflection.
- **Define** the index of refraction of a material.
- **Explain** why a prism separates white light into different colors.

Why It's Important

The images you see every day are due to the behavior of light waves.

Review Vocabulary

visible light: an electromagnetic wave with wavelengths between about 400 and 700 billionths of a meter

New Vocabulary

- opaque
- translucent
- transparent
- index of refraction

The Interaction of Light and Matter

Look around your darkened room at night. You know that some of the objects are brightly colored, but they look gray or black in the dim light. Turn on the light, and you can see all the objects in the room clearly, including their colors. What you see depends on the amount of light in the room and the color of the objects. For you to see an object, it must reflect or emit some light that reaches your eyes.

Absorption, Transmission, and Reflection Objects can absorb light, reflect light, and transmit light—allow light to pass through them. The type of matter in an object determines the amount of light it absorbs, reflects, and transmits. For example, the **opaque** material in the top candleholder in **Figure 8** only absorbs and reflects light—no light passes through it. As a result, you cannot see the candle inside. Materials that allow some light to pass through them, like the material of the middle candleholder in **Figure 8,** are described as **translucent.** You cannot see clearly through translucent materials.

Transparent materials, such as the bottom candleholder in **Figure 8,** transmit almost all the light striking them, so you can see objects clearly through them. Only a small amount of light is absorbed and reflected by transparent materials.

Figure 8 These candleholders interact with light differently.

A Opaque

B Translucent

C Transparent

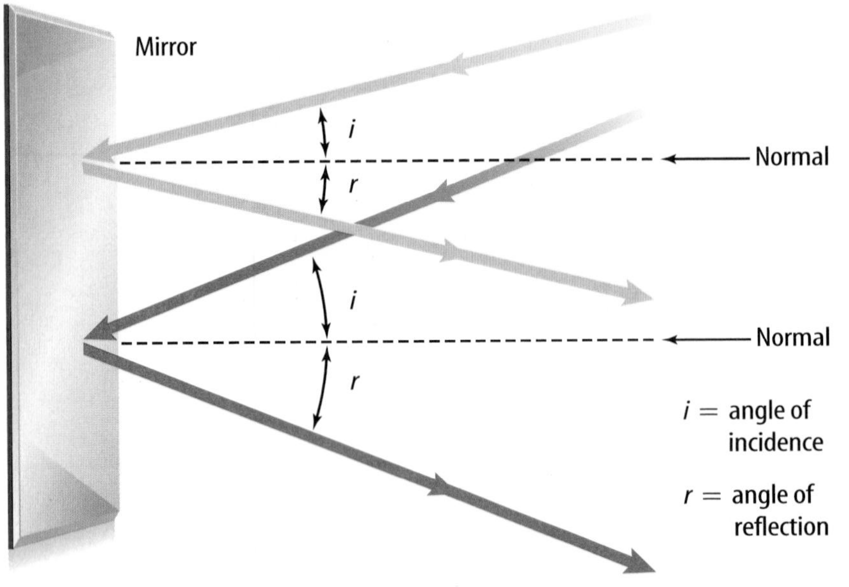

Figure 9 According to the law of reflection, light is reflected so that the angle of incidence always equals the angle of reflection.

Reflection of Light

Have you glanced in a mirror today? You see your reflection in the mirror when light is reflected off you, strikes the mirror, and is reflected off the mirror into your eye. Because light behaves as a wave, it obeys the law of reflection, as shown in **Figure 9.** Recall that, according to the law of reflection, the angle at which a light wave strikes a surface is the same as the angle at which it is reflected. Light reflected from any surface—a mirror or a sheet of paper—obeys this law.

Regular and Diffuse Reflection Why can you see your reflection in a store window but not in a sheet of paper? The answer has to do with the smoothness of the surfaces. A smooth, even surface such as a pane of glass produces a sharp image by reflecting parallel light waves in only one direction. Reflection of light waves from a smooth surface is regular reflection. To cause a regular reflection, the roughness of a surface must be less than the wavelengths it reflects. A sheet of paper has an uneven surface that causes incoming parallel light waves to be reflected in many directions, as shown in **Figure 10.** Reflection of light from a rough surface is diffuse reflection.

Scattering Diffuse reflection is a type of scattering that occurs when light waves traveling in one direction are made to travel in many different directions. Scattering also occurs when light waves traveling through the air reflect off small particles. An example is light scattering off the small water droplets that make up a cloud. Scattering causes the cloud to appear white, even though the droplets are transparent.

Reading Check *What is scattering of light?*

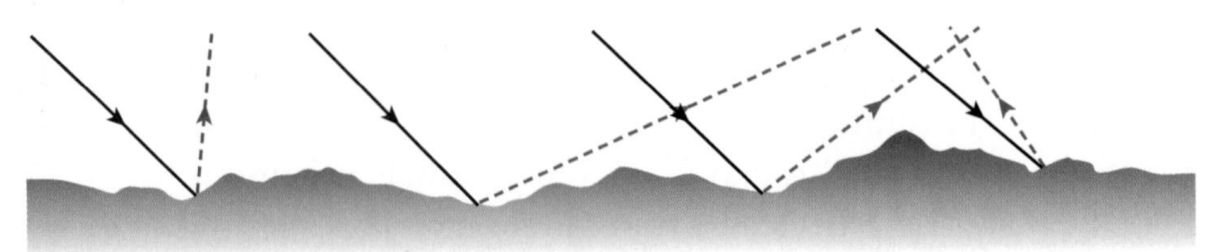

Figure 10 A sheet of paper has an uneven surface that produces a diffuse reflection.
Explain *Use the law of reflection to explain why a rough surface causes parallel light waves to be reflected in many directions.*

Refraction of Light

What occurs when a light wave passes from one material to another—from air to water, for example? Recall that refraction is caused by a change in the speed of a wave when it passes from one material to another. If the light wave is traveling at an angle to the boundary between the materials and the speed of light is different in the two materials, the wave will be bent, or refracted.

Reading Check *Why does refraction occur?*

Figure 11 The spoon looks broken because light waves are refracted as they change speed when they pass from the water to the air.

The Index of Refraction The amount of bending that takes place depends on the speeds of light in both materials. The greater the change in speed, the more the light will be bent as it passes at an angle from one material to the other. **Figure 11** shows an example of refraction. Every material has an **index of refraction** which is the ratio of the speed of light in a vacuum to the speed of light in the material. The index of refraction indicates how much the speed of light is reduced in the material compared to its speed in empty space.

The larger the index of refraction, the more light is slowed down in the material. For example, because glass has a larger index of refraction than air, light moves more slowly in glass than air. The index of refraction is usually largest for solids and smallest for gases.

Prisms A sparkling glass prism hangs in a sunny window, refracting the sunlight and projecting a colorful pattern onto the walls of the room. How does the refraction of light create these colors? The spectrum of colors is produced because the speed of light in a material also depends on the wavelength of the light. In a glass prism, light waves with longer wavelengths, such as red light waves, are slowed less than light waves with shorter wavelengths, such as blue light waves.

Figure 12 shows what occurs when white light passes through a prism. White light, such as sunlight, is made up of light waves with range of wavelengths from red to blue. The triangular prism refracts the light twice—once when it enters the prism and again when it leaves the prism and reenters the air. Because the longer wavelengths of light are slowed less than the shorter wavelengths are, red light is bent the least. As a result of these different amounts of bending, the different colors are separated when they emerge from the prism.

Figure 12 Refraction causes a prism to separate a beam of white light into different colors.

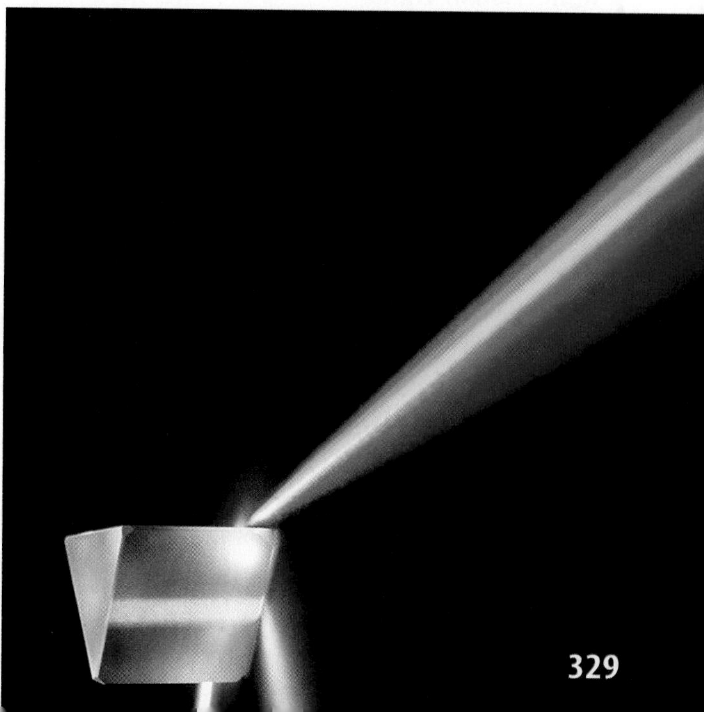

Figure 13 Mirages result when air near the ground is much warmer or cooler than the air above. This causes light waves reflected from an object to refract, creating one or more additional images.

Mirages When you're traveling in a car, you might see what looks like a pool of water on the road ahead. As you get closer, the water seems to disappear. You've seen a mirage—an image of a distant object produced by the refraction of light through air layers of different densities. Mirages result when the air at ground level is much warmer or cooler than the air above, as shown in **Figure 13.** The density of air increases as air cools and light waves move slower in cooler air than in warmer air. As a result, light waves are refracted as they pass through air layers with different temperatures.

section 2 review

Summary

Light and Matter

- The amount of light that is absorbed, reflected, or transmitted depends on the material making up the object.

Reflection of Light

- Light waves always obey the law of reflection: the angle of incidence equals the angle of reflection.
- Regular reflection causes parallel light waves to be reflected in only one direction.
- Diffuse reflection causes parallel light waves to be reflected in many directions.
- Scattering occurs when light waves are reflected and travel in many different directions.

Refraction of Light

- Refraction occurs when a light wave changes speed in moving from one material to another.
- The index of refraction of a material indicates how much light slows down in the material.

Self Check

1. **Compare and contrast** opaque, transparent, and translucent materials. Give one example of each.
2. **Discuss** why you can see your reflection in a smooth piece of aluminum foil but not in a crumpled ball of foil.
3. **Infer** what happens to light waves that are reflected off dust particles in the air.
4. **Explain** what happens to white light when it passes through a prism.
5. **Think Critically** Suppose a material has the same index of refraction as water. How would a light wave change direction as it traveled from water into this material?

Applying Math

6. **Find an Angle** A light wave strikes a mirror at an angle of 42° from the surface of the mirror. What angle does the reflected wave make with the normal?
7. **Find an Angle** A light wave reflects from a mirror at 27° from the normal. What was the angle between the mirror and the incident wave?

More Section Review gpescience.com

section 3

Mirrors, Lenses, and the Eye

Reading Guide

***What* You'll Learn**

- **Describe** how images are formed by three types of mirrors.
- **Explain** how convex and concave lenses form images.
- **Explain** how the human eye enables you to see.
- **Describe** how lenses are used to correct vision problems.

***Why* It's Important**

Mirrors and lenses are used in optical instruments such as cameras and telescopes.

Review Vocabulary

reflection: the process of changing direction after striking a surface

New Vocabulary

- plane mirror
- concave mirror
- convex mirror
- convex lens
- concave lens

Light Rays

Light sources send out light waves in all directions. These waves spread out like ripples on the surface of water spread out from the point of impact of a pebble. You can think of the light coming from the source as being many narrow beams of light traveling in all directions. Each narrow light beam is called a light ray.

Mirrors

A mirror is any surface that produces a regular reflection. A pool of still water, a metal pan, and even the back of a shiny spoon can be mirrors. Mirrors can be flat, curved inward, or curved outward.

Plane Mirrors A flat, smooth mirror is a **plane mirror.** When you look in a plane mirror, your image appears upright. If you stand 1 m from the mirror, your image appears 1 m behind the mirror, or 2 m from you. In fact, your image is what a friend standing 2 m from you would see. **Figure 14** shows how your image is formed. First, light rays from a light source strike you. Every point that is struck by the light rays reflects these rays so they travel outward in all directions. If your friend were looking at you, these reflected light rays coming from you would enter her eyes so she could see you. However, if a mirror is placed between you and your friend, light rays are reflected from the mirror back to your eyes.

Figure 14 Seeing an image of yourself in a mirror involves two sets of reflections: light rays are reflected from you and then are reflected by the mirror.

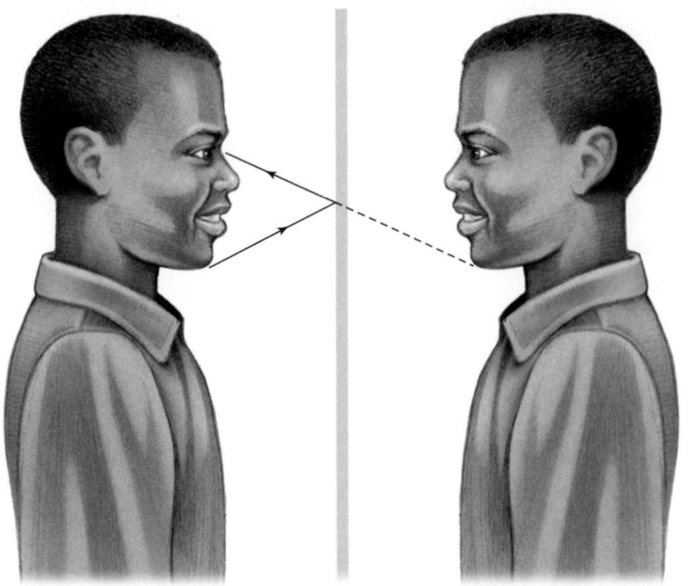

Figure 15 Your brain interprets the light rays reflected by the mirror as coming from a point behind the mirror.
Infer *how the size of your image in a plane mirror depends on your distance from the mirror.*

Virtual and Real Images The formation of an image by a plane mirror is shown in **Figure 15.** Your brain assumes that light rays travel in a straight line. As a result, rays that enter your eyes seem to come from behind the mirror. This makes the image seem to be behind the mirror. However, no light rays actually pass through the place where the image seems to be located. This type of image is called a virtual image. Plane mirrors always form virtual images. If light rays from an object pass through the location of the image, the image is called a real image. Curved mirrors can form both real and virtual images.

Concave Mirrors If the surface of a mirror is curved inward, it is called a **concave mirror,** as shown in **Figure 16.** The optical axis is an imaginary straight line drawn perpendicular to the surface of the mirror at its center. Every light ray traveling parallel to the optical axis is reflected through a point on the optical axis called the focal point. The distance from the center of the mirror to the focal point is called the focal length.

The image formed by a concave mirror depends on the location of the object relative to the focal point. **Figure 16** shows a candle located more than twice the focal length from the mirror. Although many light rays come from each point on the candle, only two rays from the same point are shown. Ray A passes through the focal point and then reflects off the mirror moving parallel to the optical axis. Ray B travels parallel to the optical axis before reflecting off the mirror and passing through the focal point. The point where the two rays meet is the location of the image of the original point on the candle. The image formed is a real image because light rays from the candle pass through the place where the image is located.

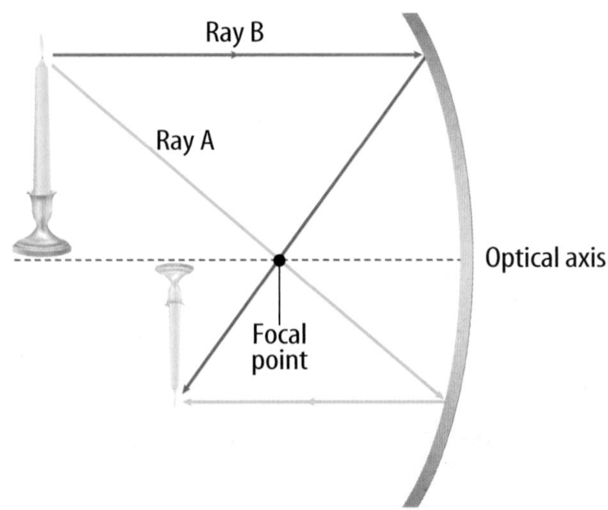

Figure 16 Rays A and B start from the same place on the candle, travel in different directions, and meet again on the reflected image.
Diagram *how two other points on the image of the candle are formed.*

Images Produced by a Concave Mirror When an object is between one and two focal lengths from a concave mirror, the image is real, inverted, and larger than the object. An object closer than one focal length from a concave mirror produces a virtual image that is upright and larger than the object. No image is produced if the object is located at the focal point.

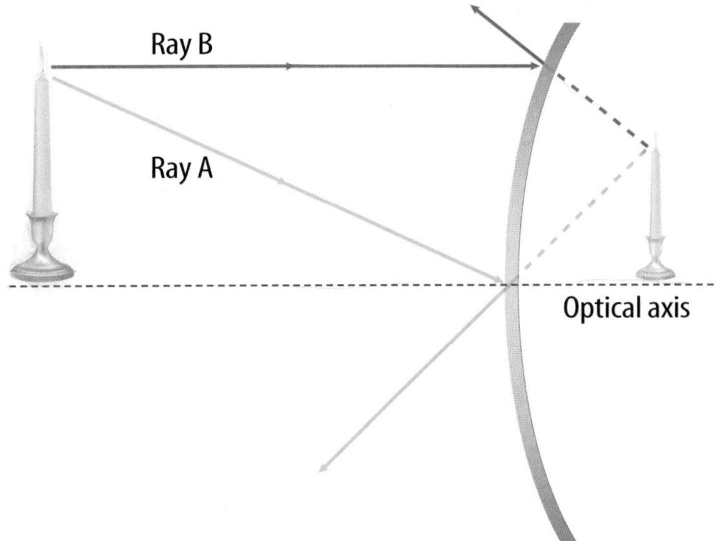

Figure 17 A convex mirror forms a reduced, upright, virtual image.

Convex Mirrors When you are in a store, look up toward one of the back corners or at the end of an aisle to see if a large, rounded mirror is mounted there. You can see a large area of the store in such a mirror. A mirror that curves outward like the back of a spoon is called a **convex mirror.** Light rays that hit a convex mirror diverge, or spread apart, after they are reflected. **Figure 17** shows how the rays from an object are reflected by a convex mirror to form an image. The reflected rays diverge and never meet, so a convex mirror forms only a virtual image. The image also is upright and smaller than the actual object is.

Reading Check *Describe the image formed by a convex mirror.*

Lenses

What do your eyes have in common with cameras, eyeglasses, and microscopes? Each of these contains at least one lens. A lens is a transparent object with at least one curved surface that causes light rays to refract. The image that a lens forms depends on the shape of the lens. Like curved mirrors, lenses can be convex or concave.

Convex Lenses A **convex lens** is thicker in the middle than at the edges. Its optical axis is an imaginary straight line that is perpendicular to the surface of the lens at its thickest point. When light rays approach a convex lens traveling parallel to its optical axis, the rays are refracted toward the center of the lens, as in **Figure 18.** Light rays traveling along the optical axis are not bent at all. All light rays traveling parallel to the optical axis are refracted so they pass through a single point, which is the focal point of the lens. The less curved the sides of the lens, the less light rays are bent. As a result, lenses with flatter sides have longer focal lengths. **Figure 19** on the next page shows that convex lenses can form either real or virtual images.

Figure 18 Convex lenses are thicker in the middle than at the edges. Light rays that are parallel to the optical axis are refracted so that they pass through the focal point. A light ray that passes through the center of the lens is not refracted.

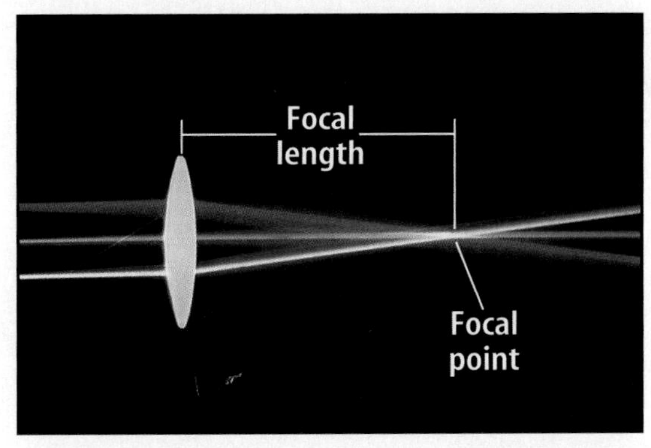

NATIONAL GEOGRAPHIC

VISUALIZING IMAGES FORMED BY A CONVEX LENS

Figure 19

A convex lens can form images that are real or virtual, enlarged or reduced, and upright or inverted. Just as for a concave mirror, the type of image formed by a convex lens depends on the location of the object relative to the focal point of the lens.

If the object is more than two focal lengths from the lens, the image formed is real, inverted, and smaller than the object. As the object moves farther from the lens, the image becomes smaller.

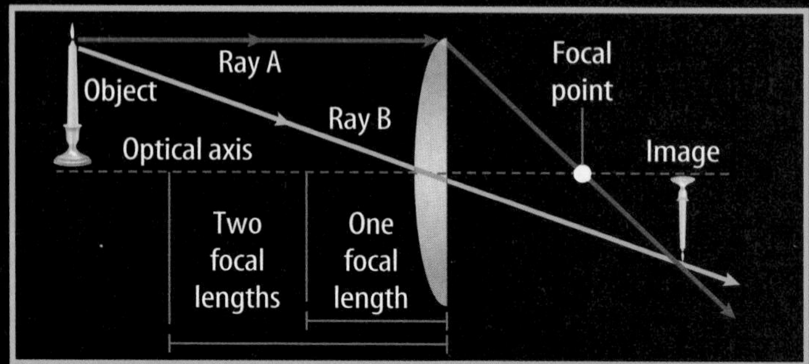

If the object is between one and two focal lengths from the lens, the image formed is real, inverted, and larger than the object. As the object moves closer to the focal point, the image becomes larger.

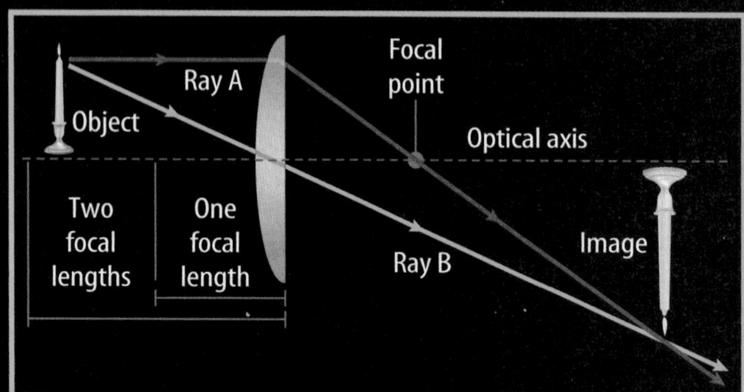

If the object is less than one focal length from the lens, the image is virtual, upright, and larger than the object. The image is virtual because the light rays from the object diverge after they pass through the lens. The image becomes smaller as the object moves closer to the lens.

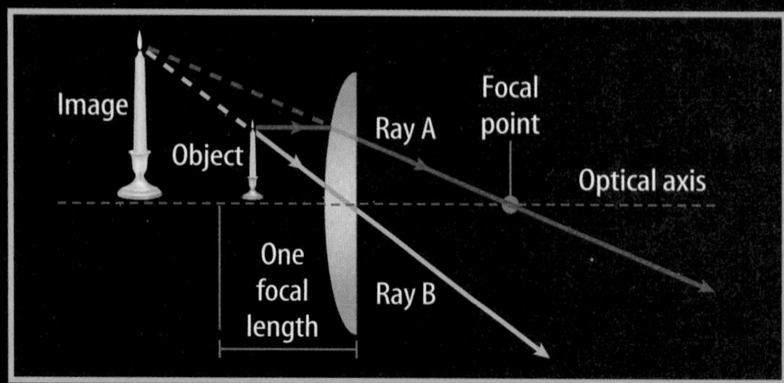

Concave Lenses A **concave lens** is thinner in the middle and thicker at the edges. As shown in **Figure 20,** light rays that pass through a concave lens bend away from the optical axis. The rays spread out and never meet, so a real image is never formed. The image is always virtual, upright, and smaller than the object is. Concave lenses are used in some types of eyeglasses and some telescopes. Concave lenses usually are used in combination with other lenses.

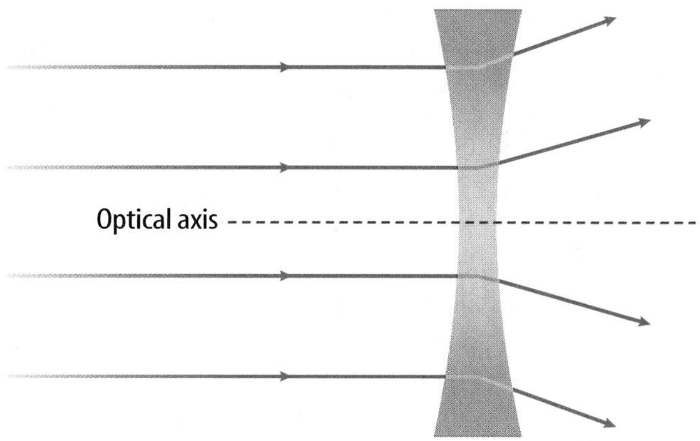

Figure 20 A concave lens refracts light rays so they spread out away from the optical axis. **Classify** *Is a concave lens most like a concave mirror or a convex mirror?*

Reading Check *What type of image is formed by a concave lens?*

The Human Eye

What determines how well you can see the words on this page? If you don't need eyeglasses, the structure of your eye gives you the ability to focus on these words and other objects around you. Look at **Figure 21.** Light enters your eye through a transparent covering on your eyeball called the cornea. The cornea causes light rays to bend so that they converge. The light then passes through an opening called the pupil. Behind the pupil is a flexible convex lens. Muscles attached to the lens change its shape to help focus light, forming a sharp image on your retina.

The retina is the inner lining of your eye. The retina contains light-sensitive cells that convert an image into electrical signals. These signals then are carried along the optic nerve to your brain to be interpreted. With this complex structure, the human eye is capable of seeing clearly in bright and dim conditions, focusing on both near and far objects, and also detecting colors.

Eye Doctors Optometrists and ophthalmologists treat many different eye problems. They often use advanced technology to diagnose and correct malfunctioning parts of the eye. Research an eye problem and how it is treated. Write a paragraph describing what parts of the eye are affected by the problem and how doctors correct it.

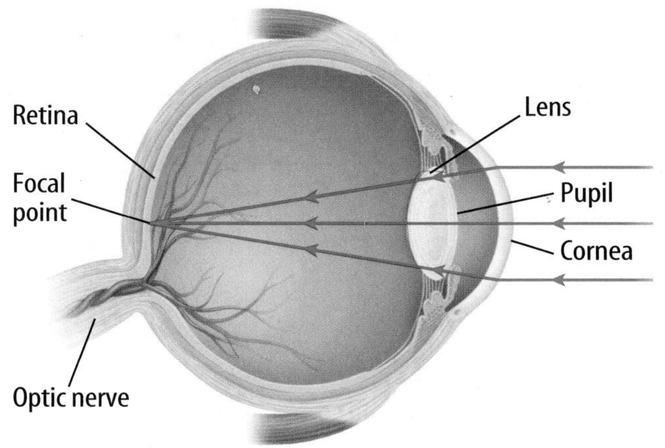

Figure 21 The cornea and lens in your eye bend light rays so that a sharp image is formed on the retina.

Experimenting with Focal Lengths

Procedure

1. Complete the safety form.
2. Fill a glass **test tube** with **water** and seal it with a **lid or stopper.**
3. Type or print *SULFUR DIOXIDE* in capital letters on a piece of **paper.**
4. Set the test tube horizontally over the words. Record your observations.
5. Hold the tube at several heights above the words. Record your observations.

Analysis

1. Describe how the words changed as the height of the test tube increased.
2. Classify the image at each height as real or virtual.

Brightness and Intensity The human eye can adjust to the brightness of the light that strikes it. If you step outside on a sunny day, you may have to shade your eyes against the intensity of the sunlight. Light intensity is the amount of light energy that strikes a certain area each second. Brightness is the human perception of light intensity. Sunlight seems bright because a large amount of light energy strikes your retina each second. Your eyes respond to bright light by decreasing the size of your pupil. This reduces the amount of light that enters your eye and strikes the retina.

Intensity depends on your distance from a light source. If you move away from a bare lightbulb, you can see how the intensity of the light rapidly decreases. Because light rays spread out from the source, less light energy strikes your retina the farther away you are from the bulb.

Correcting Vision Problems

If you have good vision, you should be able to see objects clearly when they are 25 cm or farther away from your eyes. A sharp image of an object should be formed on your retina. However, for many people, the image is blurry or formed in the wrong place, causing vision problems.

Farsightedness If you can see distant objects clearly but can't bring nearby objects into focus, then you are farsighted. In this case, the eyeball might be too short or the lens isn't curved enough to form a sharp image of nearby objects on the retina, as shown in **Figure 22.** To correct the problem, convex lenses can be used to bend incoming light rays so they converge before they enter the eye.

Figure 22 Farsightedness can be corrected by convex lenses.

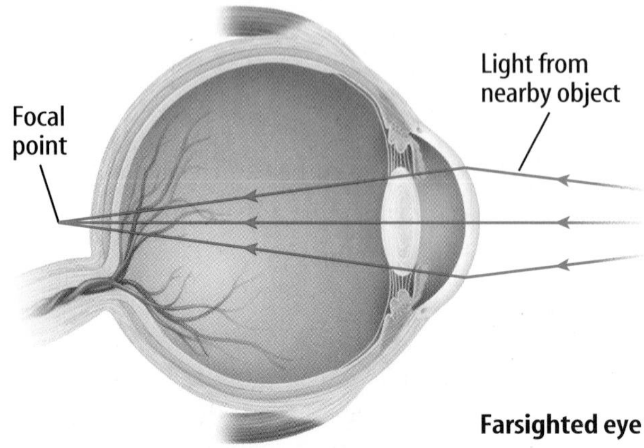

The focal length of a farsighted eye is too long to form a sharp image of nearby objects on the retina.

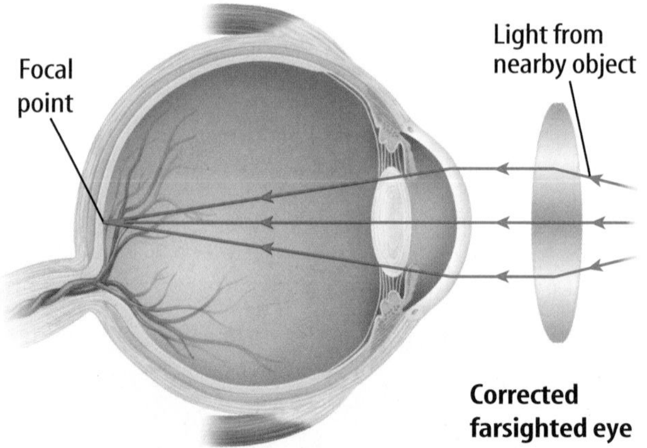

A convex lens in front of a farsighted eye enables a sharp image of nearby objects to be formed on the retina.

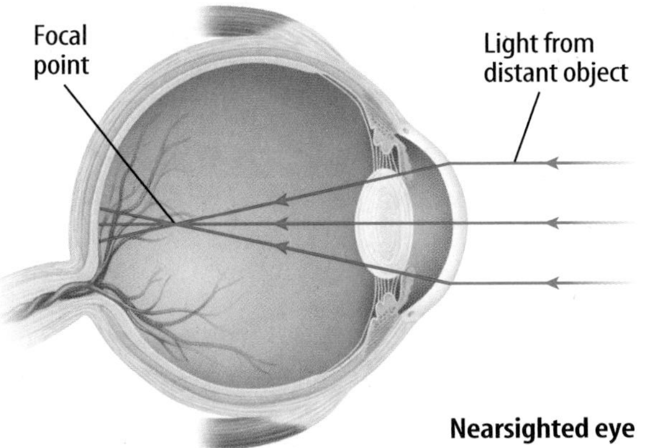

When a nearsighted person looks at distant objects, a sharp image is formed in front of the retina.

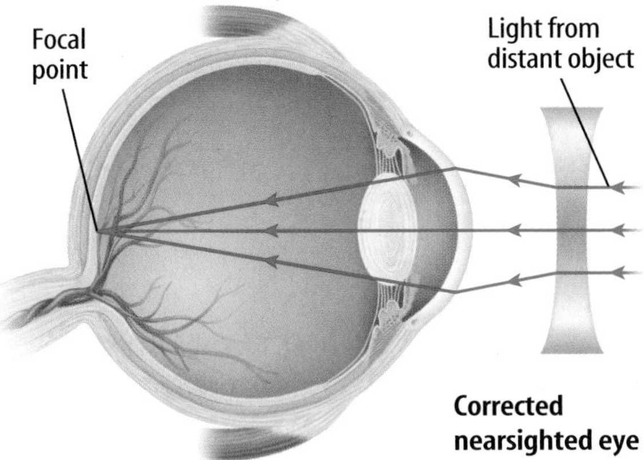

A concave lens in front of a nearsighted eye makes light rays diverge so a sharp image is formed on the retina.

Figure 23 Nearsightedness can be corrected with concave lenses.

Nearsightedness If you have nearsighted friends, you know that they can see only nearby objects clearly. Their eyes cannot form a sharp image on the retina of an object that is far away. Instead, the image is formed in front of the retina, as shown in **Figure 23.** To correct this problem, a nearsighted person can wear concave lenses. **Figure 23** shows how a concave lens causes incoming light rays from distant objects to diverge before they reach the eye. Then the rays can be focused by the eye to form a sharp image on the retina.

section 3 review

Summary

Mirrors

- A plane mirror forms upright, virtual images.
- The image formed by a concave mirror depends on the location of the object relative to the focal point.
- Convex mirrors always produce virtual, upright images that are smaller than the object.

Lenses

- The image formed by a convex lens depends on the distance of the object from the lens.
- Concave lenses always form virtual, upright images that are smaller than the object.

The Eye and Vision

- The eye contains a lens that changes shape to produce sharp images on the retina.
- Farsightedness occurs when the eye cannot form a sharp image of nearby objects. Nearsightedness occurs when the eye cannot form sharp images of distant objects.

Self Check

1. **Diagram** how light rays from an object are reflected by a convex mirror to form an image.
2. **Infer** how the size of the image changes if an object that is less than one focal length from a concave mirror moves closer to the mirror.
3. **Explain** how the focal length of a convex lens changes as the sides of the lens become less curved.
4. **Compare** the image of an object less than one focal length from a convex lens with the image of an object more than two focal lengths from the lens.
5. **Think Critically** Describe the image formed by a light source placed at the focal point of a convex lens.

Applying Math

6. **Calculate Object Distance** If you looked through a convex lens with a focal length of 15 cm and saw a real, inverted, enlarged image, what is the maximum distance between the lens and the object?

More Section Review gpescience.com

Reflections of Reflections

How can you see the back of your head? You can use two mirrors to view a reflection of a reflection of the back of your head.

Real-World Problem

How many images are seen in two mirrors?

Goal

- **Infer** how the number of images depends on the angle between two mirrors.

Materials

plane mirrors (2) protractor
masking tape paper clip

Safety Precautions

Handle glass mirrors and paper clips carefully.

Procedure

1. Complete the safety form.
2. Lay one mirror on top of the other with the mirror surfaces together. Tape together one side so they can open and close. Use tape to label them *L* and *R* for *left* and *right.*
3. Open the mirrors and stand them on a sheet of paper at an angle of 72°.

Images and Wedges Seen in the Mirrors

Angle of Mirrors	Number of Paper Clip Images		Number of Wedges
	R	L	
72°			
90°			
120°			

Do not write in this book.

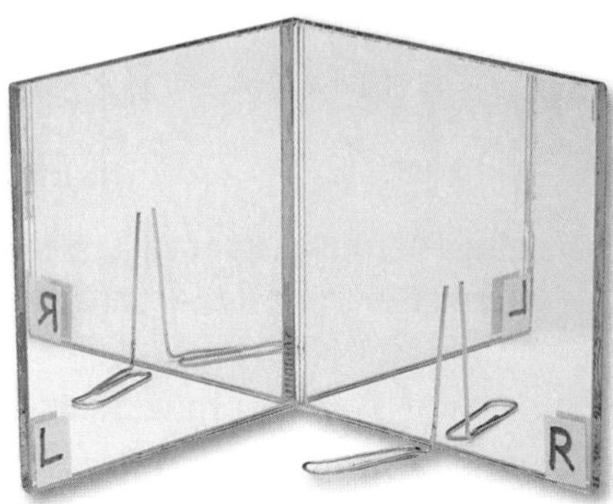

4. Bend one leg of a paper clip up 90° and place it close to the front of the *R* mirror.
5. Count the number of images of the paper clip you see in the *R* and *L* mirrors. Record these numbers in your data table.
6. The mirrors create an image of a circle divided into wedges. Record the number of wedges.
7. Hold the *R* mirror still and slowly move the *L* mirror to 90°. Record the numbers of images of the clip and wedges in the circle. Repeat, this time opening the mirrors to 120°.

Conclude and Apply

1. **Infer** the relationship between the number of wedges and paper clip images.
2. **Determine** the angle that would divide a circle into six wedges. Hypothesize how many images would be produced. Test your hypothesis.

Communicating Your Data

Demonstrate for younger students the relationship between the angle of the mirrors and the number of reflections.

section 4

Light and Color

Reading Guide

What You'll Learn

- **Explain** how you see color.
- **Describe** the difference between light color and pigment color.
- **Predict** what happens when different colors are mixed.

Why It's Important

From traffic lights to great works of art, color is an important part of your world.

Review Vocabulary

retina: inner layer of the eye containing cells that convert light images into electrical signals

New Vocabulary

- pigment

Why Objects Have Color

Why do some apples appear red, while others look green or yellow? An object's color depends on the wavelengths of light it reflects. Recall that white light is a blend of all colors of visible light. When a red apple is struck by white light, it reflects red light back to your eyes and absorbs all of the other colors. **Figure 24** shows white light striking a green leaf. Only the wavelengths corresponding to green light are reflected to your eyes.

Although some objects appear to be black, black isn't a color. Black is the absence of visible light. Objects that appear black absorb all colors of light and reflect little or no light back to your eyes. White objects appear to be white because they reflect all colors of visible light.

Reading Check *Why does a white object appear white?*

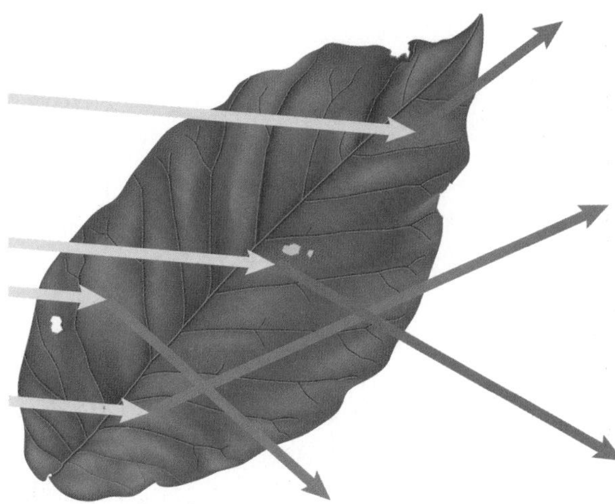

Figure 24 This leaf absorbs all wavelengths of visible light except the wavelengths you see as green.

Colored Filters Wearing tinted glasses changes the color of almost everything you see. If the lenses are yellow, the world takes on a golden glow. If they are rose colored, everything looks rosy. Something similar would occur if you placed a colored, clear plastic sheet over this white page. The paper would appear to be the same color as the plastic. The plastic sheet and the tinted lenses are filters. A filter is a transparent material that transmits one or more colors of light but absorbs all others. The color of a filter is the color of the light that it transmits.

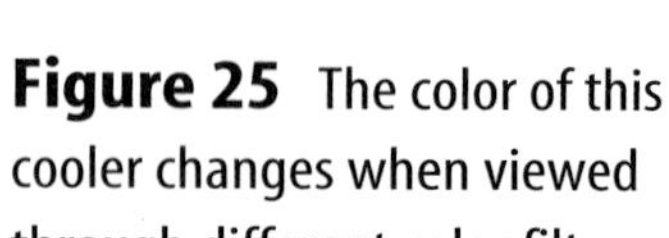

Figure 25 The color of this cooler changes when viewed through different color filters.

Looking Through Colored Filters **Figure 25** shows how the color of an object can change when you look at it through various colored filters. The cooler looks blue under white light because it reflects only the wavelengths of blue light that strike it. If you look at the cooler through a blue filter, the cooler still looks blue because the blue filter transmits the reflected blue light. However, if you look at the cooler through a red filter, the cooler seems black because the red filter blocks the blue light reflected by the cooler.

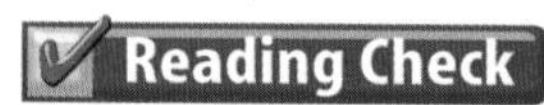

Why does a blue object appear black when viewed through a red filter?

Seeing Color

As you approach a busy intersection, the color of the traffic light changes from green to yellow to red. On the cross street, the color changes from red to green. At a busy intersection, traffic safety depends on your ability to detect color changes rapidly. How do you see colors?

Light and the Eye In a healthy eye, light enters the eye through the cornea, is focused by the lens, and finally forms an image on the retina. The retina is made up of two types of cells that absorb light, as shown in **Figure 26.** When these cells absorb light energy, chemical reactions convert light energy into nerve impulses that are transmitted to the brain. One type of cell in the retina, called a cone, enables you to distinguish colors and detailed shapes of objects. Cones need bright light to generate nerve impulses, so they do not operate in dim light. As a result, even brightly colored objects might look gray or black in dim light.

Topic: Color Blindness
Visit gpescience.com for Web links to information about the causes of color blindness.

Activity Research the causes and types of color blindness. Find out how cones are related to the ability to distinguish colors.

Figure 26 Light enters the eye and focuses on the retina. The two types of light-detecting cells that make up the retina are called rods and cones.

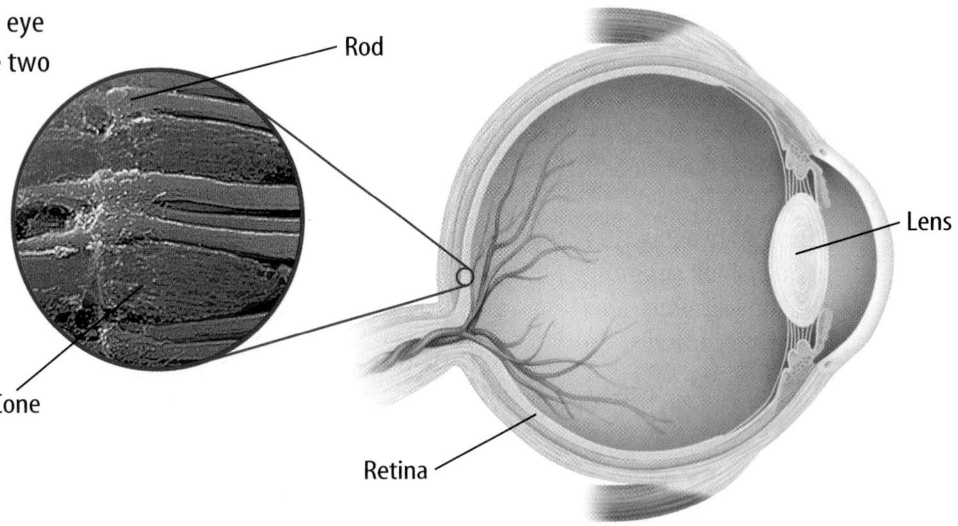

Rod

Lens

Cone

Retina

Cones and Rods Your eyes have three types of cones, each of which responds to a different range of wavelengths. Red cones respond to mostly red and yellow light. Green cones respond to mostly yellow and green light. Blue cones respond to mostly blue and violet light. The second type of cell in the retina, called a rod, is sensitive to dim light and enables you to see at night. However, rod cells do not enable you to see colors.

INTEGRATE Life Science

Interpreting Color Why does a banana look yellow? The light reflected by the banana causes the cone cells that are sensitive to red and green light to send signals to your brain. Your brain could get the same signal if a mixture of red light and green light reached your eyes. This mixture also would cause your red and green cones to respond, and you would see the color yellow. As a result, light with wavelengths corresponding to yellow light and light that is a mixture of red and green light both cause the color yellow to be seen. What happens when you look at a white shirt? You see white when all wavelengths of visible light enters your eyes. Then all three sets of cones cells send signals to the brain. The combination of these signals cause you to see the shirt as white.

Color Blindness If one or more of your sets of cone cells do not function properly, you might not be able to distinguish certain colors. This condition is called color blindness or color deficiency. About eight percent of men and one-half percent of women have some form of color blindness. The most common form of color blindness makes it difficult to distinguish between red and green. **Figure 27** shows an image that is used in a test for red–green color blindness. Because red and green are used in traffic signals, drivers and pedestrians must be able to identify them.

Figure 27 Color blindness is an inherited sex-linked condition in which certain sets of cones in the retina do not function properly. **Identify** *the number that you see in the dots.*

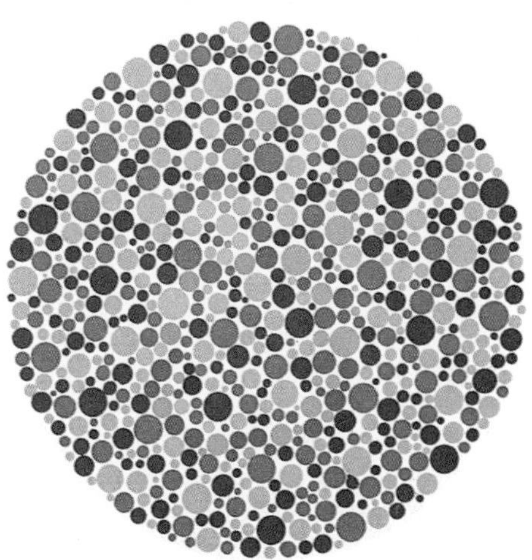

Color for Photosynthesis Plant pigments determine the wavelengths of light for photosynthesis. Leaves usually look green because of the pigment chlorophyll. Chlorophyll absorbs most wavelengths of visible light except green, which it reflects. However, not all plants are green. Research different plant pigments to find out how they allow plant species to survive in diverse habitats.

Mixing Colors

If you have ever browsed through a paint store, you have probably seen displays where customers can select paint samples of almost every imaginable color. The colors are a result of mixtures of pigments. For example, you might have mixed blue and yellow paint to produce green paint. A **pigment** is a colored material that is used to change the color of other substances. The color of a pigment results from the different wavelengths of light that the pigment reflects.

Mixing Colored Lights From the glowing orange of a sunset to the deep blue of a mountain lake, all the colors you see can be made by mixing three colors of light. These three colors—red, green, and blue—are the primary colors of light. They correspond to the three different types of cones in the retina of your eye. When mixed together in equal amounts, they produce white light, as **Figure 28** shows. Mixing the primary colors in different proportions can produce the colors you see.

What are the primary colors of light?

Paint Pigments If you were to mix equal amounts of red, green, and blue paint, would you get white paint? If mixing colors of paint were like mixing colors of light, you would, but mixing paint is different. Paints are made with pigments. Pigments produce color as a result of the wavelengths of light they reflect. Paint pigments usually are made of chemical compounds such as titanium oxide, a bright white pigment, and lead chromate, which is used for painting yellow lines on highways.

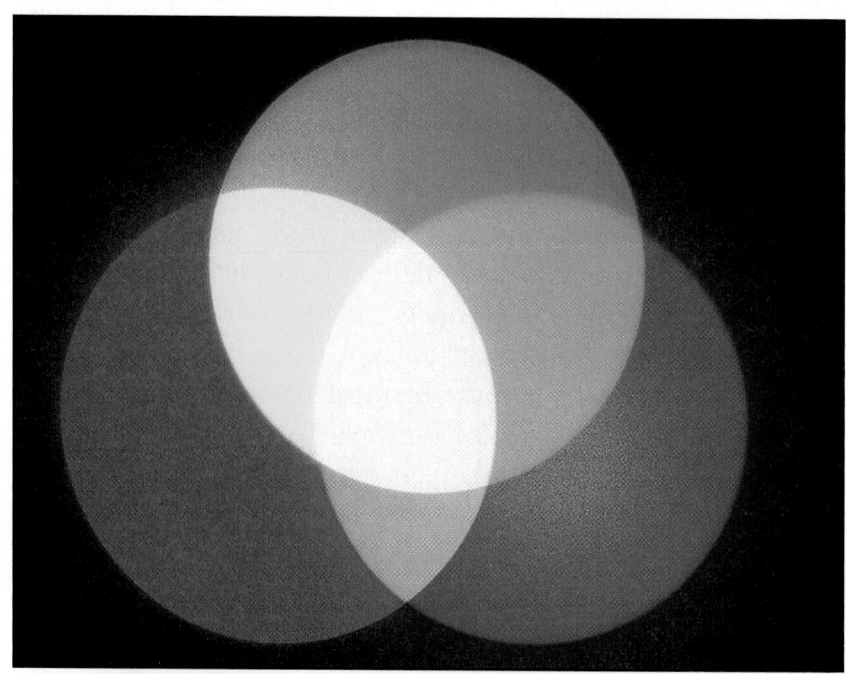

Figure 28 White light is produced when the three primary colors of light are mixed in equal amounts.

Mixing Pigments You can make any pigment color by mixing different amounts of the three primary pigments—magenta (bluish red), cyan (greenish blue), and yellow. In fact, color printers use these pigments, as well as black ink, to make full-color prints like the pages in this book. A primary pigment's color depends on the wavelengths of the light that it reflects. Actually, pigments both absorb and reflect a range of colors in sending a single color message to your eye. For example, in white light, yellow pigment appears yellow because it reflects red and green light but absorbs the other wavelengths of visible light. The color of a mixture of two primary pigments is determined by the primary colors of light that both pigments reflect.

Look at **Figure 29.** The area in the center where the colors all overlap appears to be black because the three blended primary pigments absorb all the primary colors of light. Recall that the primary colors of light combine to produce white light. They are called additive colors. However, when the primary pigment colors are combined, they absorb all wavelengths of visible light and produce black. Because black results from the absence of reflected light, the primary pigments are called subtractive colors.

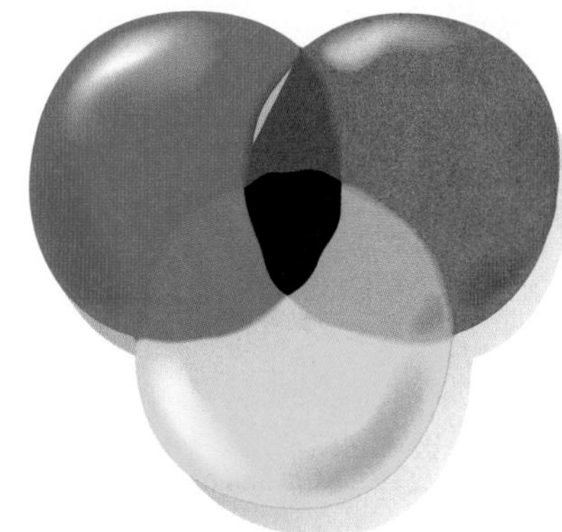

Figure 29 The three primary pigment colors appear to be black when they are mixed.
Describe *how the primary pigment colors are similar to the primary colors of light.*

section 4 review

Summary

Why Objects Have Color

- The color of an object is determined by the wavelengths of light it reflects.
- The color of a filter is the color of the light the filter transmits.

Seeing Color

- Rod and cone cells are light-sensitive cells found in the retina of the human eye.
- Rod cells are sensitive to dim light. Cone cells enable colors to be seen.
- There are three types of cone cells. One type responds to red light, another to green light, and another to blue light.

Mixing Colors

- Red, green, and blue are the primary light colors. Any color of light can be created by mixing these primary light colors.
- Any pigment color can be formed by mixing the primary pigment colors—magenta, cyan, and yellow.

Self Check

1. **Identity** what colors are reflected and what colors are absorbed if a white light shines on a red shirt.
2. **Discuss** how the primary colors of light differ from the primary pigment colors.
3. **Explain** why a red apple would appear black if you looked at it through a blue filter.
4. **Determine** why a white fence appears to be white instead of multicolored if all colors are present in white light.
5. **Think Critically** Light reflected from an object passes through a green filter, then a red filter, and finally a blue filter. What color will the object seem to be?

Applying Math

6. **Use Percentages** In the human eye there are about 120,000,000 rods. If 90,000,000 rods trigger at once, what percentage of the total number of rods triggered?
7. **Convert Units** The wavelengths of light are measured in nanometers (nm) which equals 0.000001 mm. Find the wavelength in mm of a light wave that has a wavelength of 690 nm.

More Section Review gpescience.com

Design Your Own

Blocking Noise Pollution

Goals

- **Design** an experiment that tests the effectiveness of various types of barriers and materials for blocking out noise pollution.
- **Test** different types of materials and barriers to determine the best noise blocks.

Possible Materials

radio, CD player, horn, drum, or other loud noise source
shrubs, trees, concrete walls, brick walls, stone walls, wooden fences, parked cars, or hanging laundry
sound meter
meterstick or metric tape measure

Real-World Problem

What loud noises do you enjoy, and which ones do you find annoying? Most people enjoy a music concert performed by their favorite artist, or the booming displays of fireworks on the Fourth of July. Most people enjoy these loud sounds for short periods of time. However, certain other loud noises, such as traffic, sirens, and loud talking, can be annoying. Constant, annoying noises are called noise pollution. What can be done to reduce noise pollution? What types of barriers will best block out noise pollution?

Form a Hypothesis

Based on your experiences with loud noises, form a hypothesis that predicts the effectiveness of different types of barriers at blocking out noise pollution.

Test Your Hypothesis

Make a Plan

1. Complete the safety form.
2. **Decide** what type of barriers or materials you will test.
3. **Describe** exactly how you will use these materials.

4. **Identify** the controls and variables you will use in your experiment.
5. **List** the steps you will use and describe each step precisely.
6. **Prepare** a data table in your Science Journal to record your measurements.
7. **Organize** the steps of your experiment in logical order.

Follow Your Plan

1. Ask your teacher to approve your plan and data table before you start.
2. **Conduct** your experiment as planned.
3. **Test** each barrier two or three times.
4. **Record** the results from each test in your data table in your Science Journal.

Analyze Your Data

1. **Identify** the barriers that most effectively reduced noise pollution.
2. **Identify** the barriers that least effectively reduced noise pollution.
3. **Compare** the effective barriers and identify common characteristics that might explain why they reduced noise pollution.
4. **Compare** the natural barriers you tested with the artificial barriers. Which type of barrier best reduced noise pollution?
5. **Compare** the different types of materials the barriers were made of. Which type of material best reduced noise pollution?

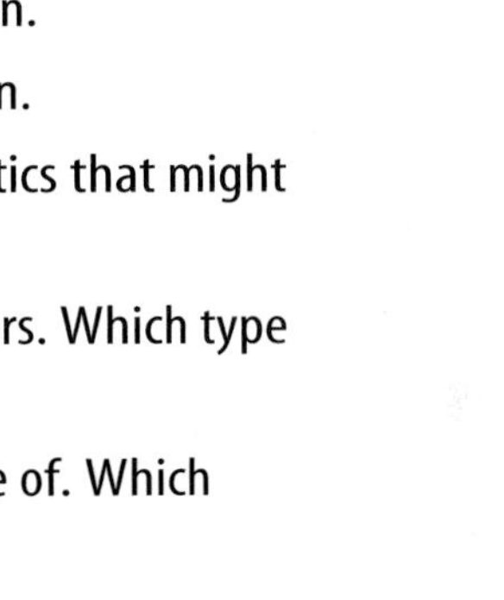

Conclude and Apply

1. **Evaluate** whether your results support your hypothesis.
2. **Predict** how your results would differ if you used a louder source of noise such as a siren.
3. **Infer** from your results how people living near a busy street could reduce noise pollution.
4. **Identify** major sources of noise pollution in or near your home. How could they be reduced?
5. **Research** how noise pollution can be unhealthy.

Draw a poster illustrating how builders and landscapers could use certain materials to better insulate a home or office from excess noise pollution.

Science and Language Arts

A Haiku Garden:

The Four Seasons in Poems and Prints

by Stephen Addiss with Fumiko and Akira Yamamoto

Withered by winter
the sound of the wind—
one-color world

Basho

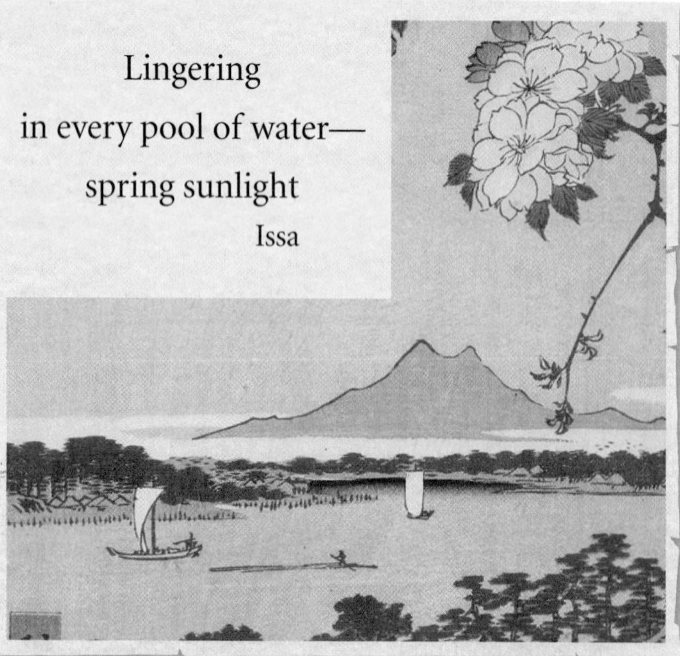

Lingering
in every pool of water—
spring sunlight

Issa

Understanding Literature

Japanese Haiku A haiku is a verse that consists of three lines and 17 syllables in the Japanese language. The first and third lines have five syllables each, and the middle line has seven syllables. Why is imagination important in reading haiku?

Respond to the Reading

1. How do the illustrations help the reader better understand the poems?
2. What do you think is meant by the word *lingering* in the Haiku about spring sunlight?
3. **Linking Science and Writing** Write one haiku about summer and another about fall. In one poem, use color to help you describe the season. In the other, use light or some property of light to help describe the season.

Research has determined that there is a connection between color and mood. Warm colors have longer wavelengths and can be more stimulating. Cool colors, which have shorter wavelengths, tend to have a calming or soothing effect on people. Light and color have long been used as literary symbols. Does the use of color change what you imagine when you read the haiku?

Reviewing Main Ideas

Section 1 Sound

1. Sound waves are compressional waves that travel only in matter.
2. The speed of sound in a material depends on the type of material as well as its temperature.
3. Loudness depends on a sound wave's intensity. Pitch depends on a sound wave's frequency.
4. Sound waves are used in echolocation, sonar, and medical imaging.

Section 2 Reflection and Refraction of Light

1. Light can be absorbed, reflected, or transmitted by a material.
2. When light waves are reflected, they obey the law of reflection—the angle of incidence equals the angle of reflection.

3. A light wave is refracted, or bent, when it changes speed as it travels at an angle from one material to another.

Section 3 Mirrors, Lenses, and the Eye

1. Plane mirrors form upright, virtual images.
2. Images formed by concave mirrors and convex lenses depend on the location of the object relative to the focal point.
3. Convex mirrors and concave lenses form virtual, upright images that are smaller than the object.
4. The lens in the human eye changes shape to produce a sharp image on the retina.

Section 4 Light and Color

1. You see color when light is reflected off objects and into your eyes.
2. Cone cells in the retina are light-sensitive cells that enable you to distinguish colors.
3. Red, blue, and green are the three primary colors of light and can be mixed to form all other colors.
4. The primary pigment colors are magenta, cyan, and yellow.

FOLDABLES™ Use the Foldable you made at the beginning of the chapter to help you review sound and light.

chapter 11 Review

Using Vocabulary

concave lens p. 335	loudness p. 322
concave mirror p. 332	opaque p. 327
convex lens p. 333	pigment p. 342
convex mirror p. 333	pitch p. 323
decibel p. 322	plane mirror p. 331
Doppler effect p. 323	translucent p. 327
index of refraction p. 329	transparent p. 327
intensity p. 322	

Complete each statement with the correct vocabulary word or phrase.

1. A change in frequency due to a moving sound source is due to the ________.
2. A flat, smooth surface that reflects light and forms an image is a(n) ________.
3. An object is ________ if you can see through it clearly.
4. The ________ of a material indicates how much the speed of light in the material changes as light passes through.
5. You can change the color of white paint by adding a(n) ________ to it.
6. A(n) ________ is a reflecting surface that curves outward like the back of a spoon.
7. The human perception of the intensity of a sound wave is ________.

Checking Concepts

Choose the word or phrase that best answers each question.

8. Which of the following best describes image formation by a plane mirror?
   A) A real image is formed in front of the mirror.
   B) A real image is formed behind the mirror.
   C) A virtual image is formed in front of the mirror.
   D) A virtual image is formed behind the mirror.
9. For a sound with low pitch, what else is also always low?
   A) amplitude **C)** wavelength
   B) frequency **D)** wave velocity
10. Which of the following occurs when a sound source moves away from you?
    A) The sound's velocity decreases.
    B) The sound's loudness increases.
    C) The sound's frequency decreases.
    D) The sound's frequency increases.

Use the table below to answer question 11.

Speed of Sound in Air

Temperature (°C)	Sound Speed (m/s)
0	331.4
10	337.4
20	343.4

11. Based on the data in the table above, which of the following would be the speed of sound in air at 30°C?
    A) 340.4 m/s **C)** 349.4 m/s
    B) 346.4 m/s **D)** 353.4 m/s
12. What happens to a light ray traveling parallel to the optical axis of a convex lens that passes through the lens?
    A) It travels parallel to the optical axis.
    B) It passes through the focal point.
    C) It is bent away from the optical axis.
    D) It forms a virtual image.
13. Light waves of which color are bent most when passing through a prism?
    A) blue **C)** violet
    B) red **D)** yellow
14. Which way does a concave lens bend light?
    A) toward its optical axis
    B) toward its center
    C) toward its edges
    D) toward its focal point

Interpreting Graphics

Use the illustration below to answer question 15.

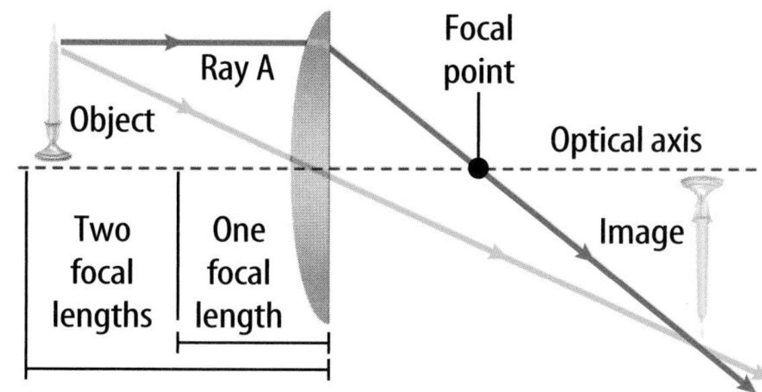

15. **Describe** how the position of the candle changed if the image of the candle moved away from the focal point.

Thinking Critically

16. **Apply** Acoustic scientists sometimes do research in rooms that absorb all sound waves. How could such a room be used to study how bats find their food?

17. **Compare** White light passes through a translucent pane of glass and shines on a shirt. Both the translucent glass and the shirt appear green. Compare the colors of light that are absorbed and transmitted by the glass and the shirt.

18. **Infer** Most mammals, including dogs and cats, can't see colors. Infer how the retina of a cat's eye might be different from the retina of a human eye.

19. **Determine** whether a convex lens could form an image that is enlarged, real, and upright.

20. **Infer** A car comes to a railroad crossing. The driver hears a train's whistle and its pitch becomes lower. What can be assumed about how the train is moving?

21. **Compare and contrast** the reflection of light from a white wall with a rough surface with the reflection of light from a mirror.

22. **Infer** why a convex mirror and a concave lens can never produce a real image.

23. **Predict** what color a white shirt would appear to be if the light reflected from the shirt passed through a red filter and then through a green filter.

24. **Explain** why windows might begin to rattle when an airplane flies overhead.

25. **Communicate** Some people enjoy using snowmobiles. Others object to the noise that they make. Write a proposal for a policy that seems fair to both groups for the use of snowmobiles in a state park.

Applying Math

Use the wave speed equation, $v = f\lambda$, to answer questions 26 and 27.

26. **Calculate Frequency** A sonar pulse has a wavelength of 3.0 cm and a speed in water of 1,500 m/s. Find its frequency.

27. **Calculate Wavelength** What is the wavelength of a sound wave with a frequency of 440 Hz if the speed of sound in air is 340 m/s?

28. **Calculate Angle of Incidence** A light ray is reflected from a mirror. If the angle between the incident ray and the reflected ray is 136°, what is the angle of incidence?

29. **Determine Object Distance** You hold an object in front of a concave mirror with a focal length of 30 cm. If you do not see a reflected image, how far from the mirror is the object?

30. **Calculate Speed** The speed of light in a vacuum is 300,000 km/s. If the index of refraction of water is 1.33, what is the speed of light in water?

Standardized Test Practice

Record your answers on the answer sheet provided by your teacher or on a sheet of paper.

Multiple Choice

1. Which of the following describes the image formed by a convex mirror?

A. enlarged

B. inverted

C. real

D. virtual

Use the figure below to answer question 2.

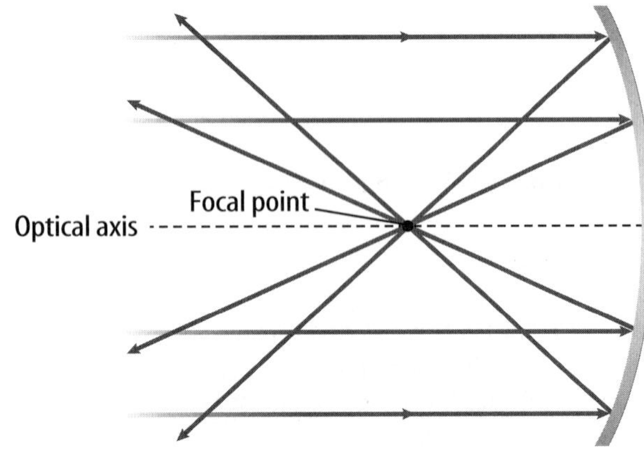

2. Which of the following describes a light ray that passes through the focal point and then is reflected by the mirror?

A. It travels parallel to the optical axis.

B. It forms a real image.

C. It is reflected back through the focal point.

D. It forms a virtual image.

3. Why does an apple look red?

A. It reflects red light.

B. It absorbs red light.

C. It reflects green and blue light.

D. It reflects all colors of light except red.

Use the figure below to answer questions 4 and 5.

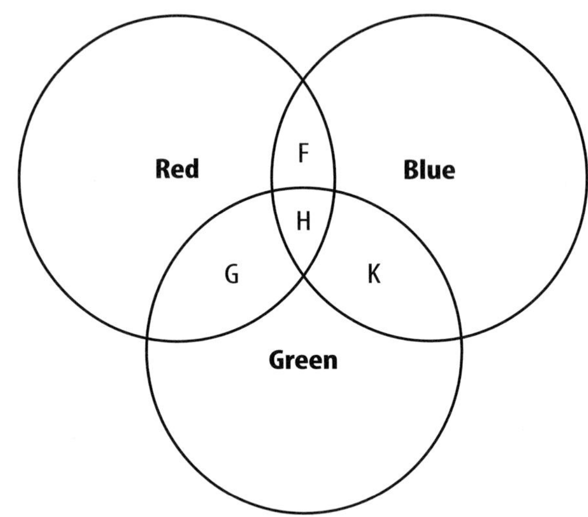

4. When the primary colors of light are added together, what color appears in area H?

A. cyan

B. magenta

C. white

D. yellow

5. When the primary colors of light are added together, what color appears in area G?

A. cyan

B. magenta

C. white

D. yellow

6. Which of the following increases as the frequency of a sound increases?

A. amplitude

B. intensity

C. loudness

D. pitch

7. Which of the following parts of the eye enable you to see colors in bright light?
   A. cone cells
   B. cornea
   C. lens
   D. rod cells

Gridded Response

Use the graph below to answer questions 8 and 9.

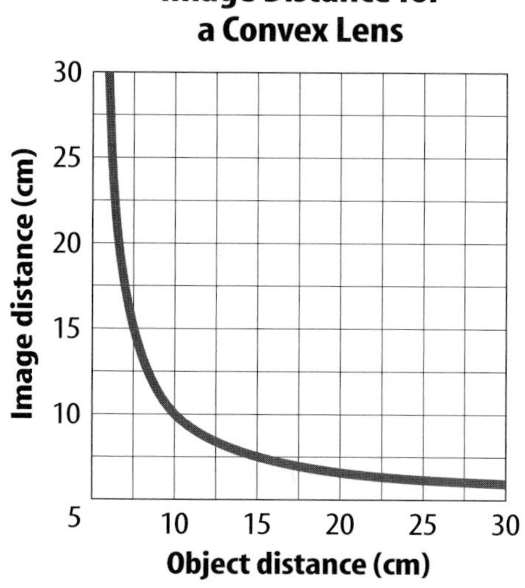

8. Determine how far, in centimeters, the image is from the lens when the object is 15 centimeters from the lens.

9. How far, in centimeters, is the object from the lens when the image distance and object distance are equal?

Test-Taking Tip

Note Units Read carefully and make note of the units used in any measurement.

Short Response

10. A shirt has stripes that are black and white in white light. If green light shines on this shirt, what colors will the black and white stripes be?

11. If you are on a moving train, what happens to the pitch of a crossing bell as you approach the crossing and then move away from the crossing?

Extended Response

Use the figure below to answer question 12.

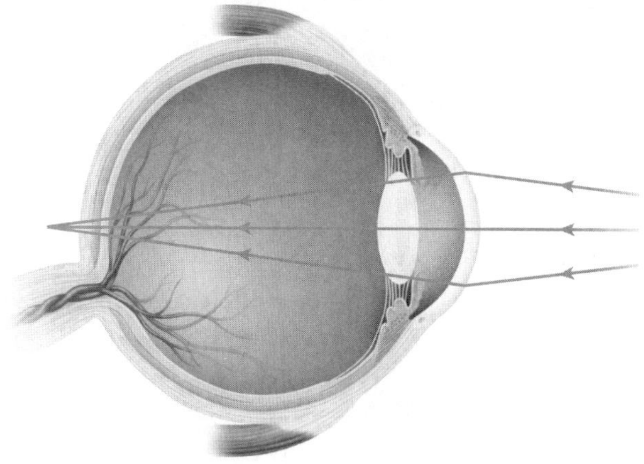

12. Describe the vision problem show by the figure and explain how this vision problem can be corrected.

13. Explain why the human eye can see colors better in bright light than in dim light.

chapter 12

Earth's Internal Processes

chapter preview

sections

Italian Fireworks

Sicily's Mt. Etna is the largest volcano in Europe. Molten rock from deep in Earth rose through a weak spot in the crust and erupted many times onto Earth's surface. Some of the exposed lava at the base of Mount Etna is nearly 300,000 years old.

Science Journal Research the most recent eruption and write a report about how it affected the surrounding environment.

Start-Up Activities

Global Jigsaw Puzzle

Alfred Wegener, a German scientist, noticed that the shapes of continental coastlines appeared as though they could match up. He suggested that the continents once were together as one giant landmass. Use a map of the world to test this idea. Can you see what he saw?

1. Cut out continents from a copy of a world map provided by your teacher.
2. Try to arrange continents so that they fit together.
3. Infer why the fit might not be perfect.
4. **Think Critically** What changes in procedure might demonstrate a better fit for the continents?

Preview this chapter's content and activities at gpescience.com

Systems Make the following Foldable to help you organize information about types of plate boundaries.

STEP 1 **Fold** one piece of paper lengthwise into thirds.

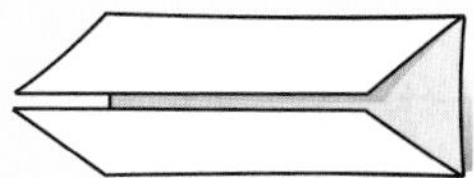

STEP 2 **Fold** the paper widthwise into fourths.

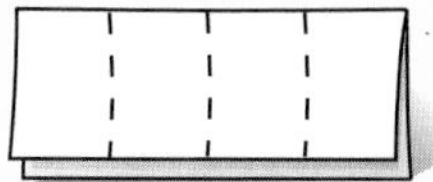

STEP 3 **Unfold,** lay the paper lengthwise, and draw lines along the folds. **Label** your table as shown.

Plate Boundary Type	Description	Illustration
Divergent		
Convergent		
Transform		

Making a Table As you read the chapter, complete the table describing and illustrating divergent, convergent, and transform plate boundaries.

section 1

Evolution of Earth's Crust

Reading Guide

What You'll Learn
- **Explain** supporting evidence for the continental drift hypothesis.
- **Discuss** the failings of the continental drift hypothesis.

Why It's Important
Wegner's continental drift hypothesis led to a unifying theory of Earth sciences known as plate tectonics.

Review Vocabulary
hypothesis: statement proposed to explain an observation or answer a question

New Vocabulary
- mid-ocean ridge
- rift valley
- divergent boundary
- convergent boundary
- subduction
- transform boundary

Continental Drift

In the early twentieth century, there was no single theory of how Earth processes interrelated. Much geologic study was done locally because transportation and communication were expensive. Based upon their observations, geologists developed theories that emphasized vertical changes, for example, an erosion process that leveled high places, and a mountain-building process that lifted them up again.

Then in 1915, Alfred Wegener (VEG nur) proposed a hypothesis that suggested that Earth's continents once were part of a large super-continent, shown in **Figure 1,** called Pangaea (pan GEE uh). Then, about 200 million years ago, the super-continent broke into pieces that drifted over the surface of Earth like rafts on water. This revolutionary idea of horizontal movement met with great resistance among his peers. He was unable to find the force capable of moving continents. It wasn't until after his death in 1930, that scientific advances finally justified his hypothesis.

Figure 1 This illustration is an artist's conception of what Pangaea may have looked like 200 million years ago.

Matching Coastlines The most apparent match of continents is the eastern coastline of South America with the western coastline of Africa. If you use your imagination, you can see that the coastline of northwestern Africa fits nicely with that of the eastern United States. When South America and Africa are joined together, their southern tips fit very well into the Weddell Sea of Antarctica.

Figure 2 Weathering of the continental edges does not affect the continental shelves (light blue).

Wegener had to show that the continents were actually joined. He used the analogy of a torn newspaper being repaired. Not only did you have to match the shapes, but also join the lines of print. And the print had to match in terms of its content as well. What kind of content could this refer to in coastal regions of the continents? Wegener argued that you could match rock types, fossils, erosion features, and mountain ranges. If you found similar formations and structures on each continent then the continents could have been joined together in that place.

Wegener's opponents pointed out that the coastlines are constantly wearing away due to wave action. How could someone compare the present coastlines? Years later, during the revival of the hypothesis, oceanographers were able to show, using sonar, that the edges of the continental shelves matched very well, as shown in **Figure 2.**

Matching Fossils Wegener could not use the remains of just any ancient living thing to support the existence of Pangaea. For instance, animals that could fly or swim could appear in the fossil record in widely separated places due to their mobility, not because the places were necessarily joined. Large land animals provided better evidence because they could not have crossed oceans. Animals such as *Lystrosaurus* or *Cynognathus,* large animals that preceded the dinosaurs, supported a contiguous landmass. *Glossopteris,* a large fern with large, heavy spores also supported the idea of Pangaea. You can see in **Figure 3** that these living things were widely distributed.

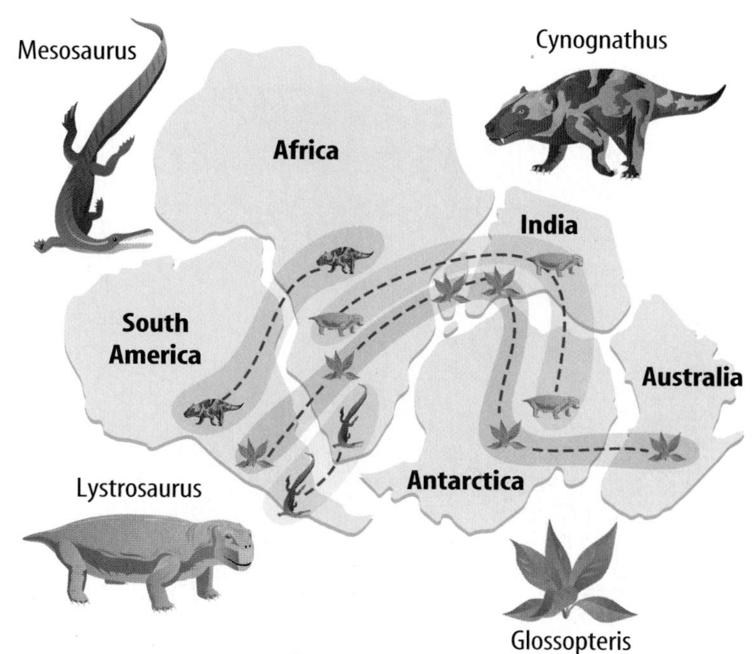

Figure 3 Wegener chose fossils of animals that could not swim or fly to prove Pangaea's existence. **Explain** *Why would being able to fly or swim eliminate a fossil organism from Wegener's proof?*

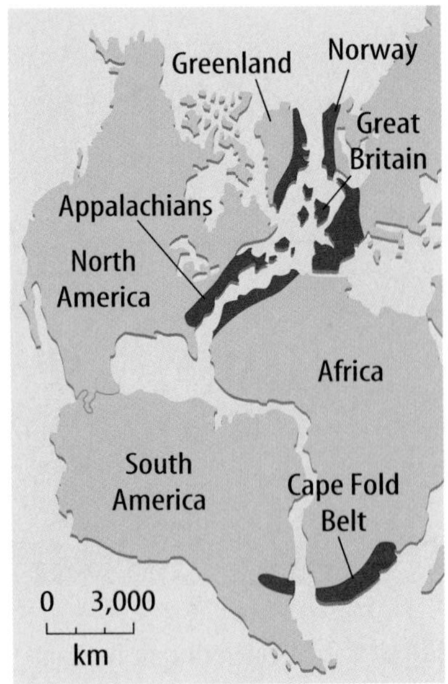

Figure 4 Wegener's hypothesis showed mountains on several continents were once part of the same range.

Matching Rocks and Mountains Mountain ranges were shown to be continuous in Pangaea, as shown in **Figure 4.** Once Pangaea broke apart, the mountain ranges became separated. For decades, geologists studied and attempted to explain the origin of these mountains as separate ranges; Wegener showed them to be one mountain range. Wegener was able to show that continents that were joined shared unique rocks and minerals.

Wegener's hypothesis was not accepted by his contemporaries because he was unable to conceive of a force or mechanism that could drive continents apart. Wegener reasoned that Earth's rotation, the gravitational pull of the Sun and the Moon, and centrifugal force could move continents. Physicists quickly showed that even combining these forces would not be sufficient.

Reading Check *Why didn't Wegener's contemporaries accept his hypothesis?*

Wegener used the analogy of continents moving over Earth's surface as ships moving through water. Skeptics argued that this ship would push a wall of water ahead of it and leave a wake. The continents didn't leave a wake. Instead of the continents pushing up a wall of water, they were deformed. How could this be if they were thicker and stronger?

Seafloor Spreading Hypothesis

After World War II, Dr. Harry Hess revived Wegener's ideas. He used sonar, intended to detect submarines, to obtain accurate maps of the seafloor. Using sonar data, astonishing three-dimensional seafloor models were created in 1960. Soon it was apparent that a **mid-ocean ridge** system, or MOR, was continuous and wrapped around Earth. A MOR is shown in **Figure 5.**

Hess proposed a hypothesis of seafloor spreading, or divergence. He suggested that magma from the mantle is forced upward because of its low density. This causes the crust to crack (fault) and move apart. The faulting causes twin mountain ranges with a down-dropped **rift valley** between. This continuous process allows new rock to form as magma fills in from below. See **Figure 6.**

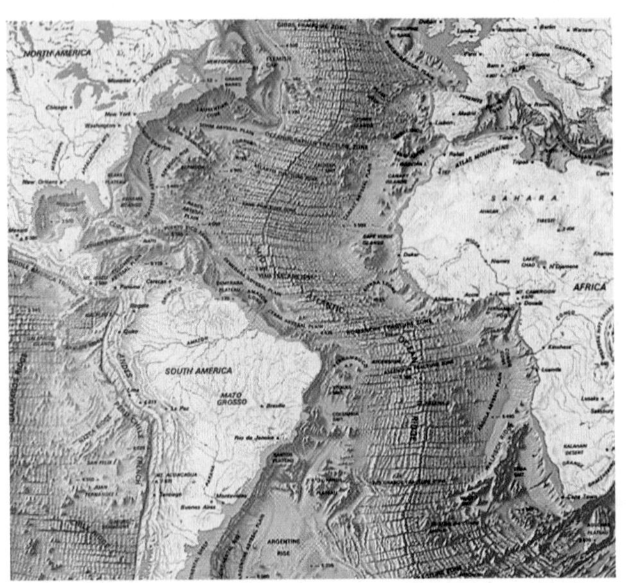

Figure 5 Discovery of a mid-ocean ridge led Hess to hypothesize seafloor spreading.

Ages of Sediment and Rocks In the early 1960s, massive programs for drilling into the seafloor began. Extracted cores of seafloor showed that sediments are thicker on top of seafloor basalt near the continents. MOR sediments, however, are thin. Cores of both sediments found that near the continents the oldest sediments are at the bottom and young sediments are at the top. MOR sediments are all of recent age. When the ages of rocks are measured, the continental rocks are billions of years old, while seafloor rocks are less than 200 million years of age. Rocks of the oceanic crust increase in age as their location extends from the MOR, and at the MOR they are new.

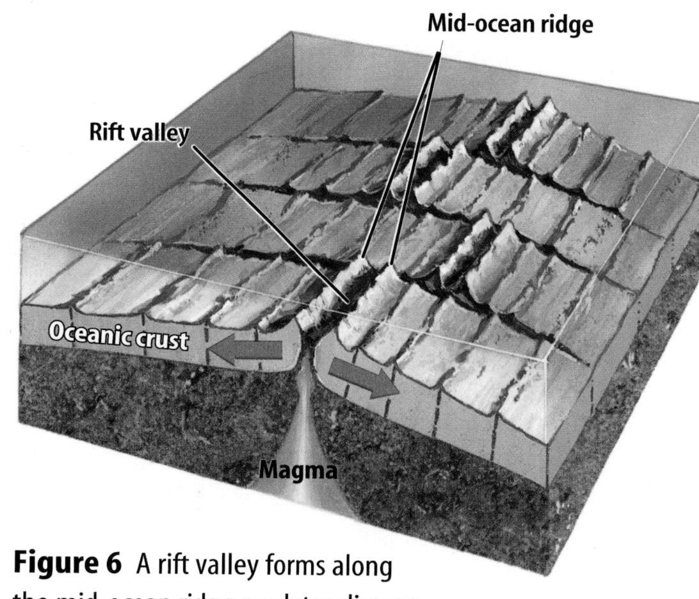

Figure 6 A rift valley forms along the mid-ocean ridge as plates diverge.

Applying Math — Solve One-Step Equations

SPREADING DISTANCES The spreading rate along the mid-ocean ridge varies. In the Atlantic Ocean, it averages about 2.5 cm/year. About how far, in kilometers, would the Atlantic seafloor have widened after 100 million years?

IDENTIFY known values and unknown values

Identify the known values:

time = 100,000,000 years

spreading rate = 2.5 cm/year

Identify the unknown values:

spreading distance

SOLVE the problem

Substitute the known values into the equation:

distance = rate × time

distance = 2.5 cm/year × 100,000,000 years = 250,000,000 cm

1 km/100,000 cm × 250,000,000 cm = 2,500 km

CHECK the answer

Does your answer seem reasonable? How wide is the Atlantic Ocean?

Practice Problems

If the East Pacific Rise spreads at 12 cm/year, how wide will it be in 10 million years?

For more practice problems, go to page 879 and visit Math Practice at gpescience.com.

Sonar In the process of refining sonar's capabilities, discoveries were made that had peacetime benefits. For example, sonar often is equipped on boats to locate schools of fish. A related technique called ultrasound is used in medicine. Research current uses and applications of sonar and ultrasound.

Magnetic Polarity of Rocks Studies show that Earth's magnetic field repeatedly reverses itself, meaning that the magnetic north pole becomes the south pole. Vine, Matthews, Wilson, et al discovered bands of reversed polarity in the seafloor rocks similar to those found on the continents. As magma crystals form, they take on the polarity of Earth at the time they form. The pattern is identical on both sides of the MOR.

Theory of Plate Tectonics

Originating in the 1960s, the theory of plate tectonics is relatively new. After seafloor spreading demonstrated that Earth's crust moved horizontally on a global scale, many investigators were determined to understand such a whole-Earth system of movement. This system consists of about a dozen major plates and many minor ones. Plates are composed of a rigid layer of uppermost mantle and a layer of either oceanic or continental crust above. Some plates are composed only of oceanic crust, and some are composed of part oceanic and part continental crust. J. T. Wilson is credited with describing the cycle of repeated opening and closing of ocean basins through Earth's history.

There are three main kinds of plate motions. These are best visualized by considering how plates interact along plate boundaries, where they meet. Plates can move apart, move together, or slide past one another. Although often visualized as narrow boundaries, scientists now consider many boundaries to be wide zones of interaction.

Figure 7 Large lakes and volcanic mountains are characteristics of a continental rift valley.

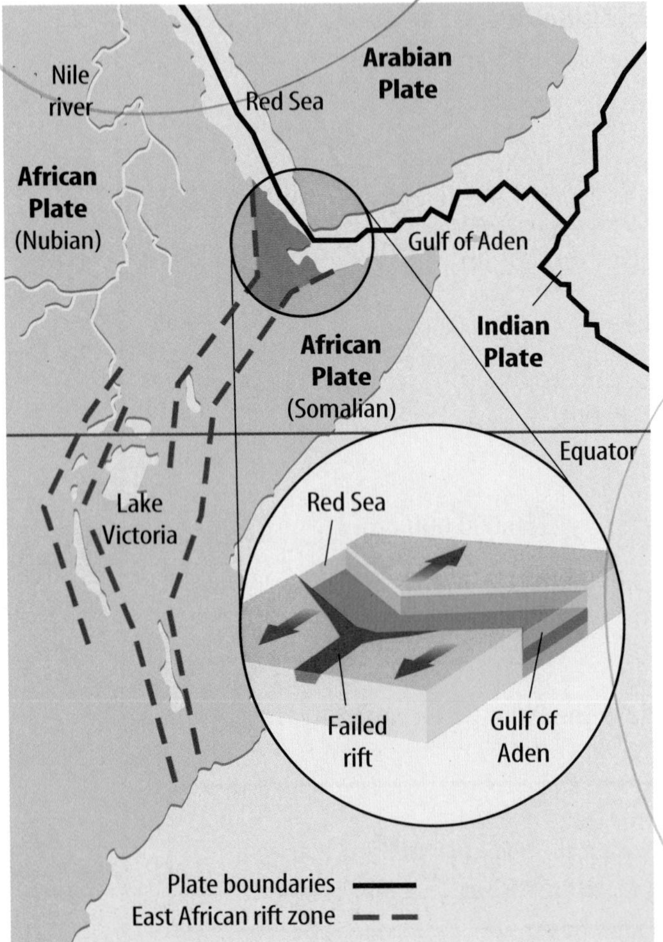

Divergent Plate Boundaries You learned that at a mid-ocean ridge (MOR), magma rises along a faulted rift valley, spreads, and cools to form new oceanic crust. This spreading apart is what happens at **divergent boundaries.** An MOR represents divergence that is well-developed and that has resulted in the production of major ocean basins. In some locations on Earth today, divergent boundaries exist as rift valleys, where no mature ocean basins exist yet, such as in East Africa, shown in **Figure 7.**

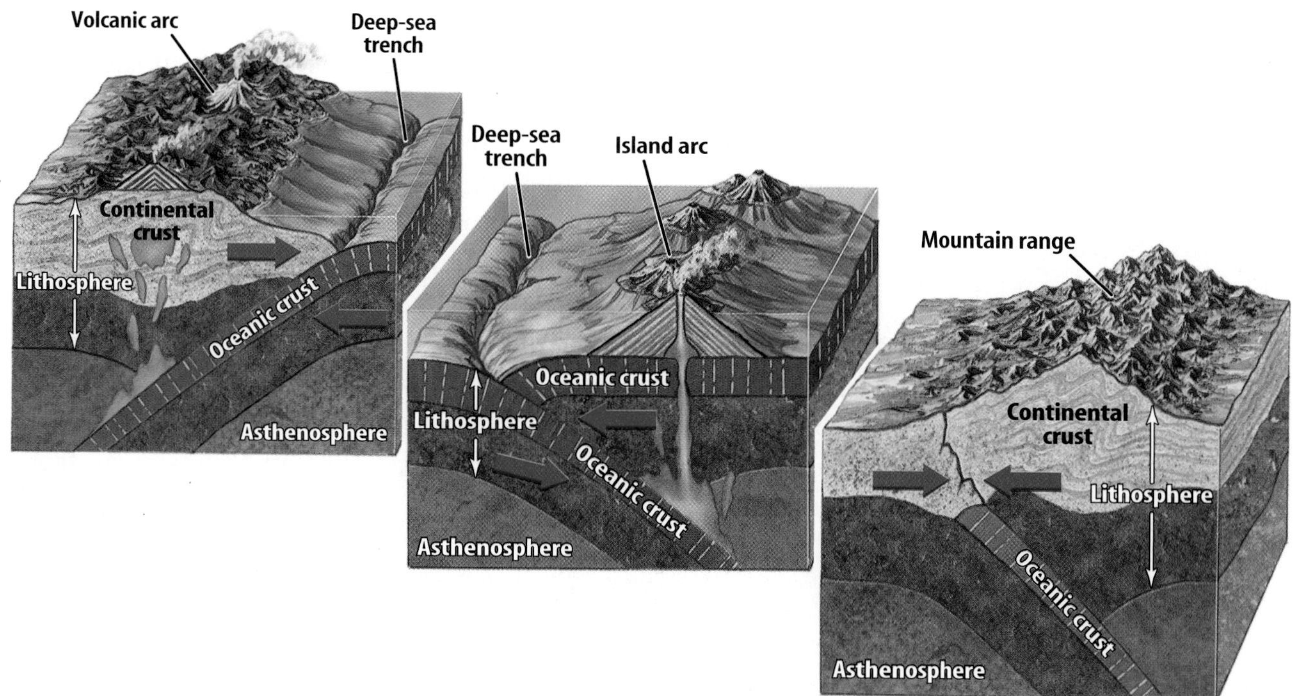

Figure 8 When plates collide, the more dense plate is subducted. The resulting features include volcanoes, mountains, and deep trenches.

Convergent Plate Boundaries Where plates collide, they come together to form **convergent boundaries.** In some cases, less-dense, thick continental lithosphere moves toward denser, thin oceanic lithosphere. This results in the oceanic side bending and being forced downward beneath the continental slab in a process called **subduction.** Heat along a subduction zone partially melts rock at depth and produces magma, which rises toward the surface. This magma feeds a volcanic arc that parallels this zone, shown in **Figure 8.** The region of collision also has a deep-sea trench that parallels the zone. The Andes mountain range in South America is an example.

Convergent plate boundaries also exist between two slabs of oceanic lithosphere. In this case, the oceanic lithosphere that is colder, and therefore denser, subducts. Magma erupted here produces chains of volcanic islands called island arcs. Japan is an example of an ocean-ocean convergent boundary, also shown in **Figure 8.** As plates converge, stress builds, which could be released as tsunami-causing earthquakes.

Along some convergent plate boundaries, two continental slabs of low density collide and tend not to subduct. Because of this resistance to subduction, the plates collide and buckle upward to form a high range of folded mountains. Volcanic activity is noticeably absent and there is no trench. The Himalaya of Asia are an example of folded mountains that occur where continental lithosphere collides.

What is formed when two continental plates converge?

Topic: Tsunamis
Visit gpescience.com for Web links to information about tsunamis.

Activity Research the most recent tsunami and its destruction. Record the epicenter and magnitude of the earthquake that caused the tsunami and post these data on a world map. Write a short report that describes the tsunami's impact on humans and the environment.

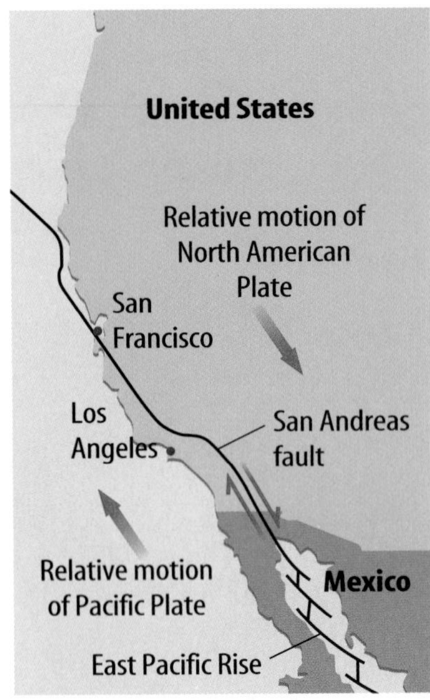

Figure 9 Friction between plates moving side by side causes cracks and breaks in the edges of the plates. This is the site of brief, but rapid energy release called an earthquake.

Transform Plate Boundaries Some boundaries among plates exist as large faults, or cracks, along which mostly horizontal movement is taking place, as shown in **Figure 9.** In this case, no new lithosphere is forming, as along a divergent boundary. In addition, old lithosphere is not being recycled, as along a subduction zone. The main result of **transform boundaries** is horizontal motion of lithosphere.

Transform faults are extremely important where they cut perpendicular to the MOR. These fault systems allow movement away from ridge crests to occur, as shown in **Figure 10.** If you observe arrows that indicate net motion along these transform faults, you will notice that this net motion trends away from the MOR.

What drives the plates?

Research indicates that plates are driven by a combination of forces. One such force is ridge push at the MOR. Because divergent boundaries are higher at the center of the ridge, gravity forces material down the slopes of the MOR.

When a plate subducts back into Earth at some convergent boundaries, the process of slab pull is thought to operate. You probably have experienced an analogy to slab pull when you found your bed covers on the floor in the morning. During the night, as you tossed and turned, the covers began to move off of the bed. Eventually, enough of the covers were over the side that gravity took over and pulled the rest of the covers to the floor. Subducting plates may act in much the same way, as portions of descending plates are pulling the rest of a plate down with them.

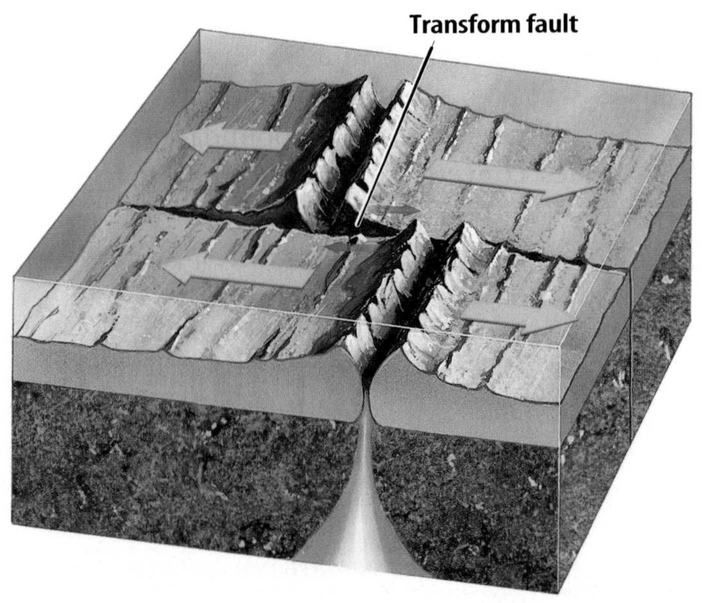

Figure 10 A transform fault cuts through the MOR, offsetting the mountain range.

Friction between a plate and mantle material below the plate probably is of major importance in relative plate motion. For example, plates that drag continental material along with them are noticeably slower than are purely oceanic plates. Scientists think that continental lithosphere has deep roots that cause more frictional force than would be expected at the base of oceanic lithosphere.

Reading Check *What role does friction play in plate motion?*

Thermal Energy Internal convection of mantle material is the driving force for all mechanisms of plate motion. In turn, the main source of thermal energy that keeps Earth materials convecting comes from the decay of radioactive elements in Earth. Increased temperature due to pressure and frictional heating produced as part of the mechanism itself probably are important. Conversion of secondary earthquake waves in the outer core may yet be another source of energy.

section 1 review

Summary

Continental Drift

- Wegener proposed that former supercontinent Pangaea broke up into pieces, which drifted to their present positions.
- Evidence favoring continental drift includes matching shorelines of continents and correlating rocks, fossils, and mountain ranges of those continents.

Seafloor Spreading Hypothesis

- Hess suggested that seafloor was created and spread apart at the mid-ocean ridge.
- Moving away from the MOR, rocks are older and sediments are thicker.
- Magnetic reversals preserved in seafloor rocks are symmetrically distributed on either side of the MOR.

Theory of Plate Tectonics

- Earth's rigid, outermost layers are composed of a dozen or so major plates and many smaller ones.
- New lithosphere is created at the MOR and recycled at convergent boundaries. Convective flow within the mantle drives the plates.

Self-Check

1. **Explain** the processes of convergence and divergence.
2. **Describe** the key features of a divergent boundary.
3. **Compare and contrast** the three types of convergent plate boundaries.
4. **Describe** the possible driving mechanisms in the plate tectonic theory.
5. **Think Critically** Predict what would happen if Earth's plates stopped moving.
6. **Think Critically** What would have to occur to stop Earth's plate movement?

Applying Math

7. If two plates diverge at a rate of 1.3 cm/year, how much farther apart will the plates be after 200 million years?
8. How many times faster are plates moving at 7.3 cm/year than those moving at 1.3 cm/year?
9. The average distance across an ocean is 16,000 km. Two continents on either side of the ocean are converging at a rate of 10 cm/year. How long will it take for them to collide?

More Section Review gpescience.com

section 2

Earthquakes

Reading Guide

What You'll Learn

- **Describe** the causes and characteristics of earthquakes.
- **Explain** how seismic waves affect Earth's surface.
- **Describe** how seismic waves are used to infer Earth's internal structure.

Why It's Important

Earthquakes kill people and destroy property. Understanding earthquakes may help minimize their effects.

Review Vocabulary

friction: force that opposes the sliding motion between two touching surfaces

New Vocabulary

- fault
- elastic rebound
- focus
- epicenter

Global Earthquake Distribution

For decades, scientists have known that earthquakes are not distributed randomly, but rather, they occur in well-defined zones. These zones coincide with the edges of lithospheric plates. In fact, seismic data originating from earthquakes helped to decipher the structure of Earth's ocean floor and to infer the structure and motion of Earth's plates. **Figure 11** shows the distribution of large earthquakes.

Figure 11 Most earthquakes occur along the edges of plates. **Identify** *some other places earthquakes occur.*

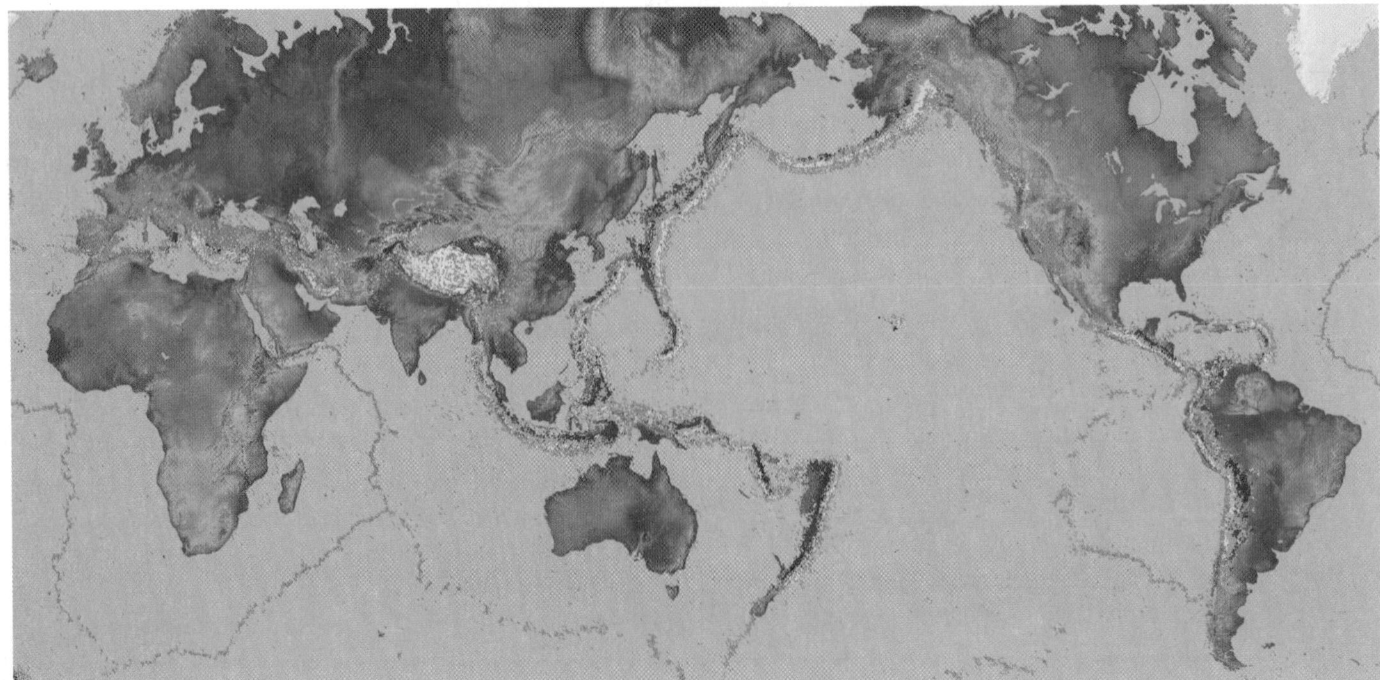

Depth of Focus Patterns develop when data about the focus depths of earthquakes are plotted on a world map. Recall that divergent boundaries are associated with transform faulting that allows plates to move in opposite directions. All of this faulting creates a narrow band of numerous, shallow earthquakes. In contrast, convergent boundaries have broad zones of earthquakes with the shallowest foci near the surface at the point of convergence, and the deepest foci located under volcanoes or mountains created in the collision area, as shown in **Figure 12.**

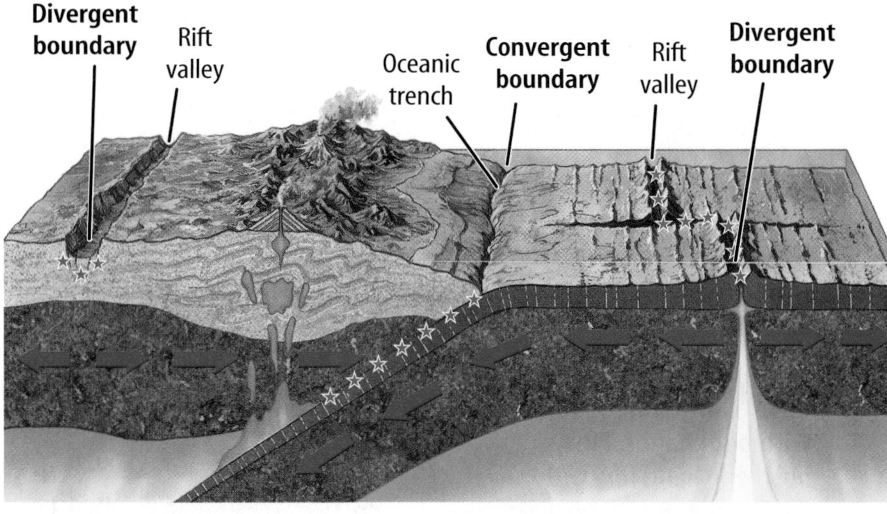

Figure 12 The depth of the earthquake focus (indicated by the stars) is related to the activity causing the earthquake.

Causes of Earthquakes

An earthquake is any seismic vibration of Earth caused by the rapid release of energy. Earthquake events can be either natural or human-caused. Passing trains or large trucks and explosions can cause Earth to vibrate. As shown in **Figure 13,** sudden, virtually unpredictable, natural earthquakes that result in major destruction are greatly feared.

Deformation Earth's crust is composed of rigid, rocky material. Engineers would describe it as brittle. When a stress is applied to a brittle material it shows little sign of strain, or deformation, until it suddenly breaks. A strain is the manner of deformation in response to a stress. Stress is the force per unit area that acts on a material. Stresses can be of four types: (1) compressive stress, in which a mass is squeezed or shortened, (2) a tension stress, in which the mass is stretched or lengthened, (3) a shear stress, in which different parts of a mass are moved in opposite directions along a plane, or (4) torsion stress, in which a mass is subjected to twisting.

Figure 13 This damage was caused when the buildings were shaken off their foundations.
Explain *How might this damage have been prevented?*

Demonstrating Four Types of Stress

Procedure

1. With palms facing down at all times and your hands in contact with each other, clasp a **large bar of taffy** with both hands. First, push one hand forward 2 cm while simultaneously pulling the other backward 2 cm. Return your hands to the original position.
2. Still holding your hands in contact, twist your hands in opposite directions and return them to the original position.
3. Next move your hands about 4 cm apart.
4. Finally push your hands back together to the original position.

Analysis

1. Which type of stress did you demonstrate in each of steps 1–4?
2. Describe the kinds of deformation you would expect to result from each of the four stresses.

Elastic Deformation Elastic deformation occurs when a material deforms as a stress is applied, but snaps back to its origin shape when the stress is removed. Plastic deformation occurs when a material deforms, or changes shape, as a stress is applied and remains in the new shape when the stress is released. Modeling clay behaves plastically. You would expect all rocks to show brittle deformation, which means breaking in response to stress. But rocks at depth, where temperatures are high enough, display plastic behavior. For example, you can break off an edge of a wax candle when it is cold and brittle, but the wax bends more under stress—without breaking—when it warms up.

Energy Release Strain energy builds up along cracks in Earth's crust in response to stress. When this strain energy is released suddenly, it causes rock to lurch to a new position. A **fault** is a crack along which movement has taken place. If no movement takes place, the crack is a fracture. Earthquake-producing faults occur in broad zones in which rock is deformed in a brittle manner during the fault movements. These zones can be tens of meters wide. The sudden energy release that goes with fault movement is called **elastic rebound.** Elastic rebound causes seismic vibrations, or earthquakes, like when you drag a table across the floor and the legs catch and release making a rumbling sound.

Earthquake Waves

Earthquake waves travel out in all directions from a point where strain energy is released. This point is the **focus,** or point of origin, of an earthquake. The point on Earth's surface directly above the focus is the **epicenter.** When you throw a stone into water you see concentric rings of waves move out across the surface from the point of impact. Earthquake waves are much the same, except they move out from the focus in all directions, like a sphere of waves. These ideas are shown in **Figure 14.**

Earthquake waves can be sorted broadly into two major types. Body waves travel through Earth. Surface waves travel across Earth's surface.

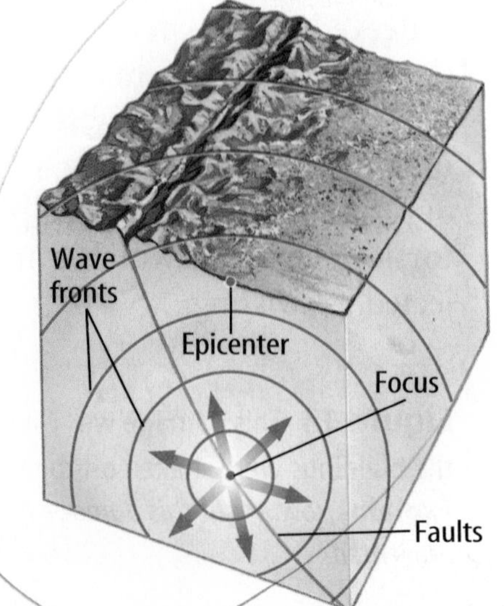

Figure 14 Waves moving out from the focus may travel through the mantle. These may be picked up by seismographs on the opposite side of Earth.

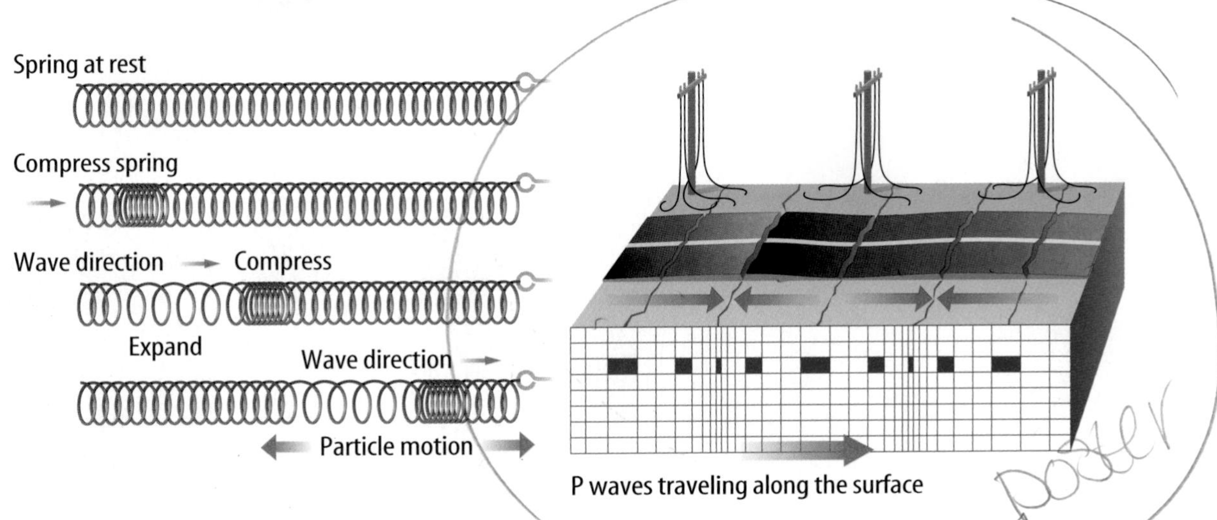

Body Waves One type of body wave is called a primary wave. Primary waves, also called P-waves, cause particles in a material to undergo a push-pull type motion as shown in **Figure 15.** Because this motion is in the direction of wave travel, the wave energy is transferred very quickly. The particles do not permanently change location. If there is matter around where particles can bump into each other, then primary waves can move through it. Much like sound waves, P-waves travel through all kinds of matter.

Secondary waves (S-waves) are body waves that travel more slowly than primary waves. They are sometimes called shear waves, because of the relative motion of particles as energy is transferred. S-waves cause particles to move perpendicular to the direction of wave travel. The farther body waves travel from an earthquake focus, the farther behind the S-waves get. It is this lag in time between the arrival of the first P-waves and the first S-waves that is important in locating epicenters.

S-waves only can travel through solids. When one particle moves, it moves its neighbors along with it. In gases and liquids the bonds are weak, or there are no bonds, so particles can move independently. The motions of S-waves are shown in **Figure 16.**

Figure 15 P-waves are compressional waves like those moving in a coiled spring.
Describe *How are P-waves like sound waves?*

Figure 16 S-waves cause the earth to undulate creating surface damage.

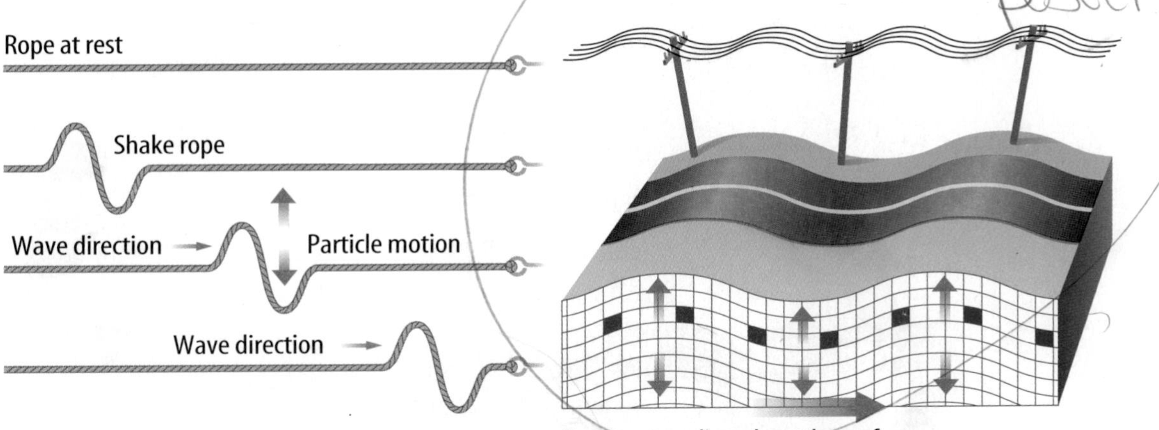

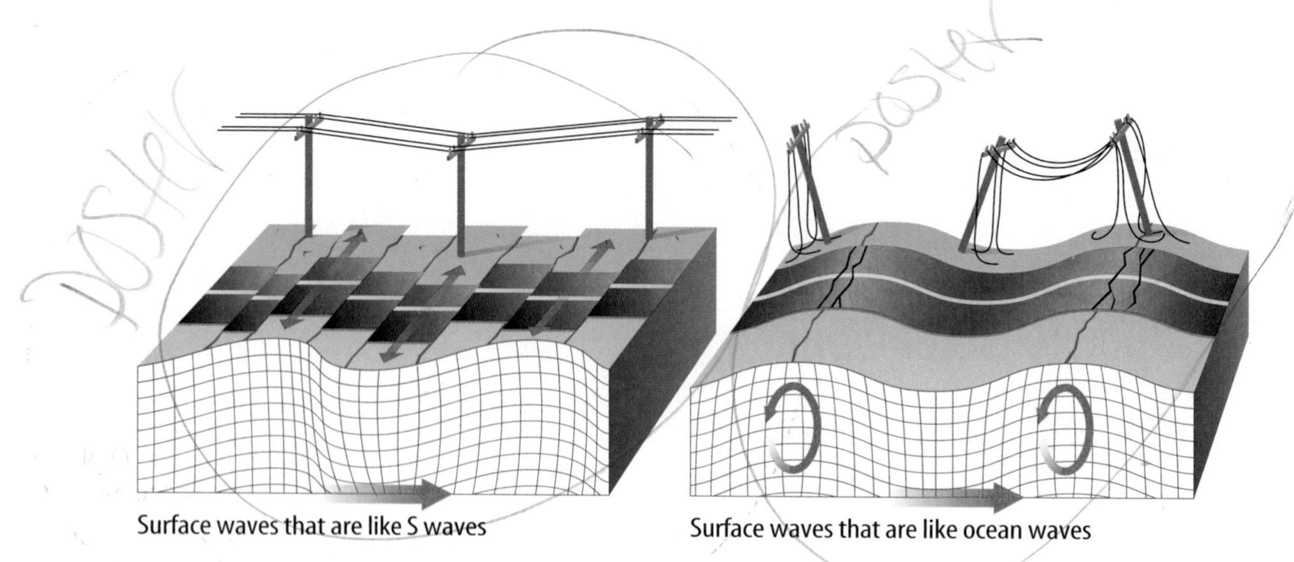

Figure 17 During an earthquake the surface can roll like the ocean and shift side to side at the same time.

Surface Waves Surface waves move in a more complex manner, often causing a rolling motion much like ocean waves. As surface waves travel through material, they can exhibit an up and down rolling motion, and also a side-to-side motion that parallels Earth's surface. Foundations of human-built structures often are susceptible to the side-to-side rocking that might result from surface waves. These surface wave motions are illustrated in **Figure 17.**

Table 1 Estimates of Earthquake Magnitude and Frequency

Richter Magnitude Range	Description Index	Estimated Occurrence per Year
< 2.0	recorded, but not generally felt	600,000
2.0–2.9	potentially felt	300,000
3.0–3.9	felt by some	49,000
4.0–4.9	felt by most	6200
5.0–5.9	damaging	800
6.0–6.9	destructive in densely populated areas	266
7.0–7.9	potential to inflict major damage	18
8.0 and above	potential to destroy communities near epicenter	1.4

Earthquake Measurement

Two measurement schemes that have been used to characterize earthquakes are the Modified Mercalli intensity scale and the Richter magnitude scale. Intensity is a measure of ground shaking and the damage that it causes. The Modified Mercalli scale, **Table 2,** ranks earthquakes in a range from I–XII, XII being the worst, and uses eyewitness observations and post-earthquake assessments to assign an intensity value. The Richter magnitude scale, Richter scale for short, uses the amplitude of the largest earthquake wave. Richter magnitude is intended to give a measure of the energy released during the earthquake. **Figure 18** shows a seismogram and how it is used to determine a Richter value. **Table 1** shows the global frequency of different magnitude earthquakes.

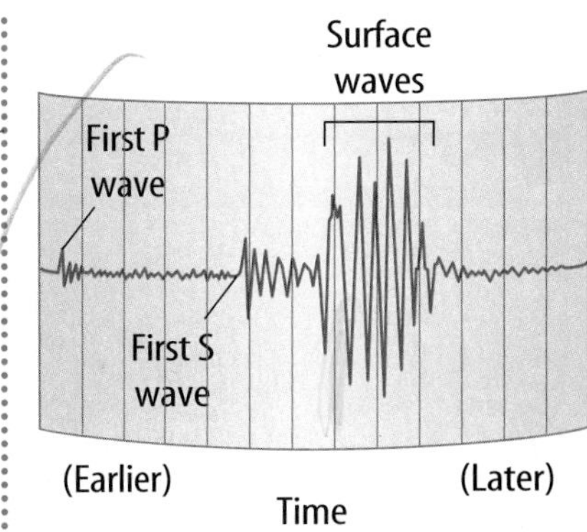

Figure 18 A seismograph is an instrument used to measure earthquake waves. A seismogram is a tracing of the seismograph's pen.

Table 2 The Mercalli Scale of Earthquake Intensity

Level	Description
I	Rarely felt by people.
II	Felt by resting people indoors; some hanging objects may swing.
III	Felt indoors by several. Vibration like passing of a light truck.
IV	Felt indoors by many. Vibration like passing of a heavy truck. Standing autos rock. Windows, dishes and doors rattle. Walls and frames may creak.
V	Felt by nearly everyone indoors and outdoors. Small unstable objects upset. Some dishes and glassware broken. Swaying of tall objects noticed.
VI	Felt by all. Walking is unsteady, many run outdoors. Windows, dishes, and glassware broken. Furniture overturned and plaster may crack.
VII	Difficult to stand. Noticed by drivers of autos. Furniture and chimneys broken. Well built buildings hardly damaged. Poor structures considerable damage.
VIII	People frightened. Ordinary buildings slightly damaged. Driving of autos affected. Tree limbs fractured. Damage to tall objects. Cracks in wet ground.
IX	General panic. Damage great in substantial buildings. Some houses thrown off foundations. Underground pipes broken. Serious ground cracks.
X	Most masonry and frame structures destroyed. Serious damage to dams, dikes, embankments. Water splashed out of rivers, canals, lakes. Rails bent.
XI	Few structures remain standing. Bridges destroyed. Broad fissures in the ground. Slumps and landslides. Rails bent generally.
XII	Damage nearly total. Waves seen on ground surfaces. Lines of sight and level distorted. Objects thrown into the air. Large rock masses displaced.

NATIONAL GEOGRAPHIC

VISUALIZING EARTHQUAKE PROOF BUILDINGS

Figure 19

Because the most severe damage from an earthquake is not caused when a structure shakes, but when it falls down, engineers are developing ways to make buildings safer. Their job is to prevent the energy of an earthquake from damaging a building's structure.

What happens when the ground moves back and forth under a building? The first floor moves back and forth, but the energy is not transferred to the whole building. The result is that the bottom of the structure collapses.

One way to keep the building stable is to design a system that allows the whole structure to move as a unit. Base isolation systems use bearings that separate the building from the ground. These bearings can be made of large rubber pads or giant metal springs and are placed between the ground and the building support beams. The stretchy rubber or metal spring absorbs the earthquake energy.

Another way to protect buildings in an earthquake is active damping. Large blocks of metal or concrete, weighing many tons slide back and forth as the building sways. The pendulum motion of the damper absorbs the energy of the earthquake, and reduces the movement of the building.

Buildings also can be protected by using structures that can bend. These diagonal braces, called unbonded braces, are made of steel and concrete. The steel beam is shaped so that it can bend back and forth without breaking. The building moves but it does not collapse.

Levels of Destruction The level of destruction by earthquakes is extremely variable. Research has shown that poor building methods are the largest contributors to earthquake damage and loss of life. In countries where there are poorly constructed buildings, it is not uncommon for tens of thousands of people to die in a single earthquake event. It is possible to use high-technology building methods to make structures earthquake resistant, but not earthquake proof. A large proportion of earthquake damage is secondary, such as damage by landslides, fires, and tsunamis. Active earthquake zones are well established, but predicting precise times for earthquakes in those zones is not yet possible.

Earthquake Proofing Although no building can be made entirely earthquake proof, scientists and engineers are finding ways to reduce the damage to structures during mild or moderate earthquakes. Much damage occurs when older structures are shaken off their foundations, so securing a building to its foundation is important. Large masses that can move with the earthquake absorb energy to make a building more secure. **Figure 19** shows some other possible methods for reducing the effects of earthquakes and making buildings safer.

section 2 review

Summary

Global Earthquake Distribution

- The majority of earthquakes occur at varying depths, and in zones that define the locations of plate boundaries.

Causes of Earthquakes

- Earthquake waves are the result of elastic rebound in faults.

Earthquake Waves

- Body waves move throughout Earth.
- Surface waves move along Earth's surface. Their motions cause the majority of earthquake damage.

Earthquake Measurement

- The Modified Mercalli scale is a subjective damage scale that indicates earthquake intensity.
- The Richter magnitude scale measures amplitudes of waves generated by an earthquake. Energy released by an earthquake are estimated from amplitude data.

Self Check

1. **Describe** the elastic rebound process.
2. **Contrast** primary and secondary seismic waves.
3. **Compare and contrast** the Richter scale with the Mercalli scale.
4. **Summarize** the patterns of global earthquake distribution.
5. **Think Critically** Why couldn't you use the Mercalli scale to measure an undersea earthquake?

Applying Math

6. **Calculate** If a primary earthquake wave travels at a rate of about 6 km/s through continental crust, how long will it take it to reach a seismic station located 1,200 km away?
7. **Calculate** If the secondary wave travels at 10 km/s to the same station in question 6, how much longer will it take for the secondary wave to arrive after the primary wave?

section 3

Earth's Interior

Reading Guide

What You'll Learn

- **Explain** how geologists infer the structure of Earth's deep interior.
- **Describe** Earth's internal structure and composition.

Why It's Important

Using waves to indirectly determine the internal structure of an object has many applications in the sciences and engineering.

Review Vocabulary

refraction: the bending of a wave as it changes speed in moving from one medium to another

New Vocabulary

- discontinuity
- shadow zone
- asthenosphere

What's inside?

How is it possible to know anything about the interior of Earth? In 1961 scientists drilled a 200 m deep hole into the oceanic crust trying to reach the Mohorovicic discontinuity. The project was discontinued after Phase I. It is 6,371 km to the center of Earth. By human standards, this attempt was barely a scratch on Earth's surface. Imagine adding this feature to the diagram in **Figure 20.**

Seismologists, geologists who use seismic earthquake waves to interpret characteristics of Earth, conceived of the idea to use these waves to gather data. It is similar to a doctor using sound waves to see inside a human body. As energy passes through matter it is scattered, absorbed, or unaffected. Observation of seismic waves allows scientists to infer images of Earth's interior.

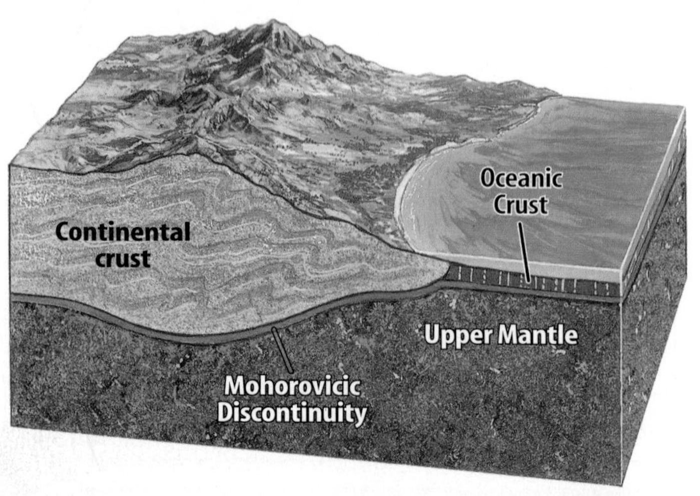

Figure 20 Also known as the Moho layer, the red line below was the target of the 1961 Mohole Project.
Infer *why they attempted to drill through oceanic crust instead of continental crust.*

Uniform Earth? If Earth was uniform in structure and composition, and you knew how fast earthquake waves traveled through its material, then it would be easy to calculate when earthquake waves should be detected on its opposite side. Observations show that seismic waves arrive at different times than expected. In order for seismic waves to change speed, Earth must not be uniform throughout. If Earth is not uniform, then is there a pattern to its interior structure?

Earthquake Observations

As seismic wave recording stations began to spread over Earth, new discoveries were made as seismic wave data from earthquakes were interpreted. Observations of refracted waves show that the waves do indeed bend as they encounter sharp changes in density. A boundary that marks a density change between layers is called a **discontinuity.** One such discontinuity separates the crust from uppermost mantle, and is known as the Mohorovicic (moh huh ROH vee chihch) discontinuity, or Moho, illustrated in **Figure 20.**

Shadow Zones Observations show that, from a given epicenter, P-waves and S-waves travel through Earth for 105 degrees of arc in all directions. Between 105 and 140 degrees from the epicenter, nothing is recorded. This "dead zone" is termed the **shadow zone.** From 140 degrees to 180 degrees (directly opposite the epicenter), only P-waves are recorded.

Shadow zones reveal two interesting facts about Earth's interior. Because S-waves seem not to appear on seismographs located beyond 105 degrees from an epicenter, scientists think that there is a layer of Earth that is absorbing them. In fact, these waves are thought to be converted to P-like waves in the outer core. Recall that S-waves only travel through solids. This suggests that the outer core is in a liquid state. See **Figure 21.**

If you move an energy source all around an object, you would eventually be able to compile many different views and describe the three-dimensional shape of the object. This process is called tomography and is the way Magnetic Resonance Imaging (MRI) can show doctors the inside images of the human body. With earthquakes happening all over the globe, seismologists have thousands of point sources of energy, and can construct a tomographic view of the core.

Figure 21 S-waves only travel through solids so they cannot penetrate the liquid outer core, creating shadow zones.

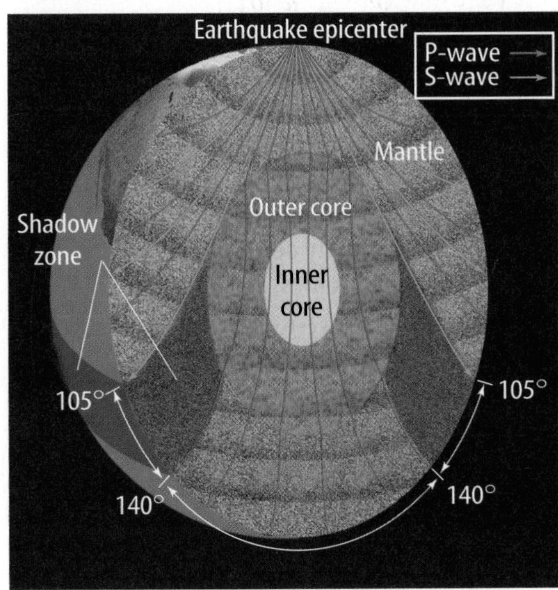

Solid Inner Core The fact that P-waves pass through the core, but are refracted along the way, indicates that the inner core is denser than the outer core and solid. The state of a particular material depends on both pressure caused by the weight of overlying material and temperature. When pressure dominates, atoms are squeezed together tightly and exist in the solid state. If temperatures are high enough, atoms move apart enough to exist in the liquid state, even at extreme pressures.

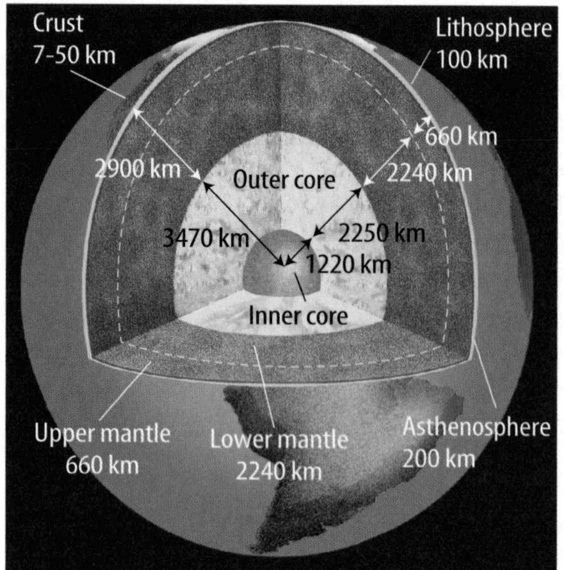

Figure 22 Layering of Earth is caused by heat and pressure. The most dense materials are at the center and the less dense materials are near the crust.

Composition of Earth's Layers

Earth's internal layers, illustrated in **Figure 22,** generally become denser with depth. The crust and uppermost mantle, which together form the lithosphere, are made of rocky material—mostly silicates. The **asthenosphere** is a weaker, plasticlike layer upon which Earth's lithospheric plates move. Much like the lithosphere, mantle below the asthenosphere also is composed of silicates, but the minerals present have different structures in response to conditions of higher pressure. The cores are made mostly of metallic material, such as iron and nickel with noticeable amounts of oxygen and sulfur also present. The core apparently has a composition similar to some iron meteorites that have struck Earth throughout its history.

Astronomers hypothesize that early Earth may have formed from meteorite-like material that was forced together by gravity and heated to melting. Some of the material then was able to migrate toward the core. Over billions of years, Earth's matter has melted and differentiated. The densest materials settled toward the core, and relatively low-density materials floated toward the surface. This differentiation due to gravity is thought to have taken place in all of the planets.

section 3 review

Summary

What's inside?

- Earthquake-generated seismic waves provide information about Earth's deep interior.

Earthquake Observations

- Earth's interior has a layered structure.
- Earth's layers become denser with depth.
- Changes in density occur at layer boundaries called discontinuities.

Composition of Earth's Layers

- Layers of crust and mantle are rocky, and composed mainly of silicates.
- The cores have a high density, metallic composition.
- Composition of Earth closely resembles the composition of meteorites.

Self Check

1. **Describe** the evidence used for subdividing Earth's interior into layers.
2. **Explain** the following points, using seismic evidence for your argument:
   a. Earth has a non-uniform density.
   b. Earth has a layered structure.
   c. Earth has a liquid outer core.
3. **Compare and contrast** the inner and outer cores of Earth.
4. **Think Critically** Explain why it is impossible to ever really know what materials compose Earth's interior.

Applying Math

5. **Calculate** What percent of the mantle are the upper mantle, lower mantle, and the asthenosphere?

More Section Review gpescience.com

section 4

Volcanoes

Reading Guide

What You'll Learn

- **Describe** the types and causes of different types of volcanic eruptions.
- **Explain** the pattern of occurrence of volcanoes and its link to plate tectonics.

Why It's Important

Volcanic eruptions have an impact on the composition of the atmosphere and on climate.

Review Vocabulary

melting point: temperature at which a solid begins to liquefy

New Vocabulary

- viscosity
- cinder cone volcano
- shield volcano
- composite volcano

Origin of Magma

Recall that faults are weaknesses in Earth's crust along which movement takes place. This movement results in a local decrease in pressure, called decompression. With less pressure, the melting point of rock material decreases, but the temperature can remain the same. Hot, nearly molten rock in Earth's asthenosphere, considered an important source for molten rock material, can change to a liquid by decompression melting. Rising magma can become more fluid as it decompresses, particularly if its gas content is high.

Any molten rock material has a lower density than that of its solid counterpart. Because of this density difference, a buoyant force acts on magma that forms from rock surrounding it. Rising magma may reach Earth's surface if pressure conditions allow and the rock has conduits through which it can flow.

Imagine hot magma rising through the crust, creating brittle deformation near the surface in the form of fractures or faults. The cracks in turn cause a drop in pressure, and more paths are available through which magma can move toward the surface, as shown in **Figure 23.** This causes more deformation until magma reaches Earth's surface as a volcanic eruption.

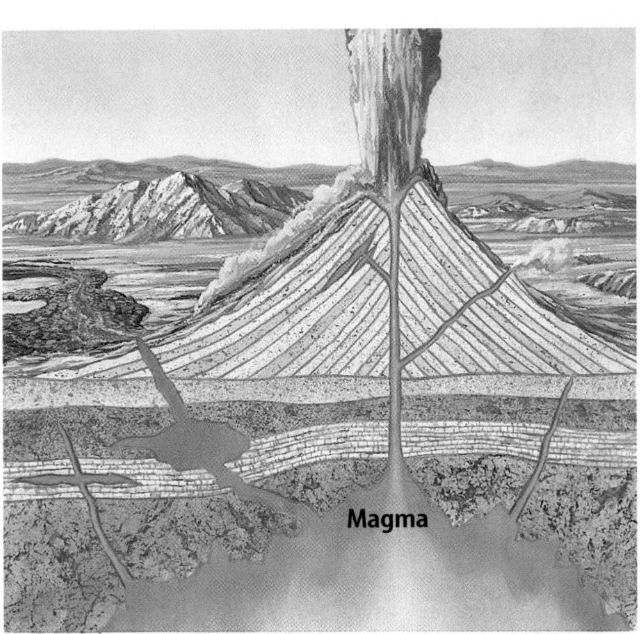

Figure 23 Less dense, liquid rock rises to the surface through cracks and fissures creating a volcano.

Figure 24 Volcanoes are common in subduction zones. Friction, conduction, and convection may all play a role in creating fissures through which a volcano may erupt.

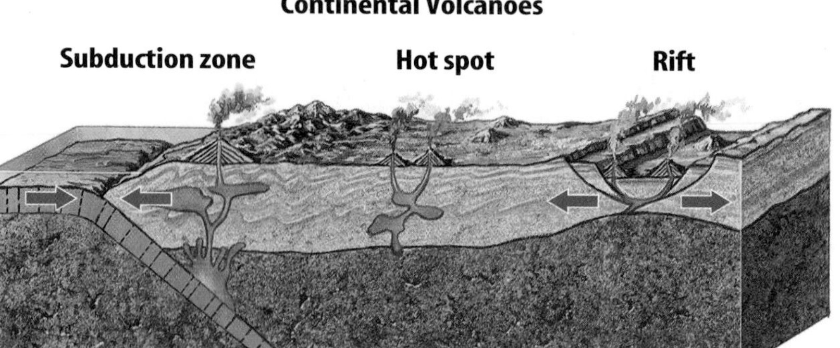

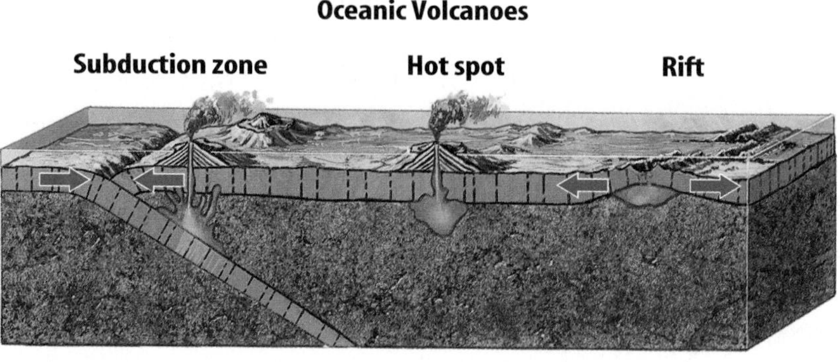

Magma on the Surface Two major physical settings on Earth produce most lava flows at the surface. Eruptions most commonly are found near boundaries that separate tectonic plates, above mantle plumes or hot spots on continents or in the ocean basins. **Figure 24** illustrates these volcanic settings.

Topic: Huge Eruptions
Visit gpescience.com for Web links to information about the world's most powerful volcanic eruptions.

Activity In your Science Journal, list information, including dates and locations, of ten of the world's most powerful volcanic eruptions. On a copy of a world map, plot the locations of these events. Is there a pattern to their occurrence in place or in time?

Eruptive Products

Volcanoes expel a wide variety of materials. These materials can be sorted first by their state of matter. Volcanoes erupt lava, gases, and chunks of solid material.

Solids All solid materials expelled by a volcano are collectively called pyroclasts. Often, lava is ejected into the air as globules. These globules cool and solidify as they fall to Earth. The smallest particles cool very quickly and form volcanic ash. Larger globules form streamlined, volcanic bombs. In addition to chunks that cool as they fall, there often are chunks of already solid material ripped away from the conduit of the volcano as material travels through it. These chunks of rock are termed volcanic blocks. The larger the size of a pyroclastic particle, the closer it will fall to the volcano. Blocks fall back to ground on a volcano's flanks. Ash can be picked up by wind and blown hundreds or even thousands of kilometers away.

Gases Volcanoes release a broad variety of superheated gases, the most common of which is water vapor. In addition carbon dioxide and gases composed of sulfur compounds are expelled high into the atmosphere. There is strong evidence that volcanoes are major contributors of greenhouse gases that can affect climate long after an eruption is over.

Liquids Magma from a volcano or fissure may remain a liquid, at least initially, and flow across the Earth's surface as lava. Lavas can vary considerably in composition, which in turn affects their physical properties.

Viscosity is a measure of the resistance of a fluid to flow. The temperature of molten rock material influences its viscosity. You have experience with temperature control on viscosity when you try to pour cold pancake syrup. It has a high viscosity when it first comes out of the refrigerator. But, let it warm up and it flows more easily because its viscosity decreases. Other factors that affect viscosity and flow are gas content and composition.

Low-viscosity lavas are generally basaltic in composition. Basaltic lavas are low in silica (SiO_2) content, and high in certain chemical elements such as calcium, magnesium, and iron. Basaltic lavas flow from fissures—such as along the MOR and zones of continental rifting, and also from hot spot volcanoes. They tend to flow easily and form huge volcanic forms such as shield volcanoes and flood basalts, both of which cover large areas on Earth's surface.

If lavas have large quantities of gas dissolved in them, then the viscosity is lowered. High gas quantities allow magma to forcefully migrate through rock, sometimes spewing out explosively as a lava fountain that behaves much like a geyser!

Eruptive Styles

Volcanoes can erupt in many different ways, depending on viscosity. Thick, sticky, high-silica magmas are so viscous that they tend not to erupt, causing internal pressure within a volcano to rise. When Earth's crust fails under such high pressure conditions, a violently explosive eruption occurs. This style of eruption is characterized by abundant pyroclasts. In contrast, the runny, low-silica, high-temperature basaltic lavas are so low in viscosity that they erupt quite easily and often produce quiet eruptions of freely flowing lava. Eruptive style is strongly linked to temperature and composition, factors that are hard to measure until after an eruption. Temperature and composition of a magma that ultimately erupts as lava can be linked to the type of plate boundary associated with it.

Modeling Lava Viscosity

Procedure

1. Mix a small batch of batter from **pancake mix** according to directions.
2. Mix a second small batch, but use 25% more **milk.**
3. Hold your finger under a **small funnel** and fill the funnel with the first batch of batter.
4. Have a partner hold a **watch with a second hand.** Remove your finger and time how long it takes the funnel to empty. Record your data.
5. Clean the funnel and repeat step 4 using the second batch of batter.
6. Gently heat a **small pan** on a **hot plate.**
7. Pour part of the first batch of batter on the pan and observe.
8. Repeat step 7 using the second batch of batter.

Analysis

1. Which batter made the flattest pancakes? Explain why.
2. Compare and contrast the properties of the batters.
3. Which batter modeled low-silica lava?

Try at Home

Figure 25 Many volcanoes occur on Earth along plate boundaries, over hot spots, or in rift valleys.

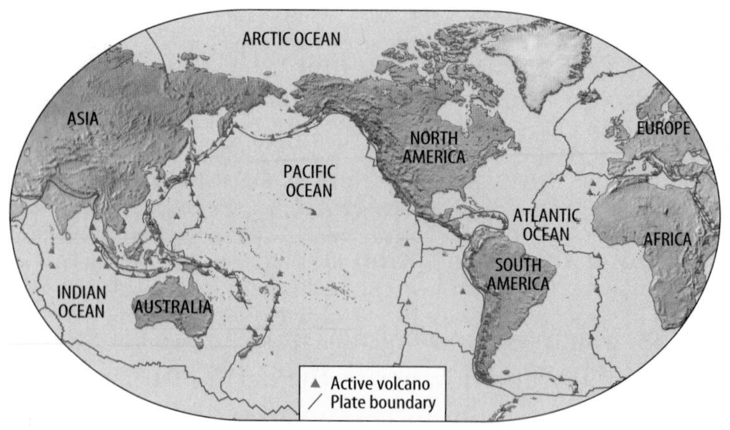

Plate Boundary Setting Look at **Figure 25.** Most of Earth's volcanoes are located along the Ring of Fire, which rims the Pacific Ocean. They lie in subduction zones where continental and oceanic materials are being mixed and partially melted. This plate motion and the associated melting create a wide variety of magma types that can potentially erupt. Large earthquakes and violent volcanic eruptions often are located along these ocean-continent and ocean-ocean convergent boundaries.

Divergent plate boundaries also are volcanically active, but most of the activity is underwater, along the MOR, and goes unnoticed by most people. There are places were divergence takes place on land and you could witness its associated volcanic activity. Iceland and the East African Rift Valley are examples of land areas that are part of divergent boundaries. Lava erupted in these settings is generally low-viscosity and basaltic in composition.

Hot Spots Hot spots are volcanically active sites that arise in places where large quantities of magma move to the surface in large, column-like plumes. Scientists think that plumes are positioned according to internal convection patterns within the mantle, and some may originate at the core-mantle boundary. It seems that hot spots do not move much, but the plates move over them. When a hot spot occurs under an oceanic plate, this stable source of hot magma forms volcanic island chains. The Hawaiian islands are such a chain. Yellowstone National Park is an example of a hot spot under a continental plate. When a volcano moves off the hot spot it becomes inactive.

Hot spot volcanic eruptions produce lava somewhat similar to that formed along divergent boundaries, but they are not an exact match. These lavas tend to contain greater abundances of alkali metals such as potassium and sodium. Like MOR lavas, hot spots, which can occur far from a plate boundary, tend to generate fluid, basaltic lavas. But their compositions can change as magmas penetrate rock material of changing composition, like in continental crust.

Topic: Mt. Pinatubo
Visit gpescience.com for Web links to information the 1991 eruption of Mt. Pinatubo in the Philippines.

Activity Construct a time line for this eruption. Begin with precursors, which are events that indicate volcanic activity is imminent, and end with long-term effects after the eruption.

Types of Volcanoes

Volcanoes are classified according to their size, shape, and the materials that compose them. Recall that eruptive materials that form a volcano are related to the physical properties of its magma source. The temperature, composition, and gas content of magma are important controls on the type of volcanic structure that forms during an eruption. **Table 3** summarizes the characteristics of main types of volcanoes.

Cinder Cone Volcanoes When eruption of gas-rich magma takes place, eruptive products often are spewed into the air explosively as large chunks. These large pyroclastic materials may pile up near the exit hole, or vent. When the primary eruptive products are large fragments of solid material, **cinder cone volcanoes** form. They tend to be small, with most cones having heights in the hundreds of meters range. When cinder cones occur on the flanks of larger volcanoes, they are called parasitic cones. An example of a volcano with parasitic cones is Mount Kilimanjaro in the African rift valley.

Reading Check *What are the characteristics of cinder cones?*

Shield Volcanoes Because they form from high-temperature, fluid, basaltic lava, **shield volcanoes** erupt with abundant lava flows that can move for kilometers over Earth's surface before stopping. Shield volcanoes are broad, flat structures made up of layer upon layer of lava. Think of pancake batter. If the batter is cold or thick, it piles up and you get thick pancakes. Add more milk and make a runny batter, and it flows easily across the skillet and makes thin pancakes. Volcanism in Hawaii produces shield volcanoes.

Volcano Eruptions
Volcanic eruptions can throw tons of ash into the atmosphere where winds carry the ash around the world. The ash blocks sunlight and affects plant growth. Scientists measure fossilized tree rings to infer the affect ancient volcanoes had on Earth. Research the amount of ash produced by Mount Pinatubo and how quickly it circled the globe.

Table 3 Comparison of Melt Properties

Composition	Silica Content	Gas Content	Viscosity	Volcano Type
Basaltic	lowest	least (1-2%)	lowest	shield, fissure eruptions (such as MOR)
Andesitic	intermediate	intermediate (3-4%)	intermediate	composite
Rhyolitic	highest	highest (4-6%)	highest	volcanic dome

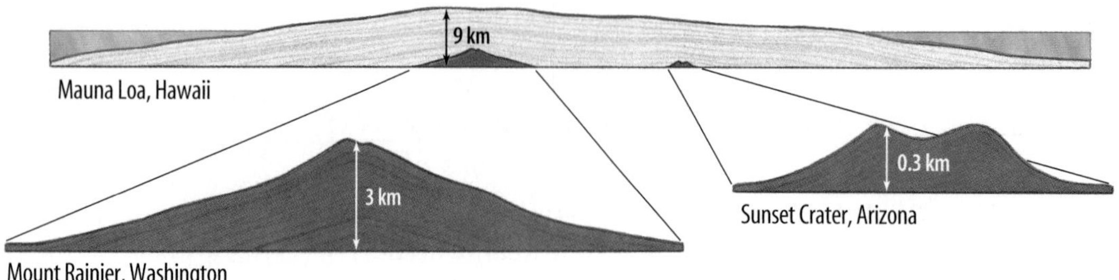

Figure 26 Shield volcanoes, like Mauna Loa, have created some of the largest mountains on Earth. Cinder cones, like Sunset Crater, are the smallest. Some of the famous volcanoes are the composites, like Mount Rainier.

Composite Volcanoes When volcanoes occur along convergent plate boundaries, they tend to have magmas that are richer in silica content than those formed at hot spots or divergent boundaries. This is because as subduction takes place, water and sediment are forced down to regions of higher temperature. Partial melting of materials, in which the silica-rich portion of rock and sediment melts first, produces viscous magma. This produces volcanoes formed from alternating explosive events that produce pyroclastic materials, and lava flows. These **composite volcanoes,** composed of alternating layers, are large, often thousands of meters high and tens of kilometers across the base. **Figure 26** shows all three types.

section  review

Summary

Origin of Magma

- Magma originates as molten rock material below the surface and erupts at Earth's surface as lava.
- When the density of magma is lower than surrounding solid rock, it is forced toward Earth's surface.

Eruptive Products

- Eruptive products can be solids, liquids, or gases.

Eruptive Styles

- The style of eruption, whether quiet or explosive, is related to its plate tectonic setting.

Types of Volcanoes

- Cinder cones are small, but they erupt violently.
- Shield volcanoes are very large and mostly expel free-flowing lava quietly.
- Composite volcanoes are large and tend to erupt violently.

Self Check

1. **Explain** why most volcanoes are found at plate boundaries.
2. **Compare and contrast** the physical settings for composite volcanoes, cinder cones, and shield volcanoes.
3. **Describe** causes for variation in eruptive style for volcanoes.
4. **Explain** how magma that originates at depth can erupt as lava at the surface.
5. **Describe** how island chains form over a hot spot.
6. **Think Critically** List some possible consequences if volcanic activity on Earth were to slow down or stop.

Applying Math

7. If a cinder cone is 540 m high and has a base diameter of 3 km, what is the volume of the volcanic cone in cubic meters? Use the formula $V_{cone} = (r^2 h \pi)/3$.
8. The dome in the caldera of the volcano has a height of 12 meters and a diameter of 50 m. What is its volume?

More Section Review gpescience.com

A Case for Pacific Plate Motion

To measure motion you have to have a starting and an ending point. You must also know the time it took to get from start to end. Volcanic activity associated with a hot spot beneath Hawaii gives geologists exactly that.

Real-World Problem

How can scientists show that Earth's plates are moving?

Goal

- **Infer** a rate of movement for the Hawaiian Islands over a hot spot

Materials

ruler
calculator
scale map of the Hawaiian Islands

Procedure

1. Make a data table like the one shown below.
2. **Measure and record** the distances between the island sets in the data table. Use the map scale to convert measurements to km.
3. Refer to average ages given for each island on the map. Calculate and record the age differences for each set of islands in the data table. Use the hot spot beneath the island of Hawaii as a starting reference point.
4. **Calculate** the rate of motion in km/year. Assume that the hot spot is stationary and that the Pacific plate is moving over it.

Distance/Time Data for Hawaiian Islands

From/To	Distance (km)	Time (years)	Rate (km/year)
Hawaii to Maui	161		
Maui to Molokai			
Molokai to Oahu	Do not write in this book.		
Oahu to Kauai			

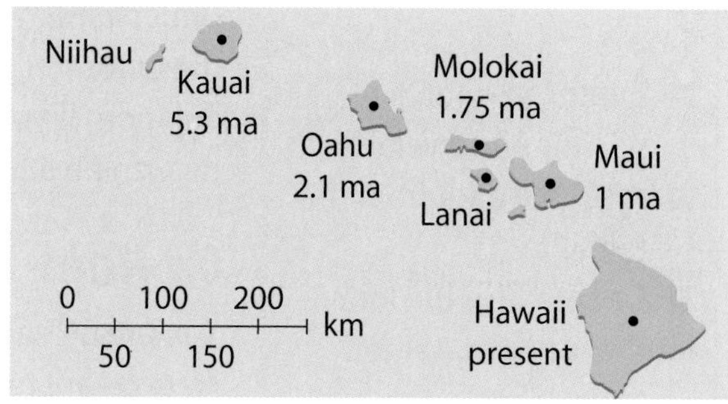

Conclude and Apply

1. **Evaluate** how meaningful your calculated rate numbers are. Determine a better rate unit and convert your km/year rates to these new units.
2. **Infer** why the rates are not consistent using what you know about plate movement.
3. **Describe** the overall motion of the Pacific Plate based on your data.
4. **Observe** a map of the Pacific Ocean and infer the location of a divergent zone that could be "pushing" the Pacific plate.

Communicating Your Data

Share your findings with the class and discuss alternative interpretations.

Earthquake! Earthquake! Where's the earthquake?

Goals

- **Examine** a table of seismic wave velocities.
- **Analyze** data from the table.
- **Determine** the location of an epicenter.

Possible Materials

plain white paper
graph paper
compass
metric ruler

Safety Precaution

Understanding earthquakes begins with locating them. To determine the epicenter of an earthquake, scientists use a method called triangulation. If you know the locations of three points on a map, you can determine a fourth point **Diagram 1** shows how triangulation works.

Real-World Problem

You are on vacation in City A and experience an earthquake. The radio stations are broadcasting information about the earthquake. Your home is in City B. Is your home near the epicenter?

Procedure

1. Draw a 20-cm $\times$ 15-cm rectangle on a piece of plain white paper. Orient the rectangle so that the 20-cm edge is vertical. This rectangle will serve as your map.
2. Using a scale of 1 cm = 200 km, draw a distance scale just below the rectangle on the white paper. Place an arrow parallel to one of the 20-cm vertical edges and label it *North*.
3. Within the rectangle, place City A 400 km from the north edge and 400 km from the west edge. Locate City B 800 km from the north edge and 800 km from the east edge; and City C 1,200 km from the south edge and 1,200 km from the west edge of the rectangle. Your map will look similar to **Diagram 2.**
4. The earthquake happened at 08:37:00 PST. Copy and complete **Table A** on your own paper. Subtract the P-wave and S-wave arrival times to find the time differences for cities B and C. Use the travel time graph and the time differences to complete the last two columns.

Table A Earthquake Arrival Time Data for Cities A, B, and C

City	P-wave Arrival	S-wave Arrival	Time Difference (min/sec)	Distance to Epicenter	Distance on Map
A	08:40:00	08:43:00	3 min/0 sec		
B	08:41:15	08:45:00			
C	08:39:40	08:42:10			

5. **Estimate** the distance from each city from Table B below.
6. Use the Distance on Map measurement for City A to set the compass width, and then draw the distance circle around City A on your map. Your circle may go off the map, but don't worry. The epicenter is somewhere on the arc you can draw on the map. Repeat this process for City B and City C.

Table B Seismic Wave Arrival Times

Point	Time difference between P-waves and S-waves (min)	Distance (km) traveled by waves
M	1.3	500
O	2.2	1,000
P	3.6	2,000
Q	4.5	3,000
R	5.4	4,000

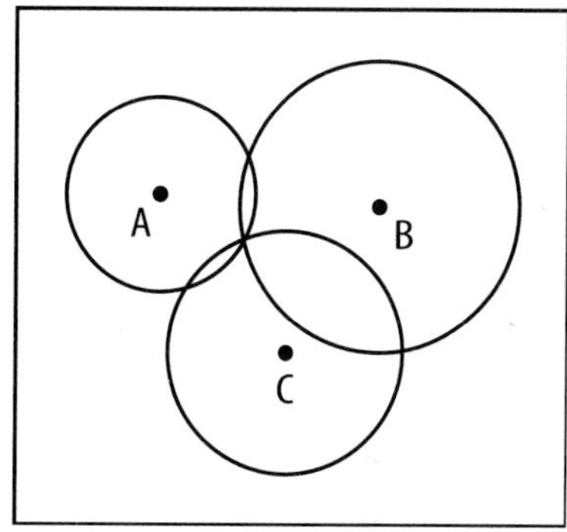

Diagram 1

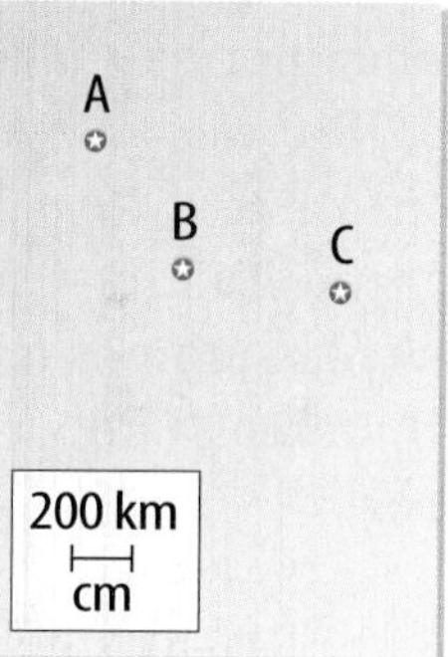

Diagram 2

Analyze Your Data

1. **Estimate** the time difference for P-waves and S-waves coming from 2,500 km.
2. **Estimate** the distance for a P-wave and S-wave pair that measure a time difference of 3.3 minutes.
3. **Determine** how much closer to the epicenter City A is than City B. City A received waves 3.0 minutes apart. City B received waves 4.2 minutes apart.

Conclude and Apply

1. **Explain** the relationship between body wave travel time differences and the distance to an earthquake epicenter.
2. **Evaluate** the potential danger to your home City B. How far is your city from the epicenter?
3. **Infer** the arrival time difference for S-waves and P-waves at or very close to the epicenter.
4. **List** the possible sources of error in the methods you used to determine an epicenter. How can you minimize errors?

Describe the location of the epicenter in terms of distances from a reference point on your map. **Compare** your descriptions with others.

TIME SCIENCE AND HISTORY

SCIENCE CAN CHANGE THE COURSE OF HISTORY!

Volcano Weather

In 1815, people worldwide noticed unusually brightly colored sunsets. Then in 1816, the weather in many parts of the world was colder than normal. In North America and Europe, 1816 was known as "the year without a summer." Parts of New England had damaging frosts in July and August. The fantastic sunsets and the cold summer resulted from the massive eruption of Tambora, a volcano in Indonesia.

Giant Weather Makers

Communication was slow in the early nineteenth century, so scientists immediately did not connect Tambora's eruption to weather change. Today, many researchers study the effects of volcanic eruptions on weather. When Mount Pinatubo in the Philippines erupted in June 1991, weather stations around the world—and even in space—recorded its effects.

Mt. Pinatubo's eruption was smaller than Tambora's but was the second largest eruption in over 100 years. It blew ash and gases into the stratosphere, where they were carried by wind around the world. Because the suspended particles and droplets absorbed and reflected sunlight, average temperatures temporarily dropped by about 0.5°C in many places. Other weather-related affects included the increased strength of hurricanes in the Atlantic and Pacific Oceans and as well as flooding rains in the U.S. Midwest.

Looking Back at Volcanoes

The effect of volcanoes on climate is so important that some scientists even study volcanic eruptions from thousands of years ago. They compare the dates of these ancient eruptions with records of unusual weather. It has been found that crop failures and disease epidemics often occurred soon after large eruptions. Because changes in weather can affect the growth of trees, the growth rings of ancient trees also are evidence that support the data from written records. These ancient records help scientist predict the weather impacts of future volcanic eruptions.

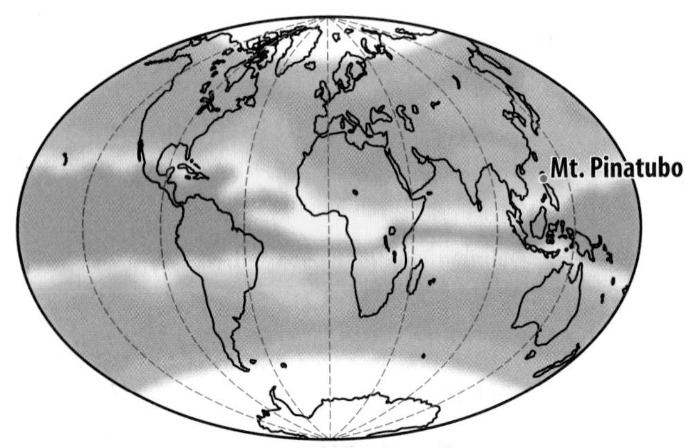

Experiment Lay two sheets of black construction paper on a sunny windowsill and place a thermometer on each sheet. Tape a large piece of gauze or cheesecloth to the window so that it shades one of the sheets. Record the change in temperature for one half hour.

TIME

For more information, visit gpescience.com

Reviewing Main Ideas

Section 1 Evolution of Earth's Crust

1. Earth's crust and uppermost mantle, together called the lithosphere, consist of about seven large and numerous smaller plates.
2. Plate motion over the asthenosphere is influenced by forces that include ridge push, frictional drag, and slab pull.
3. Mantle convection drives the system of plates. Thermal energy sources for convection include heat left over from Earth's formation, and decay of radioactive isotopes.

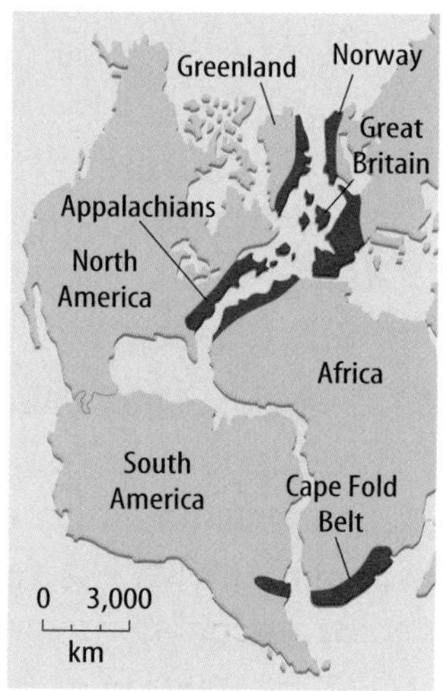

4. Plates meet along divergent, convergent, and transform boundaries.

Section 2 Earthquakes

1. Earthquakes are vibrations in Earth caused by the sudden release of energy.
2. During an earthquake caused by movement along a fault, strain energy is released in the form of seismic waves.
3. Most earthquake activity is confined to regions near plate boundaries.
4. Short-term prediction of earthquakes is not yet possible, but earthquakes can be rated by their intensities and magnitudes.

Section 3 Earth's Interior

1. The interior of Earth consists of four main layers: crust, mantle, inner and outer core.
2. The crust and mantle are largely composed of solid silicates. The asthenosphere is a layer of mantle in a semi-solid state.
3. The two innermost layers, the inner and outer core, are mainly composed of iron, nickel, oxygen and sulfur. The outer core is in a liquid state, and the inner core is solid.
4. The crust is the thinnest layer, while the mantle is the thickest.

Section 4 Volcanoes

1. Most volcanic activity on Earth occurs in the vicinity of plate boundaries or above mantle plumes.
2. The viscosity of the magma feeding volcanoes controls the volcanoes' eruptive characteristics.
3. Major controls on magma viscosity are: temperature, chemical composition, and gas content.
4. Cinder cones, shield volcanoes, and composite volcanoes are three principle volcanic landforms.

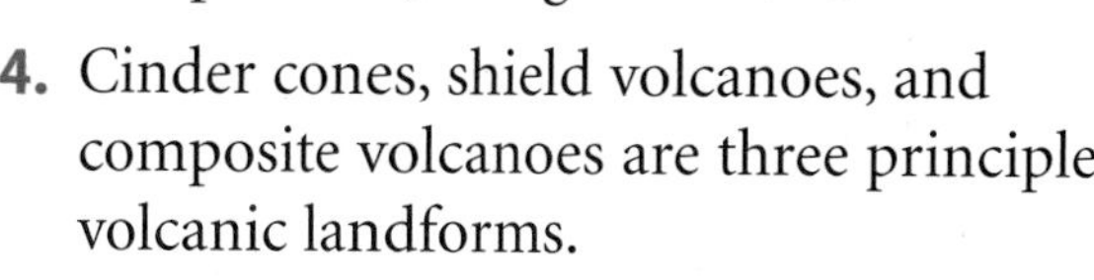

FOLDABLES Use the Foldable that you made at the beginning of this chapter to help you review Earth's internal process.

chapter 12 Review

Using Vocabulary

asthenosphere p. 372	fault p. 364
cinder cone volcano p. 377	focus p. 364
composite volcano p. 378	mid-ocean ridge p. 356
convergent boundary p. 359	rift valley p. 356
discontinuity p. 371	shadow zone p. 371
divergent boundary p. 358	shield volcano p. 377
elastic rebound p. 364	subduction p. 359
epicenter p. 364	transform boundary p. 360
	viscosity p. 375

Complete each sentence with the correct vocabulary word or words.

1. The ________ is the point of origin of an earthquake.
2. A zone of cracking in Earth's crust along which movement takes place is a ________.
3. A long, linear feature within a divergent plate boundary is a(n) ________.
4. A feature consisting of a relatively small pile of pyroclastic materials is a(n) ________.
5. A boundary marking an abrupt change in density is a(n) ________.

Checking Concepts

Choose the word or phrase that best answers the question.

6. What process causes a material to break due to excess stress?
   A) plastic deformation.
   B) elastic rebound.
   C) elastic deformation.
   D) brittle deformation.
7. Which best identifies a shield volcano?
   A) sticky, silica-rich magmas
   B) great height compared to width
   C) forms above hot spots
   D) found mostly on continents
8. What characteristic was first used to identify Earth's layers?
   A) temperature. **C)** density.
   B) composition. **D)** thickness.
9. Which pair of plate tectonic boundaries is best characterized by mostly shallow-focus earthquakes?
   A) divergent and transform
   B) divergent and continent-ocean convergent
   C) continent-continent and continent-ocean convergent
   D) transform and ocean-ocean convergent zones
10. Which feature is common to and only found in diverging regions?
    A) trenches **C)** volcanic arcs
    B) rift valleys **D)** island arcs
11. Which is **NOT** evidence used by Wegener to support the continental drift hypothesis?
    A) matching magnetic patterns symmetrical to the Mid-Ocean Ridge
    B) matching of continental margins
    C) correlation of fossils among the continents
    D) mountain-range matching among the continents

Use the illustration below to answer question 12.

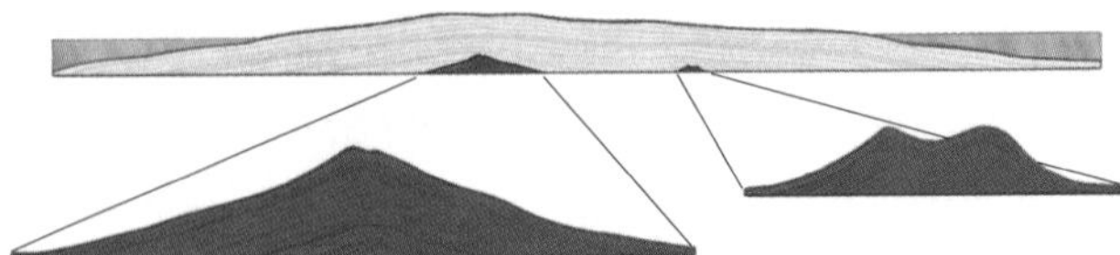

12. Which volcano type has small height, small diameter and consists mostly of pyroclasts?
    A) Hawaiian volcano
    B) cinder cone volcano
    C) composite volcano
    D) shield volcano

Vocabulary PuzzleMaker gpescience.com

13. Which earthquake waves travel through matter with a push-pull motion?
   A) secondary waves
   B) surface waves
   C) primary waves
   D) body waves

Interpreting Graphics

14. Copy and complete the concept map below summarizing characteristics of divergent, convergent and transform plate boundaries.

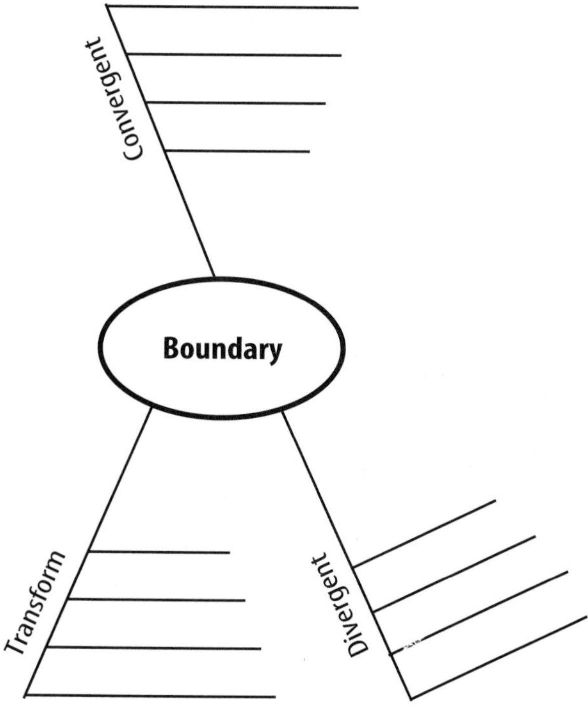

15. Copy and complete the table summarizing three types of volcanoes.

Magma Properties and Volcano Types

General Composition	Relative silica (SiO_2) Content	Relative Viscosity	Volcano Type
		lowest	
Andesitic	intermediate		
	highest		Lava Dome

Thinking Critically

16. **Explain** why the island of Kauai, often referred to as the "Garden Isle," has thicker soils and is better able to sustain agriculture than other Hawaiian Islands.

17. **Infer** what general depth of focus earthquakes are likely to occur in the Himalaya.

18. **Explain** how the processes associated with plate tectonics maintain Earth's recycling of materials.

Applying Math

19. **Calculating Lava Thickness** Suppose that hot spot volcanism produces an average of 76,000 m^3 of lava per day and does so for 340 days. This lava flows across a region that has an area of 10 km^2. How thick will the resulting lava flow be?

20. Located on the island of Hawaii, Mauna Loa is the largest volcano on Earth and rises 17 km above its base. Using the thickness you found in question 19 as the annual thickness, how many years did it take for Mauna Loa to build from the sea bed?

21. Mauna Loa rises 4 km above sea level. How many years did it take to grow from sea level to its present height?

22. An active volcano called Loihi lies just off the coast of the island of Hawaii. It is called a seamount because it still lies below the surface. If it grows at the same rate you used in question 21 and lies 1250 m below the surface, how long will it take for Loihi to break the ocean surface?

Standardized Test Practice

Record your answers on the answer sheet provided by your teacher or on a sheet of paper.

Multiple Choice

1. Who first proposed the hypothesis of continental drift?

A. Matthews

B. Vine

C. Wegener

D. Wilson

Use the image below to answer question 2.

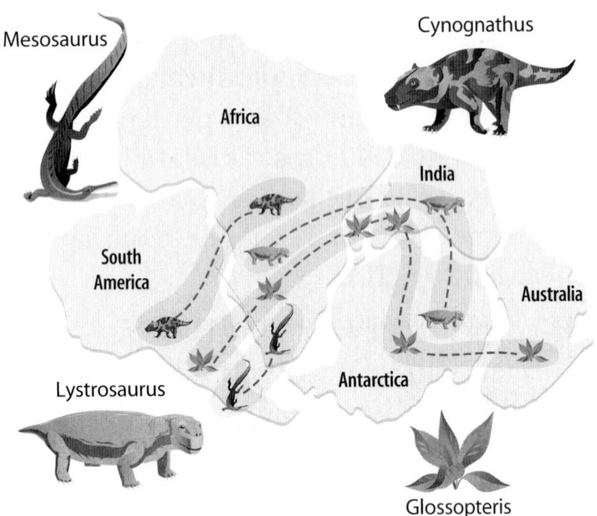

2. Which is the fern, pictured above, that provided support for Pangaea?

A. Antarctica

B. Glossopteris

C. Mesosaurus

D. Lystrosaurus

Test-Taking Tip

For each question, double-check that you are filling in the correct answer bubble for the question number you are completing.

3. What did oceanographers show, using sonar, which helped revive continental drift?

A. edges of the continental shelves matched

B. edges of the continents matched

C. fossils match similar fossils

D. rocks match similar rocks

4. Which forms when a less-dense plate converges on a denser plate?

A. mid-ocean ridge

B. rift valley

C. subduction zone

D. transform boundary

Use the image below to answer question 5.

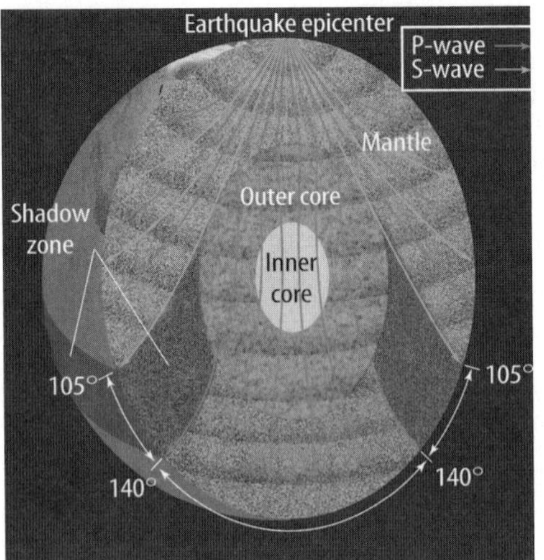

5. The shadow zone, illustrated above, is caused by which layer of Earth?

A. crust

B. mantle

C. outer core

D. inner core

Gridded Response

6. If a divergent boundary separates at a rate of 2.5 cm/year, how much farther apart would the two plates be after 230 years?

7. Compared to the two plates in question 7, how much farther apart would two plates be after 230 years if the divergent boundary between them separated at 15 cm/year?

Short Response

Use the illustration below to answer question 8.

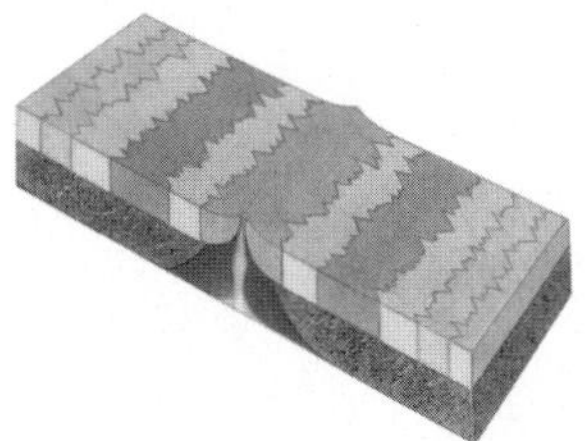

8. The illustration shows reverse polarity bands on the seafloor. How does magnetic polarity of rocks support seafloor spreading?

9. What are the three different ways in which tectonic plates can move?

10. What is different about what happens to plate edges at the different types of convergent boundaries?

11. How did Dr. Harry Hess obtain detailed maps of the ocean floor?

12. What causes a rift valley to form at divergent boundaries?

13. What is true about the thickness and age of sediments at different locations on the ocean floor that supports seafloor spreading?

14. What is the driving force for all mechanisms of tectonic plate movement?

15. What is the difference between a crack in Earth's crust and a fault?

16. What causes the energy of an earthquake?

17. How are the focus and epicenter of an earthquake related?

Extended Response

Use the image below to answer question 18.

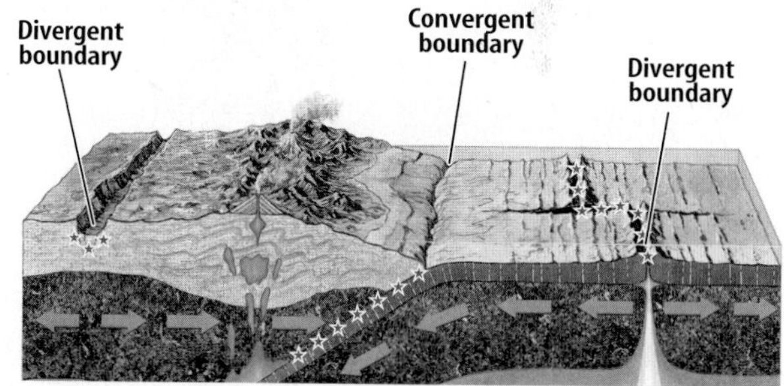

18. **Part A** What is true about the depth of earthquake foci and the location of tectonic plate boundaries?

Part B Why are deep-focus earthquakes noted only at ocean-ocean convergent boundaries and ocean-continental convergent boundaries?

unit 4

Forms of Energy

How Are Clouds & Toasters Connected?

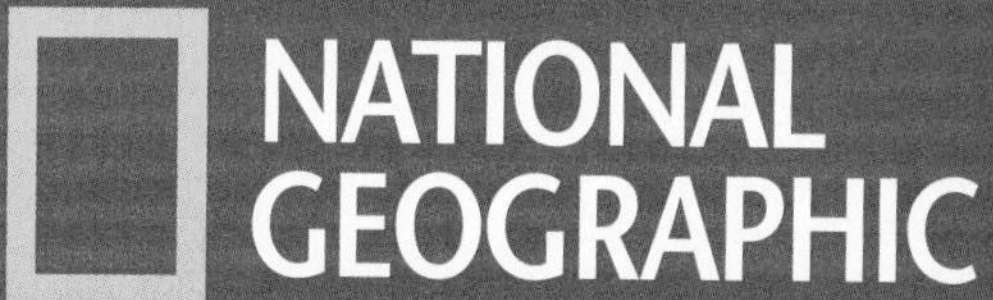

In the late 1800s, a mysterious form of radiation called X rays was discovered. One French physicist wondered whether uranium would give off X rays after being exposed to sunlight. He figured that if X rays were emitted, they would make a bright spot on a wrapped photographic plate. But the weather turned cloudy, so the physicist placed the uranium and the photographic plate together in a drawer. Later, on a hunch, he developed the plate and found that the uranium had made a bright spot anyway. The uranium was giving off some kind of radiation even without being exposed to sunlight! Scientists soon determined that the atoms of uranium are radioactive—that is, they give off particles and energy from their nuclei. In today's nuclear power plants, this energy is harnessed and converted into electricity. This electricity provides some of the power used in homes to operate everything from lamps to toasters.

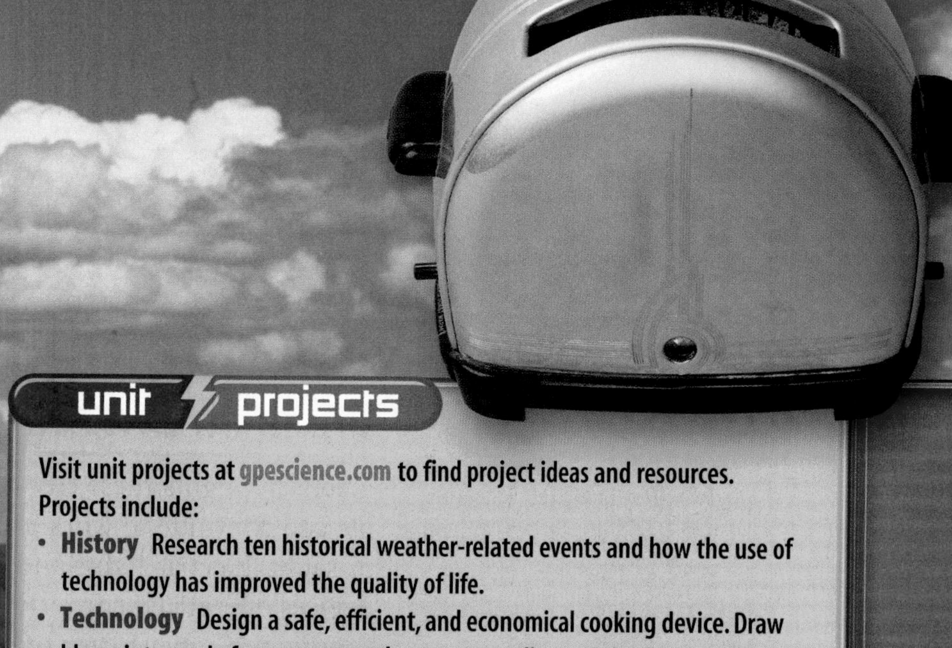

unit projects

Visit unit projects at gpescience.com to find project ideas and resources. Projects include:

- **History** Research ten historical weather-related events and how the use of technology has improved the quality of life.
- **Technology** Design a safe, efficient, and economical cooking device. Draw blueprints, apply for a patent, and test your appliance. Submit your recipe, appliance, and food to the class.
- **Model** Construct a unit review game that demonstrates electricity or magnetism. Review information with the answer key should be provided in a well-designed, marketable package.

WebQuest Using *Maglev Trains* WebQuest, research, design, build, and test your own version of a maglev train, then present your model to the class.

chapter 13

Electricity

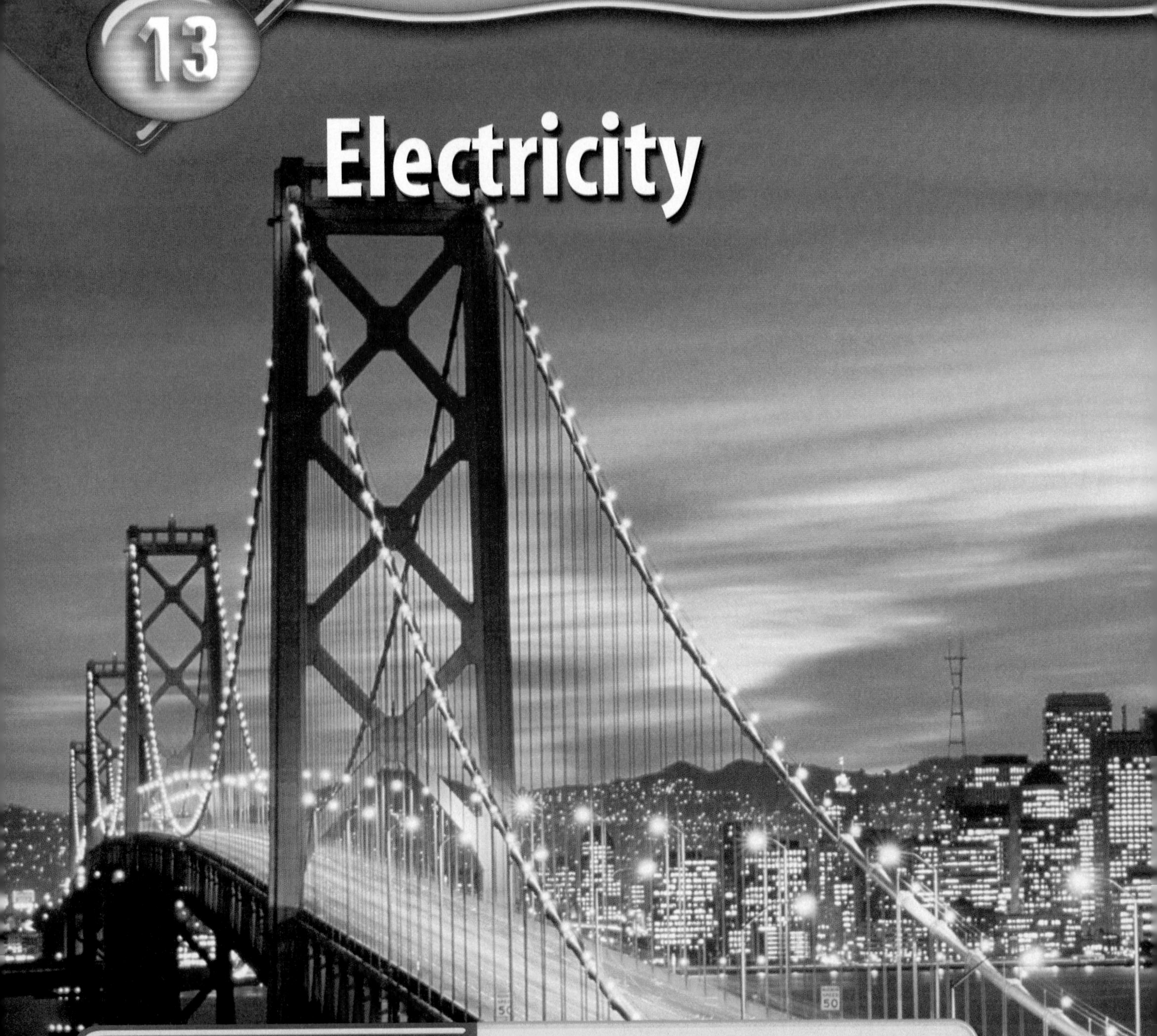

chapter preview

sections

1 Electric Charge

2 Electric Current
Lab *Identifying Conductors and Insulators*

3 Electrical Energy
Lab *Comparing Series and Parallel Circuits*

Virtual Lab *How are current, voltage and resistance related?*

Shine on Brightly

Electricity lights these city lights so that people can continue to work and have fun, day or night. Electric lights and other electric devices operate by converting electrical energy into other forms of energy.

Science Journal For five electric devices, list the form of energy that electrical energy is converted into by each device.

Start-Up Activities

Electric Circuits

No lights! No CD players! No computers, video games or TVs! Without electricity, many of the things that make your life enjoyable wouldn't exist. For these devices to operate, electric current must flow in the electric circuits that are part of the device. Under what conditions does electric current flow in an electric circuit?

1. Complete the safety form.
2. Obtain a battery, a flashlight bulb, and some wire.
3. Connect the materials so that the lightbulb lights.
4. Draw diagrams of all the ways that you were able to light the bulb.
5. Record a few of the ways that didn't work.
6. Can you light the bulb using only one wire and the battery?
7. **Think Critically** Write a paragraph describing the requirements to light the bulb. Write out a procedure for lighting the bulb and have a classmate follow your procedure.

Electricity Make the following Foldable to help you organize information about electricity.

STEP 1 **Fold** a sheet of paper vertically from top to bottom. Make the top edge about 2 cm shorter than the bottom edge.

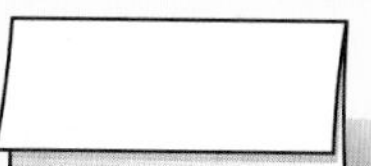

STEP 2 **Turn** lengthwise and fold into thirds.

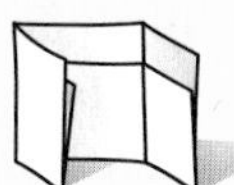

STEP 3 **Unfold and cut** only the top layer along both folds to make three tabs.

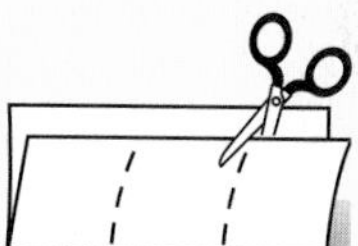

STEP 4 **Label** the Foldable as shown.

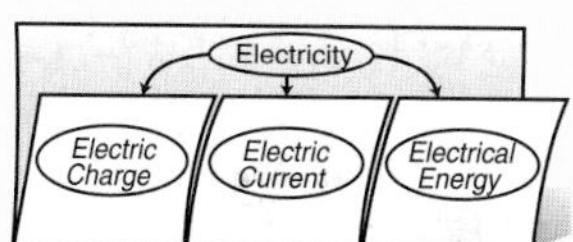

Organize Information As you read the chapter, organize the information you find about electric charge, electric current, and electrical energy under the appropriate tab.

Preview this chapter's content and activities at gpescience.com

section 1

Electric Charge

Reading Guide

What You'll Learn

- **Describe** how electric charges exert forces on each other.
- **Compare** the strengths of electric and gravitational forces.
- **Distinguish** between conductors and insulators.
- **Explain** how objects become electrically charged.

Why It's Important

The electrical energy that all electrical devices use comes from the forces that electric charges exert on each other.

Review Vocabulary

atom: the smallest particle of an element

New Vocabulary

- static electricity
- law of conservation of charge
- conductor
- insulator
- charging by contact
- charging by induction

Positive and Negative Charge

Why does walking across a carpeted floor and then touching something sometimes result in a shock? The answer has to do with electric charge. Atoms contain particles called protons, neutrons, and electrons, as shown in **Figure 1.** Protons and electrons have electric charge, and neutrons have no electric charge.

There are two types of electric charge. Protons have positive electric charge and electrons have negative electric charge. The amount of positive charge on a proton equals the amount of negative charge on an electron. If an atom contains equal numbers of protons and electrons, the positive and negative charges cancel out and the atom has no net electric charge. Objects with no net charge are said to be electrically neutral.

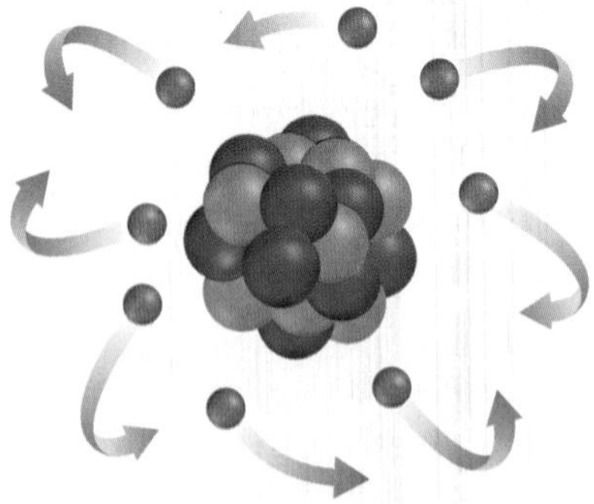

Figure 1 The center of an atom contains protons (orange) and neutrons (blue). Electrons (red) swarm around the atom's center.

Transferring Charge Electrons are bound more tightly to some atoms and molecules. For example, compared to the electrons in atoms in the carpet, electrons are bound more tightly to the atoms in the soles of your shoes. **Figure 2** shows that when you walk on carpet, electrons are transferred from the carpet to the soles of your shoes. The soles of your shoes have an excess of electrons and become negatively charged. The carpet has lost electrons and has an excess of positive charge. The carpet has become positively charged. The accumulation of excess electric charge on an object is called **static electricity.**

Before the shoe scuffs against the carpet, both the sole of the shoe and the carpet are electrically neutral.

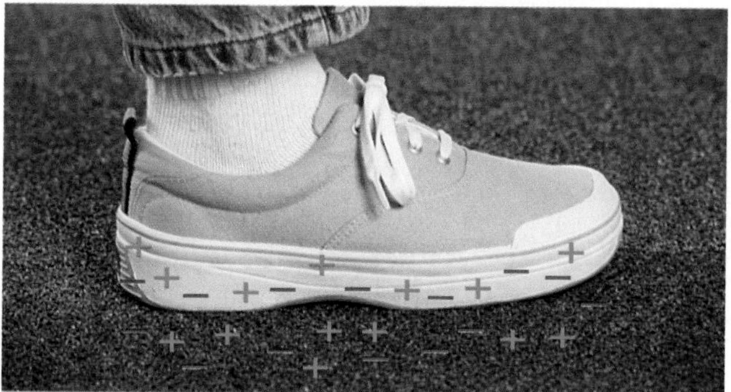

As the shoe scuffs against the carpet, electrons are transferred from the carpet to the sole of the shoe.

Figure 2 Atoms in the shoe's sole hold their electrons more tightly than atoms in the carpet hold their electrons.
Explain *how the sole gained charge even though charge can't be created or destroyed.*

Conservation of Charge When an object becomes charged, charge is neither created nor destroyed. Usually it is electrons that have moved from one object to another. According to the **law of conservation of charge,** charge can be transferred from object to object, but it cannot be created or destroyed. Whenever an object becomes charged, electric charges have moved from one place to another.

Reading Check *How does an object become charged?*

Charges Exert Forces Have you noticed how clothes sometimes cling together when removed from the dryer? These clothes cling together because of the forces electric charges exert on each other. **Figure 3** shows that unlike charges attract each other and like charges repel each other. The force between electric charges also depends on the distance between charges. The force decreases as the charges get farther apart.

Just as for two electric charges, the force between any two objects that are electrically charged decreases as the objects get farther apart. This force also depends on the amount of charge on each object. As the amount of charge on either object increases, the electrical force also increases.

As clothes tumble in a dryer, the atoms in some clothes gain electrons and become negatively charged. Meanwhile the atoms in other clothes lose electrons and become positively charged. Clothes that are oppositely charged attract each other and stick together.

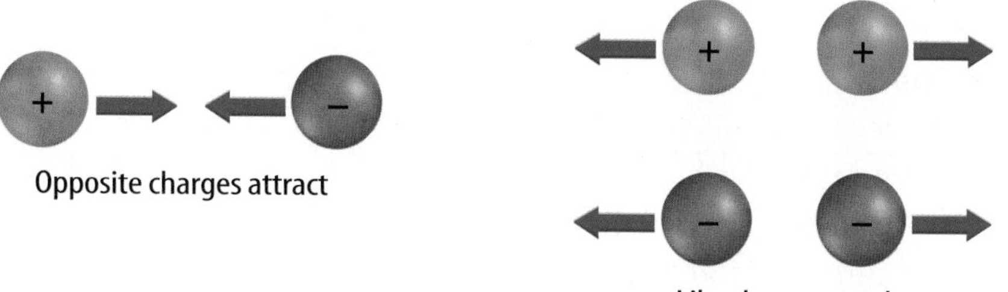
Opposite charges attract
Like charges repel

Figure 3 Positive and negative charges exert forces on each other.

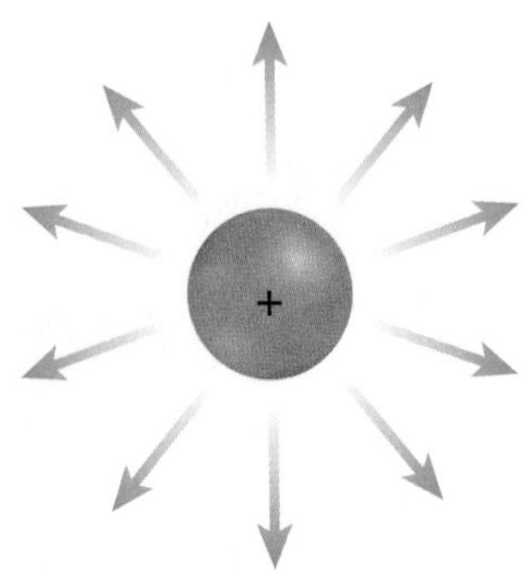

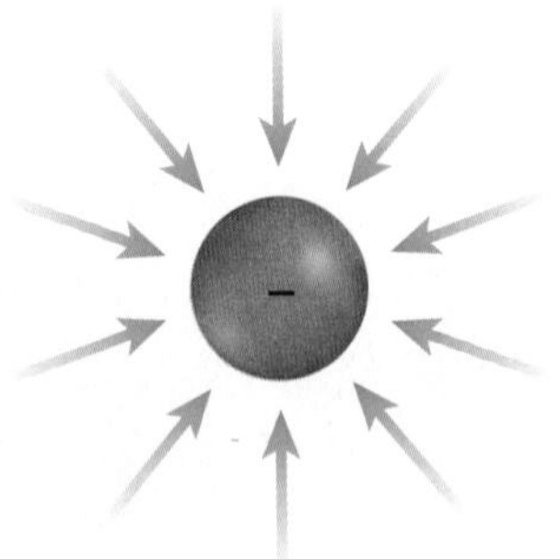

Electric Fields You might have seen bits of paper fly up and stick to a charged balloon. The bits of paper do not need to touch the charged balloon for an electric force to act on them. If the balloon and the paper are not touching, what causes the paper to move?

Figure 4 Surrounding every electric charge is an electric field that exerts forces on other electric charges. The arrows point in the direction a positive charge would move.

An electric field surrounds every electric charge, as shown in **Figure 4,** and exerts the force that causes other electric charges to be attracted or repelled. Any charge that is placed in an electric field will be pushed or pulled by the field. Electric fields are represented by arrows that show how the electric field would make a positive charge move.

The Strength of Electric Forces The force of gravity between you and Earth seems to be strong. Yet, compared with electric forces, the force of gravity is much weaker. For example, the attractive electric force between a proton and an electron in a hydrogen atom is about a thousand trillion trillion trillion times larger, or 10^{39} times larger, than the attractive gravitational force between the two particles.

In fact, all atoms are held together by electric forces between protons and electrons that are tremendously larger than the gravitational forces between the same particles. The chemical bonds that form between atoms in molecules also are due to the electric forces between the atoms. These electric forces are much larger than the gravitational forces between the atoms.

Reading Check *Compare the strength of electric and gravitational forces between protons and electrons.*

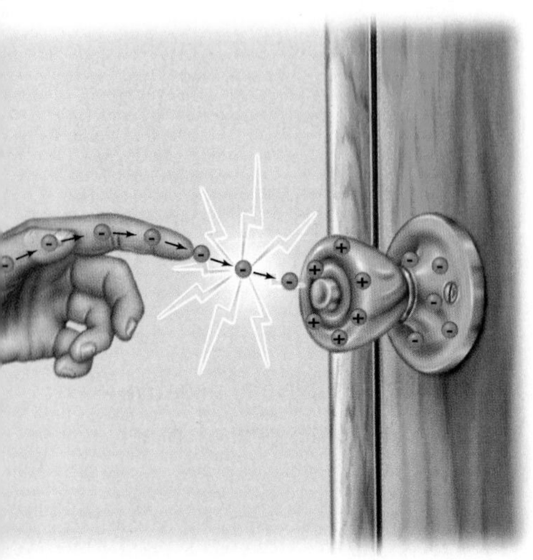

Figure 5 As you walk across a carpeted floor, excess electrons can accumulate on your body. When you reach for a metal doorknob, electrons flow from your hand to the doorknob and you see a spark.

Some Observable Forces are due to Electric Forces Many of the forces that act on objects are due to the electric forces between atoms and molecules. All atoms contain electrically charged electrons and protons. When atoms or molecules get close enough, they can exert electric forces on each other that can be attractive or repulsive. For example, when you push on a door, the atoms on the surface of your hand get close to the atoms on the surface of the door. These atoms are close enough that they exert electric forces on each other. The electric forces the atoms in your hand exert on the atoms in the door cause the door to move. The frictional force between two surfaces in contact is due to the attractive electric forces between the atoms on the two surfaces. These forces cause the surfaces to stick together.

Conductors and Insulators

If you reach for a metal doorknob after walking across a carpet, you might see a spark. The spark is caused by electrons moving from your hand to the doorknob, as shown in **Figure 5.** Recall that electrons were transferred from the carpet to your shoes. How did these electrons move from your shoes to your hand?

Conductors A material in which electrons are able to move easily is a **conductor.** Electrons on your shoes repel each other and some are pushed onto your skin. Because your skin is a better conductor than your shoes, the electrons spread over your skin, including your hand.

The best electrical conductors are metals. The atoms in metals have electrons that are able to move easily through the material. Electric wires usually are made of copper because copper metal is one of the best conductors.

Insulators A material in which electrons are not able to move easily is an **insulator.** Electrons are held tightly to atoms in insulators. Most plastics are insulators. The plastic coating around electric wires, such as the one shown in **Figure 6,** prevents a dangerous electric shock when you touch the wire. Other good insulators are wood, rubber, and glass.

Charging Objects

You might have noticed socks clinging to each other after they have been tumbling in a clothes dryer. Rubbing two materials together can result in a transfer of electrons. Then one material is left with a positive charge and the other with an equal amount of negative charge. The process of transferring charge by touching or rubbing is called **charging by contact.**

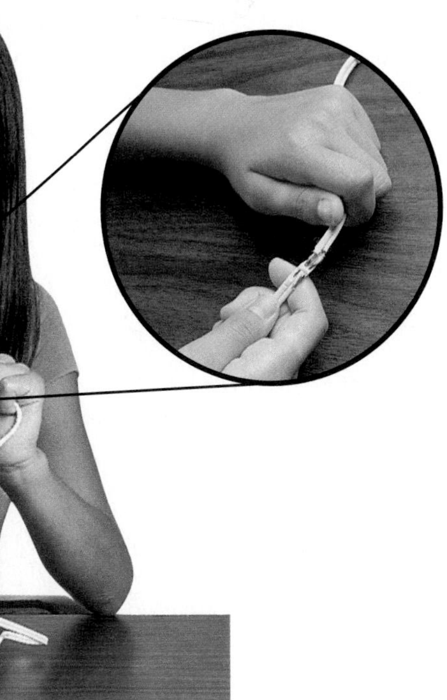

Figure 6 The plastic coating around wires is an insulator. A damaged electrical cord is hazardous when the conducting wire is exposed.
Explain *how the coating on a wire prevents an electric shock.*

Figure 7 The balloon on the left is neutral. The balloon on the right is negatively charged. It produces a positively charged area on the sleeve by repelling electrons. **Determine** *the direction of the electric force acting on the balloon.*

Charging at a Distance

Because electrical forces act at a distance, charged objects brought near a neutral object will cause electrons to rearrange their positions on the neutral object. Suppose you charge a balloon by rubbing it with a cloth. If you bring the negatively charged balloon near your sleeve, the extra electrons on the balloon repel the electrons in the sleeve. The electrons near the sleeve's surface move away from the balloon, leaving a positively charged area on the surface of the sleeve, as shown in **Figure 7.** As a result, the negatively charged balloon attracts the positively charged area of the sleeve. The rearrangement of electrons on a neutral object caused by a nearby charged object is called **charging by induction.** The sweater was charged by induction. The balloon will now cling to the sweater, being held there by an electrical force.

Lightning Have you ever seen lightning strike Earth? Lightning is a large static discharge. A static discharge is a transfer of charge between two objects because of a buildup of static electricity. A thundercloud is a mighty generator of static electricity. As air masses move and swirl in the cloud, areas of positive and negative charge build up. Eventually, enough charge builds up to cause a static discharge between the cloud and the ground. As the electric charges move through air, they collide with atoms and molecules. These collisions cause the atoms and molecules in air to emit light. You see this light as a spark, as shown in **Figure 8.**

Thunder Not only does lightning produce a brilliant flash of light, it also generates powerful sound waves. The electrical energy in a lightning bolt rips electrons off atoms in the air and produces great amounts of thermal energy. The surrounding air temperature can rise to about 25,000°C—several times hotter than the Sun's surface. The hot air in the lightning bolt's path expands rapidly, producing sound waves that you hear as thunder.

The sudden discharge of so much energy can be dangerous. It is estimated that Earth is struck by lightning about 100 times every second. Lightning strikes can cause power outages, injury, loss of life, and fires.

Topic: Lightning
Visit gpescience.com for Web links to information about lightning strikes.

Activity Make a table listing tips on how people can protect themselves from lightning.

NATIONAL GEOGRAPHIC

VISUALIZING LIGHTNING

Figure 8

Storm clouds can form when humid, Sun-warmed air rises to meet a colder air layer. As these air masses churn together, the stage is set for the explosive electrical display we call lightning. Lightning strikes when negative charges at the bottom of a storm cloud are attracted to positive charges on the ground.

A Convection currents in the storm cloud cause charge separation. The top of the cloud becomes positively charged, the bottom negatively charged.

B Negative charges on the bottom of the cloud induce a positive charge on the ground below the cloud by repelling negative charges in the ground.

C When the bottom of the cloud has accumulated enough negative charges, the attraction of the positive charges below causes electrons in the bottom of the cloud to move toward the ground.

D When the electrons get close to the ground, they attract positive charges that surge upward, completing the connection between cloud and ground. This is the spark you see as a lightning flash.

INTRACLOUD LIGHTNING never strikes Earth and can occur ten times more often in a storm than cloud-to-ground lightning.

Figure 9 A lightning rod directs the charge from a lightning bolt safely to the ground.

Grounding The sensitive electronics in a computer can be harmed by large static discharges. A discharge can occur any time that charge builds up in one area. Providing a path for charge to reach Earth prevents any charge from building up. Earth is a large, neutral object that is also a conductor of charge. Any object connected to Earth by a good conductor will transfer any excess electric charge to Earth. Connecting an object to Earth with a conductor is called grounding. For example, to prevent damage by lightning, buildings often have a metal lightning rod that provides a conducting path from the highest point on the building to the ground, as shown in **Figure 9.**

Plumbing fixtures, such as metal faucets, sinks, and pipes, often provide a convenient ground connection. Look around. Do you see anything that might act as a path to the ground?

Investigating Charged Objects

Procedure

1. Complete the safety form.
2. Fold over about 1 cm on the end of a **roll of transparent tape** to make a handle. Tear off a strip of tape about 10 cm long.
3. Stick the strip to a clean, dry, smooth surface, such as a countertop. Make an identical strip and stick it directly on top of the first.
4. Pull both pieces off the counter together and pull them apart. Bring the nonsticky sides of both tapes together. What happens?
5. Stick the two strips of tape side by side on the smooth surface. Pull them off and bring the nonsticky sides near each other again.

Analysis

1. What happened when you brought the pieces close together the first time? How were they charged? What might have caused this?
2. What happened when you brought the pieces together the second time? How were they charged? What did you do that might have changed the behavior?

Try at Home

Detecting Electric Charge

The presence of electric charges can be detected by an electroscope. One kind of electroscope is made of two thin, metal leaves attached to a metal rod with a knob at the top. The leaves are allowed to hang freely from the metal rod. When the device is not charged, the leaves hang straight down, as shown in **Figure 10A.**

Suppose a negatively charged rod touches the knob. Because the metal is a good conductor, electrons travel down the rod into the leaves. Both leaves become negatively charged as they gain electrons, as shown in **Figure 10B.** Because the leaves have similar charges, they repel each other.

If a glass rod is rubbed with silk, electrons move away from the atoms in the glass rod and build up on the silk. The glass rod becomes positively charged.

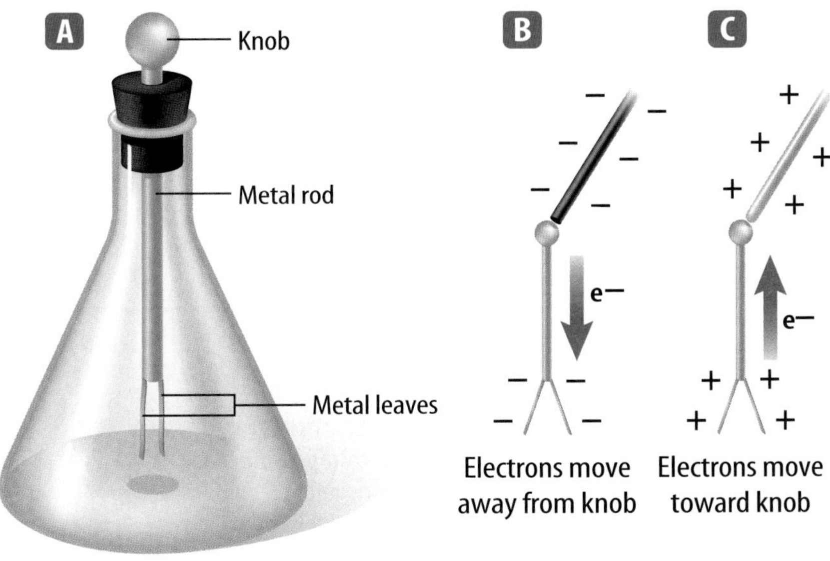

Figure 10 Notice the position of the leaves on the electroscope when they are **A** uncharged, **B** negatively charged, and **C** positively charged.
Infer *How can you tell whether an electroscope is positively or negatively charged?*

When the positively charged glass rod is brought into contact with the metal knob of an uncharged electroscope, electrons flow out of the metal leaves and onto the rod. The leaves repel each other because each leaf becomes positively charged as it loses electrons, as shown in **Figure 10C.**

section 1 review

Summary

Positive and Negative Charge

- There are two types of electric charge: positive charge and negative charge.
- Electric charges can be transferred between objects but cannot be created or destroyed.
- Like charges repel and unlike charges attract.
- An electric charge is surrounded by an electric field that exerts forces on other charges.

Electrical Conductors and Insulators

- A conductor contains electrons that can move easily. The best conductors are metals.
- The electrons in an electrical insulator do not move easily. Rubber, glass, and most plastics are examples of insulators.

Charging Objects

- Electric charge can be transferred between objects by bringing them into contact.
- Charging by induction occurs when the electric field around a charged object rearranges electrons in a nearby neutral object.

Self Check

1. **Define** static electricity.
2. **Describe** how lightning is produced.
3. **Explain** why electrically neutral objects can become electrically charged even though charge cannot be created or destroyed.
4. **Predict** what would happen if you touched the knob of a positively charged electroscope with another positively charged object.
5. **Think Critically** Humid air is a better electrical conductor than dry air. Explain why you're more likely to receive a shock after walking across a carpet when the air is dry than when the air is humid.

Applying Math

6. **Determine Lightning Strikes** Earth is struck by lightning 100 times each second. How many times is Earth struck by lightning in one day?
7. **Calculate Electric Force** A balloon with a mass of 0.020 kg is charged by rubbing and then is stuck to the ceiling. If the acceleration of gravity is 9.8 m/s^2, what is the electrical force on the balloon?

Electric Current

Reading Guide

What **You'll Learn**

- **Describe** how voltage difference causes current to flow.
- **Explain** how batteries produce a voltage difference in a circuit.
- **List** the factors that affect an object's electrical resistance.
- **Define** Ohm's law.

Why **It's Important**

You control electric current every time you change the volume on a TV, stereo, or CD player.

Review Vocabulary

pressure: amount of force exerted per unit area

New Vocabulary

- electric current
- voltage difference
- circuit
- resistance
- Ohm's law

Figure 11 Electric forces in a material cause electric current to flow, just as forces in the water cause water to flow.

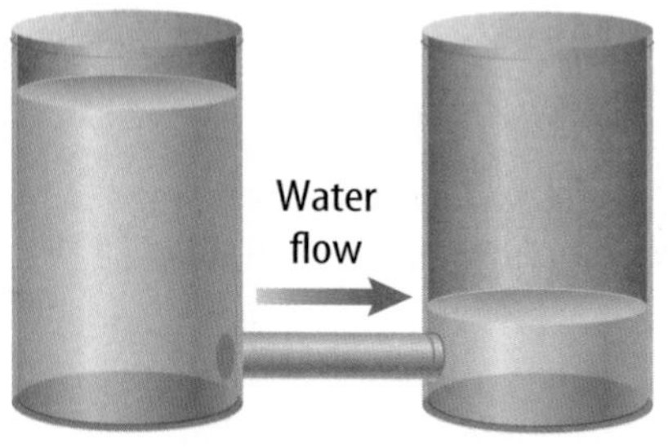

The force that causes water to flow is related to a pressure difference.

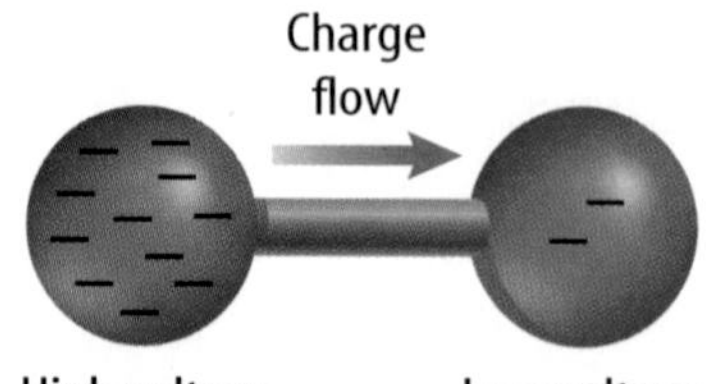

The force that causes a current to flow is related to a voltage difference.

Current and Voltage Difference

When a spark jumps between your hand and a metal doorknob, electric charges move quickly from one place to another. The net movement of electric charges in a single direction is an **electric current.** In a metal wire, or any material, electrons are in constant motion in all directions. As a result, there is no net movement of electrons in one direction. However, when an electric current flows, electrons continue their random movement, but they also drift in the direction that the electron current flows.

Electric current is measured in amperes. One ampere is equal to 6,250 million billion electrons flowing past a point every second.

Reading Check *What is electric current?*

Voltage Difference The movement of an electron in an electric current is similar to a ball bouncing down a flight of stairs. Even though the ball changes direction when it strikes a stair, the net motion of the ball is downward. The downward motion of the ball is caused by the force of gravity. When a current flows, the net movement of electric charges is caused by an electric force acting on the charges.

In some ways, the electric force that causes charges to flow is similar to the force acting on the water in a pipe. Water flows from higher pressure to lower pressure, as shown in **Figure 11.** In a similar way, electric charge flows when there is a voltage difference. A **voltage difference** is related to the force that causes electric charges to flow. Voltage difference is measured in volts.

Figure 12 Water or electric current will flow continually only through a closed loop. If any part of the loop is broken or disconnected, the flow stops.

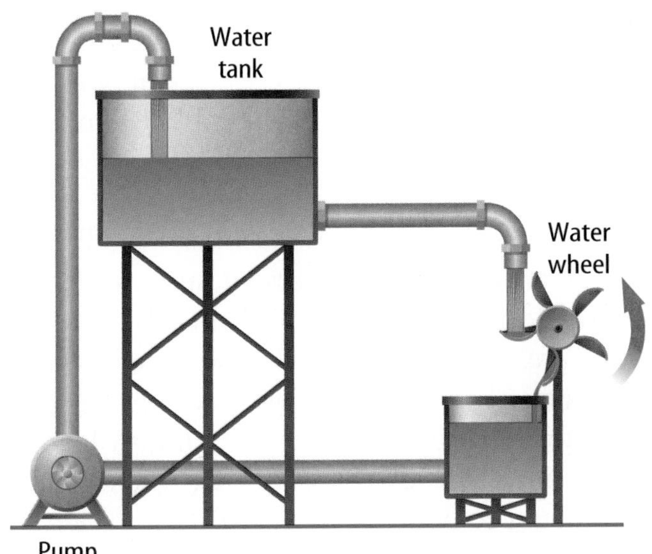

A pump provides the pressure difference that keeps water flowing.

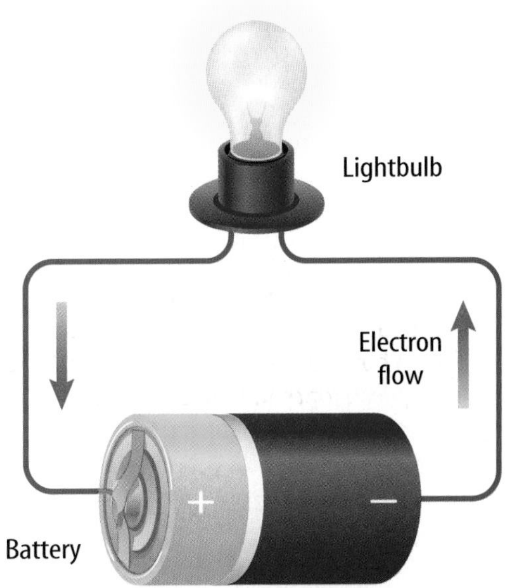

A battery provides the voltage difference that keeps electric current flowing.

Electric Circuits One way to have flowing water perform work is shown in **Figure 12.** Water flows out of the tank and falls on a paddle wheel, causing it to rotate. A pump then provides a pressure difference that lifts the water back up into the tank. The constant flow of water would stop if the pump stopped working. The flow of water also would stop if one of the pipes broke. Then water no longer could flow in a closed loop, and the paddle wheel would stop rotating.

Figure 12 also shows an electric current doing work by lighting a lightbulb. Just as the water current stops flowing if there is no longer a closed loop to flow through, the electric current stops if there is no longer a closed path to follow. A closed path that electric current follows is a **circuit.** If the circuit in **Figure 12** is broken by removing the battery, the lightbulb, or one of the wires, current will not flow.

Batteries

To keep water flowing continually in the water circuit in **Figure 12,** a pump is used to provide a pressure difference. In a similar way, to keep an electric current continually flowing in the electric circuit in **Figure 12,** a voltage difference needs to be maintained in the circuit. A battery can provide the voltage difference that is needed to keep current flowing in a circuit. Current flows as long as there is a closed path that connects one battery terminal to the other battery terminal.

Investigating Battery Addition

Procedure

1. Complete the safety form.
2. Connect two bulbs and one D-cell battery in a loop to make a circuit.
3. Add another battery by connecting the negative terminal of one battery to the positive terminal of the other.
4. Compare the brightness of the bulbs in the two circuits.

Analysis

1. Add the voltage difference for each battery to find the voltage difference in the circuit in step 2.
2. If the brightness of a bulb increases as the current increases, how are the current and the voltage difference in a circuit related?

Dry-Cell Batteries You probably are most familiar with dry-cell batteries. A cell consists of two electrodes surrounded by a material called an electrolyte. The electrolyte enables charges to move from one electrode to the other. Look at the dry cell shown in **Figure 13.** One electrode is the carbon rod, and the other is the zinc container. The electrolyte is a moist paste containing several chemicals. The cell is called a dry cell because the electrolyte is a moist paste, not a liquid solution.

INTEGRATE Chemistry When the two terminals of a dry-cell battery are connected in a circuit, such as in a flashlight, a reaction involving zinc and several chemicals in the paste occurs. Electrons are transferred between some of the compounds in this chemical reaction. As a result, the carbon rod becomes positive, forming the positive (+) terminal. Electrons accumulate on the zinc, making it the negative (−) terminal.

The voltage difference between these two terminals causes current to flow through a closed circuit. You make a battery when you connect two or more cells together to produce a higher voltage difference.

Wet-Cell Batteries Another commonly used type of battery is the wet-cell battery. A wet cell, like the one shown in **Figure 13,** contains two connected plates made of different metals or metallic compounds in a conducting solution. A wet-cell battery contains several wet cells connected together.

Figure 13 Chemical reactions in batteries produce a voltage difference between the positive and negative terminals. **Identify** *when these chemical reactions occur.*

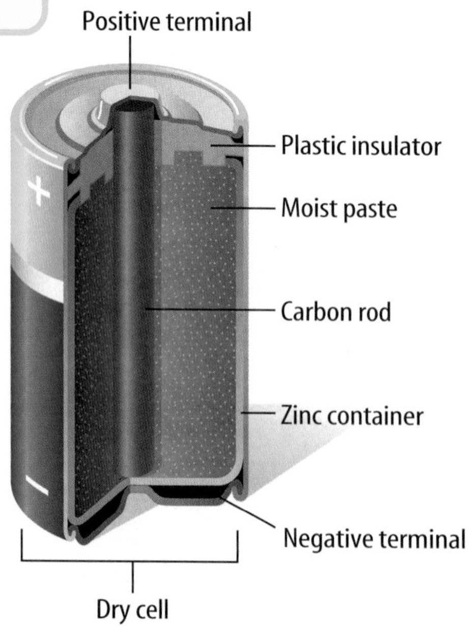

In this dry cell, chemical reactions in the moist paste transfer electrons to the zinc container.

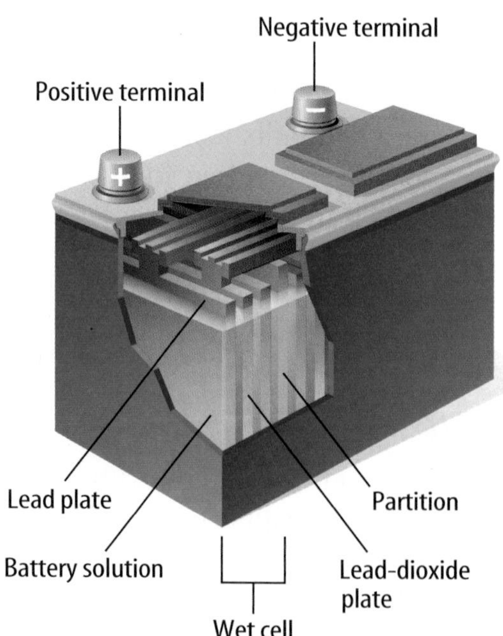

In this wet cell, chemical reactions transfer electrons from the lead plates to the lead-dioxide plates.

Lead-Acid Batteries Most car batteries are lead-acid batteries, like the wet-cell battery shown in **Figure 13.** A lead-acid battery contains a series of six wet cells made up of lead and lead dioxide plates in a sulfuric acid solution. The chemical reaction in each cell provides a voltage difference of about 2 V, giving a total voltage difference of 12 V. As a car is driven, the alternator recharges the battery by sending current through the battery in the opposite direction to reverse the chemical reaction.

A voltage difference is provided at electrical outlets, such as a wall socket. This voltage difference usually is higher than the voltage difference provided by batteries. Most types of household devices are designed to use the voltage difference supplied by a wall socket. In the United States, the voltage difference across the two holes in a wall socket is usually 120 V. Some wall sockets supply 240 V, which is required by appliances such as electric ranges and electric clothes dryers.

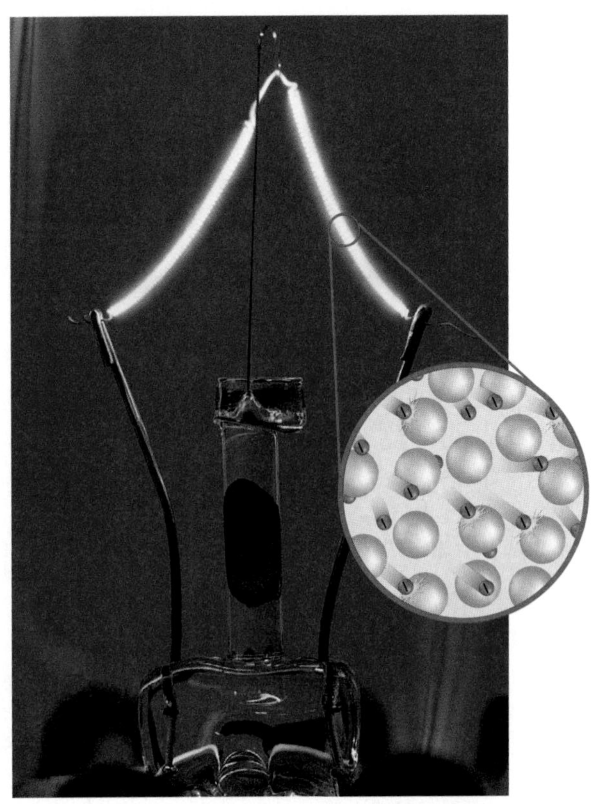

Figure 14 As electrons move through the filament in a lightbulb, they bump into metal atoms. Due to the collisions, the metal heats up and starts to glow.
Describe *the energy conversions that occur in a lightbulb filament.*

Resistance

Flashlights use dry-cell batteries to provide the electric current that lights a lightbulb. What makes a lightbulb glow? Look at the lightbulb in **Figure 14.** Part of the circuit through the bulb is a thin wire called a filament. As the electrons flow through the filament, they bump into the metal atoms that make up the filament. In these collisions, some of the electrical energy of the electrons is converted into thermal energy. Eventually, the metal filament becomes hot enough to glow, producing radiant energy that can light up a dark room.

Resisting the Flow of Current Electric current loses energy as it moves through the filament because the filament resists the flow of electrons. **Resistance** is the tendency of a material to oppose the flow of electrons, changing electrical energy into thermal energy and light. With the exception of some substances that become superconductors at low temperatures, all materials have some electrical resistance. Electrical conductors have much less resistance than insulators. Resistance is measured in ohms (Ω).

Copper is an excellent conductor and has low resistance to the flow of electrons. Copper is used in household wiring because only a small amount of electrical energy is converted to thermal energy as current flows in copper wires.

Temperature, Length, and Thickness The electric resistance of most materials usually increases as the temperature of the material increases. The resistance of an object such as a wire also depends on the length and diameter of the wire. The resistance of a wire, or any conductor, increases as the wire becomes longer. The resistance also increases as the wire becomes thinner.

In a 60-W lightbulb, the filament is a piece of tungsten wire made into a short coil a few centimeters long. The uncoiled wire is about 2 m long and only about 0.25 mm thick. Even though tungsten metal is a good conductor, by making the wire thin and long, the resistance of the filament is made large enough to cause the bulb to glow.

Reading Check *How does changing the length and thickness of a wire affect its resistance?*

The Current in a Simple Circuit

A simple electric circuit contains a source of voltage difference, such as a battery, a device that has resistance, such as a lightbulb, and conductors that connect the device to the battery terminals. When the wires are connected to the battery terminals, current flows in the closed path. An example of a simple circuit is shown in **Figure 15.**

The voltage difference, current, and resistance in a circuit are related. If the voltage difference doesn't change, decreasing the resistance increases the current in the circuit, as shown in **Figure 15.** Also, if the resistance doesn't change, increasing the voltage difference increases the current.

Figure 15 The amount of current flowing through a circuit is related to the amount of resistance in the circuit.

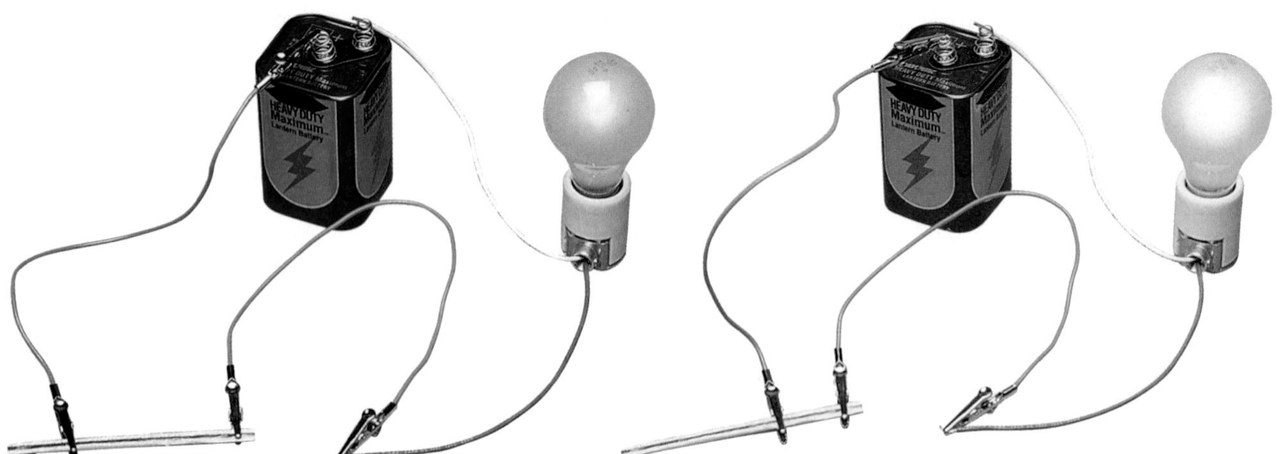

When the clips on the graphite rod are farther apart, the resistance of the rod in the circuit is larger. As a result, less current flows in the circuit and the lightbulb is dim.

When the clips on the graphite rod are closer together, the resistance of the rod in the circuit is less. As a result, more current flows in the circuit and the lightbulb is brighter.

Ohm's Law The relationship among voltage difference, current, and resistance in a circuit is known as Ohm's law. According to **Ohm's law,** the current in a circuit equals the voltage difference divided by the resistance. Ohm's law can be written as the following equation, where I stands for electric current.

Ohm's Law

$$\text{current (in amperes)} = \frac{\text{voltage difference (in volts)}}{\text{resistance (in ohms)}}$$

$$I = \frac{V}{R}$$

Ohm's law provides a way to measure the resistance of objects and materials. First, the equation above is written as:

$$R = \frac{V}{I}$$

An object is connected to a source of voltage difference and the current flowing in the circuit is measured. The object's resistance then equals the voltage difference divided by the measured current.

Current and the Human Body When an electric shock occurs, an electric current moves through some part of the body. The damage caused by an electric shock depends on how large the current is. Research the effects of current on the human body. Make a table showing the effects on the body at different amounts of current.

section 2 review

Summary

Current and Voltage Difference

- Electric current is the net movement of electric charge in a single direction.
- A voltage difference is related to the force that causes charges to flow.
- A circuit is a closed, conducting path.

Batteries

- Chemical reactions in a battery produce a voltage difference between the positive and negative battery terminals.
- Two commonly used types of batteries are dry-cell batteries and wet-cell batteries.

Resistance and Ohm's Law

- Resistance is the tendency of a material to oppose the flow of electrons.
- Ohm's law relates the current, I, resistance, R, and voltage difference, V, in a circuit:

$$I = \frac{V}{R}$$

Self Check

1. **Compare and contrast** a current traveling through a circuit with a static discharge.
2. **Explain** how a carbon-zinc dry cell produces a voltage difference between the positive and negative terminals.
3. **Identify** two ways to increase the current in a simple circuit.
4. **Compare and contrast** the flow of water in a pipe and the flow of electrons in a wire.
5. **Think Critically** Explain how the resistance of a lightbulb filament changes after the light has been turned on.

Applying Math

6. **Calculate** the voltage difference in a circuit with a resistance of 25 ohms if the current is 0.5 A.
7. **Calculate Resistance** A current of 0.5 A flows in a 60-W lightbulb when the voltage difference between the ends of the filament is 120 V. What is the resistance of the filament?

Identifying Conductors and Insulators

When an insulator is connected in a circuit, only a small current flows. As a result, a lightbulb connected in a circuit with an insulator usually will not glow. In this lab, you will use the brightness of a lightbulb to identify conductors and insulators.

Real-World Problem

What materials are conductors and what materials are insulators?

Goals

- **Identify** conductors and insulators.
- **Describe** the common characteristics of conductors and insulators.

Materials

battery
flashlight bulb
bulb holder
insulated wire

Safety Precautions

Procedure

1. Complete the safety form before you begin.
2. Set up an open circuit as shown above.
3. Touch the free bare ends of the wires to various objects around the room. Test at least 12 items.
4. Copy the table below. In your table, record which materials make the lightbulb glow and which don't.

Material Tested with Lightbulb Circuit	
Lightbulb Glows	Lightbulb Doesn't Glow
Do not write in this book.	

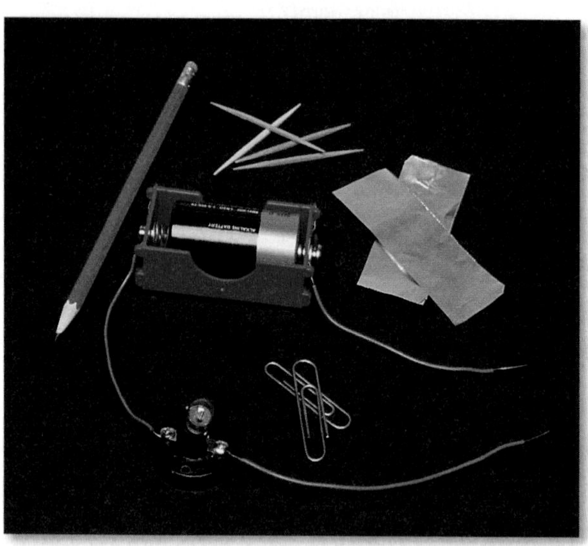

Conclude and Apply

1. Is there a pattern to your data?
2. Do all or most of the materials that light the lightbulb have something in common?
3. Do all or most of the materials that don't light the lightbulb have something in common?
4. **Explain** why one material might allow the lightbulb to light and another might prevent the lightbulb from lighting.
5. **Predict** what other materials will allow the lightbulb to light and what will prevent the lightbulb from lighting.
6. **Classify** all the materials you have tested as conductors or insulators.

Communicating Your Data

Compare your conclusions with those of other students in your class. **For more help, refer to the Science Skill Handbook.**

section 3

Electrical Energy

Reading Guide

What You'll Learn

- **Describe** the difference between series and parallel circuits.
- **Recognize** the function of circuit breakers and fuses.
- **Calculate** electrical power.
- **Calculate** the electrical energy used by a device.

Why It's Important

When you use an electric appliance, such as a hair dryer or a toaster oven, you pay for the electrical energy you use.

Review Vocabulary

energy: the ability to cause change

New Vocabulary

- series circuit
- parallel circuit
- electrical power

Series and Parallel Circuits

Look around. How many electrical devices such as lights, clocks, stereos, and televisions do you see that are plugged into electrical outlets? Circuits usually include three components. One is a source of voltage difference that can be provided by a battery or an electrical outlet. Another is one or more devices that use electrical energy. Circuits also include conductors such as wires that connect the devices to the source of voltage difference to form a closed path.

Think about using a hair dryer. The dryer must be plugged into an electrical outlet to operate. A generator at a power plant produces a voltage difference across the outlet, causing charges to move when the circuit is complete. The dryer and the circuit in the house contain conducting wires to carry current. The hair dryer turns the electrical energy into thermal energy and mechanical energy. When you unplug the hair dryer or turn off its switch, you open the circuit and break the path of the current. To use electrical energy, a complete circuit must be made. There are two kinds of circuits.

Series Circuits One kind of circuit is called a series circuit. In a **series circuit,** the current has only one loop to flow through, as shown in **Figure 16.** Series circuits are used in flashlights and some holiday lights.

Reading Check *How many loops are in a series circuit?*

Figure 16 A series circuit provides only one path for the current to follow.
Infer *What happens to the brightness of each bulb as more bulbs are added?*

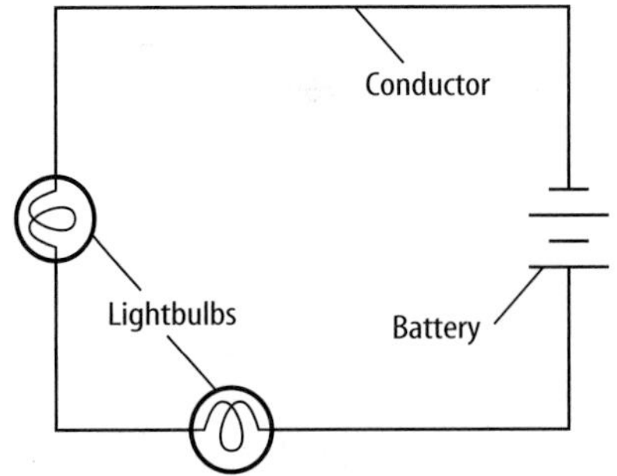

Electrician The installation of electrical wiring in any building usually requires an electrician. Electricians must have a thorough understanding of electricity. They also must constantly be aware of the safety issues involved in working with electricity. Write a paragraph describing other skills that electricians must have.

Open Circuits If you have ever decorated a window or a tree with a string of lights, you might have had the frustrating experience of trying to find one burned-out bulb. How can one faulty bulb cause the whole string to go out? Because the parts of a series circuit are wired one after another, the amount of current is the same through every part. When any part of a series circuit is disconnected, no current flows through the circuit. This is called an open circuit. The burned-out bulb causes an open circuit in the string of lights.

Parallel Circuits What would happen if your home were wired in a series circuit and you turned off one light? This would cause an open circuit, and all the other lights and appliances in your home would go out, too. This is why houses are wired with parallel circuits. **Parallel circuits** contain two or more branches for current to move through. Look at the parallel circuit in **Figure 17.** The current can flow through both or either of the branches. Because all branches connect the same two points of the circuit, the voltage difference is the same in each branch. Then, according to Ohm's law, more current flows through the branches that have lower resistance.

Parallel circuits have several advantages. When one branch of the circuit is opened, such as when you turn a light off, the current continues to flow through the other branches. Houses, automobiles, and most electrical systems use parallel wiring so individual parts can be turned off without affecting the entire circuit.

Figure 17 In parallel circuits, the current follows more than one path.
Describe *how the voltage difference will compare in each branch.*

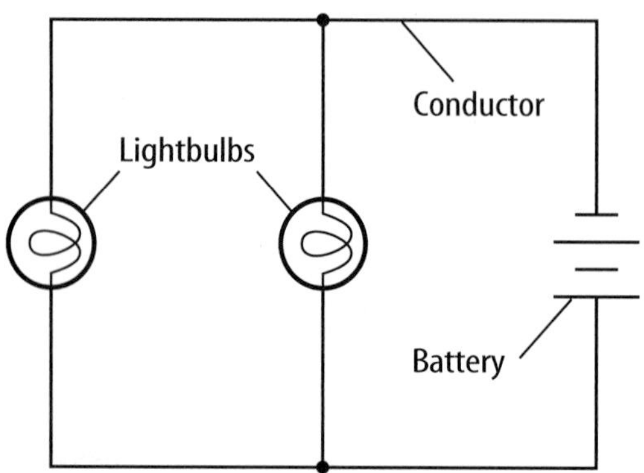

Figure 18 The wiring in a house must allow for the individual use of various appliances and fixtures. **Identify** *the type of circuit that is most common in household wiring.*

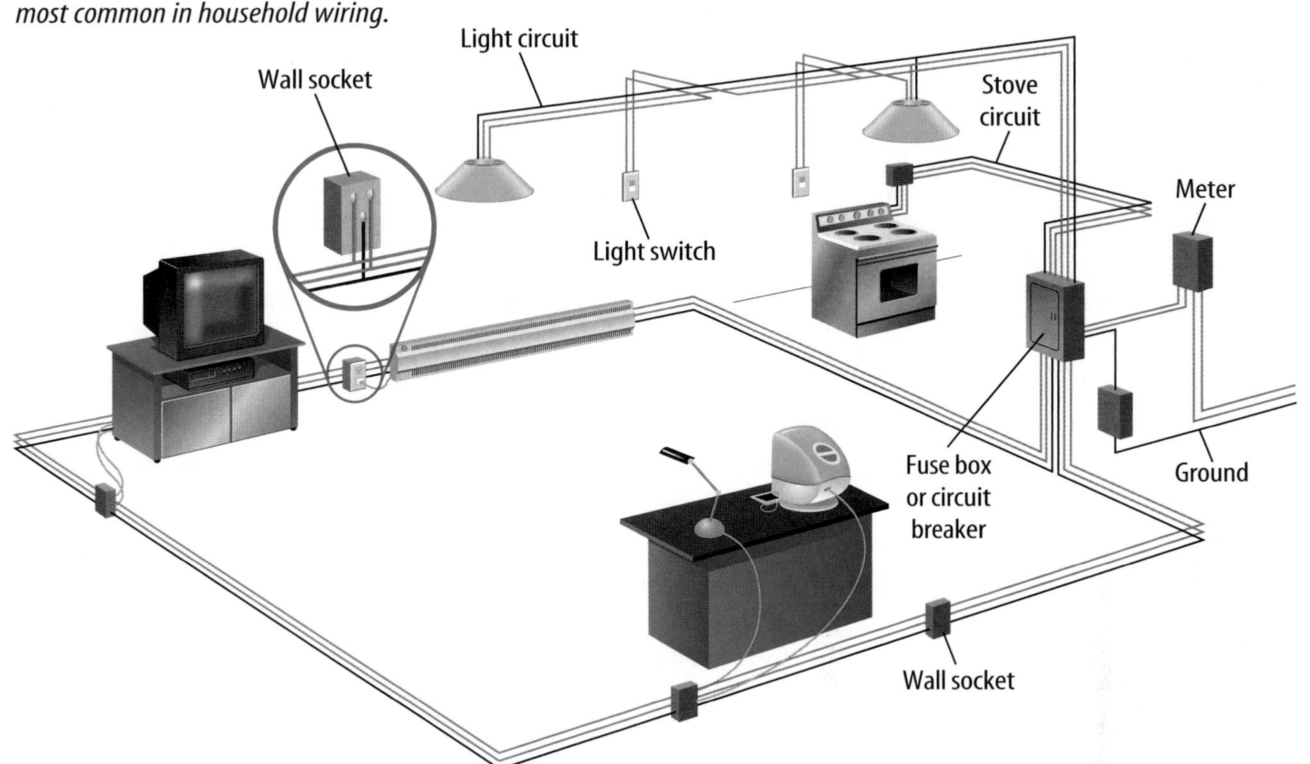

Household Circuits

Count how many different things in your home require electrical energy. You don't see the wires because most of them are hidden behind the walls, ceilings, and floors. This wiring is mostly a combination of parallel circuits connected in an organized and logical network. **Figure 18** shows how electrical energy enters a home and is distributed. In the United States, the voltage difference in most of the branches is 120 V. In some branches that are used for electric stoves or electric clothes dryers, the voltage difference is 240 V. The main switch and circuit breaker or fuse box serve as an electrical headquarters for your home. Parallel circuits branch out from the breaker or fuse box to wall sockets, major appliances, and lights.

In a house, many appliances draw current from the same circuit. If more appliances are connected, more current will flow through the wires. As the amount of current increases, so does the amount of heat produced in the wires. If the wires get too hot, the insulation can melt and the bare wires can cause a fire. To protect against overheating of the wires, all household circuits contain either a fuse or a circuit breaker.

Fuses When you hear that somebody has "blown a fuse," it means that the person has lost his or her temper. This expression comes from the function of an electrical fuse, shown in **Figure 19A,** which contains a small piece of metal that melts if the current becomes too high. When it melts, it causes a break in the circuit, stopping the flow of current through the overloaded circuit. To enable current to flow again in the circuit, you must replace the blown fuse with a new one. However, before you replace the blown fuse, you should turn off or unplug some of the appliances. Too many appliances in use at the same time is the most likely cause of the overheating of the circuit.

Figure 19 Two useful devices to prevent electric circuits from overheating are A fuses and B circuit breakers.
Evaluate *which device, a fuse or a circuit breaker, would be more convenient to have in the home.*

Circuit Breaker A circuit breaker, shown in **Figure 19B,** is another device that prevents a circuit from overheating and causing a fire. A circuit breaker contains a piece of metal that bends when the current in it is so large that it gets hot. The bending causes a switch to flip and open the circuit, stopping the flow of current. Circuit breakers usually can be reset by pushing the switch to its "on" position. Again, before you reset a circuit breaker, you should turn off or unplug some of the appliances from the overloaded circuit. Otherwise, the circuit breaker will switch off again.

Reading Check ***What is the purpose of fuses and circuit breakers in household circuits?***

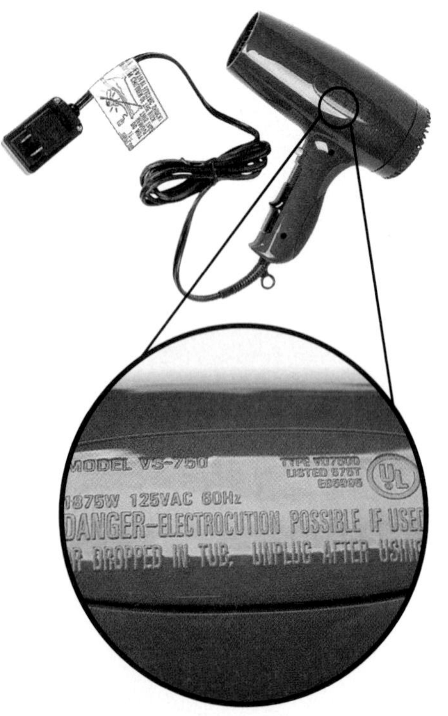

Figure 20 All appliances come with a power rating.

Electrical Power

The reason why electricity is so useful is that electrical energy is converted easily to other types of energy. For example, electrical energy is converted to mechanical energy as the blades of a fan rotate to cool you. Electrical energy is converted to light energy in lightbulbs. A hair dryer changes electrical energy into thermal energy. The rate at which electrical energy is converted to another form of energy is the **electrical power.**

The electrical power used by appliances varies. Appliances often are labeled with a power rating that describes how much power the appliance uses, as shown in **Figure 20.** Appliances that have electric heating elements, such as ovens and hair dryers, usually use more electrical power than other appliances.

Calculating Electrical Power The electrical power used depends on the voltage difference and the current. Electrical power can be calculated from the following equation.

Electrical Power Equation

electrical power (in watts) = current (in amperes) × voltage difference (in volts)

$$P = IV$$

The unit for power is the watt (W). Because the watt is a small unit of power, electrical power is often expressed in kilowatts (kW). One kilowatt equals 1,000 watts.

Applying Math — Solve a Simple Equation

POWER USED BY A CLOTHES DRYER The current in an electric clothes dryer is 15 A when it is plugged into a 240-V outlet. How much power does the clothes dryer use?

IDENTIFY known values and the unknown value

Identify the known values:

current in the clothes dryer is 15 A → means → $I = 15\ \text{A}$

plugged into a 240-V outlet → means → $V = 240\ \text{V}$

Identify the unknown value:

how much power does the clothes dryer use → means → $P = ?\ \text{W}$

SOLVE the problem

Substitute the known values $I = 15$ A and $V = 240$ V into the electrical power equation:

$$P = IV = (15\ \text{A})(240\ \text{V}) = 3{,}600\ \text{W} = 3.6\ \text{kW}$$

CHECK the answer

Does your answer seem reasonable? Check your answer by dividing the power you calculated by the current given in the problem. The result should be the voltage difference given in the problem.

Practice Problems

1. A toaster oven is plugged into an outlet that provides a voltage difference of 120 V. What power does the oven use if the current is 10 A?
2. A VCR that is not playing still uses 10.0 W of power. What is the current if the VCR is plugged into a 120-V electrical outlet?
3. A flashlight bulb uses 2.4 W of power when the current in the bulb is 0.8 A. What is the voltage difference?

For more practice problems go to page 879, and visit Math Practice at gpescience.com.

Topic: Energy

Visit gpescience.com for Web links to information about the cost of electrical energy around the country.

Activity Using a blank map of the United States, create a key showing the relative energy costs in different states in different colors. With a partner, color in the states to create a visual map of energy costs. Give your map a title.

Electrical Energy Using electrical power costs money. However, electric companies charge by the amount of electrical energy used rather than by the electrical power used. Electrical energy usually is measured in units of kilowatt-hours (kWh) and can be calculated from this equation:

Electrical Energy Equation

electrical energy (in kWh) = **electrical power** (in kW) × time (in hours)

$$E = Pt$$

In the above equation, electrical power is in units of kW and the time is the number of hours that the electrical power is used.

Applying Math — Solve a Simple Equation

ELECTRICAL ENERGY USED BY A MICROWAVE OVEN A microwave oven with a power rating of 1,200 W is used for 0.25 h. How much electrical energy is used by the microwave oven?

IDENTIFY known values and the unknown value

Identify the known values:

power rating of 1,200 W → means → $P = 1{,}200 \text{ W} = 1.2 \text{ kW}$

is used for 0.25 h → means → $t = 0.25 \text{ h}$

Identify the unknown value:

how much electrical energy is used → means → $E = ? \text{ kWh}$

SOLVE the problem

Substitute the known values $p = 1.2$ kW and $t = 0.25$ h into the electrical energy equation:

$$E = Pt = (1.2 \text{ kW})(0.25 \text{ h}) = 0.30 \text{ kWh}$$

CHECK your answer

Does your answer seem reasonable? Check your answer by dividing the electrical energy you calculated by the power given in the problem. The results should be the time given in the problem.

Practice Problems

1. A refrigerator operates on average for 10.0 h a day. If the power rating of the refrigerator is 700 W, how much electrical energy does the refrigerator use in one day?
2. A TV with a power rating of 200 W uses 0.8 kWh in one day. For how many hours was the TV on during this day?

For more practice problems go to page 879, and visit Math Practice at gpescience.com.

The Cost of Using Electrical Energy

The cost of using an appliance can be computed by multiplying the electrical energy used by the amount the power company charges for each kWh. For example, if a 100-W lightbulb is left on for 5 h, the amount of electrical energy used is

$$E = Pt = (0.1 \text{ kW})(5 \text{ h}) = 0.5 \text{ kWh}$$

If the power company charges \$0.10 per kWh, the cost of using the bulb for 5 h is

$$\text{cost} = (\text{kWh used})(\text{cost per kWh})$$
$$= (0.5 \text{ kWh})(\$0.10/\text{kWh}) = \$0.05$$

The cost of using some household appliances is given in **Table 1,** where the cost per kWh is assumed to be \$0.09/kWh.

Table 1 Cost of Using Home Appliances

Appliance	Hair Dryer	Stereo	Color Television
Power rating	1,000	100	200
Hours used daily	0.25	2.0	4.0
kWh used monthly	7.5	6.0	24.0
Cost per kWh	\$0.09	\$0.09	\$0.09
Monthly cost	\$0.68	\$0.54	\$2.16

section 3 review

Summary

Series and Parallel Circuits

- A series circuit has only one path that current can follow. A parallel circuit has two or more branches that current can follow.
- Household wiring usually consists of a number of connected parallel circuits.
- Fuses and circuit breakers are used to prevent wires from overheating when the current in the wires becomes too large.

Electrical Power

- Electrical power is the rate at which electrical energy is converted into other forms of energy.
- Electrical power can be calculated by multiplying the current by the voltage difference:

$$P = IV$$

Electrical Energy

- The electrical energy used can be calculated by multiplying the power by the time:

$$E = Pt$$

- Electric power companies charge customers for the amount of electrical energy they use.

Self Check

1. **Explain** how electrical power and electrical energy are related.
2. **Discuss** why fuses and circuit breakers are used in household circuits.
3. **Explain** what determines the current in each branch of a parallel circuit.
4. **Explain** whether or not a fuse or circuit breaker should be connected in parallel to the circuit it is protecting.
5. **Think Critically** A parallel circuit consisting of four branches is connected to a battery. Explain how the amount of current that flows out of the battery is related to the amount of current in the branches of the circuit.

Applying Math

6. **Calculate** the current flowing into a desktop computer plugged into a 120-V outlet if the power used is 180 W.
7. **Calculate Electrical Power** A circuit breaker is tripped when the current in the circuit is greater than 15 A. If the voltage difference is 120 V, what is the power being used when the circuit breaker is tripped?
8. **Calculate** the monthly cost of using a 700-W refrigerator that runs for 10 h a day if the cost per kWh is \$0.09.

Design Your Own

Comparing Series and Parallel Circuits

Goals

- **Design** and construct series and parallel circuits.
- **Compare** and contrast the behaviors of series and parallel circuits.

Possible Materials

6-V dry-cell battery
small lights with sockets (3)
aluminum foil
paper clips
tape
scissors
paper

Safety Precautions

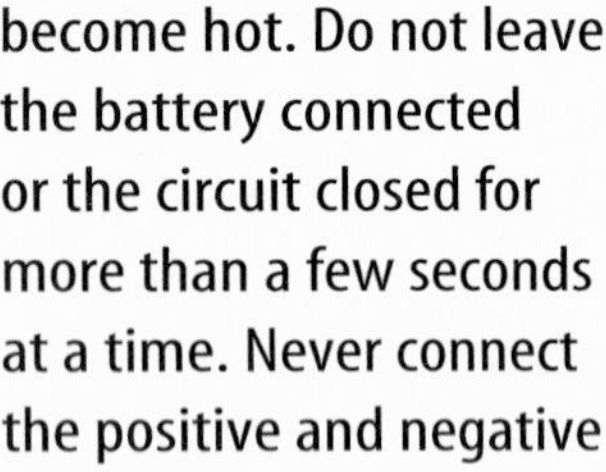

Some parts of circuits can become hot. Do not leave the battery connected or the circuit closed for more than a few seconds at a time. Never connect the positive and negative terminals of the dry-cell battery directly without including at least one bulb in the circuit.

Real-World Problem

Imagine what a bedroom might be like if it were wired in series. For an alarm clock to keep time and wake you in the morning, your lights and anything else that uses electricity would have to be on. Fortunately, most outlets in homes are wired in parallel circuits on separate branches of the main circuit. How do the behaviors of series and parallel circuits compare?

Form a Hypothesis

Predict what will happen to the other bulbs when one bulb is unscrewed from a series circuit and from a parallel circuit. Explain your prediction. Also, form a hypothesis to explain in which circuit the lights shine the brightest.

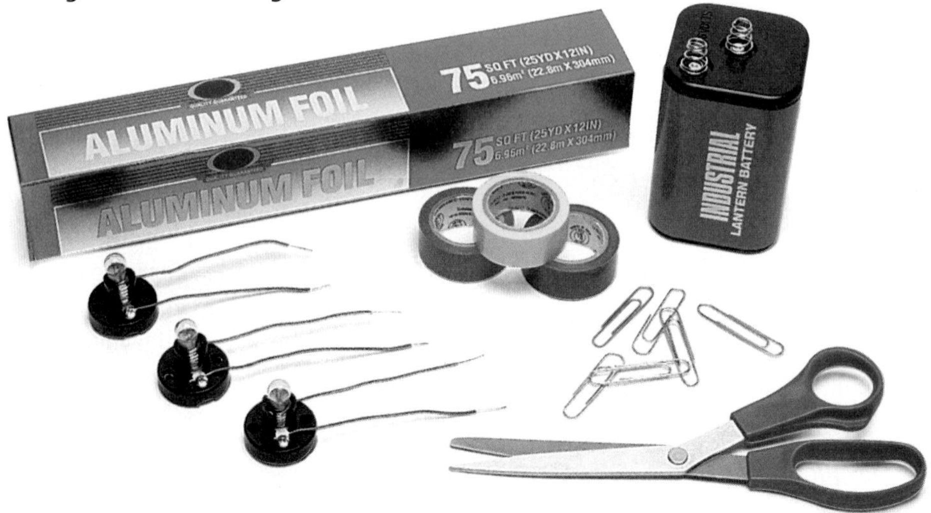

Test Your Hypothesis

Make a Plan

1. Complete the safety form before you begin.
2. As a group, agree upon and write the hypothesis statement.
3. Work together to determine and write the steps you will take to test your hypothesis. Include a list of the materials you will need.
4. How will your circuits be arranged? On a piece of paper, draw a large parallel circuit of three lights and the dry-cell battery as shown. On the other side, draw another circuit with the three bulbs arranged in series.
5. Make conducting wires by taping a piece of transparent tape to a sheet of aluminum foil and folding the foil over twice to cover the tape. Cut these to any length that works in your design.

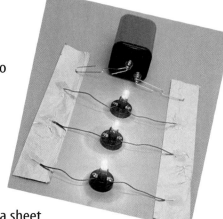

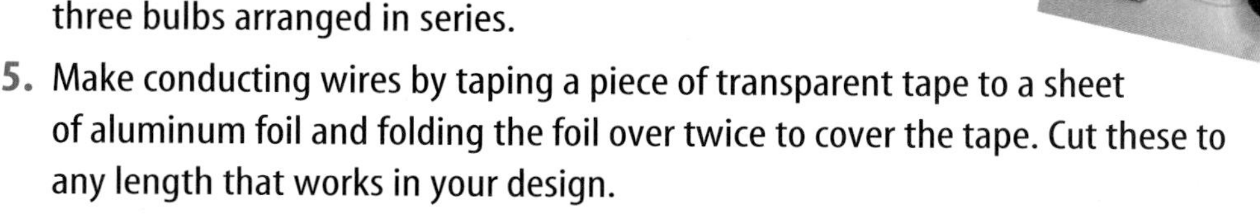

Follow Your Plan

1. Make sure your teacher approves your plan before you start.
2. Carry out the experiment. **WARNING:** *Leave the circuit on for only a few seconds at a time to avoid overheating.*
3. As you do the experiment, record your predictions and your observations in your Science Journal.

Analyze Your Data

1. **Predict** what will happen in the series circuit when a bulb is unscrewed at one end. What will happen in the parallel circuit?
2. **Compare** the brightness of the lights in the different circuits. Explain.
3. **Predict** what will happen to the brightness of the bulbs in the series circuit if you complete it with two bulbs instead of three bulbs. Test your prediction. How do the results demonstrate Ohm's law?

Conclude and Apply

1. Did the results support your hypothesis? Explain by using your observations.
2. Where in the parallel circuit would you place a switch to control all three lights? Where would you place a switch to control only one light? Test your predictions.

Prepare a poster to highlight the differences between a parallel and a series circuit. Include possible practical applications of both types of circuits. **For more help, refer to the** Science Skill Handbook.

Science *and* Language Arts

Invisible Man

by Ralph Ellison

I am an invisible man. No, I am not a spook like those who haunted Edgar Allan Poe; nor am I one of your Hollywood-movie ectoplasms.[1] I am a man of substance, of flesh and bone, fiber and liquids—and I might even be said to possess a mind. I am invisible, understand, simply because people refuse to see me.... Nor is my invisibility exactly a matter of biochemical accident to my epidermis.[2] That invisibility to which I refer occurs because of a peculiar disposition ... of those with whom I come in contact....

.... Now don't jump to the conclusion that because I call my home a "hole" it is damp and cold like a grave.... Mine is a warm hole.

My hole is warm and full of light. Yes, *full* of light. I doubt if there is a brighter spot in all New York than this hole of mine.... Perhaps you'll think it strange that an invisible man should need light, desire light, love light. Because maybe it is exactly because I *am invisible.* Light confirms my reality, gives birth to my form.... I myself, after existing some twenty years, did not become alive until I discovered my invisibility.

... In my hole in the basement there are exactly 1,369 lights. I've wired the entire ceiling, every inch of it.... Though invisible, I am in the great American tradition of tinkers. That makes me kin to Ford, Edison and Franklin.

Ralph Ellison

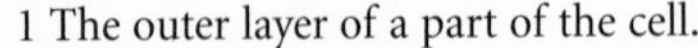

1 The outer layer of a part of the cell.

2 The outer layer of skin.

Understanding Literature

Prologue A prologue is an introduction to a novel, play, or other work of literature. Often a prologue contains useful information about events to come in the story. Foreshadowing is the use of clues by the author to prepare readers for future events or recurring themes.

Respond to the Reading

1. What clues does the narrator give that he is not really invisible?
2. Why does the narrator believe he is in the "great American tradition of tinkers"?
3. **Linking Science and Writing** Write a prologue to a fictional book describing Edison's invention of the lightbulb. Recall that a prologue is not a summary of the book. Rather, it can state general themes that the work of literature will address or set the stage or describe the setting of the story.

INTEGRATE Physics If all 1,369 light-bulbs were wired together in a series circuit, the electrical resistance in the circuit would be high. According to Ohm's law, the current in the circuit would be low and the bulbs wouldn't glow. If all the bulbs were wired in a parallel circuit, so much current would flow in the circuit that the connecting wires would melt. For the bulbs to light, the narrator must have wired them in many independent circuits.

Reviewing Main Ideas

Section 1 Electric Charge

1. There are two types of electric charge: positive charge and negative charge.
2. Electric charges exert forces on each other. Like charges repel and unlike charges attract.
3. Electric charges can be transferred from one object to another but cannot be created or destroyed.
4. Electrons can move easily in an electrical conductor. Electrons do not move easily in an insulator.
5. Objects can be charged by contact or by induction. Charging by induction occurs when a charged object is brought near an electrically neutral object.

Section 2 Electric Current

1. Electric current is the net movement of electric charges in a single direction. A voltage difference causes an electric current to flow.
2. A circuit is a closed path along which charges can move. Current will flow continually only along a circuit that is unbroken.
3. Chemical reactions in a battery produce a voltage difference between the positive and negative terminals of the battery.
4. Electrical resistance is the tendency of a material to oppose the flow of electric current.
5. In an electric circuit, the voltage difference, current, and resistance are related by Ohm's law:

$$I = \frac{V}{R}$$

Section 3 Electrical Energy

1. Current has only one path in a series circuit and more than one path in a parallel circuit.
2. Circuit breakers and fuses prevent excessive current from flowing in a circuit.
3. Electrical power is the rate at which electrical energy is used and can be calculated with $P = IV$.
4. The electrical energy used by a device can be calculated with the equation $E = Pt$.

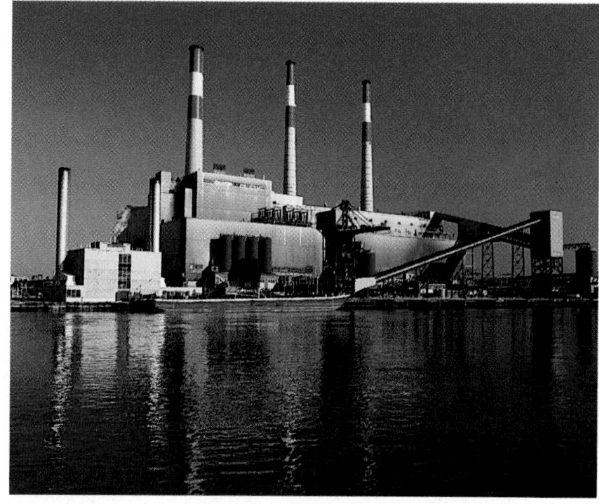

FOLDABLES™ Use the Foldable you made at the beginning of the chapter to help you review electric charge, electric current, and electrical energy.

chapter 13 Review

Using Vocabulary

charging by contact p. 395
charging by induction p. 396
circuit p. 401
conductor p. 395
electric current p. 400
electrical power p. 410
insulator p. 395
law of conservation of charge p. 393
Ohm's law p. 405
parallel circuit p. 408
resistance p. 403
series circuit p. 407
static electricity p. 392
voltage difference p. 400

Complete each statement using a word(s) from the vocabulary list above.

1. A(n) _________ is a circuit with only one path for current to follow.
2. An accumulation of excess electric charge is _________.
3. The electric force that makes current flow in a circuit is related to the _________.
4. According to _________, electric charge cannot be created or destroyed.
5. _________ is the result of electrons colliding with atoms as current flows in a material.
6. Charging a balloon by rubbing it on wool is an example of _________.

Checking Concepts

Choose the word or phrase that best answers the question.

7. Which of the following is a conductor?
   A) glass **C)** tungsten
   B) wood **D)** plastic
8. Resistance in wires causes electrical energy to be converted into which form of energy?
   A) chemical energy
   B) nuclear energy
   C) thermal energy
   D) sound
9. The electric force between two charged objects depends on which of the following?
   A) their masses and their separation
   B) their speeds
   C) their charge and their separation
   D) their masses and their charge
10. An object becomes positively charged when which of the following occurs?
   A) loses electrons **C)** gains electrons
   B) loses protons **D)** gains neutrons
11. Which of the following does NOT provide a voltage difference in a circuit?
   A) wet cell **C)** electrical outlet
   B) wires **D)** dry cell
12. A commonly used unit for electrical energy is which of the following?
   A) kilowatt-hour **C)** ohm
   B) ampere **D)** newton
13. Which of the following is the rate at which appliances use electrical energy?
   A) power **C)** resistance
   B) current **D)** speed

Interpreting Graphics

Use the table below to answer question 14.

Current in Electric Circuits

Circuit	Current (A)
A	2.3
B	0.6
C	0.2
D	1.8

14. The table shows the current in circuits that were each connected to a 6-V dry cell. Calculate the resistance of each circuit. Graph the current versus the resistance of each circuit. Describe the shape of the line on your graph.

15. Copy and complete the following concept map on electric current.

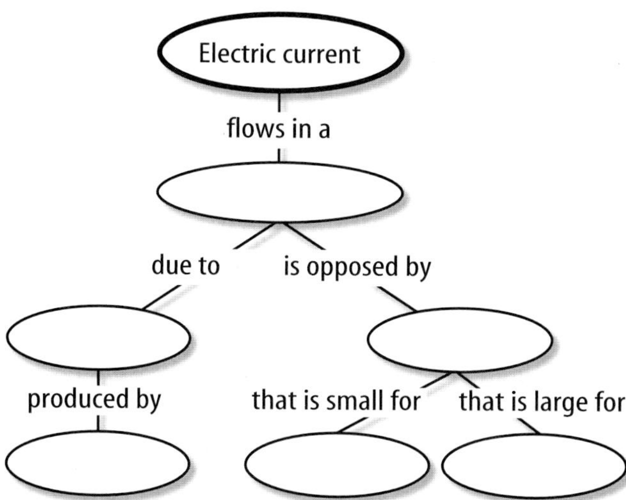

Thinking Critically

16. **Identify and Manipulate Variables** Design an experiment to test the effect on current and voltage differences in a circuit when two identical batteries are connected in series. What is your hypothesis? What are the variables and controls?

17. **Explain** A metal rod is charged by induction when a negatively charged plastic rod is brought nearby. Explain how the net charge on the metal rod has changed.

18. **Predict** You walk across a carpet on a dry day and touch a glass doorknob. Predict whether or not you would receive an electric shock. Explain your reasoning.

19. **Explain** The electric force between electric charges is much larger than the gravitational force between the charges. Why, then, is the gravitational force between Earth and the Moon much larger than the electric force between Earth and the Moon?

20. **Diagram** Draw a circuit diagram showing how a stereo, a TV, and a computer can be connected to a single source of voltage difference, such that turning off one appliance does not turn off all the others. Include a circuit breaker in your diagram that will protect all the appliances.

Applying Math

21. **Calculate Current** Using the information in the circuit diagram below, compute the current flowing in the circuit.

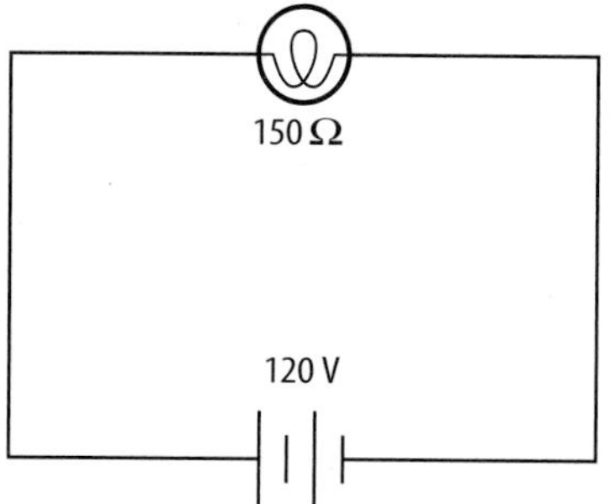

22. **Calculate Current** A toy car with a resistance of 20 Ω is connected to a 3-V battery. How much current flows in the car?

23. **Calculate Electrical Energy** The current flowing in an appliance connected to a 120-V source is 2 A. How many kilowatt-hours of electrical energy does the appliance use in 4 h?

24. **Calculate Electrical Energy Cost** A self-cleaning oven uses 5,400 W when cleaning itself. If it takes 1.5 h to clean, how many kilowatt-hours of electricity are used? At a cost of $0.09 per kWh, what does it cost to clean the oven?

25. **Calculate Power** A calculator uses a 9-V battery and draws 0.1 A of current. How much power does it use?

chapter 13 Standardized Test Practice

Record your answers on the answer sheet provided by your teacher or on a sheet of paper.

Multiple Choice

1. Which of the following is true about two adjacent electric charges?

A. If both are positive, they attract.

B. If both are negative, they attract.

C. If one is positive and one is negative, they attract.

D. If one is positive and one is negative, they repel.

The figure below shows a negatively charged electroscope. Use the figure to answer questions 2 and 3.

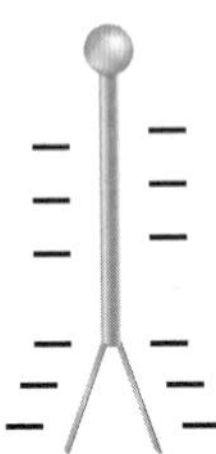

2. If a negatively charged rod is brought close to, but not touching, the knob, the two leaves will

A. move closer together.

B. move farther apart.

C. not move at all.

D. become positively charged.

3. If a positively charged rod touches the knob, the two leaves will

A. move closer together.

B. move farther apart.

C. not move at all.

D. become positively charged.

4. When an object becomes charged by induction, which of the following best describes the net charge on the object?

A. The net charge increases.

B. The net charge decreases.

C. The net charge doesn't change.

D. The net charge is negative.

Use the figure below to answer questions 5, 6, and 7.

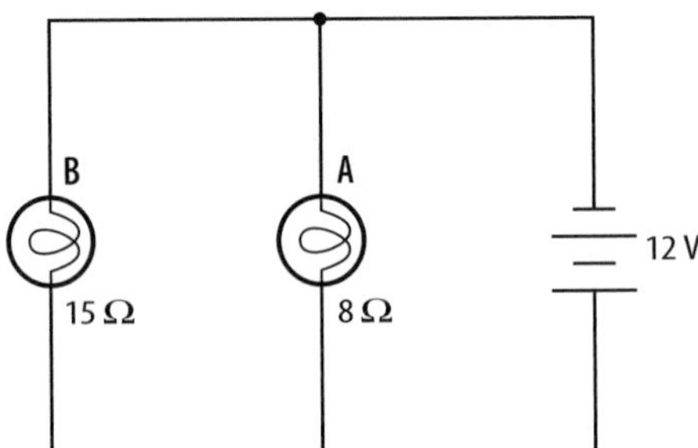

5. Which of the following is the same for each lightbulb?

A. current in the filament

B. voltage difference

C. electric resistance

D. charging by induction

6. Which of the following is the current that flows through lightbulb B?

A. 1.25 A

B. 0.67 A

C. 0.8 A

D. 1.5 A

7. Which of the following is the electrical power used by lightbulb A?

   A. 8 W

   B. 18 W

   C. 12 W

   D. 15 W

Gridded Response

Use the figure below to answer questions 8 and 9.

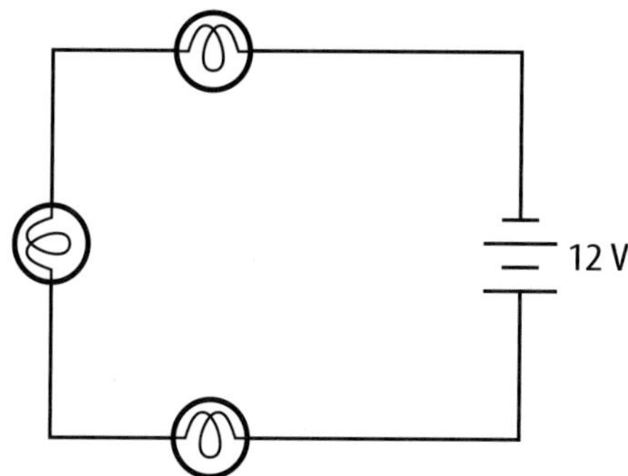

8. If the current in the circuit is 1.0 A, what is the total resistance of the circuit in ohms?

9. If the current in the circuit is 1.0 A, how much electrical energy, in kWh, is used by the circuit in 2.5 h?

Short Response

10. Two balloons are rubbed with a piece of wool. Describe what happens as the balloons are brought close together.

11. The current flowing through a lightbulb is 2.5 A. The lamp is connected to a battery supplying a voltage difference of 12 V. What is the power used by the lightbulb?

Extended Response

Use the table below to answer question 12.

Resistance of Copper Wire	
Length (m)	**Resistance (Ω)**
10	0.8
20	1.6
25	2.0
30	2.4

12. Suppose you were asked to estimate the resistance of a 5-m length of wire based on data in the table. What additional information would be needed to make your estimate more accurate?

13. The filament in a 75-W lightbulb has a smaller electric resistance than the filament in a 40-W lightbulb. Describe two ways the filament in the 75-W bulb could be different from the filament in the 40-W lightbulb.

Test-Taking Tip

Be sure you understand all symbols on a table or graph before attempting to answer any questions about the table or graph.

chapter 14

Magnetism

chapter preview

sections

A Natural Light Show

Have you ever seen an aurora? This light display occurs when blasts of charged particles from the Sun are captured by Earth's magnetic field. Atoms in the upper atmosphere emit the aurora light when they collide with charged particles that result from the Sun's blasts.

Science Journal List three things you know about magnets.

Start-Up Activities

The Strength of Magnets

Did you know that magnets are used in TV sets, computers, stereo speakers, electric motors, and many other devices? Magnets also help create images of the inside of the human body. Even Earth acts like a giant bar magnet. How do magnets work?

1. Complete the safety form.
2. Hold a bar magnet horizontally and put a paper clip on one end. Touch a second paper clip to the end of the first one. Continue adding paper clips until none will stick to one end of the chain. Copy the data table below and record the number of paper clips the magnet holds. Remove the paper clips from the magnet.
3. Repeat step 2 three more times. First, start the chain about 2 cm from the end of the magnet. Second, start the chain near the center of the magnet. Third, start the chain at the other end of the magnet.
4. **Think Critically** Infer which part of the magnet exerts the strongest attraction. Compare the attraction at the center of the magnet with the attraction at the ends.

Magnet and Paper Clip Data

	Paper Clip Chain (number of clips)
Trial 1 (end)	
Trial 2 (2 cm)	
Trial 3 (center)	
Trial 4 (other end)	

Using Magnets Many devices you use contain magnets that help convert one form of energy to another. Make the following Foldable to help you understand how magnets are used to transform electrical and mechanical energy.

STEP 1 **Fold** a sheet of paper in half lengthwise.

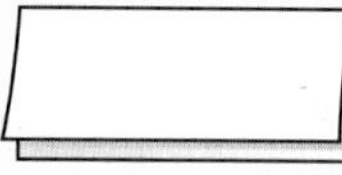

STEP 2 **Fold** the paper down about 2 cm from the top.

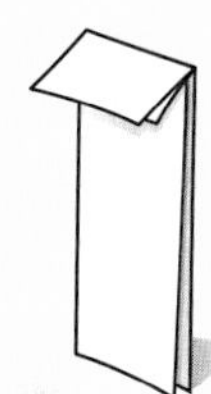

STEP 3 **Open and draw** lines along the top fold. **Label** as shown.

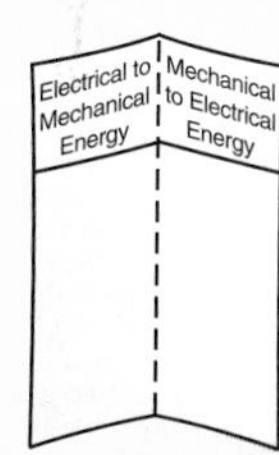

Summarize As you read the chapter, summarize how magnets are used to convert electrical energy to mechanical energy in the left column, and how magnets are used to convert mechanical energy to electrical energy in the right column.

Preview this chapter's content and activities at gpescience.com

section 1

Magnetism

Reading Guide

What You'll Learn

- **Explain** how a magnet exerts a force.
- **Describe** the properties of temporary and permanent magnets.
- **Explain** why some materials are magnetic and some are not.
- **Model** magnetic behavior using magnetic domains.

Why It's Important

Without the forces exerted by magnets, you could not use televisions, computers, CD players, or even refrigerators.

Review Vocabulary

electric field: surrounds an electric charge and exerts a force on other electric charges

New Vocabulary

- magnetism
- magnetic field
- magnetic pole
- magnetic domain

Magnets

More than 2,000 years ago, Greeks discovered deposits of a mineral that was a natural magnet. They noticed that chunks of this mineral could attract pieces of iron. This mineral was found in a region of Turkey that then was known as Magnesia, so the Greeks called the mineral magnetic. The mineral is now called magnetite. In the twelfth century, Chinese sailors used magnetite to make compasses that improved navigation. Since then, many devices have been developed that rely on magnets to operate. Today, the word **magnetism** refers to the properties and interactions of magnets. **Figure 1** shows a device you might be familiar with that uses magnets and magnetism.

Figure 1 Magnets can be found in many devices you use every day, such as TVs, video games, and telephones. Headphones and CD players also contain magnets.

Magnetic Force You probably have played with magnets and noticed that two magnets exert a force on each other. Depending on which ends of the magnets are close together, magnets either repel or attract each other. You might have noticed that the interaction between two magnets can be felt even before the magnets touch. The strength of the force between two magnets increases as magnets move closer together and decreases as the the magnets move farther apart.

Reading Check *What does the force between two magnets depend on?*

Figure 2 A magnet is suruonded by a magnetic field.

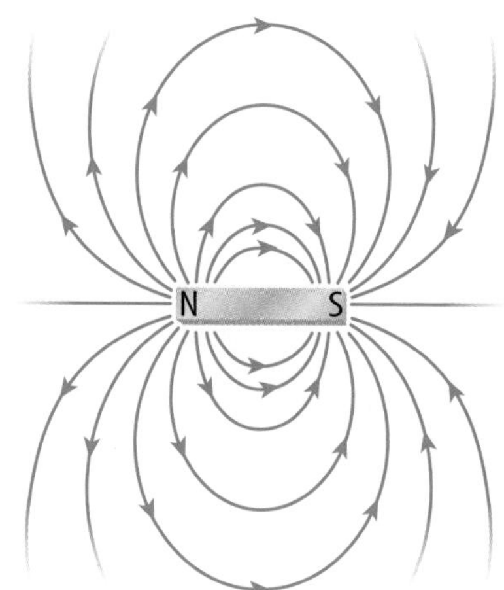

A magnet's magnetic field is represented by magnetic field lines.

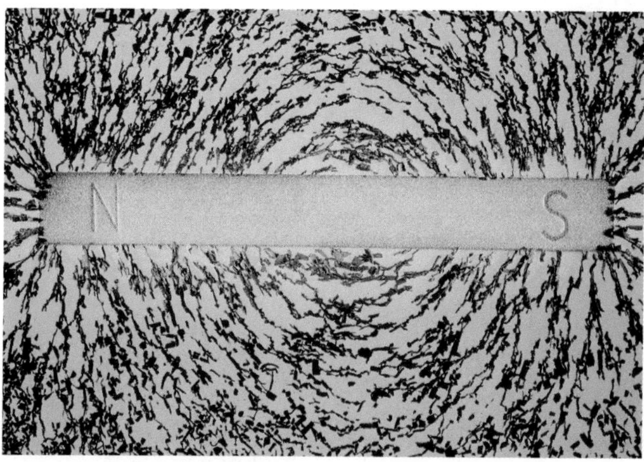

Iron filings sprinkled around a magnet line up along the magnetic field lines.

Magnetic Field A magnet is surrounded by a magnetic field. A **magnetic field** exerts a force on other magnets and objects made of magnetic materials. The magnetic field is strongest close to the magnet and weaker farther away. The magnetic field can be represented by lines of force, or magnetic field lines. **Figure 2** shows the magnetic field lines surrounding a bar magnet. A magnetic field also has a direction. The direction of the magnetic field around a bar magnet is shown by the arrows on the left side of **Figure 2.**

Magnetic Poles Look again at **Figure 2.** Do you notice that the magnetic field lines are closest together at the ends of the bar magnet? These regions, called the **magnetic poles,** are where the magnetic force exerted by the magnet is strongest. All magnets have a north pole and a south pole. For a bar magnet, the north and south poles are at the opposite ends.

Figure 3 shows the north and south poles of magnets with more complicated shapes. The two ends of a horseshoe-shaped magnet are the north and south poles. A magnet shaped like a disk has opposite poles on the top and bottom of the disk. Magnetic field lines always connect the north pole and the south pole of a magnet.

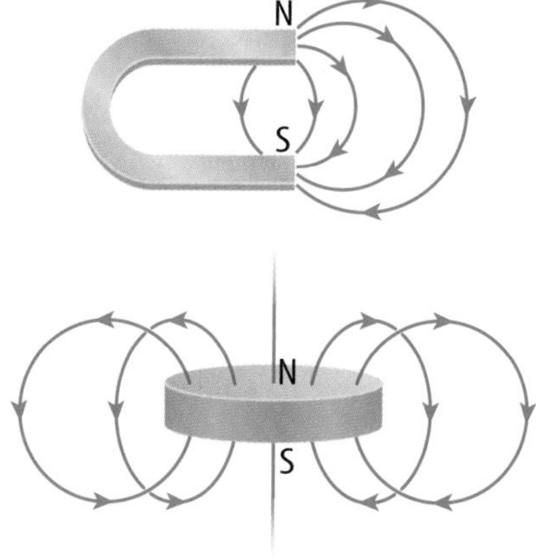

Figure 3 The magnetic field lines around horseshoe and disk magnets begin at each magnet's north pole and end at the south pole.
Identify *where the magnetic field is strongest.*

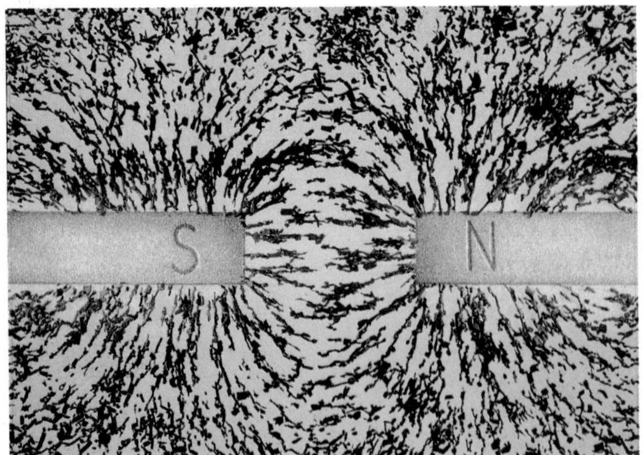

Unlike poles closest together

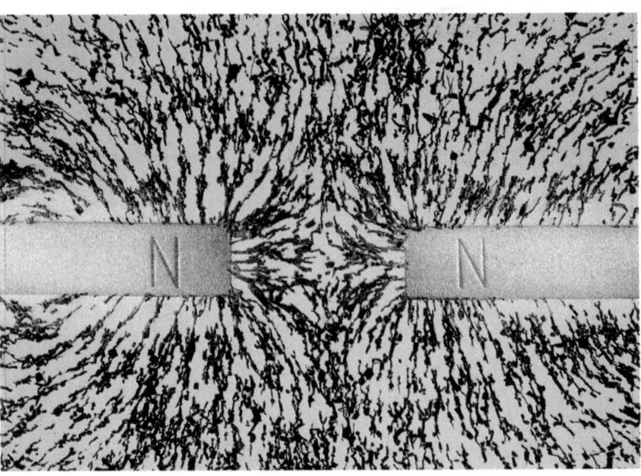

Like poles closest together

Figure 4 Two magnets can attract or repel each other, depending on which poles are closest together.

How Magnets Interact Two magnets can either attract or repel each other. Two north poles or two south poles of two magnets repel each other. However, north poles and south poles always attract each other. Like magnetic poles repel each other and unlike poles attract each other. When two magnets are brought close to each other, their magnetic fields combine to produce a new magnetic field. **Figure 4** shows the magnetic field that results when like poles and unlike poles of bar magnets are brought close to each other.

Reading Check *How do magnetic poles interact with each other?*

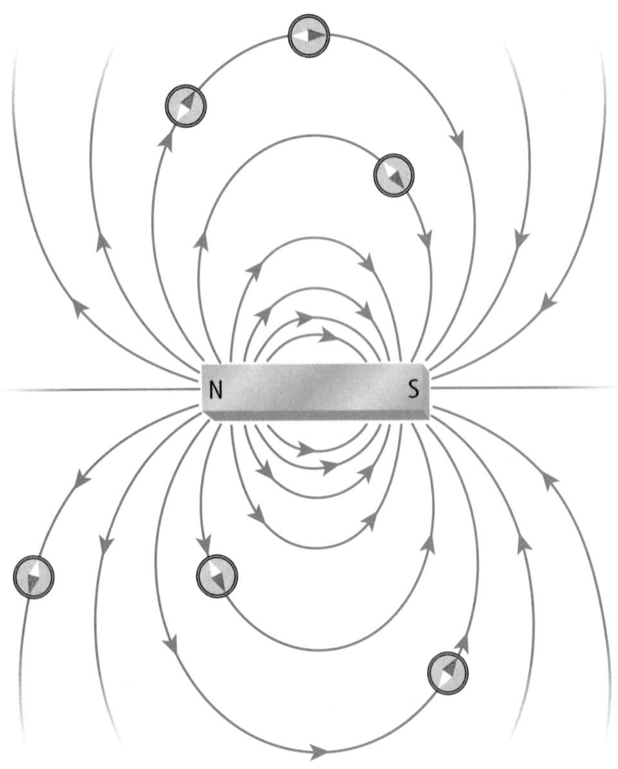

Magnetic Field Direction When a compass is brought near a bar magnet, the compass needle rotates. The compass needle is a small bar magnet with a north pole and a south pole. The force exerted on the compass needle by the magnetic field causes the needle to rotate. The compass needle rotates until it lines up with the magnetic field lines, as shown in **Figure 5.** The north pole of a compass points in the direction of the magnetic field. This direction is always away from a north magnetic pole and toward a south magnetic pole.

Figure 5 Compass needles placed around a bar magnet line up along magnetic field lines. The north poles of the compass needles are shaded red.

Earth's Magnetic Field A compass can help determine direction because the north pole of the compass needle points north. This is because Earth acts like a giant bar magnet and is surrounded by a magnetic field that extends into space. Just as with a bar magnet, the compass needle aligns with Earth's magnetic field lines, as shown in **Figure 6.**

Earth's Magnetic Poles The north pole of a magnet is defined as the end of the magnet that points toward the geographic north. Sometimes the north pole and south pole of a magnet are called the north-seeking pole and the south-seeking pole. Because opposite magnetic poles attract, the north pole of a compass is being attracted by a south magnetic pole. Earth is like a bar magnet with its south magnetic pole near its geographic north pole.

Currently, Earth's south magnetic pole is located in northern Canada, about 1,500 km from the geographic north pole. However, Earth's magnetic poles move slowly with time. Sometimes Earth's magnetic poles switch places so that Earth's south magnetic pole is in the southern hemisphere near the geographic south pole. Measurements of magnetism in rocks show that Earth's magnetic poles have changed places over 150 times in the past 70 million years.

No one is sure what produces Earth's magnetic field. Earth's inner core is made of a solid ball of iron and nickel, surrounded by a liquid layer of molten iron and nickel. According to one theory, circulation of the molten iron and nickel in Earth's outer core produces Earth's magnetic field.

Observing Magnetic Interference

Procedure

1. Complete the safety form.
2. Clamp a **bar magnet** to a **ring stand.** Tie a **thread** around one end of a **paper clip** and stick the paper clip to one pole of the magnet.
3. Anchor the other end of the thread under a **book** on the table. Slowly pull the thread until the paper clip is suspended below the magnet but not touching it.
4. Without touching the paper clip, slip a piece of paper between the magnet and the paper clip. Does the paper clip fall?
5. Try other materials, such as **aluminum foil, fabric,** and a **butter knife.**

Analysis

1. Which materials caused the paper clip to fall? Why?
2. Which materials did not cause the paper clip to fall? Why?

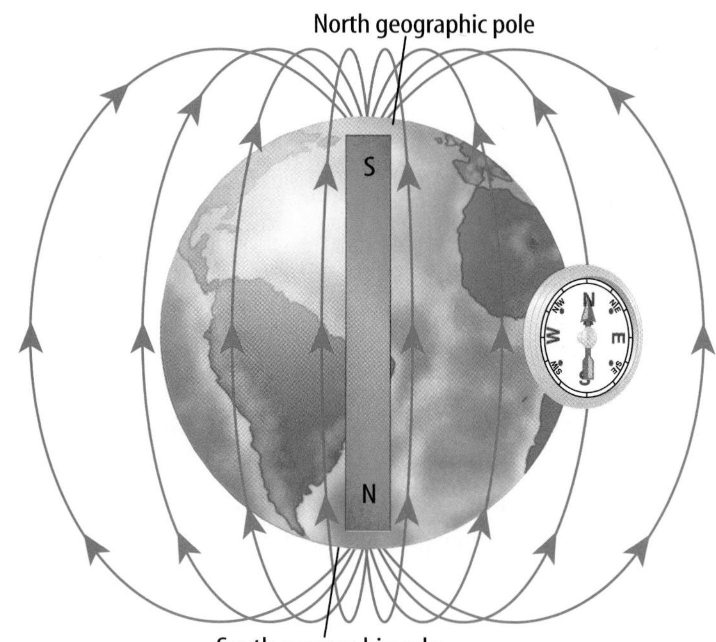

Figure 6 A compass needle aligns with the magnetic field lines of Earth's magnetic field.
Predict *Which way would a compass needle point if Earth's magnetic poles switched places?*

Magnets in Organisms Some organisms may use Earth's magnetic field to help them find their way around. Some species of birds, insects, and bacteria have been shown to contain small amounts of the mineral magnetite. Research how one species uses Earth's magnetic field, and report your findings to your class.

Magnetic Materials

You might have noticed that a magnet will not attract all metal objects. For example, a magnet will not attract pieces of aluminum foil. Only a few metals, such as iron, cobalt, and nickel, are attracted to magnets or can be made into permanent magnets. What makes these elements magnetic? Remember that every atom contains electrons. Electrons have magnetic properties. In the atoms of most elements, the magnetic properties of the electrons cancel out. However, in the atoms of iron, cobalt, and nickel, these magnetic properties don't cancel out. Each atom in these elements behaves like a small magnet and has its own magnetic field.

Even though these atoms have their own magnetic fields, objects made from these metals are not always magnets. For example, if you hold an iron nail close to a refrigerator door and let go, it falls to the floor. However, you can make the nail behave like a magnet temporarily.

Applying Science

How can magnetic parts of a junk car be salvaged?

Every year over 10 million cars containing plastics, glass, rubber, and various metals are scrapped. Magnets are often used to help retrieve some of these materials from scrapped cars. The materials can then be reused, saving both natural resources and energy. Once the junk car has been fed into a shredder, large magnets can easily separate many of its metal parts from its nonmetal parts. How much of the car does a magnet actually help separate? Use your ability to interpret a circle graph to find out.

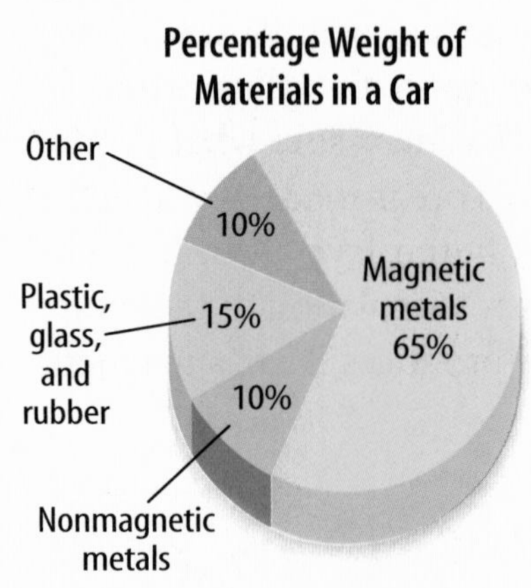

Identifying the Problem

The graph at the right shows the average percent by weight of the different materials in a car. Included in the magnetic metals are steel and iron. The nonmagnetic metals refer to aluminum, copper, lead, zinc, and magnesium. According to the graph, how much of the car can a magnet separate for recycling?

Solving the Problem

1. What percent of the car's weight will a magnet recover?
2. A certain scrapped car has a mass of 1,500 kg. What is the mass of the materials in this car that cannot be recovered using a magnet?
3. If the average mass of a scrapped car is 1,500 kg and 10 million cars are scrapped each year, what is the total mass of iron and steel that could be recovered from scrapped cars each year?

Magnetic Domains—A Model for Magnetism In iron, cobalt, nickel, and some other magnetic materials, the magnetic field created by each atom exerts a force on the other nearby atoms. Because of these forces, large groups of atoms align their magnetic poles so that almost all like poles point in the same direction. The groups of atoms with aligned magnetic poles are called **magnetic domains.** Each domain contains an enormous number of atoms, yet the domains are too small to be seen with the unaided eye. Because the magnetic poles of the individual atoms in a domain are aligned, the domain itself behaves like a magnet with a north pole and a south pole.

Lining Up Domains An iron nail contains an enormous number of these magnetic domains, so why doesn't the nail behave like a magnet? Even though each domain behaves like a magnet, the poles of the domains are arranged randomly and point in different directions, as shown in **Figure 7.** As a result, the magnetic fields from all the domains cancel each other out.

If you place a magnet against the same nail, the atoms in the domains orient themselves in the direction of the nearby magnetic field, as shown on the right in **Figure 7.** The like poles of the domains point in the same direction and no longer cancel each other out. The nail itself now acts as a magnet. However, when the external magnetic field is removed, the constant motion and vibration of the atoms bump the magnetic domains out of their alignment. The magnetic domains in the nail return to a random arrangement. For this reason, the nail is only a temporary magnet. Paper clips and other objects containing iron also can become temporary magnets.

Making Your Own Compass

Procedure

WARNING: *Use care when handling sharp objects.*

1. Complete the safety form.
2. Cut off the bottom of a **plastic foam cup** to make a polystyrene disk.
3. Magnetize a **sewing needle** by continuously stroking the needle in the same direction with a magnet for 1 min.
4. **Tape** the needle to the center of the foam disk.
5. Fill a **plate** with **water** and float the disk, needle-side up, in the water.
6. Bring the magnet close to the foam disk.

Analysis

1. How did the needle and disk move when placed in the water? Explain.
2. How did the needle and disk move when the magnet was brought near? Explain.

Try at Home

Figure 7 Magnetic materials contain magnetic domains.

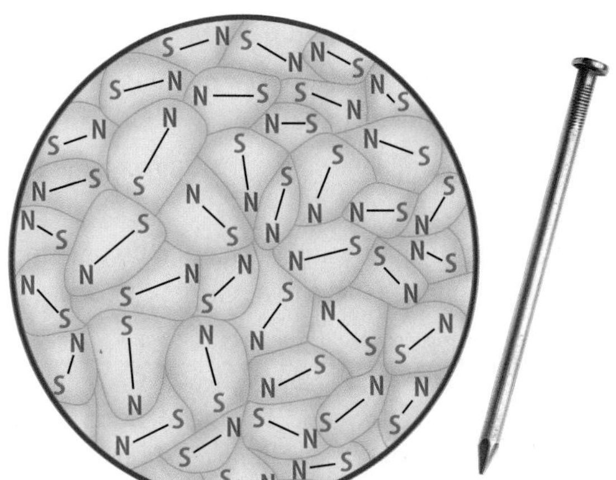

A normal iron nail is made up of microscopic domains that are arranged randomly.

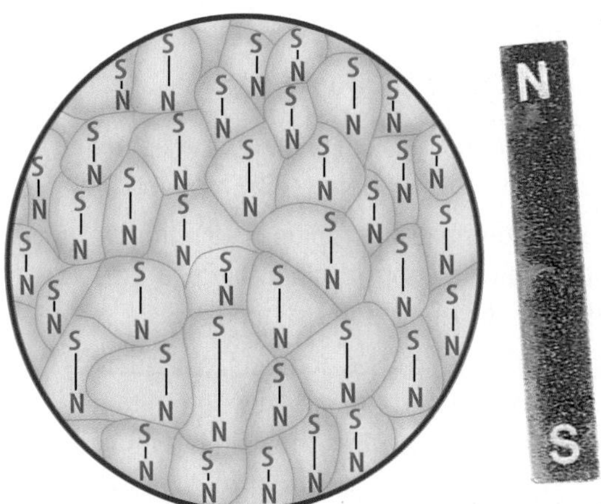

The domains will align themselves along the magnetic field lines of a nearby magnet.

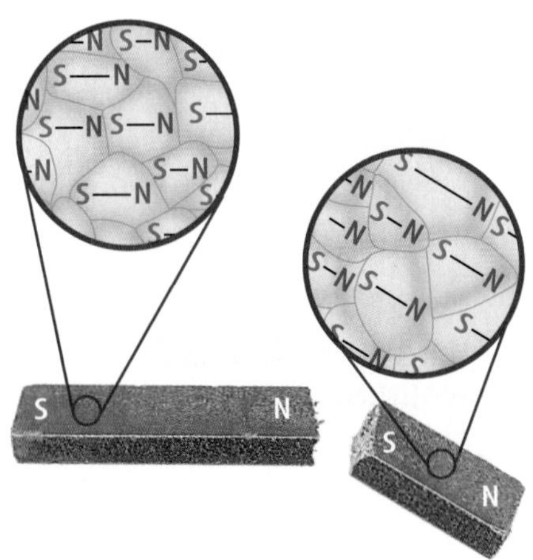

Figure 8 Each piece of a broken magnet still has a north and a south pole.

Permanent Magnets A permanent magnet can be made by placing a magnetic material, such as iron, in a strong magnetic field. The strong magnetic field causes the magnetic domains in the material to line up. The magnetic fields of these aligned domains combine and create a strong magnetic field inside the material. This field prevents the constant motion of the atoms from bumping the domains out of alignment. The material is then a permanent magnet.

Even permanent magnets can lose their magnetic behavior if they are heated. Heating causes atoms in a magnet to move faster. If a permanent magnet is heated enough, its atoms may be moving fast enough to jostle the domains out of alignment. Then the permanent magnet loses its magnetic field and is no longer a magnet.

Can a pole be isolated? What happens when a magnet is broken in two? Can one piece be a north pole and one piece be a south pole? Look at the domain model of the broken magnet in **Figure 8.** Recall that even individual atoms of magnetic materials act as tiny magnets. Because every magnet is made of many aligned smaller magnets, even the smallest pieces have both a north pole and a south pole.

section 1 review

Summary

Magnets

- A magnet is surrounded by a magnetic field that exerts a force on magnetic materials.
- A magnet has a north pole and a south pole.
- Like magnetic poles repel and unlike poles attract.

Magnetic Materials

- Iron, cobalt, and nickel are magnetic elements because their atoms behave like magnets.
- Magnetic domains are regions in a material that contain an enormous number of atoms with their magnetic poles aligned.
- A magnetic field causes domains to align. In a temporary magnet, the domains return to a random alignment when the field is removed.
- In a permanent magnet, a stong magnetic field aligns domains, and they remain aligned when the field is removed.

Self Check

1. **Describe** what happens when you move two unlike magnetic poles closer together. Draw a diagram to illustrate your answer.
2. **Describe** how a compass needle moves when it is placed in a magnetic field.
3. **Explain** why only certain materials are magnetic.
4. **Predict** how the properties of a bar magnet would change if it were broken in half.
5. **Explain** how heating a bar magnet would change its magnetic field.
6. **Think Critically** Use the magnetic domain model to explain why a magnet sticks to a refrigerator door.

Applying Math

7. **Calculate Number of Domains** The magnetic domains in a magnet have an average volume of 0.0001 mm^3. If the magnet has dimensions of 50 mm by 10 mm by 4 mm, how many domains does the magnet contain?

More Section Review gpescience.com

Electricity and Magnetism

Reading Guide

What You'll Learn

- **Describe** the magnetic field produced by an electric current.
- **Explain** how an electromagnet produces a magnetic field.
- **Describe** how electromagnets are used.
- **Explain** how an electric motor operates.

Why It's Important

Electric motors contained in many of the devices you use every day operate because electric currents produce magnetic fields.

Review Vocabulary

electric current: the flow of electric charges in a wire or any conductor

New Vocabulary

- electromagnet
- solenoid
- galvanometer
- electric motor

Electric Current and Magnetism

In 1820, Hans Christian Oersted, a Danish physics teacher, found that electricity and magnetism are related. While doing a demonstration involving electric current, he happened to have a compass near an electric circuit. He noticed that the flow of the electric current affected the direction the compass needle pointed. Oersted hypothesized that the electric current must have produced a magnetic field around the wire, and that the direction of the field changed with the direction of the current.

Moving Charges and Magnetic Fields Oersted's hypothesis that an electric current creates a magnetic field was correct. Moving charges, such as those in an electric current, produce magnetic fields. Around a current-carrying wire the magnetic field lines form circles, as shown in **Figure 9.** The direction of the magnetic field around the wire reverses when the direction of the current in the wire reverses. As the current in the wire increases, the strength of the magnetic field increases. As the distance from the wire increases, the strength of the magnetic field decreases.

Figure 9 When electric current flows through a wire, a magnetic field forms around the wire. The direction of the magnetic field depends on the direction of the current in the wire.

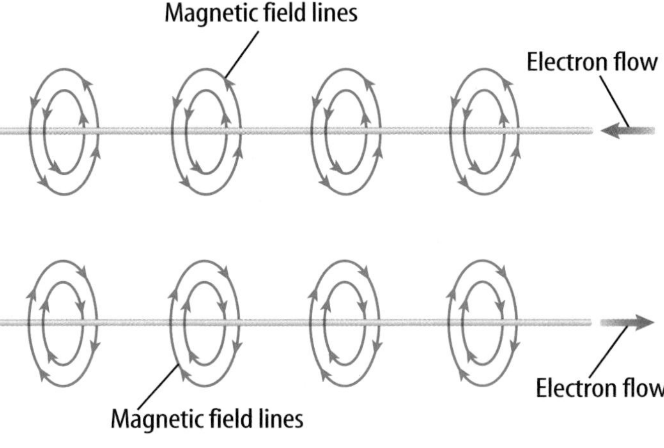

Electromagnets

The magnetic field that surrounds a current-carrying wire can be made much stronger in an electromagnet. An **electromagnet** is a temporary magnet made by wrapping a wire coil carrying a current around an iron core. When a current flows through a wire loop, such as the one shown in **Figure 10A,** the magnetic field inside the loop is stronger than the field around a straight wire. A single wire wrapped into a cylindrical wire coil is called a **solenoid.** The magnetic field inside a solenoid is stronger than the field in a single loop. The magnetic field around each loop in the solenoid combines to form the field shown in **Figure 10B.**

If the solenoid is wrapped around an iron core, an electromagnet is formed, as shown in **Figure 10C.** The solenoid's magnetic field magnetizes the iron core. As a result, the field inside the solenoid with the iron core can be more than 1,000 times greater than the field inside the solenoid without the iron core.

Figure 10 An electromagnet is made from a current-carrying wire.

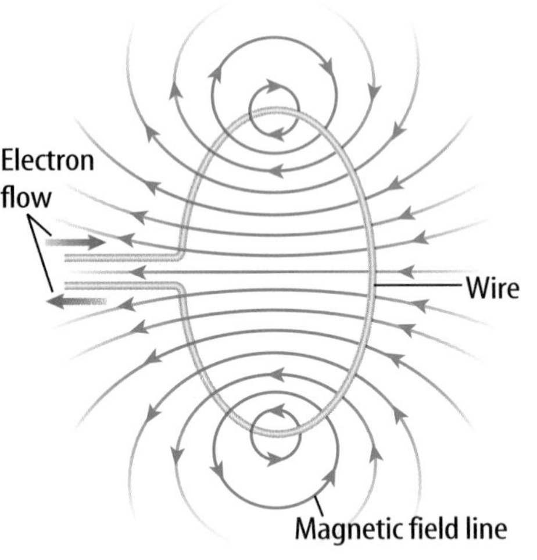

A The magnetic fields around different parts of the wire loop combine to form the field inside the loop.

Properties of Electromagnets Electromagnets are temporary magnets because a magnetic field is present only when current is flowing in the solenoid. The strength of the magnetic field can be increased by adding more turns of wire to the solenoid or by increasing the current passing through the wire.

An electromagnet behaves like any other magnet when current flows through the solenoid. One end of the electromagnet is a north pole and the other end is a south pole. If it is placed in a magnetic field, an electromagnet will align itself along the magnetic field lines, just as a compass needle will. An electromagnet also will attract magnetic materials and be attracted or repelled by other magnets. What makes an electromagnet so useful is that its magnetic properties can be controlled by changing the electric current flowing through the solenoid.

When current flows in an electromagnet and it moves toward or away from another magnet, electric energy is converted into mechanical energy to do work. Electromagnets do work in various devices, such as stereo speakers and electric motors.

B When many loops of current-carrying wire are formed into a solenoid, the magnetic field is increased inside the solenoid. The solenoid has a north pole and a south pole.
Predict *how the field would change if the current reversed direction.*

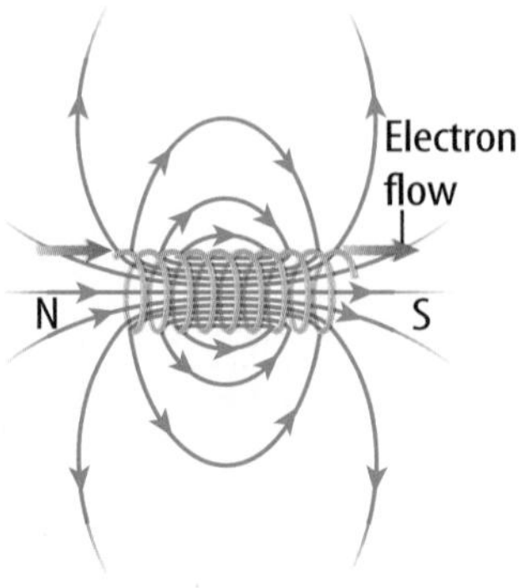

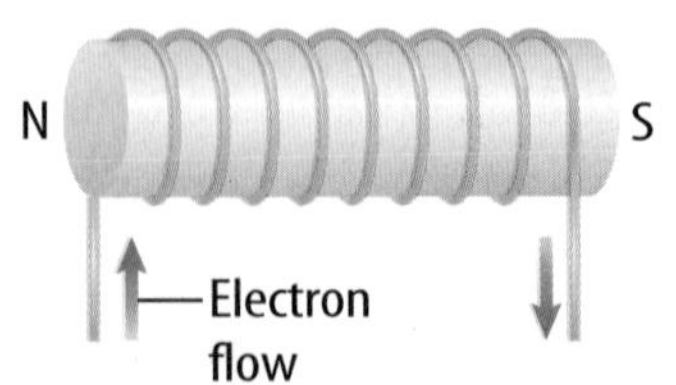

C A solenoid wrapped around an iron core forms an electromagnet.

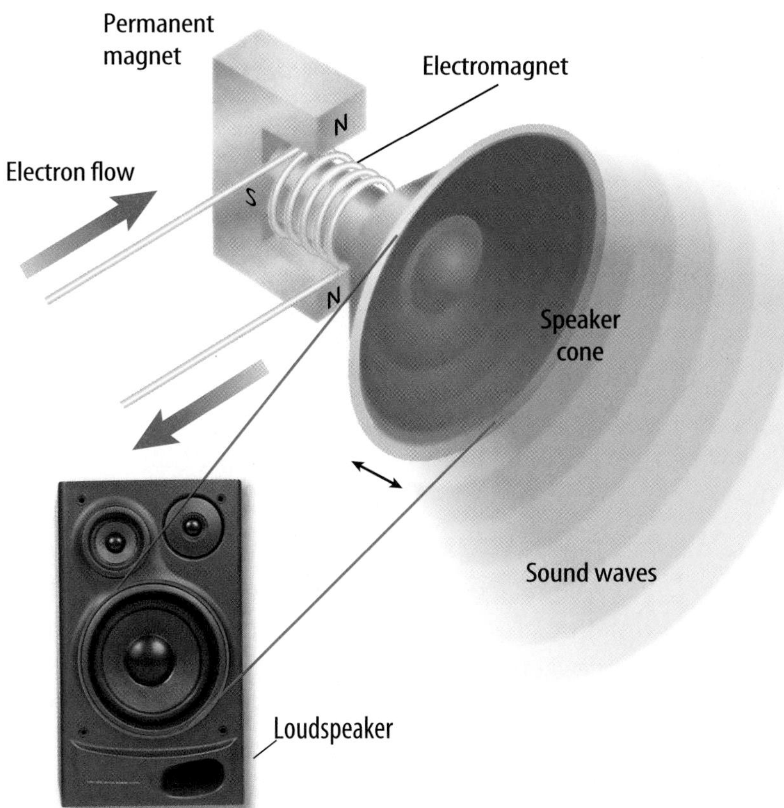

Figure 11 The electromagnet in a speaker converts electrical energy into mechanical energy to produce sound.
Explain *why a speaker needs a permanent magnet to produce sound.*

Using Electromagnets to Make Sound How does musical information stored on a CD become sound you can hear? The sound is produced by a loudspeaker that contains an electromagnet connected to a flexible speaker cone that is usually made from paper, plastic, or metal. The electromagnet changes electrical energy to mechanical energy that vibrates the speaker cone to produce sound, as shown on **Figure 11.**

Reading Check *How does a stereo speaker use an electromagnet to produce sound?*

When you listen to a CD, the CD player produces a voltage that changes according to the musical information on the CD. This varying voltage produces a varying electric current in the electromagnet connected to the speaker cone. Both the amount and the direction of the electric current change, depending on the information on the CD. The varying electric current causes both the strength and the direction of the magnetic field in the electromagnet to change. The electromagnet is surrounded by a permanent, fixed magnet. The changing direction of the magnetic field in the electromagnet causes the electromagnet to be attracted to or repelled by the permanent magnet. This makes the electromagnet move back and forth, causing the speaker cone to vibrate and reproduce the sound that was recorded on the CD.

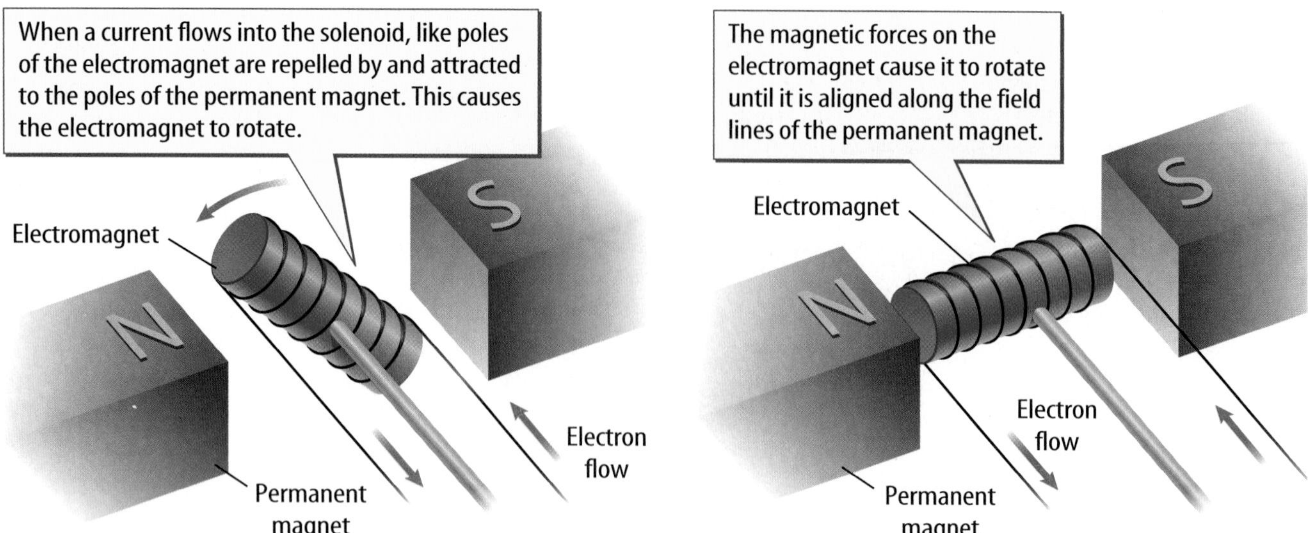
When a current flows into the solenoid, like poles of the electromagnet are repelled by and attracted to the poles of the permanent magnet. This causes the electromagnet to rotate.
Electromagnet
N
S
Permanent magnet
Electron flow
The magnetic forces on the electromagnet cause it to rotate until it is aligned along the field lines of the permanent magnet.
Electromagnet
N
S
Electron flow
Permanent magnet

Figure 12 An electromagnet can be made to rotate in a magnetic field.

Making an Electromagnet Rotate The forces exerted on an electromagnet by another magnet can be used to make the electromagnet rotate. **Figure 12** shows an electromagnet suspended between the poles of a permanent magnet. The poles of the electromagnet are repelled by the like poles and attracted by the unlike poles of the permanent magnet. When the electromagnet is in the position shown on the left side of **Figure 12,** there is a downward force on the left side and an upward force on the right side of the electromagnet forces. These forces cause the electromagnet to rotate as shown.

Reading Check *How can a permanent magnet cause an electromagnet to rotate?*

The electromagnet continues to rotate until its poles are next to the opposite poles of the permanent magnet, as shown on the right side of **Figure 12.** In this position, the forces on the north and south poles of the electromagnet are in opposite directions. The net force on the electromagnet is now zero, and the electromagnet stops rotating.

One way to change the forces that make the electromagnet rotate is to change the current in the electromagnet. Increasing the current increases the strength of the forces that the two magnets exert on each other.

Galvanometers You've probably noticed the gauges in the dashboard of a car. One gauge shows the amount of gasoline left in the tank, and another shows the engine temperature. How does a change in the amount of gasoline in a tank or the water temperature in the engine make a needle move in a gauge on the dashboard? These gauges are **galvanometers,** which are devices that use an electromagnet to measure electric current.

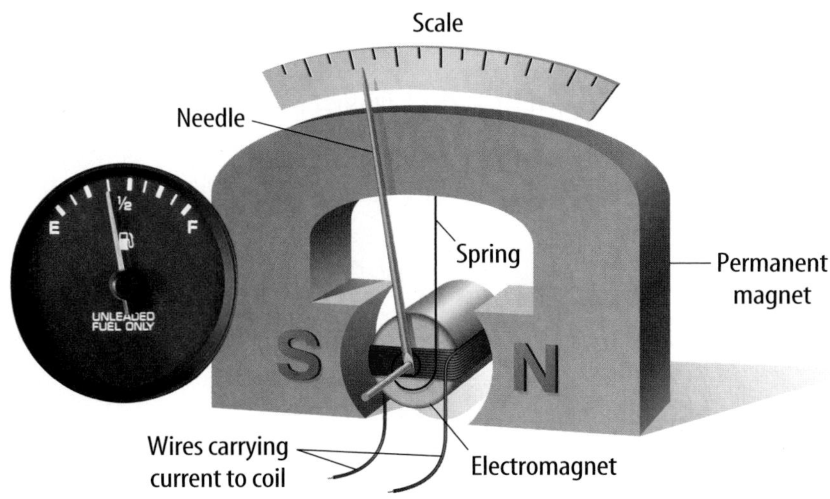

Figure 13 The rotation of the needle in a galvanometer depends on the amount of current flowing in the electromagnet. The current flowing into the galvanometer in a car's fuel gauge changes as the amount of fuel changes.

Using Galvanometers An example of a galvanometer is shown in **Figure 13.** In a galvanometer, the electromagnet is connected to a small spring. The electromagnet rotates until the force exerted by the spring is balanced by the magnetic forces on the electromagnet. Changing the current in the electromagnet causes the needle to rotate to different positions on the scale.

For example, a car's fuel gauge uses a galvanometer. A float in the fuel tank is attached to a sensor that sends a current to the fuel-gauge galvanometer. As the level of the float in the tank changes, the current sent by the sensor changes. The changing current in the galvanometer causes the needle to rotate by different amounts. The gauge is calibrated so that the current sent when the tank is full causes the needle to rotate to the full mark on the scale.

Electric Motors

On sizzling summer days, do you ever use an electric fan to keep cool? A fan uses an **electric motor,** which is a device that changes electrical energy into mechanical energy. The motor in a fan turns the fan blades, moving air past your skin to make you feel cooler.

Electric motors are used in all types of industry, agriculture, and transportation, including airplanes and automobiles. If you were to look carefully, you probably could find electric motors in every room of your home. Almost every appliance in which something moves contains an electric motor. Electric motors are used in devices such as VCRs, CD players, and computers, and appliances such as those shown in **Figure 14.**

Figure 14 All the devices shown here contain electric motors.
List *three additional devices that contain electric motors.*

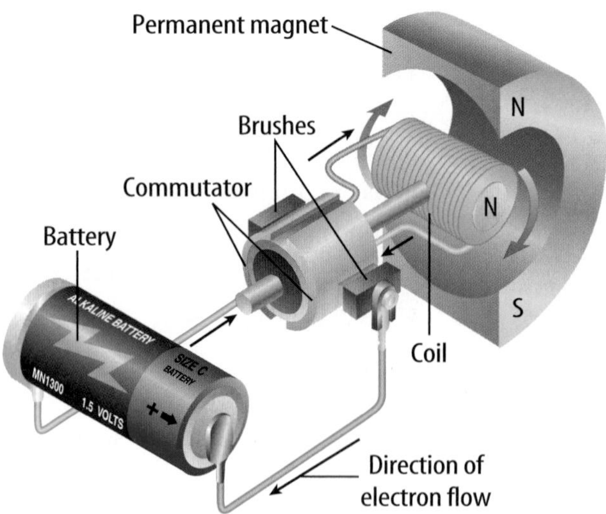

Step 1 When a current flows in the coil, the magnetic forces between the permanent magnet and the coil cause the coil to rotate.

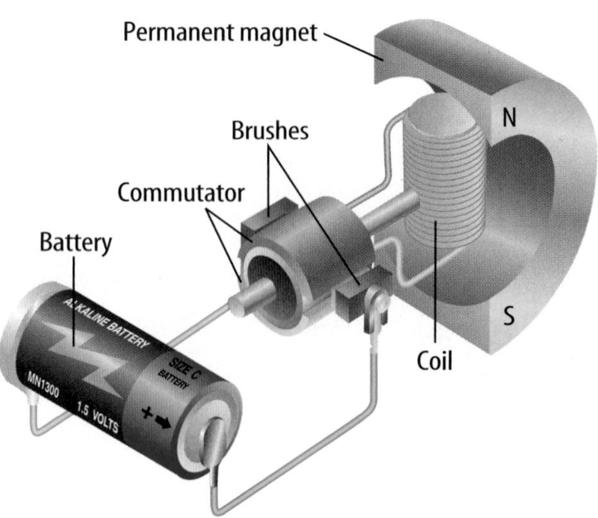

Step 2 In this position, the brushes are not in contact with the commutator and no current flows in the coil. The inertia of the coil keeps it rotating.

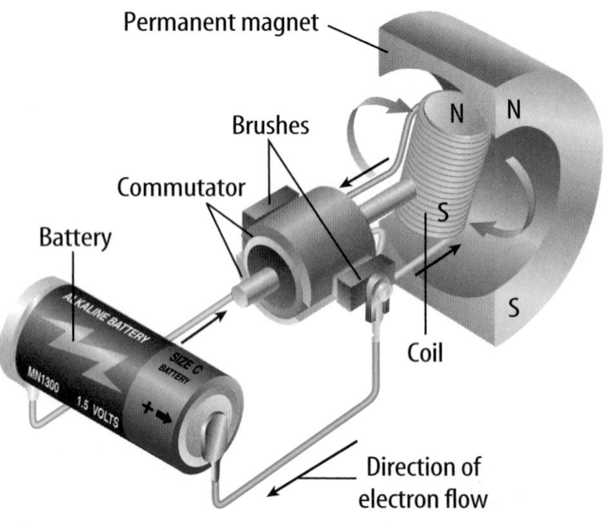

Step 3 The commutator reverses the direction of the current in the coil. This flips the north and south poles of the magnetic field around the coil.

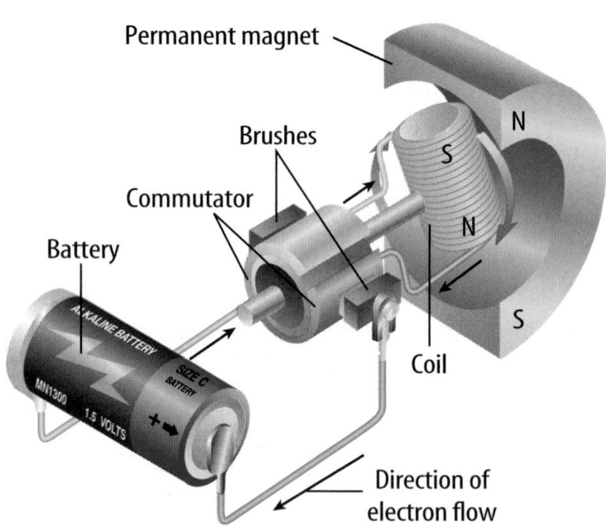

Step 4 The coil rotates until its poles are opposite the poles of the permanent magnet. The commutator reverses the current, and the coil keeps rotating.

Figure 15 In a simple electric motor, a coil rotates between the poles of a permanent magnet. To keep the coil rotating, the current must change direction twice during each rotation.

A Simple Electric Motor A diagram of the simplest type of electric motor is shown in **Figure 15.** The main parts of a simple electric motor include a wire coil, a permanent magnet, and a source of electric current, such as a battery. The battery produces the current that makes the coil an electromagnet. A simple electric motor also includes components called brushes and a commutator. The brushes are conducting pads connected to the battery. The brushes make contact with the commutator, which is a conducting metal ring that is split. Each half of the commutator is connected to one end of the coil so that the commutator rotates with the coil. The brushes and the commutator form a closed electric circuit between the battery and the coil.

Making the Motor Spin When current flows in the coil, the forces between the coil and the permanent magnet cause the coil to rotate, as shown in step 1 of **Figure 15.** The coil continues to rotate until it reaches the position shown in step 2. Then, the brushes no longer make contact with the commutator, and no current flows in the coil. As a result, there are no magnetic forces exerted on the coil. However, the inertia of the coil causes it to continue rotating.

In step 3, the coil has rotated so that the brushes again are in contact with the commutator. However, the halves of the commutator that are in contact with the positive and negative battery terminals have switched. This causes the current in the commutator to reverse direction. Now the top of the electromagnet is a north magnetic pole and the bottom is a south pole. These poles are repelled by the nearby like poles of the permanent magnet, and the magnet continues to rotate.

In step 4, the coil rotates until its poles are next to the opposite poles of the permanent magnet. Then the commutator again reverses the direction of the current, enabling the coil to keep rotating. In this way, the coil is kept rotating as long as the battery remains connected to the commutator.

Topic: Electric Motors
Visit gpescience.com for Web links to information about electric motors.

Activity Using the information provided at these links, construct a simple electric motor.

section 2 review

Summary

Electric Current and Magnetic Fields

- A magnetic field surrounds a moving electric charge.
- The strength of the magnetic field surrounding a current-carrying wire depends on the amount of current.

Electromagnets

- An electromagnet is a temporary magnet consisting of a current-carrying wire wrapped around an iron core.
- The magnetic properties of an electromagnet can be controlled by changing the current in the coil.
- A galvanometer uses an electromagnet to measure electric current.

Electric Motors

- In a simple electric motor, an electromagnet rotates between the poles of a permanent magnet.

Self Check

1. **Explain** why, if the same current flows in a wire coil and a single wire loop, the magnetic field inside the coil is stronger than the field inside the loop.
2. **Describe** two ways you could change the strength of the magnetic field produced by an electromagnet.
3. **Predict** how the magnetic field produced by an electromagnet would change if the iron core were replaced by an aluminum core.
4. **Explain** why it is necessary to continually reverse the direction of current flow in the coil of an electric motor.
5. **Think Critically** A bar magnet is repelled when an electromagnet is brought close to it. Describe how the bar magnet would have moved if the current in the electromagnet had been reversed.

Applying Math

6. **Use a Ratio** The magnetic field strength around a wire at a distance of 1 cm is twice as large as at a distance of 2 cm. How does the field strength at 0.5 cm compare to the field strength at 1 cm?

section 3

Producing Electric Current

Reading Guide

What You'll Learn

- **Define** electromagnetic induction.
- **Describe** how a generator produces an electric current.
- **Distinguish** between alternating current and direct current.
- **Explain** how a transformer can change the voltage of an alternating current.

Why It's Important

Electromagnetic induction enables power plants to generate the electric current an appliance uses when you plug it into an electric outlet.

Review Vocabulary

voltage difference: a measure of the electrical energy provided by charges as they flow in a circuit

New Vocabulary

- electromagnetic induction
- generator
- turbine
- direct current (DC)
- alternating current (AC)
- transformer

From Mechanical to Electrical Energy

Working independently in 1831, Michael Faraday in Great Britain and Joseph Henry in the United States both found that moving a loop of wire through a magnetic field causes an electric current to flow in the wire. They also found that moving a magnet through a loop of wire produces a current. In both cases, the mechanical energy associated with the motion of the wire loop or the magnet is converted into electrical energy associated with the current in the wire. The magnet and wire loop must be moving relative to each other for an electric current to be produced. This causes the magnetic field inside the loop to change with time. In addition, if the current in a wire changes with time, the changing magnetic field around the wire can also induce a current in a nearby coil. The generation of a current by a changing magnetic field is **electromagnetic induction.**

Generators Most of the electrical energy you use every day is provided by generators. A **generator** uses electromagnetic induction to transform mechanical energy into electrical energy. **Figure 16** shows one way a generator converts mechanical energy to electrical energy. The mechanical energy is provided by turning the handle on the generator.

An example of a simple generator is shown in **Figure 17.** In this type of generator, a current is produced in the coil as the coil rotates between the poles of a permanent magnet.

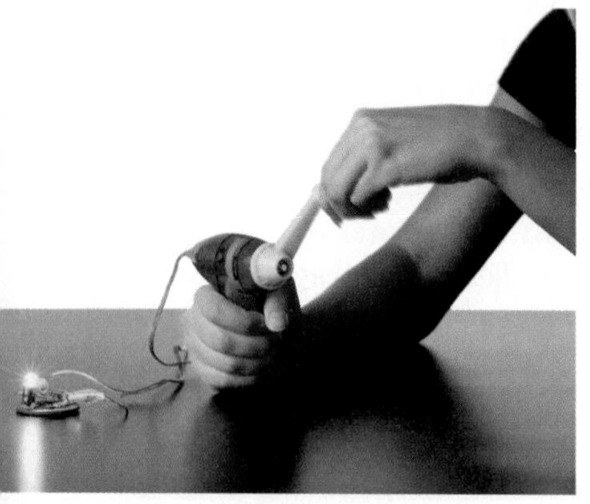

Figure 16 The coil in a generator is rotated by an outside source of mechanical energy. Here the student supplies the mechanical energy that is converted into electrical energy by the generator.

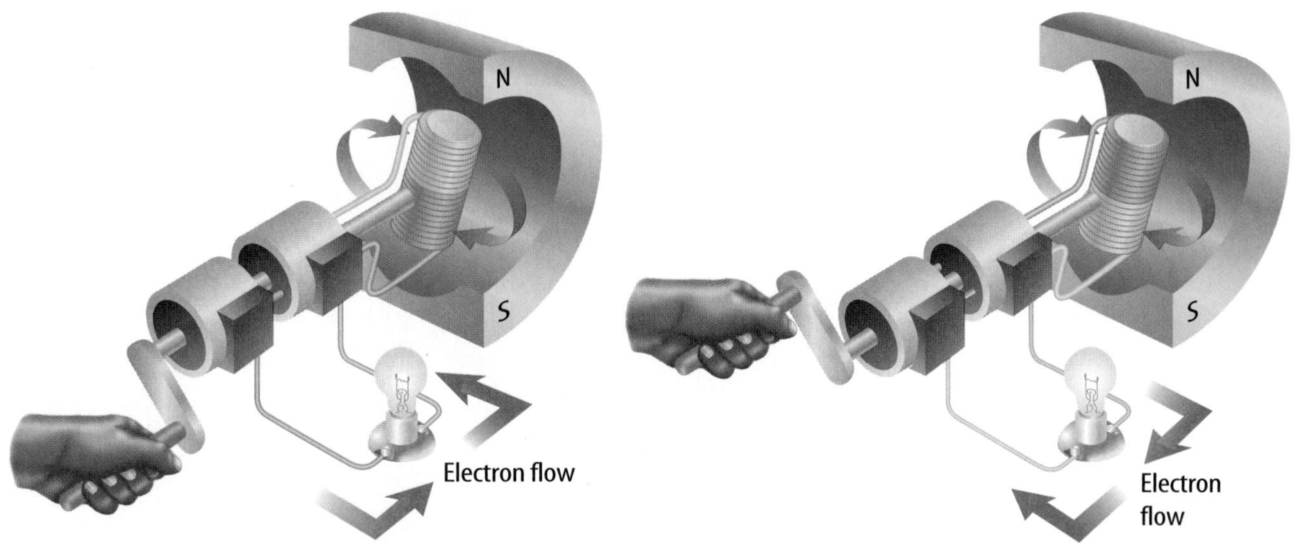

Figure 17 The current in the coil changes direction each time the ends of the coil move past the poles of the permanent magnet.
Explain *how the frequency of the changing current can be controlled.*

Switching Direction As the generator's wire coil rotates through the magnetic field of the permanent magnet, current flows through the coil. After the wire coil makes one-half of a revolution, the ends of the coil are moving past the opposite poles of the permanent magnet. This causes the current to change direction. In a generator, as the coil keeps rotating, the current that is produced periodically changes direction. The direction of the current in the coil changes twice with each revolution, as **Figure 17** shows. The frequency with which the current changes direction can be controlled by regulating the rotation rate of the generator. In the United States, current is produced by generators that rotate 60 times a second, or 3,600 revolutions per minute.

Reading Check *For each revolution of the coil, how many times does the current change direction?*

The Electromagnetic Force There is a relationship between electricity and magnetism. A changing magnetic field can cause electric charges to move in a wire. Also, a moving electric charge produces a magnetic field. The atoms of magnetic materials, such as iron, are magnets because of the motion of electric charges in the atoms. As a result, the magnetic field around a permanent magnet is due to electric charges that are in motion.

This connection occurs because the electric force and the magnetic force are two different aspects of the same force. This force is the electromagnetic force and is one of the fundamental forces in nature. An electromagnetic force exists between all objects that have electric charge. Just like the magnetic and the electric force, the electromagnetic force can be attractive or repulsive.

Figure 18 Each of these generators at Hoover Dam can produce over 100,000 kW of electric power. In these generators, a rotating magnet induces an electric current in a stationary wire coil.

Power-Plant Operator Many daily activities require electricity. Power-plant operators control the machinery that generates electricity. Operators must have a high school diploma. College-level courses may be helpful. Research to find employers in your area that hire power plant operators.

Generating Electricity for Your Home You probably do not have a generator in your home that supplies the electrical energy you use at home. This electrical energy comes from a power plant with huge generators like the one in **Figure 18.** The coils in these generators have many coils of wire wrapped around huge iron cores. The rotating magnets are connected to a **turbine,** which is a large wheel that rotates when pushed by water, wind, or steam.

Some power plants first produce thermal energy by burning fossil fuels or using the heat produced by nuclear reactions. This thermal energy is used to heat water and produce steam. Thermal energy is then converted to mechanical energy as the steam pushes the turbine blades. The generator then changes the mechanical energy of the rotating turbine into the electrical energy you use. In some areas, fields of windmills, such as those in **Figure 18,** can be used to capture the mechanical energy in wind to turn generators. Other power plants use the mechanical energy in falling water to rotate the turbines. Both generators and electric motors use magnets to produce energy conversions between electrical and mechanical energy. **Figure 20** summarizes the differences between electric motors and generators.

Figure 19 The propeller on each of these windmills is connected to an electric generator. The rotating propeller rotates a coil or a permanent magnet.

NATIONAL GEOGRAPHIC VISUALIZING MOTORS AND GENERATORS

Figure 20

Electric motors power many everyday machines, from CD players to vacuum cleaners. Generators produce the electricity these motors need to run. Both motors and generators use electromagnets, but in different ways. The table below compares motors and generators.

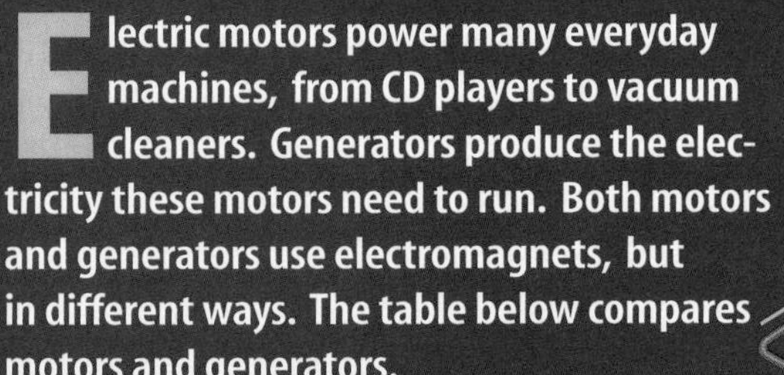

	Electric Motor	Generator
What does it do?	Changes electricity into movement	Changes movement into electricity 
What makes its electromagnetic coil rotate?	Attractive and repulsive forces between the coil and the permanent magnet 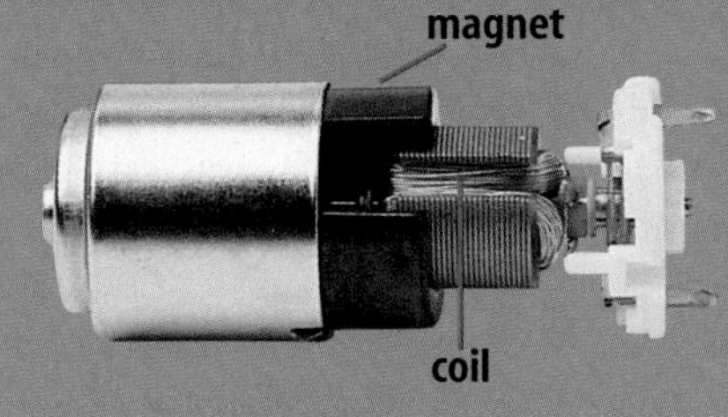	An outside source of mechanical energy 
What is the source of the current that flows in its coil?	An outside power source 	Electromagnetic induction from moving the coil through the field of the permanent magnet 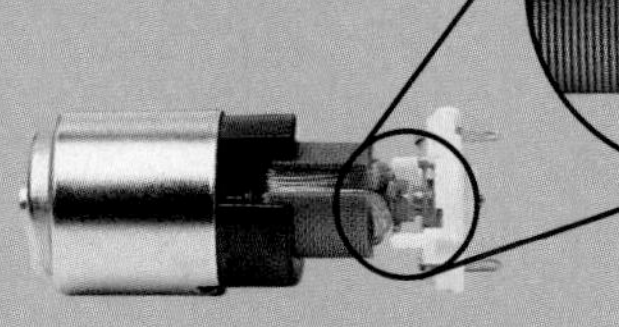
How often does the current in the coil change direction?	Twice during each rotation of the coil	Twice during each rotation of the coil

Topic: Transformers

Visit gpescience.com for links to information about how transformers transmit electric current.

Activity In England, voltage from an outlet is greater than required for U.S. appliances. Using information found at these links, diagram a transformer that would allow a U.S. appliance to operate in England.

Direct and Alternating Currents

Modern society relies heavily on electricity. Just how much you rely on electricity becomes obvious during a power outage. Out of habit, you might walk into a room and flip on the light switch. You might try to turn on a radio or television or check the clock to see what time it is. Because power outages sometimes occur, some electrical devices, such as the radio in **Figure 21,** use batteries as a backup source of electrical energy. However, the current produced by a battery is different from the current produced by an electric generator.

A battery produces a direct current. **Direct current** (DC) flows only in one direction through a wire. When you plug your CD player or any other appliance into a wall outlet, you are using alternating current. **Alternating current** (AC) reverses the direction of the current in a regular pattern. In North America, generators produce alternating current at a frequency of 60 cycles per second, or 60 Hz. The electric current produced by a generator changes direction twice during each cycle, or each rotation, of the coil. This means that a 60-Hz alternating current changes direction 120 times each second.

Electronic devices that use batteries as a backup energy source usually require direct current to operate. When the device is plugged into a wall outlet, electronic components inside the device convert the alternating current to direct current and also reduce the voltage of the alternating current.

Figure 21 Some devices, such as this radio, can use either direct or alternating current. Electronic components in these devices change alternating current from an electric outlet to direct current.

Transmitting Electrical Energy

The alternating current produced by an electric power plant carries electrical energy that is transmitted along electric transmission lines. However, when electrical energy is transmitted along power lines, some of the electrical energy is converted into heat because of the electrical resistance of the wires. The heat produced in the power lines warms the wires and the surrounding air and can't be used to power electrical devices. Also, the electrical resistance and heat production increase as the wires get longer. As a result, large amounts of heat can be produced when electrical energy is transmitted over long distances.

One way to reduce the heat produced in a power line is to transmit the electrical energy at high voltages, typically around 150,000 V. However, electrical energy at such high voltage cannot enter your home safely, nor can it be used in home appliances. Instead, a transformer is used to decrease the voltage.

Transformers

A **transformer** is a device that increases or decreases the voltage of an alternating current. A transformer is made of a primary coil and a secondary coil. These wire coils are wrapped around the same iron core, as shown in **Figure 22.** As an alternating current passes through the primary coil, the coil's magnetic field magnetizes the iron core. The magnetic field in the primary coil changes direction as the current in the primary coil changes direction. This produces a magnetic field in the iron core that changes direction at the same frequency. The changing magnetic field in the iron core then induces an alternating current with the same frequency in the secondary coil.

The voltage in the primary coil is the input voltage, and the voltage in the secondary coil is the output voltage. The output voltage divided by the input voltage equals the number of turns in the secondary coil divided by the number of turns in the primary coil.

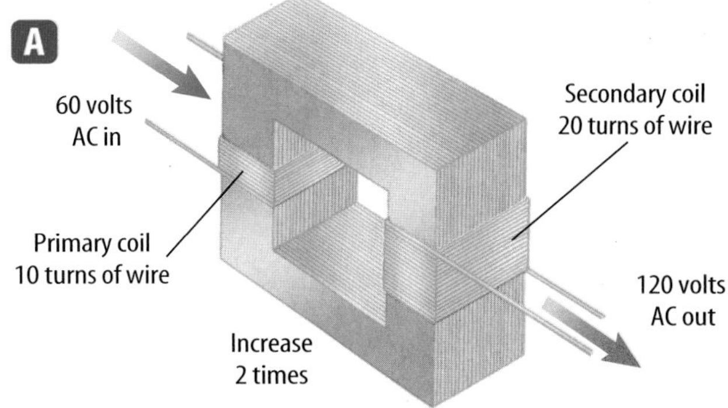

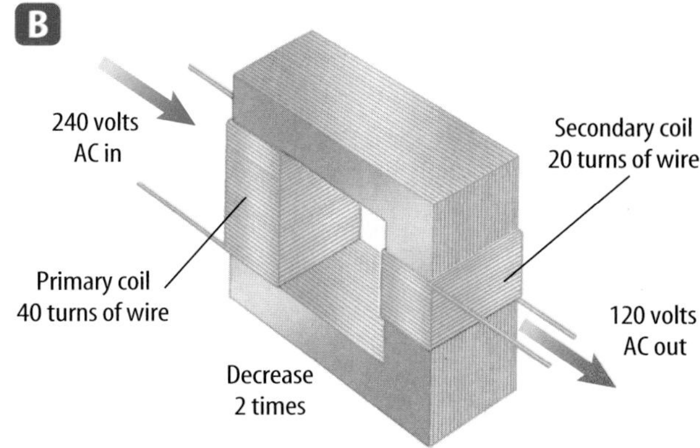

Figure 22 Transformers can increase or decrease voltage.
A A step-up transformer increases voltage. The secondary coil has more turns than the primary coil does. **B** A step-down transformer decreases voltage. The secondary coil has fewer turns than the primary coil does.
Infer *whether a transformer could change the voltage of a direct current.*

Reading Check *How does a transformer produce an alternating current in the secondary coil?*

Step-Up Transformer A transformer that increases the voltage so that the output voltage is greater than the input voltage is a step-up transformer. In a step-up transformer, the number of wire turns on the secondary coil is greater than the number of turns on the primary coil. For example, the secondary coil of the step-up transformer in **Figure 22A** has twice as many turns as the primary coil has. The ratio of the output voltage to the input voltage is 2:1, and the output voltage is twice as large as the input voltage. For this transformer, an input voltage of 60 V in the primary coil is increased to 120 V in the secondary coil.

Step-Down Transformer A transformer that decreases the voltage so that the output voltage is less than the input voltage is a step-down transformer. In a step-down transformer, the number of wire turns on the secondary coil is less than the number of turns on the primary coil. In **Figure 22B** the secondary coil has half as many turns as the primary coil has, so the ratio of the output voltage to the input voltage is 1:2. The input voltage of 240 V in the primary coil is reduced to a voltage of 120 V in the secondary coil.

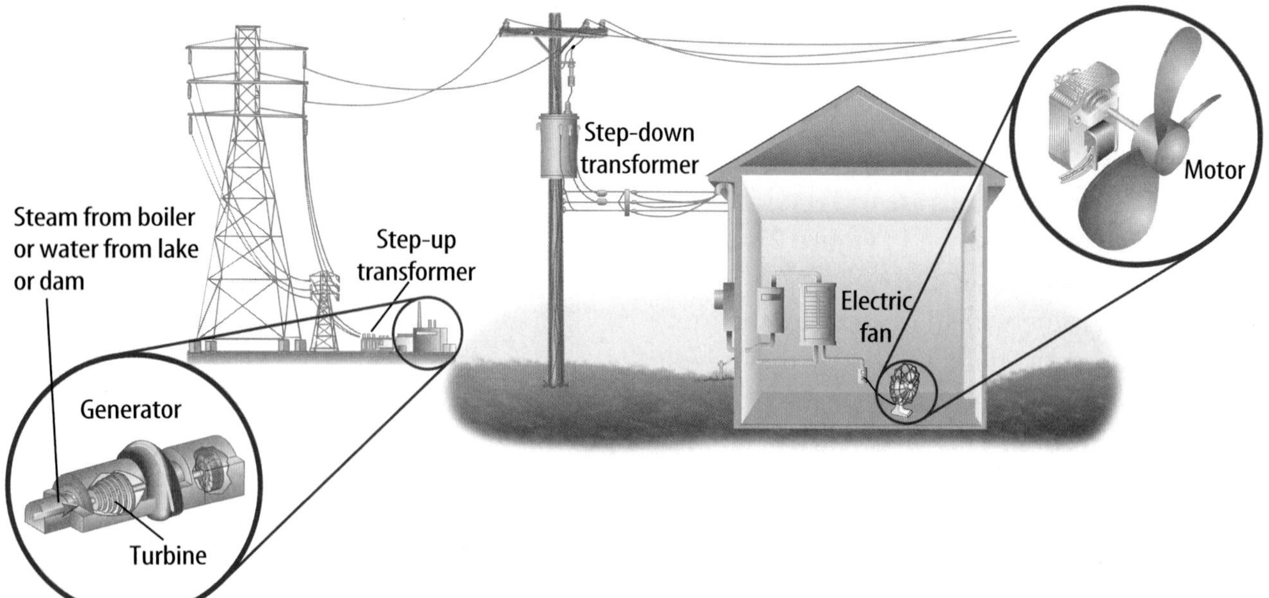

Figure 23 Many steps are involved in the creation, transportation, and use of the electric current in your home.
Identify *the steps that involve electromagnetic induction.*

Transmitting Alternating Current Power plants commonly produce alternating current because the voltage can be increased or decreased with transformers. Although step-up transformers and step-down transformers change the voltage at which electrical energy is transmitted, they do not change the amount of electrical energy transmitted. **Figure 23** shows how step-up and step-down transformers are used in transmitting electrical energy from power plants to your home.

section 3 review

Summary

From Mechanical to Electrical Energy

- An electric current is produced by moving a wire loop through a magnetic field or a magnet through a wire loop.
- A generator can produce an electric current by rotating a wire coil in a magnetic field.

Direct and Alternating Currents

- A direct current flows in one direction. An alternating current changes direction in a regular pattern.

Transformers

- A transformer changes the voltage of an alternating current. The voltage can be increased or decreased.
- The changing magnetic field in the primary coil of a transformer induces an alternating current in the secondary coil.

Self Check

1. **Describe** the energy conversions that occur when water falls on a paddle wheel connected to a generator that is connected to electric lights.
2. **Compare and contrast** a generator with an electric motor.
3. **Explain** why the output voltage from a transformer is zero if a direct current flows through the primary coil.
4. **Explain** why electric current produced by power plants is transmitted as alternating current.
5. **Think Critically** A magnet is pushed into the center of a wire loop and then stops. What is the current in the wire loop after the magnet stops moving? Explain.

Applying Math

6. **Use a Ratio** A transformer has 1,000 turns of wire in the primary coil and 50 turns in the secondary coil. If the input voltage is 2,400 V, what is the output voltage?

More Section Review gpescience.com

El💡ctricity and Magnetism

Huge generators in power plants produce electricity by moving magnets past coils of wire. How does this produce an electric current?

Real-World Problem

How can a magnet and a wire coil be used to produce an electric current?

Goals

- **Observe** how a magnet and a wire coil can produce an electric current in a wire.
- **Compare** the currents created by moving the magnet and the wire coil in different ways.

Materials

cardboard tube
bar magnet
insulated wire
scissors
galvanometer or ammeter

Safety Precautions 

Procedure

1. Complete the safety form.
2. Wrap the wire around the cardboard tube to make a coil of about 20 turns. Remove the tube from the coil.
3. Use the scissors to cut and remove 2 cm of insulation from each end of the wire.
4. Connect the ends of the wire to a galvanometer or ammeter. Record the reading on your meter.
5. Insert one end of the magnet into the coil and then pull it out. Record the current. Move the magnet at different speeds inside the coil and record the current.

6. Watch the meter and move the bar magnet in different ways around the outside of the coil. Record your observations.
7. Repeat steps 5 and 6, keeping the magnet stationary and moving the wire coil.

Conclude and Apply

1. How was the largest current generated?
2. Does the current generated always flow in the same direction? How do you know?
3. **Predict** what would happen if you used a coil made with fewer turns of wire.
4. **Infer** whether a current would have been generated if the cardboard tube had been left in the coil. Why or why not? Try it.

Communicating Your Data

Compare the currents generated by different members of the class. What was the value of the largest current that was generated? How was this current generated?

Design Your Own

Controlling Electromagnets

Goals
- **Measure** relative strengths of electromagnets.
- **Determine** which factors affect the strength of an electromagnet.

Possible Materials
22-gauge insulated wire
16-penny iron nail
aluminum rod or nail
6-V DC power supply
three 1.5-V D cells
steel paper clips
magnetic compass
duct tape (to hold D cells together)

Safety Precautions

WARNING: *Do not leave the electromagnet connected for long periods of time because the battery will run down. Magnets will get hot with only a few turns of wire. Use caution in handling them when current is flowing through the coil. Do not apply voltages higher than 6 V to the electromagnets.*

Real-World Problem

You use electromagnets every day when you use stereo speakers, power door locks, and many other devices. To make these devices work properly, the strength of the magnetic field surrounding an electromagnet must be controlled. How can the magnetic field produced by an electromagnet be made stronger or weaker? Think about the components that form an electromagnet. Make a hypothesis about how changing these components would affect the strength of the electromagnet's magnetic field.

Form a Hypothesis

As a group, write down the components of an electromagnet that might affect the strength of its magnetic field.

Test Your Hypothesis

Make a Plan

1. Write your hypothesis about the best way to control the magnetic field strength of an electromagnet.
2. **Decide** how you will assemble and test the electromagnets. Which features will you change to determine the effect on the strength of the magnetic fields? How many changes will you need to try? How many electromagnets do you need to build?
3. **Decide** how you are going to test the strength of your electromagnets. Several ways are possible with the materials listed. Which way would be the most sensitive? Be prepared to change test methods if necessary.

4. Write your plan of investigation. Make sure your plan tests only one variable at a time.

Follow Your Plan

1. Before you begin to build and test the electromagnets, make sure your teacher approves of your plan.
2. Carry out your planned investigation.
3. Record your results.

Analyze Your Data

1. **Make a table** showing how the strength of your electromagnet depends on changes you made in its construction or operation.
2. **Examine** the trends shown by your data. Are there any data points that seem out of line? How can you account for them?

Testing Electromagnets

Trial	Electromagnet Construction Features	Strength of Electromagnet
	Do not write in this book.	

Conclude and Apply

1. **Describe** how the electromagnet's magnetic-field strength depended on its construction or operation.
2. **Identify** the features of the electromagnet's construction that had the greatest effect on its magnetic-field strength. Which do you think would be easiest to control?
3. **Explain** how you could use your electromagnet to make a switch. Would it work with both AC and DC?
4. **Evaluate** whether or not your results support your hypothesis. Explain.

Compare your group's result with those of other groups. Did any other group use a different method to test the strength of the magnet? Did you get the same results?

TIME SCIENCE AND HISTORY

SCIENCE CAN CHANGE THE COURSE OF HISTORY!

Body Art

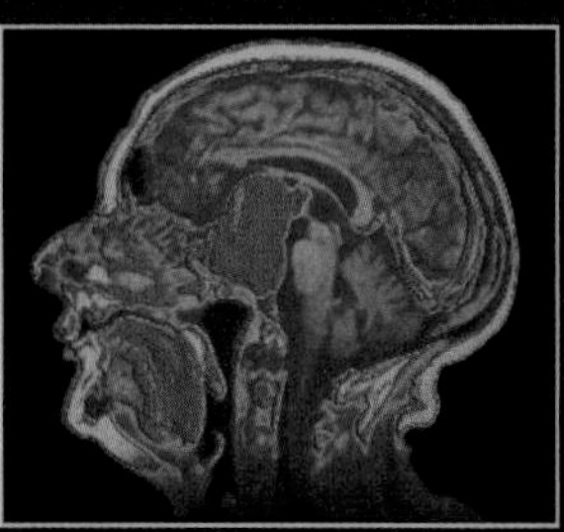

The surgeon turns the computer screen so the patient can see it. Pointing to a colorful image of the patient's brain, she reassures the worried patient. "This MRI shows exactly where your tumor is. We can remove it with little danger to you."

MRI for the Soft Stuff

MRI stands for "magnetic resonance imaging." It's a way to take 3-D pictures of the inside of your body. Before the 1980s, doctors could x-ray solid tissue such as bones, but had no way to see soft tissue such as the brain. Well, they had one way—surgery, which sometimes caused injury and infection, risking a patient's health.

MRI uses a strong magnet and radio waves. Tissues in your body contain water molecules that are made of oxygen and hydrogen atoms.

The nucleus of a hydrogen atom is a proton, which behaves like a tiny magnet. A strong magnetic field inside the MRI tube makes these proton magnets line up in the direction of the field. Radio waves are then applied to the body. The protons absorb some of the radio-wave energy, and flip their direction.

When the radio waves are turned off, the protons realign themselves with the magnetic field and emit the energy they absorbed. Different tissues in the body absorb and emit different amounts of energy. The emitted energy is detected, and a computer uses this information to form images of the body.

Your brain is getting bigger!

MRI has turned into an important research tool. For example, researchers using MRI have found that the brain grows rapidly through adolescence. Before this research, people thought that the brain stopped growing in childhood. MRI has proven that adolescents are getting bigger brains all the time.

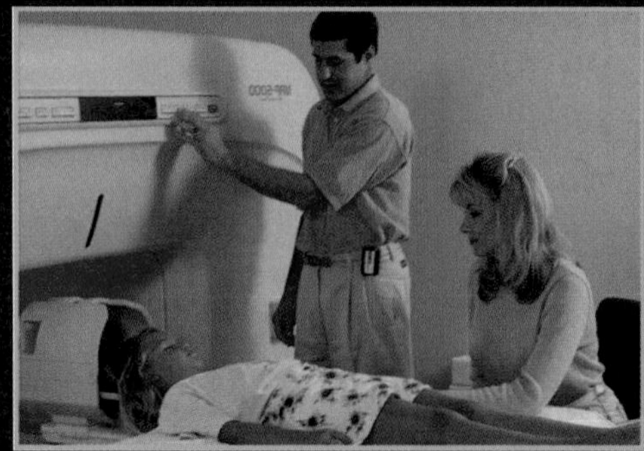

This patient is about to be placed in an MRI machine.

Interview As an oral history project, interview a retired physician or surgeon. Ask him or her to discuss with you how tools such as MRI changed during his or her career. Make a list of the tools and how they have helped improve medicine.

TIME

For more information, visit gpescience.com

Reviewing Main Ideas

Section 1 Magnetism

1. A magnetic field surrounds a magnet and exerts a magnetic force.
2. All magnets have two poles: a south pole and a north pole.

3. Opposite poles of magnets attract; like poles repel.
4. Groups of atoms with aligned magnetic poles are called magnetic domains.

Section 2 Electricity and Magnetism

1. An electric current flowing through a wire produces a magnetic field.
2. An electric current passing through a coil of wire can produce a magnetic field inside the coil. The coil becomes an electromagnet. One end of the coil is the north pole, and the other end is the south pole.

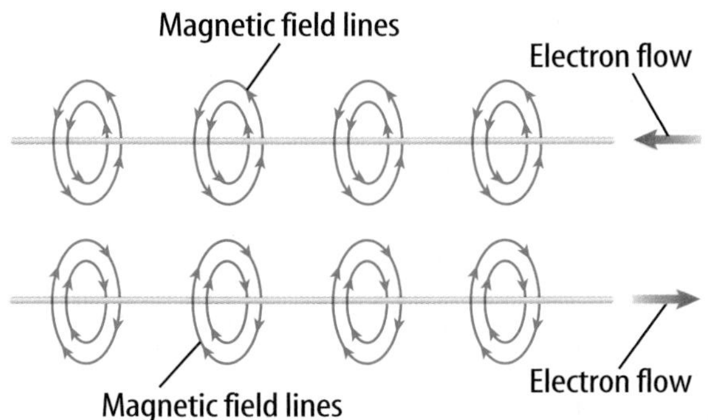

3. The magnetic field around an electromagnet depends on the current and the number of coils.
4. An electric motor contains a rotating electromagnet that converts electrical energy to mechanical energy.

Section 3 Producing Electric Current

1. By moving a magnet near a wire, you can create an electric current in the wire. This is called electromagnetic induction.

2. A generator produces electric current by rotating a coil of wire in a magnetic field. Generators at the base of this dam convert the kinetic energy in falling water into electrical energy.
3. Direct current flows in one direction through a wire; alternating current reverses the direction of current flow in a regular pattern.
4. The number of turns of wire in the primary and secondary coils of a transformer determines whether it increases or decreases voltage.

FOLDABLES Use the Foldable that you made at the beginning of this chapter to help you review magnets and magnetism.

chapter 14 Review

Using Vocabulary

alternating current (AC) p. 442	generator p. 438
direct current (DC) p. 442	magnetic domains p. 429
electric motor p. 435	magnetic field p. 425
electromagnet p. 432	magnetic pole p. 425
electromagnetic induction p. 438	magnetism p. 424
galvanometer p. 434	solenoid p. 432
	transformer p. 443
	turbine p. 440

Complete each statement with the correct vocabulary word or phrase.

1. A(n) ________ can be used to change the voltage of an alternating current.
2. A(n) ________ is the region where the magnetic field of a magnet is strongest.
3. ________ does not change direction.
4. The properties and interactions of magnets are called ________.
5. A(n) ________ can rotate in a magnetic field when a current passes through it.
6. The magnetic poles of atoms are aligned in a(n) ________.
7. A device that uses an electromagnet to measure electric current is a(n) ________.

Checking Concepts

Choose the word or phrase that best answers the question.

8. Where is the magnetic force exerted by a magnet strongest?
   A) both poles **C)** north pole
   B) south pole **D)** center
9. Which change occurs in an electric motor?
   A) electrical energy to mechanical energy
   B) thermal energy to wind energy
   C) mechanical energy to electrical energy
   D) wind energy to electrical energy
10. What happens to the magnetic force as the distance between two magnetic poles decreases?
    A) remains constant **C)** increases
    B) decreases **D)** becomes zero
11. Which of the following best describes what type of magnetic poles the domains at the north pole of a bar magnet have?
    A) north magnetic poles only
    B) south magnetic poles only
    C) no magnetic poles
    D) north and south magnetic poles
12. Which of the following would not change the strength of an electromagnet?
    A) increasing the amount of current
    B) changing the current's direction
    C) inserting an iron core inside the coil
    D) increasing the number of loops
13. Which of the following is NOT part of a generator?
    A) turbine **C)** electromagnet
    B) battery **D)** permanent magnet
14. Which of the following describes the direction of the electric current in AC?
    A) is constant **C)** changes regularly
    B) is direct **D)** changes irregularly

Interpreting Graphics

15. Copy and complete the Venn diagram below. Include the functions, part names, and power sources for these devices.

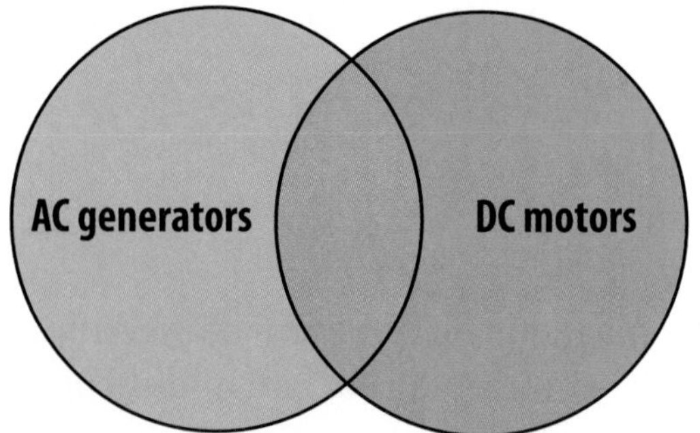

Vocabulary PuzzleMaker gpescience.com

Use the diagram below to answer questions 16 and 17.

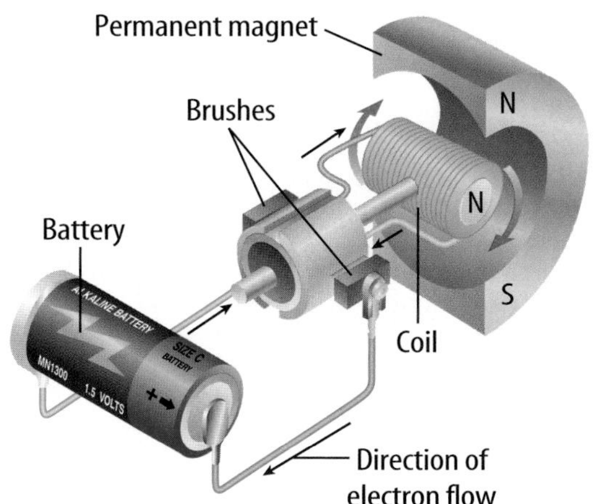

16. **Describe** Using the diagram, describe the function of the permanent magnet, the electromagnet, and the current source in a simple electric motor.

17. **Describe** the sequence of steps that occur in an electric motor that forces the coil to spin. Include the role of the commutator in your description.

Use the graph below to answer questions 18–20.

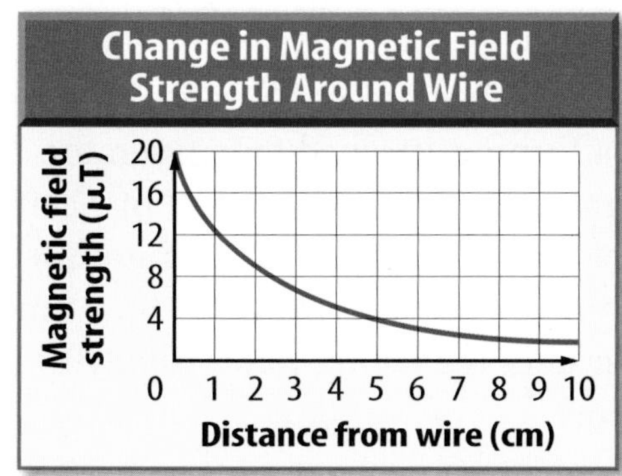

18. **Explain** How much larger is the magnetic field strength 1 cm from the wire compared to 5 cm from the wire?

19. **Explain** Does the magnetic field strength decrease more rapidly with distance closer to the wire or farther from the wire? Explain.

20. **Solve** Using the graph, estimate the magnetic field strength 11 cm from the wire.

Thinking Critically

21. **Infer** how you could you use a horseshoe magnet to find the north direction.

22. **Explain** In Europe, generators produce alternating current at a frequency of 50 Hz. Would the electric appliances you use in North America work if you plugged them into an outlet in Europe? Why or why not?

23. **Predict** Two generators are identical except for the loops of wire that rotate through their magnetic fields. One has twice as many turns of wire as the other one does. Which generator will produce the most electric current? Why?

24. **Explain** why a bar magnet will attract an iron nail to either its north pole or its south pole, but it will attract another magnet to only one of its poles.

25. **Compare and contrast** electromagnetic induction and the formation of electromagnets.

Applying Math

26. **Calculate** A step-down transformer reduces a 2,400-V current to 120 V. If the primary coil has 500 turns of wire, how many turns of wire are there on the secondary coil?

27. **Use a Ratio** To produce a spark, a spark plug requires a current of about 12,000 V. A car's engine uses a type of transformer called an induction coil to change the input voltage from 12 V to 12,000 V. In the induction coil, what is the ratio of the number of wire turns on the primary coil to the number of turns on the secondary coil?

Standardized Test Practice

Record your answers on the answer sheet provided by your teacher or on a sheet of paper.

Multiple Choice

A transformer has 50 turns of wire on the primary coil and 10 turns of wire on the secondary coil. The effect of changing the input voltage is summarized in the table below.

Use the table below to answer questions 1 and 2.

Voltage and Current in a Transformer				
Trial	Input Voltage (V)	Primary Coil Current (A)	Output Voltage (V)	Secondary Coil Current (A)
1	5	0.1	1	0.5
2	10	0.2	2	1.0
3	20	0.2	4	1.0
4	50	0.5	10	2.5

1. What is the ratio of the input voltage to the output voltage for this transformer?

A. 1:2.5

B. 4:1

C. 5:1

D. 1:5

2. What is the ratio of the secondary-coil current to the primary-coil current?

A. the ratio of the secondary-coil wire turns to the primary-coil wire turns

B. the ratio of the output voltage to the input voltage

C. the ratio of the primary-coil wire turns to the secondary-coil wire turns

D. It always equals one.

Use the figure below to answer question 3.

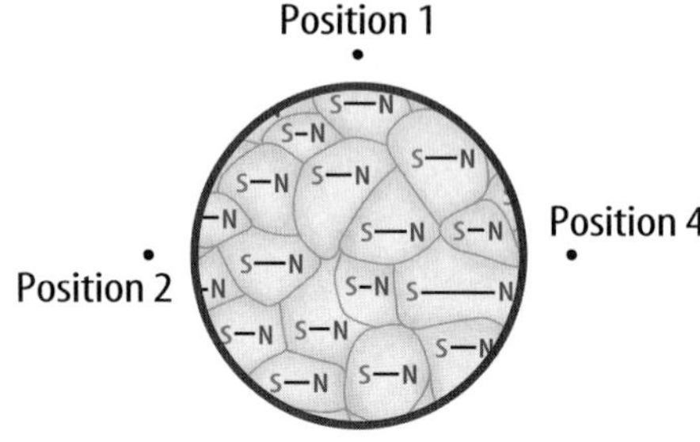

3. A steel paper clip is sitting on a desk. The figure above shows the magnetic domains in a section of the paper clip after the north pole of a magnet has been moved close to it. According to the diagram, the magnet's north pole is at which position?

A. position 1

B. position 2

C. position 3

D. position 4

4. A hydroelectric power plant uses water to spin a turbine attached to a generator. The generator produces 30,000 kW of electric power. If the turbine and generator are a combined 75 percent efficient, how much power does the falling water supply to the turbine?

A. 50,000 kW

B. 40,000 kW

C. 22,500 kW

D. 20,000 kW

Gridded Response

Use the figure below to answer questions 5 and 6.

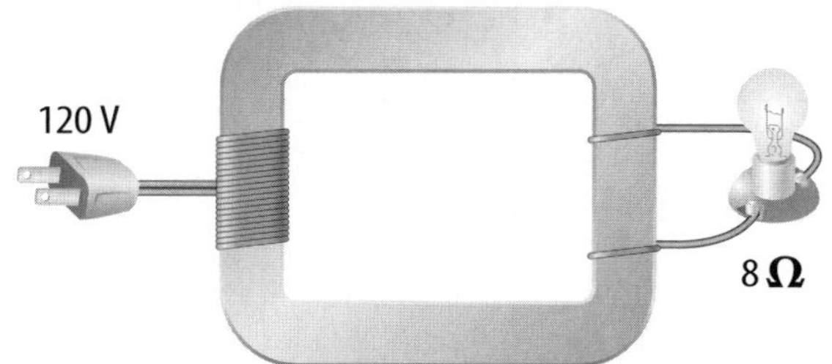

5. A step-down transformer is plugged into a 120-V electric outlet and a lightbulb is connected to the output coil. The transformer has 20 turns on the primary coil and 2 turns on the secondary coil. What is the voltage, in volts, at the output coil?

6. If the lightbulb has a resistance of 8 Ω, what is the current, in amperes, in the lightbulb?

Short Response

7. A bicycle has a small electric generator that is used to light a headlight. The generator is made to spin by rubbing against a wheel. Will the bicycle coast farther on a level surface if the light is turned on or turned off?

8. A student connects a battery to a step-up transformer to boost the voltage. Explain why a small electric motor does not spin when it is connected to the secondary coil of the transformer.

Extended Response

Use the figure below to answer question 9.

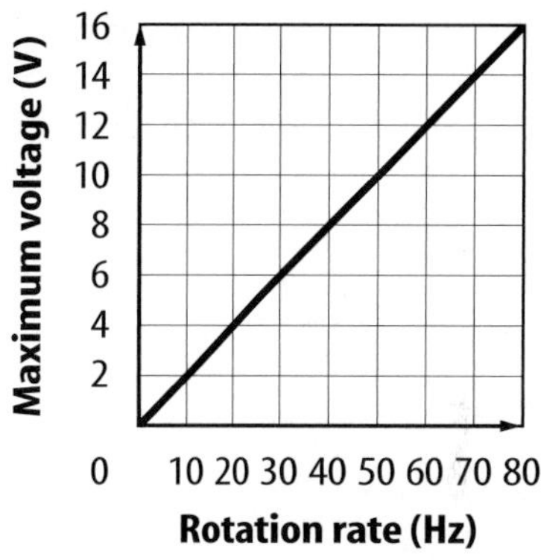

9. The graphic above shows how the voltage produced by a generator depends on the rotation rate of the coil. Explain whether this generator could produce household AC current that is 120 V at 60 Hz.

10. Compare and contrast the behavior and properties of positive and negative electric charges with north and south magnetic poles.

Test-Taking Tip

Read All the Information If a question includes a text passage and a graphic, carefully read the information in the text passage and the graphic before answering the question.

Question 1 Review the information in the text above the table and the information in the table.

chapter 15

Electromagnetic Radiation

chapter preview

sections

1 What are electromagnetic waves?

2 The Electromagnetic Spectrum
Lab *The Shape of Satellite Dishes*

3 Radio Communication
Lab *Radio Frequencies*

Virtual Lab *What is the electromagnetic spectrum?*

How's the reception?

These giant 25-m dishes aren't picking up TV signals, unless they're coming from distant stars and galaxies. They are part of a group of 27 antennas that detect radio waves, which are electromagnetic waves. However, all objects emit electromagnetic waves, not just stars and galaxies.

Science Journal List six objects around you that emit light or feel warm.

Start-Up Activities

Can electromagnetic waves change materials?

You often hear about the danger of the Sun's ultraviolet rays, which can damage the cells of your skin. When the exposure isn't too great, your cells can repair themselves, but too much at one time can cause a painful sunburn. Repeated overexposure to the Sun over many years can damage cells and cause skin cancer. In the lab below, observe how energy carried by ultraviolet waves can cause changes in other materials.

1. Complete the safety form.
2. Cut a sheet of red construction paper in half.
3. Place one piece outside in direct sunlight. Place the other in a shaded location.
4. Keep the construction paper in full sunlight for at least 45 min. If possible, allow it to stay there for 3 h or more before taking it down. Be sure the other piece remains in the shade.
5. **Think Critically** In your Science Journal, describe any differences you notice in the two pieces of construction paper. Comment on your results.

Electromagnetic Waves Make the following Foldable to help you understand electromagnetic waves.

STEP 1 **Fold** a sheet of paper vertically in half from top to bottom.

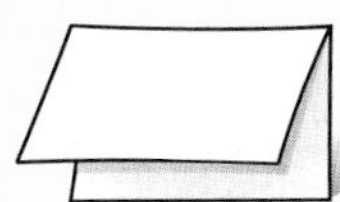

STEP 2 **Fold** in half from side to side with the fold at the top.

STEP 3 **Unfold** the paper once. **Cut** only the fold of the top flap to make two tabs.

STEP 3 **Write** on the front tabs as shown.

How do electromagnetic waves travel through space?

How do electromagnetic waves transfer energy to matter?

Identify Questions As you read the chapter, write answers to the questions on the back of the appropriate tabs.

Preview this chapter's content and activities at gpescience.com

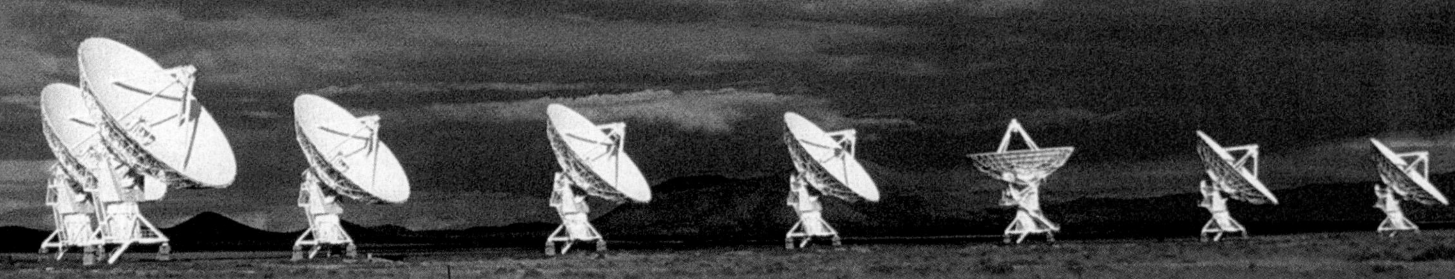

section 1

What are electromagnetic waves?

Reading Guide

What You'll Learn

- **Describe** how electric and magnetic fields form electromagnetic waves.
- **Explain** how vibrating charges produce electromagnetic waves.
- **Describe** properties of electromagnetic waves.

Why It's Important

You, and all the objects and materials around you, are radiating electromagnetic waves.

Review Vocabulary

hertz: the SI unit of frequency, abbreviated Hz; 1 Hz equals one vibration per second

New Vocabulary

- electromagnetic wave
- radiant energy
- photon

Waves in Space

Stay calm. Do not panic. As you are reading this sentence, no matter where you are, you are surrounded by electromagnetic waves. Even though you can't feel them, some of these waves are traveling right through your body. They enable you to see. They make your skin feel warm. You use electromagnetic waves when you watch television, talk on a cordless phone, or prepare popcorn in a microwave oven.

Figure 1 Water waves and sound waves require matter to move through. Energy is transferred from one particle to the next as the wave travels through the matter.

Sound and Water Waves Waves are produced by something that vibrates, and they transmit energy from one place to another. Look at the sound wave and the water wave in **Figure 1.** Both waves are moving through matter. The sound wave is moving through air and the water wave through water. These waves travel because energy is transferred from particle to particle. Without matter to transfer the energy, waves cannot move.

Electromagnetic Waves However, electromagnetic waves do not require matter to transfer energy. **Electromagnetic waves** are made by vibrating electric charges and can travel through space where matter is not present. Instead of transferring energy from particle to particle, electromagnetic waves travel by transferring energy between vibrating electric and magnetic fields.

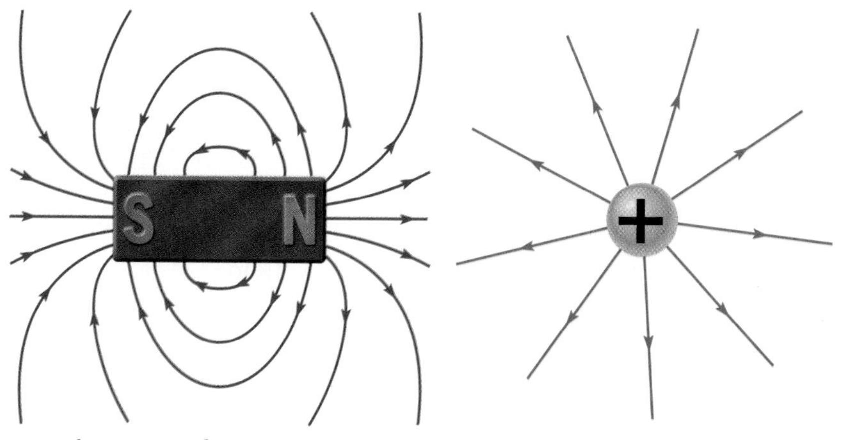

A magnetic field surrounds all magnets.

An electric field surrounds all charges.

Figure 2 Fields enable magnets and charges to exert forces at a distance. These fields extend throughout space.
Explain *how you could detect a magnetic field.*

Electric and Magnetic Fields

When you bring a magnet near a metal paper clip, the paper clip moves toward the magnet and sticks to it. The paper clip moves because the magnet exerts a force on it. The magnet exerts this force without having to touch the paper clip because all magnets are surrounded by a magnetic field, as shown in **Figure 2.** Magnetic fields exist around magnets even if the space around the magnets contains no matter.

Just as magnets are surrounded by magnetic fields, electric charges are surrounded by electric fields, also shown in **Figure 2.** An electric field enables charges to exert forces on each other even when they are far apart. Just as a magnetic field around a magnet can exist in empty space, an electric field exists around an electric charge even if the space around it contains no matter.

Magnetic Fields and Moving Charges Electric charges also can be surrounded by magnetic fields. It is the motion of electrons that generates the magnetic field. An electric current, the net flow of electrons in one direction, is always surrounded by a magnetic field. In fact, any moving electric charge is surrounded by a magnetic field, as well as an electric field. For example, an electric current flowing through a wire is surrounded by a magnetic field, as illustrated in **Figure 3.**

Figure 3 Electrons moving in a wire are surrounded by a magnetic field.
Describe *how you would confirm that a magnetic field exists around a current-carrying wire.*

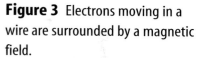

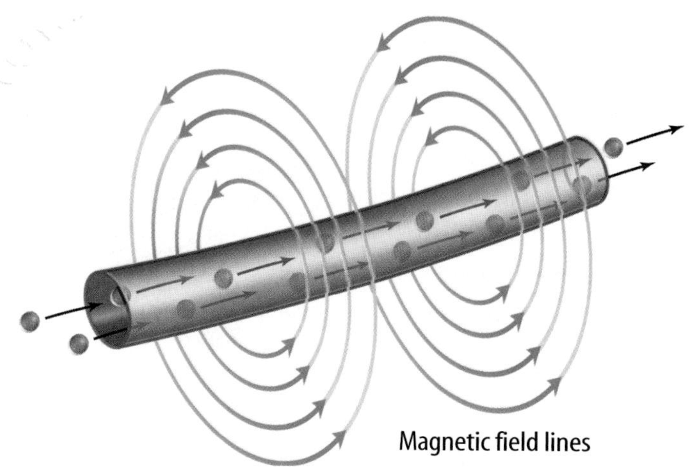

Investigating Electromagnetic Waves

Procedure

1. Complete the safety form.
2. Point your **television remote control** in different directions and observe whether it will still control the **television.**
3. Place various materials in front of the infrared receiver on the television and observe whether the remote still controls the television. Some materials you might try are **glass, a book, your hand, paper,** and a **metal pan.**

Analysis

1. Was it necessary for the remote control to be pointing exactly toward the receiver to control the television? Explain.
2. Did the remote control work when the various materials were placed between it and the receiver? Explain.

Try at Home

Changing Electric and Magnetic Fields A changing magnetic field creates a changing electric field. For example, in a transformer, changing electric current in the primary coil produces a changing magnetic field. This changing magnetic field then creates a changing electric field in the secondary coil that produces current in the coil. The reverse is also true: a changing electric field creates a changing magnetic field.

Making Electromagnetic Waves

Waves such as sound waves are produced when something vibrates. Electromagnetic waves also are produced when something vibrates—an electric charge that moves back and forth.

Reading Check *What produces an electromagnetic wave?*

When an electric charge vibrates, the electric field around it changes. Because the electric charge is in motion, it also has a magnetic field around it. This magnetic field also changes as the charge vibrates. As a result, the vibrating electric charge is surrounded by changing electric and magnetic fields.

How do the vibrating electric and magnetic fields around the charge become a wave that travels through space? The changing electric field around the charge creates a changing magnetic field. This changing magnetic field then creates a changing electric field. This process continues, with the magnetic and electric fields continually creating each other. These vibrating electric and magnetic fields are perpendicular to each other and travel outward from the moving charge, as shown in **Figure 4.** Because the electric and magnetic fields vibrate at right angles to the direction the wave travels, an electromagnetic wave is a transverse wave.

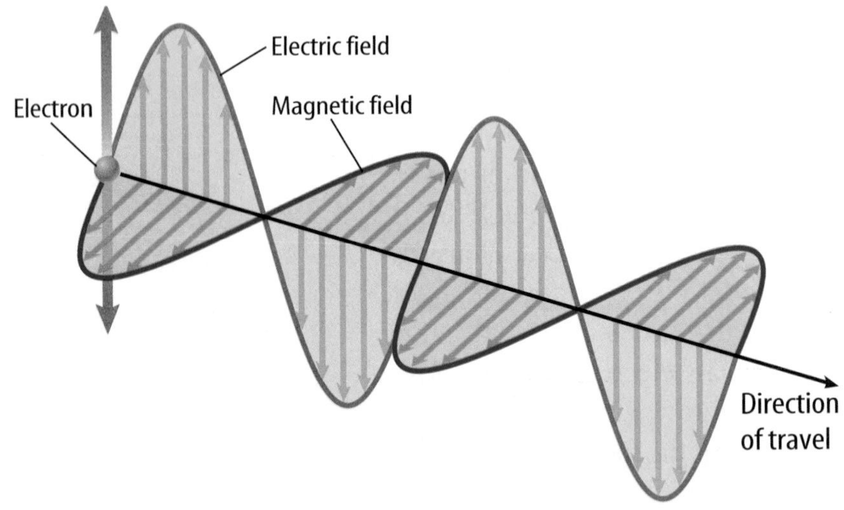

Figure 4 A vibrating electric charge creates an electromagnetic wave that travels outward in all directions from the charge. The wave in only one direction is shown here.
Determine *whether an electromagnetic wave is a transverse wave or a compressional wave.*

Properties of Electromagnetic Waves

All matter contains charged particles that are always in motion. As a result, all objects emit electromagnetic waves. The wavelengths of the emitted waves become shorter as the temperature of the material increases. As an electromagnetic wave moves, its electric and magnetic fields encounter objects. These vibrating fields can exert forces on charged particles and magnetic materials, causing them to move. For example, electromagnetic waves from the Sun cause electrons in your skin to vibrate and gain energy, as shown in **Figure 5.** The energy transferred by an electromagnetic wave is **radiant energy.** Radiant energy makes a fire feel warm and enables you to see.

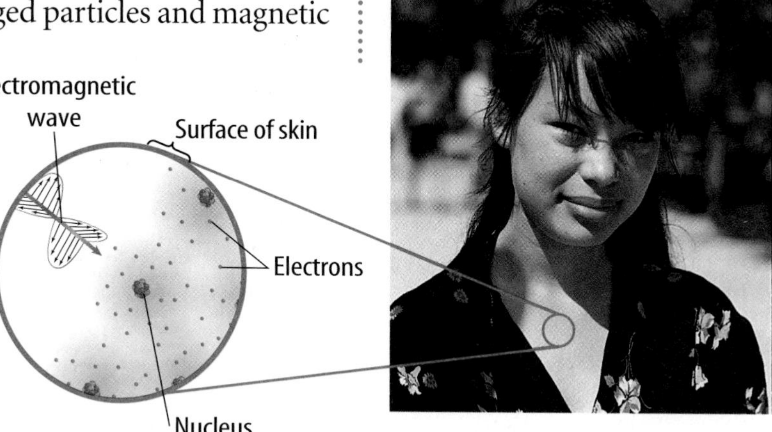

Figure 5 As an electromagnetic wave strikes your skin, electrons in your skin gain energy from the vibrating electric and magnetic fields.

Applying Math — Using Scientific Notation

THE SPEED OF LIGHT IN WATER The speed of light in water is 226,000 km/s. Write this number in scientific notation.

IDENTIFY known values and the unknown value

Identify the known values:

The speed of light in water is 226,000 km/s.

Identify the unknown value:

the number 226,000 written in scientific notation

SOLVE the problem

A number written in scientific notation has the form $M \times 10^N$. N is the number of places the decimal point in the number has to be moved so that the number M that results has only one digit to the left of the decimal point.

Write the number in scientific notation:	$226{,}000. \times 10^N$
Move the decimal point so there is only one digit to its left:	2.26000×10^N
The decimal point was moved 5 places, so N equals 5:	2.26000×10^5
Delete remaining zeroes at the end of the number.	2.26×10^5

CHECK your answer

Add zeroes to the end of the number and move the decimal point in the opposite direction five places. The result should be the original number.

Practice Problems

Write the following numbers in scientific notation: 433; 812,000,000; 73,000,000,000; 84,500.

For more practice problems go to page 879, and visit Math Practice at gpescience.com.

Material	Speed (km/s)
Vacuum	300,000
Air	slightly less than 300,000
Water	226,000
Glass	200,000
Diamond	124,000

Table 1 Speed of Visible Light

Wave Speed All electromagnetic waves travel at 300,000 km/s in the vacuum of space. Because light is an electromagnetic wave, the speed of electromagnetic waves in space is usually called the speed of light. The speed of light is nature's speed limit—nothing travels faster than the speed of light. In matter, the speed of electromagnetic waves depends on the material they travel through. Electromagnetic waves usually travel most slowly in solids and fastest in gases. **Table 1** lists the speed of visible light in various materials.

Reading Check *What is the speed of light?*

Wavelength and Frequency Like all waves, electromagnetic waves can be described by their wavelengths and frequencies. The wavelength of an electromagnetic wave is the distance from one crest to another, as shown in **Figure 6.**

The frequency of any wave is the number of wavelengths that pass a point in 1 s. The frequency of an electromagnetic wave also equals the frequency of the vibrating charge that produces the wave. This frequency is the number of vibrations, or back-and-forth movements, of the charge in one second. The frequency and wavelength of an electromagnetic wave are related. As the frequency increases, the wavelength becomes smaller.

Waves and Particles

The difference between a wave and a particle might seem obvious—a wave is a disturbance that carries energy, and a particle is a piece of matter. However, in reality, the difference is not so clear.

Waves as Particles In 1887, Heinrich Hertz found that by shining light on a metal, electrons were ejected from the metal. Hertz found that whether or not electrons were ejected depended on the frequency of the light and not on the amplitude. Because the energy carried by a wave depends on its amplitude and not its frequency, this result was mysterious. Years later, Albert Einstein provided an explanation: electromagnetic waves can behave as a particle, called a **photon,** whose energy depends on the frequency of the waves.

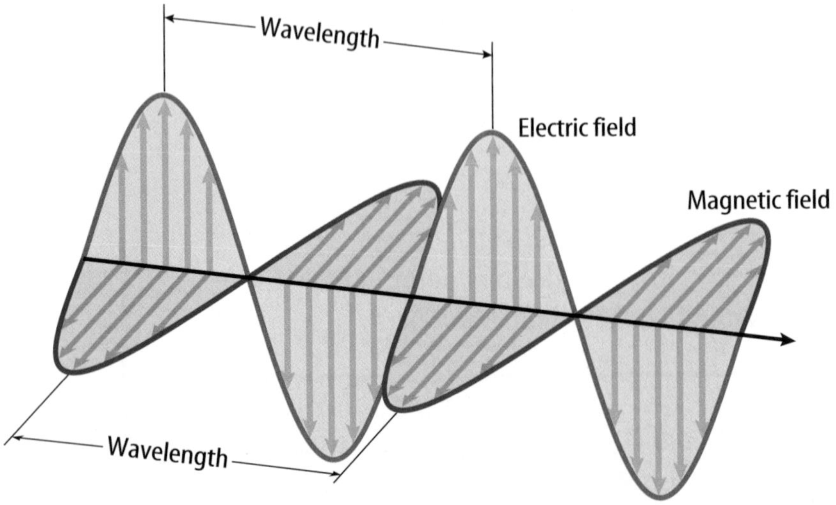

Wavelength
Electric field
Magnetic field
Wavelength

Figure 6 The wavelength of an electromagnetic wave is the distance between the crests of the vibrating electric field or magnetic field.

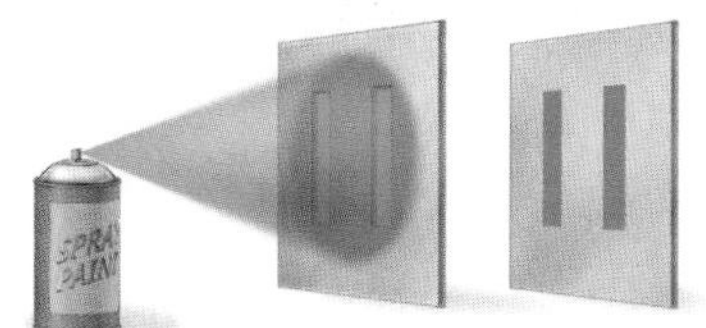

Particles of paint sprayed through two slits coat only the area behind the slits.

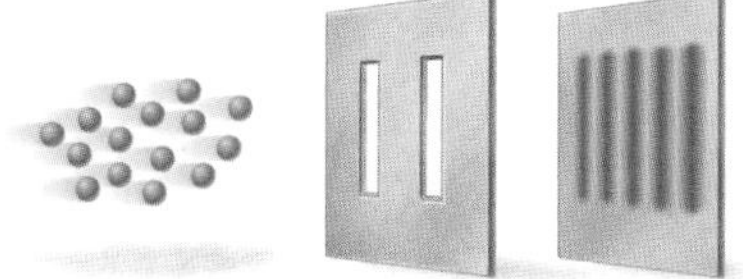

Electrons fired at two closely spaced openings form a wavelike interference pattern.

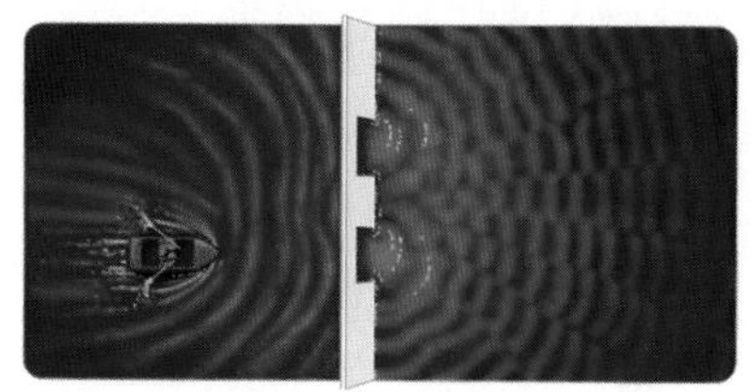

Water waves produce an interference pattern after passing through two openings.

Figure 7 When electrons are sent through two narrow slits, they behave as a wave.

Particles as Waves Because electromagnetic waves could behave as a particle, other scientists wondered whether matter could behave as a wave. If a beam of electrons were sprayed at two tiny slits, you might expect that the electrons would strike only the area behind the slits, like the spray paint in **Figure 7.** Instead, it was found that the electrons formed an interference pattern. This type of pattern is produced by waves when they pass through two slits and interfere with each other, as the water waves do in **Figure 7.** This experiment showed that electrons can behave like waves. It is now known that all particles, not only electrons, can behave like waves.

section 1 review

Summary

Making Electromagnetic Waves

- Moving electric charges are surrounded by an electric field and a magnetic field.
- A vibrating electric charge produces an electromagnetic wave.
- An electromagnetic wave consists of vibrating electric and magnetic fields that are perpendicular to each other and travel outward from the vibrating electric charge.

Properties of Electromagnetic Waves

- Electromagnetic waves carry radiant energy.
- In empty space, electromagnetic waves travel at 300,000 km/s, the speed of light.
- Electromagnetic waves travel more slowly in matter, with a speed that depends on the material.

Waves and Particles

- Electromagnetic waves can behave as particles that are called photons.
- In some circumstances, particles, such as electrons, can behave as waves.

Self Check

1. **Explain** why an electromagnetic wave is a transverse wave and not a compressional wave.
2. **Compare** the frequency of an electromagnetic wave with the frequency of the vibrating charge that produces the wave.
3. **Describe** how electromagnetic waves transfer radiant energy to matter.
4. **Explain** how an electromagnetic wave can travel through space that contains no matter.
5. **Think Critically** Suppose a moving electric charge were surrounded only by an electric field. Infer whether or not a vibrating electric charge would produce an electromagnetic wave.

Applying Math

6. **Calculate Time** How many minutes does it take an electromagnetic wave to travel 150,000,000 km?
7. **Use Scientific Notation** Calculate the distance an electromagnetic wave in space would travel in one day. Express your answer in scientific notation.

section 2

The Electromagnetic Spectrum

Reading Guide

What You'll Learn

- **Describe** the waves in the different regions of the electromagnetic spectrum.
- **Compare** the properties of different electromagnetic waves.
- **Identify** uses for different types of electromagnetic waves.

Why It's Important

Waves in different regions of the electromagnetic spectrum are used every day in many ways.

Review Vocabulary

spectrum: a continuous sequence arranged by a particular property

New Vocabulary

- visible light
- radio wave
- microwave
- infrared wave
- ultraviolet wave
- X ray
- gamma ray

A Range of Frequencies

Electromagnetic waves can have a wide variety of frequencies. They might vibrate once each second or trillions of times each second. The entire range of electromagnetic wave frequencies is known as the electromagnetic spectrum, shown in **Figure 8.** Various portions of the electromagnetic spectrum interact with matter differently. As a result, they are given different names. The electromagnetic waves that humans can detect with their eyes, called **visible light,** are a small portion of the entire electromagnetic spectrum. However, various devices have been developed to detect the other frequencies. For example, the antenna of your radio detects radio waves.

Figure 8 Electromagnetic waves are described by different names, depending on their frequencies and wavelengths.

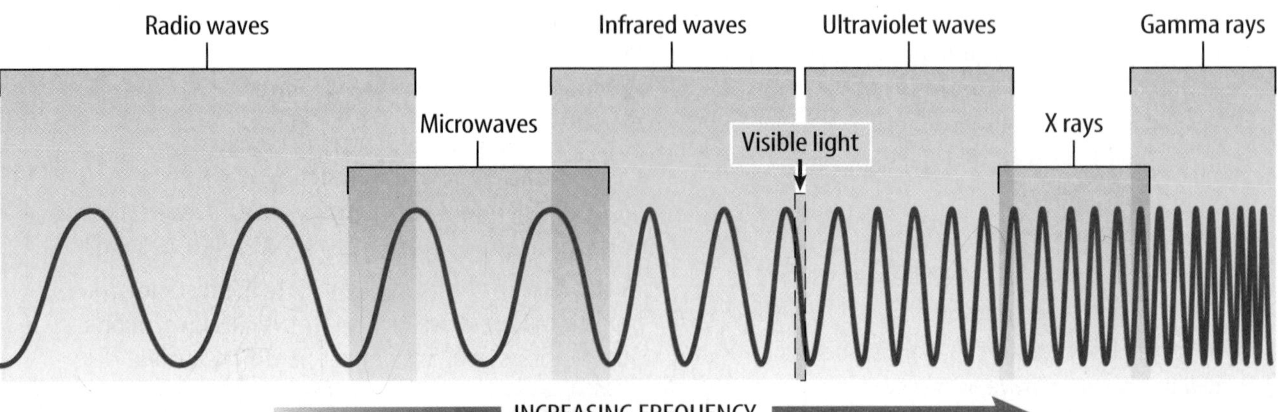

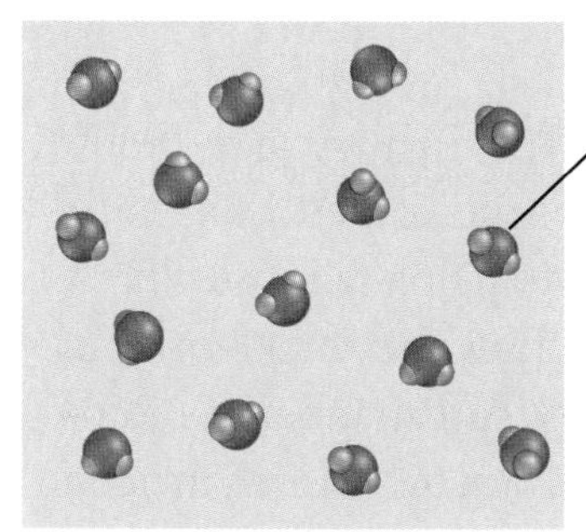
Water molecules
Normally, water molecules are randomly arranged.

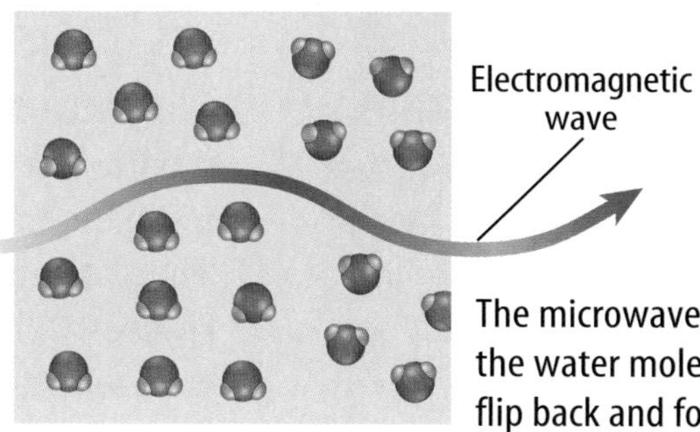
Electromagnetic wave
The microwaves cause the water molecules to flip back and forth.

Figure 9 Microwave ovens use electromagnetic waves to heat food.

Radio Waves

Stop and look around you. Even though you can't see them, radio waves are moving everywhere you look. Some radio waves carry an audio signal from a radio station to a radio. However, even though these radio waves carry information that a radio uses to create sound, you can't hear radio waves. You hear a sound wave when the compressions and rarefactions the sound wave produces reach your ears. A radio wave does not produce compressions and rarefactions as it travels through air.

Microwaves **Radio waves** are low-frequency electromagnetic waves with wavelengths longer than about 1 mm. Radio waves with wavelengths of less than 30 cm are called **microwaves.** Microwaves with wavelengths of about 20 cm to 1 cm are widely used for communication, such as for cellular telephones and satellite signals. You are probably most familiar with microwaves because of their use in microwave ovens.

Reading Check *What is the difference between a microwave and a radio wave?*

Microwave ovens heat food when microwaves interact with water molecules in food, as shown in **Figure 9.** Each water molecule is positively charged on one side and negatively charged on the other side. The vibrating electric field inside a microwave oven causes water molecules in food to rotate back and forth billions of times each second. This rotation causes a type of friction between water molecules that generates thermal energy. It is the thermal energy produced by the interactions between the water molecules that causes your food to cook.

Radar Another use for radio waves is to find the position and movement of objects by a method called radar. Radar stands for **RA**dio **D**etecting **A**nd **R**anging. With radar, radio waves are transmitted toward an object. By measuring the time required for the waves to bounce off the object and return to a receiving antenna, the location of the object can be found. Law-enforcement officers use radar to measure how fast a vehicle is moving. Radar also is used for tracking the movement of aircraft, watercraft, and spacecraft.

Mini LAB

Heating with Microwaves

Procedure

1. Complete the safety form.
2. Obtain two small **beakers or baby-food jars.** Place 50 mL of **dry sand** in each. To one of the jars, add 20 mL of **room-temperature water** and stir well.
3. Record the temperature of the sand in each jar.
4. Together, **microwave** both jars of sand for 10 s and immediately record the temperature again.

Analysis

1. Compare the initial and final temperatures of the wet and dry sand.
2. Infer why there was a difference.

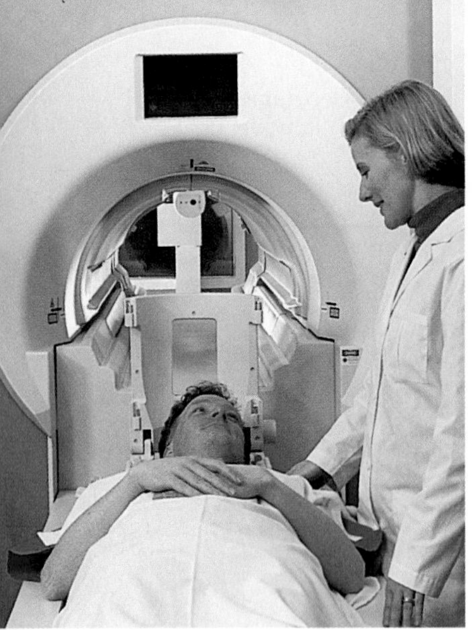

Figure 10 Magnetic resonance imaging technology uses radio waves as an alternative to X-ray imaging.

Magnetic Resonance Imaging (MRI) In the early 1980s, medical researchers developed a technique called magnetic resonance imaging, which uses radio waves to help diagnose illness. The patient lies inside a large cylinder, such as the one shown in **Figure 10.** Housed in the cylinder is a powerful magnet, a radio-wave emitter, and a radio-wave detector. Protons in hydrogen atoms in bones and soft tissue behave like magnets and align with the strong magnetic field. Energy from radio-waves causes some of the protons to flip their alignment. As the protons flip, they release radiant energy. A radio receiver detects this released energy. The amount of energy a proton releases depends on the type of tissue it is part of. The released energy detected by the radio receiver is used to create a map of the different tissues. A picture of the inside of the patient's body is produced painlessly.

Infrared Waves

Most of the warm air in a fireplace moves up the chimney, yet when you stand in front of a fireplace, you feel the warmth of the blazing fire. Why do you feel the heat? The warmth you feel is thermal energy transmitted to you by **infrared waves,** which are a type of electromagnetic wave with wavelengths between about one millimeter and 750 billionths of a meter.

You use infrared waves every day. A remote control emits infrared waves to control your television. A computer uses infrared waves to read CD-ROMs. In fact, every object emits infrared waves. Hotter objects emit more infrared waves than cooler objects emit. The wavelengths emitted also become shorter as the temperature increases. Infrared detectors can form images of objects from the infrared radiation they emit. Infrared sensors on satellites can produce infrared images that can help identify the vegetation over a region. **Figure 11** shows how cities appear different from surrounding vegetation in infrared satellite imagery.

Figure 11 Infrared images and visible light images can provide different types of information.

This visible light image of the region around San Francisco Bay in California was taken from an aircraft at an altitude of 20,000 m.

This infrared image of the same area was taken from a satellite. In this image, vegetation is red and buildings are gray.

Visible Light

Visible light is the range of electromagnetic waves that you can detect with your eyes. Light differs from radio waves and infrared waves only by its frequency and wavelength. Visible light has wavelengths around 750 billionths to 400 billionths of a meter. Your eyes contain substances that react differently to various wavelengths of visible light, so you see different colors. These colors range from long-wavelength red to short-wavelength blue. If all the colors are present, you see the light as white.

CT Scans In certain situations, doctors will perform a computerized tomography (CT) scan on a patient instead of taking a traditional X ray. Research to find out more about CT scans. Compare and contrast CT scans with X rays. What are the advantages and disadvantages of a CT scan? Write a paragraph about your findings in your Science Journal.

Ultraviolet Waves

Ultraviolet waves are electromagnetic waves with wavelengths from about 400 billionths to 10 billionths of a meter. Ultraviolet waves are energetic enough to enter skin cells. Overexposure to ultraviolet rays can cause skin damage and cancer. Most of the ultraviolet radiation that reaches Earth's surface is longer-wavelength UVA rays. The shorter-wavelength UVB rays cause sunburn, and both UVA and UVB rays can cause skin cancers and skin damage such as wrinkling. Although too much exposure to the Sun's ultraviolet waves is damaging, some exposure is necessary. Ultraviolet light striking the skin enables your body to make vitamin D, which is needed for healthy bones and teeth.

Useful UVs A useful property of ultraviolet waves is their ability to kill bacteria on objects such as food and medical supplies. When ultraviolet light enters a cell, it damages protein and DNA molecules. For some single-celled organisms, damage can mean death, which can be a benefit to human health. Ultraviolet waves are also useful because they make some materials fluoresce. Fluorescent materials absorb ultraviolet waves and reemit the energy as visible light. As shown in **Figure 12,** police detectives sometimes use fluorescent powder to show fingerprints when solving crimes.

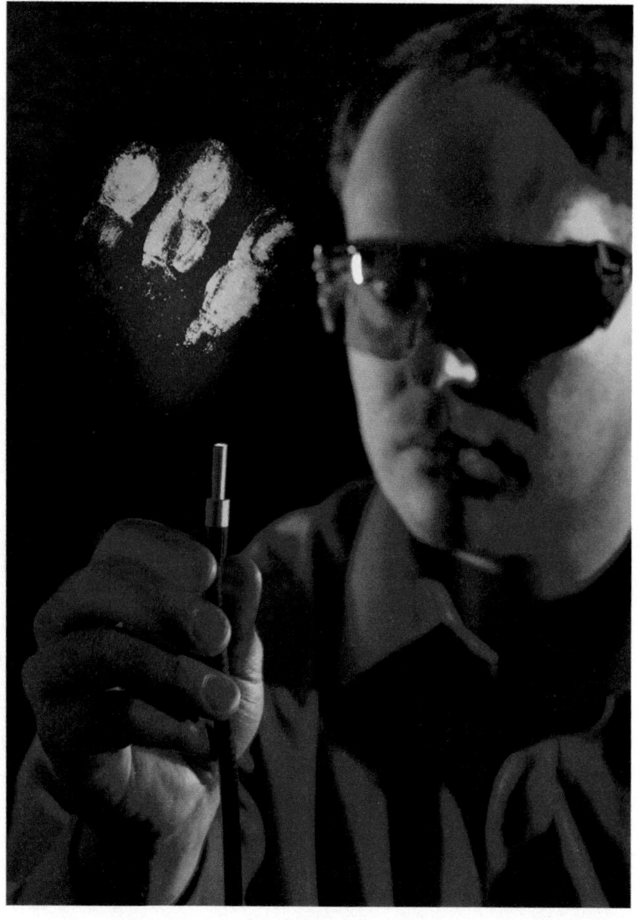

Figure 12 The police detective in this picture is shining ultraviolet light on a fingerprint dusted with fluorescent powder.

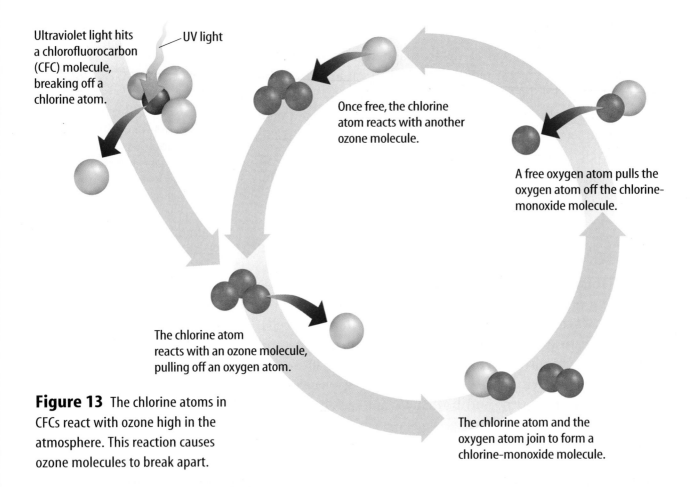

Ultraviolet light hits a chlorofluorocarbon (CFC) molecule, breaking off a chlorine atom.

UV light

Once free, the chlorine atom reacts with another ozone molecule.

A free oxygen atom pulls the oxygen atom off the chlorine-monoxide molecule.

The chlorine atom reacts with an ozone molecule, pulling off an oxygen atom.

The chlorine atom and the oxygen atom join to form a chlorine-monoxide molecule.

Figure 13 The chlorine atoms in CFCs react with ozone high in the atmosphere. This reaction causes ozone molecules to break apart.

The Ozone Layer About 20 to 50 km above Earth's surface in the stratosphere is a region called the ozone layer. Ozone is a molecule composed of three oxygen atoms. It is continually being formed and destroyed by ultraviolet waves high in the atmosphere. The ozone layer is vital to life on Earth because it absorbs most of the Sun's harmful ultraviolet waves. However, over the past few decades, the amount of ozone in the ozone layer has decreased. Averaged globally, the decrease is about three percent, but it is greater at higher latitudes.

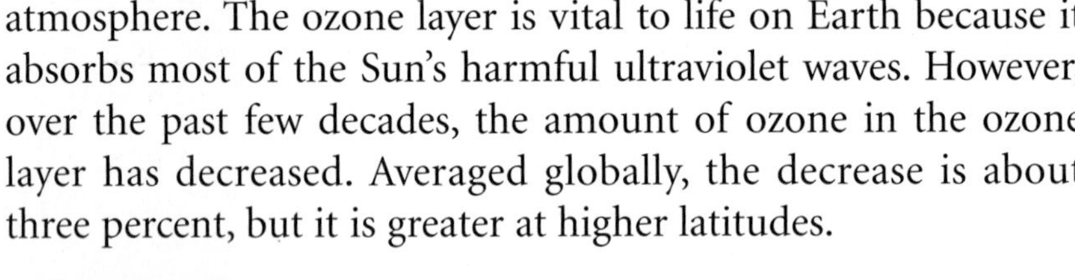

Reading Check *Why is the ozone layer vital to life on Earth?*

INTEGRATE Environment

The decrease in ozone is caused by the presence of certain chemicals, such as CFCs, high in Earth's atmosphere. CFCs are chemicals called chlorofluorocarbons that have been widely used in air conditioners, refrigerators, and cleaning fluids. When CFC molecules reach the ozone layer, they react chemically with ozone molecules, as shown in **Figure 13.** One chlorine atom from a CFC molecule can break apart thousands of ozone molecules. As a result, many countries are reducing the use of CFCs and other ozone-depleting chemicals.

X Rays and Gamma Rays

The electromagnetic waves with the shortest wavelengths and highest frequencies are X rays and gamma rays. Both X rays and gamma rays are high-energy electromagnetic waves. **X rays** have wavelengths between about 10 billionths of a meter and 10 trillionths of a meter. Doctors and dentists use low doses of X rays to form images of internal organs, bones, and teeth, like the image shown in **Figure 14.** X rays also are used in airport screening devices to examine the contents of luggage.

Electromagnetic waves with wavelengths shorter than about 10 trillionths of a meter are **gamma rays.** These are the highest-energy electromagnetic waves and can penetrate through several centimeters of lead. Gamma rays are produced by processes that occur in atomic nuclei. Both X rays and gamma rays are used in a technique called radiation therapy to kill diseased cells in the human body. A beam of X rays or gamma rays can damage the biological molecules in living cells, causing both healthy and diseased cells to die. However, by carefully controlling the amount of X-ray or gamma-ray radiation received by the diseased area, the damage to healthy cells can be reduced.

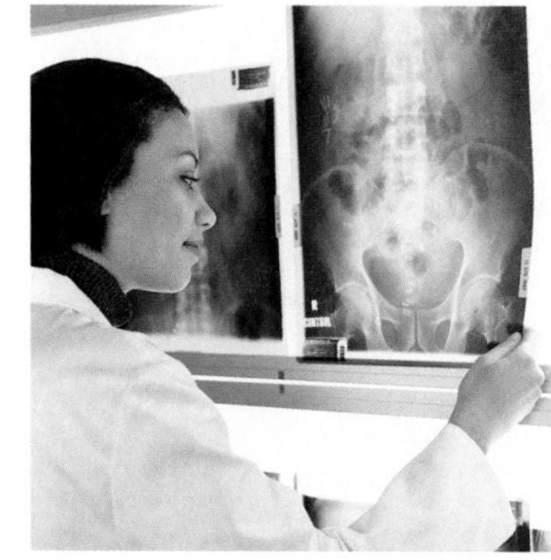

Figure 14 Bones are more dense than surrounding tissues and absorb more X rays. The image of a bone on an X ray is the shadow cast by the bone as X rays pass through the soft tissue.

section 2 review

Summary

Radio Waves and Infrared Waves

- Radio waves are electromagnetic waves with wavelengths longer than about 1 mm.
- Microwaves are radio waves with wavelengths between about 30 cm and 1 mm.
- Infrared waves have wavelengths between about 1 mm and 750 billionths of a meter.

Visible Light and Ultraviolet Waves

- Visible light waves have wavelengths between about 750 and 400 billionths of a meter.
- Ultraviolet waves have wavelengths between about 400 and 10 billionths of a meter.
- Most of the harmful ultraviolet waves emitted by the Sun are absorbed by the ozone layer.

X Rays and Gamma Rays

- X rays and gamma rays are the most energetic electromagnetic waves.
- Gamma rays have wavelengths less than 10 trillionths of a meter and are produced in the nuclei of atoms.

Self Check

1. **Explain** A mug of water is heated in a microwave oven. Explain why the water gets hotter than the mug.
2. **Describe** why you can see visible light waves but not other electromagnetic waves.
3. **List** the beneficial effects and the harmful effects of human exposure to ultraviolet rays.
4. **Identify** three objects in a home that produce electromagnetic waves and describe how the electromagnetic waves are used.
5. **Think Critically** What could an infrared image of their house reveal to homeowners?

Applying Math

6. **Use Scientific Notation** Express the range of wavelengths corresponding to visible light, ultraviolet waves, and X rays in scientific notation.
7. **Convert Units** A nanometer, abbreviated nm, equals one billionth of a meter, or 10^{-9} m. Express the range of wavelengths corresponding to visible light, ultraviolet waves, and X rays in nanometers.

The Shape of Satellite Dishes

Communications satellites transmit signals with a narrow beam pointed toward a particular area of Earth. To detect this signal, receivers are typically large, parabolic dishes.

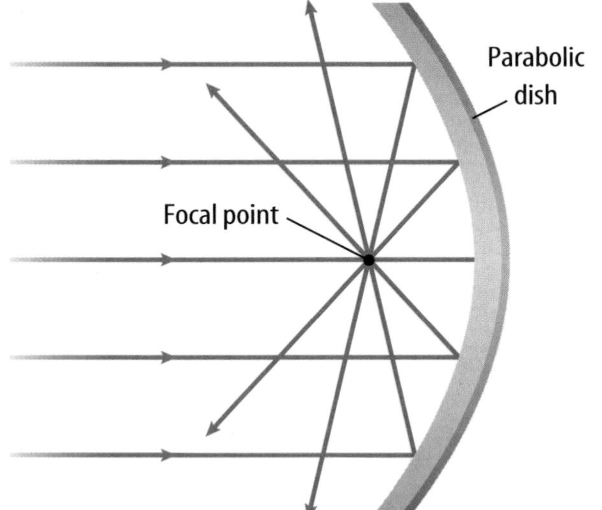

Real-World Problem

How does the shape of a satellite dish improve reception?

Goals

- **Make** a model of a satellite reflecting dish.
- **Observe** how the shape of the dish affects reception.

Materials

flashlight
several books
aluminum foil
small bowl
**large, metal spoon*
**Alternate materials*

Safety Precautions

Procedure

1. Complete the safety form.
2. Cover one side of a book with aluminum foil. Be careful not to wrinkle the foil.
3. Line the inside of the bowl with foil, also keeping it as smooth as possible.
4. Place some of the books on a table. Put the flashlight on top of the books so that its beam of light will shine several centimeters above and across the table.
5. Hold the foil-covered book on its side at a right angle to the top of the table. The foil-covered side should face the beam of light.
6. **Observe** the intensity of the light on the foil.
7. Repeat steps 5 and 6, replacing the foil-covered book with the bowl.

Conclude and Apply

1. **Compare** the brightness of the light reflected from the two surfaces.
2. **Explain** why the light you see from the curved surface is brighter.
3. **Infer** why bowl-shaped dishes are used to receive signals from satellites.

Communicating Your Data

Compare your conclusions with those of other students in your class. **For more help, refer to the** Science Skill Handbook.

section 3

Radio Communication

Reading Guide

What You'll Learn

- **Explain** how modulating carrier waves enables information to be transmitted by radio waves.
- **Distinguish** between amplitude modulation and frequency modulation.
- **Identify** various ways of communicating using radio waves.

Why It's Important

Every day, you use radio waves to communicate.

Review Vocabulary

modulate: to vary the amplitude or frequency of a wave to transmit information

New Vocabulary

- carrier wave
- cathode-ray tube
- transceiver
- Global Positioning System (GPS)

Radio Transmission

When you listen to the radio, you hear music and words that are produced at a distant location. The music and words are sent to your radio by radio waves. The metal antenna of your radio detects radio waves. As the electromagnetic waves pass by your radio's antenna, the electrons in the metal vibrate, as shown in **Figure 15.** These vibrating electrons produce a changing electric current that contains the information about the music and words. An amplifier boosts the current and sends it to speakers, causing them to vibrate. The vibrating speakers create sound waves that travel to your ears. Your brain interprets these sound waves as music and words

Dividing the Radio Spectrum Each radio station is assigned to broadcast at one particular radio frequency. Turning the tuning knob on your radio allows you to select a particular frequency to listen to. The specific frequency of the electromagnetic wave that a radio station is assigned is called the **carrier wave.**

The radio station must do more than simply transmit a carrier wave. The station has to send information about the sounds that you are to receive. This information is sent by modifying the carrier wave. The carrier wave is modified to carry information in one of two ways, as shown in **Figure 16.**

Figure 15 Radio waves exert a force on the electrons in an antenna, causing the electrons to vibrate.

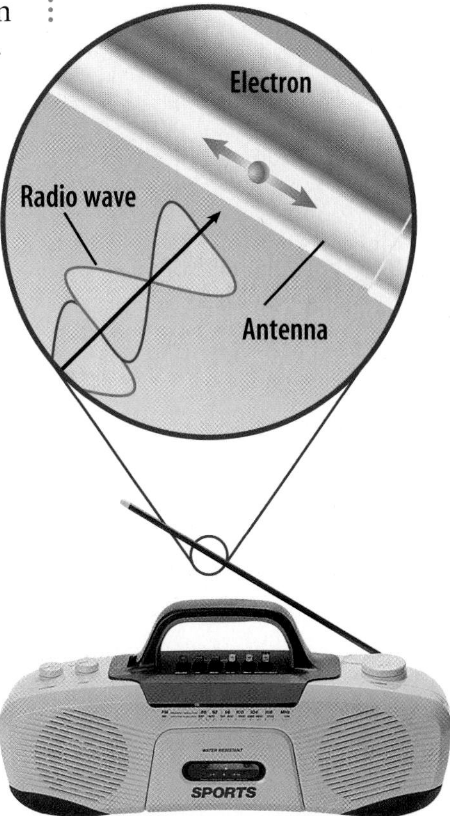

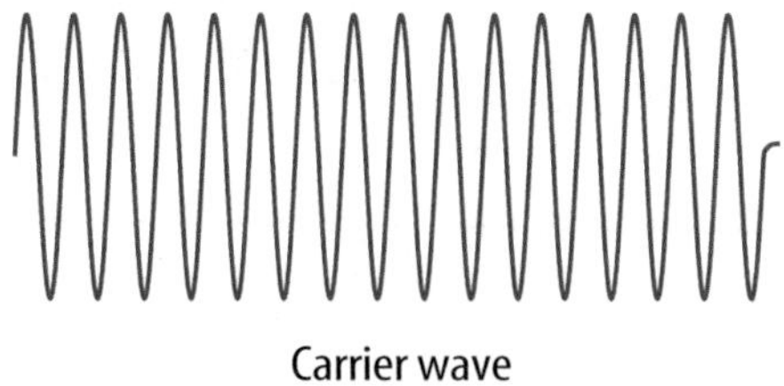
Carrier wave

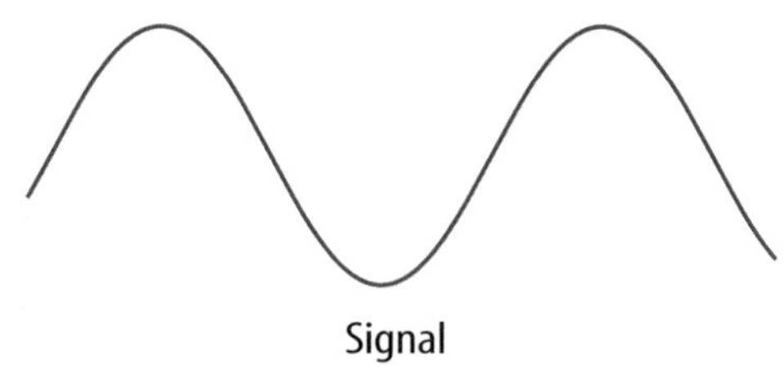
Signal

Amplitude modulation

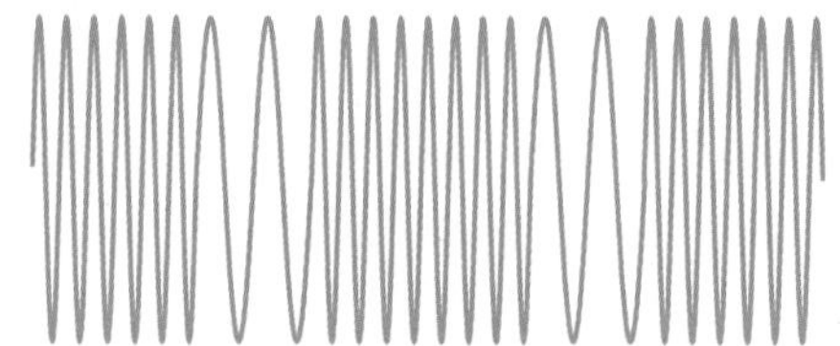
Frequency modulation

Figure 16 A carrier wave broadcast by a radio station can be altered in one of two ways to transmit a signal: amplitude modulation (AM) or frequency modulation (FM).

AM Radio An AM radio station broadcasts information by varying the amplitude of the carrier wave, as shown in **Figure 16.** Your radio detects the variations in amplitude of the carrier wave and produces a changing electric current from these variations. The changing electric current makes the speaker vibrate. AM carrier-wave frequencies range from 540,000 to 1,600,000 Hz.

FM Radio Electronic signals are transmitted by FM radio stations by varying the frequency of the carrier wave, as in **Figure 16.** Your radio detects the changes in frequency of the carrier wave. Because the strength of the FM waves is kept fixed, FM signals tend to be more clear than AM signals. FM carrier frequencies range from 88 million to 108 million Hz. This is much higher than AM frequencies, as shown in **Figure 17. Figure 18** shows how radio signals are broadcast.

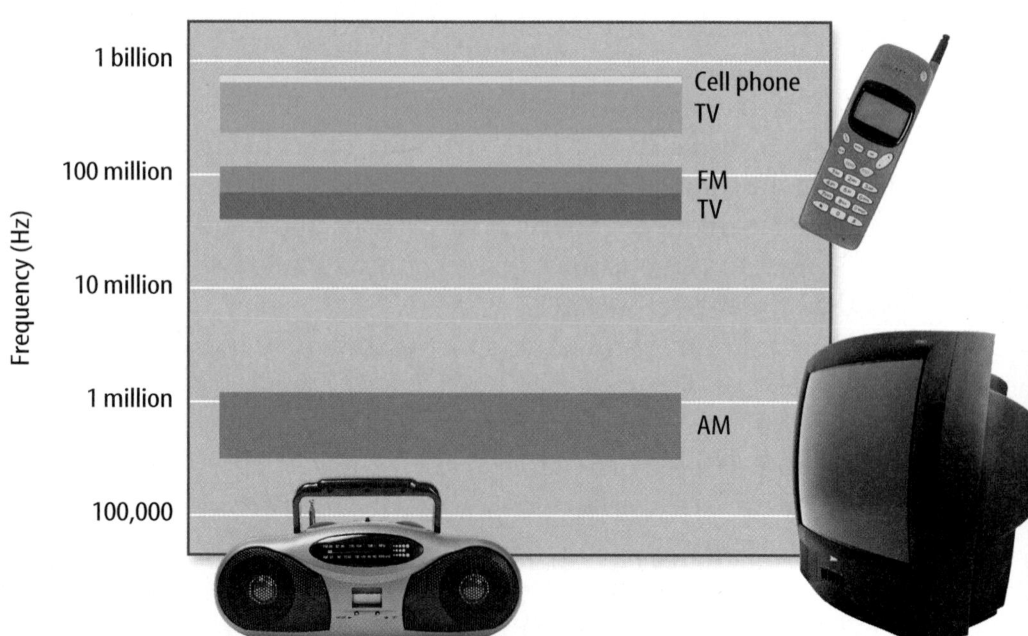

Figure 17 Cell phones, TVs, and radios broadcast at frequencies that range from more than 500,000 Hz to almost 1 billion Hz.

NATIONAL GEOGRAPHIC VISUALIZING RADIO BROADCASTS

Figure 18

You flick a switch, turn the dial, and music from your favorite radio station fills the room. Although it seems like magic, sounds are transmitted over great distances by converting sound waves to electromagnetic waves and back again, as shown here.

A At the radio station, musical instruments and voices create sound waves by causing air molecules to vibrate. Microphones convert these sound waves to a varying electric current, or electronic signal.

B This signal then is added to the station's carrier wave. If the station is an AM station, the electronic signal modifies the amplitude of the carrier wave. If the station is an FM station, the electronic signal modifies the frequency of the carrier wave.

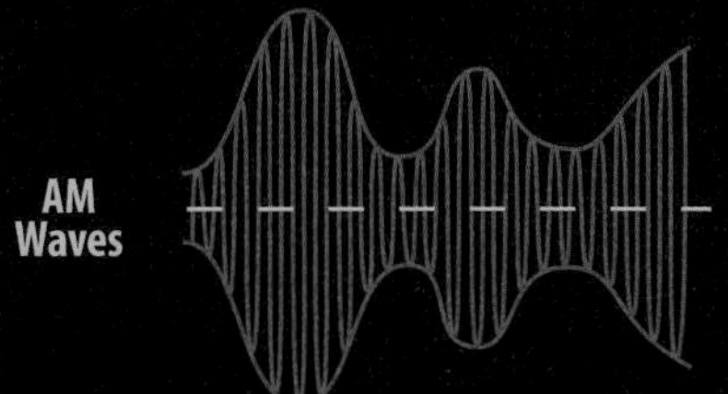

FM Waves

C The modified carrier wave is used to vibrate electrons in the station's antenna. These vibrating electrons create a radio wave that travels out in all directions at the speed of light.

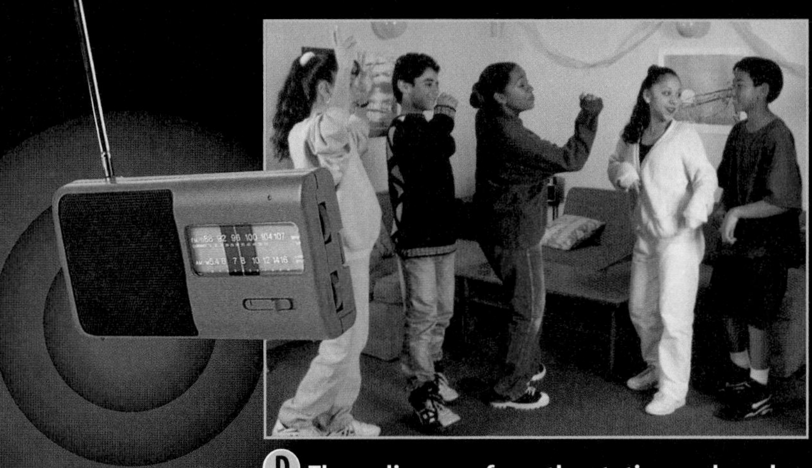

D The radio wave from the station makes electrons in your radio's antenna vibrate. This creates an electric current. If your radio is tuned to the station's frequency, the carrier wave is removed from the original electronic signal. This signal then makes the radio's speaker vibrate, creating sound waves that you hear as music.

Astronomers Do you ever look up at the stars at night and wonder how they were formed? With so many stars and so many galaxies, life might be possible on other planets. Research ways that astronomers use electromagnetic waves to investigate the universe. Choose one project astronomers currently are working on that interests you, and write about it in your Science Journal. Discuss the benefits of a career in astronomy.

Television

What would people hundreds of years ago have thought if they had seen a television? Televisions might seem like magic, but not if you know how they work. Television and radio transmissions are similar. At the television station, sounds and images are changed into electronic signals. These signals are broadcast by carrier waves. The audio part of television is sent by FM radio waves. Information about the color and brightness is sent at the same time by AM signals.

Cathode-Ray Tubes In many television sets, images are displayed on a cathode-ray tube (CRT), as shown in **Figure 19.** A **cathode-ray tube** is a sealed vacuum tube in which one or more beams of electrons are produced. The CRT in a color TV produces three electron beams that are focused by a magnetic field and strike a coated screen. The screen is speckled with more than 100,000 rectangular spots that are of three types. One type glows red, another glows green, and the third type glows blue when electrons strike it. The spots are grouped together, with a red, green, and blue spot in each group.

An image is created when the three electron beams of the CRT sweep back and forth across the screen. Each electron beam controls the brightness of each type of spot, according to the information in the video signal from the TV station. By varying the brightness of each spot in a group, the three spots together can form any color so that you see a full-color image.

Reading Check *What is a cathode-ray tube?*

Figure 19 Cathode-ray tubes produce the images you see on television. The inside surface of a television screen is covered by groups of spots that glow red, green, or blue when struck by an electron beam.

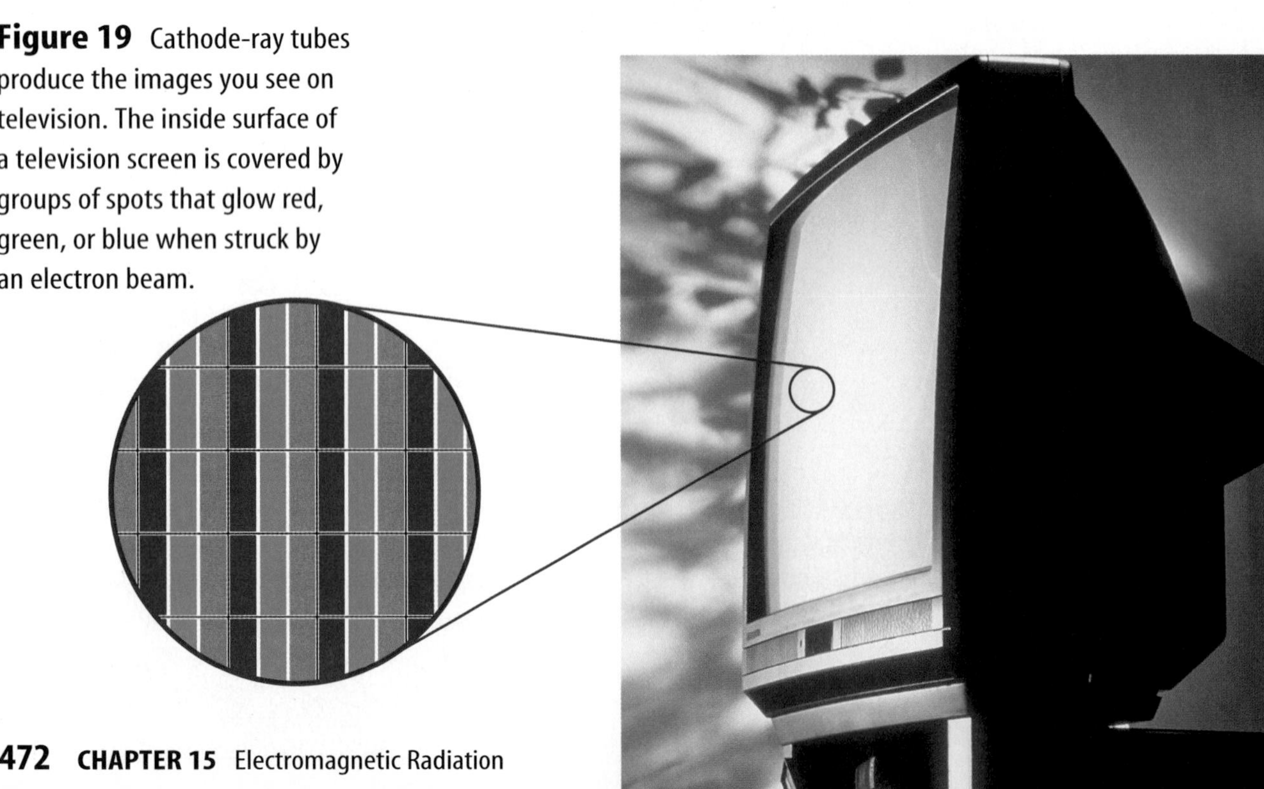

Telephones

Until about 1950, human operators were needed to connect telephone calls between people. Just 20 years ago, you never would have seen someone walking down the street talking on a telephone. Today, cell phones are seen everywhere. When you speak into a telephone, a microphone converts sound waves into an electrical signal. In cell phones, this current is used to create radio waves that are transmitted to and from a microwave tower, as shown in **Figure 20.** A cell phone uses one radio signal for sending information to a tower at a base station. It uses another signal for receiving information from the base station. The base stations are several kilometers apart. The area each one covers is called a cell. If you move from one cell to another while using a cell phone, an automated control station transfers your signal to the new cell.

Figure 20 The antenna at the top of a microwave tower receives signals from nearby cell phones. **Determine** *whether any microwave towers are located near your school or home. Describe their locations.*

Reading Check *What are the cells in a cell-phone system?*

Cordless Telephones Like a cellular telephone, a cordless telephone is a transceiver. A **transceiver** transmits one radio signal and receives another radio signal from a base unit. Having two signals at different frequencies allows you to talk and listen at the same time. Cordless telephones work much like cell phones. With a cordless telephone, however, you must be close to the base unit. Another drawback is that when someone nearby is using a cordless telephone, you could hear that conversation on your phone if the frequencies match. For this reason, many cordless phones have a channel button. This allows you to switch your call to another frequency.

Pagers Another method of transmitting signals is a pager, which allows messages to be sent to a small radio receiver. A caller leaves a message at a central terminal by entering a callback number through a telephone keypad or by entering a text message from a computer. At the terminal, the message is changed into an electronic signal and transmitted by radio waves. Each pager is given a unique number for identification. This identification number is sent along with the message. Your pager receives all messages that are transmitted in the area at its assigned frequency. However, your pager responds only to messages with its particular identification number. Newer pagers can send data as well as receive them.

Topic: Radio Wave Technology

Visit gpescience.com for Web links to information about advances in radio-wave technology.

Activity List the advances you find, and write about the significance of each one in your Science Journal.

Figure 21 Communications satellites, like the one shown here, use solar panels to provide the electrical energy they need to communicate with receivers on Earth. The solar panels are the structures on either side of the central body of the satellite.

Communications Satellites

Since satellites were first developed, thousands have been launched into Earth's orbit. Many of these, like the one in **Figure 21,** are used for communication. A station broadcasts a high-frequency microwave signal to the satellite. The satellite receives the signal, amplifies it, and transmits it to a particular region on Earth. To avoid interference, the frequency broadcast by the satellite is different from the frequency broadcast from Earth.

Topic: Satellite Communication
Visit gpescience.com for Web links to information about ways satellites are used for communication.

Activity Write a paragraph describing the advantages of placing a communications satellite in a geosynchronous orbit. Include a diagram.

Satellite Telephone Systems If you have a mobile telephone, you can make a phone call when sailing across the ocean. To make a call on a mobile telephone, the telephone transmits radio waves directly to a satellite. The satellite relays the signal to a ground station, and the call is passed on to the telephone network. Satellite links work well for one-way transmissions, but two-way communications can have an annoying delay caused by the large distance the signals must travel to and from the satellite.

Television Satellites The satellite-reception dishes that you sometimes see in yards or attached to houses are receivers for television satellite signals. Satellite television is used as an alternative to ground-based transmission. Television satellites use microwaves rather than the longer-wavelength radio waves used for normal television broadcasts. Short-wavelength microwaves travel more easily through the atmo-sphere. The ground receiver dishes are rounded to help focus the microwaves onto an antenna.

The Global Positioning System

Getting lost while hiking is not uncommon, but if you are carrying a Global Positioning System receiver, this is much less likely to happen. The **Global Positioning System (GPS)** is a system of satellites, ground monitoring stations, and receivers that determine a person's or object's exact location at or above Earth's surface. The 24 satellites necessary for 24-hour, around-the-world coverage became fully operational in 1995. GPS satellites are owned and operated by the United States Department of Defense, but the microwave signals they send out can be used by anyone. As shown in **Figure 22,** signals from four satellites are needed to determine the location of an object using a GPS receiver. Today, GPS receivers are used in airplanes, ships, cars, and even by hikers.

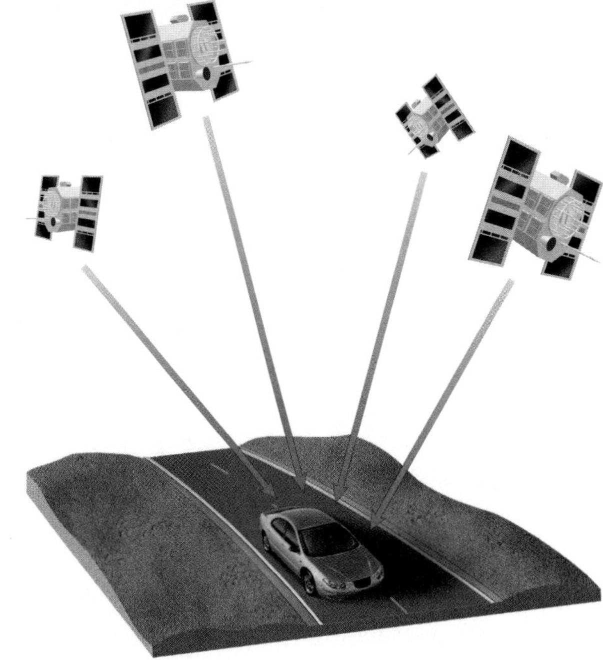

Figure 22 A GPS receiver uses signals from orbiting satellites to determine the receiver's location.

section 3 review

Summary

Radio Transmission

- Radio stations transmit electromagnetic waves that receivers convert to sound waves.
- Each AM radio station is assigned a carrier-wave frequency and varies the amplitude of the carrier waves to transmit a signal.
- Each FM radio station is assigned a carrier-wave frequency and varies the frequency of the carrier waves to transmit a signal.

Television

- TV sets use cathode-ray tubes to convert electronic signals from TV stations into both sounds and images.

Telephones

- Telephones contain transceivers that convert sound waves into electrical signals and also convert electrical signals into sound waves.
- Wires, microwave towers, and satellites are used to transmit and receive telephone signals.

Global Positioning System (GPS)

- The GPS uses a system of satellites to determine a person's or object's exact position.

Self Check

1. **Explain** the difference between AM and FM radio. Make a sketch of how a carrier wave is modulated in AM and FM radio.
2. **Define** a cathode-ray tube, and explain how it is used in a television.
3. **Describe** what happens if you are talking on a cell phone while riding in a car and you travel from one cell to another cell.
4. **Explain** some of the uses of the Global Positioning System. Why might emergency vehicles be equipped with GPS receivers?
5. **Think Critically** Why do cordless telephones stop working if you move too far from the base unit?

Applying Math

6. **Calculate a Ratio** A group of red, green, and blue spots on a TV screen is a pixel. A standard TV has 460 pixels horizontally and 360 pixels vertically. A high-definition TV has 1,920 horizontal and 1,080 vertical pixels. What is the ratio of the number of pixels in a high-definition TV to the number in a standard TV?

More Section Review gpescience.com

Use the Internet

Radio Frequencies

Goals

- **Research** which frequencies are used by different radio stations.
- **Observe** the reception of your favorite radio station.
- **Make** a chart of your findings and communicate them to other students.

Data Source

Internet Lab

Visit gpescience.com for more information on radio frequencies, different frequencies of radio stations around the country, and the ranges of AM and FM broadcasts.

Real-World Problem

The signals from many radio stations broadcasting at different frequencies are hitting your radio's antenna at the same time. When you tune to your favorite station, the electronics inside your radio amplify the signal at the frequency broadcast by the station. The signal from your favorite station is broadcast from a transmission site that may be several miles away.

You may have noticed that if you're listening to a radio station while driving in a car, sometimes the station gets fuzzy and you hear another station at the same time. Sometimes you lose the station completely. How far can you drive before this happens? Does the distance vary depending on the station you listen to? What are the ranges of radio stations? Form a hypothesis about how far you think a radio station can transmit. Which type of signal, AM or FM, has a greater range? Form a hypothesis about the range of your favorite radio station.

Make a Plan

1. **Research** what frequencies are used by AM and FM radio stations in your area and other areas around the country.
2. **Determine** these stations' broadcast locations.
3. **Determine** the broadcast range of radio stations in your area.
4. **Observe** how frequencies differ. What is the maximum difference between frequencies for FM stations in your area? AM stations?

Follow Your Plan

1. **Make** sure your teacher approves your plan before you start.
2. Visit the Web site shown below for links to different radio stations.
3. **Compare** the different frequencies of the stations and the locations of the broadcasts.
4. **Determine** the range of radio stations in your area and the power of their broadcast signals in watts.
5. **Record** your data in your Science Journal.

Analyze Your Data

1. **Make** a map of the radio stations in your area. Do the ranges of AM stations differ from those of FM stations?
2. **Make** a map of different radio stations around the country. Do you see any patterns in the frequencies for stations that are located near each other?
3. **Write** a description that compares how close the frequencies of AM stations are and how close the frequencies of FM stations are. Also compare the power of their broadcast signals and their ranges.
4. **Share** your data by posting it at the Web site shown below.

Conclude and Apply

1. **Compare** your findings with those of your classmates and other data that are posted at the Web site shown below. Do all AM stations and FM stations have different ranges?
2. **Observe** your map of the country. How close can stations with similar frequencies be? Do AM and FM stations appear to be different in this respect?
3. **Infer** The power of a broadcast signal also determines its range. How does the power (wattage) of the signals affect your analysis of your data?

Communicating Your Data

Find this lab using the Web site below. Post your data in the table provided. **Compare** your data with those of other students. Then combine your data with theirs and make a map for your class that shows all of the data.

Internet Lab
gpescience.com

Riding a Beam of Light

Einstein and the Special Theory of Relativity

Catch a Wave

At age 16, Albert Einstein wondered, "What would it be like to ride a beam of light?" He imagined what might happen if he turned on a flashlight while riding a light beam. Because the flashlight already would be traveling at the speed of light, would light from the flashlight travel at twice the speed of light?

What's so special?

Einstein thought about this problem, and in 1905, he published the special theory of relativity. This theory states that the speed of light will be the same when measured by any observer that moves with a constant speed. The measured speed of light will not depend on the speed of the observer or on how fast the source of light is moving. Einstein answered the question he had asked himself when he was 16. He found the universal speed limit that can't be broken.

It Doesn't Add Up

According to Einstein, electromagnetic waves such as light waves behave very differently from other waves. For example, sound waves from the siren of an ambulance moving toward you move faster than they would if the ambulance were not moving. The speed of the ambulance adds to the speed of the sound waves. However, for light waves, the speed of a light source doesn't add to the speed of light.

Very Strange, but True

Einstein's special theory of relativity makes other strange predictions. According to this theory, no object can travel faster than the speed of light. Another prediction is that the measured length of a moving object is shorter than when the object is at rest. Also, moving clocks should run more slowly than when they are at rest. These predictions have been confirmed by experiments. Measurements have shown, for example, that a moving clock does run more slowly.

Communicate **Research the life of Albert Einstein and make a time line showing important events in his life. Also include on your time line major historical events that occurred during Einstein's lifetime.**

TIME
For more information, visit gpescience.com

Reviewing Main Ideas

Section 1 What are electromagnetic waves?

1. Electromagnetic waves consist of vibrating electric and magnetic fields, and are produced by vibrating electric charges.

2. Electromagnetic waves carry radiant energy and can travel through a vacuum or through matter.
3. Electromagnetic waves sometimes behave like particles called photons.

Section 2 The Electromagnetic Spectrum

1. Electromagnetic waves with the longest wavelengths are called radio waves. Radio waves have wavelengths greater than about 1 mm. Microwaves are radio waves with wavelengths between about 30 cm and 1 mm.
2. Infrared waves have wavelengths between about 1 mm and 750 billionths of a meter. Warmer objects emit more infrared waves than cooler objects do.
3. Visible light rays have wavelengths between about 750 and 400 billionths of a meter. Substances in your eyes react with visible light to enable you to see.
4. Ultraviolet waves have frequencies between about 400 and 10 billionths of a meter. Excessive exposure to ultraviolet waves can damage human skin.
5. X rays and gamma rays are high-energy electromagnetic waves with wavelengths less than 10 billionths of a meter. X rays are used in medical imaging.

Section 3 Radio Communication

1. Modulated radio waves are used often for communication. AM and FM are two forms of carrier-wave modulation.
2. Television signals are transmitted as a combination of AM and FM waves.
3. Cellular telephones, cordless telephones, and pagers use radio waves to transmit signals. Communications satellites are used to relay telephone and television signals over long distances.
4. The Global Positioning System (GPS) enables the exact position of a person or object on Earth to be determined.

FOLDABLES™ Use the Foldable that you made at the beginning of this chapter to help you review electromagnetic radiation.

chapter 15 Review

Using Vocabulary

carrier wave p. 469
cathode-ray tube p. 472
electromagnetic wave p. 456
gamma ray p. 467
Global Positioning System (GPS) p. 475
infrared wave p. 464
microwave p. 463
photon p. 460
radiant energy p. 459
radio wave p. 463
transceiver p. 473
ultraviolet wave p. 465
visible light p. 462
X ray p. 467

Complete each statement using the correct word or phrase from the vocabulary list above.

1. ________ are the type of electromagnetic waves often used for communication.
2. A remote control uses ________ to communicate with a television set.
3. Electromagnetic waves transmit ________.
4. If you stay outdoors too long, your skin might be burned by exposure to ________ from the Sun.
5. A radio station broadcasts radio waves called ________, which have the specific frequency assigned to the station.
6. The image on a television screen is produced by a ________.
7. Transverse waves that are produced by vibrating electric charges and consist of vibrating electric and magnetic fields are ________.

Checking Concepts

Choose the word or phrase that best answers each question.

8. Which type of electromagnetic wave is the most energetic?
   - **A)** gamma rays
   - **B)** ultraviolet waves
   - **C)** infrared waves
   - **D)** microwaves

Use the figure below to answer question 9.

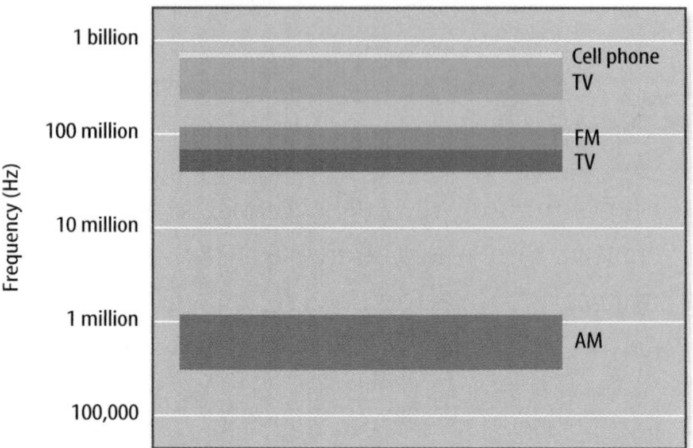

9. What signal has the longest wavelength?
   - **A)** cell phone
   - **B)** TV
   - **C)** FM
   - **D)** AM
10. Which type of electromagnetic wave enables skin cells to produce vitamin D?
    - **A)** visible light
    - **B)** ultraviolet waves
    - **C)** infrared waves
    - **D)** X rays
11. Which of the following describes X rays?
    - **A)** short wavelength, high frequency
    - **B)** short wavelength, low frequency
    - **C)** long wavelength, high frequency
    - **D)** long wavelength, low frequency
12. Which of the following is changing in an AM radio wave?
    - **A)** speed
    - **B)** frequency
    - **C)** amplitude
    - **D)** wavelength
13. Which type of electromagnetic wave has wavelengths greater than about 1 mm?
    - **A)** X rays
    - **B)** radio waves
    - **C)** gamma rays
    - **D)** ultraviolet waves
14. Which of these colors of visible light has the shortest wavelength?
    - **A)** blue
    - **B)** green
    - **C)** red
    - **D)** white

Interpreting Graphics

15. Copy and complete the following table about the electromagnetic spectrum.

Uses of Electromagnetic Waves	
Type of Electromagnetic Waves	Examples of How Electromagnetic Waves Are Used
	radio, TV transmission
Infrared waves	Do not write in this book.
Visible light	vision
	fluorescent materials
X rays	
	destroying harmful cells

16. Copy and complete the following events chain about the destruction of ozone molecules in the ozone layer by CFC molecules.

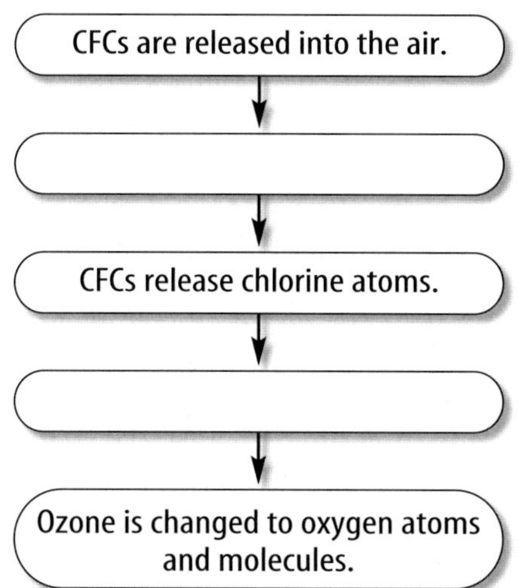
CFCs are released into the air.
CFCs release chlorine atoms.
Ozone is changed to oxygen atoms and molecules.

Thinking Critically

17. **Explain** why X rays are used in medical imaging.
18. **Predict** whether an electromagnetic wave would travel through space if its electric and magnetic fields were not changing with time. Explain your reasoning.
19. **Infer** Electromagnetic waves consist of vibrating electric and magnetic fields. A magnetic field can move a compass needle. Why doesn't a compass needle move when visible light strikes the compass?
20. **Classify** Look around your home, school, and community. Make a list of the different devices that use electromagnetic waves. Beside each device, write the type of electromagnetic wave the device uses.
21. **Form a hypothesis** to explain why communications satellites don't use ultraviolet waves to receive information and transmit signals to Earth's surface.
22. **Compare** the energy of photons corresponding to infrared waves with the energy of photons corresponding to ultraviolet waves.
23. **Determine** whether or not all electromagnetic waves always travel at the speed of light. Explain.

Applying Math

24. **Use Fractions** When visible light waves travel in ethyl alcohol, their speed is three-fourths of the speed of light in air. What is the speed of light in ethyl alcohol?
25. **Use Scientific Notation** The speed of light in a vacuum has been determined to be 299,792,458 m/s. Express this number to four significant digits using scientific notation.
26. **Calculate Wavelength** A radio wave has a frequency of 540,000 Hz and travels at a speed of 300,000 km/s. Use the wave speed equation to calculate the wavelength of the radio wave. Express your answer in meters.

Standardized Test Practice

Record your answers on the answer sheet provided by your teacher or on a sheet of paper.

Multiple Choice

1. Which of the following produces electromagnetic waves?
   A. vibrating charge
   B. constant electric field
   C. static charge
   D. constant magnetic field

Use the illustration below to answer questions 2 and 3.

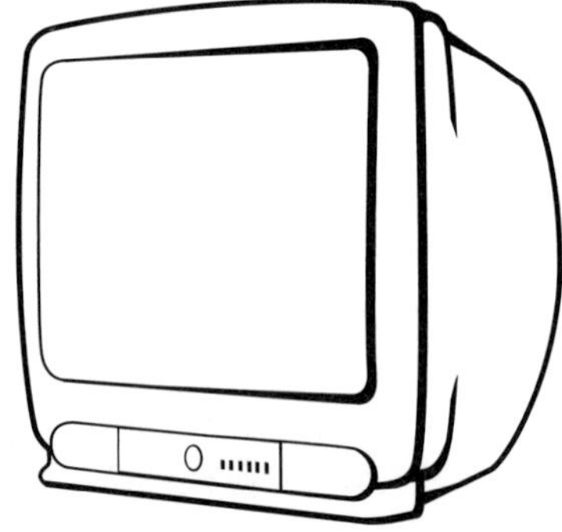

2. A television image is produced by three electron beams. What device inside a television set produces the electron beams?
   A. transceiver
   B. transmitter
   C. antenna
   D. cathode-ray tube

Test-Taking Tip

Marking on Tests Be sure to ask if it is okay to write on the test booklet when taking the test, but make sure you mark all answers on your answer sheet.

3. What colors are the three types of glowing spots that form the different colors on the television screen?
   A. red, yellow, blue
   B. red, green, blue
   C. cyan, magenta, yellow
   D. cyan, magenta, blue

4. Which of the following explains how interference is avoided between the signals communications satellites receive and the signals they broadcast?
   A. The signals travel at different speeds.
   B. The signals have different amplitudes.
   C. The signals have different frequencies.
   D. The signals are only magnetic.

Use the table below to answer questions 5 and 6.

Regions of the Electromagnetic Spectrum		
Infrared waves	Radio waves	Gamma rays
X rays	Visible light	Ultraviolet waves

5. If you arranged the list of electromagnetic waves shown above in order from shortest to longest wavelength, which would be first on the list?
   A. radio waves
   B. X rays
   C. gamma rays
   D. visible light

6. Which region of the electromagnetic spectrum listed in the table above includes microwaves?

   A. gamma rays

   B. radio waves

   C. ultraviolet waves

   D. infrared waves

7. The warmth you feel when you stand in front of a fire is thermal energy transmitted to you by what type of electromagnetic waves?

   A. X rays

   B. microwaves

   C. ultraviolet waves

   D. infrared waves

Gridded Response

8. Electromagnetic radiation travels through space at 3.00×10^5 km/s. The moon is, on average, 382,000 km from Earth. How many seconds would it take radio waves broadcast from Earth to reach the moon?

Short Response

Use the illustrations below to answer question 9.

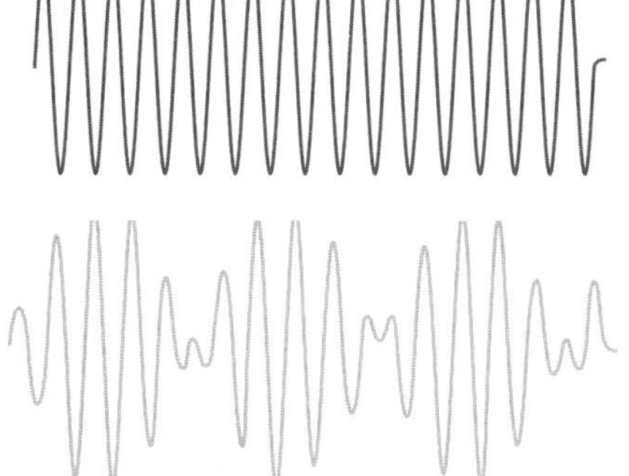

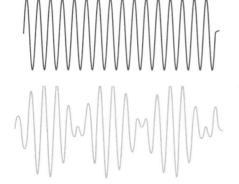

9. The illustrations above show two radio waves broadcast by a radio station. The upper, unmodulated wave is the carrier wave. The lower figure shows the same wave that has been modulated to carry sound information. What type of modulation does it show?

10. Even on a cloudy day, you can get sunburned outside. However, inside a glass greenhouse, you won't get sunburned. Based on this fact, which type of electromagnetic waves will pass through clouds but not glass?

Extended Response

11. Explain how an electromagnetic wave that strikes a material transfers radiant energy to the atoms in the material.

Use the illustration below to answer question 12.

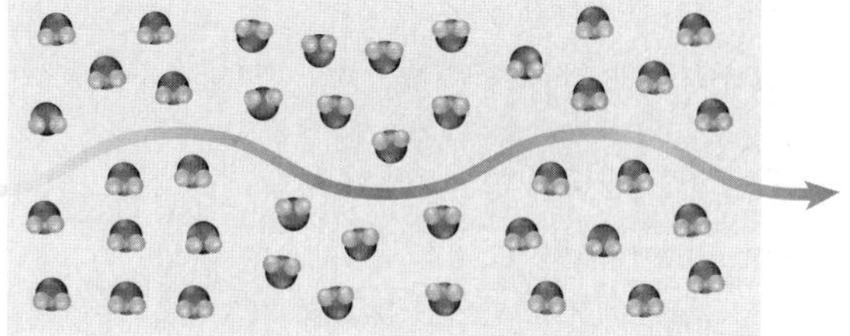

12. The illustration above shows a microwave interacting with water molecules in food.

    Part A How does the electric field in microwaves affect water molecules?

    Part B Describe how thermal energy is produced by microwaves interacting with water molecules inside food.

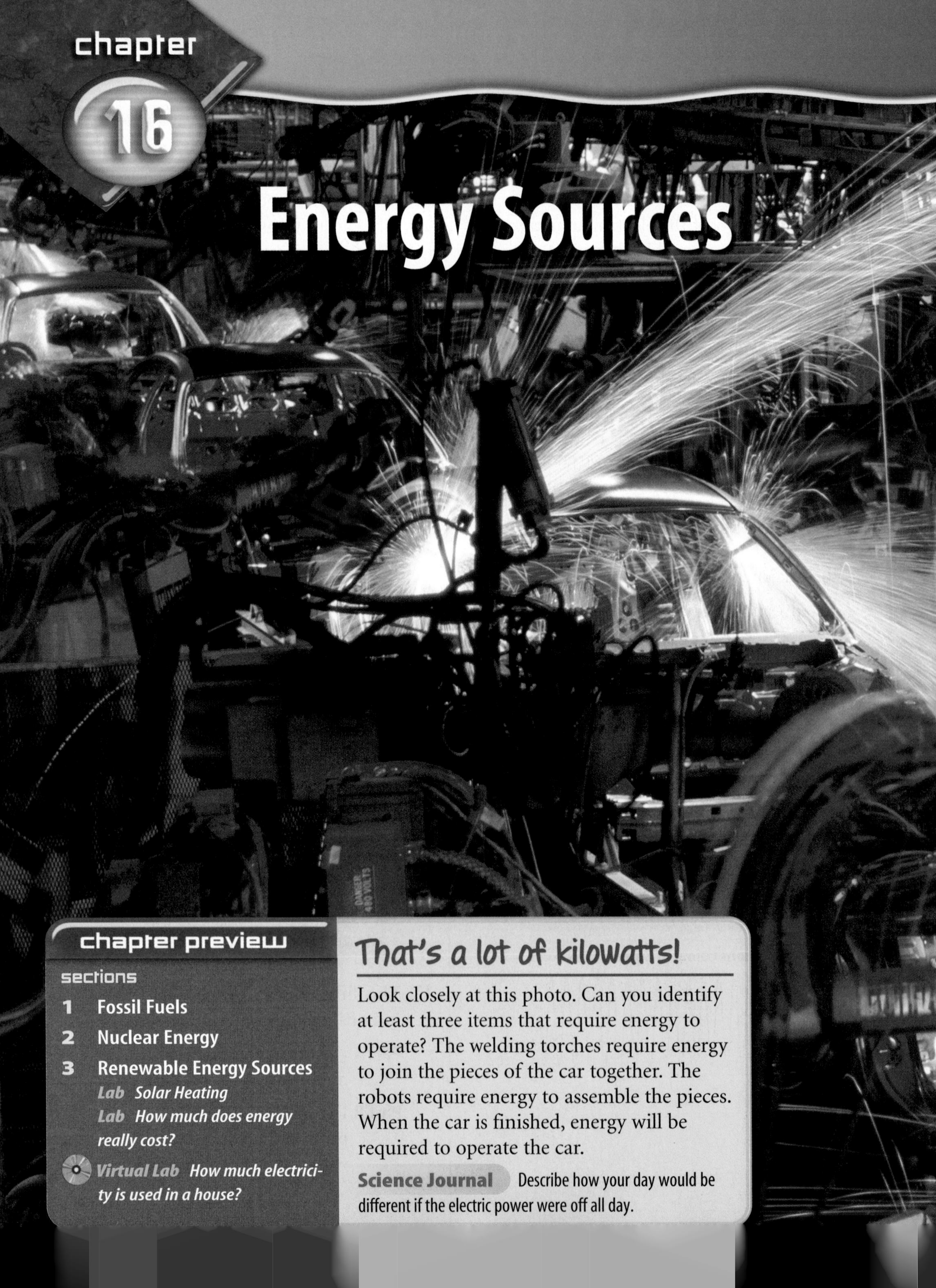

chapter 16

Energy Sources

chapter preview

sections

1 Fossil Fuels
2 Nuclear Energy
3 Renewable Energy Sources
Lab Solar Heating
Lab How much does energy really cost?
Virtual Lab How much electricity is used in a house?

That's a lot of kilowatts!

Look closely at this photo. Can you identify at least three items that require energy to operate? The welding torches require energy to join the pieces of the car together. The robots require energy to assemble the pieces. When the car is finished, energy will be required to operate the car.

Science Journal Describe how your day would be different if the electric power were off all day.

Start-Up Activities

Heating with Solar Energy

The Sun constantly bathes our planet with enormous amounts of energy. This energy can be captured and used to make electricity, heat homes, and provide hot water. How can the Sun's energy be used to heat water?

1. Complete the safety form.
2. Use scissors to poke a small hole in the center of each of two plastic coffee can lids.
3. Fill a coffee can that has been painted black with water at room temperature. Snap on the lid and push a thermometer through the hole in the lid. Record the temperature.
4. Repeat step 3 using the coffee can that has been painted white.
5. Place both cans in direct sunlight. After 15 min, record the temperature of the water in both cans again.
6. **Think Critically** Write a paragraph explaining why the temperature change differed between the two cans.

Preview this chapter's content and activities at gpescience.com

Energy Sources There are many sources of energy. Make the following Foldable to help you organize information about various types of energy sources.

STEP 1 **Fold** a sheet of paper in half lengthwise. Make the back edge about 5 cm longer than the front edge.

STEP 2 **Turn** the paper so the fold is on the bottom. Then **fold** it into thirds.

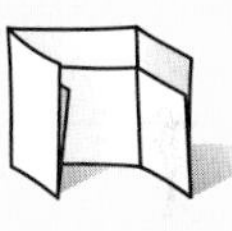

STEP 3 **Unfold and cut** only the top layer along both folds to make three tabs.

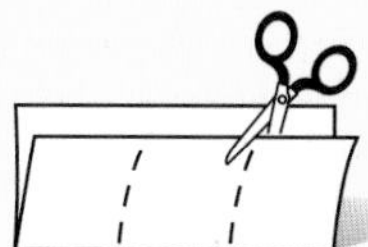

STEP 4 **Label** the Foldable as shown.

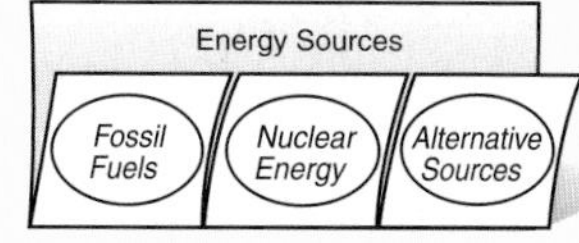

Summarize As you read this chapter, summarize important information about each type of energy source under the appropriate tab.

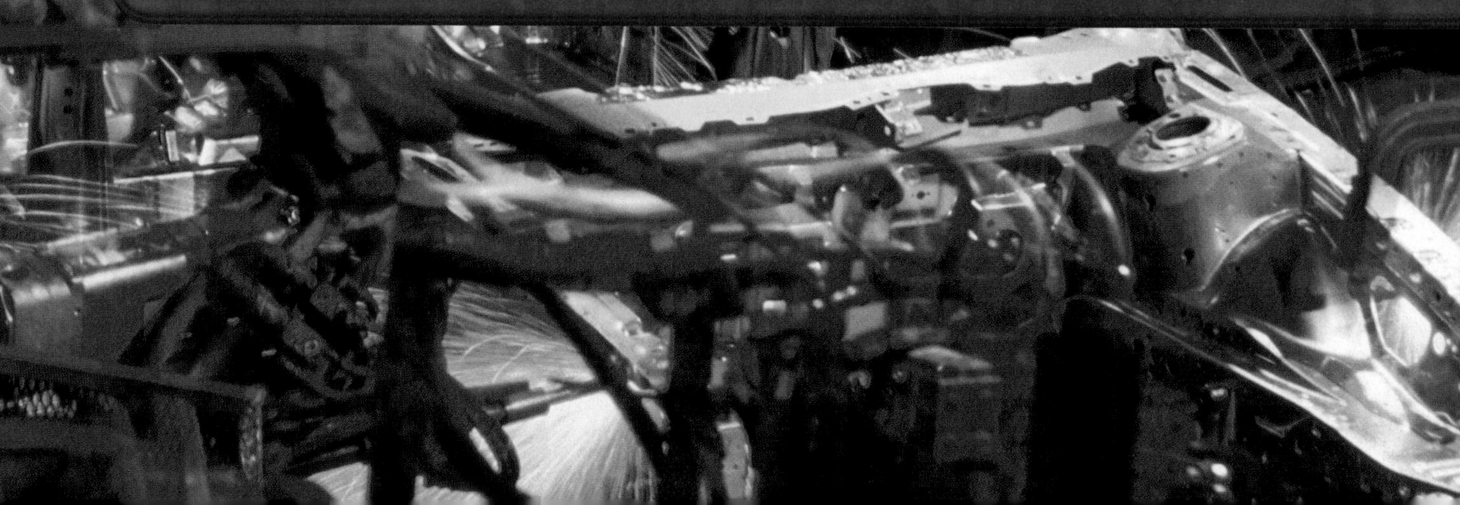

section 1

Fossil Fuels

Reading Guide

What You'll Learn

- **Discuss** properties and uses of fossil fuels.
- **Explain** how fossil fuels are formed.
- **Describe** how the chemical energy in fossil fuels is converted into electrical energy.

Why It's Important

Fossil fuels are used to generate most of the energy you use every day.

Review Vocabulary

chemical potential energy: the energy stored in the chemical bonds between atoms in molecules

New Vocabulary

- fossil fuel
- petroleum
- nonrenewable resource

Using Energy

How many different ways have you used energy today? Today you might have ridden in a car or bus, or used a hair dryer or a toaster. If you did, you used energy. Furnaces and stoves use thermal energy to heat buildings and cook food. Air conditioners use electrical energy to move thermal energy outdoors. Cars and other vehicles use mechanical energy to carry people and materials from one part of the country to another.

Transforming Energy According to the law of conservation of energy, energy cannot be created or destroyed. Energy can only be transformed, or converted, from one form to another. To use energy means to transform one form of energy to another form of energy that can perform a useful function. For example, energy is used when the chemical energy in fuels is transformed into thermal energy that heats your home.

Sometimes energy is transformed into a form that isn't useful. For example, power lines, like those shown in **Figure 1,** carry electrical energy. When electric current flows in power lines, about 10 percent of the electrical energy is transformed into thermal energy. This thermal energy flows into the air surrounding the power lines, and can no longer be used.

Figure 1 Power lines like these carry the electrical energy you use every day.

Energy Use in the United States More energy is used in the United States than in any other country in the world. **Figure 2** shows energy usage in the United States. About 20 percent of the energy is used in homes for heating and cooling, to run appliances, and to provide lighting and hot water. About 27 percent is used for transportation powering vehicles such as cars, trucks, and aircraft. Another 16 percent is used by businesses to heat, cool, and light stores, shops, and office buildings. Finally, about 37 percent of this energy is used by industry and agriculture to manufacture products and produce food. **Figure 2** also shows the main sources of the energy used in the United States. Almost 85 percent of the energy used in the United States comes from burning petroleum, natural gas, and coal. Nuclear power plants provide about eight percent of the energy used in the United States.

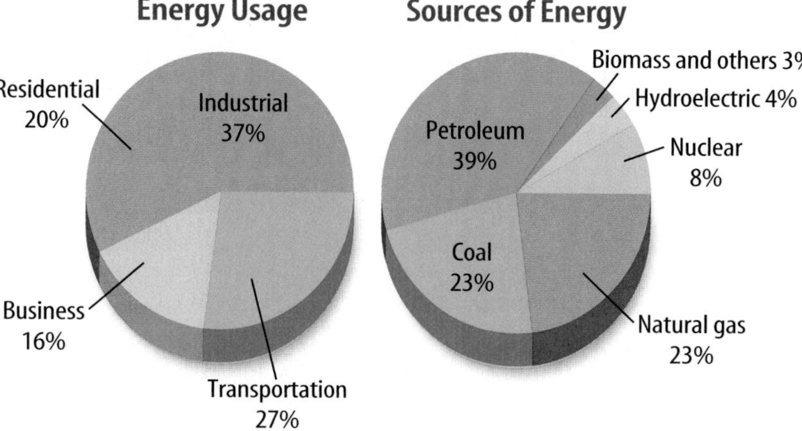

Figure 2 These circle graphs show where energy is used in the United States and sources of this energy.

Making Fossil Fuels

In one hour of freeway driving a car might use several gallons of gasoline. It may be hard to believe that it took millions of years to make the fuels that are used to produce electricity, provide heat, and transport people and materials. **Figure 4** on the next page shows how coal, petroleum, and natural gas are formed by the decay of ancient plants and animals. Fuels such as petroleum, or oil, natural gas, and coal are called **fossil fuels** because they are formed from the decaying remains of ancient plants and animals.

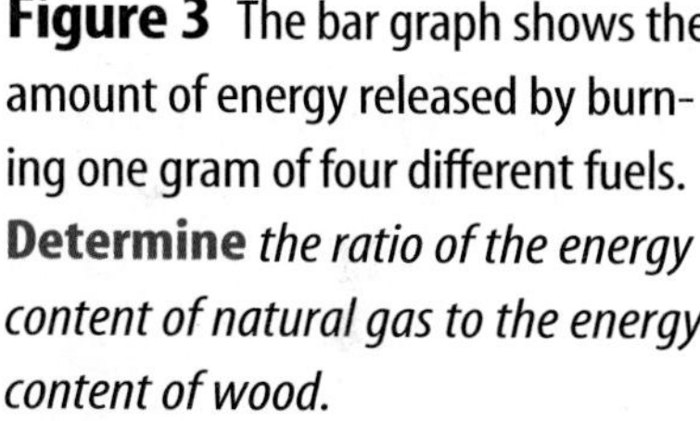

Figure 3 The bar graph shows the amount of energy released by burning one gram of four different fuels. **Determine** *the ratio of the energy content of natural gas to the energy content of wood.*

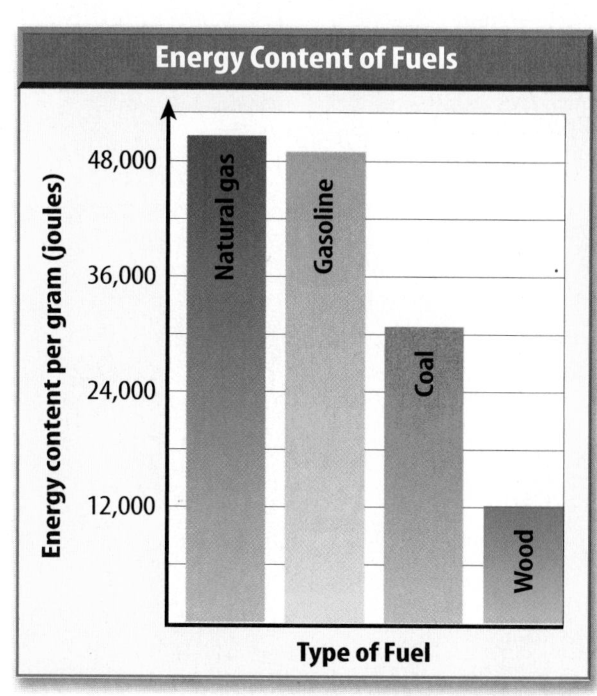

Concentrated Energy Sources When fossil fuels are burned, carbon and hydrogen atoms combine with oxygen molecules in the air to form carbon dioxide and water molecules. This process converts the chemical potential energy that is stored in the chemical bonds between atoms to heat and light. Compared to other fuels such as wood, the chemical energy that is stored in fossil fuels is more concentrated. For example, burning 1 kg of coal releases two to three times as much energy as burning 1 kg of wood. **Figure 3** compares the amount of energy that is produced by burning different fossil fuels.

NATIONAL GEOGRAPHIC VISUALIZING THE FORMATION OF FOSSIL FUELS

Figure 4

Oil and natural gas form when organic matter on the ocean floor, gradually buried under additional layers of sediment, is chemically changed by heat and crushing pressure. The oil and gas may bubble to the surface or become trapped beneath a dense rock layer. Coal forms when peat—partially decomposed vegetation—is compressed by overlying sediments and transformed first into lignite (soft brown coal) and then into harder, bituminous coal. These two processes are shown below.

HOW OIL AND NATURAL GAS ARE FORMED

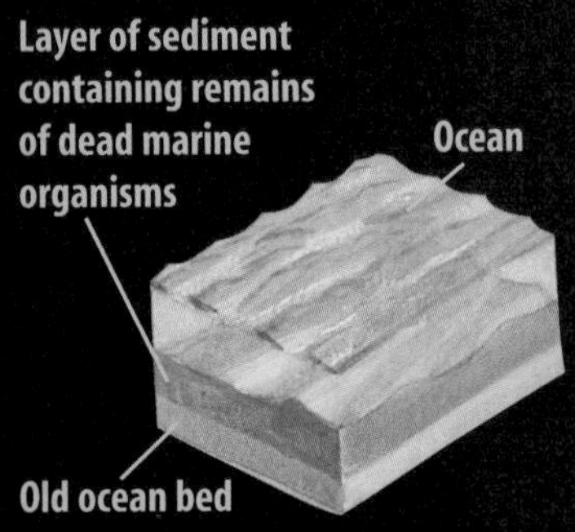

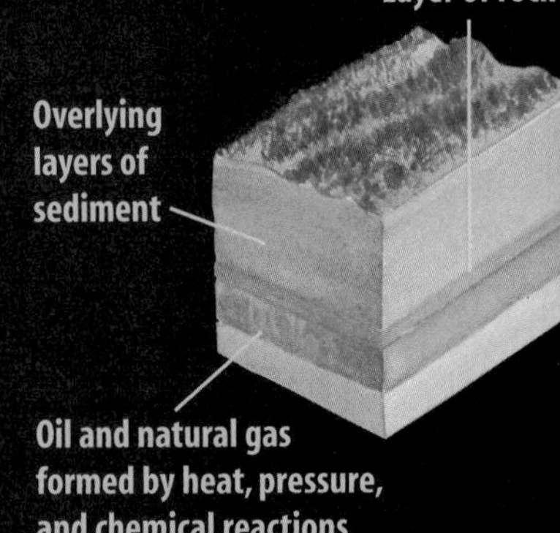

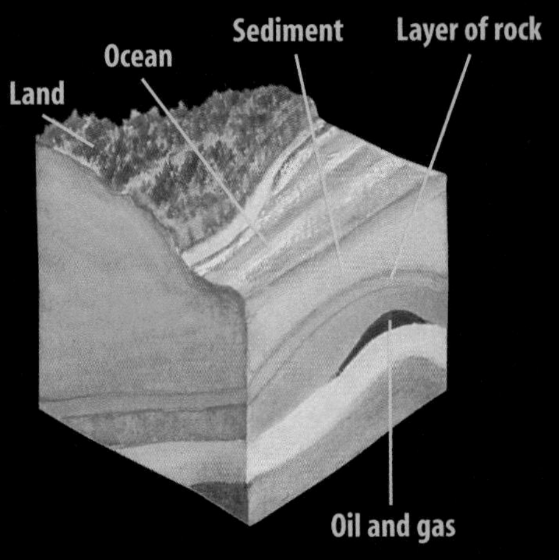

HOW COAL IS FORMED

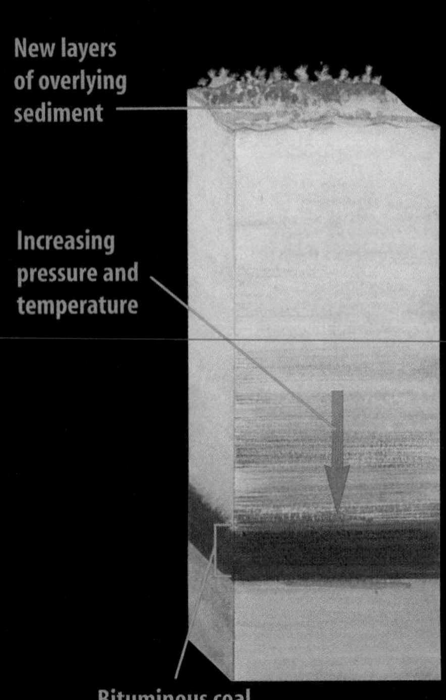

Petroleum

Millions of gallons of petroleum, or crude oil, are pumped every day from wells deep in Earth's crust. **Petroleum** is a highly flammable liquid formed by decayed ancient organisms, such as microscopic plankton and algae. Petroleum is a mixture of thousands of chemical compounds. Most of these compounds are hydrocarbons, which means their molecules contain only carbon atoms and hydrogen atoms.

Separating Hydrocarbons The different hydrocarbon molecules found in petroleum have different numbers and arrangements of carbon and hydrogen atoms. The composition and structure of hydrocarbons determines their properties.

The many different compounds that are found in petroleum are separated in a process called fractional distillation. This separation occurs in the tall towers of oil-refinery plants. First, crude oil is pumped into the bottom of the tower and heated. The chemical compounds in the crude oil boil and vaporize according to their individual boiling points. Materials with the lowest boiling points rise to the top of the tower as vapor and are collected. Hydrocarbons with high boiling points, such as asphalt and some types of waxes, remain liquid and are drained off through the bottom of the tower.

Reading Check *What is fractional distillation used for?*

Other Uses for Petroleum Not all of the products obtained from petroleum are burned to produce energy. About 15 percent of the petroleum-based substances that are used in the United States go toward nonfuel uses. Look around at the materials in your home or classroom. Do you see any plastics? In addition to fuels, plastics and synthetic fabrics are made from the hydrocarbons found in crude petroleum. Also, lubricants such as grease and motor oil, as well as the asphalt used in surfacing roads, are obtained from petroleum. Some synthetic materials produced from petroleum are shown in **Figure 5.**

Designing an Efficient Water Heater

Procedure

1. Complete the safety form.
2. Measure and record the mass of a **candle.**
3. Measure 50 mL of **water** into a **beaker.** Record the temperature of the water.
4. Use the lighted candle to increase the temperature of the water by 10°C. Put out the candle and measure its mass.
5. Repeat steps 2 to 4 with an **aluminum foil** chimney surrounding the candle.

Analysis

1. Compare the mass change in the two trials. Which heater was more efficient?
2. Explain why adding the chimney changed the efficiency of the water heater.

Figure 5 The objects shown here are made from chemical compounds found in petroleum.
Identify *four objects in your classroom that are made from petroleum.*

Figure 6 Coal mines usually are located deep underground.

Natural Gas

The chemical processes that produce petroleum as ancient organisms decay also produce gaseous compounds called natural gas. These compounds rise to the top of the petroleum deposit and are trapped there. Natural gas is composed mostly of methane, CH_4, but it also contains other hydrocarbon gases such as propane, C_3H_8, and butane, C_4H_{10}. Natural gas is burned to provide energy for cooking, heating, and manufacturing. About one fourth of the energy consumed in the United States comes from burning natural gas. There's a good chance that your home has a stove, furnace, hot-water heater, or clothes drier that uses natural gas.

Natural gas contains more energy per kilogram than petroleum or coal does. It also burns more cleanly than other fossil fuels, produces fewer pollutants, and leaves no residue such as ash.

Coal

Coal is a solid fossil fuel that is found in mines underground, such as the one shown in **Figure 6.** In the first half of the twentieth century, most houses in the United States were heated by burning coal. In fact, during this time, coal provided more than half of the energy that was used in the United States. Now, almost two-thirds of the energy used comes from petroleum and natural gas, and only about one-fourth comes from coal. About 90 percent of all the coal that is used in the United States is burned by power plants to generate electricity.

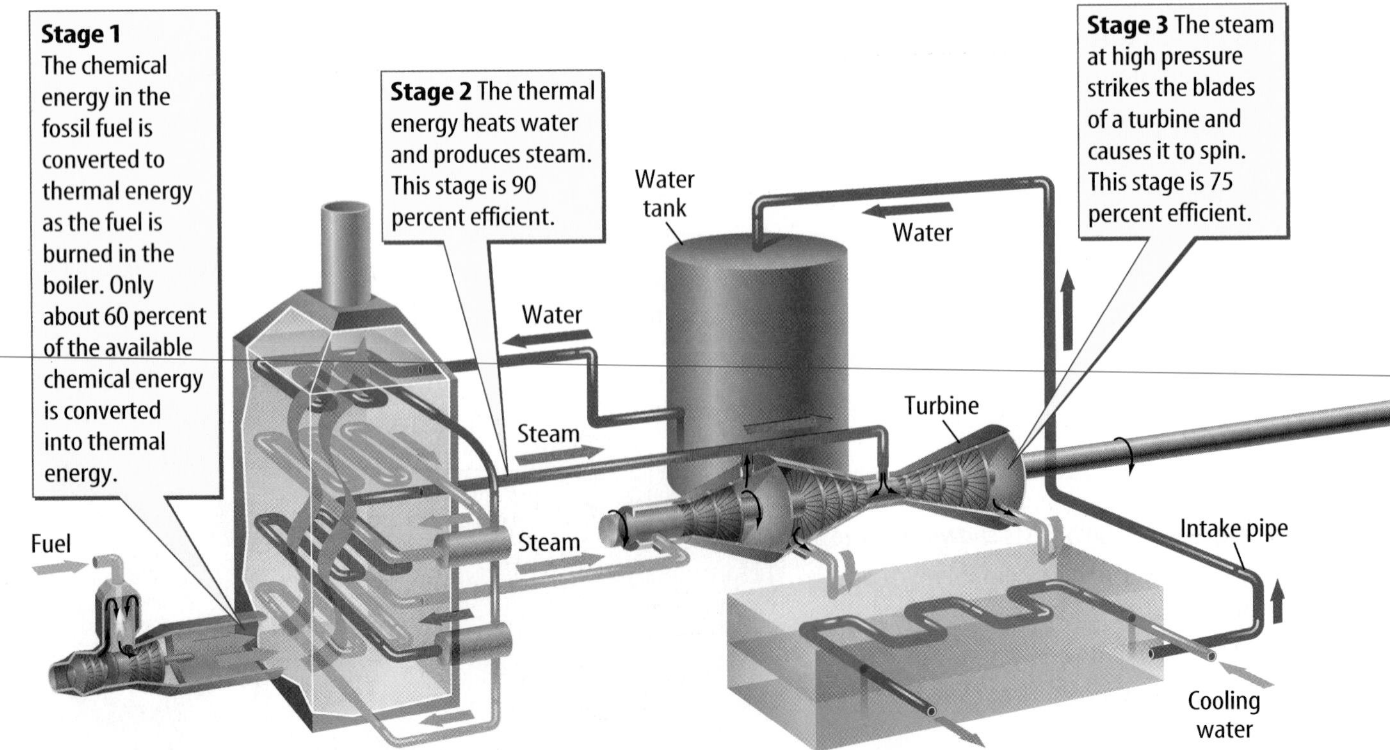

Stage 1 The chemical energy in the fossil fuel is converted to thermal energy as the fuel is burned in the boiler. Only about 60 percent of the available chemical energy is converted into thermal energy.
Stage 2 The thermal energy heats water and produces steam. This stage is 90 percent efficient.
Stage 3 The steam at high pressure strikes the blades of a turbine and causes it to spin. This stage is 75 percent efficient.
Water tank
Water
Water
Steam
Steam
Turbine
Fuel
Intake pipe
Cooling water

Origin of Coal Coal mines were once the sites of ancient swamps. Coal formed from the organic material that was deposited as the plants that lived in these swamps died. Worldwide, the amount of coal that is potentially available is estimated to be 20 to 40 times greater than the supply of petroleum.

Coal also is a complex mixture of hydrocarbons and other chemical compounds. Compared to petroleum and natural gas, coal contains more impurities, such as sulfur and nitrogen compounds. As a result, more pollutants, such as sulfur dioxide and nitrogen oxides, are produced when coal is burned.

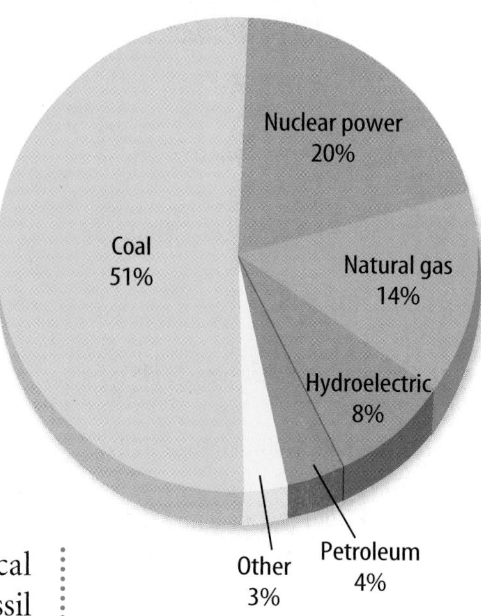

Figure 7 This circle graph shows the percentage of electricity generated in the United States that comes from various energy sources.

Generating Electricity

Figure 7 shows that almost 70 percent of the electrical energy used in the United States is produced by burning fossil fuels. How is the chemical energy contained in fossil fuels converted to electrical energy in an electric power station?

The process is shown in **Figure 8.** In the first stage, fuel is burned in a boiler or combustion chamber, and it releases thermal energy. In the second stage, this thermal energy heats water and produces steam under high pressure. In the third stage, the steam strikes the blades of a turbine, causing it to spin. The shaft of the turbine is connected to an electric generator. In the fourth stage, electric current is produced when the spinning turbine shaft rotates magnets inside the generator. In the final stage, the electric current is transmitted to homes, schools, and businesses through power lines.

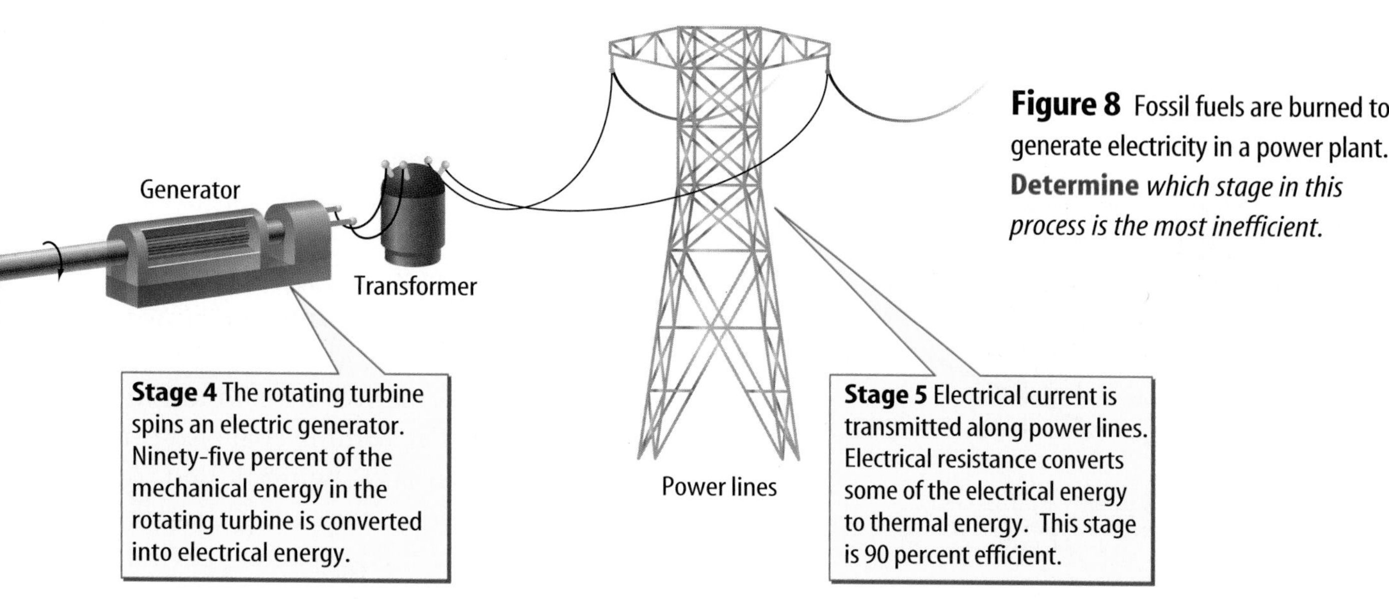

Generator
Transformer
Power lines
Stage 4 The rotating turbine spins an electric generator. Ninety-five percent of the mechanical energy in the rotating turbine is converted into electrical energy.
Stage 5 Electrical current is transmitted along power lines. Electrical resistance converts some of the electrical energy to thermal energy. This stage is 90 percent efficient.

Figure 8 Fossil fuels are burned to generate electricity in a power plant. **Determine** *which stage in this process is the most inefficient.*

Table 1 Efficiency of Fossil Fuel Conversion

Process	Efficiency (%)
Chemical to thermal energy	60
Conversion of water to steam	90
Steam-turning turbine	75
Turbine spins electric generator	95
Transmission through power lines	90
Overall efficiency	35

Efficiency of Power Plants

When fossil fuels are burned in power plants, not all of the chemical energy stored in the fuels is converted into electrical energy. In each energy transformation, some energy is converted into thermal energy that cannot be used. As a result, no stage of the process is 100 percent efficient.

The efficiency of each stage of the process in a fossil-fuel burning power plant is given in **Table 1.** The overall efficiency is found by multiplying the efficiencies together, and is only about 35 percent. This means that only about 35 percent of the chemical energy contained in fossil fuels is converted into electrical energy by power plants. The other 65 percent is converted into thermal energy that is transferred to the environment.

The Costs of Using Fossil Fuels

Although fossil fuels are a useful source of energy for generating electricity and providing the power for transportation, their use has some undesirable side effects. When petroleum products and coal are burned, smoke is given off that contains small particles called particulates. These particulates cause breathing problems for some people. Burning fossil fuels also releases carbon dioxide. **Figure 9** shows how the carbon dioxide concentration in the atmosphere has increased from 1960 to 2000. One consequence of increasing the atmospheric carbon dioxide concentration could be to cause Earth's surface temperature to increase.

Figure 9 The carbon dioxide concentration in Earth's atmosphere has been measured at Mauna Loa in Hawaii. From 1960 to 2000, the carbon dioxide concentration has increased by about 16 percent.

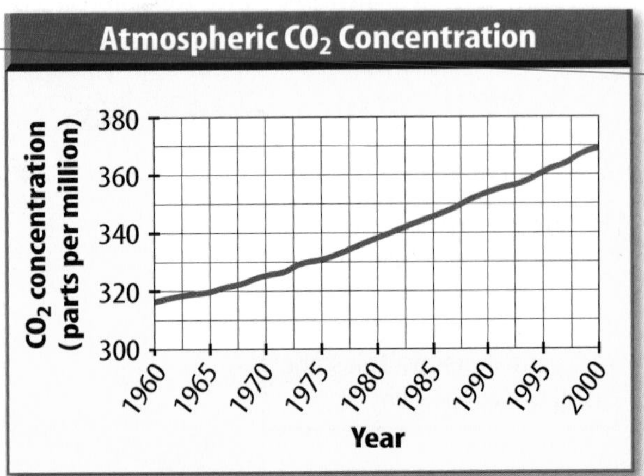

Using Coal The most abundant fossil fuel is coal, but coal contains even more impurities than oil or natural gas. Many electric power plants that burn coal remove some of these pollutants before they are released into the atmosphere. Removing sulfur dioxide, for example, helps to prevent the formation of compounds that might cause acid rain. Mining coal also can be dangerous. Miners risk being killed or injured, and some suffer from lung diseases caused by breathing coal dust over long periods of time.

Nonrenewable Resources

All fossil fuels are **nonrenewable resources,** which means they are resources that cannot be replaced by natural processes as quickly as they are used. Therefore, fossil fuel reserves are decreasing at the same time that population and industrial demands are increasing. **Figure 10** shows how the production of oil might decline over the next 50 years as oil reserves are used up. As the production of energy from fossil fuels continues, the remaining reserves of fossil fuels will decrease. Fossil fuels will become more difficult to obtain, causing them to become more costly in the future.

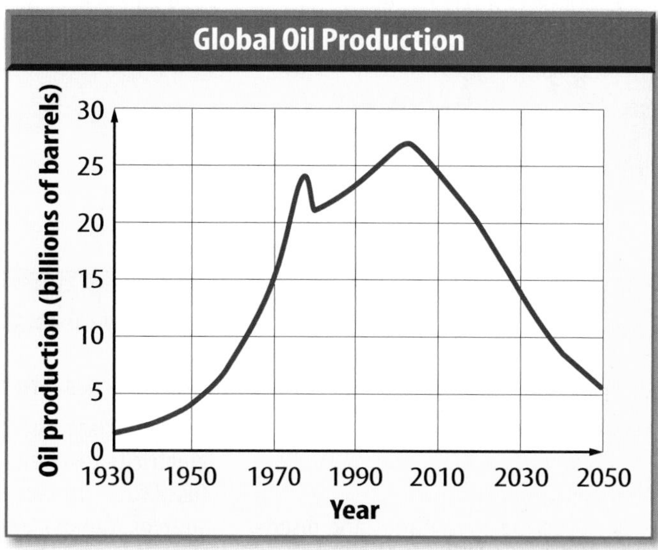

Figure 10 Some predictions show that worldwide oil production will peak by 2005 and then decline rapidly over the following 50 years.

Conserving Fossil Fuels

Even as reserves of fossil fuels decrease and they become more costly, the demand for energy continues to increase as the world's population increases. One way to meet these energy demands would be to reduce the use of fossil fuels and obtain energy from other sources.

section 1 review

Summary

Using Energy

- Energy cannot be created or destroyed, but can only be transformed from one form to another.

Fossil Fuels

- Petroleum, natural gas, and coal are fossil fuels formed by the decay of ancient plants and animals.
- Petroleum is a mixture of thousands of chemical compounds, most of which are hydrocarbons.
- About 90 percent of all coal used in the United States is burned by power plants to produce electricity.

Generating Electricity

- Power plants burn fossil fuels to produce steam that spins turbines attached to electric generators.

Self Check

1. **Describe** the advantages and disadvantages of using fossil fuels to generate electricity.
2. **Explain** how the different chemical compounds in crude oil are separated.
3. **Describe** how fossil fuels are formed.
4. **Name** three materials that are derived from the chemical compounds in petroleum.
5. **Think Critically** If fossil fuels are still forming, why are they considered to be a nonrenewable resource?

Applying Math

6. **Interpret a Graph** According to the graph in **Figure 9,** by how many parts per million did the concentration of atmospheric carbon dioxide increase from 1960 to 2000?
7. **Use a Table** In **Table 1,** if the efficiency of converting chemical to thermal energy was 90 percent, what would the overall efficiency be?

section 2

Nuclear Energy

Reading Guide

What You'll Learn

- **Explain** how a nuclear reactor converts nuclear energy to thermal energy.
- **Describe** advantages and disadvantages of using nuclear energy to produce electricity.
- **Discuss** nuclear fusion as a possible energy source.

Why It's Important

Using nuclear energy to produce electricity can help reduce the use of fossil fuels. However, like all energy sources, the use of nuclear energy has advantages and disadvantages.

Review Vocabulary

nuclear fission: the process of splitting an atomic nucleus into two or more nuclei with smaller masses

New Vocabulary

- nuclear reactor
- nuclear waste

Using Nuclear Energy

Over the past several decades, electric power plants have been developed that generate electricity without burning fossil fuels. Some of these power plants, such as the one shown in **Figure 11,** convert nuclear energy to electrical energy. Energy is released when the nucleus of an atom breaks apart. In this process, called nuclear fission, an extremely small amount of mass is converted into an enormous amount of energy. Today almost 20 percent of all the electricity produced in the United States comes from nuclear power plants. Overall, nuclear power plants produce about eight percent of all the energy consumed in the United States. In 2003, there were 104 nuclear reactors producing electricity at 65 nuclear power plants in the United States.

Figure 11 A nuclear power plant generates electricity using the energy released in nuclear fission. Each of the domes contain a nuclear reactor. A cooling tower is on the left.

Nuclear Reactors

A **nuclear reactor** uses the energy from controlled nuclear reactions to generate electricity. Although nuclear reactors vary in design, all have some parts in common, as shown in **Figure 12.** They contain a fuel that can be made to undergo nuclear fission; they contain control rods that are used to control the nuclear reactions; and they have a cooling system that keeps the reactor from being damaged by the heat produced. The actual fission of the radioactive fuel occurs in a relatively small part of the reactor known as the core.

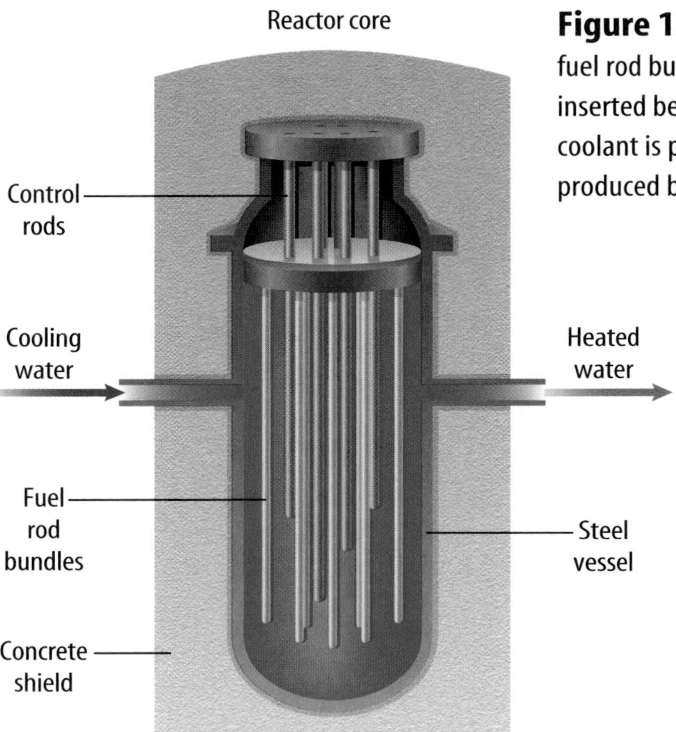

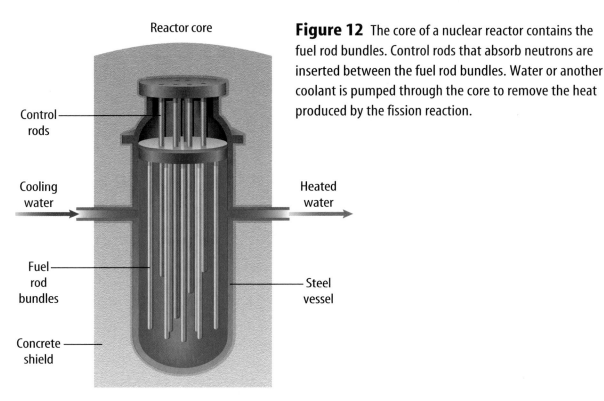

Figure 12 The core of a nuclear reactor contains the fuel rod bundles. Control rods that absorb neutrons are inserted between the fuel rod bundles. Water or another coolant is pumped through the core to remove the heat produced by the fission reaction.

Nuclear Fuel Only certain elements have nuclei that can undergo fission. Naturally occurring uranium contains an isotope, U-235, whose nucleus can split apart. As a result, the fuel that is used in a nuclear reactor is usually uranium dioxide. Naturally occurring uranium contains only about 0.7 percent of the U-235 isotope. In a reactor, the uranium usually is enriched so that it contains three percent to five percent U-235.

The Reactor Core The reactor core contains uranium dioxide fuel in the form of tiny pellets like the ones in **Figure 13.** The pellets are about the size of a pencil eraser and are placed end to end in a tube. The tubes are then bundled and covered with a metal alloy, as shown in **Figure 13.** The core of a typical reactor contains about a hundred thousand kilograms of uranium in hundreds of fuel rods. For every kilogram of uranium that undergoes fission in the core, 1 g of matter is converted into energy. The energy released by this gram of matter is equivalent to the energy released by burning more than 3 million kg of coal.

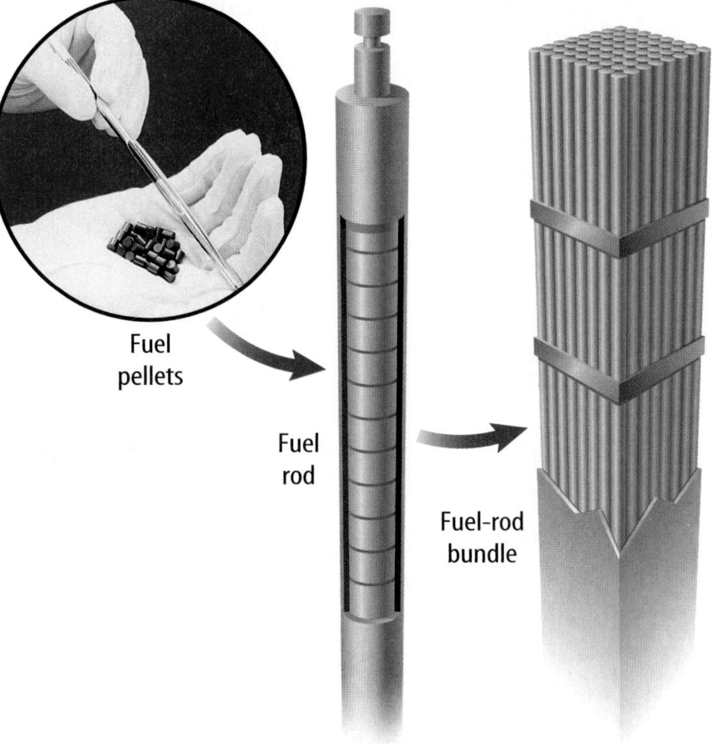

Figure 13 Nuclear fuel pellets are stacked together to form fuel rods. The fuel rods are bundled together, and the bundle is covered with a metal alloy.

Figure 14 When a neutron strikes the nucleus of a U-235 atom, the nucleus splits apart into two smaller nuclei. In the process two or three neutrons also are emitted. The smaller nuclei are called fission products. **Explain** *what happens to the neutrons that are released in this reaction.*

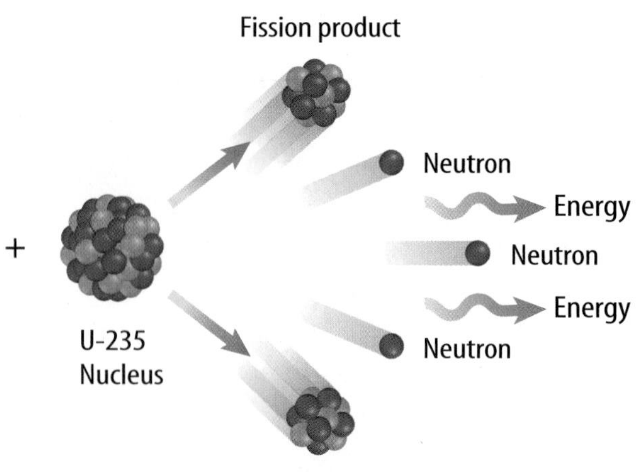

Nuclear Fission How does the nuclear reaction proceed in the reactor core? Neutrons that are produced by the decay of U-235 nuclei are absorbed by other U-235 nuclei. When a U-235 nucleus absorbs a neutron, it splits into two smaller nuclei and two or three additional neutrons, as shown in **Figure 14.** These neutrons strike other U-235 nuclei, causing them to release two or three more neutrons each when they split apart.

Because every uranium atom that splits apart releases neutrons that cause other uranium atoms to split apart, this process is called a nuclear chain reaction. In the chain reaction involving the fission of uranium nuclei, the number of nuclei that are split can more than double at each stage of the process. As a result, an enormous number of nuclei can be split after only a small number of stages. For example, if the number of nuclei involved doubles at each stage, after only 50 stages more than a quadrillion nuclei might be split.

Nuclear chain reactions take place in a matter of milliseconds. If the process isn't controlled, the chain reaction will release energy explosively rather than releasing energy at a constant rate.

Uranium-Lead Dating Uranium is used to determine the age of rocks. As uranium decays into lead at a constant rate, the age of a rock can be found by comparing the amount of uranium to the amount of lead produced. Uranium-lead dating is used by scientists to date rocks as old as 4.6 billion years. Research other methods used to determine the age of rocks.

Reading Check *What is a nuclear chain reaction?*

Controlling the Chain Reaction To control the chain reaction, some of the neutrons that are released when U-235 splits apart must be prevented from striking other U-235 nuclei. These neutrons are absorbed by rods containing boron or cadmium that are inserted into the reactor core. Moving these control rods deeper into the reactor causes them to absorb more neutrons and slow down the chain reaction. Eventually, only one of the neutrons released in the fission of each of the U-235 nuclei strikes another U-235 nucleus, and energy is released at a constant rate.

Nuclear Power Plants

Nuclear fission reactors produce electricity in much the same way that conventional power plants do. **Figure 15** shows how a nuclear reactor produces electricity. The thermal energy released in nuclear fission is used to heat water and produce steam. This steam then is used to drive a turbine that rotates an electric generator. To transfer thermal energy from the reactor core to heat water and produce steam, the core is immersed in a fluid coolant. The coolant absorbs heat from the core and is pumped through a heat exchanger. There thermal energy is transferred from the coolant and boils water to produce steam. The overall efficiency of nuclear power plants is about 35 percent, similar to that of fossil fuel power plants.

Ukraine The worst nuclear accident in history occurred at the Chernobyl nuclear power plant in the Ukraine in 1986. Many people in the area suffered from radiation sickness. Use a map or atlas to find the location of the Ukraine. Write a description of the location in your Science Journal.

The Risks of Nuclear Power

Producing energy from nuclear fission has advantages. Nuclear power plants do not produce the air pollutants that are released by fossil-fuel burning power plants. Also, nuclear power plants don't produce carbon dioxide.

The nuclear generation of electricity also has disadvantages. The mining of the uranium can cause environmental damage. Water that is used as a coolant in the reactor core must cool before it is released into streams and rivers. Otherwise, the excess heat could harm fish and other animals and plants in the water.

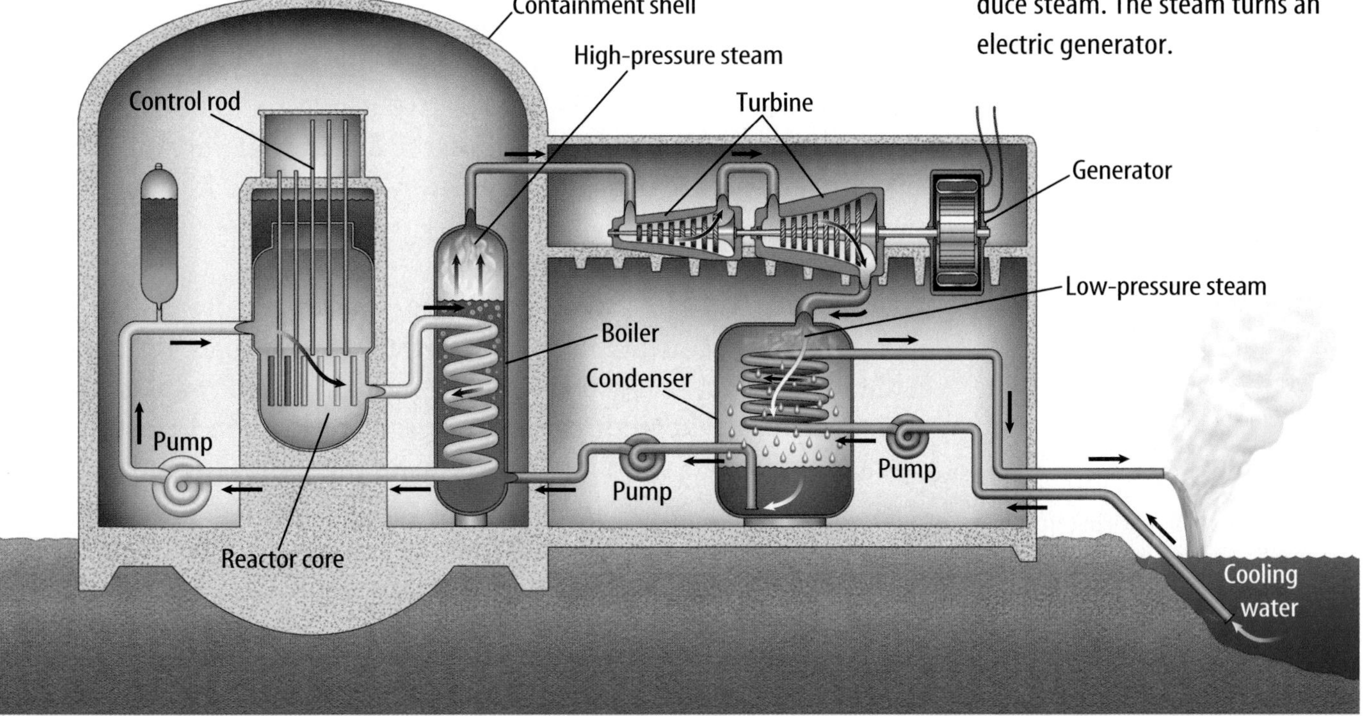

Figure 15 A nuclear power plant uses the heat produced by nuclear fission in its core to produce steam. The steam turns an electric generator.

Figure 16 An explosion occurred at the Chernobyl reactor in the Ukraine after graphite control rods caught fire. The explosion shattered the reactor's roof.

The Release of Radioactivity One of the most serious risks of nuclear power is the escape of harmful radiation from power plants. The fuel rods contain radioactive elements with various half-lives. Some of these elements could cause damage to living organisms if they were released from the reactor core. Nuclear reactors have elaborate systems of safeguards, strict safety precautions, and highly trained workers in order to prevent accidents. In spite of this, accidents have occurred.

For example, in 1986 in Chernobyl, Ukraine, an accident occurred when a reactor core overheated during a safety test. Materials in the core caught fire and caused a chemical explosion that blew a hole in the reactor, as shown in **Figure 16.** This resulted in the release of radioactive materials that were carried by winds and deposited over a large area. As a result of the accident, 28 people died of acute radiation sickness. In the United States, power plants are designed to prevent accidents such as the one that occurred at Chernobyl.

The Disposal of Nuclear Waste

After about three years, not enough fissionable U-235 is left in the fuel pellets in the reactor core to sustain the chain reaction. The spent fuel contains radioactive fission products in addition to the remaining uranium. These fuel pellets are a form of nuclear waste. **Nuclear waste** is any radioactive by-product that results when radioactive materials are used. Nuclear wastes are classified as low-level waste and high-level waste.

Topic: Storing Nuclear Wastes
Visit gpescience.com for Web links to information about storing nuclear wastes.

Activity Obtain a map or sketch an outline of the United States. Mark the locations of the nuclear waste sites that you found. What do these locations have in common? Why do you think these locations were chosen over other sites that were closer to the nuclear waste generating sites?

Low-Level Waste Low-level nuclear wastes usually contain a small amount of radioactive material. They usually do not contain radioactive materials with long half-lives. Products of some medical and industrial processes are low-level wastes, including items of clothing used in handling radioactive materials. Low-level wastes also include used air filters from nuclear power plants and discarded smoke detectors. Low-level wastes usually are sealed in containers and buried in trenches 30 m deep at special locations. When dilute enough, low-level waste sometimes is released into the air or water.

Figure 17 Spent nuclear fuel rods are placed underwater after they are removed from the reactor core. The water absorbs the nuclear radiation and prevents it from escaping into the environment.

High-Level Waste High-level nuclear waste is generated in nuclear power plants and by nuclear weapons programs. After spent fuel is removed from a reactor, it is stored in a deep pool of water, as shown in **Figure 17.** Many of the radioactive materials in high-level nuclear waste have short half-lives. However, the spent fuel also contains materials that will remain radioactive for tens of thousands of years. For this reason, the waste must be disposed of in extremely durable and stable containers.

One method proposed for the disposal of high-level waste is to seal the waste in ceramic glass, which is placed in protective metal-alloy containers. The containers then are buried hundreds of meters below ground in stable rock formations or salt deposits. It is hoped that this will keep the material from contaminating the environment for thousands of years.

What is the difference between low-level and high-level nuclear wastes?

Applying Science

Can a contaminated radioactive site be reclaimed?

In the early 1900s, with the discovery of radium, extensive mining for the element began in the Denver, Colorado, area. Radium is a radioactive element that was used to make watch dials and instrument panels that glowed in the dark. After World War I, the radium industry collapsed. The area was left contaminated with 97,000 tons of radioactive soil and debris containing heavy metals and radium, which is now known to cause cancer. The soil was used as fill, foundation material, left in place, or mishandled.

Radium
88
Ra
(226)

Identifying the Problem

In the 1980s, one area became known as the Denver Radium Superfund Site and was cleaned up by the Environmental Protection Agency. The land then was reclaimed by a local commercial establishment.

Solving the Problem

1. The contaminated soil was placed in one area and a protective cap was placed over it. This area also was restricted from being used for residential homes. Explain why it is important for the protective cap to be maintained and why homes could not be built in this area.
2. The advantages of cleaning up this site are economical, environmental, and social. Give an example of each.

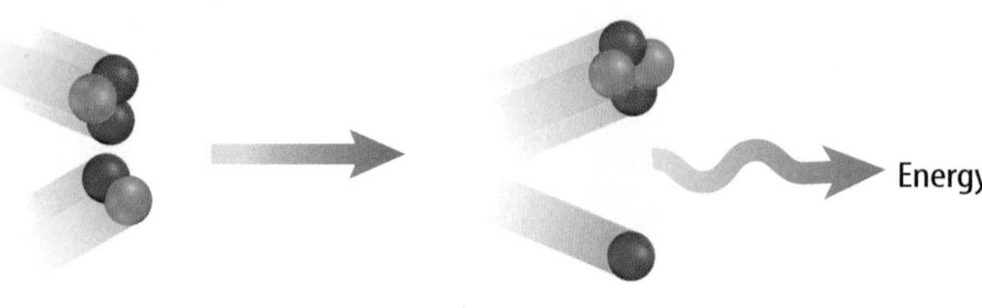

Figure 18 In nuclear fusion, two smaller nuclei join together to form a larger nucleus. Energy is released in the process. In the reaction shown here, two isotopes of hydrogen come together to form a helium nucleus.
Identify *the source of the energy released in a fusion reaction.*

Nuclear Fusion

The Sun gives off a tremendous amount of energy through a process called thermonuclear fusion. Thermonuclear fusion is the joining together of small nuclei at high temperatures, as shown in **Figure 18.** In this process, a small amount of mass is converted into energy. Fusion is the most concentrated energy source known.

An advantage of producing energy using nuclear fusion is that the process uses hydrogen as fuel. Hydrogen is abundant on Earth. Another advantage is that the product of the reaction is helium. Helium is not radioactive and is chemically nonreactive.

One disadvantage of fusion is that it occurs only at temperatures of millions of degrees Celsius. Research reactors often consume more energy to reach and maintain these temperatures than they produce. Another problem is how to contain a reaction that occurs at such extreme conditions. Until solutions to these and other problems are found, the use of nuclear fusion as an energy source is not practical.

section 2 review

Summary

Using Nuclear Energy

- Nuclear power plants produce about eight percent of the energy used each year in the United States.

Nuclear Power Plants

- Nuclear reactors use the energy released in the fission of U-235 to produce electricity.
- The energy released in the fission reaction is used to make steam. The steam drives a turbine that rotates an electric generator.

The Risks of Nuclear Energy

- Organisms could be damaged if radiation is released from the reactor.
- Nuclear waste is the radioactive by-product produced by using radioactive materials.
- Nuclear power generation produces high-level nuclear wastes.

Self Check

1. **Explain** why a chain reaction occurs when uranium-235 undergoes fission.
2. **Describe** how the chain reaction in a nuclear reactor is controlled.
3. **Compare** the advantages and disadvantages of nuclear power plants and those that burn fossil fuels.
4. **Describe** the advantages and disadvantages of using nuclear fusion reactions as a source of energy.
5. **Think Critically** A research project produced 10 g of nuclear waste with a short half-life. How would you classify this waste, and how would it be disposed of?

Applying Math

6. **Use Percentages** Naturally occurring uranium contains 0.72 percent of the isotope uranium-235. What is the mass of uranium-235 in 2,000 kg of naturally-occurring uranium?

More Section Review gpescience.com

section 3

Renewable Energy Sources

Reading Guide

What You'll Learn

- **Analyze** the need for alternate energy sources.
- **Describe** alternate methods for generating electricity.
- **Compare** the advantages and disadvantages of various alternate energy sources.

Why It's Important

Alternative energy sources need to be explored because the primary sources of energy used in the United States are nonrenewable.

Review Vocabulary

radiant energy: the energy carried by an electromagnetic wave

New Vocabulary

- renewable resource
- photovoltaic cell
- hydroelectricity
- geothermal energy
- biomass

Energy Options

The demand for energy continues to increase, but supplies of fossil fuels are decreasing. Using more nuclear reactors to produce electricity will produce more high-level nuclear waste that has to be disposed of safely. As a result, other sources of energy that can meet Earth's increasing energy demands are being developed. Some alternative energy sources are considered to be renewable resources. A **renewable resource** is an energy source that is replaced nearly as quickly as it is used.

Energy from the Sun

The average amount of solar energy that falls on the United States in one day is more than the total amount of energy used in the United States in one year. Because only about one billionth of the Sun's energy falls on Earth, and because the Sun is expected to continue producing energy for several billion years, solar energy cannot be used up. Solar energy is a renewable resource.

Many devices use solar energy for power including solar-powered calculators similar to the one in **Figure 19.** These devices use a **photovoltaic cell** that converts radiant energy from the Sun directly into electrical energy. Photovoltaic cells also are called solar cells.

Figure 19 This calculator uses a solar cell to produce the electricity it needs to operate.

Figure 20 Solar cells convert radiant energy from the Sun to electricity.
Identify *two devices that use solar cells for power.*

When sunlight strikes a solar cell, electrons are ejected from the electron-rich semiconductor. These electrons can travel in a closed circuit back to the electron-poor semiconductor.

A solar cell is made of two layers of semiconductor material.

Sunlight

Antireflective coating

Glass cover

Metal contact

Current

Electron-rich semiconductor

Electron-poor semiconductor

Metal contact

Current

Using Solar Power at Home

Procedure

1. Complete the safety form.
2. Cut a **piece of cloth** into two pieces about 10 cm on each side.
3. Wet both pieces and wring them out so they are equally damp.
4. Spread the pieces out to dry outside. One piece should be in sunlight and the other should be in shade.
5. Record the time it takes for each piece to dry.

Analysis

1. Which piece of cloth received more solar energy?
2. Explain why the drying time for a piece of cloth depends on the amount of solar energy it receives.

Try at Home

How Solar Cells Work Solar cells are made of two layers of semiconductor materials sandwiched between two layers of conducting metal, as shown in **Figure 20.** One layer of semiconductor is rich in electrons, while the other layer is electron poor. When sunlight strikes the surface of the solar cell, electrons flow through an electrical circuit from the electron-rich semiconductor to the electron-poor material. This process of converting radiant energy from the Sun directly to electrical energy is only about 7 percent to 11 percent efficient.

Using Solar Energy Producing large amounts of electrical energy using solar cells is more expensive than producing electrical energy using fossil fuels. However, in remote areas where electric distribution lines are not available, the use of solar cells is a practical way of providing electrical power.

Currently, the most promising solar technologies are those that concentrate the solar power into a receiver. One such system is called the parabolic trough. The trough focuses the sunlight on a tube that contains a heat-absorbing fluid such as synthetic oil or liquid salt. The heated fluid is circulated through a boiler where it generates steam to turn a turbine, generating electricity.

The world's largest concentrating solar power plant is located in the Mojave Desert in California. This facility consists of nine units that generate over 350 megawatts of power. These nine units can generate enough electrical power to meet the needs of approximately 500,000 people. These units use natural gas as a backup power source for generating electricity at night and on cloudy days when solar energy is unavailable.

Energy from Water

Just as the expansion of steam can turn an electric generator, rapidly moving water can as well. The gravitational potential energy of the water can be increased if the water is retained by a high dam. This potential energy is released when the water flows through tunnels near the base of the dam. **Figure 21** shows how the rushing water spins a turbine, which rotates the shaft of an electric generator to produce electricity. Dams built for this purpose are called hydroelectric dams.

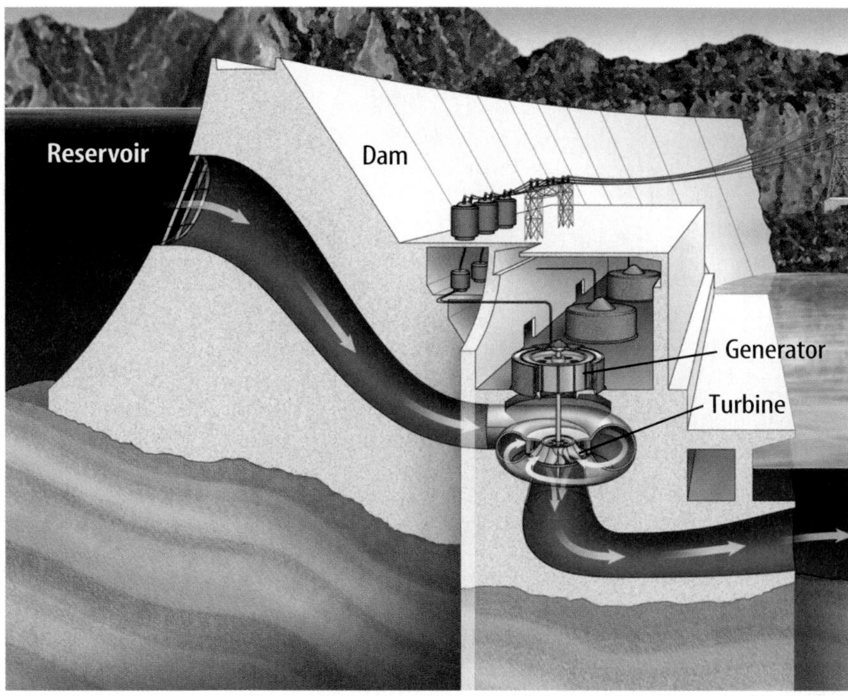

Figure 21 The potential energy in water stored behind the dam is converted to electrical energy in a hydroelectric power plant.
Diagram *the energy conversions that occur as a hydroelectric dam produces electrical energy.*

Using Hydroelectricity Electricity produced from the energy of moving water is called **hydroelectricity.** Currently about 8 percent of the electrical energy used in the United States is produced by hydroelectric power plants. Hydroelectric power plants are an efficient way to produce electricity with almost no pollution. Because no exchange of heat is involved in producing steam to spin a turbine, hydroelectric power plants are almost twice as efficient as fossil fuel or nuclear power plants.

Reading Check *Why are hydroelectric power plants more efficient than fossil fuel power plants?*

Another advantage is that the bodies of water held back by dams can form lakes that can provide water for drinking and crop irrigation. These lakes also can be used for boating and swimming. Also, after the initial cost of building a dam and a power plant, the electricity is relatively cheap.

However, artificial dams can disturb the balance of natural ecosystems. Some species of fish that live in the ocean migrate back to the rivers in which they were hatched to breed. This migration can be blocked by dams, which causes a decline in the fish population. Fish ladders, such as those shown in **Figure 22,** have been designed to enable fish to migrate upstream past some dams. Also, some water sources suitable for a hydroelectric power plant are located far from the regions needing power.

Figure 22 Fish ladders enable fish to migrate upstream past dams.

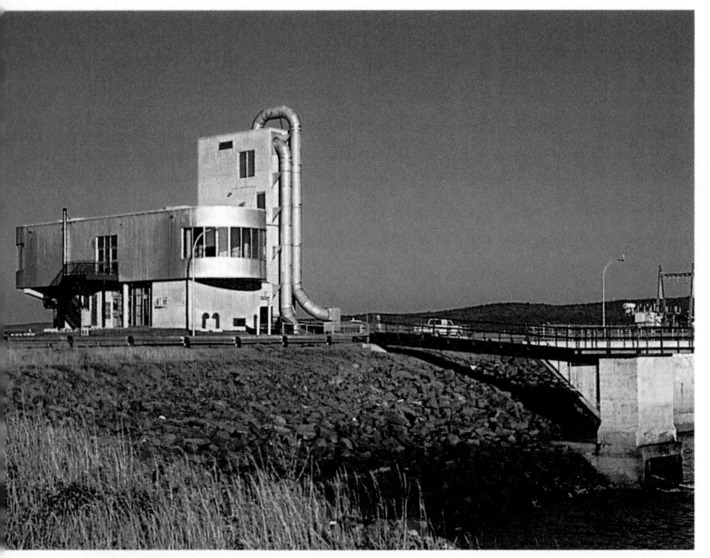

Energy from the Tides

The gravity of the Moon and Sun causes bulges in Earth's oceans. As Earth rotates, the two bulges of ocean water move westward. Each day, the level of the ocean on a coast rises and falls continually. Hydroelectric power can be generated by these ocean tides. As the tide comes in, the moving water spins a turbine that generates electricity. The water is then trapped behind a dam. At low tide the water behind the dam flows back out to the ocean, spinning the turbines and generating electric power.

Tidal energy is nearly pollution free. The efficiency of a tidal power plant is similar to that of a conventional hydroelectric power plant. However, only a few places on Earth have large enough differences between high and low tides for tidal energy to be a useful energy source. The only tidal power station in use in North America is at Annapolis Royal, Nova Scotia, shown in **Figure 23.** Tidal energy probably will be a limited source of energy in the future.

Figure 23 This tidal energy plant at Annapolis Royal, Nova Scotia, generates 20 megawatts of electric power.

Harnessing the Wind

You might have seen a windmill on a farm or pictures of windmills in a book. These windmills use the energy of the wind to pump water. Windmills also can use the energy of the wind to generate electricity. Wind spins a propeller that is connected to an electric generator. Windmill farms, like the one shown in **Figure 24,** may contain several hundred windmills.

Figure 24 Wind energy is converted to electricity as the spinning propeller turns a generator.

However, only a few places on Earth consistently have enough wind to rely on wind power to meet energy needs. Also, windmills are only about 20 percent efficient on average. Research is underway to improve the design of wind generators and increase their efficiency. Other disadvantages of wind energy are that windmills can be noisy and change the appearance of a landscape. Also, they can disrupt the migration patterns of some birds. However, wind generators do not consume any nonrenewable natural resources, and they do not pollute the atmosphere or water.

Energy from Inside Earth

INTEGRATE Earth Science Earth is not completely solid. Heat is generated within Earth by the decay of radioactive elements. This heat is called geothermal heat. Geothermal heat causes the rock beneath Earth's crust to soften and melt. This hot molten rock is called magma. The thermal energy that is contained in hot magma is called **geothermal energy.**

In some places, Earth's crust has cracks or thin spots that allow magma to rise near the surface. Active volcanoes, for example, permit hot gases and magma from deep within Earth to escape. Perhaps you have seen a geyser, like Old Faithful in Yellowstone National Park, shooting steam and hot water. The water that shoots from the geyser is heated by magma close to Earth's surface. In some areas, this hot water can be pumped into houses to provide heat.

What two natural phenomena are caused by geothermal heat?

Topic: Geothermal Energy
Visit gpescience.com for Web links to information about geothermal energy.

Activity Using the information that you find, write a paragraph describing why geothermal power plants are located where they are.

Geothermal Power Plants

Geothermal energy also can be used to generate electricity, as shown in **Figure 25.** Where magma is close to the surface, the surrounding rocks are also hot. A well is drilled and water is pumped into the ground, where it makes contact with the hot rock and changes into steam. The steam then returns to the surface, where it is used to rotate turbines that spin electric generators.

The efficiency of geothermal power plants is about 16 percent. Although geothermal power plants can release some gases containing sulfur compounds, pumping the water created by the condensed steam back into Earth can help reduce this pollution. However, the use of geothermal energy is limited to areas where magma is relatively close to the surface.

Figure 25 A geothermal power plant converts geothermal energy to electrical energy.

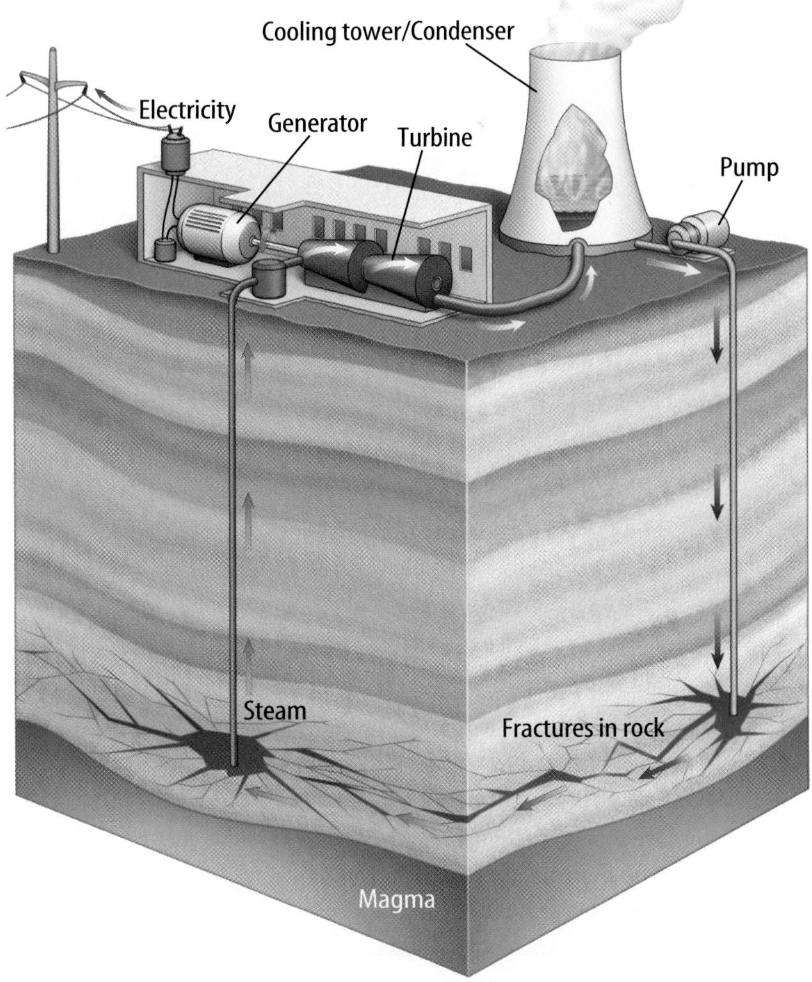

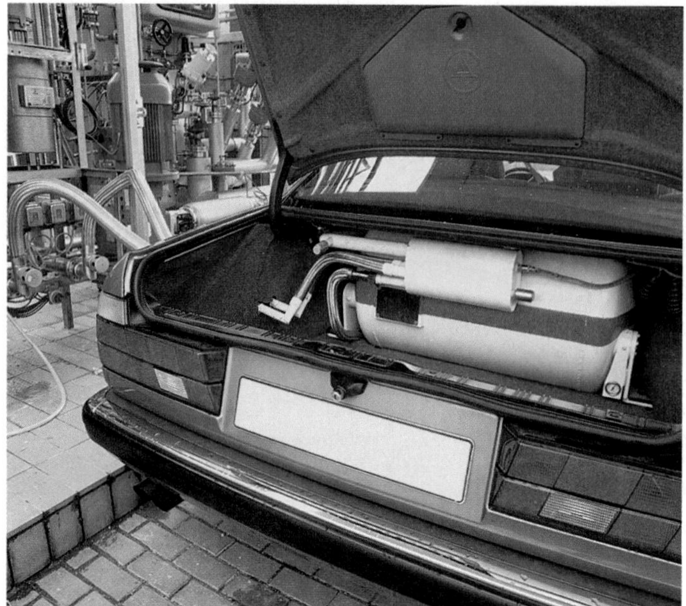

Figure 26 Hydrogen may one day replace gasoline as a fuel for automobiles. Burning hydrogen produces water vapor, instead of carbon dioxide.

Alternative Fuels

The use of fossil fuels would be greatly reduced if cars could run on other fuels or sources of energy. For example, cars have been developed that use electrical energy supplied by batteries as a power source. Hybrid cars use both electric motors and gasoline engines. Hydrogen gas is another possible alternative fuel. It produces only water vapor when it burns and creates no pollution. **Figure 26** shows a car that is equipped to use hydrogen as fuel.

Biomass Fuels Could any other materials be used to heat water and produce electricity like fossil fuels and nuclear fission? Biomass can be burned in the presence of oxygen to convert the stored chemical energy to thermal energy. **Biomass** is renewable organic matter, such as wood, sugarcane fibers, rice hulls, and animal manure. Converting biomass is probably the oldest use of natural resources for meeting human energy needs.

section 3 review

Summary

Energy Options

- The development of alternative energy sources can help reduce the use of fossil fuels.

Solar Energy

- Photovoltaic cells, or solar cells, convert radiant energy from the Sun into electrical energy.
- Producing large amounts of energy from solar cells is more expensive than using fossil fuels.

Other Renewable Energy Sources

- Hydroelectric power plants convert the potential energy in water to electrical energy.
- Tidal energy, wind energy, and geothermal energy can be converted into electrical energy, but are useable only in certain locations.
- Alternative fuels such as hydrogen could be used to power cars, and biomass can be burned to provide heat.

Self Check

1. **Explain** the need to develop and use alternative energy sources.
2. **Describe** three ways that solar energy can be used.
3. **Explain** how the generation of electricity by hydroelectric, tidal, and wind sources are similar to each other.
4. **Explain** why geothermal energy is unlikely to become a major energy source.
5. **Think Critically** What single energy source do most energy alternatives depend on, either directly or indirectly?

Applying Math

6. **Use Percentages** A house uses solar cells that generate 6.0 kW of electrical power to supply some of its energy needs. If the solar panels supply the house with 40 percent of the power it needs, how much power does the house use?

More Section Review gpescience.com

Solar Heating

Energy from the Sun is absorbed by Earth and makes its temperature warmer. In a similar way, solar energy also is absorbed by solar collectors to heat water and buildings.

Real-World Problem

Does the rate at which an object absorbs solar energy depend on the color of the object?

Goals

- **Demonstrate** solar heating.
- **Compare** the effectiveness of heating items of different colors.
- **Graph** your results.

Materials

small cardboard boxes
black, white, and colored paper
tape or glue
thermometer
watch with a second hand

Safety Precautions

Procedure

1. Complete the safety form.
2. Cover at least three small boxes with colored paper. The colors should include black and white as well as at least one other color.
3. Copy the data table into your Science Journal. Replace *Other color* with the color you use.
4. Place the three objects on a windowsill or other sunny spot and note the starting time.
5. **Measure and record** the temperature inside each box at 2-min intervals for at least 10 min.

Temperature Due to Different Colors

Color	2 min	4 min	6 min	8 min	10 min
Black					
White	Do not write in this book.				
Other color					

Conclude and Apply

1. **Graph** your data using a line graph.
2. **Describe** the shapes of the lines on your graph. What color heated up the fastest? Which heated up the slowest?
3. **Explain** why the colored boxes heated at different rates.
4. **Infer** Suppose you wanted to heat a tub of water using solar energy. Based on the results of this activity, what color would you want the tub to be? Explain.
5. **Explain** why you might want to wear a white or light-colored shirt on a hot, sunny, summer day.

Communicating Your Data

Compare your results with those of other students in your class. Discuss any differences found in your graphs, particularly if different colors were used by different groups.

Use the Internet

How much does energy really co$t?

Goals

- **Identify** three energy sources that people use.
- **Determine** the cost of the energy produced by each source.
- **Describe** the environmental impact of each source.

Data Source

Internet Lab

Visit gpescience.com for more information about energy sources and for data collected by other students.

Real-World Problem

You know that it costs money to produce energy. Using energy also can have an impact on the environment. For example, coal costs less than some other fuels. However, combustion is a chemical reaction that can produce pollutants, and burning coal produces more pollution than burning other fossil fuels, such as natural gas. Even energy sources, such as hydroelectric power, that don't produce pollution can have an impact on the environment. What are some of the environmental impacts of the evergy sources used in the United States? How can these environmental impacts be compared to the cost of the energy produced?

Make a Plan

1. **Research** the various sources of energy used in different areas of the United States and choose three energy sources to investigate.
2. **Research** the cost of the consumer of 1 kWh of electrical energy generated by energy sources you have choosen.
3. **Determine** the effects each of the three energy sources has on the environment.

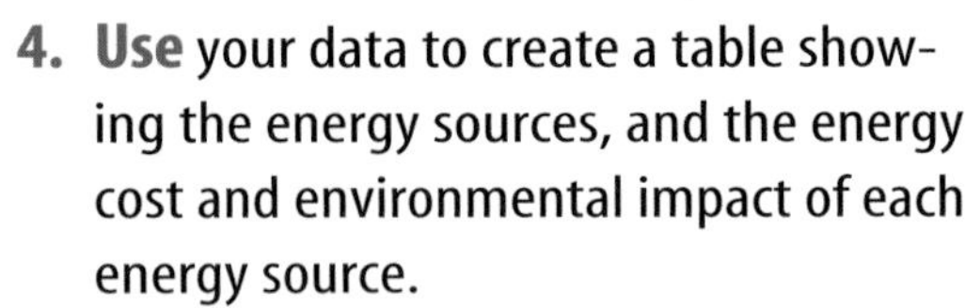

4. **Use** your data to create a table showing the energy sources, and the energy cost and environmental impact of each energy source.
5. **Decide** how you will evaluate the environmental impact of each of your energy sources.
6. **Write** a summary describing which of your three energy sources is the most cost-effective for producing energy. Consider the cost of the energy and your evaluation of the environmental impact in making your decision. Use information from your research to support your conclusions.

Follow Your Plan

1. Make sure your teacher approves your plan before you start.
2. **Record** your data in your Science Journal.

Analyze Your Data

1. Of the energy sources you investigated, which is the most expensive to use? The least expensive?
2. Which energy source do you think has the most impact on the environment? The least impact?

Energy Sources

Energy Source	Cost per kWh	Environmental Impacts
Energy source 1		
Energy source 2	Do not write in this book.	
Energy source 3		

Conclude and Apply

1. **Explain** Of the energy sources you investigated, which is the least expensive energy source? Which is the best choice to use? Why?
2. **Explain** Of the energy sources you investigated, how did the environmental impact of using that energy source influence your choice of the best energy solution?
3. **Evaluate** Which data support your decision?

Find this lab using the link below. Post your data in the table provided. **Compare** your data to those of other students.

Internet Lab
gpescience.com

Reacting to Nuclear Energy

Most people agree that thanks to energy sources, we have many things that make our quality of life better. Energy runs our cars, lights our homes, and powers our appliances. What many people don't agree on is where that energy should come from.

Almost all of the world's electric energy is produced by thermal power plants. Most of these plants burn fossil fuels—such as coal, oil, and natural gas—to produce energy. Nuclear energy is produced by fission, which is the splitting of an atom's nucleus. People in favor of nuclear energy argue that, unlike fossil fuels, nuclear energy is nonpolluting.

Opponents counter, though, that the poisonous radioactive waste created in nuclear reactors qualifies as pollution—and will be lingering in the ground and water for hundreds of thousands of years.

Supporters of nuclear energy also cite the spectacular efficiency of nuclear energy—one metric ton of nuclear fuel produces the same amount of energy as up to 3 million tons of coal. Opponents point out that uranium is in very short supply and, like fossil fuels, is likely to run out in the next 100 years.

Opponents worry that as utilities come under less government regulation, safety standards will be ignored in the interest of profit. This could result in more accidents like the one that occurred at Chernobyl in the Ukraine. There, an explosion in the reactor core released radiation over a wide area.

Supporters counter that it will never be in the best interests of those running nuclear plants to relax safety standards since those safety standards are the best safeguard of workers' health. They cite the overall good safety record of nuclear power plants.

This site at Yucca Mountain, Nevada, is the location of a proposed high-level nuclear waste storage facility. Here, radioactive materials would be buried for tens of thousands of years.

Debate **Form three teams and have each team defend one of the views presented here. If you need more information, go to the link at the right. "Debrief" after the debate. Did the arguments change your understanding of the issues?**

TIME
For more information, visit gpescience.com

Reviewing Main Ideas

Section 1 Fossil Fuels

1. Fossil fuels include oil, natural gas, and coal. They formed from the buried remains of plants and animals.
2. Fossil fuels can be burned to supply energy for generating electricity. Petroleum also is used to make plastics and synthetic fabrics.

3. Fossil fuels are nonrenewable energy resources. They can be replaced, but it takes millions of years.

Section 2 Nuclear Energy

1. A nuclear reactor transforms the nuclear energy from a controlled nuclear chain reaction into electrical energy.
2. Nuclear wastes must be contained and disposed of carefully so radiation from nuclear decay will not leak into the environment. These low-level nuclear wastes are buried to protect living organisms.

3. Nuclear fusion releases energy when two nuclei combine. Fusion only occurs at high temperatures that are difficult to produce in a laboratory.

Section 3 Renewable Energy Sources

1. Alternative energy resources can be used to supplement or replace nonrenewable energy resources.
2. Other sources of energy for generating electricity include hydroelectricity and solar, wind, tidal, and geothermal energy. Each source has its advantages and disadvantages. Also, some of these sources can damage the environment.
3. Although some alternative energy sources produce less pollution than fossil fuels do and are renewable, their use often is limited to the regions where the energy source is available. For example, tides can be used to generate electricity in coastal regions only.

4. It may be possible to use hydrogen as a fuel for automobiles and other vehicles. Biomass, such as wood and other renewable organic matter, has been used as fuel for thousands of years.

FOLDABLES™ Use the Foldable that you made at the beginning of the chapter to help you review energy sources.

chapter 16 Review

Using Vocabulary

biomass p. 506	nuclear reactor p. 494
fossil fuel p. 487	nuclear waste p. 498
geothermal energy p. 505	petroleum p. 489
hydroelectricity p. 503	photovoltaic cell p. 501
nonrenewable resource p. 493	renewable resource p. 501

Complete each statement using a term from the vocabulary list above.

1. A(n) ________ uses the Sun to generate electricity.
2. ________ makes use of thermal energy inside the Earth.
3. Energy produced by the rise and fall of ocean levels is a(n) ________.
4. ________ includes the following: oil, natural gas, and coal.
5. Fossil fuels are a(n) ________ because they are being used up faster than they are being made.
6. A special caution should be taken in disposing of ________.

Checking Concepts

Choose the word or phrase that best answers the question.

7. Why are fossil fuels considered to be nonrenewable resources?
   A) They are no longer being produced.
   B) They are in short supply.
   C) They are not being produced as fast as they're being used.
   D) They contain hydrocarbons.

8. To generate electricity, nuclear power plants produce which of the following?
   A) steam **C)** plutonium
   B) carbon dioxide **D)** water

9. What is a major disadvantage of using nuclear fusion reactors?
   A) use of hydrogen as fuel
   B) less radioactivity produced
   C) extremely high temperatures required
   D) use of only small nuclei

10. How are spent nuclear fuel rods usually disposed of?
    A) burying them in a community landfill
    B) storing them in a deep pool of water
    C) burying them at the reactor site
    D) releasing them into the air

Use the graph below to answer question 11.

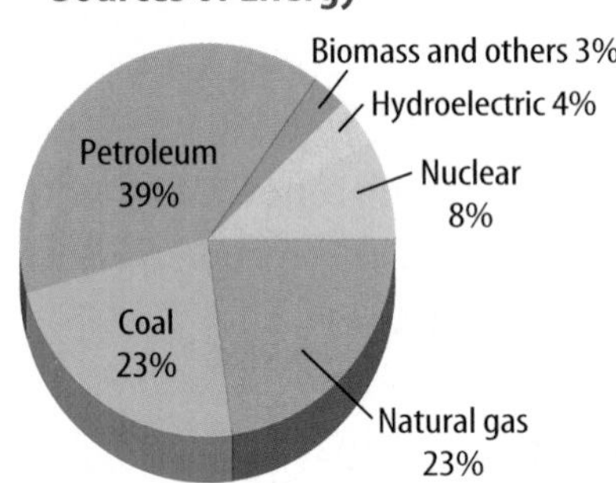

11. How much energy in the United States comes from burning fossil fuels?
    A) 85% **C)** 65%
    B) 35% **D)** 25%

12. What do hydrocarbons react with when fossil fuels are burned?
    A) carbon dioxide **C)** oxygen
    B) carbon monoxide **D)** water

13. Which of the following is NOT a source of nuclear waste?
    A) products of fission reactors
    B) materials with short half-lives
    C) some medical and industrial products
    D) products of coal-burning power plants

14. Which of the following is the source of almost all of Earth's energy resources?
    A) plants **C)** magma
    B) the Sun **D)** fossil fuels

Interpreting Graphics

15. Copy and complete the table below describing possible effects of changes in the normal operation of a nuclear reactor.

Reactor Problems

Cause	Effect
The cooling water is released hot.	Do not write in this book.
The control rods are removed.	
	The reactor core overheats and meltdown occurs.

16. Copy and complete this concept map.

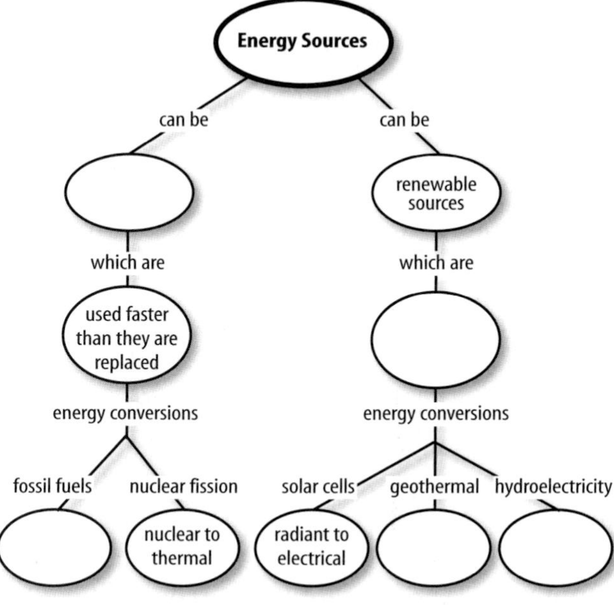

Thinking Critically

17. **Infer** why alternative energy resources aren't more widely used.

18. **Infer** whether fossil fuels should be conserved if renewable energy sources are being developed.

19. **Infer** Suppose new reserves of fossil fuels were found and a way to burn these fuels was developed that did not release pollutants and carbon dioxide into the atmosphere. Should fossil fuels still be conserved? Explain

20. **Explain** why coal is considered a nonrenewable energy source, but biomass, such as wood, is considered a renewable energy source.

21. **Make a table** listing two advantages and two disadvantages for each of the following energy sources: fossil fuels, hydroelectricity, wind turbines, nuclear fission, solar cells, and geothermal energy.

Applying Math

22. **Convert Units** Crude oil is sold on the world market in units called barrels. A barrel of crude oil contains 42 gallons. If 1 gallon is 3.8 liters, how many liters are there in a barrel of crude oil?

Use the table below to answer question 25.

High-Production Coal Mines

Coal Mine	Metric tons/year
North Antelope Rochelle	6.78×10^7
Black Thunder	6.13×10^7

23. **Use Percentages** Nine of the top coal producing mines are located in Wyoming. Production information on two of the mines is in the table above. A total of about 1.02×10^9 metric tons is produced per year in the United States. What percentage do these two coal mines contribute to the total yearly coal production in the U.S.?

Record your answers on the answer sheet provided by your teacher or on a sheet of paper.

Multiple Choice

Use the graph below to answer questions 1 and 2.

Sources of Electricity

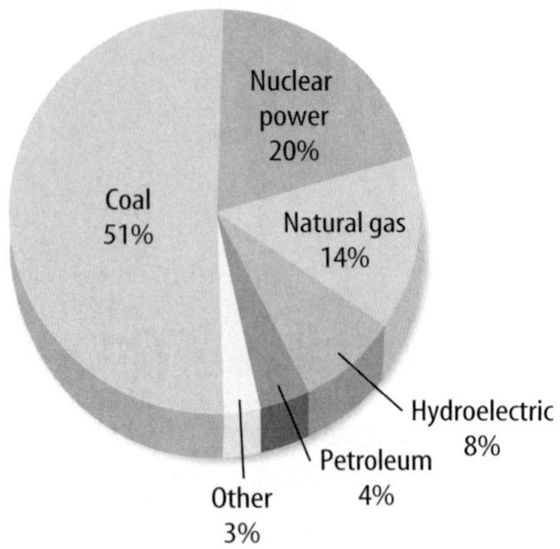

1. The graph above shows the percentage of electricity generated in the United States that comes from various energy sources. According to the graph, about what percentage comes from fossil fuels?

   A. 51%

   B. 55%

   C. 65%

   D. 69%

2. The graph shows that approximately what percentage of electricity comes from nonrenewable energy sources?

   A. 97%

   B. 89%

   C. 69%

   D. 55%

3. Which of the following is a typical efficiency for a solar cell?

   A. 10%

   B. 50%

   C. 75%

   D. 95%

4. Which of the following forms only from ancient plant material, not from ancient animal remains?

   A. coal

   B. crude oil

   C. natural gas

   D. petroleum

Use the table below to answer questions 5 and 6.

Efficiency of Fossil-fuel Conversion	
Process	**Efficiency (%)**
Chemical to thermal energy	60
Conversion of water to steam	90
Steam spins turbine	75
Turbine spins electric generator	95
Transmission through power lines	90

5. The table above shows the efficiency of different steps in the conversion of fossil fuels to electricity at a power plant. According to the table, what is the efficiency for converting chemical energy in the fossil fuels to thermal energy, and then converting water to steam?

   A. 40%

   B. 54%

   C. 25%

   D. 90%

Gridded Response

6. What is the overall efficiency shown in the table for converting chemical energy in fossil fuels to electricity?

Short Response

Use the illustration below to answer question 7.

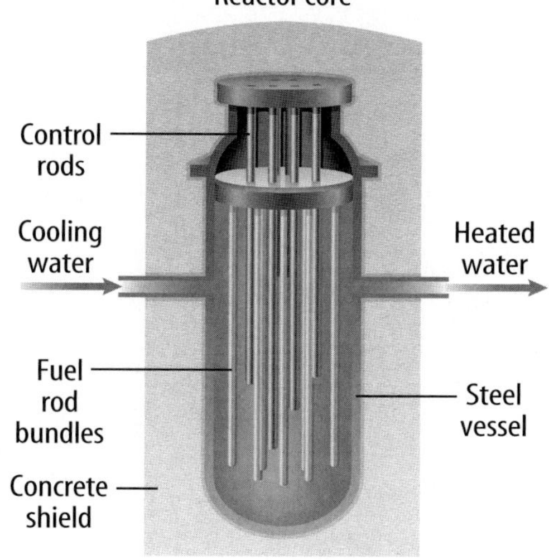

7. Describe the purpose of the control rods and explain how their placement in the reactor affects the nuclear chain reaction.

8. Fusion is the most concentrated energy source known. Why, then, is it not used at nuclear plants to make electricity?

Extended Response

Use the illustration below to answer question 9.

9. The illustration above shows a nuclear power plant that generates electricity using the energy released in nuclear fission of uranium-235. Draw a sketch showing this fission process. Describe your sketch and explain how the process results in a chain reaction.

10. Describe two advantages and two disadvantages of using fossil fuels as a source of energy.

Test-Taking Tip

Determine the Information Needed Concentrate on what the question is asking about a table, instead of all the information in the table.

Question 5 Read the question carefully to determine which rows in the table contain the information needed to answer the question.

chapter 17

Weather and Climate

chapter preview

Flash Flood!

When cool, dry air meets warm, moist air dramatic things can happen. Many desert environments are familiar with flash floods that can result from such an event. Dry river beds can suddenly become raging torrents of water, rocks, and mud.

Science Journal In your Science Journal, describe some of your observations of severe weather. Hypothesize what might cause these weather events.

Start-Up Activities

Atmospheric Pressure

Changes in atmospheric pressure are involved in producing winds and weather. You may not be aware of how much pressure the atmosphere exerts, but you can see it in this Lab.

1. Fill a glass to the brim with water.
2. Place a piece of thick paper or cardboard on top.
3. Hold the paper or cardboard securely to the brim of the glass. Turn the glass upside down.
4. Release your hand from the paper
5. **Think Critically** What keeps the paper or cardboard against the brim of the glass and the water from flowing out?

Preview this chapter's content and activities at gpescience.com

FOLDABLES™ Study Organizer

Weather and Climate Make the following Foldable to compare and contrast the characteristics of weather and climate.

STEP 1 **Fold** one sheet of paper lengthwise.

STEP 2 **Fold** into thirds.

STEP 3 **Unfold and draw** overlapping ovals. **Cut** the top sheet along the folds.

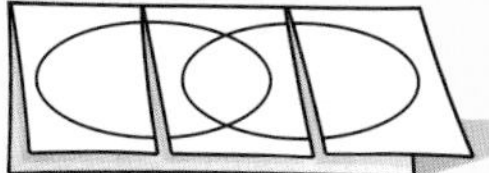

STEP 4 **Label** the ovals as shown.

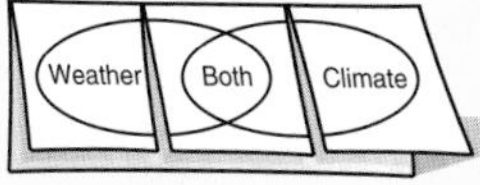

Constructing a Venn Diagram As you read the chapter, list the characteristics unique to weather under the left tab, those unique to climate under the right tab, and those elements common to both under the middle tab.

section 1

Earth's Atmosphere

Reading Guide

What You'll Learn

- **Describe** the composition of the atmosphere.
- **Explain** how the atmosphere is heated and include the role of land surface and water.
- **Describe** Earth's system of water cycling.

Why It's Important

Heat and water are essential for life on Earth.

Review Vocabulary

nucleus: a central point about which concentration or accretion takes place

New Vocabulary

- troposphere
- temperature inversion
- greenhouse effect
- latent heat

Atmospheric Composition

You probably never think about the air you breathe. Your body uses only oxygen, but air is a mixture of gases. Nitrogen is the largest component with 78 percent and oxygen is next with 21 percent by volume. Most of the remaining one percent is the inactive gas argon and water vapor in varying amounts. The small remaining portion is a mixture of trace gases, so called because they are present in such small, barely detectable amounts. For example, carbon dioxide makes up only about 0.03 percent, and the other trace gases—methane, nitrous oxide, and ozone—together make up less than 0.001 percent of the atmosphere. Still, these trace gases are critical for life on Earth.

Reading Check *What are the trace gases?*

Biological Processes Living organisms produce trace gases, except ozone. Cellular respiration by organisms produces water vapor and carbon dioxide (CO_2). Photosynthetic organisms use carbon dioxide and produce almost all the oxygen in the atmosphere. Organisms alter atmospheric CO_2 concentration at various latitudes throughout the year, as shown in **Figure 1.** Microorganisms in swamps, rice paddies, and soil produce nitrogen and methane. The microorganisms in the digestive tracts of animals such as termites, cows, and sheep produce methane.

Figure 1 In northern latitudes, carbon dioxide concentration increases during summer and decreases during winter.

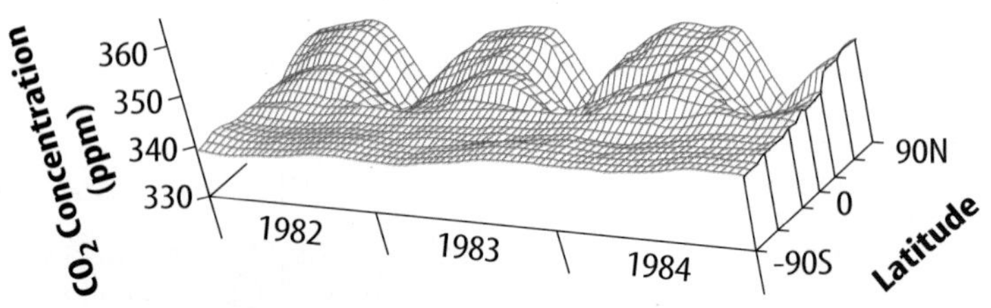

Formation of the Atmosphere Earth's early atmosphere contained mostly hydrogen and helium. These gases were lost and were replaced by gases from volcanic eruptions, including water vapor and CO_2. Oxygen from photosynthetic marine organisms accumulated in the atmosphere and intense solar radiation converted some of it into ozone. The ozone layer shielded Earth from harmful ultraviolet rays. This allowed photosynthetic organisms to emerge on land, where they produced more oxygen.

Some scientists think it took millions of years for our atmosphere to reach its current state—a delicate equilibrium between processes producing and destroying atmospheric gases. Unfortunately, human activities might be threatening this equilibrium.

Visualizing Convection

Procedure

1. Place an **ice cube** in the center of a **beaker.**
2. Add room-temperature **water** until the beaker is three-fourths full.
3. Place one drop of **food coloring** on the surface of the water.
4. Observe what happens.

Analysis

1. Describe what happened to the color.
2. Explain why this occurred.

Atmospheric Structure

Earth's atmosphere extends more than 1,000 km above Earth's surface. Most of our weather takes place within the **troposphere,** a layer extending 30 km above Earth's surface. In this layer, as shown in **Figure 2,** temperature normally decreases with height. Sometimes, however, temperature increases with height. This is called a **temperature inversion.** When this happens, the air is very stable—it resists the rising motion needed to form clouds or to disperse air pollution.

The Stratosphere Above the troposphere is the stratosphere, which is extremely dry and rich in ozone. Here, temperature always increases with height, creating a permanent temperature inversion. The place where this temperature inversion begins is called the tropopause. It acts like a lid that keeps air in the troposphere from rising into the stratosphere. The uppermost layers, the mesophere and thermosphere, are very low in density and do not affect weather.

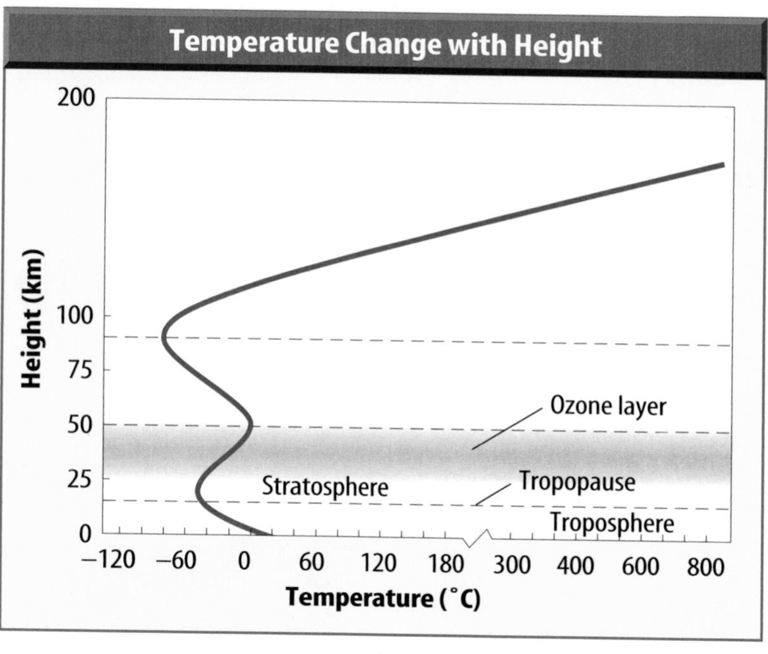

Figure 2 Temperature decreases with height in the troposphere up to the tropopause. Above the tropopause, temperature increases with height in the stratosphere.
Identify *The ozone layer is found in which atmospheric layer?*

Heating the Atmosphere

The energy that heats the atmosphere ultimately comes from the Sun. In the stratosphere, solar rays split oxygen molecules into single atoms. The oxygen atoms then react with other oxygen molecules, forming ozone. Ozone absorbs nearly all of the Sun's ultraviolet radiation. This absorption by ozone is why temperature increases with height in the stratosphere.

The remaining solar rays pass to Earth's surface where they are either absorbed or reflected back to the atmosphere. As Earth's surface is heated, it emits long-wave, infrared radiation. Trace gases, such as carbon dioxide and water vapor, absorb long wavelengths and re-emit some of them back to Earth's surface. The term **greenhouse effect** refers to this re-emission of infrared radiation back to Earth's surface, as shown in **Figure 3.**

Oxygen and nitrogen absorb little radiation and contribute little to atmospheric heating. Although water vapor and trace gases make up less than 0.03 percent of the atmosphere, they are strong absorbers and heat the atmosphere the most.

Conduction, convection, and latent heat also contribute to heating Earth's atmosphere. **Latent heat** is heat energy released or absorbed during the phase changes of water, such as evaporating water or melting snow. It is released to the atmosphere when water vapor condenses as clouds.

Figure 3 Trace gases absorb infrared radiation and are heated. These gases then re-emit infrared in all directions, sending part of it back to Earth's surface.
Infer *How does this compare to being covered by a blanket at night?*

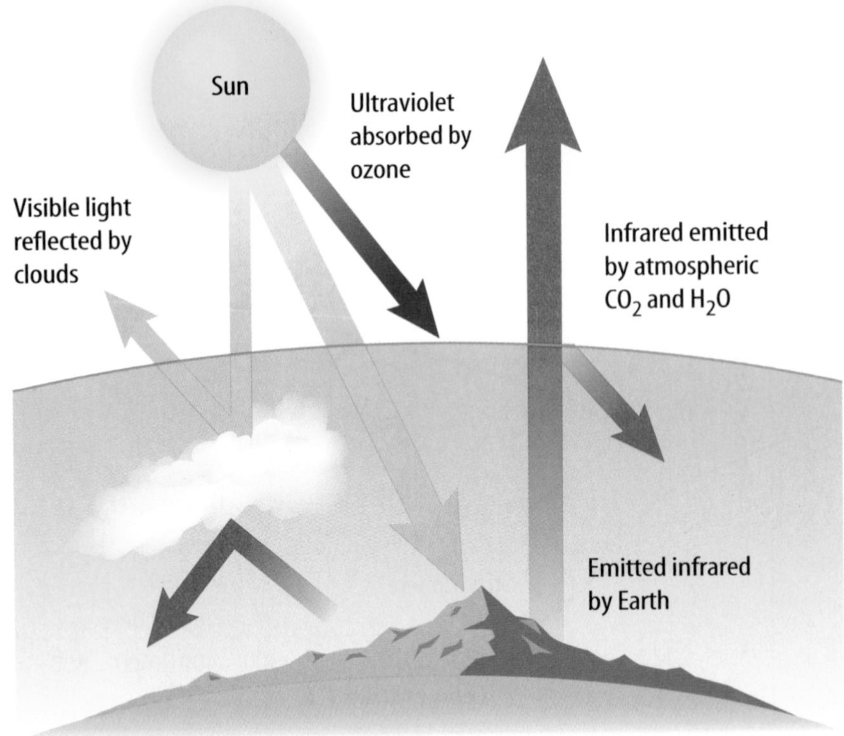

A Varied Surface

Earth's surface is not uniform and therefore, heats the atmosphere unevenly. Snow, ice, water, vegetation, and bare soil reflect different amounts of solar radiation back to space and heat at different rates when they absorb radiation. For example, dry land heats rapidly and emits much radiation to the atmosphere. In contrast, water temperature changes slowly and stores heat, releasing it at a later time. This uneven pattern of surface heating gives rise to pressure differences and wind.

Water in the Atmosphere

Uneven heating has another effect. It produces currents of air that carry water vapor aloft and form clouds. Air generally rises over warm surfaces and sinks over cold surfaces. Many birds and hang gliders soar on these warm currents, called thermals.

As air rises, it expands and cools. To form clouds, moist air must rise high enough to cool to its dew point. At this temperature, air is saturated and water vapor condenses to form cloud droplets. Small particles in the air, called condensation nuclei, trigger this process. When present in high quantities, such as in dust or polluted city air, these nuclei can trigger condensation in unsaturated air. This is how smog forms.

Reading Check *Explain how smog forms.*

Precipitation Cloud formation is the first step in the precipitation process. Two basic cloud types are the puffy cumulus type and the flat, elongated, stratus type. Cumulus clouds form from rising air parcels. If they produce rain, it is usually only brief showers. Stratus-type clouds form mainly when layers of air rise gently. They usually produce drizzle or long-lasting rain. Many in-between forms of clouds exist, depending on how stable the air is and how high in the atmosphere the clouds form. You can see some of these types in **Figure 4.** One type you might recognize is the cumulonimbus, or thunderhead. This cloud forms from unstable air and usually brings intense rain.

Cloud droplets are so small that they might be kept aloft by turbulence or evaporate before reaching Earth. For precipitation to occur, droplets must grow large. Growth can occur when droplets collide and combine. This is called warm rain. Droplets grow faster when they combine with ice crystals high in the atmosphere. This is called cold rain.

Figure 4 Clouds affect atmospheric heating by absorbing or blocking solar radiation and trapping Earth's radiation. Cloud cover also helps to reduce daytime temperature and increase nighttime temperature.

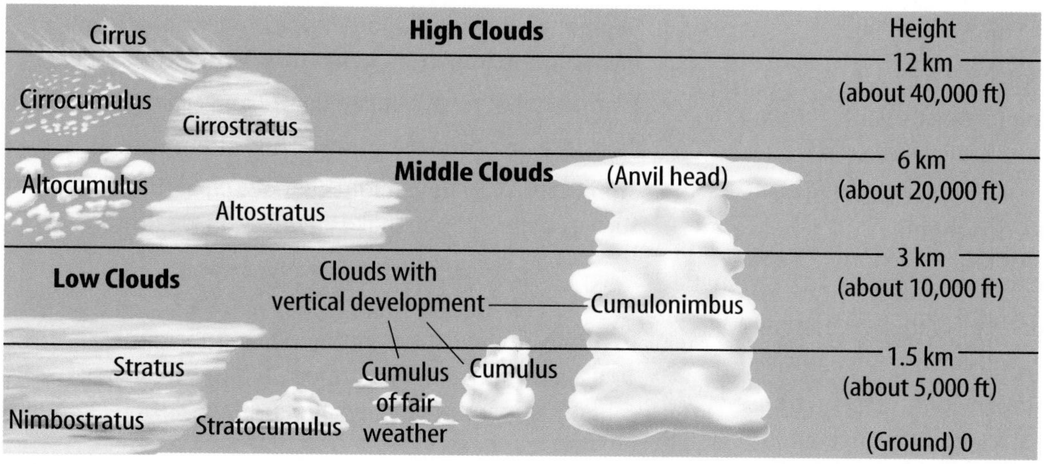

Cirrus
High Clouds
Height
12 km
(about 40,000 ft)
Cirrocumulus
Cirrostratus
Altocumulus
Middle Clouds
(Anvil head)
6 km
(about 20,000 ft)
Altostratus
Low Clouds
Clouds with vertical development
Cumulonimbus
3 km
(about 10,000 ft)
Stratus
Cumulus of fair weather
Cumulus
1.5 km
(about 5,000 ft)
Nimbostratus
Stratocumulus
(Ground) 0

Figure 5 Earth's continual cycling of water strongly affects weather and climate.

Plants Transpiration is the loss of water through pores in the leaves of plants. More than 90 percent of the water that enters a plant returns to the atmosphere through the process of transpiration. Research how humidity and air temperature affect transpiration rates in plants. Share your findings with your class.

Global Water Cycle

Precipitation, runoff, storage, and evaporation make up the global water cycle, shown in **Figure 5.** Plants are an important part of the water cycle. They affect absorption and runoff and return water to the air by evaporation from their surfaces.

People affect the water cycle in many ways. They use groundwater for irrigation or pump it from wells. They replace forests with agricultural fields and pave land to build cities. Many of these changes have reduced water quality and resources. Conserving water resources requires careful planning.

section 1 review

Summary

Atmospheric Structure

- The stratosphere and the troposphere are two lower layers of Earth's atmosphere. Most weather takes place in the troposphere.

Atmospheric Heating

- Most solar radiation first heats Earth's surface, which then heats the atmosphere.
- Characteristics of the land surface greatly influence atmospheric heating.

Water in the Atmosphere

- Cloud formation generally requires moist air, rising and cooling, and condensation nuclei.
- Water is cycled through Earth's system by precipitation, runoff, storage, and evaporation.

Self Check

1. **Describe** how temperature of the atmosphere changes with height and explain why.
2. **Explain** the greenhouse effect.
3. **Explain** why small changes in the amount of trace gasses are so important in heating Earth's atmosphere.
4. **Identify** what must happen before rain can occur.
5. **Compare and contrast** cumulus and stratus clouds.
6. **Think Critically** How might changes on Earth's surface, such as deforestation, have an effect on weather?

Applying Math

7. **Use Percentages** If the southern hemisphere contains 10 percent land and the northern hemisphere contains roughly 40 percent land, what percent of Earth is land?

More Section Review gpescience.com

Visualizing a Temperature Inversion

Normally temperature decreases with increasing altitude in the troposphere. Sometimes near the ground, a temperature inversion occurs and air becomes very stable and resists rising. This can result in fog or smog in cities. Will a liquid behave in the same manner as the atmosphere?

Real-World Problem

How can you visualize what happens during a temperature inversion?

Goals

- **Make a model** that demonstrates a temperature inversion.
- **Apply** what you observe to explain what happens in the atmosphere during a temperature inversion.

Materials

10-mL beaker containing 2–3 mL water
500-mL beaker containing 250 mL water
1,000-mL beaker containing 750 mL water
food coloring
long-stem dropper
freezer
ladle
thermal mitt
hot plate

Safety Precautions

Procedure

1. Chill 750 mL of water in a 1,000-mL beaker to near freezing.
2. Add one to two drops of food coloring to the water in the 10-mL beaker and let it stand at room temperature.
3. Heat 250 mL of water in the 500-mL beaker to near boiling.
4. Hold the ladle at the surface of the chilled water. Very slowly pour all the heated water into the ladle and allow it to slowly flow out of the ladle onto the surface of the chilled water. You should have a bottom layer of cold water and a top layer of hot water.
5. Use the long-stem dropper to inject a few drops of the colored water from step 2 into the cold water in the 10-mL beaker.

Conclude and Apply

1. **Describe** what happened to the colored water.
2. **Explain** why this happened in terms of the temperatures of the water layers.
3. **Infer** how this is related to temperature inversions in the atmosphere.

Communicating Your Data

Write a brief paragraph describing your experiment and your observations. Include a labeled diagram showing the temperature layers and explain what happened to the colored water.

section 2

Weather

Reading Guide

What You'll Learn

- **Explain** what causes Earth's major wind and pressure systems.
- **Describe** typical daily weather patterns around lows and fronts.
- **Describe** the most important storm systems and forms of severe weather.

Why It's Important

Weather affects the health, life, and livelihood of everyone on Earth.

Review Vocabulary

gradient: the rate of change of a quantity with distance

New Vocabulary

- westerlies
- jet stream
- subtropical high
- weather front

Atmospheric Pressure

Although you usually are unaware of it, the atmosphere presses down on you with a pressure equivalent to one kilogram per square centimeter. This pressure is caused by gas molecules moving and colliding with each other and any surfaces they touch. Because the number of air molecules decreases as altitude increases, pressure always decreases with altitude. This is why air is said to be thinner in the upper atmosphere. The number of air molecules, including oxygen, decreases in proportion to pressure. This is why jet aircraft cabins are pressurized and why climbers can get mountain sickness at altitudes over 3,000 m.

Global Winds and Pressure Systems Weather patterns result from complex global patterns of wind and pressure. **Figure 6** shows a simplified picture of the Earth's major pressure belts that give rise to major wind belts. The most important of these are the **westerlies**—winds that blow from the west in the middle latitudes—and the trade winds, which blow from the east, in the tropics.

Two factors produce these global patterns—unequal heating between the equator and poles and the rotation of Earth. Warm air rising near the equator and sinking over the poles creates general north-south wind circulation. Earth's rotation produces an east-west deflection of this general circulation pattern.

Figure 6 In the northern hemisphere, four major pressure belts produce three major wind belts. **Infer** *How does pressure affect wind direction?*

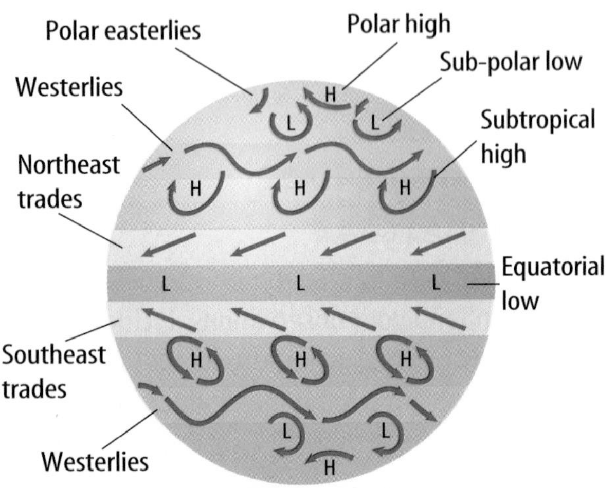

Jet Streams Imbedded in these wind systems are fast and powerful **jet streams** that control many weather processes, such as storm development. Most important for the United States is the polar front jet stream, a wind maximum in the westerlies located about 12 km above the surface. Its speeds can exceed 500 km/h. Major storm tracks follow it as it moves north and south with the seasons.

High and Low Pressure Systems

The large-scale weather systems that have the most effect on the United States are the subpolar lows, westerlies, and the subtropical highs. **Subtropical highs** are relatively stable belts of high pressure near latitudes of 30°. In contrast, sub-polar lows and the westerlies tend to meander as smaller cells of high and low pressure develop. The lows generally develop from a disturbance in the polar front jet and move eastward with the jet stream.

Specific patterns of weather are associated with high and low pressure cells because of the way air flows around them. In the northern hemisphere, winds blow counterclockwise around lows and clockwise around highs, as shown in **Figure 7.** In the southern hemisphere, the directions reverse. Lows are associated with rainfall and storms, and highs with calm winds and clear skies.

INTEGRATE Physics

Coriolis Effect Airflow around low or high pressure areas results from the net forces acting on the air. First, the pressure gradient pulls the air toward low pressure. Then an apparent force, called the Coriolis effect, deflects the air to the right in the northern hemisphere. When these forces are balanced, air flows perpendicular to lines of equal pressure. Near the surface, friction slows air and modifies its direction, turning it slightly toward low-pressure centers and slightly away from high-pressure centers. This causes air to rise in the center of lows and to sink in the center of highs.

Modeling the Coriolis Effect

Procedure

1. Put a **large, round piece of cardboard** or paper on a **turnable surface** such as a turntable.
2. Hold a **ruler** just above the diameter of the cardboard.
3. Ask someone to turn the surface while you draw a line across the turning cardboard with your **pen or pencil** against the ruler.
4. Repeat step 3 but turn the surface in the opposite direction.

Analysis

1. Compare the outlines drawn on the cardboard.
2. Explain how these outlines represent the Coriolis effect.

Try at Home

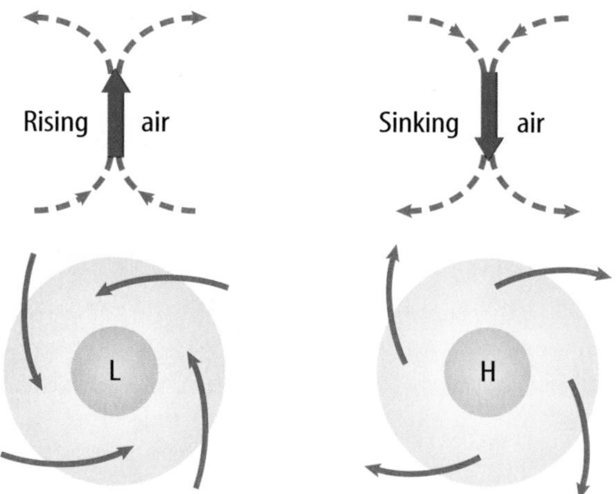

Figure 7 At Earth's surface in the northern hemisphere, air moves counterclockwise toward low pressure and clockwise from high pressure.
Infer *How does air move when it is high above Earth's surface, above each pressure system?*

Air Masses and Weather Fronts

Weather around low-pressure cells is produced by interaction of air masses—large units of air with relatively uniform moisture and temperature. These form when air remains stationary for a time, such as in regions of high pressure. The air then takes on the characteristics of the surface. Air masses can be polar or tropical and continental or maritime. Continental air originates over land. It is relatively dry and can be extremely cold or extremely warm. Maritime air masses are moist because they originate over the oceans. The maritime air masses affecting the United States come from the Atlantic Ocean, the Pacific Ocean, or the Gulf of Mexico.

Air masses interact in zones called **weather fronts,** as shown in **Figure 8,** which are associated with low pressure systems. Warm and cold fronts create different types of precipitation. In a warm front, warm air rises gently above the cold air, usually forming layered, stratus-type clouds or fog—a cloud with its base on the ground. Most layered clouds produce only drizzle or steady rain. In a cold front, cold air pushes the warm air aloft in a random and chaotic fashion forming cumulus clouds. These often produce showers and thunderstorms.

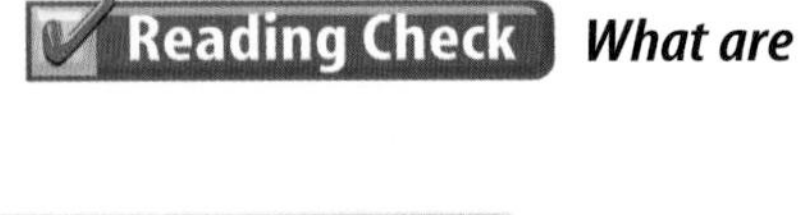

Reading Check *What are weather fronts?*

Figure 8 The symbols shown below each of these weather fronts are used by meteorologists to represent the respective fronts on weather maps.

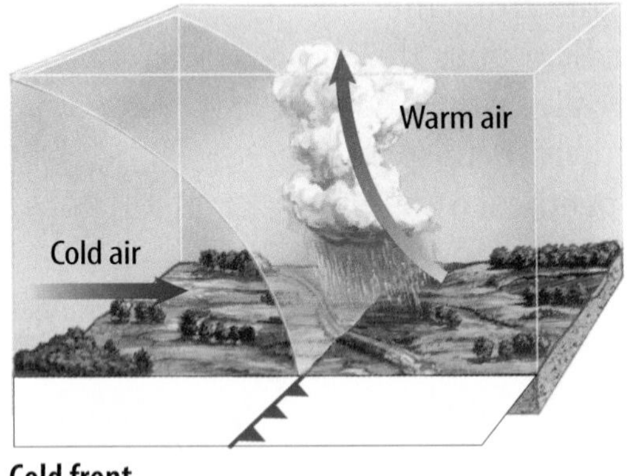

Cold front

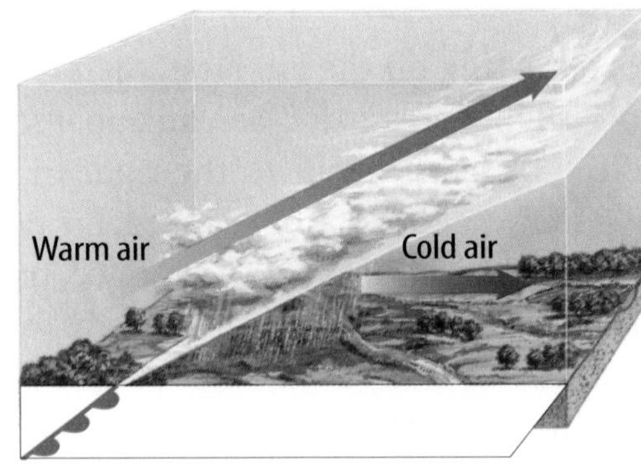

Warm front

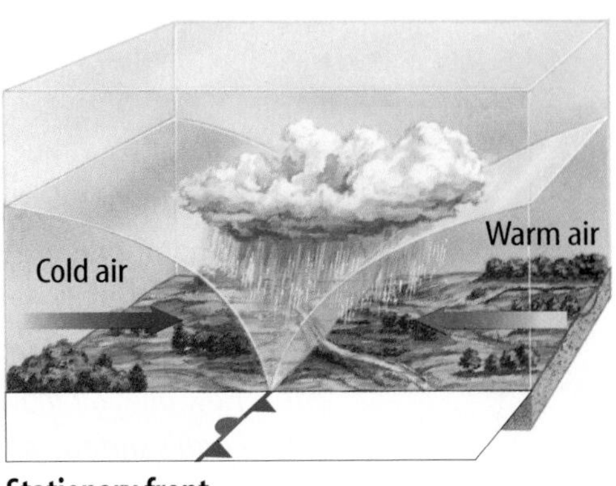

Stationary front

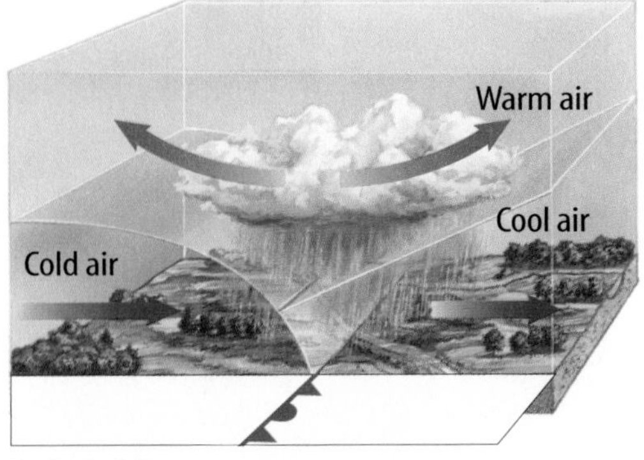

Occluded front

Severe Weather

The continental United States is prone to severe weather because of the extreme temperatures of warm and cold air masses and the availability of moisture from tropical oceans. Such conditions lead to severe thunderstorms, hurricanes, tornadoes, and violent wind storms called downbursts.

Thunderstorms Recall that cumulonimbus clouds formed from unstable air produce thunderstorms. A typical cumulonimbus cloud has ice crystals near its top, as shown in **Figure 9.** Sometimes these ice crystals act as nuclei to trigger further growth of cloud droplets, and turbulence adds layers of ice during many cycles of sinking and rising. This forms hail. Hailstones can grow to the size of softballs and can cause extensive damage to crops and structures.

Downdrafts and Squalls The force of the falling precipitation in a thunderstorm may pull with it cold air bursts from higher in the cloud. This is why the air often feels cool after a thunderstorm. This sinking current of cold air is called a downdraft. When a downdraft hits the surface with particularly strong force, it spreads out in a series of windy gusts called squalls. In arid regions, squalls produce dust storms.

Downbursts Cold air downdrafts can produce even more severe forms of weather. One example of an extreme form of wind shear is a downburst. Here, cold air descends from a thunderstorm and hits the ground. When it hits the ground, it bursts outward like the spokes on a wheel. The rapid change in wind speed and/or direction that a downburst causes can be dangerous for aircraft during both take-off and landing. The winds that result from downbursts can be as high as 260 km/h. Fortunately, automated warning systems now alert pilots to look for signs of downbursts when approaching an airport.

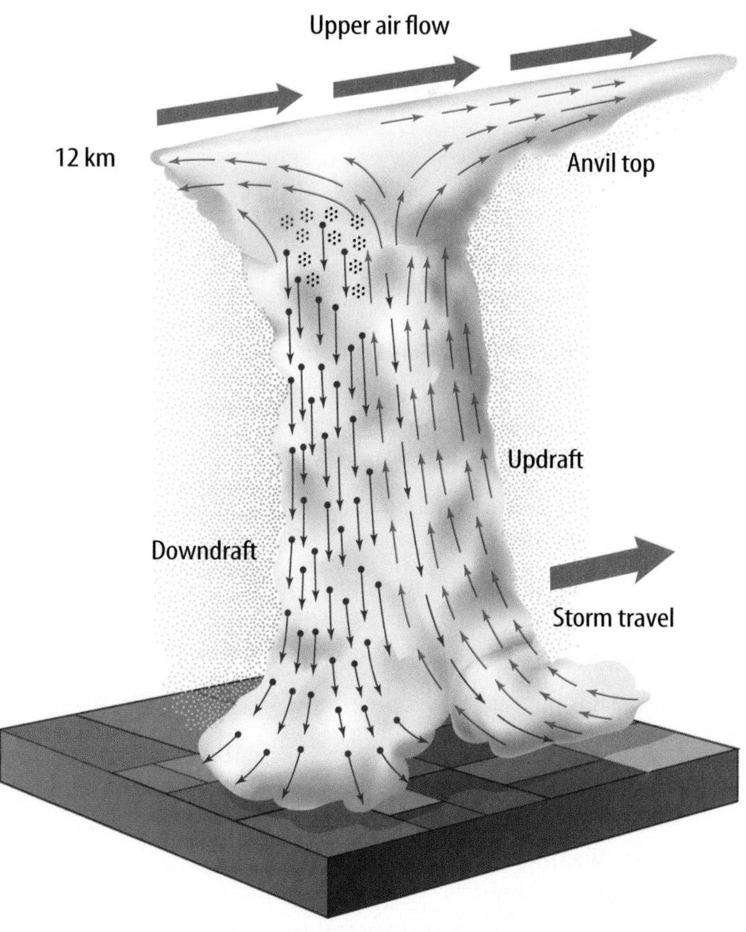

Figure 9 When hailstones fall, they are repeatedly caught in updrafts, coated with moisture, and frozen. This gives them a layered, onionlike appearance.
Infer *Why does this cumulonimbus cloud have an anvil-shaped top?*

Topic: Hurricane Tracking
Visit gpescience.com for Web links to information about hurricanes and hurricane tracking.

Activity Research six hurricanes. Make a table that lists how each was tracked as a tropical depression and then as a hurricane.

Tornadoes and Hurricanes Two types of violent wind storms that differ greatly in their origins and effects are tornadoes and hurricanes. Tornadoes are intense, short-lived, localized storms in the mid-latitudes. They originate in cumulonimbus clouds under special conditions. Typically, tornadoes that occur in the United States form when dry air from the deserts of Mexico and the southwest overrides warm, moist air from the Gulf of Mexico. This happens frequently in the Great Plains, the lower midwest, and parts of the south, as shown in **Figure 10.** In the south, they often accompany hurricanes.

A twisting, funnel-shaped tornado cloud can move across land at a speed of around 50 km/h creating a path 150 m wide and 10 km long. Intense, circular winds in the funnel can reach speeds up to 400 km/h. The extreme low pressure at the center can result in more damage than that from the wind.

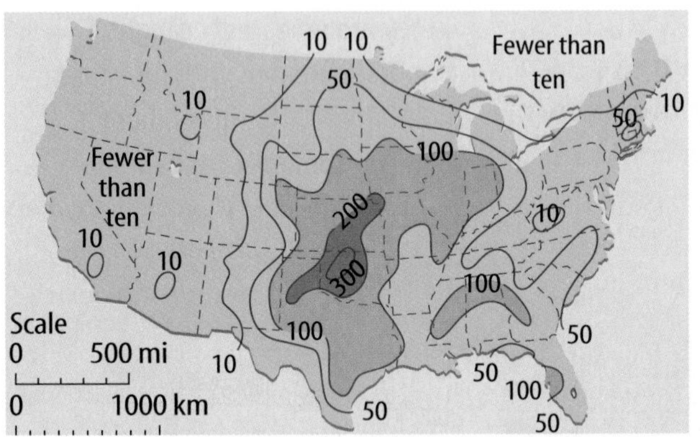

Figure 10 Texas, Oklahoma, and Kansas frequently experience tornadoes.

Hurricanes are tropical storms that cover vast areas and last for days. Those affecting the United States often form as tropical depressions over the warm waters of the southern Atlantic off the coast of Africa. When winds exceed 118 km/h, the storms are called hurricanes. A typical hurricane consists of vast cloud bands that spiral inward toward the clear center, called the eye. Scientists often fly into the eye to study the storm. Western Pacific hurricanes are called typhoons.

section 2 review

Summary

Global Wind and Pressure Systems

- Four major pressure systems produce three major wind systems in each hemisphere.

High and Low Pressure Systems

- Air flows counterclockwise around lows and clockwise around highs in the northern hemisphere.
- Air rises in the center of lows and sinks in the center of highs.

Weather Fronts and Severe Weather

- Air masses interact at weather fronts.
- Warm fronts are associated with stratus-type clouds and cold fronts with cumulus-type clouds.
- Severe weather includes hurricanes, tornadoes, and downbursts.

Self Check

1. **Explain** how Earth's rotation affects winds.
2. **Compare and contrast** tornadoes and hurricanes.
3. **Infer** the wind directions around a high in the southern hemisphere.
4. **Describe** common differences between continental air masses and maritime air masses.
5. **Compare and contrast** warm fronts and cold fronts.
6. **Think Critically** If the polar front jet stream were to move southward over the U.S., what other weather systems are likely to be affected?

Applying Math

7. **Use Percentage** A tornado watch was issued on 25 days during one year in a midwest city. What percent of the year does this represent?

More Section Review gpescience.com

section 3

Climate

Reading Guide

What You'll Learn

- **Describe** what determines climate.
- **Explain** how latitude, oceans, and other factors affect the climate of a region.
- **Classify** climate systems.
- **Describe** climate distribution over the United States.

Why It's Important

Climate affects the way you live.

Review Vocabulary

boreal: relating to northern regions

New Vocabulary

- biosphere
- continental climate
- maritime climate
- lee rain shadow
- sea breeze

Climate and Weather

What is the climate where you live? Traditionally, climate means the long-term average of weather conditions—wind, temperature, precipitation, moisture, and other aspects of weather. Climate also describes the annual variations of these conditions and their extremes.

Averages of data collected monthly over 30 years or longer are used to define climatic normals. These normals do not describe usual weather conditions of an area, but are only averages of conditions measured at one local site. For example, the conditions in a city might vary from what is measured at an airport weather station outside the city.

Climate System Climate is best considered as part of the whole Earth system. This biogeophysical system can be visualized as five spheres that interact to create the environments in which we live, as shown in **Figure 11.** The atmosphere includes the air around us. The **biosphere** is everything organic, including plants, animals, and humans. The hydrosphere is liquid water in oceans, lakes, rivers, soil, and underground. The cryosphere is frozen water in snow, ice, and glaciers. Finally, the lithosphere is the solid Earth, including its soil, rocks, and mantle.

Figure 11 All of the five spheres shown below interact. Each sphere causes changes in, and is changed by, the others.
Research *What does the prefix* cryo- *mean?*

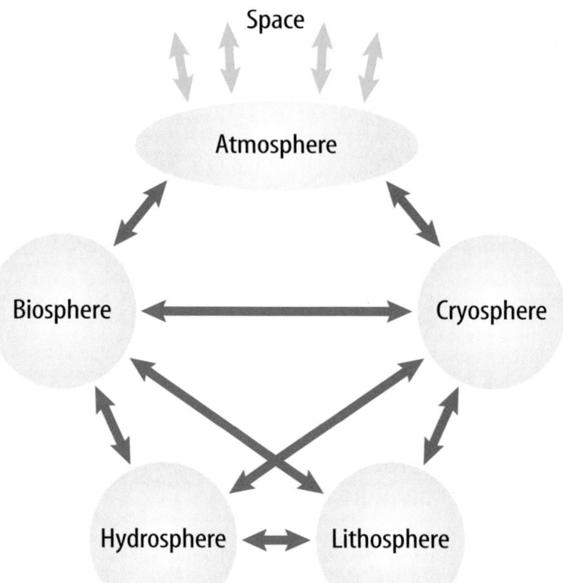

Topic: World Climates
Visit gpescience.com for Web links to data on mean annual temperature and rainfall of world cities.

Activity Use a world atlas, globe, or large classroom map to find a city corresponding to each of the eight climate divisions shown in **Figure 17.**

Sphere Interactions Gases, water, soluble materials, energy, and particulates are exchanged among these spheres. Each sphere affects all the other spheres. For example, volcanic eruptions transfer gases and particles from the lithosphere to the atmosphere. The atmosphere provides and regulates the amount of water in the hydrosphere and cryosphere, and provides water, carbon, and oxygen for the biosphere. Through winds, atmosphere causes erosion, creates soil, and absorbs and emits energy from the Sun. When you consider all these complex interactions, you can see why it's better to define climate as the average weather conditions, their variability and causes, and inter-relationships of many individual systems within the global Earth system. One example how such systems are inter-related is illustrated in **Figure 12.**

What causes climate?

Latitude is the primary factor that determines climate at a given location. The amount of radiation received from the Sun and the prevailing circulation features depend on latitude. Other factors are location near high mountains or on the east or west sides of a continent and distance from major bodies of water.

Causes of Mean Temperature The amount of solar radiation received and surface temperatures vary greatly from the equator to the poles. In the winter, the amount of solar radiation varies, because of the low angle at which it strikes Earth. As a result, temperatures decrease rapidly with increasing latitude. The high reflectance, or albedo, of snow and ice in high latitudes adds to this decrease. In summer, temperature decrease is less pronounced as sunlight strikes at a higher angle and periods of daylight are longer. The temperature patterns over the United States illustrate this, as shown in **Figure 12.** The strong temperature gradients in winter and spring help to create storms and severe weather.

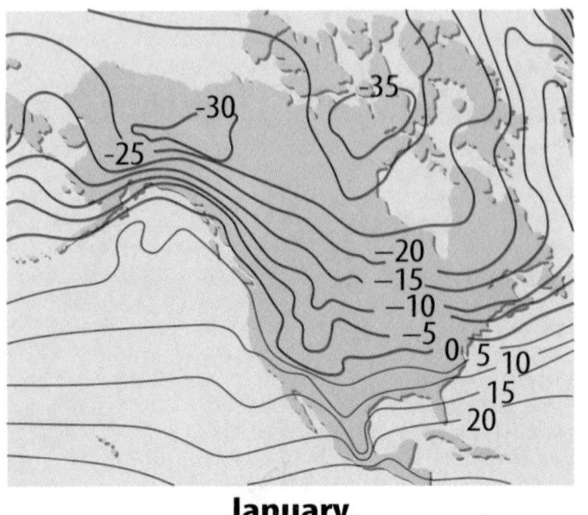

January

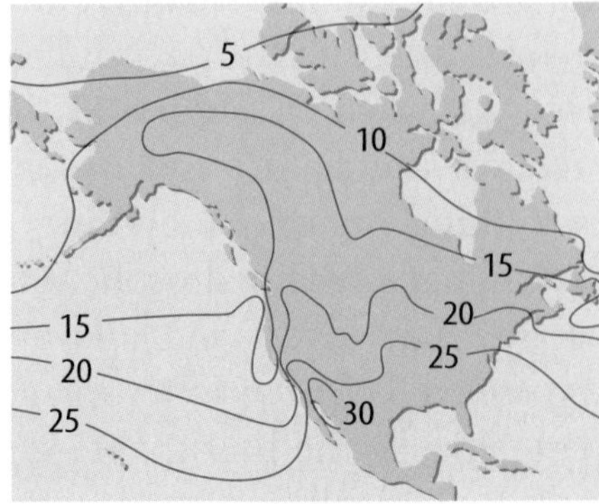

July

Figure 12 Lines drawn on a weather map that connect points of equal temperature are called isotherms.
Infer *Why are warmer temperatures shifted northward in July?*

Figure 13

The Namib desert illustrates the interconnectedness of climate, geology, plants, and animals. It is created by high pressure, wind, and cold ocean currents along the west coast of southern Africa.

Transverse dunes form perpendicular to the wind.

The desert is constantly reshaped by the wind. Transverse dunes show that the wind blows in one preferred direction. Dunes also begin when isolated plants trap sand. However, most deserts are not covered with sand dunes—they are made of stony pavement left behind when wind strips the desert surface. Winds and water remove the finer surface materials, leaving stones behind. This process is called deflation.

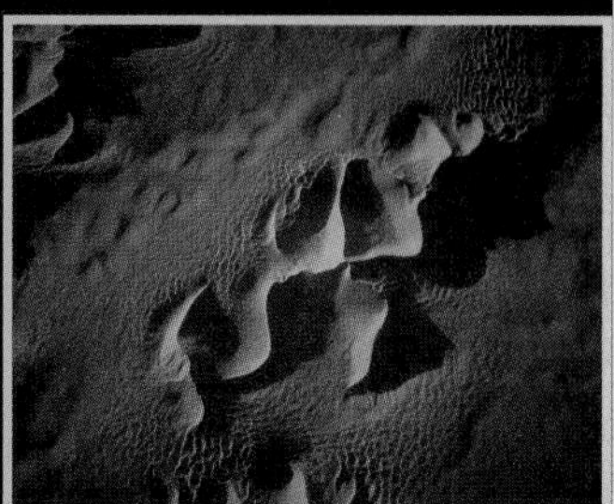

Star-shaped dunes form when winds are variable.

Dense coastal fog forms when the air above cold ocean currents interacts with moist desert air. Although deserts may appear to be dry, the air above ground contains lots of moisture. Condensation from fog produces fog-water, which drips into desert pavement and dissolves rocks, forming new sediment. It also forms small pools below the stony surface, which support plant growth. Desert plants have special adaptations such as a dense, compact form and thick, waxy coats. This helps protect them from both heat and animals.

Stone pavement is more common in a desert than dunes.

The Welwitschia plant grows in this barren land by absorbing fog-water through its elongated leaves. It is a relic of millions of years ago and is found only in the Namib.

Beetles capture fog-water by basking upside down or forming trenches at the bottom of dunes. They also survive beneath the leaves of Welwitschia plants where temperatures may be 30°C cooler than the surround bare ground.

Ocean and Land Influence Oceans and ocean currents modify the basic climate. Areas with little direct ocean influence are called **continental climates** and have steep temperature gradients. A climate with strong ocean influence is called a **maritime climate.** Maritime climates are milder—summers are cooler, winters are warmer, and daily temperatures vary less. The continental climate of Peoria, Illinois, and the maritime climate of San Francisco, California, illustrate the contrast. Although both are near 40° N latitude, the temperature difference between the warmest and coldest months is about 8° C in San Francisco and 30° C in Peoria. Maritime effects are strong enough to keep winters at arctic Spitsbergen, Norway at 78° N latitude, warmer than in International Falls, Minnesota at 48° N latitude.

Precipitation Wind and pressure patterns determine precipitation. Humid climates are associated with low-pressure areas in the tropics and the middle latitudes. Rainfall is greater near the equator because the trade winds of both hemispheres converge there, increasing the rising motion. Arid climates are common where high pressure prevails, and aridity is particularly intense in the subtropics on the eastern sides of the subtropical highs.

Superimposed upon this general pattern are differences related to location on a continent, as shown for North America in **Figure 14.** The west coast lies east of the subtropical high, which brings cold water currents and stable air. This creates the dry climates of California and the southwest. The east coast lies west of the subtropical high, where southerly winds bring warm, unstable air from the Gulf of Mexico that increases precipitation. Air masses can also influence the amount of precipitation an area receives. When continental polar air masses, which originate in the cold, dry artic regions, move across the Midwestern states in the winter, cold, dry weather results.

Another factor affecting precipitation is the prevailing winds. Because westerlies prevail in the middle latitudes, the maritime influence is stronger on the west coast. This explains why San Francisco has a strongly developed maritime climate, but Boston has a continental climate.

Figure 14 Lines drawn on a map that connect points of equal precipitation are called isohyets or isohyetal lines.
Research *What does the prefix* iso- *mean?*

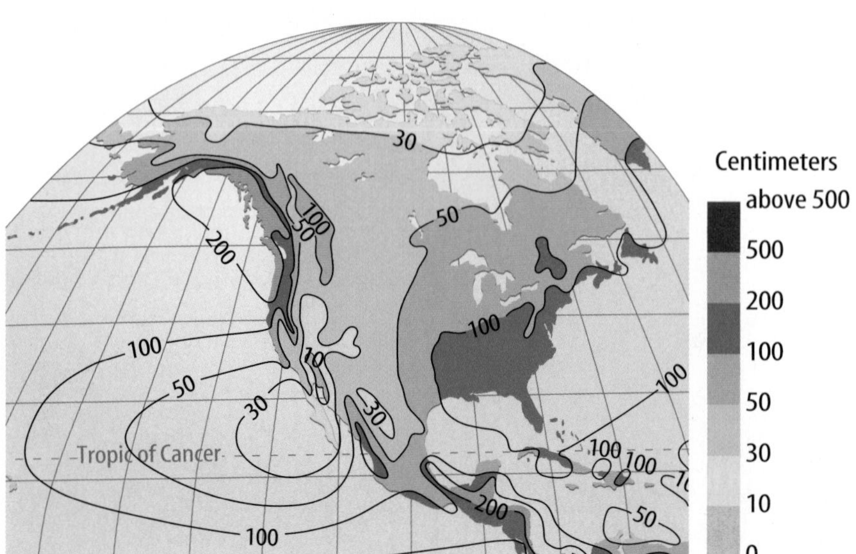

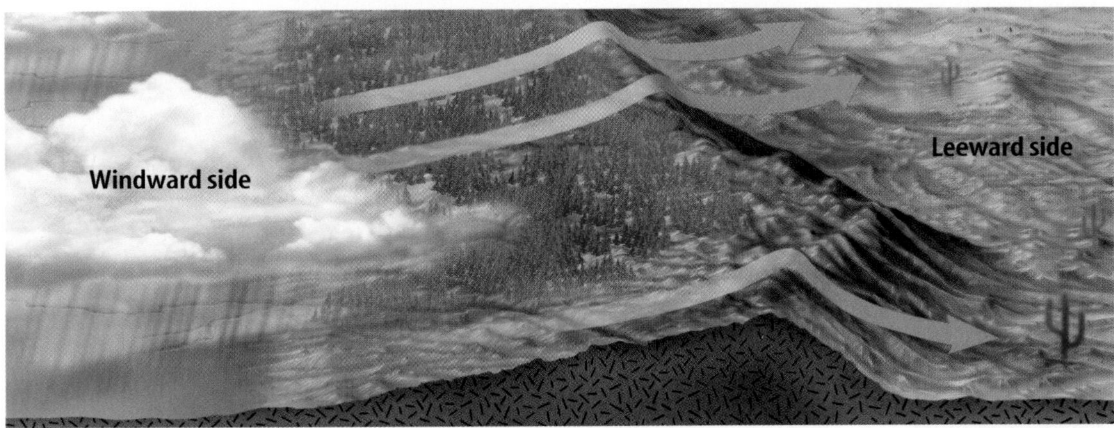

Influence of Mountains The Rocky Mountains and the Sierra Nevada also influence the climate of the west. They act as barriers in the wind, blocking weather systems and altering patterns of precipitation. When the wind blows perpendicular to one side of a mountain range, a **lee rain shadow** forms on the opposite side, as shown in **Figure 15.** The Great Plains lie in the lee shadow of the Rocky Mountains and are relatively dry.

Figure 15 The lee rain shadow of the Sierra Nevada and Rocky Mountains accounts for much of the decreased precipitation shown in Figure 14.

Influence of Water Coasts and lakeshores can affect regional climates in several ways. In winter, lake-effect snow often results in regions around the Great Lakes, like those shown in **Figure 16.** As cold, continental air from the north passes over warmer lake water, the air mass gains heat and moisture. When the air mass reaches the colder land to the south or east of a lake, heavy precipitation, in the form of snow, occurs.

Another example of how coasts affect regional climate is a sea breeze. A **sea breeze** (or lake breeze) blows from the water toward the land in the afternoon, when the land is warmer than the water. Warm air rises over the land creating low pressure that allows cool, dense air to blow from the sea toward land. The reverse happens at night when the land is cooler than the water. A land breeze occurs when cool, dense air over land creates high pressure causing the air to blow from the land toward the sea.

Figure 16 People often experience similar weather events near large bodies of water. Cities east of the Great Lakes experience lake-effect snow.

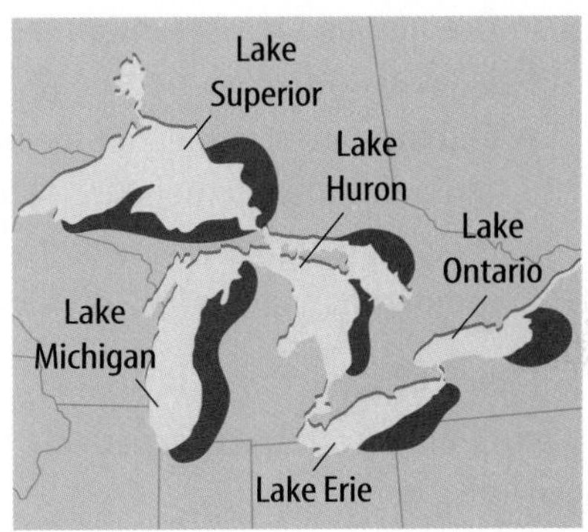

Climate Scale Many small-scale variations are superimposed upon the large-scale climate patterns. Some are regional or local and others, termed microclimates, are variations within small distances. For example, cities create a condition called the heat island effect. Building and pavement materials heat more rapidly than bare land. Vehicles and industry produce pollution that retains heat. Air rises over a heat island, pulling in air from the surrounding countryside. On some clear, calm nights, downtown San Francisco can be as much as 8°C warmer than the surrounding rural areas.

Cold Arid	*tundra*
Cold Dry winter	*boreal evergreen forest*
Warm Arid	*desert*
Warm Semi-arid	*grassland*
Wet winter Dry summer	*Mediterranean forest*
Wet summer Dry winter	*temperate woodland*
Warm Wet	*subtropical deciduous forest*
Warm Wet	*tropical deciduous forest*

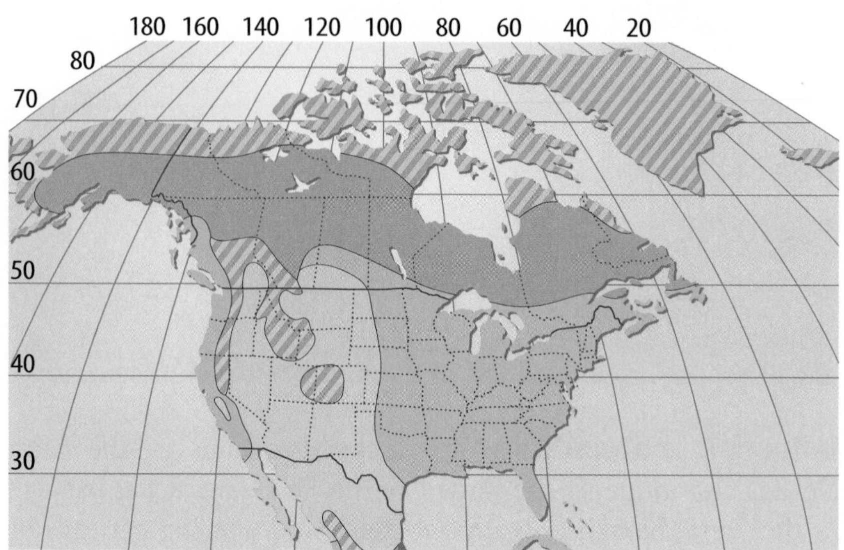

Figure 17 Climate zone influences the types of vegetation that will grow there.
Infer *What are two major factors that characterize climate types?*

Types of Climates

Geographer Glenn Trewartha and meteorologist Lyle Horn designed a system to classify climates, shown in **Figure 17.** It has three major divisions—cold or boreal, arid and semi-arid, and climates with adequate heat and precipitation. The last includes temperate, subtropical, and tropical climates. These divisions are closely correlated with vegetation.

section 3 review

Summary

Climate and Weather

- Climate is the net result of interactions involving all aspects of the biogeophysical system called Earth.

What causes climate?

- The main causes of climate are the distribution of solar radiation and the location of pressure and wind systems.
- Oceans, land masses, mountains, and large lakes also affect climate.
- Climates are classified on the basis of temperature and moisture availability.
- Prevailing high pressure or the rain shadow of mountains can produce dry climates.

Types of climate

- Climates can be classified into three major divisions.

Self Check

1. **Explain** why a climatic normal does not tell you exactly what to expect where you live.
2. **Identify** the five elements of the biogeophysical system called Earth.
3. **Explain** the association between climate and latitude.
4. **Compare and contrast** continental and maritime climates.
5. **Think Critically** Would you expect the daily temperature range to be greater in a maritime climate or a continental climate? Explain.

Applying Math

6. **Calculate Range** The coolest average summer temperature in the United States is 2°C at Barrow, Alaska, and the warmest is 37°C at Death Valley, California. Calculate the range of average summer temperatures in the United States.

More Section Review gpescience.com

Earth's Changing Climates

Reading Guide

What You'll Learn

- **Explain** how climate changes seasonally.
- **Explain** causes of climate change.
- **Describe** how we influence climate.
- **Explain** El Niño's effect on weather.

Why It's Important

Changing climates affect all life on Earth.

Review Vocabulary

trace: a minute, barely detectable amount present

New Vocabulary

- global warming
- El Niño
- La Niña

Seasonal Changes

Seasonal changes occur as Earth completes a revolution around the Sun. The hemisphere tilted toward the Sun experiences summer while the hemisphere tilted away from the Sun experiences winter. During the summer, a hemisphere receives more intense solar radiation and temperatures rise. During winter, the intensity and amount of solar radiation decreases and temperatures drop. Seasonal changes are magnified in the mid-latitudes by the temperature contrast between land and oceans. The oceans are generally colder than land in the summer, but warmer than land in the winter.

Figure 18 The last glacial maximum was 18,000 years ago when Earth's temperature was 6°C cooler than present.

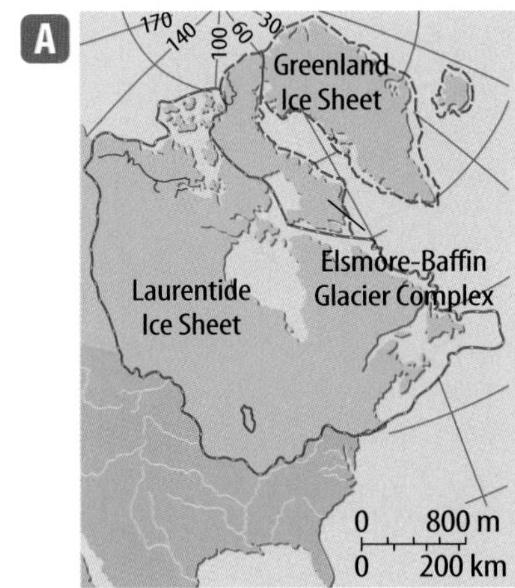

Long-term Changes

Cycles of glaciations, called ice ages, represent long-term climatic change. The peak of the last ice age was 18,000–22,000 years ago when global was about 6°C cooler than present. Glaciers covered much of the middle and high latitudes, as shown in **Figure 18A.** Deserts expanded in the tropics and rainforests all but disappeared.

However, by 5,000 years ago, most of the glacial ice had melted, rainforests returned, and grasslands spread into low latitude deserts. World climate reached its current pattern only about 3,000 years ago, but even since then, large variations have occurred. In the Medieval period, parts of the northern hemisphere were warm and much of the North Atlantic was free of ice, enabling the Vikings to sail to Greenland and establish colonies. A few centuries later, a cooler period called the Little Ice Age occurred, as shown in **Figure 18B.**

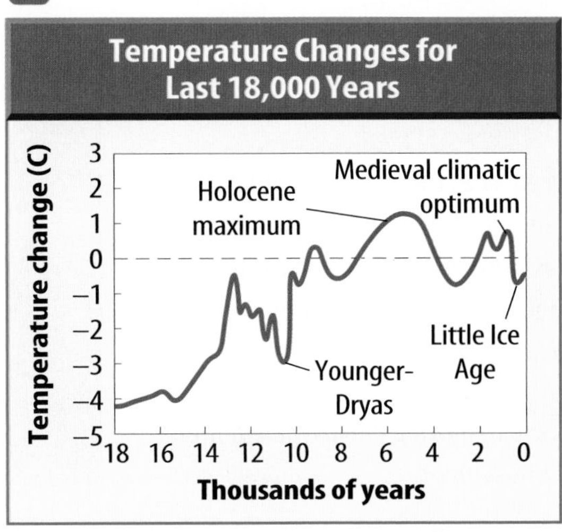

Dinosaur Extinction
The impact of a large asteroid 65 million years ago on Mexico's Yucatán peninsula may have caused the extinction of dinosaurs and other species. The dust from such an impact would have caused months of darkness and drastically lowered temperatures. Animals like dinosaurs that needed large supplies of food would have suffered most. Research what evidence scientists have for such an impact and report your findings to your class.

Causes of Climate Change Numerous factors influence climate and act on diverse time scales. Over millions of years, factors such as mountain building and continental movement are important. Over years and decades, ocean currents, temperatures, and snow and ice cover play a big role. Climate is the net result of all factors on all timescales.

Variations in the receipt of solar radiation are important on a scale of hundreds to thousands of years. Changes related to Earth's orbit are the most important factors, producing changes that determine the rhythmic cycles of glaciation. These include changes in the tilt of Earth's axis of rotation, the shape of its orbit, and the timing of the seasons with respect to distance from the Sun. The tilt, for example, is now 23.5°, but has varied between 21.5° and 24.5°. These changes alter the amount and distribution of solar radiation that reaches Earth.

Sunspots similarly affect the amount of radiation received by Earth. During one period from 1645 to 1715 sunspot activity was very low. This period is correlated with long winters and extreme cold temperatures in Western Europe, known as the Little Ice Age. Sunspots are important on historical time scales. Volcanoes also play a role on this scale. They spew out vast amounts of dust that can block sunlight for years. For example, the eruption of Mt. Tambora in Indonesia in 1815 created the cold weather conditions that gave 1816 the name "The Year Without a Summer."

The Human Factor

Human activities, such as the burning of fossil fuels, manufacturing processes, deforestation, draining of wetlands, and intensive agriculture, have influenced Earth's atmosphere significantly. These activities modify the surface heating and the water and carbon cycles. They also increase the atmospheric concentrations of trace gases, dust and air pollution.

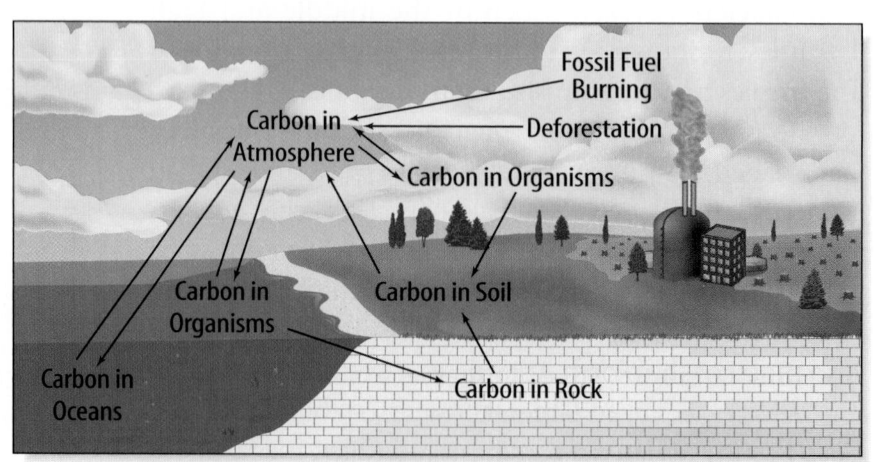

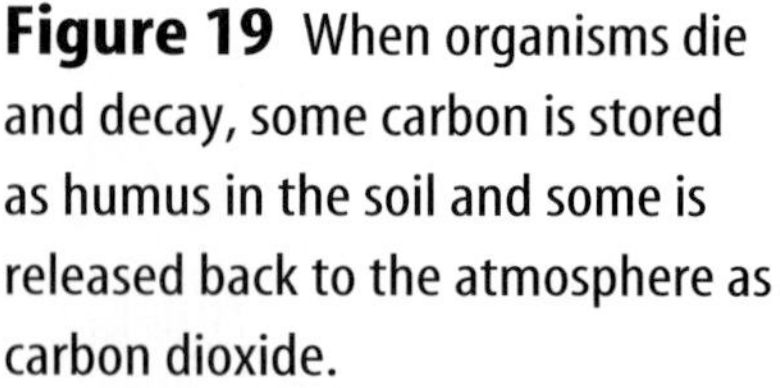
Figure 19 When organisms die and decay, some carbon is stored as humus in the soil and some is released back to the atmosphere as carbon dioxide.
Describe *how trees affect levels of carbon dioxide and oxygen in the atmosphere.*

The Carbon Cycle The carbon cycle, as shown in **Figure 19,** follows the exchange of carbon among the ocean, land, and atmosphere. Carbon is the basis of all organic matter. The changes in the carbon cycle are particularly important both meteorologically and ecologically. The carbon cycle is affected in two ways by deforestation and loss of vegetation—less carbon dioxide is absorbed from the atmosphere during photosynthesis, and decaying and burning wood adds carbon dioxide to the atmosphere.

Applying Math — Calculate

CARBON DIOXIDE (CO_2) CONCENTRATION Concentration of CO_2 in the atmosphere has increased at a constant rate of about 4.2 percent per decade over at least the last 30 years. Assuming this rate remains constant, you can predict its concentration in the future by using the following formula:

CO_2 (year + 10) = CO_2 (year) + CO_2 (year) × rate of CO_2 increase

IDENTIFY known values and unknown values

Identify the known values:

The concentration of CO_2 in 2000 was 369 ppm

Rate of CO_2 increase between 1990 and 2000 was 4.2 percent

Identify the unknown value:

What will be the concentration of carbon dioxide in 2010?

Concentration of CO_2 = ? ppm

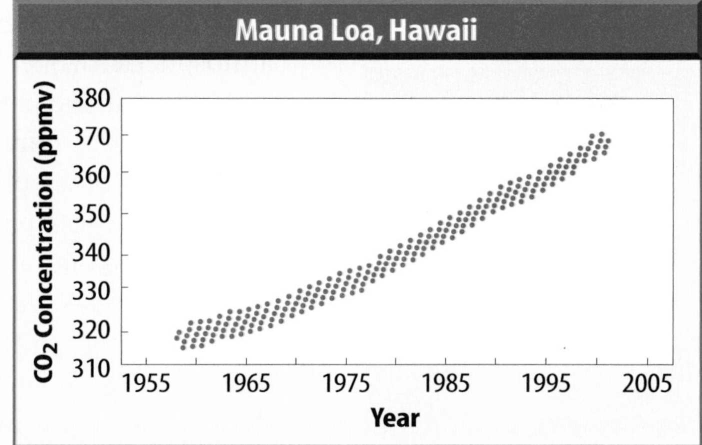

SOLVE the problem

Substitute the known values into the equation:

Concentration in 2010 = 369 ppm + 369 × 0.042

Concentration in 2010 = 384.5

CHECK your answer

Multiply 369 × .042 and subtract from 384.5 to get the value in 2000.

Practice Problems

1. Use the same method to find carbon dioxide concentration in 2020.
2. Assuming that conservation efforts reduce the rate of carbon dioxide increase to 2.6 percent per decade, predict the concentration in 2010?

For more practice problems, go to page 879 and visit Math Practice at gpescience.com.

Trace Gases Today's atmosphere contains on average about 380 ppm of carbon dioxide. This is an increase of 66 ppm, or 21 percent, since measurements began in 1957. This level is far greater still than at the beginning of the nineteenth century, when levels were stable at 280 ppm. Human activities have also increased the concentration of other trace gases—methane by more than 100 percent, nitrous oxide by about 10 percent. All of these gases are important in heating the atmosphere.

Global Warming **Global warming** is an increase in the average global temperature of Earth. Global temperatures have increased over the last century by about 1°C. This may seem small, but the entire global temperature increase since the last ice age is only 6°C. Our understanding of global warming is incomplete, but evidence strongly suggests that the increase in trace gases is an important component.

Figure 20 The concentration of ozone over Antarctic has dramatically decreased between 1979 and 2003. Concentrations in the hole are about eighty percent lower than what would be there naturally. **Explain** *why scientists are concerned about decreasing concentration of ozone in the stratosphere.*

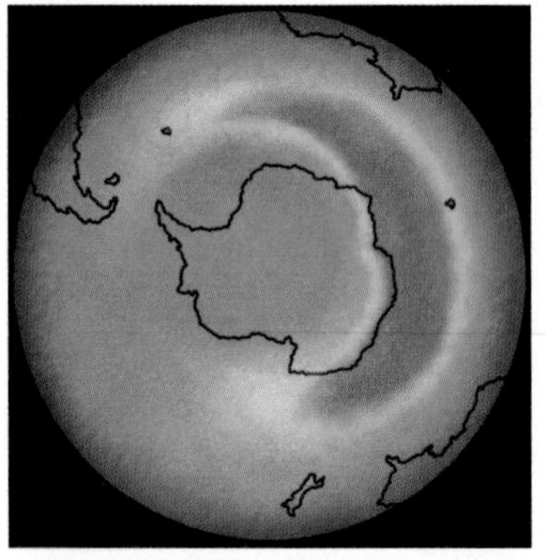

Oct. 1979

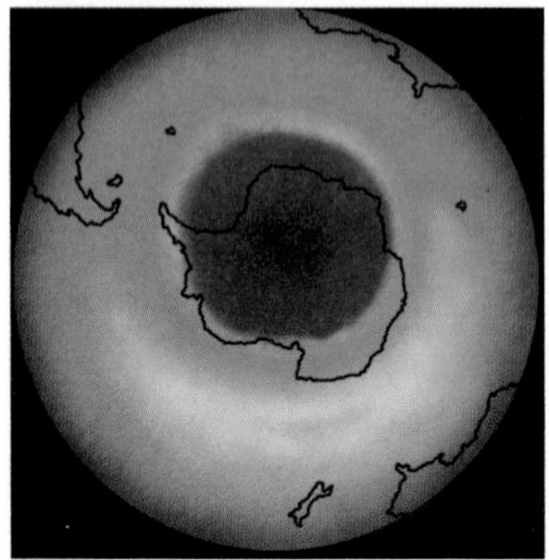

Oct. 2003

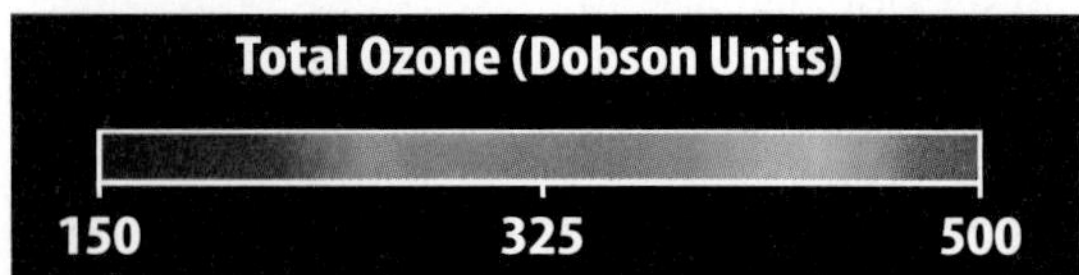

The Ozone Hole As early as the 1970s, scientists were concerned that synthetic chemical compounds could destroy atmospheric ozone. They worried about exhaust from supersonic aircraft and chlorine and fluorine compounds, such as the CFCs (chlorofluorocarbons) used in refrigeration, aerosol sprays, and other processes. Since ultraviolet radiation breaks down DNA, less protection from solar ultraviolet radiation potentially could affect the quality of life on Earth. In 1985, British scientists found a hole in the ozone layer over Antarctica.

The change in the ozone layer between 1979 and 2003 is shown in **Figure 20.** When study of the air in this hole showed that it was largely man-made, an international agreement was made to limit the use of CFCs. Recent studies indicate that actions taken as a result of the agreement are having an effect on levels of chlorine in the atmosphere, which are decreasing each year.

The Land Surface Humans change the land surface by draining swamps, plowing fields, and building cities. Extensive studies have suggested that these processes might affect local or regional climate, but the issue is still controversial. The climate of cities is different than the climate of the surrounding countryside. Effects on larger scales are not as clear.

One concern is desertification. Desertification is the end product of many types of changes that make once-productive land unusable. Human activities such as overgrazing of livestock, deforestation, and irrigation of crops may contribute to the process of desertification in some areas of the world.

El Niño and La Niña

El Niño is a climatic event that involves the atmosphere and oceans. Normally, the trade winds blow warm surface water westward toward a low pressure area in the western Pacific. The warm surface water is replaced by cold, nutrient-rich water that is upwelled from below the surface. When the tradewinds weaken, surface pressure patterns break down and the flow of warm water is reversed. Nutrient-rich cold water is no longer upwelled and warm, nutrient-poor water remains at the surface.

Fewer fish and other marine life can be supported by the nutrient-poor water. Rainfall in the western Pacific decreases, where as heavy rain and flooding can occur on the normally dry coast of Peru. El Niño can dramatically alter global weather patterns. For example, a strong El Niño can lead to flooding and mudslides in California, as shown in **Figure 21,** and droughts in India, Australia, and parts of Africa.

The opposite of El Niño is **La Niña,** which occurs when tradewinds in the Pacific are unusually strong and equatorial oceanic surface temperatures are colder than normal. La Niña can cause drought in the southern United States and excess rainfall in the northwest.

Figure 21 Parts of the Pacific Coast Highway in California have been disrupted by erosion and mudslides caused by El Niño.

section 4 review

Summary

Seasonal Changes

- The tilt of Earth's axis changes the amount of solar radiation received by the northern and southern hemispheres throughout the year, causing the seasons.

Long-term Climate Changes

- Cycles of glaciation occur on the scale of tens of thousands of years.
- Sunspots and volcanoes influence climate on shorter timescales.

The Human Factor

- Human activities have increased the concentration of trace gases and decreased stratospheric ozone.

El Niño and La Niña

- El Niño is a warming of the Pacific Ocean with worldwide effects. La Niña is the opposite of El Niño.

Self Check

1. **Identify** ways that humans may be affecting climate.
2. **Explain** why the contrast between land and water changes with the seasons.
3. **Explain** why it is difficult to identify any single cause of an observed climate change.
4. **Think Critically** What can be done to stop global warming? What can you do to help?

Applying Math

5. **Use Percentages** The eruption of Mount Pinatubo in 1991 put large amounts of dust into the atmosphere. Scientists say this decreased the average global temperature about 0.8 percent during the next year. How would this affect an area with an average temperature of 25°C before the eruption?
6. **Calculate** Areas of the southeastern United States may receive 15 percent of their rain during hurricanes. If average rainfall for this area ranges from 100–200 cm, how much of this might be from hurricanes?

Design Your Own

Investigating Microclimate

Goals

- **Investigate** how environmental variables respond to various microclimates.

Possible Materials

directional compass
thermometer
rain gauge
meter stick
small plastic cups
**graduated cylinder*
**Alternate materials*

Safety Precautions

Real World Problem

While we can talk about global climate or regional climate, we also can discuss climate on the scale of a few meters. We call climate at this scale a microclimate. For example, because cold, heavy air drains downhill, valley fog sometimes forms in moist, low-lying areas. Along the California coast, coastal fog collects on the needles of redwood trees, drips to the ground, and is absorbed by the tree's shallow root system. This microclimate helps the coastal redwoods survive, nestled in the fog belt along the California coast. In this lab, you will investigate local differences in microclimate.

Form a Hypothesis

Choose an aspect of climate, such as precipitation, light, or temperature. Form a hypothesis to explain how this aspect influences local environments. Predict how these factors will vary in response to differences in your microclimate variable.

Using Scientific Methods

Test Your Hypothesis

Make a Plan

1. Decide what microclimate factor you will investigate, and what environmental factor you will monitor for effects.
2. How will you measure the selected microclimate variable? What equipment will you need? How often will you make measurements?
3. Choose sites for making measurements. Will these sites vary in microclimate? Is the environmental variable you chose present in these places?
4. Decide how you will measure your environmental variable. What equipment will you need? How often will you make measurements? Should the microclimate measurement happen at the same time?
5. Prepare a data table to record your measurements. Will you record the time of the measurements? Should you record the weather at the time of measurements?
6. Before you begin, list the steps of your procedure. Include all materials needed for each step. Does your procedure give you the data necessary to test your hypothesis?

Follow Your Plan

1. Be sure that your teacher approves your plan before you start.
2. Carry out your experiment as planned. Be sure to record all data in the appropriate places. Follow all appropriate safety precautions.
3. Record all observations and data in your Science Journal.

Analyze Your Data

1. **Make a graph** of your data. Put the microclimate variable on the *y*-axis, and the environmental variable on the *x*-axis.
2. **Discuss** any trends you see in the data based on your graph.
3. **Infer** any effects that the weather had on microclimate variables. How did this affect your environmental variable?

Conclude and Apply

1. **Discuss** your predicted environmental response to microclimate. Did your results support your hypothesis?
2. **Predict** how changes in global climate would affect the microclimate and response variables that you studied.

Compare your results with the data of other students who measured the same microclimate variable. Present the combined data to the class. **For more help, refer to the Science Skills Handbook.**

Science and Language Arts

The Grapes of Wrath

by John Steinbeck

In *The Grapes of Wrath*, Steinbeck tells the story of the fictional Joad family, who lived in Oklahoma in the 1930s. Like thousands of families living in the drought-stricken dust bowl of the central United States, the Joads were tenant farmers who lost their farm. Such families piled their belongings into rickety trucks and migrated to California seeking work as fruit pickers.

Although the drought alone caused much hardship, it was gigantic dust storms that finally destroyed their crops. Steinbeck describes such a storm in the following words:

"The wind grew stronger, whisked under stones, carried up straws and old leaves, and even little clods, marking its course as it sailed across the fields. The air and the sky darkened and through them the sun shone redly, and there was a raw sting in the air. During a night the wind raced faster over the land, dug cunningly among the rootlets of the corn, and the corn fought the wind with its weakened leaves until the roots were freed by the prying wind and then each stalk settled wearily sideways toward the earth and pointed the direction of the wind."

When the storm finally ended, the buildings, fences, and trees were blanketed with thick layers of dust—a grim reminder of how the Joads' lives had been changed forever.

Understanding Literature

Historical Novels Authors often use historical events as inspiration. When a piece of fiction combines historical events or characters with fictional plot and dialogue, it becomes a historical novel. Well-written historical fiction can aid the reader in understanding how people actually experienced an important time in history.

Respond to the Reading

1. How does the dust storm affect the crops?
2. What might cause such a dust storm?
3. **Linking Science and Reading** Write a paragraph describing the effects of a hurricane or tornado.

The dust bowl resulted from interactions between natural elements, such as climate, plants, and soil, and human elements, such as farming practices and economics. When the dust bowl occurred, vegetation was already sparse. This reduced friction and allowed the wind to gain speed. With no moisture to bind soil particles together, the soil eroded. Dust from these storms was carried up to 3,000 km away from its original source. The dust storms were a type of squall that sometimes bring rain to these semi-arid plains.

Reviewing Main Ideas

Section 1 Earth's Atmosphere

1. Earth's atmosphere is 78 percent nitrogen, 21 percent oxygen, one percent argon, and includes small amounts of trace gases.
2. The stratosphere is the upper layer of the atmosphere where temperature always increases with height. The troposphere is the lower layer where most weather occurs.
3. The troposphere is heated primarily by Earth's surface after it absorbs radiation from the Sun.
4. Land absorbs and emits heat efficiently, but water resists temperature change.
5. Clouds form when warm air carrying water vapor rises until it is cool enough to condense.

Section 2 Weather

1. Major pressure belts and wind belts are caused by unequal heating between the equator and the poles and modifications resulting from Earth's rotation.

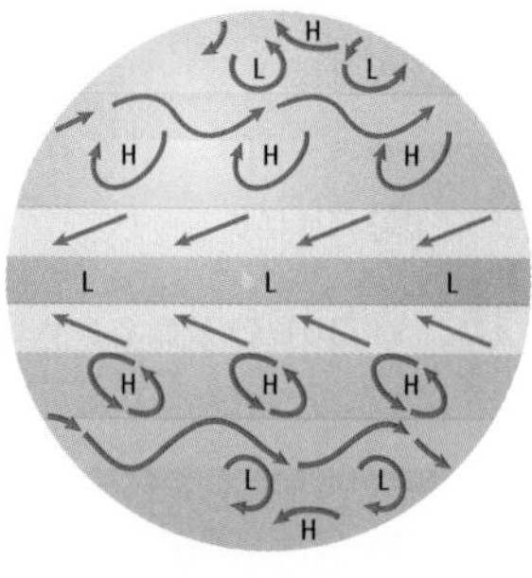

2. Specific weather patterns are associated with high and low pressure cells.
3. Air masses are large blocks of air with similar properties of moisture and temperature throughout. They interact in zones called weather fronts.
4. Severe weather includes thunderstorms, downbursts, tornadoes, and hurricanes.

Section 3 Climate

1. Climate refers to the mean weather conditions and their annual variations in an area.
2. Climate is part of an Earth system that includes the atmosphere, biosphere, hydrosphere, cryosphere, and lithosphere.
3. Latitude is the most important factor in determining climate.
4. Continents, mountains, and oceans influence climate on a large scale, and small-scale variations are termed microclimates.

Section 4 Earth's Changing Climates

1. Long-term changes include cycles of glaciation. Causes of climate change include changes in Earth's orbit, solar activity, volcanism, and human intervention.
2. Global warming has been documented. The reasons for it are complex and not fully understood.
3. Increased concentrations of trace gases and damage to the ozone layer are probably caused by human activities.

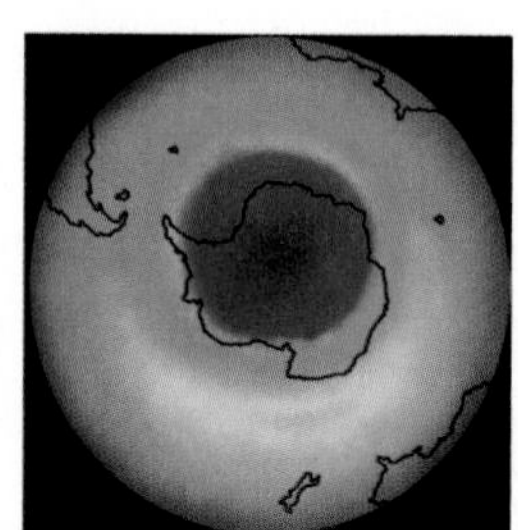

4. El Niño and La Niña affect ocean currents and coastal winds, causing serious droughts and flooding in some areas.

FOLDABLES Use the Foldable that you made at the beginning of this chapter to review what you have learned about weather and climate.

Using Vocabulary

biosphere p. 529	maritime climate p. 532
continental climate p. 532	sea breeze p. 533
El Niño p. 539	subtropical high p. 525
global warming p. 538	temperature inversion p. 519
greenhouse effect p. 520	troposphere p. 519
jet stream p. 525	weather front p. 526
La Niña p. 539	westerlies p. 524
latent heat p. 520	
lee rain shadow p. 533	

Match the correct vocabulary word(s) with each definition given below.

1. area of interaction between air masses
2. climate with a strong ocean influence
3. energy used to evaporate water
4. layer of the atmosphere where most weather occurs
5. global weather event(s) that involve oceans and the atmosphere
6. most important of three major wind belts
7. fast, powerful air current that affects many weather processes
8. area of reduced precipitation on one side of a mountain range
9. a region of very stable air that resists rising needed to form clouds and dispel pollution
10. warming of the atmosphere involving heat absorption by trace gases

Checking Concepts

Choose the word or phrase that best answers the question.

11. Which has great ranges in temperature and little ocean influence?
    A) continental climate
    B) El Niño
    C) La Niña
    D) maritime climate

12. Which is the percent of solar radiation reflected from the surface?
    A) albedo
    B) Coriolis effect
    C) greenhouse effect
    D) urban heat island

13. Which is the most important factor in determining climate at a given location?
    A) altitude **C)** latitude
    B) continents **D)** mountains

Use the illustration below to answer question 14.

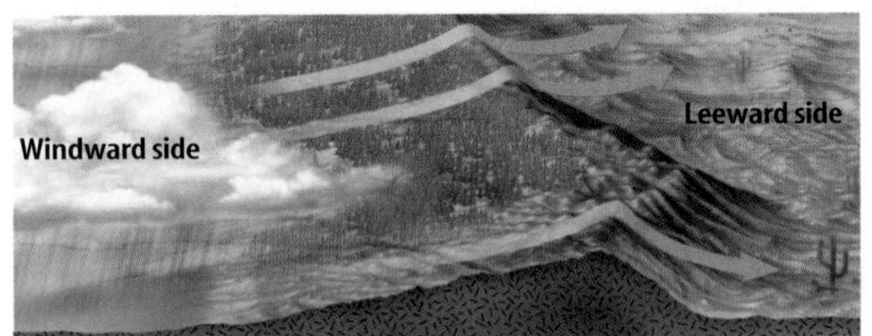

14. Which influence on regional climate is shown above?
    A) continental location
    B) lake effect
    C) lee rain shadow
    D) maritime location

15. Which triggers droplet formation in clouds?
    A) evaporation **C)** ozone
    B) dust **D)** thermals

16. What type of weather is most closely associated with a warm front?
    A) drizzle or steady rain
    B) downbursts or windshear
    C) hurricanes or tornadoes
    D) thunderstorms

17. Which is a trace gas?
    A) argon **C)** oxygen
    B) nitrogen **D)** ozone

18. Which surface reflects solar radiation the most?
    A) bare soil **C)** snow fields
    B) forest **D)** ocean

Interpreting Graphics

19. **Illustrate** Earth's major pressure belts and wind belts, with labels showing the westerlies and the trade winds.

20. **Make a table** comparing characteristics of continental and maritime climates.

Use the data in the following table to answer question 21.

City Temperatures

Cities at Degrees N Latitude	Average July Temperature (°C)	Average January Temperature (°C)
72	20	−45
41	21	9
45	29	−2
40	31	−10

21. **Identify** Based on the data in the table above, identify the latitudes of the cities that have maritime and continental climates.

Thinking Critically

22. **Explain** why the continental United States is prone to severe weather, such as tornadoes and hurricanes.

23. **Infer** why scientists are so concerned about finding a hole in the ozone layer over Antarctica.

24. **Form a hypothesis** about how we might determine whether human activities have played a role in global warming.

25. **Identify** some ways that plants and animals might adapt to changing climate patterns. *Hint: Think of how organisms adapt to extreme conditions.*

Use the illustration below to answer question 26.

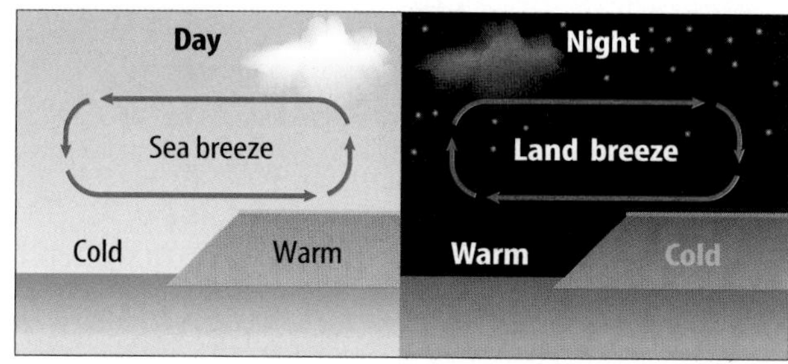

26. **Explain** why a sea breeze blows toward the land in the afternoon, and a land breeze blows toward the water in the evening.

27. **Infer** how global warming might increase the frequency of severe weather, such as hurricanes.

28. **Describe** the carbon cycle and explain how human activities might interfere with it.

29. **Explain** how water droplets form in clouds and the conditions needed to make them fall as some form of precipitation.

30. **Identify** some of the ways that human activities have interfered with the water cycle and describe some results of these activities.

31. **Explain** how El Niño affects global weather patterns.

Applying Math

32. **Convert Units** The high temperature of a summer day in the United States might be 81°F. What would this temperature be in France where they use Celsius temperatures? *Hint: Use the formula* $°C = (°F - 32) \times 5/9$.

33. **Compare Ratios** City A had 175 cloudy days in 2004. In 2004, 56 percent of city B's days were cloudy. Which city had more cloudy days? Show the calculations you did to find your answer.

chapter 17 Standardized Test Practice

Record your answers on the answer sheet provided by your teacher or on a sheet of paper.

Multiple Choice

1. Which makes up most of Earth's atmosphere?

   A. carbon dioxide

   B. carbon monoxide

   C. nitrogen

   D. oxygen

Use the figure below to answer question 2.

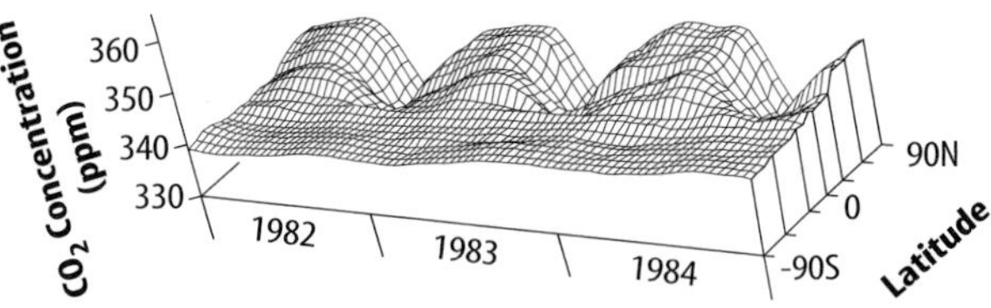

2. Which is used by photosynthetic organisms to produce almost all of Earth's atmospheric oxygen?

   A. carbon monoxide

   B. carbon dioxide

   C. methane

   D. nitrogen

3. Which accumulated in the stratosphere over millions of years and shielded Earth from ultraviolet rays from the Sun?

   A. argon

   B. nitrogen

   C. oxygen

   D. ozone

4. Which type of air mass would most likely be moist and warm?

   A. continental polar

   B. continental tropical

   C. maritime polar

   D. maritime tropical

5. Which is a series of windy gusts formed when a downdraft hits Earth's surface with particularly strong force?

   A. hail

   B. squall

   C. thunderstorm

   D. tornado

Use the diagram below to answer question 6.

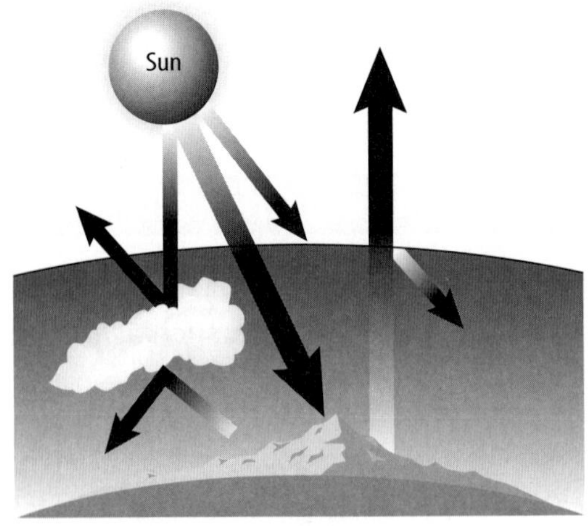

6. Which is caused by absorption of long wavelength radiation by trace gases?

   A. condensation

   B. greenhouse effect

   C. latent heat

   D. precipitation

Use the table below to answer question 7.

Land Surface	Albedo
Bare soil	25
Snow	90
Desert	50
Dense forest	10

7. Over which surface is the least amount of solar radiation absorbed?

A. bare soil

B. dense forest

C. desert

D. snow

8. Which are fast, powerful flows of wind imbedded in global wind systems?

A. jet streams

B. sea breezes

C. subtropical highs

D. westerlies

Gridded Response

9. If the average concentration of carbon dioxide in Earth's atmosphere at the beginning of the nineteenth century was 280 ppm and it presently is 380 ppm, what percentage of today's concentration existed at the beginning of the nineteenth century?

Test-Taking Tip

Review Never leave any answer blank.

Short Response

10. Describe the major differences between the conditions that existed 18,000 years ago from the conditions of the present. Include descriptions of the climatic, geological, and biological conditions.

11. Explain how the stratospheric ozone layer is formed. Also explain its importance in regard to the quality of life on Earth.

12. What causes uneven patterns of heating on Earth's surface?

13. Which factors cause climate?

Extended Response

Use diagram below to answer question 14.

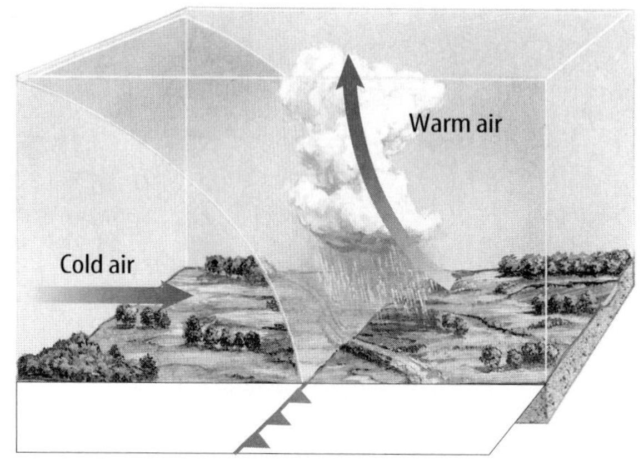

14. PART A What are weather fronts?

PART B How do air masses interact at a cold front?

unit 5

Matter and Earth Materials

How Are Playing Cards & the Periodic Table Connected?

By 1860, scientists knew of about 60 elements. However, they had yet to clearly organize their knowledge. A Russian scientist named Dmitri Mendeleev changed that. Mendeleev loved to play solitaire, a type of card game in which playing cards are arranged into patterns according to their properties. One day, Mendeleev decided to make a set of cards on which he wrote the names and properties of the known elements. Then he began to arrange the cards into rows. The result was a table in which certain chemical properties could be seen to occur periodically—that is, to occur in a repeating pattern. In 1869, Mendeleev published his periodic table (seen here in a more advanced version). He left blank spaces in the table where the pattern seemed to call for elements that were not yet known. Over the next several decades, other scientists refined the table, and new elements were added. Modern periodic tables—like the one probably hanging in your classroom—still follow the basic pattern laid out by Mendeleev.

unit projects

Visit unit projects at gpescience.com to find project ideas and resources.
Projects include:

- **History** Discover the diverse uses of lasers. Compile a class spider map of laser use in different professions.
- **Technology** Investigate core sampling, its use, and costs as you prepare a multimedia advertisement for research geologists.
- **Model** Design and construct a review game to include questions, answers, directions, playing pieces, and a creative box.

WebQuest *Art of Neon* is an investigation of the noble gases and how they are inserted into glass tubes to be used in art and signage.

chapter 18

Classification of Matter

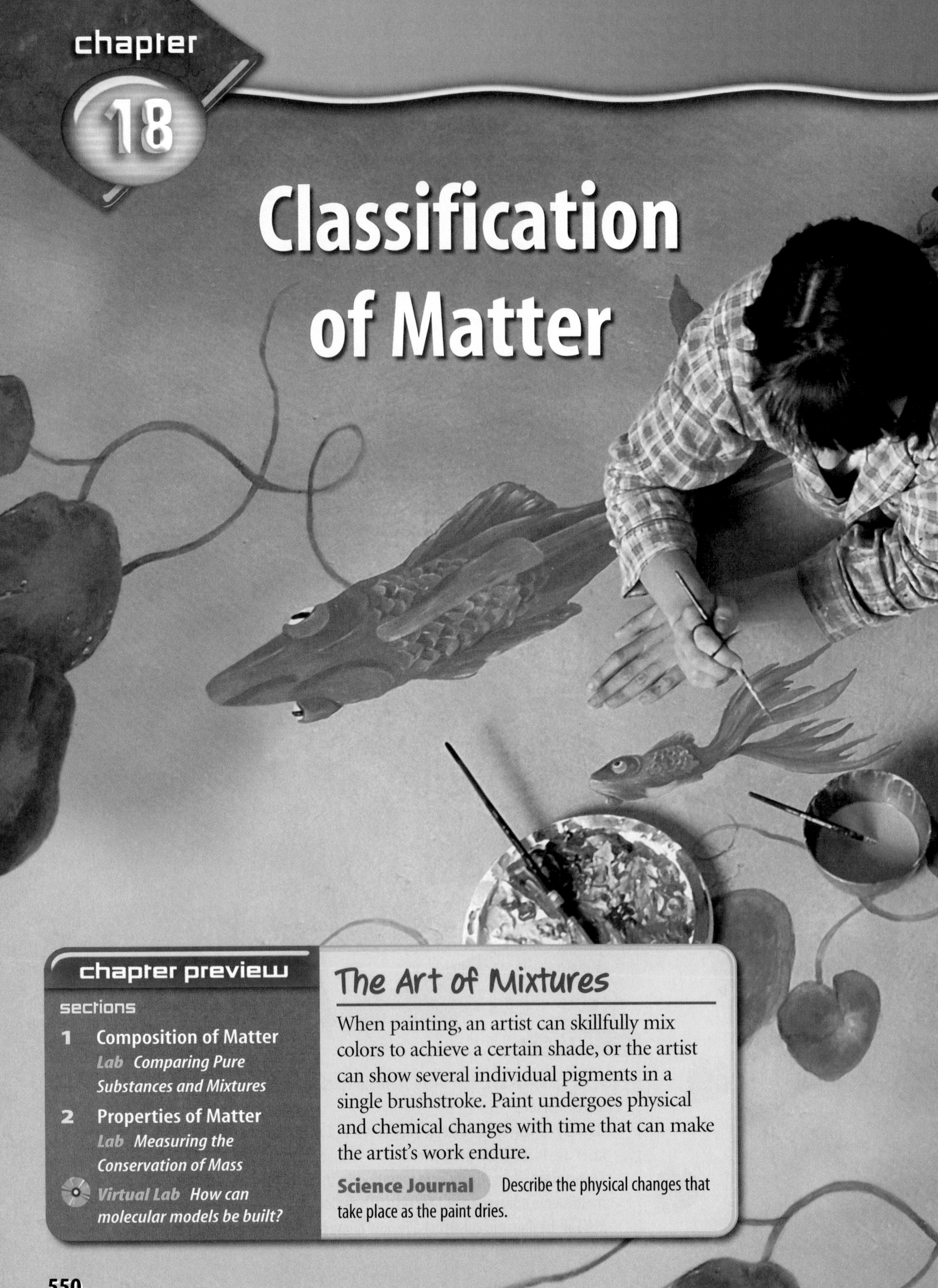

chapter preview

sections

1 Composition of Matter
Lab Comparing Pure Substances and Mixtures

2 Properties of Matter
Lab Measuring the Conservation of Mass

Virtual Lab How can molecular models be built?

The Art of Mixtures

When painting, an artist can skillfully mix colors to achieve a certain shade, or the artist can show several individual pigments in a single brushstroke. Paint undergoes physical and chemical changes with time that can make the artist's work endure.

Science Journal Describe the physical changes that take place as the paint dries.

Start-Up Activities

Demonstrate the Distillation of Water

Matter is classified according to the different properties that it exhibits. These differences in properties allow drinking water to be obtained from seawater. These properties could be very important to you if you were stranded on a desert island and needed drinking water. Purified water can be obtained through a process called distillation.

1. Complete the safety form.
2. Place 75 mL of water in a 200-mL beaker and add 20 drops of blue food coloring.
3. Place the beaker on a hot plate.
4. Add ice to an evaporating dish until the dish is half full. Place the evaporating dish on the beaker as shown in the photo.
5. Turn on the hot plate and slowly bring the water and food-coloring solution to a boil.
6. After boiling the solution for five minutes, carefully remove the evaporating dish using heat-resistant gloves. Touch the drops of liquid on the bottom of the dish to a piece of white paper.
7. Observe the liquid on the paper.
8. **Think Critically** In your Science Journal, write a paragraph explaining where the liquid came from. What was in the beaker that is not in the liquid on the paper?

Classification of Matter Make the following Foldable to ensure that you understand the content by defining the vocabulary terms from this chapter.

STEP 1 **Fold** a sheet of paper vertically from side to side. Make the front edge about 1.25 cm shorter than the back edge.

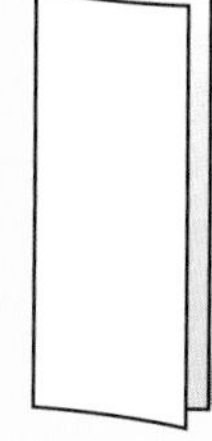

STEP 2 **Turn** lengthwise, with the fold on top, and **fold** into thirds.

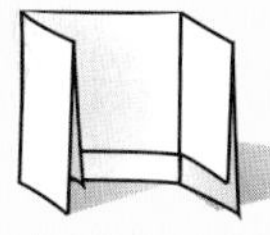

STEP 3 **Unfold and cut** only the top layer along both folds to make three tabs.

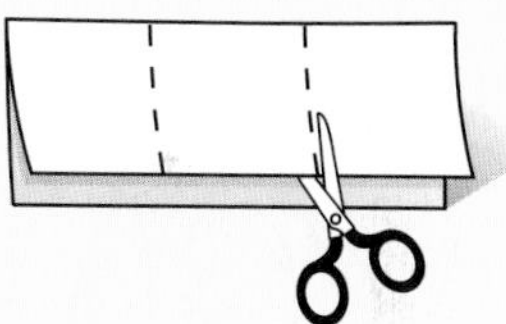

STEP 4 **Label** the tabs *Elements, Compounds,* and *Mixtures* as shown.

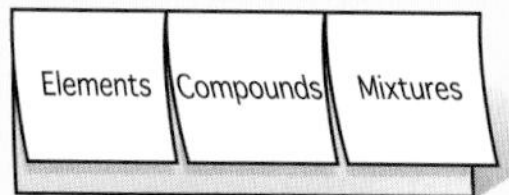

Define As you read the chapter, define each term and list examples of each under the appropriate tab.

Preview this chapter's content and activities at gpescience.com

section 1

Composition of Matter

Reading Guide

What You'll Learn

- **Define** substances and mixtures.
- **Identify** elements and compounds.
- **Compare and contrast** solutions, colloids, and suspensions.

Why It's Important

You will have a better knowledge of your world when you understand the concepts of elements and compounds.

Review Vocabulary

property: characteristic or essential quality

New Vocabulary

- substance
- element
- compound
- heterogeneous mixture
- homogeneous mixture
- solution
- colloid
- Tyndall effect
- suspension

Figure 1 All the atoms of an element are alike.

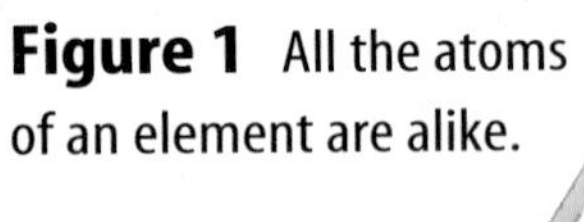

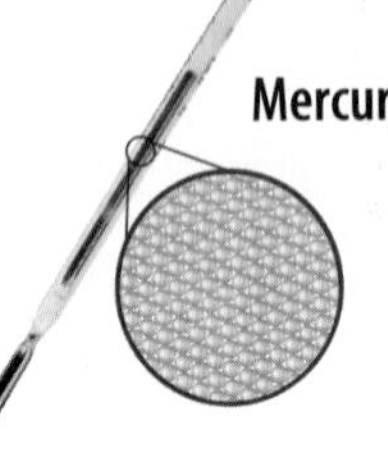

Mercury

Copper

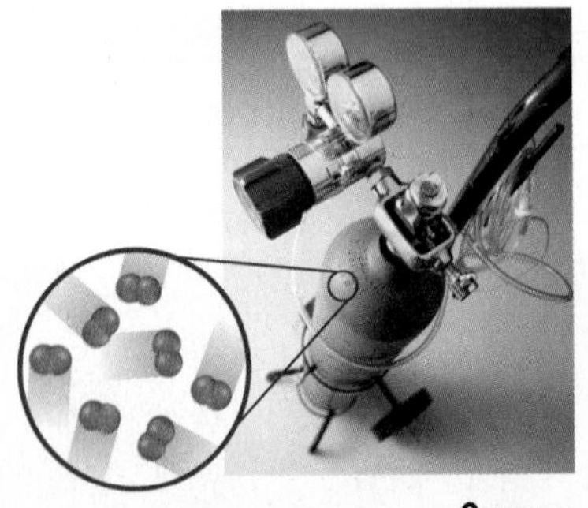

Oxygen

Pure Substances

Have you ever seen a photo hanging on a wall that looked just like a real painting? Did you have to touch it to find out? If so, the rough or smooth surface told you whether it was a painting or a photo. Every material has its own properties. The properties of materials can be used to classify them into general categories.

Every material is made of a pure substance or a mixture of substances. A pure **substance** is a type of matter with a fixed composition. A substance can be either an element or a compound. Some substances you might recognize are helium, aluminum, water, and salt.

Elements All substances are built from atoms. If all the atoms in a substance have the same identity, that substance is an **element.** The graphite in your pencil point and the copper coating of most pennies are examples of elements. In graphite, all a atoms are carbon atoms, and in a copper sample, all the atoms are copper atoms. The metal substance beneath the copper in a penny is another element—zinc. About 90 elements are found on Earth. More than 20 others have been made in laboratories, but most of these are unstable and exist only for short periods of time. Some elements you might recognize are shown in **Figure 1.** Some less-common elements and their properties are shown in **Figure 2.**

NATIONAL GEOGRAPHIC

VISUALIZING ELEMENTS

Figure 2

Most of us think of gold as a shiny yellow metal used to make jewelry. However, it is also an element that is used in more unexpected ways, such as in spacecraft parts. On the other hand, some less-common elements, such as americium (am uh REE see um), are used in everyday objects. Some elements and their uses are shown here.

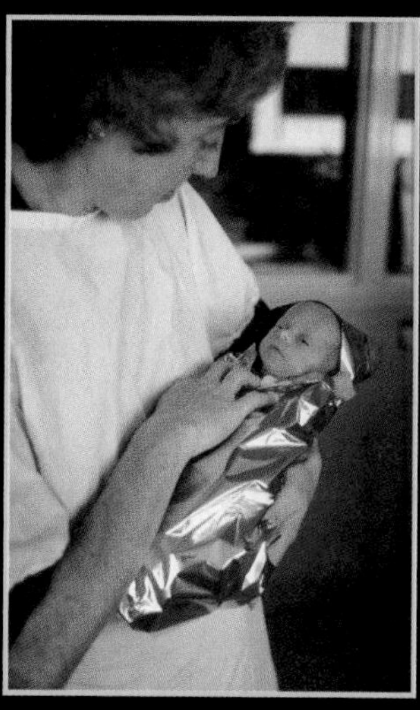

▲ ALUMINUM Aluminum is an excellent reflector of heat. Here, an aluminum plastic laminate is used to retain the body heat of a newborn baby.

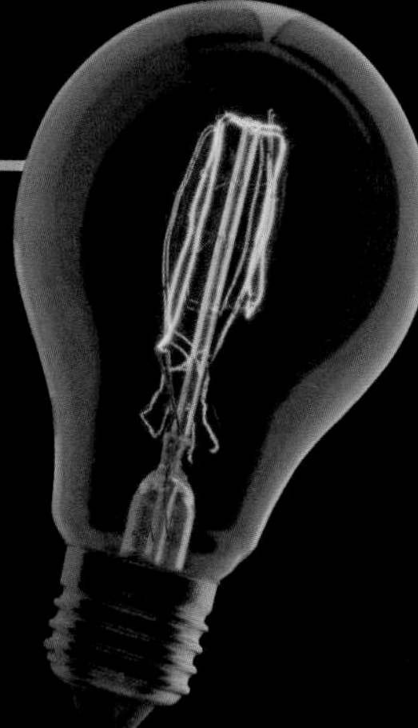

▲ TUNGSTEN Although tungsten can be combined with steel to form a very durable metal, in its pure form it is soft enough to be stretched to form the filament of a lightbulb. Tungsten has the highest melting point of any metal.

▲ TITANIUM (tie TAY nee um) Parts of the exterior of the Guggenheim Museum in Bilbao, Spain, are made of titanium panels. Strong and lightweight, titanium is also used for body implants.

▲ GOLD Gold's resistance to corrosion and its ability to reflect infrared radiation make it an excellent coating for space vehicles. The electronic box on the six-wheel *Sojourner Rover,* above, part of NASA's Pathfinder 1997 mission to Mars, is coated with gold.

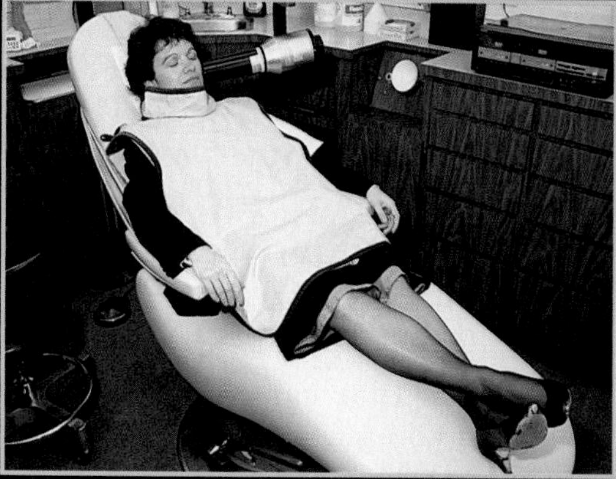

▲ LEAD Because lead has a high density, it is a good barrier to radiation. Dentists drape lead aprons on patients before taking X rays of the patient's teeth to reduce radiation exposure.

◀ AMERICIUM Named after America, where it was first produced, americium is a component of this smoke detector. It is a radioactive metal that must be handled with care to avoid contact.

Compounds When two or more different elements combine, the substance formed is called a compound. A **compound** is a pure substance in which the atoms of two or more elements are combined in a fixed proportion. For example, water is a compound in which two atoms of the element hydrogen combine with one atom of the element oxygen. Chalk contains calcium, carbon, and oxygen in the proportion of one atom each of calcium and carbon to three atoms of oxygen.

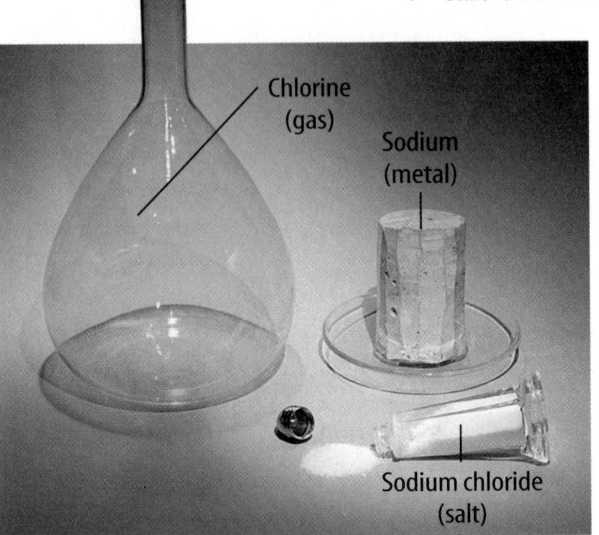

Figure 3 Chlorine gas and sodium metal combine dramatically in the ratio of one to one to form sodium chloride.

Reading Check *How are elements and compounds related?*

Can you imagine yourself putting something made from a silvery metal and a greenish-yellow, poisonous gas on your food? You might have shaken some on your food today—table salt is a chemical compound that fits this description. Even though it looks like white crystals and adds flavor to food, its components—sodium and chlorine—are neither white nor salty, as shown in **Figure 3.** Like salt, other compounds usually look different from the elements in them.

Molecules A particle consisting of two or more atoms that are bonded together is called a molecule. Oxygen in the air, as an example, is a diatomic (two-atom) molecule. A molecule is a basic unit of a molecular compound. The simple sugars you eat; the proteins in your body; and the wool and cotton fibers in your clothes all consist of molecules formed from bonded atoms.

Figure 4 The number of mixtures that can be created by combining substances is unlimited.

Mixtures

Are pizza and a soft drink one of your favorite lunches? If so, you enjoy two foods that are classified as mixtures, but two different kinds of mixtures. A mixture, such as the pizza or soft drink shown in **Figure 4,** is a material made up of two or more substances that can be easily separated by physical means.

Heterogeneous Mixtures Unlike compounds, mixtures do not always contain the same proportions of the substances that make them up. For example, a pizza chef doesn't measure precisely how much of each topping is sprinkled on a pizza. You easily can see most of the toppings on a pizza. A mixture in which different materials can be distinguished easily is called a **heterogeneous** (he tuh ruh JEE nee us) **mixture.** Granite, concrete, and dry soup mixes are other heterogeneous mixtures you can recognize.

You might be wearing another heterogeneous mixture—clothing made of a permanent-press fabric like the one shown in **Figure 5A.** Such a fabric contains fibers of two materials—polyester and cotton. The amounts of polyester and cotton can vary from one article of clothing to another, as shown by the label. Though you might not be able to distinguish the two fibers just by looking at them, you probably could if you used a microscope, as shown in **Figure 5B.**

Most of the substances you come in contact with every day are heterogeneous mixtures. Some components are easy to see, such as the ingredients in pizza, but others are not. In fact, a component you see can be a mixture itself. For example, the cheese in pizza is also a mixture, but you cannot see the individual components. Cheese contains many compounds, such as milk proteins, butterfat, colorings, and other food additives.

Classifying Materials

Procedure

1. In your **Science Journal,** create a data table with four columns and 11 rows. Label the columns *Elements, Compounds, Mixtures,* and *Explanation.*
2. Choose ten items in your home and classify each as an element, a compound, or a mixture. List them in your table in the appropriate columns.
3. In the *Explanation* column, explain why you classified the items as you did.

Analysis

1. How many elements did you find?
2. Which column has the greatest number of items? Is this the same for all students?

Try at Home

MENS L
LARGE
90% COTTON
10% POLYESTER
SEE REVERSE FOR CARE
HOMBRES G
GRANDE
90% ALGODON
10% POLIESTER

Figure 5 Heterogeneous mixtures can be hard to detect.

A You can't tell at a glance that this fabric is a mixture of cotton and polyester.

Cotton fiber

Polyester fiber

B With a microscope, however, the difference between the two fibers is clear; the polyester fiber is perfectly smooth, and the cotton is rough.

Magnification: 600×

Homogeneous Mixtures Remember that soft drink you like with your pizza? Some soft drinks contain water, sugar, flavoring, coloring, and carbon-dioxide gas.

Soft drinks in sealed bottles are examples of homogeneous mixtures. A **homogeneous** (hoh muh JEE nee us) **mixture** contains two or more gaseous, liquid, or solid substances blended evenly throughout. However, a soft drink in which you can see bubbles of carbon-dioxide gas and ice cubes is a heterogeneous mixture.

Vinegar is another homogeneous mixture. It appears to be clear, even though it is made up of particles of acetic acid mixed with water. Another name for a homogeneous mixture such as vinegar and a bottled soft drink is a solution. A **solution** is a homogeneous mixture of particles so small that they cannot be seen with a microscope and will never settle to the bottom of their container. Solutions remain constantly and uniformly mixed. The differences between substances and mixtures are summarized in **Figure 6.**

Reading Check *What kind of mixture is a solution?*

Colloids Milk is an example of a specific kind of mixture called a colloid. Like a heterogeneous mixture, milk contains water, fats, and proteins in varying proportions. Like a solution, its components won't settle when it is left standing. A **colloid** (KAH loyd) is a type of mixture with particles that are larger than those in solutions but not heavy enough to settle out. The word *colloid* comes from a Greek word for *glue.* The first colloids studied were in gelatin, a source of some types of glue.

Paint is an example of a liquid with suspended colloid particles. Gases and solids can contain colloidal particles, too. For example, fog consists of particles of liquid water suspended in air, and smoke contains solids suspended in air.

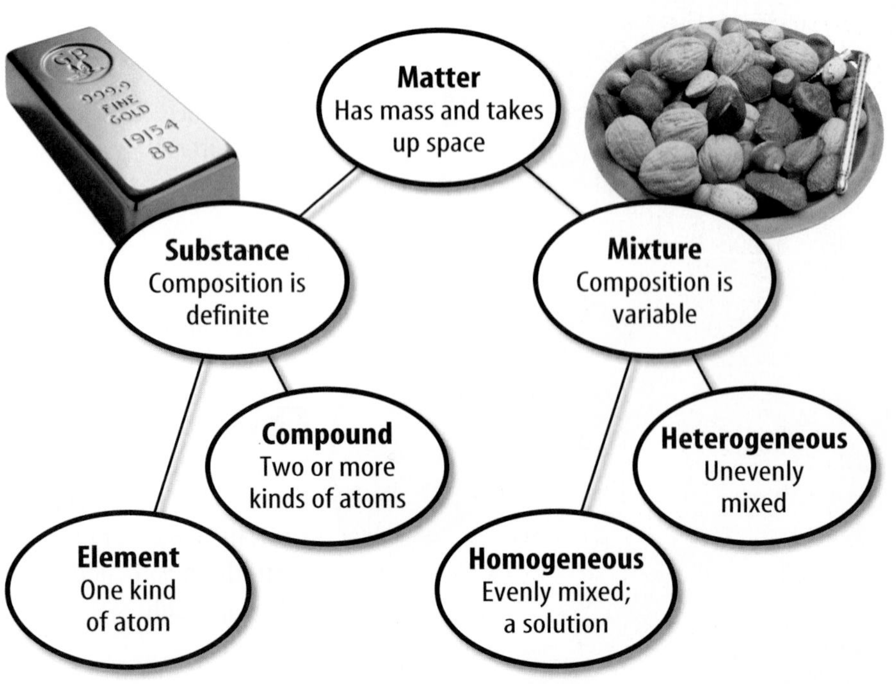

Figure 6 All matter can be divided into substances and mixtures.

Figure 7 Fog is a colloid composed of water droplets suspended in air.

The light from the headlights is scattered by the fog.

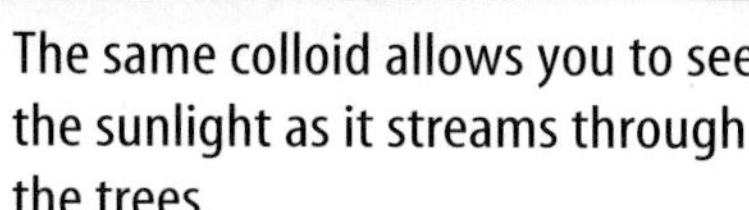

The same colloid allows you to see the sunlight as it streams through the trees.

Detecting Colloids One way to distinguish a colloid from a solution is by its appearance. Fog appears white because its particles are large enough to scatter light as shown in **Figure 7.** Sometimes it is not so obvious that a liquid is a colloid. For example, some shampoos and gelatins are colloids called gels that appear almost clear. You can tell for certain if a liquid is a colloid by passing a beam of light through it, as shown in **Figure 8.** A light beam cannot be seen as it passes through a solution, but can be seen readily as it passes through a colloid. This occurs because the particles in the colloid are large enough to scatter light, but those in the solution are not. This scattering of light by colloidal particles is called the **Tyndall effect.**

How can you distinguish a colloid from a solution?

Topic: John Tyndall
Visit gpescience.com for Web links to information about the discoverer of the Tyndall effect.

Activity Prepare a book cover describing the various activities of John Tyndall.

Figure 8 Because of the Tyndall effect, a light beam is scattered by the colloid suspension on the right, but it passes invisibly through the solution on the left.

Figure 9 The mud deposited by the Mississippi River is thought to be more than 10,000 m thick. **Calculate** *in kilometers how thick the mud in the Mississippi River is.*

Table 1 Comparing Solutions, Colloids, and Suspensions

Description	Solutions	Colloids	Suspensions
Settle upon standing?	no	no	yes
Separate using filter paper?	no	no	yes
Particle size	0.1–1 nm	1–1,000 nm	>1,000 nm
Scatter light?	no	yes	yes

Suspensions Some mixtures are neither solutions nor colloids. One example is muddy pond water. If pond water stands long enough, some mud particles will fall to the bottom, and the water clears. Pond water is a **suspension,** which is a heterogeneous mixture containing a liquid in which visible particles settle. **Table 1** summarizes the properties of different types of mixtures.

INTEGRATE Earth Science River deltas are large scale examples of how a suspension settles. Rivers flow swiftly through narrow channels, picking up soil and debris along the way. As a river widens, it flows more slowly. Suspended particles settle and form a delta at the mouth, as shown in **Figure 9.**

section 1 review

Summary

Pure Substances

- An element is a substance in which all atoms have the same identity.
- A compound is a substance that has two or more elements combined in a fixed proportion.

Mixtures

- A heterogeneous mixture is a mixture in which different materials can be distinguished easily.
- A homogeneous mixture contains two or more gaseous, liquid, or solid substances that are blended evenly throughout.
- Mixtures can be heterogeneous, homogeneous, colloids, or suspensions.

Self Check

1. **Describe** How is a compound similar to a homogeneous mixture? How is it different?
2. **Distinguish** between a substance and a mixture. Give two examples of each.
3. **Describe** the differences between colloids and suspensions.
4. **Think Critically** Why do the words "Shake well before using" indicate that a fruit juice is a suspension?

Applying Math

5. **Use Numbers** A polyester fabric is a mixture of 90 percent cotton and 10 percent polyester. If the fabric has a mass of 500 g, what is the mass of cotton in the fabric?

More Section Review gpescience.com

Comparing Pure Substances and Mixtures

Everything you see that is made of matter is either a pure substance or a mixture. Some things, such as iron nails and aluminum foil, are elements; others, such as water and salt, are compounds. Steel, lemonade, concrete, and a bowl of fruit are all examples of mixtures.

Properties Table

Substance	Matter Classification	Color	Magnetic Attraction	Density
	Do not write in this book.			

Real-World Problem

How do the properties of different pure substances and mixtures compare?

Goals

- **Classify** matter.
- **Calculate** density.
- **Compare** the properties of elements, compounds, and mixtures.

Materials

salt
iron filings
pepper water
sugar water
graduated cylinder
250-mL beakers (5)
magnet
balance
funnel
filter paper
magnifying lens

Safety Precautions

Procedure

1. Complete the safety form before you begin.
2. Copy the data table into your Science Journal. Record all data in your table.
3. **Determine** the color of each sample.
4. **Classify** each material as an element, compound, or mixture.
5. **Determine** if each sample is magnetic .
6. **Determine** the mass of a 10-mL sample of each material. Calculate the density of each sample.
7. Try to separate the particles of the pepper-water mixture and the sugar water.
8. **Infer** if the different particles of each sample can be separated by using physical properties.

Conclude and Apply

1. **Describe** the separation methods of the mixture samples.
2. **Infer** how you can identify a substance as an element.
3. **Calculate** the density of each material.

Communicating Your Data

Identify experimental errors that might have occurred during the collection of your data.

section 2

Properties of Matter

Reading Guide

What You'll Learn

- **Identify** substances using physical properties.
- **Compare and contrast** physical and chemical changes.
- **Identify** chemical changes.
- **Determine** how the law of conservation of mass applies to chemical changes.

Why It's Important

Understanding chemical and physical properties can help you use materials properly.

Review Vocabulary

state of matter: one of three physical forms of matter: solid, liquid, or gas

New Vocabulary

- physical property
- physical change
- distillation
- chemical property
- chemical change
- law of conservation of mass

Figure 10 Appearance is the most obvious physical property. **Describe** *the appearance of these items.*

Physical Properties

You can stretch a rubber band, but you can't stretch a piece of string very much, if at all. You can bend a piece of wire, but you can't easily bend a matchstick. In each case, the materials change shape, but the identity of the substances—rubber, string, wire, and wood—does not change. The abilities to stretch and bend are physical properties. Any characteristic of a material that you can observe without changing the identity of the substances that make up the material is a **physical property.** Examples of other physical properties are color, shape, size, density, melting point, and boiling point. What physical properties can you use to describe the items in **Figure 10?**

Appearance How would you describe a tennis ball? You could begin by describing its shape, color, and state of matter. You might describe the tennis ball as a brightly colored, hollow sphere. You can measure some physical properties, too. For instance, you could measure the diameter of the ball. What physical property of the ball is measured with a balance?

To describe a soft drink in a cup, you could start by calling it a liquid with a brown color. You could measure its volume and temperature. Each of these characteristics is a physical property of that soft drink.

Figure 11 The best way to separate substances depends on their physical properties. Size is the property used to separate poppy seeds from sunflower seeds in this example.

Behavior Some physical properties describe the behavior of a material or a substance. As you might recall, objects that contain iron, such as a safety pin, are attracted by a magnet. Attraction to a magnet (magnetism) is a physical property of the substance iron. Every substance has a specific combination of physical properties that make it useful for certain tasks. Some metals, such as copper, can be drawn out into wires. Others, such as gold, can be pounded into sheets as thin as 0.1 micrometers (μm), about 4-millionths of an inch. This property of gold makes it useful for decorating picture frames and other objects. Gold that has been beaten or flattened in this way is called gold leaf.

Think again about your soft drink. If you spill it, the drink will spread out over the table or floor. If you knock over a jar of molasses, however, it does not flow as easily. The ability to flow is a physical property of liquids.

Magnetism easily separates iron from sand.

Using Physical Properties to Separate Removing the seeds from a watermelon can be easily done based on the physical properties of the seeds compared to the rest of the fruit. **Figure 11** shows a mixture of poppy seeds and sunflower seeds. You can identify the two kinds of seeds by differences in color, shape, and size. By sifting the mixture, you could separate the poppy seeds from the sunflower seeds quickly because their sizes differ.

Now look at the mixture of iron filings and sand shown in **Figure 11.** You probably wouldn't be able to sift out the iron filings because they are similar in size to the sand particles. What you could do is pass a magnet through the mixture. The magnet will attract only the iron filings and pull them from the sand. This is an example of how a physical property, such as magnetic attraction, can be used to separate substances in a mixture. Something like this is done to separate iron for recycling.

What physical properties could be used to separate two different substances?

Recycling and Physical Properties Recycling conserves natural resources. In some large recycling projects, aluminum metal must be separated from scrap iron. What physical properties of the two metals could be used to separate them?

Separating Ink

Procedure

1. Complete the safety form.
2. Cut a circle out of a **clean, white coffee filter** with a 9-cm diameter. Draw a spot using a **nonpermanent black marker** in the center of the filter.
3. Place the filter on a **plate.** Using a **dropper,** place ten drops of water on the spot.

Analysis

Make a labeled drawing of your results in your **Science Journal.** What type of change did you observe? Explain.

Physical Changes

If you break a piece of chewing gum, you change some of its physical properties—its size and shape. However, you do not change the identity of the materials that make up the gum.

The Identity Remains the Same When a substance freezes, boils, evaporates, or condenses, it undergoes physical changes. A change in size, shape, or state of matter is called a **physical change.** Such a change might involve energy changes, but the kind of substance—the identity of the element or compound—does not change. Because all substances have distinct properties, such as densities, specific heats, and boiling and melting points, which are constant, these properties can be used to help identify them when a particular mixture contains substances that are not yet identified.

Reading Check ***Does a change in state mean that a new substance has formed? Explain.***

Iron is a substance that can change states if it absorbs or releases enough energy; at high temperatures, it melts. However, in both the solid and liquid state, iron has physical properties that identify it as iron. Color changes can accompany a physical change, too. For example, when iron is heated, it first glows red. Then, if it is heated to a higher temperature, it turns white, as shown in **Figure 12.**

Using Physical Change to Separate A cool drink of water is something most people take for granted, but in some parts of the world, drinkable water is scarce. Not enough drinkable water can be obtained from wells. Many such areas that lie close to the ocean obtain drinking water by using the physical properties of water to separate it from the salt. One separation method, which uses the property of boiling point, is distillation.

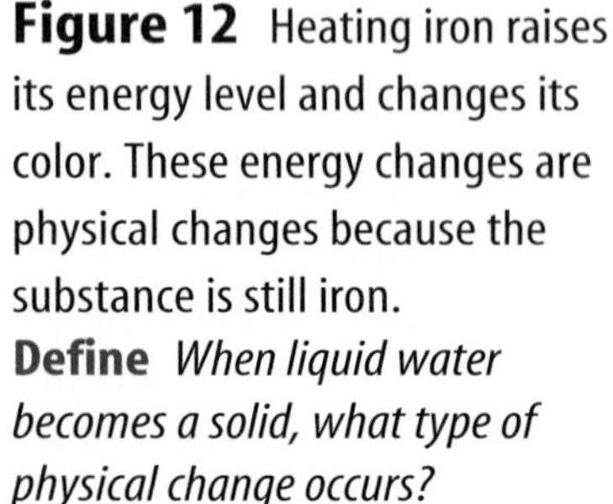

Figure 12 Heating iron raises its energy level and changes its color. These energy changes are physical changes because the substance is still iron.
Define *When liquid water becomes a solid, what type of physical change occurs?*

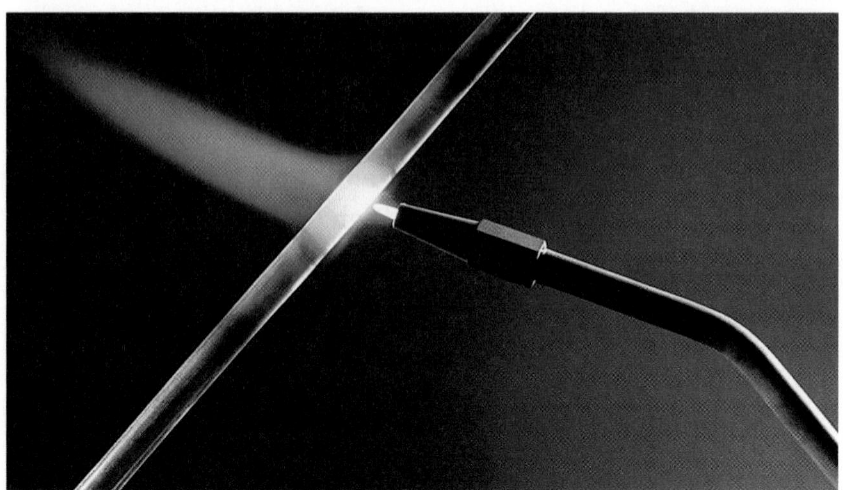

Distillation The process used for separating substances in a mixture by evaporating a liquid and recondensing its vapor is **distillation.** It usually is done in the laboratory using an apparatus similar to that shown in **Figure 13.** The liquid vaporizes and condenses, leaving the solid material behind.

Two liquids having different boiling points can be separated in a similar way. The mixture is heated slowly until it begins to boil. Vapors of the liquid with the lowest boiling point form first and are condensed and collected. Then, the temperature is increased until the second liquid boils, condenses, and is collected. Distillation is used often in industry. Natural oils such as mint are distilled.

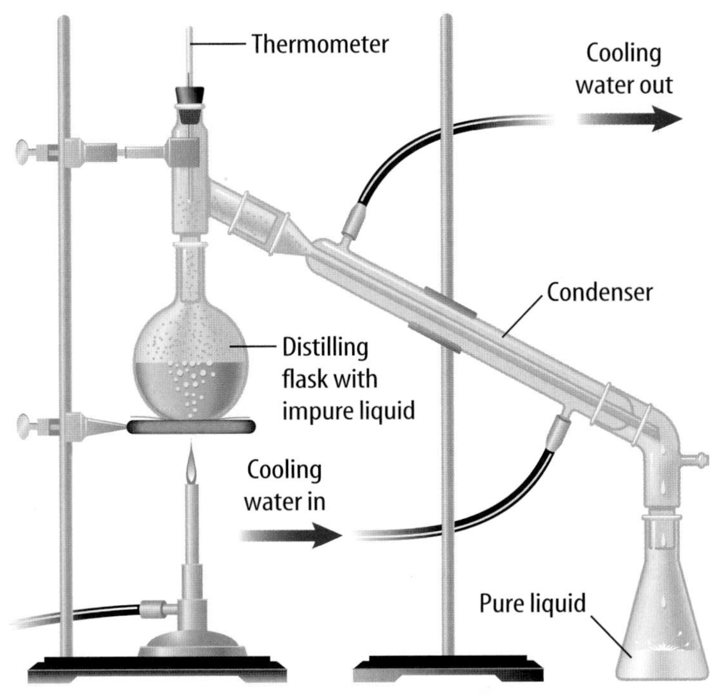

Figure 13 Distillation can easily separate liquids from solids dissolved in them. The liquid is heated until it vaporizes and moves up the column. Then, as it touches the water-cooled surface of the condenser, it becomes liquid again.

Chemical Properties and Changes

You probably have seen warnings on cans of paint thinners and lighter fluids for charcoal grills that say these liquids are flammable (FLA muh buhl). The tendency of a substance to burn, or its flammability, is an example of a chemical property. Burning produces a new substance during a chemical change. A **chemical property** is a characteristic of a substance that indicates whether it can undergo a certain chemical change. Many substances used around the home, such as lighter fluids, are flammable. Knowing which ones are flammable helps you to use them safely.

A less-dramatic chemical change can affect some medicines. Look at **Figure 14.** You probably have seen bottles like this in a pharmacy. Many medicines are stored in dark bottles because they contain compounds that can change chemically if they are exposed to light.

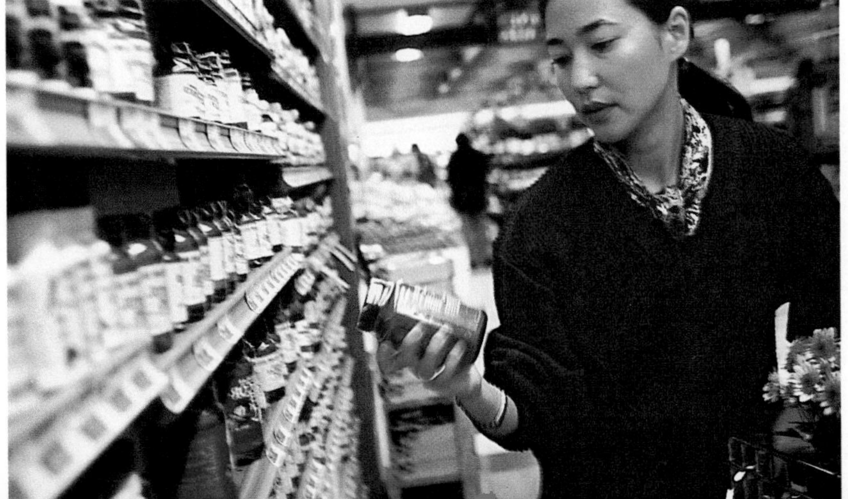

Figure 14 The brown color of these bottles tells you that these vitamins may react to light. Reaction to light is a chemical property.

Alchemy In the Middle Ages, alchemy was an early form of chemistry devoted to the study of changing baser metals into gold and also to finding the elixir of perpetual youth. Based on what we know now about the properties of metals and biology, it is easy to understand why this field of study is no longer practiced.

Detecting Chemical Changes

If you leave a pan of chili cooking unattended on the stove for too long, your nose soon tells you that something is wrong. Instead of a spicy aroma, you detect an unpleasant smell that alerts you that something is burning. This burnt odor is a clue that a new substance has formed.

The Identity Changes The smell of rotten eggs and the formation of rust on bikes or car fenders are signs that chemical changes have taken place. A change of one substance to another is a **chemical change.** The foaming of an antacid tablet in a glass of water and the smell in the air after a thunderstorm are other signs of new substances being produced. In some chemical changes, a rapid release of energy—detected as change of heat, light, and/or sound production—is a clue that changes are occurring.

Reading Check *What is a chemical change?*

A clue such as heat, cooling, or the formation of bubbles or solids in a liquid is a helpful indicator that a chemical reaction is taking place. However, the only sure proof is that a new substance is produced. Consider the following example. The heat, light, and sound produced when hydrogen gas combines with oxygen in a rocket engine are clear evidence that a chemical reaction has taken place. But no clues announce the reaction that takes place when iron combines with oxygen to form rust because the reaction takes place so slowly. The only clue that iron has changed into a new substance is the presence of rust. Burning and rusting are chemical changes because new substances form. You sometimes can follow the progress of a chemical reaction visually. For example, you can see lead nitrate forming in **Figure 15.**

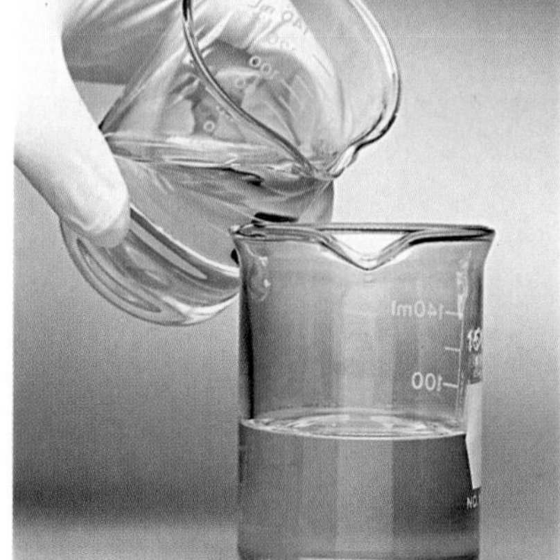

Figure 15 The solid forming from two liquids is another sign that a chemical reaction has taken place.
Identify *clues that show a reaction has taken place.*

Using Chemical Change to Separate You might separate substances using a chemical change when cleaning tarnished silver. Tarnish is a chemical reaction between silver metal and sulfur compounds in the air that results in silver sulfide. It can be changed back into silver with a chemical reaction. This chemical reversal back to silver takes place when the tarnished item is placed in a warm water bath with baking soda and aluminum foil. You don't usually separate substances using chemical changes in the home. In industry and chemical laboratories, however, this kind of separation is common. For example, many metals are separated from their ores and then purified using chemical changes.

Chemical Reactions and Lightning In a thunderstorm, light and sound tell you that changes have occurred. The pungent smells of ozone indicates that a chemical reaction also took place. Lightning converts oxygen gas, O_2, into ozone, O_3. Ozone is unstable and soon breaks up, forming oxygen again.

Weathering—Chemical or Physical Change?

The forces of nature continuously shape Earth's surface. Rocks split, deep canyons are carved out, sand dunes shift, and curious limestone formations decorate caves. Do you think these changes, often referred to as weathering, are physical or chemical? The answer is both. Geologists, who use the same criteria that you have learned in this chapter, say that some weathering changes are physical and some are chemical.

Physical Large rocks can split when water seeps into small cracks, freezes, and expands. However, the smaller pieces of newly exposed rock still have the same properties as the original sample. This is a physical change. Streams can cut through softer rock, forming canyons, and can smooth and sculpt harder rock, as shown on the left in **Figure 16.** In each case, the stream carries rock particles far downstream before depositing them. Because the particles are unchanged, the change is a physical one.

Flowing water shaped and smoothed these rocks in a physical process.

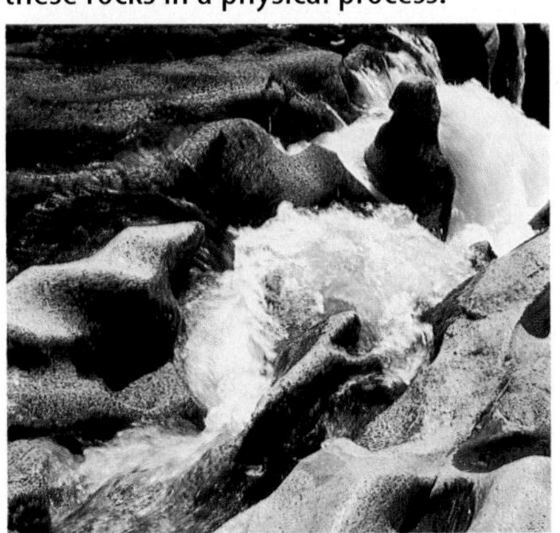

Both chemical and physical changes shaped the famous White Cliffs of Dover lining the English Channel.

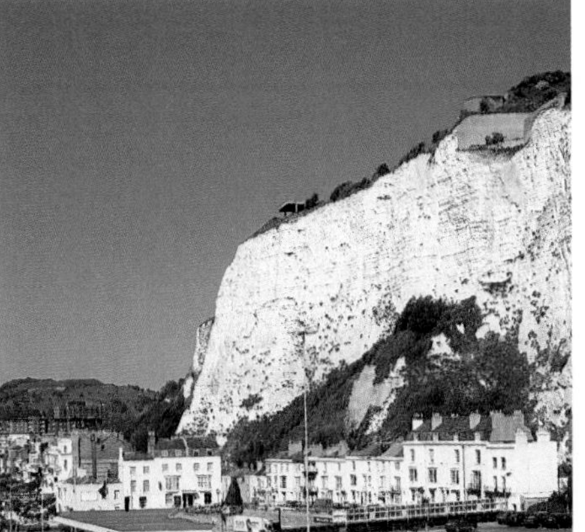

Figure 16 Weathering can involve physical or chemical change.

Topic: Caves

Visit gpescience.com for Web links to information about cave formations.

Activity Describe the formation of stalactites and stalagmites. Explain the differences between the two and whether they are the result of a physical and/or chemical process.

Chemical In other cases, the change is chemical. For example, solid calcium carbonate, a compound that mostly makes up limestone, does not dissolve easily in water. However, when the water is even slightly acidic, as it is when it contains some dissolved carbon dioxide, calcium carbonate reacts. It changes into a new substance, calcium hydrogen carbonate, which does dissolve in water. This change in limestone is a chemical change because the identity of the calcium carbonate changes. The White Cliffs of Dover, shown on the right in **Figure 16,** are made of limestone and undergo such chemical changes, as well as physical changes. A similar chemical change produces caves and the icicle-shaped rock formations that often are found in them.

Applying Math Calculate

THE LAW OF CONSERVATION OF MASS When a chemical reaction takes place, the total mass of reactants equals the total mass of products. If 18 g of hydrogen reacts completely with 633 g of chlorine, how many grams of HCl are formed? $H_2 + Cl_2 \rightarrow 2HCl$

IDENTIFY known values and the unknown value

Identify the known values:

mass of H_2 = 18 g

mass of Cl_2 = 633 g

mass of reactants = mass of products

Identify the unknown value:

mass of HCl means ? g

SOLVE the problem

Solve for the mass of HCl. $g\ H_2 + g\ Cl_2 = g\ HCl$

Substitute the known values. $18\ g + 633\ g = 651\ g\ HCl$

CHECK the answer

Does your answer seem reasonable? Check your answer by subtracting the mass of H_2 from the mass of HCl. Do you obtain the mass of the Cl_2? If so, the answer is correct.

Practice Problems

1. In the following reaction, 24 g of CH_4 (methane) reacts completely with 96 g of O_2 to form 66 g of CO_2. How many grams of H_2O are formed? $CH_4 + 2O_2 \rightarrow CO_2 + 2H_2O$
2. In the following equation, 54.0 g of Al reacts completely with 409.2 g of $ZnCl_2$ to form 196.2 g of Zn metal. How many grams of $AlCl_3$ are formed? $2Al + 3ZnCl_2 \rightarrow 3Zn + 2AlCl_3$

For more practice problems go to page 879, and visit Math Practice at gpescience.com.

The Conservation of Mass

Figure 17 This reaction appears to be destroying these logs. When it is over, only ashes will remain. Yet you know that no mass is lost in a chemical reaction. **Explain** *why this is so.*

Wood is combustible, or burnable. As you just learned, this is a chemical property. Suppose you burn a large log in a fireplace, as shown in **Figure 17,** until nothing is left but a small pile of ashes. Smoke, heat, gas, and light are given off, and the changes in the appearance of the log confirm that a chemical change took place. You might think that matter was lost during this change because the pile of ashes looks much smaller than the log did. In fact, the mass of the ashes is less than that of the log. Suppose that you could collect all the oxygen in the air that was combined with the log during the burning and all the smoke, heat, light, and gases that escaped from the burning log and measure their masses, too. Then you would find that no mass was lost after all.

Not only is no mass lost during burning, mass is not gained or lost during any chemical change. In other words, matter is neither created nor destroyed during a chemical change. According to the **law of conservation of mass,** the mass of all substances that are present before a chemical change equals the mass of all the substances that remain after the change.

Reading Check *Explain what is meant by the law of conservation of mass.*

section 2 review

Summary

Physical Properties

- You can observe physical properties without changing the identity of a substance.

Physical Changes

- Change in the size, shape, or state of matter is a physical change.

Chemical Properties and Changes

- A chemical property is a characteristic of a substance that indicates whether it can undergo a certain chemical change.
- A change of one substance to another is a chemical change.
- Many metals are separated from their ores and purified using chemical changes.

Self Check

1. **Explain** why evaporation of water is a physical change and not a chemical change.
2. **List** four physical properties you could use to describe a liquid.
3. **Describe** why flammability is a chemical property rather than a physical property.
4. **Explain** how the law of conservation of mass applies to chemical changes.
5. **Think Critically** How might you demonstrate the law of conservation of mass for melting ice and distilling water?

Applying Math

6. **Calculate** In the following equation, 417.96 g of Bi (bismuth) reacts completely with 200 g of F (fluorine). How many grams of BiF_3 (bismuth fluoride) are formed? $2\ Bi + 3\ F_2 \longrightarrow 2\ BiF_3$

Design Your Own

Measuring the Conservation of Mass

Goals

- **Measure** the total mass of water and antacid tablets before and after the tablets are dissolved in the water.
- **Compare** the total mass of water and tablets before and after the tablets are dissolved in the water.
- **Infer** whether or not the law of conservation of mass applies to antacid tablets dissolving in water

Possible Materials

antacid tablets
empty plastic drink bottle
balloon
beaker
water
spatula
balance
mortar
pestle
funnel

Safety Precautions

Do not eat the antacid tablet.

Real-World Problem

Have you ever watched burning logs in a fireplace? If you have, you might have noticed many large logs being burned in the hearth during an evening. At the end of the night, nothing more than a pile of ash remains. In this lab, your group will design an experiment to demonstrate the law of conservation of mass. Is the mass of antacid tablets conserved after they are dissolved?

Form a Hypothesis

Based on your understanding of mass conservation, form a hypothesis that predicts the total mass of antacid tablets and water before and after the tablets are dissolved.

Test Your Hypothesis

Make a Plan

1. Complete the safety form before you begin.
2. As a group, agree upon and write the hypothesis statement.
3. Decide upon any safety equipment you need or safety procedures you need to devise to ensure the safety of your group during the experiment.

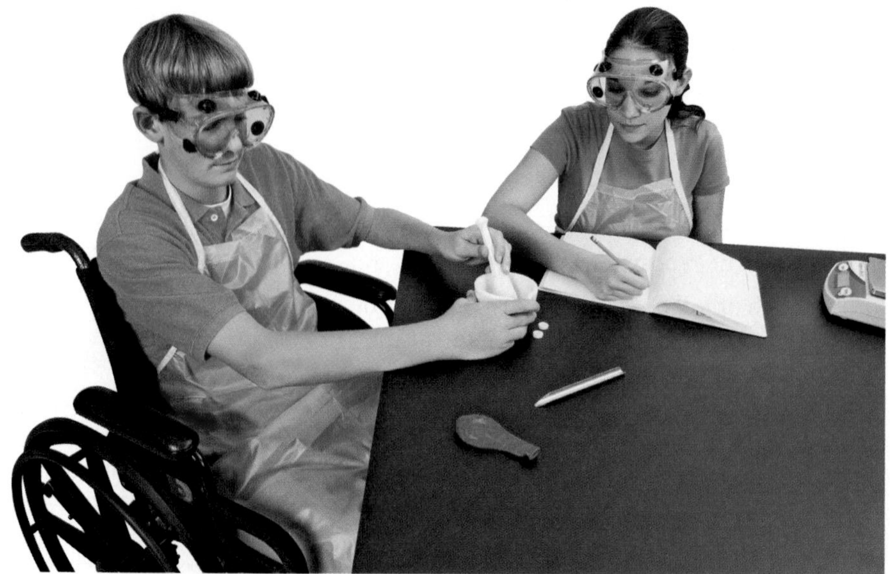

4. **Plan** an experiment to test your hypothesis. List the steps of your experiment.
5. **List** the materials you need to test your hypothesis.
6. Have one group member reread your entire experiment aloud to the group to make certain you have all the necessary materials and that your experimental steps can be easily followed.

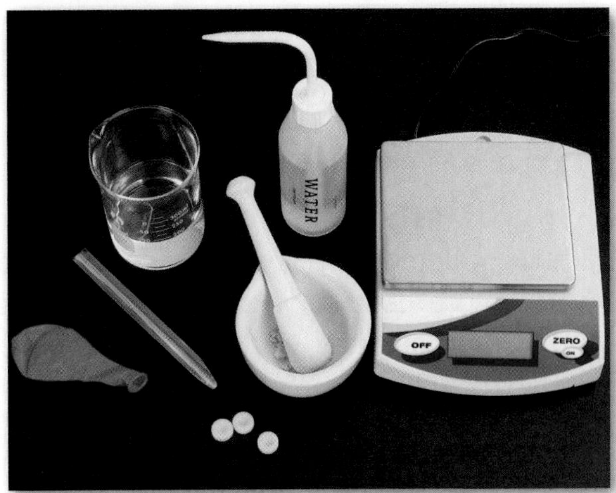

Follow Your Plan

1. Make sure your teacher approves your plan.
2. Copy the data table into your Science Journal to record measurements.
3. While doing the experiment, write your observations, and complete the data table in your Science Journal.

Data Table

Trial Number	Mass Before Reaction	Mass After Reaction
Do not write in this book.		

Analyze Your Data

1. **Describe** the effects of the chemical reaction between the antacid powder and the water.
2. **Compare** the total mass of the substances before the reaction to their mass after the reaction.
3. **Graph** the mass of the substances before and after the reaction using a bar graph.
4. **Explain** if your data supported your hypothesis.

Conclude and Apply

1. **Identify** possible experimental errors in your experiment design.
2. **Infer** how mass might have been lost or added between the initial and final weighing of the substances.
3. **Infer** what the expected mass of the substances should be after the chemical reaction occurred.

Communicating Your Data

Compare the data your group collected with the data collected by the other groups and discuss possible reasons for differences in the data.

SCIENCE Stats

Intriguing Elements

Did you know...

... Silver-white cobalt, which usually is combined with other elements in nature, is used to create rich paint pigments. It can be used to form powerful magnets, treat cancer patients, build jet engines, and prevent disease in sheep.

... Gold is the most ductile (stretchable) of all the elements. Just 29 g of gold—about ten wedding bands—can be pulled into a wire 100 km long. That's long enough to stretch from Toledo, Ohio, to Detroit, Michigan, and beyond.

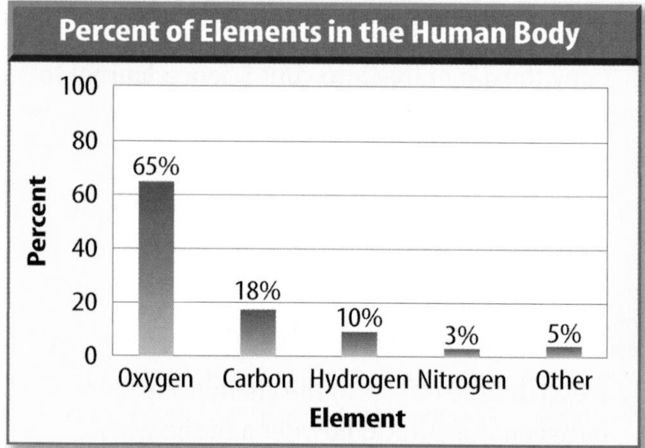

... Zinc makes chewing gum taste better. Up to 0.3 mg of zinc acetate can be added per 1000 mg of chewing gum to provide a tart, zingy flavor.

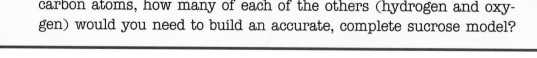

Applying Math

1. Zinc acetate is approximately 35 percent zinc. How many grams of chewing gum would be needed to provide a total of 10.0 mg of zinc?
2. If you wanted to produce a gold wire with the same diameter as mentioned in the example above, how many grams would be needed to make a wire one meter in length?
3. Table sugar has the chemical formula $C_{12}H_{22}O_{11}$. If you were going to build a scale model of sucrose (table sugar) and you had 36 model carbon atoms, how many of each of the others (hydrogen and oxygen) would you need to build an accurate, complete sucrose model?

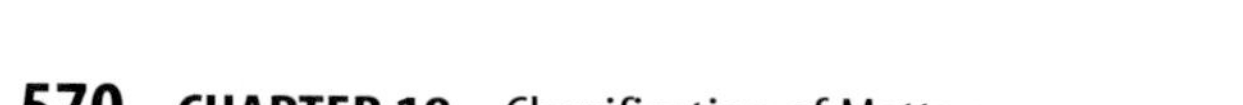

Reviewing Main Ideas

Section 1 Composition of Matter

1. Elements and compounds are substances. A mixture is composed of two or more substances.
2. You can distinguish between the different materials in a heterogeneous mixture using either your unaided eye or a microscope.
3. Colloids and suspensions are two types of heterogeneous mixtures. The particles in a suspension will settle eventually. Particles of a colloid will not. Milk is an example of a colloid.
4. In a homogeneous mixture, the particles are distributed evenly and are not visible, even when using a microscope. Homogeneous mixtures can be composed of solids, liquids, or gases.
5. A solution is another name for a homogeneous mixture that remains constantly and uniformly mixed.

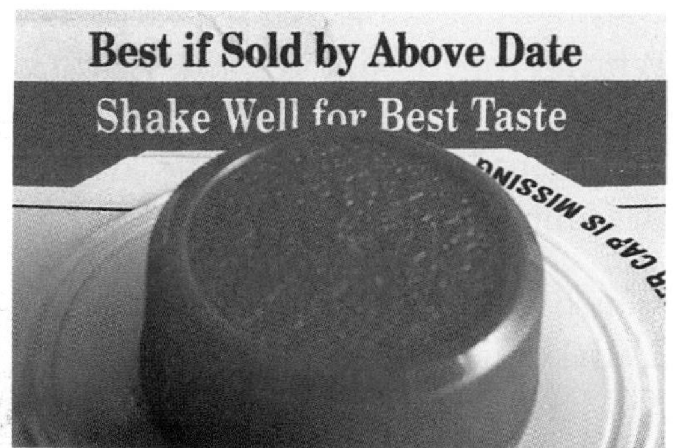

This is a clear indication that this mixture is a suspension.

Section 2 Properties of Matter

1. Physical properties are characteristics of materials that you can observe without changing the identities of the substances.
2. Chemical properties indicate what chemical changes substances can undergo. Many medicines are stored in dark bottles because they react with light.
3. In physical changes, the identities of substances remain unchanged.
4. In chemical changes, the identities of substances change: new substances are formed. A visible chemical change takes place when rust is cleaned with bleach.

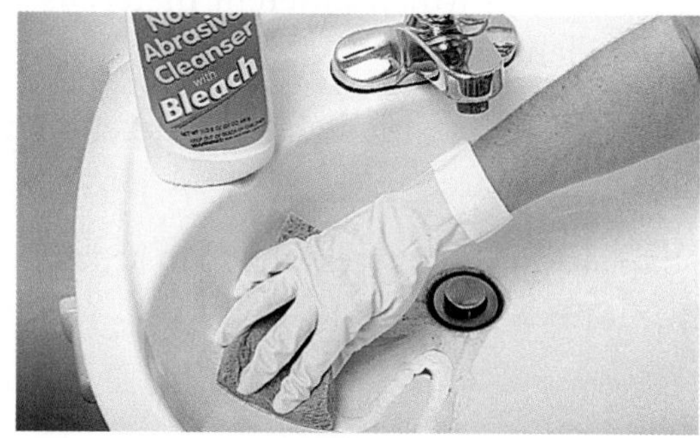

5. The law of conservation of mass states that during any chemical change, matter is neither created nor destroyed.

FOLDABLES Use the Foldable that you made at the beginning of the chapter to help you review the classification of matter.

chapter 18 Review

Using Vocabulary

chemical change p. 564
chemical property p. 563
colloid p. 556
compound p. 554
distillation p. 563
element p. 552
heterogeneous mixture p. 555
homogeneous mixture p. 556
law of conservation of mass p. 567
physical change p. 562
physical property p. 560
solution p. 556
substance p. 552
suspension p. 558
Tyndall effect p. 557

Complete each sentence with the correct vocabulary word or words.

1. Substances formed from atoms of two or more elements are called ________.
2. A(n) ________ is a heterogeneous mixture in which visible particles settle.
3. Freezing, boiling, and evaporation are all examples of ________.
4. According to the ________, matter is neither created nor destroyed during a chemical change.
5. A mixture in which different materials are easily identified is ________.
6. Compounds are made from different atoms of two or more ________.
7. Distillation is a process that can separate two liquids using ________.

Checking Concepts

Choose the word or phrase that best answers the question.

8. Bending a copper wire is an example of what property?
   A) chemical **C)** element
   B) conservation **D)** physical
9. Which is **NOT** an element?
   A) carbon **C)** oxygen
   B) hydrogen **D)** water
10. Which is an example of a chemical change?
   A) boiling **C)** evaporation
   B) burning **D)** melting
11. What is gelatin?
   A) colloid **C)** substance
   B) compound **D)** suspension
12. Which is an example of a visible sunbeam?
   A) an element **C)** a compound
   B) a solution **D)** the Tyndall effect
13. How would you classify the color of a rose?
   A) chemical change
   B) chemical property
   C) physical change
   D) physical property
14. How would you describe the process of evaporating water from seawater?
   A) chemical change
   B) chemical property
   C) physical change
   D) physical property
15. Which is a substance?
   A) colloid **C)** mixture
   B) element **D)** solution
16. Which property can be used to help identify an unknown substance?
   A) specific heat **C)** temperature
   B) combustion **D)** Tyndall effect
17. Which statement about the law of conservation of mass is **TRUE**?
   A) Mass is created only.
   B) Mass is created and destroyed.
   C) Mass is destroyed only.
   D) Mass is neither created nor destroyed.

Vocabulary PuzzleMaker gpescience.com

Interpreting Graphics

18. Copy and complete the concept map below about matter.

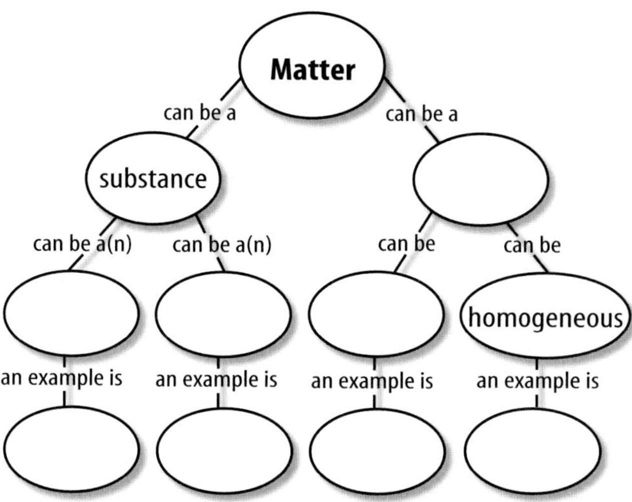

Use the table below to answer question 19.

Common Colloids

Colloid	Example
Solid in a liquid	Gelatin
Solid in a gas	
Gas in a solid	Do not
Solid in a liquid	write in this book.
Liquid in a gas	

19. **Identify** how different colloids can involve different states. For example, gelatin is formed from solid particles in a liquid. Complete the table using these colloids: *smoke, marshmallow, fog,* and *paint.*

Thinking Critically

20. **Compare and contrast** elements and compounds. Give two examples of each.

21. **Explain** Carbon and the gases hydrogen and oxygen combine to form sugar. How do you know sugar is a compound?

22. **Use** a nail rusting in air to explain the law of conservation of mass.

23. **Use Variables, Constants, and Controls** Marcos took a 100-cm^3 sample of a suspension, shook it well, and poured equal amounts into four different test tubes. He placed one test tube in a rack, one in hot water, one in warm water, and the fourth in ice water. He then observed the time it took for each suspension to settle. What was the variable in the experiment? What was one constant?

24. **Concept Map** Make a network tree to show types of liquid mixtures. Include these terms: *homogeneous mixtures, heterogeneous mixtures, solutions, colloids,* and *suspensions.*

Applying Math

Use the graph below to answer question 25.

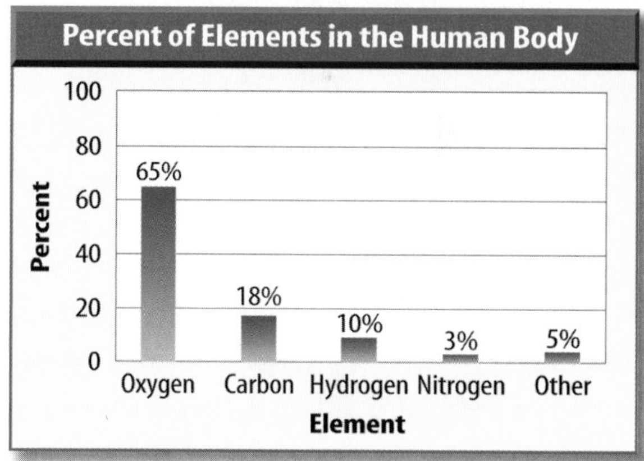

25. **Use Proportions** If you had 50 atoms of hydrogen, how many atoms each of oxygen, carbon, and nitrogen would you need to have the elements in the same proportion as they are in your body?

26. **Use Numbers** In the following equation, 243.5 g of Sb (antimony) reacts completely with 1000 g of I_2 (iodine) to form 1004.9 g of SbI_3 (antimony triiodide). How many grams of I_2 were consumed in the reaction? $2\ Sb + 3\ I_2 \rightarrow 2\ SbI_3$

chapter 18 Standardized Test Practice

Record your answers on the answer sheet provided by your teacher or on a sheet of paper.

Multiple Choice

1. $CaCO_3$ is an example of which type of material?

   A. colloid

   B. compound

   C. element

   D. mixture

Use the graph to answer question 2.

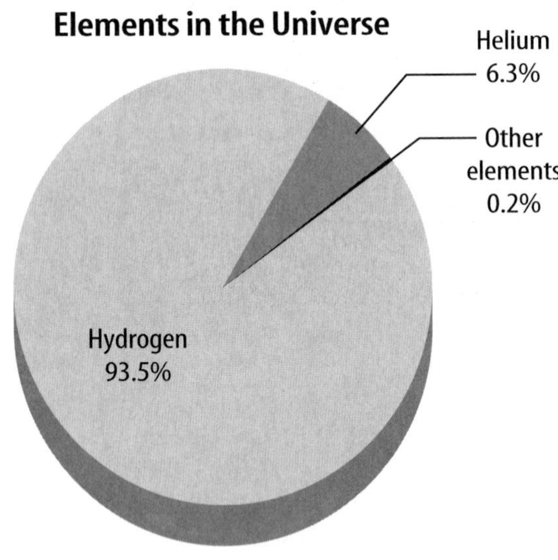

2. What percentage do the elements hydrogen and helium account for in the universe?

   A. 100 percent

   B. 99.9 percent

   C. 99.8 percent

   D. 98 percent

Use the graph below to answer question 3.

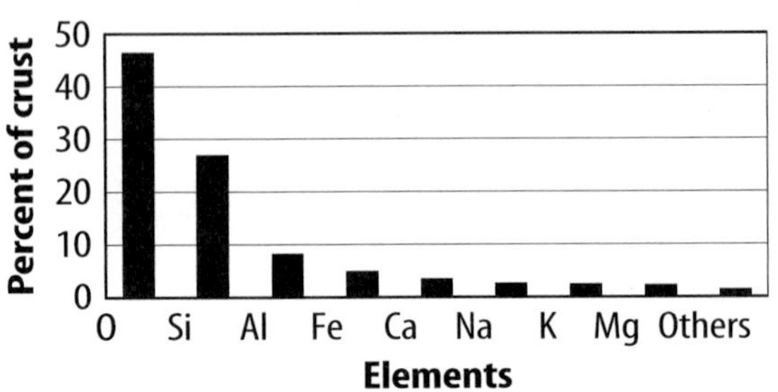

3. Which element makes up 8 percent of Earth's crust?

   A. aluminum

   B. iron

   C. oxygen

   D. silicon

4. Which element has the physical property of magnetism?

   A. calcium

   B. iron

   C. oxygen

   D. sodium

5. Which statement best decribes the law of conservation of mass?

   A. The mass of the products is always greater than the mass of the materials that react in a chemical change.

   B. The mass of the products is always less than the mass of the materials that react in a chemical change.

   C. A certain mass of material must be present for a reaction to occur.

   D. Matter is neither lost nor gained during a chemical change.

6. The most plentiful element in the universe readily burns in air. What is this chemical property called?

   A. flammability

   B. ductility

   C. density

   D. boiling point

Gridded Response

7. In the following reaction, 24 g of CH_4 (methane) react completely with 96 g of O_2 to form 66 g of CO_2. How many grams of H_2O are formed?

8. A reaction of 22.85 g of sodium hydroxide (NaOH) with 20.82 g of hydrogen chloride (HCl) gives off 10.29 g of water (H_2O). What mass of sodium chloride (NaCl) is formed in this reaction?

Short Response

Use the illustration below to answer questions 9.

9. What physical properties could be used to identify these minerals?

10. You are given a mixture of iron filings, sand, and salt. Describe how to separate this mixture.

Extended Response

11. Describe how a compound is a combination of elements in a fixed proportion. Give two examples.

Use the illustration below to answer question 12.

12. Design an experiment that shows that this type of chemical change is governed by the law of conservation of mass.

Test-Taking Tip

Directions and Instructions Listen carefully to the instructions from your teacher. Read the directions and each question carefully.

chapter 19

Properties of Atoms and the Periodic Table

chapter preview

sections

Atoms Compose All Things, Great and Small

Everything in this photo and the universe is composed of tiny particles called atoms. You will learn about atoms and their components: protons, neutrons, electrons, and quarks.

Science Journal In your Science Journal, write a few paragraphs about what you know about atoms.

Start-Up Activities

Inferring What You Can't Observe

How do scientists study atoms when they cannot see them? In situations such as these, techniques must be developed to find clues to answer questions. Do the lab below to see how clues might be gathered.

1. Complete the safety form.
2. Take an envelope and several items from your teacher.
3. Place an assortment of items in the envelope and seal it.
4. Trade envelopes with another group.
5. Without opening the envelope, try to figure out the types and number of items that are in the envelope. Record a hypothesis about the contents of the envelope in your Science Journal.
6. After you record your hypothesis, open the envelope and see what is inside.
7. **Think Critically** Describe the contents of your envelope. Was your hypothesis correct?

Preview this chapter's content and activities at gpescience.com

Atoms You have probably studied atoms before. Make the following Foldable to help identify what you already know, what you want to know, and what you learn about atoms.

STEP 1 **Fold** a sheet of paper vertically from side to side. Make the front edge about 1.25 cm shorter than the back edge.

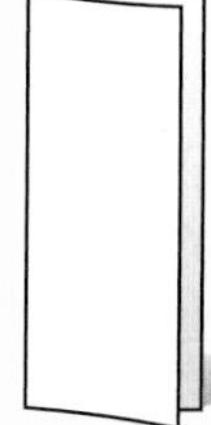

STEP 2 **Turn** lengthwise and **fold** into thirds.

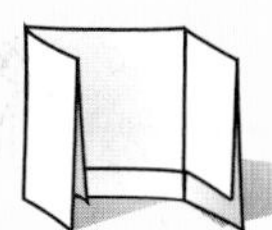

STEP 3 **Unfold and cut** only the top layer along both folds to make three tabs.

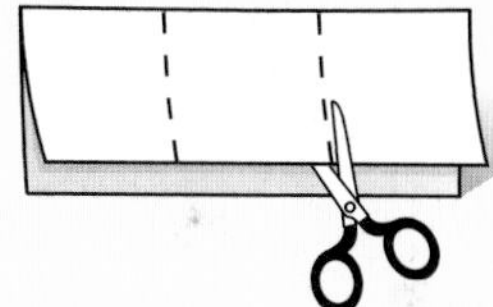

STEP 4 **Label** each tab as shown.

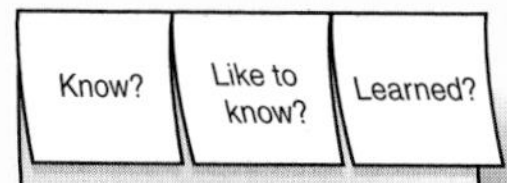

Identify Questions Before you read the chapter, write what you already know about atoms under the left tab of your Foldable, and write questions about what you'd like to know under the center tab. After you read the chapter, list what you learned under the right tab.

section 1

Structure of the Atom

Reading Guide

What You'll Learn

- **Identify** the names and symbols of common elements.
- **Identify** quarks as subatomic particles of matter.
- **Describe** the electron cloud model of the atom.
- **Explain** how electrons are arranged in an atom.

Why It's Important

Everything that you see, touch, and breathe is composed of tiny atoms.

Review Vocabulary

element: substance with atoms that are all alike

New Vocabulary

- atom
- nucleus
- proton
- neutron
- electron
- quark
- electron cloud

Scientific Shorthand

Do you have a nickname? Do you use abbreviations for long words or the names of states? Scientists also do this. In fact, scientists have developed their own shorthand for dealing with long, complicated names.

Do the letters C, Al, Ne, and Ag mean anything to you? Each letter or pair of letters is a chemical symbol, which is a short or abbreviated way to write the name of an element. Chemical symbols, such as those in **Table 1,** consist of one capital letter or a capital letter plus one or two lowercase letters. For some elements, the symbol is the first letter of the element's name. For other elements, the symbol is the first letter of the name plus another letter from its name. Some symbols are derived from Latin. For instance, *argentum* is Latin for "silver." Elements have been named in a variety of ways. Some elements are named to honor scientists, for places, or for their properties. Other elements are named using rules established by an international committee. Regardless of the origin of the names, scientists derived this international system for convenience. It is much easier to write H for hydrogen, O for oxygen, and H_2O for dihydrogen oxide (water) than to write out the names. Because scientists worldwide use this system, everyone understands what the symbols mean.

Table 1 Symbols of Some Elements

Element	Symbol	Element	Symbol
Aluminum	Al	Iron	Fe
Calcium	Ca	Mercury	Hg
Carbon	C	Nitrogen	N
Chlorine	Cl	Oxygen	O
Gold	Au	Potassium	K
Hydrogen	H	Sodium	Na

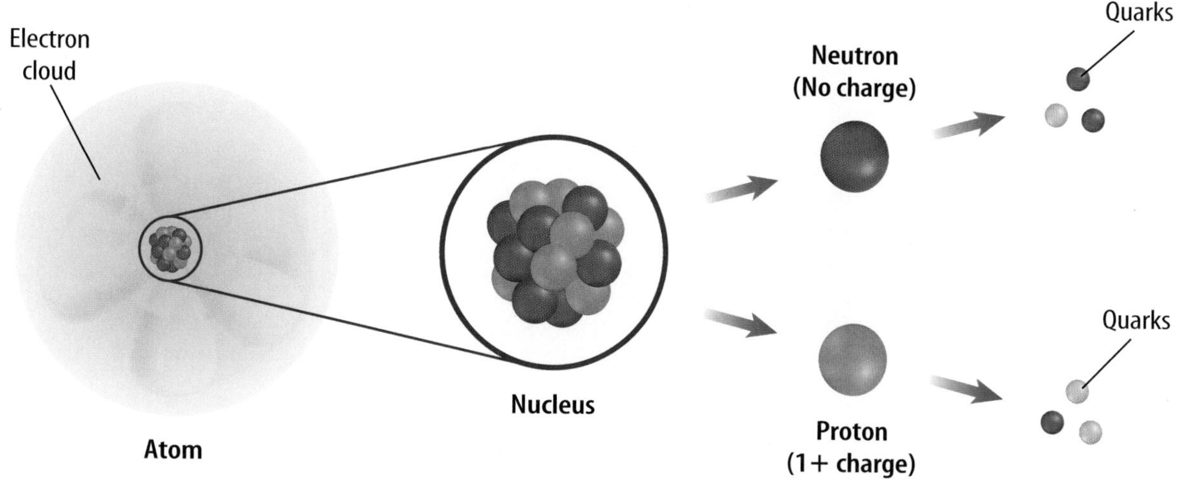

Figure 1 The nucleus of the atom contains protons and neutrons that are composed of quarks. The proton has a positive charge and the neutron has no charge. A cloud of negatively charged electrons surrounds the nucleus of the atom.
Explain *how atoms of different elements differ.*

Atomic Components

An element is matter that is composed of one type of **atom,** which is the smallest piece of matter that still retains the property of the element. For example, the element silver is composed of only silver atoms, and the element hydrogen is composed of only hydrogen atoms. Atoms are composed of particles called protons, neutrons, and electrons, as shown in **Figure 1.** Protons and neutrons are found in a small, positively-charged center of the atom called the **nucleus,** which is surrounded by a cloud containing electrons. **Protons** are particles with an electrical charge of 1+. **Neutrons** are neutral particles that do not have an electrical charge. **Electrons** are particles with an electrical charge of 1−. Atoms of different elements differ in the number of protons they contain.

Reading Check *What are the particles that make up the atom and where are they located?*

Topic: Particle Research
Visit gpescience.com for Web links to information about particle research at Fermi National Accelerator Laboratory.

Activity Write a paragraph describing the information that you found at the site.

Quarks: Even Smaller Particles

Are the protons, electrons, and neutrons that make up atoms the smallest particles that exist? Scientists hypothesize that electrons are not composed of smaller particles and are one of the most basic types of particles. Protons and neutrons, however, are made up of smaller particles called **quarks.** So far, scientists have confirmed the existence of six uniquely different quarks. Scientists theorize that an arrangement of three quarks held together with the strong nuclear force produces a proton. Another arrangement of three quarks produces a neutron. The search for the composition of protons and neutrons is an ongoing effort.

Figure 2 The Tevatron is a huge machine. The aerial photograph of Fermi National Accelerator Laboratory shows the circular outline of the Tevatron particle accelerator. The close-up photograph of the Tevatron gives you a better view of the tunnel.

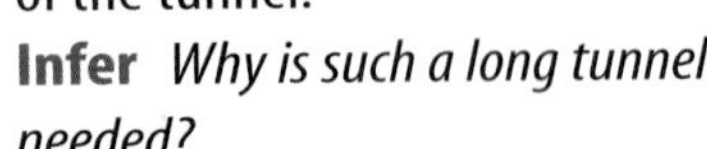

Infer *Why is such a long tunnel needed?*

Finding Quarks To study quarks, scientists accelerate charged particles to tremendous speeds and then force them to collide with—or smash into—protons. This collision causes the proton to break apart. The Fermi National Accelerator Laboratory, a research laboratory in Batavia, Illinois, houses a machine that can generate the forces that are required to make protons collide. This machine, the Tevatron, shown in **Figure 2,** is approximately 6.4 km in circumference. Electric and magnetic fields are used to accelerate, focus, and make the fast-moving particles collide.

The particles that result from the collision can be detected by various collection devices. Often, scientists use multiple collection devices to collect the greatest possible amount of information about the particles created in a collision. Just as police investigators can reconstruct traffic accidents from tire marks and other clues at the scene, scientists are able to examine and gather information about the particles, as shown in **Figure 3.** Scientists use inferences to identify the subatomic particles and to discover information about each particle's inner structure.

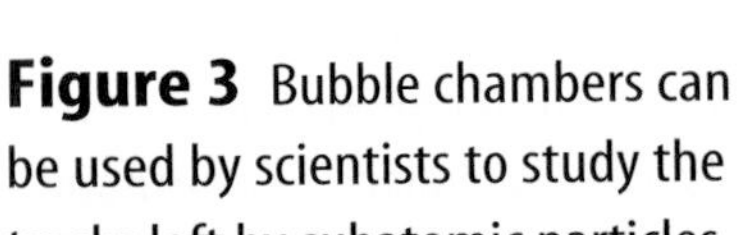

Figure 3 Bubble chambers can be used by scientists to study the tracks left by subatomic particles.

The Sixth Quark Finding evidence for the existence of quarks was not an easy task. Scientists found five quarks and hypothesized that a sixth quark existed. However, it took a team of nearly 450 scientists from around the world several years to find the sixth quark. The tracks of the sixth quark were hard to detect because only about one-billionth of a percent of the proton collisions performed showed the presence of a sixth quark, typically referred to as the *top* quark.

Models—Tools for Scientists

Scientists and engineers use models to represent things that are difficult to visualize or picture in the mind. You might have seen models of buildings, the solar system, or airplanes. These are scaled-down models. Scaled-down models allow you to see either something too large to see all at once or something that has not been built yet. Scaled-up models are often used to visualize things that are too small to see. To give you an idea of how small the atom is, it would take about 24,400 atoms stacked one on top of the other to equal the thickness of a sheet of aluminum foil. To study the atom, scientists have developed scaled-up models that they can use to visualize how the atom is constructed. For the model to be useful, it must support all of the information that is known about matter and the behavior of atoms. As more information about the atom is collected, scientists change their models to include the new information.

Reading Check *Explain how models can simplify science.*

The Changing Atomic Model You know now that all matter is composed of atoms, but this was not always known. Around 400 B.C., Democritus proposed the idea that atoms make up all substances. However, another famous Greek philosopher, Aristotle, disputed Democritus's theory and proposed that matter is uniform throughout and is not composed of smaller particles. Aristotle's incorrect theory was accepted for about 2000 years. In the 1800s, John Dalton, an English scientist, was able to offer proof that atoms exist.

Dalton's model of the atom, a solid sphere shown in **Figure 4,** was an early model of the atom. As you can see in **Figure 5,** the model has changed somewhat over time. Dalton's modernization of Aristotle's idea of the atom provided a physical explanation for chemical reactions. Scientists could then express these reactions in quantitative terms using chemical symbols and equations.

Modeling an Aluminum Atom

Procedure

1. Complete the safety form.
2. Arrange thirteen 3-cm circles cut from **orange paper** and fourteen 3-cm circles cut from **blue paper** on a **flat surface** to represent the nucleus of an atom. Each orange circle represents one proton, and each blue circle represents one neutron.
3. Position two holes punched from **red paper** about 20 cm from your nucleus.
4. Position eight punched holes about 40 cm from your nucleus.
5. Position three punched holes about 60 cm from your nucleus.

Analysis

1. How many protons, neutrons, and electrons does an aluminum atom have?
2. Explain how your circles model an aluminum atom.
3. Explain why your model does not accurately represent the true size and distances in an aluminum atom.

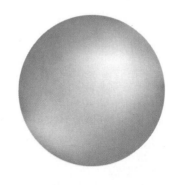

Figure 4 John Dalton's atomic model was a simple sphere.

NATIONAL GEOGRAPHIC VISUALIZING THE ATOMIC MODEL

Figure 5

The ancient Greek philosopher Democritus proposed that elements consist of tiny, solid particles that cannot be subdivided (A). He called these particles *atomos,* meaning "uncuttable." This concept of the atom's structure remained largely unchallenged until the 1900s, when researchers began to discover through experiments that atoms are composed of still smaller particles. In the early 1900s, a number of models for atomic structure were proposed (B-D). The currently accepted model (E) evolved from these ideas and the work of many other scientists.

A DEMOCRITUS'S UNCUTTABLE ATOM

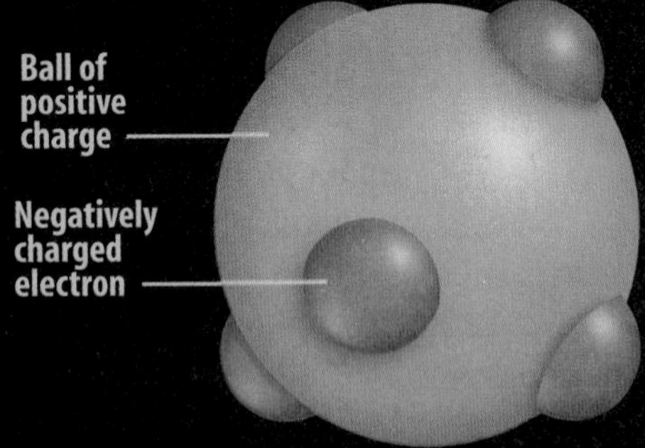

B THOMSON MODEL, 1904 English physicist Joseph John Thomson inferred from his experiments that atoms contain small, negatively charged particles. He thought these "electrons" (in red) were evenly embedded throughout a positively charged sphere, much like chocolate chips in a ball of cookie dough.

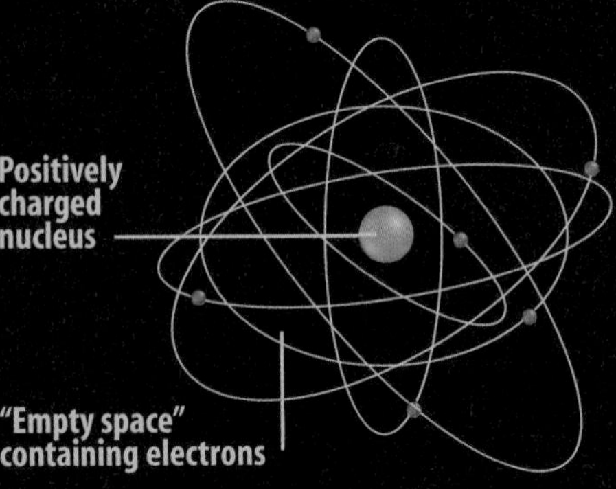

C RUTHERFORD MODEL, 1911 Another British physicist, Ernest Rutherford, proposed that almost all the mass of an atom—and all its positive charges—is concentrated in a central atomic nucleus surrounded by electrons.

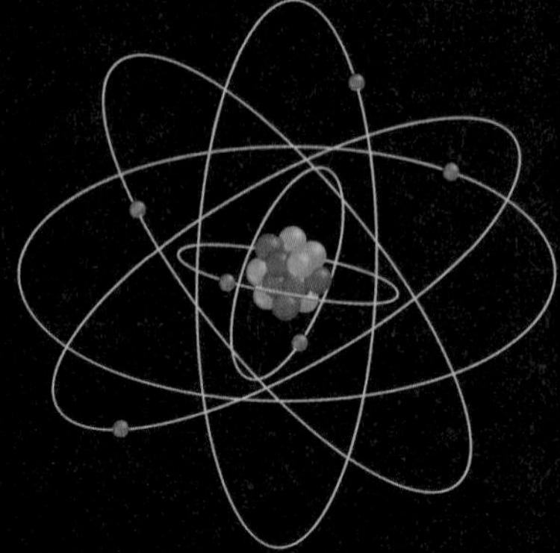

D BOHR MODEL, 1913 Danish physicist Niels Bohr hypothesized that electrons travel in fixed orbits around the atom's nucleus. James Chadwick, a student of Rutherford, concluded that the nucleus contains positive protons and neutral neutrons.

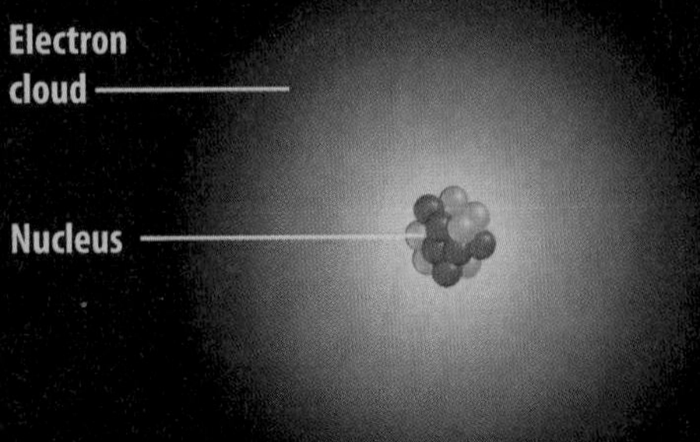

E ELECTRON CLOUD MODEL, CURRENT According to the currently accepted model of atomic structure, electrons do not follow fixed orbits but tend to occur more frequently in certain areas around the nucleus at any given time.

The Electron Cloud Model By 1926, scientists had developed the electron cloud model of the atom that is in use today. An **electron cloud** is the area around the nucleus of an atom where its electrons are most likely found. The electron cloud is 100,000 times larger than the diameter of the nucleus. In contrast, each electron in the cloud is much smaller than a single proton.

Because an electron's mass is small and the electron is moving so quickly around the nucleus, it is impossible to describe its exact location in an atom. Picture the spokes on a moving bicycle wheel. They are moving so quickly that you can't pinpoint any single spoke. All you see is a blur that contains all of the spokes somewhere within it. In the same way, an electron cloud is a blur containing all of the electrons of the atom somewhere within it. **Figure 6** illustrates what the electron cloud might look like.

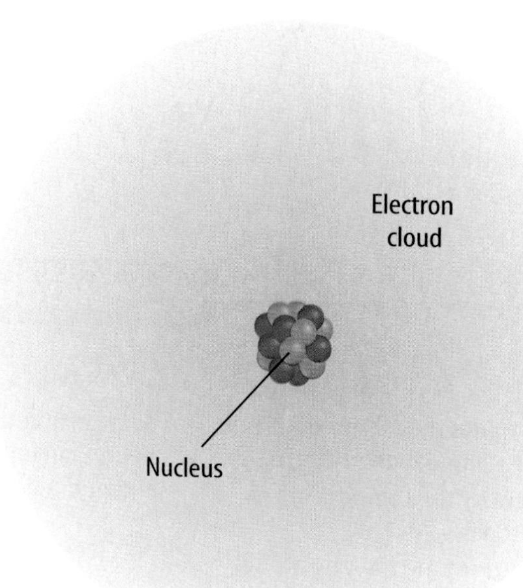

Figure 6 The electrons are located in an electron cloud surrounding the nucleus of the atom.

section 1 review

Summary

Scientific Shorthand

- Scientists use chemical symbols as shorthand when writing the names of elements.

Atomic Components

- Atoms are composed of small particles that have known charges.
- Atoms of different elements differ in the number of protons they contain.

Quarks: Even Smaller Particles

- So far, scientists have confirmed the existence of six different quarks.

Models: Tools for Scientists

- Models are used by scientists to simplify the study of concepts and things.
- The current atomic model is an accumulation of over 200 years of knowledge.
- The electron cloud model is the current atomic model.

Self Check

1. **List** the chemical symbols for the elements carbon, aluminum, hydrogen, oxygen, and sodium.
2. **Identify** the names, charges, and locations of three kinds of particles that make up an atom.
3. **Identify** the smallest particle of matter. How were these particles discovered?
4. **Describe** the electron cloud model of the atom.
5. **Think Critically** Explain how a rotating electric fan might be used to model the atom. Explain how the rotating fan is unlike an atom.

Applying Math

6. **Use Numbers** The mass of a proton is estimated to be 1.6726×10^{-24} g, and the mass of an electron is estimated to be 9.1093×10^{-28} g. How many times larger is the mass of a proton than the mass of an electron?
7. **Calculate** What is the difference between the mass of a proton and the mass of an electron?

section 2

Masses of Atoms

Reading Guide

What You'll Learn

- **Compute** the atomic mass and mass number of an atom.
- **Identify** the components of isotopes.
- **Interpret** the average atomic mass of an element.

Why It's Important

Some elements naturally exist in more than one form—radioactive and nonradioactive.

Review Vocabulary

mass: amount of matter in an object

New Vocabulary

- atomic number
- mass number
- isotope
- average atomic mass

Atomic Mass

The nucleus contains most of the mass of an atom because protons and neutrons are far more massive than electrons. The mass of a proton is about the same as that of a neutron—approximately 1.6726×10^{-24} g, as shown in **Table 2.** The mass of each is approximately 1,836 times greater than the mass of an electron. An electron's mass is so small that it is considered negligible when finding the mass of an atom.

Table 2 Subatomic Particle Masses

Particle	Mass (g)
Proton	1.6726×10^{-24}
Neutron	1.6749×10^{-24}
Electron	9.1093×10^{-28}

If you were asked to estimate the height of your school building, you probably wouldn't give an answer in kilometers. The number would be too cumbersome to use. Considering the scale of the building, you would more likely give the height in a smaller unit, meters. When thinking about the small masses of atoms, scientists found that even grams were not small enough to use for measurement. Scientists need a unit that results in more manageable numbers. The unit of measurement used for atomic particles is the atomic mass unit (amu). The mass of a proton or a neutron is almost equal to 1 amu. This is not coincidence: the unit was defined that way. The atomic mass unit is defined as one-twelfth the mass of a carbon atom containing six protons and six neutrons. Remember that the mass of the carbon atom is contained almost entirely in the mass of the protons and neutrons that are located in the nucleus. Therefore, each of the 12 particles in the nucleus must have a mass nearly equal to 1 amu.

Reading Check *Where is the majority of the mass of an atom located?*

Table 3 Mass Numbers of Some Atoms

Element	Symbol	Atomic Number	Protons	Neutrons	Mass Number	Average Atomic Mass*
Boron	B	5	5	6	11	10.81 amu
Carbon	C	6	6	6	12	12.01 amu
Oxygen	O	8	8	8	16	16.00 amu
Sodium	Na	11	11	12	23	22.99 amu
Copper	Cu	29	29	34	63	63.55 amu

*The atomic mass units are rounded to two decimal places.

Protons Identify an Element You learned earlier that atoms of different elements are different because they have different numbers of protons. In fact, the number of protons tells you what type of atom you have and vice versa. For example, every carbon atom has six protons. Also, all atoms with six protons are carbon atoms. Atoms with eight protons are oxygen atoms. The number of protons in an atom is equal to a number called the **atomic number.** The atomic number of carbon is 6. Therefore, if you are given any one of the following—the name of the element, the number of protons in the element, or the atomic number of the element—you can determine the other two.

Reading Check *Which element is an atom with six protons in the nucleus?*

Mass Number The **mass number** of an atom is the sum of the number of protons and the number of neutrons in the nucleus of the atom. Look at **Table 3** to see that this is true.

If you know the mass number and the atomic number of an atom, you can calculate the number of neutrons. The number of neutrons is equal to the atomic number subtracted from the mass number.

$$\text{number of neutrons} = \text{mass number} - \text{atomic number}$$

Atoms of the same element with different numbers of neutrons can have different properties. For example, carbon with a mass number equal to 12, or carbon-12, is the most common form of carbon. Carbon-14 is present on Earth in much smaller quantities. Carbon-14 is radioactive, while carbon-12 is not.

Carbon Dating Living organisms on Earth contain carbon. Carbon-12 makes up 99 percent of this carbon. Carbon-13 and carbon-14 make up the other one percent. Which isotopes are archaeologists most interested in when they determine the age of carbon-containing remains? Explain your answer in your Science Journal.

Isotopes

Not all the atoms of an element have the same number of neutrons. Atoms of the same element that have different numbers of neutrons are called **isotopes.** Suppose you have a sample of the element boron. Naturally occurring atoms of boron have mass numbers of 10 or 11. How many neutrons are in a boron atom? It depends upon the isotope of boron to which you are referring. Obtain the number of protons in boron from the periodic table. Then use the formula on the previous page to calculate the number of neutrons in each boron isotope. You can determine that boron can have five or six neutrons.

Uranium-238 has 92 protons. How many neutrons does it have?

Applying Science

How can radioactive isotopes help tell time?

Atoms can be used to measure the age of bones or rock formations that are millions of years old. The time it takes for half of the radioactive atoms in a piece of rock or bone to change into another element is called its half-life. Scientists use the half-lives of radioactive isotopes to measure geologic time.

Half-Lives of Radioactive Isotopes

Radioactive Element	Changes to This Element	Half-Life
Uranium-238	lead-206	4,460 million years
Potassium-40	argon-40, calcium-40	1,260 million years
Rubidium-87	strontium-87	48,800 million years
Carbon-14	nitrogen-14	5,715 years

Identifying the Problem

The table above lists the half-lives of some radioactive isotopes and into which elements they change. For example, it would take 5,715 years for half of the carbon-14 atoms in a rock to change into atoms of nitrogen-14. After another 5,715 years, half of the remaining carbon-14 atoms will change, and so on. You can use these radioactive clocks to measure different periods of time.

Solving the Problem

1. How many years would it take half of the rubidium-87 atoms in a piece of rock to change into strontium-87? How many years would it take for 75 percent of the atoms to change?
2. After a long period, only 25 percent of the atoms in a rock remained uranium-238. How many years old would you predict the rock to be? The other 75 percent of the atoms are now which radioactive element?

Identifying Isotopes Models of two isotopes of boron are shown in **Figure 7.** Because the numbers of neutrons in the isotopes are different, the mass numbers are also different. You use the name of the element followed by the mass number of the isotope to identify each isotope: boron-10 and boron-11. Because most elements have more than one isotope, each element has an average atomic mass. The **average atomic mass** of an element is the weighted-average mass of the mixture of its isotopes. For example, four out of five atoms of boron are boron-11, and one out of five is boron-10. To find the weighted-average, or the average atomic mass of boron, you would use the following equation:

$$\frac{4}{5}(11 \text{ amu}) + \frac{1}{5}(10 \text{ amu}) = 10.8 \text{ amu}$$

The average atomic mass of the element boron is 10.8 amu. Note that the average atomic mass of boron is close to the mass of its most abundant isotope, boron-11.

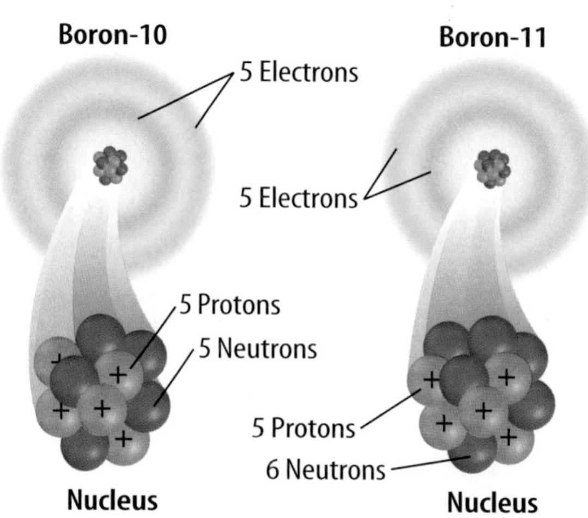

Figure 7 Boron-10 and boron-11 are two isotopes of boron. These two isotopes differ by one neutron.
Explain *why these atoms are isotopes.*

section 2 review

Summary

Atomic Mass

- The nucleus contains most of the mass of an atom.
- The masses of a proton and neutron are approximately equal.
- The mass of an electron is considered negligible when finding the mass of an atom.
- The unit of measurement for atomic particles is the atomic mass unit (amu).
- The carbon-12 isotope was used to define the atomic mass unit.
- The number of protons identifies an element.

Isotopes

- Atoms of the same element with different numbers of neutrons are called isotopes.
- The average atomic mass of an element is the weighted-average mass of the mixture of isotopes.

Self Check

1. **Identify** the mass number and atomic number of a chlorine atom that has 17 protons and 18 neutrons.
2. **Explain** how the isotopes of an element are alike and how are they different.
3. **Explain** why the atomic mass of an element is an average mass.
4. **Explain** how you would calculate the number of neutrons in potassium-40.
5. **Think Critically** Chlorine has an average atomic mass of 35.45 amu. The two naturally occurring isotopes of chlorine are chlorine-35 and chlorine-37. Why does this indicate that most chlorine atoms contain 18 neutrons?

Applying Math

6. **Use Numbers** If a hydrogen atom has two neutrons and one proton, what is its mass number?
7. **Use Tables** Use the information in **Table 2** to find the mass in kilograms of each subatomic particle.

section 3

The Periodic Table

Reading Guide

What You'll Learn

- **Explain** the composition of the periodic table.
- **Use** the periodic table to obtain information.
- **Explain** what the terms *metal, nonmetal,* and *metalloid* mean.

Why It's Important

The periodic table is an organized list of the elements that compose all living and nonliving things known to exist in the universe.

Review Vocabulary

chemical property: any characteristic of a substance that indicates whether it can undergo a certain chemical change

New Vocabulary

- periodic table
- group
- electron dot diagram
- period

Organizing the Elements

On a clear evening, you can see one of the various phases of the Moon. Each month, the Moon seems to grow larger, then smaller, in a repeating pattern. This type of change is periodic. *Periodic* means "repeated in a pattern." The days of the week are periodic because they repeat themselves every seven days. The calendar is a periodic table of days and months.

Figure 8 Mendeleev discovered that the elements have a periodic pattern in their chemical properties.

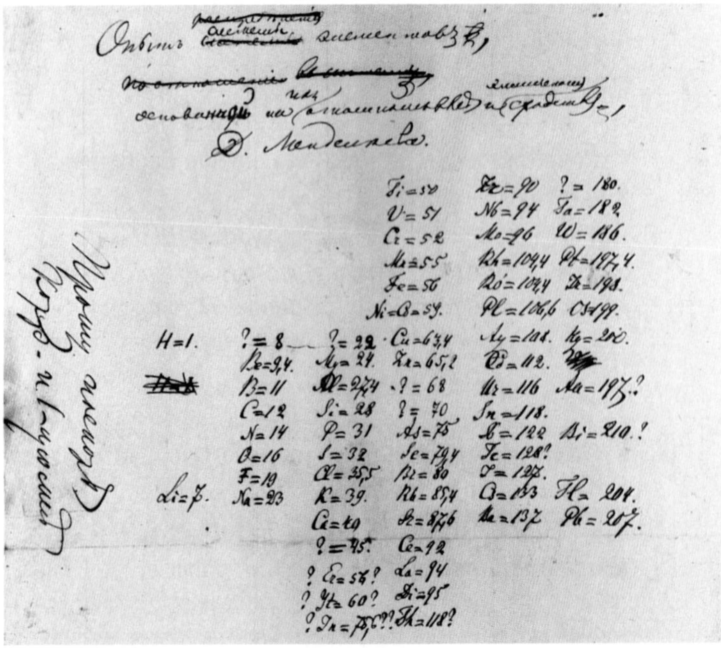

In the late 1800s, Dmitri Mendeleev, a Russian chemist, searched for a way to organize the elements. When he arranged all the elements known at that time in order of increasing atomic masses, he discovered a pattern. **Figure 8** shows Mendeleev's early periodic chart. Chemical properties found in lighter elements could be shown to repeat in heavier elements. Because the pattern repeated, it was considered to be periodic. Today, this arrangement is called the periodic table of elements. In the **periodic table,** the elements are arranged by increasing atomic number and by changes in physical and chemical properties.

Table 4 Mendeleev's Predictions

Predicted Properties of Ekasilicon (Es)	Actual Properties of Germanium (Ge)
Existence Predicted: 1871	*Actual Discovery: 1886*
Atomic mass = 72	Atomic mass = 72.61
High melting point	Melting point = 938°C
Density = 5.5 g/cm^3	Density = 5.323 g/cm^3
Dark gray metal	Gray metal
Density of EsO_2 = 4.7 g/cm^3	Density of GeO_2 = 4.23 g/cm^3

Mendeleev's Predictions Mendeleev had to leave blank spaces in his periodic table to keep the elements properly lined up according to their chemical properties. He looked at the properties and atomic masses of the elements surrounding these blank spaces. From this information, he was able to predict the properties and the mass numbers of new elements that had not yet been discovered. **Table 4** shows Mendeleev's predicted properties for germanium, which he called ekasilicon. His predictions proved to be accurate. Scientists later discovered these missing elements and found that their properties were extremely close to what Mendeleev had predicted.

Reading Check *How did Mendeleev organize his periodic table?*

Improving the Periodic Table Although Mendeleev's arrangement of elements was successful, it did need some changes. On Mendeleev's table, the atomic mass gradually increased from left to right. If you look at the modern periodic table, shown in **Table 5,** you will see several examples, such as cobalt and nickel, in which the mass decreases from left to right. You also might notice that the atomic number always increases from left to right. In 1913, the work of Henry G. J. Moseley, a young English scientist, led to the arrangement of elements based on their increasing atomic numbers instead of an arrangement based on atomic masses. This new arrangement seemed to correct the problems that had occurred in the old table. The current periodic table uses Moseley's arrangement of the elements.

Reading Check *How is the modern periodic table arranged?*

Organizing a Personal Periodic Table

Procedure

1. Complete the safety form.
2. Collect as many of the following items as you can find: **feather, penny, container of water, pencil, dime, strand of hair, container of milk, container of orange juice, square of cotton cloth, nickel, crayon, quarter, container of soda, golf ball, sheet of paper, baseball, marble, leaf, paper clip.**
3. Organize these items into several columns based on their similarities to create your own periodic table.

Analysis

1. Explain the system you used to group your items.
2. Were there any items on the list that did not fit into any of your columns?
3. Infer how your activity modeled Mendeleev's work in developing the periodic table of the elements.

Try at Home

Table 5 Periodic Table of the Elements

Columns of elements are called groups. Elements in the same group have similar chemical properties.

Element	Atomic number	Symbol	Atomic mass	State of matter
Hydrogen	1	H	1.008	Gas

Legend: Gas; Liquid; Solid; Synthetic

The first three symbols tell you the state of matter of the element at room temperature. The fourth symbol identifies elements that are not present in significant amounts on Earth. Useful amounts are made synthetically.

Period	1	2	3	4	5	6	7	8	9
1	Hydrogen 1 (Gas) **H** 1.008								
2	Lithium 3 (Solid) **Li** 6.941	Beryllium 4 (Solid) **Be** 9.012							
3	Sodium 11 (Solid) **Na** 22.990	Magnesium 12 (Solid) **Mg** 24.305							
4	Potassium 19 (Solid) **K** 39.098	Calcium 20 (Solid) **Ca** 40.078	Scandium 21 (Solid) **Sc** 44.956	Titanium 22 (Solid) **Ti** 47.867	Vanadium 23 (Solid) **V** 50.942	Chromium 24 (Solid) **Cr** 51.996	Manganese 25 (Solid) **Mn** 54.938	Iron 26 (Solid) **Fe** 55.845	Cobalt 27 (Solid) **Co** 58.933
5	Rubidium 37 (Solid) **Rb** 85.468	Strontium 38 (Solid) **Sr** 87.62	Yttrium 39 (Solid) **Y** 88.906	Zirconium 40 (Solid) **Zr** 91.224	Niobium 41 (Solid) **Nb** 92.906	Molybdenum 42 (Solid) **Mo** 95.94	Technetium 43 (Synthetic) **Tc** (98)	Ruthenium 44 (Solid) **Ru** 101.07	Rhodium 45 (Solid) **Rh** 102.906
6	Cesium 55 (Solid) **Cs** 132.905	Barium 56 (Solid) **Ba** 137.327	Lanthanum 57 (Solid) **La** 138.906	Hafnium 72 (Solid) **Hf** 178.49	Tantalum 73 (Solid) **Ta** 180.948	Tungsten 74 (Solid) **W** 183.84	Rhenium 75 (Solid) **Re** 186.207	Osmium 76 (Solid) **Os** 190.23	Iridium 77 (Solid) **Ir** 192.217
7	Francium 87 (Solid) **Fr** (223)	Radium 88 (Solid) **Ra** (226)	Actinium 89 (Solid) **Ac** (227)	Rutherfordium 104 (Synthetic) **Rf** (261)	Dubnium 105 (Synthetic) **Db** (262)	Seaborgium 106 (Synthetic) **Sg** (266)	Bohrium 107 (Synthetic) **Bh** (264)	Hassium 108 (Synthetic) **Hs** (277)	Meitnerium 109 (Synthetic) **Mt** (268)

The number in parentheses is the mass number of the longest-lived isotope for that element.

Rows of elements are called periods. Atomic number increases across a period.

The arrow shows where these elements would fit into the periodic table. They are moved to the bottom of the table to save space.

Lanthanide series	Cerium 58 (Solid) **Ce** 140.116	Praseodymium 59 (Solid) **Pr** 140.908	Neodymium 60 (Solid) **Nd** 144.24	Promethium 61 (Synthetic) **Pm** (145)	Samarium 62 (Solid) **Sm** 150.36
Actinide series	Thorium 90 (Solid) **Th** 232.038	Protactinium 91 (Solid) **Pa** 231.036	Uranium 92 (Solid) **U** 238.029	Neptunium 93 (Synthetic) **Np** (237)	Plutonium 94 (Synthetic) **Pu** (244)

Metal

Metalloid

Nonmetal

The color of an element's block tells you if the element is a metal, nonmetal, or metalloid.

Science Online

Topic: Periodic Table Updates

Visit gpescience.com for updates to the periodic table.

10	11	12	13	14	15	16	17	18
								Helium 2 **He** 4.003
			Boron 5 **B** 10.811	Carbon 6 **C** 12.011	Nitrogen 7 **N** 14.007	Oxygen 8 **O** 15.999	Fluorine 9 **F** 18.998	Neon 10 **Ne** 20.180
			Aluminum 13 **Al** 26.982	Silicon 14 **Si** 28.086	Phosphorus 15 **P** 30.974	Sulfur 16 **S** 32.065	Chlorine 17 **Cl** 35.453	Argon 18 **Ar** 39.948
Nickel 28 **Ni** 58.693	Copper 29 **Cu** 63.546	Zinc 30 **Zn** 65.409	Gallium 31 **Ga** 69.723	Germanium 32 **Ge** 72.64	Arsenic 33 **As** 74.922	Selenium 34 **Se** 78.96	Bromine 35 **Br** 79.904	Krypton 36 **Kr** 83.798
Palladium 46 **Pd** 106.42	Silver 47 **Ag** 107.868	Cadmium 48 **Cd** 112.411	Indium 49 **In** 114.818	Tin 50 **Sn** 118.710	Antimony 51 **Sb** 121.760	Tellurium 52 **Te** 127.60	Iodine 53 **I** 126.904	Xenon 54 **Xe** 131.293
Platinum 78 **Pt** 195.078	Gold 79 **Au** 196.967	Mercury 80 **Hg** 200.59	Thallium 81 **Tl** 204.383	Lead 82 **Pb** 207.2	Bismuth 83 **Bi** 208.980	Polonium 84 **Po** (209)	Astatine 85 **At** (210)	Radon 86 **Rn** (222)
Darmstadtium 110 **Ds** (281)	Roentgenium 111 **Rg** (272)	Ununbium * 112 **Uub** (285)		Ununquadium * 114 **Uuq** (289)				

* The names and symbols for elements 112 and 114 are temporary. Final names will be selected when the elements' discoveries are verified.

Europium 63 **Eu** 151.964	Gadolinium 64 **Gd** 157.25	Terbium 65 **Tb** 158.925	Dysprosium 66 **Dy** 162.500	Holmium 67 **Ho** 164.930	Erbium 68 **Er** 167.259	Thulium 69 **Tm** 168.934	Ytterbium 70 **Yb** 173.04	Lutetium 71 **Lu** 174.967
Americium 95 **Am** (243)	Curium 96 **Cm** (247)	Berkelium 97 **Bk** (247)	Californium 98 **Cf** (251)	Einsteinium 99 **Es** (252)	Fermium 100 **Fm** (257)	Mendelevium 101 **Md** (258)	Nobelium 102 **No** (259)	Lawrencium 103 **Lr** (262)

Research Physicist The study of nuclear interactions is shared by chemists and physicists. Research physicists use their knowledge of the physical laws of nature to explain the behavior of the atom and its composition. Explain in your Science Journal why physicists would study the amount of energy that electrons contain.

The Atom and the Periodic Table

Objects often are sorted or classified according to the properties they have in common. This also is done in the periodic table. The vertical columns in the periodic table are called **groups,** or families, and are numbered 1 through 18. Elements in each group have similar properties. For example, in Group 11, copper, silver, and gold have similar properties. Each is a shiny metal and a good conductor of electricity and heat. What is responsible for the similar properties? To answer this question, look at the structure of the atom.

Electron Cloud Structure You have learned about the number and location of protons and neutrons in an atom. But where are the electrons located? How many are there? In a neutral atom, the number of electrons is equal to the number of protons. Therefore, a carbon atom, with an atomic number of 6, has six protons and six electrons. These electrons are located in the electron cloud surrounding the nucleus.

Scientists have found that electrons within the electron cloud have different amounts of energy. Scientists model the energy differences of the electrons by placing the electrons in energy levels, as in **Figure 9.** Energy levels nearer the nucleus have lower energy than those levels that are farther away. Electrons fill these energy levels from the inner levels (closer to the nucleus) to the outer levels (farther from the nucleus).

Elements that are in the same group have the same number of electrons in their outer energy levels. It is the number of electrons in the outer energy level that determines the chemical properties of an element. It is important to understand the link between the location on the periodic table, chemical properties, and the structure of the atom.

Figure 9 Energy levels in atoms can be represented by a flight of stairs. Each stair step away from the nucleus represents an increase in the amount of energy within the electrons.
Explain *what determines the chemical properties of an element.*

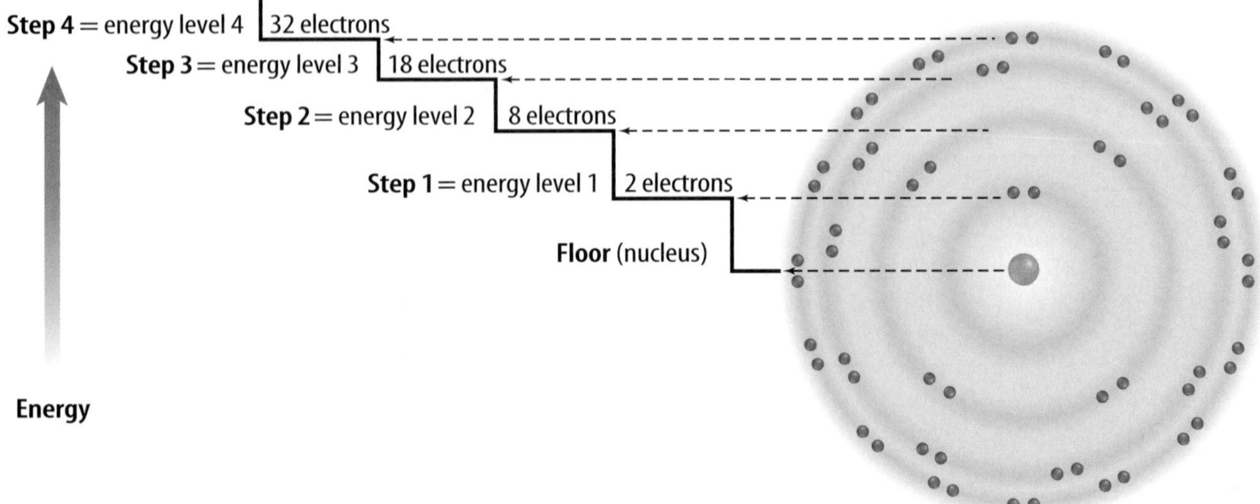

Energy Levels Energy levels are named using the numbers one to seven. The maximum number of electrons that can be contained in each of the first four levels is shown in **Figure 9.** For example, energy level one can contain a maximum of two electrons. Energy level two can contain a maximum of eight electrons. Notice that energy levels three and four contain several electrons. A complete and stable outer energy level will contain eight electrons. In elements in periods three and higher, additional electrons can be added to inner energy levels, although the outer energy level contains only eight electrons.

Topic: Atomic Energy Level Structure
Visit gpescience.com for Web links to information about the structure of atomic energy levels.

Activity Draw a diagram that details how the energy levels are structured.

Rows on the Table Remember that an atomic number found on the periodic table is equal to the number of electrons in an atom. Look at **Figure 10.** The first row has hydrogen with one electron and helium with two electrons both in energy level one. Because energy level one is the outermost level containing an electron, hydrogen has one outer electron. Helium has two outer electrons. Recall from **Figure 9** that energy level one can hold only two electrons. Therefore, helium has a full or complete outer energy level.

The second row begins with lithium, which has three electrons, two in energy level one and one in energy level two. Lithium has one outer electron. Lithium is followed by beryllium, with two outer electrons, boron with three, and so on until you reach neon, with eight outer electrons. Again, looking at **Figure 9,** energy level two can hold only eight electrons. Therefore, neon has a complete outer energy level. Do you notice how the row in the periodic table ends when an outer energy level is filled? In the third row of elements, the electrons begin filling energy level three. The row ends with argon, which has a full outer energy level of eight electrons.

How many electrons are needed to fill the outer energy level of sulfur?

Figure 10 One proton and one electron are added to each element as you go across a period in the periodic table.
Explain *what the elements in the last column share in relation to their outer energy levels.*

Hydrogen 1 **H**							Helium 2 **He**
Lithium 3 **Li**	Beryllium 4 **Be**	Boron 5 **B**	Carbon 6 **C**	Nitrogen 7 **N**	Oxygen 8 **O**	Fluorine 9 **F**	Neon 10 **Ne**
Sodium 11 **Na**	Magnesium 12 **Mg**	Aluminum 13 **Al**	Silicon 14 **Si**	Phosphorus 15 **P**	Sulfur 16 **S**	Chlorine 17 **Cl**	Argon 18 **Ar**

Figure 11 The elements in Group 1 have one electron in their outer energy level. This electron dot diagram represents that one electron.
Explain *why the outer electrons of an element are important.*

H·

Li·

Na·

K·

Rb·

Cs·

Fr·

Electron Dot Diagrams Did you notice that hydrogen, lithium, and sodium each have one electron in their outer energy levels? Elements that are in the same group have the same number of electrons in their outer energy levels. These outer electrons are so important in determining the chemical properties of an element that a special way to represent them has been developed. American chemist G. N. Lewis created this method while teaching a college chemistry class. An **electron dot diagram** uses the symbol of the element and dots to represent the electrons in the outer energy level. **Figure 11** shows the electron dot diagram for Group 1 elements. The electron configuration of an atom determines how that atom reacts with other atoms. Electron dot diagrams also are used to show how the electrons in the outer energy levels are bonded when elements combine to form compounds.

Same Group, Similar Properties The elements in Group 17, the halogens, have electron dot diagrams similar to chlorine, shown in **Figure 12.** All halogens have seven electrons in their outer energy levels. Because all of the members of a group on the periodic table have the same number of electrons in their outer energy levels, group members will undergo chemical reactions in similar ways.

A common property of the halogens is the ability to form compounds readily with elements in Group 1. Group 1 elements each have only one electron in their outer energy levels. **Figure 12** shows an example of a compound formed by one such reaction. The Group 1 element, sodium, reacts easily with the Group 17 element, chlorine. The result is the compound sodium chloride, or NaCl, ordinary table salt.

Not all elements will combine readily with other elements. The elements in Group 18 have complete outer energy levels. This special configuration makes Group 18 elements relatively unreactive. You will learn more about why and how bonds form between elements in later chapters.

Reading Check *Why do elements in a group undergo similar chemical reactions?*

Figure 12 Electron dot diagrams show the electrons in an element's outer energy level.

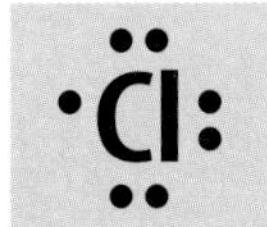

The electron dot diagram for Group 17 consists of three sets of paired dots and one single dot.

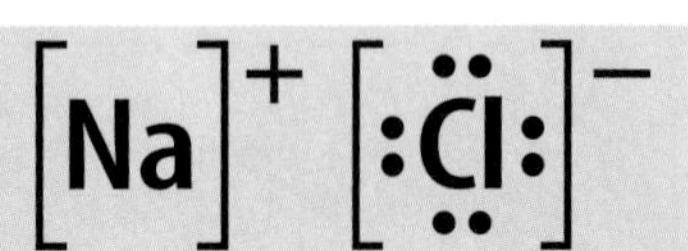

Sodium combines with chlorine to give each element a complete outer energy level in the resulting compound.

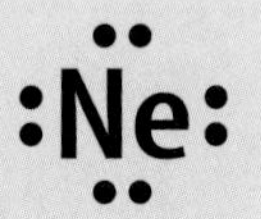

Neon, a member of Group 18, has a full outer energy level. Neon has eight electrons in its outer energy level, making it unreactive.

Regions on the Periodic Table

The periodic table has several regions with specific names. The horizontal rows of elements on the periodic table are called **periods.** The elements increase by one proton and one electron as you go from left to right in a period.

All of the elements in the blue squares in **Figure 13** are metals. Iron, zinc, and copper are examples of metals. Most metals exist as solids at room temperature. They are shiny, can be drawn into wires, can be pounded into sheets, and are good conductors of heat and electricity.

Those elements on the right side of the periodic table, in yellow, are classified as nonmetals. Oxygen, bromine, and carbon are examples of nonmetals. Most nonmetals are gases, are brittle, and are poor conductors of heat and electricity at room temperature. The elements in green are metalloids or semimetals. They have some properties of both metals and nonmetals. Boron and silicon are examples of metalloids.

Topic: New Elements
Visit gpescience.com for Web links to information about newly synthesized elements.

Activity Write an article explaining how several new elements were synthesized and who synthesized them.

What are the properties of the elements located on the left side of the periodic table?

A Growing Family Scientists around the world are continuing their research into the synthesis of elements. In 1994, scientists at the Heavy-Ion Research Laboratory in Darmstadt, Germany, discovered element 111. As of 1998, only one isotope of element 111 had been found. This isotope had a life span of 0.002 s. In 1996, element 112 was discovered at the same laboratory. As of 1998, only one isotope of element 112 had been found. The life span of this isotope was 0.00048 s. Both of these elements are produced in the laboratory by joining smaller atoms into a single atom. The search for elements with higher atomic numbers continues. Scientists think they have synthesized elements 114 and 116. However, the discovery of these elements has not yet been confirmed.

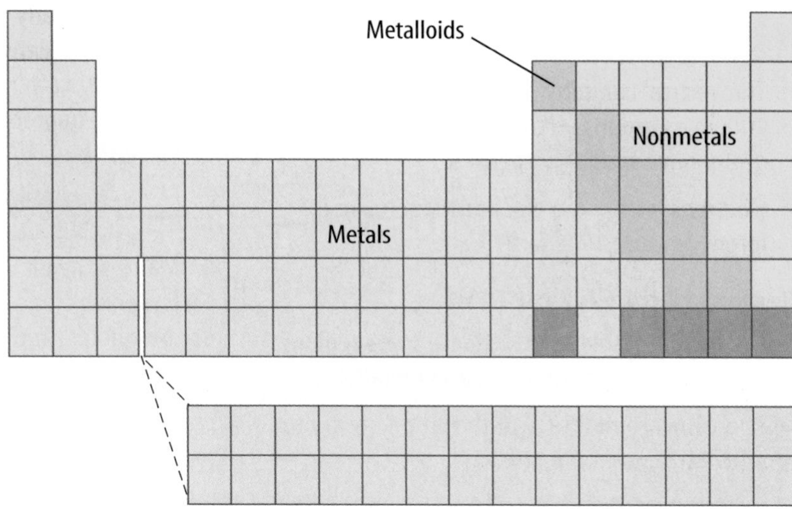

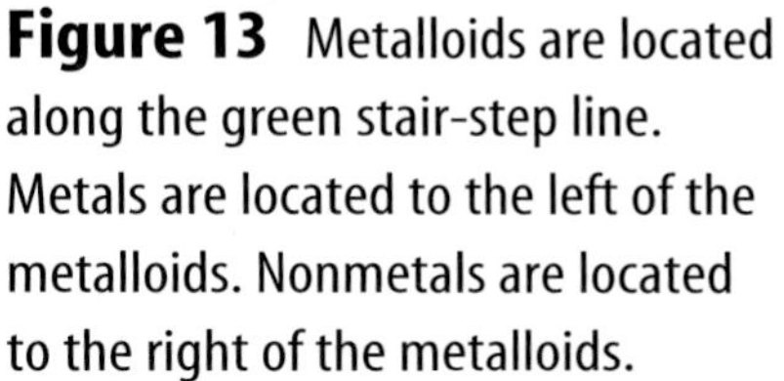
Figure 13 Metalloids are located along the green stair-step line. Metals are located to the left of the metalloids. Nonmetals are located to the right of the metalloids.

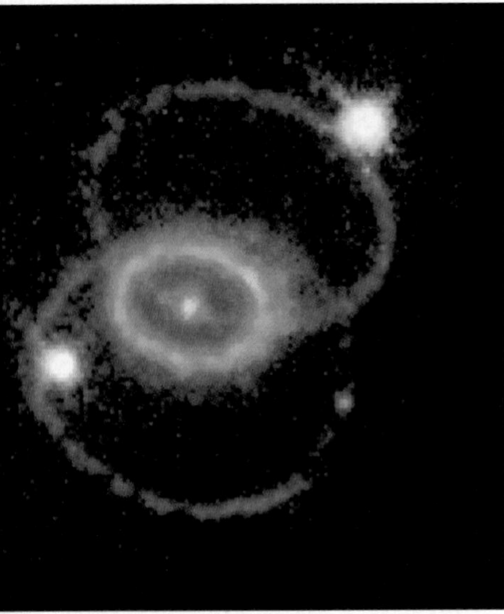

Figure 14 Scientists think that some elements are found in nature only within stars.

Elements in the Universe

Using the technology that is available today, scientists are finding the same elements throughout the universe. They have been able to study only a small portion of the universe because it is so vast. Many scientists believe that hydrogen and helium are the building blocks of other elements. Atoms join together within stars to produce elements with atomic numbers greater than 1 or 2, the atomic numbers of hydrogen and helium. Exploding stars, or supernovas, shown in **Figure 14,** give scientists evidence to support this theory. When stars reach the supernova stage, which is characterized by a massive explosion, a mixture of elements, including the heavy elements such as iron, are flung into the galaxy. Many scientists believe that supernovas have spread the elements that are found throughout the universe. Some of these elements are found only in trace amounts in Earth's crust as a result of uranium decay. Others have been found only in stars.

section 3 review

Summary

Organizing the Elements

- Mendeleev organized the elements using increasing atomic mass and chemical and physical properties.
- Mendeleev left blank spaces in his table to allow for elements that were yet undiscovered.
- Moseley corrected the problems in the periodic table by arranging the elements in order of increasing atomic number.

The Atom and the Periodic Table

- The vertical columns in the periodic table are known as groups or families. Elements in a group have similar properties.
- Electrons within the electron cloud have different amounts of energy.

Regions of the Periodic Table

- The periodic table is divided into these regions: periods, metals, nonmetals, and metalloids.
- Scientists around the world continue to try to synthesize new elements.

Self Check

1. **Identify** Use the periodic table to find the name, atomic number, and average atomic mass of the following elements: N, Ca, Kr, and W.
2. **List** the period and group in which each of these elements is found: nitrogen, sodium, iodine, and mercury.
3. **Classify** each of these elements as a metal, a nonmetal, or a metalloid and give the full name of each: K, Si, Ba, and S.
4. **Think Critically** The Mendeleev and Mosely periodic tables have gaps for the as-then-undiscovered elements. Why do you think the table used by Mosely was more accurate at predicting where new elements would be placed?

Applying Math

5. **Make a Graph** Construct a circle graph showing the percentage of elements classified as metals, metalloids, and nonmetals. Use markers or colored pencils to distinguish clearly between each section on the graph. Record your calculations in your Science Journal.

More Section Review gpescience.com

A Periodic Table of Foods

Real-World Problem

Mendeleev's task of organizing a collection of loosely related items probably seemed daunting at first. How will using your favorite foods to create your own periodic table be similar to the task that Mendeleev had?

Goals

- **Organize** 20 of your favorite foods into a periodic table of foods.
- **Analyze** and **evaluate** your periodic table for similar characteristics among groups or family members on your table.
- **Infer** where new foods would be placed on your table.

Materials

11 × 17-in paper
12- or 18-in ruler
colored pencils or markers

Safety Precautions

Procedure

1. Complete the safety form before you begin.
2. **List** 20 of your favorite foods and drinks.
3. **Describe** basic characteristics of each food and drink item. For example, you might describe the primary ingredient, nutritional value, taste, and color or identify the food group of each item.
4. **Create** a data table to organize the information that you collect.
5. Using your data table, construct a periodic table of foods. Determine which characteristics you will use to group your items. Create families (columns) of food and drink items that share similar characteristics.

For example, potato chips, pretzels, and crackers could be combined into a family of salty tasting foods. Create as many groups as you need. You do not need to have the same number of items in every family.

Conclude and Apply

1. **Evaluate** the characteristics you used to make the groups on your periodic table. Do the characteristics of each group adequately describe all the family members? Do the characteristics of each group distinguish its family members from the family members of the other groups?
2. **Analyze** the reasons why some items did not fit easily into a group.
3. **Infer** why chemists have not created a periodic table of compounds.

Communicating Your Data

Construct a bulletin board of the periodic tables of foods created by the class. Compare and contrast the tables.

Use the Internet

What's in a name?

Goals

- **Research** the names and symbols of various elements.
- **Study** the methods that are used to name elements and how they have changed through time.
- **Organize** your data by making your own periodic table.
- **Study** the history of certain elements and their discoveries.
- **Create** a table of your findings and communicate them to other students.

Data Source

Internet Lab

Visit **gpescience.com** for more information about naming elements, elements' symbols, and the discovery of new elements, and for data collected by other students.

Real-World Problem

The symbols used for different elements sometimes are easy to figure out. After all, it makes sense for the symbol for carbon to be C and the symbol for nitrogen to be N. However, some symbols aren't as easy to figure out. For example, the element silver has the symbol Ag. This symbol comes from the Latin word for *silver, argentum*. How are symbols and names chosen for elements?

Make a Plan

1. Make a list of particular elements you wish to study.
2. **Identify** the symbols for these elements.
3. **Research** the discovery of these elements. Do their names match their symbols? Were they named after a property of the element, a person, their place of discovery, or a system of nomenclature? What was that system?

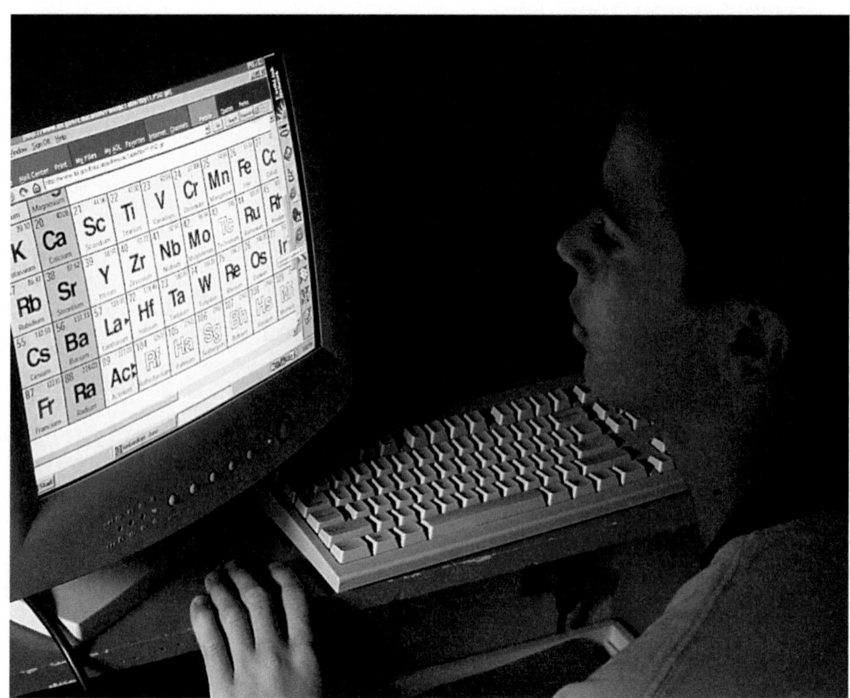

Using Scientific Methods

Follow Your Plan

1. Make sure your teacher approves your plan before you start.
2. Visit the Web site provided for links to different sites about elements, their history, and how they were named.
3. **Research** these elements.
4. Carefully record your data in your Science Journal.

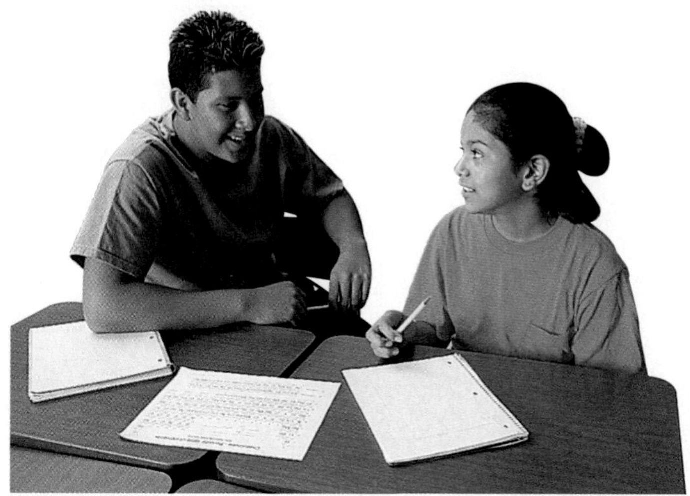

Analyze Your Data

1. **Record** in your Science Journal how the symbols for your elements were chosen. What were your elements named after?
2. **Construct** an element key that includes the research information that you found for your elements.
3. Make a chart of your class's findings. Sort the chart by year of discovery for each element.
4. How are the names and symbols for newly discovered elements chosen? Make a chart that shows how the newly discovered elements will be named.

Conclude and Apply

1. **Compare** your findings to those of your classmates. Did anyone's data differ for the same element? Were all the elements in the periodic table covered?
2. **Explain** the system that is used to name the newly discovered elements today.
3. **Explain** Some elements were assigned symbols based on their name in another language. Do these examples occur for elements discovered today or long ago?

Find this lab using the link below. Post your data in the table provided. **Compare** your data to those of other students.

Internet Lab
gpescience.com

TIME

SCIENCE AND HISTORY

SCIENCE CAN CHANGE THE COURSE OF HISTORY!

A CHILLING STORY

A scientist inspects an ice core sample from the Greenland Ice Sheet. The samples are stored in a freezer at −36°C.

Picture this: It's 1361. A ship from Norway arrives at a Norwegian settlement in Greenland. The ship's crew members hope to trade its cargo with the people living there. The crew members get off the ship. They look around. The settlement is deserted. More than 1,000 people had vanished!

New evidence has shed some light on the mysterious disappearance of the Norse settlers. The evidence came from a place on the Greenland Ice Sheet over 600 km away from the settlement. This part of Greenland is so cold that snow never melts. As new snow falls, the existing snow is buried and turns to ice.

By drilling deep into this ice, scientists can recover an ice core. The core is made up of ice formed from snowfalls way, way back in time.

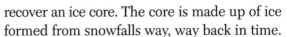

Air bubbles and dirt trapped in ice provide clues to Earth's past climate.

By measuring the ratio of oxygen isotopes in the ice core, scientists can estimate Greenland's past air temperatures. The cores provide a detailed climate history going back over 80,000 years. Individual ice layers can be dated much like tree rings to determine their ages. The air bubbles trapped within each layer are used to learn about climate variations. Dust and pollen trapped in the ice also yield clues to ancient climates.

A Little Ice Age

Based on their analysis, scientists think the Norse settlers moved to Greenland during an unusually warm period. Then, in the 1300s, the climate started to cool and a period known as the Little Ice Age began. The ways the Norse people hunted and farmed were inadequate for survival in this long chill. Because they couldn't adapt to their colder surroundings, the settlers died out.

Research Report **Evidence seems to show that Earth is warming. Rising temperatures could affect our lives. Research global warming to find out how Earth may change. Share your report with the class.**

TIME

For more information, visit gpescience.com

Reviewing Main Ideas

Section 1 Structure of the Atom

1. A chemical symbol is a shorthand way of writing the name of an element.
2. An atom consists of a nucleus made up of protons and neutrons surrounded by an electron cloud, as shown in the figure to the right.

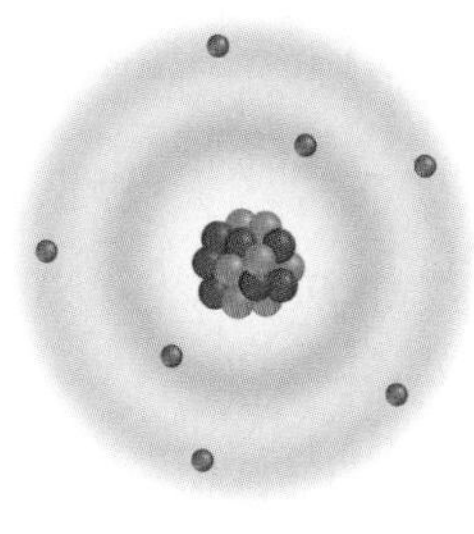

3. Quarks are particles of matter that make up protons and neutrons.
4. The model of the atom changes over time. As new information is discovered, scientists incorporate it into the model.

Section 2 Masses of Atoms

1. The number of neutrons in an atom can be computed by subtracting the atomic number from the mass number.
2. The isotopes of an element are atoms of the same element that have different numbers of neutrons. The figure below shows the isotopes of hydrogen.

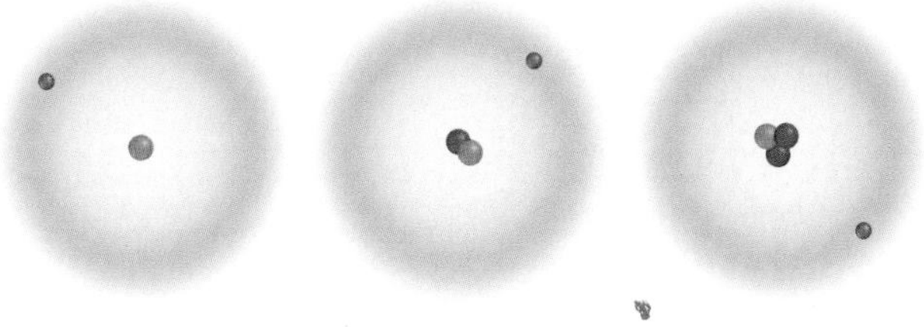

3. The average atomic mass of an element is the weighted-average mass of the mixture of its isotopes. Isotopes are named by using the element name, followed by a dash, and its mass number.

Section 3 The Periodic Table

1. In the periodic table, the elements are arranged by increasing atomic number, resulting in periodic changes in properties. Knowing that the number of protons, electrons, and atomic number are equal gives you a partial composition of an atom.
2. In the periodic table, the elements are arranged in 18 vertical columns, or groups, and seven horizontal rows, or periods.
3. Metals are found at the left of the periodic table, nonmetals at the right, and metalloids along the line that separates the metals from the nonmetals, as shown below.

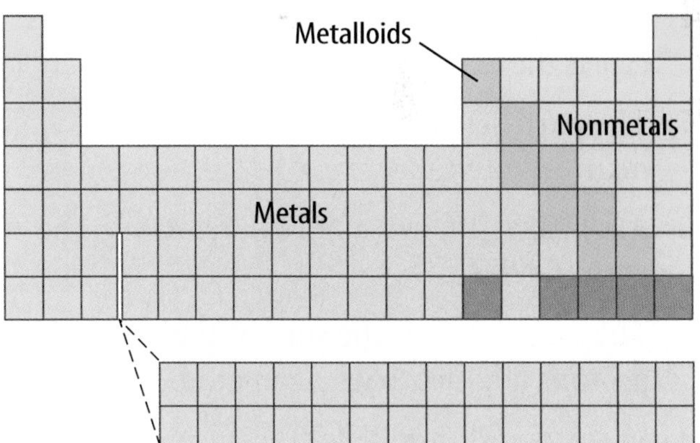

4. Elements are placed on the periodic table in order of increasing atomic number. A new row on the periodic table begins when the outer energy level of an element is filled.

FOLDABLES™ Use the Foldable that you made at the beginning of the chapter to help you review properties of atoms and the periodic table.

chapter 19 Review

Using Vocabulary

atom p. 579
atomic number p. 585
average atomic mass p. 587
electron p. 579
electron cloud p. 583
electron dot diagram p. 594
group p. 592
isotope p. 586
mass number p. 585
neutron p. 579
nucleus p. 579
period p. 595
periodic table p. 588
proton p. 579
quark p. 579

Complete the statement with the correct vocabulary word or phrase.

1. Mendeleev created an organized table of elements called the ________.
2. Two elements with the same number of protons but a different number of neutrons are called ________.
3. ________ is the weighted-average mass of all the known isotopes for an element.
4. The positively charged center of an atom is called the ________.
5. The particles that make up protons and neutrons are called ________.
6. A(n) ________ is a horizontal row in the periodic table.
7. The ________ is the sum of the number of protons and neutrons in an atom.
8. In the current model of the atom, the electrons are located in the ________.

Checking Concepts

Choose the word or phrase that best answers the question.

9. In which state of matter are most of the elements to the left of the stair-step line in the periodic table?
   A) gas **C)** plasma
   B) liquid **D)** solid

10. Which term describes a pattern that repeats?
   A) isotopic **C)** periodic
   B) metallic **D)** transition

11. Which element has similar properties to those of neon?
   A) aluminum **C)** arsenic
   B) argon **D)** silver

12. Which term describes boron?
   A) metal **C)** noble gas
   B) metalloid **D)** nonmetal

13. How many outer-level electrons do lithium and potassium have?
   A) one **C)** three
   B) two **D)** four

14. Which is **NOT** found in the nucleus of an atom?
   A) electron **C)** proton
   B) neutron **D)** quark

15. The atomic number of rhenium is 75. The atomic mass of one of its isotopes is 186. How many neutrons are in an atom of this isotope?
   A) 75 **C)** 186
   B) 111 **D)** 261

Use the photo below to answer question 16.

16. Which scientist developed this atomic model?
   A) Bohr **C)** Rutherford
   B) Dalton **D)** Thomson

Vocabulary PuzzleMaker gpescience.com

Interpreting Graphics

17. **Construct** As a star dies, it becomes more dense. Its temperature rises to a point at which He nuclei are combined with other nuclei. When this happens, the atomic numbers of the other nuclei are increased by 2 because each gains the two protons contained in the He nucleus. For example, Cr fuses with He to become Fe. Copy and complete the concept map showing the first four steps in He fusion.

He
+He
Be
+He
+He
+He

18. Copy and complete the concept map below.

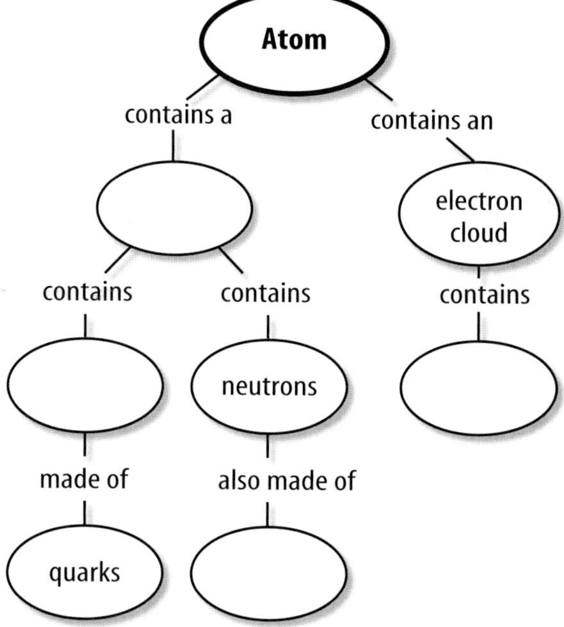

Thinking Critically

19. **Infer** Lead and mercury are two pollutants in the environment. From information about them in the periodic table, determine why they are called heavy metals.

20. **Explain** why it is necessary to change models as new information becomes available.

21. **Infer** Why did scientists choose carbon as a base for the atomic mass unit? Which isotope of carbon did they use?

22. **Infer** Ge and Si are used in making semiconductors. Are these two elements in the same group or the same period?

23. **Explain** Using the periodic table, predict how many outer level electrons will be in elements 114, 116, and 118. Explain your answer.

24. **Infer** Ca is used by the body to make bones and teeth. Sr-90 is radioactive. Ca is safe for people, and Sr-90 is hazardous. Why is Sr-90 hazardous to people?

Applying Math

25. **Solve One-Step Equations** The atomic number of yttrium is 39. The atomic mass of one of its isotopes is 89. How many neutrons are in an atom of this isotope?

Use the table below to answer question 26.

Electrons per Energy Level

Energy Level	Maximum Number of Electrons
1	2
2	Do not write in this book.
3	
4	

26. **Use Tables** Use the information in **Figure 9** to determine how many electrons should be in the second, third, and fourth energy levels for argon, atomic number 18. Copy and complete the table above with the number of electrons for each energy level.

chapter 19 Standardized Test Practice

Record your answers on the answer sheet provided by your teacher or on a sheet of paper.

Multiple Choice

1. Which group of elements on the periodic table do not combine readily with other elements?

   A. Group 1

   B. Group 2

   C. Group 17

   D. Group 18

Use the illustration below to answer question 2.

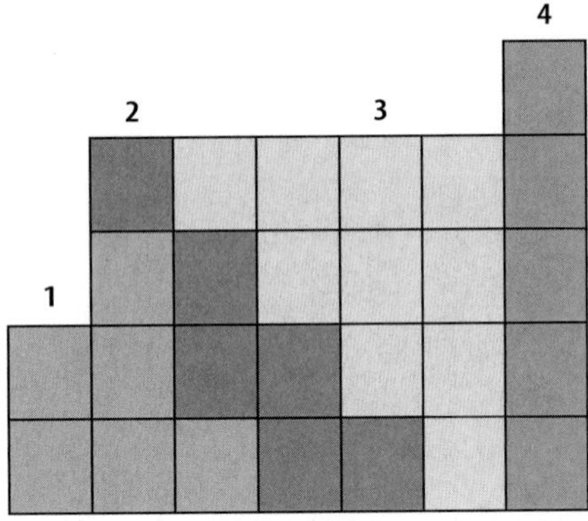

2. Which of the regions shown on the periodic table contains mostly elements that are gases at room temperature?

   A. region 1

   B. region 2

   C. region 3

   D. region 4

Test-Taking Tip

Have Breakfast The morning of the test, eat a healthy breakfast with a balanced amount of protein and carbohydrates.

3. Which scientist proposed the idea that atoms make up all substances?

   A. Aristotle

   B. Dalton

   C. Democritus

   D. Galileo

Use the table below to answer question 4.

Element	Electrons in a Neutral Atom	Electrons in Outer Energy Level
Carbon	6	4
Oxygen	8	6
Neon	10	8
Sodium	11	1
Chlorine	17	7

4. Which element would you expect to be located in Group 1 of the periodic table?

   A. oxygen

   B. neon

   C. sodium

   D. chlorine

5. The element nickel has five naturally occurring isotopes. Which of the following describes the relationship of these isotopes?

   A. same mass, same atomic number

   B. same mass, different atomic number

   C. different mass, same atomic number

   D. different mass, different atomic number

6. Atoms of different elements are different because they have different numbers of what type of particle?

   A. electrons

   B. photons

   C. protons

   D. neutrons

Gridded Response

7. According to the periodic table, an atom of lead has an atomic number of 82. How many neutrons does lead-207 have?

8. About three out of four chlorine atoms are chlorine-35, and about one out of four is chlorine-37. What is the average atomic mass of chlorine?

Short Response

Use the illustration below to answer question 9.

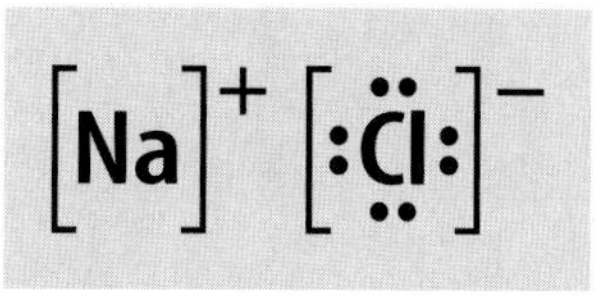

$[Na]^{+}\ [:\ddot{\underset{\cdot\cdot}{Cl}}:]^{-}$

9. The electron dot diagram above shows how a sodium atom, Na, combines with a chlorine atom, Cl, to form sodium chloride. What do the + and the − symbols indicate in the diagram?

10. Why isn't the mass of the electron included in the mass of an atom on the periodic table?

Extended Response

Use the illustration below to answer questions 11 and 12.

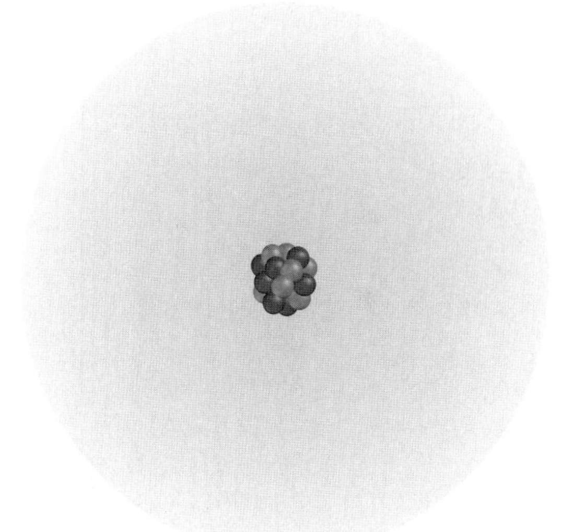

11. The illustration above shows the currently accepted model of atomic structure. Describe this model.

12. Compare and contrast the model shown above with Bohr's model of an atom.

13. How can you use the periodic table to determine the average number of neutrons an element has, even though the number of neutrons is not listed?

14. Describe how Dalton's modernization of the ancient Greeks' ideas of elements, atoms, and compounds provided a basis for understanding chemical reactions. Give an example.

chapter 20

Earth Materials

chapter preview

sections

Mountain of Oxygen

We often think of oxygen as being in the air we breathe, but we also walk on it everyday. By weight, almost half of Earth's crust is oxygen. Imagine half the weight of this mountain being made of oxygen!

Science Journal In your Science Journal, list ten of the most important materials you can think of and where you think they come from.

Start-Up Activities

Earth Materials at Home and School

You come in contact with Earth materials every day. Many of the objects around you are Earth materials, or items processed using minerals or rocks as raw materials.

1. Obtain several items from your teacher.
2. Make a three-column table on a separate sheet of paper.
3. In the first column, list each item.
4. In the second column, decide whether each item listed is a natural Earth material or a product made from Earth materials.
5. In the third column, indicate whether the item is found at home, school, or both places.
6. **Think Critically** Select three of the items in your first column and imagine that they no longer exist. Explain how you could cope without these materials. What other materials could be used instead?

Preview this chapter's content and activities at gpescience.com

Textures Make the following Foldable to help you organize the types of textures in igneous, sedimentary, and metamorphic rocks.

STEP 1 **Fold** a sheet of paper in half lengthwise. Make the back edge about 2 inches longer than the front edge.

STEP 2 **Turn** the paper so the fold is on the bottom. Then **fold** it into thirds.

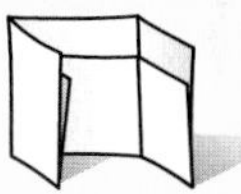

STEP 3 **Unfold and cut** only the top layer along both folds to make three tabs.

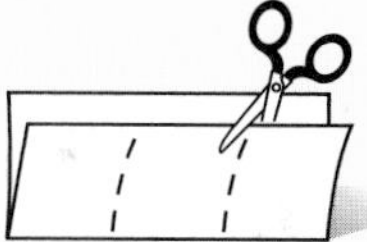

STEP 4 **Label** the Foldable as shown

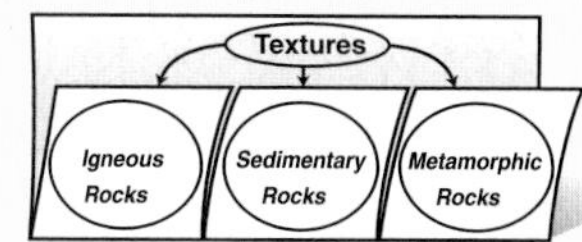

Making a Concept Map As you read the chapter, write about and illustrate the texture types shown by each group of rocks represented in your concept map.

section 1

Minerals

Reading Guide

What You'll Learn

- **Describe** characteristics of all minerals.
- **Identify** common minerals using physical properties.
- **Compare and contrast** some conditions of mineral formation.

Why It's Important

Minerals are a source of raw materials used to support civilization.

Review Vocabulary

ionic bond: attraction formed between oppositely charged ions in an ionic compound

New Vocabulary

- mineral
- cleavage
- fracture
- hardness
- streak
- magma

Common Elements

You're familiar with the periodic table of elements. Of the first 92 elements, 90 are found in Earth, and of these, only a small number combine to make up most of the common minerals in Earth's crust. **Table 1** lists these important elements. Most of the chemical formulas of common minerals contain some of these elements.

Composition of Earth's Crust The crust is the outermost layer of Earth. It includes all continental material and the material that forms the ocean bottom. The crust extends down tens of kilometers beneath the continents, but it is much thinner where it makes up the ocean bottom and material below, as shown in **Figure 1.**

Table 1 Major Elements in Earth's Crust

Element	Percent (by mass)	Element	Percent (by mass)
Oxygen	46.6%	Sodium	2.8%
Silicon	27.7%	Potassium	2.6%
Aluminum	8.1%	Magnesium	2.1%
Iron	5.0%	All others	1.5%
Calcium	3.6%		

Chemical Element Building Blocks Imagine Earth's chemical parts being like a set of interlocking blocks. In the set you are given lots of common types of blocks from which you can make thousands of forms. But, you are given only a few of the special blocks. The bulk of the forms you build consist of the common blocks. Elements and minerals in Earth's crust are a lot like the block set. For example, most minerals of the crust, whether beneath the land surface or ocean water, contain abundant oxygen and silicon.

What's a mineral?

Remember that atoms of different elements can bond chemically to form compounds. A **mineral** is a naturally occurring element or compound that is inorganic, solid, and has a crystalline structure. In addition, minerals have predictable chemical compositions that are indicated as single elements or chemical formulas. For example, the native element gold is a mineral with chemical formula Au. Similarly, the compound fluorite is a mineral that contains Ca and F in a 1:2 ratio, resulting in chemical formula CaF_2.

Physical Properties

A mineral has a characteristic set of physical properties, but some of these properties can differ from sample to sample. For example, some minerals come in several different colors. This color variation may be due to chemical impurities in the mineral. This is true for the gem varieties of the mineral corundum, Al_2O_3. Ruby is red corundum, and sapphire is blue corundum. Ruby contains more of the element chromium, which substitutes for Al in the crystalline structure. You should take care when observing mineral properties, particularly when studying their colors.

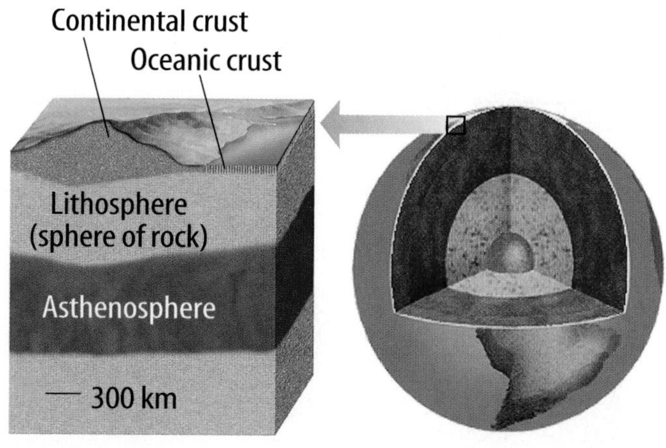

Figure 1 The crust and a thin layer of the upper mantle make up the lithosphere, which means rock sphere. The lithosphere floats on a dense, plasticlike layer called the asthenosphere.

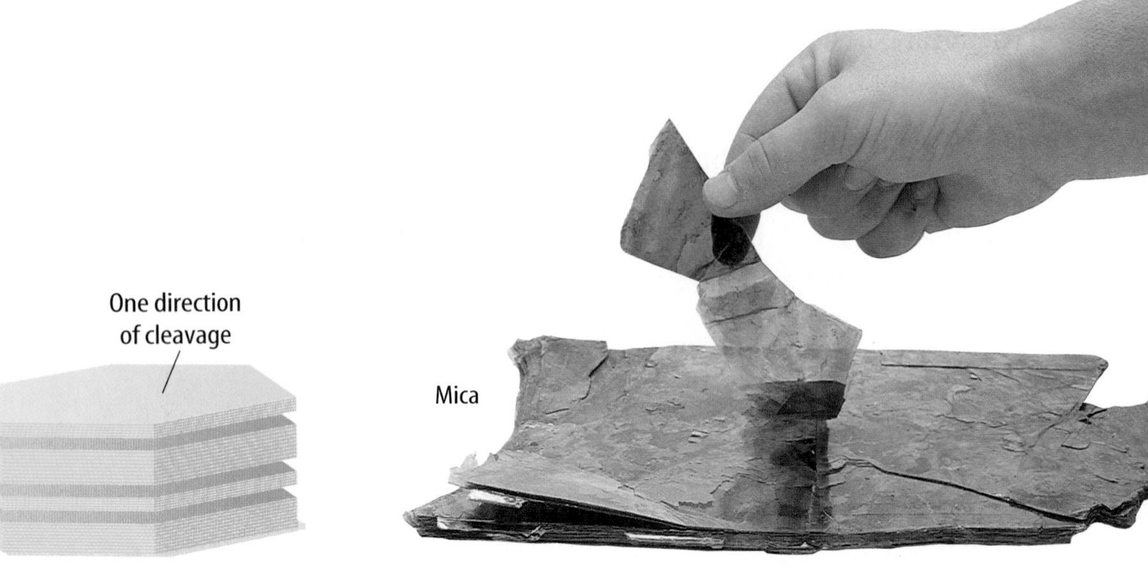

Figure 2 Because mica has one direction of cleavage, it can be separated in layers.

Atom Arrangement Some physical properties are controlled by the orderly arrangement of atoms in a mineral's structure. This orderly pattern is what makes a mineral crystalline. The arrangement of atoms and the bonds between them can reflect the way a mineral breaks, how hard it is, and what type of crystal shape it has.

How Minerals Break When minerals break along planes that cut across relatively weak chemical bonds, a smooth, flat surface is created. The ability of a mineral to do this is the physical property called **cleavage.** All parallel cleavage planes define a single direction of cleavage. The mica family of minerals exhibits only one direction of cleavage, as shown in **Figure 2.** Other minerals may show two or more different directions of cleavage, as in **Figure 3.** If this happens, those different planar directions have to meet each other at an angle.

Some minerals do not split along well-defined flat surfaces. In such cases, a mineral will break unevenly. This type of irregular break is called **fracture.** Quartz is an example of a common mineral that does not cleave, but exhibits fracture. This physical property is a key aspect in distinguishing quartz from other minerals.

Reading Check *What distinguishes quartz from other minerals?*

Figure 3 Feldspar is an example of a mineral with two planes of cleavage.

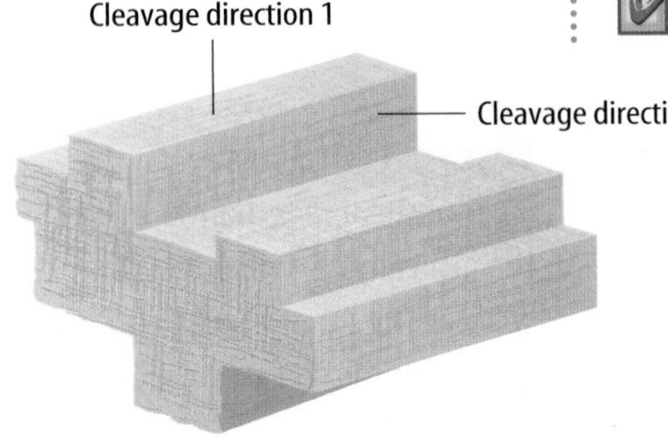

Hardness Bonds connecting atoms in materials often have different strengths. When you scratch a mineral with another material you will be breaking bonds and leaving a scratch. The physical property that measures resistance to scratching is called **hardness.** You can compare the hardness of different materials by applying scratch tests and using Mohs Scale, as shown in **Table 2.** When a hardness test is performed by rubbing two objects together, the softer of the two will wear away.

Table 2 Mohs Scale of Hardness

1. Talc	6. Feldspar
2. Gypsum (fingernail 2.2)	7. Quartz (streak plate 7.0)
3. Calcite (copper penny 3.2)	8. Topaz
4. Fluorite	9. Corundum
5. Apatite (glass plate 5.5)	10. Diamond

Today, you might be using a pencil that takes advantage of the softness of a mineral. The graphite in the core of some pencils has one plane of cleavage and very weak bonds. The weak bonds make the lead soft. As you write with the pencil, thin plates of graphite are left on your paper and the pencil becomes dull.

Luster and Streak The way a mineral reflects light is the physical property known as luster. Two main types of luster, metallic and nonmetallic, help subdivide minerals and often give a clue about their compositions. Just as you might guess, metallic luster minerals reflect light in a way that a metal surface might, such as the shiny chrome on a car or bicycle. Nonmetallic luster, shown in **Figure 4,** includes descriptions of all other types, such as those minerals that shine like glass or appear earthy or waxy.

The color of a mineral in powdered form is called **streak.** The streak of a mineral may be the same color as the mineral specimen, but when a mineral shows different colors or types of luster, the streak powder color generally stays the same, which helps identify the mineral. A streak test is performed by rubbing a mineral on a white porcelain tile.

Figure 4 The mineral on the left produces a glassy luster; the mineral on the right produces a metallic luster.

Identifying Salt's Crystal System

Procedure

1. Sprinkle a small quantity of **salt** onto a **dark surface.**
2. Using a **magnifying lens,** observe the individual grains of salt.

Analysis

1. Make a sketch showing the shapes of several salt grains.
2. Why are some of the grains more elongated than others?
3. To which crystal system does salt belong? Support your answer with details from the text.

Try at Home

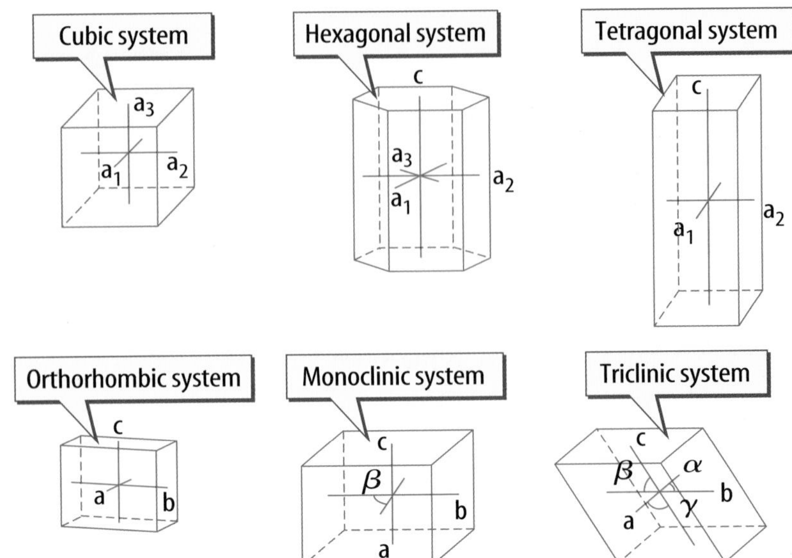

Figure 5 Minerals can be classified using six basic crystal forms.

Crystal Shape The orderly internal arrangement of atoms in a mineral often is indicated by its external crystal shape. In fact, some minerals take on a geometric shape that is a larger version of their internal atomic arrangement. Even though many unique shapes are exhibited by mineral crystals in nature, each can be sorted into one of the six crystal systems, as shown in **Figure 5.** The types of symmetry shown by the crystal are key elements in determining the crystal system to which a mineral belongs.

Mineral Formation

A mineral crystal grows as atoms are added to its surfaces, edges, or corners. The types of atoms that are added depend on what atoms are available in the growing crystal's surroundings. Growth also is controlled by how fast atoms can migrate to the crystal and by the temperature and pressure conditions of the surroundings. Some ways that minerals form include precipitation from hot, water-rich fluids, solidification from molten rock material, and evaporation of water rich in dissolved salts at low temperatures near Earth's surface.

Figure 6 Hydrothermal minerals form on the rims of hot springs.

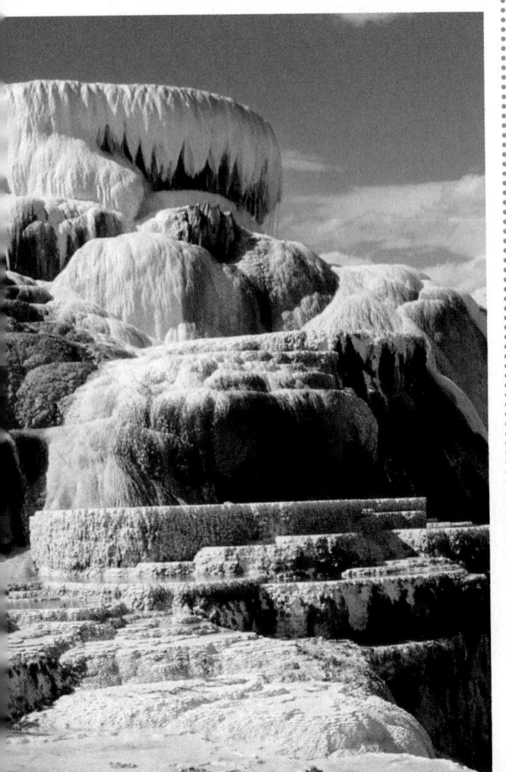

Minerals from Hot Water Some minerals are produced from hot-water solutions rich in dissolved mineral matter. You know that sugar dissolves fastest in hot water. As hot water cools, its atoms slow down and atoms of dissolved material are able to form chemical bonds. Minerals often form around the edges of hot springs, as shown in **Figure 6.** When hot water passes through cracks in cooler rock, minerals may form within the cracks. The cracks become lined and filled with mineral matter. Sometimes, veins or halos of concentrated minerals like gold, silver, or copper are produced in fractures.

Minerals from Magma Molten rock material found inside Earth is called **magma.** As magma cools, atoms slow down and begin to arrange into an orderly structure. When the temperature of magma drops well below the solidification temperature of a mineral, crystals of that particular mineral may form and grow.

Reading Check *How do minerals form in magma?*

Minerals from Evaporation Minerals also form from water at Earth's surface. When water slowly evaporates, concentrated dissolved mineral material may be left behind to form crystals. You might have seen films of material left behind in a pan that once was filled with water. Crusty material is left on the surface of the pan after the water evaporates.

Si-O Bond The silicon-oxygen tetrahedron is the base unit for the silicates in the nonliving world. Carbon chains are fundamental units for living things. Research carbon chains and compare and contrast them with the Si-O tetrahedrons.

Mineral Groups

About 3,800 minerals have been identified in nature. Some minerals are so common that they are called rock-forming minerals. Recall that few elements are needed to make up almost the entire crust of Earth. Similarly, a few important mineral groups make up most of Earth's crust.

Silicates The atomic arrangement and composition of minerals allow them to be sorted into groups. Among these, the most important by volume of Earth's crust are the silicate groups.

Most minerals contain silica. Silica is a common term for a compound that contains silicon plus oxygen or silicon dioxide (SiO_2). The mineral quartz is pure silica that has crystallized. In silicate minerals, the elements silicon and oxygen bond together to form a geometric structure called a tetrahedron, as shown in **Table 3.** Other metal atoms can attach to the oxygen atoms. Most silicate minerals contain silicon, oxygen, and one or more other elements.

Table 3 Silicate Minerals

Mineral		Photos	Silicate Structure
Olivine			O—Si Single tetrahedron
Pyroxene group (Augite)			Single chains
Amphibole group (Hornblende)			Double chains
Micas	Biotite		Sheets
	Muscovite		
Feldspars	Potassium feldspar		Three-dimensional networks
	Plagioclase		
Quartz			

Topic: Economic Minerals

Visit gpescience.com for Web links to information about uses for minerals.

Activity Make a table for at least ten minerals. Your table headings should include *Common Minerals, Occurrences,* and *Uses or Products.*

Silicate Structures You probably have watched buildings go up in your neighborhood. All the buildings are put together using similar construction materials. What really makes them different is how these materials are put together. Silicate mineral groups are similar. Silicates all contain silicon-oxygen tetrahedrons that are linked together in different ways.

The simplest silicate structures have silicon-oxygen tetrahedrons that are not linked together. By joining silicon-oxygen tetrahedrons together, chains, sheets, and three-dimensional framework structures can form, as shown in **Table 3.** Keep in mind that each group of silicates has a different structure, but most contain the silicon-oxygen tetrahedron as a basic structural unit.

Minerals of the Crust Several important silicate groups form most of Earth's crust. Quartz and some feldspar group minerals, shown in **Table 3,** are relatively low in density and form at relatively low temperatures compared to other silicate minerals. Together, quartz and feldspar group silicates make up most of Earth's continental crust.

In contrast, Earth's oceanic crust is denser and contains a larger percentage of silicates whose tetrahedrons are not linked together as much. For example, chain silicates such as the pyroxene group, and single tetrahedron silicates of the olivine group, along with plagioclase feldspars, make up the majority of crust beneath the oceans.

Figure 7 Calcite, shown above, is a common mineral found in limestone. Hematite, below, is an ore used to produce iron and steel.

Important Non-silicates Many important mineral groups are not silicates. These include the carbonates, oxides, halides, sulfides, sulfates, and native metals. The non-silicate groups are a source of many valuable ore minerals and building materials. To be an ore, a mineral must occur in large enough quantities to be economically recoverable. One example of an important ore of iron is the mineral hematite, an iron oxide. Aluminum comes in part from bauxite, a mixture of aluminum oxides. People eat salt processed from the halide mineral halite, and use calcite (a carbonate) to make cement. Common non-silicate minerals are shown in **Figure 7.**

Reading Check *To which mineral group does hematite belong?*

Mineral Uses

For centuries, humans have relied on minerals for their everyday needs, for example, humans have always needed salt. Civilizations have advanced themselves through the use of their mineral wealth. Think of the exploration and conflicts that took place over gold. Europeans spent vast sums of money and risked many lives to search for gold.

People use minerals either directly as objects of wealth, or as raw materials to make things. Those regions fortunate enough to have large quantities of hematite, iron ore, advanced rapidly. Think of all of the things made from iron and steel. Without iron there would be no machinery as we know it today for manufacturing goods.

Not all minerals need to provide metals to be valuable. Nonmetallic minerals are valuable as well. For example, quartz is used to make glass and glass fibers. Glass fiber is used to make fiber optical cables. Most sands and gravels are largely quartz. When mixed with cement (calcite), quartz is used to make concrete. Think of all the things that are made of concrete.

Salt Mining Because of its use as a preservative for food in ancient times, salt was a valuable resource. Salt has been collected by evaporating seawater and by mining deposits from below Earth's surface. Research the word *salary* in terms of its connection with salt.

section 1 review

Summary

Common Elements

- Few elements form the bulk of materials in Earth's crust.

Mineral Properties

- A mineral is an inorganic solid with a predictable chemical composition.
- All minerals occur in nature and have orderly internal atomic arrangements.
- Physical properties of minerals may reflect their chemical composition and orderly atomic arrangement.

Mineral Formation and Mineral Groups

- Minerals crystallize mainly from fluids and molten rock material, as atoms come together to form crystalline structures.
- Minerals can be sorted into two broad categories: silicates and non-silicates.

Mineral Uses

- Civilization depends on minerals for a wide range of uses, from direct wealth to making products.

Self Check

1. **Determine** whether diamonds produced in laboratories are considered to be minerals.
2. **Describe** two ways that minerals form.
3. **Compare and contrast** mineral cleavage and mineral fracture.
4. **Explain** why it is useful to test more than one physical property when attempting to identify a mineral.
5. **Think Critically** What does the fact that mower blades become dull tell you about the hardness of the blades?

Applying Math

6. **Calculate Area** Imagine you have a halite cube that measures 3 cm on each edge. What is the total surface area of the cube?
7. **Calculate Area** Imagine cleaving the halite cube in question 6 exactly in half, perpendicular to one of its faces. What is the total surface area now?
8. **Analyzing Variables** What is the relationship between particle size and surface area for a given volume of material?

More Section Review gpescience.com

LAB

Be a Mineral Detective

Detectives must gather facts and physical evidence to deduce the events that took place during a crime. Much like detectives, geologists gather physical evidence to better understand Earth processes. First, minerals are identified, and then their histories sometimes can be interpreted.

Real-World Problem

How is it possible to distinguish similar-looking materials from each other?

Goals

- **Observe and record** physical properties of minerals.
- **Determine** mineral names using your observations and identification keys.

Materials

mineral samples
unglazed porcelain tile
Mohs hardness scale
Reference Handbook, "Minerals"
penny
nail
glass plate

Procedure

1. Set up a data table on your paper. Label the columns using, at least: *Mineral Number, Color, Hardness, Streak, Cleavage, Fracture, Other Observations* (like smell, feel, or heft), and *Mineral Name.*
2. Obtain numbered mineral specimens from your teacher. Observe each mineral and accurately record the data based on your tests for physical properties. Be descriptive and broad in your observations.

   Hints: When determining hardness, test several different spots on the sample. Remember the hardness of your fingernail is a little more than 2; penny, a bit more than 3; steel nail, 5–6; glass, about 5–5.5. Minerals can have both cleavage and fracture at the same time. Cleavage is expressed as how many unique planar directions are present, and the angles formed when more than one direction occurs.
3. Using the Reference Handbook, "Minerals" at the back of your book, attempt to identify each unknown mineral. Your teacher may tell you when you are incorrect and allow you try again.

Conclude and Apply

1. **Explain** why certain minerals seemed to be difficult to identify.
2. **Explain** why certain minerals seemed to be easy to identify.
3. **Determine** which physical properties were most reliable for identification overall.
4. **Determine** which physical properties seemed to be least reliable.

Communicating Your Data

Compare your results with your classmates. Discuss what kind of mineral evidence a detective might obtain from a crime scene and how it might be used.

section 2

Igneous Rocks

Reading Guide

What You'll Learn
- **Describe** the types of materials present in most rocks.
- **Explain** how and where igneous rocks form.
- **Classify** igneous rocks.

Why It's Important
Igneous rocks make up the majority of Earth's crust and mantle and contain clues about processes that form them.

Review Vocabulary
mixture: combination of two or more substances that can be physically separated

New Vocabulary
- rock
- texture
- intrusive igneous rock
- extrusive igneous rock

What's a rock?

A **rock** is a naturally formed consolidated mixture containing minerals, rock fragments, or volcanic glass. Most rocks contain crystals of minerals that may or may not be well-formed. Rocks are identified by their composition and texture. **Texture** is a description that includes the size and arrangement of the rock's components. The rock-making process is a continuous cycle. Part of this cycle involves rock formed from magma inside Earth.

Reading Check *What types of materials compose rocks?*

Intrusive Igneous Rocks

Igneous rocks are those that form from molten rock material called magma. Geologists have termed igneous rocks from inside Earth plutonic, after Pluto, the Greek god of the underworld. Such rocks also are called **intrusive igneous rocks** because they form within, or push into, regions of Earth's crust, as shown in **Figure 8.**

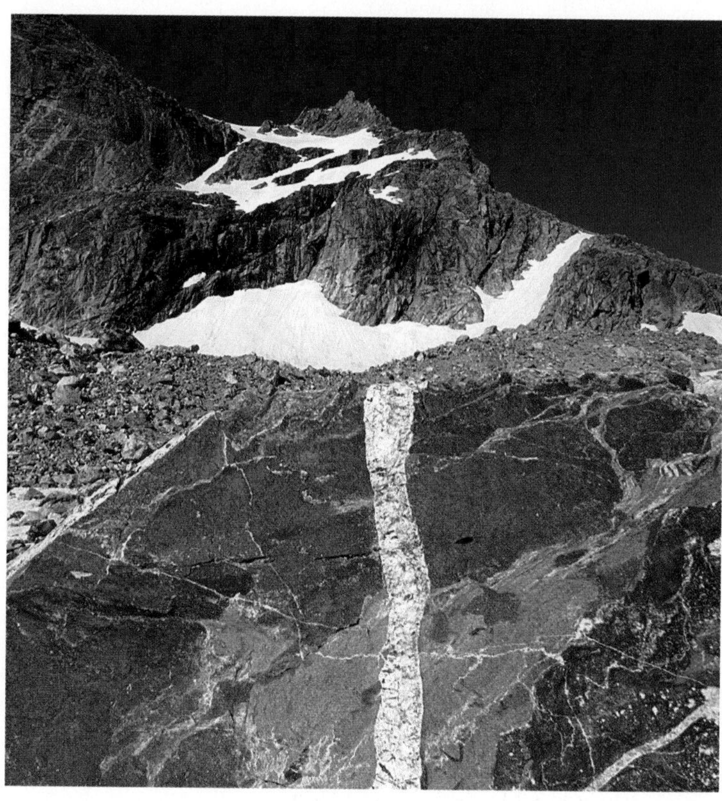

Figure 8 These light-colored veins of intrusive igneous rock formed when molten rock material cut through surrounding rock layers.

Nature of Magma As it passes through rock, magma might cause partial melting of the rock it intrudes. Geologists have learned that minerals melt at different temperatures, so some will melt when exposed to the thermal energy of the magma. This interaction between magma and surrounding rock can cause changes in the rock, contamination of the magma, or both.

You're familiar with the freezing of liquid water to form ice at 0°C on Earth's surface. Minerals freeze at very high temperatures, ranging from several hundred to more than 1,000°C. However, magmas generally must be cooled below the solidification temperature of a mineral in order for it to begin crystallizing.

Intrusive Igneous Rock Composition As crystals solidify in cooling magma, they use up certain atoms. High-temperature magmas tend to crystallize olivine and pyroxene group minerals and plagioclase feldspars first. These minerals are dense and tend to settle toward the bottom of the magma chamber. Late-forming, less dense minerals tend to solidify at lower temperatures and float to the top of the magma chamber. These relationships are shown in **Figure 9.**

Figure 9 The Bowen's reaction series, shown below, illustrates the sequence in which minerals crystallize from magma at different temperatures.
Identify *two pairs of minerals that are likely to be found in the same rock.*

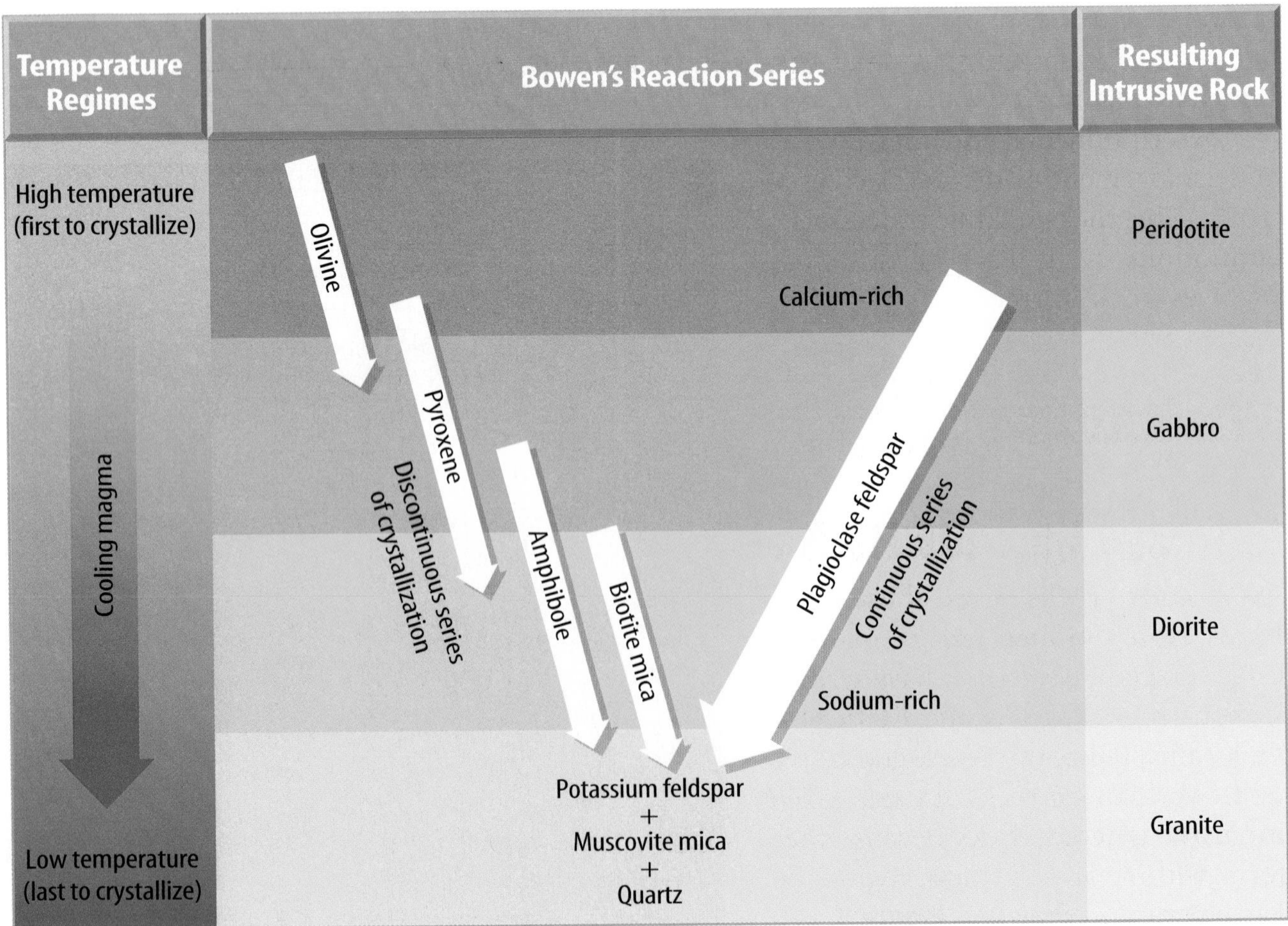

Figure 10 Intrusive igneous rocks often can be distinguished from extrusive igneous rocks because of their large grain size.

The composition of intrusive igneous rocks gives you clues as to where in Earth they formed. Igneous rocks with abundant quartz generally are associated with continental crust, while those with little or no quartz generally are associated with deep locations in continental crust or with oceanic crust. The minerals found in the rock depend on what atoms were available in the magma. If a particular element is not present in the magma, then minerals requiring that element can't be formed.

Reading Check *Where might you find iqneous rocks with little or no quartz?*

Intrusive Igneous Rock Texture Recall that texture describes the size and arrangement of rock components. In intrusive igneous rocks, grain size, which means the size of individual mineral crystals, gives you clues as to how fast magma cooled to form it. Magma that cools slowly, over many thousands or millions of years, allows atoms time to migrate about and form large crystals. Large grains are those big enough to see with the unaided eye. Such materials are termed coarse-grained, like the rock shown in **Figure 10.**

Classification of Intrusive Igneous Rocks It is worthwhile to consider a few igneous rock compositions that represent end-members of a large range in possible rock compositions. Rocks that are quartz-rich and contain potassium feldspar and plagioclase feldspar are called granite. Rocks with no quartz and abundant plagioclase feldspar and pyroxene are called gabbro. The low density of granite allows it to be a dominant rock in continental crust. Coarse-grained diorite is an igneous rock between granite and gabbro in composition. Peridotite is denser than gabbro, is composed mainly of olivine and pyroxene, and is thought to compose the bulk of Earth's uppermost mantle.

Extrusive Igneous Rocks

When magma reaches Earth's surface, it is called lava. **Extrusive igneous rocks** are those that cool from lava that has erupted at Earth's surface, as shown in **Figure 11.** These rocks may have the same compositions as intrusive igneous rocks, but they always will have different textures. Rapid cooling causes small, microscopic grains to be present, or no grains at all. Volcanic glass represents an extreme in that it cools almost instantaneously, or is quenched, in air or water.

Magma often becomes contaminated with material that is melted from surrounding rocks as it passes through them on its way to Earth's surface. The greater the distance the magma travels, the more likely it will be contaminated. For example, magma traveling several kilometers to the surface is more likely to change in composition. Magma traveling a much shorter distance might remain unchanged. Another factor is the amount of time the magma was in contact with the surrounding rock material. If the magma is rapidly erupted at Earth's surface, it is less likely to have had enough time to melt the surrounding rock material. Composition of the surrounding rock material will also affect the extrusive magma. If the surrounding rock material is similar to the magma, contamination is less likely.

Figure 11 This type of lava, called pahoehoe, has a smooth, ropy surface.

Extrusive Igneous Rock Composition The compositions of extrusive igneous rocks is much like the compositions of their intrusive cousins. A magma rich in silica (SiO_2) forms granite if it cools slowly, and rhyolite if it cools rapidly. Similarly, gabbro's fine-grained volcanic counterpart is basalt, which is a common rock in Earth's oceanic crust. An extrusive rock intermediate in the amount of SiO_2 is andesite, the volcanic equivalent of intrusive diorite. Many intrusive igneous rock compositions have extrusive igneous rock equivalents.

Applying Math — Use Percentages

EXTRUSIVE IGNEOUS ROCK CLASSIFICATION The amount of silica (SiO_2) in an extrusive rock formed when magma cools is used as a classification system. Balsatic rock contains 45 percent to 55 percent SiO_2, andesitic rock contains 55 percent to 65 percent SiO_2, and rhyolitic rock contains 65 percent to 75 percent SiO_2. A 151-g rock is found to contain 89 g of SiO_2. Based on the percentage of SiO_2 in this rock, how would you classify it?

IDENTIFY known values and the unknown value

Identify the known values:

151-g rock — means → $m_{rock} = 151$ g

89 g of SiO_2 — means → $m_{silica} = 89$ g

Identify the unknown value:

percentage of SiO_2 — means → percent SiO_2 = ? %

SOLVE the problem

Divide the mass of SiO_2 by the total mass of the rock and multiply by 100%:

$$\text{percent } SiO_2 = \frac{m_{silica}}{m_{rock}} \times 100\% = \frac{89\text{ g}}{151\text{g}} \times 100\% = 59\%$$

The rock should be classified as andesitic.

CHECK your answer

Does your answer seem reasonable? Check your answer by multiplying the percentage you calculated (expressed as a decimal) by the mass of the rock. The answer should be the mass of silica given in the problem.

Practice Problems

1. What is the greatest mass of SiO_2 that could be found in a 212-g rhyolitic rock?
2. A 120-kg balsatic boulder found in Hawaii contains 47 percent SiO_2. How many kilograms of SiO_2 are in the boulder?

For more practice problems, go to page 879 and visit Math Practice at gpescience.com.

Topic: Igneous Rock Textures

Visit gpescience.com for Web links to information about igneous rock textures.

Activity Select gabbro and granite from plutonic microtextures. Make a detailed, colored sketch of each of the photomicrographs of these rocks. Label your drawings. Repeat the activity by selecting nepheline basalt and rhyolite from the volcanic microtextures file.

Extrusive Igneous Rock Textures Rock has a limited ability to transfer thermal energy and so acts like an insulator, allowing magma to cool slowly. Should magma erupt into air or water, thermal energy loss is much faster. When magma nears the surface it meets up with much cooler surroundings. Depending on composition, magmas have temperatures ranging from about 650°C to 1,200°C. The surface of Earth might be 35°C on a hot day.

Recall that rapid cooling results in small crystals or no crystals at all and slow cooling results in large crystals, as shown in **Figure 12.** If cooling starts off slowly below the surface with large crystals, but then finishes at a faster rate to form small or no crystals, the extrusive rock is called a porphyry. If you observed a volcanic porphyry, you would see larger crystals embedded in a mass of small crystals or glass. The term *porphyritic* is used as an adjective to describe this texture. You might then have a porphyritic rhyolite or porphyritic basalt.

What conditions cause a porphyritic texture to form?

	Granitic	Basaltic
Intrusive	Granite	Gabbro
Extrusive	Rhyolite	Basalt

Figure 12 Igneous rocks often are described as being basaltic or granitic. Basaltic rocks are dense and rich in magnesium and iron, but low in silica. Granitic rocks usually are less dense and are silica-rich, containing higher quantities of quartz.
Compare and contrast *the textures and colors of extrusive igneous rocks and their intrusive counterparts.*

Volcanic eruptions can result in many different extrusive rock textures and forms as shown in **Figure 13.** Lava can be ejected into the air as large, streamlined blobs that solidify in flight called volcanic bombs, or smaller droplets known as lapilli. Solids of varying size from large blocks to ash can be expelled from a volcano.

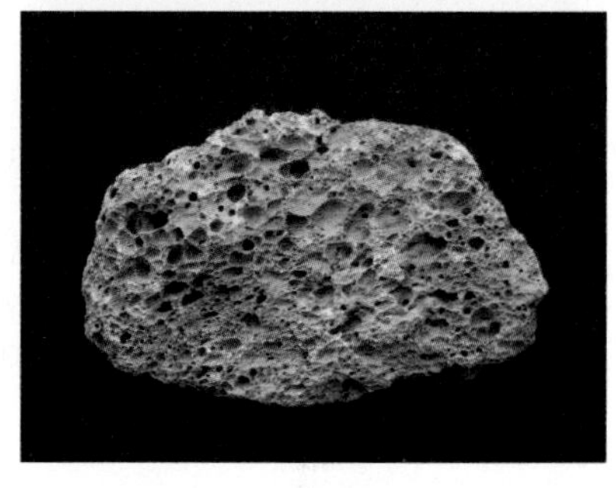

Figure 13 Some volcanic bombs develop a streamlined shape when they are hurled rapidly through the air during a volcanic explosion.

Effect of Gases Should magma contain dissolved gases, the gases expand in response to decreasing pressure at the surface, escape, and break apart or disrupt the lava into ash flows. In addition, lava that flows across the surface also can have many tiny holes, somewhat like Swiss cheese. Such a texture is called vesicular, and it mainly forms near the top surface of a flow where gases escape. The extrusive rock pumice results when the holes are left behind. Some pumice has so many gas-bubble holes that its density is reduced and it can float in water.

section 2 review

Summary

What's a rock?

- Rocks are mixtures of a wide variety of natural materials.
- The composition and texture of a rock indicate how it formed.

Intrusive Igneous Rocks

- Intrusive igneous rocks form from magma below Earth's surface, cool slowly, and are coarse-grained.
- The composition of granite is similar to the composition of continental crust as a whole.

Extrusive Igneous Rocks

- Some extrusive igneous rocks form from lava at Earth's surface, cool rapidly, and are fine-grained or glassy.
- Other extrusive rocks are consolidated solid materials expelled from volcanoes.
- Oceanic crust is composed primarily of the extrusive igneous rock basalt.

Self Check

1. **List** the materials that can form a rock.
2. **Describe** how the texture of a porphyritic andesite indicates its cooling history.
3. **Discuss** what happens to the dissolved gases in magma as the magma ascends toward the surface of Earth.
4. **Explain** why you would not expect to observe an igneous rock with quartz and olivine in it.
5. **Think Critically** Lava flows that are relatively low in SiO_2 tend to have low viscosities, or flow easily. Name an igneous rock that you would expect to form from a low-viscosity lava flow. Explain your answer.

Applying Math

6. **Calculate Granite Volume** Granite occurs in very large masses called batholiths. The Sierra Nevada in California is a batholith made up of many separate intrusions. Suppose the mountain range averaged 8.33 km in thickness, extended 700 km north to south, and averaged 91.66 km wide. Approximately how many cubic kilometers of granite are in that mountain range?

section 3

Sedimentary Rocks

Reading Guide

What You'll Learn

- **Explain** how and where sedimentary rocks form.
- **Discuss** how sedimentary rocks are used to interpret Earth's history.
- **Describe** how sedimentary rocks are classified.

Why It's Important

Knowledge of sedimentary rocks helps you understand where resources such as fossil fuels and groundwater come from.

Review Vocabulary

precipitate: an insoluble compound that comes out of solution during a chemical reaction

New Vocabulary

- clast
- porosity
- cementation

Figure 14 Rocks that are tumbled more than 3,000 km along the bottom of the Colorado River, shown below, can be broken into fine particles before they reach the Gulf of California.

Rocks from Surface Materials

Recall that rock is a consolidated mixture of minerals. Some of these minerals could be in bits and pieces of other rocks. Such small bits and pieces are called **clasts.** The word *clast* is from the Greek *klastos* which means "broken."

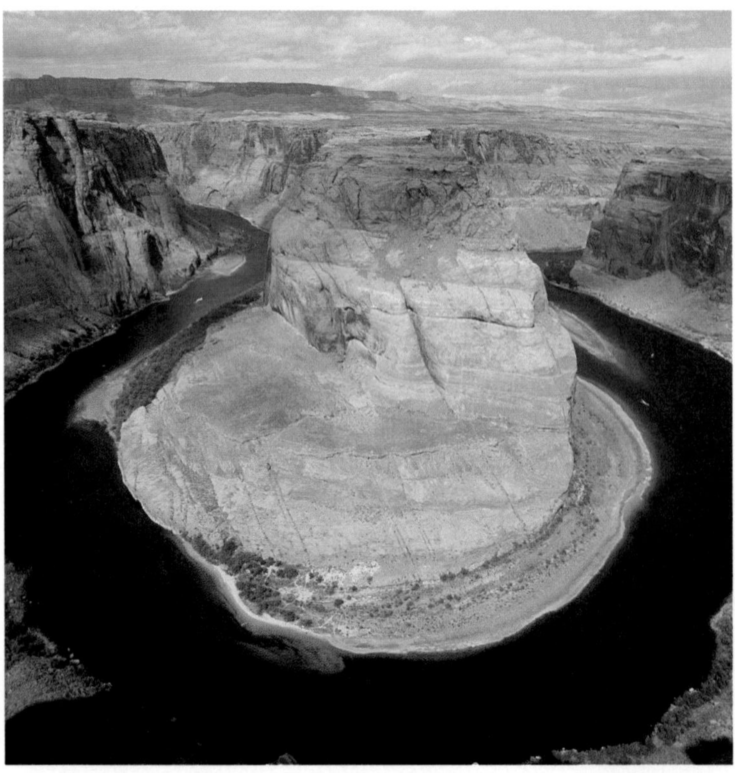

Surface Attack What could reduce rocks of the crust to smaller sizes? Rocks inside Earth are protected from surface conditions. They are squeezed and heated and are reasonably stable under those physical conditions. In contrast, rock exposed at the surface is attacked by the weather. This action over a long time breaks rocks down into smaller-sized pieces and loosens them from their original positions. Loose chunks of rock, in turn, are attacked more easily by water and air, and are able to smash into each other. It is this slow, but constant smashing, grinding, and dissolving of clasts that can take place in rivers and streams, as shown in **Figure 14,** and in other environments.

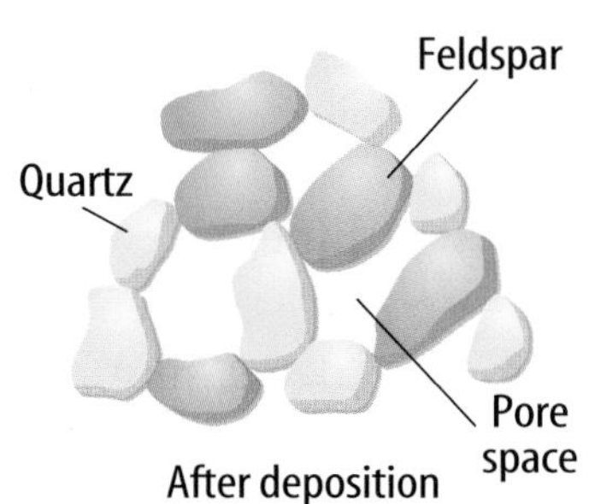

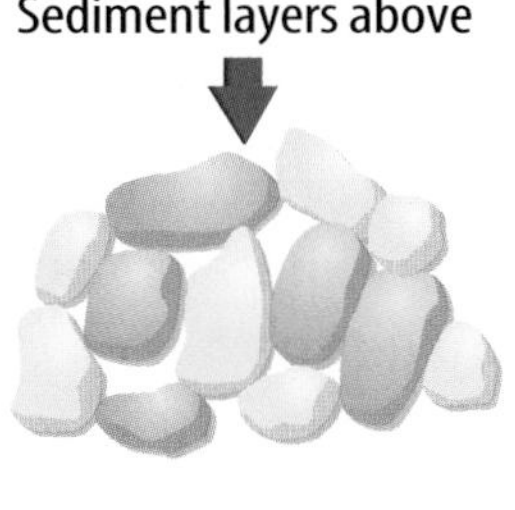

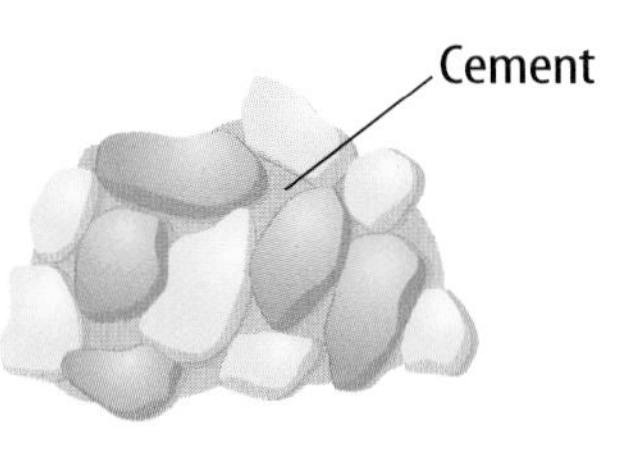

Figure 15 Sandstone can form when sand grains are deposited, compacted, and cemented together.

Transportation and Deposition Mechanical weathering processes break rocks into smaller clasts, making it possible for the clasts to be eroded or removed from their original locations. When clasts are transported to new locations, they often become rounded before being deposited.

When clasts are loose on Earth's surface, they don't fit together perfectly. For example, a pile of baseballs will have plenty of space between the balls. This empty space is called **porosity.** Now imagine a pile of clasts, some rounded as they were transported by a river. Although smaller than the baseballs, the pile of clasts also will have some porosity. Sedimentary rock starts off with some porosity. When the clasts are clay minerals, which are sheet silicates that fit together tightly when compressed, porosity is minimal. It is in the porosity that water, oil, and natural gas are stored within Earth.

Buried Clasts Eventually, clasts can become consolidated into sedimentary rock. The making of clasts by weathering is a surface process, while the making of sedimentary rock occurs below the surface. What processes below the surface could hold clasts together tightly enough to make a rock?

When buried by more sediment deposited above them, clasts can be smashed together with such great force that they become compressed and stick together. The force is pressure created by gravity pulling down on all the material that overlies the clasts. Pressure increases the deeper in Earth that the clasts are, increasing the chance that the clasts will stick together. This process is called compaction.

Reading Check *What force causes compaction?*

Water moving between clasts carries dissolved minerals that can act as cement. Common minerals that are cement materials include quartz, calcite, hematite, and clay minerals. Under the right conditions these minerals slowly precipitate out of water and begin to fill spaces between clasts. This process is called **cementation.** Most of the time both compaction and cementation work together to make sedimentary rock. **Figure 15** shows how sedimentary rock becomes consolidated.

Detrital Sedimentary Rocks

Detritus, from the Latin *deterere*, which means "to lessen or wear away," is another name given to clasts. Sedimentary rocks that are made mostly of clasts are called detrital sedimentary rocks. Clasts can come in many sizes. In order of decreasing size, clasts are known as gravel, sand, silt, or clay. Most people use size names loosely, but to geologists, each size name has a specific meaning, as shown in **Figure 16.** These size names only tell the dimension of a particle, not its composition.

Detrital Sedimentary Rock Textures If someone asked you to define sand, what would you say? Some people might think sand is made only of quartz, or the common silica sand available in hardware stores. In fact, a sand-sized particle could be made of any kind of material. Some kinds of materials, such as quartz and feldspar, hold up better against the effects of weather than others, such as olivine and pyroxene, and are more common in sedimentary rocks.

Why would anyone care about the exact size of a clast? Geologists have found that size works well as a clue to the kind of environment in which a rock formed. One clast of gravel is large and heavy compared to one clast of sand. It takes more force, or energy, to lift or move gravel than it does to lift or move sand. The larger particles in a mix of clasts often can't be lifted by air or water, and instead bounce or scoot along Earth's surface. If you sampled a muddy river you might be surprised to find gravel or sand on the bottom of the river, but a jar of river water might contain millions of tiny silt and clay particles. When water is calm, most clast sizes settle out. As water speeds up, the smallest clasts are lifted first, then larger clasts. During a flood, when the water is moving swiftly, all but the very largest particles are moved. The reverse happens as the water slows. Largest particles settle out first, followed by smaller particles.

Detrital Sedimentary Rocks

Texture (grain size)	Sediment Name	Rock Name
Coarse (over 2mm) 	Gravel (Rounded fragments)	Conglomerate
	Gravel (Angular fragments)	Breccia
Medium (1/16 to 2 mm) 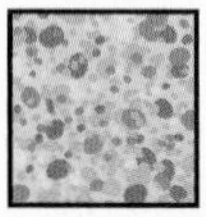	Sand	Sandstone
Fine (1/16 to 1/256 mm) 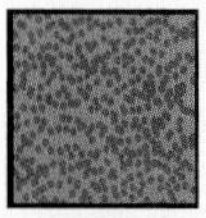	Mud	Siltstone
Very fine (less than 1/256 mm) 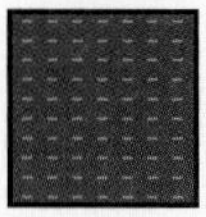	Mud	Shale

Figure 16 Sand is defined by size, not by composition. Sand doesn't even have to be made from rock material; it can be made from shells.

Detrital Sedimentary Rock Compositions Detrital sedimentary rock composition depends on sources of rock material that were eroded, transported, and eventually deposited. Original rock material can be dislodged and carried by streams or water currents, blown by the wind, or rafted along by moving ice. The clasts can come from all the regions that these carriers moved over or through. The number of possible combinations of different kinds of clasts is large. However, the size of clasts is controlled by the ability of the carrier to move them.

Some minerals tend to be more common in detrital sediments because they are harder or more resistant to being dissolved. Quartz is such a mineral, which is why you might have thought clasts are always made of quartz. Geologists examine sedimentary rock compositions and try to reconstruct what happened to form them. You can try to reconstruct a sedimentary rock's history by imagining all the places that clasts could have come from and how they got there.

What factors control the composition of a detrital sedimentary rock?

Detrital Rock Classification Just as igneous rocks are classified according to composition and texture, similar observations are used to classify detrital sedimentary rocks. Mineral composition is extremely variable, so adjectives are used to modify the general name of the rock. The general rock name is determined by clast size. The names of detrital sedimentary rocks are given in **Table 4.**

Clast size also provides clues to help determine the deposition environment of the sediment that formed the detrital rock. Suppose you have a sample of siltstone that contains shells of marine animals. You might conclude that the sediments were deposited in a shallow, calm area along the coast. The calm water allows small particles like silt to settle out. The shells give you a clue that the environment was near the ocean. Perhaps the sample also has fossils in the form of marine worm tubes. The fossils are traces of organisms that might have lived in soft mud.

Table 4 Common Clast Sizes Used to Name Sedimentary Rocks

Common Clast Size	Rock Name
Gravel or larger	Conglomerate
Sand	Sandstone
Silt	Siltstone
Clay	Shale

INTEGRATE Chemistry

Cementation Though sedimentary rocks may form in arid, desert conditions or from compaction by glaciers, most sedimentary rocks are created from being deposited in water. Dissolved minerals in the water may precipitate onto the surfaces of the clastic particles and become the cement to hold grains of sand, silt and clay together. The minerals may also precipitate to form layered rock crystals as well. Make a supersaturated solution and allow the liquid to evaporate. Do the crystals form in layers?

Modeling Evaporites

Procedure

1. Add to a **sauce pan** 1 L of **tap water** and 34 g (about 6 level teaspoons) of **salt** and stir with a **spoon.**
2. Place a **lid** on the pan and heat it on a **hotplate** to boiling. Droplets of a solution should be forming underneath the lid. This solution will be hot. Use **thermal mitts** to carefully remove the lid and drain off the liquid from the lid into a **saucer.** Replace the lid.
3. Repeat step 2 several times until you have several teaspoons worth of liquid from the lid.
4. Carefully pour enough of the pan's contents to half-fill a **second saucer.** Label each saucer to identify which water came from the lid and which came from the pan.
5. Allow both saucers to sit undisturbed until all of the water evaporates from each.

Analysis

1. What do you observe in the saucers after all of the water has evaporated?
2. Compare and contrast any substance you observe with some salt grains directly from the salt container. Use a magnifying lens to observe the substance in each saucer.

Chemical Sedimentary Rocks

Some sedimentary rocks form through the activity of chemicals dissolved in water. If the water contains more dissolved material than it can hold, then some of that dissolved material begins to form crystals that settle out of the water. These crystals build up to form new rock. There are two ways that this process gets started.

Precipitation If water receives more dissolved materials than it can hold in solution, then the excess must precipitate out as microscopic crystals. This crystallization of excess dissolved material is called precipitation. In a small percentage of limestones called oolites, calcium carbonate precipitates directly from sea water around preexisting particles, such as grains of silt or sand.

Evaporation The other option is for some water to evaporate. This leaves an oversupply of dissolved matter and again crystals are forced to form. Rock salt and rock gypsum are common chemical rocks formed in this way. They are often are called evaporites because evaporation of water initiates their formation.

Biochemical Sedimentary Rocks

If sedimentary rocks contain the remains of living organisms they are called biochemical sedimentary rocks. Most of Earth's limestone is composed, at least partially, of the remains of marine organisms that had hard parts made of calcium carbonate. The shells of the organisms were cemented together by calcite mud. If you found a rock composed almost entirely of shells and shell fragments, you would have found limestone. Coquina is a variety of limestone that is formed mostly from broken shell fragments. The shells accumulated on the seafloor and were later cemented together with calcium carbonate. Most organisms whose shells contributed to the limestone-making process lived in warm, shallow seas. Animals without shells rarely contributed enough debris to make rock. More often, their soft remains deteriorated or were scavenged before they became part of a sedimentary rock.

Reading Check *What chemical compound is limestone made of and where does it come from?*

Another common rock that originates from the remains of organisms is coal. The starting material for coal is mostly plant matter. Coal is sedimentary rock composed almost entirely of the carbon that remains after plant material is compressed underground.

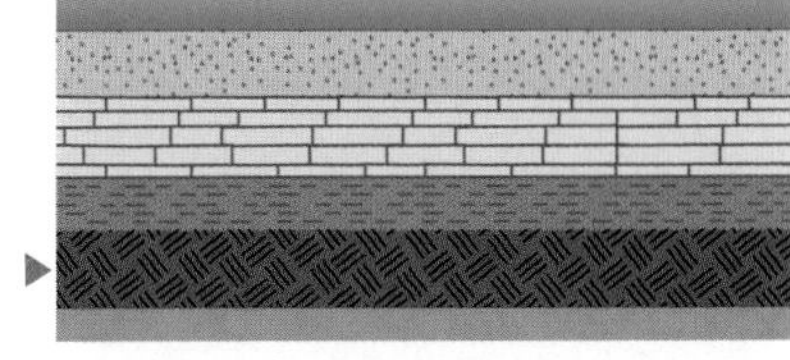

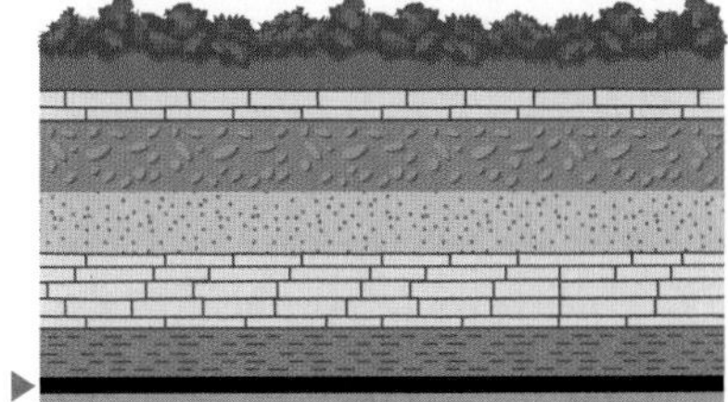

Figure 17 As sediment accumulates above a layer of peat, shown at the upper triangle, the peat becomes more compressed. Continual compression drives out water and other compounds, leaving behind a form of carbon called coal, shown at the lower triangle.

Coal usually develops from peat, a brown, lightweight deposit of moss and other plant matter. Shallow swamps or bogs in a temperate or tropical climate are likely environments of deposition. Partial decay of the plant matter uses up oxygen in the swamp water. This stops further decay and the remaining organic matter is preserved. Peat is transformed into coal largely by compaction after it has been buried by sediment.

Look at **Figure 17.** Coal goes through a series of changes as it forms from peat. Each stage of compaction drives out more impurities and leaves behind a more concentrated form of carbon. Coal often lies within repeated sequences of sandstone, shale, and limestone, which indicate alternating continental and marine conditions. This implies a low-lying environment near the sea, such as lagoons, large deltas, and swampy coastal plains. By studying coal fields, changes in Earth's climate, geological activity, and life forms may be traced and compared.

section 3 review

Summary

Rocks from Surface Materials

- Sedimentary rocks have three main types that reflect their origins: Detrital, chemical, and biochemical.

Detrital Sedimentary Rocks

- Detrital sedimentary rocks can be composed of a wide variety of clasts.
- Clasts vary widely in size.
- Many possible combinations of clast composition and clast size produce a large number of detrital sedimentary rocks.

Chemical Sedimentary Rocks

- Chemical sedimentary rocks form as minerals precipitate from water rich in dissolved mineral matter.

Biochemical Sedimentary Rocks

- Biochemical sedimentary rocks form from the remains of organisms.

Self Check

1. **Describe** the requirements for a material to be classified as a sedimentary rock.
2. **Compare and contrast** the formation of an evaporite rock to one of detrital origin.
3. **Explain** how clast size affects transportation of different-sized clasts. Also, explain the order in which different-sized clasts are deposited.
4. **Think Critically** Suppose a stream floods over its banks. If you dug a hole through the layers of sediment deposited near the edge of the channel after the water receded, what would you likely see?

Applying Math

5. **Calculating Sedimentation Rate** Imagine that thin layers of sediment are being added to a lake that is 10 m deep. Suppose the average yearly influx of sediment is 2.5 mm. About how long would it take to fill in the lake with sediment?

section 4

Metamorphic Rocks and the Rock Cycle

Reading Guide

What You'll Learn

- **Identify** physical conditions that cause metamorphism.
- **Explain** where metamorphism occurs.
- **Classify** metamorphic rocks.

Why It's Important

An awareness of change leads to an improved ability to adapt to change.

Review Vocabulary

chemical reaction: process in which one or more substances are changed into new substances

New Vocabulary

- foliated
- rock cycle

Metamorphic Rocks

In the last section you learned that sedimentary rocks show great variety with many different combinations of composition and texture. The same is true for metamorphic rocks. Metamorphic rocks, like the one shown in **Figure 18,** are rocks that have been changed by some combination of thermal energy, pressure, and chemical activity. As these agents begin to act on a preexisting rock, its atoms rearrange and sometimes form new minerals. The word metamorphic comes from two stems, "meta" (to change) and "morph" (form). Any igneous, sedimentary, or metamorphic rock is subject to change through metamorphism.

Figure 18 Sharp folds sometimes display intense transformations in metamorphic rocks.

Conditions that form metamorphic rocks are somewhere between the sedimentary and igneous environments. In other words, consolidation of sediment to form sedimentary rock occurs at lower pressures and temperatures than typical metamorphic conditions. On the other hand, igneous rock formation requires rock to melt and solidify. Igneous processes operate at higher temperatures than metamorphism which occurs while rock remains solid.

Metamorphic Rock Composition

INTEGRATE Chemistry The agents of metamorphism are thermal energy, pressure, and chemical activity. Water and carbon dioxide are common active chemicals that exist as a fluid and chemically react with rock. This fluid can come from molecular water or carbonate locked in the rock itself, or can be introduced from a nearby intrusion of magma.

Reading Check *What are three causal agents of metamorphism?*

Changing Minerals Clay minerals, micas, and amphiboles are examples of minerals that contain water in their crystal structures. Clay minerals tend to form micas with increasing metamorphic conditions. If metamorphic conditions of pressure and temperature continue to increase, the water can be driven away completely. Minerals, such as garnet, that contain some of the same elements found in clays (aluminum and silicon) can result. In this way, some new minerals form by dehydration at higher temperature and pressure.

Environments of Metamorphism So, what's causing these changes? Deep burial or regional movements of large parts of Earth's crust and uppermost mantle are possible causes. These large-scale changes, called regional metamorphism, produce huge areas of changed rock. Another smaller-scale cause is local contact of any preexisting rock with magma. This is called contact metamorphism. You produce contact metamorphism when you place food in a hot pan coated with oil and it quickly begins to burn. One important difference is that in cooking, chemical reactions take place rapidly, while metamorphic changes occur over thousands or millions of years.

Metamorphic Rock Textures

Metamorphic processes create many rock textures, but a few general types can help you understand the processes that formed them. **Foliated** textures in metamorphic rocks have lots of layers or bands. Nonfoliated metamorphic textures include rocks whose grains are in more random orientations. Some nonfoliated textures resemble igneous rock textures, in which grains are locked tightly together. **Figure 19** shows examples of these textures.

Figure 19 *Folio* means "leaf." Foliated texture has the appearance of layered leaves or pages of a book. The texture of the first metamorphic rock shown below is foliated; the second is nonfoliated.

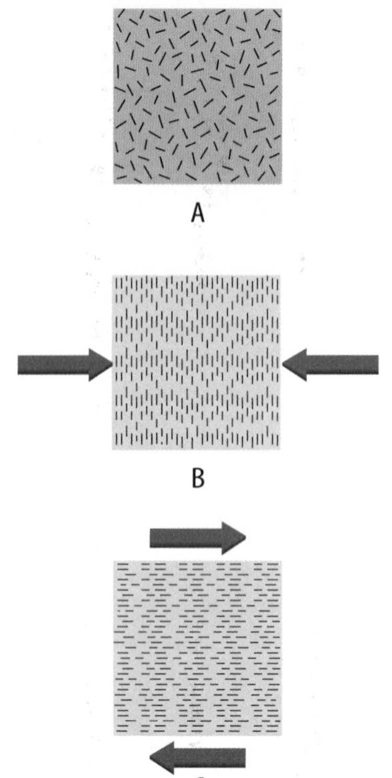

Figure 20 Direction of applied force affects orientation of mineral grains.

A Mineral grains are randomly oriented when no directed force is involved.

B Orientation of mineral grains is perpendicular to the direction of pressure caused by compression.

C Mineral grains are parallel to the direction of shearing force.

Topic: Metamorphic Rock Textures

Visit gpescience.com for Web links to information about metamorphic rock textures.

Activity: Select from metamorphic microtextures: Gneiss, marble and foliation. Make a detailed colored sketch of each of the photomicrographs showing these textures. Label your drawings.

Foliated Rocks The most common sedimentary rocks in Earth's crust are mudrocks such as shale and siltstone. These rocks contain abundant clay minerals, and when metamorphosed, the clays change to minerals in the mica group, such as muscovite. The mica group minerals also change in grain size with increasing degrees of metamorphism.

When squeezed and heated, layers of mica line up in a direction that is perpendicular to the direction of compression or parallel to the direction of shearing, as shown in **Figure 20.** For example, slate is a fine-grained, foliated metamorphic rock that splits easily along flat, parallel planes. It typically formed when shale was compressed, causing mica grains to orient perpendicular to the compression.

Reading Check *How does compression affect the orientation of mica grains in metamorphic rocks?*

Sometimes it's easiest to consider foliated rocks as a progression of increasing metamorphic conditions. The smallest-grain sizes in foliated textures occur in slate, which forms thin layers and exhibits rock cleavage. The rock cleavage comes from the parallel alignment of mineral grains within the rock. If the metamorphic conditions increase, some of the mineral grains in the slate grow and produce phyllite. With continued grain growth, a shiny mass of mica sheets called schist (SHIHST) forms. Finally, at the highest level of change a rock called gneiss (NISE) is produced. Gneiss rock textures often are banded, and gneisses generally represent the limit between metamorphic and igneous conditions.

Nonfoliated Rocks Similar in texture to intrusive igneous rocks, nonfoliated metamorphic rocks tend to have random crystal orientation and uniform grain size. In general, mineral grains tend to grow as the grade of metamorphism increases. Imagine a limestone with microscopic grains. Under the influence of thermal energy, pressure, and fluids, such a limestone recrystallizes in large beautiful grains of calcite that are readily visible in a hand specimen. This resulting rock is marble, and it is essentially chemically equivalent to the limestone.

Metamorphic Rock Classification

Much like other rock types, metamorphic rocks can be classified based on texture and composition. Beginning with texture, you can determine whether a metamorphic rock is foliated or nonfoliated. Remember that in any classification, there are exceptions, as illustrated in **Figure 21.**

NATIONAL GEOGRAPHIC **VISUALIZING SHOCK METAMORPHIC ROCK**

Figure 21

Earth materials undergo change, even when buried far below the surface. Deep in the crust, high temperature and pressure conditions alter the chemical composition and structure of rocks in predictable ways over time, producing metamorphic rocks. Metamorphism, however, also can occur rapidly, violently, and without warning.

Scientists estimate that the force created by a meteorite impact could be nine million times greater than atmospheric pressure. This type of collision would send enormous shock waves throughout surrounding rock strata, rearranging atoms into new crystal structures stable at higher temperature, and pressure conditions.

Large meteorite impacts leave scars that can remain visible for millions of years. Taken from the space shuttle, this image shows the *International Space Station* above the Manicouagan impact crater in Quebec, Canada. The meteorite that created this crater struck Earth 214 million years ago, hurling debris high into the atmosphere and leaving evidence in the rock record worldwide.

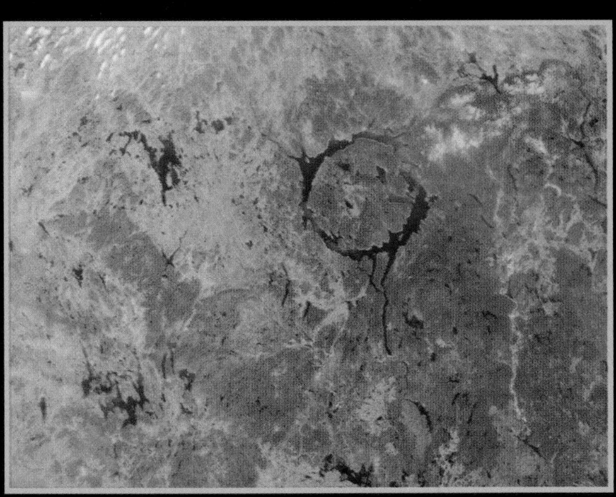

With a diameter of 100 km, the Manicouagan crater is one of the largest impact structures preserved on Earth. The force of the collision caused rock melting and rock metamorphism to a depth of about 9 km. Mineral grains in metamorphic rocks at the site are deformed extensively, evidence that an intense compression wave shocked the strata.

Mineral Composition Next, consider the mineral composition. Mineral composition provides clues about the original rock type before metamorphism, and also indicates to what degree a rock has been metamorphosed. The original rock material, often called the parent rock, and the conditions of metamorphism together control the resulting metamorphic rock. So, in studying rocks from the field, geologists must work backwards. Field samples are the results of many geologic conditions, most of which cannot be accurately reproduced in the laboratory. Some conditions have been reproduced, however, and scientists can use mineral composition to estimate the pressure, temperature, and chemical activity that resulted in the formation of a metamorphic rock.

The Rock Cycle

Illustrations often simplify complex ideas. A diagram of the rock cycle, shown in **Figure 22,** summarizes processes that you have learned about in this chapter. Processes of the **rock cycle** include any chemical and physical conditions that continuously form and change rocks. Notice that some processes happen at Earth's surface, while others happen deep below the surface.

Figure 22 The rock cycle is a continual process in which rocks change from one form to another.

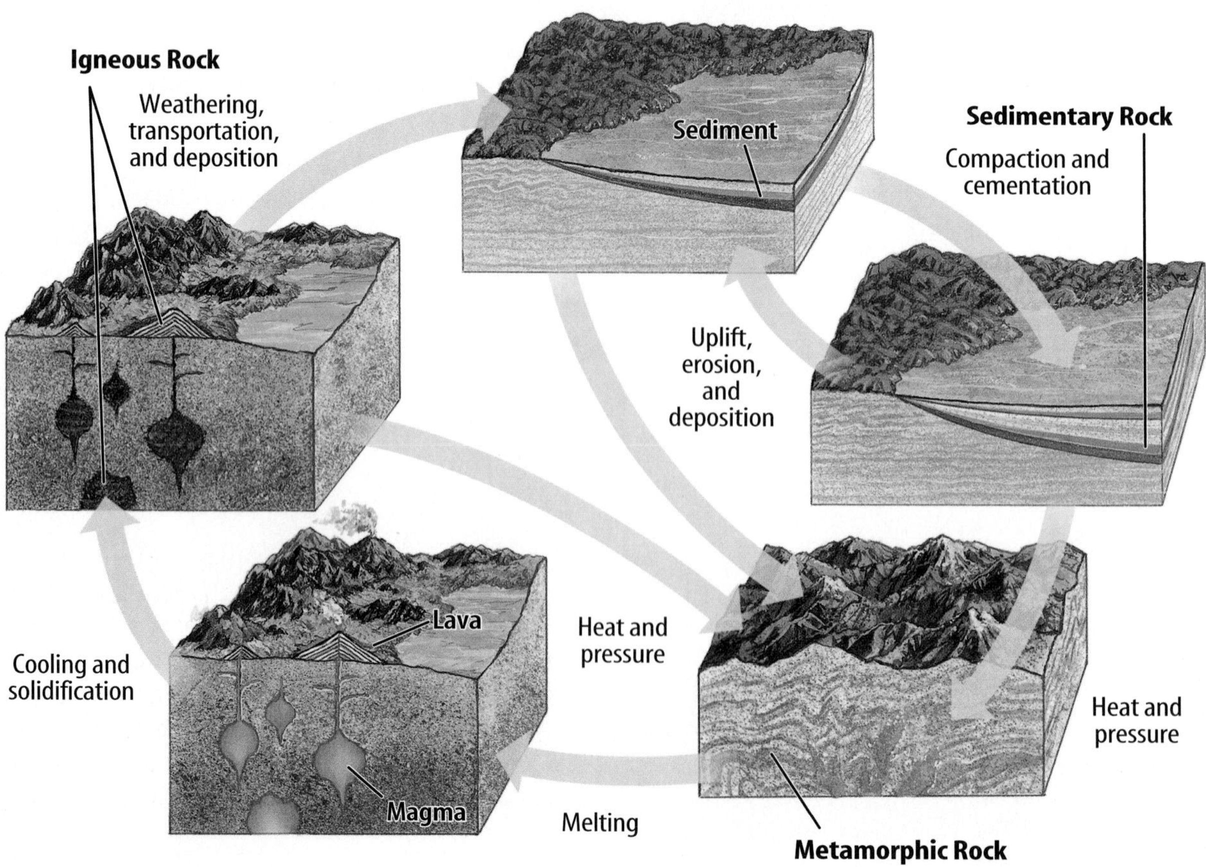

As rock material moves through stages of the cycle, all matter is conserved. Minerals from granite are found in the sediments that were eroded from those igneous rocks. Those same minerals could be present in sedimentary rocks formed later. Atoms that have been mobilized during weathering or metamorphic chemical reactions are not destroyed, but can end up as dissolved materials in solutions. Mineral crystals later can precipitate from those solutions. The atoms are never destroyed, only rearranged into new forms that are stable under the present conditions.

Also note that many different paths, indicated by arrows in **Figure 22,** can be taken by rock material as it changes within the cycle. Understand that these processes may take long periods of time, occurring over millions of years. Other processes happen suddenly, such as an erupting volcano, but this does not change the fact that there is no beginning and no end to the rock cycle.

Reading Check *Describe several examples of how matter is conserved in the rock cycle.*

section  review

Summary

Metamorphic Rocks

- Metamorphic rocks can form from any preexisting rocks.

Metamorphic Rock Composition

- Agents of metamorphism can cause new textures and new minerals to form.

Metamorphic Rock Textures

- Metamorphic rocks are foliated or non-foliated.
- As metamorphic conditions increase in magnitude, mineral grains may grow larger.

Metamorphic Rock Classification

- Metamorphic rocks are classified based on mineral composition and texture.
- As with all rock classifications, some metamorphic rocks are exceptional and not easily categorized.

The Rock Cycle

- Although rocks are subject to many changes, matter is conserved through processes of the rock cycle.

Self Check

1. **Explain** the requirements for a material to be called a metamorphic rock.
2. **Compare and contrast** foliated and non-foliated textures.
3. **List** the following sedimentary rocks: limestone and shale. Provide the respective common names of the metamorphic rocks that each one becomes.
4. **Identify** which rock in question 3 is foliated.
5. **Critical Thinking** A marble contains occasional layers of the mineral muscovite. What possible parent rock composition could have been metamorphosed to produce this rock?

Applying Math

6. **Calculate Mass** Suppose a sculptor wants to use marble with an average density of about 2.7 g/cm^3. The block of marble is one meter on every side. What is the mass of the block? If the sculptor knows from experience that the finished product will only use 45% of the block, and the rest is waste, then what will the mass of the statue be?

Identifying Rocks

Goals

- **Observe** physical properties of minerals in rocks to determine their compositions.
- **Classify** each rock according to general type: Igneous, sedimentary, or metamorphic.
- **Identify** each rock based on its composition and texture.

Materials

rock samples
nail
penny (pre-1986)
magnifying lens or particle size chart
dilute HCl
paper towels
glass plate
* binocular microscope
unglazed porcelain tile
Resource Handbook, "Rocks"

**Alternate material*

Safety Precautions

Real-World Problem

You've learned to identify most everything around you. This is an automatic process most of the time. You learned by repetition and by asking what things are called. By the time you were five years old you could name many different kinds of objects. You could identify the people in your life. You might not have had the same experience in identifying rocks—at least not yet. In this lab, you will identify rocks using a more formal approach. How can you determine the composition and texture of common rocks? What can a rock's composition and texture indicate to you about how it was formed?

Procedure

1. Construct a horizontal data table with ten columns and four rows.
2. Label the tops of columns 1 through 10, in the top row, as follows:
   1) Sample number
   2) General rock type (igneous, sedimentary, metamorphic)
   3) Crystals (yes/no)
   4) Grain size (large/small/none)
   5) Gas-bubble holes (yes/no)
   6) Fossils (yes/no)
   7) Particle size (gravel/sand/silt/clay)
   8) Foliations (yes/no)
   9) Additional observations
   10) Rock name

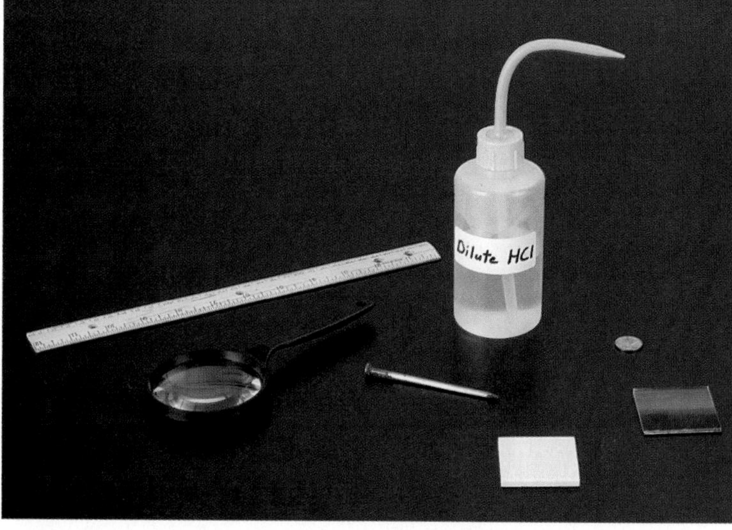

3. Obtain three rocks provided by your teacher.
4. **Observe** each rock. First, try to determine if it is igneous, sedimentary, or metamorphic. Then record everything else you observe. Your data table columns will serve as a checklist to help organize your observations.

WARNING: *If you choose to use the acid test for calcite identification, use only one drop of HCl. A positive reaction to acid is the presence of tiny bubbles rising through the drop. Blot the liquid off the sample with a paper towel before leaving it. Be careful to keep acid away from skin and clothing.*

5. **Identify** the rock with the help of Resource Handbook, "Rocks," at the back of your textbook.
6. **Repeat** steps 4 and 5 to check that your results are replicated. Keep an open mind.

Analyze Your Data

1. **Identify** the texture observations that helped you distinguish igneous rocks from sedimentary rocks.
2. **Determine** which rock samples required the fewest tests to identify them and explain why.

Conclude and Apply

1. **Evaluate** rocks according to difficulty for identifying and classifying.
2. **Explain** which observations were useful for the largest number of rock samples.
3. **Evaluate** your data table in terms of including important observations. Explain whether there were any observations you left out of your data table that are critical for distinguishing rocks.

Present your findings to the class. Discuss any observations that were different from those of your classmates. Discuss the column headings you included in your data table and defend your choices.

Earth Materials at Work

Kaolinite, a type of clay, is a common mineral on Earth.

What is clay, and where does it come from?

What do a commode, a brick, and a coffee mug have in common? While these items have widely different uses, they share an important component—a material called clay. Clay is found in many locations worldwide. Because of its unique properties and availability, clay has been used by civilizations throughout history to make items including building materials, pottery, and tablets for recording information.

Clay forms through weathering, crystallization, and deposition of Earth materials. The word *clay* really represents an entire group of many different minerals. These minerals have particles with a diameter of 0.004 mm or less. When wet, most clay minerals can be molded into a variety of shapes. When water is driven off under high temperature conditions, the material may become as hard as stone.

Clay deposits typically are found in sedimentary rock layers close to Earth's surface. As a result, clay mines are generally shallow, with depths of less than 150 m. Companies in the United States often reclaim the landscape that they mine. First, the topsoil and uppermost clay layer are removed and stored at another location. After clay is mined from an area, leftover Earth materials are returned to the pit. This layer is covered with topsoil, into which vegetation is planted.

Ceramic materials have diverse applications in manufacturing, building, art, and everyday life. Products made from clay minerals include floor tiles, fine china, and dental porcelain. The import, export, and use of clay materials contribute significantly to the world economy.

Clay minerals are used to make many items of pottery and dinnerware.

Brainstorm **Ceramic products can be found at school, at home, and in stores. Make a list of ceramic products that you encounter or use on a regular basis.**

TIME
For more information, visit gpescience.com

Reviewing Main Ideas

Section 1 Minerals

1. Only eight elements compose more than 98 percent, by mass, of Earth's crust.
2. Minerals are naturally occurring, inorganic solids that have a crystalline structure and predictable chemical composition.

3. Major mineral groups that form the bulk of Earth's crust are known as the rock-forming minerals.
4. Every mineral has characteristic physical properties that help in its identification.

Section 2 Igneous Rocks

1. Igneous rocks must form from solidification of magma or lava.
2. Igneous rocks can be separated into two categories: Intrusive—those that form in Earth—and extrusive—those that form at Earth's surface.

3. Igneous rocks are classified based on their texture and composition.
4. Igneous rocks with granite-like compositions are associated with continental crust, while igneous rocks with basalt-like compositions are associated with oceanic crust.

Section 3 Sedimentary Rocks

1. Sedimentary rocks are consolidated mixtures of minerals, rock fragments, organic debris, or other natural materials.
2. Sedimentary rocks can contain or be made entirely of the remains of organisms.

3. Detrital sedimentary rocks form through the processes of cementation and compaction.
4. Sedimentary rocks are classified based on clast size and composition.

Section 4 Metamorphic Rocks and the Rock Cycle

1. Metamorphic rocks are made through the effects of heat, pressure, and chemical activity on preexisting rocks.
2. Metamorphic rocks can be divided into two categories: Foliated—having layers or bands—and nonfoliated—with grains in random orientations.
3. Parent rock material and conditions of metamorphism control the composition of a metamorphic rock.
4. The rock cycle describes all processes in Earth that continuously change Earth materials.

FOLDABLES™ Use the Foldable that you made at the beginning of this chapter to help you review Earth materials.

Using Vocabulary

cementation p. 625	intrusive igneous rock p. 617
clast p. 624	magma p. 613
cleavage p. 610	mineral p. 609
extrusive igneous rock p. 620	porosity p. 625
foliated p. 631	rock p. 617
fracture p. 610	rock cycle p. 634
hardness p. 611	streak p. 611
	texture p. 617

Match the correct vocabulary terms with each description below.

1. molten rock
2. grain size and arrangement
3. coarse grained
4. scratch resistance
5. uneven break
6. consolidated minerals
7. particle of rock
8. layers or bands
9. empty space
10. fills porosity
11. planar break
12. natural, inorganic, crystalline solid

Checking Concepts

Choose the word or phase that best answers the question.

13. What is the most common group of minerals?
    A) carbonates
    B) silicates
    C) sulfates
    D) oxides

14. What is a factor that controls grain size in igneous rocks?
    A) porosity
    B) cleavage
    C) cooling rate
    D) temperature

15. What do all the silicate minerals have?
    A) the same hardness
    B) the same crystal structure
    C) silicon and oxygen
    D) silicon and carbon as common elements

16. Through what process does chemical sedimentary rock usually form?
    A) cementation of sand
    B) compaction of clasts
    C) evaporation of melt
    D) precipitation of crystals

17. Which of the following features are extrusive igneous rocks least likely to have?
    A) fossils
    B) small grains
    C) vesicular texture
    D) porphyrtic texture

Use the photo below to answer question 18.

18. Which of the following descriptions best describes the large-grained igneous rock shown above?
    A) extrusive and basaltic
    B) extrusive and granitic
    C) intrusive and basaltic
    D) intrusive and granitic

Vocabulary PuzzleMaker gpescience.com

Interpreting Graphics

19. Copy and complete the concept map on igneous rocks.

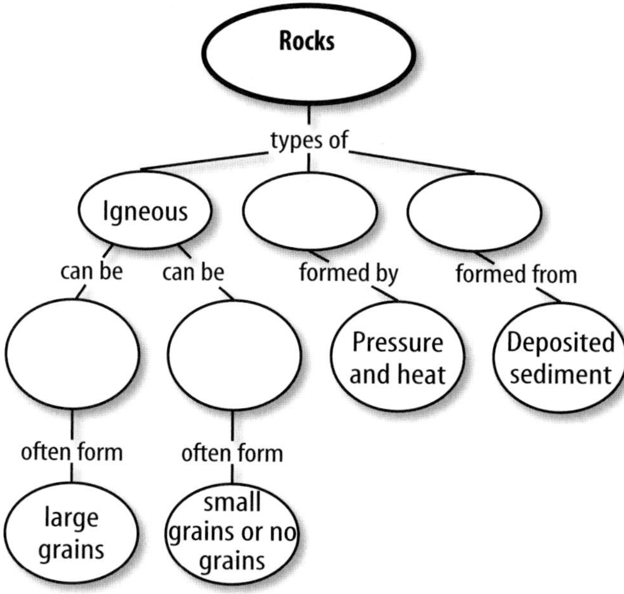

20. Copy the following table on a separate sheet of paper. Use Bowen's reaction series to fill in the blanks.

Crystalization Temperature Ranges for Mineral 1 and Mineral 2

Temperature Range	Mineral 1	Mineral 2	Rock Name
Low	mica		granite or rhyolite
Intermediate/ High	pyroxene	calcium plagioclase	

Thinking Critically

21. **Explain** how cooling rate affects the texture of an igneous rock.

22. **List** the characteristics that help you identify sedimentary rocks.

23. **Determine** the identity of each rock from the following descriptions.
    a. quartz-rich, coarse interlocking grains, with noticeable amounts of potassium feldspar
    b. foliated mass of mica group minerals with some garnet
    c. nearly pure mass of fossil shells cemented by calcite
    d. dense black, fine-grained, with some gas bubble holes
    e. light colored, very low density, frothy (vesicular), glassy

Use the photo below to answer question 24.

24. Describe the kind of force and the direction it was most likely applied to produce this foliated metamorphic rock.

Applying Math

25. If a miner were extracting rock that contained 0.015% gold by mass, how many kilograms of rock must be processed to obtain one kilogram of gold?

26. The body of ore described in question 25 is estimated to have an average density of 6.0 g/cm^3 and contain 1.5×105 m^3 of ore. How many kilograms of gold potentially could come from this body of ore?

Standardized Test Practice

Record your answers on the answer sheet provided by your teacher or on a sheet of paper.

Multiple Choice

1. Which is the most common element in Earth's crust?
   A. aluminum
   B. oxygen
   C. potassium
   D. silicon

Use the image below to answer question 2.

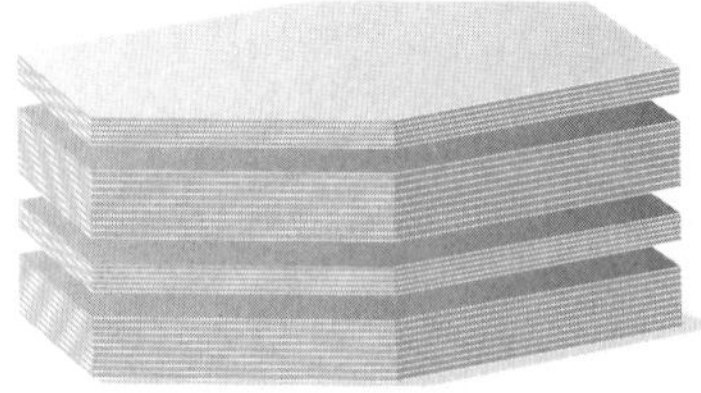

2. Which property of a mineral, used in its identification, is shown?
   A. cleavage
   B. fracture
   C. hardness
   D. luster

3. Which is a measure of a mineral's resistance to scratching?
   A. cleavage
   B. hardness
   C. luster
   D. streak

4. Why is graphite used in the core of some pencils?
   A. Its hardness is high enough to leave a scratch on paper.
   B. Its hardness is low enough to leave a streak on paper.
   C. Its metallic luster gives it a dark color.
   D. Its metallic luster makes it shine on paper.

5. Which mineral is very important to the production of machinery for manufacturing goods?
   A. feldspar
   B. halite
   C. hematite
   D. quartz

Use the photo below to answer question 6.

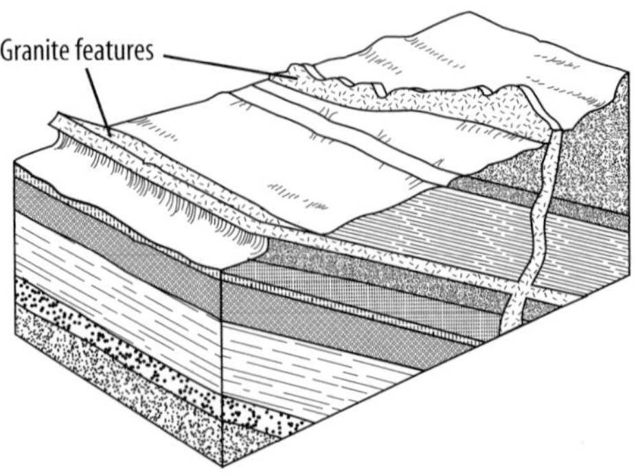

6. The formation of which type of rock produced the granite features shown above?
   A. foliated metamorphic rock
   B. extrusive igneous rock
   C. intrusive igneous rock
   D. nonfoliated metamorphic rock

7. Which is a dark-colored igneous rock with small grains?

   A. basalt

   B. gabbro

   C. granite

   D. rhyolite

Gridded Response

8. If the percent by mass of silicon in Earth's crust is 27.7 percent and the percent by mass of iron is 5.0 percent, how much more silicon is present in Earth's crust than iron?

Use the circle graph below to answer question 9.

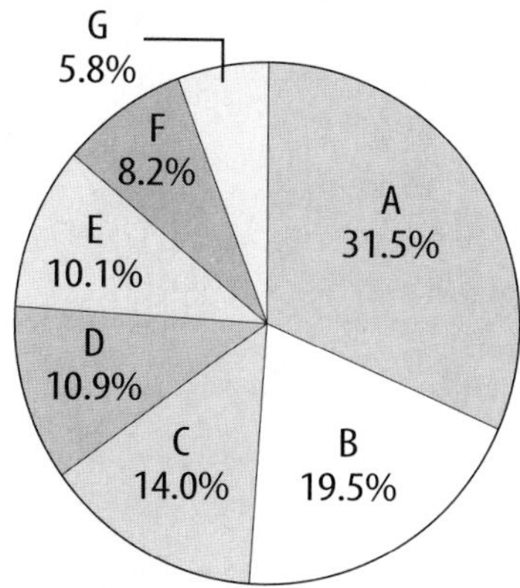

9. The percent by mass of elements in Earth's crust are: 46.6% oxygen; 27.7% silicon; 8.1% aluminum; 5.0% iron; 3.6% calcium; 2.8% sodium; 2.6% potassium; 2.1% magnesium; and 1.5% all others. When oxygen and silicon are not considered, the relative abundance of the remaining elements in Earth's crust are shown in the circle graph above. Which material represents the abundance of calcium?

Short Response

10. How does coal form? Describe the initial material, deposition environment, and geological conditions that lead to its formation.

11. Describe the rock cycle. Discuss what can happen to rock material within the cycle and the amount of time it might take for rocks to be transformed.

12. How does a mineral crystal grow during mineral formation?

13. What are some of the factors that affect the way a crystal grows?

Extended Response

Use the illustration below to answer question 14.

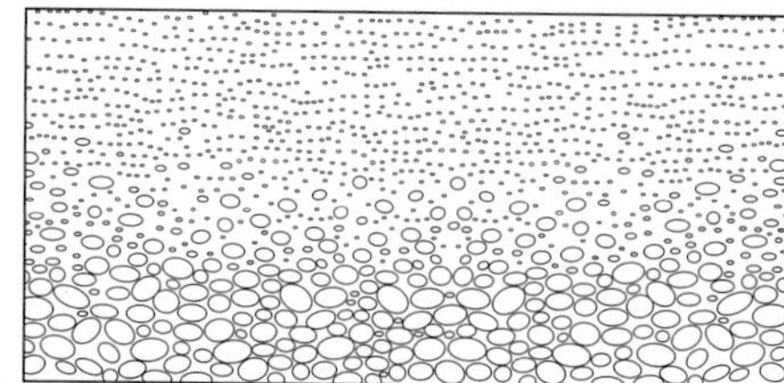

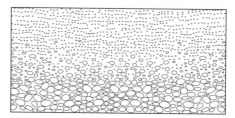

14. **PART A** Describe how the sediment particles are arranged in the illustration above.

    PART B Describe the type of environment in which this arrangement of sediment particles might occur.

Test-Taking Tip

Review Double check your answers before turning in the test.

chapter 21

Earth's Changing Surface

chapter preview

sections

1. Weathering and Soil
   Lab Graphing Soil Textures
2. Shaping the Landscape
3. Groundwater
4. Geologic Time
   Lab It's about time—relative, that is.

Carving A Path

Glaciers have been carving Earth's surface for millions of years. When this glacier eventually melts, it will leave characteristic clues behind that scientists can interpret.

Science Journal In your Science Journal, describe some clues to look for that could indicate a glacier had been there.

Start-Up Activities

Have you got the time?

We usually think of time in terms of hours, days, and years. Earth processes operate in much longer time units. Imagine that Earth's age—about 4.5 billion years—were equal to an average human lifespan of 75 years. Then one million years of Earth's existence would represent about six days of a human life.

1. Obtain a 5-m length of adding machine tape and a meterstick.
2. Draw a horizontal line across one end of the tape and label it *Present*.
3. Using a scale of 1 mm = 1 million years, draw and label lines across the tape behind *Present* to mark the beginning of each of the following divisions of time:

   Cenozoic Era: 66 million years ago (mya)

   Mesozoic Era: 248 mya

   Paleozoic Era and beginning of Phanerozoic Eon: 540 mya

   Proterozoic Eon: 2,500 mya

   Archean Eon: 3,800 mya
4. **Think Critically** Eons are the longest subdivisions of geologic time. An eon is subdivided into eras. Infer why eons and eras are different lengths of time. What kind of events might end one time interval and begin a new one?

Surface Features Make the following Foldable to help you organize features that result from erosion and deposition on Earth's surface.

STEP 1 **Fold** one piece of paper lengthwise into thirds.

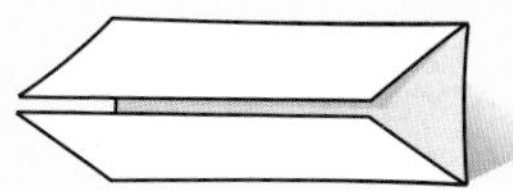

STEP 2 **Fold** the paper widthwise into fourths.

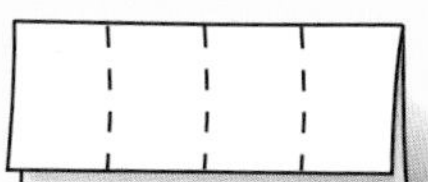

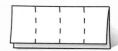

STEP 3 **Unfold,** lay the paper lengthwise, and draw lines along the folds.

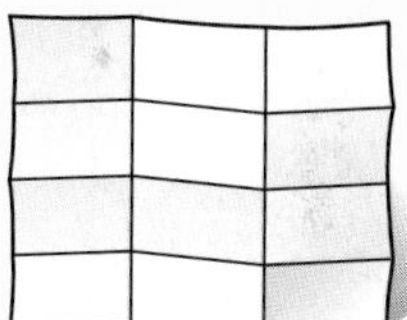

STEP 4 **Label** your table as shown.

Surface-Changing Agent	Erosion	Deposition
Wind		
Water		
Glaciers		

Making a Table As you read the chapter, complete the table describing erosional and despositional features.

Preview this chapter's content and activities at gpescience.com

section 1

Weathering and Soil

Reading Guide

What You'll Learn

- **Explain** the role of weathering in recycling Earth materials.
- **Discuss** the processes of mechanical and chemical weathering.
- **Describe** the process of and controls on soil formation.
- **Discuss** some methods for controlling soil loss.

Why It's Important

The processes of weathering and soil formation are critical in forming a base to support plant growth and ecosystems.

Review Vocabulary

sediment: particles of rock deposited on Earth's surface that may become sedimentary rocks

New Vocabulary

- weathering
- soil

Weathering

In nature, matter is recycled. Chemical, physical, and biological processes form interconnected cycles that operate to change materials. Because we tend to study each process one at a time, you might get the impression that they stand alone. This is not true. A real understanding of nature comes from learning how the cycles interconnect. Weathering is an example of physical and chemical processes that involve the interaction of air, water, and rock over time.

Natural materials become unstable and break down when exposed to conditions on Earth's surface. **Figure 1** illustrates the results of rock weathering. The minerals in this rock formed under higher temperatures and pressures compared to those on the surface. When these minerals were subjected to the conditions at the surface, they broke down. **Weathering** is the process of physical or chemical breakdown of a material at or near Earth's surface. Factors that influence weathering include the agent, such as water or air, the nature of the material being weathered, climate, and time. Weathering varies from region to region because different weathering conditions are present.

Figure 1 The corners and edges of rocks are more rapidly weathered than smooth rock faces. This often produces rounded forms.

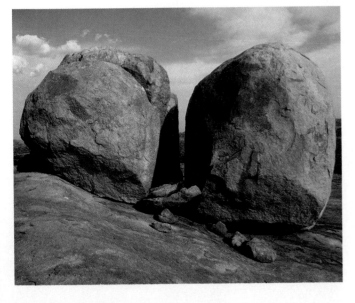

Reading Check *What causes weathering?*

Everyday Weathering You can notice weathering going on all around you. Paint on a house eventually will crack or fade in color. Roadways eventually develop cracks and potholes. Some metal items on cars might start to rust. All of these take place at different rates depending on where you live and whether or not you have taken measures to prevent weathering.

Mechanical Weathering

You can think of mechanical weathering as turning big pieces into little pieces. For this to take place, a force must be applied to the material. Forces that can cause mechanical weathering include impact, expansion or contraction of materials, and biological effects. Variables that control mechanical weathering include the nature of the material being weathered, climate, and time. Of these variables, time is the least noticeable. Extremely slow changes often go unnoticed until their long-term accumulated effects get our attention.

Prying Rock Apart If a rock contains natural fractures, water can enter them. When water freezes, it expands. This forces the cracks apart and ultimately forces the rock apart. This type of weathering is called frost wedging. Any place that has sufficient surface water and temperatures that change from freezing to above freezing can experience this type of mechanical weathering.

Growing plant roots are an example of a biological force that can cause mechanical weathering. As roots grow, they increase in diameter and exert forces that widen cracks.

Rocks can also act as impact agents. Large rocks falling from a cliff or tumbling down a slope can break other rocks into smaller pieces. **Figure 2** illustrates what happens when a large object breaks apart. As a large particle breaks into smaller particles, the amount of surface area increases. More surface area compared to volume increases the likelihood that a material will be attacked by chemical agents.

Observing a Disappearing Seasoning

Procedure

1. Sprinkle a teaspoon of **sugar** on a **paper plate.** Using the back of a **metal spoon,** crush the sugar into the finest powder you can. Do the same for a teaspoon of **salt.**
2. Half-fill **four clear drinking glasses** with **tap water.** Allow them to stand for several minutes.
3. Have a **watch with a second hand** ready. Drop a teaspoon of uncrushed sugar into a glass of water and stir. Record how long it takes the sugar to completely dissolve.
4. Repeat step 3 using the powdered sugar, the powdered salt, and a teaspoon of uncrushed salt.

Analysis

1. Which substance dissolved fastest? Why?
2. How does particle size affect dissolving rate?
3. List some variables other than particle size that could affect dissolving rate.

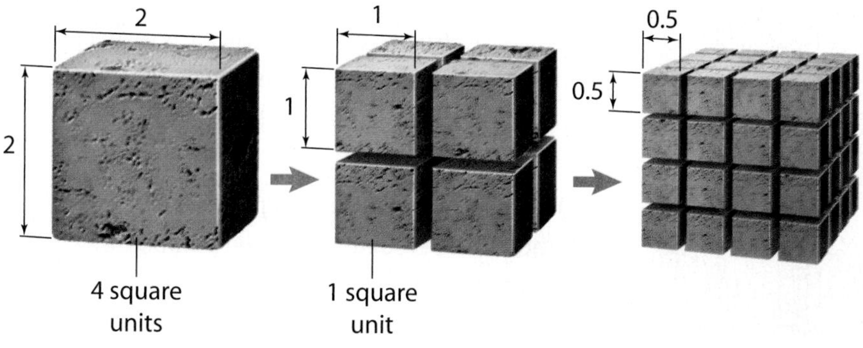

Figure 2 The cube on the right has exposed, inner surfaces that the cube on the left does not have.

Illegible Rocks The Egyptians carved tall, slender granite obelisks. One is displayed in New York City and another in London. Over time, these two obelisks have become almost smooth and the hieroglyphs barely can be seen. Form a hypothesis stating why the obelisks changed.

Chemical Weathering

The process of chemical weathering forms new compounds and releases elements into the environment. Water and oxygen are the key agents of chemical weathering, along with naturally occurring acids. Elements released by chemical weathering enrich soil and nourish plants. Without weathering of rocks there would be no soil to support land plants, terrestrial ecosystems, and humans.

Reading Check *List three key agents of chemical weathering.*

Matter on the Move Cycles of chemical weathering affect all surface environments from land to sea. If streams carry elements released on land into the ocean, then those elements become raw materials that support marine ecosystems. For example, marine organisms that build shells require dissolved products such as calcium ions (Ca^{2+}) and bicarbonate ions (HCO_3^-) to make their shells. These ions come from the weathering of limestone on land and from the dissolving of other seashells.

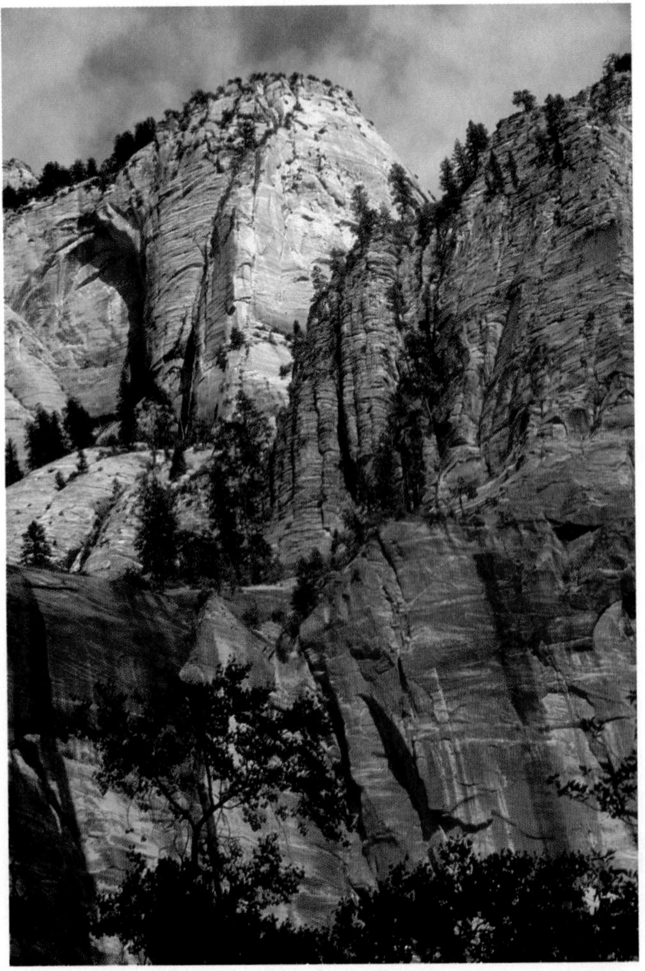

Rock Rust Oxidation is a common chemical weathering process. Iron that is present in minerals is released through weathering and combines with oxygen to form iron oxides. You might know that when something made of iron or steel is exposed to air and water, it quickly begins to rust. The mineral hematite is natural "rust" produced through oxidation reactions. Hematite is mined to obtain iron for steel production.

Recall that natural materials tend to convert to a more stable form at Earth's surface. Unstable iron released by weathering forms stable hematite. Hematite is easily spotted in rocks because it stains them a brownish-red color, as shown in **Figure 3.** Other metals also form oxide minerals. Examples include copper, which oxidizes to a greenish-colored patina or silver, which oxidizes to a black coating.

Figure 3 When iron-bearing rocks are weathered and eroded, they often produce reddish sediment. Iron oxidation also can give some soils a reddish color.

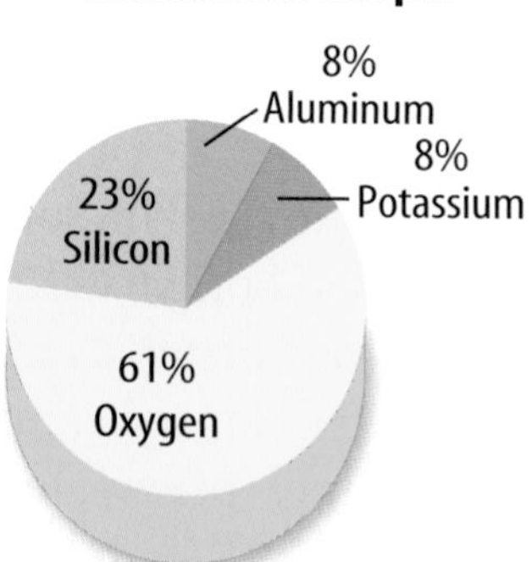

Elements in Kaolinite

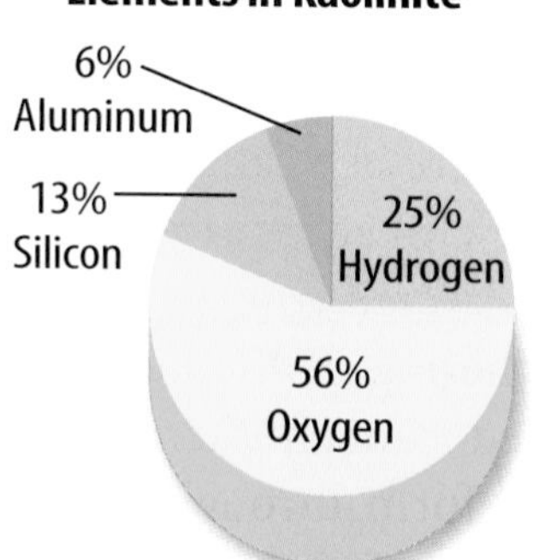

Figure 4 Chemical weathering changes the chemical composition of minerals and rocks.
Describe *how kaolinite is different from feldspar.*

Feldspar reacts with carbonic acid.

The mineral kaolinite is formed.

Feldspar Weathering An important chemical weathering process for continental rocks is the weathering of feldspar minerals into clay minerals.

This reaction starts with a common rock-forming mineral shown in **Figure 4,** and ends with minerals that are common in continental sediment and soil. The silica end product often forms quartz cement in sedimentary rocks. The kaolinite is a clay mineral that helps soil hold water and nutrients. Chemical elements freed by weathering now are available for use by plants. Similar chemical weathering reactions of other minerals provide particles necessary for roots to anchor themselves and nutrient elements to support plant growth.

Differential Weathering Different rock formations tend to weather at different rates. The types of minerals, cementing agent, and the presence of defects, such as fractures, influence how a rock formation weathers. As a result, rocks that are the most resistant to weathering remain in the landscape. They are products of differential weathering and erosion.

Together, erosion and deposition of sediment create a variety of landforms. They also represent important processes of the rock cycle. Landforms, such as those in **Figure 5,** are geologists' clues to the processes taking place on Earth's surface. Some of these processes include floods, landslides, and beach erosion.

Figure 5 As water seeks a path of least resistance, it acts as a differential weathering agent. This type of weathering creates many interesting landforms.

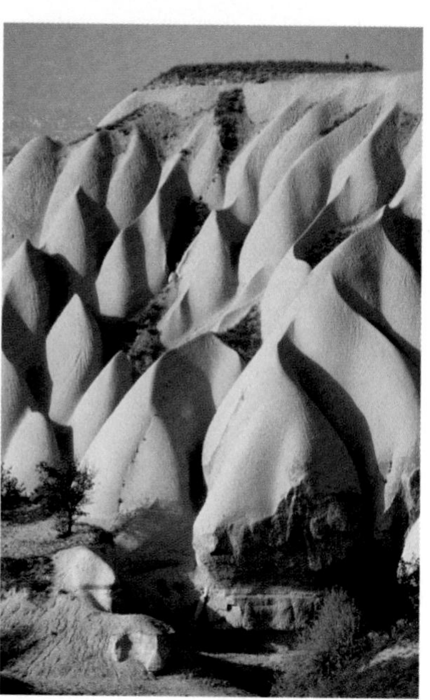

Soil

One natural resource that is vital to life on Earth is soil. **Soil** is a mixture of weathered rock, organic matter, water, and air that is capable of supporting plant life. Much like human skin, soil is a thin, layered covering of its parent material. Soil originates from weathering of the bedrock beneath it or from materials transported from another location. The raw materials for soil continuously form through the weathering of bedrock below and the addition of organic matter from above.

Reading Check *Soil is made up of what?*

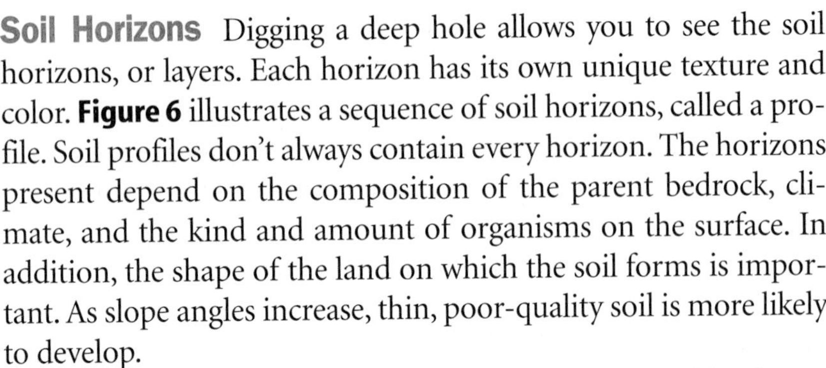

Soil Horizons Digging a deep hole allows you to see the soil horizons, or layers. Each horizon has its own unique texture and color. **Figure 6** illustrates a sequence of soil horizons, called a profile. Soil profiles don't always contain every horizon. The horizons present depend on the composition of the parent bedrock, climate, and the kind and amount of organisms on the surface. In addition, the shape of the land on which the soil forms is important. As slope angles increase, thin, poor-quality soil is more likely to develop.

Figure 6 This illustration provides a generalized view of a soil profile. Soil horizons vary in color and thickness. Some horizons don't appear in certain soils. Notice that the E horizon is lighter in color than the B horizon.
Explain *why the E horizon is light in color.*

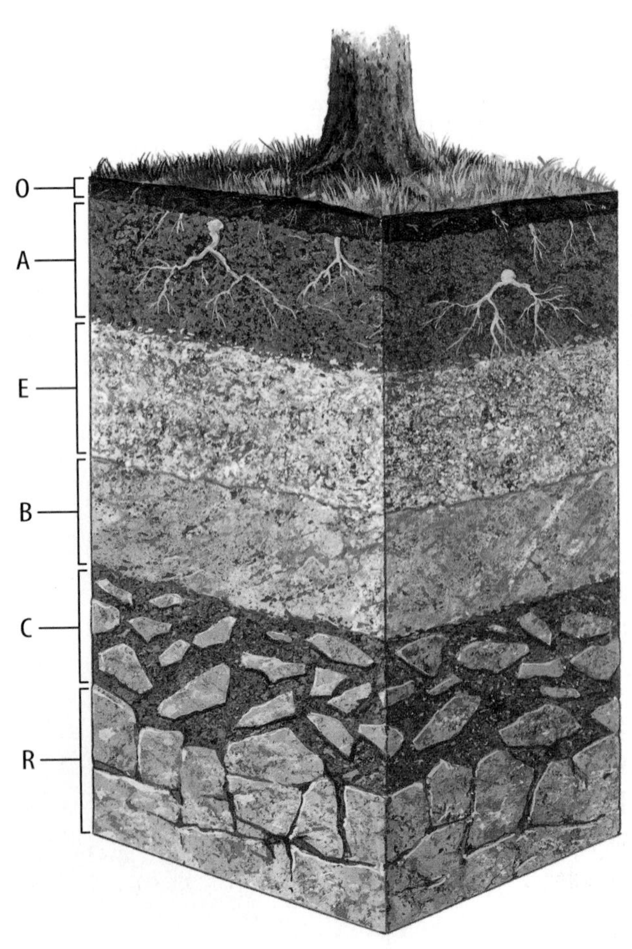

Figure 6 shows a complete set of horizons, designated as O, A, E, B, C, and R from the surface to bedrock. The O horizon, for organic, and the A horizon often are referred to as topsoil. Plants get most of their nutrients from the topsoil.

The E horizon, for eluviation, is a zone in which finer sediments and soluble materials are transported downward. The process of dissolving soluble elements and transporting them deeper into the soil is called leaching. This is much like a coffee maker that leaches flavors from ground coffee. The B horizon collects the materials from above and is usually darker than the E horizon. Sometimes so much material accumulates that water can't easily penetrate this zone. This formation is referred to as hardpan. The E horizon and the B horizon often are referred to as subsoil. Roots from large trees sometimes penetrate these horizons to get water and are able to anchor themselves in the subsoil.

Collectively, horizons O, A, E, and B make up the true soil. The C horizon is partially crumbled and weathered bedrock. The R horizon is unweathered bedrock.

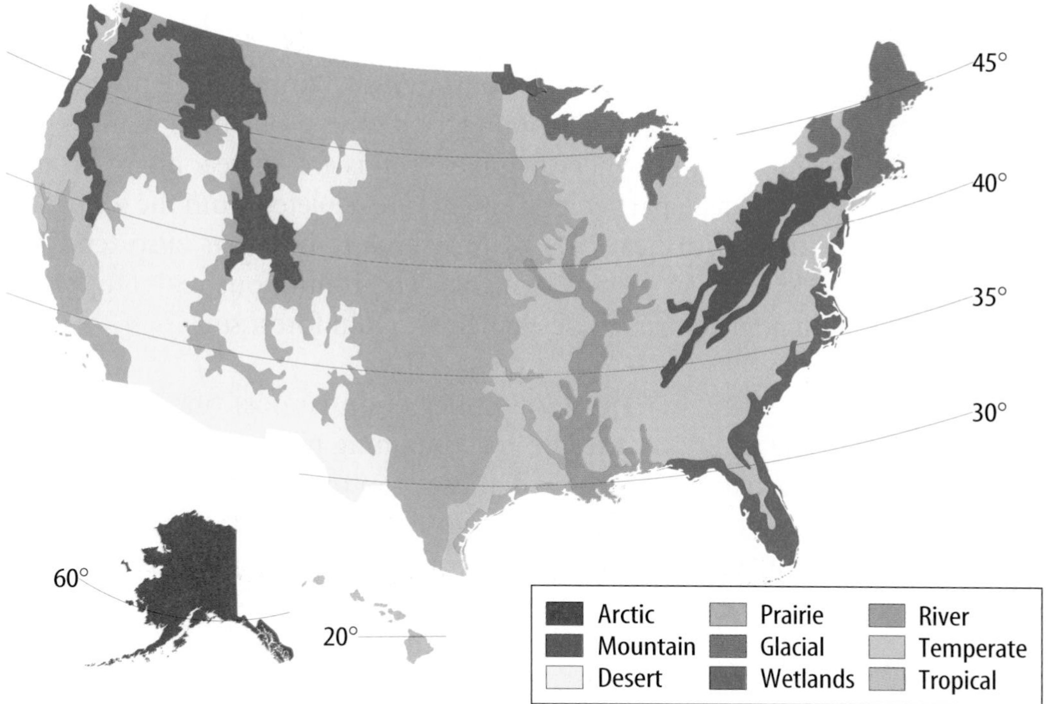

Figure 7 Glacial soil in the northeastern U.S. often is composed of poorly-sorted glacial till. The eastern U.S. receives large amounts of precipitation which contribute to wetland soil.

Soil Types There are many different types of soil. Climate often is used as a basis to characterize many different types of soil, as illustrated in **Figure 7.** They are separated according to composition and physical properties. The amount of precipitation and the temperature range of a region, together with the type of parent material, affect the soil that forms. Vegetation, if present, also is important for providing a supply of organic matter. For example, topsoils that develop beneath the forests of the eastern U.S. are acidic, sandy in texture and light in color. This contrasts with soils that develop beneath the grasses of the western U.S. These often are enriched in calcite and are whitish in color. In tropical environments, the soils are strongly leached. They are rich in insoluble aluminum, iron oxides, and clay minerals. Finally, in the arctic or desert regions, where little or no vegetation grows, there is no soil. Without organic material, a true soil cannot develop.

Parent Material Some soils form in the same place as the parent bedrock. Other soils form from materials transported from distant sources. Rivers and glaciers are important transporting agents for soils. Some of the most fertile soils form when rivers flood and deposit new sediment on floodplains or when glaciers transport and deposit sediment. The ancient Egyptians relied for thousands of years on the annual flooding of the Nile River to deposit nutrient-rich sediment. During the most recent ice advance, glaciers in North America transported materials that later would become some of the world's most fertile soils.

Figure 8 The tie-ridging method makes fields look like waffles.

Soil Conservation

Plants and animals continuously remove elements from soil. Soil depletion is a serious agricultural problem in many regions. For example, when the same crop is grown year after year in the same field, particular elements are depleted from the soil. Soils in areas with heavy rainfall, such as rain forests, also tend to be deficient in certain elements. The continuous recycling of plant matter returns these elements to rain forest soils.

Adding Nutrients Soil depletion is most often corrected by the addition of fertilizers containing nitrogen, phosphorous, or potassium. Farmers often use crop rotation to help preserve the quality of soils. This is because what one crop removes from the soil, another crop may return. For example, some crops, such as legumes, return nutrients to the soil. Crop rotation also helps prevent erosion and reduces the risk of diseases and attack by pests. Another alternative is to allow the soil to rest, called going fallow, by not planting crops. This gives the soil time to replenish some of its lost nutrients.

Preventing Soil Loss Most soil lost to erosion occurs because the vegetative cover has been removed or because the land is overly steep. Modern farmers use contour plowing, which runs the furrows around a hill instead of up and down a slope, to reduce erosion. In the tie-ridging method, shown in **Figure 8,** crops are planted on ridges and rainfall collects in basins. Slowing the downhill flow of water helps prevent erosion.

section 1 review

Summary

Mechanical and Chemical Weathering

- Weathering is the breakdown of materials that are not stable under the physical conditions at Earth's surface.
- Physical breakdown of surface materials is accomplished by mechanical weathering.
- Chemical reactions at Earth's surface change mineral composition and release elements.

Soil and Soil Conservation

- Soil is a mixture of organic matter, weathered rock material, water, and air.
- Measures are taken to prevent soil depletion and soil erosion.

Self Check

1. **Explain** how mechanical and chemical weathering processes are related.
2. **Describe** how the controls on soil formation affect soil.
3. **Distinguish** between soil depletion and soil erosion.
4. **Think Critically** Outline the soil conservation measures you expect from any contractor building new roads in your region.

Applying Math

5. **Calculate** Suppose that soil erosion in your area averages 2.65 cm per year. The average soil profile is 3 m thick, and 45 percent of that is topsoil. Estimate how long it will take for the topsoil to erode.

More Section Review gpescience.com

Graphing Soil Textures

Three particle sizes characterize the textures of most soils—clay, silt, and sand. The relative amounts of these three variables plot on a triangular graph. In a triangular graph, a corner represents 100 percent of a variable and its opposite base represents 0 percent of that variable.

Real-World Problem

Different plants require different soil textures. How can you classify soil according to texture?

Goals

- **Graph** data on a triangular diagram.
- **Interpret** graphed data to classify soils.

Materials

soil texture triangular diagram

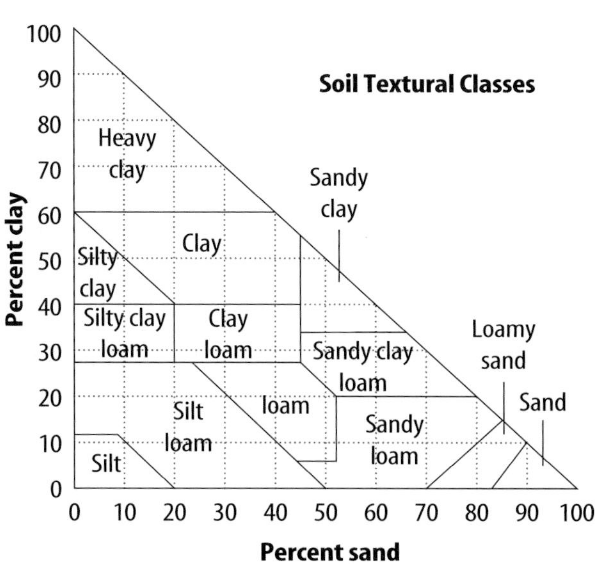

Procedure

1. Obtain a copy of a soil texture triangular diagram from your teacher.
2. Find the line representing 40 percent clay.
3. Find the line representing 50 percent sand.
4. Plot the intersection of these two lines on your copy of the graph.
5. The remainder of the sample is silt. Your first plotted point represents a soil texture that is 40 percent clay, 50 percent sand, and 10 percent silt, and is called sandy clay.
6. Plot and label the following textures and determine the percentage of silt. Record your answers in a data table.

Soil Texture Data

Sample #	% Clay	% Sand	% Silt
1	20	40	
2	10	70	
3	15	15	
4	60	20	
5	25	65	

Conclude and Apply

1. **Describe** the composition of any soil that plots along the base of the triangle.
2. **List** the texture names for each sample.
3. **Evaluate** each sample for its ability to retain or drain water.
4. **Predict** which soil samples would be best for root crops such as potatoes or carrots.

Communicating Your Data

Prepare a brief oral presentation outlining how to use the soil texture triangle.

section 2

Shaping the Landscape

Reading Guide

What You'll Learn

- **Explain** how agents of erosion operate.
- **Describe** kinds of landforms created by erosion.
- **Identify** landforms created by deposition of eroded material.

Why It's Important

Some erosion events, such as floods and landslides, have catastrophic effects on lives and property.

Review Vocabulary

physical change: any change in size, shape, or state in which the identity of a substance remains the same

New Vocabulary

- erosion
- sediment transport
- deposition
- drainage basin
- longshore current

Erosion, Transport, and Deposition

You've learned about forces capable of building mountains. Forces with origins within Earth can create uplifted landforms through folding, faulting, or volcanic activity. Although they might seem like permanent features, mountains and other landforms are subject to continuous change. Gravity is a factor in all change. Some of these changes also involve the action of agents at the surface of Earth, such as air, water, or ice. The shape of the land you see is what is left over after material has been removed or added by these agents.

Erosion is the removal of surface material through the processes of weathering. **Sediment transport** is the movement of eroded materials from one place to another by water, wind, and/or glaciers. When a transporting agent drops its load of eroded material, **deposition** occurs. The characteristics of landforms created on the surface, such as the eskers and moraines shown in **Figure 9,** can help identify the type of agent that formed it. Analyzing the sediment in a surface deposit can reveal clues about its point of origin. For this reason surface sediment often is used to piece together the plate tectonic history of a region.

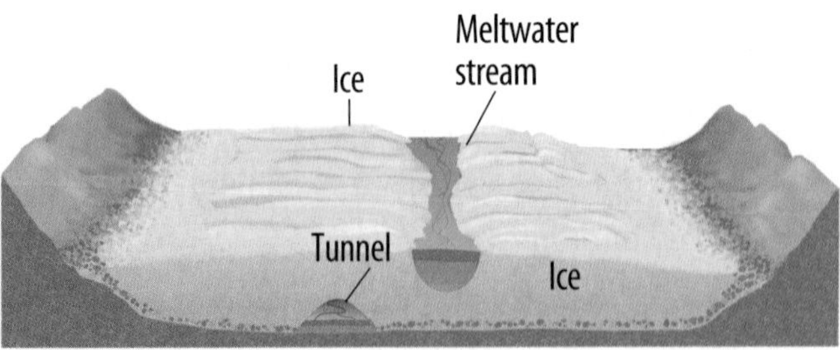

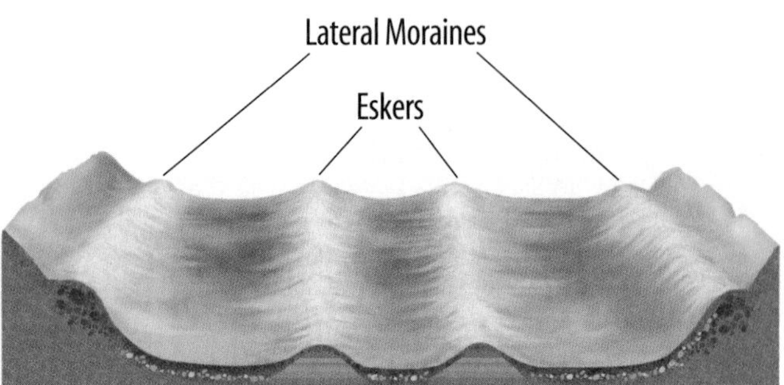

Figure 9 When valley glaciers melt they deposit sediment in characteristic landforms. Moraines consist of poorly sorted glacial till and eskers consist of layers of sediment.

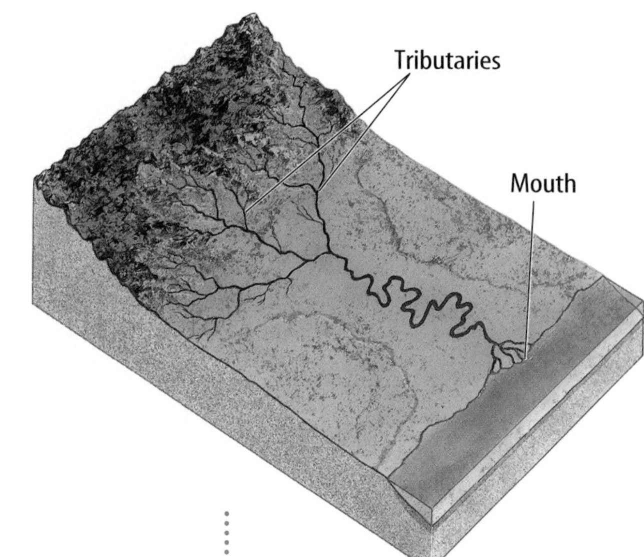

Figure 10 Gradient, or slope, affects the shape of a stream and its containing valley. **Compare and contrast** *how a steep gradient and a gentle gradient affect the shape of a stream.*

Running Water

An important agent of erosion that exerts a downward force on slopes is running water. In steep areas, running water cuts down through sediment and rock, creating V-shaped valleys. As the steepness of a slope decreases, more force is exerted sideways. Lateral cutting broadens river valleys and allows for floodplains to develop. The mouth of a river is where it meets the body of water or land surface into which it ultimately flows. Near the mouth, a river might wind back and forth, or meander, across its valley. **Figure 10** shows how characteristics of a river can change as it flows from its source to its mouth.

Drainage Basins In most river systems, small streams called tributaries flow into larger streams, which in turn flow into even larger streams. This process of branching is much like a tree with small branches that can be traced to larger branches, and then to a trunk. Most major river systems are well-known, such as the Mississippi River. All of the land area that gathers water for a major river is the river's **drainage basin.** Drainage basins are larger in area for major rivers and include all of the drainage basins of its tributaries. The Mississippi River drainage basin extends from the Appalachian Mountains westward to the Rocky Mountains. Both of these mountain ranges represent drainage divides. A drainage divide is a boundary line separating distinct drainage basins. **Figure 11** illustrates the Mississippi River drainage basin.

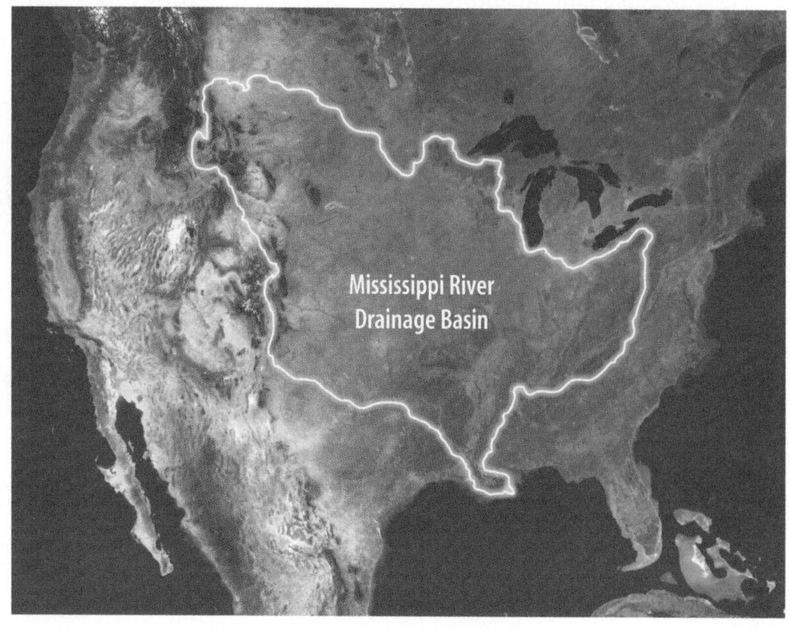

Figure 11 The Mississippi River drainage basin is the largest drainage basin in the United States.

Figure 12 This portion of the Colorado River is cutting a deep, V-shaped valley as the Colorado Plateau is being uplifted.

Modeling Runoff

Procedure

1. Obtain a clean, flat, smooth **baking sheet.**
2. Turn on a faucet and adjust the flow of **water** to a slow, smooth stream about the diameter of a pencil lead.
3. Place the baking sheet at a steep angle under the stream of water. Hold the sheet so that the water runs off along the length of the sheet. Record your observations.
4. Make several more observations reducing the steepness each time until your last observation is with the stream perpendicular to the sheet. Record all observations.

Analysis

What happened to the stream flow pattern as you decreased the angle of the sheet? Explain.

Try at Home

Channel Development As surface water flows downhill under the influence of gravity, water erodes the surface, creating its own path or following existing paths called channels. V-shaped valleys are formed from rivers cutting down through sediment and rock, eroding material from a slope. Steep canyons can form when downcutting is rapid. This is often a result of a recent uplift of land, as shown in **Figure 12.**

What is the shape of a valley that is cut by a young stream?

Stream Deposits Maybe you've seen accumulations or bars of sediment in stream channels. When running water slows down, it drops part of the sediment load it is carrying, largest particles first. Generally, bars form on the inside of bends in a river channel, where water speeds are slowest. But in regions where sediment is plentiful, bars might interrupt the flow of water right in the middle of the channel, as shown in **Figure 13.**

Figure 13 Streams with many bars within their channels are called braided streams.

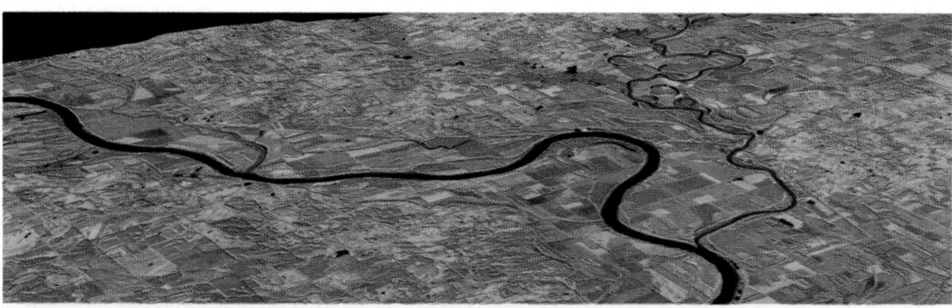

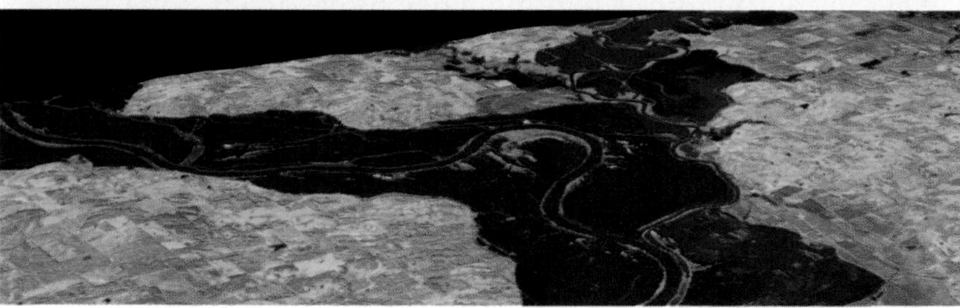

Figure 14 The Missouri River, shown above, sometimes overflows its banks and spills onto its floodplain, shown below.

Topic: Bayou
Visit gpescience.com for Web links to information about the Louisiana Bayous.

Activity A bayou can be a creek, secondary watercourse, or any of various, usually marshy or sluggish bodies of water. Research the origin of the word *bayou*. Locate the Louisiana Bayous on a United States map.

Floodplains After a river floods, and as floodwaters subside, slow, and drop their sediment, floodplains form along the valley sides. These floodplains are part of a river, but they are only submerged during floods. **Figure 14** shows a satellite image of a portion of the Missouri river before and during the 1993 flood. While they can be catastrophic to property, a beneficial effect of such a flood event is to rejuvenate soil quality.

Deltas At the mouth of a stream that empties into a body of water, a fan-shaped sediment deposit called a delta will form. Deltas often have branching channels, like roots of a tree, called distributaries. Distributaries distribute water and mud into the water body they enter, as shown in **Figure 15.** Similar depositional features, called alluvial fans, form where the mouth of a stream enters dry land.

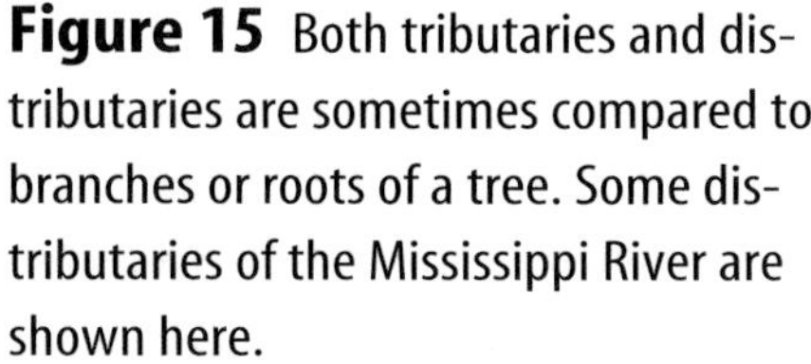

Figure 15 Both tributaries and distributaries are sometimes compared to branches or roots of a tree. Some distributaries of the Mississippi River are shown here.
Compare and contrast *tributaries and distributaries.*

Topic: Isostatic Rebound
Visit gpescience.com for Web links to information about the effect of the weight of glacial ice on Earth's crust.

Activity In your Science Journal, describe evidence that supports the fact that Earth's crust rebounded upward at a location where the weight of glacial ice was removed.

Glaciers

A region where winter snowfall exceeds summer melt has the potential to form a glacier. This excess snowfall must take place year after year for hundreds to thousands of years. Compaction within a thick mass of snow eventually can change it to glacial ice. Valley glaciers form at the top, or head, of river valleys in mountainous regions. They move down slope, usually following the path of the preexisting valleys as shown in **Figure 16.** Continental glaciers, which occupy huge land areas, cover all but the highest peaks. Ice moves away in all directions from where it accumulates.

Reading Check *What is a requirement for a glacier to form?*

Imagine what happens when you pour pancake batter onto a cooking surface. It spreads out under its own weight. The more you pour the more it spreads. As ice accumulates, the pressure at its base increases and causes it to flow outward. The polar ice sheets of Greenland and Antarctica move in this manner.

Glaciers become effective agents of erosion as they move either down slope as in valley glaciers or under their own weight as in continental glaciers. As the ice moves, it acts like a giant bulldozer scraping, gouging, and plucking soil and surface rocks.

Figure 16 Valley glaciers create a variety of alpine, or mountainous, features. Along their paths, they accumulate soil and surface rocks that become abrasive material used to transform the landscape.
Compare and contrast *trunk valleys and hanging valleys.*

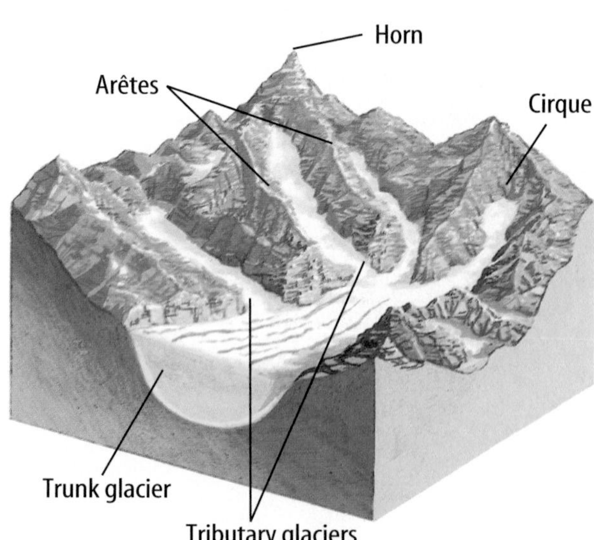

Bowl-shaped basins called cirques form by erosion at the start of a valley glacier. Arêtes form where two adjacent valley glaciers meet and erode a long, sharp ridge. Horns are sharpened peaks formed by glacial action at three or more cirques.

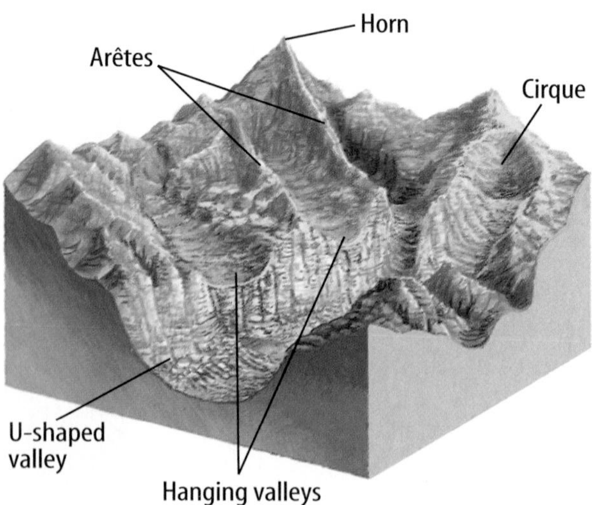

U-shaped valleys result when valley glaciers move through regions once occupied by streams. A tributary glacial valley whose mouth is high above the floor of the main, trunk valley is called a hanging valley. The discordance between the different valley floors is due to the greater erosive power of the trunk glacier.

Erosional Features Valley and continental glaciers leave behind deep grooves or striations, which give clues about the direction the ice was moving. Continental glaciers gouged out the Great Lakes and the finger lakes of New York. Valley glaciers convert V-shaped stream valleys into distinctive U-shaped valleys. The downcutting force of valley glaciers differs from those of streams. Rather than simply cutting downward, glacial ice exerts pressure to the sides as well. This creates the distinct and easily recognized U-shaped valley.

Reading Check *What is the shape of a valley that is cut by glaciers?*

This downward cutting force of ice is a function of mass. Tributary glaciers with less mass cut less deeply. Larger trunk glaciers have more mass and erode more deeply. This produces tributary valleys that enter the main valley at different elevations, creating hanging valleys. Stream divides are eroded into sharp ridges called arêtes. Where ice first accumulates, a bowl-shaped depression called a cirque is formed. The formation of several cirques around a mountain results in the formation of a horn. **Figure 16** illustrates some glacial landforms that form in mountainous areas.

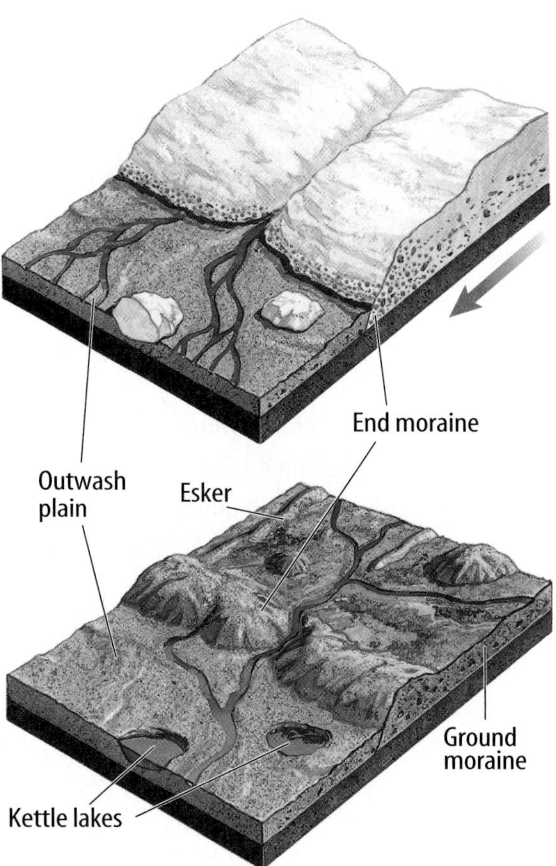

Figure 17 End moraines are composed of poorly sorted soil and rocks that were scraped up by the glacier, moved ahead of it, and then deposited where it melted. Eskers are composed of finer soil and rock material that was deposited within the glacier.

Glacier Deposits Ice cannot sort material like moving water does. So, when continental and valley glaciers melt, they dump all the rock material they are carrying. This material, known as till, is composed of random sediment sizes ranging from tiny clay particles to house-sized boulders. The large ridges of till that accumulate at the edge of a glacier are called moraines. An end moraine forms at the front of the ice, and lateral moraines form at the side of the ice. If two valley glaciers join together down slope, their lateral moraines get sandwiched together to form medial moraines. A ground moraine forms underneath melting continental ice sheets.

Some of the smaller-sized particles ranging from clay to gravel are carried away by meltwater and form layered deposits on outwash plains. Deposits of outwash material are layered and sorted by size because of the action of water in transporting and depositing them. Glacial deposits in flat outwash plains often become fertile soils that support plant growth. **Figure 17** shows examples of landforms deposited by melting ice.

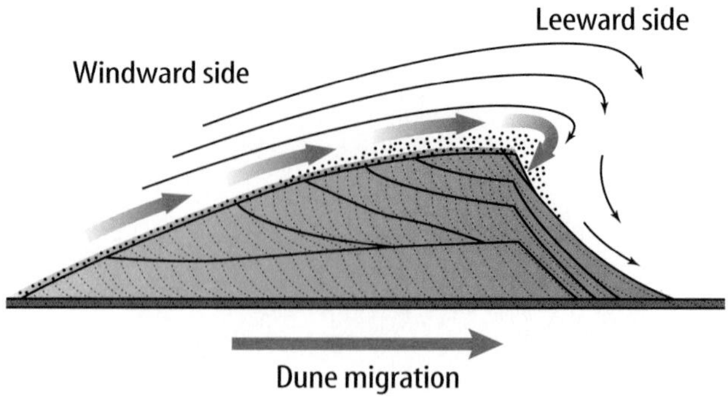

Figure 18 Erosion on the windward side and deposition on the leeward side can cause sand dunes to appear to move.
Infer *Why does deposition occur on the leeward side of a dune?*

Wind

You might think that wind is the most important agent of erosion in deserts, but this is not so. Running water erodes more, especially during infrequent flash floods. Compared to water and ice, wind lacks the ability to pick up and carry large particles. However, wind can act as a sand blaster, blowing sediment around and polishing and smoothing landforms. In places where sand and smaller particles are abundant and not held in place by vegetation, wind can be a very effective agent of erosion.

Erosion by Wind The removal of small particles by wind, leaving heavier particles behind, is called deflation. Deserts often are pictured as endless sand dunes, but this is misleading. Many deserts have a rocky, rough texture caused by deflation. The remaining surface is known as desert pavement. Blowouts, or deflation hollows, are common landforms where wind is the dominant agent of erosion. These landforms are shallow depressions where wind has scooped out surface material, leaving high places behind.

Wind Deposits As wind velocity decreases, the load of sediment is dropped. The shapes and sizes of landforms that form depend on how constant the wind velocity is and on the supply of sediment. Fine silt from glacial outwash plains or deserts is picked up and deposited as thick, unlayered deposits called loess (LUS). Dunes form as the wind moves sand-sized particles into distinct forms, as shown in **Figure 18.**

Wave Action

The zone where land meets open water is an area where extreme forces operate. One cubic meter of water has a large mass. The force generated by breaking waves is large and provides the energy to change surface features.

Currents from Wave Action When waves approach a shoreline at an angle, they change directions, or are refracted. They wash up at a slight angle to the beach. The backwash returns to the sea perpendicular to the shoreline. Waves churn up beach sediment, which is carried back toward the sea with the backwash, then moved toward shore at an angle again. This process results in net sediment transport parallel to the shoreline, as illustrated in **Figure 19.** The resulting movement of water parallel to the shoreline is called a **longshore current.**

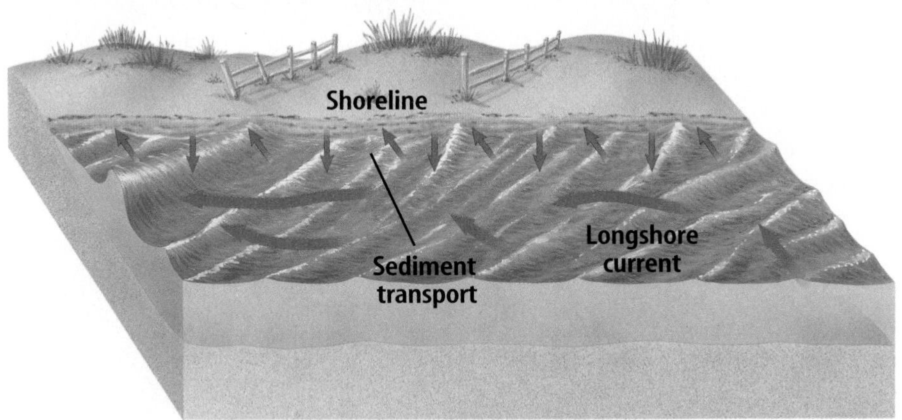

Figure 19 Longshore currents run parallel to the shoreline. They can completely remove some beaches by erosion and later deposit the sediment, forming new beaches farther down-current.

Erosion from Wave Action The action of waves loaded with sediment works like sandpaper and causes abrasion. This results in rounded pebbles and cobbles on the beach. Coastal cliffs often are eroded at their bases by wave action. This results in periodic collapse of cliff material into the sea. This material then is battered by waves and eventually reduced to smaller sizes that can be transported by the longshore current. Shorelines that are rock-bound with steep cliffs tend to experience differential weathering. A wide variety of landforms, such as sea stacks, sea arches, and headlands, are characteristic of rocky shorelines.

Reading Check *How does beach sand become rounded?*

Deposition by Wave Action Shorelines commonly lack erosional landforms when they have a thick sediment cover. Instead, the sediment on these beaches is constantly reworked producing a variety of depositional landforms. Longshore currents move sediment along the shore. Where this current slows, sediments are deposited as offshore sand ridges or bars. Sandbars run parallel to the shore. A sandbar that seals off a bay from the open ocean is called a baymouth bar. Spits are sand bars that project into the water from land and curve back toward land in a hook shape. Cape Cod is a well-known spit off the coast of Massachusetts. During severe storms, water levels can rise and create sand deposits known as barrier islands that can serve as protection for the mainland. The shallow lagoons between the islands and the mainland are nurseries for many forms of marine life.

Mass Wasting

When erosion occurs primarily as a result of gravity, whether triggered by an influx of water, by earthquakes, or by human activity, mass wasting occurs. Mass wasting takes place where slopes are overly steep or lacking vegetation. Events usually begin when support at the base of a slope is removed and the material above moves downhill.

Figure 20 In 1985 in Colombia, the Nevado del Ruiz volcano erupted. Meltwater from its glacier-covered peak created a mudflow that took 23,000 lives. In 1991 in the Philippines, Mount Pinatubo erupted and covered its slopes with ash and pyroclastic flows. Heavy rains mixed with the volcanic debris, creating a lahar, or mudflow, that covered much of Bacolor Village, shown above.

Erosion by Mass Wasting Erosion caused by mass wasting is dependent upon the type of event and the type of materials involved. For example, slumps, which are common in soil along hillsides, produce concave, upward scars from which material has detached and moved downward. Landslides or rockslides also can produce distinct scars on slopes. Mudflows move rapidly and are as dense and damaging as fluid cement. These most often occur following heavy rainfall or snowmelt. **Figure 20** illustrates an example of a mass wasting event.

Deposition by Mass Wasting Mass wasting events tend to dump their material in disorganized masses. This material replaces the undermined material at the base of the landform. When sufficient material builds up, the slope becomes stabilized and mass wasting stops. Geologists often can identify ancient landslides by their disorganized structure.

section 2 review

Summary

Erosion, Transport, and Deposition

- Features resulting from erosion, transport, and deposition of sediment give clues about the natural processes that formed them.

Running Water and Glaciers

- Regions along streams have specific characteristics depending on whether they are near the head or near the mouth of the stream.
- Glaciers create characteristic landforms.

Wind, Wave Action, and Mass Wasting

- Wind erosion creates landforms such as blowouts and dunes.
- Wave action creates longshore currents.
- Mass wasting occurs when material moves downslope mainly due to gravity.

Self Check

1. **Define** the general process of erosion.
2. **Describe** the process of sediment transport by running water, glaciers, wind, wave action, and mass wasting.
3. **List** at least two landforms associated with each of processes listed in question 2.
4. **Think Critically** How can you explain a sediment deposit that contains unlayered and randomly sorted material at the bottom and layered, well-sorted material at the top?

Applying Math

5. **Calculate** If a glacier moves 0.5 m per day, how far does it move in 100 years?

More Section Review gpescience.com

section 3

Groundwater

Reading Guide

What You'll Learn

- **Explain** the importance of groundwater.
- **Describe** how groundwater is stored and obtained.
- **List** some concerns about future groundwater use.

Why It's Important

Groundwater is a main source of drinking water for society.

Review Vocabulary

porosity: the percentage of open space in a material

New Vocabulary

- infiltration
- water table
- permeability
- aquifer

The Water Cycle

The oceans contain 97.2 percent of Earth's water. That leaves only 2.8 percent as freshwater, three-quarters of which is tied up in glacial ice. This means less than 1.0 percent of Earth's water is available for use.

Water Cycle Freshwater supplies are constantly being replenished through the water cycle. For example, large amounts of freshwater enter the water cycle by evaporation of seawater. A lesser amount enters the cycle through transpiration—the plant process in which water enters roots and is released through leaves as water vapor. The energy to power both evaporation and transpiration comes from the Sun. Water vapor in the air forms clouds, which in turn provide precipitation. Precipitation is freshwater, but it often is contaminated with pollutants washed from the air. Most precipitation goes back into the ocean. The water that falls back on land can evaporate, runoff as stream flow, or infiltrate Earth. **Infiltration** is the process by which water enters Earth to become groundwater below the surface. **Figure 21** illustrates how these processes are interconnected through the water cycle.

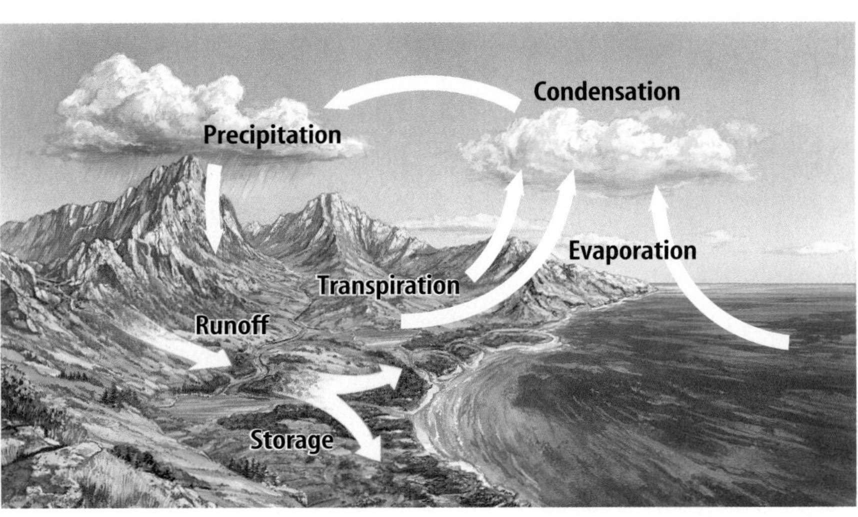

Figure 21 The water cycle continually redistributes Earth's water.

Groundwater

All the water that is naturally stored underground is called groundwater, as illustrated in **Figure 22.** When water moves underground through infiltration into the pore spaces of sediment and rock, groundwater is replenished. The region near the surface where water can infiltrate freely is the unsaturated zone. The region below that, where water completely fills the pore space, is the saturated zone. The **water table** is the boundary separating these two zones. Factors affecting how easily water can infiltrate and recharge groundwater supplies include the slope of the land, the nature of the surface material, and the type and amount of vegetation.

Reading Check *What is infiltration?*

Groundwater Storage A material must be capable of absorbing water in order for it to allow infiltration. If a material is not already saturated and enough time is available for water to pass through it, infiltration will occur. Unsaturated, porous sand or gravel will generally allow rapid infiltration. Surfaces made of clay or other impermeable material will generally allow little water to infiltrate. Most rainfall or snow melt runs off steep surfaces or evaporates. Plants take in significant amounts of infiltrated water before the water can move deep enough to be stored. In some regions, the supply of groundwater has been stored for millions of years, but humans are withdrawing it faster than it can naturally be replaced. This is particularly bad in arid climates where rainfall is low.

A rock unit that can transmit water through its pore space is called an **aquifer.** Sandstone and limestone often make good aquifers. A rock that slows or stops infiltration is an aquitard. A shale or clay layer makes a good aquitard. Small, local water tables can happen where an aquitard prevents downward infiltration. In this case water is said to be perched, or positioned above an aquitard.

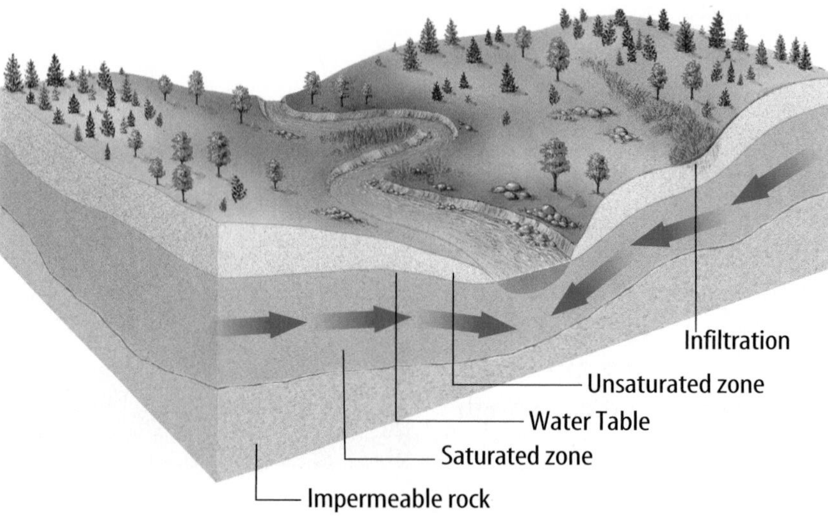

Figure 22 Groundwater is stored in porous rock called aquifers. It is often recharged, or replenished, by infiltration in wetlands.

Porosity and Permeability Water infiltrates through pore spaces, which are open spaces in sediments or rock. The combined volume of these spaces defines the material's porosity. If porosity is low or if the pores don't interconnect, then water will not flow easily, if at all. Such a condition is called low permeability. **Permeability,** illustrated in **Figure 23,** is a measure of how well a fluid can pass through a material. Water can continue to move through sediment and rock as long as there are interconnected pores available. This can cause the water table to rise and the zone of saturation to grow. When the water table rises to the surface, water can no longer infiltrate and run-off or flooding from more rainfall is likely.

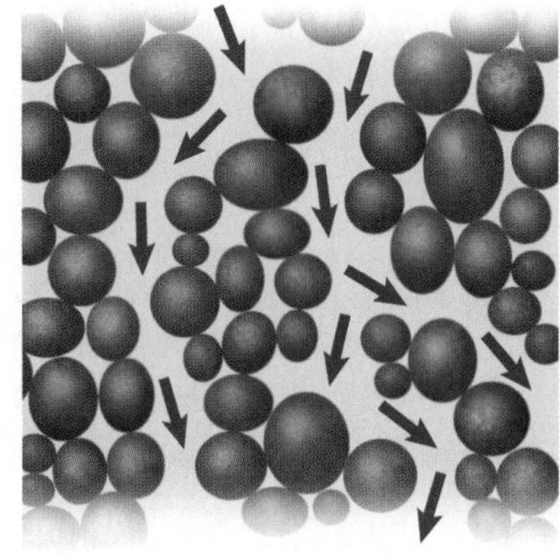

Figure 23 Fluids pass through interconnected pores but cannot pass through isolated pores.

Applying Math — Use Geometry

POROSITY VOLUME Suppose that an aquifer with dimensions of about 450 km in length and 200 km in width is 75 m thick. It has an average porosity of 7 percent. About how much water can the aquifer potentially hold?

IDENTIFY known values and the unknown value

Identify the known values:

Length of aquifer (L) = 450 km or 450,000 m

Width of aquifer (W) = 200 km, or 200,000 m

Thickness of aquifer (T) = 75 m

Porosity = 7% = 0.07

Identify the unknown value: the porosity volume

SOLVE the problem

Volume of aquifer = L × W × T = 450,000 m × 200,000 m × 75 m

$(4.5 \times 10^5 m) \times (2.0 \times 10^5 \ m) \times (75 \ m) = 6.8 \times 10^{12} \ m^3$

Total volume of porosity = volume × porosity (%)

$6.8 \times 10^{12} \ m^3 \times .07 = 4.7 \times 10^{11} \ m^3$

CHECK the answer

Does your answer seem reasonable? Check your answer by dividing it by the porosity, thickness, and width values. The result should be the length of the aquifer in meters.

Practice Problems

The Dakota Sandstone is 925 km long, 560 km wide, averages 4,200 m in thickness, and has an average porosity of 9.3 percent. What is its water storage volume?

For more practice problems, go to page 879 and visit Math Practice at gpescience.com.

VISUALIZING THE ROLE OF GROUNDWATER IN SHAPING EARTH'S SURFACE

NATIONAL GEOGRAPHIC

Figure 24

Some of the rainwater that falls on the ground flows into rivers and lakes. However, much of the water flows between soil particles and travels underground. Below ground, acidic water gradually dissolves limestone. Landforms that result from acidic water dissolving underground limestone are known as karst topography. Indications of karst areas include springs, caves, and sinkholes.

Spring waters flow from the ground where the zone of saturation meets Earth's surface. Depending on the level of the water table, springs can be either permanent or intermittent features of the landscape.

Caves form as acidic groundwater flows through bedrock and dissolves limestone. When the water table drops, open channels remain as caves. Limestone caves can extend many kilometers underground and include chambers that are tens of meters high.

Sinkholes can form when the roof of a cave collapses. Sinkholes can range from small, sloping depressions where water can slowly work its way underground, to large chasms where the flow of a stream or river can be transferred to an underground cave.

Water Resources

Life is dependent on the availability of freshwater. Because of this, supplies are monitored and attempts are made to preserve water resources. Humans have a responsibility to protect freshwater no matter what its source is.

Obtaining Groundwater Some groundwater comes naturally from springs, which are where the water table intersects the ground surface. Alternatively, wells are holes dug or drilled into Earth. When a drill hole encounters the water table, water can be pumped to the surface. As water is removed from the zone of saturation, the water table is lowered. You might expect the water table's surface to remain horizontal as water is taken away. This does not happen near the well because the groundwater flow path changes.

Much like surface water, groundwater flows downhill. The natural flow path of groundwater has both horizontal and vertical components, as illustrated in **Figure 24.** Also, groundwater flow rates usually are slow. As water is withdrawn, the direction of water flow shifts downward toward the well, depressing the water table around it into what is called a cone of depression. Depending on the rate of pumping, the cone of depression can affect water availability from other local wells, or force wells to be drilled deeper into an aquifer to obtain water. **Figure 25** illustrates this situation.

Water Under Pressure Wells drilled into aquifers that are under sufficient natural pressure to force water up into a well are called artesian wells. They form when an aquifer is sandwiched between two aquitards. One example might be sandstone positioned between two shale layers. The aquitards keep the water confined to the aquifer. When the aquifer is sloped, gravity on the water provides pressure on the low side of the slope. These wells usually are free flowing at the surface.

Not In My Back Yard (NIMBY) In the past, abandoned mines and parts of uninhabited valleys were used as landfill sites. Trash was thrown into these landfills. When it was discovered that many hazardous materials in trash were leaching into nearby water supplies, something had to be done. Research problems associated with locating sites for landfills and current improvements in landfill technology. Hold a class debate about the pros and cons of landfills.

Topic: Hard v. Soft Water
Visit gpescience.com for Web links to information about the difference between hard and soft water and where they occur.

Activity Create a map showing locations of hard and soft water across the United States. Explain what makes water hard or soft and how the effects can be reduced.

Figure 25 As areas become more populated, more water wells often are drilled. This lowers the water table, causing some existing water wells to go dry.
Infer *Why is a cone of depression created in the water table at the base of a water well?*

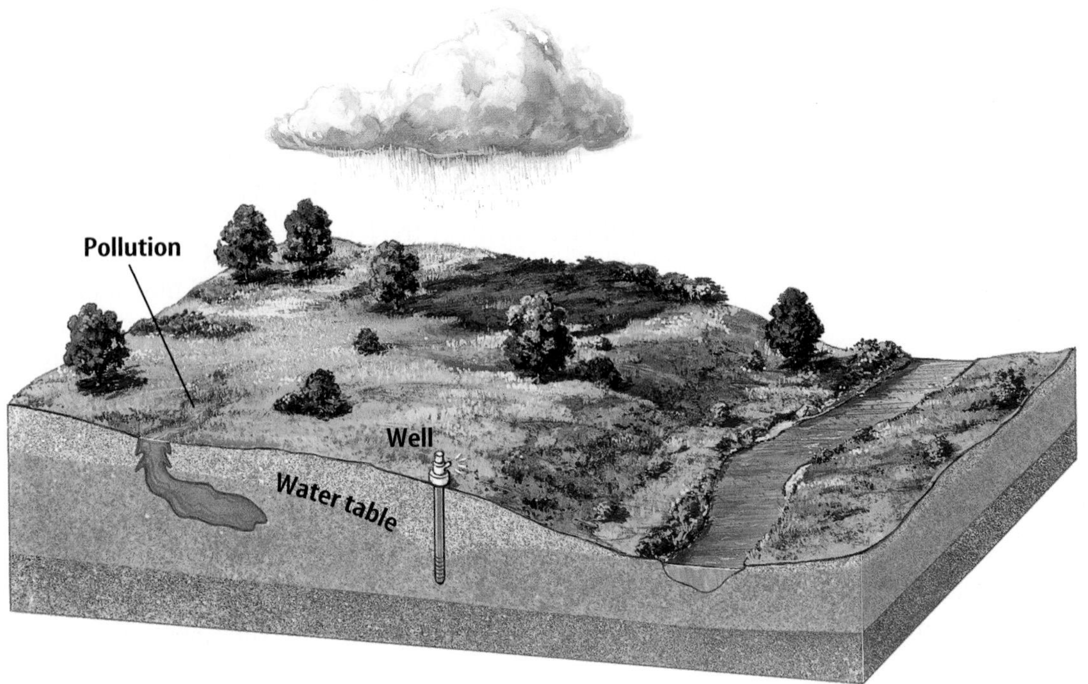

Figure 26 Pollution can enter groundwater through infiltration. Cleaning polluted groundwater can be difficult and costly.
Infer *Why is it important to prevent groundwater resources from being polluted?*

Pollution and Groundwater Resources Pollutants that are spilled or dumped on the ground and even air pollutants, washed from the air by precipitation, enter groundwater through infiltration, as shown in **Figure 26.** Particles of soil can filter larger contaminants and prevent them from moving deeper. But, natural pollutants such as arsenic can contaminate infiltrated water. Such groundwater is not useful as a freshwater source.

Reading Check *How can groundwater become polluted?*

section 3 review

Summary

Water Cycle

- Only about three percent of Earth's water sources are freshwater. Freshwater is renewed through water cycle processes.
- Supplies of groundwater are replenished when surface water infiltrates into Earth.

Groundwater

- The slope of the land, nature of the surface material, and the kind and amount of vegetation control infiltration.
- Porosity, permeability, and gravity control how water will move through materials.

Water Resources

- Both surface and groundwater resources are susceptible to human and natural contamination.

Self Check

1. **Explain** the importance of groundwater.
2. **Discuss** the role of porosity and permeability in the transmission of water through rock.
3. **Distinguish** among the unsaturated zone, the saturated zone, and the water table.
4. **Think Critically** In a rapidly growing neighborhood, streets, shopping centers, and homes are being built. Discuss the possible long-term effects of such growth on groundwater infiltration and supplies.

Applying Math

5. **Calculate** Suppose a community of 1,500 uses water at a rate of 1,000 L per person per day. This community relies on an aquifer that contains 1.5 billion L of water. How many years will this water supply last?

More Section Review gpescience.com

Geologic Time

Reading Guide

What You'll Learn

- **Distinguish** among geologic eras, periods, and epochs.
- **Explain** how scientists decide when a geologic time period ends and another begins.
- **Explain** the methods and logic used in relative dating.
- **Interpret** a geologic cross section.
- **Describe** the methods and logic used in absolute dating.

Why It's Important

An understanding of geologic time allows you to put current geologic events into a realistic perspective.

Review Vocabulary

radioactivity: the process that occurs when a nucleus decays and emits alpha, beta, and gamma radiation

New Vocabulary

- absolute dating
- relative dating
- uniformitarianism
- principle of superposition
- unconformity
- fossil

Time

In your daily life you have become accustomed to the length of a year, a day, or an hour. These units of time are ancient and were established by observing astronomical events. **Figure 27** shows a globe illustrating Earth's time zones. They are determined by the position of Earth with respect to the Sun during one cycle of Earth's rotation.

Figure 27 Earth is divided into 24 time zones.

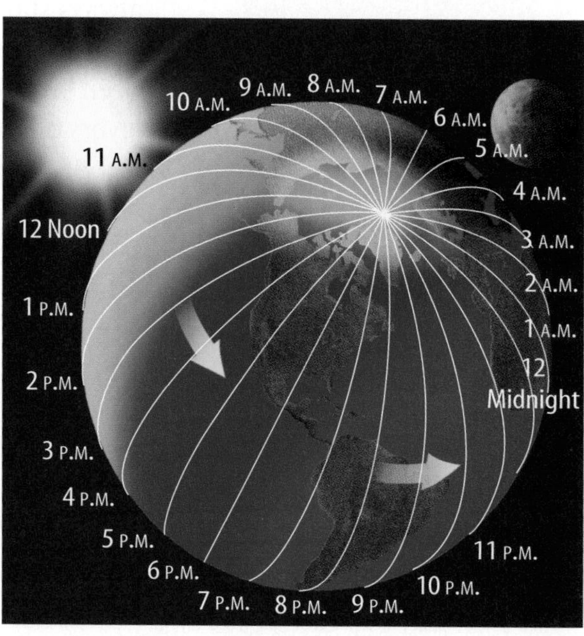

Absolute and Relative Dating You can give your age in years, months, days, and even hours. To do this you must have a starting reference—your birth date. Birth records often give the exact hour and minute of birth. This is an example of absolute dating. **Absolute dating** is the process of assigning a precise numerical age to an organism, object, or event. Technology allows scientists to add precision to their findings involving geologic time. **Relative dating** is the process of placing objects or events in their proper sequence in time. When you date events relatively, you know which came first, second, and so on. For example, students could be asked to line up in order of birth date, with oldest first. The key to relative dating is that the order is controlled by logical relationships, which might or might not involve the use of numbers.

It's how old? James Hutton was instrumental in changing the way scientists think about Earth history. Research Hutton's life to discover how a gentleman farmer became such an important scientific figure.

Uniformitarianism Geologic time is so vast that when James Hutton introduced his ideas for relative dating of rocks in his native Scotland and proposed that they might be millions of years old, he was seriously doubted. Hutton proposed the idea of **uniformitarianism,** which states that the laws of nature operate today as they have in the past. The only changes might be in their rate or scale. By studying how geologic events take place today, scientists can infer how events took place in the past. Uniformitarianism often is stated "the present is the key to the past." The idea of uniformitarianism is supported by many observations.

Principles of Relative Dating

To accurately date an event requires a set of logical rules that help determine the order of events. In the mid-1600s, Nicolaus Steno, a Danish physician, stated four principles that became the basis for understanding geologic events and the order in which they occurred. Suppose you were throwing papers into a wastebasket all day long, and at the end of the day you discovered you needed to find a paper from early that morning. You would know to look at the bottom of the basket, because the first things tossed away, or oldest papers, are at the bottom. In much the same way, the **principle of superposition** states that the oldest rocks in an undisturbed sequence of rock layers will be at the bottom of the sequence.

Reading Check *What is the principle of superposition?*

Original Horizontality In formulating the principle of original horizontality, Steno recognized that sedimentary layers always start off as horizontal layers. If you saw sedimentary layers that were not horizontal, you should know that they had been disturbed. You could infer that they began as horizontal layers of sediment, were formed into rock, and then were moved out of the horizontal position by some force.

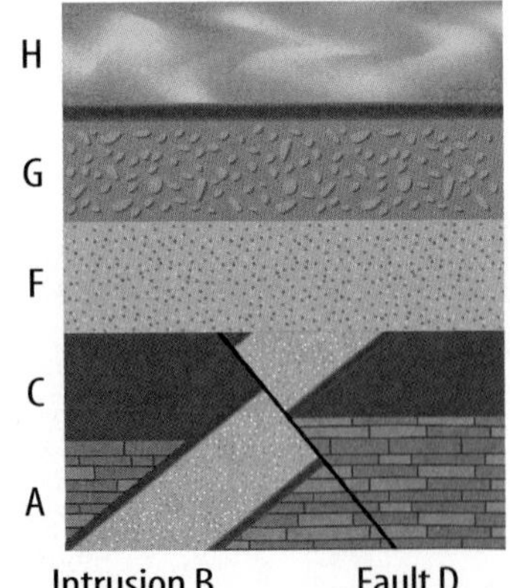

Figure 28 The principle of cross cutting relationships often can be used to determine the chronological order of some of Earth's geologic events. This illustration shows an intrusion of igneous rock that was separated later by a fault.

Overlapping Features Another point of logic that helps establish a relative age is the principle of cross cutting relationships. This principle states that any rock formation or fault is younger than the rock or feature that it cuts through. How could an igneous rock intrusion squeeze through something that is not there? It seems apparent that the rocks affected by the intrusion in **Figure 28** had to be there first in order to be affected. Likewise, if a piece of rock were to become embedded in part of another rock, the piece had to exist before the rock into which it was incorporated. These rock pieces are called inclusions.

Unconformities Rock sequences are like Earth's diary. If you keep a diary there might be days when you make no new entries. Also, there might be days when you make entries, but tear them out later and destroy them. Deposition is similar. Rocks that occur in their proper time sequence might have gaps within the sequence. There are times when no sediments are being deposited that can be observed later. Also, there are times when erosion destroys existing rock layers. **Unconformities** represent gaps in the rock record during which erosion occurred or deposition was absent. **Figure 29** shows a sequence of rocks that contains several unconformities.

Reading Check *What is an unconformity?*

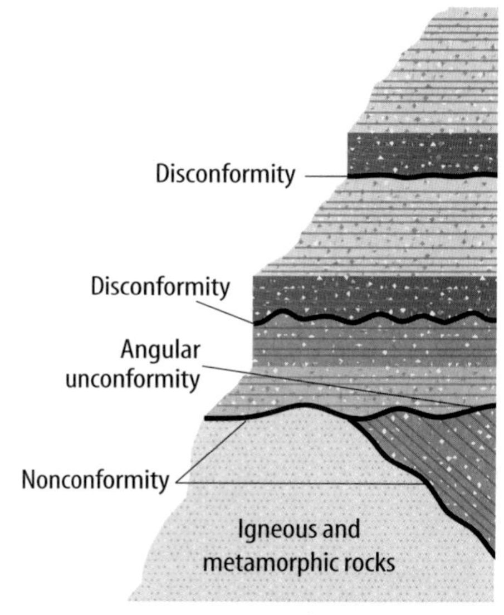

Figure 29 Three types of unconformities are shown above: disconformity, angular unconformity, and nonconformity. A disconformity is an erosional surface between horizontal rock layers. An angular unconformity is an erosional surface between rock layer segments that intercept at an angle. A nonconformity is an erosional surface between sedimentary rock and igneous or metamorphic rock.

Fossils

The remains or traces of organisms found in the geologic rock record are called **fossils.** Fossils can be direct remains, such as an actual bone or shell. Sometimes fossil remains, such as the inside of a shell, act as molds that form casts when filled with sediment that later hardens to rock. In a process called replacement, water containing dissolved mineral material might replace original shell or bone material with different minerals such as pyrite or quartz. In other cases, only a trace impression such as a footprint or burrow is left for paleontologists to observe.

Correlation While working as a canal engineer in England at the end of the eighteenth century, William Smith proposed the principle of faunal succession. Smith noticed that fossils appeared in the same order in the rock layers dug out to build the canals, no matter where he found them. For example, if fossil Q were found in a sedimentary rock layer above fossil F, then Q also appeared above F in a different geographic region. If you apply the law of superposition to this observation, you can conclude that fossil Q is younger than fossil F.

The process of matching distinctive rock units from different regions is called correlation. **Figure 30** illustrates how this is done. Paleontologists have been successful in correlating fossil-bearing rock units on a global scale. Fossils that are most useful as global time markers often are from organisms that were widespread geographically, but lived in only a narrow, well-defined period of time. Such fossils are called index fossils.

Not all fossil organisms existed over a wide area, however. Some fossils are not useful as time markers because they appear the same over millions of years. In addition, recall that geologic processes such as mountain building and erosion can destroy fossils, making correlation difficult.

Figure 30 Sometimes the relative age of rock units can be found by matching distinctive fossil-bearing rock units from different locations.

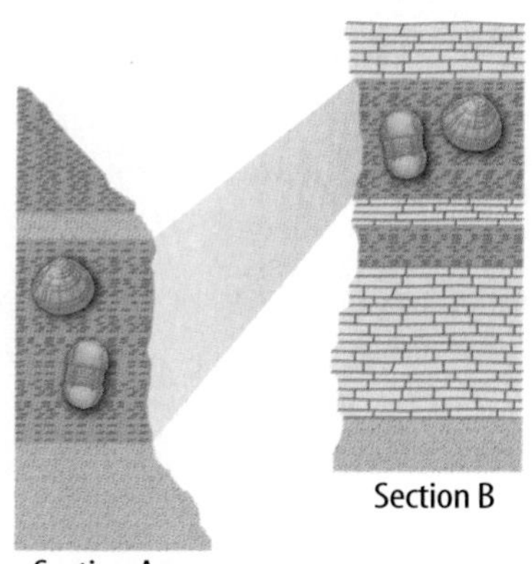

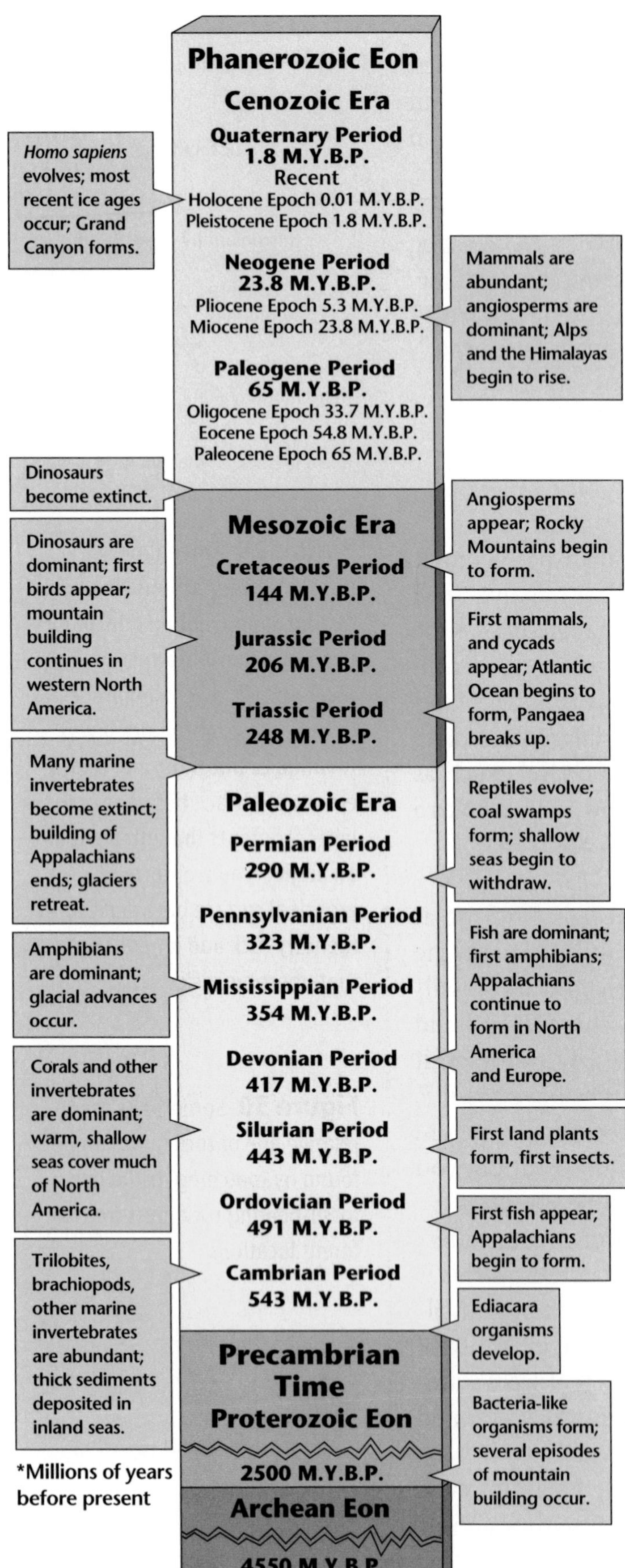

Geologic Time Scale Historians speak of the Industrial Revolution, and musicians refer to the Classical Period. These time periods began and ended with some event that marked a noticeable change in society and music. Similarly, paleontologists observe changes in the fossil record and associate them with boundaries of time units, shown in **Figure 31.** At these boundaries, fossils of certain life forms are no longer present and new life forms begin to appear. Boundaries also might be established by evidence of a catastrophic geologic event. Such events might have caused environmental changes that led to extinctions or changes in life forms.

Absolute Dating

Some atoms in minerals of a newly-formed igneous rock are like a clock that has been reset. A quantity of unstable radioactive element—the parent isotope—starts its decay at the time of mineral formation and gradually decays to a stable daughter isotope. Because radioactive isotopes decay in predictable ways, scientists can analyze the ratio of parent to daughter isotopes in a mineral to calculate the time that has elapsed since that mineral formed.

Radioactive Decay Some types of atoms are unstable and decay by casting off, or emitting, parts of their nuclei. Decay mechanisms produce new isotopes—some produce daughters with different atomic numbers, some involve a change in atomic mass, and some involve a change in both.

Figure 31 *Phanerozoic* means "visible life."
Calculate *What fraction of Earth's history does the Phanerozoic eon represent?*

Table 1 Commonly Used Isotopes for Dating Earth Materials

Radioactive Parent	Stable Daughter	Half-Life
Carbon-14	Nitrogen-14	5,730 years
Uranium-235	Lead-207	713 million years
Potassium-40	Argon-40	1.3 billion years
Uranium-238	Lead-206	4.5 billion years
Thorium-232	Lead-208	14.1 billion years
Rubidium-87	Strontium-87	49.0 billion years

INTEGRATE Physics

Half-Life Every radioactive isotope has a half-life. The half-life is the time it takes for one-half of a radioactive parent sample to decay to its stable daughter. After one half-life, the ratio of parent atoms to daughter atoms is 50:50, or 1:1. For example, if the half-life of a parent isotope is 500,000 years and a rock is analyzed and the observed daughter/parent ratio is 1:1, one half-life has elapsed and the age of that rock is 500,000 years. During the next half-life, there is decay of one-half of the remaining parent isotope, resulting in one-fourth parent to three-fourths daughter, or 25:75.

Reading Check *What is half-life?*

Useful Isotopes All of the isotopes in **Table 1** except carbon-14 are common in many types of rocks. Carbon-14 is useful for dating organic Earth materials such as wood and peat.

Generally, absolute ages are determined from unaltered minerals in igneous rocks. This is because sedimentary and metamorphic processes involve water or other fluids. When they interact with the rock it can cause the amount of parent isotopes, daughter isotopes, or both to change. When this happens, the age that is determined will be too young or old.

To see how geologists combine relative dating techniques with absolute dating techniques, look at **Figure 32.** You could conclude from the rock layers at Location A that Fossil C is between 2.1 and 2.4 million years old, Fossil A is older than 2.4 million years, and Fossil E is younger than 2.1 million years. These conclusions use radiometric dating for the age values and superposition for the order of events. What age relations can you determine for Fossils C, B, E and F at Location B?

Figure 32 Relative dating and absolute dating techniques can be combined to help determine the ages of some fossils.
Infer *Why is Fossil C considered to be between 2.1 and 2.4 million years old?*

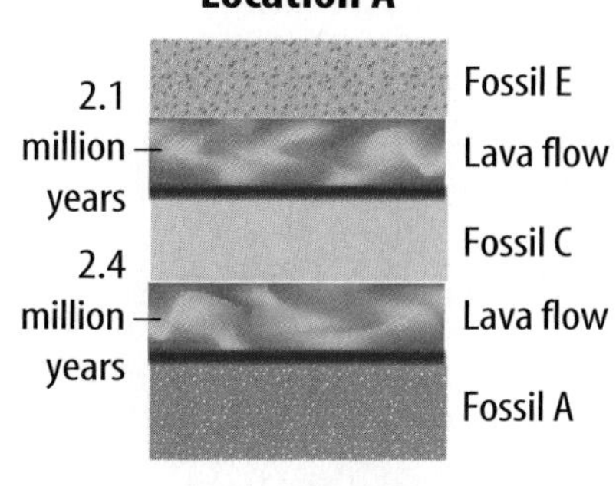

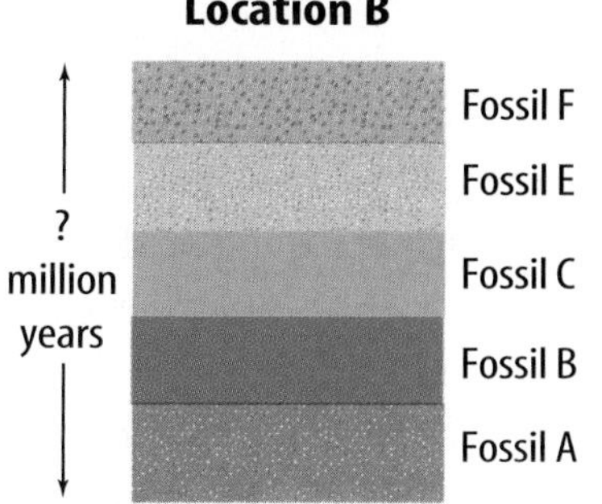

Geologic Maps

Geologic cross-sections, like the simplified sketch in **Figure 32,** represent a vertical slice taken through Earth. Geologic maps, on the other hand, show the horizontal surface distribution of various rock formations. Most geologic maps also include a legend that indicates the ages of the rock formations at the surface. Geologic maps are two-dimensional models of Earth's crust for a given region. Geologists plot the location, orientation, and boundaries of rocks found at the surface. These data are obtained by direct field observation, through drill cores, aerial photography, and satellite observations.

Reading Check *What is a geologic map?*

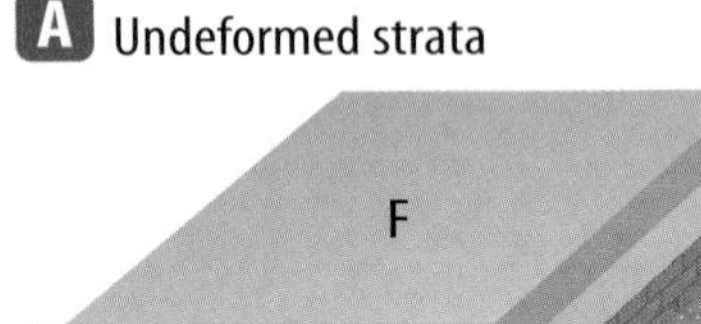
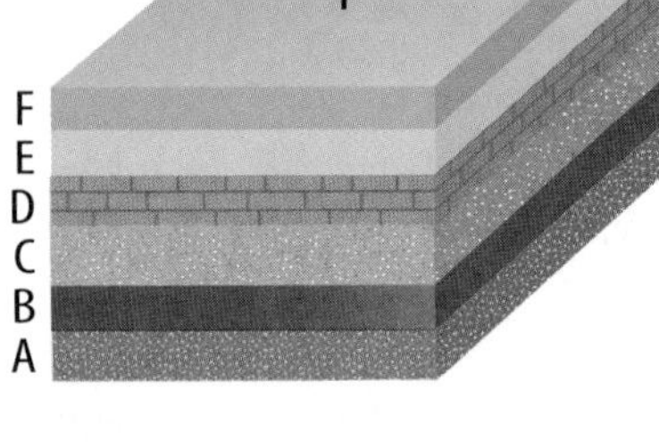

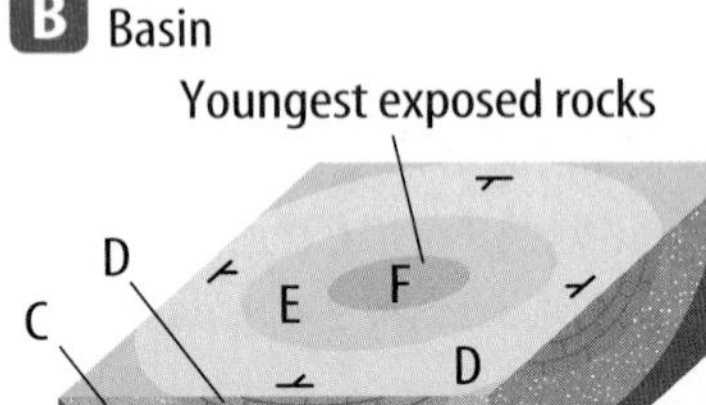
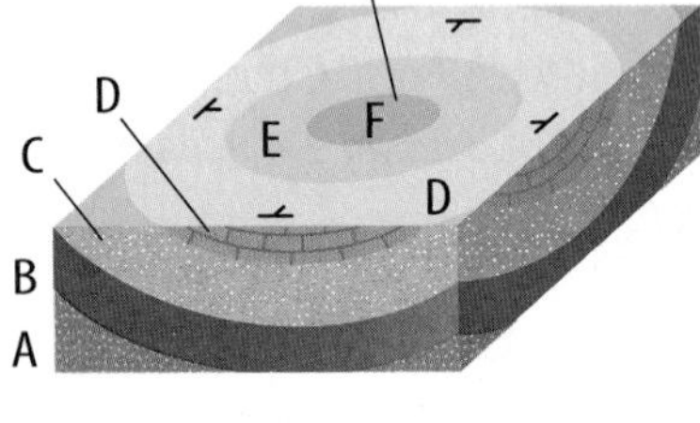

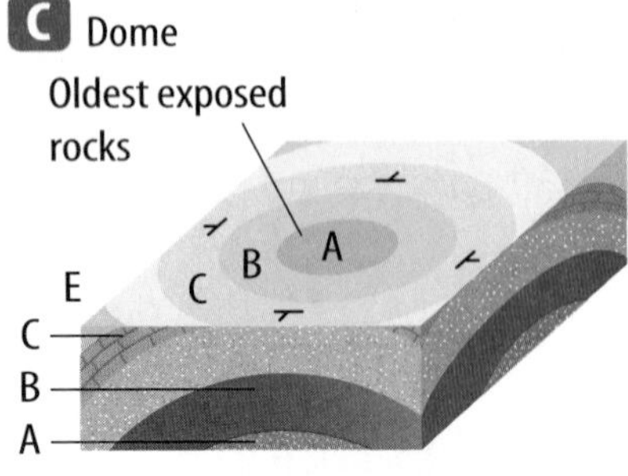

Figure 33 Tectonic processes can cause Earth's plates to shift, collide, and separate. Sometimes this leads to the deformation of horizontal rock layers.
Infer *What direction(s) of force might be needed to create a dome?*

Basins and Domes Recall that rock layers can be deformed by tectonic processes that result in the formation of faults and folds. One example of a deformed structure is a basin. Recall that the principle of original horizontality states that sediments are laid down in horizontal layers as shown in **Figure 33A.** Compressional forces due to plate tectonic activity cause the layers to fold and to dip downward. **Figure 33B** shows the surface map expression of a basin. In basins, the youngest rocks are exposed in the center and tilt or dip inward.

Domes are essentially the opposite of basins. In this case, the rock layers are uplifted at the center of the structure. The surface appearance of a dome, shown in map view in **Figure 33C,** is similar to a basin. In this case, the oldest rocks are at the center and dip outward. Visualize this by warping up a stack of papers into a hill. Next, imagine sawing off the top of the paper hill. The sawing would expose the bottom layers of paper in the middle. In the case of rock formations, the sawing represents erosion, and the lower paper layers represent the oldest rock layers.

Folds Folding of the crust can also be recognized on geologic maps. Recall that compressional stresses tend to squeeze and shorten materials. If the forces on part of Earth's crust are mainly compression forces, then rocks subjected to the compression might wrinkle up or fold. Hold a sheet of paper flat with both hands and push the sides together. The paper wrinkles either up or down. Folds of rock appear much the same way. An upward wrinkle of rock formations is an anticline and a downward wrinkle is a syncline, shown in **Figure 34A.** Try the paper trick again, but hold one hand still and push with the other. You get a fold that flops over horizontally. This is called a recumbent fold.

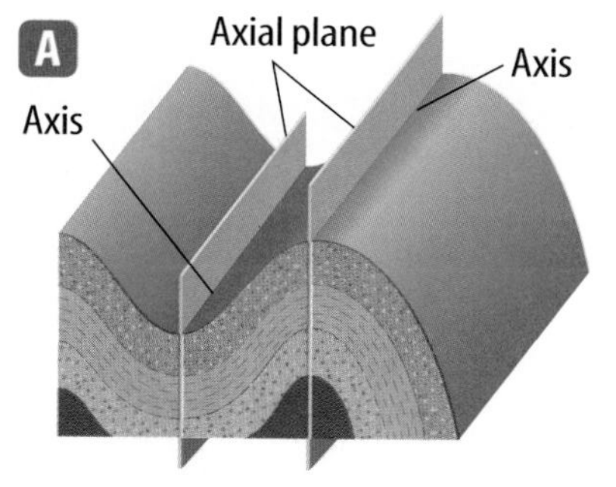

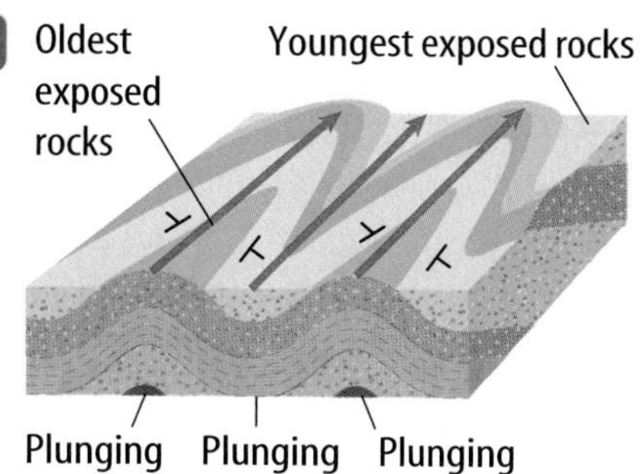

Figure 34 Two basic types of folds are found in rock layers.
A Synclines form a trough and anticlines form an arch.
B If these folds are tilted, they form plunging folds. On a geologic map, plunging folds appear as V-shaped or horseshoe patterns.

Plunging Folds The axis of a fold is an imaginary line where Earth's surface meets the axial plane. The axial plane divides the fold in half, and is usually perpendicular to the compression force. If the forces are applied unevenly the axis of the fold might plunge back into the Earth, producing a surface pattern of rock like the one shown in **Figure 34B.**

These structural features can give you an idea of how geologists illustrate what the surface of the crust looks like in given locations. Practice viewing the structures in cross section and map view, and apply relative dating logic to the structures. You will soon be able to reconstruct the events that formed the features, and tell a story about their history.

section 4 review

Summary

Time

- Uniformitarianism states the processes occurring today are the same as they have been in the past. Rate and extent may have changed.

Principles of Relative Dating

- Steno's principles helped to develop an understanding of geologic events.

Fossils

- Fossils are the remains or traces of once-living organisms. Correlation is the matching of rock outcrops from one region to another.

Absolute Dating

- Analyzing the radioactive decay of radioactive isotopes helps to establish numerican dates for objects and events.

Geologic Maps

- Geologic maps are two-dimensional models of a region of Earth's crust.

Self Check

1. **Discuss** how events in Earth history determine boundaries for geologic time units.
2. **Compare and contrast** absolute and relative dating.
3. **Describe** three lines of logic used in relative dating. Draw and label respective diagrams.
4. **Think Critically** You're hiking down a canyon that has been deeply eroded, exposing numerous folds. One fold is recumbent, with layers turned completely over. Describe the problems this structure could cause for relative dating, if only a small portion of its rock layers were exposed.

Applying Skills

5. **Calculate an Age** Suppose the ratio of radioactive parent to stable daughter isotope in a mineral within an igneous rock is 12.5:87.5. The half-life of the parent isotope is 7.13×10^8 years. How old is the rock sample?

It's about time—relative, that is.

Goals

- **Observe** geologic cross sections.
- **Interpret** events that caused the observed features.
- **Infer** a probable chronology for the events.

Materials

copies of geologic cross sections

Real-World Problem

When interpreting the geologic history of a region, geologists follow the principles of relative dating. As with all scientific disciplines, they might get help by exchanging information with other scientists. Some rocks in an area of interest might have been dated using radioactive isotopes or index fossils. How can geologists determine a chronological order of events that took place in a section of Earth's crust over many millions of years?

Procedure

1. Familiarize yourself with the symbols in the legend that signify different rock types and other cross section features.
2. Examine the cross sections. On your own paper, list the order of events that took place for each cross section, listing the oldest first. Note that the number of steps is not simply the number of rock layers in the cross section.

Legend

 Limestone

 Sandstone

 Shale

 Conglomerate

 Extrusive igneous rock lava flow

 Contact metamorphism

 Intrusive igneous rock

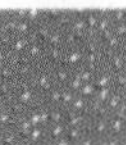 Intrusive igneous rock

Tiltin Hilton

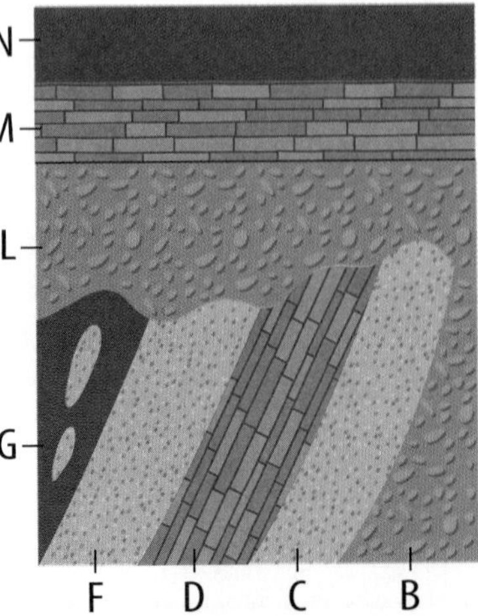

Criss cross

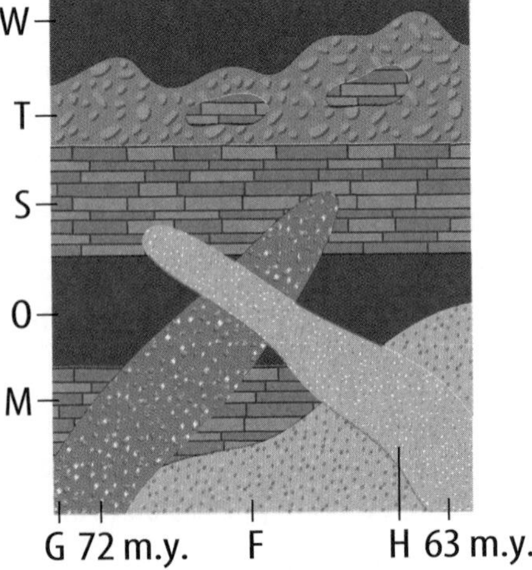

3. Be sure to include all of the events or actions that changed or helped form the Earth materials shown in each of the four cross sections. For example, unconformities, faulting, folding, or other changes also need to be placed in proper order.

Analyze Your Data

1. **Analyze** each cross section and determine whether there is more than one possible chronological interpretation.
2. **Identify** features that are most useful in supporting relative age relationships.

Conclude and Apply

1. **Evaluate** cross sections that apparently have more than one possible explanation for a geologic history.
2. **Explain** whether or not the absolute ages for the intrusions in cross section *Crisscross* agree with the relative ages of the intrusions.
3. How would your answer have changed for cross section *Crisscross* if you did not have absolute age data?
4. **Summarize** your reasoning for establishing a relative age for fault X in the cross section *No Fault of Mine*.

No fault of mine

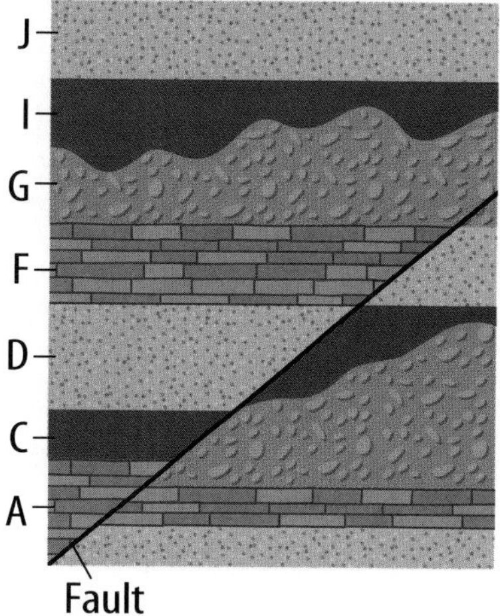

Intruder

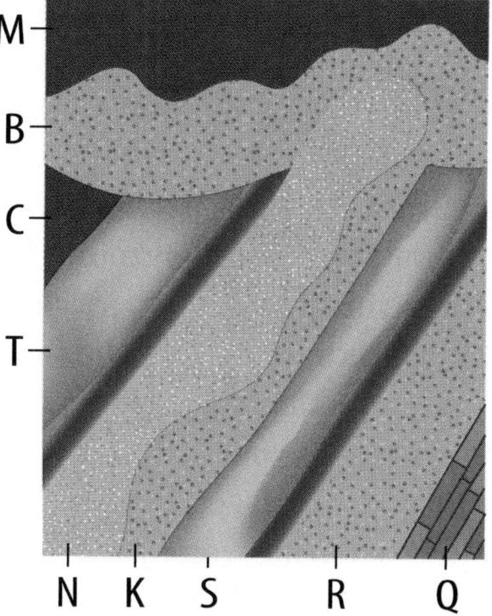

Communicating Your Data

Prepare an oral defense for your interpretation of one of the cross sections. Answer questions from classmates who might have made different conclusions about the same cross section.

Tapping into the Underground

Freshwater is one of our most important natural resources. Although it may seem that there is an unlimited supply of water—just turn on the faucet and there it is—water supplies can run out. Much of Earth's freshwater is stored in aquifers, inside the cracks and pores of rock and soil. Some of this water flows to the surface and is continually replaced by rainfall that filters through the ground. Other aquifers lie far below the surface, holding water that fell as rain thousands of years ago.

Humans pump water from underground for many uses including drinking, bathing, irrigating crops, and for manufacturing. Because the time needed for water to move into an aquifer can be years or even centuries, these uses consume water faster than it can be replaced by natural processes. When that happens, water is no longer a renewable resource and groundwater shortages become likely.

If water is pumped from a well faster than it can be replaced, the area around it becomes depeleted. This results in wells must be drilled deeper and deeper. Eventually the water flow becomes too slow to be useful. Groundwater shortages don't just affect the amount of water, but they can also affect its quality. Under normal conditions, the direction of underground water flow tends to be the same as the slope of the surface. As the supply is depleted, water flows from all directions toward the well. In places near the ocean, salt water flowing into the aquifer can make the water useless.

Water conservation—using less water—slows the depletion of the aquifer. This helps the supply refill naturally. Many communities are taking steps to avoid groundwater shortages. In some places, wastewater is cleaned and recycled for other uses to reduce the amount of underground water needed. Another way to use recycled wastewater is ro refill the groundwater supply. Treated wastewater is either pumped back into the ground or allowed to filter back into the aquifer from constructed wetlands.

Interview **Contact a local hydrologist for an interview. Before the interview, do research to determine the sources of water for your community and what happens to treated wastewater. Make a list of interview questions. Conduct the interview and write a summary.**

TIME

For more information, visit gpescience.com

Reviewing Main Ideas

Section 1 Weathering and Soil

1. Mechanical weathering reduces the size of materials. Chemical weathering creates new materials or dissolves material through chemical reactions.
2. Differential weathering causes rock that is resistant to particular climate conditions to stand out in the landscape.
3. Soil is composed of weathered rock, decaying organic matter, water, and air. Soil type depends on the nature of the material from which it forms, climate type, and location.

4. Erosion and chemical depletion of soil are global problems.

Section 2 Shaping the Landscape

1. Erosion is a set of processes that physically remove material at Earth's surface.
2. All the land area that collects water for a major river is a drainage basin.
3. In order to develop, glaciers require long periods of time in which winter snowfall exceeds summer melt. Glaciers create erosional and depositional features.
4. Erosion by wind generally is limited to particle sizes that are sand-sized or smaller.
5. Erosion by wave action on rocky coastlines leaves resistant landforms behind, such as sea stacks and arches. Beach erosion and deposition is caused partly by movement of sediment by longshore currents.
6. Mass wasting is any form of erosion driven by gravity, such as landslides and mudflows.

Section 3 Groundwater

1. Groundwater accumulates by the infiltration of surface water. Infiltration is controlled by slope of land, type of surface material and type and amount of vegetation.
2. Groundwater is stored in aquifers. The movement of groundwater in an aquifer is influenced by gravity, the structure of the aquifer, and by the porosity and permeability of the aquifer.

3. Groundwater might return to the surface naturally by springs or be brought to the surface by human-made wells.

Section 4 Geologic Time

1. Relative dating principles are used to determine the order of geologic events. Radiometric dating uses the decay of radioactive isotopes to establish a numerical age for materials.
2. Geologic maps illustrate the horizontal distribution of rock formations.

FOLDABLES™ Use the Foldable that you made at the beginning of this chapter to help you review Earth's erosional and depositional features.

chapter 21 Review

Using Vocabulary

absolute dating p. 669
aquifer p. 664
deposition p. 654
drainage basin p. 655
erosion p. 654
fossil p. 671
infiltration p. 663
longshore current p. 660
permeability p. 664
principle of superposition p. 670
relative dating p. 669
sediment transport p. 654
soil p. 650
unconformity p. 671
uniformitarianism p. 670
water table p. 664
weathering p. 646

Match the correct vocabulary word(s) with each definition given below.

1. a rock layer that stores and transmits water
2. a means of establishing the order of events
3. region that collects water for a stream system
4. concept that the laws of nature act today as they have in the past
5. a measure of the ability of a fluid to pass through a material
6. processes of chemical or physical breakdown of a substance
7. Earth's surface covering consisting of weathered rock, decaying organic matter, water, and air
8. boundary separating the saturated zone and the unsaturated zone
9. a surface representing a period of erosion or non-deposition

Checking Concepts

10. What is a process that leaves resistant rock remaining in the landscape?
    A) deposition
    B) differential weathering
    C) sediment transport
    D) infiltration

Use the illustration below to answer questions 11–12.

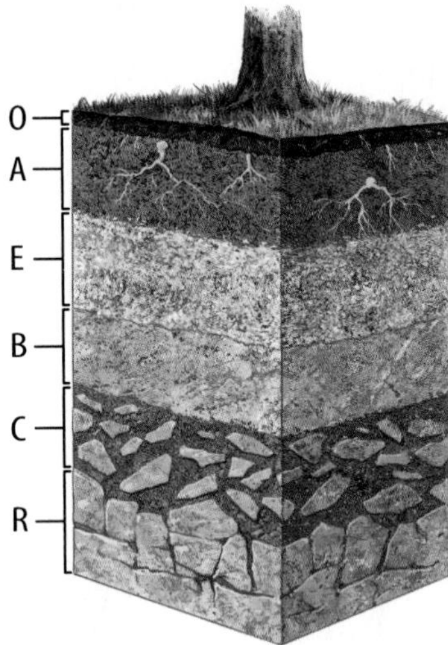

11. Which soil horizon contains the greatest amount of organic matter?
    A) A horizon
    B) B horizon
    C) C horizon
    D) O horizon

12. Which soil horizon represents weathered bedrock?
    A) A horizon
    B) B horizon
    C) C horizon
    D) O horizon

13. Which of the following is NOT a typical soil texture component?
    A) gravel
    B) sand
    C) silt
    D) clay

14. Which of the following is formed by glaciers?
    A) V-shaped valleys
    B) moraines
    C) flood plain
    D) deltas

15. Select the material that slows or stops movement of ground water.
A) aquifer
B) aquitard
C) water table
D) saturated zone

16. Select the relative principle that states: *The oldest rock layer is on the bottom.*
A) original horizontality
B) cross cutting relationships
C) faunal succession
D) superposition

17. Which time unit of the geological time scale is the shortest?
A) eon
B) epoch
C) period
D) era

Interpreting Graphic

18. Copy and complete the table below.

Radioactive Decay		
Half-lives	% Parent	% Daughter
	50	50
2		
3		
4		

Thinking Critically

19. **Explain** why a soluble rock such as limestone can stand out as cliffs in the western United States.

20. **Infer** how water from a natural spring could become contaminated.

21. **Explain** the difference between a confined aquifer and an unconfined aquifer.

22. **Infer** why cities that draw their water from a drainage basin need to understand how surrounding land is used.

23. **Infer** why rocks or fossils that are found lying on the ground are not useful for relative dating.

Use the illustration below to answer question 24.

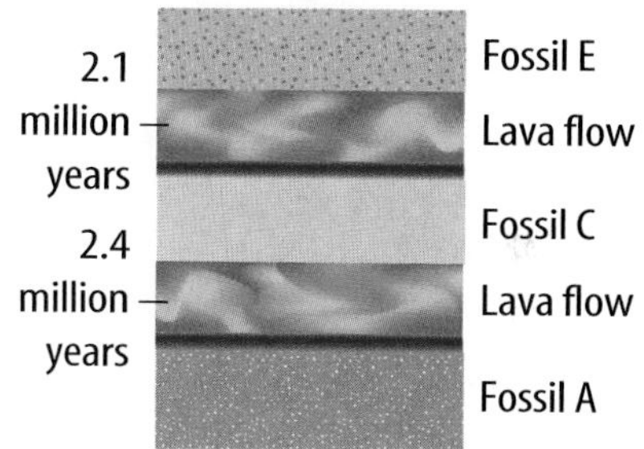

24. **Describe** how absolute dates from ancient lava flows could be used to determine the absolute ages of fossils.

25. **Apply** The half-life of carbon-14 is 5730 years. Explain why carbon-14 is useful for dating objects that are 75,000 years old and younger.

26. **Apply** Explain why glacial deposits made of till are often poorly drained and swampy.

Applying Math

27. Suppose an aquifer is known to be 200 km long and 75 km wide. If its average saturated thickness is 40 m and it has an average porosity of 3 percent, estimate how many cubic kilometers of water could potentially be stored in the aquifer.

28. An igneous rock sample was dated using radioactive potassium. Measurement indicates that 1/32 of the original parent potassium is left in the sample. The half-life of potassium is 1.3 billion years. How old is this igneous rock?

chapter 21 Standardized Test Practice

Record your answers on the answer sheet provided by your teacher or on a sheet of paper.

Multiple Choice

Use the illustration below to answer question 1.

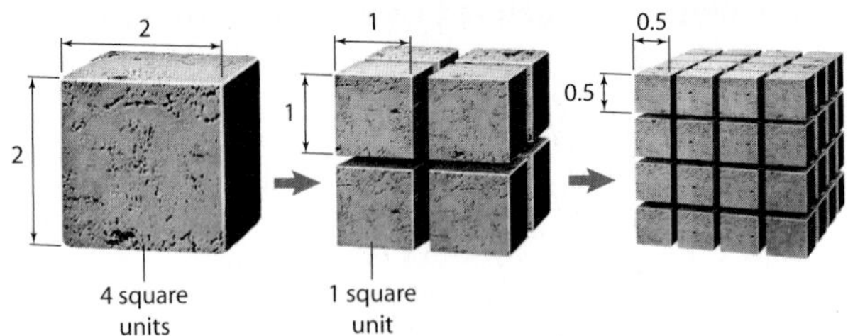

1. How does mechanical weathering affect rock material?
   A. alters chemical composition
   B. increases rock surface area
   C. increases rock volume
   D. slows chemical weathering

2. Which is a mixture of weathered rock, organic matter, water, and air?
   A. feldspar
   B. parent material
   C. rock rust
   D. soil

3. Which is the dropping of sediment load?
   A. deposition
   B. erosion
   C. transporting
   D. weathering

Test-Taking Tip

Remember Never leave any answer blank.

4. Which develops as force within a stream is exerted sideways?
   A. floodplain
   B. mouth
   C. source
   D. tributary

Use the illustration below to answer question 5.

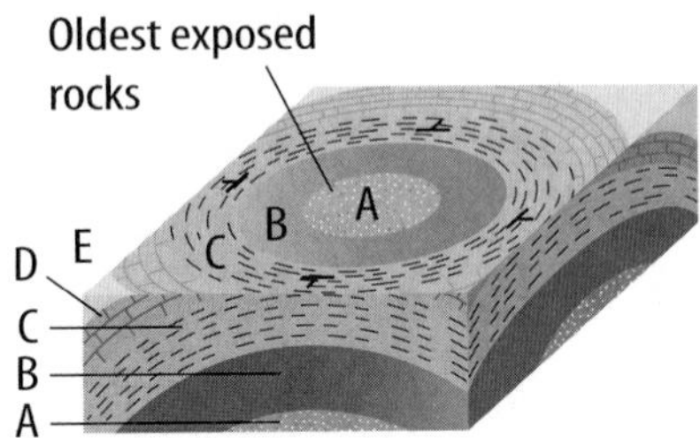

5. Which structural form is illustrated above?
   A. anticline
   B. basin
   C. dome
   D. syncline

6. Which is where a stream's water flows into another body of water?
   A. floodplain
   B. mouth
   C. source
   D. tributary

7. Which forms at the front of a glacier?
   A. end moraine
   B. ground moraine
   C. lateral moraine
   D. medial moraine

8. Which forms when sand is moved by wind?

   A. currents

   B. delta

   C. dunes

   D. loess

Gridded Response

Use the illustration below to answer question 9.

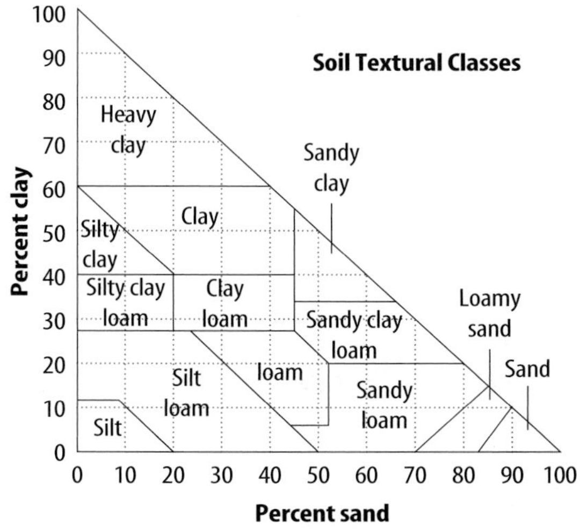

9. What is the greatest percentage of sand that can be in a heavy clay soil?

10. If the half-life of an isotope is 7,000 years and the amount of that isotope present in an igneous rock is only one-fourth of the original amount, how old is the rock?

11. Soil erosion in an area is 3.0 cm per year. The soil profile is 2.4 m thick and 25 percent of that is topsoil. How long will it take for the topsoil to erode away?

Short Response

12. How does water work to cause mechanical weathering called frost wedging?

13. How does slope affect the soil horizon?

14. Why is vegetation important to the type of soil that forms?

15. What is the drainage basin of a river?

16. How does a delta form?

17. How does a bar form on the inside of a bend in a stream?

18. How does a glacier erode the land's surface?

Extended Response

Use the art below to answer question 19.

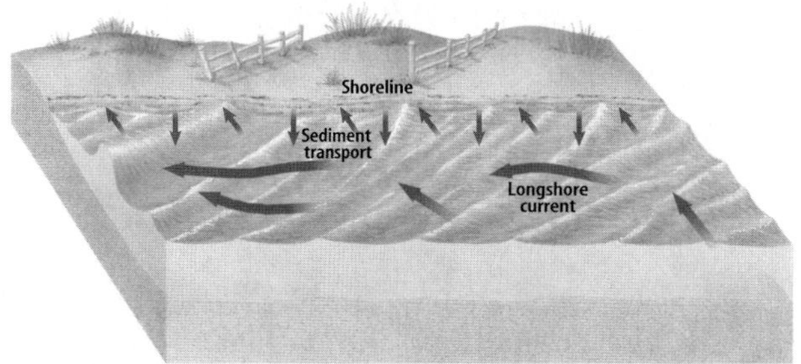

19. **PART A** How does a longshore current form?

    PART B Explain how a longshore current affects erosion and deposition.

unit 6 Interactions of Matter

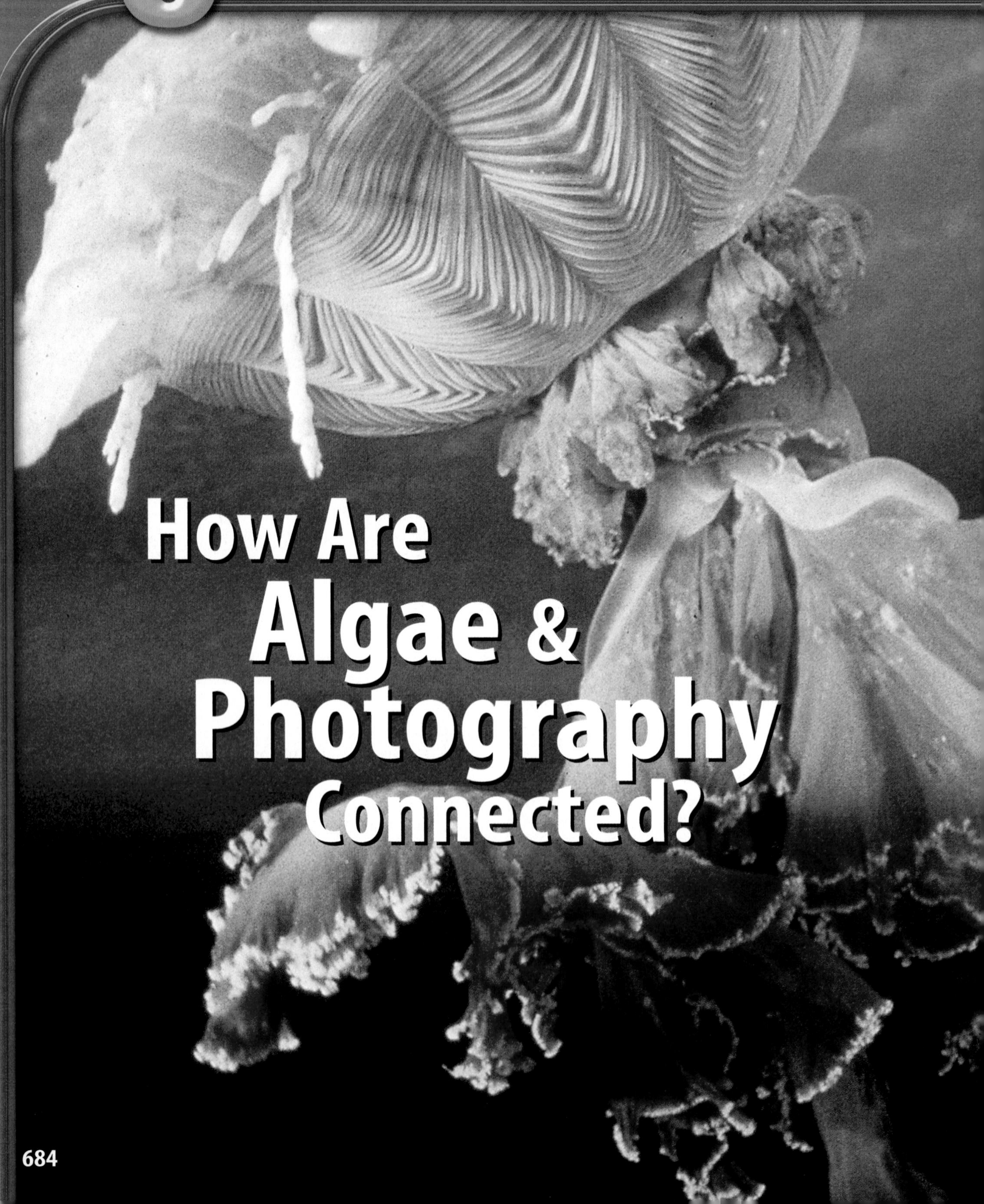

How Are Algae & Photography Connected?

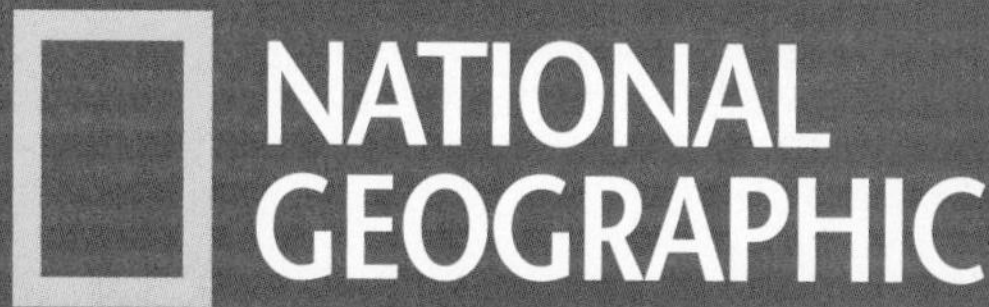

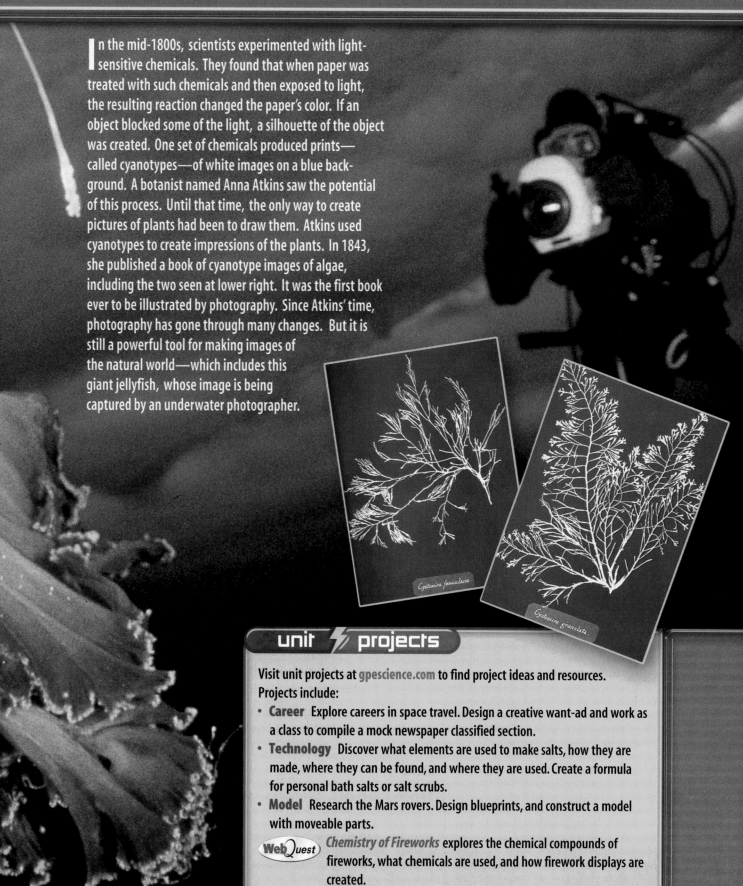

In the mid-1800s, scientists experimented with light-sensitive chemicals. They found that when paper was treated with such chemicals and then exposed to light, the resulting reaction changed the paper's color. If an object blocked some of the light, a silhouette of the object was created. One set of chemicals produced prints—called cyanotypes—of white images on a blue background. A botanist named Anna Atkins saw the potential of this process. Until that time, the only way to create pictures of plants had been to draw them. Atkins used cyanotypes to create impressions of the plants. In 1843, she published a book of cyanotype images of algae, including the two seen at lower right. It was the first book ever to be illustrated by photography. Since Atkins' time, photography has gone through many changes. But it is still a powerful tool for making images of the natural world—which includes this giant jellyfish, whose image is being captured by an underwater photographer.

unit projects

Visit unit projects at gpescience.com to find project ideas and resources. Projects include:

- **Career** Explore careers in space travel. Design a creative want-ad and work as a class to compile a mock newspaper classified section.
- **Technology** Discover what elements are used to make salts, how they are made, where they can be found, and where they are used. Create a formula for personal bath salts or salt scrubs.
- **Model** Research the Mars rovers. Design blueprints, and construct a model with moveable parts.

WebQuest *Chemistry of Fireworks* explores the chemical compounds of fireworks, what chemicals are used, and how firework displays are created.

chapter 22

Chemical Bonds

chapter preview

sections

Elements Form Chemical Bonds

Just as these skydivers are linked together to make a stable formation, the atoms in elements can link together with chemical bonds to form a compound. You will read about how chemical bonds form and learn how to write chemical formulas and equations.

Science Journal Describe how glue is similar to chemical bonds.

Start-Up Activities

Chemical Bonds and Mixing

You probably have noticed that some liquids such as oil-and-vinegar salad dressings will not stay mixed after the bottle is shaken. However, rubbing alcohol and water will mix. The compounds that make up the two liquids are different. This lab will demonstrate the influence the types of chemical bonds have on how the compounds mix.

1. Complete the safety form.
2. Pour 20 mL of water into a 100-mL graduated cylinder.
3. Pour 20 mL of vegetable oil into the same cylinder. Vigorously swirl the two liquids together, and observe for several minutes.
4. Add two drops of food dye and observe.
5. After several minutes, slowly pour 30 mL of rubbing alcohol into the cylinder.
6. Add two more drops of food dye and observe.
7. **Think Critically** In your Science Journal, write a paragraph describing how the different liquids mixed. Would your final results be different if you added the liquids in a different order? Explain.

Chemical Formulas Every compound has a chemical formula that tells exactly which elements are present in that compound and exactly how many atoms of each element are present in that compound. Make the following Foldable to help identify the chemical formulas in this chapter.

STEP 1 **Fold** a sheet of notebook paper vertically from side to side.

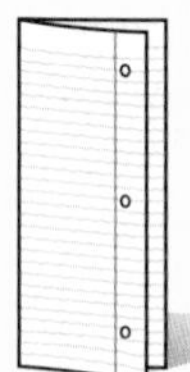

STEP 2 **Cut** along every third line of only the top layer to form tabs.

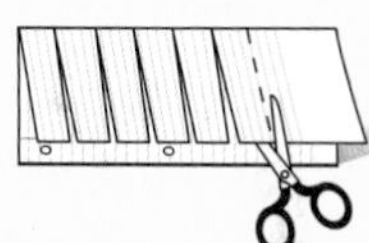

Read and Write Go through the chapter, find ten chemical formulas, and write them on the front of the tabs. As you read the chapter, write what compound each formula represents under the appropriate tab.

Preview this chapter's content and activities at gpescience.com

section 1

Stability in Bonding

Reading Guide

What You'll Learn

- **Describe** how a compound differs from its component elements.
- **Explain** what a chemical formula represents.
- **Explain** that the electric forces between electrons and protons, which are oppositely charged, are essential to forming compounds.
- **State** a reason why chemical bonding occurs.

Why It's Important

The millions of different kinds of matter around us are a result of chemical bonds.

Review Vocabulary

compound: substance formed from two or more elements in which the exact combination and proportion of elements is always the same

New Vocabulary

- chemical formula
- ion

Combined Elements

Have you noticed the color of the Statue of Liberty? It is green. The Statue of Liberty is made of copper, which is an element. Uncombined, elemental copper is a bright, shiny copper color. When copper is exposed to the weather over a long period of time, the metal combines with substances in the atmosphere and changes color.

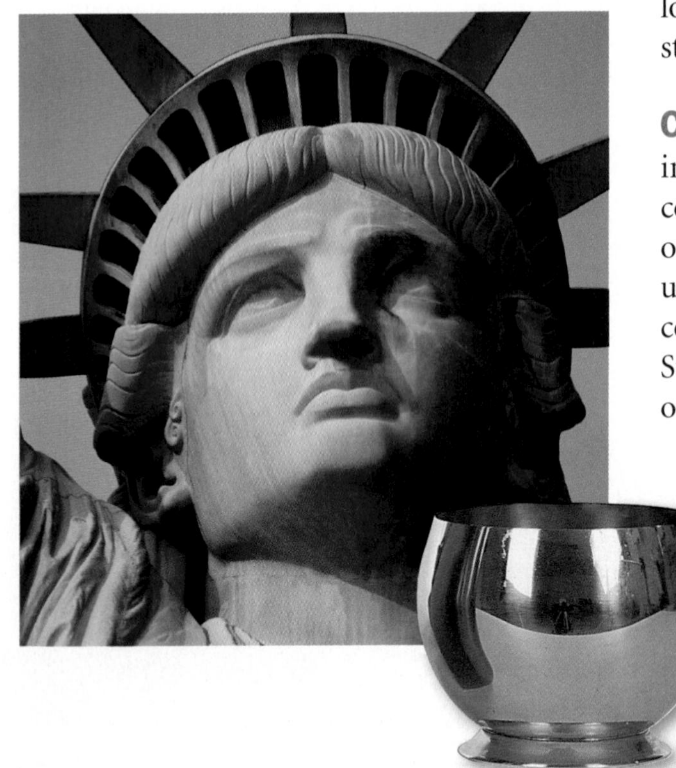

Figure 1 The difference between the elemental copper metal and the copper compound formed on the Statue of Liberty is striking.

Compounds Some of the matter around you is in the form of uncombined elements such as copper, sulfur, and oxygen. However, like many other sets of elements, these three elements can unite chemically to form a compound when the conditions are right. The green coating on the Statue of Liberty and some old pennies is a result of this chemical change. One compound in this coating, seen in contrast with the elemental copper in **Figure 1,** is a compound called copper sulfate. Copper sulfate isn't shiny and copper colored like elemental copper. Nor is it a pale-yellow solid like sulfur or a colorless, odorless gas like oxygen. It has its own unique properties.

New Properties An observation you will make is that the compound formed when elements combine often has chemical and physical properties that aren't anything like those of the individual elements. Sodium chloride, for example, shown in **Figure 2,** is a compound made from the elements sodium and chlorine. Sodium is a shiny, soft, silvery metal that reacts violently with water. Chlorine is a poisonous greenish yellow gas. Would you have guessed that these elements combine to make ordinary table salt?

Figure 2 Sodium is a soft, silvery metal that combines with chlorine, a greenish yellow gas, to form sodium chloride, which is a white crystalline solid.
Describe *how the properties of table salt are different from those of sodium and chlorine.*

Formulas

The chemical symbols Na and Cl represent the elements sodium and chlorine. When written as NaCl, the symbols make up a formula, or chemical shorthand, for the compound sodium chloride. A **chemical formula** tells what elements a compound contains and the exact number of the atoms of each element in a unit of that compound. The compound that you are probably most familiar with is H_2O, commonly known as water. This formula contains the symbols H for the element hydrogen and O for the element oxygen. Notice the number 2 written as a subscript after the H for hydrogen. *Subscript* means "written below." A subscript written after a symbol tells how many atoms of that element are in a unit of the compound. If a symbol has no subscript, the unit contains only one atom of that element. A unit of H_2O contains two hydrogen atoms and one oxygen atom.

Look at the formulas for each compound listed in **Table 1.** What elements combine to form each compound? How many atoms of each element are required to form each of the compounds?

Table 1 Some Familiar Compounds

Familiar Name	Chemical Name	Formula
Sand	Silicon dioxide	SiO_2
Milk of magnesia	Magnesium hydroxide	$Mg(OH)_2$
Cane sugar	Sucrose	$C_{12}H_{22}O_{11}$
Lime	Calcium oxide	CaO
Vinegar	Acetic acid	CH_3COOH
Laughing gas	Dinitrogen oxide	N_2O
Grain alcohol	Ethanol	C_2H_5OH
Battery acid	Sulfuric acid	H_2SO_4
Stomach acid	Hydrochloric acid	HCl

Reading Check *Describe what a chemical formula tells you.*

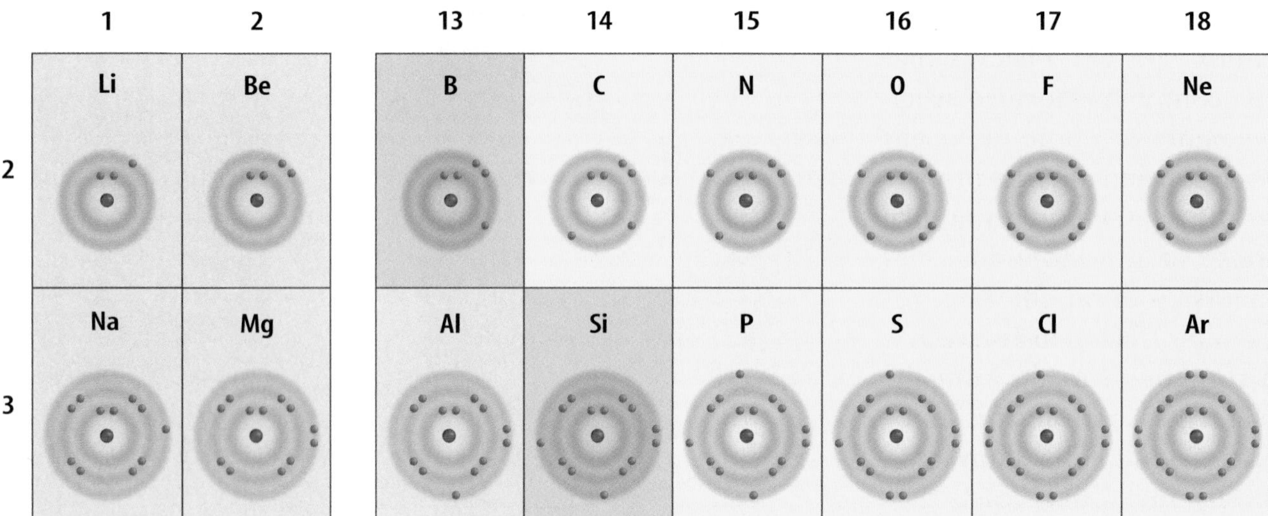

Figure 3 The number of electrons in each group's outer level increases across the table until the noble gases in Group 18, where each has a complete outer energy level.

Atomic Stability

Why do atoms form compounds? The electric forces between electrons and protons, which are oppositely charged, hold atoms and molecules together, and thus they are the forces that cause compounds to form. The periodic table on the inside back cover of your book lists the known elements. However, the six noble gases in Group 18 do not form compounds, or do so with difficulty. Atoms of noble gases are unusually stable. Compounds of these atoms rarely form because they are almost always less stable than the original atoms.

The Unique Noble Gases To understand the stability of the noble gases, it is helpful to look at electron dot diagrams across a period. Electron dot diagrams show only the electrons in the outer energy level of an atom. They contain the chemical symbol for the element surrounded by dots representing its outer electrons. How do you know how many dots to make? For Groups 1 and 2 and 13 through 18, you can use the periodic table or the portion of it shown in **Figure 3.** Elements in Group 1 each have one outer electron. Those in Group 2 each have two. Those in Group 13 each have three, those in Group 14, four, and so on to Group 18, the noble gases, which each have eight.

Figure 4 Electron dot diagrams of noble gases show that they each have a stable outer energy level.

He	Kr
Ne	Xe
Ar	Rn

Chemical Stability An atom is chemically stable when its outermost energy level has the maximum number of electrons. The outer energy levels of helium and hydrogen are stable with two electrons. The outer energy levels of all the other elements are stable when they contain eight electrons. **Figure 4** shows electron dot diagrams of some of the noble gases. Notice that eight dots surround Kr, Ne, Xe, Ar, and Rn, and two dots surround He.

Energy Levels and Other Elements How do the dot diagrams represent other elements, and how does this relate to their ability to make compounds? Hydrogen and helium, the elements in period 1 of the periodic table, can hold a maximum of two electrons in their outer energy levels. Hydrogen contains one electron in its lone energy level. A dot diagram for hydrogen has a single dot next to its symbol. This means that hydrogen's outer energy level is not full. It is more stable when it is part of a compound.

In contrast, helium's outer energy level contains two electrons. Its dot diagram has two dots—a pair of electrons—next to its symbol. Helium has a full outer energy level and is chemically stable. Helium rarely forms compounds and the element is a commonly used gas.

When you look at the elements in Groups 13 through 17, you see that none of the elements has a stable energy level. Each group contains too few electrons for a stable level of eight electrons.

Reading Check *Why is it rare for helium to form a compound?*

Topic: Dot Diagrams
Visit gpescience.com for Web links to information about using dot diagrams to represent outer energy level electrons.

Activity Draw a dot diagram of methane, CH_4.

Outer Levels—Getting Their Fill As you just learned, hydrogen is an element that does not have a full outer energy level. How does hydrogen, or any other element, become stable? Atoms with partially stable outer energy levels can lose, gain, or share electrons to obtain stable outer energy levels. They do this by combining with other atoms that also have partially complete outer energy levels. As a result, each becomes stable. **Figure 5** shows electron dot diagrams for sodium and chlorine. When they combine, sodium loses one electron and chlorine gains one electron. You can see from the electron dot diagram that chlorine now has a stable outer energy level, similar to a noble gas. But what about sodium?

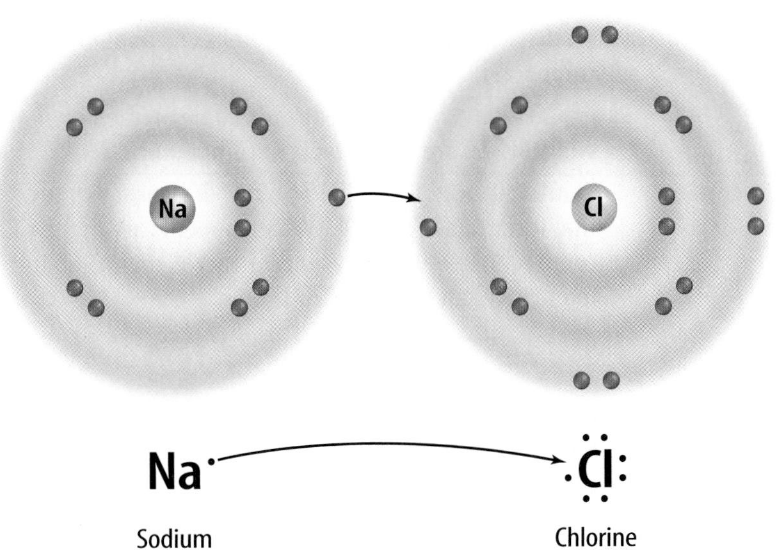

Figure 5 Sodium, as a Group 1 element, will lose one electron. It will have the same number of electrons as neon. Chlorine, which belongs to Group 17, will gain one electron. It will have the same number of electrons as argon. **Identify** *the group number where neon and argon can be found.*

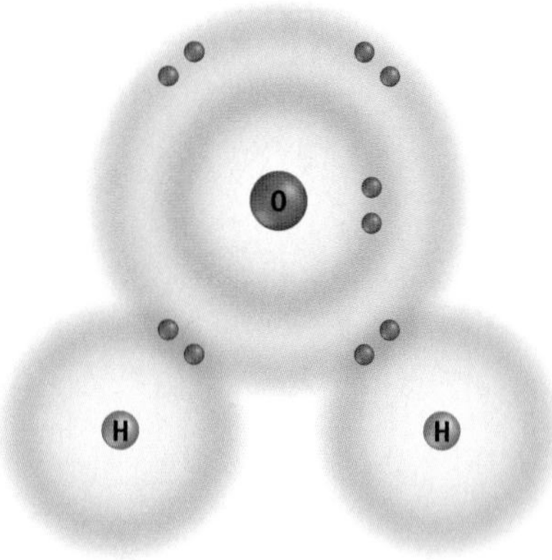

Figure 6 In water, hydrogen and oxygen each contribute one electron to each hydrogen-oxygen bond. The atoms share those electrons instead of giving them up.

Stability Is Reached Sodium had only one electron in its outer energy level, which it lost when it combined with chlorine to form sodium chloride. However, look back to the next, outermost energy level of sodium. This is now the new outer energy level, and it is stable with eight electrons. When the outer electron of sodium is removed, a complete inner energy level becomes the new outer energy level. Sodium and chlorine are stable now because of the exchange of an electron.

In the compound water, each hydrogen atom needs one electron to have a stable outer energy level. The oxygen atom needs two electrons for its outer level to be stable with eight electrons. Hydrogen and oxygen become stable and form bonds in a different way from sodium and chlorine. Instead of gaining or losing electrons, they share them. **Figure 6** shows how hydrogen and oxygen share electrons to achieve a more stable arrangement of electrons.

Atoms, too, lose or gain to meet a goal—a stable energy level. They do not lose or gain an advantage. Instead, they lose or gain electrons. An atom that has lost or gained electrons is called an ion. An **ion** is a charged particle because it now has either more or fewer electrons than protons. The positive and negative charges are not balanced. It is the electric forces between oppositely charged particles, such as ions, that hold compounds together.

section 1 review

Summary

Combined Elements

- When elements combine, the new compound has unique properties that are different from the original properties of the elements.

Formulas

- Chemical symbols and numbers are shorthand for the elements and their amounts in chemical formulas.

Atomic Stability

- The elements of Group 18, the noble gases, rarely combine with other elements.
- Electron dot diagrams show the electrons in the outer energy level of an atom.
- Most atoms need eight electrons to complete their outer energy levels.

Self Check

1. **Compare and contrast** the properties of the individual elements that combine to make salt with the compound salt.
2. **Identify** what the formula BaF_2 tells you about this compound.
3. **Explain** why some elements are stable on their own while others are more stable in compounds.
4. **Think Critically** The label on a box of cleanser states that it contains CH_3COOH. What elements are in this compound? How many atoms of each element can be found in a unit of CH_3COOH?

Applying Math

5. **Use Percentages** Given that the molecular weight of $Mg(OH)_2$, magnesium hydroxide, is 58.32 g, what percentage of this compound is oxygen?

Atomic Trading Cards

Perhaps you have seen or collected trading cards of famous athletes. Usually, each card has a picture of the athlete on one side with important statistics related to the sport on the back. Atoms also can be identified by their properties and statistics.

Real-World Problem

How can a visible model show how energy levels fill when atoms combine?

Goals

- **Display** the electrons of elements according to their energy levels.
- **Compare and classify** elements according to their outer energy levels.

Materials

4-in × 6-in index cards
periodic table

Procedure

1. You will get an assigned element from your teacher. Write the following information for your element on your index card: name, symbol, group number, atomic number, atomic mass, and metal, nonmetal, or metalloid.
2. On the other side of your index card, show the number of protons and neutrons in the nucleus (e.g., 6*p* for six protons and 6*n* for six neutrons for carbon).
3. Draw circles around the nucleus to represent the energy levels of your element. The number of circles you will need is the same as the period the element is in on the periodic table.
4. Draw dots on each circle to represent the electrons in each energy level. Remember, elements in row 1 become stable with two outer electrons, while levels two and three become stable with eight electrons.
5. Look at the picture side only of four or five of your classmates' cards. Identify the element and the group to which it belongs.

Conclude and Apply

1. As you classify the elements according to their group number, what pattern do you see in the number of electrons in the outer energy levels?
2. Atoms that give up electrons combine with atoms that gain electrons to form compounds. In your Science Journal, predict some pairs of elements that combine in this way.

Communicating Your Data

Make a graph that relates the groups to the number of electrons in their outer energy levels. **For more help, refer to the** Science Skill Handbook.

Types of Bonds

Reading Guide

What You'll Learn

- **Describe** ionic bonds and covalent bonds.
- **Identify** the particles produced by ionic bonding and by covalent bonding.
- **Distinguish** between a nonpolar covalent bond and a polar covalent bond.

Why It's Important

Bond type determines how compounds mix and interact with other compounds.

Review Vocabulary

atom: the smallest particle of an element that still retains the properties of the element

New Vocabulary

- chemical bond
- ionic bond
- covalent bond
- molecule
- polar molecule
- nonpolar molecule

Gain or Loss of Electrons

You and a friend decide to go to the movies. When you arrive at the theater, you discover that you do not have enough money to buy a ticket. Your friend has enough money for both tickets and loans you the money. Now you both have enough money to go to the movies.

Recall that atoms also can loan electrons to other atoms so that both can reach a stable energy level. When atoms gain, lose, or share electrons, an attraction forms that pulls the atoms together to form a compound. This attraction is called a chemical bond. A **chemical bond** is the force that holds atoms together in a compound. A compound has different physical and chemical properties from those of the atoms that make up the compound.

Some of the most common compounds are made by the loss and gain of just one electron. These compounds contain an element from Group 1 on the periodic table and an element from Group 17. Some examples are sodium chloride, commonly known as table salt, and potassium iodide, an ingredient in iodized salt.

Figure 7 A goiter, an enlargement of the thyroid gland in the neck, can be caused by iodine deficiency.

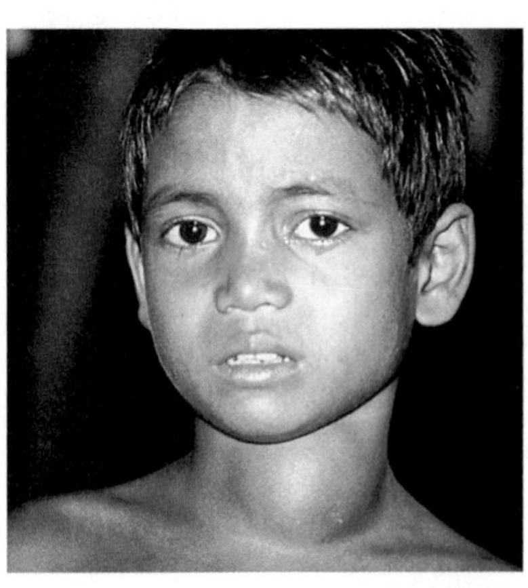

Why do people need iodized salt? A lack of iodine causes a wide range of problems in the human body. The most obvious is an enlarged thyroid gland, as shown in **Figure 7,** but the problems can include mental retardation, neurological disorders, and physical problems.

A Bond Forms What happens when potassium and iodine atoms collide? A neutral atom of potassium has one electron in its outer level. This is not a stable outer energy level. When potassium forms a compound with iodine, potassium loses one electron from its fourth level, and the third level becomes the complete outer level. However, the atom is no longer neutral. The potassium atom has become an ion. When a potassium atom loses an electron, the atom becomes positively charged because there is one electron less in the atom than there are protons in the nucleus. The 1+ charge is shown as a superscript written after the element's symbol, K^+, to indicate its charge. *Superscript* means "written above."

The iodine atom in this reaction undergoes change as well. An iodine atom has seven electrons in its outer energy level. Recall that a stable outer energy level contains eight electrons. During the reaction with potassium, the iodide atom gains an electron, leaving its outer energy level with eight electrons. This atom is no longer neutral because it gained an extra negative particle. It now has a charge of 1− and is called an iodide ion, written as I^-. The compound formed between potassium and iodine is called potassium iodide. The dot diagrams for the process are shown in **Figure 8.**

Reading Check *What part of an ion's symbol indicates its charge?*

Another way to look at the electron in the outer shell of a potassium atom is as an advertisement to other atoms saying, "Available: One electron to lend." The iodine atom would have the message, "Wanted: One electron to borrow." When the two atoms get together, each becomes a stable ion. Notice that the resulting compound has a neutral charge because the positive and negative charges of the ions cancel each other.

Ions and Nerve Cells Ions are important in many processes in your body. The movement of muscles is just one of these processes. Muscle movement would be impossible without the movement of ions in and out of nerve cells. Research the type of ions used by the nerve cells.

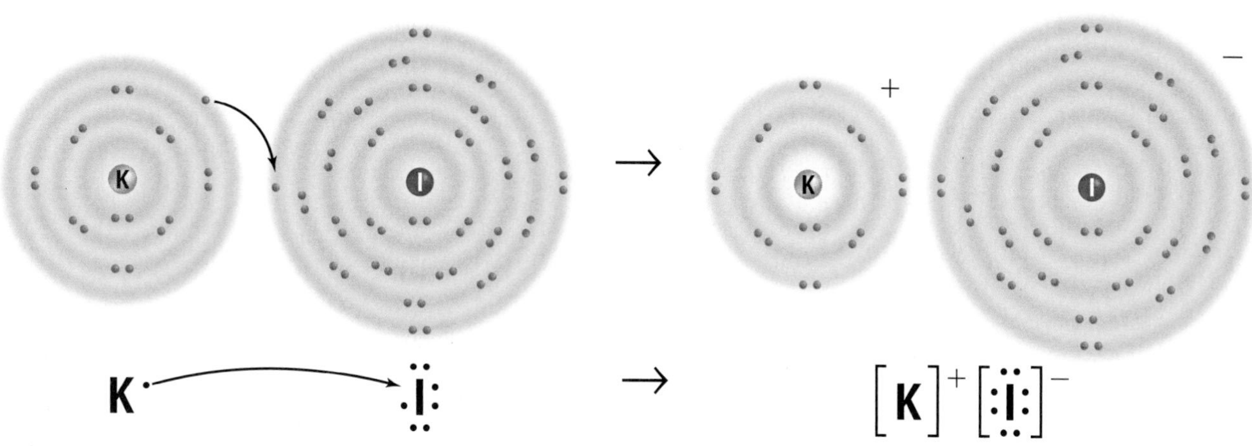

Figure 8 Potassium and iodine must perform a transfer of one electron. Potassium and iodine end up with stable outer energy levels. **Infer** *why the size of the ion changes with the change in electron distribution.*

Figure 9 This sparkler contains iron, which burns in air to produce an ionic compound that contains iron and oxygen.

The Ionic Bond

When ions attract in this way, a bond is formed. An **ionic bond** is the force of attraction between the opposite charges of the ions in an ionic compound. In an ionic bond, a transfer of one or more electrons takes place. With this transfer of electrons to form an ionic compound, a large amount of energy is released, as shown in **Figure 9**.

Now that you have seen how an ionic bond forms when one electron is involved, see how it works when more than one electron is involved. The formation of magnesium chloride, $MgCl_2$, is another example of ionic bonding. When magnesium reacts with chlorine, a magnesium atom loses two electrons and becomes a positively charged ion, Mg^{2+}. At the same time, two chlorine atoms gain one electron each and become negatively charged chloride ions, Cl^-. In this case, a magnesium atom has two electrons to lend, but a single chlorine atom needs to borrow only one electron. Therefore, it takes two chlorine atoms, as shown in **Figure 10,** to take the two electrons from the magnesium ion to form the compound magnesium chloride.

Zero Net Charge The result of this bond is a neutral compound. The compound as a whole is neutral because the sum of the charges on the ions is zero. The positive charge of the magnesium ion is equal to the negative charge of the two chloride ions. In other words, when different atoms form an ionic compound, electrons move from one atom to a different atom, but the overall number of protons and electrons of the combined atoms remains equal and unchanged. Therefore, the compound is neutral.

Metals and nonmetals usually combine by forming ionic bonds. Looking at the periodic table, you will see that the elements that bond ionically are often across the table from each other. Ionic compounds are often crystalline solids with high melting points.

Figure 10 A magnesium atom gives an electron to each of two chlorine atoms to form $MgCl_2$.

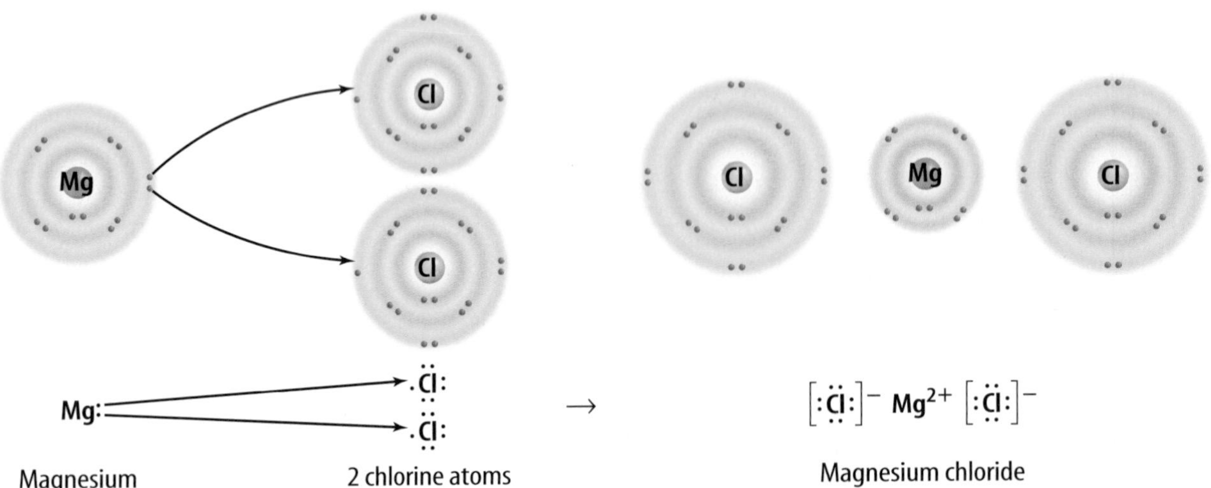

Sharing Electrons

Some atoms of nonmetals are unlikely to lose or gain electrons. For example, the elements in Group 14 of the periodic table have four electrons in their outer levels. They would have to either gain or lose four electrons to have a stable outer level. Losing four electrons takes a great deal of energy. Each time an electron is removed, the nucleus holds the remaining electrons even more tightly. These atoms become more chemically stable by sharing electrons, rather than by losing or gaining electrons.

The attraction that forms between atoms when they share electrons is known as a **covalent bond.** A neutral particle that forms as a result of electron sharing is called a **molecule,** as shown in **Figure 11.**

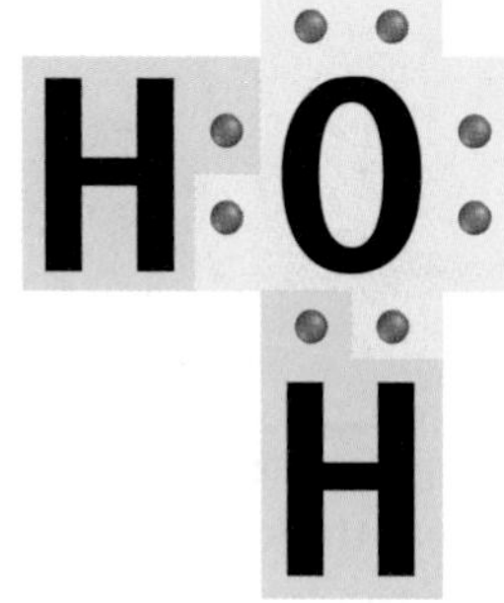

Figure 11 Each of the pairs of electrons between the two hydrogens and the oxygen is shared as each atom contributes one electron to the pair to make the bond.

Single Covalent Bonds A single covalent bond is made up of two shared electrons. Usually, one of the shared electrons comes from one atom in the bond and the other comes from the other atom in the bond. A water molecule contains two single bonds. In each bond, a hydrogen atom contributes one electron to the bond, and the oxygen atom contributes the other. The two electrons are shared, forming a single bond. The result of this type of bonding is a stable outer energy level for each atom in the molecule.

Multiple Bonds A covalent bond also can contain more than one pair of shared electrons. An example of this is the bond in nitrogen (N_2), shown in **Figure 12.** A nitrogen atom has five electrons in its outer energy level and needs three more electrons to become stable. It does this by sharing three of its electrons with another nitrogen atom. The other nitrogen atom also shares three of its electrons. When each atom contributes three electrons, they share six electrons, or three pairs of electrons. Each pair of electrons represents a bond. Therefore, three pairs of electrons represent three bonds, or a triple bond. Each nitrogen atom is stable with eight electrons in its outer energy level. In a similar way, a bond that contains two shared pairs of electrons is a double bond. Carbon dioxide is an example of a molecule with double bonds.

Covalent bonds form between nonmetallic elements. These elements are close together in the upper right-hand corner of the periodic table. Many covalent compounds are liquids or gases at room temperature.

Figure 12 The dot diagram shows that the two nitrogen atoms in nitrogen gas share six electrons. **Explain** *which of these gases would require the most energy to react with another element to form a compound, H_2 or N_2.*

Observing a Bond Type

Procedure

1. Complete the safety form.
2. Turn on the faucet to produce a thin **stream of water.**
3. Rub an inflated **balloon** with **wool or fur.**
4. Bring the balloon near the stream of water, and describe what you see.

Analysis

1. Explain your observations.
2. Relate the attraction between the balloon and the water to the attraction between the north and south poles of two magnets. Why might water act like a magnet?

Try at Home

Unequal Sharing Electrons are not always shared equally between atoms in a covalent bond. The strength of the attraction of each atom to its electrons is related to the size of the atom, the charge of the nucleus, and the total number of electrons the atom contains. Part of the strength of attraction has to do with how far the electron being shared is from the nucleus. For example, a magnet has a stronger pull when it is right next to a piece of metal rather than several centimeters away. The other part of the strength of attraction has to do with the size of the positive charge in the nucleus. Using a magnet as an example again, a strong magnet will hold the metal more firmly than a weak magnet will.

One example of this unequal sharing is found in a molecule of hydrogen chloride, HCl, which is shown in **Figure 13.** In water, HCl is hydrochloric acid, which is used in laboratories and in industry to clean metal, and is found in your stomach, where it helps to digest food. Chlorine atoms have a stronger attraction for the shared electrons than hydrogen atoms do. As a result, the shared electrons in hydrogen chloride will spend more time near the chlorine atom than near the hydrogen atom. The chlorine atom has a partial negative charge, represented by a lowercase Greek symbol *delta* followed by a negative superscript, δ^-. The hydrogen atom has a partial positive charge, represented by a δ^+.

Reading Check *What determines the strength of attraction of an atom?*

Tug-of-War You might think of a covalent bond as the rope in a tug-of-war, and the shared electrons as the knot in the center of the rope. **Figure 14** illustrates this concept. Each atom in the molecule attracts the electrons that they share. However, sometimes the atoms aren't the same size.

The same thing happens in tug-of-war. Sometimes one team has more people or stronger participants than the other. When this is true, the knot in the middle of the rope ends up closer to the stronger team. Similarly, the electrons being shared in a molecule are held more closely to the atoms with the stronger pull or larger nucleus.

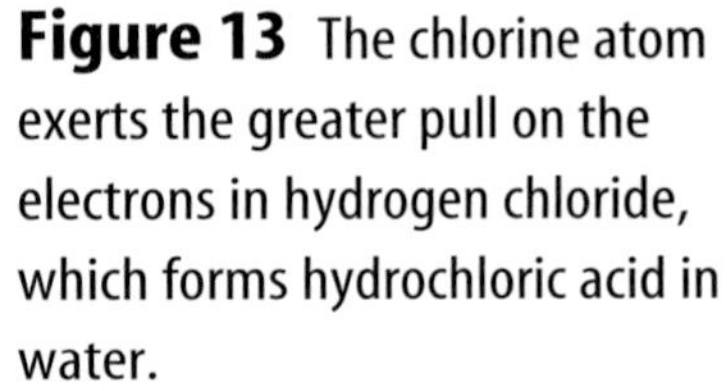

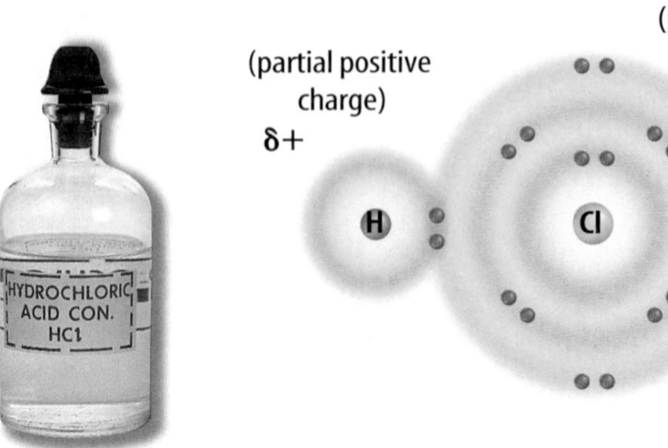

Figure 13 The chlorine atom exerts the greater pull on the electrons in hydrogen chloride, which forms hydrochloric acid in water.
Explain *why the chlorine atom has a greater pull than the hydrogen atom.*

NATIONAL GEOGRAPHIC **VISUALIZING POLAR MOLECULES**

Figure 14

When playing tug-of-war, if there are more—or stronger—team members on one end of the rope than the other, there is an unequal balance of power. The stronger team can pull harder on the rope and has the advantage. A similar situation exists in polar molecules, in which electrons are attracted more strongly by one type of atom in the molecule than another. Because of this unequal sharing of electrons, polar molecules have a slightly negative end and a slightly positive end, as shown below.

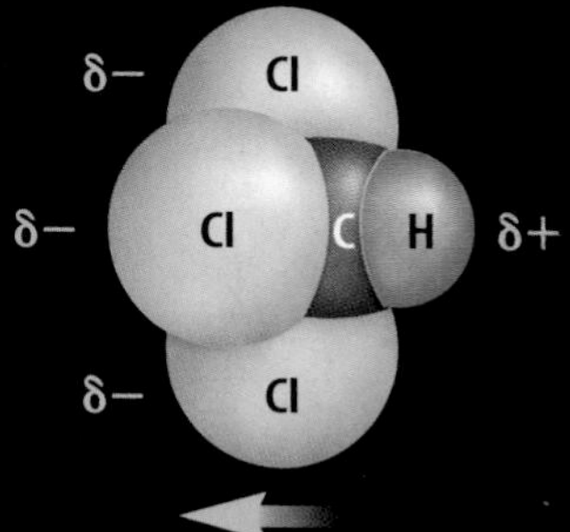

CHLOROFORM In a molecule of chloroform ($CHCl_3$), or trichloromethane (tri klor oh ME thayn), the three chlorine atoms attract electrons more strongly than the hydrogen atom does, creating a partial negative charge on the chlorine end of the molecule and a partial positive charge on the hydrogen end. This polar molecule is a clear, sweet-smelling liquid once widely used as an anesthetic in human and veterinary surgery.

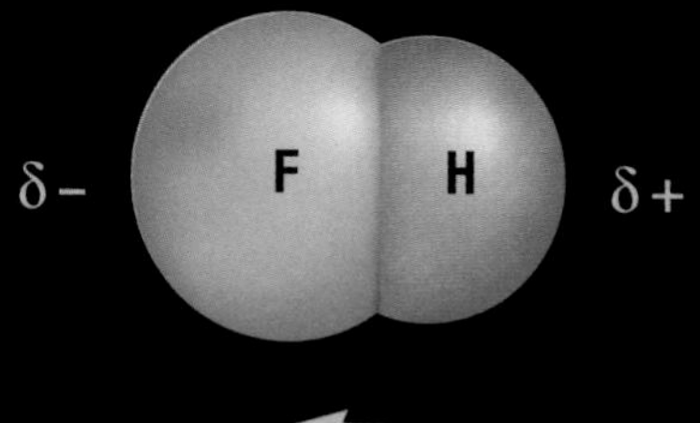

HYDROGEN FLUORIDE Hydrogen and fluorine react to form hydrogen fluoride (HF). In an HF molecule, the two atoms are bound together by a pair of electrons, one contributed by each atom. However, the electrons are not shared equally because the fluorine atom attracts them more strongly than the hydrogen atom does. The result is a polar molecule with a slightly positive charge near the hydrogen end and a slightly negative charge near the fluorine end.

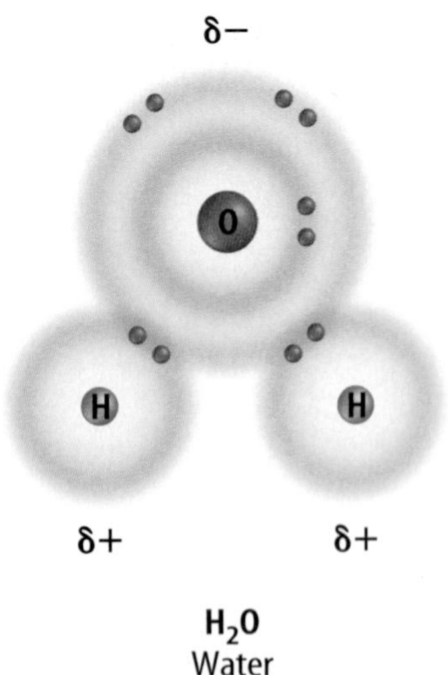

Figure 15 The polarity of water is responsible for many of its unique properties.
Explain *the cause of polarity.*

Polar or Nonpolar? For the molecule involved in this electron tug-of-war, there is another consequence. Again, look at the molecule of hydrogen chloride. The unequal sharing of electrons gives each chlorine atom a slight negative charge and each hydrogen atom a slight positive charge. The atom holding the electron more closely always will have a slightly negative charge. The charge is balanced but not equally distributed. This type of molecule is called polar. The term *polar* means "having opposite ends." A **polar molecule** is one that has a slightly positive end and a slightly negative end, although the overall molecule is neutral. Water is an example of a polar molecule, as shown in **Figure 15.**

Reading Check *What is a polar molecule?*

Two atoms that are exactly alike can share their electrons equally, forming a nonpolar molecule. A **nonpolar molecule** is one in which electrons are shared equally in bonds. Such a molecule does not have oppositely charged ends. This is true of molecules made from two identical atoms or molecules that are symmetric, such as CCl_4.

Properties of Compounds

Recall that atoms can form two types of bonds: covalent and ionic. A compound whose atoms are held together by covalent bonds is a covalent compound.

Sugar is a covalent compound. A compound that is composed of ions is an ionic compound. Table salt, NaCl, is an ionic compound. Looking at both substances, you would think they would have similar physical and chemical properties.

Both sugar and salt, as shown in **Figure 16,** are used to change the taste of foods. You add sugar to lemonade or iced tea to make the drink sweet. Table salt is used to enhance the flavors of foods, such as meats and vegetables. Sugar and salt have different physical and chemical properties.

Figure 16 Table sugar is an example of a covalent compound that is a crystalline solid soluble in water. Ionic compounds, such as table salt, are also soluble in water.

Figure 17 Candle wax and propane are covalent compounds that are insoluble in water.

Covalent and Ionic Properties The chemical and physical properties of covalent compounds and ionic compounds are different. This difference in properties, as shown in **Table 2,** is due to differences in attractive forces of the bonds.

Covalent Compounds The covalent bonds between atoms in molecules are strong. However, the attraction between individual molecules is weak. It is the weak forces between individual molecules that are responsible for the properties of covalent compounds.

Melting and boiling points of covalent compounds are relatively lower when compared to ionic compounds. Sugar will melt at approximately 185°C, whereas table salt will melt at 801°C. Covalent compounds, which will form soft solids, have poor electrical and thermal conductivity. Candles and propane gas shown in **Figure 17** are covalent compounds.

Table 2 Comparison of Covalent and Ionic Compounds

	Covalent Compounds	Ionic Compounds
Bond Type	Electron Sharing	Electron Transfer
Melting and Boiling Points	Lower	Higher
Electrical Conductivity	Poor	Good
State at Room Temperature	Solid, liquid, or gas	Solid
Forces Between Particles	Strong bonds between atoms; weak attraction between molecules	Strong attraction between positive and negative ions

Figure 18 Some common ionic compounds used in everyday life are potassium chloride as a salt substitute; potassium iodide added to table salt; and sodium fluoride added to toothpaste.

Ionic Compounds The ionic bonds between ions are relatively strong. This accounts for the high melting and boiling points of ionic compounds such as table salt. Examples of common ionic compounds are shown in **Figure 18.** Ionic compounds are hard and brittle. When in a liquid or aqueous state, ionic compounds can conduct electric current.

Strong attractive forces hold the ions in place. Ionic compounds are stable due to the attraction between unlike charges. The ions are drawn together and energy is released. Formation of ionic compounds is always exothermic.

Reading Check *Why are ionic compounds brittle?*

section 2 review

Summary

Gain or Loss of Electrons

- An ion is a charged particle that has either fewer or more electrons than protons, resulting in a negative or positive charge.

Ionic Bond

- An ionic bond is the force or attraction between opposite charges of ions in an ionic bond.
- An ionic compound is neutral because the sum of the ion charges is zero.

Sharing Electrons

- Some atoms, such as those in Group 4, share electrons instead of losing or gaining them.
- Covalent bonds can form single, double, or triple bonds.
- In a polar molecule, the electrons are shared unequally in the bond. This results in slightly charged ends.
- Electrons are shared equally in a nonpolar molecule.

Self Check

1. **Explain** why an atom makes an ionic bond only with certain other atoms.
2. **Compare and contrast** the possession of electrons in ionic and covalent bonds.
3. **Name** the types of particles formed by covalent bonds.
4. **Concept Map** Using the following terms, make a network-tree concept map of chemical bonding: *ionic, covalent, ions, positive ions, negative ions, molecules, polar,* and *nonpolar.* Define the terms *atom* and *molecule.*
5. **Think Critically** Choose two elements that are likely to form an ionic bond: O, Ne, S, Ca, K. Next, select two elements that would likely form a covalent bond. Explain.

Applying Math

6. **Solve One-Step Equations** Aluminum oxide, Al_2O_3, can be produced during space shuttle launches. Show that the sum of the positive and negative charges in a unit of Al_2O_3 equals zero.

More Section Review gpescience.com

section 3

Writing Formulas and Naming Compounds

Reading Guide

What You'll Learn

- **Explain** how to determine oxidation numbers.
- **Write** formulas and names for ionic compounds.
- **Write** formulas and names for covalent compounds.

Why It's Important

The name and the formula convey information about the compound.

Review Vocabulary

anion: a negatively charged ion

New Vocabulary

- binary compound
- oxidation number
- polyatomic ion
- hydrate

Binary Ionic Compounds

Does the table in **Figure 19** look like it has anything to do with chemistry? It is an early table of the elements made by alchemists, scientists who tried to make gold from other elements. The alchemists used symbols to write the formulas of substances. The first formulas of compounds that you will write are for binary ionic compounds. A **binary compound** is one that is composed of two elements. Potassium iodide, the salt additive mentioned in Section 2, is a binary ionic compound. Before you can write a formula, you must have all the needed information at your fingertips. What will you need to know?

Oxidation Numbers You need to know which elements are involved and what number of electrons they lose, gain, or share to become stable. Section 1 discussed the relationship between an element's position on the periodic table and the number of electrons it gains or loses. Because all elements in a given group have the same number of electrons in their outer energy levels, they must gain or lose the same number of electrons. Metals always lose electrons and nonmetals always gain electrons when they form ions. The charge on the ion is known as the **oxidation number** of the atom.

For ionic compounds, the oxidation number is the same as the charge on the ion. For example, a sodium ion has a charge of 1+ and an oxidation number of 1+. A chloride ion has a charge of 1− and an oxidation number of 1−.

Figure 19 This old chart of the elements used pictorial symbols to represent elements.

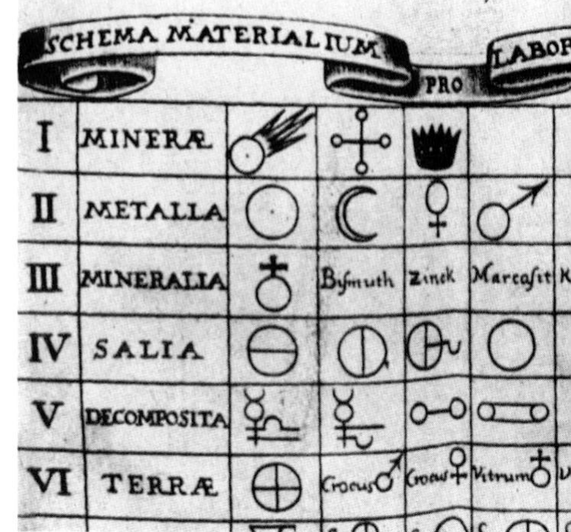

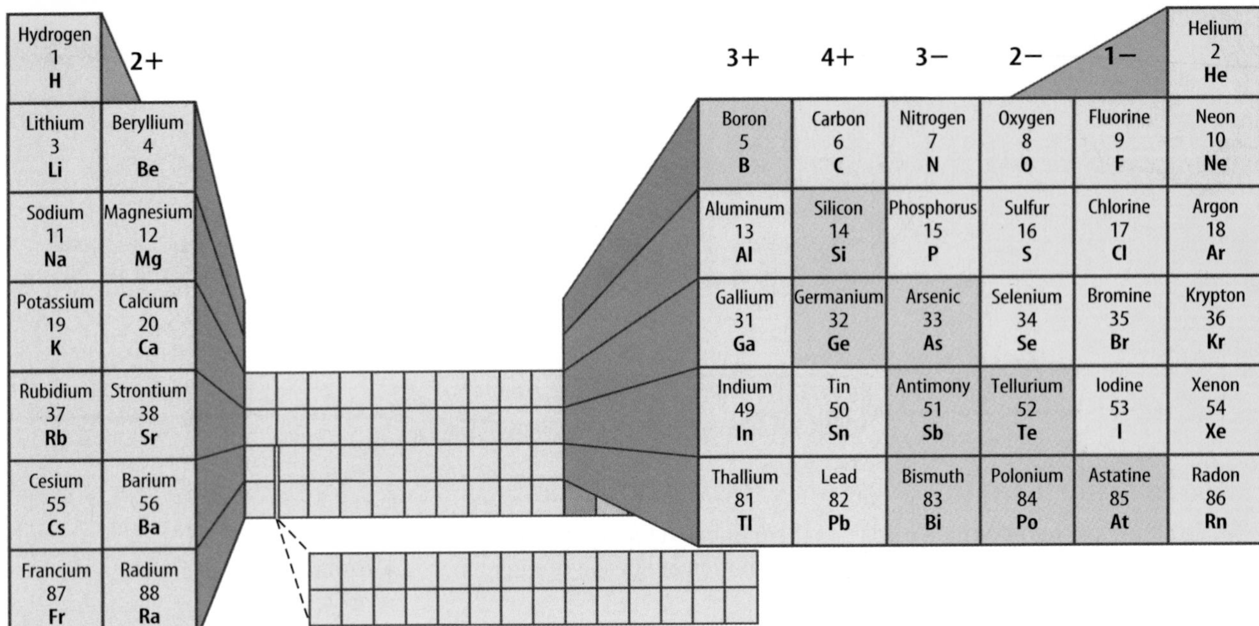

Figure 20 The number at the top of each column is the most common oxidation number of elements in that group.
Define *oxidation number.*

Oxidation Numbers and the Periodic Table The numbers with positive or negative signs in **Figure 20** are the oxidation numbers for the elements in the columns below them. Notice how they fit with the periodic-table groupings.

The elements in **Table 3** can have more than one oxidation number. When naming these compounds, the oxidation number is expressed in the name with a roman numeral. For example, the oxidation number of iron in iron(III) oxide is 3+.

Reading Check *What is the oxidation number for barium, Ba?*

Table 3 Special Ions

Name	Oxidation Number
Copper(I)	1+
Copper(II)	2+
Iron(II)	2+
Iron(III)	3+
Chromium(II)	2+
Chromium(III)	3+
Lead(II)	2+
Lead(IV)	4+

Ionic Compounds Are Neutral When writing formulas, it is important to remember that although the individual ions in a compound carry charges, the compound itself is neutral. A formula must have the same number of positive ions and negative ions so the charges balance. For example, sodium chloride is made up of a sodium ion with a 1+ charge and a chloride ion with a 1− charge.

However, what if you have a compound such as calcium fluoride? A calcium ion has a charge of 2+ and a fluoride ion has a charge of 1−. In this case, you need to have two fluoride ions for every calcium ion for the charges to cancel and the compound to be neutral with the formula CaF_2.

Some compounds require more thought. Aluminum oxide contains an ion with a 3+ charge and an ion with a 2− charge. You must find the least common multiple of 3 and 2 to determine how many of each ion you need. You need two aluminum ions and three oxygen ions to have a 6+ charge and, a 6− charge and, therefore, the neutral compound Al_2O_3.

Writing Formulas After you've learned how to find the oxidation numbers and their least common multiple, you can write formulas for ionic compounds. Write the formula for an ionic compound containing sodium and oxygen by using the following rules in this order.

1. Write the symbol of the element that has the positive oxidation number or charge. Sodium, a Group 1 element, has an oxidation number of 1+.
2. Write the symbol of the element with the negative number. Nonmetals other than hydrogen have negative oxidation numbers. Oxygen has an oxidation number of 2−.
3. To have a neutral compound, the positive charges have to balance the negative charges. It takes two sodium ions to balance the one oxygen ion. Thus, the formula becomes Na_2O.

Jöns Jakob Berzelius As a Swedish chemist, Berzelius is credited with developing the concepts of ions and ionic compounds. Because of his many interests and discoveries in chemistry, Berzelius, like Antoine Lavoisier, is known as one of the founders of modern chemistry.

Applying Math — Writing Chemical Formulas

DETERMINING A FORMULA What is the formula for lithium nitride?

IDENTIFY known values

Identify the known values:

Symbol and oxidation number of the positive element:

Lithium $\xrightarrow{\text{means}}$ Li^{1+}

Symbol and oxidation number of the negative element:

Nitrogen $\xrightarrow{\text{means}}$ N^{3-}

SOLVE the problem

To have a neutral compound, it takes three Li^{1+} ions to balance the one N^{3-} ion.

$$3Li^{1+} = 1N^{3-} \xrightarrow{\text{means}} Li_3N_1 \text{ or } Li_3N$$

Reduce the subscripts to the smallest whole numbers that retain the ratios of ions.

CHECK the answer

Does your answer seem reasonable? Check your answer by determining whether your compound is neutral.

Practice Problems

1. What is the formula for lead (IV) phosphide?
2. What is the formula for iron (III) oxide?

For more practice problems go to page 879, and visit Math Practice at gpescience.com.

Table 4 Elements in Binary Compounds	
Element	*-ide* Name
Oxygen	oxide
Phosphorus	phosphide
Nitrogen	nitride
Sulfur	sulfide

Writing Names You can name a binary ionic compound from its formula by using these rules.

1. Write the name of the positive ion.
2. Using **Table 3,** check to see if the positive ion is capable of forming more than one oxidation number. If the ion has only one possible oxidation number, proceed to step 3. If it has more than one, determine the oxidation number of the ion from the formula of the compound. To do this, keep in mind that the overall charge of the compound is zero and the negative ion has only one possible charge. Write the charge of the positive ion using roman numerals in parentheses after the ion's name.
3. Write the root name of the negative ion. The root is the first part of the element's name. For chlorine, the root is *chlor-*.
4. Add the ending *-ide* to the root. **Table 4** lists several elements and their *-ide* counterparts. For example, sulfur in a binary compound becomes sulfide.

Subscripts do not become part of the name for ionic compounds. However, subscripts can be used to help determine the charges of these metals that have more than one positive charge.

Applying Science

Can you name binary ionic compounds?

What would a chemist name the compound CuCl?

Identifying the Problem

There are four simple steps in naming binary ionic compounds.

1. Write the name of the positive ion in the compound. In CuCl, the name of the positive ion is copper.
2. Check **Table 3** to determine whether copper is one of the elements that can have more than one oxidation number. Looking at **Table 3,** you can see that copper can have a 1+ or a 2+ oxidation number. You need to determine which to use. Looking at the compound, you see that there is one copper atom and one chlorine atom. You know that the overall charge of the compound is zero and that chlorine only forms a 1− ion. For the charge of the compound to be zero, the charge of the copper ion must be 1+. Write this charge using roman numerals in parentheses after the element's name, copper(I).
3. Write the root name of the negative ion. The negative ion is chlorine and its root is *chlor-*.
4. Add the ending *-ide* to the root, *chloride.*
5. The full name of the compound CuCl is copper(I) chloride.

Solving the Problem

1. What is the name of CuO?
2. What is the name of $AlCl_3$?

Compounds with Polyatomic Ions

Not all ionic compounds are binary. Baking soda—used in cooking, as a medicine, and for brushing your teeth—has the formula $NaHCO_3$. This is an example of an ionic compound that is not binary. Some ionic compounds are composed of more than two elements. They contain polyatomic ions. The prefix *poly-* means "many," so the term *polyatomic* means "having many atoms." A **polyatomic ion** is a positively or negatively charged, covalently bonded group of atoms. Thus, the polyatomic ions as a whole contain two or more elements. The polyatomic ion in baking soda is the bicarbonate or hydrogen carbonate ion, HCO_3^-.

Table 5 Polyatomic Ions

Charge	Name	Formula
1+	ammonium	NH_4^+
1–	acetate chlorate hydroxide nitrate	$C_2H_3O_2^-$ ClO_3^- OH^- NO_3^-
2–	carbonate sulfate	CO_3^{2-} SO_4^{2-}
3–	phosphate	PO_4^{3-}

Writing Names Several polyatomic ions are listed in **Table 5.** To name a compound that contains one of these ions, first write the name of the positive ion. Use **Table 5** to find the name of a polyatomic ion. Then write the name of the negative ion. For example, K_2SO_4 is potassium sulfate. What is the name of $Sr(OH)_2$? Begin by writing the name of the positive ion, strontium. Then find the name of the polyatomic ion, OH^-, which is hydroxide. Thus, the name is strontium hydroxide.

Writing Formulas To write formulas for these compounds, follow the rules for binary compounds, with one addition. When more than one polyatomic ion is needed, write parentheses around the polyatomic ion before adding the subscript. How would you write the formula of barium chlorate?

First, identify the symbol of the positive ion. Barium has the symbol Ba and forms a 2+ ion, Ba^{2+}. Next, identify the negative chlorate ion. **Table 5** shows that it is ClO_3^-. Finally, you need to balance the charges of the ions to make the compound neutral. It will take two chlorate ions with a 1– charge to balance the 2+ charge of the barium ion. Because the chlorate ion is polyatomic, you use parentheses before adding the subscript. The formula is $Ba(ClO_3)_2$. Another example of naming complex compounds is shown in **Figure 21.**

Figure 21 Naming Complex Compounds

How would a scientist write the chemical formula for ammonium phospate?
To write the formula, answer the following questions:

1. What is the positive ion and its charge?
   The positive ion is NH_4^{1+} and its charge is 1^+.
2. What is the negative ion and its charge?
   The negative ion is PO_4^{3-} and its charge is 3–.
3. Balance the charges to make the compound neutral.
   a) three NH_4^{1+} ions (+3) balance one PO_4^{3-} (3–)

   b) The charge of one ion (without the sign) becomes the subscript of the other. Add parentheses for subscripts greater than 1.
   NH_4^{1+} PO_4^{3-} gives $(NH_4)_3PO_4$

The chemical formula for ammonium phosphate is $(NH_4)_3PO_4$.

Making a Hydrate

Procedure

1. Complete the safety form.
2. Mix 150 g of **plaster of paris** with 75 mL of **water** in a small **bowl.**
3. Let the plaster dry overnight and then take the hardened plaster out of the bowl.
4. Lightly tap the plaster with a **rubber hammer.**
5. Heat the plaster with a **hair dryer** on the hottest setting and observe.
6. Place a towel over the sample, then lightly tap the plaster with the hammer after heating it.

Analysis

1. What happened to the plaster when you tapped it before and after heating it?
2. What did you observe happening to the plaster as you heated it? Explain.

Compounds with Added Water

Some ionic compounds have water molecules as part of their structure. These compounds are called hydrates. A **hydrate** is a compound that has water chemically attached to its ions and written into its chemical formula.

Common Hydrates The term *hydrate* comes from a word that means "water." When a solution of cobalt chloride evaporates, pink crystals that contain six water molecules for each unit of cobalt chloride are formed. The formula for this compound is $CoCl_2 \cdot 6H_2O$ and is called cobalt chloride hexahydrate.

You can remove water from these crystals by heating them. The resulting blue compound is called anhydrous, which means "without water." When anhydrous (blue) $CoCl_2$ is exposed to water, even from the air, it will revert back to its hydrated state.

The plaster of paris shown in **Figure 22** also forms a hydrate when water is added. It becomes calcium sulfate dihydrate, which is also known as gypsum. The water that was added to the powder became a part of the compound.

To write the formula for a hydrate, write the formula for the compound and then place a dot followed by the number of water molecules. The dot in the formula represents a ratio of a compound to water molecules. For example, calcium sulfate dihydrate, $CaSO_4 \cdot 2H_2O$, is the formula for the hydrate of calcium sulfate that contains two molecules of water.

Naming Binary Covalent Compounds

Covalent compounds are those formed between elements that are nonmetals. Some pairs of nonmetals can form more than one compound with each other. For example, nitrogen and oxygen can form N_2O, NO, NO_2 and N_2O_5. In the system you have learned so far, each of these compounds would be called nitrogen oxide. You would not know from that name what the composition of the compound is.

Figure 22 The presence of water changes this powder into a material that can be used to create art.
Identify *the formula for this powder prior to the addition of water.*

Using Prefixes Scientists use the Greek prefixes in **Table 6** to indicate how many atoms of each element are in a binary covalent compound. The nitrogen and oxygen compounds N_2O, NO, NO_2, and N_2O_5 would be named dinitrogen oxide, nitrogen oxide, nitrogen dioxide, and dinitrogen pentoxide. Notice that the last vowel of the prefix is dropped when the second element begins with a vowel, as in pentoxide. Often, the prefix *mono-* is omitted, although it is used for emphasis in some cases. Carbon monoxide is one example.

Reading Check ***What prefix would be used for seven atoms of one element in a covalent compound?***

These same prefixes are used when naming the hydrates previously discussed. The main ionic compound is named the regular way, but the number of water molecules in the hydrate is indicated by the Greek prefix.

You have learned how to write formulas of binary ionic compounds and of compounds containing polyatomic ions. Using oxidation numbers to write formulas, you can predict the ratio in which atoms of elements might combine to form compounds. You also have seen how hydrates have water molecules as part of their structures and formulas. Finally, you have learned how to use prefixes in naming binary covalent compounds. As you continue to study, you will see many uses of formulas.

Table 6 Prefixes for Covalent Compounds

Number of Atoms	Prefix
1	*mono-*
2	*di-*
3	*tri-*
4	*tetra-*
5	*penta-*
6	*hexa-*
7	*hepta-*
8	*octa-*

section 3 review

Summary

Binary Ionic Compounds

- A binary compound is one composed of two elements.
- The oxidation number tells how many electrons an atom has gained, lost, or shared to become stable.
- The net charge of a compound is zero.

Compounds with Complex Ions

- A polyatomic ion is a positively or negatively charged, covalently bonded group of atoms.
- A hydrate is a compound that has water chemically attached to its ions.
- Greek prefixes are used to indicate how many atoms of each element are in a binary covalent compound.

Self Check

1. **Use Formulas** Write formulas for the following compounds: potassium iodide, magnesium hydroxide, aluminum sulfate, and chlorine heptoxide.
2. **Use Formulas** Write the names of these compounds: KCl, Cr_2O_3, $Ba(ClO_3)_2$, NH_4Cl, and PCl_3.
3. **Name** $Mg_3(PO_4)_2 \cdot 4H_2O$ and write the formula for calcium nitrate trihydrate.
4. **Think Critically** Explain why sodium and potassium will or will not react to form a bond with each other.

Applying Math

5. **Solve One-Step Equations** The overall charge on the polyatomic sulfate ion, found in some acids, is 2^-. Its formula is SO_4^{2-}. If the oxygen ion has a 2^- oxidation number, determine the oxidation number of sulfur in this polyatomic ion.

Model and Invent

Modeling Chemical Bonding

Goals

- **Infer** chemical formulas by making models of outer electron levels.
- **Compare and contrast** models of ionic and covalent bonding.
- **Draw and label** diagrams to illustrate chemical bonding.

Possible Materials

modified egg carton
beans
**pennies*
**buttons*
**Alternate materials*

Safety Precautions

Real-World Problem

Chemical bonding is one of the most important concepts in chemistry. Bonding is what makes different compounds. Even though atoms have a nucleus containing protons and neutrons, the part of the atom that is most important to chemists is outside the nucleus. How are electrons involved in chemical bonding?

Make a Model

1. Complete the safety form before you begin.
2. Obtain a modified egg carton and beans from your teacher. The egg carton represents the first and second energy levels of an atom. The beans represent electrons.
3. **Decide** which elements you can model by the number of beans you have. Consider how the periodic table can be used to determine the number of outer electrons in an element.
4. **Model** all the elements that you can create using the modified egg carton and beans.
5. **Draw** the models of elements that you created in your Science Journal.
6. **Determine** if the elements form ionic or covalent bonding. Record your conclusions in your Science Journal.

Using Scientific Methods

Test Your Model

1. Look for element combinations that could represent chemical formulas for compounds and molecules. Some formulas may require more than one atom to form some types of elements and compounds.
2. Work with other students to make models of compounds and molecules produced by chemical bonding.
3. **Draw** the models of the compounds and molecules you made in your Science Journal. Record the type of bonding used to make the compounds and molecules.

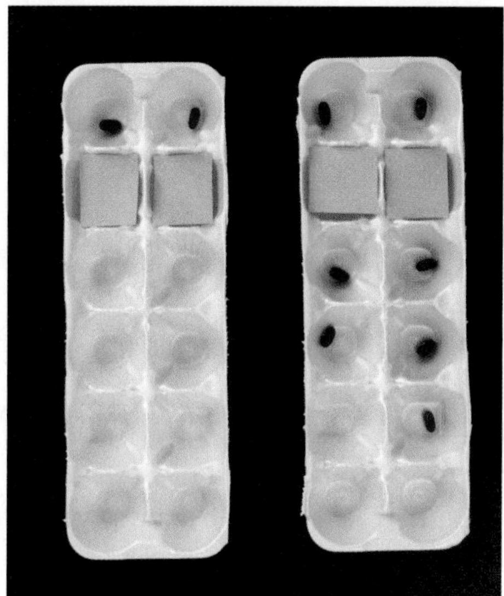

Analyze Your Data

1. **Explain** how you can use the periodic table to identify the element that represents your model.
2. **Explain** whether elements in the metal groups on the periodic table have more or fewer electrons in their outer energy levels than the nonmetals.
3. The combinations of elements could represent chemical formulas. Explain why some formulas require more than one atom of an element.

Conclude and Apply

1. **Describe** why your model could be used to show examples of both ionic and covalent bonding.
2. **Predict** what element your model would be if you received ten beans. Explain your reasoning.

Communicating Your Data

Compare your conclusions with other students' conclusions. What elements did their models represent? Explain how their elements differed in ability to form ionic and covalent bonds.

Oops! Accidents in SCIENCE

SOMETIMES GREAT DISCOVERIES HAPPEN BY ACCIDENT!

A Sticky Subject

In 1942, a research team was working on creating a new kind of glass. The group was working with some cyanoacrylate monomers (si uh noh A kruh layt • MAH nuh muhrz) that showed promise, but a problem kept coming up. Everything the monomers touched stuck to everything else!

Cyanoacrylate is the chemical name for instant, supertype glues. The researchers were so focused on finding a different type of glass that at the time, none of them recognized that they were working with an important new adhesive.

In 1952, a member of the research team, working on new materials for jet plane canopies, made a similar complaint. The ethyl cyanoacrylate the team was working with again made everything stick together. This time, the insight stuck to the scientists like, well, like GLUE! "I began gluing everything I could lay my hands on—glass plates, rubber stoppers, metal spatulas, wood, paper, plastic. Everything stuck to everything, almost instantly, and with bonds I could not break apart," recalls the head of the research group.

Stick to It

Most adhesives, commonly called glues, are long chains of bonded molecules called polymers. Cyanoacrylate, however, exists as monomers—single molecules with double bonds. And it stays that way until it hits anything with moisture in it, such as air. Yes, even the small amount of moisture in air and on the surfaces of most materials is enough to dissolve the double bonds in the monomers of cyanoacrylate, making them join together in long chains. The chains bond to surfaces as they polymerize.

The discovery of cyanoacrylates had an immediate impact on the automobile and airplane industries. And it soon "held" a spot in almost every household toolbox. Since the 1990s, cyanoacrylate glues have also found a place in doctors' offices. A doctor can apply a thin layer of instant glue instead of putting stitches in a cut. This specially made medical glue was approved by the U.S. Food and Drug Administration in 1998.

Take Note Visit a store and make a table of different kinds of glues. List their common names, their chemical names, what they are made of, how long it takes them to set, and the types of surfaces for which they are recommended. Note any safety precautions.

For more information, visit gpescience.com

Reviewing Main Ideas

Section 1 Stability in Bonding

1. The properties of compounds are generally different from the properties of the elements they contain.
2. A chemical formula for a compound indicates the composition of a unit of the compound. This model of a water molecule shows the shape of the molecule and the relative sizes of the atoms.

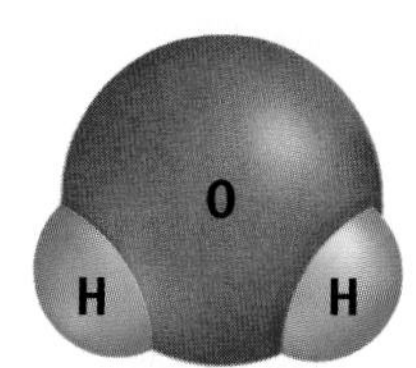

3. Chemical bonding occurs because atoms of most elements become more stable by gaining, losing, or sharing electrons to obtain stable outer energy levels.

Section 2 Types of Bonds

1. Covalent bonds are formed by the sharing of electrons. Ionic bonds between atoms are formed by the attraction between ions. Below is an example of an ionically bonded compound.

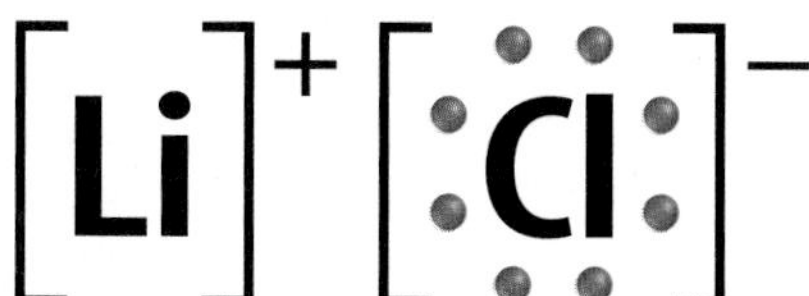

2. Ionic bonding occurs between charged particles called ions and produces ionic compounds. Covalent bonding produces units called molecules and occurs between nonmetallic elements.
3. The unequal sharing of electrons produces compounds that contain polar bonds, and the equal sharing of electrons produces nonpolar compounds.

Section 3 Writing Formulas and Naming Compounds

1. An oxidation number indicates how many electrons an atom has gained, lost, or shared when bonding with other atoms.
2. In the formula of an ionic compound, the element or ion with the positive oxidation number is written first, followed by the one with the negative oxidation number.
3. The name of a binary compound is derived from the names of the two elements that compose the compound. Salt is an example of a binary compound, sodium chloride.

4. A hydrate is a compound that has water chemically attached to its ions and written into its formula.
5. Greek prefixes are used in the names of covalent compounds. These indicate the number of each atom present.

FOLDABLES™ Use the Foldable that you made at the beginning of this chapter to help you review chemical bonds.

chapter 22 Review

Using Vocabulary

binary compound p. 703
chemical bond p. 694
chemical formula p. 689
covalent bond p. 697
hydrate p. 708
ion p. 692
ionic bond p. 696
molecule p. 697
nonpolar molecule p. 700
oxidation number p. 703
polar molecule p. 700
polyatomic ion p. 707

Match each phrase with the correct vocabulary word or phrase.

1. a charged group of atoms
2. a compound composed of two elements
3. a molecule with partially charged areas
4. a positively or negatively charged particle
5. a chemical bond between oppositely charged ions
6. a bond formed from shared electrons
7. a crystalline substance that contains water
8. a particle made of covalently bonded atoms
9. shows an element's combining ability
10. tells which elements are in a compound and their ratios.

Checking Concepts

Choose the word or phrase that best answers the question.

11. Which elements are least likely to react with other elements?
    A) metals
    B) noble gases
    C) nonmetals
    D) transition elements

12. What is the name of CuO?
    A) copper oxide
    B) copper(I) oxide
    C) copper(II) oxide
    D) copper(III) oxide

13. Which formula represents a nonpolar molecule?
    A) N_2
    B) H_2O
    C) NaCl
    D) HCl

14. How many electrons are in the outer energy levels of Group 17 elements?
    A) 1
    B) 2
    C) 7
    D) 17

15. Which is a binary ionic compound?
    A) O_2
    B) NaF
    C) H_2SO_4
    D) $Cu(NO_3)_2$

16. Which is an anhydrous compound?
    A) H_2O
    B) $CaSO_4$
    C) $CuSO_4 \cdot 5H_2O$
    D) $CaSO_4 \cdot 2H_2O$

17. Which atom has gained an electron?
    A) negative ion
    B) positive ion
    C) polar molecule
    D) nonpolar molecule

18. Which is a covalent compound?
    A) sodium chloride
    B) calcium fluoride
    C) calcium chloride
    D) sulfur dioxide

Interpreting Graphics

19. Copy and complete the concept map.

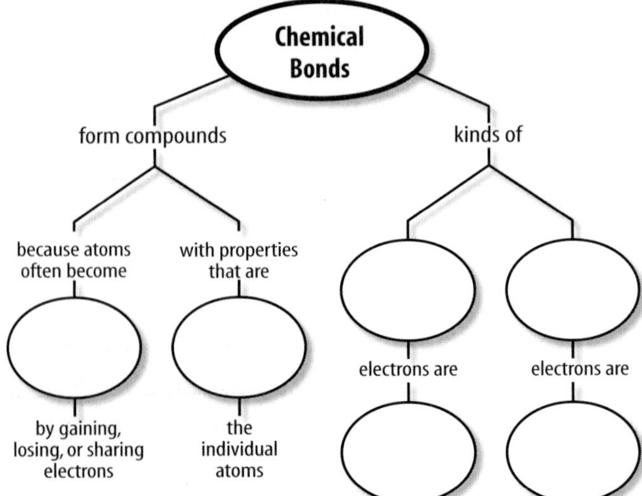

Vocabulary PuzzleMaker gpescience.com

20. **Identify** and write the name and formula for the compound illustrated to the right.

Use the table below to answer question 21.

Which compounds exist?

Formula	Possible Compounds
SF_6	AlF_6 or TeF_6
K_2SO_4	Na_2SO_4 or Ba_2SO_4
CO_2	CCl_2 or CS_2
$CaCO_3$	OCO_3 or $BaCO_3$

21. **Predict** Elements from one family (vertical column) of the periodic table generally combine with elements from another family and polyatomic ions in the same ratio. For example, one calcium atom combines with two chlorine atoms to give $CaCl_2$ (calcium chloride), as it does with two fluorine atoms to give CaF_2 (calcium fluoride). Using the periodic table as a guide, predict which of the two compounds on the right side of the table above is more likely to exist based upon the formula on the left side.

Thinking Critically

22. **Draw** Anhydrous magnesium chloride is used to make wood fireproof. Draw a dot diagram of magnesium chloride.

23. **Use Formulas** Artificial diamonds are made using thallium carbonate. If thallium has an oxidation number of 1+, what is the formula for the compound?

24. **Compare and contrast** polar and nonpolar molecules.

25. **Write** Baking soda, which is sodium hydrogen carbonate, and vinegar, which contains hydrogen acetate, can be used as household cleaners. Write the chemical formulas for these two compounds.

26. **Draw Conclusions** Ammonia gas and water react to form household ammonia, which contains NH_4^+ and OH^- ions. The formula for water is H_2O. What is the formula for ammonia gas?

27. **Draw Conclusions** The name of a compound called copper(II) sulfate is written on a bottle. What is the charge of the copper ion? What is the charge of the sulfate ion?

28. **Explain** what electric forces between oppositely charged electrons and protons have to do with chemical reactions.

29. **Name Compounds** Write the chemical name for the following compounds.
A) Fe_2S_3
B) $Cu(ClO_3)_2$
C) $Ca(PO_4)_2$
D) $(NH_4)_2SO_4$

30. **Model** One common form of phosphorus, white phosphorus, has the formula P_4 and is formed by four covalently bonded phosphorus atoms. Make a model of this molecule, showing that all four atoms are now chemically stable.

31. **Determine** the chemical formulas for the following compounds.
A) potassium chloride
B) calcium carbonate
C) copper sulfate
D) sodium oxide

Applying Math

32. **Use Numbers** What is the oxidation number of Fe in the compound Fe_2S_3?
A) 1^+
B) 2^+
C) 3^+
D) 4^+

chapter 22 Standardized Test Practice

Record your answers on the answer sheet provided by your teacher or on a sheet of paper.

Multiple Choice

1. Which statement about this molecule is **TRUE**?
   A. This is a nonpolar molecule.
   B. The electrons are shared equally in the bonds of this molecule.
   C. This molecule does not have oppositely charged ends.
   D. This is a polar molecule.

Use the figure below to answer question 2.

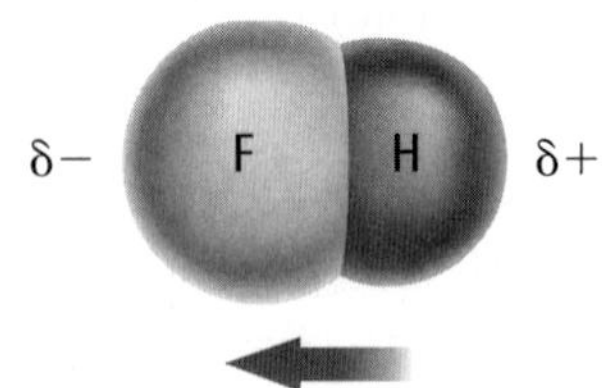

2. What type of bond holds the atoms of this molecule together?
   A. covalent
   B. ionic
   C. triple
   D. double

3. What do the Group 7A elements become when they react with Group 1A elements?
   A. negative ions
   B. neutral
   C. positive ions
   D. polyatomic ions

4. What is the name of $KC_2H_3O_2$?
   A. potassium carbide
   B. potassium acetate
   C. potassium hydroxide
   D. potassium oxide

5. What is the chemical formula for lead(II) oxide?
   A. PbO
   B. Pb_2O
   C. PbO_2
   D. Pb_2O_2

Use the illustration below to answer question 6.

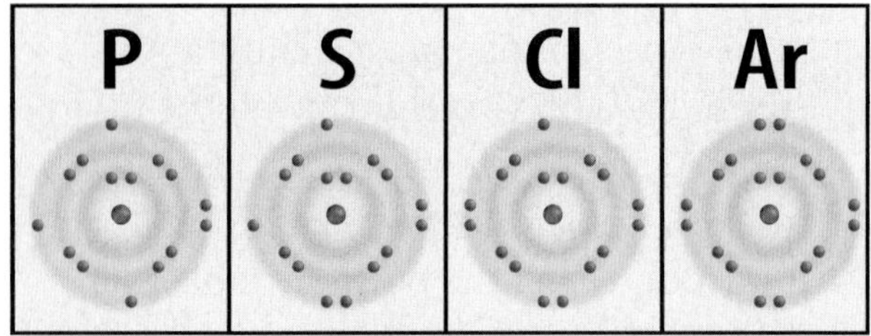

6. How many electrons are required to complete the outer energy level of a phosphorous atom?
   A. one
   B. two
   C. three
   D. four

Test-Taking Tip

Comprehension Be sure you understand the question before you read the answer choices. Make special note of words like NOT or EXCEPT. Read and consider all the answer choices before you mark your answer sheet.

7. Which element is **NOT** part of the compound NH_4NO_3?

   A. nitrogen

   B. nickel

   C. oxygen

   D. hydrogen

Gridded Response

8. When an atom is chemically stable, how many electrons are in its outer energy level?

9. How many electrons are there in an argon atom?

Short Response

Use the illustration below to answer question 10.

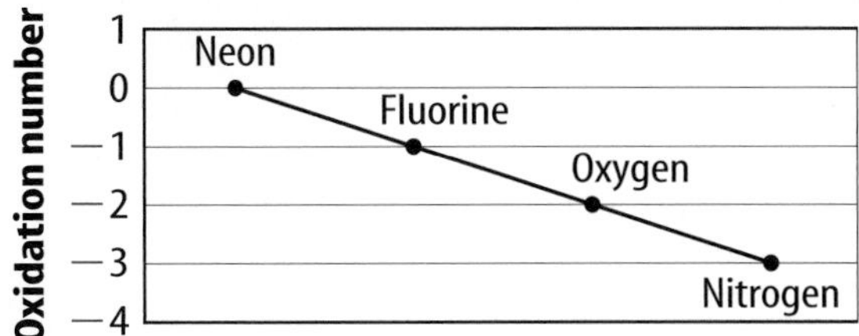

10. Compare the oxidation numbers of nitrogen and fluorine. Why do they differ?

11. The bonding of atoms and molecules is the result of oppositely charged electrons and protons being held together by electric forces within the atoms. Using this information, explain the bonding of NaCl.

Extended Response

12. KCl is an example of ionic bonding. HCl is an example of covalent bonding. Describe the difference in the bonds in terms of electrons and outer energy levels.

Use the illustration below to answer question 13.

$\cdot\ddot{\underset{\cdot}{N}}\cdot + \cdot\ddot{\underset{\cdot}{N}}\cdot \rightarrow :N\vdots\vdots N:$

13. Describe the bond holding the nitrogen atoms together in this molecule.

14. Nitrogen occurs naturally as a diatomic molecule because N_2 molecules are more stable than nitrogen atoms. H_2, O_2, F_2, Cl_2, Br_2, and I_2 are other diatomic molecules. Draw dot diagrams for three of these molecules.

15. Explain why elements in Groups 4A, which have four electrons in their outer energy levels, are unlikely to lose all of the electrons in their outer energy levels.

16. What factors affect how strongly an atom is attracted to its electrons?

17. Create a chart that compares the properties of polar and nonpolar molecules. Your chart should include several examples of each type of molecule.

18. What is the difference between nitrogen oxide and dinitrogen pentoxide? Why are prefixes used to name these compounds?

chapter 23

Chemical Reactions

chapter preview

sections

All-American Chemistry

Few things are as American as fireworks on the Fourth of July. The explosions of color and deafening booms get huge reactions from crowds across the country. These sights and sounds are the results of chemical reactions involving various substances with oxygen.

Science Journal Describe several cause-and-effect types of events that might happen in your refrigerator. Later, decide which of the events are chemical reactions.

Start-Up Activities

Rusting—A Chemical Reaction

Like exploding fireworks, rusting is a chemical reaction in which iron metal combines with oxygen. Other metals combine with oxygen, too—some more readily than others. In this lab, you will compare how iron and aluminum react with oxygen.

1. Complete the safety form.
2. Place a clean iron or steel nail in a dish prepared by your teacher.
3. Place a clean aluminum nail in a second dish. These dishes contain agar gel and an indicator that detects a reaction with oxygen.
4. Observe both nails after one hour. Record any changes around the nails in your Science Journal.
5. Carefully examine both of the dishes the next day.
6. **Think Critically** Record any differences you noticed between the two dishes. Predict if a reaction occurred. How can you tell? What might have caused the differences you observed between the two nails. Explain.

FOLDABLES™ Study Organizer

Chemical Reactions Make the following Foldable to help you classify chemical reactions.

STEP 1 **Fold** a sheet of paper in half lengthwise.

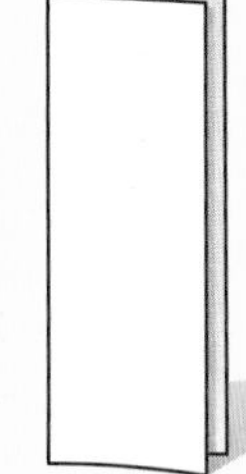

STEP 2 **Mark** four lines evenly spaced at even intervals down the page.

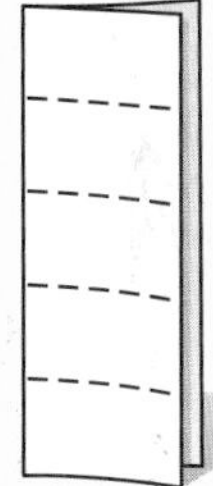

STEP 3 **Cut** only the top layer along the four marks to make five tabs. Label the tabs as shown.

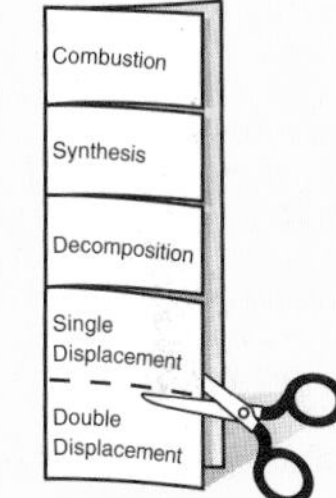

Classify As you read, record examples of each type of reaction from the book, then review the chapter and list other examples mentioned in the text or from classroom discussions.

Preview this chapter's content and activities at gpescience.com

section 1

Chemical Changes

Reading Guide

What You'll Learn

- **Identify** the reactants and products in a chemical reaction.
- **Determine** how a chemical reaction satisfies the law of conservation of mass.
- **Determine** how chemists express chemical changes using equations.

Why It's Important

Chemical reactions cook our food, warm our homes, and provide energy for our bodies.

Review Vocabulary

chemical change: change of one substance into a new substance

New Vocabulary

- chemical reaction
- reactant
- product
- chemical equation

Describing Chemical Reactions

Dark mysterious mixtures react, gas bubbles up and expands, and powerful aromas waft through the air. Where are you? You are in the kitchen baking a chocolate cake. Nowhere in the house do so many chemical reactions take place as in the kitchen.

Chemical reactions are taking place all around you and even within you. A **chemical reaction** is a change in which one or more substances are converted into new substances. The substances that react are called **reactants.** The new substances produced are called **products.**

Chemical and Nuclear Reactions When chemical reactions occur, new compounds form when bonds between atoms in the reactants break and new bonds form. Recall that chemical bonds form when outer electrons, called valence electrons, are shared between atoms or are transferred from one atom to another. As a result, only the outer electrons of atoms are involved in chemical reactions. The nucleus of an atom is not affected by a chemical reaction. An atomic nucleus changes only when nuclear decay or a nuclear reaction, such as nuclear fission or fusion, occurs. The energy released by a nuclear reaction is millions of times greater than the energy released by a chemical reaction. **Figure 1** summarizes the difference between nuclear and chemical reactions.

Figure 1 Different parts of the atom are used in chemical and nuclear reactions.
Identify *how chemical bonds are formed.*

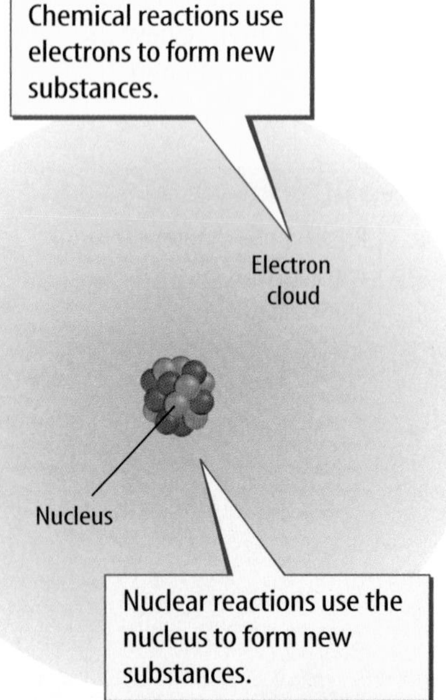
Chemical reactions use electrons to form new substances.
Electron cloud
Nucleus
Nuclear reactions use the nucleus to form new substances.

Conservation of Mass

By the 1770s, chemistry was changing from the art of alchemy to a true science. Instead of being satisfied with a superficial explanation of unknown events, scientists began to study chemical reactions more thoroughly. Through such study, the French chemist Antoine Lavoisier established that the total mass of the products always equals the total mass of the reactants. This principle is illustrated in **Figure 2.**

The Father of Modern Chemistry When Lavoisier demonstrated the law of conservation of mass, he set the field of chemistry on its modern path. In fact, Lavoisier is known today as the father of modern chemistry for his more accurate explanation of the conservation of mass and for describing a common type of chemical reaction called combustion, which you will learn about later in this chapter. Lavoisier also pioneered early experimentation on the biological phenomena of respiration and metabolism that contributed early milestones in the study of biochemistry, medicine, and even sports medicine.

Reading Check *How did Lavoisier's contributions earn him the title of Father of Modern Chemistry?*

Nomenclature Antoine Lavoisier's work led him to the conclusion that language terminology would be critical to communicate novel scientific ideas. Lavoisier began to develop the system of naming substances based on their composition that we still use today. In 1787, Lavoisier and several colleagues published *Méthode de Nomenclature Chimique* as one of the first sets of nomenclature guidelines.

Figure 2 The mass of the candles and oxygen before burning is exactly equal to the mass of the remaining candle and gaseous products.
Describe *the principle that is represented in this figure.*

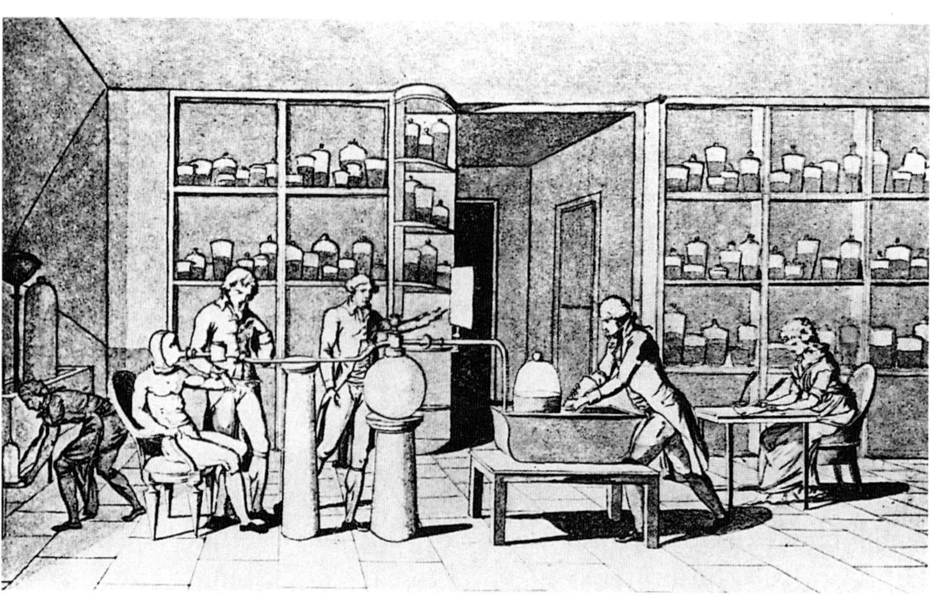

Figure 3 Antoine Lavoisier's wife, Marie-Anne, drew this view of Lavoisier in his laboratory performing studies on oxygen. She depicted herself at the right taking notes.

Lavoisier's Contribution One of the questions that motivated Lavoisier was the mystery of exactly what happened when substances changed form. He began to answer this question by experimenting with mercury. In one experiment, Lavoisier placed a carefully measured mass of solid mercury(II) oxide, which he knew as mercury calx, into a sealed container. When he heated this container, he noted a dramatic change. The red powder had been transformed into a silvery liquid that he recognized as mercury metal, and a gas was produced. When he determined the mass of the liquid mercury and gas, their combined masses were exactly the same as the mass of the red powder he had started with.

mercury(II) oxide		oxygen	plus	mercury
10.0 g	=	0.7 g	+	9.3 g

Lavoisier also established that the gas produced by heating mercury(II) oxide, which we call oxygen, was a component of air. He did this by heating mercury metal with air and saw that a portion of the air combined to give red mercury(II) oxide. He studied the effect of this gas on living animals, including himself. Hundreds of experiments carried out in his laboratory, as shown in **Figure 3,** confirmed that in a chemical reaction, matter is not created or destroyed, but is conserved. This principle became known as the law of conservation of mass. This means that the total starting mass of all reactants equals the total final mass of all products.

Reading Check ***What does the law of conservation of mass state?***

Topic: Antoine Lavoisier
Visit gpescience.com for Web links to information about Antoine Lavoisier and his contributions to chemistry.

Activity In your Science Journal, write a brief biography of Antoine Lavoisier that includes some of his non-scientific activities and political interests, as well as his scientific contributions.

Writing Equations

If you wanted to describe the chemical reaction shown in **Figure 4,** you might write something like this:

Nickel(II) chloride, dissolved in water, plus sodium hydroxide, dissolved in water, produces solid nickel(II) hydroxide plus sodium chloride, dissolved in water.

This series of words is rather cumbersome, but all of the information is important. The same is true of descriptions of most chemical reactions. Many words are needed to state all the important information. As a result, scientists have developed a shorthand method to describe chemical reactions. A **chemical equation** is a way to describe a chemical reaction using chemical formulas and other symbols. Some of the symbols used in chemical equations are listed in **Table 1.**

The chemical equation for the reaction described above in words and shown in **Figure 4** looks like this:

$$NiCl_2(aq) + 2NaOH(aq) \longrightarrow Ni(OH)_2(s) + 2NaCl(aq)$$

On the left side of the equation are the reactants, nickel(II) chloride and sodium hydroxide. On the right side of the equation are the products, nickel(II) hydroxide and sodium chloride.

It is much easier to tell what is happening by writing the information in this form. Later, you will learn how chemical equations make it easier to calculate the quantities of reactants that are needed and the quantities of products that are formed.

Table 1 Symbols Used in Chemical Equations

Symbol	Meaning
$\rightarrow$	produces or yields
+	plus
(s)	solid
(l)	liquid
(g)	gas
(aq)	aqueous, a substance is dissolved in water
heat $\rightarrow$	the reactants are heated
light $\rightarrow$	the reactants are exposed to light
elec. $\rightarrow$	an electric current is applied to the reactants

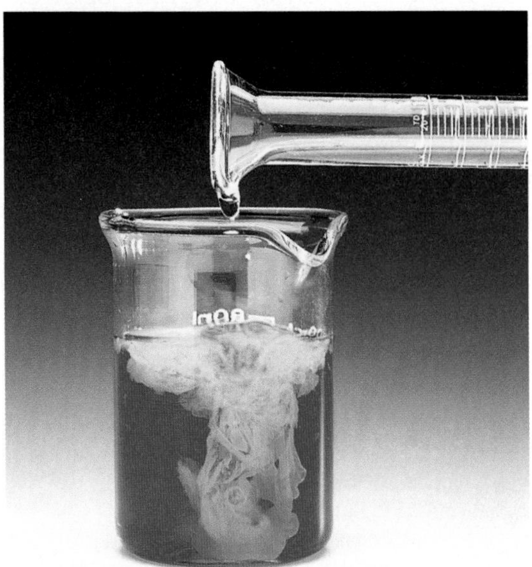

Figure 4 A white precipitate of nickel(II) hydroxide forms when sodium hydroxide is added to a green solution of nickel(II) chloride. Sodium chloride, the other product formed, is in solution.

Designing a Team Equation

Procedure

1. Obtain **15 index cards** and mark each as follows: five with *Guard,* five with *Forward,* and five with *Center.*
2. Group the cards to form as many complete basketball teams as possible. Each team needs two guards, two forwards, and one center.

Analysis

1. Write the formula for a team. Write the formation of a team as an equation. Use coefficients in front of each type of player needed for a team.
2. How is this equation like a chemical equation? Why can't you use the remaining cards?
3. How do the remaining cards illustrate the law of conservation of matter in this example?

Try at Home

Unit Managers

What do the numbers to the left of the formulas for reactants and products mean? Remember that according to the law of conservation of mass, matter is neither made nor lost during chemical reactions. Atoms are rearranged but never lost or destroyed. These numbers, called coefficients, represent the number of units of each substance taking part in a reaction. Coefficients can be thought of as unit managers.

Reading Check ***What is the function of coefficients in a chemical equation?***

Imagine that you are responsible for making sandwiches for a picnic. You have been told to make a certain number of three kinds of sandwiches, and that no substitutions can be made. You would have to figure out exactly how much food to buy so that you had enough without any food left over. You might need two loaves of bread, four packages of turkey, four packages of cheese, two heads of lettuce, and ten tomatoes. With these supplies you could make exactly the right number of each kind of sandwich.

In a way, your sandwich-making effort is like a chemical reaction. The reactants are your bread, turkey, cheese, lettuce, and tomatoes. The number of units of each ingredient are similar to the coefficients of the reactants in an equation. The sandwiches are similar to the products, and the numbers of each kind of sandwich are similar to the coefficients.

Knowing the number of units of reactants enables chemists to add the correct amounts of reactants to a reaction. Also, these units, or coefficients, tell them exactly how much product will form. An example of this is the reaction of one unit of $NiCl_2$ with two units of NaOH to produce one unit of $Ni(OH)_2$ and two units of NaCl. You can see these units in **Figure 5.**

Figure 5 Each coefficient in the equation represents the number of units of each type in this reaction.

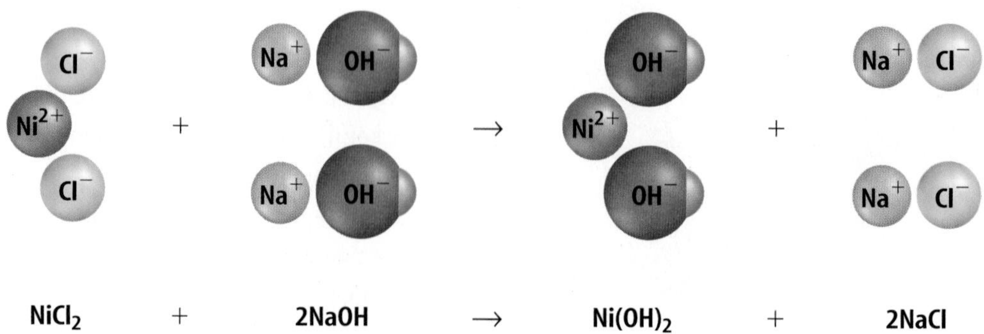

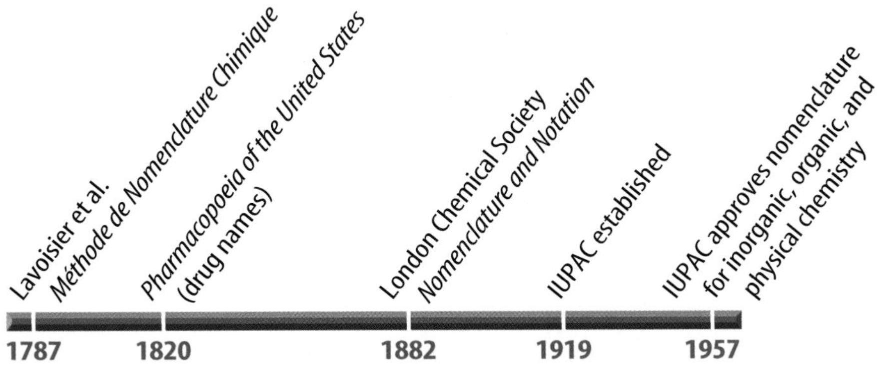

Figure 6 This time line of nomenclature development and publications does not end in 1957. In fact, today there are nomenclature organizations for almost every branch of scientific study, and the rules and guidelines for naming substances continue to evolve.

INTEGRATE History

IUPAC Antoine Lavoisier and his colleagues understood the importance of using a nomenclature system that helps scientists communicate their data. Since that time, the guidelines have continued to evolve with scientific discovery. In 1919, the International Union of Applied Chemistry (IUPAC) was formed. The primary mission of the IUPAC is to coordinate guidelines for naming chemical compounds systematically. **Figure 6** illustrates some of the early events in nomenclature development.

In addition to chemical nomenclature, IUPAC also is responsible for standardizing chemistry methods, evaluating atomic weights and environmental programs, along with many other areas relating to chemistry.

section 1 review

Summary

Describing Chemical Reactions

- A chemical reaction is a process that involves one or more reactants changing into one or more products.
- In chemical reactions and nuclear reactions, different parts of the atom form new substances.

Conservation of Mass

- A basic principle of chemistry is that matter, during a chemical change, can neither be created nor destroyed.

Writing Equations

- Chemical equations describe the change of reactants to products and obey the law of conservation of mass.

Unit Managers

- Coefficients represent how many units of each substance are involved in a chemical reaction.

Self Check

1. **Identify** the reactants and the products in the following chemical equation.
   $Cd(NO_3)_2(aq) + H_2S(g) \longrightarrow CdS(s) + 2HNO_3(aq)$
2. **Identify** the state of matter of each substance in the following reaction.
   $Zn(s) + 2HCl(aq) \longrightarrow H_2(g) + ZnCl_2(aq)$
3. **Compare and contrast** a chemical reaction and a nuclear reaction.
4. **Explain** the importance of the law of conservation of mass.
5. **Think Critically** How would global communication between scientists have been changed if IUPAC had not been established?

Applying Math

6. **Solve One-Step Equations** When making soap, if 890 g of a specific fat react completely with 120 g of sodium hydroxide, the products formed are soap and 92 g of glycerin. Calculate the mass of soap formed to satisfy the law of conservation of mass.

More Section Review gpescience.com

section 2

Chemical Equations

Reading Guide

What You'll Learn

- **Identify** what is meant by a balanced chemical equation.
- **Determine** how to write balanced chemical equations.

Why It's Important

Chemical equations are the language used to describe chemical change, which allow scientists to develop products for our world.

Review Vocabulary

subscript: in a chemical formula, a number below and to the right of a symbol indicating number of atoms

New Vocabulary

- balanced chemical equation

Figure 7 Mercury metal forms when mercury oxide is heated. Because mercury is poisonous, this reaction is never performed in a classroom laboratory.

Balanced Equations

Lavoisier's mercury(II) oxide reaction, shown in **Figure 7,** can be written as:

$$HgO(s) \xrightarrow{\text{heat}} Hg(l) + O_2(g)$$

Notice that the number of mercury atoms is the same on both sides of the equation but that the number of oxygen atoms is not the same. One oxygen atom appears on the reactant side of the equation and two appear on the product side.

Atoms	HgO	→	Hg	+	O_2
Hg	1		1		
O	1				2

But according to the law of conservation of mass, one oxygen atom cannot just become two. Nor can you simply add the subscript 2 and write HgO_2 instead of HgO. The formulas HgO_2 and HgO do not represent the same compound. In fact, HgO_2 does not exist. The formulas in a chemical equation must accurately represent the compounds that react.

Fixing this equation requires a process called balancing. Balancing an equation doesn't change what happens in a reaction—it simply changes the way the reaction is represented. The balancing process involves changing coefficients in a reaction to achieve a **balanced chemical equation,** which has the same number of atoms of each element on both sides of the equation.

Choosing Coefficients Finding out which coefficients to use to balance an equation is often a trial-and-error process. In the equation for Lavoisier's experiment, the number of mercury atoms is balanced, but one oxygen atom is on the left and two are on the right. If you put a coefficient of 2 before the HgO on the left, the oxygen atoms will be balanced, but the mercury atoms become unbalanced. To balance the equation, also put a 2 in front of mercury on the right. The equation is now balanced.

Atoms	2HgO	→	2Hg	+	O_2
Hg	2		2		
O	2				2

Balancing Equations Magnesium burns with such a brilliant white light that it is often used in emergency flares as shown in **Figure 8.** Burning leaves a white powder called magnesium oxide. To write a balanced chemical equation for this and most other reactions, follow these four steps.

Step 1 Write a chemical equation for the reaction using formulas and symbols. Recall that oxygen is a diatomic molecule.

$$Mg(s) + O_2(g) \rightarrow MgO(s)$$

Step 2 Count the atoms in reactants and products.

Atoms	Mg	+	O_2	→	MgO
Mg	1				1
O			2		1

The magnesium atoms are balanced, but the oxygen atoms are not. Therefore, this equation isn't balanced.

Step 3 Choose coefficients that balance the equation. Remember, never change subscripts of a correct formula to balance an equation. Try putting a coefficient of 2 before MgO.

$$Mg(s) + O_2(g) \rightarrow 2MgO(s)$$

Step 4 Recheck the numbers of each atom on each side of the equation and adjust coefficients again if necessary. Now two Mg atoms are on the right side and only one is on the left side. So a coefficient of 2 is needed for Mg to balance the equation.

$$2Mg(s) + O_2(g) \rightarrow 2MgO(s)$$

Reading Check *How can you balance a chemical equation using coefficients?*

Topic: Balancing Chemical Equations

Visit gpescience.com for Web links to information about balancing chemical equations.

Activity Using the Web links, locate a website that offers practice problems for balancing chemical equations. Copy several of the unbalanced equations in your Science Journal and try to balance them. Check your work against the answers on the web site when you are done.

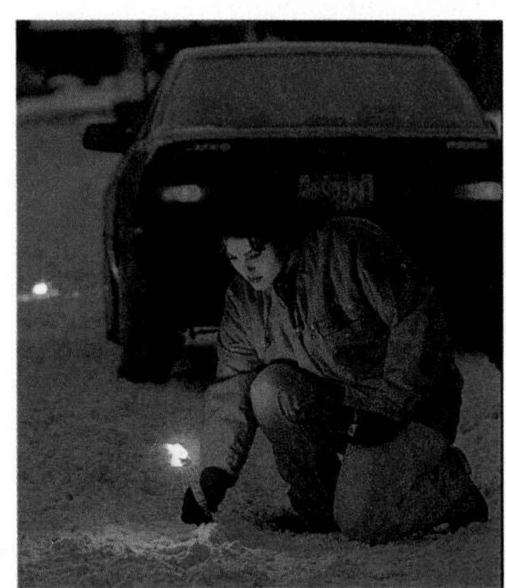

Figure 8 Magnesium combines with oxygen, giving an intense white light.

Applying Math — Use Coefficients

BARIUM SULFATE REACTION A sample of barium sulfate is placed on a piece of paper, which is then ignited. Barium sulfate reacts with the carbon from the burned paper producing barium sulfide and carbon monoxide. Write a balanced chemical equation for this reaction.

IDENTIFY known values

We know the substances that are involved in the reaction. From this, we can write a chemical equation showing reactants and products.

$$BaSO_4(s) + C(s) \longrightarrow BaS(s) + CO(g)$$

SOLVE the problem

The chemical equation above is not balanced. There are more oxygen atoms on the left side of the equation than there are on the right side. This must be corrected while keeping all other atom counts in balance. Begin to balance the equation by first counting and listing the atoms before and after the reaction.

Kind of Atom	Number of Atoms Before Reaction	Number of Atoms After Reaction
Ba	1	1
S	1	1
O	4	1
C	1	1

Next, adjust the coefficients until all atoms are balanced on the left and right sides of the arrow. Try putting a 4 in front of CO. Now you have 4 oxygen atoms on the right, which balances on both sides, but the carbon atoms become unbalanced. To fix this, add a 4 in front of the C in the reactants. The balanced equation looks like this:

$$BaSO_4(s) + 4C(s) \longrightarrow BaS(s) + 4CO(g)$$

CHECK your answer

Count the number of atoms on each side of the equation and verify that they are equal.

Practice Problems

1. Balance this equation: $NaOH(aq) + CaBr_2(aq) \longrightarrow Ca(OH)_2(s) + NaBr(aq)$.
2. HCl is slowly added to aqueous Na_2CO_3 forming NaCl, H_2O, and CO_2. Follow the steps above to write a balanced equation for this reaction.

For more practice problems, go to page 879, and visit Math Practice at gpescience.com .

Polish Your Skill When lithium metal is treated with water, hydrogen gas and lithium hydroxide are produced, as shown in **Figure 9.**

Step 1 Write the chemical equation.

$$Li(s) + H_2O \rightarrow LiOH(aq) + H_2(g)$$

Step 2 Check for balance by counting the atoms.

Atoms	Li	+	H_2O	→	LiOH	+	H_2
Li	1				1		
H			2		1		2
O			1		1		

This equation is not balanced. There are three hydrogen atoms on the right and only two on the left. Complete steps 3 and 4 to balance the equation. After each step, count the atoms of each element. When equal numbers of atoms of each element are on both sides, the equation is balanced. The balanced chemical equation looks like this:

$$2Li + 2H_2O \rightarrow 2LiOH + H_2$$

The coefficients and subscripts are multiplied together, not added, to determine the correct number of atoms for each element.

Figure 9 When lithium metal is added to water, it reacts, producing a solution of lithium hydroxide and bubbles of hydrogen gas.

section 2 review

Summary

Balanced Equations

- A chemical equation is a way to indicate reactants and products and relative amounts of each.
- A balanced chemical equation tells the exact number of atoms involved in the reaction.
- Balanced chemical equations must satisfy the law of conservation of matter; no atoms of reactant or product can be lost from one side to the other.
- Coefficients are used to achieve balance in a chemical equation.
- Chemical equations cannot be balanced by adjusting the subscript numerals in compound names because doing so would change the compounds.

Self Check

1. **Describe** two reasons for balancing chemical equations.
2. **Balance** this chemical equation: $Fe(s) + O_2(g) \rightarrow FeO(s)$.
3. **Explain** why oxygen gas must always be written as O_2 in a chemical equation.
4. **Infer** What coefficient is assumed if no coefficient is written before a formula in a chemical equation?
5. **Think Critically** Explain why the sum of the coefficients on the reactant side of a balanced equation does not have to equal the sum of the coefficients on the product side of the equation.

Applying Math

6. **Use Numbers** Balance the equation for the reaction $Fe(s) + Cl_2(g) \rightarrow FeCl_3(s)$.

section 3

Classifying Chemical Reactions

Reading Guide

What You'll Learn

- **Identify** the five general types of chemical reactions.
- **Identify** redox reactions.
- **Predict** which metals will replace other metals in compounds.

Why It's Important

Classifying chemical reactions helps to understand what is happening and predict the outcome of reactions.

Review Vocabulary

states of matter: the physical forms in which all matter naturally exists, most commonly solid, liquid, and gas

New Vocabulary

- combustion reaction
- synthesis reaction
- decomposition reaction
- single-displacement reaction
- double-displacement reaction

Types of Reactions

You might have noticed that there are all sorts of chemical reactions. In fact, there are literally millions of chemical reactions that occur every day, and scientists have described many of them and continue to describe more. With all these reactions, it would be impossible to use the information without first having some type of organization. With this in mind, chemists have defined five main categories of chemical reactions: combustion, synthesis, decomposition, single displacement, and double displacement.

Figure 10 Rust has accumulated on the *Titanic* since it sank in 1912. **Identify** *the general formula for a synthesis reaction.*

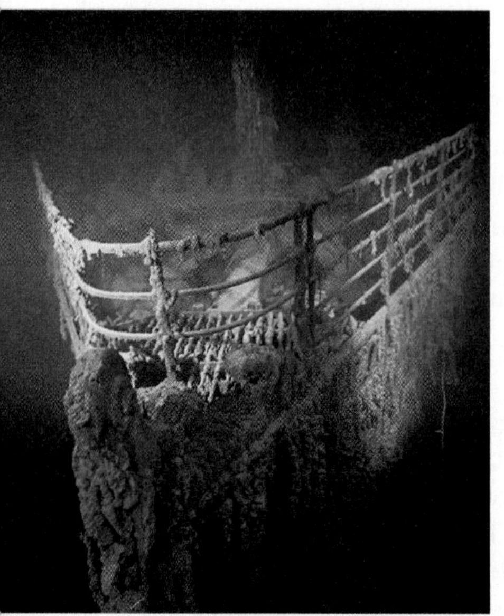

Combustion Reactions If you have ever observed something burning, you have observed a combustion reaction. As mentioned previously, Lavoisier was one of the first scientists to accurately describe combustion. He deduced that the process of burning (combustion) involves the combination of a substance with oxygen. A **combustion reaction** occurs when a substance reacts with oxygen to produce energy in the form of heat and light. Combustion reactions also produce one or more products that contain the elements in the reactants. For example, the reaction between carbon and oxygen produces carbon dioxide. Many combustion reactions also will fit into other categories of reactions. For example, the reaction between carbon and oxygen also is a synthesis reaction.

Synthesis Reactions One of the easiest reaction types to recognize is a synthesis reaction. In a **synthesis reaction,** two or more substances combine to form another substance. The generalized formula for this reaction type is as follows: A + B → AB.

The reaction in which hydrogen burns in oxygen to form water is an example of a synthesis reaction.

$$2H_2(g) + O_2(g) \rightarrow 2H_2O(g)$$

This reaction is used to power some types of rockets. Another synthesis reaction is the combination of oxygen with iron in the presence of water to form hydrated iron(II) oxide or rust. This reaction is shown in **Figure 10.**

Decomposition Reactions A decomposition reaction is just the reverse of a synthesis. Instead of two substances coming together to form a third, a **decomposition reaction** occurs when one substance breaks down, or decomposes, into two or more substances. The general formula for this type of reaction can be expressed as follows: AB → A + B.

Most decomposition reactions require the use of heat, light, or electricity. An electric current passed through water produces hydrogen and oxygen as shown in **Figure 11.**

$$2H_2O(l) \xrightarrow{\text{elec.}} 2H_2(g) + O_2(g)$$

Figure 11 Water decomposes into hydrogen and oxygen when an electric current is passed through it. A small amount of sulfuric acid is added to increase conductivity. Notice the proportions of the gases collected. **Describe** *how this is related to the coefficients of the products in the equation.*

Single Displacement When one element replaces another element in a compound, it is called a **single-displacement reaction.** Single-displacement reactions are described by the general equation A + BC → AC + B. Here you can see that atom A displaces atom B to produce a new molecule AC. A single displacment reaction is illustrated in **Figure 12,** where a copper wire is put into a solution of silver nitrate. Because copper is a more active metal than silver, it replaces the silver, forming a blue copper(II) nitrate solution. The silver, which is not soluble, forms on the wire.

$$Cu(s) + 2AgNO_3(aq) \rightarrow Cu(NO_3)_2\,(aq) + 2Ag(s)$$

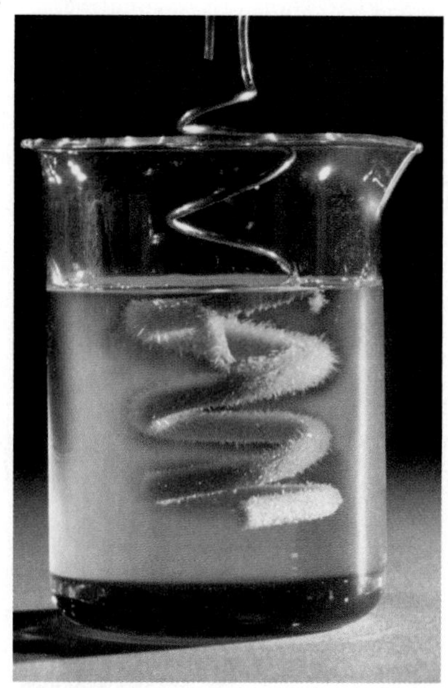

Figure 12 Copper in a wire replaces silver in silver nitrate, forming a blue-tinted solution of copper(II) nitrate.

Reading Check *Describe a single-displacement reaction.*

Figure 13 This figure shows the activity series of metals. A metal will replace any other metal that is less active.

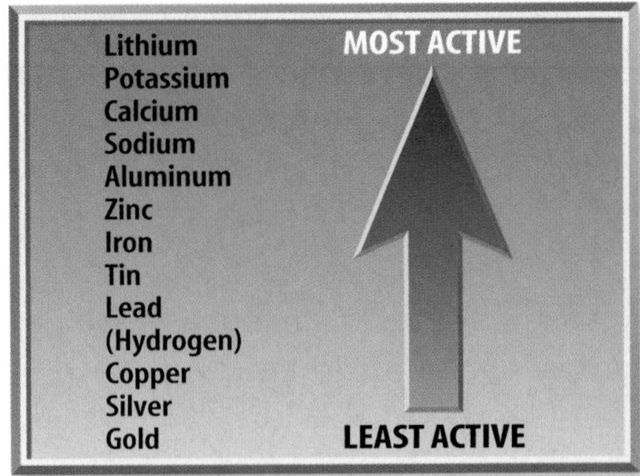

The Activity Series Sometimes single-displacement reactions can cause problems. For example, if iron-containing vegetables such as spinach are cooked in aluminum pans, aluminum can displace iron from the vegetable. This causes a black deposit of iron to form on the sides of the pan. For this reason, it is better to use stainless steel or enamel cookware when cooking spinach.

We can predict which metal will replace another using the diagram shown in **Figure 13,** which lists metals according to how reactive they are. A metal can replace any metal below it on the list but not above it. Notice that copper, silver, and gold are the least active metals on the list. That is why these elements often occur as deposits of the relatively pure element. For example, gold is sometimes found as veins in quartz rock. Copper is found in pure lumps known as native copper. Other metals can occur as compounds.

Double Displacement In a **double-displacement reaction,** the positive ion of one compound replaces the positive ion of the other to form two new compounds. A double-displacement reaction takes place if a precipitate, water, or a gas forms when two ionic compounds in solution are combined. A precipitate is an insoluble compound that comes out of solution during this type of reaction. The generalized formula for this type of reaction is as follows: $AB + CD \rightarrow AD + CB$.

Reading Check *What type of reaction produces a precipitate?*

The reaction of barium nitrate with potassium sulfate is an example of this type of reaction. A precipitate—barium sulfate—forms, as shown in **Figure 14.** The chemical equation is as follows:

$$Ba(NO_3)_2(aq) + K_2SO_4(aq) \rightarrow BaSO_4(s) + 2KNO_3(aq)$$

These are a few examples of chemical reactions classified into types. Many more reactions of each type occur around you.

Figure 14 Solid barium sulfate is formed from the reaction of two solutions.
Observe *Has a chemical change occurred in this photo? How can you tell?*

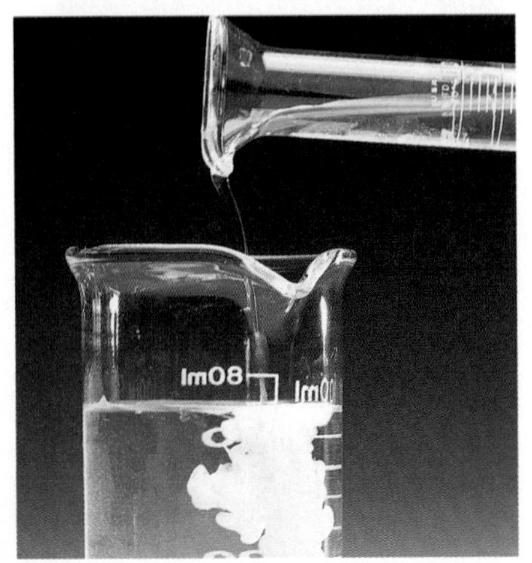

Oxidation-Reduction Reactions One characteristic that is common to many chemical reactions is the tendency of the substances to lose or gain electrons. Chemists use the term oxidation to describe the loss of electrons and the term reduction to describe the gain of electrons. Chemical reactions involving electron transfer of this sort often involve oxygen, which is very reactive, pulling electrons from metallic elements. Corrosion of metal is a visible result, as shown in **Figure 15.**

The cause and effect of oxidation and reduction can be taken one step further by describing the substances after the electron transfer. The substance that gains an electron or electrons obviously becomes more negative, so we say it is reduced. On the other hand, the substance that loses an electron or electrons then becomes more positive, and we say it is oxidized. The electrons that were pulled from one atom were gained by another atom in a chemical reaction called reduction. Reduction is the partner to oxidation; the two always work as a pair, which is commonly referred to as redox.

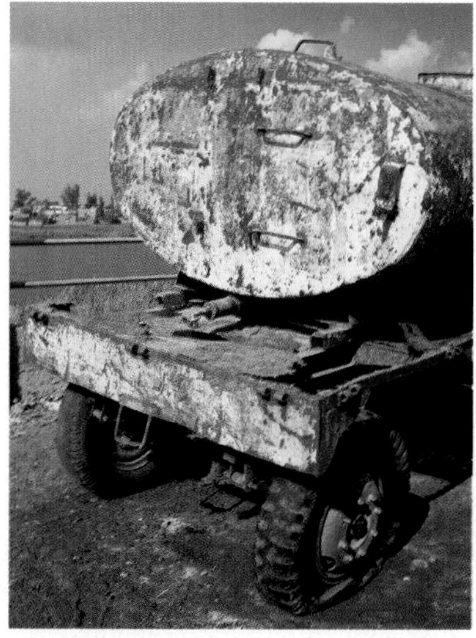

Figure 15 One of the results of all of these electrons moving from one place to another might show up on the metal body of a tanker.

section 3 review

Summary

Types of Reactions

- Lavoisier was one of the first scientists to accurately describe a combustion reaction.
- For single-displacement reactions, we can predict which metal will replace another by comparing the activity characteristic of each.
- Some reactions produce a solid called a precipitate when two ionic substances are combined.
- Chemical reactions are organized into five basic classes: combustion, synthesis, decomposition, single displacement, and double displacement.

Oxidation-Reduction Reactions

- Oxidation is the loss of electrons and reduction is the corresponding gain of electrons.
- Redox reactions often result in corrosion and rust.
- A substance that gains elections is reduced, and a substance that loses elections is oxidized.

Self Check

1. **Classify** each of the following reactions:
   a. $CaO(s) + H_2O \longrightarrow Ca(OH)_2\,(aq)$
   b. $Fe(s) + CuSO_4(aq) \longrightarrow FeSO_4(aq) + Cu(s)$
   c. $NH_4NO_3(s) \longrightarrow N_2O(g) + 2H_2O(g)$
2. **Describe** what happens in a combustion reaction.
3. **Explain** the difference between synthesis and decomposition reactions.
4. **Determine,** using **Figure 13,** if zinc will displace gold in a chemical reaction and explain why or why not.
5. **Think Critically** In the reaction $2\,Na + Cl_2 \longrightarrow 2NaCl$, which atom is reduced and which atom is oxidized?

Applying Math

6. **Use Proportions** The following chemical equation is balanced, but the coefficients used are larger than necessary. Rewrite this balanced equation using the smallest coefficients.
   $9Fe(s) + 12H_2O(g) \longrightarrow 3Fe_3O_4(s) + 12H_2(g)$
7. **Use Coefficients** Sulfur trioxide, (SO_3), a pollutant released by coal-burning plants, can react with water in the atmosphere to produce sulfuric acid, H_2SO_4. Write a balanced equation for this reaction.

section 4

Reaction Rates and Energy

Reading Guide

What You'll Learn

- **Identify** the source of energy changes in chemical reactions.
- **Compare and contrast** exergonic and endergonic reactions.
- **Examine** the effects of catalysts and inhibitors on the speed of chemical reactions.

Why It's Important

Chemical reactions provide energy to cook your food, keep you warm, and transform the food you eat into substances you need to live and grow.

Review Vocabulary

chemical bond: the force that holds two atoms together

New Vocabulary

- activation energy
- endothermic reaction
- exothermic reaction
- rate of reaction
- catalyst
- inhibitor

Chemical Reactions and Energy

Often a crowd gathers to watch a building being demolished using dynamite. In a few breathtaking seconds, tremendous structures of steel and cement that took a year or more to build are reduced to rubble and a large cloud of dust. A dynamite explosion, as shown in **Figure 16,** is an example of a rapid chemical reaction.

All chemical reactions release or absorb energy. This energy can take many forms, such as heat, light, sound, or electricity. The heat produced by a wood fire and the light emitted by a glow stick are two examples of reactions that release energy.

Figure 16 When its usefulness is over, a building is sometimes demolished using dynamite. Dynamite charges must be placed carefully so that the building collapses inward, where it cannot harm people or property.

Conservation of Energy in Chemical Reactions

According to the law of conservation of energy, energy cannot be created or destroyed, but can only change form. In compounds, chemical potential energy is stored in chemical bonds between atoms. In some chemical reactions, chemical potential energy is changed to other forms of energy, such as heat or light, and is released. In other chemical reactions, forms of energy such as heat or light are converted to chemical potential energy and stored in bonds that form, and energy is absorbed. In all chemical reactions, energy is never created or destroyed, but only changes form. All reactions follow the laws of conservation of mass and energy.

Activation Energy

As you learned earlier, atoms and molecules have to bump into each other before a product can be formed. In order to form new bonds, atoms have to be close together. In addition to being close, the reactants require a certain amount of energy in order to allow the reaction to start. This minimum amount of energy needed to start a reaction is called **activation energy.** If there is not enough energy, the reaction will not start. Spilled gasoline does not ignite unless there is an energy source. Once ignited, the gasoline releases enough energy to keep going. Activation energy, which differs from reaction to reaction, is required for both exothermic and endothermic reactions.

Controlling Body Heat Some animals, such as birds, are classified as endotherms because their body temperature remains constant. An animal whose body temperature varies with its environment, such as a reptile, is classified as an ectotherm. What other animals that can be classified as ectotherms?

Endergonic Reactions

Sometimes a chemical reaction requires more energy to break bonds than is released when new ones are formed. These reactions are called endergonic reactions. The energy absorbed can be in the form of light, heat, or electricity.

Electricity is often used to supply energy to endergonic reactions. For example, electroplating deposits a coating of metal onto a surface.

Heat Absorption When the energy needed is in the form of heat, the reaction is called an **endothermic reaction.** The term *endothermic* is not just related to chemical reactions. It also can describe physical changes. The process of dissolving a salt in water is a physical change. If you ever had to soak a swollen ankle in an Epsom salt solution, you probably noticed that when you mixed the Epsom salt in water, the solution became cold. The dissolving of Epsom salt absorbs heat. Thus, it is a physical change that is endothermic.

Some reactions are so endothermic that they can cause water to freeze. One such endothermic reaction is that of barium hydroxide ($Ba(OH)_2$) and ammonium chloride (NH_4Cl) in water, shown in **Figure 17.**

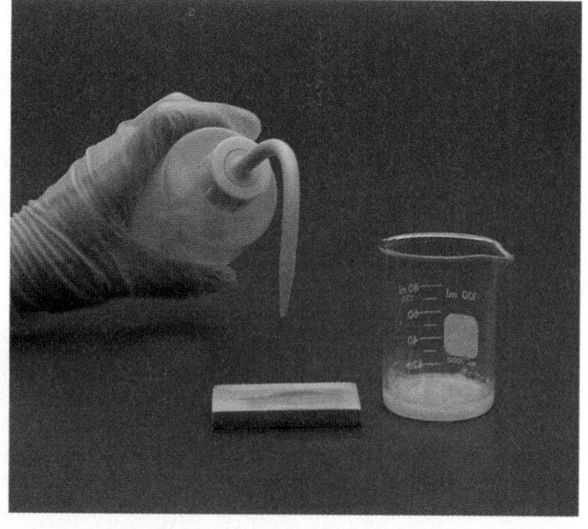

Figure 17 As an endothermic reaction happens, such as the reaction of barium hydroxide and ammonium chloride, energy from the surrounding environment is absorbed, causing a cooling effect. Here, the reaction absorbs so much heat that a drop of water freezes and the beaker holding the reaction sticks to the wood.

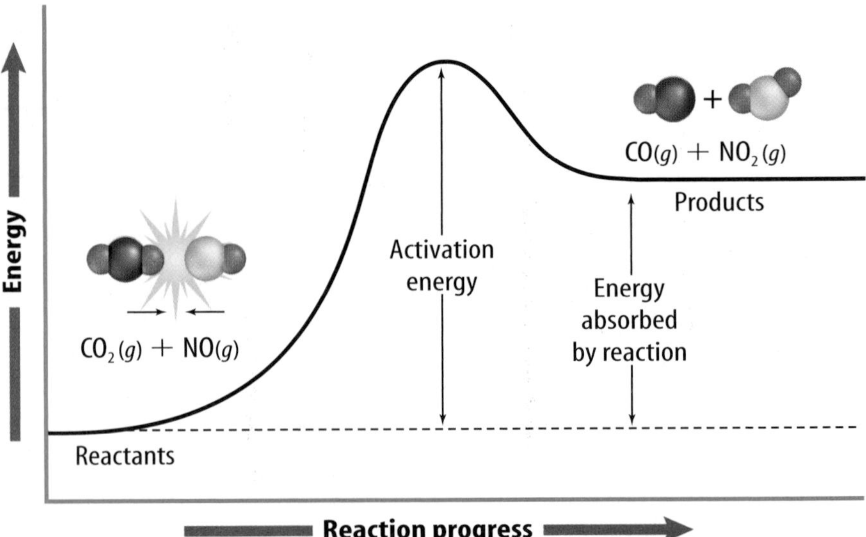

Figure 18 With an endothermic reaction, the reactants have a lower energy level than the products. The reactants must overcome the activation energy barrier in order to form new products.

Endothermic Reactions With an endothermic reaction, the chemical reaction will not take place unless energy is added. A constant source of energy must be added to keep the reaction going. The products have more stored energy than the reactants.

Figure 18 shows an energy diagram for the reaction of carbon dioxide (CO_2) and nitrogen monoxide (NO). With an endothermic reaction, the reactants have a lower energy level than the products. In order for the products to form, an input of energy is needed for the reactants to overcome the activation energy barrier.

Exergonic Reactions

You have seen many reactions that release energy. Chemical reactions that release energy are called exergonic (ek sur GAH nihk) reactions. In these reactions, less energy is required to break the original bonds than is released when new bonds are formed. As a result, some form of energy, such as light or heat, is given off by the reaction. The familiar glow from the reaction inside a glow stick, shown in **Figure 19,** is an example of an exergonic reaction, which produces visible light. In other reactions, however, energy is released as heat. This is the case with some heat packs that are used to treat muscle aches and other medical conditions. Another release of energy is used to power rockets, as shown in **Figure 20.**

Figure 19 Glow sticks contain three different chemicals—an ester and a dye in the outer section and hydrogen peroxide in a center glass tube. Bending the stick breaks the tube and mixes the three components. The energy released is in the form of visible light.

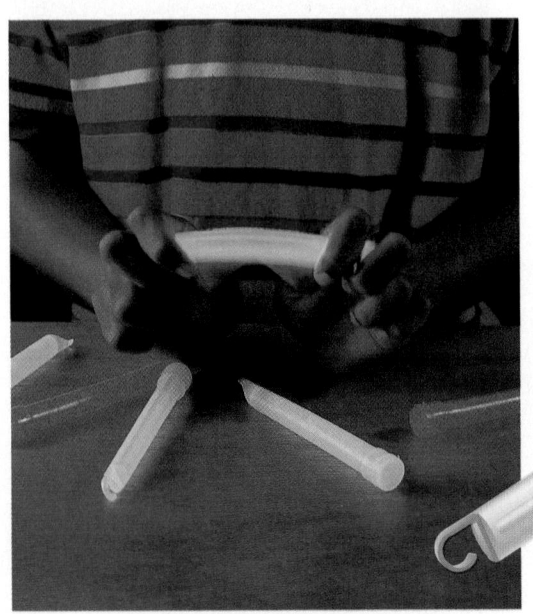

NATIONAL GEOGRAPHIC

VISUALIZING CHEMICAL ENERGY

Figure 20

Rockets burn fuel to provide the thrust necessary to propel them upward. In 1926, engineer Robert Goddard used gasoline and liquid oxygen to propel the first ever liquid-fueled rocket. Although many people at the time ridiculed Goddard's space travel theories, his rockets eventually served as models for those that have gone to the Moon and beyond. A selection of rockets—including Goddard's—is shown here. The number below each craft indicates the amount of thrust—expressed in newtons (N)—produced during launch.

◀ SPACE SHUTTLE The main engines produce enormous amounts of energy by combining liquid hydrogen and oxygen. Coupled with solid rocket boosters, they produce over 32.5 million newtons (N) of thrust to lift the system's 2 million kg off the ground.

▶ JUPITER C This rocket launched the first United States satellite in 1958. It used a fuel called hydyne plus liquid oxygen.

▼ GODDARD'S MODEL ROCKET Although his first rocket rose only 12.6 m, Goddard successfully launched 35 rockets in his lifetime. The highest reached an altitude of 2.7 km.

▼ LUNAR MODULE Smaller rocket engines, like those used by the Lunar Module to leave the Moon, use hydrazine-peroxide fuels. The number shown below indicates the fixed thrust from one of the module's two engines; the other engine's thrust was adjustable.

400 N

369,350 N

32,500,000 N

15,920 N

Creating a Colorful Chemical Reaction

Procedure

1. Complete the safety form.
2. Sprinkle a few crystals of **copper(II) bromide** into the **test tube.** Record the color of the crystals in your Science Journal.
3. Pour 5 mL of **water** into the test tube. Record your observations.
4. Slowly add more water and observe and record what happens.

Analysis

1. What color were the crystals after water was added?
2. What color were they when you added more water?
3. What caused this color change?

Exothermic Reactions When the energy given off in a reaction is primarily in the form of heat, the reaction is called an **exothermic reaction.** The burning of wood and the explosion of dynamite are exothermic reactions. Iron rusting is also exothermic, but, under typical conditions, the reaction proceeds so slowly that it's difficult to detect any temperature change.

Reading Check *Why is a log fire considered to be an exothermic reaction?*

Exothermic reactions provide most of the power used in homes and industries. Fossil fuels that contain carbon, such as coal, petroleum, and natural gas, combine with oxygen to yield carbon dioxide gas and energy. Unfortunately impurities in these fuels, such as sulfur, burn as well, producing pollutants such as sulfur dioxide. Sulfur dioxide combines with water in the atmosphere, producing acid rain.

Energy Release The energy diagram for an exothermic reaction is the reverse of an endothermic reaction. With an exothermic reaction, the products have less stored energy than the reactants. As shown in **Figure 21,** the reactants, carbon monoxide (CO) and nitrogen dioxide (NO_2) have a higher energy level than the products. The molecules have enough energy to overcome the activation energy barrier.

Figure 21 In an exothermic reaction, molecules have enough energy to overcome the activation energy barrier. Energy is released with the formation of new products.

Energy

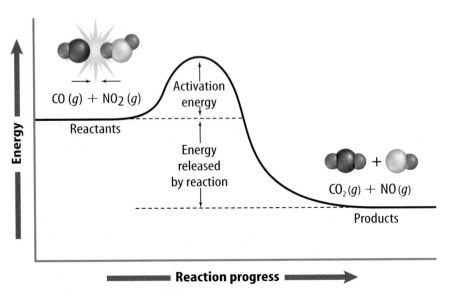

$CO(g) + NO_2(g)$

Reactants

Activation energy

Energy released by reaction

$CO_2(g) + NO(g)$

Products

Reaction progress

Chemical Reaction Rates

According to the kinetic theory of matter, atoms and molecules are always moving. In order for a chemical reaction to occur, the atoms and molecules that are the reactants have to bump into each other or collide. The **rate of reaction** is the speed at which reactants are consumed and products are produced in a given reaction.

Reaction rate is important in the manufacturing industry because the faster the product can be made, the less it usually costs. Sometimes a fast reaction rate is undesirable, such as the rate of reaction that causes food spoilage. In this case, the slower the reaction rate, the longer the food will stay edible. What conditions control the reaction rate, and how can the rate be changed?

Temperature Energy is needed by atoms and molecules to break old bonds and to form new ones. One way to increase the activation energy is to add heat or increase the temperature. With an increase in temperature, atoms and molecules move faster and kinetic energy increases. With faster moving atoms and molecules, more molecules have kinetic energy greater than activation energy. The atoms and molecules now will have enough energy to break old bonds at higher temperature, which will increase the reaction rate. A chemical reaction will go faster at higher temperature and slower at lower temperature as shown in **Figure 22.**

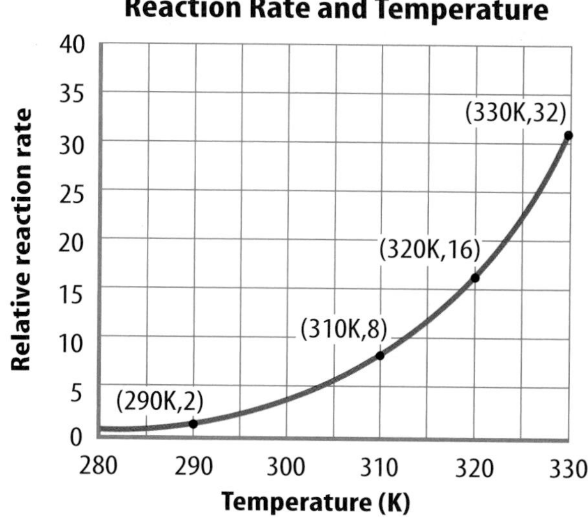

Figure 22 Increasing the temperature of a reaction increases the frequency of collisions, which will increase the rate of reaction.

Concentration When you walk through the hallways at school, you are more likely to bump into another student if the hallways are crowded. A similar situation is shown in **Figure 23.** The closer atoms and molecules are to each other, the greater the chance of collision. The amount of substance present in a certain volume is called its concentration. Increasing the concentration of a substance increases the reaction rate.

Surface Area Only atoms or molecules in the outer layer of a substance can collide with other reactants. When a substance is finely divided, it has a larger surface area than when it was whole. Increasing the surface area increases the chance for collisions, which will increase the reaction rate. A powdered reactant will produce a faster reaction rate than a lump of the same reactant. For instance, grain elevator operators have to be careful of explosions and fire. The dust from the grains in the bin has a larger surface area than the whole grains. With an ignition spark, the dust will catch fire rapidly and cause an explosion.

Agitation If you are making lemonade, the water, sugar, and lemon juice are mixed in order to get the product. Agitation or stirring is a physical process that allows reactants to mix. A low stirring rate will slow the reaction due to fewer collisions. Chemical reactions can be controlled by agitation.

Pressure Another way to influence the reaction rate is with pressure. By increasing the pressure of gases, molecules have less room to move about and the concentration of the reactants increases. This will boost the chance of collisions, which means the reaction rate increases. Decreasing the pressure means fewer collisions, and lower reaction rate.

Collisions are more frequent in a concentrated solution.

Collisions are less frequent in a dilute solution.

Figure 23 People are more likely to collide in crowds. Molecules behave similarly.

Catalysts Metals, such as platinum and palladium, are used as catalysts in the exhaust systems of automobiles. What reactions do you think they catalyze?

Catalysts and Inhibitors Some reactions proceed too slowly to be useful. To speed them up, a catalyst reaction can be added. A **catalyst** is substance that speeds up a chemical reaction without being permanently changed itself. When you add a catalyst to a reaction, the mass of the product that is formed remains the same, but it will form more rapidly. A catalyst lowers the activation energy of the reaction. With the lowering of the activation energy, collisions among molecules now will become more effective, thus increasing the rate of reaction.

Reading Check *Explain how a catalyst speeds up a reaction.*

At times, it is worthwhile to prevent certain reaction from occurring. Substances that are used to slow down a chemical reaction are called **inhibitors.** The food preservations BHT and BHA are inhibitors that prevent spoilage of certain foods, such as cereals and crackers.

One thing to remember when thinking about catalysts and inhibitors is that they do not change the amount of product produced. They only change the rate of production. Catalysts increase the rate and inhibitors decease the rate. Other factors, including concentration, pressure, and temperature, also affect the rate of reaction and must be considered when catalyzing or inhibiting a reaction.

section 4 review

Summary

Conservation of Energy

- In a chemical reaction, the total amount of energy is the same. Energy is neither created nor destroyed.

Chemical Reactions and Energy

- Chemical reactions release or absorb energy, as chemical bonds are broken and formed.
- An endothermic reaction absorbs heat.
- An exothermic reaction releases heat.

Chemical Reaction Rates

- Temperature, concentration, and pressure can affect reaction rates.
- In an exothermic reaction, reactants have more energy than the products.
- In an endothermic reaction, the reactants have less energy than the products.

Self Check

1. **Explain** whether the term *endothermic* can be used to describe physical changes.
2. **Describe** how temperature can affect reaction rate.
3. **Illustrate** As an exothermic reaction, methane and oxygen will react to form carbon dioxide and water. Draw the activation energy chart showing where the reactants and products belong.
4. **Explain** why crackers containing BHT stay fresh longer than those without it.
5. **Think Critically** To develop a product that warms people's hands, would you choose an exothermic or endothermic reaction to use? Why?

Applying Math

6. **Calculate** If an endothermic reaction begins at 26°C and loses 2°C per minute, how long will it take to reach 0°C?
7. **Use Graphs** Create a graph of the data in question 6. After 5 min, what is the temperature of the reaction?

More Section Review gpescience.com

To Glow or Not to Glow

Many chemical reactions release energy as light. The light seen from a light stick is the result of a chemical reaction.

Real-World Problem

How does changing the water temperature affect the amount of light produced from a light stick?

Goals

- **Observe** the effect of temperature on a light stick.
- **Explain** how temperature affects the rate of reaction.

Possible Materials

light sticks (3)
400-mL beakers (4)
water
ice
hot plate
thermometer
graduated cylinder

Safety Precautions

Procedure

1. Complete the safety form before you begin.
2. Write a hypothesis in your Science Journal about how the light stick will be affected by temperature.
3. Prepare a hot water bath by pouring 200-mL of hot water into a beaker. Record the temperature of the hot water.
4. Prepare a 200-mL ice water bath and a 200-mL room temperature water bath. Record the temperature of the water in each bath.
5. Bend a light stick until the inner capsule snaps. Shake the light stick for 10 seconds. Place the light stick into the hot water bath. Record your observations.
6. Repeat step 5 with the remaining light sticks using the ice water bath and the room temperature bath, instead. Record your observations.

Conclude and Apply

1. **Summarize** your observations.
2. **Explain** why the amount of light released is different above and below the water level.
3. **Evaluate** your hypothesis.
4. **Infer** how temperature affects the light intensity.

Compare your results with those of your classmates. **For more help refer to the** Science Skill Handbook.

Investigating Reaction Conditions

Goals
- **Evaluate** the effect of concentration on the rate of a chemical reaction.
- **Examine** the effect of temperature on the rate of a chemical reaction.

Materials
water
vinegar solution
baking soda
balloons (3)
plastic 0.5-liter soft-drink bottles (3)
test tube
150-mL beakers (3)
100-mL graduated cylinder
marker
tape measure
stop watch
**clock with second hand*

**Alternate material*

Safety Precautions

Real-World Problem

Many people believe that you cannot perform chemical reactions without expensive equipment or costly chemicals. But this isn't true; chemical reactions happen everywhere. All you need is a food store to find many substances that can produce exciting chemical reactions. What factors determine how much product is produced in a chemical reaction or how fast a reaction occurs?

Procedure

1. Complete the safety form before you begin.
2. Copy the data tables into your Science Journal.
3. Prepare the following solutions: a 50% vinegar solution by mixing 50 mL of vinegar with 50 mL of water; a 30% vinegar solution by mixing 30 mL vinegar with 70 mL of water and a 10% vinegar solution by mixing 10 mL of vinegar with 90 mL of water. Pour these solutions into labeled 0.5-L plastic bottles.
4. Mark a small test tube about 1–2 cm from its bottom. Fill the test tube to the line with baking soda. Pour the baking soda into one balloon. Repeat, adding baking soda to the other two balloons.

5. Place the mouth of one balloon over the mouth of one 0.5-L bottle. Do not let any of the baking soda fall into the solution.
6. Lift each balloon to allow the baking soda to fall into each vinegar solution. Observe time of reaction and measure how much each balloon inflates. Record your observations in your data table.
7. Rinse the plastic bottles with water. Prepare three 30% vinegar solutions. Bottle A will have cold water added. Bottle B will have room temperature water added. Bottle C will have hot water added.
8. Prepare the baking soda in the balloons as in step 3. Add the baking soda to the solution. Time how long it takes for the reaction to finish. Measure how much each balloon inflates. Record your observations in the data table.

Concentration Table

Concentration	50%	30%	10%
Observations	Do not write in this book.		

Temperature Table

Temperature	Cold Temp	Room Temp	Hot Temp
Observations	Do not write in this book.		

Analyze Your Data

1. **Describe** how increasing the concentration of a solution affects the rate of a chemical reaction.
2. **Summarize** how temperature affects the rate of a chemical reaction.
3. **Explain** why the balloons became inflated.

Conclude and Apply

1. **Infer** how a change in pressure can affect the rate of chemical reaction of this system.
2. **Explain** how temperature and balloon inflation are related.

Communicating Your Data

Compare your results with other students in your class.

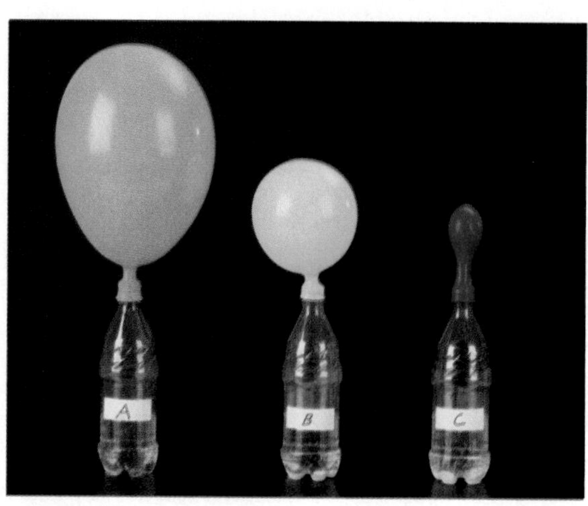

A Clumsy Move Pays Off

Hilaire de Chardonnet

Great scientific discoveries can happen in some very unlikely ways. Most people might not think that an accidental spill left uncleaned would become significant, but that's exactly what led a chemist named Hilaire de Chardonnet (hee LAYR • duh • shar doh NAY) to his discovery. In 1878, Chardonnet accidentally knocked over some nitrate chemicals. He put off cleaning up the mess and ended up inventing artificial silk.

Silk is produced naturally by silkworms. In the mid-1800s, though, silkworms were dying from disease and the silk industry was suffering. Businesses were going under and people were put out of work. Many scientists were working to develop a solution to this problem. Chardonnet had been searching for a silk substitute for years—he just didn't plan to find it by knocking it over!

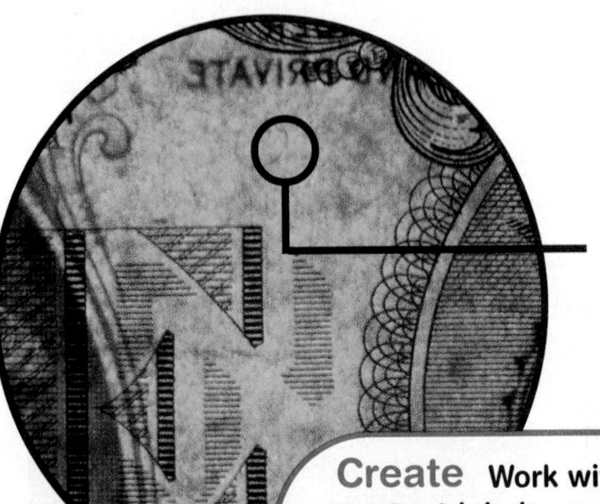

A Messy Discovery

Chardonnet was in his darkroom developing photographs when the accidental spill took place. He decided to clean up the spill later and finish what he was working on. By the time he returned to wipe up the spill, the chemical solution had turned into a thick, gooey mess. When he pulled the cleaning cloth away, the goop formed long, thin strands of fiber that stuck to the cloth. The chemicals had reacted with the cellulose in the wooden table and liquefied it. The strands of fiber looked just like the raw silk made by silkworms.

Within six years, Chardonnet had developed a way to make the fibers into an artificial silk. Other scientists extended his work, developing a fiber called rayon. Today's rayon is made from sodium hydroxide mixed with wood fibers, which is then stranded and woven into cloth.

Rayon has another real-world application. To help prevent counterfeiting, dollars are printed on paper that contains red and blue rayon fibers. If you can scratch off the red or blue, that means it's ink and your bill is counterfeit. If you can pick out the red or blue fiber with a needle, it's a real bill.

Create Work with a partner to examine the fabric content labels on the inside collars of your clothes. Research the materials, then make a data table that identifies their characteristics.

Oops! For more information, visit gpescience.com

Reviewing Main Ideas

Section 1 Chemical Changes

1. In a chemical reaction, one or more substances are changed to new substances.
2. The substances that react are called reactants, and the new substances formed are called products. Charcoal, the reactant shown below, is almost pure carbon.

3. The law of conservation of mass states that in chemical reactions, matter is neither created nor destroyed, just rearranged.
4. The law of conservation of energy states that energy is neither created nor destroyed.

Section 2 Chemical Equations

1. Balanced chemical equations give the exact number of atoms involved in the reaction.
2. A balanced chemical equation has the same number of atoms of each element on both sides of the equation. This satisfies the law of conservation of mass.
3. When balancing chemical equations, change only the coefficients of the formulas, never the subscripts. To change a subscript would change the compound.

Section 3 Classifying Chemical Reactions

1. In synthesis reactions, two or more substances combine to form another substance.
2. In single-displacement reactions, one element replaces another in a compound.
3. In double-displacement reactions, ions in two compounds switch places, often forming a gas or insoluble compound.
4. Using the activity series chart, scientists can determine which metal can replace another metal.

Section 4 Reaction Rates and Energy

1. When energy is released in the form of heat, the reaction is exothermic. An endothermic reaction requires heat to get started. This flame is releasing light and heat energy.

2. Temperature, concentration, and the presence of a catalyst can change the rate of reaction.
3. Activation energy is the minimum amount of energy needed to start a reaction.
4. Reactions may be sped up by adding catalysts, such as a nickel catalyst. A reaction can be slowed down by adding inhibitors.

FOLDABLES™ Use the Foldable that you made at the beginning of this chapter to help you review chemical reactions.

chapter 23 Review

Using Vocabulary

activation energy p. 735
balanced chemical equation p. 726
catalyst p. 740
chemical equation p. 723
chemical reaction p. 720
combustion reaction p. 730
decomposition reaction p. 731
double-displacement reaction p. 732
endothermic reaction p. 735
exothermic reaction p. 738
inhibitor p. 740
product p. 720
rate of reaction p. 738
reactant p. 720
single-displacement reaction p. 731
synthesis reaction p. 731

For each set of vocabulary words below, explain the relationship that exists.

1. synthesis reaction—decomposition reaction
2. reactant—product
3. catalyst—inhibitor
4. exothermic reaction—endothermic reaction
5. chemical reaction—product
6. single-displacement reaction—double-displacement reaction
7. chemical reaction—synthesis reaction

Checking Concepts

Choose the word or phrase that best answers the question.

8. Oxygen gas is always written as O_2 in chemical equations. What term is used to describe the "2" in this formula?
   A) catalyst
   B) coefficient
   C) product
   D) subscript

9. In an endothermic reaction, which has the lower energy?
   A) catalyst
   B) inhibitor
   C) product
   D) reactant

10. What law is based on the experiments of Lavoisier?
    A) atomic theory
    B) collision theory
    C) conservation of energy
    D) conservation of mass

11. What must an element be in order to replace another element in a compound?
    A) less inhibiting
    B) less reactive
    C) more inhibiting
    D) more reactive

12. How do you indicate that a substance in an equation is a solid?
    A) (l)
    B) (g)
    C) (s)
    D) (aq)

13. Classify the reaction:
    $2H_2(g) + O_2(g) \rightarrow 2H_2O(g)$.
    A) combustion
    B) decomposition
    C) single displacement
    D) synthesis

14. What compound is the food additive BHA?
    A) catalyst
    B) inhibitor
    C) oxidized
    D) reduced

15. Which symbol indicates that a substance is dissolved in water when writing an equation?
    A) (aq)
    B) (s)
    C) (g)
    D) (l)

16. What word would you use to describe HgO in the reaction that Lavoisier used to show conservation of mass?
    A) catalyst
    B) inhibitor
    C) product
    D) reactant

17. When hydrogen burns, what is oxygen's role?
    A) catalyst
    B) inhibitor
    C) product
    D) reactant

Vocabulary PuzzleMaker gpescience.com

Interpreting Graphics

18. Copy and complete the concept map using the following terms: *oxidized, redox reactions, lost, reduced, oxidation, gained,* and *reduction.*

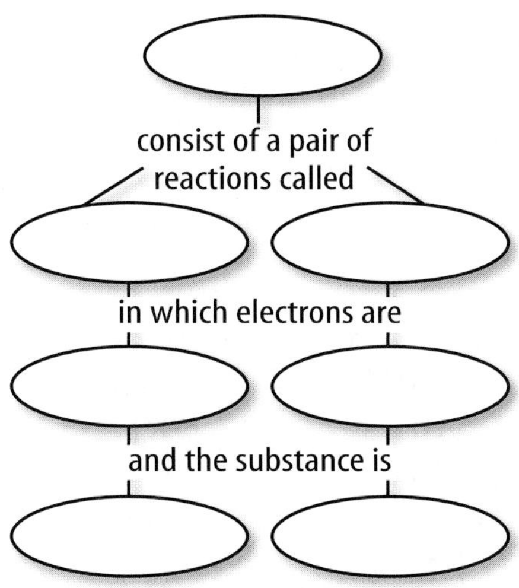

19. **Sequence** Sometimes a bond formed in a chemical reaction is weak and the product breaks apart as it forms. This is shown by a double arrow in chemical equations. Copy and complete the concept map, using the words *product(s)* and *reactant(s).* In the blank in the center, fill in the formulas for the substances appearing in the reversible reaction.

$$H_2(g) + I_2(g) \rightleftharpoons 2HI(g)$$

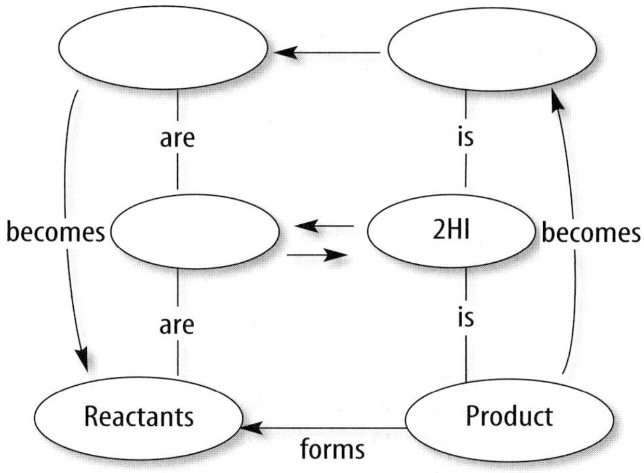

Thinking Critically

20. **Write a balanced chemical equation** for the reaction of propane $C_3H_8(g)$ burning in oxygen to form carbon dioxide and water vapor.

21. **Interpret** the balanced chemical equation from question 23 to explain the law of conservation of mass.

22. **Hypothesize** Zn is placed in a solution of $Cu(NO_3)_2$ and Cu is placed in a $Zn(NO_3)_2$ solution. In which of these will a reaction occur?

23. **Predict** what kind of energy process happens when lye, $NaOH(s)$, is put in water and the water gets hot.

24. **Recognize Cause and Effect** Sucrose, or table sugar, is a disaccharide. This means that sucrose is composed of two simple sugars chemically bonded together. Sucrose can be separated into its components by heating it in an aqueous sulfuric acid solution. Research what products are formed by breaking up sucrose. What role does the acid play?

25. **Classify** Make an outline with the general heading *Chemical Reactions.* Include the five types of reactions, with a description and example of each.

Applying Math

26. **Use Numbers** When 46 g of sodium were exposed to dry air, 62 g of sodium oxide formed. How many grams of oxygen from the air were used?

27. **Calculate Mass** Chromium is produced by reacting its oxide with aluminum. If 76 g of Cr_2O_3 and 27 g of Al completely react to form 51 g of Al_2O_3, how many grams of Cr are formed?

Standardized Test Practice

Record your answers on the answer sheet provided by your teacher or on a sheet of paper.

Multiple Choice

Use the illustration below to answer question 1.

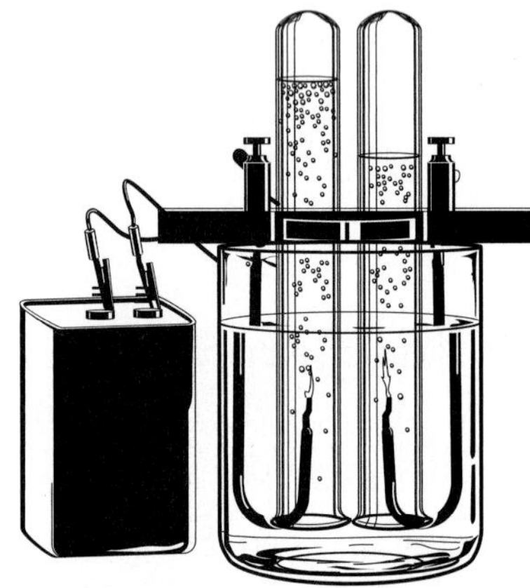

1. The illustration above shows a chemical reaction in which water decomposes into hydrogen gas and oxygen gas when an electric current is passed through it. Which of the following is the correct chemical equation for this reaction?

   A. $H_2O(l) \rightarrow H_2(g) + O(g)$

   B. $H_2O(l) \rightarrow 2H(g) + O(g)$

   C. $2H_2O(l) \rightarrow 2H_2(g) + 2O(g)$

   D. $2H_2O(l) \rightarrow 2H_2(g) + O_2(g)$

2. Which substance is the precipitate in the following reaction?

   $Ba(NO_3)_2(aq) + K_2SO_4(aq) \rightarrow BaSO_4(s) + 2KNO_3(aq)$

   A. $Ba(NO_3)_2$

   B. K_2SO_4

   C. $BaSO_4$

   D. KNO_3

3. Which reaction is endothermic?

   A. iron rusting

   B. burning wood

   C. exploding dynamite

   D. mixing Epsom salt in water

Use the figure below to answer question 4.

4. Which metal would most likely replace lead in a solution?

   A. potassium

   B. copper

   C. silver

   D. gold

5. Which type of reaction is the opposite of a synthesis reaction?

   A. displacement

   B. reversible

   C. combustion

   D. decomposition

Gridded Response

6. A reaction of 24 g of sodium hydroxide with 20 g of hydrogen chloride gives off 12 g of water. What mass of sodium chloride is formed in this reaction?

7. From an experiment to separate water into hydrogen and oxygen, a student collected 10.0 g of hydrogen and 79.4 g of oxygen. How much water was originally involved in the process?

Short Response

8. What is a synthesis reaction?

Use the illustration to answer question 9.

9. The illustration above shows the reaction of aqueous nickel(II) chloride, $NiCl_2$, and aqueous sodium hydroxide, NaOH, to form solid nickel(II) hydroxide, $Ni\ (OH)_2$, and aqueous sodium chloride, NaCl. State the conversation of mass as it applies to this chemical reaction.

10. Food preservatives are a type of inhibitor. Explain why this is useful in foods.

Extended Response

11. Explain what is wrong with the following balanced equation:

$4Al(s) + 6O(g) \rightarrow 2Al_2O_3(s)$

What is the correct form of the equation?

Use the illustration to answer question 12.

12. The drawing above illustrates a chemical reaction between magnesium, Mg, and oxygen gas, O_2. This reaction is exergonic and exothermic. Explain what these terms mean and how you can tell that a chemical reaction is exergonic or exothermic.

13. The reaction of magnesium and oxygen gas forms magnesium oxide, MgO. Write the chemical equation for this reaction and explain the process you use to balance the equation.

Test-Taking Tip

Missing Information Questions often will ask about missing information. Notice what is missing as well as what is given.

chapter 24

Solutions, Acids, and Bases

chapter preview

sections

Mixed-Up Chemistry

This diver is swimming in a solution. Seas are solutions that contain dissolved solids, such as calcium and magnesium, and dissolved gases, such as carbon dioxide, in water. In this chapter, you will learn about solutions, acids, and bases.

Science Journal Are all liquids solutions, and are all solutions liquids? Answer this question in your Science Journal. Check your answer later and revise it if you've learned - differently.

Start-Up Activities

Magic Solutions

To prevent spies from learning military secrets, scientists developed a special ink that is invisible on paper at room temperature. When the paper is warmed by a candle or lightbulb, the message appears. In this lab, you will create an invisible message.

1. Complete the safety form.
2. Write a message or draw a picture on a white sheet of paper using a cotton swab dipped in either phenolphthalein indicator or universal indicator. In your Science Journal, record which indicator you used.
3. Allow the paper to dry thoroughly.
4. Dip a clean cotton ball in household ammonia (a base) or vinegar (an acid) and rub it across your paper. Record which solution you used and your observations in your Science Journal.
5. Compare your results with those of others.
6. **Think Critically** In your Science Journal, infer how the magic solution works.

Preview this chapter's content and activities at gpescience.com

Solutions, Acids, and Bases Make the following Foldable to help you identify the main characteristics of solutions, acids, and bases.

STEP 1 **Fold** a sheet of paper vertically from side to side. Make the front edge about 1.25 cm longer than the back edge.

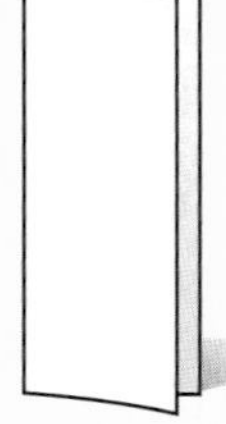

STEP 2 **Turn** lengthwise and **fold** into thirds.

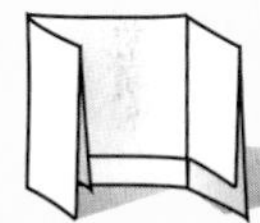

STEP 3 **Unfold and cut** only the top layer along both folds to make three tabs.

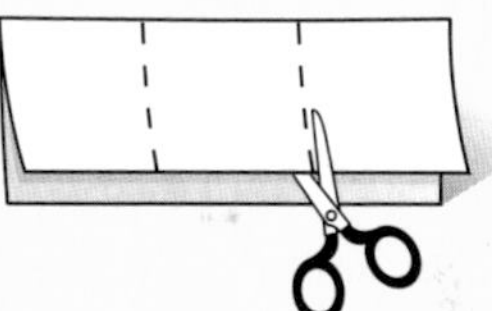

STEP 4 **Label** each tab.

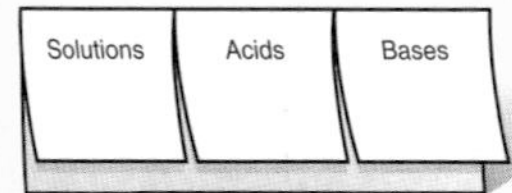

Reading for Main Ideas As you read the chapter, list the characteristics of solutions, acids, and bases.

section 1

How Solutions Form

Reading Guide

What You'll Learn

- **Identify** the components of a solution.
- **Explain** how things dissolve.
- **Identify** the factors that affect the rates at which solids and gases dissolve in liquids.

Why It's Important

Solutions play an important role in your everyday life. Examples of everyday solutions are sports drinks, dishwashing detergents, and shampoos.

Review Vocabulary

homogeneous mixture: solid, liquid, or gas that contains two or more substances blended evenly throughout

New Vocabulary

- solution
- solute
- solvent
- aqueous solution

What is a solution?

Hummingbirds are fascinating creatures. They can hover for long periods while they sip nectar from flowers through their long beaks. To attract hummingbirds, many people use feeder bottles containing a red liquid, as shown in **Figure 1.** The liquid is a solution of sugar and red food coloring in water.

Suppose you are making some hummingbird food. When you add sugar to water and stir, the sugar crystals disappear. When you add a few drops of red food coloring and stir, the color spreads evenly throughout the sugar water. Why does this happen?

Hummingbird food is one of many solutions. A **solution** is a mixture that has the same composition, color, density, and even taste throughout. The reason you no longer see the sugar crystals and the reason the red dye spreads out evenly is that they have formed a completely homogeneous mixture. The sugar crystals broke up into sugar molecules, the red dye into its molecules, and both mixed evenly among the water molecules.

Figure 1 The liquid solution in the hummingbird feeder contains sugar, food coloring, and other substances.

Solutes and Solvents

In a solution, one substance is dissolved in another. In the hummingbird solution, sugar which is the substance being dissolved, is the **solute.** Water, which is the substance doing the dissolving, is the **solvent.** When a solid dissolves in a liquid, the solid is the solute and the liquid is the solvent. In salt water, salt is the solute and water is the solvent. In carbonated soft drinks, carbon dioxide gas is one of the solutes and water is the solvent. When a liquid dissolves in another liquid, the substance present in the larger amount is usually called the solvent. A solution in which water is the solvent is called an **aqueous** (A kwee us) **solution.**

Reading Check *How do you know which substance is the solute in a solution?*

Nonliquid Solutions Solutions also can be gaseous or even solid. Examples are shown in **Figure 1** and **Figure 2.** Did you know that the air you breathe is a solution? In fact, all mixtures of gases are solutions. Air is a solution of 78 percent nitrogen, 21 percent oxygen, and small amounts of other gases such as argon, carbon dioxide, and hydrogen. The sterling silver and brass used in musical instruments are an example of a solid solution. The sterling silver contains 92.5 percent silver and 7.5 percent copper. The brass is a solution of copper and zinc metals. Solid solutions are known as alloys. They are made by melting the metal solute and solvent together. Most coins, as shown in **Figure 3,** are alloys.

Surface-Coating Chemist
A scientist who develops products such as house paints and polyurethane wood finishers is called a surface-coating chemist. Paints, varnishes, and synthetic clear coatings are solutions. These products are designed to protect the surface of an object from the environment. Make a list of five items that have a protective coating that was applied as a solution.

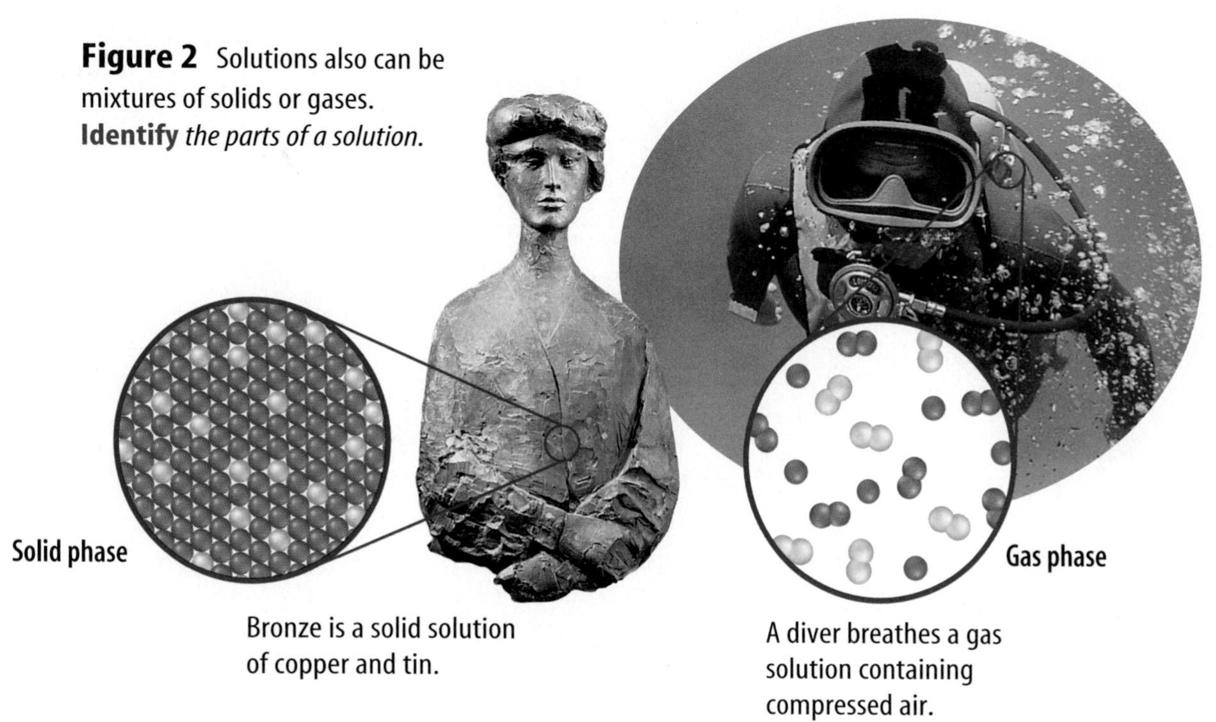

Solid phase
Gas phase
Bronze is a solid solution of copper and tin.
A diver breathes a gas solution containing compressed air.

Figure 2 Solutions also can be mixtures of solids or gases.
Identify *the parts of a solution.*

NATIONAL GEOGRAPHIC VISUALIZING METAL ALLOYS

Figure 3

Have you ever accidentally put a non-United States coin into a vending machine? Of course, the vending machine didn't accept it. If a vending machine is that selective, how can it be fooled by two coins that look and feel very different? This is exactly the case with the silver Susan B. Anthony dollar and the new golden Sacagawea dollar. Vending machines can't tell them apart.

◀ Vending machines recognize coins by size, weight, and electrical conductivity. The size and weight of the Susan B. Anthony coin were easy to copy. Copying the coin's electrical conductivity was more difficult.

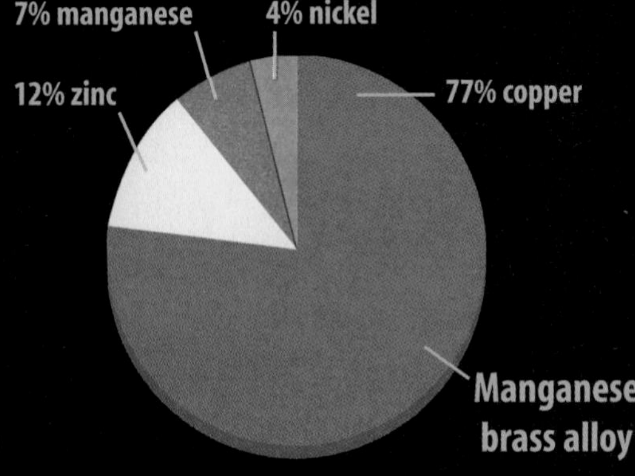

▲ Over 30,000 samples of coin coatings were tested to find an alloy and thickness that would copy the conductivity of the Susan B. Anthony dollar. The final composition of the alloy is shown in the graph above. The key ingredient? Manganese.

Manganese brass alloy

Copper core

Manganese brass alloy

▲ The dollar's copper core is half the coin's thickness. It is sandwiched between two layers of manganese brass alloy.

How Substances Dissolve

Fruit drinks and sports drinks are examples of solutions made by dissolving solids in liquids. Like hummingbird food, both contain sugar as well as other substances that add color and flavor. How do solids such as sugar dissolve in water?

The dissolving of a solid in a liquid occurs at the surface of the solid. To understand how water solutions form, keep in mind two things you have learned about water. Like the particles of any substance, water molecules are constantly moving. Also, water molecules are polar, which means they have a positive area and a negative area. Molecules of sugar also are polar.

How It Happens Molecules of sugar dissolving in water are shown in **Figure 4.** First, water molecules cluster around sugar molecules with their negative ends attracted to the positive ends of the sugar. Then, the water molecules pull the sugar molecules into solution. Finally, the water molecules and the sugar molecules mix evenly, forming a solution.

Reading Check *How do water molecules help sugar molecules dissolve?*

The process described in **Figure 4** repeats as layer after layer of sugar molecules move away from the crystal, until all the molecules are evenly spread out. The same three steps occur for most solid solutes dissolving in a liquid solvent.

Dissolving Liquids and Gases The same process takes place when a gas dissolves in a liquid. Particles of liquids and gases move much more freely than do particles of solids. When gases dissolve in gases or when liquids dissolve in liquids, this movement spreads solutes evenly throughout the solvent, resulting in a homogenous solution.

Dissolving Solids in Solids How can you mix solids to make alloys? Although solid particles do move a little, this movement is not enough to spread them evenly throughout the mixture. The solid metals are first melted and then mixed together. In this liquid state, the metal atoms can spread out evenly and will remain mixed when cooled.

Figure 4 Dissolving sugar in water can be thought of as a three-step process.

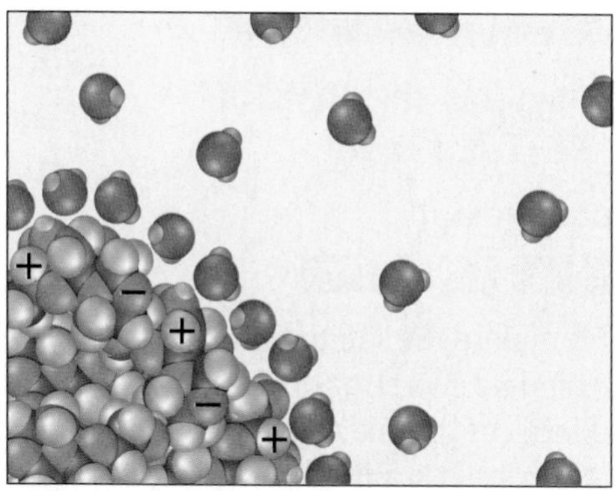

Step 1 Moving water molecules cluster around the sugar molecules as their negative ends are attracted to the positive ends of the sugar molecules.

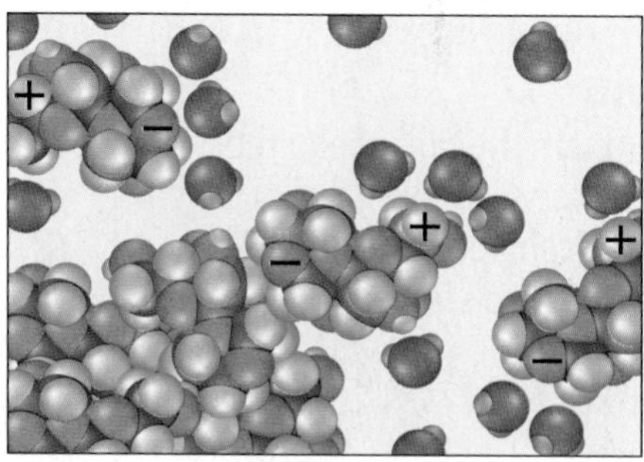

Step 2 Water molecules pull the sugar molecules into solution.

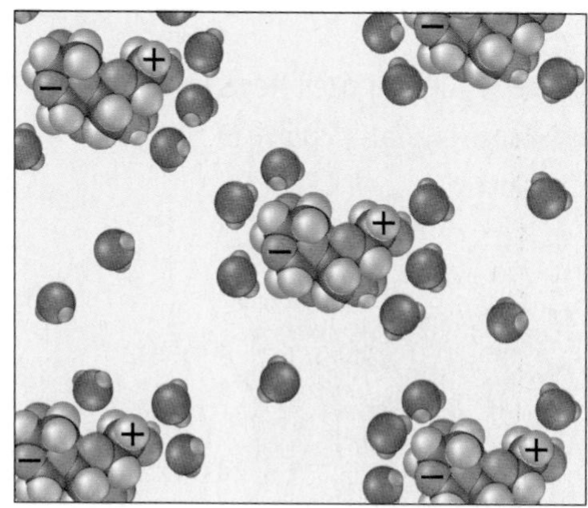

Step 3 Water molecules and sugar molecules spread out to form a homogeneous mixture.

Observing the Effect of Surface Area

Procedure

1. Complete the safety form.
2. Grind up two **sugar cubes.**
3. Place the ground sugar particles into a **medium-sized glass** and place two **unground sugar cubes** into a **similar glass.**
4. Add an equal amount of **water** at room temperature to each glass.

Analysis

1. Compare the times required to dissolve each.
2. What can you conclude about the dissolving rate and surface area?

Try at Home

Rate of Dissolving

When two substances form a solution, the dissolving occurs at different rates. Sometimes the rate at which a solute dissolves into a solvent is fast, while other times it is slow. There are several things you can do to speed up the rate of dissolving—stirring, reducing crystal size, and increasing temperature.

Stirring How does stirring speed up the dissolving process? Think about how you make a drink from a powdered mix. After you add the mix to water, you stir it. Stirring a solution speeds up the dissolving process because it brings more fresh solvent into contact with more solute. The fresh solvent attracts the particles of solute, causing the solid solute to dissolve faster.

Crystal Size Another way to speed the dissolving of a solid in a liquid is to grind large crystals into smaller ones. Suppose you want to use a 5-g crystal of rock candy to sweeten your water. If you put the whole crystal into a glass of water, it might take several minutes to dissolve, even with stirring. However, if you first grind the crystal of rock candy into a powder, it will dissolve in the same amount of water in a few seconds.

Why does breaking up a solid cause it to dissolve faster? Breaking the solid into many smaller pieces greatly increases its surface area, as you can see in **Figure 5.** Because dissolving takes place at the surface of the solid, increasing the surface area allows more solvent to come into contact with more solid solute. Therefore, the speed of the dissolving process increases.

Figure 5 Crystal size affects solubility. Large crystals dissolve in water slowly.

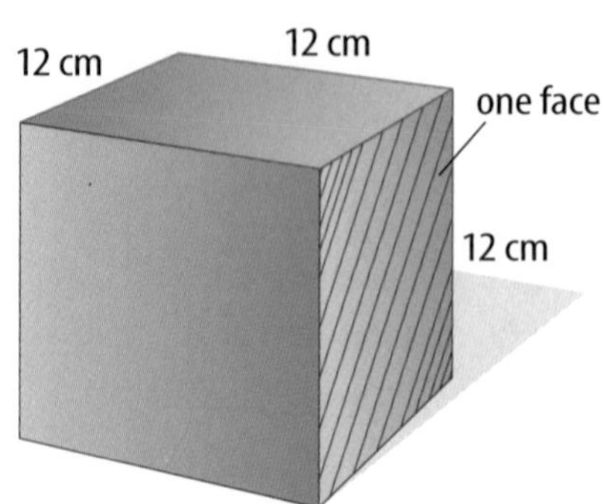

A face of a cube is the outer surface that has four edges.

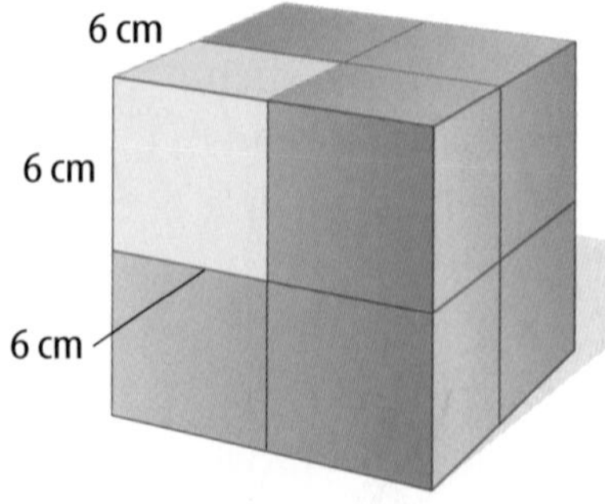

Pull apart the cube into smaller cubes of equal size. You now have 8 cubes and 48 faces.

If you divide the cube into smaller cubes that are 1 cm on a side, you will have 1,728 cubes and 10,368 faces.

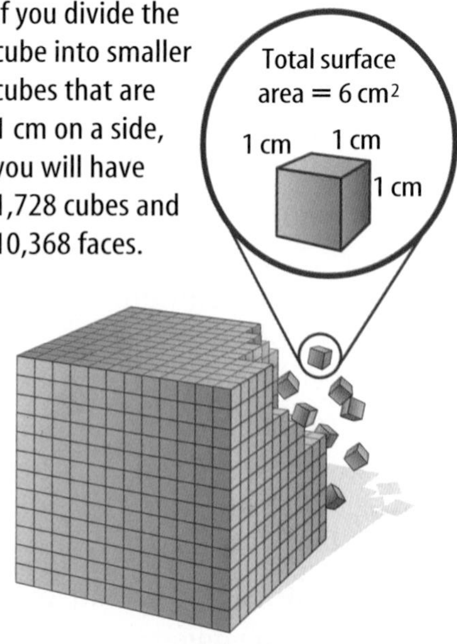

Applying Math Calculate

SURFACE AREA The length, height, and width of a cube are each 1 cm. If the cube is cut in half to form two rectangles, how much new surface area has been created?

IDENTIFY **the known values and the unknown value**

Identify the known values:

- The cube has dimensions of $l = h = w = 1$ cm.
- The rectangular solid has a width $w = 0.5$ cm.
- The rectangular solid has a length and height $l = h = 1$ cm.

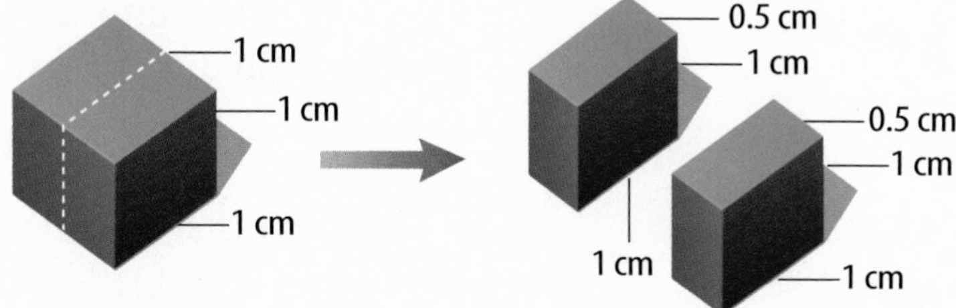

- The cube and the rectangular solids each have six faces: front and back $(h \times w)$; left and right $(h \times l)$; and top and bottom $(w \times l)$. The total surface area of the cube or the rectangular solid is the sum of these areas, or $2(h \times w) + 2(h \times l) + 2(w \times l)$.

Identify the unknown value:

Find the total surface area of the two rectangular solids.

SOLVE **the problem**

The surface area of the cube is:
$2(1 \text{ cm} \times 1 \text{ cm}) + 2(1 \text{ cm} \times 1 \text{ cm}) + 2(1 \text{ cm} \times 1 \text{ cm}) = 6 \text{ cm}^2$

The surface area of the rectangular solid is:
$2(1 \text{ cm} \times 0.5 \text{ cm}) + 2(1 \text{ cm} \times 1 \text{ cm}) + 2(0.5 \text{ cm} \times 1 \text{ cm}) = 4 \text{ cm}^2$

Because there are two rectangular solids, their total surface area is:
$4 \text{ cm}^2 + 4 \text{ cm}^2 = 8 \text{ cm}^2$

To find out how much new surface area has been created, compare the two results:
$8 \text{ cm}^2 - 6 \text{ cm}^2 = 2 \text{ cm}^2$

CHECK **the answer**

Does your answer seem reasonable? Consider what you've done by splitting the cube in two. You have created two new faces, each with a surface area of 1 cm^2; therefore, the answer 2 cm^2 is correct.

Practice Problems

1. A cube of salt with a length, height, and width of 5 cm is attached along a face to another cube of salt with the same dimensions. What is the combined surface area of the new rectangular solid?
2. How much surface area has been lost?

For more practice problems go to page 879, and visit Math Practice at gpescience.com.

Temperature In addition to stirring and decreasing particle size, a third way to increase the rate at which most solids dissolve is to increase the temperature of the solvent. Think about making hot chocolate from a mix. You can make the sugar in the chocolate mix dissolve faster by putting it in hot water instead of cold water. Increasing the temperature of a solvent speeds up the movement of its particles. This increase causes more solvent particles to bump into the solute. As a result, solute particles break loose and dissolve faster.

Controlling the Process Think about how the three factors you just learned about affect the rate of dissolving. Can these factors combine to further increase the rate, or perhaps control the rate of dissolving? Each technique—stirring, crushing, and heating—is known to speed up the rate of dissolving by itself. However, when two or more techniques are combined, the rate of dissolving is even faster. Consider a sugar cube placed in cold water. You know that the sugar cube eventually will dissolve. You can predict that heating the water will increase the rate by some amount. You also can predict that heat and stirring will increase the rate further. Finally, you can predict that crushing the cube combined with heating and stirring will result in the fastest rate of dissolving. Knowing how much each technique affects the rate will allow you to control the rate of dissolving more precisely.

section 1 review

Summary

What is a solution?

- A solution is a uniform mixture.
- A solution has the same composition, color, density, and taste throughout.

Solutes and Solvents

- In a solution, the solute is the substance that is being dissolved; the solvent is the substance that is doing the dissolving.

How Substances Dissolve

- The process of dissolving happens at the surface and is aided by polarity and molecular movement.

Rate of Dissolving

- Stirring, surface area, and temperature affect the rate of dissolving.

Self Check

1. **List** possible ways that phases of matter could combine to form a solution.
2. **Describe** how temperature affects the rate of dissolving.
3. **Describe** how the metal atoms in an alloy are mixed.
4. **Think Critically** Amalgams, which sometimes are used in tooth fillings, are alloys of mercury with other metals. Is an amalgam a solution? Explain.

Applying Math

5. **Find Surface Area** Calculate the surface area of a rectangular solid with dimensions $\ell = 2$ cm, $w = 1$ cm, and $h = 0.5$ cm.
6. **Calculate Percent Increase** If the length of the rectangle in question 5 is increased by 10 percent, by how much will the surface area increase?

More Section Review gpescience.com

section 2

Solubility and Concentration

Reading Guide

What You'll Learn

- **Define** the concept of solubility.
- **Identify** how to express the concentration of solutions.
- **List** and define three types of solutions.
- **Describe** the effects of pressure and temperature on the solubility of gases.

Why It's Important

Solutions such as medicine and lemonade work and taste a particular way because of the specific solution concentrations.

Review Vocabulary

substance: element or compound that cannot be broken down into simpler components

New Vocabulary

- solubility
- concentration
- saturated solution
- unsaturated solution
- supersaturated solution

How much can dissolve?

You can stir several teaspoons of salt into water, and the salt will dissolve. However, if you continue adding salt, eventually the point is reached when no more salt dissolves and the excess salt sinks to the bottom of the glass. This indicates how soluble salt is in water. **Solubility** (sol yuh BIH luh tee) is the maximum amount of a solute that can be dissolved in a given amount of solvent at a given temperature.

Reading Check *What is solubility?*

Comparing Solubilities The amount of a solute that can dissolve in a solvent depends on the nature of these substances. **Figure 6** shows two beakers with the same volume of water and two different solutes. In one beaker, 1 g of solute A dissolves completely, but additional solute does not dissolve and falls to the bottom of the beaker. On the other hand, 1 g of solute B dissolves completely, and two more grams also dissolve before solute begins to fall to the bottom. You can conclude that substance B is more soluble than substance A, if the temperature of the water is the same in both beakers. **Table 1** shows how the solubility of several substances varies at 20°C. For solutes that are gases, the pressure also must be given.

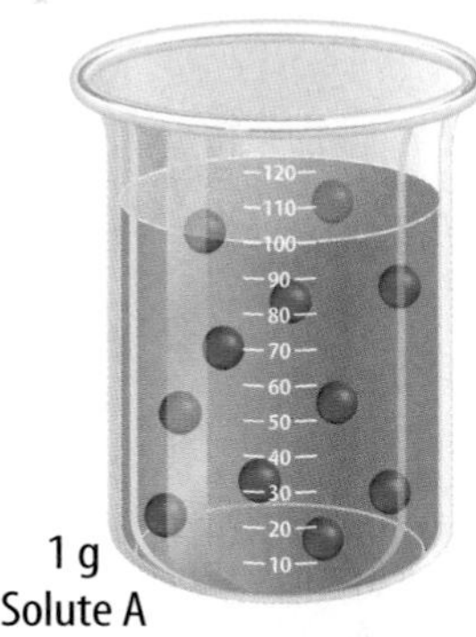

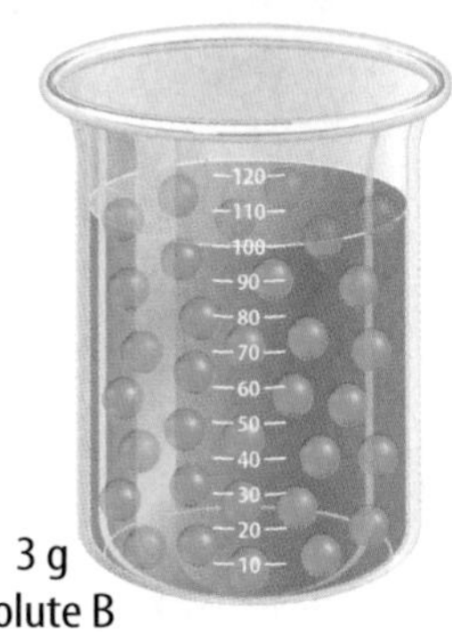

Figure 6 Substance B is more soluble in water than substance A is at the same temperature.

Substance	Solubility in g/100 g of Water
Solid Substances	
Salt (sodium chloride)	35.9
Baking soda (sodium bicarbonate)	9.6
Washing soda (sodium carbonate)	21.4
Lye (sodium hydroxide)	109.0
Sugar (sucrose)	203.9
Gaseous Substances*	
Hydrogen	0.00017
Oxygen	0.005
Carbon dioxide	0.16

Table 1 Solubility of Substances in Water at 20°C

*at normal atmospheric pressure

Concentration

Suppose you add one teaspoon of lemon juice to a glass of water to make lemonade. Your friend adds four teaspoons of lemon juice to another glass of the same size. What is the difference between these two glasses of lemonade? The difference between the two drinks is the amount of water present compared to the amount of lemon juice. The **concentration** describes how much solute is present in a solution compared to the amount of solvent. A concentrated solution is one in which a large amount of solute is dissolved in the solvent. A dilute solution is one that has a small amount of solute in the solvent.

Precise Concentrations How much real fruit juice is there in one of those boxed fruit drinks? You can read the label to find out. *Concentrated* and *dilute* are not precise terms. Concentrations of solutions can be described precisely. One way is to state the percentage by volume of the solute. The percentage by volume of the juice in the bottled drink shown in **Figure 7** is 10 percent. Adding 10 mL of juice to 90 mL of water makes 100 mL of this drink. Commonly, fruit-flavored drinks can have ten percent or less fruit juice. Generally, if two or more liquids are being mixed, the concentration is given in percentage by volume.

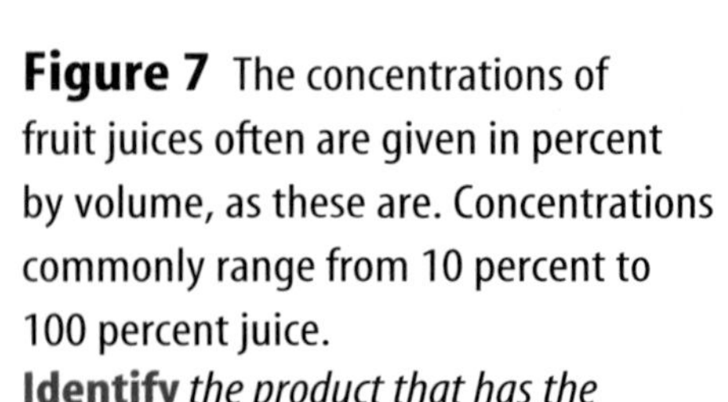

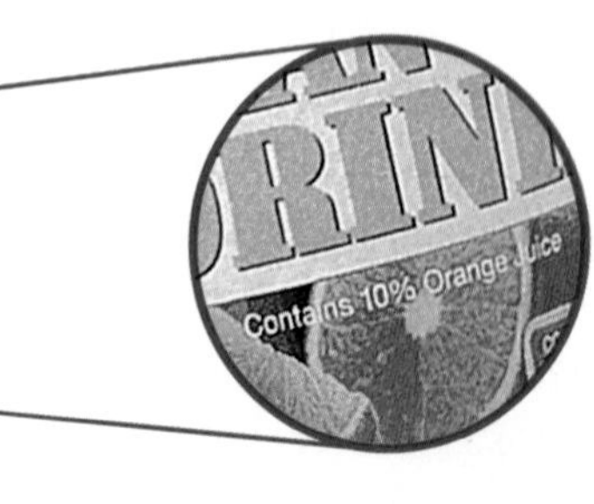

Figure 7 The concentrations of fruit juices often are given in percent by volume, as these are. Concentrations commonly range from 10 percent to 100 percent juice.
Identify *the product that has the highest concentration.*

Types of Solutions

How much solute can dissolve in a given amount of solvent? That depends on a number of factors, including the solubility of the solute. Here you will examine the types of solutions based on the amount of a solute dissolved.

Saturated Solutions If you add 35 g of copper(II) sulfate, $CuSO_4$, to 100 g of water at 20°C, only 32 g will dissolve. You have a saturated solution because no more copper(II) sulfate can dissolve. A **saturated solution** is a solution that contains all the solute it can hold at a given temperature. However, if you heat the mixture to a higher temperature, more copper(II) sulfate can dissolve. Generally, as the temperature of a liquid solvent increases, the amount of solid solute that can dissolve in it also increases. **Table 2** shows the amounts of a few solutes that can dissolve in 100 g of water at different temperatures to form saturated solutions. Some of these data also are shown on the accompanying graph.

Table 2 Solubility of Compounds in g/100 g of Water

Compound	0°C	20°C	100°C
Copper(II) sulfate	23.1	32.0	114
Potassium bromide	53.6	65.3	104
Potassium chloride	28.0	34.0	56.3
Potassium nitrate	13.9	31.6	245
Sodium chlorate	79.6	95.9	204
Sodium chloride	35.7	36.0	39.2
Sucrose (sugar)	179.2	203.9	487.2

Solubility Curves Each line on the graph from **Table 2** is called a solubility curve for a particular substance. You can use a solubility curve to determine how much solute will dissolve at any temperature given on the graph. For example, about 78 g of KBr (potassium bromide) will form a saturated solution in 100 g of water at 47°C. How much NaCl (sodium chloride) will form a saturated solution with 100 g of water at the same temperature?

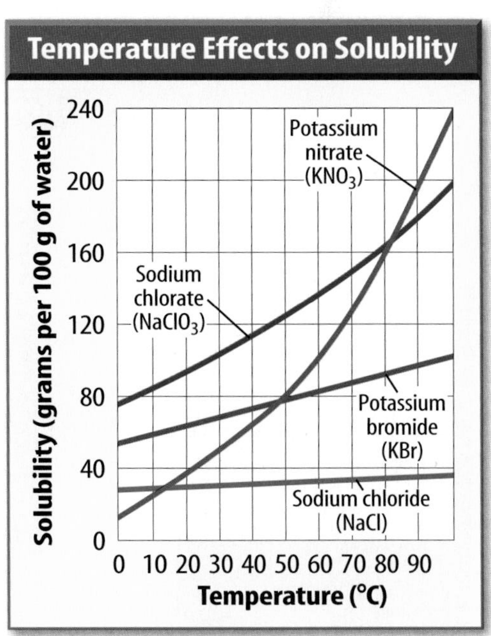

Unsaturated Solutions An **unsaturated solution** is any solution that can dissolve more solute at a given temperature. Each time a saturated solution is heated to a higher temperature, it generally becomes unsaturated. The term *unsaturated* isn't precise. If you look at **Table 2,** you'll see that at 20°C, 35.9 g of NaCl (sodium chloride) forms a saturated solution in 100 g of water. However, an unsaturated solution of NaCl could be any amount less than 35.9 g in 100 g of water at 20°C.

Reading Check *What happens to a saturated solution if it is heated?*

Supersaturated Solutions If you make a saturated solution of potassium nitrate at 100°C and then let it cool to 20°C, part of the solute comes out of solution. This is because the solvent cannot hold as much solute at the lower temperature. Most other saturated solutions behave in a similar way when cooled. However, if you cool a saturated solution of sodium acetate from 100°C to 20°C without disturbing it, no solute comes out. At this point, the solution is supersaturated. A **supersaturated solution** is one that contains more solute than a saturated solution at the same temperature. Supersaturated solutions are unstable. For example, if a seed crystal of sodium acetate is dropped into the supersaturated solution, excess sodium acetate crystallizes out, as shown in **Figure 8.**

Topic: Crystallization
Visit gpescience.com for Web links to information about crystals and crystallization.

Activity Find instructions for a safe "do-it-yourself" home crystallization experiment. Grow the crystals as directed and share the results with the class.

Solution Energy As the supersaturated solution of sodium acetate crystallizes, the solution becomes hot. Energy is given off as the sodium acetate forms crystals. Some portable heat packs use crystallization from supersaturated solutions to produce heat. After crystallization, the heat pack can be reused by heating it to again dissolve all the solute.

Substances, such as ammonium nitrate, draw energy from their surroundings to dissolve. This is what happens when a cold pack is activated to treat minor injuries or to reduce swelling. When the inner bags of ammonium nitrate and water are broken, the ammonium nitrate draws energy from the water, which causes the temperature of the water to drop, which makes the the pack cool.

Reading Check *Why is energy given off between ions and water molecules?*

Figure 8 A supersaturated solution is unstable.
Explain *why this is so.*

A seed crystal of sodium acetate is added to a supersaturated solution of sodium acetate.

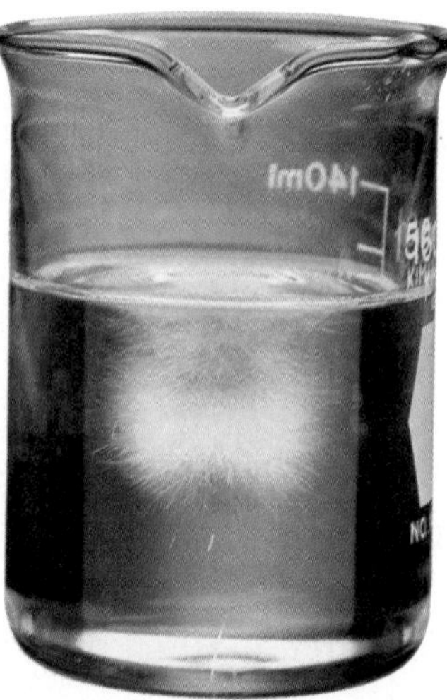

Excess solute immediately crystallizes from solution.

The crystallization reaction draws solute from the solution.

Solubility of Gases

When you shake an opened bottle of soda, it bubbles up and may squirt out. Shaking or pouring a solution of a gas in a liquid causes gas to come out of solution. Agitating the solution exposes more gas molecules to the surface, where they escape from the liquid.

Figure 9 Solutions of gases behave differently from those of solids or liquids. This soda is bottled under pressure to keep carbon dioxide in solution. When the bottle is opened, pressure is released and carbon dioxide bubbles out of solution.

Pressure Effects What might you do if you want to dissolve more gas in a liquid? One thing you can do is increase the pressure of that gas over the liquid. Soft drinks are bottled under increased pressure. This increases the amount of carbon dioxide that dissolves in the liquid. When the pressure is released, the carbon dioxide bubbles out, as shown in **Figure 9.**

Temperature Effects Another way to increase the amount of gas that dissolves in a liquid is to cool the liquid. This is just the opposite of what you do to increase the speed at which most solids dissolve in a liquid. Imagine what happens to the carbon dioxide when a bottle of soft drink is opened. Even more carbon dioxide will bubble out of a soft drink as it gets warmer.

section 2 review

Summary

How much can dissolve?

- Solubility tells how much solute can dissolve in a solvent at a particular temperature.

Concentration

- A concentrated solution has a larger amount of dissolved solute. A dilute solution has a smaller amount of dissolved solute.

Types of Solutions

- Saturated, unsaturated, and supersaturated solutions are defined by how much solute is dissolved.
- Solubility curves help predict how much solute can dissolve at a particular temperature.
- Some supersaturated solutions absorb or give off energy.

Solubility of Gases

- Pressure and temperature affect gases in solution. High pressure and low temperature allow more gas to dissolve.

Self Check

1. **Explain** Do all solutes dissolve to the same extent in the same solvent? How do you know?
2. **Interpret** from **Table 2** the mass of sugar that would have to be dissolved in 100 g of water to form a saturated solution at 20°C.
3. **Determine** which solution is more concentrated: 17 g of solute X dissolved in 100 mL of water at 23°C or 26 g of solute Z dissolved in 100 mL of water at 23°C.
4. **Identify** the type of solution you have if, at 35°C, the solute continues to dissolve as you add more.
5. **Think Critically** Explain how keeping a carbonated beverage capped helps keep it from going "flat."

Applying Math

6. **Calculate Cost** By volume, orange drink is ten percent each of orange juice and corn syrup. A 1.5-L can of the drink costs $0.95. A 1.5-L can of orange juice is $1.49, and 1.5 L of corn syrup is $1.69. Per serving, does it cost less to make your own orange drink or buy it ?

section 3

Acids, Bases, and Salts

Reading Guide

What You'll Learn

- **Compare and contrast** acids and bases and identify the characteristics they have.
- **Examine** some formulas and uses of common acids and bases.
- **Determine** how the processes of ionization and dissociation apply to acids and bases.

Why It's Important

Acids and bases are found almost everywhere—from fruit juice and gastric juice to soaps.

Review Vocabulary

electrolyte: compound that breaks apart in water, forming charged particles (ions) that can conduct electricity

New Vocabulary

- acid
- indicator
- base

Acids

What comes to mind when you hear the word *acid?* Do you think of a substance that can burn your skin or even burn a hole through a piece of metal? Do you think about sour foods such as those shown in **Figure 10?** Although some acids can burn and are dangerous to handle, most acids in foods are safe to eat. What acids have in common, however, is that they contain at least one hydrogen atom that can be removed when the acid is dissolved in water.

Figure 10 The acids in these common foods give each their distinctive sour taste.

Properties of Acids When an acid dissolves in water, some of the hydrogen is released as hydrogen ions, H^+. An **acid** is a substance that produces hydrogen ions in a water solution. It is the ability to produce these ions that gives acids their characteristic properties. When an acid dissolves in water, H^+ ions interact with water molecules to form H_3O^+ ions, which are called hydronium ions (hi DROH nee um • I ahnz).

Acids have several common properties. For one thing, all acids taste sour. The familiar, sour taste of many foods is due to acids. However, taste never should be used to test for the presence of acids. Some acids can damage tissue by producing painful burns. Acids are corrosive. Some acids react strongly with certain metals, eating away the metals and forming metallic compounds and hydrogen gas. Acids also react with indicators to produce predictable changes in color. An **indicator** is an organic compound that changes color in acids and bases. For example, the indicator litmus paper turns red in acid.

Common Acids Many foods contain acids. In addition to citric acid in citrus fruits, lactic acid is found in yogurt and buttermilk, and food, such as pickles, contain vinegar, also known as acetic acid. Your stomach uses acid to help digest your food. At least four acids (sulfuric, phosphoric, nitric, and hydrochloric) play roles in industrial applications.

Reading Check *Which four acids are important for industry?*

Table 3 lists the names and formulas of a few acids, their uses, and some properties. Three acids are used to make fertilizers; most of the nitric acid and sulfuric acid and approximately 90 percent of phosphoric acid produced are used for this purpose. Many acids can burn, but sulfuric acid can burn by removing water from your skin as easily as it takes water from sugar, as shown in **Figure 11.**

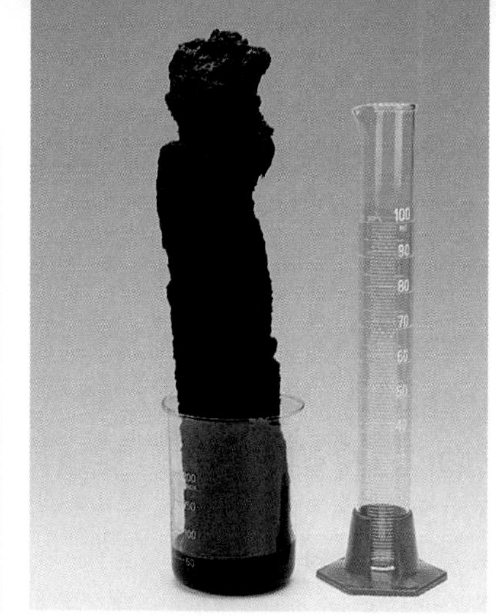

Figure 11 When sulfuric acid is added to sugar, the mixture foams, removing hydrogen and oxygen atoms as water and leaving air-filled carbon.

Table 3 Common Acids and Their Uses

Name, Formula	Use	Other Information
Acetic acid, CH_3COOH	Food preservation and preparation	When in solution with water, it is known as vinegar.
Acetylsalicylic acid, $HOOC{-}C_6H_4{-}OOCCH_3$	Pain relief, fever relief, to reduce inflammation	Known as aspirin
Ascorbic acid, $H_2C_6H_6O_6$	Antioxidant, vitamin	Called vitamin C
Carbonic acid, H_2CO_3	Carbonated drinks	Involved in cave, stalactite, and stalagmite formation
Hydrochloric acid, HCl	Digestion as gastric juice in stomach, to clean steel in a process called pickling	Commonly called muriatic acid
Nitric acid, HNO_3	To make fertilizers	Used to produce nitrogen fertilizers
Phosphoric acid, H_3PO_4	To make detergents, fertilizers, and soft drinks	Slightly sour but pleasant taste, detergents containing phosphates cause water pollution
Sulfuric acid, H_2SO_4	Car batteries, to manufacture fertilizers and other chemicals	Dehydrating agent, causes burns by removing water from cells

Observing Acid Relief

WARNING: *Do not eat antacid tablets.*

Procedure

1. Complete the safety form.
2. Add 150 mL of **water** to a **250-mL beaker.**
3. Add three drops of **1*M* HCl** and 12 drops of **universal indicator.**
4. Observe the color of the solution.
5. Add an **antacid tablet** and observe for 15 min.

Analysis

1. Describe any changes that took place in the solution.
2. Explain why these changes occurred.

Bases

You might not be as familiar with bases as you are with acids. Although you eat some foods that are acidic, you don't consume many bases. Some foods, such as egg whites, are slightly basic. Another example of basic materials is baking powder, which is found in some foods. Medicines, such as milk of magnesia and antacids, are basic, too. Still, you come in contact with many bases every day. Each time you wash your hands using soap, you are using a base. One characteristic of bases is that they feel slippery, like soapy water. Bases are important in many types of cleaning materials, as shown in **Figure 12.** Bases are important in industry, also. For example, sodium hydroxide is used in the paper industry to separate fibers of cellulose from wood pulp. The freed cellulose fibers are made into paper.

Bases can be defined in two ways. Any substance that forms hydroxide ions, OH^-, in a water solution is a **base.** In addition, a base is any substance that accepts H^+ from acids. The definitions are related, because the OH^- ions produced by some bases do accept H^+ ions.

Reading Check *What are the two definitions of a base?*

Properties of Bases One way to think about bases is as the complements, or opposites, of acids. Although acids and bases share some common features, bases have their own characteristic properties. In the pure, undissolved state, many bases are crystalline solids. In solution, bases feel slippery and have a bitter taste. Like strong acids, strong bases are corrosive, and contact with skin can result in severe burns. Taste and touch never should be used to test for the presence of a base or an acid. Finally, like acids, bases react with indicators to produce changes in color. The indicator litmus turns blue in bases.

Figure 12 Bases commonly are found in many cleaning products used around the home.
Identify *the property of bases evident in soaps.*

Figure 13 Two applications of bases are shown here.

Aluminum hydroxide is a base used in water-treatment plants. Its sticky surface collects impurities, making them easier to filter from the water.

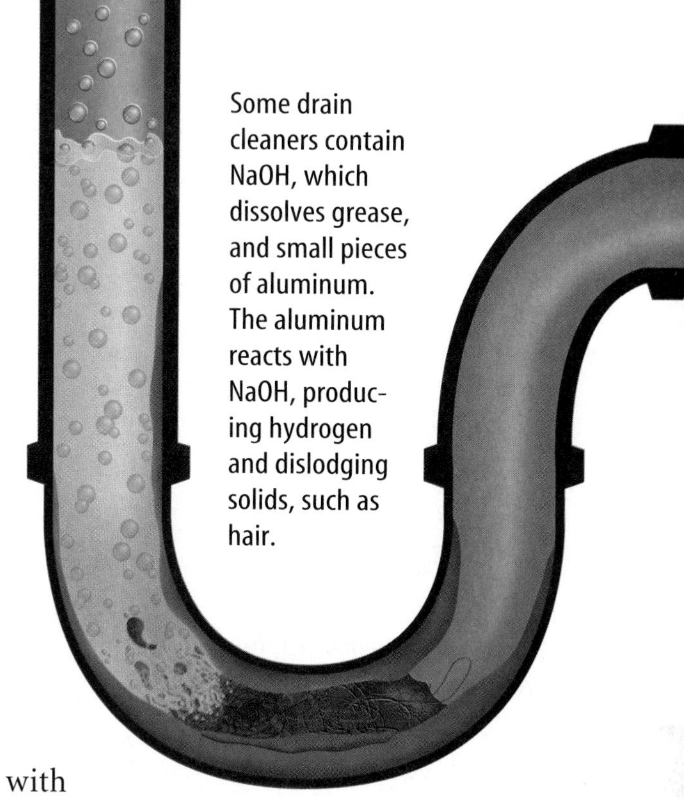

Some drain cleaners contain NaOH, which dissolves grease, and small pieces of aluminum. The aluminum reacts with NaOH, producing hydrogen and dislodging solids, such as hair.

Common Bases You probably are familiar with many common bases because they are found in cleaning products used in the home. These and some other bases are shown in **Table 4,** which also includes their uses and some information about them. **Figure 13** shows two uses of bases that you might not be familiar with.

Table 4 Common Bases and Their Uses

Name, Formula	Use	Other Information
Aluminum hydroxide, $Al(OH)_3$	Color-fast fabrics, antacid, water purification as shown in **Figure 13**	Sticky gel that collects suspended clay and dirt particles on its surface
Calcium hydroxide, $Ca(OH)_2$	Leather-making, mortar and plaster, to lessen acidity of soil	Called caustic lime
Magnesium hydroxide, $Mg(OH)_2$	Laxative, antacid	Called milk of magnesia when in water
Sodium hydroxide, NaOH	To make soap, oven cleaner, drain cleaner, textiles, and paper	Called lye and caustic soda; generates heat (exothermic) when combined with water, reacts with metals to form hydrogen
Ammonia, NH_3	Cleaners, fertilizer, to make rayon and nylon	Irritating odor that is damaging to nasal passages and lungs

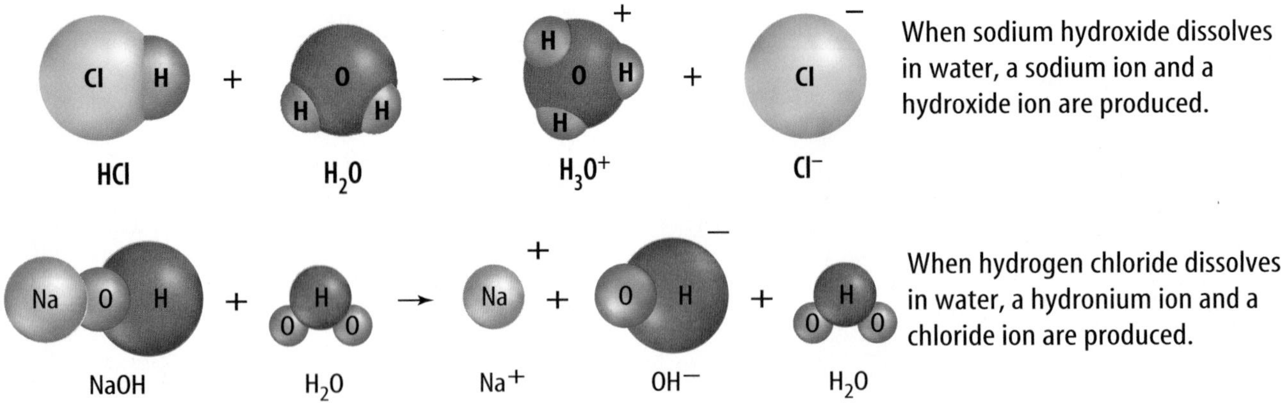

Figure 14 When acids and bases are dissolved in water, they produce ions. Acids produce hydronium ions in water. Bases produce hydroxide ions in water.

Solutions of Acids and Bases

Many of the products that rely on the chemistry of acids and bases are solutions, such as the cleaning products and food products mentioned previously. Because of its polarity, water is the main solvent in these cleaning products.

Dissociation of Acids You have learned that substances such as HCl, HNO_3, and H_2SO_4 are acids because of their ability to produce hydrogen ions (H^+) in water. When an acid dissolves in water, the negative areas of nearby water molecules attract the positive hydrogen in the acid. The acid dissociates into ions and the hydrogen atom combines with a water molecule to form hydronium ions (H_3O^+). Dissociation is the process in which an ionic solid separates into its positive and negative ions. An acid can more accurately be described as a compound that produces hydronium ions when dissolved in water, as shown in **Figure 14.**

Acidic Stings Some ants add sting to their bite by injecting a solution of formic acid. In fact, formic acid was named for ants, which make up the genus *Formica.* Still, ants are considered tasty treats by many animals. For example, one woodpecker called a flicker has saliva that is basic enough to take the sting out of ants. Research the chemical formula and physical properties of formic acid.

Dissociation of Bases Compounds that can form hydroxide ions (OH^-) in water are classified as bases. If you look at **Table 4,** you will find that most of the substances listed contain –OH in their formulas. When bases that contain –OH dissolve in water, the negative areas of nearby water molecules attract the positive ion in the base. The positive areas of nearby water molecules attract the –OH of the base. The base dissociates into a positive ion and a negative ion—a hydroxide ion (OH–). This process also is shown in **Figure 14.** Unlike in acid dissociation, water molecules do not combine with the ions formed from the base.

What ions form when acids and bases dissolve in water?

Neutralization Advertisements for antacids claim that these products neutralize the excess stomach acid that causes indigestion. Normally, gastric juice is acidic. Too much acid can produce discomfort. Antacids contain bases or other compounds containing sodium, calcium, magnesium, or aluminum that react with acids to lower acid concentration. **Figure 15** shows what happens when you ingest an antacid tablet containing sodium bicarbonate, $NaHCO_3$. The acid (HCl) is neutralized by the base ($NaHCO_3$).

Neutralization is a chemical reaction between an acid and a base that takes place in a water solution. When HCl is neutralized by NaOH, hydronium ions from the acid combine with hydroxide ions from the base to produce water.

$$H_3O^+(aq) + OH^- \rightarrow 2H_2O(l)$$

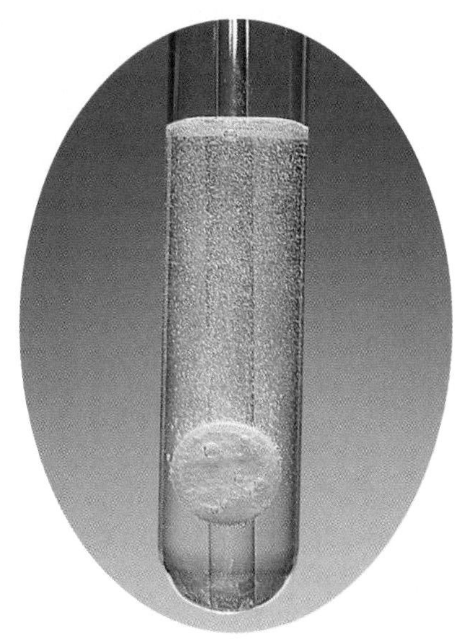

Figure 15 An antacid tablet reacts in your stomach much as it does in this dilute HCl. Many antacid tablets are made to be chewed before they are swallowed. **Explain** *how chewing the tablet would affect the rate of the reaction.*

Acid-Base Reactions The following general equation represents acid-base reactions in water. A few common salts are listed in **Table 5.**

Acid-Base Reactions

acid + base → salt + water

Salt Formation The acid-base equation accounts for only half of the ions in the solution. The remaining ions react to form a salt. A salt is a compound formed when the negative ions from an acid combine with the positive ions from a base. In the reaction between HCl and NaOH, the salt formed in water solution is sodium chloride.

$$Na^+(aq) + Cl^-(aq) \rightarrow NaCl(aq)$$

Table 5 Some Common Salts and Their Uses

Name, Formula	Common Name	Uses
Sodium chloride, NaCl	Salt	Food, manufacture of chemicals
Sodium hydrogen carbonate, $NaHCO_3$	Sodium bicarbonate, baking soda	Food, antacids
Calcium carbonate, $CaCO_3$	Calcite, chalk	Manufacture of paint and rubber tires

Figure 16 Ammonia reacts with water to produce some hydroxide ions; therefore, it is a base.

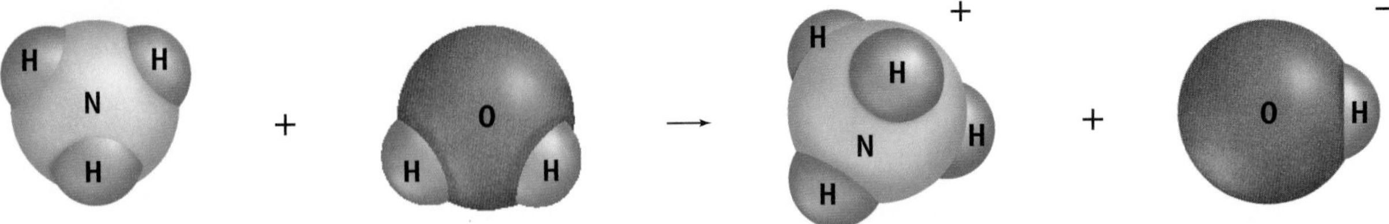

Topic: Cleaner Chemistry
Visit gpescience.com for Web links to information about the dangers of mixing ammonia cleaners with chlorine or hydrochloric acid cleaners.

Activity Visit the cleaning products and laundry sections of the grocery store. Read the labels on several products. Make a list of products that include warnings on the labels and those that do not. Share your findings with the class.

An Exception Ammonia is a base that does not contain −OH. In a water solution, dissociation takes place when the ammonia molecule attracts a hydrogen ion from a water molecule, forming an ammonium ion (NH_4^+). This leaves a hydroxide ion (OH^-), as shown in **Figure 16.**

Reading Check *How does ammonia react in a water solution?*

Ammonia is a common household cleaner. However, products containing ammonia never should be used with other cleaners that contain chlorine or sodium hypochlorite, such as some bathroom toilet-bowl cleaners and bleach. A reaction between sodium hypochlorite and ammonia produces the toxic gases hydrazine and chloramine. Breathing these gases can severely damage lung tissues and cause death.

The reactions of acids and bases are important to the chemistry of living systems, the environment, and many industrial processes.

section 3 review

Summary

Acids

- Acids, when dissolved in water, release H^+, which forms hydronium ions (H_3O^+).
- Acids are sour tasting, corrosive, and reactive with indicators.

Bases

- Bases, when dissolved in water, form OH^-.
- Bases are slippery, have a bitter taste, are corrosive, and are reactive with indicators.

Solutions of Acids and Bases

- The polar nature of water allows acids and bases to dissolve in water.
- Dissociation is the separation of substances, such as acids and bases, into ions in water.

Self Check

1. **Identify** three important acids and three important bases and describe their uses.
2. **Describe** an indicator.
3. **Predict** what metallic compound forms when sulfuric acid reacts with magnesium metal.
4. **Infer** If an acid donates H^+ and a base produces OH^-, what compound is likely to be produced when acids react with bases?
5. **Think Critically** Vinegar contains acetic acid, CH_3COOH. What salt is formed if acetic acid reacts with sodium hydroxide, NaOH?

Applying Math

6. **Calculate** the molecular weight of acetylsalicylic acid, $C_9H_8O_4$.

More Section Review gpescience.com

Strength of Acids and Bases

Reading Guide

What You'll Learn

- **Determine** what is responsible for the strength of an acid or a base.
- **Compare and contrast** strength and concentration.
- **Examine** the relationship between pH and acid or base strength.
- **Examine** electrical conductivity.

Why It's Important

Understanding the strength of acids and bases helps you use them safely.

Review Vocabulary

acid strength: the ability of an acid to dissociate completely

New Vocabulary

- strong acid
- weak acid
- strong base
- weak base
- pH

Strong and Weak Acids and Bases

Some acids must be handled with great care. For example, sulfuric acid found in car batteries can burn your skin, yet you drink acids such as citric acid in orange juice. Obviously, some acids are stronger than others. One measure of acid strength is the ability to dissociate in solution.

The strength of an acid or base depends on how many acid or base particles dissociate into ions in water. When a **strong acid** dissolves in water, almost 100 percent of the acid molecules dissociate into ions. HCl, HNO_3, and H_2SO_4 are examples of strong acids. When a **weak acid** dissolves in water, only a small fraction of the acid molecules dissociates into ions. Acetic acid and carbonic acid are examples of weak acids.

Ions in solution can conduct an electric current are called electrolytes. The more ions a solution contains, the more current it can conduct. The ability of a solution to conduct a current can be demonstrated using a lightbulb connected to a battery with leads placed in the solution, as shown in **Figure 17.** The strong acid solution conducts more current. The weak acid solution does not conduct as much current as a strong acid solution.

Figure 17 HCl is a strong acid. The bulb burns brightly. Acetic acid is a weak acid. The bulb is dimmer.

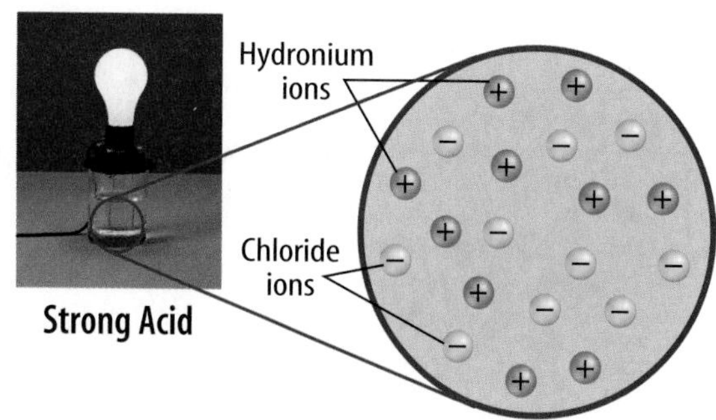

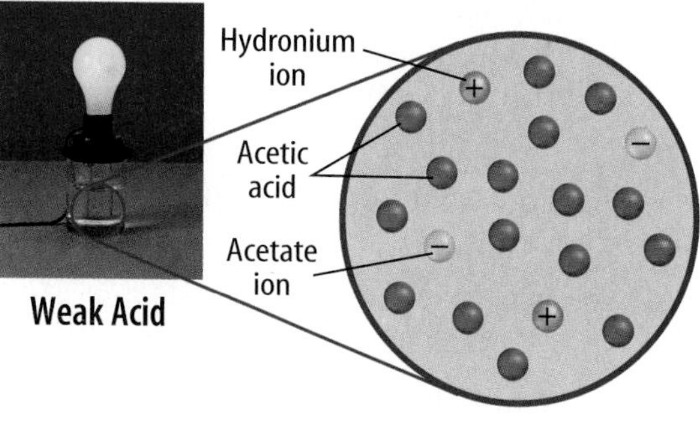

Strong and Weak Acids Equations describing dissociation are written in two ways. In strong acids, such as HCl, nearly all the acid dissociates. This is shown by writing the equation using a single arrow pointing toward the ions that are formed.

$$HCl(g) + H_2O(l) \rightarrow H_3O^+(aq) + Cl^-(aq)$$

Almost 100 percent of the particles in solution are H_3O^+ and Cl^- ions, and only a negligible number of HCl molecules are present.

Equations describing the dissociation of weak acids, such as acetic acid, are written using double arrows pointing in opposite directions. This means that only some of the CH_3COOH dissociates and the reaction does not go to completion.

$$CH_3COOH(aq) + H_2O(l) \rightleftharpoons H_3O^+(aq) + CH_3COO^-(aq)$$

In an acetic acid solution, most of the particles are CH_3COOH molecules, and only a few CH_3COO^- and H^+ ions are in solution.

Strong and Weak Bases Remember that many bases are ionic compounds that dissociate to produce ions when they dissolve. A **strong base** dissociates completely in solution. The following equation shows the dissociation of sodium hydroxide, a strong base.

$$NaOH(s) + H_2O(l) \rightarrow Na^+(aq) + OH^-(aq)$$

The dissociation of ammonia, which is a weak base, is shown using double arrows to indicate that not all the ammonia dissociates. A **weak base** is one that does not dissociate completely.

$$NH_3(aq) + H_2O(l) \rightleftharpoons NH_4^+(aq) + OH^-(aq)$$

Because ammonia produces only a few ions and most of the ammonia remains in the form of NH_3, ammonia is a weak base.

Strength and Concentration Sometimes, when talking about acids and bases, the terms *strength* and *concentration* can be confused. The terms *strong* and *weak* are used to classify acids and bases. The terms refer to the ease with which an acid or base dissociates in solution. *Strong* acids and bases dissociate completely; *weak* acids and bases dissociate only partially. In contrast, the terms *dilute* and *concentrated* are used to indicate the concentration of a solution, which is the amount of acid or base dissolved in the solution. It is possible to have dilute solutions of strong acids and bases and concentrated solutions of weak acids and bases, as shown in **Figure 18.**

Figure 18 You can have a dilute solution of a strong acid and a concentrated solution of a weak acid.

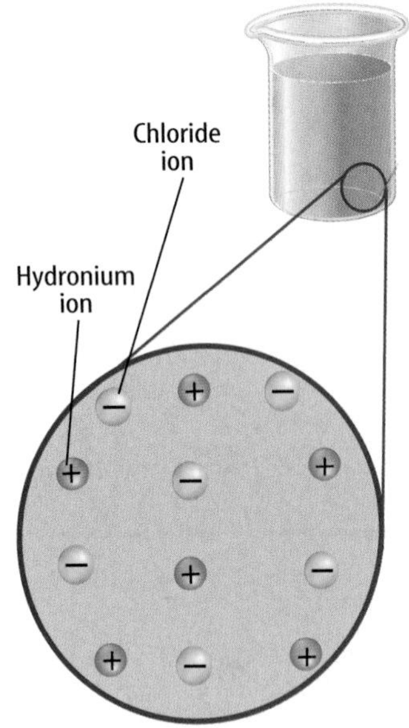

This shows a dilute solution of HCl.

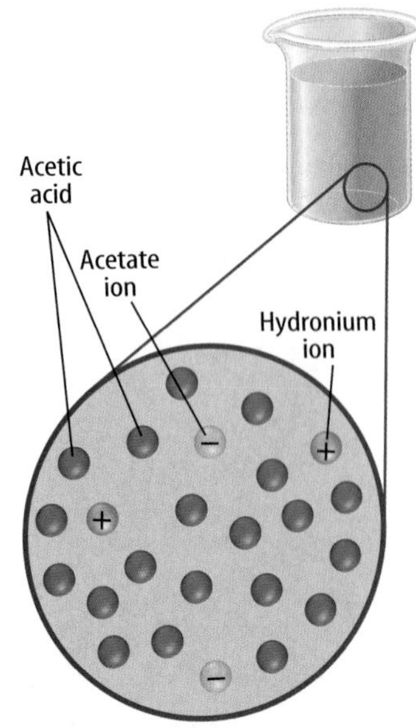

This shows a concentrated solution of acetic acid.

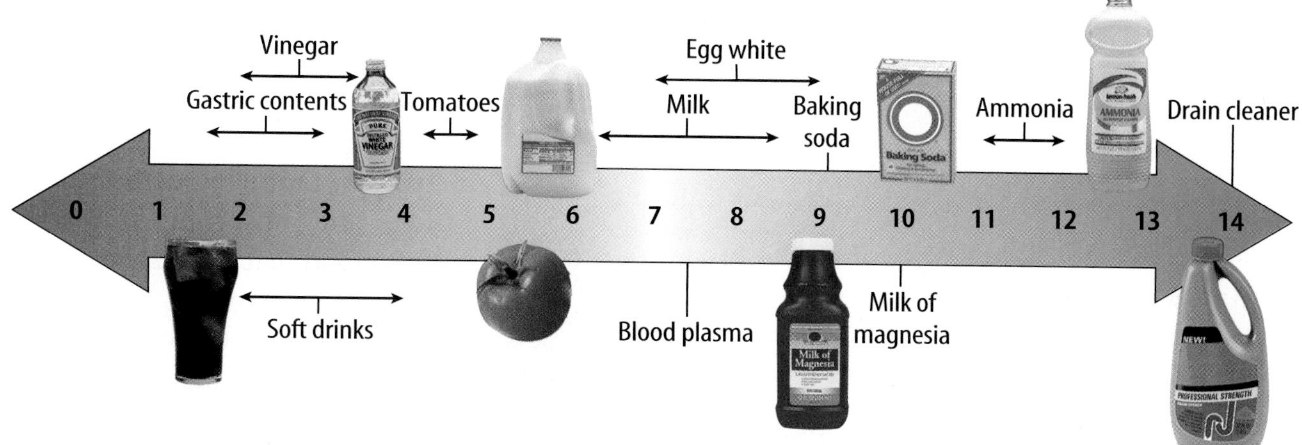

Figure 19 The pH scale helps classify solutions as acidic or basic.

pH of a Solution

If you have a swimming pool or keep tropical fish, you know that the pH of the water must be controlled. Also, many shampoo products claim to control pH, making them suitable for your type of hair. The **pH** of a solution is a measure of the concentration of H^+ ions in it. The greater the H^+ concentration is, the lower the pH is and the more acidic the solution is. The pH measures how acidic or basic a solution is. To indicate pH, a scale ranging from 0 to 14 has been devised, as shown in **Figure 19.**

Reading Check *If a solution has a pH of 8.5, does it have more hydroxide ions or hydronium ions?*

As the scale shows, solutions with a pH lower than 7 are described as acidic, and the lower the value is, the more acidic the solution is. Solutions with a pH greater than 7 are basic, and the higher the pH is, the more basic the solution is. A solution with a pH of exactly 7 indicates that the concentrations of H^+ ions and OH^- ions are equal. These solutions are considered neutral. Pure water at 25°C has a pH of 7.

One way to determine pH is by using indicator paper. This paper undergoes a color change in the presence of H_3O^+ ions and OH^- ions in solution. The final color of the pH paper is matched with colors in a chart to find the pH, as shown in **Figure 20.** Is this an accurate way to determine pH?

An instrument called a pH meter is another tool to determine the pH of a solution. This meter is operated by immersing the electrode in the solution to be tested and reading the dial. Small, battery-operated pH meters with digital readouts, as shown in **Figure 20,** are precise and convenient for use outside the laboratory when testing the pH of soils and streams.

Figure 20 The pH of a sample can be measured in several ways. Indicator paper gives an approximate value quickly, however, a pH meter is quick and more precise.

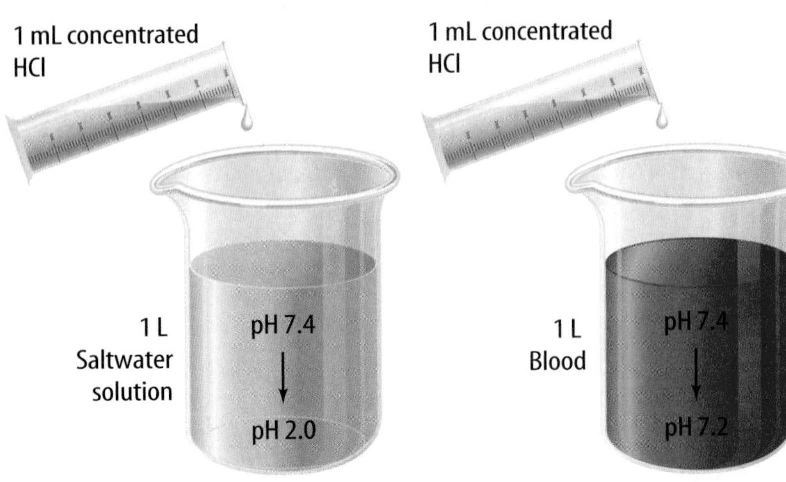

Figure 21 This experiment shows how well blood plasma acts as a buffer. Adding 1 mL of concentrated HCl to 1 L of saltwater solution changes the pH from 7.4 to 2.0. Adding the same amount of concentrated HCl to 1 L of blood plasma changes the pH from 7.4 to 7.2.

Blood pH

Your blood circulates throughout your body carrying oxygen, removing carbon dioxide, and absorbing nutrients from food that you have eaten. In order to carry out its many functions properly, the pH of blood must remain between 7.0 and 7.8. The main reason for this is that some enzymes cannot work outside this pH range. An enzyme is a protein molecule that acts as a catalyst for many reactions in the body. Yet you can eat foods that are acidic without changing the pH of your blood. How can this be? The answer is that your blood contains compounds called buffers that enable small amounts of acids or bases to be absorbed without harmful effects.

Buffers are solutions containing ions that react with additional acids or bases to minimize their effects on pH. One buffer system in blood involves bicarbonate ions, HCO_3^-. Because of these buffer systems, small amounts of even concentrated acid will not change pH much, as shown in **Figure 21.** Buffers help keep your blood close to a nearly constant pH of 7.4.

What are buffers and how are they important for health?

section 4 review

Summary

Strong and Weak Acids and Bases

- When strong acids dissolve in water, nearly all the acid particles dissociate into ions. When weak acids dissolve in water, few particles dissociate.
- When strong bases dissolve in water, nearly all base particles dissociate. When weak bases dissove, only a few particles dissociate.
- Ions in solution can conduct electricity.
- Strength refers to the ability of an acid or base to dissociate in water; concentration refers to how much acid or base is in solution.

pH of a Solution

- pH describes a substance as acidic or basic.
- Buffers are substances that minimize the effects of an acid or base on pH.

Self Check

1. **Describe** what determines the strength of an acid. A base?
2. **Explain** how to make a dilute solution of a strong acid.
3. **Explain** how electricity can be conducted by solutions.
4. **Describe** pH values of 9.1, 1.2, and 5.7 as basic, acidic, or very acidic.
5. **Think Critically** The proper pH range for a swimming pool is between 7.2 and 7.8. Some pools use two substances, $NaHCO_3$ and HCl, to maintain this range. How would you adjust the pH if you found it was 8.2? 6.9?

Applying Math

6. **Use Equations** To determine the difference in acid strength, calculate 10^n, where n = difference between pHs. How much more acidic is a solution of pH 2.4 than a solution of pH 4.4?

More Section Review gpescience.com

Determining Acidity

The science of acids and bases is not practiced only in high-tech laboratories by scientists. You can investigate the acidic concentrations of things in your own home using a simple homemade indicator solution.

Real-World Problem

How can you tell if a substance is a strong or weak acid?

Goal

- **Determine** the relative concentrations of common acidic substances.

Materials

homemade cabbage indicator (indicates both acids and bases)
grease pencil or masking tape
coffee filter
waxed paper
teaspoons (3)
alum
cream of tartar
fruit preservative

Safety Precautions

Procedure

1. Complete the safety form.
2. Use the grease pencil or masking tape and a pencil to label three areas on the waxed paper *alum, cream of tartar,* and *fruit preservative.* These areas should be about 8 cm apart.
3. Place approximately $\frac{1}{2}$ teaspoon of each of the three powders on the waxed paper where labeled. Use a separate teaspoon for each substance.
4. Cut three strips from the coffee filter, about 1 cm wide by 8 cm long.
5. Dip the end of one of the strips into the cabbage indicator solution, then lay the wet end on top of the alum.
6. Wet a second strip and lay it in on top of the cream of tartar.
7. Wet the third strip and lay on top of the fruit preservative.
8. Wait 5 min, then check the indicator strips and record your observations.

Conclude and Apply

1. **Determine** if all three substances were acids. Did the indicator strips turn a similar color?
2. **Explain** why each substance produced a different color.
3. **Propose** a possible rank of the concentrations.
4. **Predict** what you would have observed if you used sodium hydroxide instead of alum.

Communicating Your Data

Compare your results with other groups in the class. Discuss any differences in the results you obtained.

Saturated Solutions

Goal

- **Observe** the effects of temperature on the amount of solute that dissolves.

Materials

distilled water at room temperature
large test tubes
Celsius thermometer
table sugar
copper wire stirrer, bent into a spiral
test-tube holder
graduated cylinder (25-mL)
beaker (250-mL) with 150 mL of water
electric hot plate
test-tube rack
ring stand

Safety Precautions

WARNING: *Do NOT touch the test tubes or hot-plate surface when hot plate is turned on or cooling down. When heating a solution in a test tube, keep it pointed away from yourself and others. Do NOT remove goggles until cleanup, including washing hands, is completed.*

Real-World Problem

Two major factors to consider when you are dissolving a solute in water are temperature and the ratio of solute to solvent. What happens to a solution as the temperature changes? How does solubility change as temperature is increased? To be able to draw conclusions about the effect of temperature, you must keep other variables constant. For example, you must be sure to stir each solution in a similar manner.

Procedure

1. Complete the safety form.
2. Place 20 mL of distilled water in a test tube. Add 30 g of sugar.
3. Stir. Does the sugar dissolve?
4. If it dissolves completely, add another 5 g of sugar to the test tube. Does it dissolve?
5. Continue adding 5-g amounts of sugar until no more sugar dissolves.
6. Now place the beaker of water on the hot plate and hang the thermometer from the ring stand so that the bulb is immersed about halfway into the beaker, making sure it does not touch the sides or bottom. Record the starting temperature.

7. Using a test-tube holder, place the test tube into the water.
8. Gradually increase the temperature of the hot plate while stirring the solution in the test tube, until all the sugar dissolves.
9. Note the temperature at which this happens.
10. Add another 5 g of sugar and continue. Note the temperature at which this additional sugar dissolves.
11. Continue in this manner until you have at least four data points. Note the total amount of sugar that has dissolved. Record your data in the data table.

Analyze Your Data

1. **Graph** your results using a line graph. Plot grams of solute per 20 g of water on the *y*-axis and plot temperature on the *x*-axis.
2. **Interpret Data** Using your graph, estimate the solubility of sugar at 100°C and at 0°C, the boiling and freezing point of water, respectively.

Dissolving Sugar in Water

Temperature	Total Grams of Sugar Dissolved
Do not write in this book.	

Conclude and Apply

1. **Explain** how the saturation change as the temperature was increased.
2. **Compare** your results with the data in **Table 2.**

Communicating Your Data

Compare your results with those of other groups and discuss any differences noted. Why might these differences have occurred? **For more help, refer to the Science Skill Handbook.**

TIME SCIENCE AND Society

SCIENCE ISSUES THAT AFFECT YOU!

Acid Rain

Protecting Earth from the Damaging Effects of Chemically Loaded Precipitation

Acid precipitation is rain, snow, or sleet that is more acidic than unpolluted precipitation. It's caused by the burning of fossil fuels, such as coal, oil, and natural gas. In the United States, most gasoline and electricity come from fossil fuels. People burn fossil fuels each time they drive a car, heat a building, or turn on a light.

Normally, raindrops pick up particles and natural chemicals in the air. When rain falls, it mixes with the carbon dioxide in the atmosphere, giving clean rain a slightly acidic pH of 5.6. Then, natural chemicals found in the air and soil balance out the acidity, giving most lakes and streams a pH between 6.0 and 8.0. But when pollutants are introduced, these natural bases are not strong enough to neutralize these solutions. Wind can carry this acidic moisture for hundreds of miles before it falls to Earth as acid rain.

Eating Away at History

Like all acids, acid rain can corrode, or eat away at, substances. Many historical monuments, such as the Mayan temples in Mexico and the Parthenon in Greece, have been slowly but steadily damaged by acid rain. This kind of damage can be fixed, though it costs billions of dollars to ensure that ancient monuments and buildings are not destroyed.

Some Solutions

In some countries, high acid levels in lakes and streams have been lowered by adding lime to the water. Lime, a natural base, balances out the damaging chemicals. In the United States, all new cars must have catalytic converters, which help reduce the amount of exhaust pollution that vehicles give off.

You also can make a difference. Turning off the lights when you are not using them means a power plant does not have to produce as much electricity. By carpooling, using public transportation, and walking, there is less pollution from cars. The results of all these individual actions can make a huge difference in preserving our environment.

Ride a bike! It saves fuel, is nonpolluting, and helps preserve the environment.

List Go to a local park or forest. List any effects of acid rain that you see. Make a list of the things you do that use energy or cause pollution. Think about what your family can do to reduce pollution and save energy. Share your list with an adult.

TIME

For more information, visit gpescience.com

Reviewing Main Ideas

Section 1 How Solutions Form

1. A solution is a mixture that has the same composition, color, density, and taste throughout.
2. The substance being dissolved is called a solute, and the substance that does the dissolving is called a solvent.
3. The rate of dissolving can be increased by stirring, increasing surface area, or increasing temperature.
4. Under similar conditions, small particles of solute dissolve faster than large particles.

Section 2 Solubility and Concentration

1. Some compounds are more soluble than others, and this can be measured.
2. *Concentrated* and *dilute* are not precise terms used to describe concentrations of solutions.
3. Concentrations can be expressed as percent by volume.
4. An unsaturated solution can dissolve more solute, while a saturated solution, such as this tea, cannot. A supersaturated solution is made by raising the temperature of a saturated solution and adding more solute. If it is cooled carefully, the supersaturated solution will retain the dissolved solute.

Section 3 Acids, Bases, and Salts

1. An acid is a substance that produces hydrogen ions, H^+, in solution. A base produces hydroxide ions, OH^-, in solution.
2. Most foods can be classified as acidic or basic. Properties of acids and bases are due, in part, to the presence of the H^+ and OH^- ions.

3. Salts form when negative ions from an acid combine with positive ions from a base.
4. Dissociation is the process where an ionic solid separates into its positive and negative ions.

Section 4 Strength of Acids and Bases

1. The strength of an acid or base is determined by how completely it forms ions when it is in solution.
2. Strength and concentration are not the same thing. Concentration involves the relative amounts of solvent and solute in a solution, whereas strength is related to the extent to which a substance dissociates.

FOLDABLES™ Use the Foldable that you made at the beginning of this chapter to help you review the characteristics of solutions, acids, and bases.

chapter 24 Review

Using Vocabulary

acid p. 764	solution p. 752
aqueous solution p. 751	solvent p. 751
base p. 766	strong acid p. 771
concentration p. 760	strong base p. 772
indicator p. 764	supersaturated solution p. 762
pH p. 773	unsaturated solution p. 761
saturated solution p. 761	weak acid p. 771
solubility p. 759	weak base p. 772
solute p. 751	

Fill in the vocabulary word(s) that correctly completes each sentence.

1. In lemonade, sugar is the _________ and water is the _________.
2. A(n) _________ is a substance that produces hydrogen ions, H^+, in solution.
3. Adding a seed crystal may cause solute to crystallize from a(n) _________.
4. A(n) _________ produces hydroxide ions, OH^-, in solution.
5. If more of substance B dissolves in water than substance A, then substance B has a higher _________ than substance A.
6. When a(n) _________ dissolves in water, nearly all the acid molecules dissociate into ions.
7. The measure of the concentration of H^+ ions in solution is called the _________.
8. A solution in which water is the solvent is called a(n) _________ solution.

Checking Concepts

Choose the word or phrase that best answers the question.

9. What term is **NOT** appropriate to use when describing solutions?
   A) gaseous **C)** liquid
   B) heterogeneous **D)** solid

10. What is another name for sodium hydroxide (NaOH)?
   A) ammonia **C)** lye
   B) caustic lime **D)** milk of magnesia

11. When iodine is dissolved in alcohol, what term is used to describe the alcohol?
   A) alloy **C)** solution
   B) solvent **D)** solute

12. What word is used to describe a solution that is 85 percent copper and 15 percent tin?
   A) alloy **C)** saturated
   B) solvent **D)** solute

13. What is a common name for hydrochloric acid?
   A) battery acid **C)** muriatic acid
   B) citric acid **D)** vinegar

14. Which acid ionizes only partially in water?
   A) HCl **C)** HNO_3
   B) H_2SO_4 **D)** CH_3COOH

15. Carrots have a pH of 5.0. How would you describe them?
   A) acidic **C)** an indicator
   B) basic **D)** neutral

16. What is the pH of pure water at 25°C?
   A) 0 **C)** 7
   B) 5 **D)** 14

17. What can you increase to make a gas more soluble in a liquid?
   A) particle size **C)** stirring
   B) pressure **D)** temperature

18. If a solute crystallizes out of a solution when a seed crystal is added, what kind of solution is it?
   A) dilute **C)** supersaturated
   B) saturated **D)** unsaturated

Interpreting Graphics

Use the table below to answer question 19.

19. **Explain** which substance listed in the table would be most effective for neutralizing battery acid.

pH Readings

Substance	pH
Battery acid	1.5
Lemon juice	2.5
Apple	3
Milk	6.7
Seawater	8.5
Ammonia	12

Use the table below to answer question 20.

Limits of Solubility

Compound	Type of Solution	Solubility in 100 g Water at 20°C
$CuSO_4$	Do not write in this book.	32.0 g
KCl		38.0 g
KNO_3		31.6 g
$NaClO_3$		45.8 g

20. **Identify** Using the data in **Table 2,** complete the table. Use the terms *saturated, unsaturated,* and *supersaturated* to describe the type of solution.

Thinking Critically

21. **Describe** what happens to hydrogen chloride, HCl, when it is dissolved in water to form hydrochloric acid.
22. **Explain** why ammonia is considered a base, even though it contains no hydroxide ions. Is it a strong or weak base?
23. **Explain** why a concentrated acid is not necessarily a strong acid.
24. **Compare and Contrast** How would the pH of a dilute solution of HCl compare with the pH of a concentrated solution of the same acid?
25. **Explain** why the statement, "Water is the solvent in a solution," is not always true.

Applying Math

26. **Use Proportions** To make an indicator solution, a student mixes 3 mL of a concentrated solution with 97 mL of water. How much concentrate is needed to make 3 L of the indicator?

Use the graph below to answer question 27.

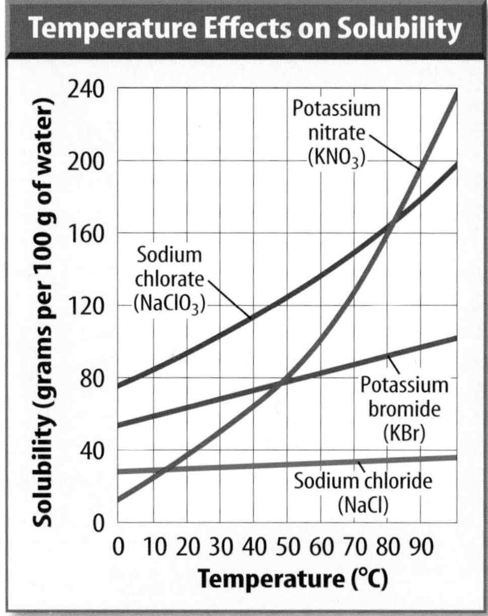

27. **Interpret Data** Determine the temperature at which a solution of 80 g of potassium nitrate (KNO_3) in 100 mL of water is saturated.
28. **Measure in SI** You dissolve 153 g of potassium nitrate in enough water to make 1 L of solution. Then, you use a graduated cylinder to measure 80 mL of solution. What mass of potassium nitrate is in the 80-mL sample?
29. **Use Numbers** How would you make a 25-percent solution by volume of an apple-juice drink?

Record your answers on the answer sheet provided by your teacher or on a sheet of paper.

Multiple Choice

The graph below shows the solubility of various salts in water.

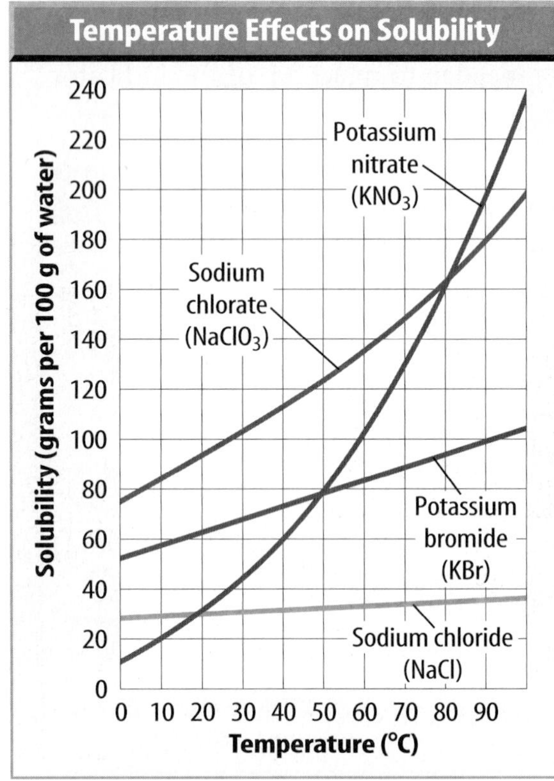

1. Which compound will make a saturated solution when added to 100 g of water?
   A. 20 g NaCl if the water is 50°C
   B. 60 g of KNO_3 if the water is 100°C
   C. 80 g $NaClO_3$ if the water is 30°C
   D. 100 g KBr if the water is 90°C

2. A label on a flask reads 130 g KNO_3/100 g of water. A thermometer reading shows a room temperature of 23°C. No crystals are seen on the bottom of the flask. How is this solution best described?
   A. concentrated
   B. dilute
   C. saturated
   D. supersaturated

3. What allows you to eat acidic foods without changing your blood pH?
   A. blood plasma
   B. buffers
   C. enzymes
   D. protein molecules

4. The following graph is a titration curve. The data indicate the changes that happened to the solution as drops of a strong base were added.

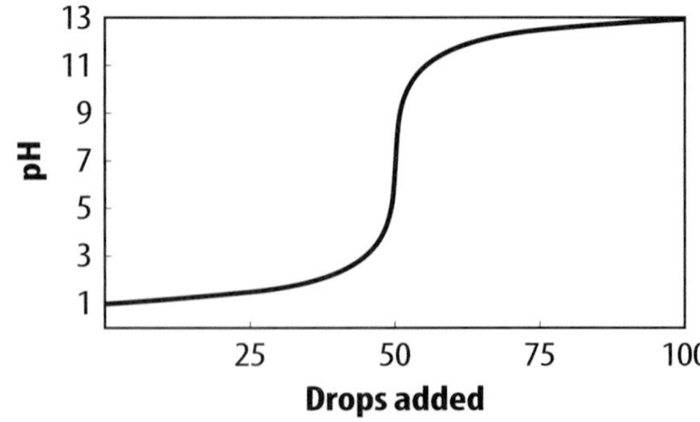

At the instant of neutralization, what is in the beaker (besides water)?
   A. acid only
   B. base only
   C. salt only
   D. equal amounts of each

5. Which chemical formula below describes a hydronium ion?

   A. H_3O^+

   B. OH^-

   C. COOH

   D. H_2O

Gridded Response

6. The solubility of potassium chloride in water is 34 g per 100 g of water at 20°C. A warm solution containing 100 g of potassium chloride in 200 g of water is cooled to 20°C. How many grams of potassium chloride will come out of solution?

7. When the pH of a solution drops from 3.0 to 1.0, the hydronium ion concentration increases by a factor of 100 fold from 0.0010. What is the concentration at pH = 1.0?

Short Response

8. The drawing below shows carbon dioxide gas dissolved in water.

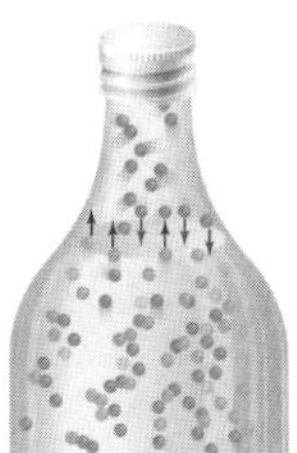

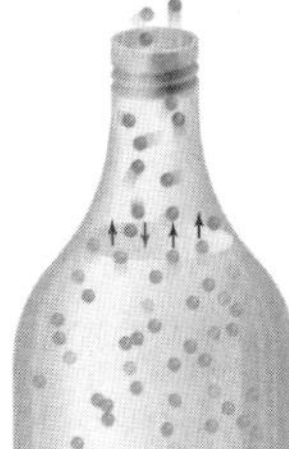

What are two ways you could make more gas dissolve in the water?

9. Name the salt that is produced by each of the following acid-base pairs.
   HCl + NaOH; HNO_3 + KOH;
   H_2SO_4 + $Ca(OH)_2$

Extended Response

10. One of the solutions has a pH of 10, the other pH of 12. In one beaker, the bulb glows more brightly than does the other.

Explain why the bulbs glow with different intensities. Use the words *strong* and *weak* in your answer.

11. You are given a clear water solution containing potassium nitrate. How could you determine whether the solution is unsaturated, saturated, or supersaturated?

12. Explain why a weak acid in solution has a higher pH than a strong acid of the same concentration.

Test-Taking Tip

Answer Bubbles Double-check that you are filling in the correct answer bubble for the question number you are working on.

chapter 25

Nuclear Changes

chapter preview

sections

1 Radioactivity
2 Nuclear Decay
3 Detecting Radioactivity
4 Nuclear Reactions
Lab *Chain Reactions*
Lab *Modeling Transmutations*

Virtual Lab *How can you simulate the radioactive half-life of an element?*

Planet Power

Although less than one billionth of the energy emitted by the Sun falls on Earth, the energy Earth receives from the Sun powers the entire planet. Without the Sun's energy, life on Earth could not exist. This energy is produced inside the Sun by nuclear fusion—a nuclear reaction in which atomic nuclei join together.

Science Journal In your Science Journal, write a paragraph describing your impressions of the Sun.

Start-Up Activities

The Size of a Nucleus

Do you realize you are made up mostly of empty space? Your body is made of atoms, and atoms are made of electrons whizzing around a small nucleus of protons and neutrons. The size of an atom is the size of the space in which the electrons move around the nucleus. In this lab, you'll find out how the size of an atom compares with the size of a nucleus.

1. Complete the safety form.
2. Pour several grains of sugar onto a sheet of dark paper.
3. Choose one of the grains of sugar to represent the nucleus of an atom. Tape this grain to the center of the sheet of paper.
4. Place the sheet of paper in a large open space.
5. Use a meterstick to measure a distance of 10 m from the sugar grain. This distance represents the radius of the electron cloud around an atom.
6. **Think Critically** In your Science Journal, compare the size of an atom to the size of a nucleus. If an electron is much smaller than a nucleus, explain why an atom contains mostly empty space.

Radioactivity and Nuclear Reactions Make the following Foldable to help you understand radioactivity and nuclear reactions.

STEP 1 **Fold** a sheet of paper in half lengthwise.

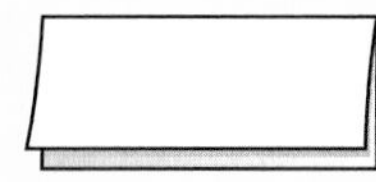

STEP 2 **Fold** paper down 2.5 cm from the top. (Hint: From the tip of your index finger to your middle knuckle is about 2.5 cm.)

STEP 3 **Open and draw** lines along the 2.5-cm fold. **Label** as shown.

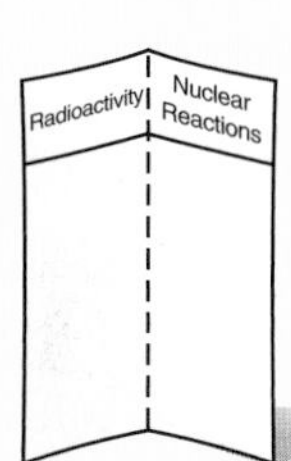

Summarize in a Table As you read the chapter, write what you learn about radioactivity in the left column and what you learn about nuclear reactions in the right column.

Preview this chapter's content and activities at gpescience.com

section 1

Radioactivity

Reading Guide

What You'll Learn

- **Describe** the structure of an atom and its nucleus.
- **Explain** what radioactivity is.
- **Compare and contrast** properties of radioactive and stable nuclei.
- **Discuss** the discovery of radioactivity.

Why It's Important

Radioactivity is everywhere because every element on the periodic table has some atomic nuclei that are radioactive.

Review Vocabulary

long-range force: a force that becomes weaker with distance, but never vanishes

New Vocabulary

- strong force
- radioactivity

The Nucleus

Every second you are being bombarded by energetic particles. Some of these particles come from unstable atoms in soil, rocks, and the atmosphere. What types of atoms are unstable? What type of particles do unstable atoms emit? The answers to these questions begin with the nucleus of an atom.

Recall that atoms are composed of protons, neutrons, and electrons. The nucleus of an atom contains the protons, which have a positive charge, and neutrons, which have no electric charge. The total amount of charge in a nucleus is determined by the number of protons, which also is called the atomic number. You might remember that an electron has a charge that is equal but opposite to a proton's charge. Atoms usually contain the same number of protons as electrons. Negatively charged electrons are electrically attracted to the positively charged nucleus and swarm around it.

Figure 1 The size of a nucleus in an atom can be compared to a marble sitting in the middle of an empty football stadium.

Protons and Neutrons in the Nucleus Protons and neutrons are packed together tightly in a nucleus. The region outside the nucleus in which the electrons are located is large compared to the size of the nucleus. As **Figure 1** shows, the nucleus occupies only a tiny fraction of the space in the atom. If an atom were enlarged so that it was 1 km in diameter, its nucleus would have a diameter of only a few centimeters. But the nucleus contains almost all the mass of the atom, because the mass of one proton or neutron is almost 2,000 times greater than the mass of an electron.

Figure 2 The particles in the nucleus are attracted to each other by the strong force.

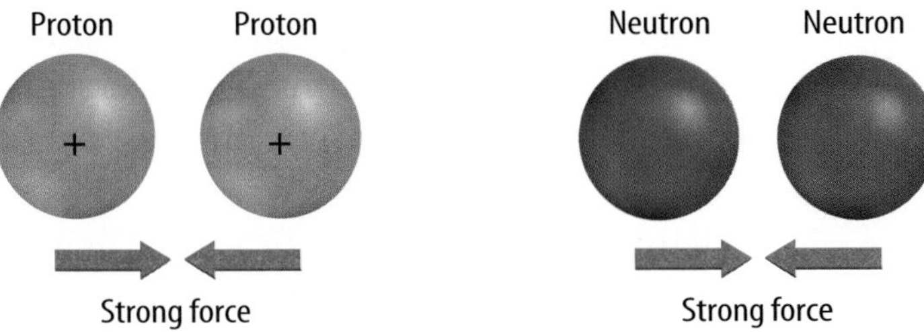

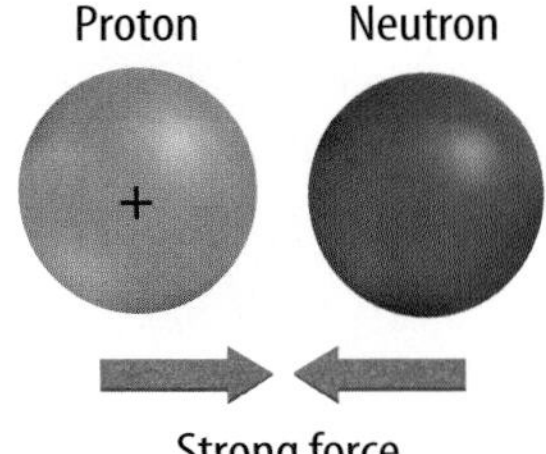

The Strong Force

How do you suppose protons and neutrons are held together so tightly in the nucleus? Positive electric charges repel each other, so why don't the protons in a nucleus push each other away? Another force, called the **strong force,** causes protons and neutrons to be attracted to each other, as shown in **Figure 2.**

The strong force is one of the four basic forces in nature and is about 100 times stronger than the electric force. The attractive forces between all the protons and neutrons in a nucleus keep the nucleus together. However, protons and neutrons have to be close together, like they are in the nucleus, to be attracted by the strong force. The strong force is a short-range force that quickly becomes extremely weak as protons and neutrons get farther apart. The electric force is a long-range force, so protons that are far apart still are repelled by the electric force, as shown in **Figure 3.**

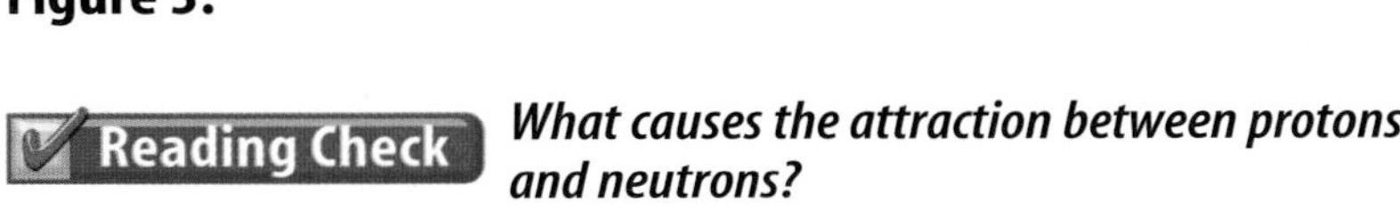

Reading Check ***What causes the attraction between protons and neutrons?***

Figure 3 The total force between two protons depends on how far apart they are. **Infer** *whether the total force between two protons could become zero.*

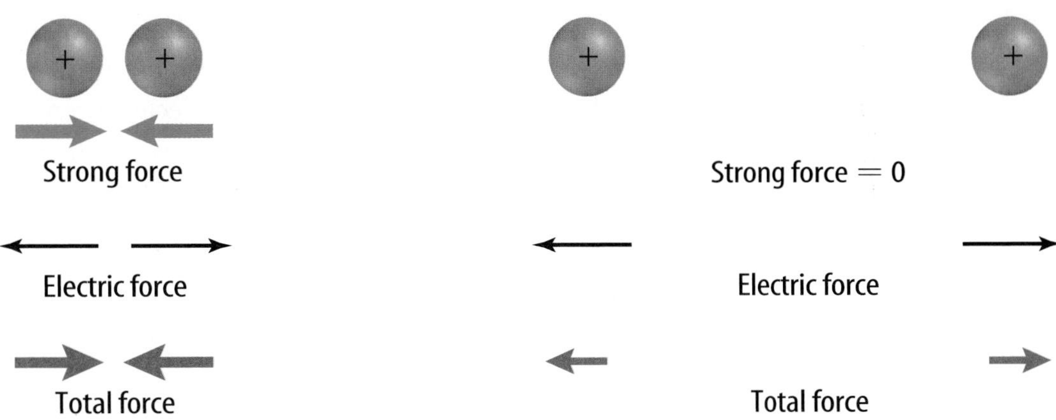

When protons are close together, they are attracted to each other. The attraction due to the short-range strong force is much stronger than the repulsion due to the long-range electric force.

When protons are too far apart to be attracted by the strong force, they still are repelled by the electric force between them. Then the total force between them is repulsive.

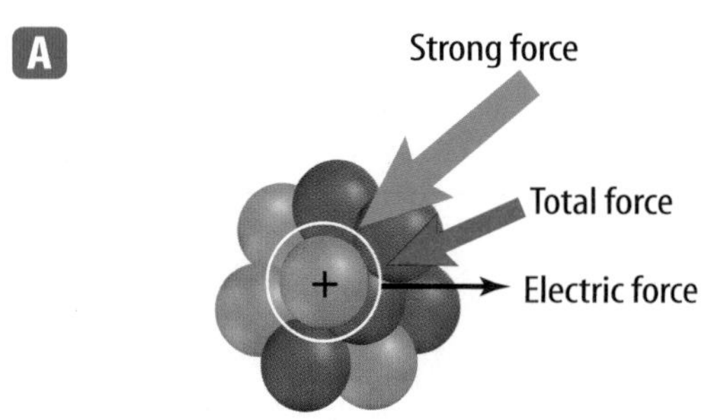

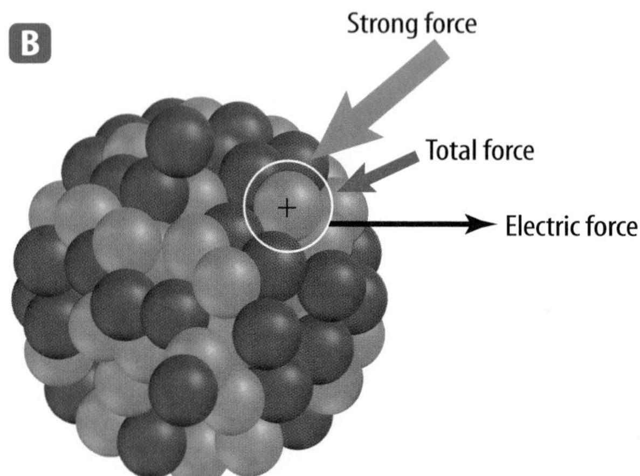

Figure 4 Protons and neutrons are held together less tightly in large nuclei. The circle shows the range of the attractive strong force. **A** Small nuclei have few protons, so the repulsive force on a proton due to the other protons is small. **B** In large nuclei, the attractive strong force is exerted only by the nearest neighbors, but all the protons exert repulsive forces. The total repulsive force is large.

Attraction and Repulsion Some atoms, such as uranium, have many protons and neutrons in their nuclei. These nuclei are held together less tightly than nuclei containing only a few protons and neutrons. To understand this, look at **Figure 4A.** If a nucleus has only a few protons and neutrons, they are all close enough together to be attracted to each other by the strong force. Because only a few protons are in the nucleus, the total electric force causing protons to repel each other is small. As a result, the overall force between the protons and the neutrons attracts the particles to each other.

Forces in a Large Nucleus However, if nuclei have many protons and neutrons, each proton or neutron is attracted to only a few neighbors by the strong force, as shown in **Figure 4B.** The other protons and neutrons are too far away. Because only the closest protons and neutrons attract each other in a large nucleus, the strong force holding them together is about the same as in a small nucleus. However, all the protons in a large nucleus exert a repulsive electric force on each other. Thus, the electric repulsive force on a proton in a large nucleus is larger than it would be in a small nucleus. Because the repulsive force increases in a large nucleus while the attractive force on each proton or neutron remains about the same, protons and neutrons are held together less tightly in a large nucleus.

Radioactivity

Many types of nuclei are held together permanently and are stable. However, there are many other types of nuclei that are unstable. These nuclei break apart, or decay, by emitting particles and energy. This process of nuclear decay is called **radioactivity.** A nucleus that decays is called a radioactive nucleus.

Nuclei that contain large numbers of protons and neutrons tend to be unstable. In fact, all nuclei that contain more than 83 protons are radioactive. However, many other nuclei that contain fewer than 83 protons also are radioactive. Even some nuclei with only one or a few protons are radioactive.

Almost all elements with more than 92 protons don't exist naturally on Earth. They have been produced only in laboratories and are called synthetic elements. These synthetic elements are unstable and decay soon after they are created.

Converting Mass into Energy When an unstable nucleus decays, energy is emitted. If energy is conserved and cannot be created or destroyed, where does this energy come from? Recall that in nuclear reactions, mass can be converted into energy. As an unstable nucleus decays, a small amount of mass is converted into energy. As a result, the mass of the initial nucleus is slightly larger than the mass of the final nucleus plus the mass of any particles that are emitted. A large amount of energy is produced by the conversion of only a small amount of mass.

Isotopes The atoms of an element all have the same number of protons in their nuclei. For example, the nuclei of all carbon atoms contain six protons. However, naturally occurring carbon nuclei can have six, seven, or eight neutrons. Nuclei that have the same number of protons but different numbers of neutrons are called isotopes. The element carbon has three isotopes that occur naturally. The atoms of all isotopes of an element have the same number of electrons and have the same chemical properties. **Figure 5** shows two isotopes of helium.

Nuclear Numbers A nucleus can be described by the number of protons and neutrons it contains. The number of protons in a nucleus is called the atomic number. Because the mass of all the protons and neutrons in a nucleus is nearly the same as the mass of the atom, the number of protons and neutrons is called the mass number.

Reading Check *What is the atomic number of a nucleus?*

A nucleus can be represented by a symbol that includes its atomic number, mass number, and the symbol of the element it belongs to. The symbol for the nucleus of the stable isotope of carbon is shown below as an example.

$$\begin{matrix}\text{mass number} \rightarrow \\ \text{atomic number} \rightarrow\end{matrix} {}^{12}_{6}\text{C} \leftarrow \text{element symbol}$$

This isotope is called carbon-12. The number of neutrons in the nucleus is the mass number minus the atomic number. So the number of neutrons in the carbon-12 nucleus is $12 - 6 = 6$. Carbon-12 has six protons and six neutrons. Now, compare the isotope carbon-12 to this radioactive isotope of carbon:

$$\begin{matrix}\text{mass number} \rightarrow \\ \text{atomic number} \rightarrow\end{matrix} {}^{14}_{6}\text{C} \leftarrow \text{element symbol}$$

The radioactive isotope is carbon-14. How many neutrons does carbon-14 have?

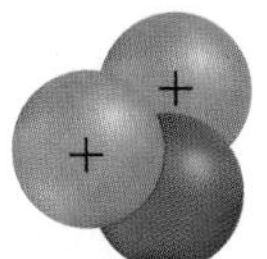

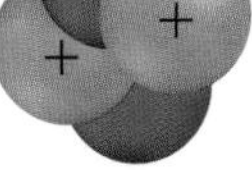

Helium-3 Helium-4

Figure 5 These two isotopes of helium each have the same number of protons, but different numbers of neutrons.
Identify *the ratio of protons to neutrons in each of these isotopes of helium.*

Modeling the Strong Force

Procedure

1. Complete the safety form.
2. Gather **15 yellow candies** to represent neutrons and **13 red** and **2 green candies** to represent protons.
3. Model a small nucleus by placing two red protons and three neutrons around a green proton so they touch.
4. Model a larger nucleus by arranging the remaining candies around the other green proton so they are touching.

Analysis

1. Compare the number of protons and neutrons touching a green proton in both models.
2. Suppose the strong force on a green proton is due to protons and neutrons that touch it. Compare the strong force in both models.

Try at Home

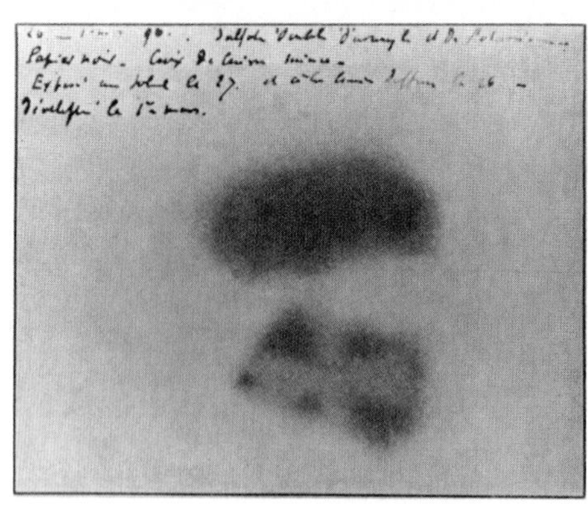

Figure 6 The dark spots on this photographic plate were made by the radiation emitted by radioactive uranium atoms. Uranium salt had been placed next to the plate by Henri Becquerel in 1896.

The Discovery of Radioactivity In 1896, Henri Becquerel left uranium salt in a desk drawer with a photographic plate. Later, when he developed the plate, shown in **Figure 6,** he found an outline of the clumps of the uranium salt. He hypothesized that the uranium salt had emitted some unknown invisible rays, or radiation, that had darkened the film.

Two years after Becquerel's discovery, Marie and Pierre Curie discovered two new elements, polonium and radium, that also were radioactive. To obtain a sample of radium large enough to be studied, they developed a process to extract radium from the mineral pitchblende. After more than three years, they were able to obtain about 0.1 g of radium from several tons of pitchblende. Years of additional processing gradually produced more radium that was made available to other researchers all over the world.

Topic: Marie Curie
Visit gpescience.com for Web links to information about the life of Marie Curie.

Activity Create a timeline showing important events in the life of Marie Curie.

section 1 review

Summary

The Strong Force

- The short-ranged strong force causes neutrons and protons to be attracted to each other.
- The long-ranged electric force causes protons to repel each other.
- The combination of the strong and electric forces causes protons and neutrons in a large nucleus to be held together less tightly than in a small nucleus.

Radioactivity and Isotopes

- Radioactivity is the process of nuclear decay.
- Isotopes of an element have the same number of protons, but different numbers of neutrons.
- The atomic number is the number of protons in a nucleus. The mass number is the number of protons and neutrons in a nucleus.

Self Check

1. **Describe** the properties of the strong force.
2. **Compare** the strong force between protons and neutrons in a small nucleus and a large nucleus.
3. **Explain** why large nuclei are unstable.
4. **Identify** the contributions of the three scientists who discovered the first radioactive elements.
5. **Think Critically** What is the ratio of protons to neutrons in lead-214? Explain whether you would expect this isotope to be radioactive or stable.

Applying Math

6. **Calculate a Ratio** What is the ratio of neutrons to protons in a nucleus of radon-222?
7. **Use Percentages** A silicon rod contains 30.21 g of silicon-28, 1.53 g of silicon-29, and 1.02 g of silicon-30. Calculate the percentage of each isotope in the rod.

More Section Review gpescience.com

section 2

Nuclear Decay

Reading Guide

What You'll Learn

- **Compare and contrast** alpha, beta, and gamma radiation.
- **Define** the half-life of a radioactive material.
- **Describe** the process of radioactive dating.

Why It's Important

Nuclear decay produces nuclear radiation that can both harm people and be useful.

Review Vocabulary

electromagnetic wave: a transverse wave consisting of vibrating electric and magnetic fields

New Vocabulary

- alpha particle
- transmutation
- beta particle
- gamma ray
- half-life

Nuclear Radiation

When an unstable nucleus decays, particles and energy called nuclear radiation are emitted from it. The three types of nuclear radiation are alpha, beta (BAY tuh), and gamma radiation. Alpha and beta radiation are particles. Gamma radiation is an electromagnetic wave.

Alpha Particles

When alpha radiation occurs, an **alpha particle**—made of two protons and two neutrons, as shown in **Table 1**—is emitted from the decaying nucleus. An alpha particle is the same as the nucleus of a helium atom. It has a charge of +2 and an atomic mass of 4. Its symbol is the same as the symbol of a helium nucleus, $^{4}_{2}He$.

Reading Check *What does an alpha particle consist of?*

Compared to beta and gamma radiation, alpha particles are much more massive. They also have the most electric charge. As a result, alpha particles lose energy more quickly when they interact with matter than the other types of nuclear radiation do. When alpha particles pass through matter, they exert an electric force on the electrons in atoms in their path. This force pulls electrons away from atoms and leaves behind charged ions. Alpha particles lose energy quickly during this process. As a result, alpha particles are the least penetrating form of nuclear radiation. Alpha particles can be stopped by a sheet of paper.

Table 1 Alpha Particles

Symbol	$^{4}_{2}He$
Mass	4
Charge	+2

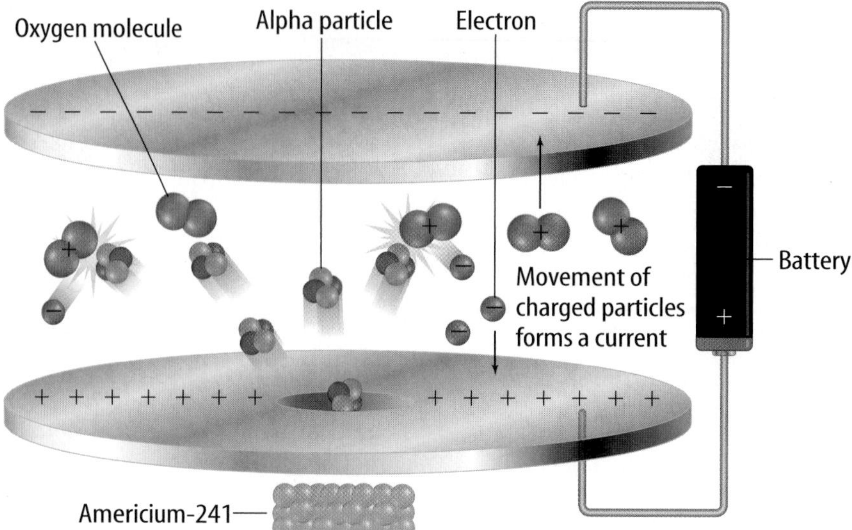

Figure 7 When alpha particles collide with molecules in the air, positively charged ions and electrons result. The ions and electrons move toward charged plates, creating a current in the smoke detector.

Damage from Alpha Particles Alpha particles can be dangerous if they are released by radioactive atoms inside the human body. Biological molecules inside your body are large and easily damaged. A single alpha particle can damage many fragile biological molecules. Damage from alpha particles can cause cells not to function properly, leading to illness and disease.

Smoke Detectors Some smoke detectors give off alpha particles that ionize the surrounding air. Normally, an electric current flows through this ionized air to form a circuit, as in **Figure 7.** But if smoke particles enter the ionized air, they will absorb the ions and electrons. The circuit is broken and the alarm goes off.

Transmutation When an atom emits an alpha particle, it has two fewer protons, so it is a different element. **Transmutation** is the process of changing one element to another through nuclear decay. In alpha decay, two protons and two neutrons are lost from the nucleus. The new element has an atomic number two less than that of the original element. The mass number of the new element is four less than the original element. **Figure 8** shows a nuclear transmutation caused by alpha decay. The charge of the original nucleus equals the sum of the charges of the nucleus and the alpha particle that are formed.

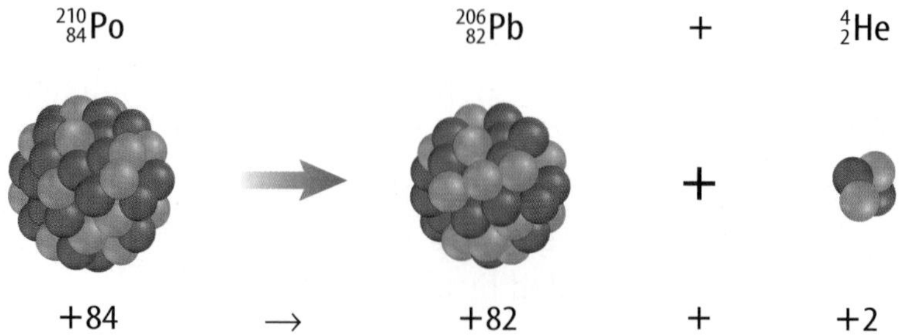

Figure 8 In this transmutation, polonium emits an alpha particle and changes into lead.
Determine *whether the charges and mass numbers of the products equal the charge and mass number of the polonium nucleus.*

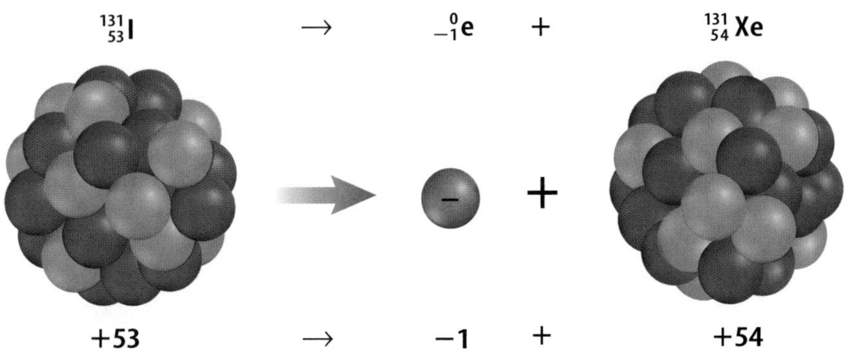

Figure 9 Nuclei that emit beta particles undergo transmutation. In beta decay shown here, iodine changes to xenon.
Compare *the total atomic number and mass number of the products with the atomic number and mass number of the iodine nucleus.*

Beta Particles

A second type of radioactive decay is called beta decay, which is summarized in **Table 2.** Sometimes in an unstable nucleus a neutron decays into a proton and emits an electron. The electron is emitted from the nucleus and is called a **beta particle.** Beta decay is caused by another basic force called the weak force.

Because the atom now has one more proton, it becomes the element with an atomic number one greater than that of the original element. Atoms that lose beta particles undergo transmutation. However, because the total number of protons and neutrons does not change during beta decay, the mass number of the new element is the same as that of the original element. **Figure 9** shows a transmutation caused by beta decay.

Damage from Beta Particles Beta particles are much faster and more penetrating than alpha particles. They can pass through paper but are stopped by a sheet of aluminum foil. Just like alpha particles, beta particles can damage cells when they are emitted by radioactive nuclei inside the human body.

Table 2 Beta Particles

Symbol	$^{0}_{-1}e$
Mass	0.0005
Charge	−1

Gamma Rays

The most penetrating form of nuclear radiation is gamma radiation. **Gamma rays** are electromagnetic waves with the highest frequencies and the shortest wavelengths in the electromagnetic spectrum. They have no mass and no charge and travel at the speed of light. They usually are emitted from a nucleus when alpha decay or beta decay occurs. The properties of gamma rays are summarized in **Table 3.**

Thick blocks of dense materials, such as lead and concrete, are required to stop gamma rays. However, gamma rays cause less damage to biological molecules as they pass through living tissue. Suppose an alpha particle and a gamma ray travel the same distance through matter. The gamma ray produces fewer ions because it has no electric charge.

Table 3 Gamma Rays

Symbol	γ
Mass	0
Charge	0

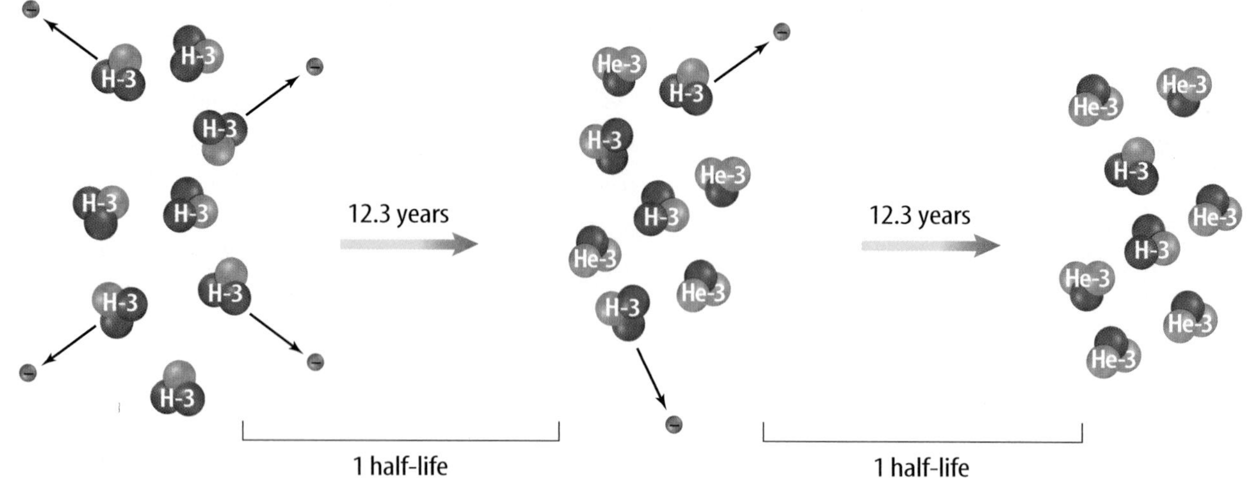

Figure 10 The half-life of $^{3}_{1}H$ is 12.3 years. During each half-life, half of the atoms in the sample decay into helium.
Infer *how many hydrogen atoms will be left in the sample after the next half-life.*

Radioactive Half-Life

If an element is radioactive, how can you tell when its atoms are going to decay? Some radioisotopes decay to stable atoms in less than a second. However, the nuclei of certain radioactive isotopes require millions of years to decay. A measure of the time required by the nuclei of an isotope to decay is called the half-life. The **half-life** of a radioactive isotope is the amount of time it takes for half the nuclei in a sample of the isotope to decay. The nucleus left after the isotope decays is called the daughter nucleus. **Figure 10** shows how the number of decaying nuclei decreases after each half-life.

Half-lives vary widely among the radioactive isotopes. For example, polonium-214 has a half-life of less than a thousandth of a second, but uranium-238 has a half-life of 4.5 billion years. The half-lives of some other radioactive elements are listed in **Table 4.**

Reading Check *What is a daughter nucleus?*

Table 4 Sample Half-Lives

Isotope	Half-Life
$^{3}_{1}H$	12.3 years
$^{212}_{82}Pb$	10.6 hr
$^{14}_{6}C$	5,730 years
$^{211}_{84}Po$	0.5 s
$^{235}_{92}U$	7.04×10^{8} years
$^{131}_{53}I$	8.04 days

Radioactive Dating

Some geologists, biologists, and archaeologists, among others, are interested in the ages of rocks and fossils found on Earth. The ages of these materials can be determined using radioactive isotopes and their half-lives. First, the amounts of the radioactive isotope and its daughter nucleus in a sample of material are measured. Then, the number of half-lives that need to pass to give the measured amounts of the isotope and its daughter nucleus is calculated. The number of half-lives is the amount of time that has passed since the isotope began to decay. Also, it is usually the amount of time that has passed since the object was formed, or the age of the object. Different isotopes are useful in dating different types of materials.

Carbon Dating The radioactive isotope carbon-14 often is used to estimate the ages of plant and animal remains. Carbon-14 has a half-life of 5,730 years and is found in molecules such as carbon dioxide. Plants use carbon dioxide when they make food, so all plants contain carbon-14. When animals eat plants, carbon-14 is added to their bodies.

The decaying carbon-14 in a plant or animal is replaced when an animal eats or when a plant makes food. As a result, the ratio of the number of carbon-14 atoms to the number of carbon-12 atoms in the organism remains nearly constant. But when an organism dies, its carbon-14 atoms decay without being replaced. The ratio of carbon-14 to carbon 12 then decreases with time. By measuring this ratio, the age of an organism's remains can be estimated. However, only material from plants and animals that lived within the past 50,000 years contains enough carbon-14 to be measured.

Uranium Dating Radioactive dating also can be used to estimate the ages of rocks. Some rocks contain uranium, which has two radioactive isotopes with long half-lives. Each of these uranium isotopes decays into a different isotope of lead. The number of these uranium isotopes and their daughter nuclei are measured. From the ratios of these amounts, the number of half-lives since the rock was formed can be calculated.

section 2 review

Summary

Nuclear Radiation

- When an unstable nucleus decays it emits nuclear radiation that can be alpha particles, beta particles, or gamma rays.
- An alpha particle consists of two protons and two neutrons.
- A beta particle is an electron and is emitted when a neutron decays into a proton.
- Gamma rays are electromagnetic waves of very high frequency that usually are emitted when alpha decay or beta decay occurs.

Half-Life and Radioactive Dating

- The half-life of a radioactive isotope is the amount of time for half the nuclei in a sample of the isotope to decay.
- The amounts of radioactive isotopes and their daughter nuclei are needed to date materials.

Self Check

1. **Infer** how the mass number and the atomic number of a nucleus change when it emits a beta particle.
2. **Determine** the daughter nucleus formed when a radon-222 nucleus emits an alpha particle.
3. **Describe** how each of the three types of radiation can be stopped.
4. **Think Critically** Sample 1 contains nuclei with a half-life of 10.6 hr and sample 2 contains an equal number of nuclei with a half-life of 0.5 s. After 3 half-lives pass for each sample, which sample contains more of the original nuclei?

Applying Math

5. **Use Percentages** What is the percentage of radioactive nuclei left after 3 half-lives pass?
6. **Use Fractions** If the half-life of iodine-131 is 8 days, how much of a 5-g sample is left after 32 days?

section 3

Detecting Radioactivity

Reading Guide

What You'll Learn

- **Describe** how radioactivity can be detected in cloud and bubble chambers.
- **Explain** how an electroscope can be used to detect radiation.
- **Explain** how a Geiger counter can measure nuclear radiation.

Why It's Important

Devices that detect and measure radioactivity are used to monitor exposure to humans.

Review Vocabulary

ion: an atom that has gained or lost electrons

New Vocabulary

- cloud chamber
- bubble chamber
- Geiger counter

Radiation Detectors

Because you can't see or feel alpha particles, beta particles, or gamma rays, you must use instruments to detect their presence. Some tools that are used to detect radioactivity rely on the fact that radiation forms ions in the matter it passes through. The tools detect these newly formed ions in several ways.

Cloud Chambers A **cloud chamber,** shown in **Figure 11,** can be used to detect alpha or beta particle radiation. A cloud chamber is filled with water or ethanol vapor. When a radioactive sample is placed in the cloud chamber, it gives off charged alpha or beta particles that travel through the water or ethanol vapor. As each charged particle travels through the chamber, it knocks electrons off the atoms in the air, creating ions. It leaves a trail of ions in the chamber. The water or ethanol vapor condenses around these ions, creating a visible path of droplets along the track of the particle. Beta particles leave long, thin trails, and alpha particles leave shorter, thicker trails.

Reading Check *Why are trails produced by alpha and beta particles seen in cloud chambers?*

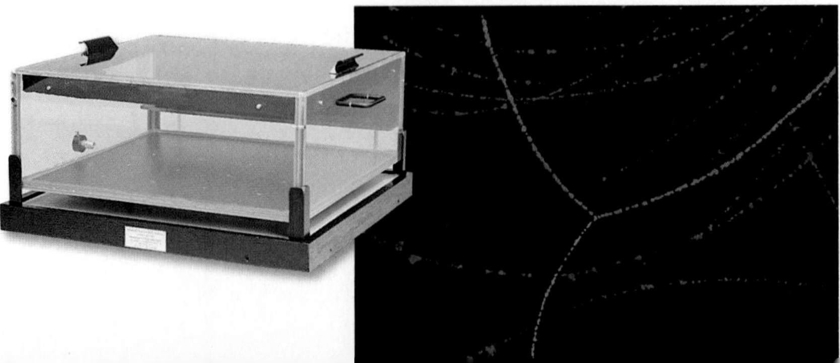

Figure 11 If a sample of radioactive material is placed in a cloud chamber, a trail of condensed vapor will form along the paths of the emitted particles.

Bubble Chambers Another way to detect and monitor the paths of nuclear particles is by using a bubble chamber. A **bubble chamber** holds a superheated liquid, which doesn't boil because the pressure in the chamber is high. When a moving particle leaves ions behind, the liquid boils along the trail. The path shows up as tracks of bubbles, like the ones in **Figure 12.**

Figure 12 Particles of nuclear radiation can be detected as they leave trails of bubbles in a bubble chamber.

Electroscopes Do you remember how an electroscope can be used to detect electric charges? When an electroscope is given a negative charge, its leaves repel each other and spread apart, as in **Figure 13A.** They will remain apart until their extra electrons have somewhere to go and discharge the electroscope. The excess charge can be neutralized if it combines with positive charges. Nuclear radiation moving through the air can remove electrons from some molecules in air, as shown in **Figure 13B,** and cause other molecules in air to gain electrons. When this occurs near the leaves of the electroscope, some positively charged molecules in the air can come in contact with the electroscope and attract the electrons from the leaves, as **Figure 13C** shows. As these negatively charged leaves lose their charges, they move together. **Figure 13D** shows this last step in the process. The same process also will occur if the electroscope leaves are positively charged. Then the electrons move from negative ions in the air to the electroscope leaves.

Figure 13 Nuclear radiation can cause an electroscope to lose its charge.

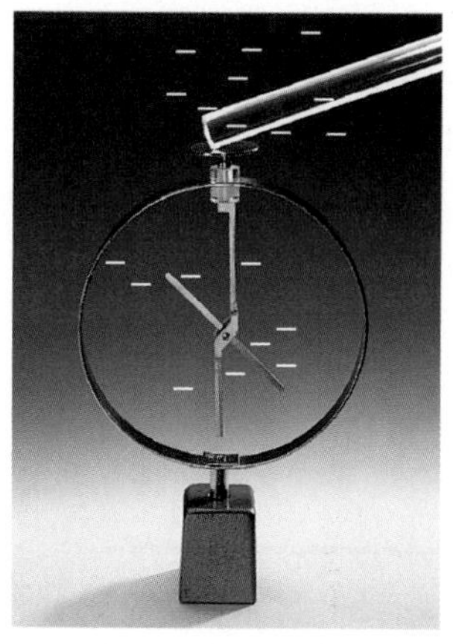

A The electroscope leaves are charged with negative charge.

B Nuclear radiation, such as alpha particles, can create positive ions.

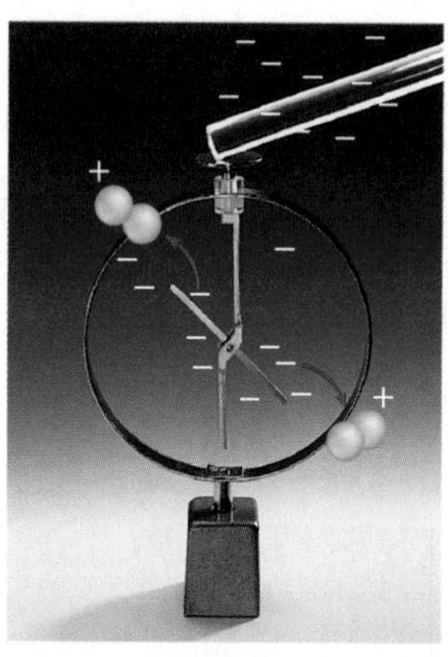

C Negative charges move from the leaves to positively charged ions.

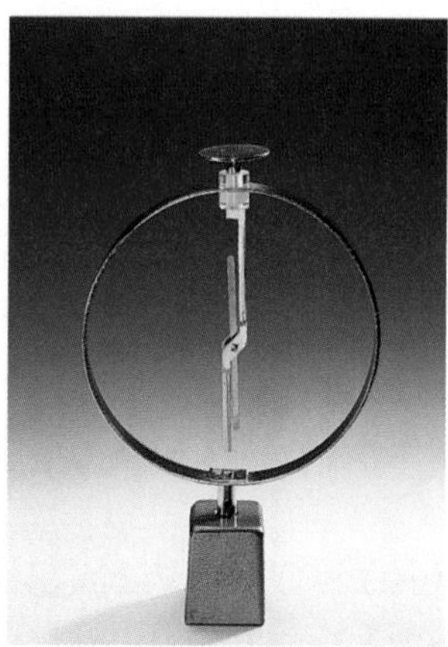

D The electroscope leaves lose their negative charge and come together.

Measuring Radiation

It is important to monitor the amount of radiation a person is being exposed to because large doses of radiation can be harmful to living tissue. A **Geiger counter** is a device that measures the amount of radiation by producing an electric current when it detects a charged particle.

Applying Math — Use Logarithms

THE AGE OF ROCKS The radioactive nucleus uranium-238 produces the daughter nucleus lead-206 with a half-life of 4.5 billion years. The age of a mineral sample can be calculated from the equation:

$$\text{age} = (1.44H)\ln\left[1 + \frac{N_L}{N_U}\right]$$

In this equation, H is the half-life of uranium 238, N_L is the number of lead-206 atoms in the sample, and N_U is the number of uranium-238 atoms in the sample. Find the age of a sample in which the ratio N_L/N_U is measured to be 0.55.

IDENTIFY known values and unknown values

Identify the known values:

the ratio of the number of uranium-238 atoms to the number of lead-206 atoms is 0.55 means $\frac{N_L}{N_U} = 0.55$

a half-life of 4.5 billion years means $H = 4.5$ billion years

Identify the unknown value:

what is the age of the sample? means age = ? years

SOLVE the problem

Substitute the known values into the equation for the age:

$$\text{age} = (1.44H)\ln\left[1 + \frac{N_L}{N_U}\right] = (1.44 \times 4.5 \text{ billion yrs}) \ln [1 + 0.55]$$
$$= (6.48 \text{ billion yrs}) \ln(1.55)$$

To calculate ln(1.55), enter 1.55 on your calculator and press the "ln" button. The result is 0.438. Substitute this value in the above equation:

$$\text{age} = (6.48 \text{ billion yrs})(0.438) = 2.84 \text{ billion yrs}$$

CHECK the answer

Does your answer seem reasonable? The ratio $N_L/N_U = 0.55$, means that less than one half-life has elapsed. The calculated age is less than one half-life.

Practice Problems

Find the age of a sample in which the ratio N_L/N_U has been measured to be 1.21.

For more practice problems go to page 879, and visit Math Practice at gpescience.com.

Figure 14 Electrons that are stripped off gas molecules in a Geiger counter move to a positively charged wire in the device. This causes current to flow in the wire. The current then is used to produce a click or a flash of light.

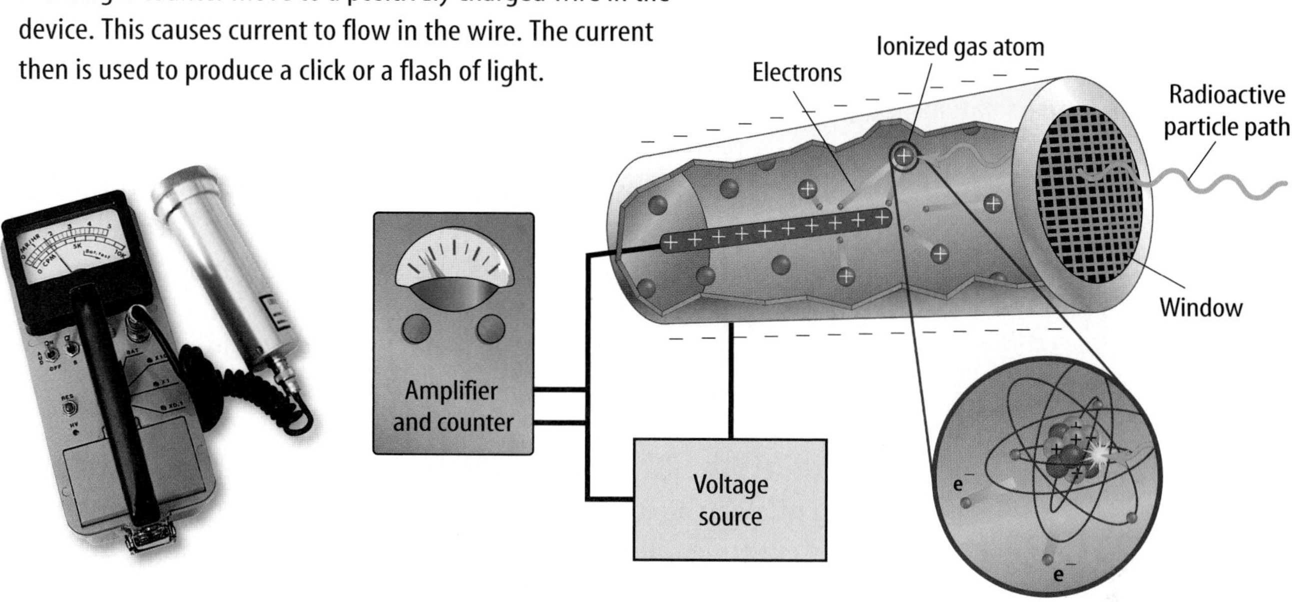

Geiger Counters A Geiger counter, shown in **Figure 14,** has a tube with a positively charged wire running through the center of a negatively charged copper cylinder. This tube is filled with gas at a low pressure. When radiation enters the tube at one end, it knocks electrons from the atoms of the gas. These electrons then knock more electrons off other atoms in the gas, and an "electron avalanche" is produced. The free electrons are attracted to the positive wire in the tube. When a large number of electrons reaches the wire, a short, intense current is produced in the wire. This current is amplified to produce a clicking sound or flashing light. The intensity of radiation present is determined by the number of clicks or flashes of light each second.

How does a Geiger counter indicate that radiation is present?

Background Radiation

It might surprise you to know that you are bathed in radiation that comes from your environment. This radiation, called background radiation, is not produced by humans. Instead it is low-level radiation emitted mainly by naturally occurring radioactive isotopes found in Earth's rocks, soils, and atmosphere. Building materials such as bricks, wood, and stones contain traces of these radioactive materials. Traces of naturally occurring radioactive isotopes are found in the food, water, and air consumed by all animals and plants. As a result, animals and plants also contain small amounts of these isotopes.

Artificial Rainmaking It may be possible to ease the health and economic hardships caused by severe droughts by artificially making rain. The formation of raindrops in a cloud is similar to the formation of droplets in a cloud chamber. Rain forms when cold droplets freeze around microscopic particles of dust, and then melt as they fall through warmer air. Research artificial rainmaking and report your findings to your class.

Sources of Background Radiation

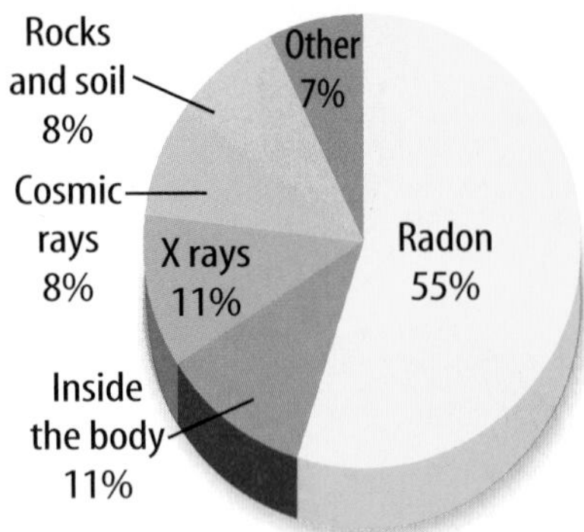

Figure 15 This circle graph shows the average amount of background radiation from different sources received by a person living in the United States.

Sources of Background Radiation Background radiation comes from several sources, as shown in **Figure 15.** The largest source comes from the decay of radon gas. Radon, which emits an alpha particle when it decays, is produced in Earth's crust by the decay of uranium-238. Radon gas can seep into houses and basements from the surrounding soil and rocks.

Some background radiation comes from high-speed nuclei, called cosmic rays, that strike Earth's atmosphere. They produce showers of particles, including alpha, beta, and gamma radiation. Most of this radiation is absorbed by the atmosphere. Higher up there is less atmosphere to absorb this radiation, so the background radiation from cosmic rays increases with altitude.

Radiation in Your Body Some of the elements that are essential for life have naturally occurring radioactive isotopes. For example, about one out of every trillion carbon atoms is carbon-14, which emits a beta particle when it decays. With each breath, you inhale about 3 million carbon-14 atoms.

The amount of background radiation a person receives can vary greatly. The amount depends on the type of rocks underground, the type of materials used to construct the person's home, and the elevation at which the person lives, among other things. However, because it comes from naturally occurring processes, background radiation never can be eliminated.

section 3 review

Summary

Radiation Detectors

- Alpha and beta particles can be detected by the trail of ions they form when they pass through a cloud chamber or a bubble chamber.
- The presence of alpha or beta particles can cause an electroscope to become discharged.
- A Geiger counter is used to measure radiation levels. It produces a clicking sound or a flash of light when alpha or beta particles enter the Geiger counter tube.

Background Radiation

- Background radiation is low-level radiations emitted mainly by radioactive isotopes in Earth's rocks, soils, and atmosphere.
- The largest source of background radiation is from the alpha decay of radon gas.

Self Check

1. **Describe** why a charged electroscope will discharge when placed near a radioactive material.
2. **Compare and contrast** cloud and bubble chambers.
3. **Describe** the process that occurs in a Geiger counter when radiation is detected.
4. **Explain** why background radiation never can be completely eliminated.
5. **Think Critically** If the radioactive isotope radon-222 has a half-life of only four days, how can radon gas be continually present inside houses?

Applying Math

6. **Use Percentages** The amount of radiation can be measured in units called millirems. If 25 millirems from cosmic rays is 8.0 percent of the average background radiation, what is the amount of the average background radiation in millirems?

More Section Review gpescience.com

Nuclear Reactions

Reading Guide

What You'll Learn

- **Explain** nuclear fission and how it can begin a chain reaction.
- **Discuss** how nuclear fusion occurs in the Sun.
- **Describe** how radioactive tracers can be used to diagnose medical problems.
- **Discuss** how nuclear reactions can help treat cancer.

Why It's Important

Almost all of the different atoms that you are made of were formed by the nuclear reactions inside ancient, distant stars.

Review Vocabulary

kinetic energy: energy of motion; increases as the mass or speed of an object increases

New Vocabulary

- nuclear fission
- chain reaction
- critical mass
- nuclear fusion
- tracer

Nuclear Fission

In the 1930s, physicist Enrico Fermi thought that by bombarding nuclei with neutrons, nuclei would absorb neutrons and heavier nuclei would be produced. However, in 1938, Otto Hahn and Fritz Strassmann found that when a neutron strikes a uranium-235 nucleus, the nucleus splits apart into smaller nuclei.

In 1939 Lise Meitner was the first to offer a theory to explain these results. She proposed that the uranium-235 nucleus is so distorted when the neutron strikes it that it divides into two smaller nuclei, as shown in **Figure 16.** The process of splitting a nucleus into several smaller nuclei is **nuclear fission.** The word *fission* means "to divide."

Reading Check *What initiates nuclear fission of a uranium-235 nucleus?*

Only large nuclei, such as the nuclei of uranium and plutonium, undergo nuclear fission. A fission reaction usually produces several individual neutrons in addition to the smaller nuclei. The total mass of the products is slightly less than the mass of the original nucleus and the neutron. This small amount of missing mass is converted to a tremendous amount of energy during the fission reaction.

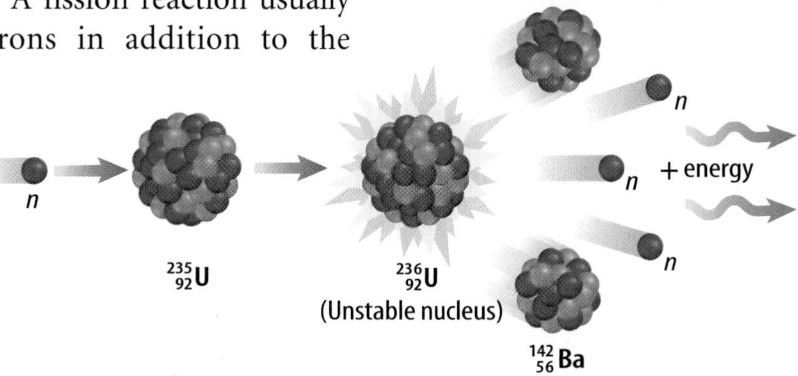

Figure 16 When a neutron hits a uranium-235 nucleus, the uranium nucleus splits into two smaller nuclei and two or three free neutrons. Energy also is released.

Figure 17 A chain reaction occurs when neutrons emitted from a split nucleus cause other nuclei to split and emit additional neutrons.

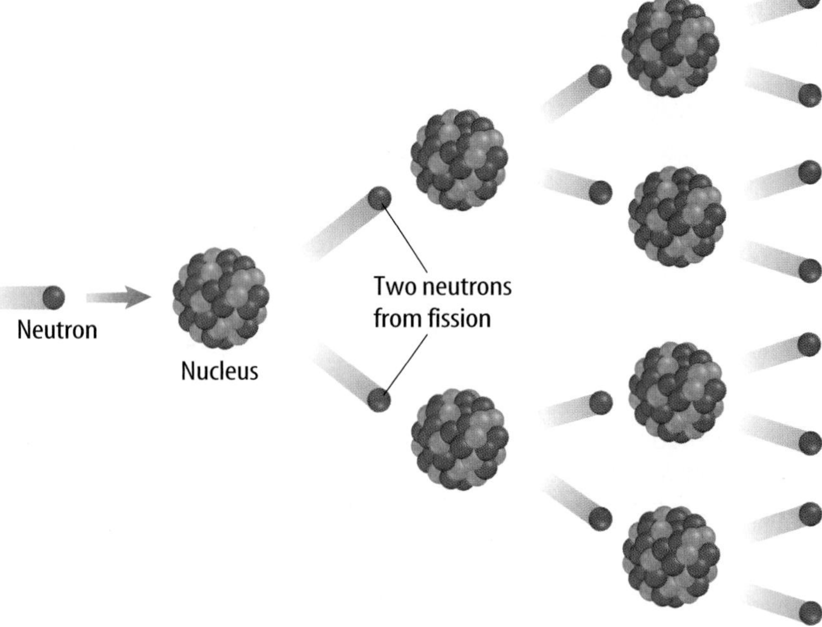

Modeling a Nuclear Reaction

Procedure

1. Complete the safety form.
2. Place **32 marbles,** each with an attached lump of **clay,** into a large **beaker.** These marbles with clay represent unstable atoms.
3. During a 1-min period, remove half of the marbles and pull off the clay. Place the removed marbles into another beaker and place the lumps of clay into a pile. Marbles without clay represent stable atoms. The clay represents waste from the reaction—smaller atoms that still might decay and give off energy.
4. Repeat this procedure four more times.

Analysis

1. What is the half-life of this reaction?
2. Explain whether the waste products could undergo nuclear fission.

Mass and Energy Albert Einstein proposed that mass and energy were related in his special theory of relativity. According to this theory, mass can be converted to energy and energy can be converted to mass. The relation between mass and energy is given by this equation:

Mass-Energy Equation

Energy (joules) = **mass** (kg) × [speed of light (m/s)]^2

$$E = mc^2$$

A small amount of mass can be converted into an enormous amount of energy. For example, if one gram of mass is converted to energy, about 100 trillion joules of energy are released.

Chain Reactions When a nuclear fission reaction occurs, the neutrons emitted can strike other nuclei in the sample and cause them to split. These reactions then release more neutrons, causing additional nuclei to split, as shown in **Figure 17.** The series of repeated fission reactions caused by the release of neutrons in each reaction is a **chain reaction.**

If the chain reaction is uncontrolled, an enormous amount of energy is released in an instant. However, a chain reaction can be controlled by adding materials that absorb neutrons. If enough neutrons are absorbed, the reaction will continue at a constant rate.

For a chain reaction to occur, a critical mass of material that can undergo fission must be present. The **critical mass** is the amount of material required so that each fission reaction produces approximately one more fission reaction. If less than the critical mass of material is present, a chain reaction will not occur.

Nuclear Fusion

Even though nuclear fission reactions release tremendous amounts of energy, even more energy can be released by nuclear fusion reactions. In a **nuclear fusion** reaction, two nuclei with small masses combine to form a nucleus of larger mass. Nuclear fission reactions release millions of times more energy than can be released by chemical reactions.

The reason nuclear reactions release so much more energy than chemical reactions is that the strong force is much stronger than the electric force. This means that the nuclear potential energy stored in atomic nuclei is much greater than the chemical potential energy stored in chemical bonds between atoms.

Topic: Fusion Reactors

Visit gpescience.com for Web links to information about the use of nuclear fusion as a future energy source.

Activity Write a paragraph describing the different types of fusion reactors that have been developed.

Temperature and Fusion For nuclear fusion to occur, positively charged nuclei must get close to each other. However, nuclei repel each other because they have the same positive electric charge. If nuclei are moving fast, they can have enough kinetic energy to overcome the repulsive electrical force between them and get close to each other.

Remember that the kinetic energy of atoms or molecules increases as their temperature increases. Only at temperatures of millions of degrees Celsius are nuclei moving so fast that they can get close enough for fusion to occur. These extremely high temperatures are found in the center of stars, such as the Sun.

Nuclear Fusion and the Sun The Sun is composed mainly of hydrogen. Most of the energy given off by the Sun is produced by a process involving the fusion of hydrogen nuclei. This process occurs in several stages, and one of the stages is shown in **Figure 18.** The net result of this process is that four hydrogen nuclei are converted into one helium nucleus. As these nuclear reactions occur, a small amount of mass is changed into an enormous amount of energy. Earth receives a small amount of this energy as thermal energy and light.

As the Sun ages, the hydrogen nuclei are used up as they are converted into helium. So far, only about one percent of the Sun's mass has been converted into energy. It is estimated that the Sun has enough hydrogen to keep this reaction going for another 5 billion years.

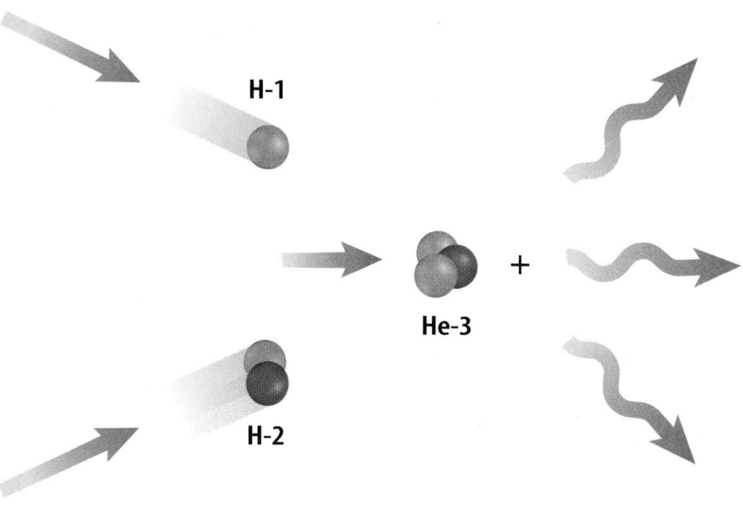

Figure 18 The fusion of hydrogen to form helium takes place in several stages in the Sun. One of these stages is shown here. An isotope of helium is produced when a proton and the hydrogen isotope H-2 undergo fusion.

Radioactive Decay Equations A uranium-235 atom can fission, or break apart, to form barium and krypton. Use a periodic table to find the atomic numbers of barium and krypton. What do they add up to? A uranium-235 atom can fission in several other ways such as producing neodymium and another element. What is the other element?

Using Nuclear Reactions in Medicine

If you were going to meet a friend in a crowded area, it would be easier to find her if she told you that she would be wearing a red hat. In a similar way, scientists can find one molecule in a large group of molecules if they know that it is "wearing" something unique. Although a molecule can't wear a red hat, if it has a radioactive atom in it, it can be found easily in a large group of molecules, or even in a living organism. Radioactive isotopes can be located by detecting the radiation they emit.

When a radioisotope is used to find or keep track of molecules in an organism, it is called a **tracer.** Scientists can use tracers to follow where a particular molecule goes in your body or to study how a particular organ functions. Tracers also are used in agriculture to monitor the uptake of nutrients and fertilizers. Examples of tracers include carbon-11, iodine-131, and sodium-24. These three radioisotopes are useful tracers because they are important in certain body processes. As a result, they accumulate inside the organism being studied.

Reading Check *How are tracers located inside the human body?*

INTEGRATE Health

Iodine Tracers in the Thyroid The thyroid gland is located in your neck and produces chemical compounds called hormones. These hormones help regulate several body processes, including growth. Because the element iodine accumulates in the thyroid, the radioisotope iodine-131 can be used to diagnose thyroid problems. As iodine-131 atoms are absorbed by the thyroid, their nuclei decay, emitting beta particles and gamma rays. The beta particles are absorbed by the surrounding tissues, but the gamma rays penetrate the skin. The emitted gamma rays can be detected and used to determine whether the thyroid is healthy, as shown in **Figure 19.** If the detected radiation is not intense, then the thyroid has not properly absorbed the iodine-131 and is not functioning properly. This could be due to the presence of a tumor. **Figure 20** shows how radioactive tracers are used to study the brain.

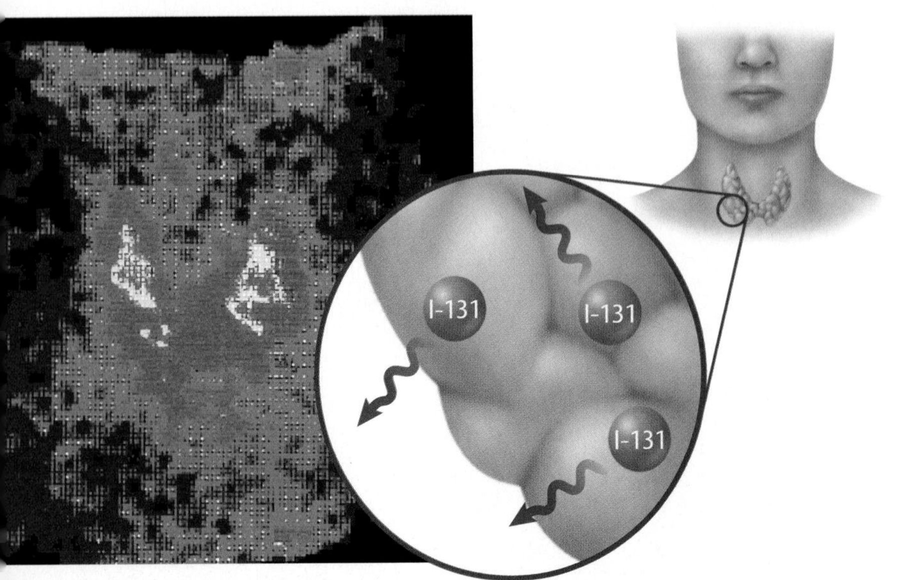

Figure 19 Radioactive iodine-131 accumulates in the thyroid gland and emits gamma rays, which can be detected to form an image of a patient's thyroid. **List** *some advantages of being able to use iodine-131 to form an image of a thyroid.*

VISUALIZING PET SCANS

Figure 20

The diagram below shows an imaging technique known as Positron Emission Tomography, or PET. Positrons are emitted from the nuclei of certain radioactive isotopes when a proton changes to a neutron. PET can form images that show the level of activity in different areas of the brain. These images can reveal tumors and regions of abnormal brain activity.

A When positrons are emitted from the nucleus of an atom, they can hit electrons from other atoms and become transformed into gamma rays.

B The radioactive isotope fluorine-18 emits positrons when it decays. Fluorine-18 atoms are chemically attached to molecules that are absorbed by brain tissue. These compounds are injected into the patient and carried by blood to the brain.

C Inside the patient's brain, the decay of the radioactive fluorine-18 nuclei emits positrons that collide with electrons. The gamma rays that are released are sensed by the detectors.

D A computer uses the information collected by the detectors to generate an image of the activity level in the brain. This image shows normal activity in the right side of the brain (red, yellow, green) but below-normal activity in the left (purple).

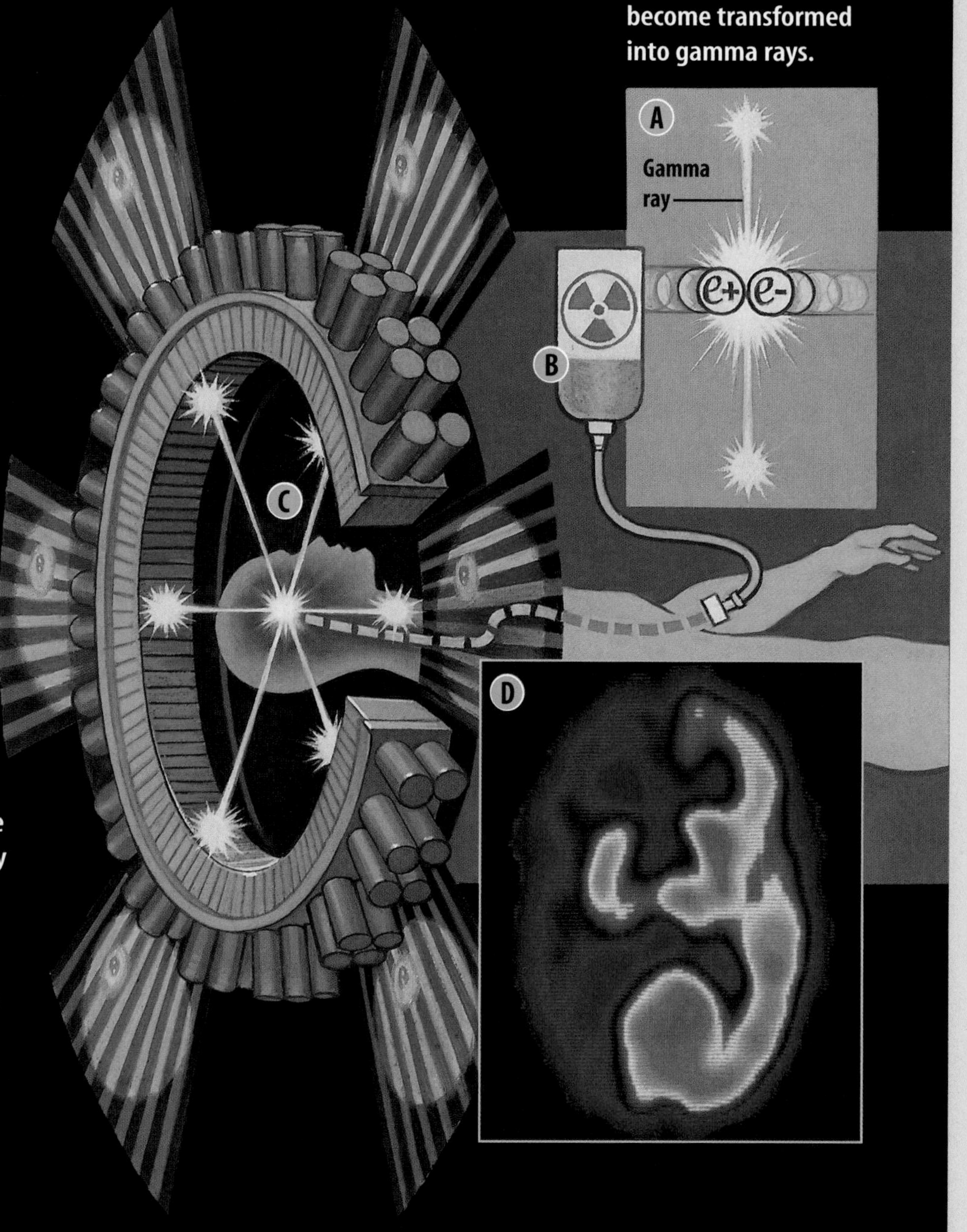

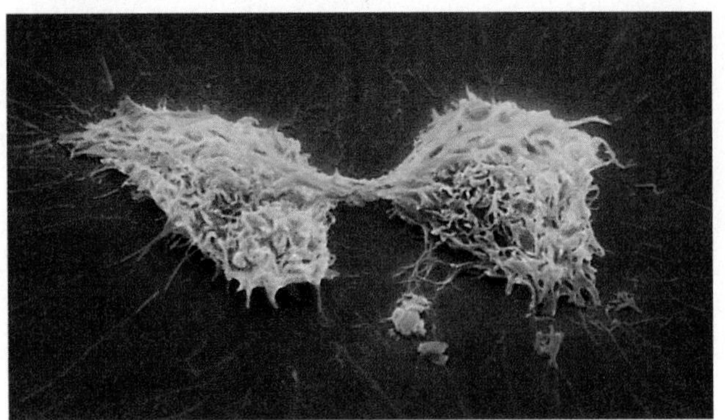

Figure 21 Cancer cells, such as the ones shown here, can be killed with carefully measured doses of radiation.

Treating Cancer with Radioactivity

When a person has cancer, a group of cells in that person's body grows out of control and can form a tumor. Radiation can be used to stop some types of cancerous cells from growing. Remember that the radiation that is given off during nuclear decay is strong enough to ionize nearby atoms. If a source of radiation is placed near cancer cells, such as those shown in **Figure 21,** atoms in the cells can be ionized. If the ionized atoms are in a critical molecule, such as the DNA or RNA of a cancer cell, then the molecule might no longer function properly. The cell then could die or stop growing.

When possible, a radioactive isotope such as gold-198 or iridium-192 is implanted within or near the tumor. Other times, tumors are treated from outside the body. Typically, an intense beam of gamma rays from the decay of cobalt-60 is focused on the tumor for a short period of time. The gamma rays pass through the body and into the tumor. How can physicians be sure that only the cancer cells will absorb radiation? Because cancer cells grow quickly, they are more susceptible to absorbing radiation and being damaged than healthy cells are. However, other cells in the body that grow quickly also are damaged, which is why cancer patients who have radiation therapy sometimes experience severe side effects.

section 4 review

Summary

Nuclear Fission

- Nuclear fission occurs when a neutron strikes a nucleus, causing it to split into smaller nuclei.
- A chain reaction requires a critical mass of fissionable material.

Nuclear Fusion

- Nuclear fusion occurs when two nuclei combine to form another nucleus.
- Nuclear fusion occurs at temperatures of millions of degrees, which occur inside the Sun.

Medical Uses of Radiation

- Radioactive isotopes are used as tracers to locate various atoms or molecules in organisms.
- Radiation emitted by radioactive isotopes is used to kill cancer cells.

Self Check

1. **Infer** whether mass is conserved in a nuclear reaction.
2. **Explain** why fusion reactions can occur inside stars.
3. **Explain** how a chain reaction can be controlled.
4. **Describe** two properties of a tracer isotope used for monitoring the functioning of an organ in the body.
5. **Think Critically** Explain why high temperatures are needed for fusion reactions to occur, but not for fission reactions to occur.

Applying Math

6. **Calculate Number of Nuclei** In a chain reaction, two neutrons are emitted by each nucleus that is split. If one nucleus is split in the first step of the reaction, how many nuclei will have been split after the fifth step?

More Section Review gpescience.com

Chain Reactions

In an uncontrolled nuclear chain reaction, the number of reactions increases as additional neutrons split more nuclei. In a controlled nuclear reaction, neutrons are absorbed, so the reaction continues at a constant rate. How could you model a controlled and an uncontrolled nuclear reaction in the classroom?

Real-World Problem

How can you model chain reactions?

Goals

- **Model** a controlled and uncontrolled chain reaction.
- **Compare** the two types of chain reactions.

Materials

dominoes
stopwatch

Safety Precautions

Procedure

1. Complete the safety form.
2. Set up a single line of dominoes standing on end so that when the first domino is pushed over, it will knock over the second and each domino will knock over the one following it.
3. Using the stopwatch, time how long it takes from the moment the first domino is pushed over until the last domino falls over. Record the time.
4. Line up the same number of dominoes in the shape of a Y, as shown above. Be sure that both dominoes at the split in the Y will get knocked down by the falling dominoes.
5. Repeat step 3.

Conclude and Apply

1. **Compare** the amount of time it took for all of the dominoes to fall in each of your two arrangements.
2. **Determine** the average number of dominoes that fell per second in both domino arrangements.
3. **Identify** which of your domino arrangements represented a controlled chain reaction and which represented an uncontrolled chain reaction.
4. **Describe** how the concept of critical mass was represented in your model of a controlled chain reaction.
5. Assuming that they had equal amounts of material, which would finish faster—a controlled or an uncontrolled nuclear chain reaction? Explain.

Communicating Your Data

Explain to friends or members of your family how a controlled nuclear chain reaction can be used in nuclear power plants to generate electricity.

Model and Invent

Modeling Transmutations

Goal
- **Model** decay of a uranium atom.

Possible Materials
brown rice
white rice
colored candies
dried beans
dried seeds
glue
poster board

Safety Precautions

WARNING: *Never eat foods used in the lab.*

Data Source
Refer to your textbook for general information about transmutation.

Real-World Problem

Imagine what would happen if the oxygen atoms around you began changing into nitrogen atoms. Without oxygen, most living organisms, including people, could not live. Fortunately, more than 99.9 percent of all oxygen atoms are stable and do not decay. Usually, when an unstable nucleus decays, an alpha or beta particle is thrown out of its nucleus, and the atom becomes a new element. This process of one element changing into another element is called transmutation. How could you create a model of a uranium-238 atom and the decay process it undergoes during transmutation? What types of materials could you use to represent the protons and neutrons in a U-238 nucleus? How could you use these materials to model transmutation?

Make a Model

1. Complete the safety form.
2. **Choose** two materials of different colors or shapes for the protons and neutrons of your nucleus model. Choose a material for the negatively charged beta particle.

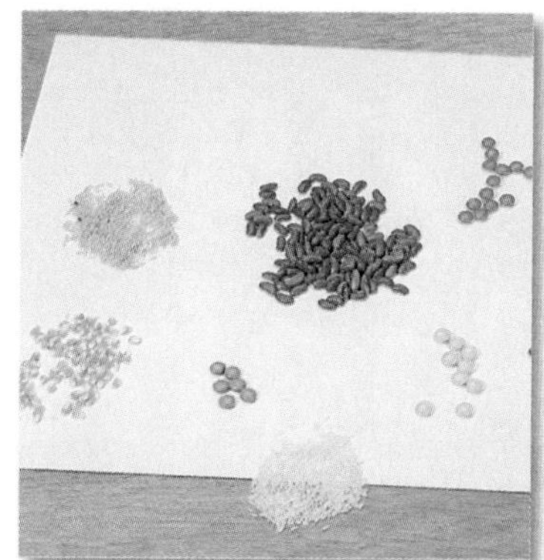

3. **Decide** how to model the transmutation process. Will you create a new nucleus model for each new element? How will you model an alpha or beta particle leaving the nucleus?
4. **Create** a transmutation chart to show the results of each transmutation step of a uranium-238 atom with the identity, atomic number, and mass number of each new element formed and the type of radiation particle emitted at each step. A uranium-238 atom will undergo the following decay steps before transmuting into a lead-206 atom: alpha decay, beta decay, beta decay, alpha decay, alpha decay, alpha decay, alpha decay, alpha decay, beta decay, beta decay, alpha decay, beta decay, beta decay, alpha decay.
5. **Describe** your model plan and transmutation chart to your teacher and ask how they can be improved.
6. **Present** your plan and chart to your class. Ask classmates to suggest improvements in both.
7. Construct your model of a uranium-238 nucleus showing the correct number of protons and neutrons.

Test Your Model

1. Using your nucleus model, demonstrate the transmutation of a uranium-238 nucleus into a lead-206 nucleus by following the decay sequence outlined in the previous section.
2. Show the emission of an alpha particle or beta particle between each transmutation step.

Analyze Your Data

1. **Compare** how alpha and beta decay change an atom's atomic number.
2. **Compare** how alpha and beta decay change the mass number of an atom.

Conclude and Apply

1. **Calculate** the ratio of neutrons to protons in lead-206 and uranium-238. In which nucleus is the ratio closer to 1.5?
2. **Identify** Alchemists living during the Middle Ages spent much time trying to turn lead into gold. Identify the decay processes needed to accomplish this task.

Show your model to the class and explain how your model represents the transmutation of U-238 into Pb-206.

TIME **SCIENCE** AND HISTORY

SCIENCE CAN CHANGE THE COURSE OF HISTORY!

The Nuclear Alchemists

The colored tracks are alpha particles emitted from a speck of radium salt placed on a special photographic plate.

For centuries, ancient alchemists tried in vain to convert common metals into gold. However, in the early 20th century, some scientists realized there was a way to convert atoms of some elements into other elements—nuclear fission.

A Startling Discovery

As the twentieth century dawned, most scientists thought atoms could not be broken apart. In 1902, New Zealand physicist Ernest Rutherford and his colleague Frederick Soddy showed that heavy elements uranium and thorium decayed into slightly lighter elements, with the production of helium gas. "Don't call it transmutation. They'll have our heads off as alchemists!" Rutherford warned Soddy. In 1908, Rutherford showed that the alpha particles emitted in radioactive decay were the same as helium nuclei.

Something's Missing

In 1938 in Germany, Otto Hahn and Fritz Strassmann found the uranium-235 nucleus would split if struck by a neutron. The process was called nuclear fission.

Enrico Fermi lead the development of the first nuclear reactor.

A year later, Austrian physicist Lise Meitner pointed out that the total mass of the particles produced when the uranium nucleus split was less than that of the original uranium nucleus. According to the special theory of relativity, this small amount of missing mass results in the release of a tremendous amount of energy when fission occurs. But is there any way this energy can be controlled?

Lise Meitner was the first to explain how nuclear fission occurs.

Controlling a Chain Reaction

Only a few years later, Italian physicist Enrico Fermi, working with colleagues in the United States, found the answer. Fermi realized that the neutrons released when fission occurs could lead to a chain reaction. However, materials that absorb neutrons could be used to control the chain reaction. In late 1942, Fermi and his colleagues built the first nuclear reactor by using cadmium rods to absorb neutrons and control the chain reaction. The tremendous energy released by nuclear fission could be controlled.

Research **Find out more about the contributions these scientists made to understanding radioactivity and the nucleus. What other discoveries did Rutherford and Fermi make?**

TIME

For more information, visit gpescience.com

Reviewing Main Ideas

Section 1 Radioactivity

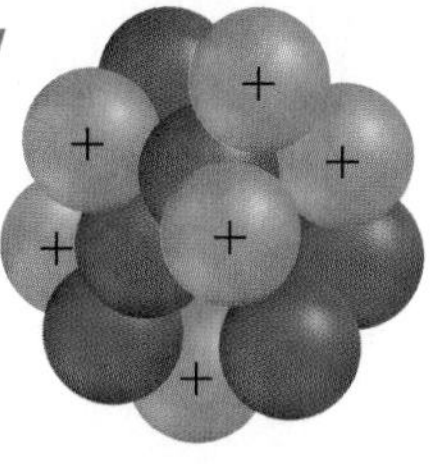
Nucleus

1. The protons and neutrons in an atomic nucleus, like the one to the right, are held together by the strong force.
2. The ratio of protons to neutrons indicates whether a nucleus will be stable or unstable. Large nuclei tend to be unstable.
3. Radioactivity is the emission of energy or particles from an unstable nucleus.
4. Radioactivity was discovered accidentally by Henri Becquerel about 100 years ago.

Section 2 Nuclear Decay

1. Unstable nuclei can decay by emitting alpha particles, beta particles, and gamma rays.
2. Alpha particles consist of two protons and two neutrons. A beta particle is an electron.
3. Gamma rays are the highest frequency electromagnetic waves.
4. Half-life is the amount of time in which half of the nuclei of a radioactive isotope will decay.
5. Because all living things contain carbon, the radioactive isotope carbon-14 can be used to date the remains of organisms that lived during the past 50,000 years, such as this skeleton.

6. Radioactive isotopes of uranium are used to date rocks.

Section 3 Detecting Radioactivity

1. Radioactivity can be detected with a cloud chamber, a bubble chamber, an electroscope, or a Geiger counter.
2. A Geiger counter measures the amount of radiation by producing electric current when it is struck by a charged particle.
3. Background radiation is low-level radiation emitted by naturally occurring isotopes found in Earth's rocks and soils, the atmosphere, and inside your body.

Section 4 Nuclear Reactions

1. When nuclear fission occurs, a nucleus splits into smaller nuclei. Neutrons and a large amount of energy are emitted.
2. Neutrons emitted when a nuclear fission reaction occurs can cause a chain reaction. A chain reaction can occur only if a critical mass of material is present.
3. Nuclear fusion occurs at high temperatures when light nuclei collide and form heavier nuclei, releasing a large amount of energy.
4. Radioactive tracers that are absorbed by specific organs can help diagnose health problems. Nuclear radiation is used to kill cancer cells.

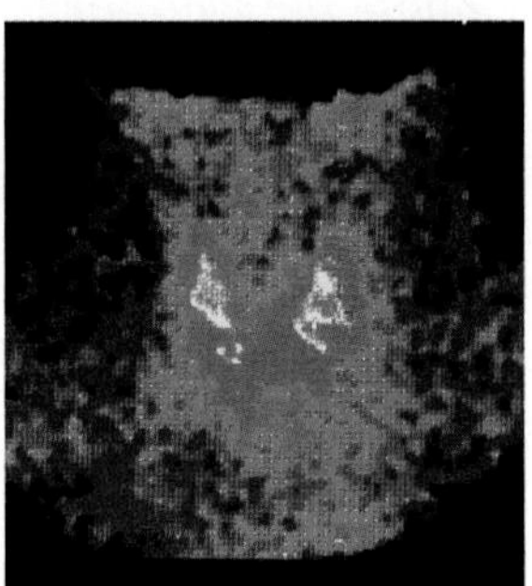

FOLDABLES™ Use the Foldable that you made at the beginning of this chapter to help you review advantages and disadvantages of using radioactive materials and nuclear reactions.

Using Vocabulary

alpha particle p. 791	half-life p. 794
beta particle p. 793	nuclear fission p. 801
bubble chamber p. 797	nuclear fusion p. 803
chain reaction p. 802	radioactivity p. 788
cloud chamber p. 796	strong force p. 787
critical mass p. 802	tracer p. 804
gamma ray p. 793	transmutation p. 792
Geiger counter p. 798	

Use what you know about the vocabulary words to explain the differences in the following sets of words. Then explain how the words are related.

1. cloud chamber—bubble chamber
2. chain reaction—critical mass
3. nuclear fission—nuclear fusion
4. radioactivity—half-life
5. alpha particle—beta particle—gamma ray
6. Geiger counter—tracer
7. nuclear fission—transmutation
8. electroscope—Geiger counter
9. strong force—radioactivity

Checking Concepts

Choose the word or phrase that best answers the question.

10. What keeps particles in a nucleus together?
    A) strong force **C)** electrical force
    B) repulsion **D)** atomic glue
11. Which device would be most useful for measuring the amount of radiation in a nuclear laboratory?
    A) a cloud chamber
    B) a Geiger counter
    C) an electroscope
    D) a bubble chamber
12. What is an electron that is produced when a neutron decays called?
    A) an alpha particle
    B) a beta particle
    C) gamma radiation
    D) a negatron
13. Which of the following describes an isotope's half-life?
    A) a constant time interval
    B) a varied time interval
    C) an increasing time interval
    D) a decreasing time interval
14. For which of the following could carbon-14 dating be used?
    A) a bone fragment
    B) a marble column
    C) dinosaur fossils
    D) rocks

Use the illustration below to answer question 15.

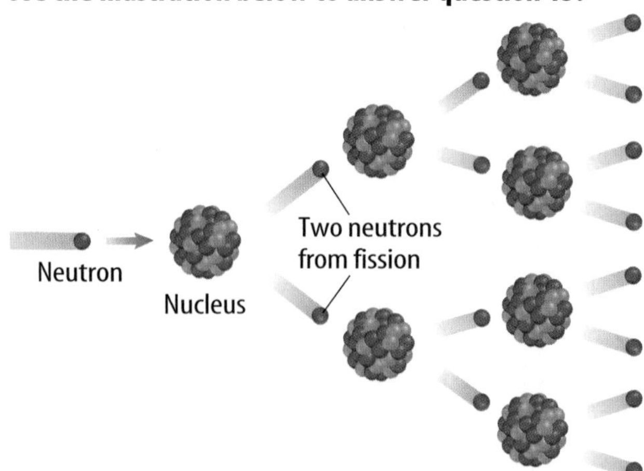

15. Which term describes an ongoing series of fission reactions such as the one pictured above?
    A) chain reaction **C)** positron emission
    B) decay reaction **D)** fusion reaction
16. Which of the following describes atoms with the same number of protons and a different number of neutrons?
    A) unstable **C)** radioactive
    B) synthetic **D)** isotopes

Interpreting Graphics

17. Copy and complete the following concept map on radioactivity.

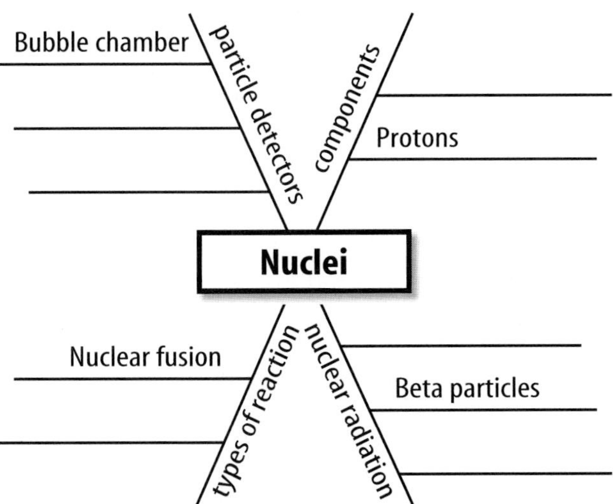

18. **Make a table** summarizing the use of radioactive isotopes or nuclear radiation in the following applications: radioactive dating, monitoring the thyroid gland, and treating cancer. Include a description of the radioactive isotope or radiation involved.

Use the data in the table below to answer question 19.

Isotope Half-Lives

Isotope	Mass Number	Half-Life
Radon-222	222	4 days
Thorium-234	234	24 days
Iodine-131	131	8 days
Bismuth-210	210	5 days
Polonium-210	210	138 days

19. **Graph** the data in the table above with the *x*-axis the mass number and the *y*-axis the half-life. Infer from your graph whether there is a relationship between the half-life and the mass number. If so, how does half-life depend on mass number?

Thinking Critically

20. **Explain** why the amount of background radiation a person receives can vary greatly from place to place.

21. **Infer** how the atomic number of a nucleus changes when the nucleus emits only gamma radiation.

22. **Identify** the properties of alpha particles that make them harmful to living cells.

23. **Determine** the type of nuclear radiation that is emitted by each of the following nuclear reactions:
    a. uranium-238 to thorium-234
    b. boron-12 to carbon-12
    c. cesium-130 to cesium-130
    d. radium-226 to radon-222

24. **Determine** how the motion of an alpha particle is affected when it passes between a positively charged electrode and a negatively charged electrode. How is the motion of a gamma ray affected?

25. **Infer** how the background radiation a person receives changes when they fly in a jet airliner.

Applying Math

26. **Use a Ratio** The mass of an alpha particle is 4.0026 mass units, and the mass of a beta particle is 0.000548 mass units. How many times larger is the mass of an alpha particle than the mass of a beta particle?

27. **Calculate Number of Half-Lives** How many half-lives have elapsed when the amount of a radioactive isotope in a sample is reduced to 3.125 percent of the original amount in the sample.

Standardized Test Practice

Record your answers on the answer sheet provided by your teacher or on a sheet of paper.

Multiple Choice

1. If a radioactive material has a half-life of 10 years, what fraction of the material will remain after 30 years?

A. one-half

B. one-third

C. one-fourth

D. one-eighth

2. Which of the following statements is true about all the isotopes of an element?

A. They have the same mass number.

B. They have different numbers of protons.

C. They have different numbers of neutrons.

D. They have the same number of neutrons.

3. How does the beta decay of a nucleus cause the nucleus to change?

A. The number of protons increases.

B. The number of neutrons increases.

C. The number of protons decreases.

D. The number of protons plus the number of neutrons decreases.

Use the illustration below to answer question 4.

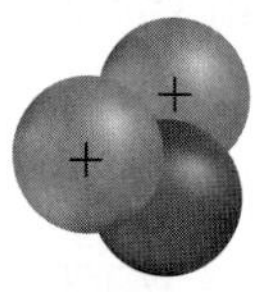

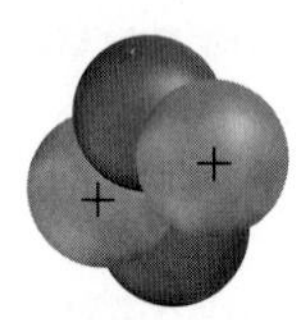

4. Which is a true statement about the two nuclei?

A. They have the same atomic number.

B. They have the same mass number.

C. They have different numbers of electrons.

D. They have different numbers of protons.

5. What is the atomic number of a nucleus equal to?

A. the number of neutrons

B. the number of protons

C. the number of neutrons and protons

D. the number of neutrons minus the number of protons

Use the illustration below to answer questions 6 and 7.

$^{210}_{84}Po \qquad ^{206}_{82}Pb + {}^{4}_{2}He$

6. What process is shown by this illustration?

A. nuclear fusion

B. chain reaction

C. transmutation

D. beta decay

Test-Taking Tip

Understand the Question Be sure you understand the question before you read the answer choices. Make special note of words like NOT or EXCEPT. Read and consider all the answer choices before you mark your answer sheet.

7. How do the total charge and total mass number of the products compare to the charge and mass number of the polonium nucleus?

   A. The charges are equal but the mass numbers are not equal.

   B. The mass numbers are equal but the charges are not equal.

   C. Neither the mass numbers or the charges are equal

   D. The mass numbers and charges are equal.

8. Radioactive isotopes of which element are used to study the brain?

   A. uranium

   B. fluorine

   C. carbon

   D. lead

Gridded Response

Use the table below to answer questions 9 and 10.

Half-Lives of Isotopes	
Isotope	**Half-life**
Carbon-14	5,730 years
Potassium-40	1.28 billion years
Iodine-131	8.04 days
Radon-222	4 days

9. How many grams of an 80-g sample of carbon-14 will be left after 17,190 years?

Short Response

10. A sample containing which radioactive isotope will have one-eighth of the isotope left after 24 days?

11. When the boron isotope boron-10 is bombarded with neutrons, it absorbs a neutron and then emits an alpha particle. Identify the isotope that is formed in this process.

Extended Response

12. Compare the strength of the strong force and the strength of the electric force on a proton in a small nucleus and in a large nucleus.

Use the illustration below to answer question 13.

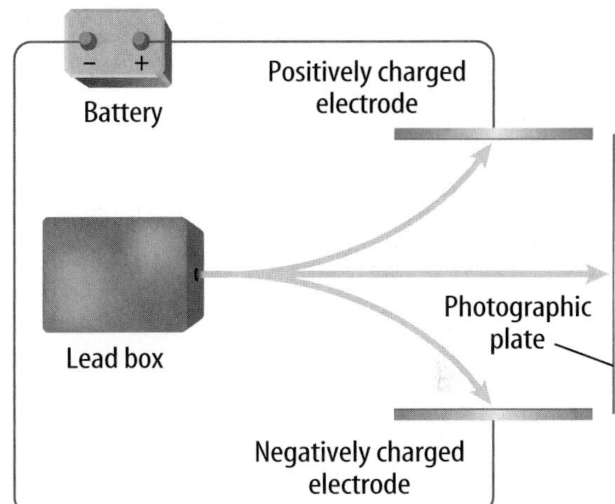

13. In the figure above, nuclear radiation is escaping from a small hole in the lead box.

   PART A Which type of nuclear radiation is deflected toward the positively-charged electrode, and why is this radiation deflected toward this electrode?

   PART B Explain why the radiation that struck the photographic plate was not deflected by the electrodes.

chapter 26

Stars and Galaxies

chapter preview

sections

1 Observing the Universe
2 Evolution of Stars
Lab *Star Cluster Age*
3 Galaxies and the Milky Way
4 Cosmology
Lab *Expansion of the Universe*

Colliding Galaxies

These two spiral galaxies, known as "the mice," have been colliding for about 160 million years. Eventually, they will merge into one giant, elliptical galaxy.

Science Journal In your Science Journal, write a paragraph about what you know about the Sun as a star.

Start-Up Activities

Stars in the Sky

Have you ever looked up at the night sky and been amazed at the number of stars you could see? But if you live in a well lighted area, you may not have seen very many stars at all. In this lab, you will explore a quick way to estimate how many stars you can see in different parts of the sky.

1. Ask an adult to help you locate an area near your home suitable for star gazing where there are not very many lights.
2. Hold a cardboard tube up to one eye and look through it at one area of the sky.
3. Count the number of stars you can see easily through the tube.
4. Look at three other areas of the sky in the same way and count those stars. Try looking to the south, west, straight up, and in a random direction.
5. Compare your data with those of two or three other students.
6. **Think Critically** In your Science Journal, report whether a similar number of stars were easily visible no matter where you looked or whether the number of stars varied. Explain any similarities or differences noted by you or other students.

Cosmology Make the following Foldable to help identify what you know, what you want to know, and what you learned about stars, galaxies, and cosmology.

STEP 1 **Fold** a vertical sheet of paper from side to side. Make the front edge about $\frac{1}{2}$ inch longer than the back edge.

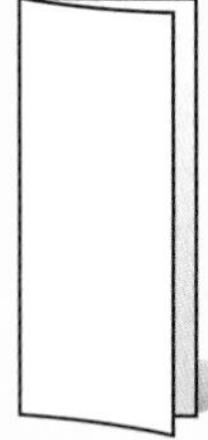

STEP 2 **Turn** lengthwise and **fold** into thirds.

STEP 3 **Unfold and cut** only the top layer along both folds to make three tabs. Label each tab as shown.

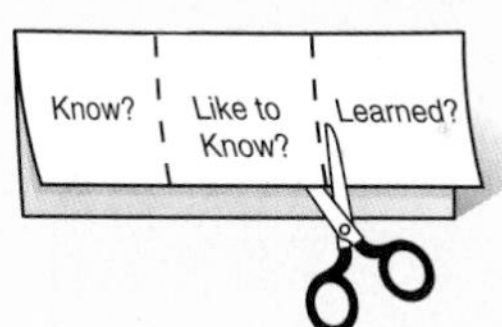

Questioning Before you read the chapter, write what you know about cosmology under the left tab of your Foldable, and write questions about what you'd like to know under the center tab. After you read the chapter, list what you learned under the right tab.

Preview this chapter's content and activities at gpescience.com

Observing the Universe

Reading Guide

What You'll Learn

- **Describe** a constellation as a pattern of stars.
- **Compare and contrast** types of optical telescopes.
- **Explain** how a radio telescope differs from an optical one.

Why It's Important

You can observe different parts of the universe from your home.

Review Vocabulary

electromagnetic spectrum: the entire range of wavelengths of electromagnetic energy

New Vocabulary

- constellation
- refracting telescope
- reflecting telescope
- radio telescope
- light-year
- spectroscope

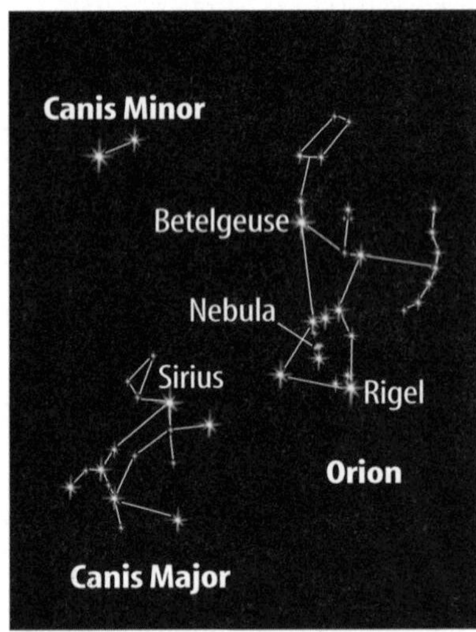

Figure 1 These three constellations contain some of the brightest stars visible from the northern hemisphere—Betelgeuse, Rigel, and Sirius.

Constellations

Have you ever watched clouds drift by on a summer day and tried to find shapes and patterns in them? One might look like a ship and another might resemble a rabbit or a bear. Long ago, people did much the same thing with stars. They named these patterns of stars after characters in stories, animals, or tools. Many of the names given to these star patterns by ancient cultures survive today and are called **constellations.** Astronomers use these constellations to locate and name stars.

From Earth, the stars in a constellation appear relatively close to one another. You can see that some of the stars are brighter than others, but you can't see how far they are from you or from each other. Usually, they lie at greatly different distances and just happen to line up and form a pattern.

In **Figure 1,** three constellations are shown with some of their brighter stars. The constellations visible in the evening sky change throughout the year.

Mythology In many cultures, Orion was a great hunter who had two hunting dogs, Canis Major (big dog) and Canis Minor (little dog). In another myth, two bears traveled around Earth's north pole. The constellations Ursa Major and Ursa Minor (big and little bears) were named for them. Indeed, they do swing around the north pole. In fact, Polaris, the polestar, is in Ursa Minor. If you have seen the Big Dipper you have seen most of Ursa Major, because the dipper is part of this constellation.

Telescopes Constellations and the stars that make them up are visible with the unaided eye. However, to see other objects in space, or to see some objects better, you need a telescope. Scientists and amateur astronomers use many different types of telescopes. Optical telescopes are used to study objects in visible light, and radio telescopes are used to study objects in the radio wavelengths. What do you think would be used to study X rays?

Optical Telescopes There are two basic types of optical telescopes. One type uses only lenses to study light and the other uses lenses and mirrors. Optical telescopes collect visible light and produce magnified images of objects. Light is collected by an objective lens or mirror. Because starlight is so distant, the light forms an image at the focus of the telescope. The focus is where light that is bent by the objective lens or reflected by the objective mirror comes together. A second lens, the eyepiece lens, then magnifies the image. The distance from the objective to the focus is the focal length of the telescope. You can find the magnifying power (M_p) of a telescope by dividing the focal length of the objective (f_o) by the focal length of the eyepiece (f_e).

$$M_p = f_o / f_e$$

Refracting Optical Telescopes A **refracting telescope** uses a convex lens, which is curved outward like the surface of a ball, as an objective, shown in **Figure 2.** When the lens curves outward on both sides, it is a double convex lens. Light passes through the objective lens. The eyepiece, which also can be a double convex lens, then magnifies the image. There is a limit to how large a refracting telescope can be. Since the objective lens can be supported only at its edges, it could sag in the middle if it is too large. When a larger telescope is needed, a reflecting telescope is used.

Reading Check *What type of optical telescope uses a lens as an objective?*

Mini LAB

Modeling a Telescope

Procedure

1. Obtain two **lenses of different focal length** from your teacher.
2. Place **double-sided tape** around the edge of one lens.
3. Place the lens along the short side of a piece of **construction paper** and roll the paper around the lens to form a tube. Tape the tube securely.
4. Repeat steps 2 and 3 with the other lens.
5. Insert the small tube into the large tube with the lenses on opposite ends.
6. Look through the eyepiece lens (small tube) and slide it in and out until you see a clear image.

Analysis

1. Observe an object with and without your telescope. Estimate how many times larger it appears through the telescope.
2. Obtain the focal lengths of your lenses from your teacher. Calculate the magnifying power of your telescope using the equation on this page. Compare this value to your estimate.

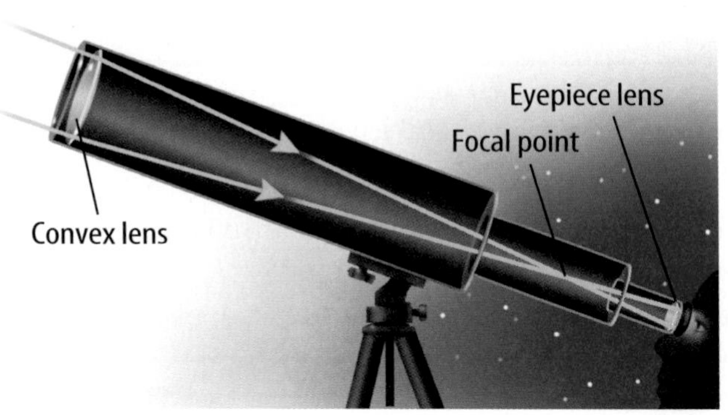

Figure 2 Two lenses are all that is needed for a refracting telescope. The tube and tripod aid in viewing.
Identify *the curvature of an objective lens.*

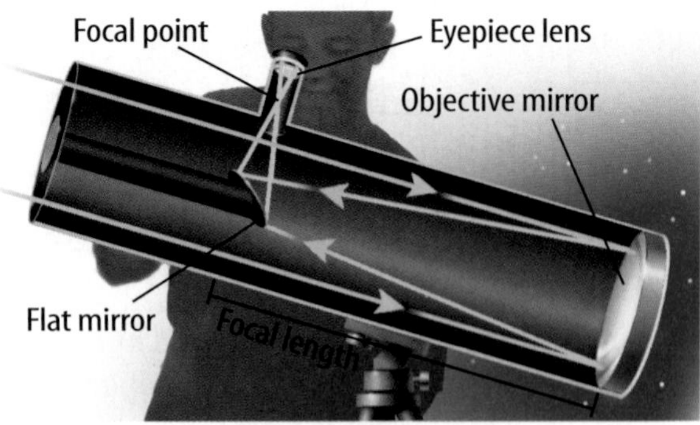

Figure 3 Reflecting telescopes use concave mirrors to gather light.

Reflecting Optical Telescopes A **reflecting telescope** uses a mirror as an objective to reflect light to the focus. **Figure 3** shows how light passes through the open end of a reflecting telescope and strikes a concave mirror at the base of the telescope. Often, a smaller mirror is used to reflect light into the eyepiece where it is magnified for viewing. However, in very large reflecting telescopes, the astronomer sits inside the telescope and looks through the eyepiece at the focus. Because mirrors can be supported from underneath a much larger telescope can be built.

Applying Math — Use Equations

THE MAGNIFYING POWER OF TELESCOPES If the focal lengths of a telescope's objective and eyepiece are known, the magnifying power can be calculated from the equation:

$$M_p = f_o/f_e$$

In this equation, M_p is the magnifying power of the telescope, f_o is the focal length of the objective, and f_e is the focal length of the eyepiece. Find the magnifying power of a telescope with a focal length of 1200 mm using eyepieces of 20 mm and 6 mm.

IDENTIFY known values and unknown values.

Identify the known values.

the focal length of the objective is 1,200 mm, this means $f_o = 1{,}200$ mm

the focal lengths of the eyepieces are 20 mm and 6 mm, this means $f_e = 20$ mm; 6 mm

Identify the unknown value

what is the magnifying power of the telescope means $M_p = ?$

SOLVE the problem

Substitute the known values into the equation for magnifying power:

$M_p = 1{,}200 \text{ mm}/20 \text{ mm} = 60$ $M_p = 1{,}200 \text{ mm}/6 \text{ mm} = 200$

Notice that the units cancel, and that magnifying power has no unit.

CHECK the answer

Does your answer seem reasonable? Do images look 60 times larger than the object when using the 20 mm eyepiece and 200 times larger when using the 6 mm eyepiece?

Practice Problems

1. Find the magnifying power of a telescope with a focal length of 2,500 mm when using eyepieces with focal lengths of 50 mm and 10 mm.
2. Find the magnifying power for a telescope with a focal length of 900 mm when using an eyepiece with a focal length of 12 mm.

For more practice problems, go to page 879 and visit Math Practice at gpescience.com.

New Telescope Design The most recent innovations in optical telescopes involve active and adaptive optics. With active optics, a computer is used to correct changes in temperature, mirror distortions, and bad viewing conditions. Adaptive optics uses a laser to probe the atmosphere and relay information to a computer about air turbulence. The computer then adjusts the telescope's mirror thousands of times per second to lessen the effects of atmospheric turbulence. The European Southern Observatory's *Very Large Telescope* in Chile, the largest optical telescope in use, uses adaptive optics.

Figure 4 27 dish antennae of the VLA are mounted on railroad tracks for rapid repositioning. The 304-m dish antenna in Arecibo, Puerto Rico is shown in the inset.

Radio Telescopes Radio waves, like visible light, are a form of electromagnetic energy emitted by stars and other objects in space. Radio waves can be detected even during the day, when the Sun's light makes it impossible to see the fainter visible light from other stars. Radio waves pass freely through Earth's atmosphere, even on completely cloudy days.

A telescope that collects and amplifies radio waves is a **radio telescope.** Because radio waves have long wavelengths, a radio telescope must be built with a very large objective, usually some form of dish antenna. Astronomers often build several radio telescopes close together and connect them to form one large telescope. The VLA (very large array), shown in **Figure 4,** is an example of this.

Topic: New Telescope Design

Visit gpescience.com for Web links to information about the development of new telescopes.

Activity Write and illustrate a paragraph about one of the new designs being considered for Earth-based telescopes or Earth-orbital telescopes.

What type of telescope is used to study radio waves?

Hubble Space Telescope Even using active and adaptive optics, the atmosphere limits what Earth-based telescopes can achieve. For this reason, astronomers use space telescopes, such as the *Hubble Space Telescope* shown in **Figure 5.**

The clear images provided by *Hubble* are changing scientists' ideas about space. One object viewed by *Hubble* is the massive galaxy cluster Abell 2218, which is about 2 billion light-years away. This cluster acts as a gravity lens that magnifies the light of even more distant galaxies. Such large distances in space are measured in a unit called a **light-year,** the distance that light travels in one year—about 9.5 trillion km. Even though it may seem confusing, remember that a light-year measures distance, not time.

Figure 5 The *Hubble Space Telescope* orbits Earth at an altitude of 610 km.

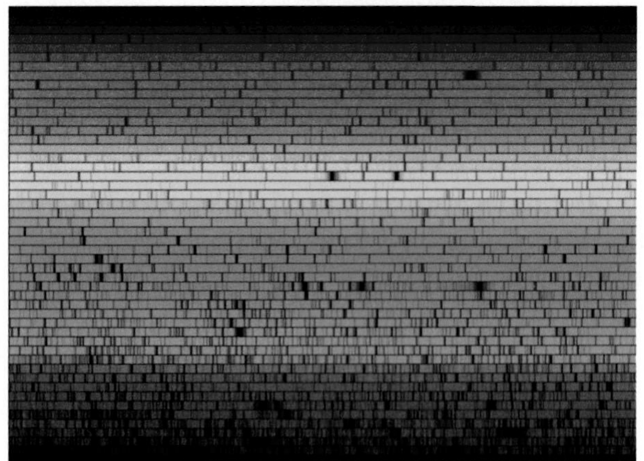

Figure 6 The dark lines in stellar spectra tell astronomers what elements are present in the stars being studied.
Explain *how a spectroscope produces a spectrum of colors.*

Spectroscopes

A **spectroscope** is a device that uses a prism or diffraction grating to disperse the light into its component wavelengths. When connected to a telescope, it disperses the light from the star or other celestial object collected by the telescope into its electromagnetic spectrum. This tells astronomers a great deal about a star. For example, they can determine its chemical composition, its surface temperature, and whether it is moving away from or toward Earth. Astronomers can even tell how fast the star is moving in relation to Earth.

Reading Check *What does a spectroscope do?*

Spectra A spectroscope disperses light into its individual wavelengths, or its spectrum. Visible light yields a spectrum of colors, including red, orange, yellow, green, blue, indigo, and violet, the colors of a rainbow. In fact, a rainbow is actually a visible spectrum of sunlight that has been dispersed by droplets of water in Earth's atmosphere. A spectrum displays all wavelengths in the light being studied, shown in **Figure 6.** You have learned that a star's spectrum indicates its surface temperature. For example, if the blue section is brightest, the star has a relatively high surface temperature. If the red section is brightest, the temperature of the star is much lower.

section 1 review

Summary

Constellations

- Constellations are patterns of stars that are used today to name and locate stars.

Telescopes

- Refracting telescopes use lenses to collect light and magnify the image.
- Reflecting telescopes use a mirror to collect light and a lens to magnify the image.
- A radio telescope collects and amplifies radio waves.

Spectroscopes

- A spectroscope disperses light into its spectrum.

Self Check

1. **Define** constellation.
2. **Compare and contrast** refractor telescopes and reflector telescopes.
3. **Describe** the objective of a radio telescope.
4. **Explain** how radio waves differ from visible light.
5. **Think Critically** Why do astronomers change eyepieces rather than objectives when they wish to increase the magnifying power of a telescope?

Applying Math

6. **Use Numbers** If the magnifying powers (M_p) of refracting and reflecting telescopes are 20 and 100, respectively, how much greater is the M_p of the reflecting telescope?

More Section Review gpescience.com

section 2

Evolution of Stars

Reading Guide

What You'll Learn

- **Explain** how stars form.
- **Classify** the stages of stellar evolution using a Hertzsprung-Russel (H-R) diagram.
- **Describe** the Sun and explain how it has and will evolve.

Why It's Important

The Sun is the star that provides energy for life on Earth.

Review Vocabulary

absolute magnitude: a measure of the amount of light given off by a star

New Vocabulary

- main sequence
- giant
- white dwarf
- solar mass
- photosphere
- sunspots

How do stars form?

Star formation begins with condensation of a large cloud of gas, ice, and dust called a nebula. These particles exert a gravitational force on each other, and the nebula contracts. Gravitational instability within the nebula causes it to break up into smaller cloud fragments. As a cloud fragment condenses its temperature increases. When the interior temperature reaches 1 million K, the center of the cloud is called a protostar. When the temperature reaches 10 million K, hydrogen fuses to form helium and a star is born.

H-R Diagram In the early 1900s, Ejnar Hertzsprung and Henry Russell studied the relationship between absolute magnitude and temperature of stars. They noticed that higher-temperature stars radiate more energy and have higher absolute magnitudes. As stars form, even while they are still protostars, they can be plotted on the Hertzsprung-Russell (H-R) diagram, like the one shown in **Figure 7.** About 90 percent of all stars fall on a line drawn from the upper left to the lower right of the H-R diagram, called the **main sequence.** The other 10 percent of stars fall elsewhere on the graph and will be discussed later.

Figure 7 The Sun is located in the center of the main sequence on this H-R diagram. It is cooler than young stars like Vega and Sirius, but warmer than the giant Betelgeuse.

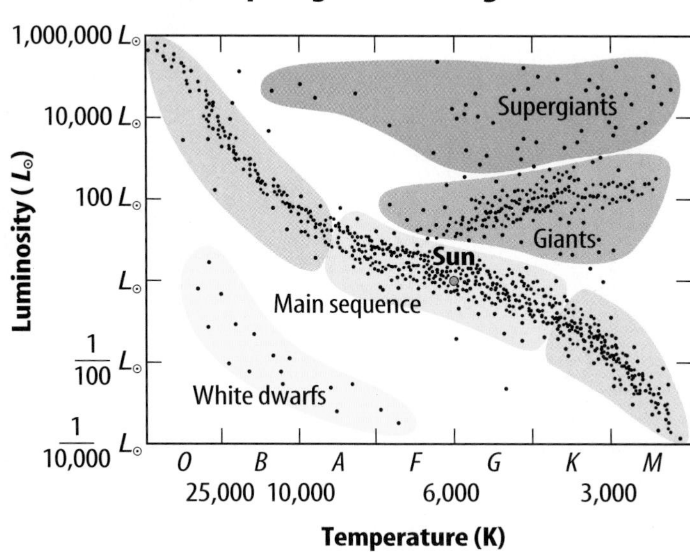

Counting Stars

Procedure

1. Draw a grid of twenty squares, with each square measuring 5 cm on a side. The final grid will be 25 cm long and 20 cm wide.
2. Randomly scatter about one **tablespoon** of **white rice** over the grid.
3. Select three squares and count the number of rice grains that are either completely within each square or partly touching its right side or bottom. Calculate the average number in each square.
4. Multiply the average by 20.

Analysis

1. How many rice grains are located within the grid?
2. How does this process model counting stars?

Try at Home

How do stars change?

A star like our Sun probably had a diameter about 100 times its present size while a protostar. As a star continues to form, it shrinks and increases in density, raising its interior temperature. Once fusion begins and the star attains stellar equilibrium, it settles onto the main sequence. In general, stellar equilibrium is the balance between outward pressure due to energy released in fusion and inward pressure due to gravity. Once this state of equilibrium is lost, the star enters the next stage of its life.

Main Sequence As long as the star's gravity balances outward pressures, the star remains on the main sequence. Stars are thought to spend most of their lives on the main sequence, which explains why this is the largest group on the H-R diagram. The Sun has been a main sequence star for about 5 billion years and will continue in this stage for about another 5 billion years.

Reading Check ***The Sun is classified as which type of star?***

When its hydrogen fuel is depleted, a star loses its equilibrium and its main sequence status. What it becomes next is determined by the total mass of the star. An average star like the Sun will become a giant, then a white dwarf, and finally a black dwarf. Stars more massive than the Sun can become supergiants and end up as neutron stars or black holes. Stars much lower in mass than the Sun, like the red dwarf star *Proxima Centauri,* could remain on the main sequence for 16 trillion years. Most stars on the main sequence are red dwarfs, and they probably make up about 80 percent of all stars in the universe.

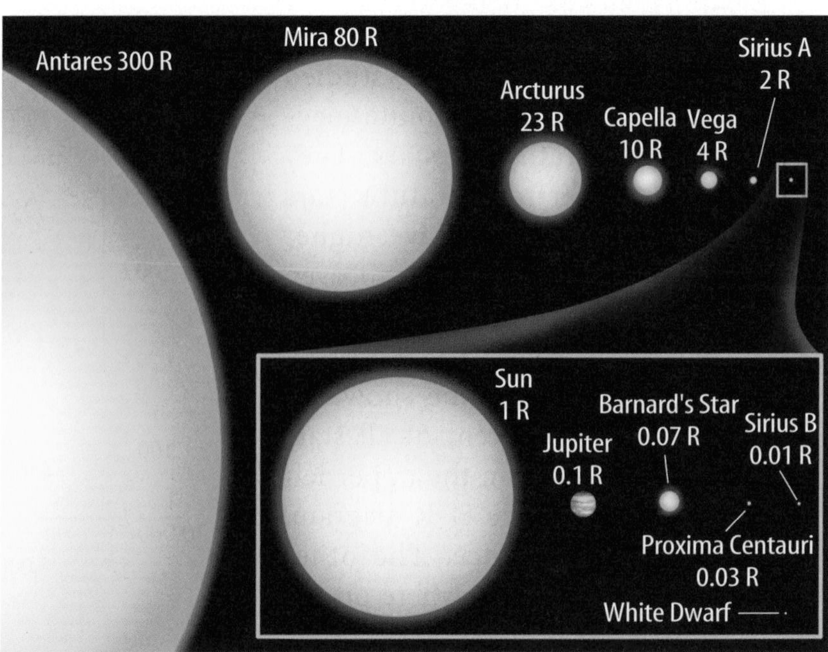

Figure 8 Only part of the giant star Antares is shown here. Super giant stars, such as Betelgeuse, are too large to be shown at all.

Giants and Dwarfs When hydrogen in a star's core is used up, its outward pressure is overcome by gravity. Its core contracts and increases in temperature. The outer layers expand and cool. In this late stage of its life cycle, an average star like our Sun is called a **giant.** In about five billion years, our Sun will become a giant like the one shown in **Figure 8.**

The giant's core continues to contract and become hotter. Eventually, the core uses up its helium and contracts even more. When temperature reaches 100 million K, helium fuses, forming carbon. Now the star is enormous and its surface is much cooler. Its outer layers escape into space leaving behind the hot, dense core that continues to contract. At this final stage in an average star's evolution, it is a **white dwarf.** A white dwarf, also shown in **Figure 8,** is about the size of Earth.

Supergiants, Neutron Stars, and Black Holes Stars that are over eight times more massive than our Sun take a different course. Their cores reach temperatures high enough to cause fusion that produces heavier and heavier elements. The star expands into a supergiant, such as Betelgeuse shown in **Figure 9.** Eventually, iron accumulates in the core. Since iron does not fuse readily, there is no outward radiation of energy to counteract the inward pull of gravity. The core collapses violently, and the outer portion of the star explodes, producing a supernova.

A supernova is a gigantic explosion in which the temperature in the collapsing core reaches 10 billion K and atomic nuclei are split into neutrons and protons. Protons merge with electrons to form neutrons, and the collapsing core becomes a neutron star. A typical neutron star is the size of a major city on Earth, but has a mass greater than the Sun's.

Very massive stars, with masses greater than 25 times that of the Sun, face a different end. In this case, the final collapse of the core continues past the neutron-star stage, forming a black hole—an object so dense that nothing can escape its gravity if it gets too close.

Supernovas The heavy elements you are made of formed during supernova explosions. Type I supernovas form from hydrogen-poor, low mass stars that have pulled in matter from a nearby red giant star. This process, called carbon detonation, causes carbon fusion almost everywhere inside the star and is thought to destroy the star completely. In contrast, Type II supernovas form from hydrogen-rich, high mass stars. They leave behind a collapsed core that can then condense further to form a neutron star or black hole.

Figure 9 Betelgeuse is located in the constellation of Orion. Its diameter is larger than the diameter of Jupiter's orbit around the Sun.
Classify *What kind of star is Betelgeuse?*

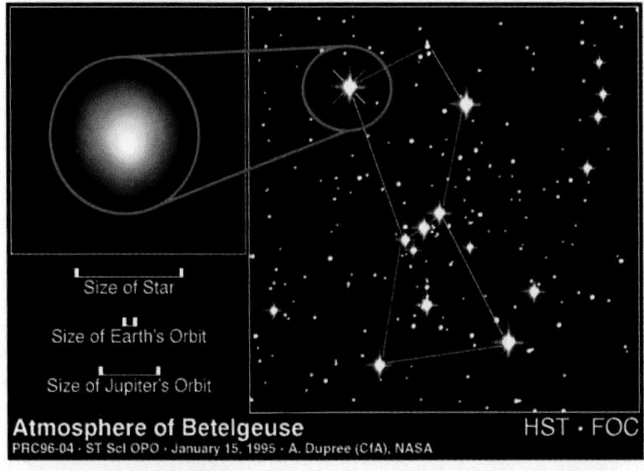

NATIONAL GEOGRAPHIC

VISUALIZING H-R DIAGRAMS OF STAR CLUSTERS

Figure 10

The Pleiades star cluster blazed into existence a mere 100 million years ago, when dinosaurs reigned on Earth and mammals were just gaining a foothold. The Pleiades, located in the back of the constellation Taurus, is among the earliest of celestial objects to be known and recorded by humans. Many myths sprang up to explain their origin. Today, astronomers study star clusters like the Pleiades because they offer valuable insights into how the universe is evolving.

Pleiades

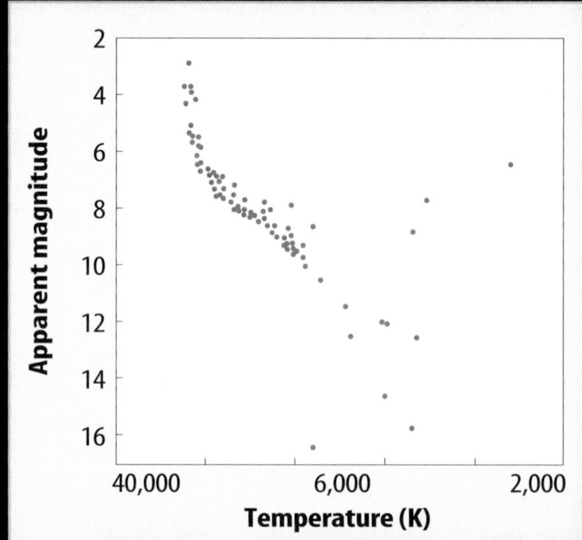

Because star clusters are believed to have formed from the same nebula, they are assumed to be about the same distance away. For this reason they are plotted using apparent magnitude instead of absolute magnitude or luminosity. Note that a lower magnitude number represents a brighter star. The H-R diagram for the Pleiades shows a well-defined main sequence. Stars in the upper left are hot, blue, and massive. As stars age, they burn all available hydrogen and eventually become red giants. The presence of hot, blue stars and the lack of red giants indicate that the Pleiades is a relatively young star cluster.

M5 is a globular cluster, located roughly 24,500 light-years from Earth. It has a compact appearance and vastly more stars than are found in open clusters like the Pleiades. Astronomers believe M5 is one of the oldest star clusters in our galaxy, with an estimated age of 13 billion years.

M5 Globular Cluster

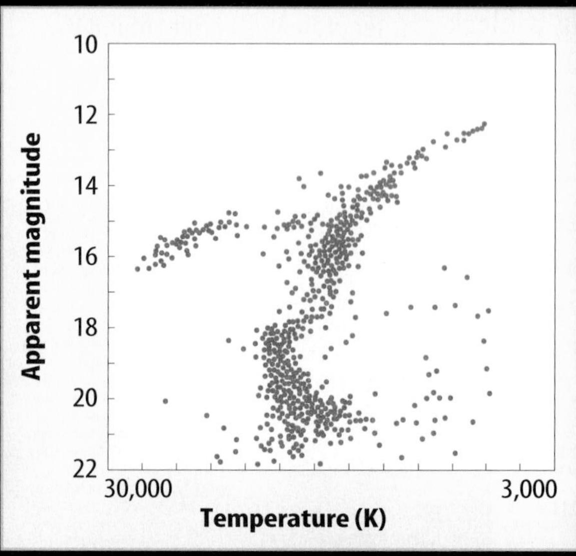

Compare the H-R diagram for M5 to the H-R diagram for the Pleiades. The stars in cluster M5, which were once hot and blue, have evolved into cooler, highly luminous red giants that appear in the upper right of the diagram. Sun-sized stars evolve from main sequence to red giants over billions of years, indicating that M5 is an ancient star cluster.

The Sun—A Main Sequence Star

The Sun is a middle-aged star and is plotted just about in the middle of the main sequence on the H-R diagram. It is average in size, temperature, and absolute magnitude. A **solar mass** is simply the mass of the Sun. For most stars, the relationship between mass and luminosity can be approximated by

$$\frac{L}{L_\odot} = \left(\frac{M}{M_\odot}\right)^{3.5}$$

where $L_\odot$ and $M_\odot$ are the luminosity and mass of the Sun. The H-R diagrams, shown in **Figure 10,** are used to compare very young stars to very old stars.

Structure of the Sun Although scientists cannot see inside the Sun, they have developed some theories about its interior. Much of what they know comes from studying its outer layers and its surface. The solar interior is composed of the core, the radiation layer, and the convection layer. The surface of the Sun is called the **photosphere.** This is the layer of the Sun that gives us light. The atmosphere above the photosphere is composed of the chromosphere and the corona. The Sun's structure is shown in **Figure 11.**

Solar Interior The innermost layer of the Sun is the core. This is where fusion occurs. The energy produced at the core may take millions of years to reach the photosphere where it is radiated into space.

The layer of the Sun just above the core is the radiation zone. In this layer, gases are completely ionized. Since no electrons remain on atoms to capture photons, this layer of the Sun is transparent to radiation. Energy formed by fusion travels easily through the radiation zone.

As you move farther outward from the Sun's core, the temperature drops and some electrons remain bound to their atoms. Here the gas becomes opaque to radiation and by the time you reach the outer edge of the radiation zone, all of the photons generated in the core have been absorbed. The energy from these photons is carried to the Sun's surface by convection through the next layer, the convection zone. Hotter gases from the bottom of this zone move upward toward the Sun's surface and cooler gases sink, setting up convection cells.

Energy for Life Plants use energy from the Sun and produce carbohydrates in a process called photosynthesis. When we burn fossil fuels, such as coal, we are using energy from the Sun that is stored in fossilized plant materials. Scientists also are trying to make direct use of solar energy by using solar cells. Research some applications of solar energy and share your results with your class.

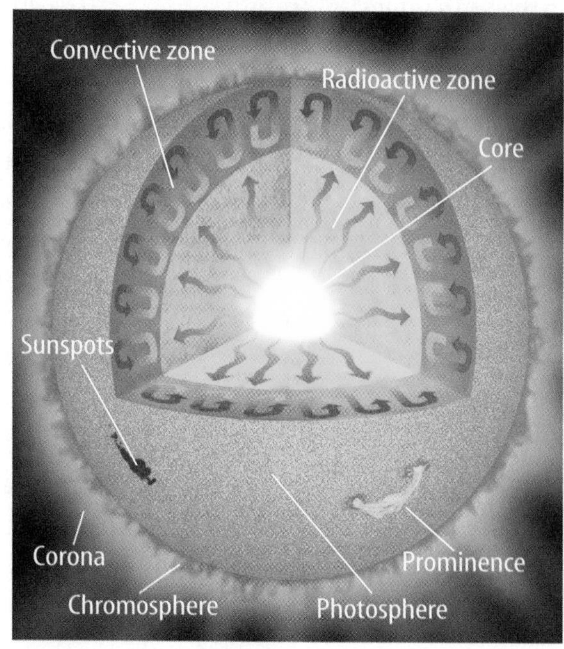

Figure 11 The Sun's interior is divided into three parts: the core, the radiation zone, and the convection zone.
List *three of the Sun's spheres or layers, starting with the surface and moving outward.*

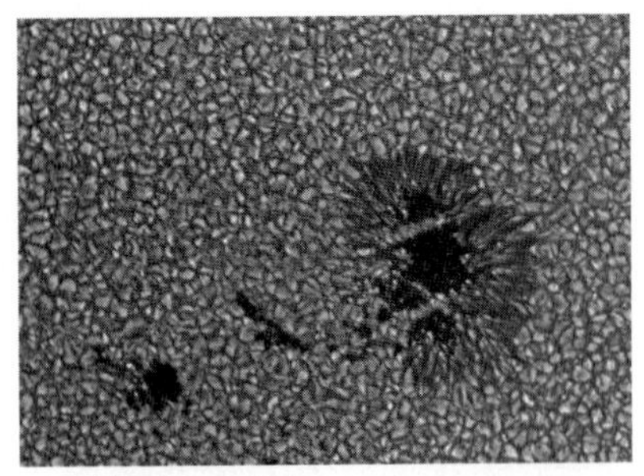

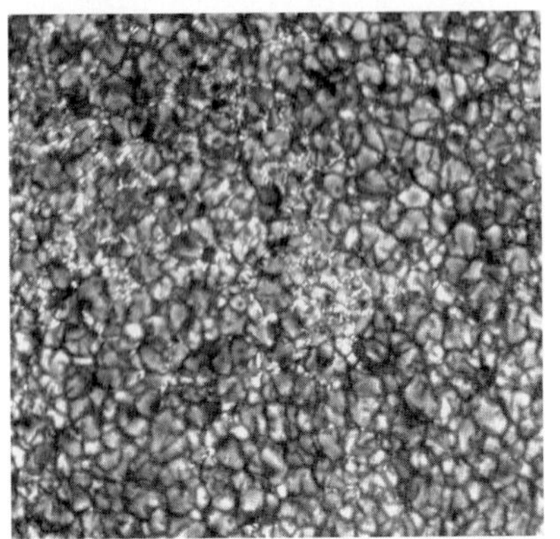

Figure 12 Sunspots, above, are often 2,000 K cooler than the surrounding granules, shown below.

Figure 13 This prominence of dense plasma appears suspended in the Sun's chromosphere as it loops back to the surface.

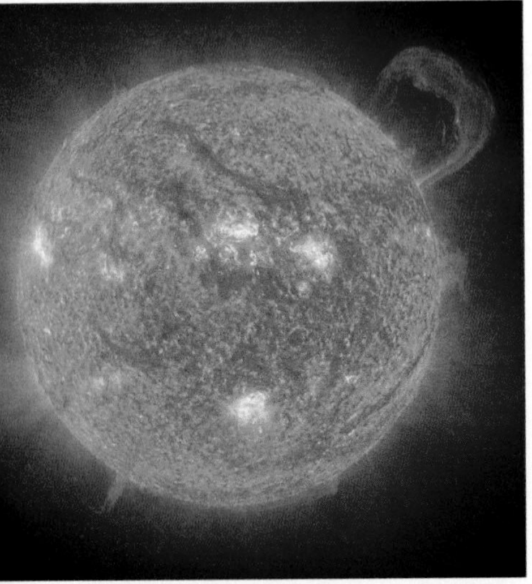

Photosphere The Sun's photosphere, or surface, is at the top of the convection zone and has a mottled appearance, called granulation. This is caused by rising hot material and sinking cooler material within the convection cells. Each granule shown in **Figure 12** is about 1,000 km across and is a direct result of convection cells in the convection zone.

Sunspots Some areas of the Sun appear darker than others. These darker areas of the Sun's photosphere, called **sunspots** are cooler than surrounding areas. Scientists can observe the movement of individual sunspots as they move with the Sun's rotation. They show that the Sun doesn't rotate as a solid body—it rotates faster at its equator than at its poles. Sunspots near the equator take about 25 days to go around the Sun, but at about 60 degrees north or south latitude, they take about 31 days. **Figure 12** shows a closeup view of some sunspots.

Reading Check *What is a sunspot?*

Sunspots aren't permanent features of the Sun. They appear and disappear over periods of days, weeks, or months. The number of sunspots changes in a fairly regular pattern called the sunspot, or solar activity, cycle. Periods of sunspot maximum occur about every 11 years, with periods of sunspot minimum occurring in between. At a sunspot minimum, sunspots appear at high latitudes. As the cycle progresses, they appear closer to the Sun's equator. By the next minimum, they again cluster at high latitudes.

Prominences and Flares Intense magnetic fields associated with sunspots can cause huge arching columns of gas called prominences to erupt, as shown in **Figure 13.** Convection in the convection zone causes magnetized gases to flow upward toward the photosphere. Sometimes the magnetic field strength is great enough that magnetic field lines shoot out from the surface near a pair of sunspots and cause a prominence of solar material to loop from one spot to the other. Some prominences blast material from the sun into space at speeds ranging from 600 km/s to more than 1,000 km/s.

Gases near a sunspot sometimes brighten suddenly, shooting gas outward at high speed in what are called solar flares. Temperatures within these flares can reach 100 million K. Particles produced in a flare possess so much energy that the Sun's magnetic field cannot hold them as it can hold prominences and they blast into space.

CMEs Sometimes large bubbles of ionized gas are emitted from the Sun. These are known as CMEs (coronal mass ejections). During sunspot minimums there is usually one CME per week, but during a sunspot maximum, there are two to three per day. When a CME is released in the direction of Earth, it appears as a halo around the Sun, as shown in **Figure 14.** As it passes Earth, the planet is exposed to a sudden shock wave of increased solar wind. Earth's atmosphere protects us, but occurrences of auroras increase. When scientists note a CME, they post an alert to watch for auroras at lower latitudes than normal.

Auroras take place when high-energy particles in CMEs and the solar wind are carried past Earth's magnetic field. This generates large electric currents that flow toward Earth's poles. These electric currents ionize gases in Earth's atmosphere. When these ions recombine with electrons and drop to a lower energy level, they produce light. This light is called the aurora borealis, or northern lights, when it occurs in the northern hemisphere. In the southern hemisphere, it is called the aurora australis. CMEs present little danger to life on Earth, but some of the highly charged solar wind material disrupts Earth's magnetosphere and interferes with orbiting satellites and radio signals.

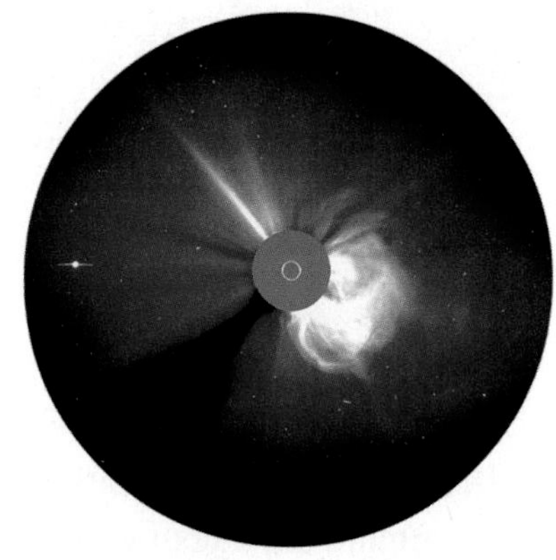

Figure 14 CMEs like this one can hurl a million tons of matter into space each second. They can damage satellites and endanger astronauts.

section 2 review

Summary

How do stars form?

- Stars form from a large cloud of gas, ice, and dust. Once the temperature inside the nebula reaches 10 million K, fusion begins.
- The H-R diagram plots temperature v. absolute magnitude of stars.

How do stars change?

- Once fusion begins, a star develops stellar equilibrium and becomes a main sequence star.
- When the hydrogen fuel is depleted, a star loses stellar equilibrium and evolves into a giant, or a supergiant.

The Sun—A Main Sequence Star

- The Sun's energy is produced by fusion.
- Sunspots are dark, cooler regions on the Sun's photosphere.
- Other solar features that can affect Earth are prominences, flares, and CMEs.

Self Check

1. **Explain** how stars form from nebulae.
2. **Describe** the different evolution stages of stars as shown on the H-R diagram.
3. **Describe** the structure of the Sun.
4. **Explain** the difference between Type I and Type II supernovas.
5. **Think Critically** What happens to stellar equilibrium to make a main sequence star evolve into a giant star?

Applying Math

6. **Use Percentages** Eighty percent of all stars are red dwarfs. Out of a random sample of 2,000 stars in the galaxy, about how many will plot on the H-R diagram as a red dwarf?
7. **Use Numbers** If the Sun will remain a main sequence star for a total of 10 billion years and Proxima Centauri will remain one for 16 trillion years, how many times longer will Proxima Centauri be a main sequence star?

More Section Review gpescience.com

Star Cluster Age

Stars are plotted on the H-R diagram based on their temperature and brightness. You can determine the age of a cluster of stars by studying the H-R diagram of the stars within the cluster. Massive main sequence stars, located at the upper left of the diagram, evolve faster than stars farther down the main sequence. Stars in the lower right of the diagram evolve more slowly. How could you use this idea to help determine the relative ages of star clusters from their H-R diagrams?

Real-World Problem

How can the relative age of star clusters be determined from H-R diagrams of each cluster?

Goals

- **Compare and contrast** H-R diagrams of several star clusters.
- **Determine** the relative ages of four star clusters.

Materials

four H-R diagrams of star clusters

Procedure

1. Study the four H-R diagrams your teacher gives you.
2. Note which types of stars remain on the main sequence and which stars have evolved off the main sequence.
3. Based on the evolution of the stars within each star cluster, decide which cluster is youngest and which is oldest. Also decide the relative ages of the other two star clusters. *HINT: The oldest cluster will have some that have evolved to white dwarfs.*

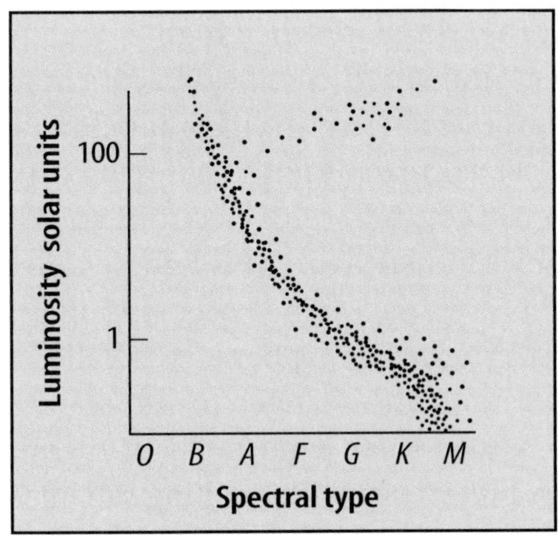

Conclude and Apply

1. **Compare and contrast** H-R diagrams of four star clusters.
2. **Determine** which star cluster was youngest and which was oldest.
3. **Explain** how you determined the ages of these two star clusters.
4. **Determine** the relative ages of the remaining two star clusters.
5. **Explain** how you determined the ages of these two star clusters.

Communicating Your Data

Explain to your friends how the relative ages of star clusters can be found by studying H-R diagrams.

section 3

Galaxies and the Milky Way

Reading Guide

What You'll Learn

- **Explain** that the same natural laws that apply in the Milky Way Galaxy also apply in other galaxies.
- **Compare** the three main types of galaxies.
- **Describe** the Milky Way galaxy and the Sun's position in it.

Why It's Important

The Milky Way galaxy is your galaxy and you can see part of it stretching across the evening sky.

Review Vocabulary

ellipse: oblong, closed curve

New Vocabulary

- galaxy
- Milky Way
- Local Group

Galaxies

One reason to study astronomy is to learn about your place in the universe. Long ago, people thought Earth was at the center of the universe. You know this isn't true, but do you know where you are in the universe?

You are on Earth, and Earth orbits the Sun. But the Sun orbits something also and it interacts with other objects in the universe. The Sun is one star among billions of stars in a galaxy. A **galaxy** is a large group of stars, gas, and dust held together by gravity, shown in **Figure 15.** Our galaxy, called the **Milky Way,** contains 400 billion stars, by most recent estimates, including the Sun. Countless other galaxies exist throughout the universe—an estimated 40 billion galaxies can be seen. Each of these galaxies contains the same elements, forces, and types of energy as our galaxy. There are three major types of galaxies: spiral, elliptical, and irregular.

Spiral Galaxies Take another look at **Figure 15.** Notice that spiral galaxies have spiral arms that wind outward from the galaxy's center. These spiral arms are made up of bright stars, dust, and gas. Our neighbor, the Andromeda galaxy, is visible to the unaided eye as a fuzzy patch in the constellation Andromeda. It is a normal spiral galaxy with its arms starting close to center. Barred spirals are another type of galaxy that have spiral arms extending from a large central bar of stars, dust, and gas that passes through the center, or hub, of the galaxy. Astronomers are not certain whether the Milky Way is normal or barred spiral.

Figure 15 Spiral galaxy 4414. Galaxies often are millions of light-years apart.
Explain *what is meant by* a million light-years.

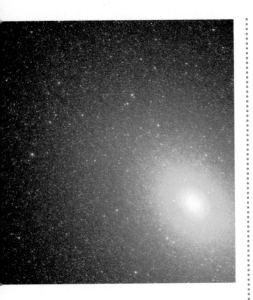

Figure 16 In ultraviolet light, this elliptical galaxy reveals a core of thousands of old, helium-burning stars.
Identify *the most common type of galaxies in the universe.*

Elliptical Galaxies Astronomers once thought that spiral galaxies were the most common galaxies because they are relatively large and easy to see. But as observations of the universe became more detailed, it became clear that most galaxies are elliptical galaxies, and most of these are dwarf galaxies. They are just too small and dim to be easily seen.

These galaxies are shaped like large, three-dimensional ellipses. Many are football-shaped, but others are spherical as shown in **Figure 16.** These giant elliptical galaxies can be over 9 million light-years across and contain trillions of stars. However, most dwarf ellipticals are only about 3,000 light-years across and contain fewer than a million stars.

Irregular Galaxies Most galaxies that aren't elliptical or spiral are considered irregular galaxies. They take many different shapes and contain 100 million to 10 billion stars, making them larger than dwarf ellipticals but smaller than spirals. Irregular galaxies are less common than spirals or ellipticals.

Two irregular galaxies called the Clouds of Magellan orbit the Milky Way. One of these, known as the Large Magellanic Cloud, is shown in **Figure 17.** Several other dwarf galaxies also are affected by the Milky Way's gravity. One of these, the elliptical Sagittarius dwarf, is being absorbed by the Milky Way. This dwarf galaxy lies about 60,000 light-years from the center of the Milky Way on the opposite side of the galaxy from us and 20,000 light-years below the galactic plane.

The Local Group Just as stars are grouped together within galaxies, galaxies are grouped into clusters. Clusters of galaxies are even grouped into superclusters. Our Milky Way galaxy belongs to a cluster called the **Local Group.** It is a relatively small cluster containing about 45 galaxies of various types and sizes, most of which are dwarf elliptical galaxies. The largest galaxy in the Local Group is the Andromeda galaxy, a spiral galaxy a little larger than ours that lies about 2.6 million light-years away. If the Andromeda galaxy and the Milky Way galaxy continue to travel through space as they are, they may collide in the distant future.

Figure 17 The irregular structure of the Large Magellanic Cloud might have been produced by interactions with the Milky Way galaxy.

How do galaxies form?

Astronomers aren't sure how galaxies originally formed. It is thought that fluctuations in density of primordial matter in the universe began to form blobs of gas that would eventually form into galaxies. These blobs might have had masses equivalent to the mass of the dwarf galaxies. In fact, one idea is that the dwarf galaxies may be remnants of these earlier blobs of matter.

Most astronomers now believe that the galaxies we see closer to us grew by absorbing or merging with other smaller objects. One bit of evidence that supports this idea is that more distant galaxies tend to be much smaller than those closer to the Milky Way. It makes sense to conclude that these smaller galaxies that existed long ago merged to make the larger, more organized galaxies of the universe today. Also, we have evidence that galaxies do collide.

Colliding Galaxies In some galaxy clusters, the galaxies are concentrated very close together. In the Virgo Cluster, for example, thousands of galaxies orbit within 10 million light-years of each other. Do these galaxies collide? If so, what happens to them and to the stars within them? It seems that little happens to the individual stars within the galaxies. There is so much open space between the stars that the individual stars of the two galaxies just move past each other.

What happens to individual stars in colliding galaxies?

However, galaxy collisions have a strong effect on the overall structure and shape of the colliding galaxies. They may lose all of their spiral shape, if they had any.

It is thought that the two galaxies shown in **Figure 18** will eventually collide. The smaller galaxy on the right does not seem to have enough energy to escape the larger one to the left. When galaxies interact by passing close to each other or by colliding, there is a burst of star formation in each. Their interstellar gas and dust clouds are shocked and squeezed, leading to star formation. This can be detected by the blue light emitted by young, hot stars.

Figure 18 In about a billion years, these two galaxies will merge. Much of their spiral structures probably will be lost.

Topic: The Milky Way Galaxy

Visit gpescience.com for Web links to information about the Milky Way galaxy.

Activity Draw a model of the Milky Way galaxy showing its possible barred structure. Include the approximate location of the Sun and information about the Sagittarius dwarf galaxy.

The Milky Way

Recall that the Milky Way galaxy contains about 400 billion stars, including our Sun. The Sun makes one complete orbit around the center of the Milky Way in about 225 million years, traveling at a speed of 220 km/s. This means that since it formed, the Sun has made a little over 22 orbits of the Milky Way.

The Milky Way is usually classified as a normal spiral galaxy. However, recent evidence suggests that it might be a barred spiral. It is difficult to know for sure because astronomers can never see our galaxy from the outside. You can't see the normal spiral or barred shape of the Milky Way because the Sun and Earth are located within one of its spiral arms. However, you can see the Milky Way stretching across the sky as a faint band of light. All of the stars you can see belong to the Milky Way galaxy.

Reading Check *What type of galaxy is the Milky Way?*

Evidence indicates that the Milky Way, like many galaxies, grows by absorbing other galaxies. It has been gobbling up the Sagittarius dwarf elliptical galaxy for 2 billion years. If we could see infrared light, we could see stars and other material from this galaxy becoming part of our section of the Milky Way, as shown in **Figure 19.** Eventually, the Milky Way probably will absorb both Clouds of Magellan and several other dwarf galaxies.

Figure 19 The swirls of red seen in this model are stars in the remnants of the Sagittarius dwarf elliptical galaxy.

Structure of the Milky Way The Milky Way galaxy, shown in **Figure 20,** measures about 100,000 light-years from one side to the other. The Sun lies about 26,000 light-years from the galactic center on the edge of one of the spiral arms. The Milky Way's disk is about 1,000 light-years thick—it would take 1,000 years to travel from top to bottom even at the speed of light. The central bulge of the Milky Way is about 10,000 light-years in diameter.

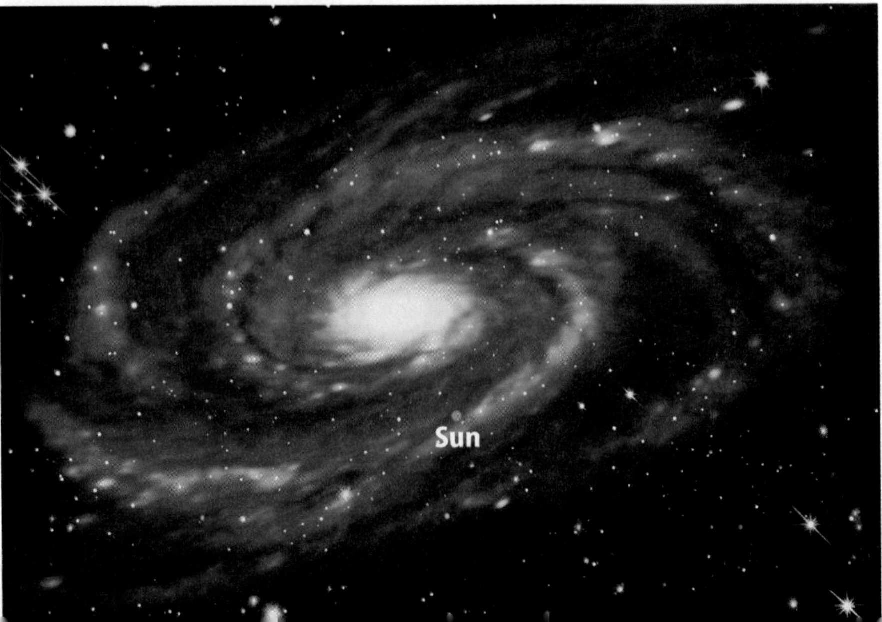

Figure 20 Star density is greatest near the center of the Milky Way, with over 1,400 stars in every cubic light-year of space. The Sun is located about halfway out on the Orion arm of the Milky Way galaxy.
Explain *what is meant by* a cubic light-year.

Spiral Arms The arms of a spiral galaxy look like pinwheels that begin near the galactic center and extend outward through the disk of the galaxy. These spiral arms contain both young stars and prestellar material, such as glowing nebulae. Young open star clusters are present too. This is the part of a spiral galaxy where stars are forming. Astronomers do not yet fully understand what causes the spiral structure to form. They speculate that it might be caused by instabilities in the gas near the galactic bulge or gravitational effects of other galaxies that are or were nearby, or might be just an extension of the shape of the galactic bulge, itself. They just don't know.

Galaxy Center What strange objects lurk in the very dense population of stars at a galaxy's core? Recent theories suggest that extremely massive black holes might exist at the cores of galaxies. The problem is that this part of a galaxy is hidden from view by material that exists in between the densely packed stars. The total energy emitted from an object called Sgr A* (saj-ay-star), located in the nucleus of the Milky Way, is equivalent to the energy that would be emitted by a million suns. The leading theory about this object, shown in **Figure 21,** is that it is a supermassive black hole, containing the mass of 3 million suns.

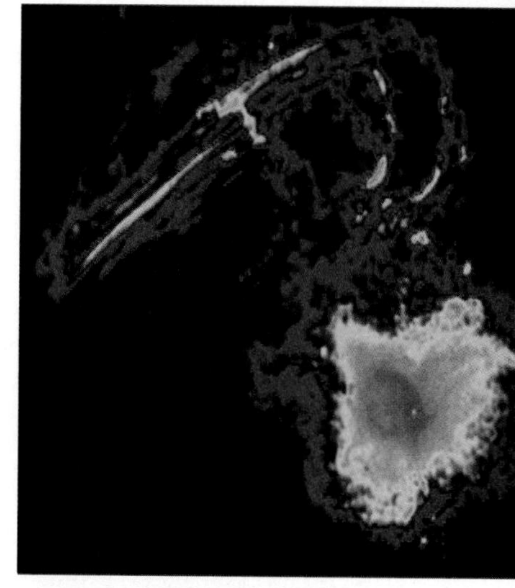

Figure 21 This bright radio source captured by the VLA is indirect evidence of a black hole at the center of our galaxy.

section 3 review

Summary

Galaxies

- A galaxy is a large group of stars, gas, and dust held together by gravity.
- The three main types of galaxies are elliptical, irregular, and spiral.

How do galaxies form?

- Astronomers believe that galaxies form by absorbing or merging with smaller objects.
- Young, hot stars form when galaxies collide, but individual stars within each galaxy are not affected much.

The Milky Way Galaxy

- The Milky Way galaxy contains about 400 billion stars and has a spiral shape.
- The Milky Way galaxy is about 100,000 light-years across and the Sun lies about 26,000 light-years from its center.
- Supermassive black holes are thought to exist in the centers of galaxies.

Self Check

1. **Compare and contrast** the Milky Way galaxy to other galaxies in the universe.
2. **Describe** the three main types of galaxies.
3. **Draw** the overall structure of the Milky Way galaxy and indicate where the Sun is located.
4. **Describe** the most common type of galaxy.
5. **Think Critically** How might the Sun be affected if the Andromeda galaxy and the Milky Way galaxy collide.

Applying Math

6. **Use Numbers** Assume there are 400 billion stars in the Milky Way galaxy, plus or minus 200 billion. Based on this estimate, what is the range of the number of stars that might exist in the Milky Way?
7. **Use Percentages** A dwarf elliptical galaxy has fewer than 1 million stars and a small irregular galaxy contains 100 million stars. What percent of the number of stars found in a small irregular galaxy are found in a dwarf elliptical galaxy?

More Section Review gpescience.com

section 4

Cosmology

Reading Guide

What You'll Learn

- **Describe** the most accepted theory about the origin of the universe.
- **Define** dark matter.
- **Explain** how dark energy may cause the universe to expand forever.

Why It's Important

Conditions that existed when the universe began led to the formation of galaxies like the Milky Way.

Review Vocabulary

universe: the space that contains all known matter and energy

New Vocabulary

- cosmology
- big bang theory
- dark matter
- dark energy

How did it begin?

The study of the universe—how it began, how it evolves, and what it is made of—is known as **cosmology.** Several models of the origin and evolution of the universe have been proposed.

One model, proposed in 1948, is called the steady state theory. In this theory, the universe has always existed and it always will. As the universe expands, new matter is created. The density of the universe remains the same or in steady state.

A second idea is called the oscillating model. In this model, the universe expanded rapidly, then slowed, and eventually contracted. This oscillating process continues back and forth through time. Some scientists currently believe that enough matter exists to cause the universe to eventually contract.

Figure 22 This map produced by the WMAP team has been called a "baby picture" of the universe. It is oval because it is a projection, just as maps of Earth can be projected as ovals. **Infer** *why this map is called a "baby picture" of the universe.*

The Big Bang Theory

The most accepted theory of how the universe formed is the **big bang theory.** It states that the universe started with a big bang, or explosion, and has been expanding ever since. The big bang is not like an explosion of matter into empty space; it is the rapid expansion of space itself.

When did it begin? A NASA-related mission, called the Wilkinson Microwave Anisotropy Probe (WMAP), produced a map of the oldest light in the universe. Based on the map, shown in **Figure 22,** and other data, the WMAP team proposed a more specific age of the universe. Their findings indicate that the universe began about 13.7 billion years ago with a big bang. The team believes its data are correct within a one percent margin of error. They measured temperature variations over the entire universe found in the cosmic microwave background radiation. This radiation is thought to have been produced about 400,000 years after the big bang when temperatures became low enough for atoms to form. Bright areas of the map are thought to indicate places that collapsed, forming the galaxies that we see today.

Reading Check *When did the universe begin?*

Science as Literature
Many famous scientists have written about science for the general public. Among them are physicists Albert Einstein and George Gamow. More recently, physicist Stephen Hawking wrote "A Brief History of Time," which describes current theories of how the universe began. Research the work of Steven Hawking and his contributions to cosmology and write a brief biography.

Expansion of the Universe

The motion of stars within the Milky Way can be detected by using the Doppler effect. For example, sound waves from a moving source are compressed as the object approaches and stretched as it recedes. Doppler shifts occur in light as well as sound. If a star approaches Earth, its wavelengths of light are compressed, causing a blueshift. If a star moves away, its wavelengths are stretched, causing a redshift. Using the Doppler shift, scientists found that some galaxies in the Local Group are moving toward, some are moving away, and others are moving with the Milky Way. A redshift is also seen in the light from distant galaxies, but this is explained differently.

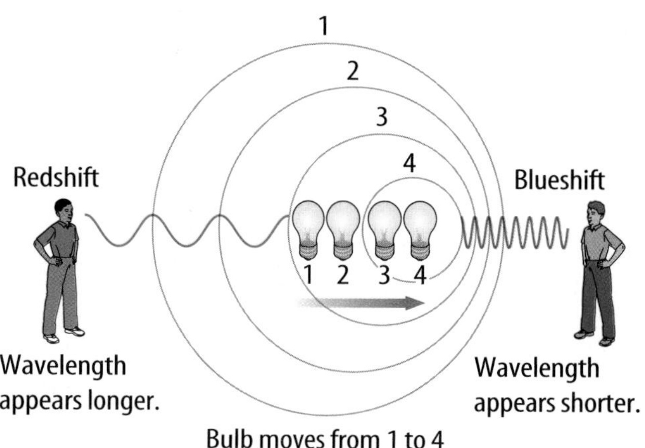

Figure 23 The observer on the right experiences a blueshift as wavelengths of light emitted by an approaching object are compressed. The observer on the left experiences a redshift as wavelengths are stretched.

Figure 24 This *Hubble* view looks deep into space and time when the universe was more chaotic and oddly shaped galaxies were common. Dark matter might have influenced galaxies to take the classic spiral and elliptical forms we see today.

Hubble Redshift The Doppler shift does not explain the shift in wavelength of light coming from distant galaxies. This shift is known as the Hubble redshift. It is caused by the stretching of space itself as the universe expands. Light waves traveling through space lengthen as space expands. The fact that this red shift is seen in the light from all galaxies outside the Local Group indicates that the entire universe is expanding.

What is the universe made of?

The way in which galaxies like the Milky Way rotate and move should depend on the amounts of mass they contain. The problem is that if only the visible or otherwise detectable mass (called regular matter) is counted, the Milky Way and other galaxies shouldn't be rotating, moving, and interacting with other galaxies the way they are. It appears that regular matter makes up only a very small amount of the known universe. Much of the mass that must be present cannot be seen. However, its effect on other galaxies can be seen, shown in **Figure 24.** This unseen and little-understood matter that affects galaxies has been named **dark matter.**

Which type of matter is considered to make up the least amount of the universe?

Dark Matter Although scientists are uncertain about what dark matter is, the concept helps them explain how the universe may have formed. Temperature variations, as shown on the WMAP map, could have led to density fluctuations in the early universe. As the universe expanded, gravity from dark matter pulled matter together in regions with higher density. Galaxies could have formed in the denser regions and voids could have formed elsewhere. This structure is seen when sections of the universe are mapped, as shown in **Figure 25.**

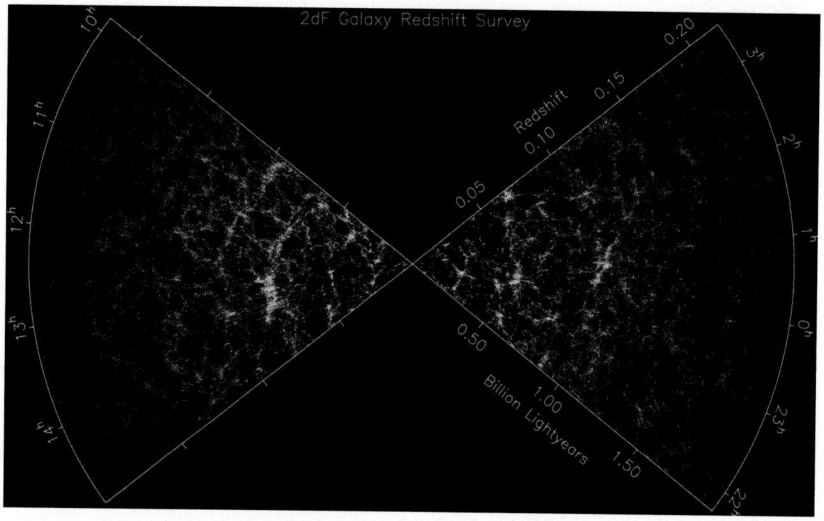

Figure 25 These maps of sections of the universe show areas densely populated with galaxies and other areas almost empty.
Identify *What are the almost-empty areas called?*

Dark Energy Data indicate that the expansion of the universe is accelerating. Explaining this acceleration is difficult. One hypothesis is that a form of energy, called **dark energy,** might be causing the acceleration. When matter was closer together in the early universe, gravity could easily overcome expansion caused by dark energy. Now, with matter farther apart, gravity is insufficient to overcome it and expansion accelerates. This does not mean, however, that dark energy is related to dark matter.

section 4 review

Summary

How did it begin?

- Cosmology is the study of how the universe began and evolves.
- The big bang theory is the most accepted theory about the beginning of the universe.
- The universe is estimated to be 13.7 billion years old.

Expansion of the Universe

- The Doppler effect indicates whether stars are coming toward Earth or moving away.
- The Hubble redshift is caused by the expansion of space and the entire universe.

What is the universe made of?

- The universe might contain dark matter and dark energy.

Self Check

1. **Define** cosmology.
2. **Describe** the big bang theory.
3. **Explain** how the expansion of space could cause a red shift in the light from distant galaxies.
4. **Describe** the Doppler effect.
5. **Think Critically** How could the presence of some repulsing force cause the universe to expand forever?

Applying Math

6. **Use Percentages** The currently accepted age of the universe is 13.7 billion years plus or minus one percent. How much time does this one percent represent?
7. **Use Percentages** What percent of the universe's age is the first 400,000 years?

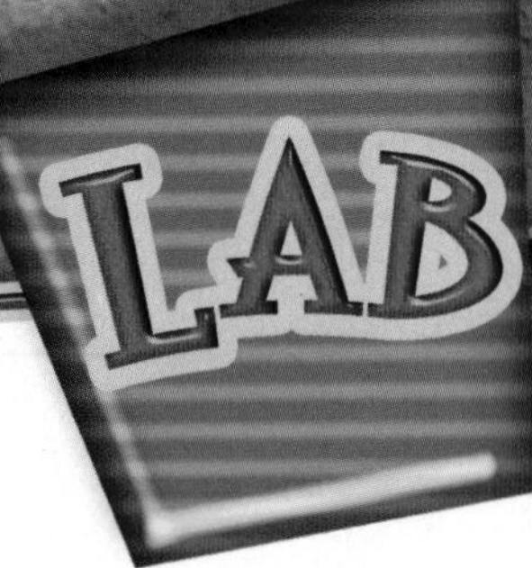

Model and Invent

Expansion of the Universe

Goals

- **Model** the stretching of the wavelengths of light caused by the expansion of the universe.
- **Measure** the amount of wavelength lengthening produced in the model.
- **Measure** increases in distances between galaxy clusters caused by the expansion of the model.

Possible Materials

round balloon
permanent marker (black or dark blue)
medium-sized binder clip (3 cm)

Safety Precautions

Real-World Problem

You have read that the universe is expanding. In fact, recent calculations indicate that it is expanding more quickly now than it did in the past. Astronomers are able to measure this expansion because it stretches the wavelengths of light coming from distant objects. Can you make a model that demonstrates how this happens?

Make a Model

1. Work in teams of two or more. **Collect** all needed materials.
2. Sketch an image on the balloon slightly inflated. **Model** the positions of galaxy clusters by placing three or four dots at four different locations and different distances apart on the surface of the balloon. Mark the locations R (for reference), A, B, and C.
3. Make a table for data.
4. Check to see if the binder clip can hold air in the balloon long enough for you to make measurements and to draw on the balloon.
5. Obtain your teacher's approval of your sketches and data table before proceeding.

Test Your Model

1. Slightly inflate your balloon so that it is not very big.
2. Use the binder clip to temporarily seal the balloon (or have your partner hold it closed). Do not tie it.
3. **Measure** the distance from your reference galaxy cluster (R) to each of the other galaxy clusters and record these distances on your data table.

Data

Location	First Distance Measurement	Second Distance Measurement	Change in Distance	First λ Measurement	Second λ Measurement	Change in λ
A						
B			Do not write in this book.			
C						

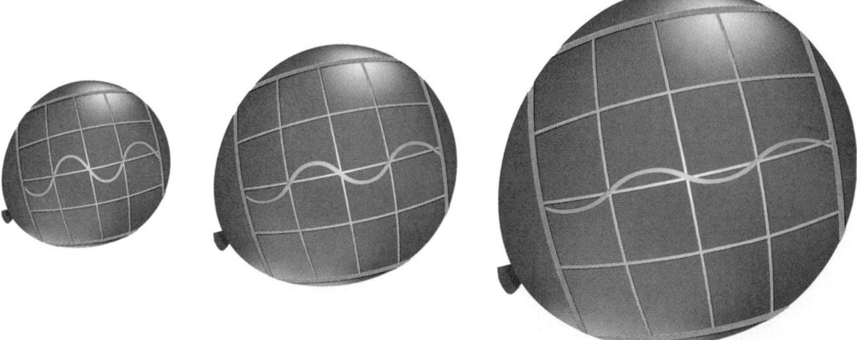

4. Draw a wavy line from your reference galaxy cluster to each of the other galaxy clusters marked to represent the wavelength (λ) of light coming from each galaxy cluster. Measure and record the wavelength of each wavy line.
5. Inflate your balloon farther (be careful not to inflate it too much). Replace the binder clip or tie the end shut.
6. **Measure** and record the distances from your reference galaxy cluster to each of the other galaxy clusters.
7. **Measure** and record the wavelength of each wavy line on your inflated model.

Analyze Your Data

1. **Calculate** the change in distance and the change in wavelength for each galaxy cluster and record it on your data table.
2. **Analyze** whether objects moved on your model or whether your entire model expanded.

Conclude and Apply

1. **Explain** any changes in distance or wavelength noted.
2. **Conclude** whether the objects moved apart because of their individual motions or because the model expanded.
3. **Infer** how measurements on your model can be related to measurements of the universe.

Enter the data from your model on a table that shows the data from all teams of students.

Science and Language Arts

Stars

Sara Teasdale

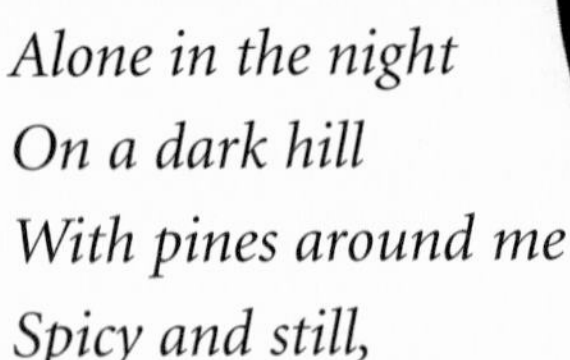

Alone in the night
On a dark hill
With pines around me
Spicy and still,

And a heaven full of stars
Over my head,
White and topaz
And misty red;

Myriads with beating
Hearts of fire
That aeons
Cannot vex or tire;

Up the dome of heaven
Like a great hill,
I watch them marching
Stately and still,

And I know that I
Am honored to be
Witness
Of so much majesty.

Understanding Literature

Imagery Imagery refers to how a literary work invites the reader to see, hear, smell, touch, or taste something in his mind. Authors often use related images throughout a work, encouraging the reader to consider the subject using more than one sense. Of what specific senses does this poet make the reader aware?

Respond to the Reading

1. How does the author use personification to make the stars in the poem seem alive?
2. Tone refers to the overall mood of a literary work. Describe the tone of this poem.
3. **Linking Science and Writing** The poet refers to the color, age, energy source, and apparent movement of stars. Write a paragraph relating what you have learned about these properties to how they are mentioned in the poem.

On a clear, dark night, it is possible to see as many as 1,500 stars with the unaided eye. However, this number is far lower because of light pollution. Light from houses, shopping areas, signs, and street lights floods the sky, making it difficult to see the stars. Inside major cities, fewer than 50 stars may be visible in the night sky.

Reviewing Main Ideas

Section 1 Observing the Universe

1. Constellations are patterns of stars that resemble things familiar to the observer.
2. Optical telescopes collect visible light and magnify viewed objects.
3. A refracting telescope uses lenses to collect light and magnify the image, and a reflecting telescope uses a mirror to collect light and a lens to magnify the image.
4. A radio telescope collects and amplifies radio waves.

Section 2 Evolution of Stars

1. Stars form from a large cloud of gas, ice, and dust, called a nebula like the one shown here. When the temperature inside the contracting nebula reaches 10 million K, fusion begins, and a star is born.

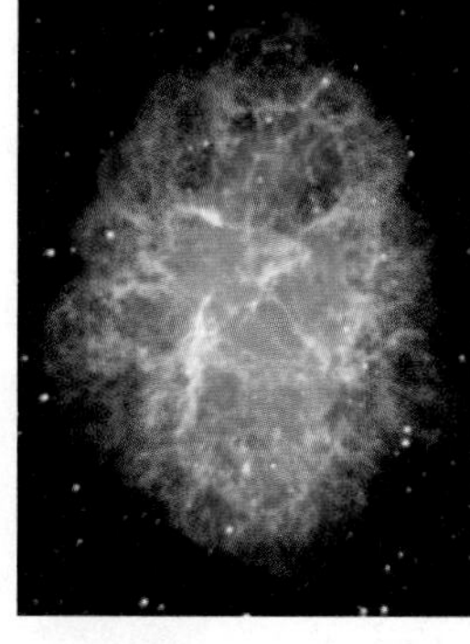

2. Stars are classified as main sequence stars, giants, and white dwarfs on the H-R diagram.
3. When a star reaches stellar equilibrium it is considered a main sequence star. When the hydrogen fuel is depleted, a star loses equilibrium and evolves into a giant or supergiant.
4. After losing its outer layers, a giant becomes a white dwarf. A supergiant can evolve into a neutron star or a black hole.
5. The Sun's energy is produced at its core by nuclear fusion.

Section 3 Galaxies and the Milky Way

1. A galaxy is a large group of stars, gas, and dust held together by gravity. The Local Group of galaxies is a cluster that contains the Milky Way Galaxy.

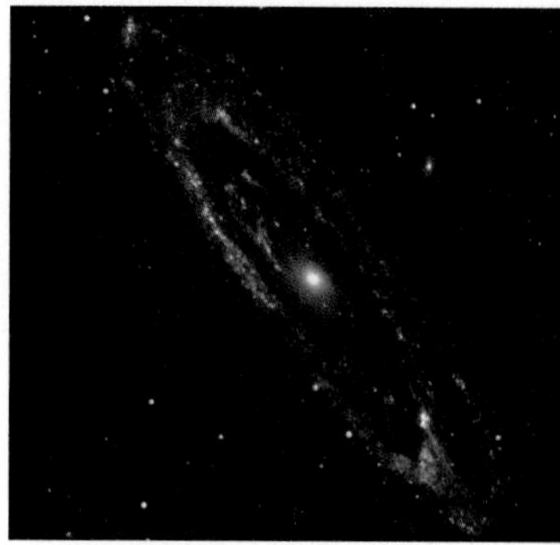

2. The three main types of galaxies are elliptical, irregular, and spiral as shown here.
3. Astronomers believe that galaxies form by absorbing or merging with smaller objects. They continue to evolve by colliding or merging with other galaxies.
4. The Milky Way galaxy is about 100,000 light-years across and the Sun lies about 26,000 light-years from its center.

Section 4 Cosmology

1. The big bang theory is the most accepted theory of how the universe began.
2. The universe is 13.7 billion years old and appears to be expanding faster now than in the past.
3. The Hubble redshift is caused by the expansion of space, not the movement of galaxies.

FOLDABLES™ Use the Foldable that you made at the beginning of this chapter to help you review what you learned about cosmology.

Using Vocabulary

big bang theory p. 837	Milky Way p. 831
constellation p. 818	photosphere p. 827
cosmology p. 836	radio telescope p. 821
dark energy p. 839	reflecting telescope p. 820
dark matter p. 838	refracting telescope p. 819
galaxy p. 831	solar mass p. 827
giant p. 825	spectroscope p. 822
light-year p. 821	sunspots p. 828
Local Group p. 832	white dwarf p. 825
main sequence p. 823	

Match the correct vocabulary word(s) with each definition given below.

1. patterns of stars
2. an optical telescope that uses a mirror to collect light
3. the distance light travels in one year
4. plotted from the upper left to the lower right on the H-R diagram
5. star in which the core contracts and outer layers expand and cool
6. layer of the Sun from which light is emitted
7. dark, cooler areas on the Sun's photosphere
8. large group of stars, gas, and dust held together by gravity
9. spiral galaxy that contains the solar system
10. study of the evolution of the universe

Checking Concepts

Choose the word or phrase that best answers the question.

11. Which telescope uses lenses to collect light and form an image?
    A) adaptive **C)** reflecting
    B) radio **D)** refracting

12. Which form of energy are optical telescopes used to study?
    A) infrared radiation
    B) radio waves
    C) visible light
    D) X rays

13. Which magnifies the image in a telescope?
    A) eyepiece **C)** focus
    B) focal length **D)** objective

14. Which is the most common type of star in the universe?
    A) giant **C)** red dwarf
    B) neutron **D)** white dwarf

15. Which forms from a star that is over 25 times the mass of the Sun?
    A) black hole **C)** neutron star
    B) giant star **D)** white dwarf

16. Which is a feature of the Sun that can reach 100 million K?
    A) CME **C)** prominence
    B) flare **D)** sunspot

Use the illustration below to answer question 17.

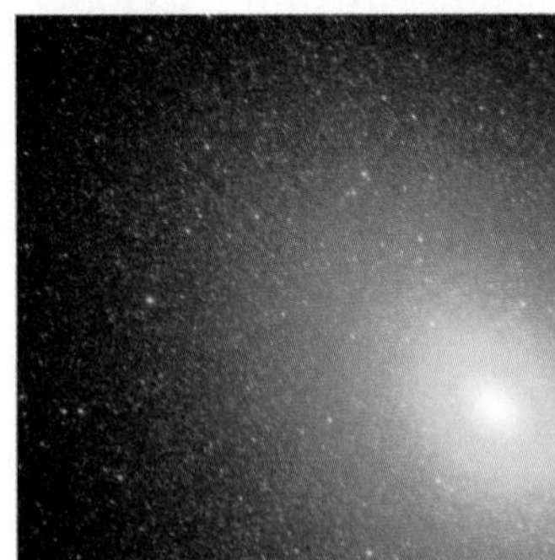

17. Which type of galaxy is most common?
    A) barred spiral
    B) dwarf elliptical
    C) irregular
    D) spiral

18. Which is the most accepted theory of how the universe formed?
    A) big bang **C)** oscillating
    B) collision **D)** steady state

Vocabulary PuzzleMaker gpescience.com

19. Which adds to the gravity of a galaxy, but cannot be seen or detected?
 A) dark energy
 B) dark matter
 C) regular energy
 D) regular matter

Interpreting Graphics

20. **Make a table** summarizing the absolute brightness (magnitude) and temperatures of stars on this H-R diagram.

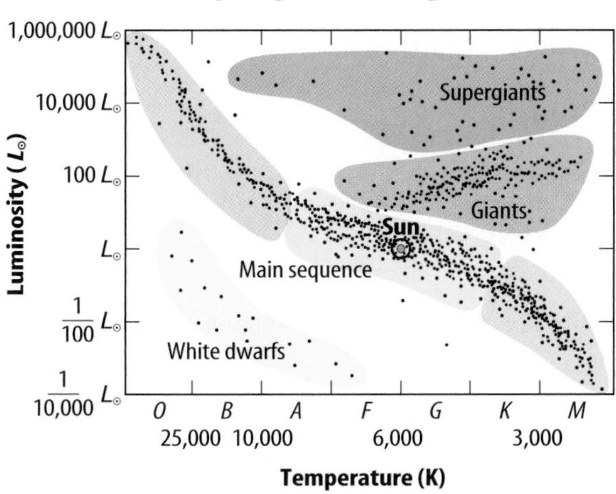

21. **Draw and label** scale models of the following stars: the Sun; Antares, 500 times larger than the Sun; and Sirius B 0.01 times the Sun's diameter.

22. **Draw and label** a concept map showing the life history of a star like the Sun.

23. **Draw and label** the parts in a reflecting telescope.

Thinking Critically

24. **Explain** how energy created in the core of the Sun is eventually emitted from the photosphere.

25. **Compare and contrast** elliptical, irregular, and spiral galaxies.

26. **Discuss** the benefits of using a radio telescope.

27. **Explain** why high sunspot activity on the Sun can affect Earth's magnetic field.

28. **Explain** how the Sun's position in the Milky Way affects how we perceive our galaxy.

Applying Math

29. **Solve One-Step Equations** Use the equation $M_p = f_o/f_e$, to determine the magnifying power of a telescope in which the focal lengths of the objective and eyepiece are 1500 mm and 9 mm, respectively.

Use the illustration below to answer question 30.

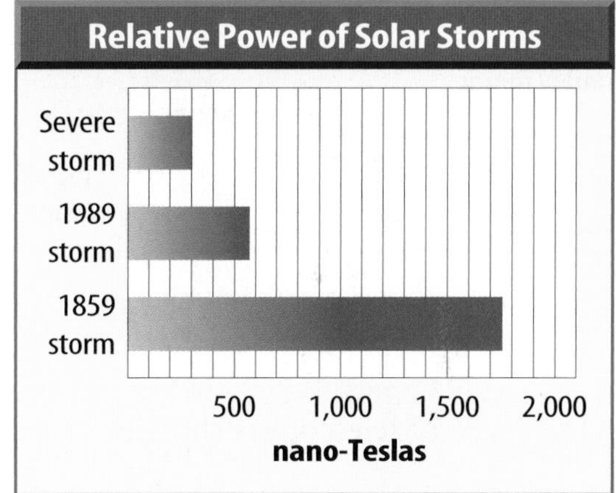

30. A solar storm that took place in 1859 is called "the perfect storm," because of its great power. The diagram shows the relative destructive power of some solar storms. Using this diagram, calculate approximately how much more destructive was the solar storm of 1859 than the one of 1989.

31. **Calculate** how long ago the light we see today left a star that is 25 light-years away. Calculate the distance in km to that star.

Standardized Test Practice

Record your answers on the answer sheet provided by your teacher or on a sheet of paper.

Multiple Choice

1. What is used as an objective in a reflecting telescope?
   - **A.** antenna
   - **B.** camera
   - **C.** lens
   - **D.** mirror

Use the illustration below to answer question 2.

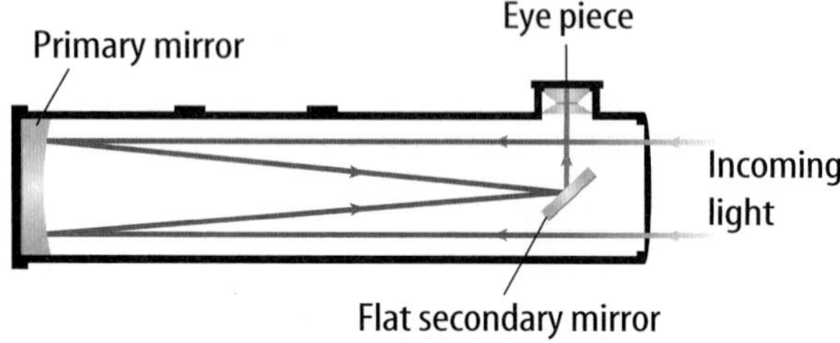

2. Which type of telescope is shown?
   - **A.** optical
   - **B.** radio
   - **C.** ultraviolet
   - **D.** X-ray

3. Which group contains most stars on the H-R diagram ?
   - **A.** giant
   - **B.** main sequence
   - **C.** supergiant
   - **D.** white dwarf

4. Which stage of stellar evolution occurs when the outer layers escape, leaving behind the hot core?
   - **A.** black hole
   - **B.** giant
   - **C.** main sequence
   - **D.** white dwarf

Use the illustration below to answer question 5.

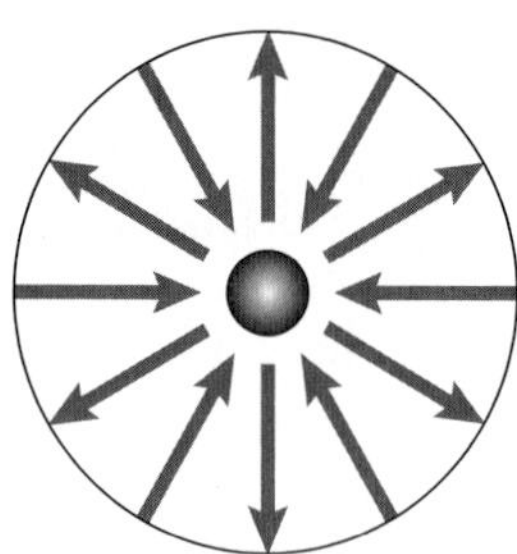

5. What occurs inside a main sequence star?
   - **A.** Energy from fusion exceeds gravity.
   - **B.** Fusion shuts down.
   - **C.** Gravity exceeds energy from fusion.
   - **D.** It attains stellar equilibrium.

Test-Taking Tip

Concentrate Stay focused during the test and don't rush, even if you notice that other students are finishing the test early.

6. Which may be responsible for the accelerating expansion of the universe?

   A. dark energy

   B. dark matter

   C. regular energy

   D. regular matter

Gridded Response

7. If the focal length of a telescope objective is 2,400 mm and the focal length of the eyepiece is 20 mm, what is the magnifying power of the telescope?

8. Use this equation, $A = \pi r^2$. What is the area in square meters of one of the four 8.2-meter reflectors in the *Very Large Telescope*?

Short Response

Use the illustration below to answer question 9.

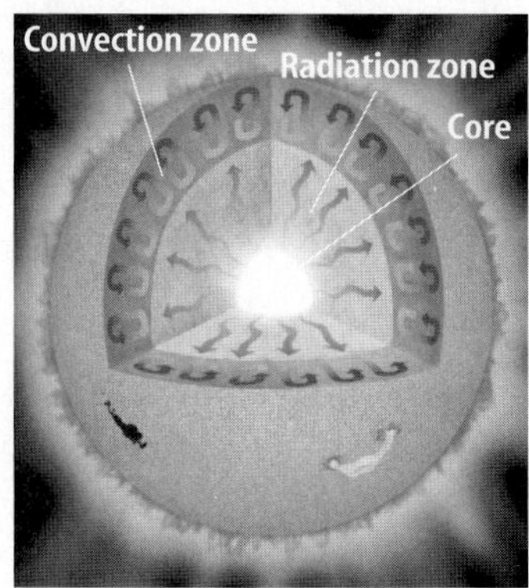

9. Describe the structure of the Sun's interior.

10. What does a spectroscope do to starlight that enables astronomers to determine the star's composition?

11. What does the H-R diagram show about the stars plotted on it?

12. What are coronal mass ejections?

13. What is the Local Group?

14. How do astronomers think that galaxies like the Milky Way formed?

15. What is cosmology?

16. Why have astronomers proposed the existence of dark energy in the universe?

Extended Response

Use the photo below to help answer question 17.

17. **PART A** What are the evolutionary stages of a star like the Sun?

    PART B What are the evolutionary stages of a star more than eight times the mass of the Sun?

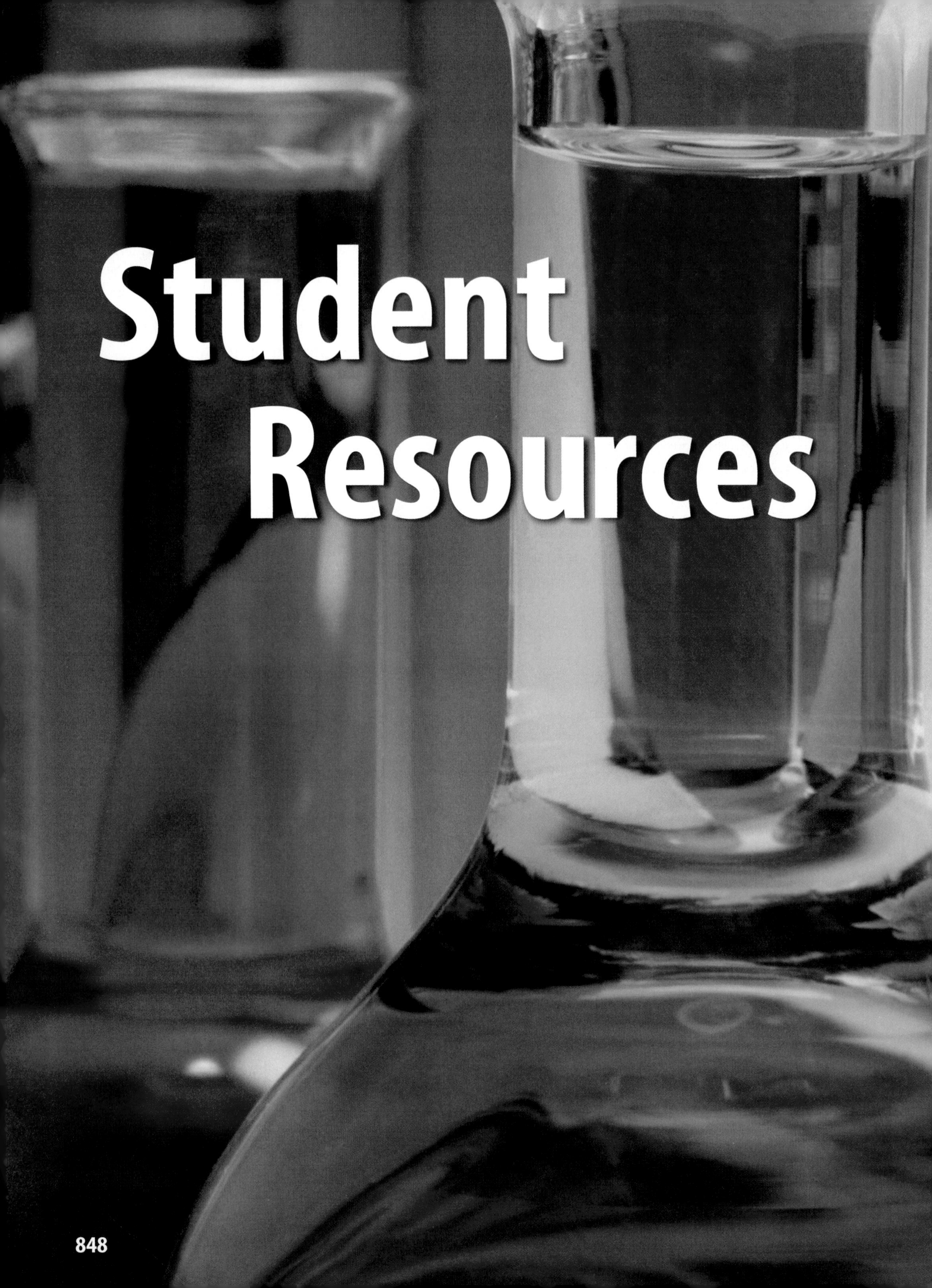

Student Resources

CONTENTS

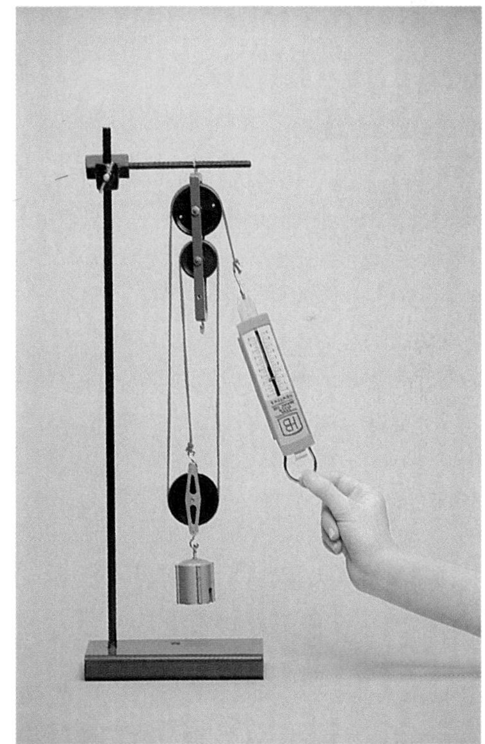

Scientific Methods

Scientists use an orderly approach called the scientific method to solve problems. This includes organizing and recording data so others can understand them. Scientists use many variations in this method when they solve problems.

Identify a Question

The first step in a scientific investigation or experiment is to identify a question to be answered or a problem to be solved. For example, you might ask which gasoline is the most efficient.

Gather and Organize Information

After you have identified your question, begin gathering and organizing information. There are many ways to gather information, such as researching in a library, interviewing those knowledgeable about the subject, testing, and working in the laboratory and field. Fieldwork is investigations and observations done outside of a laboratory.

Researching Information Before moving in a new direction, it is important to gather the information that already is known about the subject. Start by asking yourself questions to determine exactly what you need to know. Then you will look for the information in various reference sources, like the student is doing in **Figure 1.** Some sources may include textbooks, encyclopedias, government documents, professional journals, science magazines, and the Internet. Always list the sources of your information.

Figure 1 The Internet can be a valuable research tool.

Evaluate Sources of Information Not all sources of information are reliable. You should evaluate all your sources of information, and use only those you know to be dependable. For example, if you are researching ways to make homes more energy efficient, a site written by the U.S. Department of Energy would be more reliable than a site written by a company that is trying to sell a new type of weatherproofing material. Also, remember that research always is changing. Consult the most current resources available to you. For example, a 1985 resource about saving energy would not reflect the most recent findings.

Sometimes scientists use data that they did not collect themselves, or conclusions drawn by other researchers. These data must be evaluated carefully. Ask questions about how the data were obtained, if the investigation was carried out properly, and if it has been duplicated exactly with the same results. Would you reach the same conclusion from the data? Only when you have confidence in the data can you believe it is true and feel comfortable using it.

Interpret Scientific Illustrations As you research a topic in science, you will see drawings, diagrams, and photographs to help you understand what you read. Some illustrations are included to help you understand an idea that you can't see easily by yourself, like the tiny particles in an atom in **Figure 2.** A drawing helps many people to remember details more easily and provides examples that clarify difficult concepts or give additional information about the topic you are studying. Most illustrations have labels or a caption to identify or to provide more information.

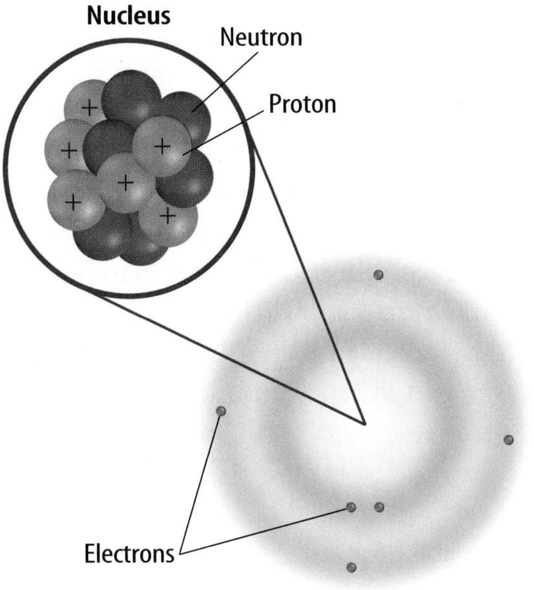

Figure 2 This drawing shows an atom of carbon with its six protons, six neutrons, and six electrons.

Concept Maps One way to organize data is to draw a diagram that shows relationships among ideas (or concepts). A concept map can help make the meanings of ideas and terms more clear, and help you understand and remember what you are studying. Concept maps are useful for breaking large concepts down into smaller parts, making learning easier.

Network Tree A type of concept map that not only shows a relationship, but how the concepts are related is a network tree, shown in **Figure 3.** In a network tree, the words are written in the ovals, while the description of the type of relationship is written across the connecting lines.

When constructing a network tree, write down the topic and all major topics on separate pieces of paper or notecards. Then arrange them in order from general to specific. Branch the related concepts from the major concept and describe the relationship on the connecting line. Continue to more specific concepts until finished.

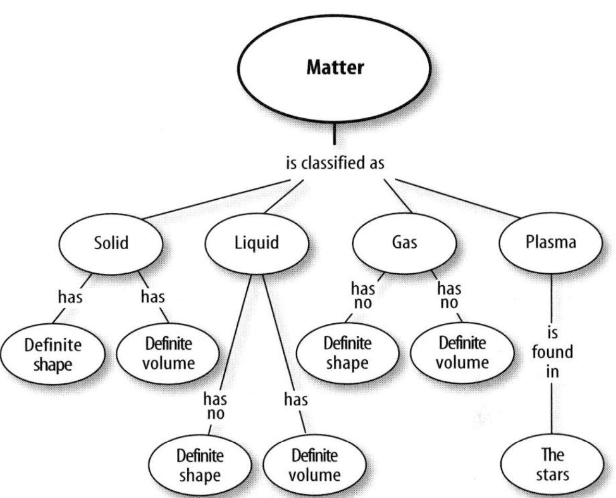

Figure 3 A network tree shows how concepts or objects are related.

Events Chain Another type of concept map is an events chain. Sometimes called a flow chart, it models the order or sequence of items. An events chain can be used to describe a sequence of events, the steps in a procedure, or the stages of a process.

When making an events chain, first find the one event that starts the chain. This event is called the initiating event. Then, find the next event and continue until the outcome is reached, as shown in **Figure 4.**

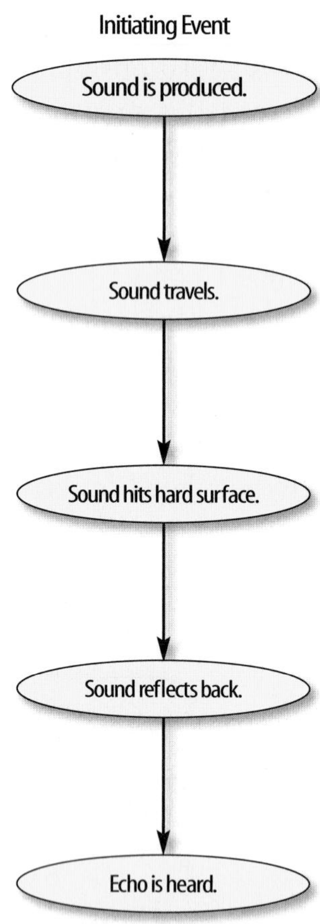

Figure 4 Events-chain concept maps show the order of steps in a process or event. This concept map shows how a sound makes an echo.

Cycle Map A specific type of events chain is a cycle map. It is used when the series of events do not produce a final outcome, but instead relate back to the beginning event, such as in **Figure 5.** Therefore, the cycle repeats itself.

To make a cycle map, first decide what event is the beginning event. This is also called the initiating event. Then list the next events in the order that they occur, with the last event relating back to the initiating event. Words can be written between the events that describe what happens from one event to the next. The number of events in a cycle map can vary, but usually contain three or more events.

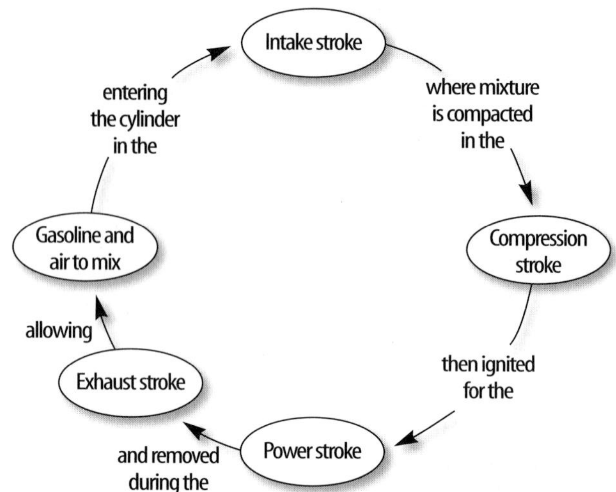

Figure 5 A cycle map shows events that occur in a cycle.

Spider Map A type of concept map that you can use for brainstorming is the spider map. When you have a central idea, you might find that you have a jumble of ideas that relate to it but are not necessarily clearly related to each other. The spider map on sound in **Figure 6** shows that if you write these ideas outside the main concept, then you can begin to separate and group unrelated terms so they become more useful.

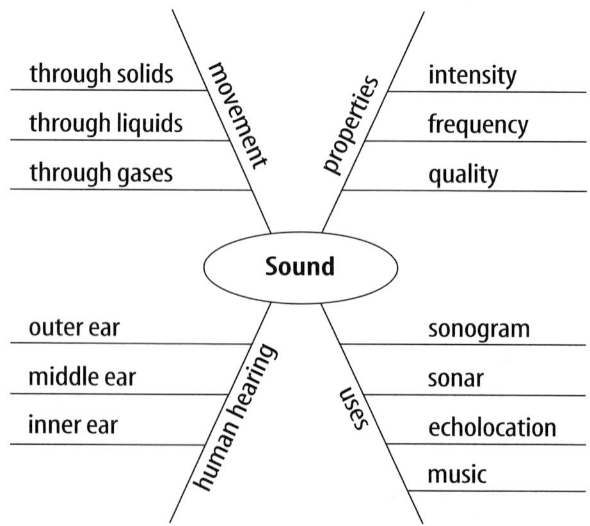

Figure 6 A spider map allows you to list ideas that relate to a central topic but not necessarily to one another.

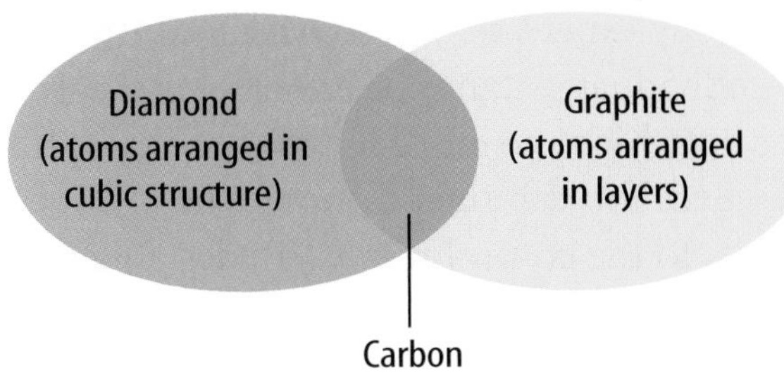

Figure 7 This Venn diagram compares and contrasts two substances made from carbon.

Venn Diagram To illustrate how two subjects compare and contrast you can use a Venn diagram. You can see the characteristics that the subjects have in common and those that they do not, shown in **Figure 7.**

To create a Venn diagram, draw two overlapping ovals that that are big enough to write in. List the characteristics unique to one subject in one oval, and the characteristics of the other subject in the other oval. The characteristics in common are listed in the overlapping section.

Make and Use Tables One way to organize information so it is easier to understand is to use a table. Tables can contain numbers, words, or both.

To make a table, list the items to be compared in the first column and the characteristics to be compared in the first row. The title should clearly indicate the content of the table, and the column or row heads should be clear. Notice that in **Table 1** the units are included.

Table 1 Recyclables Collected During Week

Day of Week	Paper (kg)	Aluminum (kg)	Glass (kg)
Monday	5.0	4.0	12.0
Wednesday	4.0	1.0	10.0
Friday	2.5	2.0	10.0

Make a Model One way to help you better understand the parts of a structure, the way a process works, or to show things too large or small for viewing is to make a model. For example, an atomic model made of a plastic-ball nucleus and pipe-cleaner electron shells can help you visualize how the parts of an atom relate to each other. Other types of models can by devised on a computer or represented by equations.

Form a Hypothesis

A possible explanation based on previous knowledge and observations is called a hypothesis. After researching gasoline types and recalling previous experiences in your family's car you form a hypothesis—our car runs more efficiently because we use premium gasoline. To be valid, a hypothesis has to be something you can test by using an investigation.

Predict When you apply a hypothesis to a specific situation, you predict something about that situation. A prediction makes a statement in advance, based on prior observation, experience, or scientific reasoning. People use predictions to make everyday decisions. Scientists test predictions by performing investigations. Based on previous observations and experiences, you might form a prediction that cars are more efficient with premium gasoline. The prediction can be tested in an investigation.

Design an Experiment A scientist needs to make many decisions before beginning an investigation. Some of these include: how to carry out the investigation, what steps to follow, how to record the data, and how the investigation will answer the question. It also is important to address any safety concerns.

Test the Hypothesis

Now that you have formed your hypothesis, you need to test it. Using an investigation, you will make observations and collect data, or information. This data might either support or not support your hypothesis. Scientists collect and organize data as numbers and descriptions.

Follow a Procedure In order to know what materials to use, as well as how and in what order to use them, you must follow a procedure. **Figure 8** shows a procedure you might follow to test your hypothesis.

Procedure

1. Use regular gasoline for two weeks.
2. Record the number of kilometers between fill-ups and the amount of gasoline used.
3. Switch to premium gasoline for two weeks.
4. Record the number of kilometers between fill-ups and the amount of gasoline used.

Figure 8 A procedure tells you what to do step by step.

Identify and Manipulate Variables and Controls In any experiment, it is important to keep everything the same except for the item you are testing. The one factor you change is called the independent variable. The change that results is the dependent variable. Make sure you have only one independent variable, to assure yourself of the cause of the changes you observe in the dependent variable. For example, in your gasoline experiment the type of fuel is the independent variable. The dependent variable is the efficiency.

Many experiments also have a control—an individual instance or experimental subject for which the independent variable is not changed. You can then compare the test results to the control results. To design a control you can have two cars of the same type. The control car uses regular gasoline for four weeks. After you are done with the test, you can compare the experimental results to the control results.

Collect Data

Whether you are carrying out an investigation or a short observational experiment, you will collect data, as shown in **Figure 9.** Scientists collect data as numbers and descriptions and organize it in specific ways.

Observe Scientists observe items and events, then record what they see. When they use only words to describe an observation, it is called qualitative data. Scientists' observations also can describe how much there is of something. These observations use numbers, as well as words, in the description and are called quantitative data. For example, if a sample of the element gold is described as being "shiny and very dense" the data are qualitative. Quantitative data on this sample of gold might include "a mass of 30 g and a density of 19.3 g/cm^3."

Figure 9 Collecting data is one way to gather information directly.

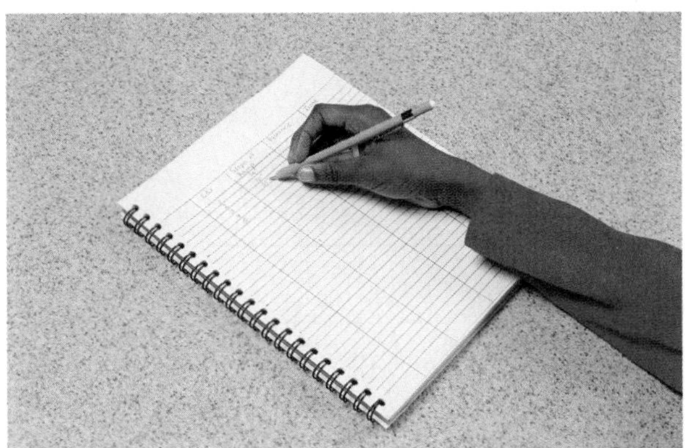

Figure 10 Record data neatly and clearly so it is easy to understand.

When you make observations you should examine the entire object or situation first, and then look carefully for details. It is important to record observations accurately and completely. Always record your observations immediately as you make them, so you do not miss details or make a mistake when recording results from memory. Never put unidentified observations on scraps of paper. Instead they should be recorded in a notebook, like the one in **Figure 10.** Write your data neatly so you can easily read it later. At each point in the experiment, record your observations and label them. That way, you will not have to determine what the figures mean when you look at your notes later. Set up any tables that you will need to use ahead of time, so you can record any observations right away. Remember to avoid bias when collecting data by not including personal thoughts when you record observations. Record only what you observe.

Estimate Scientific work also involves estimating. To estimate is to make a judgment about the size or the number of something without measuring or counting. This is important when the number or size of an object or population is too large or too difficult to accurately count or measure.

Sample Scientists may use a sample or a portion of the total number as a type of estimation. To sample is to take a small, representative portion of the objects or organisms of a population for research. By making careful observations or manipulating variables within that portion of the group, information is discovered and conclusions are drawn that might apply to the whole population. A poorly chosen sample can be unrepresentative of the whole. If you were trying to determine the rainfall in an area, it would not be best to take a rainfall sample from under a tree.

Measure You use measurements every day. Scientists also take measurements when collecting data. When taking measurements, it is important to know how to use measuring tools properly. Accuracy also is important.

Length To measure length, the distance between two points, scientists use meters. Smaller measurements might be measured in centimeters or millimeters.

Length is measured using a metric ruler or meterstick. When using a metric ruler, line up the 0-cm mark with the end of the object being measured and read the number of the unit where the object ends. Look at the metric ruler shown in **Figure 11.** The centimeter lines are the long, numbered lines, and the shorter lines are millimeter lines. In this instance, the length would be 4.50 cm.

Figure 11 This metric ruler has centimeter and millimeter divisions.

Mass The SI unit for mass is the kilogram (kg). Scientists can measure mass using units formed by adding metric prefixes to the unit gram (g), such as milligram (mg). To measure mass, you might use a triple-beam balance similar to the one shown in **Figure 12.** The balance has a pan on one side and a set of beams on the other side. Each beam has a rider that slides on the beam.

When using a triple-beam balance, place an object on the pan. Slide the largest rider along its beam until the pointer drops below zero. Then move it back one notch. Repeat the process for each rider proceeding from the larger to smaller until the pointer swings an equal distance above and below the zero point. Sum the masses on each beam to find the mass of the object. Move all riders back to zero when finished.

Instead of putting materials directly on the balance, scientists often take a tare of a container. A tare is the mass of a container into which objects or substances are placed for measuring their masses. To mass objects or substances, find the mass of a clean container. Remove the container from the pan, and place the object or substances in the container. Find the mass of the container with the materials in it. Subtract the mass of the empty container from the mass of the filled container to find the mass of the materials you are using.

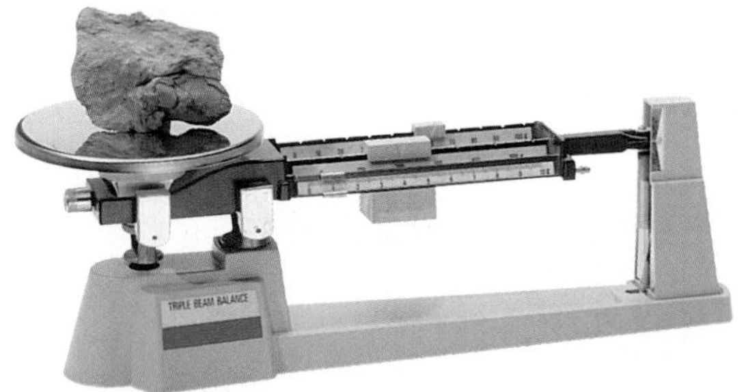

Figure 12 A triple-beam balance is used to determine the mass of an object.

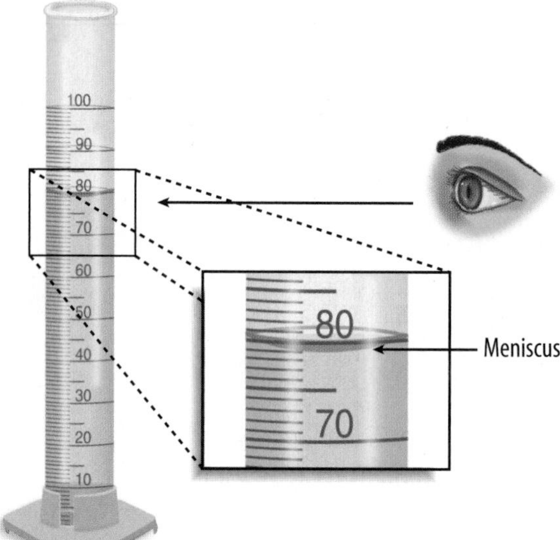

Figure 13 Graduated cylinders measure liquid volume.

Liquid Volume To measure liquids, the unit used is the liter (L). When a smaller unit is needed, scientists might use a milliliter (mL). Because a milliliter takes up the volume of a cube measuring 1 cm on each side it also can be called a cubic centimeter ($cm^3 = cm \times cm \times cm$).

You can use beakers and graduated cylinders to measure liquid volume. A graduated cylinder, shown in **Figure 13,** is marked from bottom to top in milliliters. In the lab, you might use a 10-mL graduated cylinder or a 100-mL graduated cylinder. When measuring liquids, notice that the liquid has a curved surface. Look at the surface at eye level, and measure the bottom of the curve. This is called the meniscus. The graduated cylinder in **Figure 13** contains 79.0 mL, or 79.0 cm^3, of a liquid.

Temperature Scientists often measure temperature using the Celsius scale. Pure water has a freezing point of 0°C and boiling point of 100°C. The unit of measurement is degrees Celsius. Two other scales often used are the Fahrenheit and Kelvin scales.

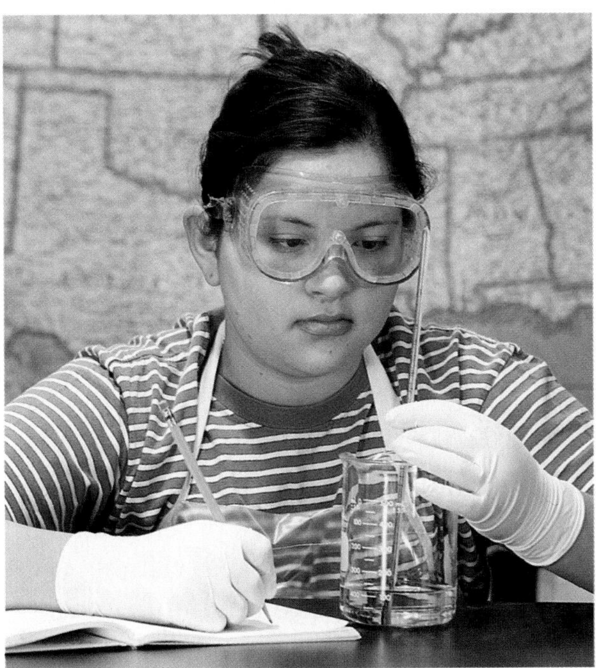

Figure 14 A thermometer measures the temperature of an object.

Scientists use a thermometer to measure temperature. Most thermometers in a laboratory are glass tubes with a bulb at the bottom end containing a liquid such as colored alcohol. The liquid rises or falls with a change in temperature. To read a glass thermometer like the thermometer in **Figure 14,** rotate it slowly until a red line appears. Read the temperature where the red line ends.

Form Operational Definitions An operational definition defines an object by how it functions, works, or behaves. For example, when you are playing hide and seek and a tree is home base, you have created an operational definition for a tree.

Objects can have more than one operational definition. For example, a ruler can be defined as a tool that measures the length of an object (how it is used). It can also be a tool with a series of marks used as a standard when measuring (how it works).

Analyze the Data

To determine the meaning of your observations and investigation results, you will need to look for patterns in the data. Then you must think critically to determine what the data mean. Scientists use several approaches when they analyze the data they have collected and recorded. Each approach is useful for identifying specific patterns.

Interpret Data The word *interpret* means "to explain the meaning of something." When analyzing data from an experiement, try to find out what the data show. Identify the control group and the test group to see whether or not changes in the independent variable have had an effect. Look for differences in the dependent variable between the control and test groups.

Classify Sorting objects or events into groups based on common features is called classifying. When classifying, first observe the objects or events to be classified. Then select one feature that is shared by some members in the group, but not by all. Place those members that share that feature in a subgroup. You can classify members into smaller and smaller subgroups based on characteristics. Remember that when you classify, you are grouping objects or events for a purpose. Keep your purpose in mind as you select the features to form groups and subgroups.

Compare and Contrast Observations can be analyzed by noting the similarities and differences between two more objects or events that you observe. When you look at objects or events to see how they are similar, you are comparing them. Contrasting is looking for differences in objects or events.

Recognize Cause and Effect A cause is a reason for an action or condition. The effect is that action or condition. When two events happen together, it is not necessarily true that one event caused the other. Scientists must design a controlled investigation to recognize the exact cause and effect.

Draw Conclusions

When scientists have analyzed the data they collected, they proceed to draw conclusions about the data. These conclusions are sometimes stated in words similar to the hypothesis that you formed earlier. They may confirm a hypothesis, or lead you to a new hypothesis.

Infer Scientists often make inferences based on their observations. An inference is an attempt to explain observations or to indicate a cause. An inference is not a fact, but a logical conclusion that needs further investigation. For example, you may infer that a fire has caused smoke. Until you investigate, however, you do not know for sure.

Apply When you draw a conclusion, you must apply those conclusions to determine whether the data support the hypothesis. If your data do not support your hypothesis, it does not mean that the hypothesis is wrong. It means only that the result of the investigation did not support the hypothesis. Maybe the experiment needs to be redesigned, or some of the initial observations on which the hypothesis was based were incomplete or biased. Perhaps more observation or research is needed to refine your hypothesis. A successful investigation does not always come out the way you originally predicted.

Avoid Bias Sometimes a scientific investigation involves making judgments. When you make a judgment, you form an opinion. It is important to be honest and not to allow any expectations of results to bias your judgments. This is important throughout the entire investigation, from researching to collecting data to drawing conclusions.

Communicate

The communication of ideas is an important part of the work of scientists. A discovery that is not reported will not advance the scientific community's understanding or knowledge. Communication among scientists also is important as a way of improving their investigations.

Scientists communicate in many ways, from writing articles in journals and magazines that explain their investigations and experiments, to announcing important discoveries on television and radio. Scientists also share ideas with colleagues on the Internet or present them as lectures, like the student is doing in **Figure 15.**

Figure 15 A student communicates to his peers about his investigation.

SAFETY SYMBOLS	HAZARD	EXAMPLES	PRECAUTION	REMEDY
DISPOSAL	Special disposal procedures need to be followed.	certain chemicals, living organisms	Do not dispose of these materials in the sink or trash can.	Dispose of wastes as directed by your teacher.
BIOLOGICAL	Organisms or other biological materials that might be harmful to humans	bacteria, fungi, blood, unpreserved tissues, plant materials	Avoid skin contact with these materials. Wear mask or gloves.	Notify your teacher if you suspect contact with material. Wash hands thoroughly.
EXTREME TEMPERATURE	Objects that can burn skin by being too cold or too hot	boiling liquids, hot plates, dry ice, liquid nitrogen	Use proper protection when handling.	Go to your teacher for first aid.
SHARP OBJECT	Use of tools or glassware that can easily puncture or slice skin	razor blades, pins, scalpels, pointed tools, dissecting probes, broken glass	Practice common-sense behavior and follow guidelines for use of the tool.	Go to your teacher for first aid.
FUME	Possible danger to respiratory tract from fumes	ammonia, acetone, nail polish remover, heated sulfur, moth balls	Make sure there is good ventilation. Never smell fumes directly. Wear a mask.	Leave foul area and notify your teacher immediately.
ELECTRICAL	Possible danger from electrical shock or burn	improper grounding, liquid spills, short circuits, exposed wires	Double-check setup with teacher. Check condition of wires and apparatus.	Do not attempt to fix electrical problems. Notify your teacher immediately.
IRRITANT	Substances that can irritate the skin or mucous membranes of the respiratory tract	pollen, moth balls, steel wool, fiberglass, potassium permanganate	Wear dust mask and gloves. Practice extra care when handling these materials.	Go to your teacher for first aid.
CHEMICAL	Chemicals can react with and destroy tissue and other materials	bleaches such as hydrogen peroxide; acids such as sulfuric acid, hydrochloric acid; bases such as ammonia, sodium hydroxide	Wear goggles, gloves, and an apron.	Immediately flush the affected area with water and notify your teacher.
TOXIC	Substance may be poisonous if touched, inhaled, or swallowed.	mercury, many metal compounds, iodine, poinsettia plant parts	Follow your teacher's instructions.	Always wash hands thoroughly after use. Go to your teacher for first aid.
FLAMMABLE	Flammable chemicals may be ignited by open flame, spark, or exposed heat.	alcohol, kerosene, potassium permanganate	Avoid open flames and heat when using flammable chemicals.	Notify your teacher immediately. Use fire safety equipment if applicable.
OPEN FLAME	Open flame in use, may cause fire.	hair, clothing, paper, synthetic materials	Tie back hair and loose clothing. Follow teacher's instruction on lighting and extinguishing flames.	Notify your teacher immediately. Use fire safety equipment if applicable.

Eye Safety
Proper eye protection should be worn at all times by anyone performing or observing science activities.

Clothing Protection
This symbol appears when substances could stain or burn clothing.

Animal Safety
This symbol appears when safety of animals and students must be ensured.

Handwashing
After the lab, wash hands with soap and water before removing goggles.

Introduction to Science Safety

Confucius, an ancient and well-known Chinese philosopher, is credited with a statement that could serve as a legacy of all types of human wisdom. It seems especially appropriate for the active learning you will experience in this science program.

"I hear and I forget, I see and I remember, I do and I understand."

This is the basis for the safety routine that will be used in all the labs in this book. It is assumed that you will use all of your senses as you "experience" the labs. However, with such experience comes the potential for injury. The purpose of this section of the book is to help keep you safe by involving you in the safety process.

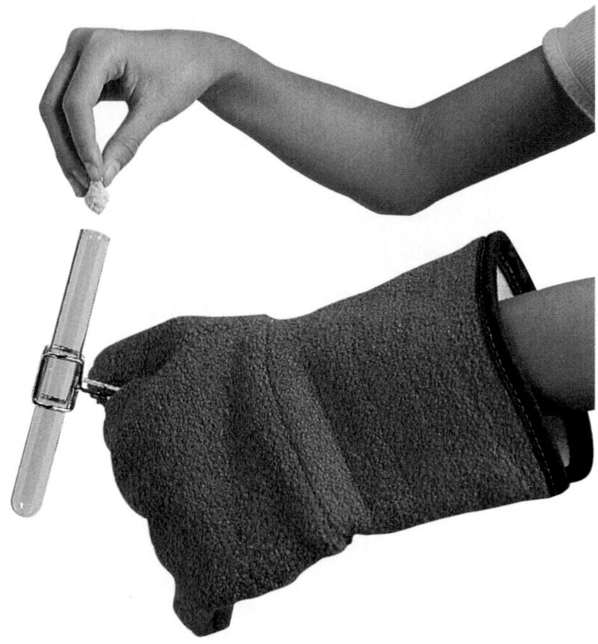

How will your teacher help?

It will be your teacher's responsibility to decide which science labs are safe and appropriate for you. Your teacher will identify the hazards involved in each activity and likely will ask for your assistance for ways to reduce the dangers as much as possible. He or she will involve you regularly in safety discussions about your understanding of the actual and potential dangers and the safety measures needed to keep everyone safe. Ideally, this will become a habit with each lab in which you take part.

Your teacher also will explain the safety features of your room as well as the most important safety equipment and routines for addressing safety issues. He or she will also require that you complete a *Student Lab-Safety Form* for each lab to make certain you are prepared to perform the lab safely. Your teacher will review your comments, make corrections, and sign or initial this form **BEFORE** you will be permitted to begin the lab.

The ultimate purpose of the safety discussions and the *Student Lab-Safety Form* will be to ***help you take some responsibility for your own safety*** and to help you to develop good habits when you prepare and perform science experiments and labs.

How can you help?

Since your teacher cannot anticipate every safety hazard that might occur and he or she cannot be everywhere in the room at the same time, you need to take some responsibility for your own safety. The following general information should apply to nearly every science lab in which you will be involved as part of this class.

Adapted from Gerlovich, et al. (2004). The Total Science Safety System D. JaKel, Inc. http://www.netins.net/showcase/jakel. Used with Permission

You must:

- Review any *Safety Symbols* in the labs and be certain that you know what they mean.
- Follow all teacher instructions for safety and make certain that you understand all the hazards related to the labs you are about to do.
- Be able to explain the purpose of the lab.
- Be able to explain, or demonstrate, all reasonable emergency procedures, such as:
  - how to evacuate the room during emergencies;
  - how to react to any chemical emergencies;
  - how to deal with fire emergencies;
  - how to perform a scientific investigation safely;
  - how to anticipate some safety concerns and be prepared to address them;
  - how to use equipment properly and safely.
- Be able to locate and use all safety equipment as directed by your teacher, such as:
  - fire extinguishers;
  - fire blankets;
  - eye protective equipment (goggles, safety glasses, face shield);
  - eyewash;
  - drench shower.
- Complete the Student Lab-Safety Form before starting any science lab.
- Ask questions about any safety concerns that you might have BEFORE starting any lab of science investigation.

Remember! Your teacher will review your comments, make corrections, and sign or initial the *Student Lab-Safety Form* **BEFORE** you will be permitted to begin the lab. A copy of this form appears below.

Student Lab-Safety Form

Student Name:________________
Date:________________
Lab Title:________________

____
Teacher Approval Initials
____
Date of Approval

In order to show your teacher that you understand the safety concerns of this lab, the following questions must be answered after the teacher explains the information to you. You must have your teacher initial this form before you can proceed with the lab.

1. How would you describe what you will be doing during this lab?

2. What are the safety concerns in this lab (explained by your teacher)?
   - ________________
   - ________________
   - ________________
   - ________________
   - ________________

3. What additional safety concerns or questions do you have?

Adapted from Gerlovich, et al. (2004). The Total Science Safety System D, JaKel, Inc. http://www.netins.net/showcase/jakel. Used with Permission

Adapted from Gerlovich, et al. (2004). The Total Science Safety System D. JaKel, Inc. http://www.netins.net/showcase/jakel. Used with Permission

Math Review

Use Fractions

A fraction compares a part to a whole. In the fraction $\frac{2}{3}$, the 2 represents the part and is the numerator. The 3 represents the whole and is the denominator.

Reduce Fractions To reduce a fraction, you must find the largest factor that is common to both the numerator and the denominator, the greatest common factor (GCF). Divide both numbers by the GCF. The fraction has then been reduced, or it is in its simplest form.

Example Twelve of the 20 chemicals in the science lab are in powder form. What fraction of the chemicals used in the lab are in powder form?

Step 1 Write the fraction.
$\frac{\text{part}}{\text{whole}} = \frac{12}{20}$

Step 2 To find the GCF of the numerator and denominator, list all of the factors of each number.
Factors of 12: 1, 2, 3, 4, 6, 12 (the numbers that divide evenly into 12)
Factors of 20: 1, 2, 4, 5, 10, 20 (the numbers that divide evenly into 20)

Step 3 List the common factors.
1, 2, 4

Step 4 Choose the greatest factor in the list.
The GCF of 12 and 20 is 4.

Step 5 Divide the numerator and denominator by the GCF.
$\frac{12 \div 4}{20 \div 4} = \frac{3}{5}$

In the lab, $\frac{3}{5}$ of the chemicals are in powder form.

Practice Problem At an amusement park, 66 of 90 rides have a height restriction. What fraction of the rides, in its simplest form, has a height restriction?

Add and Subtract Fractions To add or subtract fractions with the same denominator, add or subtract the numerators and write the sum or difference over the denominator. After finding the sum or difference, find the simplest form for your fraction.

Example 1 In the forest outside your house, $\frac{1}{8}$ of the animals are rabbits, $\frac{3}{8}$ are squirrels, and the remainder are birds and insects. How many are mammals?

Step 1 Add the numerators.
$\frac{1}{8} + \frac{3}{8} = \frac{(1 + 3)}{8} = \frac{4}{8}$

Step 2 Find the GCF.
$\frac{4}{8}$ (GCF, 4)

Step 3 Divide the numerator and denominator by the GCF.
$\frac{4 \div 4}{8 \div 4} = \frac{1}{2}$

$\frac{1}{2}$ of the animals are mammals.

Example 2 If $\frac{7}{16}$ of the Earth is covered by freshwater, and $\frac{1}{16}$ of that is in glaciers, how much freshwater is not frozen?

Step 1 Subtract the numerators.
$\frac{7}{16} - \frac{1}{16} = \frac{(7 - 1)}{16} = \frac{6}{16}$

Step 2 Find the GCF.
$\frac{6}{16}$ (GCF, 2)

Step 3 Divide the numerator and denominator by the GCF.
$\frac{6 \div 2}{16 \div 2} = \frac{3}{8}$

$\frac{3}{8}$ of the freshwater is not frozen.

Practice Problem A bicycle rider is riding at a rate of 15 km/h for $\frac{4}{9}$ of his ride, 10 km/h for $\frac{2}{9}$ of his ride, and 8 km/h for the remainder of the ride. How much of his ride is he riding at a rate greater than 8 km/h?

Unlike Denominators To add or subtract fractions with unlike denominators, first find the least common denominator (LCD). This is the smallest number that is a common multiple of both denominators. Rename each fraction with the LCD, and then add or subtract. Find the simplest form if necessary.

Example 1 A chemist makes a paste that is $\frac{1}{2}$ table salt (NaCl), $\frac{1}{3}$ sugar ($C_6H_{12}O_6$), and the remainder is water (H_2O). How much of the paste is a solid?

Step 1 Find the LCD of the fractions.

$$\frac{1}{2} + \frac{1}{3} \quad \text{(LCD, 6)}$$

Step 2 Rename each numerator and each denominator with the LCD.

Step 3 Add the numerators.

$$\frac{3}{6} + \frac{2}{6} = \frac{(3+2)}{6} = \frac{5}{6}$$

$\frac{5}{6}$ of the paste is a solid.

Example 2 The average precipitation in Grand Junction, CO, is $\frac{7}{10}$ inch in November, and $\frac{3}{5}$ inch in December. What is the total average precipitation?

Step 1 Find the LCD of the fractions.

$$\frac{7}{10} + \frac{3}{5} \quad \text{(LCD, 10)}$$

Step 2 Rename each numerator and each denominator with the LCD.

Step 3 Add the numerators.

$$\frac{7}{10} + \frac{6}{10} = \frac{(7+6)}{10} = \frac{13}{10}$$

$\frac{13}{10}$ inches total precipitation, or $1\frac{3}{10}$ inches.

Practice Problem On an electric bill, about $\frac{1}{8}$ of the energy is from solar energy and about $\frac{1}{10}$ is from wind power. How much of the total bill is from solar energy and wind power combined?

Example 3 In your body, $\frac{7}{10}$ of your muscle contractions are involuntary (cardiac and smooth muscle tissue). Smooth muscle makes $\frac{3}{15}$ of your muscle contractions. How many of your muscle contractions are made by cardiac muscle?

Step 1 Find the LCD of the fractions.

$$\frac{7}{10} - \frac{3}{15} \quad \text{(LCD, 30)}$$

Step 2 Rename each numerator and each denominator with the LCD.

$$7 \times 3 = 21, \quad 10 \times 3 = 30$$

$$3 \times 2 = 6, \quad 15 \times 2 = 30$$

Step 3 Subtract the numerators.

$$\frac{21}{30} - \frac{6}{30} = \frac{(21-6)}{30} = \frac{15}{30}$$

Step 4 Find the GCF.

$$\frac{15}{30} \quad \text{(GCF, 15)}$$

$$\frac{1}{2}$$

$\frac{1}{2}$ of all muscle contractions are cardiac muscle.

Example 4 Tony wants to make cookies that call for $\frac{3}{4}$ of a cup of flour, but he only has $\frac{1}{3}$ of a cup. How much more flour does he need?

Step 1 Find the LCD of the fractions.

$$\frac{3}{4} - \frac{1}{3} \quad \text{(LCD, 12)}$$

Step 2 Rename each numerator and each denominator with the LCD.

$$\frac{3 \times 3}{4 \times 3} = \frac{9}{12}$$

$$\frac{1 \times 4}{3 \times 4} = \frac{4}{12}$$

Step 3 Subtract the numerators.

$$\frac{9}{12} - \frac{4}{12} = \frac{(9-4)}{12} = \frac{5}{12}$$

$\frac{5}{12}$ of a cup of flour

Practice Problem Using the information provided to you in Example 3 above, determine how many muscle contractions are voluntary (skeletal muscle).

Multiply Fractions To multiply with fractions, multiply the numerators and multiply the denominators. Find the simplest form if necessary.

Example Multiply $\frac{3}{5}$ by $\frac{1}{3}$.

Step 1 Multiply the numerators and denominators.

$$\frac{3}{5} \times \frac{1}{3} = \frac{(3 \times 1)}{(5 \times 3)} = \frac{3}{15}$$

Step 2 Find the GCF.

$\frac{3}{15}$ (GCF, 3)

Step 3 Divide the numerator and denominator by the GCF.

$$\frac{3 \div 3}{15 \div 3} = \frac{1}{5}$$

$\frac{3}{5}$ multiplied by $\frac{1}{3}$ is $\frac{1}{5}$.

Practice Problem Multiply $\frac{3}{14}$ by $\frac{5}{16}$.

Find a Reciprocal Two numbers whose product is 1 are called multiplicative inverses, or reciprocals.

Example Find the reciprocal of $\frac{3}{8}$.

Step 1 Inverse the fraction by putting the denominator on top and the numerator on the bottom.

$\frac{8}{3}$

The reciprocal of $\frac{3}{8}$ is $\frac{8}{3}$.

Practice Problem Find the reciprocal of $\frac{4}{9}$.

Divide Fractions To divide one fraction by another fraction, multiply the dividend by the reciprocal of the divisor. Find the simplest form if necessary.

Example 1 Divide $\frac{1}{9}$ by $\frac{1}{3}$.

Step 1 Find the reciprocal of the divisor.

The reciprocal of $\frac{1}{3}$ is $\frac{3}{1}$.

Step 2 Multiply the dividend by the reciprocal of the divisor.

$$\frac{\frac{1}{9}}{\frac{1}{3}} = \frac{1}{9} \times \frac{3}{1} = \frac{(1 \times 3)}{(9 \times 1)} = \frac{3}{9}$$

Step 3 Find the GCF.

$\frac{3}{9}$ (GCF, 3)

Step 4 Divide the numerator and denominator by the GCF.

$$\frac{3 \div 3}{9 \div 3} = \frac{1}{3}$$

$\frac{1}{9}$ divided by $\frac{1}{3}$ is $\frac{1}{3}$.

Example 2 Divide $\frac{3}{5}$ by $\frac{1}{4}$.

Step 1 Find the reciprocal of the divisor.

The reciprocal of $\frac{1}{4}$ is $\frac{4}{1}$.

Step 2 Multiply the dividend by the reciprocal of the divisor.

$$\frac{\frac{3}{5}}{\frac{1}{4}} = \frac{3}{5} \times \frac{4}{1} = \frac{(3 \times 4)}{(5 \times 1)} = \frac{12}{5}$$

$\frac{3}{5}$ divided by $\frac{1}{4}$ is $\frac{12}{5}$ or $2\frac{2}{5}$.

Practice Problem Divide $\frac{3}{11}$ by $\frac{7}{10}$.

Use Ratios

When you compare two numbers by division, you are using a ratio. Ratios can be written 3 to 5, 3:5, or $\frac{3}{5}$. Ratios, like fractions, also can be written in simplest form.

Ratios can represent one type of probability, called odds. This is a ratio that compares the number of ways a certain outcome occurs to the number of possible outcomes. For example, if you flip a coin 100 times, what are the odds that it will come up heads? There are two possible outcomes, heads or tails, so the odds of coming up heads are 50:100. Another way to say this is that 50 out of 100 times the coin will come up heads. In its simplest form, the ratio is 1:2.

Example 1 A chemical solution contains 40 g of salt and 64 g of baking soda. What is the ratio of salt to baking soda as a fraction in simplest form?

Step 1 Write the ratio as a fraction.

$$\frac{\text{salt}}{\text{baking soda}} = \frac{40}{64}$$

Step 2 Express the fraction in simplest form.

The GCF of 40 and 64 is 8.

$$\frac{40}{64} = \frac{40 \div 8}{64 \div 8} = \frac{5}{8}$$

The ratio of salt to baking soda in the sample is 5:8.

Example 2 Sean rolls a 6-sided die 6 times. What are the odds that the side with a 3 will show?

Step 1 Write the ratio as a fraction.

$$\frac{\text{number of sides with a 3}}{\text{number of possible sides}} = \frac{1}{6}$$

Step 2 Multiply by the number of attempts.

$$\frac{1}{6} \times 6 \text{ attempts} = \frac{6}{6} \text{ attempts} = 1 \text{ attempt}$$

1 attempt out of 6 will show a 3.

Practice Problem Two metal rods measure 100 cm and 144 cm in length. What is the ratio of their lengths in simplest form?

Use Decimals

A fraction with a denominator that is a power of ten can be written as a decimal. For example, 0.27 means $\frac{27}{100}$. The decimal point separates the ones place from the tenths place.

Any fraction can be written as a decimal using division. For example, the fraction $\frac{5}{8}$ can be written as a decimal by dividing 5 by 8. Written as a decimal, it is 0.625.

Add or Subtract Decimals When adding and subtracting decimals, line up the decimal points before carrying out the operation.

Example 1 Find the sum of 47.68 and 7.80.

Step 1 Line up the decimal places when you write the numbers.

$$\begin{array}{r} 47.68 \\ +\ 7.80 \\ \hline \end{array}$$

Step 2 Add the decimals.

$$\begin{array}{r} {\scriptstyle 1\,1} \\ 47.68 \\ +\ 7.80 \\ \hline 55.48 \end{array}$$

The sum of 47.68 and 7.80 is 55.48.

Example 2 Find the difference of 42.17 and 15.85.

Step 1 Line up the decimal places when you write the number.

$$\begin{array}{r} 42.17 \\ -15.85 \\ \hline \end{array}$$

Step 2 Subtract the decimals.

$$\begin{array}{r} {\scriptstyle 3\,11\,1} \\ \not{4}\not{2}.17 \\ -15.85 \\ \hline 26.32 \end{array}$$

The difference of 42.17 and 15.85 is 26.32.

Practice Problem Find the sum of 1.245 and 3.842.

Multiply Decimals To multiply decimals, multiply the numbers like you multiply numbers without decimals. Count the decimal places in each factor. The product will have the same number of decimal places as the sum of the decimal places in the factors.

Example Multiply 2.4 by 5.9.

Step 1 Multiply the factors like two whole numbers.
$24 \times 59 = 1416$

Step 2 Find the sum of the number of decimal places in the factors. Each factor has one decimal place, for a sum of two decimal places.

Step 3 The product will have two decimal places.
14.16

The product of 2.4 and 5.9 is 14.16.

Practice Problem Multiply 4.6 by 2.2.

Divide Decimals When dividing decimals, change the divisor to a whole number. To do this, multiply both the divisor and the dividend by the same power of ten. Then place the decimal point in the quotient directly above the decimal point in the dividend. Then divide as you do with whole numbers.

Example Divide 8.84 by 3.4.

Step 1 Multiply both factors by 10.
$3.4 \times 10 = 34,\ 8.84 \times 10 = 88.4$

Step 2 Divide 88.4 by 34.

$$\begin{array}{r} 2.6 \\ 34\overline{)88.4} \\ -68 \\ \hline 20\,4 \\ -20\,4 \\ \hline 0 \end{array}$$

8.84 divided by 3.4 is 2.6.

Practice Problem Divide 75.6 by 3.6.

Use Proportions

An equation that shows that two ratios are equivalent is a proportion. The ratios $\frac{2}{4}$ and $\frac{5}{10}$ are equivalent, so they can be written as $\frac{2}{4} = \frac{5}{10}$. This equation is a proportion.

When two ratios form a proportion, the cross products are equal. To find the cross products in the proportion $\frac{2}{4} = \frac{5}{10}$, multiply the 2 and the 10, and the 4 and the 5. Therefore $2 \times 10 = 4 \times 5$, or $20 = 20$.

Because you know that both ratios are equal, you can use cross products to find a missing term in a proportion. This is known as solving the proportion.

Example The heights of a tree and a pole are proportional to the lengths of their shadows. The tree casts a shadow of 24 m when a 6-m pole casts a shadow of 4 m. What is the height of the tree?

Step 1 Write a proportion.
$$\frac{\text{height of tree}}{\text{height of pole}} = \frac{\text{length of tree's shadow}}{\text{length of pole's shadow}}$$

Step 2 Substitute the known values into the proportion. Let *h* represent the unknown value, the height of the tree.
$$\frac{h}{6} = \frac{24}{4}$$

Step 3 Find the cross products.
$h \times 4 = 6 \times 24$

Step 4 Simplify the equation.
$4h = 144$

Step 5 Divide each side by 4.
$$\frac{4h}{4} = \frac{144}{4}$$
$h = 36$

The height of the tree is 36 m.

Practice Problem The ratios of the weights of two objects on the Moon and on Earth are in proportion. A rock weighing 3 N on the Moon weighs 18 N on Earth. How much would a rock that weighs 5 N on the Moon weigh on Earth?

Use Percentages

The word *percent* means "out of one hundred." It is a ratio that compares a number to 100. Suppose you read that 77 percent of the Earth's surface is covered by water. That is the same as reading that the fraction of the Earth's surface covered by water is $\frac{77}{100}$. To express a fraction as a percent, first find the equivalent decimal for the fraction. Then, multiply the decimal by 100 and add the percent symbol.

Example Express $\frac{13}{20}$ as a percent.

Step 1 Find the equivalent decimal for the fraction.

$$\begin{array}{r} 0.65 \\ 20\overline{)13.00} \\ \underline{12\,0} \\ 1\,00 \\ \underline{1\,00} \\ 0 \end{array}$$

Step 2 Rewrite the fraction $\frac{13}{20}$ as 0.65.

Step 3 Multiply 0.65 by 100 and add the % symbol.

$0.65 \times 100 = 65 = 65\%$

So, $\frac{13}{20} = 65\%$.

This also can be solved as a proportion.

Example Express $\frac{13}{20}$ as a percent.

Step 1 Write a proportion.

$\frac{13}{20} = \frac{x}{100}$

Step 2 Find the cross products.

$1300 = 20x$

Step 3 Divide each side by 20.

$\frac{1300}{20} = \frac{20x}{20}$

$65\% = x$

Practice Problem In one year, 73 of 365 days were rainy in one city. What percent of the days in that city were rainy?

Solve One-Step Equations

A statement that two expressions are equal is an equation. For example, $A = B$ is an equation that states that A is equal to B.

An equation is solved when a variable is replaced with a value that makes both sides of the equation equal. To make both sides equal the inverse operation is used. Addition and subtraction are inverses, and multiplication and division are inverses.

Example 1 Solve the equation $x - 10 = 35$.

Step 1 Find the solution by adding 10 to each side of the equation.

$x - 10 = 35$

$x - 10 + 10 = 35 + 10$

$x = 45$

Step 2 Check the solution.

$x - 10 = 35$

$45 - 10 = 35$

$35 = 35$

Both sides of the equation are equal, so $x = 45$.

Example 2 In the formula $a = bc$, find the value of c if $a = 20$ and $b = 2$.

Step 1 Rearrange the formula so the unknown value is by itself on one side of the equation by dividing both sides by b.

$a = bc$

$\frac{a}{b} = \frac{bc}{b}$

$\frac{a}{b} = c$

Step 2 Replace the variables a and b with the values that are given.

$\frac{a}{b} = c$

$\frac{20}{2} = c$

$10 = c$

Step 3 Check the solution.

$a = bc$

$20 = 2 \times 10$

$20 = 20$

Both sides of the equation are equal, so $c = 10$ is the solution when $a = 20$ and $b = 2$.

Practice Problem In the formula $h = gd$, find the value of d if $g = 12.3$ and $h = 17.4$.

Use Statistics

The branch of mathematics that deals with collecting, analyzing, and presenting data is statistics. In statistics, there are three common ways to summarize data with a single number—the mean, the median, and the mode.

The **mean** of a set of data is the arithmetic average. It is found by adding the numbers in the data set and dividing by the number of items in the set.

The **median** is the middle number in a set of data when the data are arranged in numerical order. If there were an even number of data points, the median would be the mean of the two middle numbers.

The **mode** of a set of data is the number or item that appears most often.

Another number that often is used to describe a set of data is the range. The **range** is the difference between the largest number and the smallest number in a set of data.

A **frequency table** shows how many times each piece of data occurs, usually in a survey. **Table 2** below shows the results of a student survey on favorite color.

Table 2 Student Color Choice

Color	Tally	Frequency
Red	||||	4
Blue	~~||||~~	5
Black	||	2
Green	|||	3
Purple	~~||||~~ ||	7
Yellow	~~||||~~ |	6

Based on the frequency table data, which color is the favorite?

Example The speeds (in m/s) for a race car during five different time trials are 39, 37, 44, 36, and 44.

To find the mean:

Step 1 Find the sum of the numbers.
$39 + 37 + 44 + 36 + 44 = 200$

Step 2 Divide the sum by the number of items, which is 5.
$200 \div 5 = 40$

The mean is 40 m/s.

To find the median:

Step 1 Arrange the measures from least to greatest.
36, 37, 39, 44, 44

Step 2 Determine the middle measure.
36, 37, <u>39</u>, 44, 44

The median is 39 m/s.

To find the mode:

Step 1 Group the numbers that are the same together.
44, 44, 36, 37, 39

Step 2 Determine the number that occurs most in the set.
<u>44, 44</u>, 36, 37, 39

The mode is 44 m/s.

To find the range:

Step 1 Arrange the measures from greatest to least.
44, 44, 39, 37, 36

Step 2 Determine the greatest and least measures in the set.
<u>44</u>, 44, 39, 37, <u>36</u>

Step 3 Find the difference between the greatest and least measures.
$44 - 36 = 8$

The range is 8 m/s.

Practice Problem Find the mean, median, mode, and range for the data set 8, 4, 12, 8, 11, 14, 16.

Use Geometry

The branch of mathematics that deals with the measurement, properties, and relationships of points, lines, angles, surfaces, and solids is called geometry.

Perimeter The **perimeter** (P) is the distance around a geometric figure. To find the perimeter of a rectangle, add the length and width and multiply that sum by two, or $2(l + w)$. To find perimeters of irregular figures, add the length of the sides.

Example 1 Find the perimeter of a rectangle that is 3 m long and 5 m wide.

Step 1 You know that the perimeter is 2 times the sum of the width and length.
$P = 2(3 \text{ m} + 5 \text{ m})$

Step 2 Find the sum of the width and length.
$P = 2(8 \text{ m})$

Step 3 Multiply by 2.
$P = 16 \text{ m}$

The perimeter is 16 m.

Example 2 Find the perimeter of a shape with sides measuring 2 cm, 5 cm, 6 cm, 3 cm.

Step 1 You know that the perimeter is the sum of all the sides.
$P = 2 + 5 + 6 + 3$

Step 2 Find the sum of the sides.
$P = 2 + 5 + 6 + 3$
$P = 16$

The perimeter is 16 cm.

Practice Problem Find the perimeter of a rectangle with a length of 18 m and a width of 7 m.

Practice Problem Find the perimeter of a triangle measuring 1.6 cm by 2.4 cm by 2.4 cm.

Area of a Rectangle The **area** (A) is the number of square units needed to cover a surface. To find the area of a rectangle, multiply the length times the width, or $l \times w$. When finding area, the units also are multiplied. Area is given in square units.

Example Find the area of a rectangle with a length of 1 cm and a width of 10 cm.

Step 1 You know that the area is the length multiplied by the width.
$A = (1 \text{ cm} \times 10 \text{ cm})$

Step 2 Multiply the length by the width. Also multiply the units.
$A = 10 \text{ cm}^2$

The area is 10 cm^2.

Practice Problem Find the area of a square whose sides measure 4 m.

Area of a Triangle To find the area of a triangle, use the formula:

$$A = \frac{1}{2}(\text{base} \times \text{height})$$

The base of a triangle can be any of its sides. The height is the perpendicular distance from a base to the opposite endpoint, or vertex.

Example Find the area of a triangle with a base of 18 m and a height of 7 m.

Step 1 You know that the area is $\frac{1}{2}$ the base times the height.
$A = \frac{1}{2}(18 \text{ m} \times 7 \text{ m})$

Step 2 Multiply $\frac{1}{2}$ by the product of 18×7. Multiply the units.
$A = \frac{1}{2}(126 \text{ m}^2)$
$A = 63 \text{ m}^2$

The area is 63 m^2.

Practice Problem Find the area of a triangle with a base of 27 cm and a height of 17 cm.

Circumference of a Circle The **diameter** (d) of a circle is the distance across the circle through its center, and the **radius** (r) is the distance from the center to any point on the circle. The radius is half of the diameter. The distance around the circle is called the **circumference** (C). The formula for finding the circumference is:

$C = 2\pi r$ *or* $C = \pi d$

The circumference divided by the diameter is always equal to 3.1415926... This nonterminating and nonrepeating number is represented by the Greek letter π (pi). An approximation often used for π is 3.14.

Example 1 Find the circumference of a circle with a radius of 3 m.

Step 1 You know the formula for the circumference is 2 times the radius times π.
$C = 2\pi(3)$

Step 2 Multiply 2 times the radius.
$C = 6\pi$

Step 3 Multiply by π.
$C = 19$ m

The circumference is 19 m.

Example 2 Find the circumference of a circle with a diameter of 24.0 cm.

Step 1 You know the formula for the circumference is the diameter times π.
$C = \pi(24.0)$

Step 2 Multiply the diameter by π.
$C = 75.4$ cm

The circumference is 75.4 cm.

Practice Problem Find the circumference of a circle with a radius of 19 cm.

Area of a Circle The formula for the area of a circle is:
$A = \pi r^2$

Example 1 Find the area of a circle with a radius of 4.0 cm.

Step 1 $A = \pi(4.0)^2$

Step 2 Find the square of the radius.
$A = 16\pi$

Step 3 Multiply the square of the radius by π.
$A = 50\ \text{cm}^2$

The area of the circle is 50 cm^2.

Example 2 Find the area of a circle with a radius of 225 m.

Step 1 $A = \pi(225)^2$

Step 2 Find the square of the radius.
$A = 50625\pi$

Step 3 Multiply the square of the radius by π.
$A = 158962.5$

The area of the circle is 158,962 m^2.

Example 3 Find the area of a circle whose diameter is 20.0 mm.

Step 1 You know the formula for the area of a circle is the square of the radius times π, and that the radius is half of the diameter.
$A = \pi\left(\frac{20.0}{2}\right)^2$

Step 2 Find the radius.
$A = \pi(10.0)^2$

Step 3 Find the square of the radius.
$A = 100\pi$

Step 4 Multiply the square of the radius by π.
$A = 314\ \text{mm}^2$

The area is 314 mm^2.

Practice Problem Find the area of a circle with a radius of 16 m.

Volume The measure of space occupied by a solid is the **volume** (V). To find the volume of a rectangular solid multiply the length times width times height, or $V = l \times w \times h$. It is measured in cubic units, such as cubic centimeters (cm^3).

Example Find the volume of a rectangular solid with a length of 2.0 m, a width of 4.0 m, and a height of 3.0 m.

Step 1 You know the formula for volume is the length times the width times the height.

$V = 2.0 \text{ m} \times 4.0 \text{ m} \times 3.0 \text{ m}$

Step 2 Multiply the length times the width times the height.

$V = 24 \text{ m}^3$

The volume is 24 m^3.

Practice Problem Find the volume of a rectangular solid that is 8 m long, 4 m wide, and 4 m high.

To find the volume of other solids, multiply the area of the base times the height.

Example 1 Find the volume of a solid that has a triangular base with a length of 8.0 m and a height of 7.0 m. The height of the entire solid is 15.0 m.

Step 1 You know that the base is a triangle, and the area of a triangle is $\frac{1}{2}$ the base times the height, and the volume is the area of the base times the height.

$V = \left[\frac{1}{2}(b \times h)\right] \times 15$

Step 2 Find the area of the base.

$V = \left[\frac{1}{2}(8 \times 7)\right] \times 15$

$V = \left(\frac{1}{2} \times 56\right) \times 15$

Step 3 Multiply the area of the base by the height of the solid.

$V = 28 \times 15$

$V = 420 \text{ m}^3$

The volume is 420 m^3.

Example 2 Find the volume of a cylinder that has a base with a radius of 12.0 cm, and a height of 21.0 cm.

Step 1 You know that the base is a circle, and the area of a circle is the square of the radius times π, and the volume is the area of the base times the height.

$V = (\pi r^2) \times 21$

$V = (\pi 12^2) \times 21$

Step 2 Find the area of the base.

$V = 144\pi \times 21$

$V = 452 \times 21$

Step 3 Multiply the area of the base by the height of the solid.

$V = 9{,}490 \text{ cm}^3$

The volume is 9,500 cm^3.

Example 3 Find the volume of a cylinder that has a diameter of 15 mm and a height of 4.8 mm.

Step 1 You know that the base is a circle with an area equal to the square of the radius times π. The radius is one-half the diameter. The volume is the area of the base times the height.

$V = (\pi r^2) \times 4.8$

$V = \left[\pi\left(\frac{1}{2} \times 15\right)^2\right] \times 4.8$

$V = (\pi 7.5^2) \times 4.8$

Step 2 Find the area of the base.

$V = 56.25\pi \times 4.8$

$V \approx 176.63 \times 4.8$

Step 3 Multiply the area of the base by the height of the solid.

$V = 847.8$

The volume is 847.8 mm^3.

Practice Problem Find the volume of a cylinder with a diameter of 7 cm in the base and a height of 16 cm.

Science Applications

Measure in SI

The metric system of measurement was developed in 1795. A modern form of the metric system, called the International System (SI), was adopted in 1960 and provides the standard measurements that all scientists around the world can understand.

The SI system is convenient because unit sizes vary by powers of 10. Prefixes are used to name units. Look at **Table 3** for some common SI prefixes and their meanings.

Table 3 Common SI Prefixes

Prefix	Symbol	Meaning	
kilo-	k	1,000	thousand
hecto-	h	100	hundred
deka-	da	10	ten
deci-	d	0.1	tenth
centi-	c	0.01	hundredth
milli-	m	0.001	thousandth

Example How many grams equal one kilogram?

Step 1 Find the prefix *kilo-* in **Table 3.**

Step 2 Using **Table 3,** determine the meaning of *kilo-*. According to the table, it means 1,000. When the prefix *kilo-* is added to a unit, it means that there are 1,000 of the units in a "kilounit."

Step 3 Apply the prefix to the units in the question. The units in the question are grams. There are 1,000 grams in a kilogram.

Practice Problem Is a milligram larger or smaller than a gram? How many of the smaller units equal one larger unit? What fraction of the larger unit does one smaller unit represent?

Dimensional Analysis

Convert SI Units In science, quantities such as length, mass, and time sometimes are measured using different units. A process called dimensional analysis can be used to change one unit of measure to another. This process involves multiplying your starting quantity and units by one or more conversion factors. A conversion factor is a ratio equal to one and can be made from any two equal quantities with different units. If 1,000 mL equal 1 L then two ratios can be made.

$$\frac{1{,}000\ \text{mL}}{1\ \text{L}} = \frac{1\ \text{L}}{1{,}000\ \text{mL}} = 1$$

One can convert between units in the SI system by using the equivalents in **Table 3** to make conversion factors.

Example 1 How many cm are in 4 m?

Step 1 Write conversion factors for the units given. From **Table 3,** you know that 100 cm = 1 m. The conversion factors are

$$\frac{100\ \text{cm}}{1\ \text{m}} \quad \textit{and} \quad \frac{1\ \text{m}}{100\ \text{cm}}$$

Step 2 Decide which conversion factor to use. Select the factor that has the units you are converting from (m) in the denominator and the units you are converting to (cm) in the numerator.

$$\frac{100\ \text{cm}}{1\ \text{m}}$$

Step 3 Multiply the starting quantity and units by the conversion factor. Cancel the starting units with the units in the denominator. There are 400 cm in 4 m.

$$4\ \cancel{\text{m}} \times \frac{100\ \text{cm}}{1\ \cancel{\text{m}}} = 400\ \text{cm}$$

Practice Problem How many milligrams are in one kilogram? (Hint: You will need to use two conversion factors from **Table 3.**)

Table 4 Unit System Equivalents

Type of Measurement	Equivalent
Length	1 in = 2.54 cm 1 yd = 0.91 m 1 mi = 1.61 km
Mass and weight*	1 oz = 28.35 g 1 lb = 0.45 kg 1 ton (short) = 0.91 tonnes (metric tons) 1 lb = 4.45 N
Volume	1 in^3 = 16.39 cm^3 1 qt = 0.95 L 1 gal = 3.78 L
Area	1 in^2 = 6.45 cm^2 1 yd^2 = 0.83 m^2 1 mi^2 = 2.59 km^2 1 acre = 0.40 hectares
Temperature	$°C = \frac{(°F - 32)}{1.8}$ K = °C + 273

*Weight is measured in standard Earth gravity.

Convert Between Unit Systems **Table 4** gives a list of equivalents that can be used to convert between English and SI units.

Example If a meterstick has a length of 100 cm, how long is the meterstick in inches?

Step 1 Write the conversion factors for the units given. From **Table 4,** 1 in = 2.54 cm.

$$\frac{1 \text{ in}}{2.54 \text{ cm}} \textit{ and } \frac{2.54 \text{ cm}}{1 \text{ in}}$$

Step 2 Determine which conversion factor to use. You are converting from cm to in. Use the conversion factor with cm on the bottom.

$$\frac{1 \text{ in}}{2.54 \text{ cm}}$$

Step 3 Multiply the starting quantity and units by the conversion factor. Cancel the starting units with the units in the denominator. Round your answer to the nearest tenth.

$$100 \cancel{\text{cm}} \times \frac{1 \text{ in}}{2.54 \cancel{\text{cm}}} = 39.37 \text{ in}$$

The meterstick is about 39.4 in long.

Practice Problem A book has a mass of 5 lbs. What is the mass of the book in kg?

Practice Problem Use the equivalent for in and cm (1 in = 2.54 cm) to show how 1 in^3 = 16.39 cm^3.

Precision and Significant Digits

When you make a measurement, the value you record depends on the precision of the measuring instrument. This precision is represented by the number of significant digits recorded in the measurement. When counting the number of significant digits, all digits are counted except zeros at the end of a number with no decimal point such as 2,050, and zeros at the beginning of a decimal such as 0.03020. When adding or subtracting numbers with different precision, round the answer to the smallest number of decimal places of any number in the sum or difference. When multiplying or dividing, the answer is rounded to the smallest number of significant digits of any number being multiplied or divided.

Example The lengths 5.28 and 5.2 are measured in meters. Find the sum of these lengths and record your answer using the correct number of significant digits.

Step 1 Find the sum.

5.28 m	2 digits after the decimal
+ 5.2 m	1 digit after the decimal
10.48 m	

Step 2 Round to one digit after the decimal because the least number of digits after the decimal of the numbers being added is 1.

The sum is 10.5 m.

Practice Problem How many significant digits are in the measurement 7,071,301 m? How many significant digits are in the measurement 0.003010 g?

Practice Problem Multiply 5.28 and 5.2 using the rule for multiplying and dividing. Record the answer using the correct number of significant digits.

Scientific Notation

Many times numbers used in science are very small or very large. Because these numbers are difficult to work with scientists use scientific notation. To write numbers in scientific notation, move the decimal point until only one non-zero digit remains on the left. Then count the number of places you moved the decimal point and use that number as a power of ten. For example, the average distance from the Sun to Mars is 227,800,000,000 m. In scientific notation, this distance is 2.278×10^{11} m. Because you moved the decimal point to the left, the number is a positive power of ten.

The mass of an electron is about 0.000 000 000 000 000 000 000 000 000 000 911 kg. Expressed in scientific notation, this mass is 9.11×10^{-31} kg. Because the decimal point was moved to the right, the number is a negative power of ten.

Example Earth is 149,600,000 km from the Sun. Express this in scientific notation.

Step 1 Move the decimal point until one non-zero digit remains on the left.
1.496 000 00

Step 2 Count the number of decimal places you have moved. In this case, eight.

Step 3 Show that number as a power of ten, 10^8.

Earth is 1.496×10^8 km from the Sun.

Practice Problem How many significant digits are in 149,600,000 km? How many significant digits are in 1.496×10^8 km?

Practice Problem Parts used in a high performance car must be measured to 7×10^{-6} m. Express this number as a decimal.

Practice Problem A CD is spinning at 539 revolutions per minute. Express this number in scientific notation.

Make and Use Graphs

Data in tables can be displayed in a graph—a visual representation of data. Common graph types include line graphs, bar graphs, and circle graphs.

Line Graph A line graph shows a relationship between two variables that change continuously. The independent variable is changed and is plotted on the *x*-axis. The dependent variable is observed, and is plotted on the *y*-axis.

Example Draw a line graph of the data below from a cyclist in a long-distance race.

Table 5 Bicycle Race Data

Time (h)	Distance (km)
0	0
1	8
2	16
3	24
4	32
5	40

Step 1 Determine the *x*-axis and *y*-axis variables. Time varies independently of distance and is plotted on the *x*-axis. Distance is dependent on time and is plotted on the *y*-axis.

Step 2 Determine the scale of each axis. The *x*-axis data ranges from 0 to 5. The *y*-axis data ranges from 0 to 50.

Step 3 Using graph paper, draw and label the axes. Include units in the labels.

Step 4 Draw a point at the intersection of the time value on the *x*-axis and corresponding distance value on the *y*-axis. Connect the points and label the graph with a title, as shown in **Figure 20.**

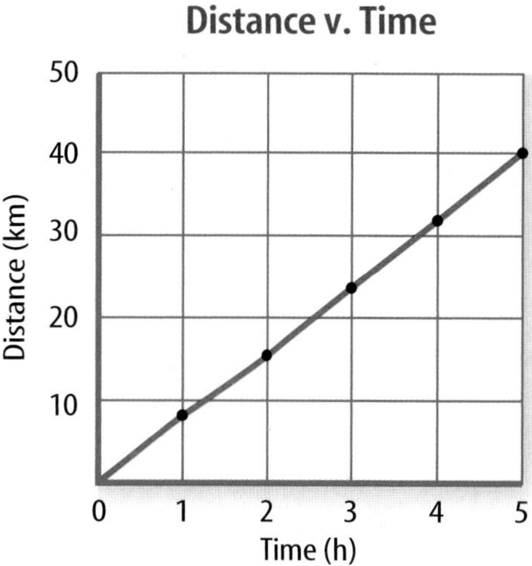

Figure 20 This line graph shows the relationship between distance and time during a bicycle ride.

Practice Problem A puppy's shoulder height is measured during the first year of her life. The following measurements were collected: (3 mo, 52 cm), (6 mo, 72 cm), (9 mo, 83 cm), (12 mo, 86 cm). Graph this data.

Find a Slope The slope of a straight line is the ratio of the vertical change, rise, to the horizontal change, run.

$$\text{Slope} = \frac{\text{vertical change (rise)}}{\text{horizontal change (run)}} = \frac{\text{change in } y}{\text{change in } x}$$

Example Find the slope of the graph in **Figure 20.**

Step 1 You know that the slope is the change in *y* divided by the change in *x*.

$$\text{Slope} = \frac{\text{change in } y}{\text{change in } x}$$

Step 2 Determine the data points you will be using. For a straight line, choose the two sets of points that are the farthest apart.

$$\text{Slope} = \frac{(40-0)\text{ km}}{(5-0)\text{ h}}$$

Step 3 Find the change in *y* and *x*.

$$\text{Slope} = \frac{40\text{ km}}{5\text{h}}$$

Step 4 Divide the change in *y* by the change in *x*.

$$\text{Slope} = \frac{8\text{ km}}{\text{h}}$$

The slope of the graph is 8 km/h.

Bar Graph To compare data that does not change continuously you might choose a bar graph. A bar graph uses bars to show the relationships between variables. The *x*-axis variable is divided into parts. The parts can be numbers such as years, or a category such as a type of animal. The *y*-axis is a number and increases continuously along the axis.

Example A recycling center collects 4.0 kg of aluminum on Monday, 1.0 kg on Wednesday, and 2.0 kg on Friday. Create a bar graph of this data.

Step 1 Select the *x*-axis and *y*-axis variables. The measured numbers (the masses of aluminum) should be placed on the *y*-axis. The variable divided into parts (collection days) is placed on the *x*-axis.

Step 2 Create a graph grid like you would for a line graph. Include labels and units.

Step 3 For each measured number, draw a vertical bar above the *x*-axis value up to the *y*-axis value. For the first data point, draw a vertical bar above Monday up to 4.0 kg.

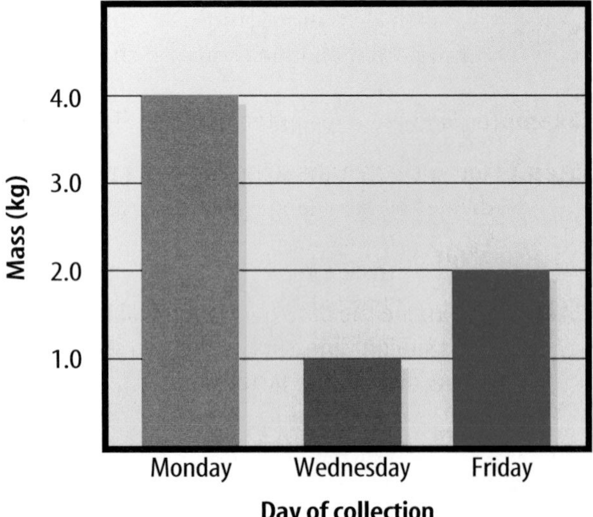

Practice Problem Draw a bar graph of the gases in air: 78% nitrogen, 21% oxygen, 1% other gases.

Circle Graph To display data as parts of a whole, you might use a circle graph. A circle graph is a circle divided into sections that represent the relative size of each piece of data. The entire circle represents 100%, half represents 50%, and so on.

Example Air is made up of 78% nitrogen, 21% oxygen, and 1% other gases. Display the composition of air in a circle graph.

Step 1 Multiply each percent by 360° and divide by 100 to find the angle of each section in the circle.

$$78\% \times \frac{360°}{100} = 280.8°$$

$$21\% \times \frac{360°}{100} = 75.6°$$

$$1\% \times \frac{360°}{100} = 3.6°$$

Step 2 Use a compass to draw a circle and to mark the center of the circle. Draw a straight line from the center to the edge of the circle.

Step 3 Use a protractor and the angles you calculated to divide the circle into parts. Place the center of the protractor over the center of the circle and line the base of the protractor over the straight line.

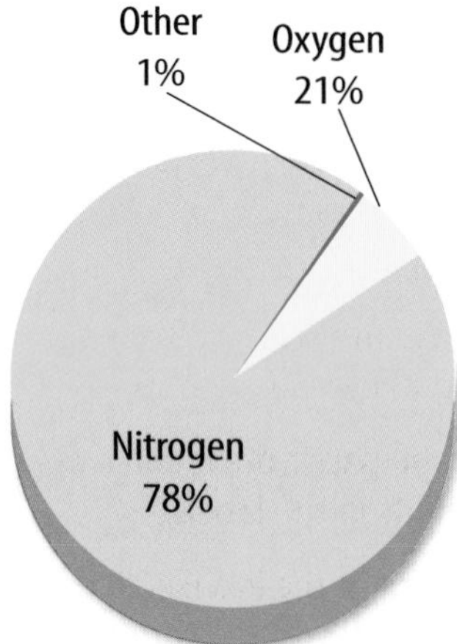

Practice Problem Draw a circle graph to represent the amount of aluminum collected during the week shown in the bar graph to the left.

Formulas

Chapter 1 The Nature of Science

density = mass/volume
Kelvin = °Celsius + 273
% Error = [(Accepted value − Experimental value)/Accepted value] × 100

Chapter 3 Motion, Acceleration, and Forces

average speed = total distance/total time
acceleration = (final velocity − initial velocity)/(final time − initial time)

Chapter 4 The Laws of Motion

acceleration = net force/mass
net force = mass × acceleration
gravitational force = mass × (acceleration due to gravity)
weight = mass × 9.8 m/s^2
momentum (p) = mass × velocity

Chapter 5 Energy

kinetic energy = $\frac{1}{2}$ mass × $(\text{velocity})^2$
gravitational potential energy (GPE) = mass × 9.8 m/s^2 × height
mechanical energy = gravitational potential energy + kinetic energy

Chapter 6 Work and Machines

work = force × distance
power = work/time
power = energy transferred/time
mechanical advantage = output force/input force
efficiency % = (output work/input work) × 100
Ideal mechanical advantage of a lever = length of input arm/length of output arm
Ideal mechanical advantage of a wheel and axle = radius of wheel/radius of axle
Ideal mechanical advantage of an inclined plane = length of slope/height of slope

Chapter 9 Heat and States of Matter

change in thermal energy = mass × change in temperature × specific heat
or
$Q = m \times (T_{final} - T_{initial}) \times C_p$
latent heat of melting = mass × heat of fusion
latent heat of vaporization = mass × heat of vaporization

Chapter 10 Waves

wave speed = wavelength × frequency
frequency = number of wavelengths/time
wavelength = wave speed/frequency

Chapter 11 Sound and Light

index of refraction = speed of light in a vacuum/speed of light in a substance or $n = c/v$

Chapter 13 Electricity

electric current = voltage difference/resistance
electric power = current × voltage difference
electric energy = power × time

EXTRA Math Problems

For help and hints with these problems, visit Math Practice at gpescience.com.

Chapter 1 The Nature of Science

1. How many centimeters are in four meters?
2. How many deciliters are in 500 mL?
3. How many liters are in 2540 cm^3?
4. A young child has a mass of 40 kg. What is the mass of the child in grams?
5. Iron has a density of 7.9 g/cm^3. What is the mass in kg of an iron statue that has a volume of 5.4 L?
6. A 2-L bottle of soda has a volume of 2000 cm^3. What is the volume of the bottle in cubic meters?
7. A big summer movie has a running time of 96 minutes. What is the movie's running time in seconds?
8. The temperature in space is approximately 3 K. What is this temperature in degrees Celsius?
9. The *x*-axis of a certain graph is distance traveled in meters and the *y*-axis is time in seconds. Two points are plotted on this graph with coordinates (2, 43) and (5, 68). What is the elapsed time between the two points?
10. A circle graph has labeled segments of: 57%, 21%, 13%, and 6%. What percentage does the unlabeled segment have?

Chapter 2 Science, Technology, and Society

11. A car can travel 14 km on 1 L of gasoline. What percentage of its fuel efficiency does another car have if it travels 10 km on 1 L of gasoline?
12. What is the fuel efficiency of a car if it gets 45 percent of the fuel efficiency of another car that travels 15 km on 1 L of gasoline?
13. If new information technology causes a three percent reduction in the 40,000,000 HIV infections worldwide, how many people would be infected?
14. A circle graph shows the comparative effects of five new technologies. The circle graph has segments labeled 45 percent, 32 percent, 12 percent, and 3 percent. What is the correct label of the fifth segment?
15. Data from a new Web site takes 17 min to download. How many seconds does it take?
16. If a new internet connection reduces a download time of 17 min by 85 percent, how much time (in minutes and seconds) will it take now?
17. Because of a new health center in a town, the cases of influenza are 25 percent fewer. If the town usually has 300 cases per month, how many cases should you expect during the next month?
18. If a person can sell 10 computers per day and a new technology allows her to triple her sales, how many computers can she expect to sell the next day?
19. If it normally takes 45 min to travel to a nearby town using an older train, but a new technologically advanced train can make the trip in one-third of the time, how long will the trip take?
20. A technologically advanced engine runs at twice the efficiency of an older engine. If the older engine allows a car to travel at a fuel efficiency of 8 km/L, what is the fuel efficiency of the advanced engine?

Chapter 3 Motion, Acceleration, and Forces

21. John rides his bike 2.3 km to school. After school, he rides an additional 1.4 km to the mall in the opposite direction. What is his total distance traveled?
22. A squirrel runs 4.8 m across a lawn, stops, then runs 2.3 m back in the opposite direction. What is the squirrel's displacement from its starting point?

23. An ant travels 75 cm in 5 s. What was the ant's speed?

24. It took you 6.5 h to drive 550 km. What was your speed?

25. A bus leaves at 9 A.M. with a group of tourists. They travel 350 km before they stop for lunch. Then they travel an additional 250 km until the end of their trip at 3 P.M. What was the average speed of the bus?

26. Halfway through a cross-country meet, a runner's speed is 4 m/s. In the last stretch, she increases her speed to 7 m/s. What is her change in speed?

27. It takes a car one minute to go from rest to 30 m/s. What is the acceleration of this car?

28. You are running at a speed of 10 km/h and hit a patch of mud. Two seconds later your speed is 8 km/h. What is your acceleration in units of m/s^2?

29. During a tug-of-war, Team A is applying a force of 5000 N while Team B is applying a force of 8000 N. What is the net force applied to the rope?

30. You are in a car traveling an average speed of 60 km/h. The total trip is 240 km. How long does the trip take?

31. You are riding in a train that is traveling at a speed of 120 km/h. How long will it take to travel 950 km?

32. A car goes from rest to a speed of 90 km/h in 10 s. What is the car's acceleration in m/s^2?

33. A cart rolling at a speed of 10 m/s comes to a stop in 2 s. What is the cart's acceleration?

Chapter 4 The Laws of Motion

34. A 85-kg mass has an acceleration of 5.5 m/s^2. What is the net force applied?

35. A 3200-N force is applied to a 160-kg mass. What is the acceleration of the mass?

36. A 2-kg object is dropped from a height of 1000 m. What is the force of air resistance on the object when it reaches terminal velocity?

37. How much force is needed to lift a 25-kg mass?

38. If you are pushing on a box with a force of 20 N and there is a force of 7 N on the box due to sliding friction, what is the net force on the box?

39. The acceleration due to gravity on the moon is about 1.6 m/s^2. If you weigh 539 N on Earth, how much would you weigh on the moon?

40. If a 5000-kg mass is moving at a speed of 43 m/s, what is its momentum?

41. How fast must a 50-kg mass travel to have a momentum of 1500 kg m/s?

42. What is the net force on a 4000-kg car that doubles its speed from 15 m/s to 30 m/s over 10 seconds?

43. A book with a mass of 1 kg is sliding on a table. If the frictional force on the book is 5 N, calculate the book's acceleration. Is it speeding up or slowing down?

44. What is the weight of a person with a mass of 80 kg?

45. A car with a mass of 1,200 kg has a speed of 30 m/s. What is the car's momentum?

Chapter 5 Energy

46. What is the kinetic energy of a 5-kg object moving at 7 m/s?

47. An object has kinetic energy of 600 J and a speed of 10 m/s. What is its mass?

48. If you throw a 0.4-kg ball at a speed of 20 m/s, what is the ball's kinetic energy?

49. A rollercoaster car moving around a high turn has 100,000 J of GPE and 23,000 J of KE. What is its mechanical energy?

50. If you have a mass of 80 kg and you are standing on a platform 3 m above the ground, what is your gravitational potential energy?

51. A car is traveling at 30 m/s with a kinetic energy of 900 kJ. What is its mass?

52. At top of a hill, a rollercoaster has 67,500 J of kinetic energy and 290,000 J of potential energy. Gradually the roller coaster comes to a stop due to friction. If the roller coaster has 30,000 J of potential when it stops, how much heat energy is generated by friction from the top of the hill until it stops?

53. A 2-kg book is moved from a shelf that is 2 m off the ground to a shelf that is 1.5 m off the ground. What is its change in GPE?

54. A system has a total mechanical energy of 350 J and kinetic energy of 220 J. What is its potential energy?

55. An object held in the air has a GPE of 470 J. The object then is dropped. Halfway down, what is the object's kinetic energy?

56. A car with a mass of 900 kg is traveling at a speed of 25 m/s. What is the kinetic energy of the car in joules?

57. What is the gravitational potential energy of a diver with a mass of 60 kg who is 10 m above the water?

58. If your weight is 500 N, and you are standing on a floor that is 20 m above the ground, what is your gravitational potential energy?

Chapter 6 Work and Machines

59. When moving a couch, you exert a force of 400 N and push it 4 m. How much work have you done?

60. How much work is needed to lift a 50-kg weight to a shelf 3 m above the floor?

61. By applying a force of 50 N, a pulley system can lift a box with a mass of 20 kg. What is the mechanical advantage of the pulley system?

62. How much energy do you save per hour if you replace a 60-watt lightbulb with a 55-watt lightbulb?

63. Suppose you supply energy to a machine at a rate of 700 W and that the machine converts 560 J into heat every second. At what rate does the machine do work?

64. You exert a force of 200 N on a machine over a distance of 0.3 m. If the machine moves an object a distance of 0.5 m, how much force does the machine exert on the object? Assume friction can be ignored.

65. What is the efficiency of a machine if you do work on the machine at a rate of 1200 W and the machine does work at a rate of 300 W?

66. What is the IMA of a seesaw with a 1.6-m effort arm and a 1.2-m resistance arm?

67. What is the IMA of a wheel with a radius of 0.35 m and an axle radius of 0.04 m?

68. An inclined plane has an IMA of 1.5 and a height of 2.0 m. How long is this inclined plane?

69. What power is used by a machine to perform 800 J of work in 25 s?

70. A person pushes a box up a ramp that is 3 m long, and 1 m high. If the box has a mass of 20 kg, and the person pushes with a force of 80 N, what is the efficiency of the ramp?

71. A first class lever has a mechanical advantage of 5. How large would a force need to be to lift a rock with a mass of 100 kg?

Chapter 7 The Earth-Moon-Sun System

72. Earth's diameter is 12,714 km from pole to pole and 12,756 km at the equator. How much less is Earth's diameter from pole to pole than that at the equator?

73. Earth's circumference is 40,075 km at the equator and 40,008 km through the two poles. How much greater is Earth's circumference around the equator?

74. Earth's average density is 5.52 g/cm^3. If the Moon's average density is 3.31 g/cm^3, what percentage of Earth's density is the Moon's density?

75. If a day on Earth lasts 23 h and 56 min, how many minutes are in one day?

76. Earth is one AU (149,600,000 km) from the Sun. If Jupiter is 5.2 AUs from the Sun, how many kilometers is Jupiter from the Sun?

77. If Earth rotates 15° each hour, how many degrees will it rotate in three hours?

78. Earth is tilted 23.5°. If the Sun is 50° above the southern horizon when it is directly over the equator, how high in the sky will the Sun be on the first day of summer?

79. If the Sun has a diameter of 1,392,000 km and is about 400 times larger than the Moon, what is the approximate diameter of the Moon?

80. If there are 29.5 days in one synodic month, how many synodic months are there in one year of 365 days?

81. The South Pole-Aitken Basin on the Moon is 12 km deep and 2,500 km wide. How many times wider is the basin than it is deep?

Chapter 8 The Solar System

82. What fraction of a complete orbit (360°) would a planet move through if its H.L. changes from 32° to 302°?

83. What fraction of a complete orbit (360°) would a planet move through if its H.L. changes from 152° to 242°?

84. Earth's atmospheric pressure is 101.3 kPa. What is atmospheric pressure on the surface of Titan if it is 1.5 times that of Earth?

85. What is the atmospheric pressure on the surface of Mars if it is 0.6 percent of Earth's?

86. If the atmospheric pressure of Triton's atmosphere is 0.002 percent of Earth's, what is the atmospheric pressure on the surface of Triton?

87. If a planet has an H.L. of 73°, what percentage of the total orbit (360°) would this represent?

88. If the diameters of Earth and Uranus are 12,756 km and 51,118 km respectively, approximately how many Earths could fit across Uranus's diameter?

89. The volume of sphere is $4/3(\pi r^3)$. If the radius of Earth is 6,378 km and the radius of Jupiter is 71,492, how many Earths would fit inside Jupiter?

90. The volume of sphere is $4/3(\pi r^3)$. If the radius of Mars is 3,394 km and the radius of Earth is 6,378 km, what percentage of Earth's volume is Mars's volume?

91. On average, Earth is 150 million km (1 AU) from the Sun. If Saturn is 9.53 AU from the Sun, what is the distance from the Sun to Saturn?

92. If Venus takes 0.62 years to orbit the Sun and Saturn takes 29.42 years to orbit the Sun, how many times will Venus orbit the Sun during each orbit of Saturn?

93. If Earth is tilted 23.5° and Uranus is tilted 97.9°, how many times greater is the axial tilt of Uranus than that of Earth?

Chapter 9 Heat and States of Matter

94. Water has a specific heat of 4184 J/(kg K). How much energy is needed to increase the temperature of a kilogram of water 5°C?

95. The temperature of a block of iron, which has a specific heat of 450 J/(kg K), increases by 3 K when 2700 J of energy are added to it. What is the mass of this block of iron?

96. How much energy is needed to heat 1 kg of sand, which has a specific heat of 664 J/(kg K), from 30°C to 50°C?

97. 1 kg of water (specific heat = 4184 J/(kg K)) is heated from freezing (0°C) to boiling (100°C). What is the change in thermal energy?

98. A concrete statue (specific heat = 600 J/(kg K)) sits in sunlight and warms up to 40°C. Overnight, it cools to 15°C and loses 90,000 J of thermal energy. What is its mass?

99. A glass of water has temperature of 70°C. What is its temperature in K?

100. A substance with a mass of 10 kg loses 106.5 kJ of heat when its temperature drops 15°C. What is this substance's specific heat?

101. Air is cooled from room temperature (25°C) to 100 K. What is the temperature change in K?

102. To remove 800 J of heat from a refrigerator, the compressor in the refrigerator does 500 J of work. How much heat is released into the surrounding room?

103. How much heat is needed to raise the temperature of 100 g of water by 50 K, if the specific heat of water is 4,184 J/kg K?

104. A sample of an unknown metal has a mass of 0.5 kg. Adding 1,985 J of heat to the metal raises its temperature by 10 K. What is the specific heat of the metal?

Chapter 10 Waves

105. What is the wavelength of a wave with a frequency of 0.4 kHz traveling at 16 m/s?

106. Two waves are traveling in the same medium with a speed of 340 m/s. What is the difference in frequency of the waves if the one has a wavelength of 5 m and the other has a wavelength of 0.2 m?

107. Transverse wave A has an amplitude of 7 cm. This wave constructively interferes with wave B. While the two waves overlap, the amplitude of the resulting wave is 10 cm. What is the amplitude of wave B?

108. What is the wavelength of a wave with a frequency of 5 Hz traveling at 15 m/s?

109. What is the velocity of a wave that has a wavelength of 6 m and a frequency of 3 Hz?

110. A ray of light hits a mirror at an angle of 35° to the normal. What is the angle of the reflected ray to the normal?

111. A wave has a wavelength of 250 cm and a frequency of 4 Hz. What is its speed?

112. A wave has a frequency of 5.6 MHz. What is the frequency of this wave in Hz?

113. A light ray strikes a mirror and is reflected. The angle between the incident and reflected rays is 86°. What is the angle of the incident ray to the normal?

114. What is the frequency of a wave with a wavelength of 7 m traveling at 21 m/s?

Chapter 11 Sound and Light

115. A sound wave with a frequency of 440 Hz travels in steel with a speed of 5200 m/s. What is the wavelength of the sound wave?

116. A wave traveling in water has a wavelength of a 750 m and a frequency of 2 Hz. How fast is this wave moving?

117. At 0°C sound travels through air with a speed of about 331 m/s and through aluminum with a speed of 4877 m/s. How many times longer is the wavelength of a sound wave in aluminum compared to the wavelength of a sound wave in air if both waves have the same frequency?

118. The speed of sound in air at 0°C is 331 m/s, and at 20°C is 344 m/s. What is the percentage change in the speed of sound at 20°C compared to 0°C?

119. The wreck of the *Titanic* is at a depth of about 3800 m. A sonar unit on a ship above the *Titanic* emits a sound wave that travels at a speed of 1500 m/s. How long does it take a sound wave reflected from the *Titanic* to return to the ocean surface?

120. A sonar unit on a ship emits a sound wave. The echo from the ocean floor is detected two seconds later. If the speed of sound in water is 1500 m/s, how deep is the ocean beneath the ship?

121. A tsunami travels across the ocean at a speed of 500 km/h. If the distance between the wave crests is 200 km, what is the frequency of the wave?

122. A light ray strikes a plane mirror. The angle between the light ray and the surface of the mirror is 25°. What angle does the reflected ray make with the normal?

123. If the index of refraction of the mineral rock salt is 1.52, and the speed of light in a vacuum is 300,000 km/s, what is the speed of light in rock salt?

124. In the human eye, there are about 7,000,000 cone cells distributed over an area of 5 cm^2. If cone cells are evenly distributed over this region, how many cone cells are distributed over an area of 2 cm^2? Express your answer in scientific notation.

125. The magnification of a mirror or lens equals the image size divided by the object size. If a plant cell with a diameter of 0.0035 mm is magnified so that the diameter of the image is 0.028 cm, what is the magnification?

126. Light enters the human eye through the pupil. In the dark, the pupil is dilated and has a diameter of about 1 cm. The Keck telescope has a mirror with a diameter of 10 m. If both the pupil and the Keck mirror are circles, what is the ratio of the area of the Keck telescope mirror to the area of a dilated human pupil?

127. A light source is placed a distance of 1.2 m from a concave mirror on the optical axis. The reflected light rays are parallel and form a light beam. What is the focal length of the mirror?

Chapter 12 Earth's Internal Processes

128. How long after an earthquake will a seismograph 3,000 km away from the epicenter record S-waves that travel at 3.6 km/s?

129. If it takes 48 min and 20 s for P-waves traveling at 6.0 km/s to reach and be recorded by a seismograph, how far away is the epicenter?

130. If S-waves lag behind P-waves by 1 min 52 s for every 1000 km of distance from the earthquake epicenter, how far away from the earthquake epicenter is a seismograph that measures a time-lag of 5 min 30 s between the arrivals of P-waves and S-waves.?

131. With a P-wave-S-wave lag time of 1 min 52 sec, how much difference between the arrival times of P-waves and S-waves would be measured at 3,000 km? At 3,500 km?

132. If surface waves travel at 3.2 km/s, when would you expect them to arrive at a seismograph that is 2,500 km distant from the earthquake epicenter?

133. If S-waves travel at 3.6 km/s and P-waves travel at 6.0 km/s through Earth's crust, what percentage of the speed of P-waves would you expect S-waves to travel at other depths inside Earth?

134. The volume of a cone is $\frac{1}{3}\pi r^2 h$. If Paricutín is 424 m high and has a base 2.8 km across, what is the volume of the cinder cone?

135. If two plates diverge at a rate of 1.3 cm/year, how much farther apart in kilometers will the two plates be after 200 million years?

136. How many times faster are plates moving at 7.3 cm/year than are plates moving at 1.3 cm/year?

137. The volume of a sphere is $4/3(\pi r^3)$. The radius of Earth is 6,378 km, and the radius of its core (including both outer and inner cores) is 3,486 km. What are the volumes of both? What percentage of the total volume of Earth is its core?

Chapter 13 Electricity

138. A circuit has a resistance of 4 Ω. What voltage difference will cause a current of 1.4 A to flow in the circuit?

139. How many amperes of current will flow in a circuit if the voltage difference is 9 V and the resistance in the circuit is 3 Ω?

140. If a voltage difference of 3 V causes a 1.5 A current to flow in a circuit, what is the resistance in the circuit?

141. The current in an appliance is 3 A and the voltage difference is 120 V. How much power is being supplied to the appliance?

142. What is the current into a microwave oven that requires 700 W of power if the voltage difference is 120 V?

143. What is the voltage difference in a circuit that uses 2420 W of power if 11 A of current flows into the circuit?

144. How much energy is used when a 110 kW appliance is used for 3 hours?

145. A television has a power rating of 210 W. If the television uses 1.68 kWh of energy, for how long has the television been on?

146. How much does it cost to light six 100-W lightbulbs for six hours if the price of electrical energy is $0.09/kWh?

147. An electric clothes dryer uses 4 kW of electric power. How long did it take to dry a load of clothes if electric power costs $0.09/kWh, and the cost of using the dryer was $0.27?

148. What is the resistance of a lightbulb that draws 0.5 amp of current when plugged into a 120-V outlet?

149. How much current flows through a 100-W lightbulb that is plugged into a 120-V outlet?

150. Eight amps of current flow through a hair dryer connected to a 120-V outlet. How much electrical power does the hair dryer use?

151. Compare the electrical energy that is used by a 100-W lightbulb that burns for 10 h, and a 1,200-W hair dryer that is used for 15 min.

Chapter 14 Magnetism

152. How many turns are in the secondary coil of a step-down transformer that reduces a voltage from 900 V to 300 V and has 15 turns in the primary coil?

153. A step-down transformer reduces voltage from 2400 V to 120 V. What is the ratio of the number of turns in the primary coil to the number of turns in the secondary coil of the transformer?

154. The current produced by an AC generator switches direction twice for each revolution of the coil. How many times does a 110-Hz alternating current switch direction each second?

155. What is the output voltage from a step-down transformer with 200 turns in the primary coil and 100 turns in the secondary coil if the input voltage was 750 V?

156. What is the output voltage from a step-up transformer with 25 turns in the primary coil and 75 turns in the secondary coil if the input voltage was 120 V?

157. How many turns are in the primary coil of a step-down transformer that reduces a voltage from 400 V to 100 V and has 80 turns in the secondary coil?

158. How many turns are in the secondary coil of a step-up transformer that increases voltage from 30 V to 150 V and has seven turns in the primary coil?

159. The coil of a 60-Hz generator makes 60 revolutions each second. How many revolutions does the coil make in five minutes?

160. If a generator coil makes 6000 revolutions in two minutes, how many revolutions does it make each second?

Chapter 15 Electromagnetic Radiation

161. Express the number 20,000 in scientific notation.

162. An electromagnetic wave has a wavelength of 0.054 m. What is the wavelength in scientific notation?

163. Earth is about 4,500,000,000 years old. Express this number in scientific notation.

164. The speed of electromagnetic waves in air is 300,000 km/s. What is the frequency of electromagnetic waves that have a wavelength of 5×10^{-3} km?

165. The speed of radio waves in water Is about 2.26×10^{5} km/s. What is the frequency of radio waves that have a wavelength of 3.0 km?

166. Radio waves with a frequency of 125,000 Hz have a wavelength of 1.84 km when traveling in ice. What is the speed of the radio waves in ice?

167. Some infrared waves have a frequency of 10,000,000,000,000 Hz. Express this frequency in scientific notation.

168. An infrared wave has a frequency of 1×10^{13} Hz and a wavelength of 3×10^{-5} m. Express this wavelength as a decimal number.

169. An AM radio station broadcasts at a frequency of 620 kHz. Express this frequency in Hz using scientific notation.

170. An FM radio station broadcasts at a frequency of 101 MHz. Express this frequency in Hz using scientific notation.

Chapter 16 Energy Sources

171. A gallon of gasoline contains about 2800 g of gasoline. If burning one gram of gasoline releases about 48 kJ of energy, how much energy is released when a gallon of gasoline is burned? (1 kJ = 1000 J)

172. An automobile engine converts the energy released by burning gasoline into mechanical energy with an efficiency of about 25%. If burning 1 kg of gasoline releases about 48,000 kJ of energy, how much mechanical energy is produced by the engine when 1 kg of gasoline is burned?

173. You heat a cup of water in a 750-W microwave oven for 40 s, and warm the water by 20°C. If it takes about 20 kJ of energy to raise the temperature of a cup of water by 20°C, what is the efficiency of the microwave oven?

174. On average, solar energy strikes Earth's surface with an intensity of about 200 W/m^2. If solar cells are 10% efficient, how large an area would have to be covered by solar cells to generate enough electrical power to light a 100-W lightbulb?

175. What is the overall efficiency of a hydroelectric plant if the process of falling water turning a turbine is 80% efficient, the turbine spinning an electric generator is 95% efficient, and the transmission through power lines is 90% efficient?

176. When a certain ${}^{235}_{92}U$ nucleus is struck by a neutron, it forms the two nuclei ${}^{91}_{36}Kr$ and ${}^{142}_{56}Ba$. How many neutrons are emitted when this occurs?

177. A nuclear reactor contains 100,000 kg of enriched uranium. About 4% of the enriched uranium is the isotope uranium-235. What is the mass of uranium-235 in the reactor core?

178. Suppose the number of uranium-235 nuclei that are split doubles at each stage of a chain reaction. If the chain reaction starts with one nucleus split in the first stage, how many nuclei will have been split after six stages?

179. From 1970 to 1995 the carbon dioxide concentration in Earth's atmosphere increased from about 325 parts per million to about 360 parts per million. What was the percentage change in the concentration of carbon dioxide?

180. About 85% of the energy used in the U.S. comes from fossil fuels. How many times greater is the amount of energy used from fossil fuel than the amount used from all other energy sources?

Chapter 17 Weather and Climate

181. If a snowy surface reflects 90 percent of the solar radiation that strikes it and bare soil reflects 30 percent, how many times more solar radiation is reflected from a snowy surface than from bare soil?

182. The amount of rainfall over five days is 4 cm, 2 cm, 0.4 cm, 0.2 cm, and 1.3 cm. What is the average rainfall per day?

183. Air pressure at Earth's surface is 101.3 kPa. What is the air pressure at 16 km if it is $\frac{1}{10}$ the air pressure at Earth's surface?

184. Much of northern Florida receives 1,300 mm of rain on average per year and the southern part of the Everglades receives 1,650 mm of rain on average per year. What percentage is the rainfall in northern Florida of the rainfall in the southern part of the Everglades?

185. If the tilt of Earth's axis has varied from 21.5° to 24.5° over time, what is the value of the range of Earth's axial tilt?

186. Air pressure at Earth's surface is 101.3 kPa. If air pressure in an average car tire is equal to two atmospheres, what is the pressure in an average car tire?

187. Most of Earth's atmosphere is within 30 km of Earth's surface. If Earth's atmosphere extends about 10,000 km above Earth's surface, what percentage of the atmosphere's depth contains most of Earth's atmosphere?

188. If a tornado travels at 50 km/h and cuts a path of destruction 10 km long, how many minutes did it take to do this?

189. If there have been eight glacial periods during the past 200,000 years, what is the rate at which these have occurred?

190. The three main gases in Earth's atmosphere are nitrogen—78 percent, oxygen—21 percent, and argon and trace gases—1 percent. If a circle graph were drawn to represent these data, how many degrees would represent each section?

Chapter 18 Classification of Matter

191. Two solutions, one with a mass of 450 g and the other with a mass of 350 g, are mixed. A chemical reaction occurs and 125 g of solid crystals are produced that settle on the bottom of the container. What is the mass of the remaining solution?

192. Carbon reacts with oxygen to form carbon dioxide according to the following equation: $C + O_2 \rightarrow CO_2$. When 120 g of carbon reacts with oxygen, 440 g of carbon dioxide are formed. How much oxygen reacted with the carbon?

193. Salt water is distilled by boiling it and condensing the vapor. After distillation, 1,164 g of water have been collected and 12 g of salt are left behind in the original container. What was the original mass of the salt water?

194. Calcium carbonate, $CaCO_3$, decomposes according to the reaction: $CaCO_3 \rightarrow CaO + CO_2$. When 250 g of $CaCO_3$ decompose completely, the mass of CaO is 56% of the mass of the products of this reaction. What is the mass of CO_2 produced?

195. Water breaks down into hydrogen gas and oxygen gas according to the reaction: $2H_2O \rightarrow 2H_2 + O_2$. In this reaction the mass of oxygen produced is eight times greater than the mass of hydrogen produced. If 36 g of water form hydrogen and oxygen gas, what is the mass of hydrogen gas produced?

196. The size of particles in a solution is about 1 nm (1 nm = 0.000000001 m). Write 0.000000001 m in scientific notation.

197. A chemical reaction produces two new substances, one with a mass of 34 g and the other with a mass of 39 g. What was the total mass of the reactants?

198. The human body is about 65% oxygen. If a person has a mass of 75.0 kg, what is the mass of oxygen in their body?

199. A 112-g serving of ice cream contains 19 g of fat. What percentage of the serving is fat?

200. The mass of the products produced by a chemical reaction is measured. The reaction is repeated five times, with the same mass of reactants used each time. The measured product masses are 50.17 g, 50.12 g, 50.17 g, 50.10 g, and 50. 14 g. What is the average of these measurements?

Chapter 19 Properties of Atoms and the Periodic Table

201. Boron has a mass number of 11 and an atomic number of 5. How many neutrons are in a boron atom?

202. A magnesium atom has 12 protons and 12 neutrons. What is its mass number?

203. Iodine-127 has a mass number of 127 and 74 neutrons. What percentage of the particles in an iodine-127 nucleus are protons?

204. How many neutrons are in an atom of phosphorus-31?

205. What is the ratio of neutrons to protons in the isotope radium-234?

206. About 80% of all magnesium atoms are magnesium-24, about 10% are magnesium-25, and about 10% are magnesium-26. What is the average atomic mass of magnesium?

207. The half-life of the radioactive isotope rubidium-87 is 48,800,000,000 years. Express this half-life in scientific notation.

208. The radioactive isotope nickel-63 has a half-life of 100 years. How much of a 10.0-g sample of nickel-63 is left after 300 years?

209. A sample of the radioactive isotope cobalt-62 is prepared. The sample has a mass of 1.00 g. After three minutes, the mass of cobalt-62 remaining is 0.25 g. What is the half-life of cobalt-62?

210. A neutral phosphorus atom has 15 electrons. How many electrons are in the third energy level?

Chapter 20 Earth Materials

211. If oxygen makes up 46.6 percent of the mass of Earth's crust and silicon makes up 27.7 percent, what is the total percent of the crust's mass for oxygen and silicon?

212. If oxygen and silicon make up 74.3 percent of the mass of Earth's crust, what percent of this percentage is silicon's percentage?

213. How many total atoms of aluminum and oxygen combined make up one molecule of corundum (Al_2O_3)?

214. What is the ratio of silicon to oxygen in a molecule of olivine ($(Mg, Fe)_2SiO_4$)?

215. Other than the common eight elements that make up Earth's crust, all other elements make up only 1.5 percent. If oxygen makes up 46.6 percent of Earth's crust, how many times greater is the amount of oxygen in Earth's crust than those other elements.

216. The Mohs scale of hardness consists of ten categories, each associated with a mineral standard. How many categories are between hardness 7–quartz and hardness 2–gypsum?

217. The volume of sphere is $4/3\ \pi r^3$. If a particle of gravel has a radius of 2 mm and a sand grain has a radius of 1 mm, how much bigger is the volume of the gravel?

218. If a sedimentary rock has a porosity of 15 percent, how much volume do the mineral grains and cement take up?

219. If you have a halite cube that measures 4 cm on each side, what is the total surface area of the cube?

220. What is the ratio of oxygen atoms to potassium atoms in one molecule of K-feldspar, ($KAlSiO_8$)?

Chapter 21 Earth's Changing Surface

221. If soil erosion averages 2.5 cm per year and the average soil profile is 3.2 m thick with 40 percent of topsoil, how long will it take for the topsoil to erode away?

222. The average soil profile in your area is 2.1 m thick. Topsoil erodes at 2.0 cm per year. What percent of the soil profile is topsoil if it erodes in 14 years?

223. Soil erosion in your area averages 3.5 cm per year. The average soil profile is 3.7 m thick and 35 percent of that is topsoil. Soil replacement through weathering is 0.2 cm per year. How long will it take for the topsoil to erode?

224. Observations taken at the mouth of a stream include a flow rate of 120 m^3/s and a suspended sediment load of 1.8 kg/m^3. How many kilograms of sediment potentially could drop out of this stream each day?

225. Observations taken at the mouth of a stream include a flow rate of 85 m^3/s and a suspended sediment load of 1.6 kg/m^3. How many kilograms of sediment potentially could drop out of this stream each day?

226. Suppose a community with a human population of 800 has a water consumption rate of 900 L per person per day. This community relies on an aquifer that is thought to contain 1.2 billion L of water. Assuming no change in average water consumption or significant recharge of the aquifer, how many years will this water supply last?

227. An aquifer is 400 km in length, 185 km in width, and 80 km thick. Sampling of the aquifer material shows it has an average porosity of 15 percent. What is the volume of porosity?

228. An aquifer with dimensions of 350 km in length, 175 km in width, and 65 km thick. Sampling of the aquifer material shows it has an average porosity of 10 percent. What is the volume of porosity?

229. If three half-lives for an isotope have passed, what fraction of the original isotope would be present in the igneous rock?

230. If the half-life of an isotope is 12,000 years and the amount of the isotope present in the rock is only $\frac{1}{16}$ of the original amount present, how old is the igneous rock containing the isotope?

Chapter 22 Chemical Bonds

231. What is the formula of the compound formed when ammonium ions, NH_4^+, and phosphate ions, PO_4^{3-}, combine?

232. Show that the sum of positive and negative charges in a unit of calcium chloride ($CaCl_2$) equals zero.

233. What is the formula for iron(III) oxide?

234. How many hydrogen atoms are in three molecules of ammonium phosphate, $(NH_4)_3PO_4$?

235. The overall charge on the polyatomic phosphate ion, PO_4^{3-}, is 3−. What is the oxidation number of phosphorus in the phosphate ion?

236. The overall charge on the polyatomic dichromate ion, $Cr_2O_7^{2-}$, is 2−. What is the oxidation number of chromium in this polyatomic ion?

237. What is the formula for lead(IV) oxide?

238. What is the formula for potassium chlorate?

239. What is the formula for carbon tetrachloride?

240. What percentage of the mass of a sulfuric acid molecule, H_2SO_4, is sulfur?

Chapter 23 Chemical Reactions

241. Lithium reacts with oxygen to form lithium oxide according to the equation: $4Li + O_2 \rightarrow 2Li_2O$. If 27.8 g of Li react completely with 32.0 g of O_2, how many grams of Li_2O are formed?

242. What coefficients balance the following equation: $\_Zn(OH)_2 + \_H_3PO_4 \rightarrow \_Zn_3(PO_4)_2 + \_H_2O$?

243. Aluminum hydroxide, $Al(OH)_3$, decomposes to form aluminum oxide, Al_2O_3, and water according to the reaction: $2Al(OH)_3 \rightarrow Al_2O_3 + 3H_2O$. If 156.0 g of $Al(OH)_3$ decompose to from 102.0 g of Al_2O_3, how many grams of H_2O are formed?

244. In the following balanced chemical reaction one of the products is represented by the symbol *X*: $BaCO_3 + C + H_2O \rightarrow Ba(OH)_2 + H_2O + 2X$. What is the formula for the compound represented by *X*?

245. When propane, C_3H_8, is burned, carbon dioxide and water vapor are produced according to the following reaction: $C_3H_8 + 5O_2 \rightarrow 3CO_2 + 4H_2O$. How much propane is burned if 160.0 g of O_2 are used and 132.0 g of CO_2 and 72.0 g of H_2O are produced?

246. Increasing the temperature usually causes the rate of a chemical reaction to increase. If the rate of a chemical reaction doubles when the temperature increases by 10°C, by what factor does the rate of reaction increase if the temperature increases by 30°C?

247. When acetylene gas, C_2H_2, is burned, carbon dioxide and water are produced. Find the coefficients that balance the chemical equation for the combustion of acetylene: $\_C_2H_2 + \_O_2 \rightarrow \_CO_2 + \_H_2O$.

248. What coefficients balances the following equation: $\_CS_2 + \_O_2 \rightarrow \_CO_2 + \_SO_2$?

249. When methane, CH_4, is burned, 50.1 kJ of energy per gram are released. When propane, C_3H_8, is burned, 45.8 kJ of energy are released. If a mixture of 1 g of methane and 1 g of propane is burned, how much energy is released per gram of mixture?

250. A chemical reaction produces 0.050 g of a product in 0.18 s. In the presence of a catalyst, the reaction produces 0.050 g of the same product in 0.007 s. How much faster is the rate of reaction in the presence of the enzyme?

Chapter 24 Solutions, Acids, and Bases

251. A cup of orange juice contains 126 mg of vitamin C and 1/2 cup of strawberries contain 42 mg of vitamin C. How many cups of strawberries contain as much vitamin C as one cup of orange juice?

252. What is the total surface area of a 2-cm cube?

253. A cube has 2-cm sides. If it is split in half, what is the total surface area of the two pieces?

254. At 20°C, the solubility in water of potassium bromide, KBr, is 65.3 g/100 mL. What is the maximum amount of potassium bromide that will dissolve in 237 mL of water?

255. At 20°C, the solubility of sodium chloride, NaCl, in water is 35.9 g/100 mL. If the maximum amount of sodium chloride is dissolved in 500 mL of water at 20°C, the mass of the dissolved sodium chloride is what percentage of the mass of the solution?

256. The difference between the pH of an acidic solution and the pH of pure water is 3. What is the pH of the solution?

257. The pH of rain that fell over a region had measured values of 4.6, 5.1, 4.8, 4.5, 4.5, 4.9, 4.7, and 4.8. What was the mean value of the measured pH?

258. If 5.5% of 473.0 mL of vinegar is acetic acid, how many milliliters of acetic acid are there?

259. Write the balanced chemical equation for the neutralization of H_2SO_4, sulfuric acid, by KOH, potassium hydroxide.

260. Write the balanced chemical equation for the neutralization of HBr, hydrobromic acid, by $Al(OH)_3$, aluminum hydroxide.

Chapter 25 Nuclear Changes

261. How many protons are in the nucleus $^{81}_{36}Kr$?

262. How many neutrons are in the nucleus $^{56}_{26}Fe$?

263. What is the ratio of neutrons to protons in the nucleus $^{241}_{95}Am$?

264. How many alpha particles are emitted when the nucleus $^{222}_{86}Rn$ decays to $^{218}_{84}Po$?

265. How many beta particles are emitted when the nucleus $^{40}_{19}K$ decays to the nucleus $^{40}_{20}Ca$?

266. An alpha particle is the same as the helium nucleus $^{4}_{2}He$. What nucleus is produced when the nucleus $^{226}_{88}Ra$ decays by emitting an alpha particle?

267. How long will it take a sample of $^{194}_{84}Po$ to decay to 1/8 of its original amount if $^{194}_{84}Po$ has a half-life of 0.7 s?

268. The half-life of $^{131}_{53}I$ is 8.04 days. How much time would be needed to reduce 1 g of $^{131}_{53}I$ to 0.25 g?

269. A sample of radioactive carbon-14 sample has decayed to 12.5% of its original amount. If the half-life of carbon-14 is 5730 years, how old is this sample?

270. A sample of $^{38}_{17}Cl$ is observed to decay to 25% of the original amount in 74.4 minutes. What is the half-life of $^{38}_{17}Cl$?

Chapter 26 Stars and Galaxies

271. If the focal length of a telescope's objective is 900 mm and the focal length of its eyepiece is 10 mm, what is the magnifying power of this telescope?

272. If the focal length of a telescope's objective is 700 mm and the magnifying power of the telescope is 40 times, what is the focal length of an eyepiece?

273. If it takes 25 days for a sunspot to travel once around the Sun, about how many days would it take to travel across the face of the Sun?

274. If a solar prominence blasts material from the Sun at a speed of 600 km/s, how long in hours and minutes would it take for material to arrive at Earth 150 million km away?

275. How many times larger in area is a 250-mm-diameter objective than a 100-mm diameter objective?

276. How many times larger in area is an 8-m-diameter objective than a 5-m-diameter objective?

277. How many times larger is an elliptical galaxy that is nine million light-years across than one that is 3,000 light-years across?

278. A telescope's objective has a focal length of 1,200 mm. It is used with two eyepieces that have focal lengths of 12 mm and 18 mm. How many times greater is the magnifying power of the telescope when the 12-mm eyepiece is used?

279. A telescope's objective has a focal length of 1500 mm. It is used with two eyepieces that have focal lengths of 20 mm and 12.5 mm. What percentage of the magnifying power obtained using the 12.5-mm eyepiece is achieved using the 20-mm eyepiece?

280. If the age of the universe was thought to be 20 billion years, but is now considered to be 13.7 billion years, how much younger is the universe thought to be now?

Rocks

Rocks		
Rock Type	**Rock Name**	**Characteristics**
Igneous (intrusive)	Granite	Large mineral grains of quartz, feldspar, hornblende, and mica. Usually light in color.
	Diorite	Large mineral grains of feldspar, hornblende, and mica. Less quartz than granite. Intermediate in color.
	Gabbro	Large mineral grains of feldspar, augite, and olivine. No quartz. Dark in color.
Igneous (extrusive)	Rhyolite	Small mineral grains of quartz, feldspar, hornblende, and mica, or no visible grains. Light in color.
	Andesite	Small mineral grains of feldspar, hornblende, and mica or no visible grains. Intermediate in color.
	Basalt	Small mineral grains of feldspar, augite, and possibly olivine or no visible grains. No quartz. Dark in color.
	Obsidian	Glassy texture. No visible grains. Volcanic glass. Fracture looks like broken glass.
	Pumice	Frothy texture. Floats in water. Usually light in color.
Sedimentary (detrital)	Conglomerate	Coarse grained. Gravel or pebble-size grains.
	Sandstone	Sand-sized grains 1/16 to 2 mm.
	Siltstone	Grains are smaller than sand but larger than clay.
	Shale	Smallest grains. Often dark in color. Usually platy.
Sedimentary (chemical or organic)	Limestone	Major mineral is calcite. Usually forms in oceans and lakes. Often contains fossils.
	Coal	Forms in swampy areas. Compacted layers of organic material, mainly plant remains.
Sedimentary (chemical)	Rock Salt	Commonly forms by the evaporation of seawater.
Metamorphic (foliated)	Gneiss	Banding due to alternate layers of different minerals, of different colors. Parent rock often is granite.
	Schist	Parallel arrangement of sheetlike minerals, mainly micas. Forms from different parent rocks.
	Phyllite	Shiny or silky appearance. May look wrinkled. Common parent rocks are shale and slate.
	Slate	Harder, denser, and shinier than shale. Common parent rock is shale.
Metamorphic (nonfoliated)	Marble	Calcite or dolomite. Common parent rock is limestone.
	Soapstone	Mainly of talc. Soft with greasy feel.
	Quartzite	Hard with interlocking quartz crystals. Common parent rock is sandstone.

Minerals

Minerals					
Mineral (formula)	Color	Streak	Hardness	Breakage Pattern	Uses and Other Properties
Graphite (C)	black to gray	black to gray	1–1.5	basal cleavage (scales)	pencil lead, lubricants for locks, rods to control some small nuclear reactions, battery poles
Galena (PbS)	gray	gray to black	2.5	cubic cleavage perfect	source of lead, used for pipes, shields for X rays, fishing equipment sinkers
Hematite (Fe_2O_3)	black or reddish-brown	reddish-brown	5.5–6.5	irregular fracture	source of iron; converted to pig iron, made into steel
Magnetite (Fe_3O_4)	black	black	6	conchoidal fracture	source of iron, attracts a magnet
Pyrite (FeS_2)	light, brassy, yellow	greenish-black	6–6.5	uneven fracture	fool's gold
Talc ($Mg_3 Si_4O_{10}(OH)_2$)	white, greenish	white	1	cleavage in one direction	used for talcum powder, sculptures, paper, and tabletops
Gypsum ($CaSO_4 \cdot 2H_2O$)	colorless, gray, white, brown	white	2	basal cleavage	used in plaster of paris and dry wall for building construction
Sphalerite (ZnS)	brown, reddish-brown, greenish	light to dark brown	3.5–4	cleavage in six directions	main ore of zinc; used in paints, dyes, and medicine
Muscovite ($KAl_3Si_3O_{10}(OH)_2$)	white, light gray, yellow, rose, green	colorless	2–2.5	basal cleavage	occurs in large, flexible plates; used as an insulator in electrical equipment, lubricant
Biotite ($K(Mg,Fe)_3(AlSi_3O_{10})(OH)_2$)	black to dark brown	colorless	2.5–3	basal cleavage	occurs in large, flexible plates
Halite ($NaCl$)	colorless, red, white, blue	colorless	2.5	cubic cleavage	salt; soluble in water; a preservative

Minerals

Minerals					
Mineral (formula)	**Color**	**Streak**	**Hardness**	**Breakage Pattern**	**Uses and Other Properties**
Calcite ($CaCO_3$)	colorless, white, pale blue	colorless, white	3	cleavage in three directions	fizzes when HCl is added; used in cements and other building materials
Dolomite ($CaMg(CO_3)_2$)	colorless, white, pink, green, gray, black	white	3.5–4	cleavage in three directions	concrete and cement; used as an ornamental building stone
Fluorite (CaF_2)	colorless, white, blue, green, red, yellow, purple	colorless	4	cleavage in four directions	used in the manufacture of optical equipment; glows under ultraviolet light
Hornblende ($(CaNa)_{2\text{-}3}(Mg,Al,Fe)_5\text{-}(Al,Si)_2Si_6O_{22}(OH)_2$)	green to black	gray to white	5–6	cleavage in two directions	will transmit light on thin edges; 6-sided cross section
Feldspar ($KAlSi_3O_8$) ($NaAlSi_3O_8$), ($CaAl_2Si_2O_8$)	colorless, white to gray, green	colorless	6	two cleavage planes meet at 90° angle	used in the manufacture of ceramics
Augite ($(Ca,Na)(Mg,Fe,Al)(Al,Si)_2O_6$)	black	colorless	6	cleavage in two directions	square or 8-sided cross section
Olivine ($(Mg,Fe)_2SiO_4$)	olive, green	none	6.5–7	conchoidal fracture	gemstones, refractory sand
Quartz (SiO_2)	colorless, various colors	none	7	conchoidal fracture	used in glass manufacture, electronic equipment, radios, computers, watches, gemstones

Physical Science Reference Tables

Standard Units

Symbol	Name	Quantity
m	meter	length
kg	kilogram	mass
Pa	pascal	pressure
K	kelvin	temperature
mol	mole	amount of a substance
J	joule	energy, work, quantity of heat
s	second	time
C	coulomb	electric charge
V	volt	electric potential
A	ampere	electric current
Ω	ohm	resistance

Physical Constants and Conversion Factors

Acceleration due to gravity	g	9.8 m/s/s or m/s^2
Avogadro's Number	N_A	6.02×10^{23} particles per mole
Electron charge	e	1.6×10^{-19} C
Electron rest mass	m_e	9.11×10^{-31} kg
Gravitation constant	G	6.67×10^{-11} N $\times$ m^2/kg^2
Mass-energy relationship		1 u (amu) = 9.3×10^2 MeV
Speed of light in a vacuum	c	3.00×108 m/s
Speed of sound at STP	331 m/s	
Standard Pressure	1 atmosphere 101.3 kPa 760 Torr or mmHg 14.7 lb/in.^{2}	

Wavelengths of Light in a Vacuum

Violet	$4.0 - 4.2 \times 10^{-7}$ m
Blue	$4.2 - 4.9 \times 10^{-7}$ m
Green	$4.9 - 5.7 \times 10^{-7}$ m
Yellow	$5.7 - 5.9 \times 10^{-7}$ m
Orange	$5.9 - 6.5 \times 10^{-7}$ m
Red	$6.5 - 7.0 \times 10^{-7}$ m

The Index of Refraction for Common Substances

($\lambda = 5.9 \times 10^{-7}$ m)

Air	1.00
Alcohol	1.36
Canada Balsam	1.53
Corn Oil	1.47
Diamond	2.42
Glass, Crown	1.52
Glass, Flint	1.61
Glycerol	1.47
Lucite	1.50
Quartz, Fused	1.46
Water	1.33

Heat Constants

	Specific Heat (average) (kJ/kg × °C) (J/g × °C)	Melting Point (°C)	Boiling Point (°C)	Heat of Fusion (kJ/kg) (J/g)	Heat of Vaporization (kJ/kg) (J/g)
Alcohol (ethyl)	2.43 (liq.)	−117	79	109	855
Aluminum	0.90 (sol.)	660	2467	396	10500
Ammonia	4.71 (liq.)	−78	−33	332	1370
Copper	0.39 (sol.)	1083	2567	205	4790
Iron	0.45 (sol.)	1535	2750	267	6290
Lead	0.13 (sol.)	328	1740	25	866
Mercury	0.14 (liq.)	−39	357	11	295
Platinum	0.13 (sol.)	1772	3827	101	229
Silver	0.24 (sol.)	962	2212	105	2370
Tungsten	0.13 (sol.)	3410	5660	192	4350
Water (solid)	2.05 (sol.)	0	–	334	–
Water (liquid)	4.18 (liq.)	–	100	–	–
Water (vapor)	2.01 (gas)	–	–	–	2260
Zinc	0.39 (sol.)	420	907	113	1770

Standard Units

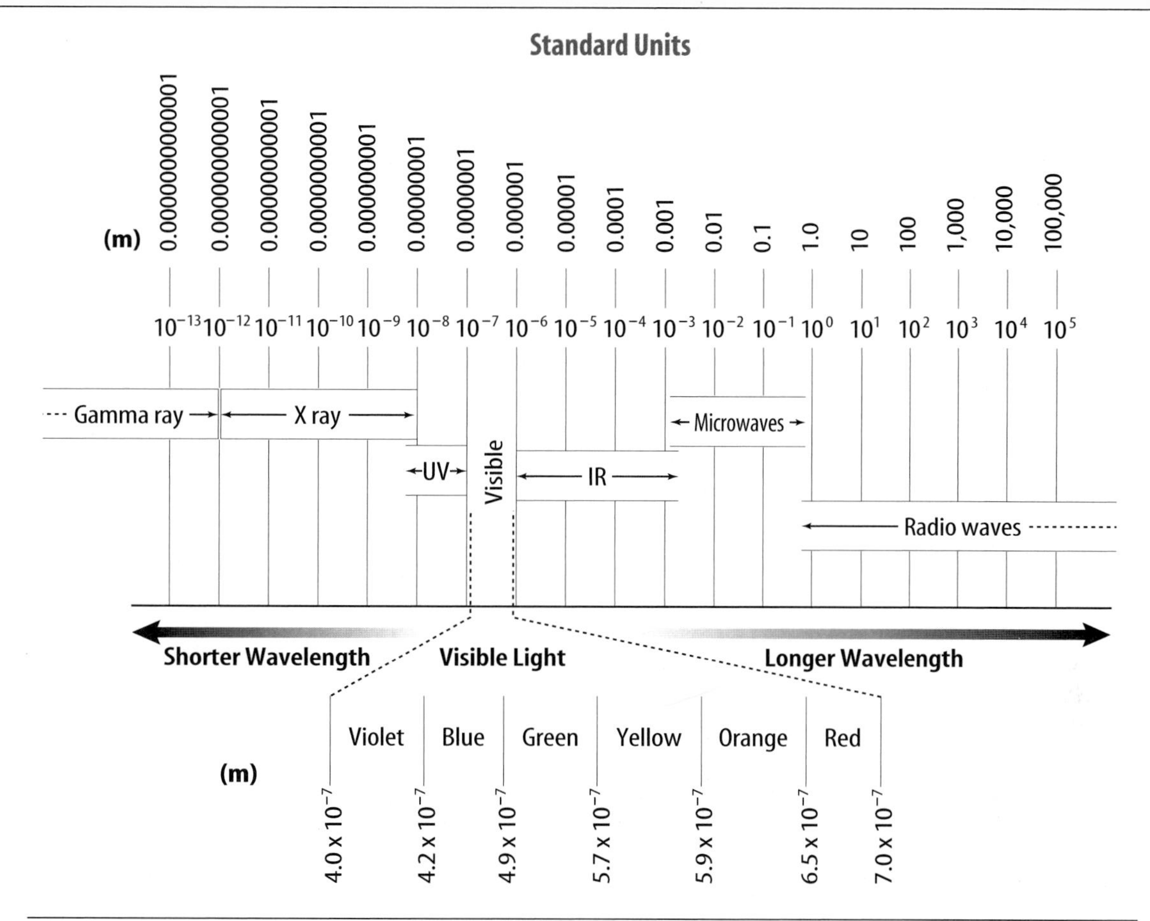

Radioactive Decay

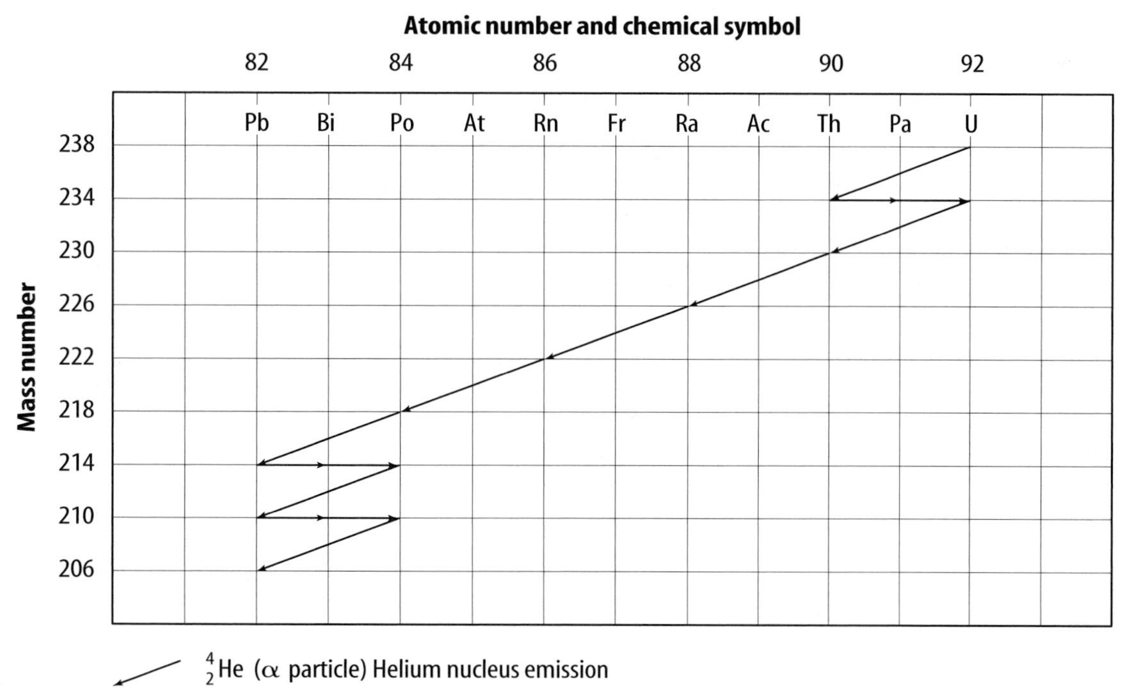

PERIODIC TABLE OF THE ELEMENTS

Columns of elements are called groups. Elements in the same group have similar chemical properties.

Gas

Liquid

Solid

Synthetic

Element — Hydrogen
Atomic number — 1
Symbol — H
Atomic mass — 1.008
State of matter

The first three symbols tell you the state of matter of the element at room temperature. The fourth symbol identifies elements that are not present in significant amounts on Earth. Useful amounts are made synthetically.

Period	1	2	3	4	5	6	7	8	9
1	Hydrogen 1 H 1.008								
2	Lithium 3 Li 6.941	Beryllium 4 Be 9.012							
3	Sodium 11 Na 22.990	Magnesium 12 Mg 24.305							
4	Potassium 19 K 39.098	Calcium 20 Ca 40.078	Scandium 21 Sc 44.956	Titanium 22 Ti 47.867	Vanadium 23 V 50.942	Chromium 24 Cr 51.996	Manganese 25 Mn 54.938	Iron 26 Fe 55.845	Cobalt 27 Co 58.933
5	Rubidium 37 Rb 85.468	Strontium 38 Sr 87.62	Yttrium 39 Y 88.906	Zirconium 40 Zr 91.224	Niobium 41 Nb 92.906	Molybdenum 42 Mo 95.94	Technetium 43 Tc (98)	Ruthenium 44 Ru 101.07	Rhodium 45 Rh 102.906
6	Cesium 55 Cs 132.905	Barium 56 Ba 137.327	Lanthanum 57 La 138.906	Hafnium 72 Hf 178.49	Tantalum 73 Ta 180.948	Tungsten 74 W 183.84	Rhenium 75 Re 186.207	Osmium 76 Os 190.23	Iridium 77 Ir 192.217
7	Francium 87 Fr (223)	Radium 88 Ra (226)	Actinium 89 Ac (227)	Rutherfordium 104 Rf (261)	Dubnium 105 Db (262)	Seaborgium 106 Sg (266)	Bohrium 107 Bh (264)	Hassium 108 Hs (277)	Meitnerium 109 Mt (268)

The number in parentheses is the mass number of the longest-lived isotope for that element.

Rows of elements are called periods. Atomic number increases across a period.

The arrow shows where these elements would fit into the periodic table. They are moved to the bottom of the table to save space.

Lanthanide series	Cerium 58 Ce 140.116	Praseodymium 59 Pr 140.908	Neodymium 60 Nd 144.24	Promethium 61 Pm (145)	Samarium 62 Sm 150.36
Actinide series	Thorium 90 Th 232.038	Protactinium 91 Pa 231.036	Uranium 92 U 238.029	Neptunium 93 Np (237)	Plutonium 94 Pu (244)

Metal
Metalloid
Nonmetal

The color of an element's block tells you if the element is a metal, nonmetal, or metalloid.

Science Online
Visit gpescience.com for updates to the periodic table.

10	11	12	13	14	15	16	17	18
								Helium 2 He 4.003
			Boron 5 B 10.811	Carbon 6 C 12.011	Nitrogen 7 N 14.007	Oxygen 8 O 15.999	Fluorine 9 F 18.998	Neon 10 Ne 20.180
			Aluminum 13 Al 26.982	Silicon 14 Si 28.086	Phosphorus 15 P 30.974	Sulfur 16 S 32.065	Chlorine 17 Cl 35.453	Argon 18 Ar 39.948
Nickel 28 Ni 58.693	Copper 29 Cu 63.546	Zinc 30 Zn 65.409	Gallium 31 Ga 69.723	Germanium 32 Ge 72.64	Arsenic 33 As 74.922	Selenium 34 Se 78.96	Bromine 35 Br 79.904	Krypton 36 Kr 83.798
Palladium 46 Pd 106.42	Silver 47 Ag 107.868	Cadmium 48 Cd 112.411	Indium 49 In 114.818	Tin 50 Sn 118.710	Antimony 51 Sb 121.760	Tellurium 52 Te 127.60	Iodine 53 I 126.904	Xenon 54 Xe 131.293
Platinum 78 Pt 195.078	Gold 79 Au 196.967	Mercury 80 Hg 200.59	Thallium 81 Tl 204.383	Lead 82 Pb 207.2	Bismuth 83 Bi 208.980	Polonium 84 Po (209)	Astatine 85 At (210)	Radon 86 Rn (222)
Darmstadtium 110 Ds (281)	Roentgenium 111 Rg (272)	Ununbium * 112 Uub (285)		Ununquadium * 114 Uuq (289)				

* The names and symbols for elements 112 and 114 are temporary. Final names will be selected when the elements' discoveries are verified.

Europium 63 Eu 151.964	Gadolinium 64 Gd 157.25	Terbium 65 Tb 158.925	Dysprosium 66 Dy 162.500	Holmium 67 Ho 164.930	Erbium 68 Er 167.259	Thulium 69 Tm 168.934	Ytterbium 70 Yb 173.04	Lutetium 71 Lu 174.967
Americium 95 Am (243)	Curium 96 Cm (247)	Berkelium 97 Bk (247)	Californium 98 Cf (251)	Einsteinium 99 Es (252)	Fermium 100 Fm (257)	Mendelevium 101 Md (258)	Nobelium 102 No (259)	Lawrencium 103 Lr (262)

Weather Map Symbols

Sample Station Model

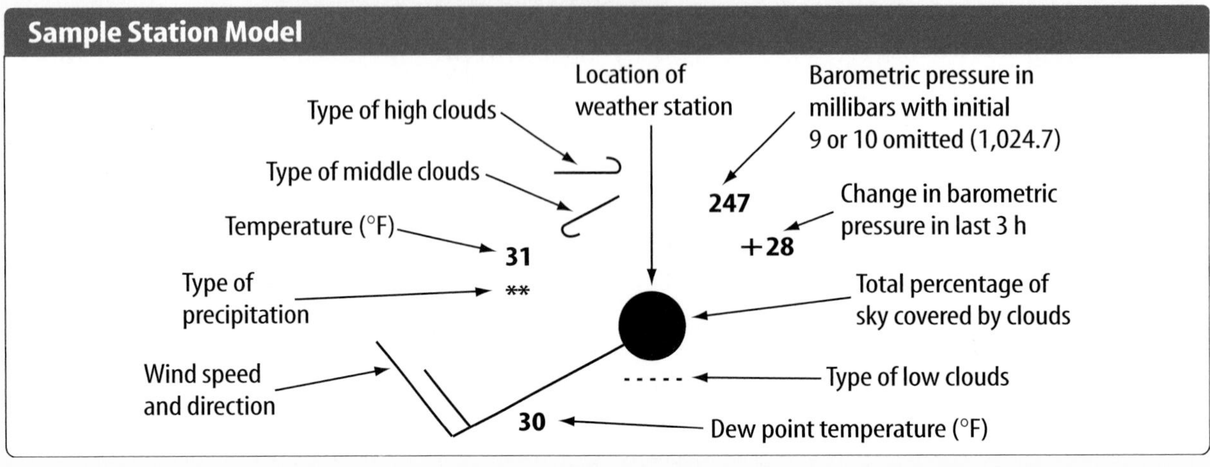

Sample Plotted Report at Each Station

Precipitation	Wind Speed and Direction	Sky Coverage	Some Types of High Clouds
Fog	0 calm	No cover	Scattered cirrus
Snow	1–2 knots	1/10 or less	Dense cirrus in patches
Rain	3–7 knots	2/10 to 3/10	Veil of cirrus covering entire sky
Thunderstorm	8–12 knots	4/10	Cirrus not covering entire sky
Drizzle	13–17 knots	–	
Showers	18–22 knots	6/10	
	23–27 knots	7/10	
	48–52 knots	Overcast with openings	
	1 knot = 1.852 km/h	Completely overcast	

Some Types of Middle Clouds	Some Types of Low Clouds	Fronts and Pressure Systems	
Thin altostratus layer	Cumulus of fair weather	(H) or High (L) or Low	Center of high- or low-pressure system
Thick altostratus layer	Stratocumulus		Cold front
Thin altostratus in patches	Fractocumulus of bad weather		Warm front
Thin altostratus in bands	Stratus of fair weather		Occluded front
			Stationary front

A multilingual science glossary at gpescience.com includes Arabic, Bengali, Chinese, English, Haitian Creole, Hmong, Korean, Portuguese, Russian, Tagalog, Urdu, and Vietnamese.

Cómo usar el glosario en español:

1. Busca el término en inglés que desees encontrar.
2. El término en español, junto con la definición, se encuentran en la columna de la derecha.

Pronunciation Key

Use the following key to help you sound out words in the glossary.

a	**back** (BAK)	**ew**	**food** (FEWD)
ay	**day** (DAY)	**yoo**	**pure** (PYOOR)
ah	**father** (FAH thur)	**yew**	**few** (FYEW)
ow	**flower** (FLOW ur)	**uh**	**comma** (CAH muh)
ar	**car** (CAR)	**u (1 con)**	**rub** (RUB)
e	**less** (LES)	**sh**	**shelf** (SHELF)
ee	**leaf** (LEEF)	**ch**	**nature** (NAY chur)
ih	**trip** (TRIHP)	**g**	**gift** (GIHFT)
i (i 1 con 1 e)	**idea** (i DEE uh)	**j**	**gem** (JEM)
oh	**go** (GOH)	**ing**	**sing** (SING)
aw	**soft** (SAWFT)	**zh**	**vision** (VIH zhun)
or	**orbit** (OR buht)	**k**	**cake** (KAYK)
oy	**coin** (COYN)	**s**	**seed, cent** (SEED, SENT)
oo	**foot** (FOOT)	**z**	**zone, raise** (ZOHN, RAYZ)

English — **A** — Español

absolute dating: process of assigning a precise numerical age to an organism, object, or event, based on an absolute reference. (p. 669)

acceleration: rate of change of velocity; can be calculated by dividing the change in the velocity by the time it takes the change to occur. (p. 76)

accuracy: compares a measurement to the real or accepted value. (p. 14)

acid: any substance that produces hydrogen ions, H^+, in a water solution. (p. 764)

activation energy: the minimum amount of energy needed to start a reaction (p. 735)

agricultural biotechnology: scientific techniques, such as genetic engineering, used to increase farm crop yields and nutritional values of foods by creating, modifying, or improving plants, animals, and microorganisms. (p. 45)

air resistance: force that opposes the motion of objects that move through the air. (p. 85)

alpha particle: particle consisting of two protons and two neutrons that is emitted from a decaying atomic nucleus. (p. 791)

alternating current (AC): electric current that reverses its direction of flow in a regular pattern. (p. 442)

amplitude: maximum distance a wave causes the particles in a medium to move from the rest position. (p. 298)

datación absoluta: proceso de asignar una edad numérica precisa a un organismo, objeto o evento, basado en una referencia absoluta. (p. 669)

aceleración: tasa de cambio de la velocidad; se calcula dividiendo el cambio en la velocidad por el tiempo que toma para que ocurra el cambio. (p. 76)

exactitud: comparación de una medición con el valor real o aceptado. (p. 14)

ácido: sustancia que produce iones de hidrógeno, H^+, en una solución de agua. (p. 764)

energía de activación: cantidad mínima de energía necesaria para comenzar una reacción. (p. 735)

biotecnología agrícola: técnicas científicas, como la ingeniería genética, utilizadas para incrementar la producción de las cosechas de las granjas y los valores nutricionales de los alimentos al crear, modificar o mejorar las plantas, animales y microorganismos. (p. 45)

resistencia del aire: fuerza que se opone al movimiento de los objetos que se mueven por el aire. (p. 85)

partícula alfa: partícula compuesta por dos protones y dos neutrones y que es emitida por un núcleo atómico en descomposición. (p. 791)

corriente alterna (CA): corriente eléctrica que invierte su dirección de flujo en un patrón regular. (p. 442)

amplitud: distancia máxima a la que una onda causa que las partículas en un medio salgan de su posición de reposo. (p. 298)

aqueous solution: a solution in which water is the solvent. (p. 753)

aquifer: rock unit that transmits water through its pore space. (p. 664)

asteroids: rocky solar system objects of widely varying size usually found between the orbits of Mars and Jupiter in an area called the asteroid belt. (p. 236)

asthenosphere: weaker, semi-solid, plasticlike layer beneath Earth's lithosphere on which lithospheric plates move. (p. 372)

astronomical unit: about 150 million km, which equals the average distance from Earth to the Sun; used to measure distances within the solar system. (p. 220)

atom: the smallest particle of an element that still retains the properties of the element. (p. 589)

atomic number: number of protons in an atom's nucleus. (p. 585)

average atomic mass: weighted-average mass of the mixture of an element's isotopes. (p. 587)

average speed: total distance an object travels divided by the total time it takes to travel that distance. (p. 72)

Solución acuosa: solución en la que el agua es el solvente. (p. 753)

Acuífero: unidad rocosa que conduce agua a través de su espacio poroso. (p. 664)

Asteroides: objetos rocosos del sistema solar que varían ampliamente en tamaño y usualmente se encuentran entre las órbitas de Marte y Júpiter en un área llamada el cinturón de asteroides. (p. 236)

astenosfera: capa débil, semisólida, similar al plástico que se encuentra por debajo de la litosfera terrestre en la cual se mueven las placas litosféricas. (p. 372)

unidad astronómica: cerca de 150 millones de km, lo cual equivale a la distancia promedio de la tierra al sol; utilizada para medir distancias entre el sistema solar. (p. 220)

átomo: la partícula más pequeña de un elemento que mantiene las propiedades del elemento. (p. 589)

número atómico: número de protones en el núcleo de un átomo. (p. 585)

masa atómica promedio: masa de peso promedio resultado de la mezcla de los isótopos de un elemento. (p. 587)

velocidad promedio: distancia que recorre un objeto dividida por el tiempo que dura en recorrer dicha distancia. (p. 72)

B

balanced chemical equation: chemical equation with the same number of atoms of each element on both sides of the equation. (p. 726)

balanced forces: forces on a object that combine to give a zero net force and do not change the motion of the object. (p. 82)

base: any substance that forms hydroxide ions, OH^-, in a water solution. (p. 766)

beta particle: electron that is emitted from a decaying atomic nucleus. (p. 793)

bias: occurs when a scientist's expectations change how the results of an experiment are viewed. (p. 10)

big bang theory: the theory that the universe started with a big bang, or explosion, and has been expanding ever since. (p. 837)

binary compound: compound that is composed of two elements. (p. 703)

biomass: renewable organic matter from plants and animals, such as wood and animal manure, that can be burned to provide heat. (p. 506)

ecuación química Fnceada: ecuación química con el mismo número de átomos de cada elemento en los dos lados de la ecuación. (p. 726)

fuerzas equilibradas: fuerzas en un objeto que se combinan para dar una fuerza neta de cero y no cambiar el movimiento del objeto. (p. 82)

base: sustancia que forma iones de hidróxido, OH^-, en una solución de agua. (p. 766)

partícula beta: electrón emitido por un núcleo atómico en descomposición. (p. 793)

predisposición: ocurre cuando las expectativas de un científico cambian la forma en que son vistos los resultados de un experimento. (p. 10)

teoría de la gran explosión: teoría que explica que el universo comenzó con una gran explosión y que se ha estado expandiendo desde entonces. (p. 837)

compuesto binario: compuesto conformado por dos elementos. (p. 703)

biomasa: materia orgánica renovable proveniente de plantas y animales, tales como madera y estiércol animal, que puede ser incinerada para producir calor. (p. 506)

biosphere: everything organic, including plants, animals, and humans. (p. 529)

bubble chamber: radiation detector, consisting of a container of superheated liquid under high pressure, that is used to detect the paths of charged particles. (p. 797)

biosfera: cualquier entidad orgánica, incluyendo plantas, animales y humanos. (p. 529)

cámara de burbujas: detector de radiación que consiste de un contenedor de un líquido sobrecalentado a alta presión, usado para detectar la trayectoria de las partículas cargadas. (p. 797)

C

carrier wave: specific frequency that a radio station is assigned and uses to broadcast signals. (p. 469)

catalyst: substance that speeds up a chemical reaction without being permanently changed itself. (p. 740)

cathode-ray tube: sealed vacuum tube that produces one or more beams of electrons that produce an image when they strike the coating on the inside of a TV screen. (p. 472)

cementation: sedimentary rock-forming process in which minerals precipitate out of water and fill the spaces between clasts. (p. 625)

centripetal acceleration: acceleration of an object toward the center of a curved or circular path. (p. 110)

centripetal force: a net force that is directed toward the center of a curved or circular path. (p. 110)

chain reaction: ongoing series of fission reactions. (p. 802)

charging by contact: process of transferring charge between objects by touching or rubbing. (p. 395)

charging by induction: process of rearranging electrons on a neutral object by bringing a charged object close to it. (p. 396)

chemical bond: force that holds atoms together in a compound. (p. 694)

chemical change: change of one substance into a new substance. (p. 564)

chemical equation: shorthand method to describe chemical reactions using chemical formulas and other symbols. (p. 723)

chemical formula: chemical shorthand that uses symbols to tell what elements are in a compound and their ratios. (p. 689)

chemical potential energy: energy stored in chemical bonds. (p. 131)

chemical property: any characteristic of a substance, such as flammability, that indicates whether it can undergo a certain chemical change. (p. 563)

chemical reaction: process in which one or more substances are changed into new substances. (p. 720)

onda transportadora: frecuencia específica que se le asigna a una estación de radio y que la usa para emitir señales. (p. 469)

catalizador: sustancia que acelera una reacción química sin cambiar el mismo permanentemente. (p. 740)

tubo de rayos catódicos: tubo vacío sellado que produce uno o más haces de electrones para producir una imagen al chocar con el revestimiento del interior de una pantalla de televisor. (p. 472)

cementación: proceso de formación de rocas sedimentarias en el que los minerales se precipitan del agua y rellenan los espacios entre los clastos. (p. 625)

aceleración centrípeta: aceleración de un objeto dirigida hacia el centro de un trayecto curvo o circular. (p. 110)

fuerza centrípeta: fuerza neta dirigida hacia el centro de un trayecto curvo o circular. (p. 110)

reacción en cadena: serie continua de reacciones de fisión. (p. 802)

carga por contacto: proceso de transferir carga entre objetos por contacto o frotaciòn. (p. 395)

carga por inducción: proceso de redistribución de los electrones en un objeto neutro acercándoles un objeto con carga. (p. 396)

enlace químico: fuerza que mantiene a los átomos juntos dentro de un compuesto. (p. 694)

cambio químico: transformación de una sustancia en una nueva sustancia. (p. 564)

ecuación química: método simplificado para describir reacciones químicas usando fórmulas químicas y otros símbolos. (p. 723)

fórmula química: nomenclatura química que usa símbolos para expresar qué elementos están en un compuesto y en qué proporción. (p. 689)

energía química potencial: energía almacenada en los enlaces químicos. (p. 131)

propiedad química: cualquier característica de una sustancia, como por ejemplo la combustibilidad, que indique si puede someterse a determinado cambio químico. (p. 563)

reacción química: proceso en el cual una o más sustancias son cambiadas por nuevas sustancias. (p. 720)

cinder cone volcano: small, violently erupting volcano formed by accumulation of large pyroclastic materials around a vent. (p. 377)

volcán de cono de ceniza: volcán pequeño, de erupción violenta formado por la acumulación de materiales piroclásticos grandes alrededor de una chimenea. (p. 377)

circuit: closed conducting loop through which an electric current can flow. (p. 401)

circuito: circuito conductor cerrado a través del cual puede fluir una corriente eléctrica. (p. 401)

clasts: bits and pieces of rock that vary widely in size and composition, are broken down by weathering, and may become consolidated into sedimentary rock. (p. 624)

clastos: toda clase de objetos rocosos que varían ampliamente en tamaño y composición, son degradados por el clima y pueden consolidarse en rocas sedimentarias. (p. 624)

cleavage: physical property of a mineral that causes it to break along planes that cut across relatively weak chemical bonds, creating a well-defined, smooth, flat surface. (p. 610)

clivaje: propiedad física de un mineral que causa que éste se rompa en planos que toman el camino más corto entre los enlaces químicos relativamente débiles, creando una superficie bien definida, suave y plana. (p. 610)

cloud chamber: radiation detector that uses water or ethanol vapor to detect the paths of charged particles. (p. 796)

cámara de vapor: detector de radiaciones que usa vapor de agua o de etanol para detectar la trayectoria de las partículas cargadas. (p. 796)

colloid (KAHL oyd): heterogeneous mixture whose particles never settle. (p. 556)

coloide: mezcla heterogénea cuyas partículas nunca se sedimentan. (p. 556)

combustion reaction: a type of chemical reaction that occurs when a substance reacts with oxygen to produce energy in the form of heat and light. (p. 730)

reacción de combustión: un tipo de reacción química que ocurre cuando una sustancia reacciona con oxígeno para producir energía en forma de calor y luz. (p. 730)

comet: mass of dust, rock particles, frozen water, methane, and ammonia that travels through space and develops a bright, distinctive tail of light as it approaches the Sun and is pushed on by the solar wind. (p. 236)

cometa: masa de polvo, partículas de roca, agua congelada, metano y amoniaco que viaja a través del espacio; desarrolla una cauda de luz brillante y distintiva a medida que se acerca al sol, y es empujada por vientos solares. (p. 236)

composite volcano: large volcano formed by alternating lava flows and violently erupting pyroclastic materials. (p. 378)

volcán compuesto: volcán grande formado por flujos alternantes de lava y materiales piroclásticos erupcionados violentamente. (p. 378)

compound: substance formed from two or more elements in which the exact combination and proportion of elements is always the same. (p. 554)

compuesto: sustancia formada por dos o más elementos en la que la combinación y proporción exacta de los elementos es siempre la misma. (p. 554)

compound machine: machine that is a combination of two or more simple machines. (p. 174)

máquina compuesta: máquina compuesta por dos o más máquinas simples. (p. 174)

compressional wave: a wave for which the matter in the medium moves back and forth along the direction that the wave travels. (p. 290)

onda de compresión: onda para la cual la materia en el medio se mueve hacia adelante y hacia atrás en la dirección en que viaja la onda. (p. 290)

computer simulation: performance-testing method using a computer to imitate the process or procedure or to gather data. (p. 55)

simulación computarizada: método de prueba del desempeño utilizando una computadora para imitar el proceso o procedimiento o para recopilar datos. (p. 55)

concave lens: a lens that is thicker at the edges than in the middle; causes light rays to diverge and forms reduced, upright, virtual images; usually used in combination with other lenses. (p. 335)

lente cóncavo: lente que es más delgado en los bordes que en el centro; hace que los rayos de luz se desvíen y forma imágenes reducidas, verticales y virtuales, generalmente utiliza en combinación con otros lentes. (p. 335)

concave mirror: a reflective surface that curves inward and can magnify objects or create beams of light. (p. 332)

espejo cóncavo: superficie reflexiva que se curva hacia el interior y que puede amplificar los objetos o crear rayos de luz. (p. 332)

concentration: how much solute is present in a solution compared to the amount of solvent. (p. 706)

concentración: cantidad de soluto que está presente en una solución en comparación con la cantidad de solvente. (p. 706)

conduction: transfer of thermal energy by collisions between particles in matter at a higher temperature and particles in matter at a lower temperature. (p. 266)

conducción: transferencia de energía térmica por colisiones entre partículas de materia a una temperatura alta y partículas de materia a una temperatura más baja. (p. 266)

conductor: material, such as copper wire, in which electrons can move easily. (p. 395)

conductor: material, como el alambre de cobre, a través del cual los electrones se pueden mover con facilidad. (p. 395)

constant: in an experiment, a variable that does not change when other variables change. (p. 9)

constante: en un experimento, una variable que no cambia cuando cambian otras variables. (p. 9)

constellation: star pattern that appears to form images, is used by astronomers to locate and name stars, and often is named for a mythological figure. (p. 818)

constelación: patrón estelar que parece formar imágenes; es usado por los astrónomos para ubicar y nombrar las estrellas, y con frecuencia se denomina con base en una figura mitológica (p. 818)

constraints: design limitations placed on products by outside factors, such as available materials, cost, and environmental impact. (p. 55)

restricciones: limitaciones de diseño impuestas en los productos por factores externos, como los materiales disponibles, el costo y el impacto ambiental. (p. 55)

continental climate: climate with little direct ocean influence and steep temperature gradients. (p. 532)

clima continental: clima con una pequeña influencia oceánica directa y gradientes abruptos de temperatura. (p. 532)

control: standard used for comparison of test results in an experiment. (p. 9)

control: estándar usado para la comparación de resultados de pruebas en un experimento. (p. 9)

control system: device or collection of devices used to monitor a system and limit system failures. (p. 56)

sistema de control: dispositivo o conjunto de dispositivos utilizados para monitorear un sistema y limitar sus fallas. (p. 56)

convection: transfer of thermal energy in a fluid from one place to another. (p. 267)

convección: transferencia de energía térmica en un fluido de un lugar a otro. (p. 267)

convergent boundary: plate tectonic boundary where lithospheric plates collide. (p. 539)

límite convergente: límite de la placa tectónica donde las placas litosféricas colisionan. (p. 539)

convex lens: a lens that is thicker in the middle than at the edges and can form real or virtual images. (p. 333)

lente convexo: lente que es más delgado en el centro que en los bordes y que puede formar imágenes reales o virtuales. (p. 333)

convex mirror: a reflective surface that curves outward and forms a reduced, upright, virtual image. (p. 333)

espejo convexo: una superficie reflexiva que se curva hacia el exterior y forma una imagen reducida, vertical y virtual. (p. 333)

cosmology: study of how the universe began, what it is made of, and how it continues to evolve. (p. 836)

cosmología: estudio del origen del universo, de su composición y de la forma como continúa evolucionando. (p. 836)

covalent bond: attraction formed between atoms when they share electrons. (p. 697)

enlace covalente: atracción formada entre átomos que comparten electrones. (p. 697)

crest: the highest points on a transverse wave. (p. 294)

cresta: los puntos más altos en una onda transversal. (p. 294)

critical mass: amount of fissionable material required so that each fission reaction produces approximately one more fission reaction. (p. 802)

masa crítica: cantidad de material fisionable requerido de manera que cada reacción de fisión produzca aproximadamente una reacción de fisión adicional. (p. 802)

D

dark energy: energy that might be causing accelerated expansion of the universe. (p. 837)

energía oscura: energía que podría estar causando la expansión acelerada del universo. (p. 837)

dark matter: unseen mass that adds to the gravity of a galaxy, but cannot be detected or seen. (p. 838)

materia oscura: masa no vista que se suma a la gravedad de una galaxia pero que no puede detectarse o verse. (p. 838)

decibel: unit for sound intensity; abbreviated dB. (p. 322)

decibel: unidad que mide la intensidad del sonido; se abrevia dB. (p. 322)

decomposition reaction: chemical reaction in which one substance breaks down into two or more substances. (p. 731)

reacción de descomposición: reacción química en la cual una sustancia se descompone en dos o más sustancias. (p. 731)

density: mass per unit volume of a material. (p. 19)

densidad: masa por unidad de volumen de un material. (p. 19)

dependent variable: factor that changes as a result of changes in the other variables. (p. 9)

variable dependiente: factor que varía como resultado de los cambios en las otras variables. (p. 9)

deposition: process in which eroded materials are dropped by their transporting agents. (p. 654)

deposición: proceso en el que los materiales erosionados son decantados a partir de sus agentes transportadores. (p. 654)

diffraction: the bending of waves around an obstacle; can also occur when waves pass through a narrow opening. (p. 304)

difracción: curvatura de las ondas alrededor de un obstáculo, la cual también puede ocurrir cuando éstas pasan a través de una abertura angosta. (p. 304)

direct current (DC): electric current that flows in only one direction. (p. 442)

corriente directa (CD): corriente eléctrica que fluye en una sola dirección. (p. 442)

discontinuity: boundary marking an abrupt density change between Earth's layers. (p. 371)

discontinuidad: límite que marca un cambio abrupto de densidad entre las capas de la Tierra. (p. 371)

displacement: distance and direction of an object's change in position from the starting point. (p. 71)

desplazamiento: distancia y dirección del cambio de posición de un objeto desde el punto inicial. (p. 71)

distance: how far an object moves. (p. 71)

distancia: qué tan lejos se mueve un objeto. (p. 71)

distillation: process that can separate two substances in a mixture by evaporating a liquid and recondensing its vapor. (p. 563)

destilación: proceso que puede separar dos sustancias de una mezcla por medio de la evaporación de un líquido y la recondensación de su vapor. (p. 563)

divergent boundary: plate tectonic boundary where lithospheric plates are moving apart. (p. 358)

límite divergente: límite de las placas tectónicas donde las placas litosféricas se alejan. (p. 358)

Doppler effect: change in pitch or frequency that occurs when a source of a sound is moving relative to a listener. (p. 330)

efecto Doppler: cambio en la altura o frecuencia que ocurre cuando una fuente de sonido se mueve en relación con un oyente. (p. 330)

double-displacement reaction: chemical reaction that produces a precipitate, water, or a gas when two ionic compounds in solution are combined. (p. 732)

reacción de doble desplazamiento: reacción química que produce un precipitado, agua o gas cuando se combinan dos compuestos iónicos en una solución. (p. 732)

drainage basin: land area that gathers water for a major river. (p. 655)

cuenca de drenaje: área terrestre que capta agua para un río grande. (p. 655)

E

Earth: third planet from the Sun; the only planet known to support life and to have water on its surface as a gas, a liquid, and a solid. (p. 225)

tierra: tercer planeta más cercano al ; el único planeta que se sabe que alberga vida y que tiene agua en su superficie en estado sólido, líquido y gaseoso. (p. 225)

ecliptic: yearly path of Earth around the Sun; also, the Sun's apparent path through the zodiac. (p. 192)

efficiency: ratio of the output work done by the machine to the input work done on the machine, expressed as a percentage. (p. 164)

elastic potential energy: energy stored when an object is compressed or stretched. (p. 131)

elastic rebound: sudden energy release that accompanies fault movement and causes earthquakes, or seismic vibrations. (p. 363)

electric current: the net movement of electric charges in a single direction, measured in amperes (A). (p. 400)

electric motor: device that converts electrical energy to mechanical energy by using the magnetic forces between an electromagnet and a permanent magnet to make a shaft rotate. (p. 435)

electric power: rate at which electrical energy is converted to another form of energy; expressed in watts (W). (p. 410)

electromagnet: temporary magnet made by wrapping a wire coil, carrying a current, around an iron core. (p. 432)

electromagnetic induction: process in which electric current is produced in a wire loop by a changing magnetic field. (p. 438)

electromagnetic waves: waves created by vibrating electric charges, can travel through a vacuum or through matter, and have a wide variety of frequencies and wavelengths. (p. 456)

electron cloud: area around the nucleus of an atom where the atom's electrons are most likely to be found. (p. 583)

electron dot diagram: uses the symbol for an element and dots representing the number of electrons in the element's outer energy level. (p. 594)

electrons: particles surrounding the center of an atom that have a charge of 1−. (p. 579)

element: substance with atoms that are all alike. (p. 552)

ellipse: elongated, closed curve with two foci; shape of Earth's orbit around the Sun. (p. 188)

El Niño: a warming of the Pacific Ocean every three to five years, which dramatically alters worldwide weather patterns, and occurs when prevailing trade winds weaken near the equator. (p. 539)

eclíptico: trayectoria anual de la tierra alrededor del Sol; también, la trayectoria aparente del Sol a través del zodíaco. (p. 192)

eficiencia: relación del trabajo efectuado por una máquina y el trabajo hecho en ésta, expresada en porcentaje. (p. 164)

energía elástica potencial: energía almacenada cuando un objeto es comprimido o estirado. (p. 131)

rebote elástico: liberación súbita de energía que acompaña al movimiento de las fallas y que causa terremotos o vibraciones sísmicas. (p. 363)

corriente eléctrica: movimiento neto de cargas eléctricas en una sola dirección, medido en amperios (A). (p. 400)

motor eléctrico: dispositivo que convierte la energía eléctrica en energía mecánica usando las fuerzas magnéticas entre un electroimán y un imán permanente para que el eje gire. (p. 435)

potencia eléctrica: proporción a la cual la energía eléctrica se convierte en otra forma de energía; se expresa en vatios (V). (p. 410)

electroimán: imán temporal que se hace envolviendo una bobina de cable que conduce una corriente, alrededor de un núcleo de hierro. (p. 432)

inducción electromagnética: proceso en el cual una corriente eléctrica es producida en un circuito cerrado de cable mediante un campo magnético cambiante. (p. 438)

ondas electromagnéticas: ondas creadas por la vibración de cargas eléctricas, que pueden viajar a través del vacío o de la materia y que tienen una amplia variedad de frecuencias y de longitudes de onda. (p. 456)

nube de electrones: área alrededor del núcleo de un átomo en donde hay más probabilidad de encontrar los electrones de los átomos. (p. 583)

diagrama de punto de electrones: usa el símbolo de un elemento y puntos que representan el número de electrones en el nivel de energía externo del elemento. (p. 594)

electrones: partículas que rodean el centro de un átomo que tienen la carga de 1−. (p. 579)

elemento: sustancia en la cual todos los átomos son iguales. (p. 552)

elipse: curva alargada y cerrada con dos distancias focales; forma de la órbita terrestre alrededor del Sol. (p. 188)

El Niño: calentamiento del Océano Pacífico cada tres a cinco años, el cual altera dramáticamente los patrones climáticos en todo del mundo y que ocurre cuando los vientos alisios se debilitan cerca del ecuador. (p. 539)

endothermic reaction: chemical reaction that requires heat energy in order to proceed. (p. 735)

energy: the ability to do work and cause a change to occur. (p. 128)

engineer: researcher who uses scientific information or ideas to solve problems or human needs and bring technology to consumers. (p. 53)

entropy: a measure of how spread out, or dispersed energy is. (p. 276)

epicenter: point on Earth's surface directly above an earthquake's focus. (p. 364)

equinox: occurs twice yearly (March and September), when the Sun is directly above Earth's equator, and the number of daylight hours equals the number of nighttime hours worldwide. (p. 195)

erosion: removal of surface material through weathering by agents such as wind and water. (p. 654)

exothermic reaction: chemical reaction in which energy is primarily given off in the form of heat. (p. 738)

experiment: organized procedure for testing a hypothesis; tests the effect of one thing on another under controlled conditions. (p. 8)

extrasolar planet: planet in orbit around another star. (p. 222)

extraterrestrial life: life on other worlds. (p. 239)

extrusive igneous rock: rock formed from rapidly cooling lava that has erupted at Earth's surface or formed from consolidated solid materials expelled from volcanoes. (p. 620)

reacción endotérmica: reacción química que requiere energía de calor para proceder. (p. 735)

energía: capacidad de realizar un trabajo y hacer que ocurra un cambio. (p. 128)

ingeniero: investigador que utiliza información científica o ideas para solucionar problemas o necesidades humanas y brindar tecnología a los consumidores. (p. 53)

entropía: una medida del grado de diseminación o dispersión de la energía. (p. 276)

epicentro: punto de la superficie terrestre directamente encima del foco de un terremoto. (p. 364)

equinoccio: fenómeno que ocurre dos veces al año (marzo y septiembre), cuando el Sol está directamente sobre el ecuador terrestre y en el que el número de horas en el día es igual al número de horas en la noche en todo el mundo. (p. 195)

erosión: remoción del material superficial a través del desgaste por agentes como el viento y el agua. (p. 654)

reacción exotérmica: reacción química en la cual la energía es inicialmente emitida en forma de calor. (p. 738)

experimento: procedimiento organizado para probar una hipótesis; prueba el efecto de una cosa sobre otra bajo condiciones controladas. (p. 8)

planeta extrasolar: planeta en órbita alrededor de otra estrella. (p. 222)

vida extraterrestre: vida en otros mundos. (p. 239)

roca ígnea extrusiva: roca formada a partir de lava que fue erupcionada a la superficie terrestre y que se enfrió rápidamente o formada a partir de materiales sólidos consolidados expulsados de los volcanes. (p. 620)

F

fault: crack in Earth's crust along which movement has taken place. (p. 364)

first law of motion: states that if the net force on an object is zero, then an object at rest remains at rest, or, if the object is moving, it continues moving with constant velocity. (p. 98)

first law of thermodynamics: states that the increase in thermal energy of a system equals the work done on the system plus the heat added to the system. (p. 274)

focus: point of origin of an earthquake. (p. 363)

foliated: type of metamorphic rock whose grains are arranged in layers or bands. (p. 631)

falla: fisura en la corteza terrestre a lo largo de la cual ha tenido lugar un movimiento. (p. 364)

primera ley del movimiento: ley que establece que si la fuerza neta sobre un objeto es cero, entonces un objeto en reposo permanece en reposo, o, si el objeto está moviéndose, éste continúa moviéndose con una velocidad constante. (p. 98)

primera ley de la termodinámica: establece que el aumento en la energía térmica de un sistema es igual al trabajo realizado sobre el sistema más el calor agregado a éste. (p. 274)

foco: punto de origen de un terremoto. (p. 363)

foliado: tipo de roca metamórfica cuyos granos están dispuestos en capas o bandas. (p. 631)

force: a push or pull exerted on an object. (p. 81)

fuerza: impulso o tracción sobre un objeto. (p. 81)

fossil fuels: oil, natural gas, and coal; formed from the decayed remains of ancient plants and animals. (p. 487)

combustibles fósiles: petróleo, gas natural y carbón formados por los restos descompuestos de plantas y animales ancestrales. (p. 487)

fossils: remains or traces of organisms found in the geologic rock record; can be direct remains such as bones, mold and cast formations, and trace impressions. (p. 671)

fósiles: restos o vestigios de organismos que se encuentran en el registro geológico; pueden ser restos directos como huesos, moldes, impresiones fosilizadas e impresiones de huellas. (p. 671)

fracture: physical property of a mineral that causes it to break into uneven pieces. (p. 610)

fractura: propiedad física de un mineral que causa que éste se rompa en piezas asimétricas. (p. 610)

frequency: the number of wavelengths that pass a fixed point each second; expressed in hertz (Hz). (p. 295)

frecuencia: el número de longitudes de onda que pasan por un punto fijo en un segundo; expresa en hertz (Hz). (p. 295)

friction: a force that opposes the sliding motion between objects that are in contact. (p. 83)

fricción: una fuerza que se opone al movimiento deslizante entre los objectos que están el contacto. (p. 83)

G

galaxy: large group of stars, gas, and dust held together by gravity; most commonly can be elliptical, spiral, and irregular. (p. 831)

galaxia: grupo grande de estrellas, gas y polvo mantenidos juntos por la gravedad; comúnmente pueden ser elípticas, espirales e irregulares. (p. 831)

galvanometer: a device that uses an electromagnet to measure electric current. (p. 434)

galvanómetro: dispositivo que usa un electroimán para medir la corriente eléctrica. (p. 434)

gamma ray: electromagnetic wave with no mass and no charge that travels at the speed of light and is usually emitted with alpha or beta particles from a decaying atomic nucleus; has a wavelength less than about ten trillionths of a meter. (pp. 467, 793)

rayo gama: onda electromagnética sin masa ni carga que viaja a la velocidad de la luz y que usualmente es emitida con partículas alfa o beta a partir de un núcleo atómico en descomposición; tiene una electromagnética con longitudes de onda menores a diez trillonésimas de metro. (pp. 467, 793)

Geiger counter: radiation detector that produces a click or a flash of light when a charged particle is detected. (p. 798)

contador Geiger: detector de radiación que produce un sonido seco o un destello de luz al detectar una partícula cargada. (p. 798)

generator: device that uses electromagnetic induction to convert mechanical energy to electrical energy. (p. 400)

generador: dispositivo que usa inducción electromagnética para convertir energía mecánica en energía eléctrica. (p. 400)

geocentric model: Earth-centered model of the solar system. (p. 218)

modelo geocéntrico: modelo del sistema solar centrado en la tierra. (p. 218)

geothermal energy: thermal energy in hot magma; can be converted by a power plant into electrical energy. (p. 505)

energía geotérmica: energía térmica en el magma caliente, la cual se puede convertir mediante una planta industrial en energía eléctrica. (p. 505)

giant: late stage in a star's life cycle that occurs when its hydrogen fuel is depleted, its core contracts, and its outer layers expand and cool. (p. 825)

gigante: última etapa en el ciclo de vida de una estrella que ocurre cuando su combustible hidrógeno es agotado, su núcleo se contrae y sus capas externas se expanden y enfrían. (p. 825)

Global Positioning System (GPS): a system of satellites and ground monitoring stations that enable a receiver to determine its location at or above Earth's surface. (p. 475)

Sistema de Posicionamiento Global (GPS): sistema de satélites y estaciones de monitoreo en tierra que permiten que un receptor determine su ubicación en o sobre la superficie terrestre. (p. 475)

global warming: increased atmospheric heating from activities such as fossil fuel burning, which have increased air pollution and concentrations of trace gases and dust and modified the water and carbon cycles. (p. 538)

graph: visual display of information or data that can provide a quick way to communicate a lot of information and allow scientists to observe patterns. (p. 22)

gravitational potential energy: energy stored by objects due to their position above Earth's surface; depends on the distance above Earth's surface and the object's mass. (p. 132)

gravity: attractive force between two objects that depends on the masses of the objects and the distance between them. (p. 104)

greenhouse effect: atmospheric warming involving heat absorption by trace gases such as carbon dioxide and water vapor. (p. 520)

group: vertical column in the periodic table. (p. 592)

calentamiento global: aumento de la temperatura atmosférica por actividades como la quema de combustibles fósiles, lo cual incrementa la polución del aire y las concentraciones de gases traza y polvo y modifica los ciclos del agua y del carbono. (p. 538)

gráfica: presentación visual de información que puede suministrar una forma rápida de comunicar gran cantidad de información y que permite que los científicos puedan observar los patrones. (p. 22)

energía gravitacional potencial: energía almacenada por objetos debido a su posición sobre la superficie terrestre, la cual depende de la distancia sobre la superficie terrestre y de la masa del objeto. (p. 132)

gravedad: fuerza de atracción entre dos objetos que depende de las masas de los objetos y de la distancia entre ellos. (p. 104)

efecto de invernadero: calentamiento atmosférico que involucra la absorción de calor por gases traza como el dióxido de carbono y el vapor de agua. (p. 520)

grupo: columna vertical en la tabla periódica. (p. 592)

H

half-life: amount of time it takes for half the nuclei in a sample of a radioactive isotope to decay. (p. 794)

hardness: physical property of a mineral that measures resistance to scratching. (p. 611)

heat: thermal energy that flows from a warmer material to a cooler material. (p. 257)

heat of fusion: amount of energy required to change a substance from the solid phase to the liquid phase. (p. 262)

heat of vaporization: the amount of energy required for the liquid at its boiling point to become a gas. (p. 262)

heliocentric model: Sun-centered model of the solar system. (p. 219)

heterogeneous (het uh ruh JEE nee us) mixture: mixture, such as mixed nuts or a dry soup mix, in which different materials are unevenly distributed and are easily identified. (p. 555)

homogeneous (hoh moh JEE nee us) mixture: solid, liquid, or gas that contains two or more substances blended evenly throughout. (p. 556)

hydrate: compound that has water chemically attached to its ions and written into its chemical formula. (p. 708)

vida media: tiempo requerido para que se descomponga la mitad de los núcleos de una muestra de isótopo radiactivo. (p. 794)

dureza: propiedad física de un mineral que mide la resistencia al desgaste. (p. 611)

calor: energía térmica que fluye de un material caliente a uno frío. (p. 257)

calor de fusión: cantidad de energía necesaria para cambiar una sustancia del estado sólido al líquido. (p. 262)

calor de vaporización: cantidad de energía necesaria para que un líquido en su punto de ebullición se convierta en gas. (p. 262)

modelo heliocéntrico: modelo del sistema solar centrado en el sol. (p. 219)

mezcla heterogénea: mezcla, tal como una mezcla de nueces o una mezcla seca para hacer sopa, en la cual diferentes materiales están distribuidos en forma desigual y se pueden identificar fácilmente. (p. 555)

mezcla homogénea: sólido, liquido, o gas que contiene dos o más sustancias mezcladas de manera uniforme en toda la mezcla. (p. 556)

hidrato: compuesto que contiene agua químicamente conectada a sus iones y representada en su fórmula química. (p. 708)

hydroelectricity: electricity produced from the energy of falling water. (p. 503)

hidroelectricidad: electricidad producida a partir de la energía generada por una caída de agua. (p. 503)

hypothesis: educated guess using what you know and what you observe. (p. 8)

hipótesis: suposición fundamentada que se basa en lo que se sabe y lo que se observa. (p. 8)

inclined plane: simple machine that consists of a sloping surface, such as a ramp, that reduces the amount of force needed to lift something by increasing the distance over which the force is applied. (p. 172)

plano inclinado: máquina simple que consiste de una superficie inclinada, tal como una rampa, que reduce la fuerza necesaria para levantar un objeto aumentando la distancia sobre la cual se aplica dicha fuerza. (p. 172)

independent variable: factor that, as it changes, affects the measure of another variable. (p. 9)

variable independiente: factor que, a medida que cambia, afecta la medida de otra variable. (p. 9)

index of refraction: property of a material indicating how much light slows down when traveling in the material. (p. 329)

índice de refracción: propiedad de un material para indicar la cantidad de luz que se frena al pasar a través del material. (p. 329)

indicator: organic compound that changes color in acids and bases. (p. 764)

indicador: compuesto orgánico que cambia de color en presencia de ácidos y bases. (p. 764)

inertia: resistance of an object to a change in its motion. (p. 99)

inercia: resistencia de un objeto a cambiar su movimiento. (p. 99)

infiltration: process by which water enters Earth to become groundwater, controlled by the slope of the land, type of surface material, and type and amount of vegetation. (p. 663)

infiltración: proceso mediante el cual el agua entra en la tierra para convertirse en agua subterránea, controlado por la pendiente del suelo, el tipo de material superficial y el tipo y cantidad de vegetación. (p. 663)

infrared waves: electromagnetic waves that have a wavelength between about 1 mm and 750 billionths of a meter. (p. 464)

ondas infrarrojas: ondas electromagnéticas que tienen una longitud de onda entre aproximadamente 1 mm y 750 billonésimas de metro. (p. 464)

inhibitor: substance that slows down a chemical reaction or prevents it from occurring by combining with a reactant. (p. 740)

inhibidor: sustancia que reduce una reacción química o previene que ocurra por una combinación con un reactivo. (p. 740)

input force: force exerted on a machine. (p. 162)

fuerza de entrada: fuerza ejercida sobre una máquina. (p. 162)

instantaneous speed: speed of an object at a given point in time; is constant for an object moving with constant speed, and changes with time for an object that is slowing down or speeding up. (p. 71)

velocidad instantánea: velocidad de un objeto en un punto dado en el tiempo; es constante para un objeto que se mueve a una velocidad constante y cambia con el tiempo en un objeto que está reduciendo o aumentando su velocidad. (p. 71)

insulator: material in which electrons are not able to move easily. (p. 395); material in which heat flows slowly. (p. 270)

aislador: material a través del cual los electrones no se pueden mover con facilidad. (p. 395); material en el cual el calor fluye lentamente. (p. 270)

intensity: amount of energy that flows through a certain area in a specific amount of time. (p. 322)

intensidad: cantidad de energía que fluye a través de cierta área en un tiempo específico. (p. 322)

interference: occurs when two or more waves overlap and combine to form a new wave. (p. 306)

interferencia: ocurre cuando dos o más ondas se sobreponen y combinan para formar una nueva onda. (p. 306)

internal combustion engine: heat engine that burns fuel inside the engine in chambers or cylinders. (p. 275)

motor de combustión interna: motor de calor que quema combustible en su interior en cámaras o cilindros. (p. 275)

intrusive igneous rock: coarse-grained rock formed from slowly cooling magma below Earth's surface. (p. 617)

roca ígnea intrusiva: roca áspera y granulosa formada a partir de un enfriamiento lento del magma debajo de la superficie terrestre. (p. 617)

ion: charged particle that has either more or fewer electrons than protons. (pp. 692, 707)

ionic bond: attraction formed between oppositely charged ions in an ionic compound. (p. 696)

isotopes: atoms of the same element that have different numbers of neutrons. (p. 586)

ion: partícula cargada que tiene ya sea más o menos electrones que protones. (pp. 692, 707)

enlace iónico: atracción formada entre iones con cargas opuestas en un compuesto iónico. (p. 696)

isótopos: átomos del mismo elemento que tienen diferente número de neutrones. (p. 586)

J

jet stream: high-speed, powerful air current that affects many weather processes, such as the development of storms. (p. 525)

joule: SI unit of energy. (p. 130)

Jupiter: largest and fifth planet from the Sun; has continuous, swirling, high-pressure gas storms, the largest of which is the Great Red Spot. (p. 231)

corriente a presión: corriente de aire poderosa y de alta velocidad que afecta a muchos procesos climáticos, como el desarrollo de las tormentas. (p. 525)

julio: unidad SI de energía. (p. 130)

Júpiter: quinto planeta más cercano al Sol, también es el más grande; tiene tormentas gaseosas arremolinadas y continuas de alta presión, la más grande de las cuales es la Gran Mancha Roja. (p. 231)

Kelvin: the SI unit for temperature, abbreviated K; a temperature change of 1 Kelvin is the same as a temperature change of 1°C. (p. 255)

kinetic energy: energy a moving object has because of its motion; depends on the mass and speed of the object. (p. 130)

kinetic theory: explanation of the behavior of molecules in matter; states that all matter is made of constantly moving particles that collide without losing energy. (p. 254)

Kelvin: la unidad del sistema internacional de medidas para la temperatura, cuya abreviatura es K; un cambio en la temperatura de 1 grado kelvin es lo mismo que un cambio en la temperatura de 1°C. (p. 255)

energía cinética: energía que tiene un cuerpo debido a su movimiento, la cual depende de la masa y velocidad del objeto. (p. 130)

teoría cinética: explicación del comportamiento de las moléculas en la materia, la cual establece que todas las sustancias están compuestas de partículas en constante movimiento que colindan sin perder energía. (p. 254)

La Niña: climatic phenomenon that occurs when Pacific trade winds are very strong and temperatures colder than normal; the opposite of El Niño. (p. 539)

latent heat: energy used to melt snow or evaporate water. (p. 520)

law of conservation of charge: states that charge can be transferred from one object to another but cannot be created or destroyed. (p. 393)

law of conservation of energy: states that energy can never be created or destroyed. (p. 139)

law of conservation of mass: states that the mass of all substances present before a chemical change equals the mass of all the substances remaining after the change. (p. 567)

La Niña: fenómeno climático que ocurre cuando los vientos alisios del pacífico son muy fuertes y la temperatura disminuye más de lo normal; lo opuesto a El Niño. (p. 539)

calor latente: energía utilizada para derretir nieve o evaporar agua. (p. 520)

ley de la conservación de carga: establece que la carga puede ser transferida entre un objeto y otro pero no puede ser creada o destruida. (p. 393)

ley de la conservación de energía: establece que la energía nunca puede ser creada ni destruida. (p. 139)

ley de conservación de la masa: establece que la masa de todas las sustancias presente antes de un cambio químico es igual a la masa de todas las sustancias resultantes después del cambio. (p. 567)

law of conservation of momentum: states that if a group of objects exerts forces only on each other, then the total momentum of the objects doesn't change (p. 117)

ley de la conservación del movimiento: ley que establece que si en un grupo de objetos se ejercen fuerzas sólo en unos sobre otros, entonces el momento total de los objetos no cambia. (p. 117)

lee rain shadow: area of reduced precipitation on one side of a mountain range. (p. 533)

sombra lluviosa a sotavento: área de precipitación reducida en un lado de un sistema montañoso. (p. 533)

lever: simple machine consisting of a bar free to pivot about a fixed point called the fulcrum. (p. 166)

palanca: máquina simple que consiste de una barra que puede girar sobre un punto fijo llamado pivote. (p. 166)

light-year: distance light travels in one year—about 9.5 trillion km. (p. 820)

año luz: distancia que viaja la luz en un año—cerca de 9 500 trillones de km. (p. 820)

Local Group: cluster of about 45 galaxies, including the Milky Way. (p. 831)

Grupo Local: agrupación de cerca de 45 galaxias, incluyendo la Vía Láctea. (p. 831)

longshore current: movement of water parallel to the shoreline caused by waves colliding with the shore at an angle, which results in net sediment transport parallel to the shoreline. (p. 659)

corriente costera: movimiento de agua paralelo a la línea costera causado por olas que colisionan con la costa en ángulo, lo cual resulta en transporte neto de sedimentos paralelo a la línea costera. (p. 659)

loudness: human perception of sound intensity. (p. 322)

volumen de sonido: percepción humana de la intensidad del sonido. (p. 322)

lunar eclipse: occurs during full moon, when the Moon enters Earth's umbra and Earth casts a curved shadow on the Moon's surface. (p. 202)

eclipse lunar: fenómeno que ocurre durante la luna llena, cuando la Luna entra en el cono de sombra de la Tierra y ésta proyecta una sombra curva sobre la superficie de la Luna. (p. 202)

M

machine: device that makes doing work. easier by increasing the force applied to an object, changing the direction of an applied force, or increasing the distance over which a force can be applied. (p. 160)

máquina: artefacto que facilita la ejecución del trabajo aumentando la fuerza que se aplica a un objeto, cambiando la dirección de una fuerza aplicada o aumentando la distancia sobre la cual se puede aplicar una fuerza. (p. 160)

magma: molten rock material found inside Earth. (p. 613)

magma: material rocoso derretido que se encuentra en el interior de la Tierra. (p. 613)

magnetic domain: group of atoms in a magnetic material with the magnetic poles of the atoms pointing in the same direction. (p. 429)

dominio magnético: grupo de átomos en un material magnético en el cual los polos magnéticos de los átomos apuntan en la misma dirección. (p. 429)

magnetic field: surrounds a magnet and exerts a force on other magnets and objects made of magnetic materials. (p. 425)

campo magnético: rodea a un imán y ejerce una fuerza sobre otros imanes y objetos hechos de materiales magnéticos. (p. 425)

magnetic pole: region on a magnet where the magnetic force exerted by a magnet is strongest; like poles repel and opposite poles attract. (p. 425)

polo magnético: zona en un imán en donde la fuerza magnética ejercida por un imán es la más fuerte; los polos iguales se repelen y los polos opuestos se atraen. (p. 425)

magnetism: the properties and interactions of magnets. (p. 424)

magnetismo: propiedades e interacciones de los imanes. (p. 424)

main sequence: section of an H-R diagram that is plotted from the upper left to the lower right and contains 90 percent of all known stars. (p. 823)

secuencia principal: sección de un diagrama H-R que es graficado desde la parte superior izquierda hasta la parte inferior derecha y contiene el 90% de todas las estrellas conocidas. (p. 823)

maria: relatively flat, dark-colored regions on the Moon's surface. (p. 203)

maría: egiones relativamente planas y de color oscuro sobre la superficie de la luna. (p. 203)

maritime climate: climate with a strong ocean influence and milder temperatures. (p. 532)

Mars: fourth planet from the Sun; called the red planet because of high concentrations of iron oxide. (p. 225)

mass: amount of matter in an object. (p. 19)

mass number: sum of the number of protons and neutrons in an atom's nucleus. (p. 585)

mechanical advantage (MA): ratio of the output force exerted by a machine to the input force applied to the machine. (p. 164)

mechanical energy: sum of the potential energy and kinetic energy in a system. (p. 136)

medium: matter in which a wave travels. (p. 289)

Mercury: closest planet to the Sun; has a larger than expected iron core. (p. 223)

meteoroid: rocky solar system object formed from pieces of comets and asteroids. (pp. 236, 633)

microwaves: radio waves with wavelengths of between about 1 m and 1 mm. (p. 463)

mid-ocean ridge (MOR): a continuous system of twin mountain ranges with a rift valley between them that extends around Earth on the seafloor; formed where two oceanic plates are forced apart due to magma rising from Earth's mantle; a source of new rock. (p. 356)

Milky Way: our spiral-shaped galaxy, which contains the solar system and about 400 billion stars, including the Sun, and measures about 100,000 light-years across. (p. 831)

mineral: naturally occurring, inorganic, crystalline solid with a predictable chemical composition and a characteristic set of physical properties. (p. 609)

model: can be used to represent an idea, object, or event that is too big, too small, too complex, or too dangerous to observe or test directly. (p. 11)

molecule: a neutral particle that forms as a result of electron sharing. (p. 697)

momentum: property of a moving object that equals its mass times its velocity. (p. 116)

moon phase: changing appearance of the Moon as viewed from Earth, depending on the relative positions of the Sun, the Moon, Earth, and the observer. (p. 199)

clima marítimo: clima con una fuerte influencia oceánica y temperaturas templadas. (p. 532)

marte: cuarto planeta más cercano al Sol; llamado el planeta rojo debido a altas concentraciones de óxido de hierro. (p. 225)

masa: cantidad de materia en un objeto. (p. 19)

número de masa: suma del número de protones y neutrones en el núcleo de un átomo. (p. 585)

ventaja mecánica (MA): relación de la fuerza ejercida por una máquina y la fuerza aplicada a dicha máquina. (p. 164)

energía mecánica: suma de la energía potencial y energía cinética en un sistema. (p. 136)

medio: materia a través de la cual viaja una onda. (p. 289)

Mercurio: planeta más cercano al sol; tiene un núcleo de hierro más grande de lo esperado. (p. 223)

meteorito: objeto rocoso del sistema solar formado de pedazos de cometas y asteroides. (pp. 236, 633)

microondas: ondas de radio con longitudes de onda entre aproximadamente 1 mm y 1 m. (p. 463)

canto del mediados de-océano (MOR): un sistema continuo de la montaña gemela se extiende con un valle de la grieta entre ellos que extienda alrededor de la tierra en el seafloor; formado donde están dos placas oceánicas separado forzado debido al magma que se levanta de la capa de la tierra; una fuente de la roca nueva. (p. 356)

Vía Láctea: nuestra galaxia en forma de espiral, la cual contiene al sistema solar y cerca de 400 000 millones de estrellas, incluyendo al Sol, y tiene una extensión de aproximadamente 100 000 años luz. (p. 831)

mineral: sólido de origen natural, inorgánico y cristalino con una composición química predecible y un conjunto característico de propiedades físicas. (p. 609)

modelo: puede ser usado para representar una idea, objeto o evento que es demasiado grande, demasiado pequeño, demasiado complejo o demasiado peligroso para ser observado o probado directamente. (p. 11)

molécula: partícula neutra que se forma al compartir electrones. (p. 697)

ímpetu: propiedad de un objeto en movimiento que es igual a su masa por su velocidad. (p. 116)

fase lunar: cambio en la apariencia de la Luna según es vista desde la Tierra, dependiendo de las posiciones relativas del Sol, la Luna, la Tierra y el observador. (p. 199)

Neptune: eighth planet from the Sun; has storms similar to Jupiter's and appears blue because of atmospheric methane. (p. 235)

net force: sum of the forces that are acting on an object. (p. 82)

neutron: neutral particle, composed of quarks, inside the nucleus of an atom. (p. 579)

nonpolar molecule: molecule that shares electrons equally and does not have oppositely charged ends. (p. 700)

nonrenewable resources: natural resource, such as fossil fuels, that cannot be replaced by natural processes as quickly as it is used. (p. 493)

nuclear fission: process of splitting an atomic nucleus into two or more nuclei with smaller masses. (p. 801)

nuclear fusion: reaction in which two or more atomic nuclei form a nucleus with a larger mass. (p. 803)

nuclear reactor: uses energy from a controlled nuclear chain reaction to generate electricity. (p. 494)

nuclear waste: radioactive by-product that results when radioactive materials are used. (p. 498)

nucleus: positively charged center of an atom that contains protons and neutrons and is surrounded by a cloud of electrons. (p. 579)

Neptuno: octavo planeta más cercano al Sol; tiene tormentas similares a las de Júpiter y parece azul debido al metano atmosférico. (p. 235)

fuerza neta: suma de fuerzas que actúan sobre un objeto. (p. 82)

neutrón: partícula neutra, compuesta por quarks, dentro del núcleo de un átomo. (p. 579)

molécula no polar: molécula que comparte equitativamente los electrones y que no tiene extremos con cargas opuestas. (p. 700)

recursos no renovables: recursos naturales, tales como combustibles fósiles, que no pueden ser reemplazados por procesos naturales tan pronto como son usados. (p. 493)

fisión nuclear: proceso de división de un núcleo atómico en dos o más núcleos con masas más pequeñas. (p. 801)

fusión nuclear: reacción en la cual dos o más núcleos atómicos forman un núcleo con mayor masa. (p. 803)

reactor nuclear: usa energía de una reacción nuclear controlada en cadena para generar electricidad. (p. 494)

desperdicio nuclear: subproducto radioactivo que resulta del uso de materiales radiactivos. (p. 498)

núcleo: centro de un átomo con carga positiva que contiene protones y neutrones y está rodeado por una nube de electrones. (p. 579)

Ohm's law: states that the current in a circuit equals the voltage difference divided by the resistance. (p. 405)

opaque: material that absorbs or reflects all light and does not transmit any light. (p. 327)

output force: force applied by a machine. (p. 162)

oxidation number: positive or negative number that indicates how many electrons an atom has gained, lost, or shared to become stable. (p. 703)

ley de Ohm: establece que la corriente en un circuito es igual a la diferencia de voltaje dividida por la resistencia. (p. 405)

opaco: material que absorbe o refleja toda la luz pero no la transmite. (p. 327)

fuerza de salida: fuerza aplicada por una máquina. (p. 162)

número de oxidación: número positivo o negativo que indica cuántos electrones ha ganado, perdido o compartido un átomo para poder ser estable. (p. 703)

P

parallel circuit: circuit in which electric current has more than one path to follow. (p. 406)

circuito paralelo: circuito en el cual la corriente eléctrica tiene más de una trayectoria para seguir. (p. 406)

period: the amount of time it takes one wavelength to pass a fixed point; expressed in seconds. (p. 295); horizontal row in the periodic table. (p. 595)

periodic table: organized list of all known elements that are arranged by increasing atomic number and by changes in chemical and physical properties. (p. 588)

permeability: a measure of the ability of a fluid to pass through a material. (p. 665)

petroleum: liquid fossil fuel formed from decayed remains of ancient organisms; can be refined into fuels and used to make plastics. (p. 489)

pH: a measure of the concentration of hydronium ions in a solution using a scale ranging from 0 to 14, with 0 being the most acidic and 14 being the most basic. (p. 773)

photon: particle that electromagnetic waves sometimes behave like; has energy that increases as the frequency of the electromagnetic wave increases. (p. 460)

photosphere: light-emitting surface of the Sun. (p. 827)

photovoltaic cell: device that converts solar energy into electricity; also called a solar cell. (p. 501)

physical change: any change in size, shape, or state of matter in which the identity of the substance remains the same. (p. 562)

physical property: any characteristic of a material, such as size or shape, that you can observe or attempt to observe without changing the identity of the material. (p. 560)

pigment: colored material that is used to change the color of other substances. (p. 342)

pilot plant: scaled-down version of real production equipment that closely models actual manufacturing conditions and is used to test a new manufacturing process. (p. 56)

pitch: how high or low a sound seems; related to the frequency of the sound waves. (p. 323)

plane mirror: flat, smooth mirror that reflects light to form upright, virtual images. (p. 331)

plasma: matter consisting of positively and negatively charged particles. (p. 261)

Pluto: smallest and ninth planet from the Sun; has a thin atmosphere and a solid, ice-rock surface. (p. 236)

polar molecule: molecule with a slightly positive end and a slightly negative end as a result of electrons being shared unequally. (p. 700)

período: la cantidad de tiempo que requiere una longitud de onda para pasar un punto fijo; expresa en segundos. (p. 295); fila horizontal en la tabla periódica. (p. 595)

tabla periódica: lista organizada de todos los elementos conocidos y que han sido ordenados de manera ascendente por número atómico y por cambios en sus propiedades químicas y físicas. (p. 588)

permeabilidad: medida de la capacidad de un fluido para pasar a través de un material. (p. 665)

petróleo: combustible fósil líquido que se forma a partir de residuos en descomposición de organismos ancestrales y que puede ser refinado para producir combustibles y usado para hacer plásticos. (p. 489)

pH: medida de la concentración de iones de hidronio en una solución, usando una escala de 0 a 14, en la cual 0 es la más ácida y 14 la más básica. (p. 773)

fotón: partícula como la cual algunas veces se comportan las ondas electromagnéticas; tiene energía que aumenta a medida que la frecuencia de la onda electromagnética aumenta. (p. 460)

fotosfera: superficie solar que emite luz. (p. 827)

células fotovoltaicas: dispositivo que convierte la energía solar en electricidad; también llamada celda solar. (p. 501)

cambio físico: cualquier cambio en tamaño, forma o estado de una sustancia en la cual la identidad de la sustancia sigue siendo la misma. (p. 562)

propiedad física: cualquier característica de un material, tal como tamaño o forma, que se puede haber observar o tratado de observar sin cambiar la identidad del material. (p. 560)

pigmento: material de color que se usa para cambiar el color de otras sustancias. (p. 342)

planta piloto: versión a escala reducida de un equipo de producción real que simula cercanamente las condiciones reales de fabricación y es utilizada para probar un proceso de fabricación nuevo. (p. 56)

altura: qué tan alto o bajo parece un sonido; tiene relación con la frecuencia de las ondas sonoras. (p. 323)

espejo plano: espejo plano y liso que refleja la luz para formar imágenes verticales y virtuales. (p. 331)

plasma: materia consistente de partículas con cargas positivas y negativas. (p. 261)

Plutón: noveno planeta y el más lejano al Sol, también el más pequeño; tiene una atmósfera delgada y una superficie sólida de hielo y roca. (p. 236)

molécula polar: molécula con un extremo ligeramente positivo y otro ligeramente negativo como resultado de un compartir desigual de los electrones. (p. 700)

polyatomic ion: positively or negatively charged, covalently bonded group of atoms. (p. 707)

ion poliatómico: grupo de átomos enlazados covalentemente, con carga positiva o negativa. (p. 707)

porosity: empty space in sedimentary rock in which water, oil, and natural gas are stored within Earth. (p. 625)

porosidad: espacio vacío en la roca sedimentaria en el que el agua, el petróleo y el gas natural son almacenados dentro de la Tierra. (p. 625)

potential energy: stored energy an object has due to its position. (p. 131)

energía potencial: energía almacenada que un objeto tiene debido a su posición. (p. 131)

power: amount of work done, or the amount of energy transferred, divided by the time required to do the work or transfer the energy; measured in watts (W). (p. 157)

potencia: cantidad de trabajo realizado o cantidad de energía transferida, dividida por el tiempo requerido para realizar el trabajo o transferir la energía; medida en vatios (V). (p. 157)

precision: describes how closely measurements are to each other and how carefully the measurements were made. (p. 14)

precisión: descripción de la cercanía de las mediciones una respecto a otra y del cuidado con el que se hacen éstas. (p. 14)

principle of superposition: states that the oldest rocks in an undisturbed sequence of rocks are at the bottom of the undisturbed sequence. (p. 670)

principio de superposición (estratigráfico): principio que establece que las rocas más antiguas en una secuencia de rocas sin perturbaciones están en la parte más baja de la secuencia no perturbada. (p. 670)

product: in a chemical reaction, the new substance that is formed. (p. 720)

producto: es la nueva sustancia que se forma en una reacción química. (p. 720)

proton: particle, composed of quarks, inside the nucleus of an atom that has a charge of 1+. (p. 579)

protón: partícula, compuesta por quarks, dentro del núcleo de un átomo que tiene una carga de 1+. (p. 579)

prototype: first full-scale model built to performance-test a new product. (p. 56)

prototipo: el primer modelo a escala completa construido para probar el desempeño de un producto nuevo. (p. 56)

pulley: simple machine that consists of a grooved wheel with a rope, chain, or cable running along the groove; can be either fixed or movable. (p. 169)

polea: máquina simple que consiste de una rueda acanalada con una cuerda, cadena o cable que se desliza por el canal y que puede ser fija o móvil. (p. 169)

Q

quarks: particles of matter that make up protons and neutrons. (p. 579)

quarks: partículas de materia que constituyen los protones y neutrones. (p. 579)

R

radiant energy: energy carried by an electromagnetic wave. (p. 459)

energía radiante: energía transportada por una onda electromagnética. (p. 459)

radiation: transfer of thermal energy by electromagnetic waves. (p. 269)

radiación: transferencia de energía térmica mediante ondas electromagnéticas. (p. 269)

radioactivity: process that occurs when a nucleus decays and emits alpha, beta, or gamma radiation. (p. 788)

radiactividad: proceso que ocurre cuando un núcleo se descompone y emite radiación alfa, beta o gama. (p. 788)

radio telescope: telescope that collects and magnifies radio waves. (p. 821)

radio telescopio: telescopio que colecta y magnifica las ondas de radio. (p. 821)

radio waves: electromagnetic waves with wavelengths longer than about 1 mm, used for communications. (p. 463)

ondas de radio: ondas electromagnéticas con longitudes de onda más largas de aproximadamente 1 mm y que se usan en las comunicaciones. (p. 463)

rarefaction: the least dense regions of a compressional wave. (p. 294)

rate of reaction: the speed at which reactants are consumed and products are produced in a given reaction. (p. 738)

reactant: in a chemical reaction, the substance that reacts. (p. 720)

reflecting telescope: optical instrument that uses a concave mirror to collect light and a lens to magnify an image. (p. 820)

refracting telescope: optical instrument that uses double convex lenses to collect light and magnify an image. (p. 819)

refraction: the bending of a wave as it changes speed in moving from one medium to another. (p. 302)

regolith: layer of debris on the Moon's surface formed by the accumulation of meteoric material. (p. 203)

relative dating: process of dating objects or events in time order, or sequence, based on logical relationships; may or may not involve the use of numbers. (p. 669)

renewable resource: energy source that is replaced almost as quickly as it is used. (p. 501)

resistance: tendency for a material to oppose electron flow and change electrical energy into thermal energy and light; measured in ohms (Ω). (p. 403)

resonance: the process by which an object is made to vibrate by absorbing energy at its natural frequencies. (p. 309)

revolution: movement of Earth in its orbit around the Sun; used to measure time in years. (p. 192)

rift valley: long, linear, dropped-down valley between twin, parallel mountain ranges produced by faulting. (p. 356)

rock: consolidated, natural mixture of minerals, volcanic glass, or rock fragments. (p. 617)

rock cycle: continuous process that forms and changes rocks on Earth's surface and deep below the surface. (p. 634)

rotation: spinning of Earth on its axis; used to measure time in days. (p. 192)

rarefacción: las regiones menos densas de una onda de compresión. (p. 294)

fasa de reacción: velocidad a la que los reactantes son consumidos y los productos son generados en una reacción dada. (p. 738)

reactante: es la sustancia que reacciona en una reacción química. (p. 720)

telescopio de reflexión: instrumento óptico que utiliza un espejo cóncavo para captar la luz y una lente para magnificar una imagen. (p. 820)

telescopio de refracción: instrumento óptico que utiliza lentes convexas dobles para captar la luz y magnificar una imagen. (p. 819)

refracción: curvatura de una onda al cambiar su velocidad al pasar de un medio a otro. (p. 302)

regolito: capa de escombros sobre la superficie lunar formada por la acumulación de material meteórico. (p. 203)

datación relativa: proceso de datar objetos o eventos en un orden o secuencia temporal, basado en relaciones lógicas; puede o no involucrar el uso de números. (p. 669)

recursos renovables: fuente de energía que es reemplazada casi tan pronto como es usada. (p. 501)

resistencia: tendencia de un material de oponerse al fluido de los electrones y convertir la energía eléctrica en energía térmica y luz; se mide en ohmios (Ω). (p. 403)

resonancia: el proceso por el cual un objeto vibra al absorber energía en sus frecuencias naturales. (p. 309)

revolución: movimiento de la Tierra en su órbita alrededor del sol, utilizado para medir el tiempo en años. (p. 192)

valle de hendidura: valle alargado, lineal y deprimido producido por una falla que se encuentra entre dos sistemas montañosos paralelos. (p. 356)

roca: mezcla natural y consolidada de minerales, obsidiana o fragmentos de roca. (p. 617)

ciclo de la roca: proceso continuo que forma y cambia las rocas sobre y debajo de la superficie terrestre. (p. 634)

rotación: giro de la Tierra sobre su eje, utilizado para medir el tiempo en días. (p. 192)

S

saturated solution: any solution that contains all the solute it can hold at a given temperature. (p. 761)

solución saturada: cualquier solución que contiene todo el soluto que puede retener a una temperatura determinada. (p. 761)

Saturn: second-largest and sixth planet from the Sun; has the most complex system of rings. (p. 234)

Saturno: sexto planeta más cercano al Sol, también el segundo más grande; tiene el sistema de anillos más complejo. (p. 234)

scientific law: statement about what happens in nature that seems to be true all the time; does not explain why or how something happens. (p. 12)

ley científica: enunciado acerca de lo que ocurre en la naturaleza, lo cual parece ser cierto en todo momento, sin explicar cómo o por qué algo ocurre. (p. 12)

scientific methods: organized set of investigation procedures that can include stating a problem, forming a hypothesis, researching and gathering information, testing a hypothesis, analyzing data, and drawing conclusions. (p. 7)

métodos científico: conjunto organizado de procedimientos de investigación que puede incluir el planteamiento de un problema, formulación de una hipótesis, investigación y recopilación de información, comprobación de la hipótesis, análisis de datos y elaboración de conclusiones. (p. 7)

screw: simple machine that consists of an inclined plane wrapped in a spiral around a cylindrical post. (p. 173)

tornillo: máquina simple que consiste de un plano inclinado envuelto en espiral alrededor de un poste cilíndrico. (p. 173)

sea breeze: cooling breeze that blows from the sea toward land during the afternoon. (p. 533)

brisa marina: brisa refrescante que sopla del mar hacia la tierra durante la tarde. (p. 533)

second law of motion: states that the acceleration of an object is in the same direction as the net force on the object, and that the acceleration equals the net force divided by the mass. (p. 102)

segunda ley de movimiento: establece que la aceleración de un objeto es en la misma dirección que la fuerza neta del objeto y que la aceleración es igual a la fuerza neta dividida por su masa. (p. 102)

second law of thermodynamics: states that is impossible for heat to flow from a cool object to a warmer object unless work is done. (p. 274)

segunda ley de la termodinámica: establece que es imposible que el calor fluya de un objeto frío a uno caliente, a menos que se realice un trabajo. (p. 274)

sediment transport: movement of eroded materials by wind, water, and/or glaciers from one location to another. (p. 654)

transporte de sedimentos: movimiento de materiales erosionados por el viento, agua o glaciares de un lugar a otro. (p. 654)

Sedna: unofficial name for object 2003 VB12, a distant planetoid with a very elliptical orbit. (p. 237)

Sedna: nombre no oficial para el objeto 2003 VB12, un planetoide distante con una órbita muy elíptica. (p. 237)

series circuit: circuit in which electric current has only one path to follow. (p. 407)

circuito en serie: circuito en el cual la corriente eléctrica tiene una sola trayectoria para seguir. (p. 407)

shadow zone: "dead zone" between 105 and 140 degrees from an earthquake's epicenter, where nothing is recorded on a seismogram. (p. 371)

zona de sombra: "zona muerta" entre 105 y 140 grados del epicentro de un terremoto, donde nada es registrado sobre un sismograma. (p. 371)

shield volcano: broad, flat volcano formed by layers of free-flowing, high-temperature, basaltic lava. (p. 377)

volcán de escudo: or capas de lava basáltica que fluyeron libremente a alta temperatura. (p. 377)

simple machine: machine that does work with only one movement—lever, pulley, wheel and axle, inclined plane, screw, and wedge. (p. 166)

máquina simple: máquina que realiza el trabajo con un solo movimiento: palanca, polea, rueda y eje, plano inclinado, tornillo y cuña. (p. 166)

single-displacement reaction: chemical reaction in which one element replaces another element in a compound. (p. 731)

reacción de un solo desplazamiento: reacción química en la cual un elemento reemplaza a otro elemento en un compuesto. (p. 731)

sliding friction: frictional force that opposes the motion of two surfaces sliding past each other. (p. 84)

fricción deslizante: fuerza de fricción que se opone al movimiento de dos superficies que se deslizan entre sí. (p. 84)

society: group of people who share similar values and beliefs. (p. 46)

sociedad: grupo de personas que comparten valores y creencias similares. (p. 46)

soil: mixture of weathered rock, decaying organic matter, water, and air that overlies bedrock and is capable of supporting plant life. (p. 650)

solar eclipse: occurs during new moon, when the Moon passes directly between the Sun and Earth, casting a shadow on part of Earth. (p. 201)

solenoid: a wire wrapped into a cylindrical coil. (p. 432)

solstice: occurs twice yearly (June and December), when the Sun reaches its greatest distance north or south of the equator. (p. 195)

solubility: maximum amount of a solute that can be dissolved in a given amount of solvent at a given temperature. (p. 759)

solute: in a solution, the substance being dissolved. (p. 753)

solution: homogeneous mixture that remains constantly and uniformly mixed and has particles that are so small they cannot be seen with a microscope. (pp. 556, 752)

solvent: in a solution, the substance in which the solute is dissolved. (p. 753)

specific heat: amount of thermal energy needed to raise the temperature of 1 kg of a material 1°C. (p. 257)

spectroscope: device that disperses light into its component wavelengths, using a prism or diffraction grating. (p. 822)

speed: distance an object travels per unit of time. (p. 71)

sphere: three-dimensional, round object whose surface is the same distance from the center in all directions. (p. 186)

standing wave: a wave pattern that forms when waves of equal wavelength and amplitude, but traveling in opposite directions, continuously interfere with each other; has points called nodes that do not move. (p. 308)

static electricity: the accumulation of excess electric charge on an object. (p. 392)

static friction: frictional force that prevents two surfaces from sliding past each other. (p. 84)

streak: color of a mineral in powdered form; helps identify a mineral and can be observed by rubbing a mineral on a white porcelain tile. (p. 611)

strong acid: any acid that dissociates almost completely in solution. (p. 771)

strong base: any base that dissociates completely in solution. (p. 772)

suelo: mezcla de rocas degradadas, materia orgánica en descomposición, agua y aire que yace sobre el lecho rocoso y que es capaz de soportar la vida vegetal. (p. 650)

eclipse solar: ocurre durante la luna nueva, cuando la Luna pasa directamente entre Sol y la Tierra, proyectando una sombra sobre una parte de la Tierra. (p. 201)

solenoide: cable envuelto en forma de bobina cilíndrica. (p. 432)

solsticio: fenómeno que ocurre dos veces al año (junio y diciembre), cuando el sol alcanza su mayor distancia al norte o al sur del ecuador. (p. 195)

solubilidad: máxima cantidad de soluto que puede ser disuelto en una cantidad dada de solvente a una temperatura determinada. (p. 759)

soluto: en una solución, la sustancia que está disuelta. (p. 753)

solución: mezcla homogénea que permanece constante y uniformemente mezclada y que tiene partículas tan pequeñas que no pueden ser vistas en un microscopio. (pp. 556, 752)

solvente: en una solución, la sustancia en la cual se disuelve el soluto. (p. 753)

calor específico: cantidad de energía térmica necesaria para aumentar un grado centígrado la temperatura de un kilogramo de material. (p. 257)

espectroscopio: dispositivo que dispersa la luz en las longitudes de onda que la componen, utilizando un prisma o rejilla de difracción. (p. 822)

velocidad: distancia que recorre un objeto por unidad de tiempo. (p. 71)

esfera: objeto redondo tridimensional cuya superficie está a la misma distancia desde el centro hacia todas las direcciones. (p. 186)

onda estacionaria: patrón de una onda que se forma cuando ondas con la misma longitud de onda y amplitud, pero que viajan en direcciones opuestas, interfieren continuamente entre sí; tiene puntos llamados nodos que no se mueven. (p. 308)

electricidad estática: la acumulación del exceso de carga eléctrica en un objeto. (p. 392)

fricción estática: fuerza que evita que dos superficies en contacto se deslicen una sobre otra. (p. 84)

veta: color de un mineral en forma de polvo; ayuda a identificar un mineral y puede observarse al frotar un mineral sobre una losa de porcelana blanca. (p. 611)

ácido fuerte: cualquier ácido que se disocie casi por completo en una solución. (p. 771)

base fuerte: cualquier base que se disocie completamente en una solución. (p. 772)

strong force: attractive force that acts between protons and neutrons in an atomic nucleus. (p. 878)

fuerza de atracción: fuerza de atracción que actúa entre protones y neutrones en un núcleo atómico. (p. 878)

subduction: occurs when lithospheric plates converge and the edge of one plate is forced downward beneath another; recycles old lithosphere. (p. 359)

subducción: fenómeno que ocurre cuando las placas litosféricas convergen y el borde de una placa es forzado a pasar por debajo del otro; recicla la litosfera antigua. (p. 359)

substance: element or compound that cannot be broken down into simpler components and maintain the properties of the original substance. (p. 552)

sustancia: elemento o compuesto que no se puede descomponer en componentes más simples y que mantiene las propiedades de la sustancia original. (p. 552)

subtropical high: persistent, relatively stationary, high-pressure system. (p. 525)

alto subtropical: sistema persistente de alta presión relativamente estacionario. (p. 525)

sunspots: cool, darker areas of the Sun's photosphere. (p. 827)

manchas solares: áreas frías y más oscuras de la fotosfera solar. (p. 827)

supersaturated solution: any solution that contains more solute than a saturated solution at the same temperature. (p. 762)

solución sobresaturada: cualquier solución que contenga más solutos que una solución saturada a la misma temperatura. (p. 762)

suspension: heterogeneous mixture containing a liquid in which visible particles settle. (p. 558)

suspensión: mezcla heterogénea que contiene un líquido en el cual las partículas visibles se sedimentan. (p. 558)

synthesis reaction: chemical reaction in which two or more substances combine to form a different substance. (p. 731)

reacción síntesis: reacción química en la cual se combinan dos o más sustancias para formar una sustancia diferente. (p. 731)

T

technology: application of science to help people. (p. 13)

tecnología: aplicación de la ciencia en beneficio de la población. (p. 13)

temperature: measure of the average kinetic energy of all the particles in an object. (p. 255)

temperatura: medida de la energía cinética promedio de todas las partículas en un objeto. (p. 255)

temperature inversion: an increase of temperature with height, resulting in very stable air that resists the rising needed for cloud formation or dispersal of air pollution. (p. 519)

inversión de temperatura: incremento de la temperatura con la altura, resultando en aire muy estable que resiste la elevación necesaria para la formación de nubes o dispersión de la polución del aire. (p. 519)

texture: describes the size and arrangement of rock components and can help indicate how a rock formed. (p. 617)

textura: patrón que describe el tamaño y disposición de los componentes rocosos y que puede ayudar a indicar cómo está formada la roca. (p. 617)

theory: explanation of things or events that is based on knowledge gained from many observations and investigations. (p. 12)

teoría: explicación de las cosas o eventos que se basa en el conocimiento obtenido a partir de numerosas observaciones e investigaciones. (p. 12)

thermal energy: sum of the kinetic and potential energy of the particles in an object; is transferred by conduction, convection, and radiation. (p. 256)

energía térmica: suma de la energía cinética y potencial de las partículas en un objeto, la cual se transfiere por conducción, convección y radiación. (p. 256)

third law of motion: states that when one object exerts a force on a second object, the second object exerts a force on the first object that is equal in strength and in the opposite direction. (p. 113)

tercera ley de movimiento: establece que cuando un objeto ejerce una fuerza sobre un segundo objeto, el segundo objeto ejerce una fuerza igual de fuerte sobre el primer objeto y en dirección opuesta. (p. 113)

tide: rise and fall in Earth's sea level, caused by a giant wave formed by the gravitational pull of the Sun and Moon. (p. 199)

marea: elevación y caída del nivel del mar en la Tierra, causadas por una ola gigante formada por la atracción gravitacional del Sol y de la Luna. (p. 199)

time zone: 15°-wide area on Earth's surface in which the time is the same. (p. 190)

zona horaria: área sobre la superficie terrestre de 15° de ancho en la que la hora es la misma. (p. 190)

tracer: radioactive isotope, such as iodine-131, that can be detected by the radiation it emits after it is absorbed by a living organism. (p. 804)

indicador radiactivo: isótopo radioactivo, tal como el yodo-131, que puede ser detectado por la radiación que emite después de ser absorbido por un organismo vivo. (p. 804)

transceiver: device that transmits one radio signal and receives another radio signal at the same time, allowing a cordless phone user to talk and listen at the same time. (p. 473)

radio transmisor-receptor: dispositivo que transmite y recibe una señal de radio al mismo tiempo, permitiendo que un usuario de un teléfono inalámbrico pueda hablar y escuchar al mismo tiempo. (p. 473)

transform boundary: plate tectonic boundary that exists as a large fault, or crack, along which lithospheric plates move in a horizontal direction. (p. 360)

límite de transformación: límite de la placa tectónica que existe como una gran falla o grieta, a lo largo del cual las placas litosféricas se mueven en dirección horizontal. (p. 360)

transformer: device that uses electromagnetic induction to increase or decrease the voltage of an alternating current. (p. 443)

transformador: dispositivo que usa inducción electromagnética para aumentar o disminuir el voltaje de una corriente alterna. (p. 443)

translucent: material that transmits some light but not enough to see objects clearly through it. (p. 327)

translúcido: material que transmite alguna luz pero no la suficiente para ver claramente los objetos a través del mismo. (p. 327)

transmutation: process of changing one element to another through radioactive decay. (p. 792)

transmutación: proceso de cambio de un elemento a otro mediante la descomposición radioactiva. (p. 792)

transparent: material that transmits almost all the light striking it so that objects can be clearly seen through it. (p. 327)

transparente: material que transmite casi toda la luz, chocándola de tal manera que los objetos pueden ser claramente vistos a través del mismo. (p. 327)

transverse wave: wave for which the matter in the medium moves back and forth at right angles to the direction the wave travels; has crests and troughs. (p. 290)

onda transversal: onda para la cual la materia en el medio se mueve hacia adelante y hacia atrás en ángulos rectos respecto a la dirección en que viaja la onda; ésta tiene cresta y depresiones. (p. 290)

troposphere: lowest layer of Earth's atmosphere where most weather occurs and temperature normally decreases with height. (p. 519)

troposfera: capa más baja de la atmósfera terrestre donde ocurre la mayoría de los fenómenos climáticos y la temperatura disminuye normalmente con la altura. (p. 519)

trough: the lowest points on a transverse wave. (p. 294)

depresión: los puntos más bajos en una onda transversal. (p. 294)

turbine: large wheel that rotates when pushed by steam, wind, or water and provides mechanical energy to a generator. (p. 440)

turbina: rueda grande que gira al ser impulsada por vapor, viento o agua y que suministra energía mecánica a un generador. (p. 440)

Tyndall effect: scattering of a light beam as it passes through a colloid. (p. 557)

efecto Tyndall: difusión de un rayo de luz al pasar a través de un coloide. (p. 557)

U

ultraviolet waves: electromagnetic waves with wavelengths between about 400 billionths and 10 billionths of a meter. (p. 465)

ondas ultravioleta: ondas electromagnéticas con longitudes de onda entre aproximadamente 10 y 400 billonésimas de metro. (p. 465)

unbalanced forces: forces that combine to produce a net force that is not equal to zero and cause the velocity of an object to change. (p. 82)

unconformity: gap in the rock record representing a period of erosion or non-deposition. (p. 671)

uniformitarianism: Hutton's concept that the laws of nature act today as they have in the past. (p. 670)

unsaturated solution: any solution that can dissolve more solute at a given temperature. (p. 671)

Uranus: seventh planet from the Sun; appears blue-green because of atmospheric methane, and its axis of rotation is tilted on its side. (p. 235)

fuerzas desbalanceadas: fuerzas que se combinan para producir una fuerza neta diferente a cero y causan que cambie la velocidad de un objeto. (p. 82)

discordancia: brecha en el registro rocoso que representa un periodo de erosión o falta de deposición. (p. 671)

uniformitarianismo: concepto de Hutton respecto a que las leyes de la naturaleza que actúan hoy día son iguales a las del pasado. (p. 670)

solución no saturada: cualquier solución que puede disolver más solutos a una temperatura determinada. (p. 671)

Urano: séptimo planeta más cercano al Sol; tiene una apariencia verde-azul debido al metano atmosférico, y cuyo eje de rotación está inclinado hacia un lado. (p. 235)

V

variable: factor that can cause a change in the results of an experiment. (p. 9)

vector: a physical quantity that is specified by both a size and a direction. (p. 71)

velocity: the speed and direction of a moving object. (p. 73)

Venus: second planet outward from the Sun; has a dense, atmosphere of mostly carbon dioxide and very high surface temperatures. (p. 224)

viscosity: a fluid's resistance to flow. (p. 375)

visible light: electromagnetic waves with wavelengths of 750 to 400 billionths of a meter that can be detected by human eyes. (p. 465)

voltage difference: related to the force that causes electric charges to flow; measured in volts (V). (p. 400)

volume: amount of space occupied by an object. (p. 18)

variable: factor que puede causar un cambio en los resultados de un experimento. (p. 9)

vector: cantidad física que está especificada por un tamaño y una dirección. (p. 71)

velocidad direccional: la rapidez y dirección de un objeto en movimiento. (p. 73)

Venus: segundo planeta más cercano al Sol; tiene una atmósfera densa compuesta principalmente de dióxido de carbono y una temperatura superficial muy alta. (p. 224)

viscosidad: resistencia de un fluido al flujo. (p. 375)

luz visible: ondas electromagnéticas con longitudes de onda entre 400 y 750 billonésimas de metro y que pueden ser detectadas por el ojo humano. (p. 465)

diferencia de voltaje: se refiere a la fuerza que causa que las cargas eléctricas fluyan; se mide en voltios (V). (p. 400)

volumen: espacio ocupado por un objeto. (p. 18)

W

water table: boundary separating the saturated zone from the unsaturated zone. (p. 664)

wave: a repeating disturbance or movement that transfers energy through matter or space. (p. 288)

wavelength: distance between one point on a wave and the nearest point just like it. (p. 295)

weak acid: any acid that only partly dissociates in solution. (p. 771)

tabla de agua: límite que separa la zona saturada de la zona insaturada. (p. 664)

onda: alteración o movimiento repetitivo que transfiere energía a través de la materia o el espacio. (p. 288)

longitud de onda: distancia entre un punto en una onda y el punto semejante más cercano. (p. 295)

ácido débil: cualquier ácido que solamente se disocie parcialmente en una solución. (p. 771)

weak base: any base that does not dissociate completely in solution. (p. 772)

weather fronts: zones in which air masses interact. (p. 526)

weathering: process of physical or chemical breakdown of materials at or near Earth's surface, influenced by factors such as the nature of the parent material, climate, and time. (p. 646)

wedge: simple machine that is an inclined plane with one or two sloping sides. (p. 173)

weight: gravitational force exerted on an object. (p. 106)

westerlies: global winds that blow from the west in the middle latitudes. (p. 524)

wheel and axle: simple machine that consists of a shaft or axle attached to the center of a larger wheel, so that the shaft and the wheel rotate together. (p. 171)

white dwarf: giant star that has lost its outer layers, leaving behind a hot, dense core that continues to contract under gravity. (p. 825)

work: transfer of energy that occurs when a force makes an object move; measured in joules. (p. 154)

base débil: cualquier base que no se disocie completamente en una solución. (p. 772)

frentes climáticos: zonas en las que interactúan las masas de aire. (p. 526)

desgaste: proceso de fraccionamiento físico o químico de materiales en o cerca de la superficie terrestre, influenciado por factores como la naturaleza del material de origen, el clima y el tiempo. (p. 646)

cuña: máquina simple que consiste de un plano inclinado con uno o dos lados en declive. (p. 173)

peso: fuerza gravitacional ejercida sobre un objeto. (p. 106)

vientos del oeste: vientos globales que soplan desde el oeste en las latitudes medias. (p. 524)

rueda y eje: máquina simple que consiste de una barra o eje sujeto al centro de una rueda de mayor tamaño de manera que el eje y la rueda giran juntos. (p. 171)

enana blanca: estrella gigante que ha perdido sus capas externas, dejando un núcleo denso y caliente que continúa contrayéndose bajo la gravedad. (p. 825)

trabajo: transferencia de energía que se produce cuando una fuerza hace mover un objeto y que se mide en julios. (p. 154)

X

X rays: electromagnetic waves with wavelengths between about 10 billionths of a meter and 10 trillionths of a meter, that are often used for medical imaging. (p. 467)

rayos X: ondas electromagnéticas con longitudes de onda entre 10 billonésimas de metro y 10 trillonésimas de metro, las cuales se utilizan con frecuencia para producir imágenes de uso médico. (p. 467)

Italic numbers = illustration/photo **Bold numbers = vocabulary term**
lab = a page on which the entry is used in a lab
act = a page on which the entry is used in an activity

A

D

Index

F

Index

Index

J

K

P

Index

Credits

Magnification Key: Magnifications listed are the magnifications at which images were originally photographed.
LM–Light Microscope
SEM–Scanning Electron Microscope
TEM–Transmission Electron Microscope

Acknowledgments: Glencoe would like to acknowledge the artists and agencies who participated in illustrating this program: Absolute Science Illustration; Andrew Evansen; Argosy; Articulate Graphics; Craig Attebery represented by Frank & Jeff Lavaty; CHK America; John Edwards and Associates; Gagliano Graphics; Pedro Julio Gonzalez represented by Melissa Turk & The Artist Network; Robert Hynes represented by Mendola Ltd.; Morgan Cain & Associates; JTH Illustration; Laurie O'Keefe; Matthew Pippin represented by Beranbaum Artist's Representative; Precision Graphics; Publisher's Art; Rolin Graphics, Inc.; Wendy Smith represented by Melissa Turk & The Artist Network; Kevin Torline represented by Berendsen and Associates, Inc.; WILDlife ART; Phil Wilson represented by Cliff Knecht Artist Representative; Zoo Botanica.

Photo Credits

cover (bkgd)Getty Images, (l)Jack Fields/Photo Researchers, (r)Alfred Pasieka/Photo Researchers; **vii** NASA/Science Source/Photo Researchers; **viii** Peticolas-Megna/Fundamental Photographs; **ix** NASA/JPL; **x** Ray Ellis/Photo Researchers; **xi** SuperStock; **xii** Photo Researchers; **xiii** Richard Megna/Fundamental Photographs; **xiv** (t)Matt Meadows, (b)NASA; **xv** Getty Images; **xvi** NASA; **xvii** Joe Lertola; **xx** Tim Courlas/Horizons; **1** James H. Karales/Peter Arnold, Inc.; **2–3** Jon Feingersh/The Stock Market; **3** Dorling Kindersley; **4–5** Roger Ressmeyer/CORBIS; **6** Getty Images; **7** Will McIntyre/Photo Researchers; **9** James L. Amos/CORBIS; **10** David Young-Wolff/PhotoEdit; **11** (l)Roger Ressmeyer/CORBIS, (r)Douglas Mesney/CORBIS; **12** J. Marshall/The Image Works; **13** (t)Jonathan Nourok/PhotoEdit, (b)Tony Freeman/PhotoEdit; **14** First Image; **15** courtesy Bureau International Des Poids et Mesures; **16** Matt Meadows; **17** (t)Amanita Pictures, (l)Runk/Schoenberger from Grant Heilman, (r)CORBIS; **20** Stephen R. Wagner; **24** First Image; **28 29** Matt Meadows; **30** Rosalie Winard; **31** (l)TSADO/NCDC/NOAA/Tom Stack & Associates, (r)David Ball/CORBIS; **32** Amanita Pictures; **35** courtesy Bureau International Des Poids et Mesures; **36–37** David Luzzi, University of Pennsylvania/Photo Researchers; **38** Bettmann/CORBIS; **39** Reuters/CORBIS; **40** Getty Images; **41** (l)Reunion des Musees Nationaux/Art Resource, NY, (r)file photo; **42** Torlief Svensson/CORBIS; **43** Getty Images; **44** (t)Getty Images, (cl)George D. Lepp/CORBIS, (cr)Richard Price/Getty Images, (bl)Lester Lefkowitz/CORBIS, (br)Joel Stettenheim/CORBIS; **46** Jill Connelly/AP/Wide World Photos; **47** Najlah Feanny/CORBIS; **48** Randy Brandon/Peter Arnold, Inc.; **49** (t)Pascal Le Segretain/CORBIS, (b)CORBIS; **52** Jim Reed/CORBIS; **53** (t)Hulton-Deutsch Collection/CORBIS, (b)Orlin Wagner/AP/Wide World Photos; **54** Roger Ressmeyer/CORBIS; **55** NASA/Science Source/Photo Researchers; **56** Toshiyuki Aizawa/Reuters/CORBIS; **57** Steve Dunwell/Index Stock Imagery; **58 59** Matt Meadows; **60** (t)AP, National Inventors Hall of Fame, (b)Rita Maas/Getty Images; **61** (t)Arthur Tilley/Getty Images, (bl)E. Dygas/Getty Images, (br)Roger Ressmeyer/CORBIS; **64** (l)Torlief Svensson/CORBIS, (r)Najlah Feanny/CORBIS; **65** (l)Ed Pritchard/Getty Images, (r)NASA/Science Source/Photo Researchers, Inc.; **66–67** John Terence Turner/FPG; **67** (l)Artville, (r)Charles L. Perrin; **68–69** Lester Lefkowitz/CORBIS; **70** Icon Images; **73** (t)J. Silver/SuperStock, (b)Robert Holmes/CORBIS; **78** (t)RDF/Visuals Unlimited, (c)Ron Kimball, (b)Richard Megna/Fundamental Photographs; **79** (l)The Image Finders, (r)Richard Hutchings; **80** Dan Feicht/Cedar Point Amusement Park; **81** Globus Brothers Studios, NYC; **82** Tim Courlas/Horizons; **84** Bob Daemmrich; **85** (t)James Sugar/Black Star/PictureQuest, (b)Michael Newman/PhotoEdit; **86** Keith Kent/Peter Arnold; **87** First Image; **88 89** Icon Images; **90** Sylvain Grandadam/Stone; **91** (t bl)Tony Freeman/PhotoEdit, (br)Keith Kent/Peter Arnold; **96–97** Tim Wright/CORBIS; **98** David Young-Wolff/PhotoEdit; **99** Neal Haynes/Rex USA, Ltd.; **100** (tl tr)Bob Daemmrich; (b)courtesy Insurance Institute for Highway Safety; **101** Donald Johnston/Stone; **103** Pictor; **105** StockTrek/CORBIS; **106** Peticolas-Megna/Fundamental Photographs; **107** NASA; **109** (tl)KS Studios, (tr)David Young-Wolff/PhotoEdit, (b)Richard Megna/Fundamental Photographs; **110** Pictor; **113** Steven Sutton/Duomo; **114** Philip Bailey/The Stock Market; **115** (l)NASA, (tr)(br)North American Rockwell, (cr)Teledyne Ryan Aeronautical; **117** (l)Jeff Smith/Fotosmith, (r)Richard Megna/Fundamental Photographs; **118** Matt Meadows; **119** Tim Courlas/Horizons; **120** (t)Tony Craddock/Photo Researchers, (b)Bettmann/CORBIS; **121** (l)Neal Haynes/Rex USA, Ltd., (r)Richard Megna/Fundamental Photographs; **126–127** Jim Cummins/CORBIS; **128** Runk/Schoenberger from Grant Heilman; **129** (l)Tony Walker/PhotoEdit, (c)Mark Burnett, (r)D. Boone/CORBIS; **133** KS Studios; **137** Walter H. Hodge/Peter Arnold, Inc.; **138** RFD/Visuals Unlimited; **142** Rudi Von Briel; **144 145** Matt Meadows; **146** (r)TIME, (l)Brompton Studios; **147** (tl)SuperStock, (bl)Jana R. Jirak/Visuals Unlimited, (r)Telegraph Colour Library/FPG; **152–153** Jakob Helbig/Getty Images; **153** Timothy Fuller; **154** Tim Courlas/Horizons; **155** (t)Tim Courlas/Horizons, (b)Michael Newman/PhotoEdit; **157** Michelle Bridwell/PhotoEdit; **159** Jules Frazier/Getty Images; **160** Michael Newman/Getty Images; **161** (t)Mark Burnett, (b)Tony Freeman/PhotoEdit; **162** Joseph P. Sinnot/Fundamental Photographs; **163** Richard Megna/Fundamental Photographs; **167** (t)Mark Burnett, (l)Tom Pantages, (r)Tony Freeman/PhotoEdit; **168** (t)Richard T. Nowitz, (c)David Madison/Stone, (b)John Henley/The Stock Market; **169** A.J. Copley/Visuals Unlimited; **171** Mark Burnett; **172** Tom Pantages; **173** (tl)file photo, (tr)Mark Burnett, (bl br)Amanita Pictures; **174** SuperStock; **175** Mark Burnett; **177** Hickson-Bender; **178** (t)courtesy D. Carr & H. Craighead, Cornell University, (b)Joe Lertola; **179** (t)Ed Lallo/Liaison Agency, (bl)file photo, (br)C. Squared Studios/PhotoDisc; **183** Michael Newman/PhotoEdit; **184–185** Richard Cummins/CORBIS; **189** NASA; **190** NASA; **196** Matt Meadows; **200** Lick Observatory; **201** Roger Ressmeyer/CORBIS; **202** NASA Kennedy Space Center; **203** NASA Goddard Space Flight Center; **205** (t)NASA/JPL/USGS, (c)NASA/CORBIS, (b)Julian Baum/Photo Researchers; **208** Bettmann/CORBIS; **210** (t)Bill Sanderson/Photo Researchers, (b)Ronald Royer/Science Photo Library; **211** (t)NASA Kennedy Space Center, (b)NASA/CORBIS; **215** NASA/GSFC; **216–217** Detlev Van Ravenswaay/SPL/Photo Researchers; **223** (t)NASA, (b)NASA/JPL/Northwestern University; **224** (t)NASA/JPL, (b)JPL/NASA; **225** NASA; **226** (t bl)JPL/NASA, (br)NASA/JPL/Malin Space Science Systems; **227** JPL/NASA; **228** (t)JPL/NASA, (b)Cornell/JPL/NASA; **229** JPL/NASA; **231** CICLOPS/University of Arizona/NASA; **232** JPL/NASA; **233** (t)CORBIS, (c)NASA-HQ-GRIN, (b)NASA-JPL; **234** (t)NASA/JPL, (b)NASA/JPL/University of Colorado; **235** (l)Goddard Space Flight Center/NASA, (r)JPL/NASA; **236** (t)Dr. R. Albrecht, ESA/ESO Space Telescope European Coordinating Facility/ NASA, (b)JPL/NASA; **237** Caltech/JPL/NASA; **238** OAR/ National Undersea Research Program (NURP); Texas A&M University; **239** Marshall Space Flight Center/NASA; **240** (t)Lunar and Planetary Institute, (b)JPL/NASA; **241** JPL/NASA; **242 243** Matt Meadows; **244** Gordon Garradd/Photo Researchers; **245** (l)HQ/GRIN/NASA, (r)JPL/NASA; **246** CICLOPS/University of Arizona/NASA; **248** JPL/NASA; **249** JPL/NASA; **250–251** Nonstock Inc.; **250** John Eastcott & Yva Momatiuk/Woodfin Camp & Associates; **252–253** Charles E. Rotkin/CORBIS; **253** Matt Meadows; **257** Aaron Haupt Photography; **261** JISAS/Lockheed/Science Photo Library/Photo Researchers; **262** Doug Martin; **263** Getty Images; **264** (t)Mark C. Burnett/Photo Researchers, (b)CORBIS; **265** Mark E. Gibson/Visuals Unlimited; **266** Zefa Visual Media/Index Stock; **267** Charles D. Winters/Photo Researchers; **268** (t)Earth Imaging/Stone, (bl)K&K Arman/ Bruce Coleman, Inc./PictureQuest, (br)Charles & Josette Lenars/CORBIS; **270** Science Photo Library/Photo Researchers; **271** Leonard Lee Rue III/Bruce Coleman, Inc./PictureQuest; **273** Steve Callahan/Visuals Unlimited; **278** Tim Courlas/Horizons; **280** (t b)NASA, (c)Roy Johnson/ Tom Stack & Associates; **281** (l)Breck P. Kent/Animals Animals, (r)Michael Newman/PhotoEdit; **286–287** Jason Childs/Getty Images; **288** James H. Karales/Peter Arnold, Inc.; **292** (t)Galen Rowell, (b)Stephen R. Wagner; **301** Richard Hutchings/Photo Researchers; **302** (t)Richard Megna/ Fundamental Photographs, (b)SIU/Visuals Unlimited; **304** (t)John S. Shelton, (b)Richard Megna/Fundamental Photographs; **305** Richard Megna/Fundamental Photographs; **308** Kodansha; **310** Doug Martin; **311** Tim Courlas/Horizons Companies; **312** Ralph White/CORBIS; **313** (tl)Stan Osolinski/The Stock Market, (tr)Bruce Heinemann/ PhotoDisc, (b)SIU/Visuals Unlimited; **318–319** AFP/ CORBIS; **322** (l)Mark A. Schneider/Visuals Unlimited, (cl)Rafael Macia/Photo Researchers, (cr)David Young-Wolff/ PhotoEdit, (r)Robert Llewellyn/SuperStock; **326** Doug Martin; **327** First Image; **329** (t)Stephen Frisch/Stock Boston/PictureQuest, (b)David Parker/Photo Researchers; **330** Charles O'Rear/CORBIS; **331** David Young-Wolff/PhotoEdit; **333** David Parker/Photo Researchers; **338** Matt Meadows; **340** First Image; **341** (t)Ralph C. Eagle, Jr./Photo Researchers, (b)Diane Hirsch/Fundamental Photographs; **342** Matt Meadows; **344** John Wang/Getty Images; **345** David Young-Wolff/PhotoEdit; **346** (t)Archivo Iconografico, S.A./CORBIS, (b)Ashmolean Museum, Oxford, UK/Bridgeman Art Library, London/New York; **347** (t)Jeff Greenberg/Visuals Unlimited, (b)Dick Poe/Visuals Unlimited; **352–353** Alfio Scigliano/CORBIS; **354** Mark Garlick/Photo Researchers; **355** NOAA/NGDC; **356** The Floor of the Oceans by Bruce C. Heezen and Marie Tharp, ©1977 by Marie Tharp. Reproduced by permission of Marie Tharp; **362** USGS, National Earthquake Information Center; **363** AP/Wide World Photos; **368** (tl)Bob Riha/Getty Images, (tr)Otto Greule Jr./Getty Images, (bl)Steven Powell, Star Seismic LLC, (br)Jerome Favre/AP/Wide World Photos; **382** Roger Ressmeyer/CORBIS; **383** Gary Braasch/CORBIS; **388–389** National Geographic Society Image Collection; **389** National Geographic Society Image Collection; **390–391** John Lawrence/Getty Images; **391** Michael Newman/PhotoEdit; **393** Geoff Butler; **395** KS Studios; **396** Tim Courlas; **397** T. Wiewandt/DRK Photo; **398** Dale Sloat/PhotoTake NYC/PictureQuest; **403** Ray Ellis/Photo Researchers; **404** Thomas Veneklasen; **406 407 408** Geoff Butler; **410** (tl)Geoff Butler, (tr)Tony Freeman/PhotoEdit, (c b)Aaron Haupt; **414 415** Geoff Butler; **416** Bernard Gotfryd/Woodfin Camp & Assoc.; **417** (t)file photo, (bl)Lester Lefkowitz/CORBIS, (br)Harold Stucker/Black Star; **422–423** Kennan Ward/CORBIS; **424** Don Smetzer/Getty Images; **425 426** Richard Megna/Fundamental Photographs; **429** (l)Stephen Frisch/Stock Boston, (r)Mark Burnett; **430** Mark Burnett; **433** Icon Images; **435** (t)Allen Zak, (b)Tim Courlas; **438** Tim Courlas; **440** (t)Tom Campbell/Getty Images, (b)Russell D. Curtis/Photo Researchers; **441** (tl)Clement Mok/PictureQuest, (tr)Joseph Palmieri/Pictor, (cl)Tim Ridley/DK Images, (cr)Robin Adshead/Military Picture Library/CORBIS, (b)Brian Gordon Green; **442** Mark Burnett; **445** Tim Courlas; **446 447** KS Studios; **448** (t)Brian Blauser/Stock Shop, (b)Laurence Dutton/Getty Images; **449** (l)Richard Megna/Fundamental Photographs, (r)Jeff Greenberg/Peter Arnold, Inc.; **454–455** Bob Burch/Index Stock Imagery; **459** Mat Jacob/The Image Works; **464** (t)Pete Saloutos/The Stock Market, (bl)NASA/GSFC/JPL, MISR and AirMISR Teams, (br)NASA/GSFC/MITI/ERSDAC/JAROS, and USA/Japan ASTER Science Team; **465** AP/Wide World Photos; **467** Telegraph Colour Library/FPG; **469** StudiOhio; **470** (t bl)First Image, (br)PhotoDisc; **471** (tl)Stephen Frisch/ Stock Boston/PictureQuest, (tr)Mark Richards/PhotoEdit/ PictureQuest, (bl)CORBIS, (bc)Dorling Kindersley, Ltd., (br)Michael Newman/PhotoEdit/PictureQuest; **472** Nino Mascardi/The Image Bank; **473** Photolink/PhotoDisc; **474** VCG/FPG; **476** Len Delessio/Index Stock; **477** Jim Wark/ Index Stock; **478** (t)Getty Images, (b)Bettman/CORBIS; **479** (t)Martin Dohrn/Science Photo Library/Photo Researchers, (c)Mickey Pfleger/Photo 20-20, (b)Joe McDonald/Tom Stack & Associates; **482** PhotoDisc; **484–485** Sandy Felsenthal/CORBIS; **486** Bill Heinsohn/Stone; **488** Robert E. Pratt; **489** Amanita Pictures; **490** (t)Digital Vision/PictureQuest, (b)Doug Martin; **494** Robert Essel/The Stock Market; **495** Photo Researchers; **498** AFP/CORBIS; **499** Tim Wright/CORBIS; **501** Amanita Pictures; **503** S.K. Patrick/Tom Stack & Associates; **504** (t)John Elk III/Stock Boston, (b)Lester Lefkowitz/The Stock Market; **506** Martin Bond/Science Photo Library/Photo Researchers; **508** Timothy Fuller; **509** Martin B. Withers/Frank Lane Picture Agency/ CORBIS; **510** Dan Lamont/CORBIS; **511** (t)Amanita Pictures, (c)Kathy Ferguson/PhotoEdit, (b)US Department of Energy/Science Photo Library/Photo Researchers; **515** Robert Essel/The Stock Market; **516–517** Byron Aughenbaugh/Getty Images; **531** (cw from top) Cartesia/Getty Iamges, Bernhard Edmaier/Photo Researchers, Bernhard Edmaier/Photo Researchers, Michele Westmorland/CORBIS, Wendy Dennis/ Visuals Unlimited, Anthony Bannister/CORBIS; **538** NASA; **539** AFP/Getty Images; **540** Bill Ross/CORBIS; **541** Philippe Colombi/Getty Images; **542** AP/Wide World Photos; **543** NASA; **548–549** Granger Collection, NY; **549** National Geographic Society Image Collection; **550–551** Michael Newman/PhotoEdit; **551** Matt Meadows; **552** (t)Siede Pries/Getty Images, (c)Ryan McVay/Getty Images, (b)Getty Images; **553** (tl)James L. Amos/CORBIS, (tr)Stockbyte/ PictureQuest, (cl)Yann Arthus-Bertrand/CORBIS, (cr)CalTech, (bl)Robert Fried/Stock Boston/PictureQuest, (br)Dorling Kindersley, Ltd.; **554** (t)Richard Megna/ Fundamental Photographs, (b)Icon Images; **555** (l c)Icon Images, (r)Andrew Syred/Science Photo Library/Photo Researchers; **556** (l)Getty Images, (r)C. Squared Studios/ Getty Images; **557** (tl)Chuck Keeler, Jr./CORBIS, (tr)Kaj R. Svensson/Science Photo Library/Photo Researchers, (b)Kip & Pat Peticolas/Fundamental Photographs; **558** Larry Mayer/

Getty News Images; **560** Matt Meadows; **561** (t)Amanita Pictures, (b)Doug Martin; **562** Richard Megna/Fundamental Photographs; **563** Jed & Kaoru Share/Getty Images; **564** Richard Megna/Fundamental Photographs; **565** (l)M. Romesser, (r)SuperStock; **567** Chris Rogers/CORBIS; **568 569** Matt Meadows; **570** (t)Dominic Oldershaw, (c)Doug Martin, (b)Frank Siteman/Stock Boston/PictureQuest; **571** (tl)Icon Images, (tr)Will Ryan/CORBIS, (bl)Tony Freeman/PhotoEdit, (br)Icon Images; **575** (tl tr)Doug Martin, (b)Chris Rogers/CORBIS; **576–577** Walter Bibikow/Index Stock Imagery, NY; **580** (tl)Science Photo Library/Photo Researchers, (tr)Hank Morgan/Photo Researchers, (b)Photo Researchers; **581** Sheila Terry/Photo Researchers; **588** Science Museum/Science & Society Picture Library; **596** CORBIS; **597** The Image Bank; **598** Tom Pantages; **599** PhotoEdit; **600** (t)Roger Ressmeyer/CORBIS, (b)Maria Stenzel/National Geographic Image Collection; **606–607** Hans Peter Merten/Robert Harding World Imagery/Getty Images; **610** (t)Doug Martin, (b)Marli Miller/Visuals Unlimited; **611** (l)Ken Lucas/Visuals Unlimited, (r)Alfred Pasieka/Photo Researchers; **612** NPS Photo; **613** (t to b) Mark A. Schneider/Photo Researchers, Carolina Biological/Visuals Unlimited, Marli Miller/Visuals Unlimited, Marli Miller/Visuals Unlimited, John R. Foster/Photo Researchers; **614** (t)Mark A. Schneider/Visuals Unlimited, (b)Martin Land/Photo Researchers; **616** (t)Icon Images, (b)Charles D. Winters/Photo Researchers; **617** Marli Miller/Visuals Unlimited; **619** Breck P. Kent/Earth Scenes; **620** Marli Miller/Visuals Unlimited; **622** (tl tr bl)Breck P. Kent/Earth Scenes, (br)Amanita Pictures; **623** (t)NPS Photo, (b)Wally Eberhart/Visuals Unlimited; **624** Darrell Gulin/CORBIS; **630** Kevin Schafer/CORBIS; **631** Breck P. Kent/Earth Scenes; **633** (t)Max Dannenbaum/Getty Images, (c)NASA, (b)NASA/GSFC/LaRC/JPL/MISR Team; **636 637** Matt Meadows; **638** (t)Albert Copley/Visuals Unlimited, (b)CORBIS; **639** (tl)Lawrence Lawry/Photo Researchers, (tr)Larry Stepanowicz/Visuals Unlimited, (b)Brad Lewis/Visuals Unlimited; **640** John R. Foster/Photo Researchers; **641** Breck P. Kent/Earth Scenes; **642** Marli Miller/Visuals Unlimited; **643** Breck P. Kent/Earth Scenes; **644–645** Tom Bean/CORBIS; **646** Robert Gill/CORBIS; **648** Owen Franken/CORBIS; **649** (tl tr)Doug Martin, (b)Adam Woolfitt/CORBIS; **652** Yann Arthus-Bertrand/CORBIS; **655** NASA/Goddard Space Flight Center/LaRC/JPL; **656** (t)Francois Gohier/Photo Researchers, (b)Martin Bond/Science Photo Library/Photo Researchers; **657** (t c)NASA/Goddard Space Flight Center Scientific Visualization Studio, (b)NASA/GSFC/METI/ERSDAC/JAROS, and U.S./Japan ASTER Science Team; **662** Yann Arthus-Bertrand/CORBIS; **666** (t)Gary Hincks/Photo Researchers, (cl)Carl & Ann Purcell/CORBIS, (cr)Stephen Alvarez/Getty Images, (b)Mark Gormus/AP/Wide World Photos; **678** (t)Peter Cade/Getty Images, (b)Ron & Patty Thomas/Getty Images; **679** (l)Barry Runk from Grant Heilman Photography, (r)Jon Turk/Visuals Unlimited; **684–685** Norbert Wu; **685** Spencer Collection, New York Public Library, Astor, Lenox & Tilden Foundation; **686–687** Tom Sanders/CORBIS; **687** Amanita Pictures; **688** (l)CORBIS, (r)John Evans; **689** (l c)Richard Megna/Fundamental Photographs, (r)Syndicated Features Limited/The Image Works; **693** First Image; **694** John Paul Kay/Peter Arnold; **696** Paul Silverman/Fundamental Photographs; **698** John Evans; **699** Patricia Lanza; **700** Amanita Pictures; **701** (l)Richard Megna/Fundamental Photographs, (r)Matt Meadows; **702** Mark Burnett; **703** Bettmann/CORBIS; **708** Amanita Pictures; **710 711** Matt Meadows; **712** Daniel Belknap; **713** Syndicated Features Limited/The Image Works; **718–719** Transglobe/Index Stock; **722** Mansell Collection/TimePix; **723** Richard Megna/Fundamental Photographs; **727** Doug Martin/Photo Researchers; **729** Richard Megna/Fundamental Photographs; **730** Emory Kristof/National Geographic Image Collection; **731** (t)Charles D. Winters/Photo Researchers, (b)Stephen Frisch/Stock Boston/PictureQuest; **732** Richard Megna/Fundamental Photographs; **733** Royalty-Free/CORBIS; **734** Tim Matsui/Liaison Agency; **735 736** Matt Meadows; **737** (t)Roger Ressmeyer/NASA/CORBIS, (l c)Bettmann/CORBIS, (r)NASA/CORBIS; **739** SuperStock; **741** Joel Sartore/Getty Images; **742 743** Matt Meadows; **744** (t)Bibliotheque du Museum D'Histoire Naturelle, (b)Brad Maushart/Graphistock; **745** (l)Charles D. Winters/Photo Researchers, (r)Joseph P. Sinnot/Fundamental Photographs; **748** Charles D. Winters/Photo Researchers; **749** (l)Richard Megna/Fundamental Photographs, (r)Firefly Productions/The Stock Market; **750–751** Stephen Frink/Index Stock Imagery; **752** Annie Griffiths Belt/CORBIS; **753** (l)Christie's Images, (r)Flip Schulke/Black Star; **754** (t)Bettmann/CORBIS, (c)Brian Gordon Green, (b)Michael Newman/PhotoEdit/PictureQuest; **760** Mark Thayer; **762** Richard Megna/Fundamental Photographs; **764 765** Matt Meadows; **766** Geoff Butler; **767** Rick Poley/Index Stock; **769** Charles D. Winters/Photo Researchers; **771** Matt Meadows; **773** (from top left) CORBIS, Elaine Shay, CORBIS, Brent Turner/BLT Productions, Icon Images, Matt Meadows, Elaine Shay, StudiOhio, (c)Dominic Oldershaw, (bl)Matt Meadows, (br)Mark Burnett/Stock Boston/PictureQuest; **775** Matt Meadows; **776** Tim Courlas/Horizons; **777** Matt Meadows; **778** TIME; **779** (l)Icon Images, (r)KS Studios; **783** Matt Meadows; **784–785** Paul Souders/Getty Images; **785** Aaron Haupt; **786** (t)First Image, (b)David Frazier/The Image Works; **790** American Institute of Physics/Emilio Segrè Visual Archives; **792** Amanita Pictures; **796** (l)courtesy Supersaturated Environments, (r)Science Photo Library/Photo Researchers; **797** (t)CERN, P. Loiez/Science Photo Library/Photo Researchers, (bl bcl bcr br)Richard Megna/Fundamental Photographs; **799** Hank Morgan/Photo Researchers; **804** Oliver Meckes/Nicole Ottawa/Photo Researchers; **805** UCLA School of Medicine; **806** J.L. Carson/Custom Medical Stock Photo; **807** Tim Courlas/Horizons Companies; **808** (l)Fermilab/Visuals Unlimited, (r)Matt Meadows; **809** Matt Meadows; **810** (t)C. Powell, P. Fowler & D. Perkins/Science Photo Library/Photo Researchers, Inc., (bl br)CORBIS; **811** (l)Martha Cooper/Peter Arnold, Inc., (r)Oliver Meckes/Nicole Ottawa/Photo Researchers; **816–817** Goddard Space Flight Center/NASA; **821** (t)NRAO/AUI/NSF, (b)NASA; **822** NOAO/Photo Researchers; **825** NASA Goddard Space Flight Center; **826** (t)Royal Observatory Edinburgh, D. Malin/AAO/Photo Researchers, (b)Celestial Image Co./Photo Researchers; **828** (t c)NASA, (b)Courtesy SOHO/Extreme Ultraviolet Imaging Telescope (EIT) Consortium/NASA; **829** SOHO/LASCO (ESA & NASA; **831** NASA; **832** (t)NASA, (b)JPL/NASA; **833** NASA; **834** Myron Jay Dorf/CORBIS; **835** National Center for Supercomputing Applications; **836** NASA; **838** NASA/ESA/S. Beckwith (STScI) and the HUDF Team; **839** The 2dF Galaxy Redshift Survey team; **842** (l)Jim Ballard/Getty Images, (r)E.O. Hoppé/CORBIS; **843** (l)NASA, (r)JPL/NASA; **844 846** NASA; **847** David A. Hardy/Photo Researchers; **848** PhotoDisc; **849** Hickson-Bender; **850** Tom Pantages; **854** Michell D. Bridwell/PhotoEdit; **855** (t)Mark Burnett, (b)Dominic Oldershaw; **856** StudiOhio; **857** Timothy Fuller; **858** Aaron Haupt Photography; **860** KS Studios; **861** Matt Meadows